Mechanical Science and Technology Update (MSTU 2018)

Journal of Physics: Conference Series
Volume 1050

Omsk, Russia
27 – 28 February 2018

ISBN: 978-1-5108-6822-9
ISSN: 1742-6588

Printed from e-media with permission by:

Curran Associates, Inc.
57 Morehouse Lane
Red Hook, NY 12571

Printed by Curran Associates, Inc. (2018)

For permission requests, please contact the Institute of Physics
at the address below.

Institute of Physics
Dirac House, Temple Back
Bristol BS1 6BE UK

Phone: 44 1 17 929 7481
Fax: 44 1 17 920 0979

techtracking@iop.org

Additional copies of this publication are available from:

Curran Associates, Inc.
57 Morehouse Lane
Red Hook, NY 12571 USA
Phone: 845-758-0400
Fax: 845-758-2633
Email: curran@proceedings.com
Web: www.proceedings.com

Table of contents

Volume 1050

Mechanical Science and Technology Update (MSTU-2018)

27–28 February 2018, Omsk, Russia

Accepted papers received: 25 June 2018
Published online: 26 July 2018

Preface

Papers

IOP Conf. Series: Journal of Physics: Conf. Series **1050** (2018) 011001 doi:10.1088/1742-6596/1050/1/011001

PREFACE

Mechanical Science and Technology Update (MSTU-2018) were held from 27-28 February in Omsk State Technical University, Omsk, Russia.

The conference materials cover the generalized issues of achievements in applied mechanics, engineering and materials science, quality management, mathematical modeling of processes, numerical solution of differential equations and development of micro-mechanical systems.

IOP Conf. Series: Journal of Physics: Conf. Series **1050** (2018) 011002 doi:10.1088/1742-6596/1050/1/011002

Peer review statement

All papers published in this volume of *Journal of Physics: Conference Series* have been peer reviewed through processes administered by the proceedings Editors. Reviews were conducted by expert referees to the professional and scientific standards expected of a proceedings journal published by IOP Publishing.

IOP Conf. Series: Journal of Physics: Conf. Series **1050** (2018) 012001 doi:10.1088/1742-6596/1050/1/012001

Work modeling processes of the hand-held impact machines

A Abramov, D Abramenkov and E Izhbuldin
Siberian Transport University, 191 Dusi Kovalchuk, Novosibirsk 630049, Russia

Corresponding author e-mail: abramov@stu.ru

Abstract. The research is devoted to the study of the processes occurring when the hand-held machine is machined by the impactor and their effect on the operator's work. To calculate the level of force impact for performing operations and to evaluate the transmitted vibration from the machine body to the hands of the operator, a model for interaction of the operator-machine-workable system, which is universal for any processed material, was created. The working cycle of the machine is divided into six main phases, for which the main energy indicators are determined. The calculation of the vibration acceleration of the machine body at different stages of the working cycle is made, the most vibro-dangerous phases of the cycle are identified, the elements of the machine are identified that need new design solutions, a layout scheme is proposed, at which the area of the vibration damper is significantly increased and the impact of the striker on the body of the machine is significantly reduced in the return stroke.

1. Introduction

Today, one of the most promising ways of connecting the building fittings is the connection in couplings of various types (crimping, threaded, and bolted). In civil engineering, it seems advisable to use fittings for diameters of up to 18 mm in diameter, which are pieces of steel cylindrical pipes. The crimping operation can be performed by various equipment, but the most compact and universal in terms of climatic and other operating conditions is a hand-operated electric tool, as evidenced by its wide prevalence on construction sites and at home. In this research, we propose the use of hand press machines with a linear electromagnetic drive as press equipment.

When the operation of a manual pulse machine, intensive shock processes occur, which are necessary to perform work operations with the workpiece, which excite the body vibrations transmitted to the operator's hands that have limitations specified in the sanitary norms [1], [2]. Therefore, it is necessary to determine the natural frequencies of machine oscillations, as a complex oscillatory system in the frequency range 5-2800 Hz, in which the vibration of the hand-operated machine must be ensured.

2. Experimental

For further calculations, we use the parameters of the developed V.A. Kargin and A.D. Abramov hand-operated impact machine UIM-3, which meets the requirements of the technological operation at the required energy level of a single impact.

The operating cycle of hand-operated impact machines with a linear electromagnetic drive is repeated in 1 second. This cycle can be divided in time into 6 stages with substantially different dynamic processes:

IOP Conf. Series: Journal of Physics: Conf. Series **1050** (2018) 012001 doi:10.1088/1742-6596/1050/1/012001

The 1st phase is a direct acceleration of the striker under the action of an electromagnetic force that occurs as a result of the current interaction in the coil of the machine with the ferromagnetic anchor bolt. Its motion is described by a sinusoidal wave, shown in Figure 1.

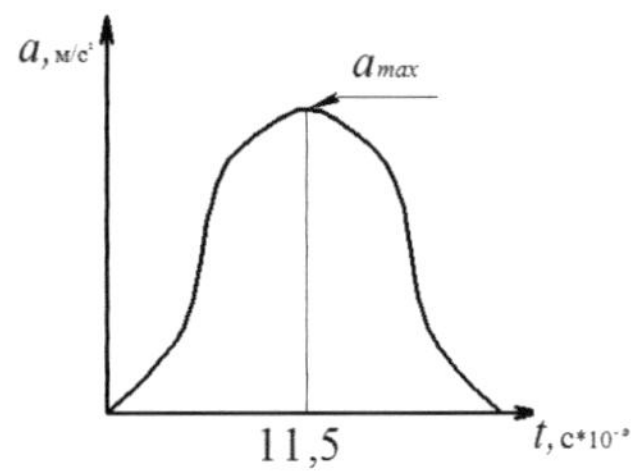

Figure 1. Characteristics of the striker acceleration.

Where a is the acceleration of the striker, the t-cycle time.

The 2d phase is the inelastic impact of the striker on the punch. The deformation of the workpiece is carried out with the joint action of the striker and punch. Since the punch and the punch move together after impact, their collision is assumed to be inelastic.

Formulas of the classical impact theory can be used if [3]:

$$\beta = \frac{t_y}{T} > 3 \div 5 \quad (1)$$

Where t_y - is the duration of the impact, $t_y = 11{,}5 * 10^{-3}c$, T- is the period of natural oscillations of the bodies, as systems with distributed parameters, in which elastic tensile-compression waves are transmitted when impact.

$$\beta = \frac{11{,}5 * 10^{-3}}{2{,}35 * 10^{-4}} = 48{,}9 > 5.$$

The impacted body (punch) before the impact is motionless, that is $v_{02} = 0$. The initial velocity of the striking body is v_{01} = 12 m / s. When the striker strikes the punch, they start joint motion, so we take the recovery factor K = 0 (absolutely inelastic impact). In this case, the joint speed of the "striker-punch" system will be:

$$v_g = \frac{m_1 v_{01}}{m_1 + m_2} \quad (2)$$

Where the mass of the striker is m_1= 0.9 kg, v_{01} = 12 m / s, punch weight m_2= 0.27 kg.

$$T_g = \frac{m_1 + m_2}{2} v_g = \frac{1{,}17}{2} * 9{,}2^2 = 49{,}1\, J$$

The 3d phase is the elastoplastic deformation of the workpiece.

To assess the energy intensity of operations and determine the boundary values of the impact energy, the lowest of which will be the minimum energy of a single impact, at which a qualitative realization of the technological operation is possible, the upper value will be the level of vibration acceleration transferred to the handle of the device, depending on the energy of a single impact and the frequency of impacts, In this paper, the process of interaction between the operator-machine system and the workpiece is simulated.

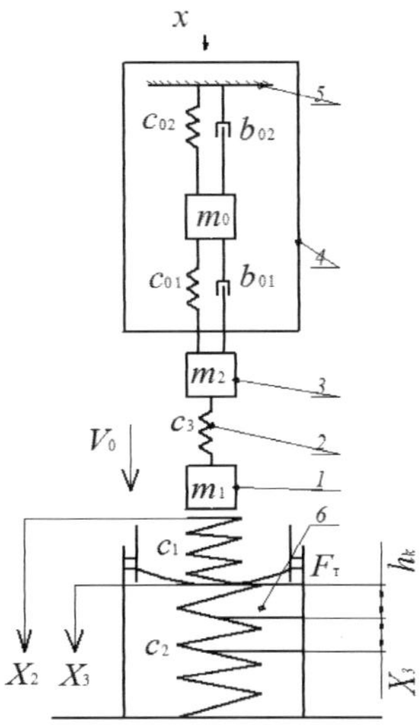

Figure 2. The model of the operator-machine system – the workpiece: 1– shock assembly; 2 – vibration isolator; 3 – body-handle; 4 – model of the operator's hand; 5 – the limiter; 6 – model of the body being treated; c_1 – the rigidity of the medium in the elastic zone; $F_т$ – resistance of environment to the beginning of introduction; c_2 – conditional stiffness of the medium in the plastic zone; ; V_0 is the pre-shock speed of the striker, $h_к$ is the sample sediment.

The special feature of this model is its versatility for all types of hand-operated impact machines and any bodies exposed to impact. The model shown in Figure 4 includes the hand of the operator holding and activating the tool, the impact machine itself, with a set of movable m_1 and fixed massesm_2, and directly the model of the workpiece, which are two springs with conditional stiffness, in the elastic c_1 and the plastic c_2 deformation zone, as well as the dry friction damper$F_т$, describing the static load characteristic. The use of such a model makes it possible to calculate the value of the workpiece draft $h_к$ according to the previously known formula for several successive cycles of dynamic loading *k:*

$$h_k = \sqrt{\left(\frac{F_т}{c_2}\right)^2 \left(\frac{c_1}{c_1+c_2}\right)^k + \frac{mV_0^2 c_1}{c_2^2}\left[1 - \left(\frac{c_1}{c_1+c_2}\right)^k\right]} - \frac{F_т}{c_2} \tag{3}$$

where c_2 is the nominal resistance of the plastic deformation part, N / m;

c_1 – conditional stiffness of the part in the zone of elastic deformation, N / m;

m – weight of the striker, kg;

V_0 – speed of the striker immediately before the contact, m / s.

The impact energy of the projected machine, which is necessary to deposit the workpiece by an amount $h_к$ for *k* impacts, can be obtained from equation (4):

$$T \geq \frac{h_k(c_2 h_k + 2F_т)}{2k} \tag{4}$$

Under the influence of the forces occurring during the working stroke of the striker and the impact of the punch and the treated body, the machine body moves. The movement spreads in three coordinate axes, but since such machines are produced and designed in the overwhelming majority of cases with pistol-type handles, long-term studies and regulatory documents state that the most critical for the machine operator is the direction of the vibrational movement of the machine body with the handle along the axis of the striker's movement. This circumstance makes it possible to consider the model of interaction between the operator and the handle of the device as two-dimensional, which simplifies further calculations.

Calculations show that with the energy of a single impact of 50 J, it is possible to obtain a qualitative connection of the reinforcement bars from 10 to 18 mm by means of crimping in a steel cylindrical bushing. Depending on the size of the rods, the tooling is replaced and the number of strokes required to complete the operation is increased.

The 4th phase is the transfer of direct shock to the body through the workpiece and the contact connection of the matrix to the body. The diagram of this interaction is shown in the figure. The

IOP Conf. Series: Journal of Physics: Conf. Series **1050** (2018) 012001 doi:10.1088/1742-6596/1050/1/012001

machine in question is low speed impact. The impact frequency is 1-2 Hz, the total number of strokes for one operation does not exceed 15.

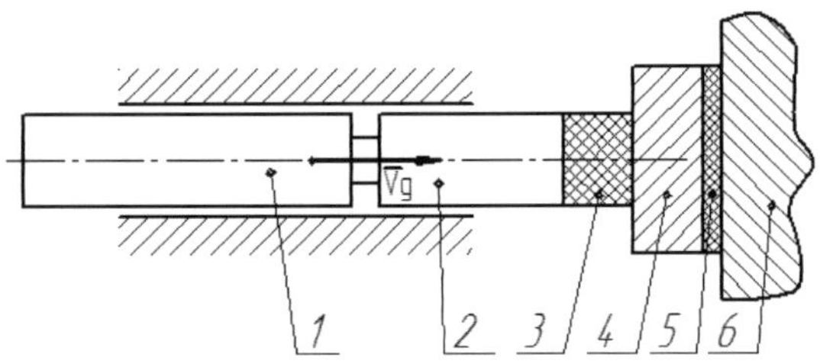

Figure 3. Contact interaction of system elements.

1 – firing pin; 2 – punch; 3 – workpiece; 4 – matrix; 5 – contact stiffness; 6 – housing.

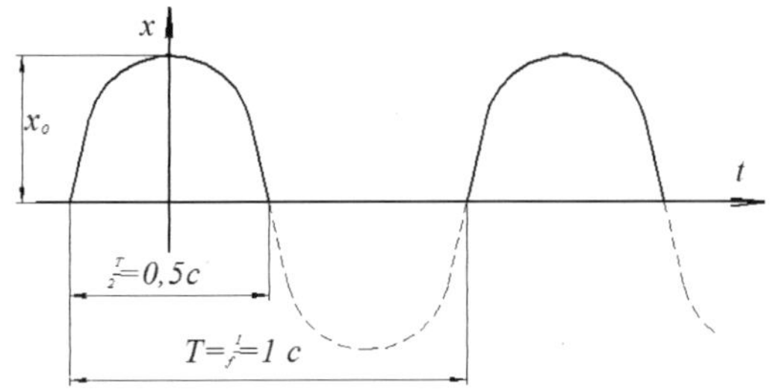

Figure 4. The operating cycle of the impact machine.

Impact pulses are approximated in the form of a half-sinusoid. Consequently, expanding in a Fourier series we obtain:

$$x(t) = \frac{2x_0}{\pi}\left(\frac{1}{2} + \frac{\pi}{4}\cos\omega t + \frac{1}{1\cdot 3}\cos 2\omega t - \frac{1}{3\cdot 5}\cos 4\omega t + \frac{1}{5\cdot 7}\cos 6\omega t - \cdots\right), \quad (5)$$

where $\omega = \frac{2\pi}{T} = 2\pi f$.

The acceleration calculation of the hull and striker joint motion at the moment of the article finishing treatment appears to be an irresolvable analytical method even for bodies of regular geometric shape, since the elastic and plastic properties of the material change with each impact. In addition, the task of selecting the impact energy for the material to be treated should virtually eliminate or minimize the acceleration of the hull at the moment of rebound by absorbing the vibrations of the treated body. Therefore, such a problem can be solved only for the case of an almost absolutely rigid body – the moment when the blows on the product continue after the crimping termination.

The case of a hard collision can be considered using the kinetic energy equation:

$$\frac{m_б V_б^2}{2} - \frac{m_к V_к^2}{2} = 0. \quad (6)$$

The calculated vibration velocity in the case of a rigid collision is 5, 1 m / s, which does not exceed the standard value of 5.270 m / s. It should be noted that the actual vibration speed will be significantly lower, since the mass of the body will be added to the mass of the workpiece, which, as a rule, exceeds the mass of the machine body.

This phase of work is controlled by the operator and can be ruled out with timely termination of work.

The 5th phase is a reverse acceleration of the striker by the return mechanism under the action of the spring.

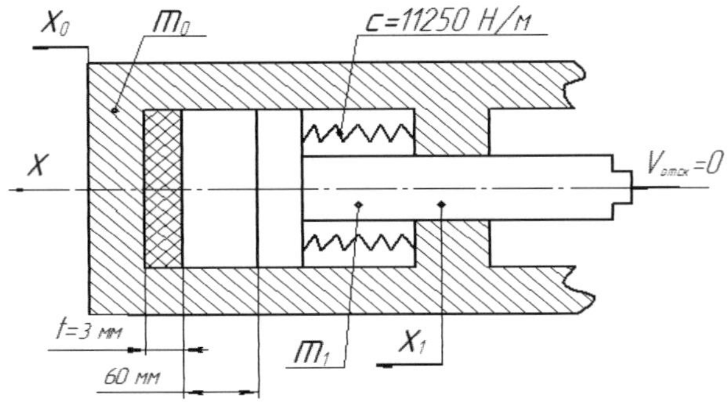

Figure 5. Calculation scheme for the return movement of the striker.

The coordinates x_n are counted from the equilibrium position.
Differential equation of motion of the striker under the action of the spring force of the return spring:

$$m_1\ddot{x}_1 + cx_1 = 0 => \ddot{x}_1 + \frac{c}{m_1}x_1 = 0 \quad (7)$$

Knowing the free move of the striker A_2 = 0,06 m, it is possible to calculate the maximum speed of the striker's movement with a return motion v_{2max}

$$v_{1max} = A_1\omega_1 = 0{,}06 * 111{,}8 = 6{,}7 \text{ м/с} \quad (8)$$

v_{1max} is the speed just before the impact of the striker with the rubber gasket – the damper of the case of the return mechanism.

The kinetic energy of the striker before impact:

$$A_{return} = \frac{m_1 v_{max}^2}{2} = \frac{0{,}9 \cdot 6^2}{2} = 20{,}2\ N \cdot \text{м}.$$

The 6th phase is the kick of the striker against the body through a rubber gasket.
To get started with this step, you need to get the numerical value of the stiffness of the gasket:

$$\frac{\Delta h}{h} = \frac{F_y}{ES} => F_y = \frac{ES}{h}\Delta h \quad (9)$$

it is possible to obtain the stiffness of the gasket, knowing the elastic modulus of the rubber

$E = 0{,}5 \cdot 10^7\ {}^{N}/_{\text{м}^2}$:

$$C_p = \frac{ES}{h} = \frac{0{,}5\cdot10^7\cdot593{,}5\cdot10^{-6}}{3\cdot10^{-3}} = 989 \cdot 10^3\ {}^{N}/_{\text{м}}. \quad (10)$$

forceN_m:

$$N_m = V_1\sqrt{\frac{C_p m_0 m_1}{m_0+m_1}} = 6{,}7\sqrt{\frac{989\cdot10^3\cdot5\cdot0{,}9}{4{,}7+0{,}9}} = 5972{,}9\ N \quad (11)$$

time of impactt_y:

$$t_y = \pi\sqrt{\frac{m_0 m_1}{C_p(m_0+m_1)}} = 3{,}14\sqrt{\frac{5\cdot0{,}9}{989\cdot10^3\cdot5{,}6}} = 8{,}9 \cdot 10^{-3}\ \text{c}. \quad (12)$$

The use of the classical theory of absolutely rigid bodies impact is possible under the following conditions:

$$\frac{t_y}{T} > 3 \div 5 \quad (13)$$

$$T = 2 \cdot 90 \cdot 10^{-3}\sqrt{\frac{7{,}9\cdot 10^3}{2{,}1\cdot 10^{11}}} = 3{,}5 \cdot 10^{-5}\text{c}.$$

$$\frac{t_y}{T} = \frac{8{,}9\cdot 10^{-3}}{3{,}5\cdot 10^{-5}} = 254{,}3 > 3.$$

Since the ratio of the time of impact to the frequency of natural oscillations satisfies the given condition, it is possible to use the classical theory of hard collision.

Next, it is necessary to calculate the vibration safety of hand-operated impact machines during the processing of products.

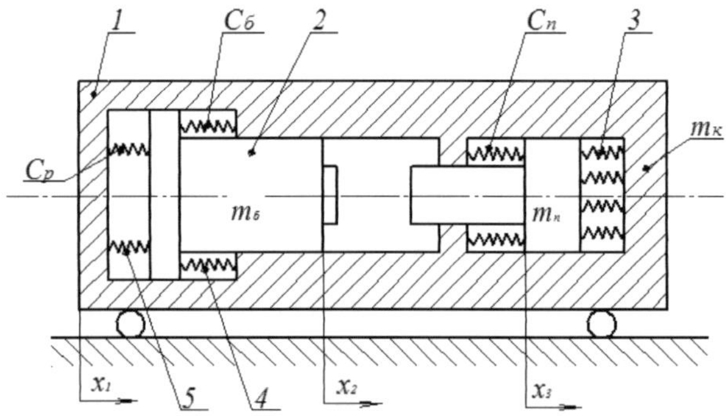

Figure 6. Schematic diagram of the system.

1– the body of the machine; 2 – kick with a return mechanism; 3 – rigidity of the workpiece, $C_{\text{ои}}$; 4 – a returnable spring; 5 – rubber gasket; C_6 – stiffness of the return spring; $C_{\text{р}}$ – the rigidity of the rubber gasket; $C_{\text{п}}$ – stiffness of the bias spring; $m_{\text{к}}$ is the mass of the shell; $m_{\text{п}}$ is the mass of the punch; m_6 is the mass of the striker.

It is necessary to determine the natural frequencies of the oscillations that occur during the shock impulse action on the workpiece. To compile the differential equations of free vibrations of a vibro-impact machine (VIM), we use the Lagrange equations of the II kind:

$$\frac{d}{dt}\left(\frac{dT}{dq_i}\right) - \frac{dT}{dq_i} + \frac{d\Pi}{dq_i} = 0 \tag{14}$$

The computational model is a mechanical system with three degrees of freedom.

The kinetic energy of the system, the inertial elements of which perform rectilinear free oscillations, is determined by the formula:

$$T = \frac{m_1\dot{x}_1^2}{2} + \frac{m_2\dot{x}_2^2}{2} + \frac{m_3\dot{x}_3^2}{2} \tag{15}$$

Potential energy of the system:

$$P = \frac{(C_{\text{р}}+C_6)(x_1-x_2)^2}{2} + \frac{(C_{\text{п}}+C_{\text{ои}})(x_1-x_3)^2}{2} \tag{16}$$

The coordinatesx_1, x_2, x_3 are counted from the equilibrium position.

Further we substitute T and P into equation (2. *). As a result, differential equations of free oscillations of the system are obtained.

$$\begin{cases} m_1\ddot{x}_1 + C_2(x_1 - x_2) + C_3(x_1 - x_3) = 0 \\ m_2\ddot{x}_2 + C_2(x_1 - x_2) = 0 \\ m_3\ddot{x}_3 + C_2(x_1 - x_3) = 0 \end{cases} \tag{17}$$

We assume, in the first approximation, that all the elastic elements in the mechanical system are linear. Then the free oscillations of inertial elementsm_1, m_2 и m_3 will be harmonic:

$$x_i = A_i \sin\omega t, , \dot{x}_i = A_i\omega\cos\omega t, \ \ddot{x}_i = -A_i\omega^2\cos\omega t$$

where i = 1, 2, 3, substituting in the differential equation (2. *):

$$\begin{cases} -m_1A_1\omega^2 + (C_2 + C_3)A_1 - C_2A_2 - C_3A_3 = 0 \\ -m_2A_2\omega^2 + C_2A_2 - C_2A_1 = 0 \\ -m_3A_3\omega^2 + C_3A_3 - C_3A_1 = 0 \end{cases} \tag{18}$$

After substituting numerical values:

$$\begin{vmatrix} (7\cdot 10^8-5\omega^2) & -322\cdot 10^4 & -7\cdot 10^8 \\ 322\cdot 10^4 & (322\cdot 10^4-\omega^2) & 0 \\ 7\cdot 10^8 & 0 & (7\cdot 10^8-0{,}3\omega^2) \end{vmatrix}=0$$

Further, expanding according to the rule of the triangle:

$$(7\cdot 10^8-5\omega^2)\,(322\cdot 10^4-\omega^2)(7\cdot 10^8-0{,}3\omega^2)=0. \quad (19)$$

We assume$\omega^2=\gamma$. After the expansion of equation (19), it takes the form of the third order algebraic equation, which is solved with the help of a computer:

$$a\gamma^3+b\gamma^2+c\gamma+d=0. \quad (20)$$

$$\omega^6-23{,}3\cdot 10^8\omega^4+32{,}7\cdot 10^{16}\omega^2-99\cdot 10^{22}=0.$$

$$\gamma^3-2{,}3\cdot 10^8\gamma^2+32{,}7\cdot 10^{16}\gamma-99\cdot 10^{22}=0.$$

f_1 = 278 Hz, f_2 = 1928 Hz, f_3 = 7431 Hz. The predicted vibration levels in these frequencies are not less than 2 times greater than the levels in neighboring frequency bands. At these frequencies, it is possible to exceed the sanitary standards of vibration acceleration in the phase of the striker impact against the body when returning to its original position, so it is necessary to improve the damping element.

3. Results and Discussion

It was found that the most successful solution is the universal low-frequency manual impact machine with a linear electromagnetic motor (LEM) for the reciprocal action of the UIM-3, the principal scheme of which is shown in Fig. 7.

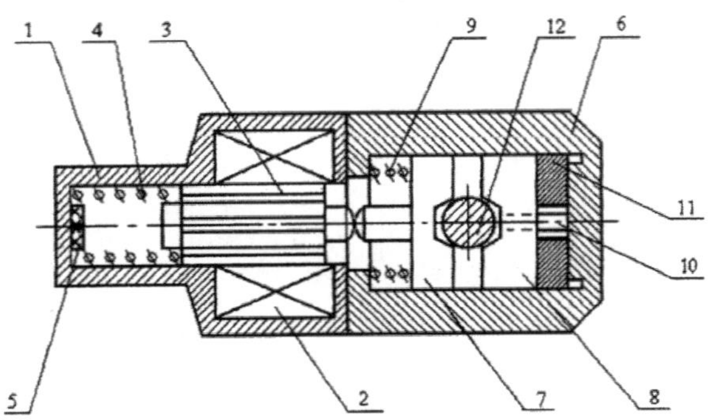

Figure 7. Diagram of the created hand-operated impact machine prototype with a spring return mechanism: 1 – body; 2 – electromagnetic coil; 3 – striker; 4 – a returnable spring; 5 – damper; 6 – matrix holder; 7 – punch; 8 – matrix; 9 – a spring; 10 – the guide; 11 – damper.

This design diagram of the machine has several drawbacks. The main ones are the low reliability of the return spring working on the principle of tension, due to the concentration of stresses in the places of its attachment to the striker and the body, as well as high vibration loading in the phase of impact of the striker against the body when returning.

To solve these problems, a new design solution [4] is proposed, based on the improvement of the layout scheme of the previously mentioned known machine, which consists in the use of a compression spring, which allows to increase by 3.2 times the area of the rubber gasket - the vibration damping damper hitting the striker against the body when returning to its original position, which in turn reduces the impact force, perceived by the body 57 times. Figure 13 shows an example of the device design being developed with a tool for crimping steel cylindrical bushings at the ends of reinforcing bars. The peculiarity of this scheme is, in addition to an improved return mechanism, a detachable matrix holder, which is a necessity for such a technological operation.

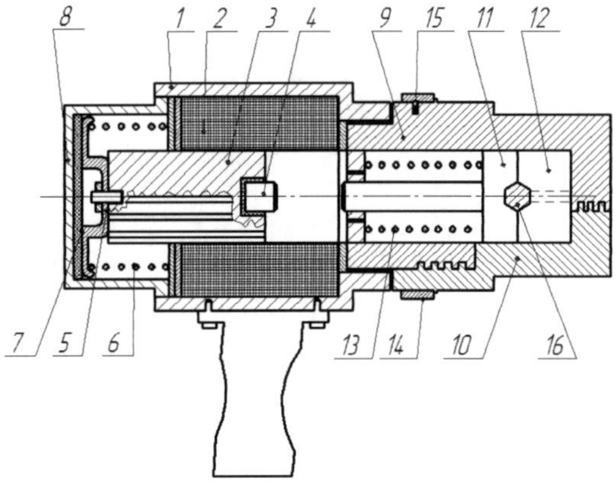

Figure 8. Schematic diagram of the device for crimping the valve.

1 – body, 2 – electromagnetic coil of the direct travel, 3 – massive body of the striker, 4 – impactor, 5 – thrust element, 6 – return spring, 7 – damper, 8 – protective casing, 9 – detachable matrix holder, 10 – removable part of the matrix holder, 11 – a punch, 12 – a matrix, 13 – a clamping spring, 14 – a fixing ring, 15 – the holding mechanism of a ring, 16 – a processed product.

4. Conclusion

Thus, the construction diagram of the striker return system drive, with a compression spring located in a separate housing, allows by increasing in the area of the damper by 3.2 times increase the reliability of the machine and reduce the vibration load on the operator to sanitary standards.

References

[1] Voss P and Jensen J G 1990 Vibration labelling of hand held concrete breakers *Int. J. of Industrial Ergonomics* vol 6, I.1 (Amsterdam: Elsevier) pp 83-87

[2] Golycheva E V, Babitsky V I and Veprik A M 2003 Dynamic correction of excitation in hand-held electro-pneumatic percussion machines *J. of Sound and Vibration*, vol 259, I 4 (Amsterdam: Elsevier) pp 829-43

[3] Aleksandrov E V and Sokolinskiy V B 1969 The applied theory and calculations of the impact systems 201 p

[4] Izhbuldin E A and Abramov A D Electric hand-held impact tool for the implementation of shock vibrating technologies in transport engineering and construction *Proceedings of Irkutsk State Technical University* 2017 vol 21 I 1 pp 32–41

A new algorithm of fast SSD calculation for motion estimation

E A Altman and E I Zakharenko

Omsk State Transport University, Omsk, 644050, Russia

Abstract. This paper considers motion estimation algorithms using a Sum of Squared Differences (SSD) metric and making an exhaustive search of motion vectors. A new fast algorithm, based on a new method of fast 2D convolution, is presented. It reduces a number of required arithmetic operations for finding all motion vectors in some area, compared to early published algorithms. The paper gives equations, underlines the presented algorithm, shows the tables comparing a new algorithm to another one in various cases. The new algorithm reduces the computational amount to 3.4%.

1 Introduction

A block matching algorithm is one of motion estimation methods. It is used for several problems of video processing, such as a video compression or an enhancement [1], [2].

The aim of this algorithm is looking to a fragment of a target frame which a best matching to a block of a source frame. The block of the source frame is a group of a neighbor pixels square form. A typical size of the block is 8×8 or 16×16 pixels. The fragment of the target frame is the frame block or an image, obtained from the frame by shifting him on half or quarter pixels. A searching area of the fragment is located around the block position in the source frame.

A block matching algorithms are classified by criteria of the matching and by methods of the searching.

As a criterion of the matching are used various metrics such as SAD (sum of absolute difference), SSD (Sum of Squared Differences) and others [3]. Commonly, SSD has the best quality of a matching evaluation, but the SAD requires less numbers of calculations [4].

There are two groups of methods for searching the best matching fragment. Methods of the first group find a fragment looking exhaustively over all possible positions of the block in the target frame [5].

Methods of the second group use a suboptimal algorithm, which presumes a smooth dependency of the used criterion on the fragment coordinate. They are based on an iterative searching procedure, have a small work time and a reasonable quality of result [6].

A choice of a group of methods depends on the problem that has to be solved. There are lots of problems, which demand the best quality of a block matching evaluation, and there are widely used problems, like motion estimation for video compression, which demand the fastest algorithms. Therefore, both groups of methods are being intensively researched.

In the paper we consider the first group. The main method of this group is brute force. There are two possibilities to perform an exhaustive searching with fewer operations, than brute force requires. The first possibility is an adaptation of fast algorithms of digital signal processing (like fast convolution) for calculation of a matching criterion (usually SSD) [7].

IOP Conf. Series: Journal of Physics: Conf. Series **1050** (2018) 012002 doi:10.1088/1742-6596/1050/1/012002

The second possibility is skipping surely bad positions of the fragment target frame. Under certain conditions applying this possibility reduces a number of arithmetic operations about 50 times (for example the New Fast Multilevel Successive Elimination Algorithm [8]).

These possibilities give the reduction of the number of operations of the same order [9]. Although elimination algorithms require less operations, fast algorithms have a regular computational scheme and often give faster implementation [10].

For the fast calculation of the matching criterion SSD, there are algorithms of fast calculation of a correlation, that based on a fast Fourier transform [4] or on number theoretic transforms [11]. This paper presences new algorithms of fast SSD calculation based on a new fast algorithm for computing the two-dimensional correlation [12].

2. Methods

The sum of squared difference (SSD) of two dimensional signals X and Y is equivalent to an Euclidean norm:

$$SSD_{i,j} = \sum_{m=0}^{N_h-1} \sum_{n=0}^{N_w-1} (X_{m,n} - Y_{m+i,n+j})^2, \quad (1)$$

where i, j – coordinates of the current candidate motion vector, Y is a candidate block in position (i, j) relative to the current block X, $N_h \times N_w$ – size of the current block X.

The double summation in the equation (1) requires a much calculations for each point of the result. To reduce the number of the operation, the expression (1) has been decomposed into three components:

$$\sum_{m=0}^{N_h-1} \sum_{n=0}^{N_w-1} X_{m,n}^2 + \quad (2)$$

$$+\sum_{m=0}^{N_h-1} \sum_{n=0}^{N_w-1} Y_{m+i,n+j}^2 - \quad (3)$$

$$-\sum_{m=0}^{N_h-1} \sum_{n=0}^{N_w-1} X_{m,n} Y_{m+i,n+j}, \quad (4)$$

The first component (2) does not depend on the arguments of the (1), thus when searching for the minimum value of this metric its calculation can be omitted.

The evaluation of the second component (3) requires M^2N^2 multiplications and $M^2(N^2-1)$ additions, where $N \times N$ – size of the current block X, $M \times M$ – coordinate range i, j. There are fast algorithms reducing number of calculations for its evaluating. They use an overlap of blocks, which appears when searching for motion vectors in adjacent points. These algorithms require $(M+N-1)N+2(M-1)\times \times(M+N-1)^2$ multiplications and $(M+N-1)(N-1)+2(M+N-1)(M-1)+M(N-1)+2M(M-1)$ additions [13].

The third component (4) is equivalent to a linear 2D correlation. It requires the largest number of operations among all component decomposition: M^4 multiplications and $M^2(M-1)^2$additions.

There are several ways for fast computation of convolution. The first group of methods uses fast algorithms for cyclic convolution based on fast Fourier transform or another number-theoretic transform.[11]. In this paper, we consider methods of the second group based on "divide and conquer" principle – they divide convolution into several smaller convolutions.

Paper [14] describes a fast algorithm for SSD computation based on a decomposition of equation (4) into 9 convolution of quarter size. Two dimensional signals X and Y are divided into odd and even sub-signals on each dimension:

$$X_{i,j} = \left[\hat{x}_{i,j}, \hat{x}_{i,j+2}, \dots \hat{x}_{i,j+N-2}\right]^T; Y_{i,j} = \left[\hat{y}_{i,j}, \hat{y}_{i,j+2}, \dots \hat{y}_{i,j+L-2}\right]. \quad (5)$$

The equation (4) may be rewritten as:

$$C_{i,j}=\begin{bmatrix}c_{i,j}\\c_{i+1,j}\\c_{i,j+1}\\c_{i+1,j+1}\end{bmatrix}=\begin{bmatrix}Y_{i,j} & Y_{i+1,j} & Y_{i,j+1} & Y_{i+1,j+1}\\Y_{i+1,j} & Y_{i+2,j} & Y_{i+1,j+1} & Y_{i+2,j+1}\\Y_{i,j+1} & Y_{i+1,j+1} & Y_{i,j+2} & Y_{i+1,j+2}\\Y_{i+1,j+1} & Y_{i+2,j+1} & Y_{i+1,j+2} & Y_{i+2,j+2}\end{bmatrix}\begin{bmatrix}X_{0,0}\\X_{1,0}\\X_{0,1}\\X_{1,1}\end{bmatrix}. \quad (6)$$

Introducing the notation:

$$A_0=X_{0,0}-X_{1,0},\ A_1=X_{0,1}-X_{1,1},\ B_{i,j}=Y_{i,j}+Y_{i+1,j}, \quad (7)$$

the equation (6) can be transformed into:

$$C_{i,j}=\begin{bmatrix}(B_{i,j}+B_{i,j+1})X_{0,0}\\(B_{i+1,v}+B_{i+1,j+1})X_{1,0}\\(B_{i,j+1}+B_{i,j+2})X_{0,1}\\(B_{i+1,j+1}+B_{i+1,j+2})X_{1,1}\end{bmatrix}+\begin{bmatrix}-1 & 0\\0 & -1\\1 & 0\\0 & 1\end{bmatrix}\begin{bmatrix}B_{i,j+1}(X_{0,0}-X_{0,1})\\B_{i+1,j+1}(X_{1,0}-X_{1,1})\end{bmatrix}+$$
$$+\begin{bmatrix}-1 & 0\\1 & 0\\0 & -1\\0 & 1\end{bmatrix}\begin{bmatrix}(Y_{i+1,j}+Y_{i+1,j+1})A_0\\(Y_{i+1,j+1}+Y_{i+1,j+2})A_1\end{bmatrix}+\begin{bmatrix}1\\-1\\-1\\1\end{bmatrix}Y_{i+1,j+1}(A_0-A_1). \quad (8)$$

The calculation of 9 convolution of quarter size requires $9/16(N^2\times M^2)$ cycles. In addition, we have to calculate input data for these convolutions (preprocessing) and to unite their output data (post processing) for obtaining the result of source convolution. The preprocessing includes $5N^2/4+2(N+M-2)\times(N+M-1)+(N+M-2)^2$ cycles. The post processing takes $10M^2/4$ cycles.

Decomposition (8) may perform recursively until a trivial convolution of size 2×2. Without pre and post processing operations, cost of convolution is divided on 16/9 times for each decomposition.

The cost of pre and post processing reduces a win obtained by the decomposition. For small size convolutions this cost is too big.

In this paper, we propose applying another decomposition method for 2D convolution for the motion estimation problem.This method has small expenses for pre and post processing and suits well to our task.

3. Results

Let's introduce the designations:

$$\begin{gathered}A_0=X_{0,0}-X_{1,0},\ A_1=X_{0,1}-X_{1,1},\\ B0_{i,j}=Y_{i,j}-Y_{i+1,j},\ B1_{i,j}=Y_{i,j+1}-Y_{i+1,j+1},\\ c_{i+1,j}=X_{0,0}Y_{i+1,j}+X_{0,1}Y_{i+1,j+1}+X_{1,0}Y_{i+2,j}+X_{1,1}Y_{i+2,j+1},\\ c_{i+1,j+1}=X_{0,0}Y_{i+1,j+1}+X_{0,1}Y_{i+1,j+2}+X_{1,0}Y_{i+2,j+1}+X_{1,1}Y_{i+2,j+2}.\end{gathered} \quad (9)$$

Using these designations we can transform the equation (6) to the next:

$$C_{i,j}=\begin{bmatrix}A_0B0_{i,j}+A_1B1_{i,j}+c_{i+1,j}\\A_0B1_{i,j}+A_1B0_{i,j}+c_{i+1,j+1}\\c_{i+1,j}\\c_{i+1,j+1}\end{bmatrix}, \quad (10)$$

IOP Conf. Series: Journal of Physics: Conf. Series **1050** (2018) 012002 doi:10.1088/1742-6596/1050/1/012002

There are 12 convolution of the quarter size in the equation (4). They require $3/4(N^2{\times}M^2)$ cycles. The preprocessing and post processing take jointly includes $N^2/2+3/2M^4$ cycles, where $N{\times}N$ – size of the current block X, $M{\times}M$ – coordinate range i, j.

Because of the low cost of pre and post processing, the algorithm based on the equation (10) is more effective than the algorithm based on the equation (8) in case of small sized convolution. Table 1 shows the numbers of operations of both algorithms for several sizes of the convolution. The size of X is $N{\times}N$ and $(2N–1){\times}(2N–1)$ – size of Y.

Table 1. The number of operations of the convolution for two methods of the decomposition.

N	The number of operations	
	9 convolution	12 convolution
2	**28**	**28**
4	501	**485**
8	**6 106**	6 709
16	**65 631**	86 189
32	**666 556**	1 073 533
64	**6 565 473**	13 170 845

We can reduce the number of operations by using the various methods on different stages of the recursive decomposition. The idea of a new fast algorithm of the convolution consists in the recursive decomposition using the best method on each stage. Samples of a benefit of the new algorithm are shown in table 2.

Table 2. The number of the operations of the various algorithms of the calculation of the convolution.

N	The number of operations	
	The fastest algorithm from the previously known	New algorithm
2	28	28
4	485	485
8	6 106	5 962
16	65 631	64 335
32	666 556	654 892

IOP Conf. Series: Journal of Physics: Conf. Series **1050** (2018) 012002 doi:10.1088/1742-6596/1050/1/012002

64	6 565 473	6 460 497

A new algorithm of a SSD calculation uses the described above algorithm of the convolution to find the third component of the equation (4). Table 3 shows the number of operations is required for the SSD calculation by various algorithms.

Table 3. The number of operations required for the SSD calculation by various algorithms.

N	The number of operations	
	The fastest algorithm from the previously known	New algorithm
2	207	207
4	2 178	2 106
8	21 003	20 355
16	201 528	195 696
32	1 926 597	1 874 109
64	18 309 546	17 837 154

4. Discussion

The presented algorithm of the motion estimation reduces a quantity of calculation in most cases. It doesn't take an additional memory or other resources. A disadvantage of this algorithm is a complexity and a bigger size of its implementation.

For a practical application of the discussed methods, the next useful step of research is developing a code generation software. In order to generate a program for a particular case it must analyze the available methods of decomposition and pick the best one. Such approach also enables taking into consideration a hardware specific.

In the sense of a theoretical research, the represented algorithm decreases the number of the operations for calculation of 2D convolution, but the question of the minimum quantity of operations is still open.

References

[1] Madisetti V 2009 *Wireless, networking, radar, sensor array processing, and nonlinear signal processing* (CRC press)

[2] Anand Deshpande P P and Rao D 2014 Motion Estimation Algorithms in Video Super-Resolution *Fifth International Conference on Recent Trends in Information, Telecommunication and Computing*

[3] Paramkusam A and Reddy V 2011 An efficient fast full search block matching algorithm with SSD criterion *India Conference (INDICON) Annual IEEE* pp 1–6

IOP Conf. Series: Journal of Physics: Conf. Series **1050** (2018) 012002 doi:10.1088/1742-6596/1050/1/012002

[4] Setkov A 2011 Matching quality investigation of Motion Estimation Algorithms *9th conference of open innovations community fruct* pp 133–140
[5] Mattoccia S, Tombari F, Di Stefano L and Pignoloni M 2007 Efficient and optimal block matching for motion estimation *Image Analysis and Processing. 14th International Conference* pp 705–710
[6] Lam C-W, Po L-M and Cheung C H 2003 A new cross-diamond search algorithm for fast block matching motion estimation *Neural Networks and Signal Processing, 2003 (Proceedings of the 2003 International Conference* vol 2), pp 1262–1265
[7] Lee C-H and Chen L-H 1997 A fast motion estimation algorithm based on the block sum pyramid *IEEE Transactions on Image Processing* **6**(11) pp 1587–1591
[8] Rao K S, Rao K N, Paramkusam A and Bhavana G 2017 An efficient multilevel motion estimation in video coding *International Journal of Mechanical Engineering and Technology* **8** pp 724–730
[9] Zakharenko E I and Altman E A 2015 Fast Full-Search Motion Estimation Method Based On Fast Fourier Transform Algorithm *AIST (Supplement)* pp 183–186
[10] Kilthau S L, Drew M S and Moller T 2002 Full search content independent block matching based on the fast fourier transform *Image Processing. International Conference* vol 1 pp I–I
[11] Toivonen T Heikkilä J and Silvén O 2002 A new algorithm for fast full search block motion estimation based on number theoretic transforms *Proceedings of the 9th International Workshop on Systems, Signals, and Image Processing* pp 90–94
[12] Altman E A and Zakharenko E I 2015 Fast algorithm for computing the two-dimensional correlation for video processing *Proceedings of TUSUR* **2**(36) pp 119–124
[13] Gao X, Duanmu C and Zou C 2000 A multilevel successive elimination algorithm for block matching motion estimation *IEEE Transactions on Image Processing* **9**(3) pp 501–504
[14] Naito Y, Miyazaki T and Kuroda I 1996 A fast full-search motion estimation method for programmable processors with a multiply-accumulator *Acoustics, Speech, and Signal Processing, 1996. ICASSP-96. IEEE International Conference* **6** pp 3221–3224

"Regular element" global SLAE of the finite element method when simulating electromagnetic processes of electric devices

E G Andreeva
Omsk State Technical University, Mira ave., 11, Omsk, 644050, Russia

E-mail: lenandr02@yandex.ru

Abstract. The article is devoted to mathematical and simulation modeling of electromagnetic processes of electric devices (ED). The development of industry, transport, power facilities, agriculture, and household services, health care leads to creation of electrical devices and systems for various functional purposes. It complicates the design (geometry) of electromagnetic and magnetic systems, improves magnetic and electrical characteristics of ED materials. The development of new software and hardware allows solving mixed (connected) tasks of simulating various physical processes in ED, electrical and electromechanical systems. However, general-purpose software complicates its engineering assimilation and implementation into production. The stage of ED numerical simulation modeling based on the finite element method when their investigating and designing is offered to be reduced and cheapened by changing the process of developing a global system of linear algebraic equations (SLAE) out of the system of elemental equations. The purpose of the work is to develop an algorithm to form global SLAE based on recurrent relations for its coefficients. The study subject is quasi-variable and magneto static vector models of ED electromagnetic processes, electro technical and electromechanical systems. Tasks are: developing of a "regular element", obtaining recurrent relations for global SLAE coefficients. Research methods are: numerical calculations based on the Galerkin's projection-grid method along with the finite element method (FEM). Results are: a seven-point "regular element" for a regular triangulation network from triangular finite elements (FE) of arbitrary shape and rectangular triangles has been developed; recurrent relations for global SLAE coefficients have been determined; a possibility of obtaining engineering techniques for investigating ED with minimal computer's dynamic memory expenses has been shown on the example of calculating a linear induction motor (LIM).

1. Introduction

ED operation both as in system and as a separate energy-transforming unit, such as electric motors of various designs, electric drives, transformers, iron separators, switching and measuring equipment, etc., depends on various factors. However, it is the electromagnetic field power action that determines the electrical device operation, as it has an electromagnet or a permanent magnet as the main assembly, therefore much attention is paid to simulating electromagnetic processes there. Knowing the field distribution in the device, its differential and integral parameters and power characteristics in both static and dynamic modes can be calculated. And it, in turn, allows determining of ED energy, technical, economic and design parameters and characteristics [1, 2]. Research of electromagnetic processes, parameters and characteristics of electrical devices are carried out, mainly with known software packages, that is 2D and 3D simulation of ED with open core [3], mixed simulation of mechanical and electrical oscillations of induction motors [4], analytical simulation of linear induction machines (linear induction motors) [5, 6], FEM-simulation for determining losses in windings of electrical machines [7].

At present, a systematic approach to the analysis of various problems, especially technical, technological, production, design and engineering in the development of mixed or connected models of devices and systems connecting nodes and subsystems of various physical nature is being applied. The use of high-performance computer technology and its development to increase resource capacity and speed operation allows solving the listed problems at a high theoretical and mathematical level [8]. ED electromagnetic processes research and simulation is carried out based on numerical

IOP Conf. Series: Journal of Physics: Conf. Series **1050** (2018) 012003 doi:10.1088/1742-6596/1050/1/012003

projection-grid methods, mainly the finite element method [9, 10]. It is the basis of many modern software packages ANSYS, ELCUT, COMSOL etc [11, 12]. Due to the software packages versatility it is hard for the engineering staff to use them in production and it also leads to rather high cost, especially when replacing with a higher package version. That is why, the author suggests another, simplified approach [1] in the finite element method to form global SLAE out of elemental algebraic systems based on the triangulation net "regular element".

2. Problem statement. ED vector models

Electromagnetic or field processes occurring in electrical devices are described with two-dimensional and three-dimensional numerical vector models based on Maxwell's equations in partial derivatives [1, 2]. Maxwell differential equations system along with boundary conditions at outer boundaries and the conjugation conditions on the inner boundaries of domains with various magnetic and electrical properties, initial conditions is ED electromagnetic field simulation and allows to determine vectors values any time any point; $\overline{E}$ is an electric field vector, $\overline{H}$ is a magnetic field vector, $\overline{B}$ is a magnetic induction vector for given initial conditions.

1. ED quasistationary or quasi-variable vector model

In determining quasistationary distribution of the electromagnetic field in the linear (the relative material permeability is μ = const) medium of investigated ED when there is no free electricity, with a sufficiently slow change in the time of currents and fields (quasi-variable or sinusoidal with frequency 50 Hz) Maxwell's equations take the form:

$$rot\,\overline{H} = \overline{J}_{st} + \overline{J}_{pr}, \quad (1)$$

$$rot\,\overline{E} = -\partial\overline{B}/\partial t, \quad (2)$$

$$\text{div}\ \overline{B} = 0, \quad (3)$$

$$\overline{B} = \mu\mu_0\overline{H}, \quad (4)$$

$$\overline{J}_{pr} = \gamma(\overline{E} + \overline{v}\times\overline{B}), \quad (5)$$

where $\overline{J}_{st}$ is an extraneous currents density vector; $\overline{J}_{pr}$ is a vector of conduction current density caused in the conducting medium by electromagnetic field change in time and this medium movement there with velocity $\overline{v}$; μ_0 is a magnetic constant; γ is specific electrical conductivity.

To calculate quasistationary vector model (1)-(5) vector magnetic potential $\overline{A}$ is used, its rotor is rot $\overline{A}=\overline{B}$. Considering Coulomb's calibration test div $\overline{A}$ = 0 an equation of quasi-variable electromagnetic field with respect to vector potential $\overline{A}$ for slow moving medium is obtained

$$\frac{1}{\mu}\nabla^2\overline{A} = -\mu_0\overline{J} + \mu_0\gamma\frac{\partial\overline{A}}{\partial t} - \mu_0\gamma\ \overline{v}\times rot\overline{A}, \quad (6)$$

and for the medium fixed in space ($\overline{v}$ = 0)

$$\frac{1}{\mu}\nabla^2\overline{A} = -\mu_0\overline{J} + \mu_0\gamma\frac{\partial\overline{A}}{\partial t}, \quad (7)$$

where $\overline{J} = \overline{J}_{st}$.

Equations (6) and (7) are parabolic and have one solution if initial conditions are set and boundary (marginal) conditions are described. These equations are solved with numerical methods, and it allows determining the distribution of the vector potential $\overline{A}$ in the modeling domain V in time period t, and then it allows determining the distribution of vectors $\overline{H}$ and $\overline{E}$; and induction $\overline{B}$ of electromagnetic field, as well as ED power, technical and economic parameters and characteristics.

2. ED magnetostatic vector model

For some electrical devices when there is no time-varying current supply Maxwell's equations have form

$$rot\,\bar{H} = \bar{J}\,,$$
$$\text{div}\ \bar{B} = 0,$$
$$\bar{B} = \mu\mu_0\bar{H}\,.$$

For linear medium, these equations are transformed to an elliptic equation or Laplace-Poisson's equation with respect to the vector potential

$$\frac{1}{\mu}\nabla^2\bar{A} = -\mu_0\bar{J}\,. \qquad (8)$$

In linear sectionally-homogeneous medium equation (8) with boundary conditions and conjugation conditions on inner boundaries of V domain is a magnetostatic vector model or a boundary problem to be solved with numerical projection-grid.

Quasistationary and magnetostatic vector models are used to study electromagnetic processes occurring in electric energy converters, such as electromagnetic and magnetic iron separators, magnetic field concentrators of various design versions, an electromagnetic motor, a linear induction pump (LIP).

3. Numerical projection-grid method or finite elements method for solving quasistationary electromagnetic boundary problems

Numerical calculation methods are used to study quasistationary magnetic fields and power characteristics of electrical devices with a complex geometry of the modeling domain and dissimilar physical properties. ED numerical models have inner boundaries of sectionally-homogeneous medium and some of them are mobile, as well as domains with distributed density of extraneous winding currents and secondary elements with eddy currents. The solution of differential equations in partial derivatives for their models is carried out with the projection-grid method or the finite element method as projection methods modifications (Ritz, Galerkin etc.) [1, 8]. Projection methods allow approximating the solution of a differential equation by a finite linear combination of basis (trial) functions. The form of a basis function and a criterion for calculating coefficients of a linear combination determine the projection method.

If a linear combination

$$\tilde{A} = \sum_{m=1}^{q} \varphi_m\, A_m$$

is an approximate solution of equation $L_f(A) = f$, where L_f is a differential operator, then according to Galerkin coefficients of a given linear combination are determined out of the orthogonality condition residual to basis functions N_m, i.e.

$$\int_V N_m(\,L_f(\tilde{A})\text{ - }f)\,dV = 0\,, \qquad (9)$$

where V is a modeling domain; m = 1 ,..., q. In case when $\varphi_m = N_m$, the result obtained by Galerkin's method coincides with the result of solution $L_f(A) = f$ at variational approach [8].

If the modeling domain is divided into finite elements (FE), where the required function A is approximated with polynomial, and finite functions are used as basic functions, that become zero everywhere except FE, then transformation of integral equations system (9) results to SLAE with a rarefied banded structure. The numerical solution of equation $L_f(A) = f$ by Galekin's method along with FEM has positive qualities of variational, projective and finite difference methods.

This article considers electric motors and power converters of electromagnetic energy with a planar magnetic system; therefore, a plane quasistationary electromagnetic field in a Cartesian coordinate system is investigated, and the modeling domain V is presented with S domain and x and y coordinates.

4. "Regular element" and algorithm of forming global SLAE

Distribution of ED quasistationary plane-parallel magnetic field is described by the Laplace-Poisson's equation system relatively to the vector magnetic potential, and also with homogeneous and inhomogeneous Neumann's or Dirichlet's boundary conditions at the outer boundary of the model. The solution of this problem is carried out by Galerkin's method - FEM, the sectionally-homogeneous modeling domain S of the electrical device is divided by a grid with q nodes per p triangular simplex-elements, vector magnetic potential is represented on everyone as follows:

$$\tilde{A}_R = (N_l\ A_l + N_j\ A_j + N_k\ A_k\ )_R = [\ N_m\ ]_R\{\ A_m\ \}_R, \qquad (10)$$

where $\{\ A_m\ \}_R$ is a column matrix of the magnetic vector potential in the computing net nodes, $[\ N_m\ ]_R$ is row matrix of basis functions $N_m = (\ a_m + b_m x + c_m y\ )/(\ 2S_R\ )$, and coefficients a_m, b_m, c_m are determined with coordinates of three nodes m = l, j, k of element R [9, 10]. The elemental SLAE is ordered with respect to q nodes of the computing triangulation net by summing the corresponding element equations (10), and thereby a global SLAE is formed

$$[\ U\ ]\{A\} = \{F\}, \qquad (11)$$

where *[U]* is a banded matrix of coefficients u_{ms} and m = 1, ..., q; s = 1, ..., q; {A} is a column matrix of nodal values of vector magnetic potential A_m, to be determined; *{F}* is a column matrix of transient components of global SLAE.

The algorithm for obtaining matrix coefficients *[U]* is simplified by using a regular triangulation net, and it automates SLAE formation, therefore minimizing the costs of the heap memory. The global system of equations (11) is solved with a direct or iterative method.

"Regular element" (figure 1), is a geometric figure, there is a calculating node in the center, it is surrounded with p_n finite elements, for example, six, i.e. p_6 = 6. A triangular element of arbitrary shape (figure 1, a) or a rectangular triangle (figure 1, b) is selected in Cartesian coordinate system as FE. FE number can be arbitrary in a "regular element". The selected seven-point pattern is the most convenient, as for FE in rectangular triangles form a part of coefficients u_{ms} is equal to zero.

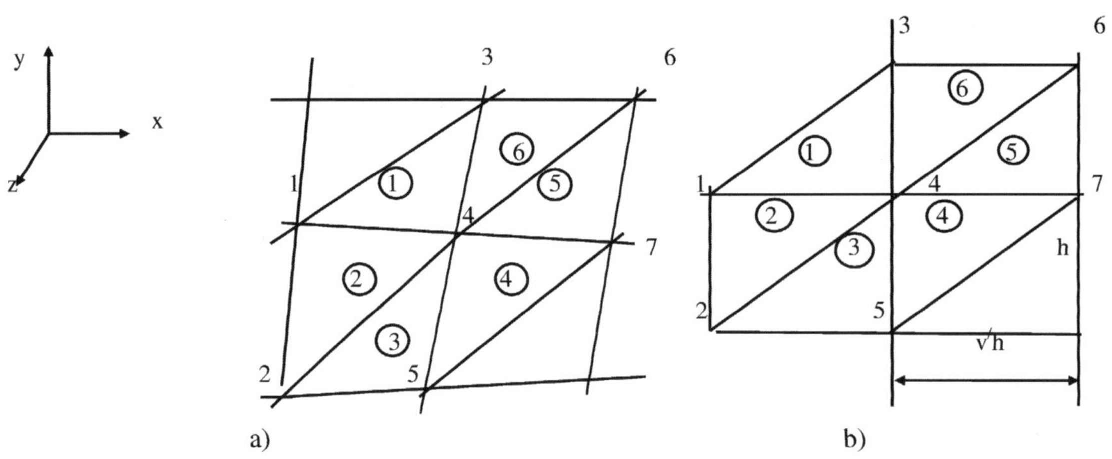

Figure 1. "Regular element" for a regular triangulation net out of triangles (FE) of arbitrary form (a) and of rectangular triangles (b)

The simulation domain is divided with the same or similar "regular elements". For nodes of the net consisting of "regular elements", the global SLAE equations are written down according to a single computing scheme.

As an example of the approach, the process of research and design of a linear induction motor (LIM) was considered [1, 5, 6]. Its technical-economic and power parameters and characteristics analysis was

carried out and it was based on the calculation of vector models of a quasi-variable electromagnetic field for complex variables that simulate values varying in time with a sinusoidal frequency 50 Hz.
LIM mathematical model is two-dimensional in the Cartesian coordinate system, it has an ideally laminated inductor ($\mu_c = \infty$). Secondary medium with specific electrical conductivity γ moves with speed $\mathbf{v} = \mathbf{i}\, v$ along a working channel with height Δ. The model is symmetric about the x axis, so only its upper part is considered. The current load is a sinusoidal time-varying extraneous current and is given with a system of pulsatory currents in q_1 slots on the inductor surface with a current density vector that has only z-component, i.e. $\boldsymbol{J}_n = \boldsymbol{k}\, \dot{j}_n\, e^{j\omega t}$

$$\dot{j}_n = \frac{I\ w}{S_n} e^{-j\psi_n} = J_{mm\,n}(\cos\psi_n - j\sin\psi_n),$$

where I is amplitude of extraneous current; S_n is slot space; w is a number of turns in the slot; Ψ_n is current phase; n = 1, 2 ,..., q_1.
For the described LIM model in rectangular Cartesian coordinates and considering the complex nature of changes in time of the vector potential, the current density vector, quasi-variable magnetic field vectors

$$\boldsymbol{A} = \boldsymbol{k}\ \dot{A} e^{j\omega t}, \qquad \dot{E} = -\frac{\partial \dot{A}}{\partial t} = -j\omega\, \dot{A},$$

scalar analogue (6) for the z-component of the magnetic vector potential can be written down

$$\frac{\partial^2 \dot{A}}{\partial x^2} + \frac{\partial^2 \dot{A}}{\partial y^2} - \mu_0 \gamma\, v \frac{\partial \dot{A}}{\partial x} - j\, \mu_0 \gamma \omega \dot{A} = -\mu_0 \sum_{n=1}^{q_1} \dot{J}_n . \qquad (12)$$

To solve equation (12) with Galerkin's projection-grid method - FEM, the computing domain is divided into p triangular simplex-elements; the function $\dot{A}$ is approximated with the polynomial

$$\tilde{A}_R = (N_l\ A_l + N_{jj}\ A_{jj} + N_k\ A_k\ )_R = [\,N_m\,]_R\{\ A_m\ \}_R, \qquad (13)$$

where $\{A_m\}_R = \{A'_m\}_R + j\{A''_m\}_R$ nodal values column vector of complex values of magnetic vector potential $\dot{A}$.
The system of linear algebraic equations (13) is transformed into global q nodes with respect to the computing net. Each node corresponds to two equations from the global system, for example, for node 4 (figure 1), they have the form

$$\sum_{m=1}^{7}(w_{m4}A'_m - v_{m4}A''_m) = -\sum_{R=1}^{6} F'_R; \qquad \sum_{m=1}^{7}(v_{m4}A'_m + w_{m4}A''_m) = -\sum_{R=1}^{6} F''_R. \quad (14)$$

Coefficients w_{m4} and v_{m4} are determined with the recurrence relations:

$$w_{14} = \frac{1}{4\mu_1 S_1}(b_1^{(1)} b_4^{(1)} + c_1^{(1)} c_4^{(1)}) + p'_1\, b_1^{(1)}/6 + p'_2\, b_1^{(2)}/6 + \frac{1}{4\mu_2 S_2}(b_1^{(2)} b_4^{(2)} + c_1^{(2)} c_4^{(2)});$$

$$v_{14} = (q'_1\, S_1 + q'_2\, S_2)/12.$$

Real and imaginary components of transient components in (14) for the elements simulating the stator winding slots are calculated as follows:

$$F'_R = -\mu_0 I w S_R \cos\Psi_n/(3S_n),$$

$$F''_R = \mu_0 I w S_R \sin\Psi_n/(3S_n).$$

Coefficients for indeterminate A'_m and A''_m in SLAE (14) for the "regular element" with the same rectangular triangles (figure 1, a) are simplified:

$$w_{14} = -1/v' - k_1\, p'\, h/6; \quad w_{74} = -1/v' + k_1\, p'\, h/6;$$

$$w_{34} = - v' - p' h/6; \; w_{54} = - v' + p' h/6;$$
$$w_{24} = - p' h/6; \; w_{64} = p' h/6; \; w_{44} = 2/v' + 2 v';$$
$$v_{14} = v_{24} = v_{34} = v_{54} = v_{64} = v_{74} = k_1 q' S/12; \quad v_{44} = k_1 q' S/2,$$

where $k_1 = 1$ for nodes on the boundary "air-secondary medium" and $k_1 = 2$ for secondary medium nodes.

Global SLAE is solved with iterative or direct methods. Then power and technical and economic parameters and characteristics can be calculated.

5. Discussion

A seven-point "regular element" for a regular triangulation net out of triangular finite elements (FE) of arbitrary shape and rectangular triangles has been developed; recurrence relations for global SLAE coefficients has been derived, possibility to obtain engineering methods for ED investigating with minimal costs of heap memory of computation, time has been presented on the example of linear induction motor (LIM) calculation. It simplifies the process of numerical simulation and research both ED and electrical and electromechanical systems, i.e. to reduce the cost of these devices engineering.

The improvement of the algorithm to form global SLAE coefficients based on the recurrence relations obtained for the "regular element" in the numerical method FEM while studying two-dimensional electromagnetic fields is aimed at the use of a three-dimensional "regular element" based on a FE parallelepiped-type for the calculation three-dimensional electromagnetic fields of ED.

6. Conclusion

The use of "regular elements" in calculation of ED magnetic field with Galerkin's method - FEM allows passing the stage of forming elemental equations and start forming global SLAE based on recurrence relations. Thus, the problem of storage and transformation (primarily for direct methods of solving SLAE) of the global matrix in the computer memory, especially for a PC, is simplified. It is important when developing software for engineering design systems for electrical devices and drives when mixed or linked problems for technical systems with subsystems of a diverse physical nature: electric, magnetic, electromechanical, thermal, magneto hydrodynamic, are to be solved.

7. References

[1] Andreeva E G, Kovalev V Z 1999 *Mathematical stimulating of the electrical complexes* (Omsk: Omsk State Technical University) p 172

[2] Tamm I E 1989 *Basics of Electricity theory* (Moscow : Nauka) p 504

[3] Andreeva E G 2017 Modeling of electrical devices with open magnetic circuit *Industrial Power Enginering* **3** pp 19-24

[4] Martinez J, Belahcen A, Detoni A 2016 A 2D magnetic and 3D mechanical coupled finite element model for the study of the dynamic vibrations in the stator of induction motors: *Mechanical Systemsand Signal Processing* **66-67** pp 640–656 https://www.elsevier.com/locate/ymssp

[5] Yamamyra S, Ito, Ishikawa Y, 1972 *Teories of the linear induction Motor and compensated Linear induction Motors* , IEEE Trans.m PAS–91 pp 1700–1710 DOI: 10.1109/TPAS.1972.2933

[6] Vol'dek A I 1970 *Induction magnetohydrodynamic machines with liquid metal working fluid* (Leningrad: Energy) p 272

[7] Mark D Ainslie, Victor M Rodriguez-Zermeno, Zhiyong Hong, Weijia Yuan, Timothy J Flack and Timothy A Coombs An improved FEM model for computing transport AC loss in coils made of RABiTS YBCO coated conductors for electric machines 2011 *Published 27 January IOP Publishing Ltd Superconductor Science and Technology* Vol 24 **4** https://doi.org/10.1088/0953-2048/24/4/045005

[8] Marchuk G I, Agoshkov V I 1981 *Introduction in projection grid methods* (Moscow: Science) p 416
[9] Segerlind L J 1984 *Applied Finite Element Analysis* 2nd edition (New York: John Wiley & Sons)
[10] Zienkiewicz O C , Morgan K 1983 *Finite Elements and Approximation* (New York Crichester Brisbane Toronto Singapore: University of Wales, Swansea, United Kigdom, A Wiley-Interscience Publication, John Wiley & Sons)
[11] ANSOFT Maxwell / URL: http://www.ansoft-axwell.narod.ru /maxwell/ Magnetostatic FieldCalculation.htm/
[12] *ELCUT Simulation of two-dimensional fields by finite element method Version 5.8 User Manual* (Saint Petersburg: TOR 2010) http://www.exponenta.ru/SOFT/OTHERS/elcut/Manual.pdf

IOP Conf. Series: Journal of Physics: Conf. Series **1050** (2018) 012004 doi:10.1088/1742-6596/1050/1/012004

Modelling the reusable space transport system for small payloads delivery

V S Aslanov, T A Ledkova

Samara National Research University, 34, Moscovskoe shosse, Samara, 443086, Russia

Abstract. The problem of delivering a payload from orbit without the consumption of jet fuel is considered. To solve this problem, it is proposed to use a reusable space transport system that consists of a satellite in a circular orbit, a weightless tether, a capsule with a payload, and a lower module. The purpose of the paper is to develop a mathematical model of the transport system, and to determine the system's parameters ensuring safe payload descent without entangling the tether. Equations of motion were obtained by Lagrange formalism. The capsule and the lower module were considered as rigid bodies, and the friction force between the capsule and the tether was taken into account. Numerical simulation showed that when the capsule approaches the lower end of the tether, the tether segment under the capsule can transfer into rotation. In turn, this will cause the capsule to stop and entangle the tether. The influence of the mass of the lower module on the dynamics of the system was investigated. The dependence of the entanglement point position on the mass of the module was identified. For safe descent from orbit, the separation of the capsule must occur before it reaches this dangerous point.

1. Introduction

The delivering a payload from orbit is one of the most topical issues of modern cosmonautics. Today, various re-entry capsules translated into the descent orbit with the help of jet engines are used. Alternative approach based on the use of space tethered systems are widely discussed in scientific literature. A detailed review of the different projects and methods of applying tethered systems in space is given in [1-3]. Tethered systems allow to deorbit a payload due to the momentum exchange between the connected by the tether bodies [4], or using the Earth's electromagnetic field [5, 6]. The projects of space tethered systems for delivering payloads based on using variable-length tethers are the most elaborated and close to practical implementation [7-10]. At the moment, three orbital transport experiments have been conducted: SEDS-1 in 1993, SEDS-2 in 1994 and YES2 in 2007 [3, 10]. In these projects variable-length tethers with the capsules at their ends were used [11].

In this paper, the reusable transport system based on the space tethered system of constant length is proposed. The payload is delivered to Earth by a capsule that slides along the tether. Similar mechanical systems with one or several capsules moving along the tether were investigated in [12-14]. In [15] the motion of a system consisting of two material points connected by weightless tether, and the load which can move freely along the tether was described. In paper [16] the problem of the motion of a mass point along a space elevator was considered. The space elevator was regarded as a weightless inextensible absolutely flexible tether. Conditions of the tether slackness were found, stationary motions were obtained and their stability was investigated. In [17] a simple analysis of a material point moving along an orbital tower was conducted. In the article [18] the dynamics of a flexible tethered system with a climber was analyzed. In [19], the dynamics of a tethered system with a movable climber is studied for the case of circular orbit. The tether tension control, which avoids the tether deflection on the final segment of the capsule descent along tether, was proposed. In contradistinction to previous works our study takes into consideration the motion of the capsule around its center of mass, and the friction force between the capsule and the tether.

2. Formulation of the problem

The paper considers a reusable transport space system, which consists of a satellite in a circular orbit, a weightless tether, a capsule with a payload, and a lower module attached to the tether end. The system is designed to deliver small payloads to the Earth without the cost of jet fuel. The payload is placed into the capsule, which slides along tether. At some distance from the lower module, the separation of the capsule from the tether occurs by dividing the capsule into two parts. Then these parts move into the descent orbit. An alternative approach involving braking and docking of the capsule with the lower module, and its further separation from the module by its automation, is also possible. The capsule delivery scheme is shown in Figure 1. The capsule deviates from the local vertical of the satellite in the direction of its orbital motion as the result of the Coriolis force action. This can cause the swing-up and rotation of the tether segment bellow the capsule, which can lead to entanglement of the tether and make it impossible to continue using the transport system. The aim of the work is to develop a mathematical model of the reusable transport space system, and to define the system parameters ensuring safe descent of the payload without the tether entanglement.

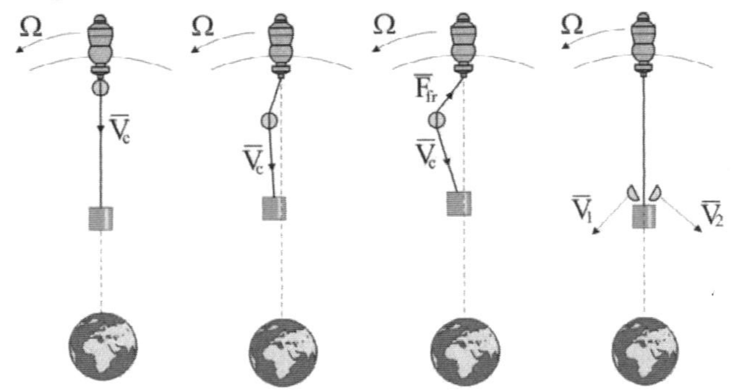

Figure 1. The scheme of delivery of small payload to Earth.

3. Mathematical model

Let's consider the tethered system, consisting of a satellite, a tether, a lower module and a sliding capsule (Figure 2). The satellite moves in a circular orbit of the radius R with constant angular velocity Ω and it is considered as a material point of mass m_1. The inextensible and weightless tether, which length is l, is attached to the satellite. It is assumed that the tether does not change its length during the transport operation. The descending capsule is a rigid body of a spherical shape. Its mass is m_2. The lower module is attached to the end of the tether. It is considered as a rigid body with mass m_3. The state of the mechanical system is described by five generalized coordinates $q_j = (\varphi_1, \varphi_2, \varphi_3, \varphi_4, s_1)$, where $\varphi_1, \varphi_2, \varphi_3$ are the angles of tether deviation the from the local vertical, φ_4- is the angle of deviation of the lower module axis from the local vertical, s_1 is the length of the tether segment from the satellite to the capsule. The length of the tether segment inside the capsule is denoted as s_2. The distance from the capsule to the lower module is s_3 , the distance from the tether attachment point to the center of mass of the lower module is s_4.

When the capsule moves along the tether, it experiences the influence of forces and torques (figure 3): gravity force G, centrifugal force of inertia Φ^e, Coriolis force Φ^c, friction force from tether F_{fr}, normal reaction from tether N and moment of reaction M_N. If the reaction force is greater than the Coriolis force, the tangential component of the capsule orbital velocity gradually decreases. As a result, the angular velocity of the capsule around the Earth and the centrifugal force also decreases.

The centrifugal force ceases to compensate for the gravitational force, and the capsule descends down the tether.

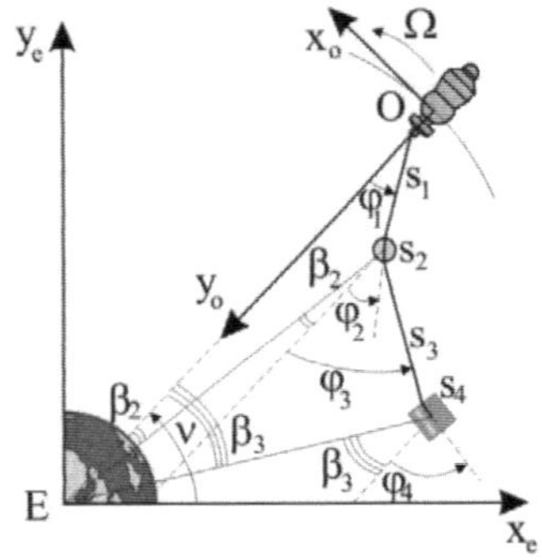

Figure 2. Space tethered system.

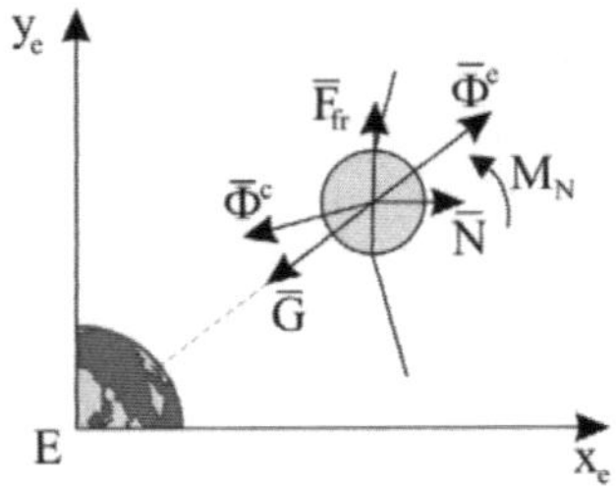

Figure 3. Forces acting on the capsule.

The equations of motion can be obtained using the Lagrange equations of the second kind. In order to calculate the Lagrangian of the mechanical system $L = T - \Pi$, the kinetic T and potential Π energy should be found. The kinetic energy is given by

$$T = \sum_{i=1}^{3} \frac{m_i V_i^2}{2} + \frac{J_{z2}(\dot{\varphi}_2 + \Omega)^2}{2} + \frac{J_{z3}(\dot{\varphi}_4 + \Omega)^2}{2}, \tag{1}$$

where m_i is the mass of the i-th body, V_i is the velocity of the center of mass of the i-th body, J_{xi}, J_{yi}, J_{zi} is the moments of inertia of the i-th body; $i = 1$ denotes the satellite, $i = 2$ - the capsule, $i = 3$ - the lower module. The potential energy can be found as

$$\Pi = -\mu \sum_{u=1}^{3} \frac{m_i}{r_i} - \frac{\mu}{2} \sum_{i=2}^{3} \frac{J_{xi} + J_{yi} + J_{zi}}{r_i^3} - \\ -\frac{3\mu}{2} \frac{J_{x2}\cos^2(\varphi_2 + \beta_2) + J_{y2}\sin^2(\varphi_2 + \beta_2) + J_{z2}}{r_2^3} - \frac{3\mu}{2} \frac{J_{x3}\cos^2(\varphi_4 + \beta_3) + J_{y3}\sin^2(\varphi_4 + \beta_3) + J_{z3}}{r_3^3}, \tag{2}$$

where $r_i = \sqrt{x_i^2 + y_i^2}$ is the distance from the center of the Earth E to the center of mass of the i-th body, angles β_1 and β_2 are shown on Figure 2 and can be calculated as

$$\cos \beta_i = \frac{x_1 x_i + y_1 y_i}{r_1 r_i},$$

x_i and y_i are the coordinates of the i-th points in the fixed coordinate reference frame $Ex_e y_e$. It is follows from the geometry (Figure 2) that

$$x_1 = R\cos\nu, \quad y_1 = R\sin\nu;$$

$$x_2 = x_1 - s_1\cos(\varphi_1 + \nu) - \frac{s_2}{2}\cos(\varphi_2 + \nu), \quad y_2 = y_1 - s_1\sin(\varphi_1 + \nu) - \frac{s_2}{2}\sin(\varphi_2 + \nu); \tag{3}$$

$$x_3 = x_1 - \sum_{i=1}^{4} s_i\cos(\varphi_i + \nu), \quad y_3 = y_1 - \sum_{i=1}^{4} s_i\sin(\varphi_i + \nu);$$

where $\nu = \Omega t$ is the true anomaly angle of the satellite.

Let us write the general form of the Lagrange equations of the second kind

$$\frac{d}{dt}\frac{\partial L}{\partial \dot{q}_j} - \frac{\partial L}{\partial q_j} = Q_j, \tag{4}$$

where Q_j is nonpotencial generalized forces. To calculate the generalized forces arising from the friction, let us write

$$Q_j = \bar{F}_{fr} \cdot \frac{\partial \bar{r}_2}{\partial q_j}, \tag{5}$$

where $F_{fr} = \mu_f N\, sign(\dot{s}_1)$ is the friction force, and μ_f is the friction coefficient. From (5) with regard to (3) we obtain:

$$Q_{\varphi_1} = 0,\ Q_{\varphi_2} = 0,\ Q_{\varphi_3} = 0,\ Q_{\varphi_4} = 0,\ Q_{s_1} = -\mu_f N\, sign(\dot{s}_1). \quad (6)$$

The tether reaction force N can be found by Newton's second law for the capsule in the projections on the axis of the inertial coordinate system (Figure 2):

$$\begin{aligned} m_2\ddot{x}_2 &= -\frac{\mu m_2 x_2}{r_2^3} + N\cos(\varphi_2 - \pi/2 + \nu) - F_{fr}\sin(\varphi_2 - \pi/2 + \nu), \\ m_2\ddot{y}_2 &= -\frac{\mu m_2 y_2}{r_2^3} + N\sin(\varphi_2 - \pi/2 + \nu) + F_{fr}\cos(\varphi_2 - \pi/2 + \nu), \end{aligned} \quad (7)$$

Substitution of the coordinates (3) in (7) and expression of N gives

$$\begin{aligned} N = m_2 \Big[& \ddot{\varphi}_1 s_1 \cos(\varphi_1 - \varphi_2) + \frac{s_2}{2}\ddot{\varphi}_2 + \ddot{s}_1 \sin(\varphi_1 - \varphi_2) - s_1(\Omega + \dot{\varphi}_1)^2 \sin(\varphi_1 - \varphi_2) + \\ & + 2\dot{s}_1(\Omega + \dot{\varphi}_1)\cos(\varphi_1 - \varphi_2) - R\Omega^2 \sin\varphi_2 + \frac{\mu}{r_2^3}\left(R\sin\varphi_2 - s_1\sin(\varphi_1 - \varphi_2)\right) \Big]. \end{aligned} \quad (8)$$

Substituting (8) in (6) and then in (4), we obtain a system of five second-order differential equations describing the motion of the system. These equations are not presented here because of their bulkiness.

4. Results of numerical simulation

Let us consider the motion of a mechanical system with the following parameters: the length of the tether is $l = 30000$ m, the radius of the satellite orbit is $R = 6645000$ m, the friction coefficient is $\mu_{fr} = 0.1$, the diameter of the capsule is $s_2 = 0.4$ m, the initial velocity of the capsule is $\dot{s}_{10} = 2.8$ m / s. The lower module is a cylinder, whose radius is $r_c = 0.5$ m and length is $d_c = 2 \cdot s_4 = 2$ m. Mass of the satellite is $m_1 = 6300$ kg, mass of the capsule is $m_2 = 50$ kg. At the initial time, the tether and the capsule are oriented along the satellite's local vertical, and the lower module is deflected by the angle $\varphi_4 = 0.1$ rad.

Integrating the equations of motion obtained in the previous section, we obtain the dependence of the deviation angles φ_j on the distance s_1 for the case when $m_3 = 100$ kg (Figure 4). During descent the capsule deviates from the local vertical in the direction of the orbital flight under the influence of the Coriolis force. This effect is well known and described, for example, in [13,18]. Calculations show that at the distance of $s_1 = 29575$ m, the capsule stops. In this moment the tether segment below the capsule rotated to a large angle, and the lower module is above the capsule. This situation can be interpreted as entanglement of the tether. A sharp drop in the capsule speed relative the tether $\dot{s}_1$ is observed with entanglement (Figure 5). This is caused by the fact that in this position, the braking of the capsule is provided not only by the friction force, but also by the tension reaction force of the lower tether segment.

Let's perform a series of numerical calculations, and investigate the influence of the mass of the lower module on the position of the entanglement point. It is seen from Figure 6 that when the mass of the lower module is increased, the tether entanglement occurs later. For example, in the case of $m_3 = 1000$ kg, it occurs when the capsule is at a distance of $s_3 = 24$ cm from the lower module (Figure 7), and in the case of $m_3 = 2000$ kg, "entanglement" occurs at a distance of $s_3 = 2$ mm, in fact, the capsule will strike the upper end of the lower module. Note, that as the mass of the lower module increases, the amplitude of oscillations of the angular $\varphi_1, \varphi_2, \varphi_3$ decreases. The tether is stronger, and a larger Coriolis force, and hence a larger relative velocity, is required to deflect the capsule.

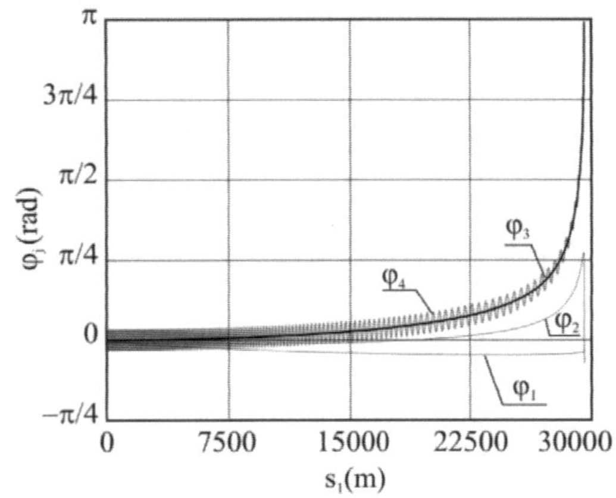

Figure 4. Dependence of angles of deviation φ_j on the distance s_1 ($m_3 = 100$ kg).

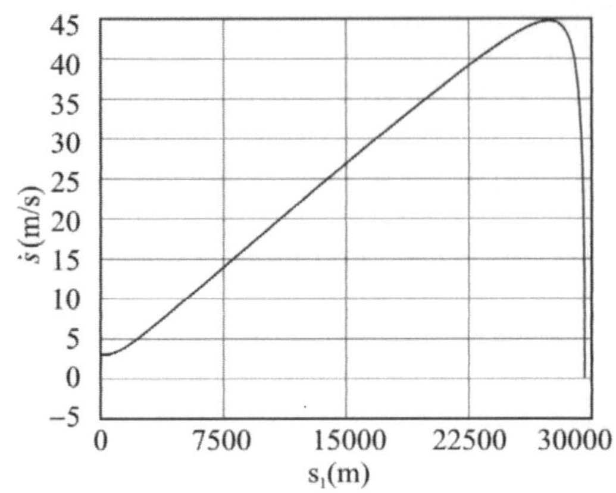

Figure 5. Dependence of the speed of the capsule $\dot{s}_1$ on the distance s_1 .

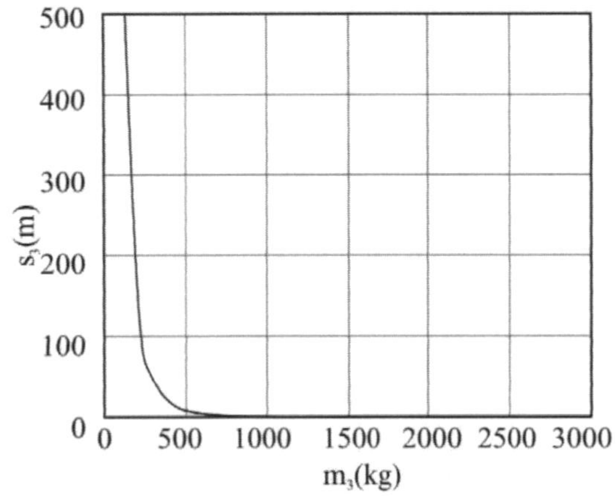

Figure 6. Dependence of the distance between the entanglement point on the mass of the lower module.

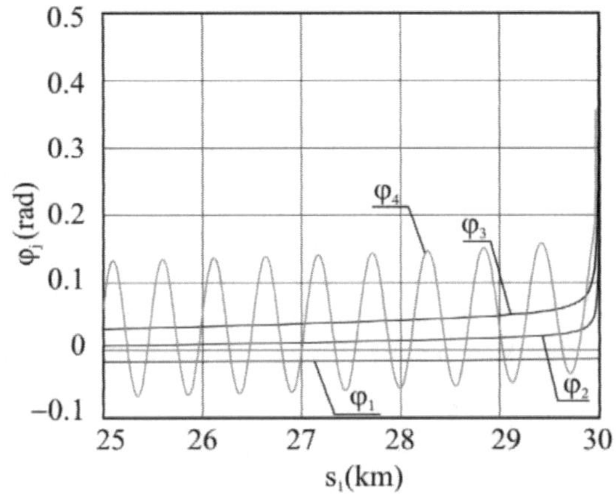

Figure 7. Dependence of angles of deviation φ_j on the distance s_1 ($m_3 = 1000$ kg).

5. The discussion of the results

The studies show that the radially oriented space tethered system, which is attached to the satellite in circular orbit, allows the descent of the capsule with a small payload without jet fuel costs. However, two problems can be noted. Firstly, the capsule develops a high velocity relative to the tether during its uncontrolled sliding (Figure 5). If the descent scheme does not involve the separation of the capsule before it reaches the lower module, then the docking with the lower module is assumed, and the capsule braking system is required. The second problem is entanglement of the tether at the final stage of the capsule motion in the case of a small relative mass of the lower module. This problem can be solved by installing jet engines in the lower module, or by separating the capsule at a great distance from the lower module. A sharp increase in the angles φ_2, φ_3 is observed at the final stage of the capsule motion even in the case of a large mass of the lower module. Therefore, the lower module must be equipped with capturing devices that allow to stop and fix the capsule before it reaches a dangerous point. After docking, the automation of the lower module should provide the separation of the capsule from the tether.

6. Conclusions

The article considers a reusable space transportation system aimed for the delivery of small payloads to the Earth and consisting of a satellite moving in a circular orbit, a weightless three-segment tether, a capsule with a payload and a lower module. The equations of motion of the system were obtained with help the Lagrange equations of the second kind. Within the framework of the model the capsule and the lower module were considered as a rigid bodies, and the friction force between the capsule and the tether was taken into account. Simulation of the uncontrolled sliding of the capsule along the tether was performed. It was shown that the motion of the capsule caused a deflection of the tether in the direction of the orbital flight. When the capsule approaches to the lower module, the segment of the tether bellow the capsule can transfer into rotation. In turn, this will cause the capsule to stop and entangle the tether. The influence of the mass of the lower module on entanglement of the tether was investigated. It was found that with the mass of the module increasing, the cable entanglement occurs later. Based on a series of numerical calculations, the graph that demonstrates the dependence of the distance to the point, at which the tether entanglement occurs, on the module mass was constructed. When the lower module has a relatively small mass, the separation of the capsule from the tethered system must occur before reaching the dangerous point of entanglement. In the case of a large mass of the lower module, a sharp increase in the angle of deflection of the cable segment between the capsule and the lower module is also observed, when the capsule approaches the lower end of the tether. Itcan lead to collision between the capsule and the lower module. Therefore, the automation of the module must ensure the capture and fixation of the capsule until it reaches the attachment point of the tether to the module.
The results obtained in the work can be used to create new transport space systems, including long tether and capsules moving along them.

Acknowledgment
The work was supported by the Russian Foundation for Basic Research, grants No. 18-01-00215 A, No. 17-41-630274 R_A.

References

[1] Beletsky V V and Levin E M 1993 *Dynamics of space tether systems* (Univelt Incorporated) p 329
[2] Aslanov V S and Ledkov A S 2012 *Dynamics of tethered satellite systems* (Elsevier) p 331
[3] Cartmell M P and McKenzie D J 2008 *A review of space tether research* vol **44** Progress in Aerospace Sciences p 1
[4] Williams P 2010 *Tether capture and momentum exchange from hyperbolic orbits* vol **47** Journal of Spacecraft and Rockets p 205
[5] Yu B S, Jin D P and Wen H A 2016 *Method for Stable Deployment of an Electrodynamic Tethered Satellite in Three-Dimensional Space* vol **744** Journal of Physics: Conference Series. IOP Publishing p 012119
[6] Hoyt R P 2002 *Stabilization of electrodynamic tethers* vol **68** AIP Conference Proceedings p 570
[7] Gorelov S K, Sof'in A P and Shcherbakov V I 2016 *The deployment of space tether systems in the delivery of lander from orbital station to Earth* vol **652** Proc. of the Mozhaisky Military Space academy p 154
[8] Aslanov V S, Ledkov A S and Stratilatov N R 2009 *The influence of the cable system dedicated to deliver freights to the Earth on the rotary motion* vol **1** (Samara: Samara State Aerospace University: Flight) p 54
[9] Kruijff M and van der Heide E J 2009 *Qualification and in-flight demonstration of a European tether deployment system on YES2* vol **64** Acta Astronautica p 882
[10] Aslanov V S 2016 *Swing principle for deployment of a tether-assisted return mission of a re-entry capsule* vol **120** Acta Astronautica p 154
[11] Zimmermann F, Schöttle U M and Messerschmid E 2005 *Optimization of the tether-assisted return mission of a guided re-entry capsule* vol **9** Aerospace Science and Technology p 713

IOP Conf. Series: Journal of Physics: Conf. Series **1050** (2018) 012004 doi:10.1088/1742-6596/1050/1/012004

[12] Cohen S and Misra A K 2015 *Static deformation of space elevator tether due to climber* vol **111** Acta Astronautica p 317

[13] Williams P and Ockels W 2010 *Climber motion optimization for the tethered space elevator* vol **66** Acta Astronautica p 1458

[14] Williams P 2009 *Dynamic multibody modeling for tethered space elevators* vol **65** Acta Astronautica p 399

[15] Rodnikov A V 2004 *Motion of a load along a cable fastened to a two-body-tethered spacecraft* vol **42** Cosmic Research p 427

[16] Beletsky V V, Ivanov M B and Otstavnoe E I 2005 *Model problem of the space elevator* vol **43** Cosmic Research p 157

[17] McInnes C R 2005 *Dynamics of a particle moving along an orbital tower* vol **31** Journal of guidance, control, and dynamics p 380

[18] Fujii H A et al 2008 *Dynamics of a flexible space tether equipped with a crawler mass* vol **31** Journal of guidance, control, and dynamics p 436

[19] Shi G, Zhu Z and Zhu Z H 2018 *Libration suppression of tethered space system with a moving climber in circular orbit* vol **91** Nonlinear Dynamics p 923

Investigation of the orientation of floccules in magnetic sump during cleaning of cooling lubricants

B A Avdeyev[1], V B Masyagin[2], R R Dema[3], O B Kalugina[3]

[1]Kerch State Maritime Technological University, Kerch, Russia
[2]Omsk State Technical University, 11, Mira ave., Omsk, 644050, Russia
[3]Nosov Magnitogorsk State Technical University, Magnitogorsk, Russia
E-mail: dirigeant@mail.ru

Abstract. The article deals with the problem of cleaning cooling lubricants involved in technological processes of parts machining. As a result of cutting, drilling, grinding, cooling lubricants are contaminated with extraneous impurities. The most dangerous for the quality of products and equipment are magnetic particles of various sizes. One of the cleaning methods is a magnetic sump, which will reduce the mass-dimensional parameters of the system on retention of a high degree of purification. An analysis of existing designs is made and it is shown how the configuration of the magnetic field affects the kinetics of the motion of impurities in cooling lubricants. Experimental dependencies of the deposition rate for different orientations of floccules formed as a result of the effect of a magnetic field on particles are found. The coefficient of drag of the floccules is found from the number of particles. It is proved that the use of only the shape factor and the equivalent diameter is not enough for calculations.

1. Introduction
When cutting, extruding, rolling, stamping, drilling parts, etc. cooling lubricants are often used, which divert heat from the working tool and the workpiece and lubricate the rubbing parts. In general, the use of cooling lubricants allows increasing the intensity of technological processes, labor productivity, and equipment productivity, and improving the quality of products. During operation, the cooling lubricants are contaminated with swarf, hard dust, extraneous oils or microorganisms. Pollution leads not only to a decrease in the efficiency and quality of materials processing and, accordingly, to frequent replacement, but also cause equipment breakdown.

2. Objectives of work
There are various ways to clean viscous media from extraneous impurities: settling tanks, sumps, hydro cyclones, separators. Sumps designed for cleaning cooling lubricants from mechanical impurities have found wide application in various sectors of the national economy due to the simplicity of the designs and their operation, reliability of work, and low material and energy costs. They are most often used as the first stage of purification from large extraneous impurities [1-3]. Due to the fact that particles possessing electrical properties are often contained in impurities, electric fields are used to improve the efficiency of purification [2]. Despite the fact that these devices have been used for a long time, many theoretical and experimental questions related to the use of electric fields have not been investigated so far. The objective of the work is to analyze the effect of the

magnetic field for reducing the size of the sumps as a consequence of increasing the extraction efficiency of magnetic impurities and the effect of floccules' orientation on the deposition process.

3. Information analysis of researches and publications

The use of a magnetic field for the purification of cooling lubricants is most effective when the amount of magnetic particles allows them to be forced into coagulation. The main force effecting the particles in the sump is gravitational. Except it, in the presence of a field, magnetic attractions and coagulations affect the particles [1]:

$$F_M = \mu_0 \chi V H \,\mathrm{grad} H \tag{1}$$

where $\mu_0 = const = 4\pi \cdot 10^{-7}$ - magnetic constant, H/m; χ - magnetic susceptibility; V - volume of the particle, m^3; H, grad$H$ - the intensity of the magnetic field and its gradient, A/m and A/m^2, respectively.

However, magnetic forces not only attract magnetic particles to zones with the greatest field strength, but also serve to coagulate them. Under the effect of magnetic (Coulomb) forces, the particles grow, forming floccules of an acicular structure:

$$F_F = \frac{4\pi M_1 M_2}{\mu_0 \mu r^2} \tag{2}$$

where M_1, M_2 - the "magnetic masses" of the particles, $m^2 \cdot kg/s^2 \cdot A$; μ - the magnetic permeability of the medium in which the particles are located; r - the distance between the particles, m.

4. Theoretical part

The magnetic field inside the working chamber increases not only the cleaning efficiency, but also reduces the mass-dimensional parameters, because the coagulated particles have a greater weight, and accordingly, they need less time to sediment. Thus, the volume of the sump and the residence time of the cooling lubricants in it decrease [4]. In figure 1, two sumps are shown: one is ordinary; the other is with a magnetic field source located at the bottom of the working chamber.

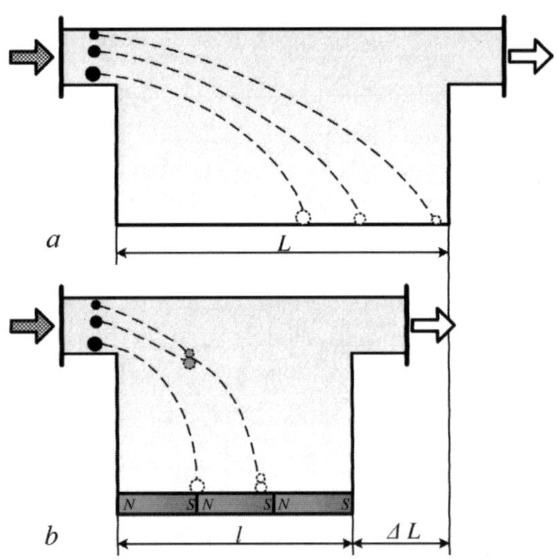

Figure 1. Sedimentation of magnetic particles in sumps: *a* - without a magnetic field, *b* - with a magnetic field

As we can see, the particles are attracted to the pole (areas with the greatest intensity), thereby reducing the distance that requires deposition to the bottom of the sump. The other two particles coagulated, which significantly reduced the distance traveled by the particle. Summarizing these two factors, we can say that the use of a magnetic field in a continuous flow sump reduces its length by ΔL while maintaining the degree of purification.

There are many designs of magnetic sumps [4-6], but three basic ones can be distinguished, shown in figure 2.

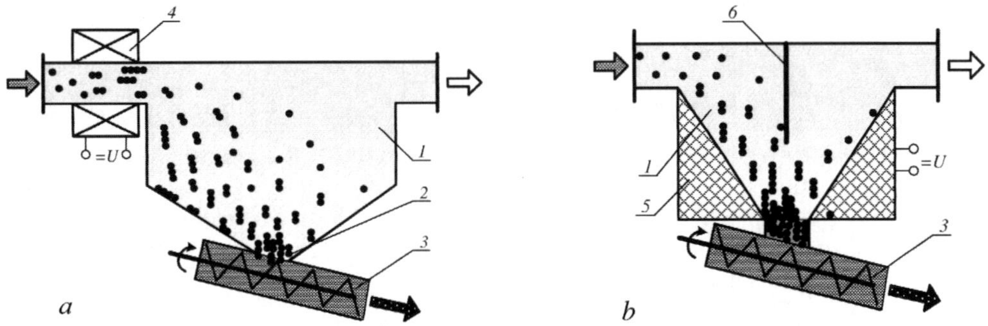

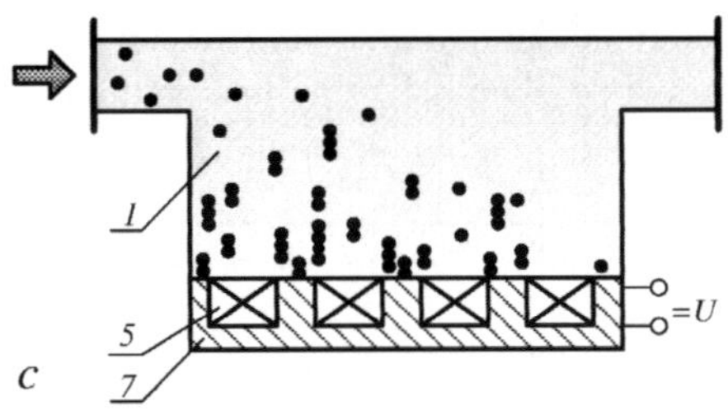

Figure 2. Basic designs of magnetic sumps: *a*) sump with electromagnetic coagulator; *b*) sedimentation tank-coagulator; *c*) precipitation tank (1 – working chamber, 2 – slurry, 3 – screw for discharge of sludge, 4 – electromagnetic coagulator, 5 – coil, 6 – baffle, 7 – ferromagnetic core)

The first type uses magnetic fields for precoagulation of the particles at the inlet. The coagulated particles are deposited under gravity to the bottom, where they are removed from the working chamber by means of various devices (in this case by means of a screw). This design is characterized by a minimum consumption of electricity, but at the same time, it has defects. The main disadvantage of a sump with an electromagnetic coagulator is that the destruction of the floccule can occur in the working chamber or under the effect of the mixing of the liquid, or in the case where the mechanical impurities belong to soft magnetic materials.

The second type differs from the previous one in that coagulation of the particles does not take place at the inlet, but in the working chamber, thereby preventing the floccules from particle decay. In addition, the coil is wound in such a way that the strength of the magnetic field increases with the approach to the bottom of the sump. Thereby not only gravity, but also magnetic force will effect on the particle. Baffle serves to ensure that the flow of cooling lubricant passes through the entire working chamber; otherwise the effect of the magnetic field on the mechanical impurities will be minimal. The disadvantages of this type of sumps are that large material and energy costs for the electromagnetic system. So this type is used extremely rarely, despite the high degree of purification.

The process of particle coagulation in a magnetic precipitation tank takes place directly in the working chamber. The coagulated particles settle at the poles of the electromagnetic system under the effect of gravity and magnetic forces. Advantage of a precipitation tank is that it also captures soft magnetic impurities. This type of magnetic sumps has become the most widespread because it provides high purification efficiency at relatively low material and energy costs. However, this design of magnetic sumps does not have a convenient slurry collector and discharge of slurry can be made either through the lid of the collector or through the sidewall.

Magnetic sumps often have baffles and non-magnetic nozzles [1, 7, 8] in order to let the entire contaminated stream pass by a magnetic field with high strength.

Calculation of pre coagulation units is rather complicated, the formed floccules have an unequal number of particles, and therefore the application of the Stoke's law, which is used in calculating the process of gravity separation (clarification), is incorrect, since it is applicable only to spherical particles [9]. A similar situation exists in the calculation of other magnetic clarifiers, so we have to use the criteria for performance evaluation of the magnetic field on the process of purifying of technical fluids, among which a special place is occupied by the cooling lubricants [10].

A formula for clarifying liquids in clarifiers, in which there are mechanical impurities (for the Stoke's law application) is well-known.

$$-\frac{dn}{dt}=\frac{nU_0}{h} \quad (3)$$

for $\tau(0)=0$, $n(0)=n_0$, h=const, U_0=const,
where n_0, n - initial and current value of particles in a clarifier, units; h – height of clarifier, m; U_0 - the particle deposition rate, m/s.

In the case of high polydispersity of particles, U_0 is calculated for each fraction (for each particle size) separately. Thus, additional theoretical and experimental studies are necessary to determine the velocity of floccules formed under the effect of a magnetic field.

5. Experimental technique

The results of an experimental study of the deposition of floccules, which were formed after the effect of a magnetic field on magnetic particles are given below. The scheme of experimental unit and the results of the research are shown in figure 3.

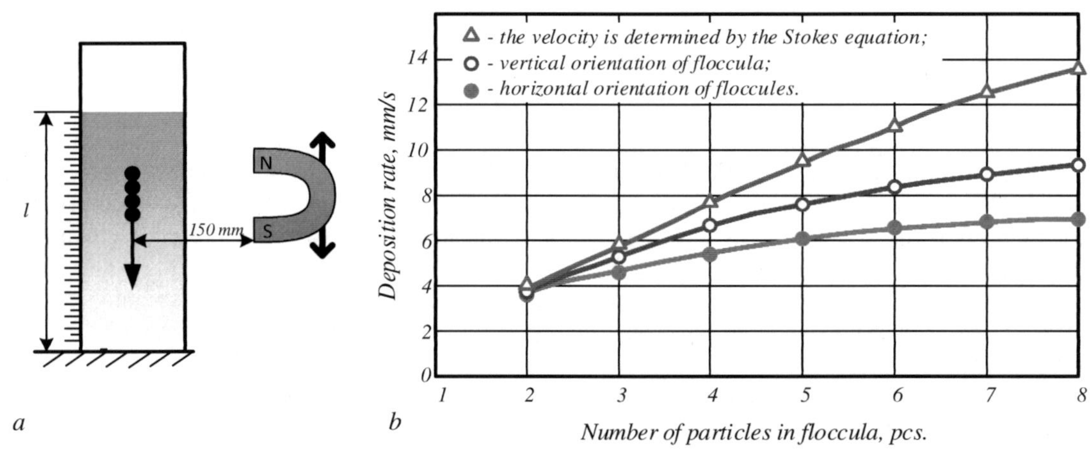

Figure 3. *a*) the scheme of unit for research of the hydraulic resistance to the motion of particles for different orientations in a viscous medium; *b*) the dependence of the rate of deposition of floccules with horizontal and vertical orientation.

Experimental conditions: spherical particles (d=1mm), medium-glycerin, permanent magnet for orientation of floccules during deposition, thermometer, stopwatch. The motion of single, double, built particles, etc. has been monitored. The permanent magnet moved simultaneously with the deposition of bodies at a distance of 150 mm from the flask, so the magnetic field had no effect on the horizontal and vertical movement of the particles. Their deposition passed only under the gravity. Each body was deposited repeatedly (from 7 to 12 times) in horizontal and vertical positions.

6. Results and discussion

Analysis of the experiment showed that the difference in precipitation rates increases with the increase in the number of particles in the floccule. A further increase in the number of particles leads to a stabilization of the difference, which is quite understandable physically due to the equalization of the drag coefficient.

The gravity was counteracted by the medium's resistance force, which is calculated by the following formula [11]:

$$F_d = \lambda S \rho \frac{U^2}{2} \tag{4}$$

where λ - the coefficient of medium resistance; S – the cross section of the particle, m^2; ρ – the density of the medium, kg/m^3; U – the velocity of the particle, m/s.

The resistance force of the medium for particles of spherical shape is calculated by the Stoke's formula [12]:

$$F_d = 3\pi\eta d_d U \tag{5}$$

where η – the dynamic viscosity of the medium, Pa·s; d_d – the diameter of the spherical particle, m.

Despite the fact that the equivalent diameter and the shape factor of the floccule for different orientations are the same, the deposition rates differ from each other, so bringing the particles to an equivalent sphere and finding the shape factor is meaningless.

The specified numerical values of speed and deposition depending on the number of particles in the floccule - vertically oriented particles deposit faster than horizontally oriented ones.

Based on formula (4), the highest deposition rate should be observed with a vertical orientation of particle, because the cross section of the floccule will be minimal (see table 1). However, the resistance of the medium to the lateral surface of the floccule is not taken into account (this is especially noticeable in the vertically oriented particle). This makes the prediction of the deposition rate from the cross section incorrect. Experimental data confirm fully the last statement.

Table 1. The values of the cross section of the floccule with its different orientations and equivalent diameter

Orientation	Value
Vertical	$\pi d^2/4$
Horizontal	$n\pi d^2/4$
Equivalent diameter	$n\pi^{2/3} d^2/4$

where d – the diameter of one particle in the floccule, m; n – the number of particles in the floccule, unit.

Let's represent the results in the form of a sign model, finding it by the rate of deposition of the floccules U_F:

$$U_F = U_0 f(n_F) \tag{6}$$

where U_0 – the deposition rate of a particle of spherical shape, m/s; n_F – the number of particles in the floccule, units

$$n_F = l_F / d_F \tag{7}$$

where l_F, d_F – length and diameter of the floccule, respectively, m; $f(n_F)$ – the functional dependency, which is found from the experiment.

$$U_{F.V} = U_0 \left(0,455 \cdot \exp(-0,263 \cdot n_F) + 0,644\right) \tag{8}$$

$$U_{F.G} = U_0 \left(0,782 \cdot \exp(-0,389 \cdot n_F) + 0,462\right) \tag{9}$$

Coefficient of determination (R^2) for the deposition rate of floccules with horizontal orientation is 0.996, with vertical one - 0.998.

Let's represent the results in the form of a sign model, having found it in terms of the coefficient of aerodynamic resistance λ_F:

$$\lambda_F = \lambda_0 f(n_F) \tag{10}$$

Generally, it can be found on the basis of the equation of the resistance and gravitational forces:

$$\lambda_F = \frac{2Mg}{SU^2} \tag{11}$$

where $M = m_i \cdot n_F$ – mass of floccule, kg.

For vertical orientation, the dependency will be as follows:

$$\lambda_F = \lambda_0 \cdot \left(2,227 \cdot \exp(-0,952 \cdot n_F) + 0,139\right) \tag{12}$$

Coefficient of determination (R^2) for the coefficient of aerodynamic resistance is 0.998. Graphically, the dependency is shown in figure 4. The values of λ_F for a horizontal orientation will be independent of the number of particles in the floccule.

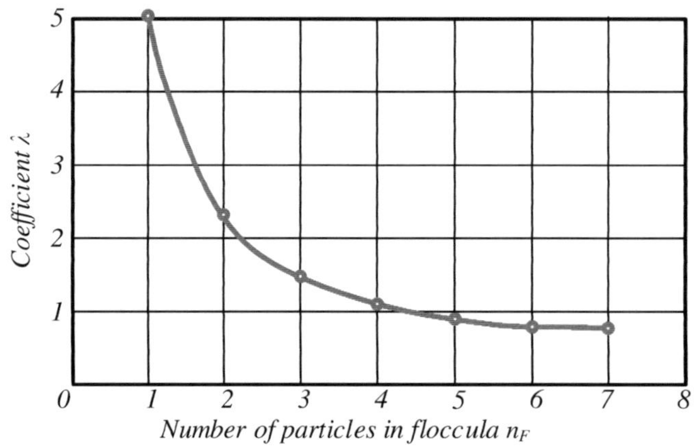

Figure 4. Dependency $\lambda = f\left(n_F\right) = f\left(l_F / d_F\right)$ with the vertical orientation of particles.

It is advisable to use the results in engineering practice for the calculation of magnetic sumps and for mathematical modeling of the kinematics of magnetic coagulation processes and particle deposition.

7. Conclusion

The article deals with the problem of cleaning cooling lubricants from magnetic impurities in magnetic sumps. The analysis of existing designs is made and the basic design versions of purifying devices of gravitational type are represented. Experimental dependencies of the deposition rate and the drag coefficient on the number of particles in the floccule are found. It is proved that the use of only the shape factor and the equivalent diameter is not enough for calculations. The results are recommended to be used in the calculation of magnetic sumps.

References

[1] Aleksandrov E E, Kravec I A, Lysikov E N and other 2006 *Increase the resource of technical systems through the use of electric and magnetic fields. Monograph* (Khar'kov: NTU KhPI) p 544

[2] Zhukov V, Masyutkin E, Avdeyev B 2017 *IOP Conf. Ser.: Mater. Sci. Eng.* **177** 012015

[3] Avdeyev B, Masyutkin E, Golikov S, Sokolov S and Gavrilov V. 2017 *IEEE Conf. of Rus. Young Res. in Electric. and Electron. Eng.* **1225**

[4] Sokolova E A, Dzhioev G A 2017 *IOP Conf. Series: Conf. Ser.: Mater. Sci. Eng.* **177** 012037

[5] Sokolova E 2017 *MATEC Web Conf.: Int. Conf. on Mod. Trends in Manuf. Technol. and Equip.* **129** 03008

[6] Augusto P A, Castelo-Grande T and Augusto P 2005 *Chem. Eng. J.* **111** 85

[7] Sokolov A A, Miroshnikov A S, Sokolova E A 2016 *Gornyi Zhurnal* **12** 83

[8] Chernyi S 2016 *Oil Industry* **2** 106

[9] Masyutkin E P, Prosvirnin V I and Avdeyev B A 2012 *Eastern-European J. Enterprise Technol.* **5/8 (59)** 52

[10] Avdeyev B 2017 *MATEC Web Conf.: Int. Conf. on Mod. Trends in Manuf. Technol. and Equip.* **129** 06012

[11] Yunus A Cengel Dr and Cimbala J M 2014 *Fluid Mechanics Fundamentals and Applications* (New York: The McGraw-Hill) p 1006
[12] Chernyi S 2016 *IOP Conf. Series: Conf. Ser.: Mater. Sci. Eng.* **124** 012048

IOP Conf. Series: Journal of Physics: Conf. Series **1050** (2018) 012006 doi:10.1088/1742-6596/1050/1/012006

Increase of steel surface operational properties by combined hardening

Yulia Bakhracheva

Volgograd State University, Prosp. Universitetsky, 100, Volgograd, 400062, Russia

E-mail: bakhracheva@volsu.ru

Abstract. . This article examines the influence of laser heat treatment of nitrocementation steel on the phase composition, structure and hardness of surface layers. It is shown that the combined heat treatment of steels – nitrocementation and laser hardening allows to provide high wear resistance of surface layers of steel.

1. Introduction

Mining, including oil and gas, requires a large amount of drilling. Drilling tools - drill bits of various types and designs readily used when conducting drilling operations. This is a consequence of the low resistance of drilling bits as well, and many other parts of the drilling equipment [1-3].

Thus, highly relevant is the development and improvement of technologies of chemical-thermal processing of drill bit steels, which provide improved abrasive wear resistance [3-7].

The impact of concentrated energy fluxes on materials currently receiving attention [8-9]. One of the directions of development of this technology is the laser heat treatment case hardening (nitrocementation) steel, which allows to increase the hardness, strength and wear resistance of the surface layers.

2. Experiment

The aim of this work is to research the influence of laser heat treatment nitrocementation layers on the phase composition, structure and hardness of surface layers of low carbon steel 18CrNi3Mo.

Nitrocementation with subsequent laser heat treatment was subjected the specimens of low carbon steel 18CrNi3Mo (0.16-0,18 % C; 3.4 % Ni; 0.8 % Cr; 0.52 % Mo; 0.43 % Mn; 0.35 % Si; 0.06 % Al; 0.008 % S; 0.012 % P; 0.015 % N; 0.01 % O; 0.01 % H).

3. Results and discussion

In nitrocementation layers with high nitrogen content, especially in obtaining layers with a thickness more than 1.2 mm, the formation of clusters of carbonitrides, oxides, troostite and bainite structures (Fig. 1, a). These structural and phase formation reduces the hardness, toughness, plastic characteristics, surface contact fatigue, wear resistance in conditions of impact-cyclic loading and under conditions of sliding friction. Application of laser heat treatment nitrozomocevina layers without melting the surface allowed to form a structure with high amounts of residual austenite (up to 80 – 100 %), crushed grain, eliminate troostite and bainite structure (Fig. 1, b), improve the hardness, ductility and wear resistance of the diffusion layer.

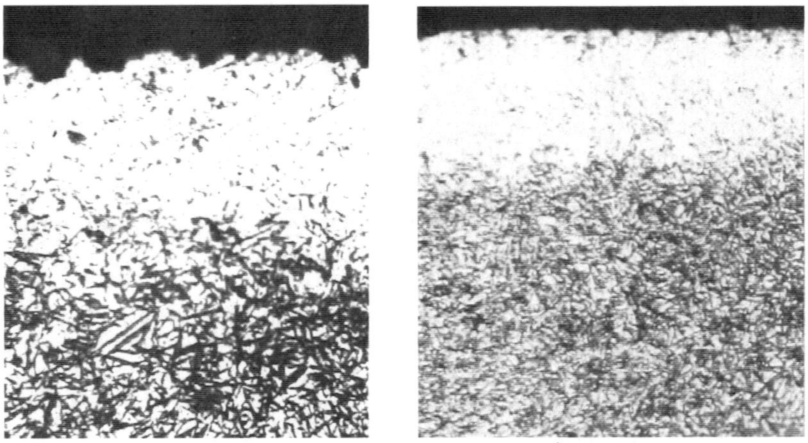
a b

Figure 1. Microstructure nitrocementation layers of steel 18CrNi3Mo (X 400)
a – before laser treatment; b – after laser treatment

In nitrocementation layers with high nitrogen content, especially in obtaining layers with a thickness more than 1.2 mm, the formation of clusters of carbonitrides, oxides, troostite and bainite structures (Fig. 1, a). These structural and phase formation reduces the hardness, toughness, plastic characteristics, surface contact fatigue, wear resistance in conditions of impact-cyclic loading and under conditions of sliding friction. Application of laser heat treatment nitrozomocevina layers without melting the surface allowed to form a structure with high amounts of residual austenite (up to 80 – 100 %), crushed grain, eliminate troostite and bainite structure (Fig. 1, b), improve the hardness, ductility and wear resistance of the diffusion layer.

Study of thin foils made of nitrocementation layers with subsequent laser heat treatment at a distance of 0.15 and 0.30 mm from the surface revealed the presence of a large number of selections of various dispersity and shape. The distribution of precipitates is uniform, their size is from 100 to 5 nm or less.

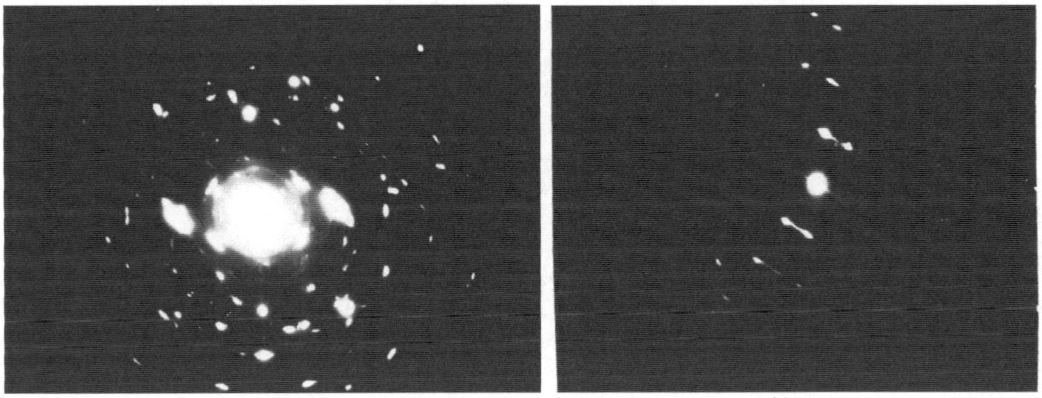
a) b)

Figure 2. Electronic diffraction (a) and microdiffraction (b) nitrocementation layers with subsequent laser heat treatment of steel 18CrNi3Mo

When the content of nitrogen in solid solution from 0.3 to 0.8 % of the identified region in the form of small ripples. Electronic diffraction and microdiffraction these plots revealed the reflexes of the matrix (Fig. 2).

Analysis of the physico-chemical and physico-mechanical characteristics nitrozomocevina layers showed that the nitrocementation followed by laser treatment has a significant influence on the properties of friction surfaces. From the presented data it is seen (Fig. 3) that the highest wear resistance have nitrocementation layers with high nitrogen content after the laser heat treatment. This is due to the particular influence of high nitrogen content on the formation before the test structure with high strength, ductility, stability and on the kinetics of changes of structure and properties of diffusion layers in the process of testing for wear. In nitrocementation layers with nitrogen content of 0.5 – 1.0 after laser hardening the structure with the content of nitrogenous austenite 80 – 90 % undergoing the process of wear of the $\gamma \rightarrow \alpha$ transformation is not more than 5 – 10 % and has a greater ability to deformation hardening due to wear (Fig. 3).

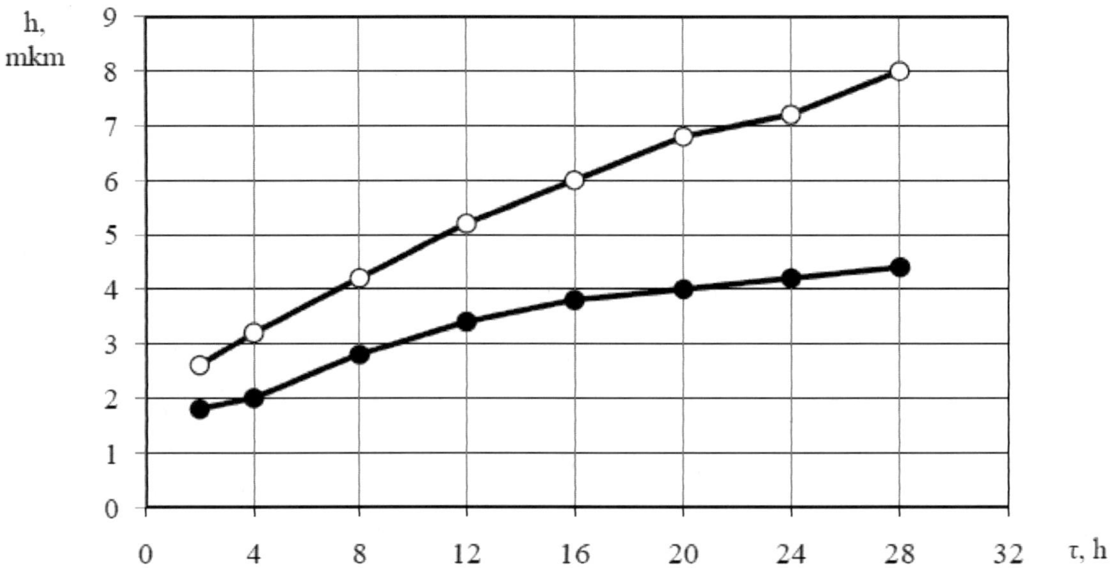

Figure 3. The average depth of the worn hole nitrocementation layers
○ – before the laser treatment; ● – after laser treatment

In nitrocementation layers with nitrogen content of 0.5 – 1.0 after laser hardening the structure with the content of nitrogenous austenite 80 – 100 % undergoing the process of wear of the $\gamma \rightarrow \alpha$ transformation is not more than 5 – 10% and has a greater ability to deformation hardening in the process of wear than the diffusion layer with nitrogen concentration of 0.3 – 0.4 %. For example, the level of hardness on the surface of the wear nitrocementation steel with a nitrogen content of 0.7 – 1.0 % treated laser reaches 1200 HV(Fig. 4).

High level of hardness while maintaining a large amount of residual austenite in nitrocementation steel with a high content of nitrogen, treated with laser, provides a higher wear resistance.

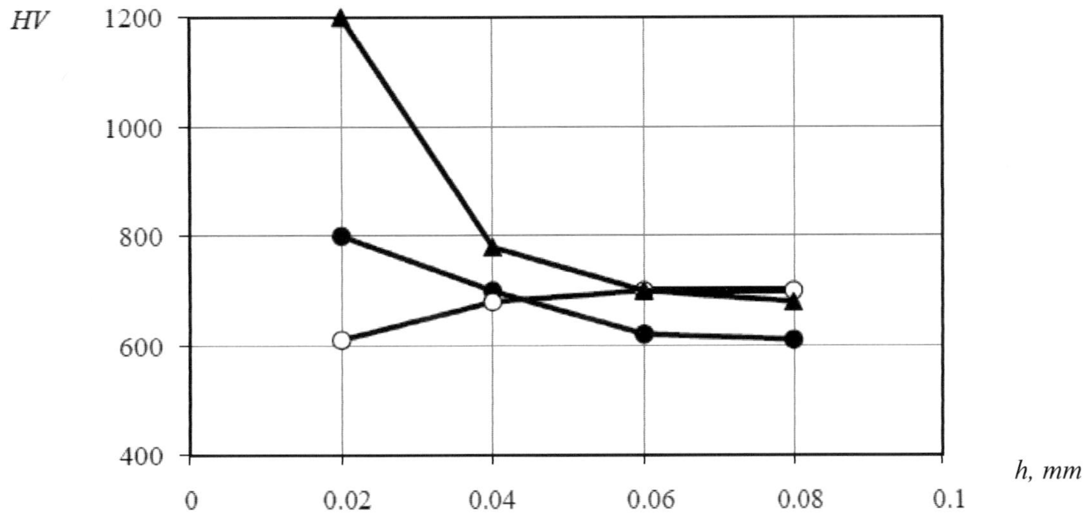

Figure 4. The distribution of hardness in thickness nitrocementation layers of steel 18CrNi3Mo
○ – nitrocementation; ● – nitrocementation and laser treatment (to test)
▲ – nitrocementation and laser treatment (post-test)

4. Conclusion

The analysis of the research shows that combined heat treatment of steels – nitrocementation with a high content of nitrogen and laser hardening allows to form a surface structure consisting of austenite strengthened by fine precipitates. This allows, along with the required depth contact strength to provide good performance properties of surface layers of metal products.

5. References

[1] Shapochkin V I et al. 2011 Effect of nitrogen content on the structure and properties of nitrocarburized steel *Metal Science and Heat Treatment* **52** pp 413–419

[2] Semenova L M et al. 2013 Laws of formation of diffusion layers and solution of the diffusion problem in temperature-cycle carbonitriding of steel *Metal Science and Heat Treatment* 55 pp 34–37

[3] Khokhlov V M 1996 Wear laws at elastic interaction *Russia Engineering Research.* **16** pp 11–12.

[4] Bakhracheva Ju S 2013 Fracture toughness prediction by means of indentation test *International Journal for Computational Civil and Structural Engineering* **9** pp 21–24

[5] Baron A, Bakhracheva J 2007 A method for impact strength estimation *Mechanika* **66** pp 31–35

[6] Bakhracheva Yu S 2014 The Method for Lif etime Estimation Through the Mechanical Properties in Tension *Science Journal of Volgograd State University. Technology and Innovations* **11** pp 27–32. DOI: 10.15688/jvolsu10.2014.2.4

[7] Bakhracheva Y S, Vasilyev A V, Petikova T N 2016 Solution to the diffusion problem in the thermocyclic nitrocementation of steel *J. Materials science forum* **870** pp 545-549 DOI: 10.4028/www.scientific.net/MSF.870.545

[8] Baron A, Bakhracheva J, Osipenko A 2007 Fracture toughness estimation by means of indentation test *Mechanika* **67** pp 33–36

[9] Vasilyev A V, et al. 2014 Valve Cam Design Using Numerical Step-by-Step Method. *Vestnik Volgogradskogo gosudarstvennogo universiteta. Seriya 10, Innovatsionnaya deyatelnost [Science Journal of Volgograd State University. Innovation Activity]* **10** pp 26–32

[10] Vasilyev A V et al. 2016 Dynamics simulation model for the internal combustion engine valve gear *Procedia engineering* **150** pp 312–317. DOI:10.1016/j.proeng.2016.07.017

IOP Conf. Series: Journal of Physics: Conf. Series **1050** (2018) 012007 doi:10.1088/1742-6596/1050/1/012007

The Dynamic Reactions of the Piston Machines Mechanical Drive Joints

P D Balakin and V N Belkov

Omsk State Technical University, 11, Mira ave., Omsk 644050, Russia

Abstract. On the basis of the mechanics fundamental provisions the high-speed cycle machines mechanical system joints backlashes are shown to be shifted. This is the impact process in which the dynamic reactions joints become greater by the values of kinetic and static reactions and it should be taken into account when carrying out the engineering calculations of machine elements. The importance of determining the impact speed as the criterion for the impact zone materials plastic deformation starting is represented. For reducing the impact process harmful influence on the joints potential, the wide use of engineering solutions concerning the automatic adjustment of developing joints backlashes is recommended.

Key-words: mechanical system, joints, backlash, impact process, dynamic reactions

1. Introduction

The joints kinetic and static reactions take into account the rings variable external and inertial loadings and in the absence of (not considering) the mechanical system joints backlashes express the nature of rings and joints loading of the ideal system movement as a first approximation, which, in most cases, would be enough for the general-purpose machines mechanical system design stage.

Having completed the system with forces and friction forces moments of joints, the second approximation of the system construct strength loading is obtained.

In movable geometric constraints of the real mechanical systems, the backlashes growing during the machine operation due to the joints surfaces non-repairable wear are normalized depending on the type of the active surfaces matching. The joints backlashes cause the additional dynamic loading of the joints elements.

Due to the cycle machines mechanical system backlashes shift, the additional dynamic loading is impacting in nature, the impact loads amounts of the high-speed systems can multiply exceed the kinetic and static reactions values, their knowledge is absolutely necessary for the engineering calculation of joins elements, as well as for the development of the growing backlashes automatic adjustment devices, impact energy dampening devices and, generally, for the development of the engineering solutions on the force impact process harmful effect reducing, which occurs in all the cycle machines mechanical systems joints.

The examination of the extremely undesirable conditions possible occurrence of the joints active surfaces plastic deformation starting and their destruction has the particular importance when defining the impact process parameters.

2. The problem statement

The problem defining the impact process parameters when shifting the single joint backlash is stated and solved using the example of the numerical calculation, thus, the similar study algorithm is represented, the impact loading comparative estimation is given, as well as the joint active surfaces plastic deformation possible start conditions examination is conducted.

3. Theory

For the impact process parameters defining, the fundamental regularities based on the momentum conservation principle are used. In order to determine the impact speed at the backlash shifting, the ideas of the complex motion decomposition into the components will be used.

The postulates of the solids impacting classical theory are used as the assumptions. The solids are considered as rigid, and if the bodies interaction is instantaneous, then the impact force is assumed to be infinitely great. In our calculations the restitution coefficient *R* is taken to be zero which corresponds to the perfectly inelastic impact conditions.

Moreover, the hypothesis having been experimentally confirmed [1, 2] that metals and alloys elastic characteristics (modulus of elasticity, Poisson ratio) do not depend on the loading speed, while the yield point, consequently the plastic deformations threshold depend on the impact speed, which limit value is absolutely possible to be calculated, is accepted in the study.

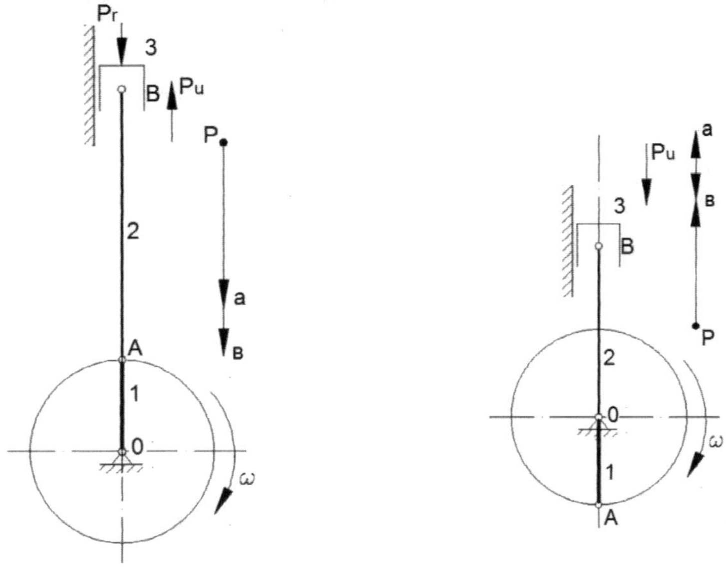

a) b)

Figure 1. To the calculation of the slider inertia forces in the extreme positions:

1 – crankshaft, 2 – connecting rod, 3 – slider.

For calculation the actual mechanical systems joints reactions, the study of the motion conversion kinematic model motion parameters of the particular reciprocating mechanism in which the reciprocating motion of one ring is realized, is considered.

The elementary crank-slider scheme used as the main mechanism of compressors, internal and external combustion engines and other machines without restrictions is chosen as an example (Fig. 1).

The scheme parameters: the slide (piston) *H* stroke 3 equals to *80 mm* , therefore, the crankshaft kinematic size is $\ell_{OA}=R=40$ *mm*; $\ell_{AB}=120$ *mm*; the crankshaft maximum rotation speed is $n_{max}=5500$ *rpm;* the slider mass m_3 is *0,5 kg*, the slider diameter is *80 mm*.

The backlash shifting only in connection of the connecting rod 2 with the slide 3 is examined, having taken the backlash size as follows $\Delta = 0.1$ *mm*.

It is evident that the backlash is shifted in all kinematic pairs including the rectilinear pair «slider-shaft» as demonstrated by hodographs and kinetic and static reactions diagrams built during the mechanism operation cycle.

IOP Conf. Series: Journal of Physics: Conf. Series **1050** (2018) 012007 doi:10.1088/1742-6596/1050/1/012007

Figure 1 shows two extreme positions of the slider 3 in which the examined joint backlash is shifted and the slider speed direction is reversed. Linear acceleration of the point B is defined according to the known vector equation:

$$\bar{a}_B = \bar{a}_A + \bar{a}_{BA}^n + \bar{a}_{BA}^C. \tag{1}$$

$\ell_{OA} = R$ and $\omega = \pi n/30$, at $n = n_{max}$, we obtain $\omega = 575$ 1/s, and $a_A = 13225$ m/s^2.

In the considered extreme position (figure 1a), $a_{BA}^n = \frac{V_{BA}^2}{\ell_{AB}}$, but $V_{BA}=V_A=\omega \cdot R=23$ m/s, then $a_{BA}^n = 4408 m/s^2$, and $a_{BA}^\tau = 0$. Having folded the equally directed accelerations $\bar{a}_A$ and $\bar{a}_{BA}^n$, a_B=17633 m/s^2 is derived.

In the mechanism position in figure 1b, the point B acceleration determines the components values difference: a_B=13225 m/s^2-4408 m/s^2=8817 m/s.

Subsequently, by using the value a_B=17633 m/s^2 and considering it to be constant in the backlash sample motion Δ, $\Delta = a_B t^2/2$ or $t^2 = 2\ \Delta/a_B$ are obtained having substituted the initial data numerical value, the impact time is derived t=0,0001 s or t=10^{-4} s, that corresponds to the values given in [1-4].

The impact speed is $V = a_B t = 1{,}7$ m/s.

The momentum conservation principle of the finite increments has the following form:

$$P_{dyn} \Delta t = m \Delta V \tag{2}$$

where P_{dyn} is the dynamic reaction or impact force when shifting the backlash $P_{dyn} = \frac{0.5 \cdot 1.7 m/s}{10^{-4} s} = 8500 H.$

For comparison: gas operating pressure at the compression level of 10 will amount to 10 atm or 1MPa, and gas pressure force P_g applied to the slide (piston) will be equal to the following: $P_g = 1\frac{H}{mm^2} \cdot 3.14(40mm)^2 = 5024H$. Thus the module gas force is less than the dynamic reaction.

The impact limiting speed when adjusting the backlash at which the impacting bodies (rings) plastic deformation start is possible, is defined.

By using the connection of the impact velocity and deformation according to the sources [1, 2] the following equation is derived:

$$V = a\varepsilon \tag{3}$$

where a is the deformation wave propagation speed; ε is the material compression relative deformation in the impact zone.

Whereas, the deformation wave propagation speed is defined by the following expression [1]

$$a = \sqrt{\frac{E}{\rho}} \tag{4}$$

where E is the elasticity modulus, for steel $2 \cdot 10^5$ *MPa*; ρ is the material density, for steel $\rho = 7.8 \cdot 10^3$ *kg/cm*^{3}; having aligned the physical values dimension, a = *5000 m/s* is obtained.

The impact velocity limit value in accordance with the equation (3) is V=*5 m/s*, and in the joint example under consideration it is V=*1.7 m/s* which represents the examined joint materials plastic deformation conditions absence, however, when increasing threefold the initial backlash *Δ* up to 0.3 mm, the joint backlash non-wearing development mechanism is started [5, 6].

4. Results discussion

The conducted study has shown the real high-speed cycle machines mechanical system joints backlashes to be adjusted and this process is an impact one at which the joints dynamic reactions have values exceeding the kinetic and static reaction ones.

Besides, the hypothesis that mechanical system rings and joints inertial loading is possible to exceed the strength loading by the values by the useful loading from the machine operating member, is confirmed. It should be considered when carrying out the machine elements design calculations.
The impact process characteristics are largely determined by the impact time, therefore the automatic adjustment techniques of the growing backlashes in operation are relevant for the high-speed machines during the implementation, such techniques have been developed and engineering solutions for their implementation exist.

5. Conclusions

- By applying the case study, the backlash adjustment process examination algorithm is represented and the backlashes shifting of the high-speed cycle machines mechanical system joints is proved to have the impact character.
- The real mechanical systems joints dynamic reactions values exceed the kinetic and static reactions ones and it should be taken into account in conducting the design calculations of machine elements.
- The joints elements plastic deformation process possible start control is necessary and relevant as the variety of construction materials is constantly expanding, and limiting impact speed knowledge is the defining criterion both in determining the initial backlash and in forecasting the operational connection resource.

6. References

[1] Ponomarev S D 1959 *The strength calculation in machine building* (Moscow: GNTI)
[2] Ponovko J G 1976 *The fundamentals of the oscillation and impact applied theory* (Leningrad: Machine building)
[3] Landau L. D., Lifschitz E. M. Mechanics. Course of Theoretical Physics. 3-rd ed. (London: Pergamon) 1976 Vol **1** 198 p
[4] Lurie A. L. Analytical mechanics. Foundations of Engineering Mechanics (Berlin: Springer) 2002 864 p
[5] Balakin P D 2017 Dynamic Reactions of the Real Mechanical Systems Joints *Dynamics of systems, mechanisms and machines* vol **5** pp 13–6
[6] Balakin P D 2016 *The Real Mechanical Systems Theory Elements* (Omsk: OmGTU) 272 p

IOP Conf. Series: Journal of Physics: Conf. Series **1050** (2018) 012008 doi:10.1088/1742-6596/1050/1/012008

Universal technique of elimination of power parametric excitation in general mechanical systems

P D Balakin, I P Zgonnik

Omsk State Technical University, 11, Mira ave., Omsk, 644050, Russia

E-mail: tmm@omgtu.ru

Abstract. A universal technique to transform general mechanical systems in rotative ones with a constant equivalent inertial characteristic is proposed. The technique is based on attaching of a kinematic chain with a program change of inertial characteristic to a reduction link. The inertial characteristic in addition to the main one forms the reduction link with a constant equivalent moment of inertia. Such modification of a mechanical drive eliminates parametric power drive excitation, reduces the level of dynamic load in links and drive joints, that results in reliability and system resource improvement.

1. Introduction

General issues of dynamics in mechanical systems are well studied in fundamental literature [1-9].

It is known that a dynamic simulation of mechanical systems motion is based on using a concept of a reduction link to which inertial characteristics of the system and external forces causing motion are reduced.

The variable equivalent inertia moment of the reduction link is a source of parametric excitation of inertial origin resulting in the increase and decrease of inertial characteristic of the reduction link in the system with a cyclic motion of links twice for one revolution of the reduction link. Thus, twice for the revolution it brings to the increase and decrease of motion velocity in the reduction link that generates the power moment of inertia forces in addition to the main flow, loading links and joints of the drive in general mechanical system.

2. Problem statement

Let us set and solve the problem of complete elimination of inertial parametric excitation in a general mechanical system by transformation of a general system with a variable value of inertial characteristic into a rotative system with a constant value of inertia equivalent moment.

The transformation of general mechanical system into the rotative one can be fulfilled by means of additional, embedded kinematic chain with a program variable reduced to a reduction link by the inertial characteristic, which combined with the equivalent moment of inertia in the main kinematic chain forms a constant equivalent moment of inertia in the modified system.

3. Theory

Lagrange equation for a reduction link of one-degree-of-freedom nondamper system with angular generalized coordinate φ and rigid links is

$$J_{eq}\frac{d\omega}{dt}+\frac{1}{2}\frac{dJ_{eq}}{d\varphi}\omega^2 = Mg^{eq} - Mc^{eq},$$

or

$$J_{eq}\ddot{\varphi}+\frac{1}{2}\frac{dJ_{eq}}{d\varphi}\dot{\varphi}^2 = Mg^{eq} - Mc^{eq}. \quad (1)$$

where J_{eq} is a variable equivalent inertia moment of a reduction link defined from the equivalent of a kinetic energy of the reduction link and a kinetic energy of movable links of the system. φ is a generalized coordinate of the system; $\omega = \dot{\varphi}$ is an angular velocity of the reduction link; t is time; $Mg^{eq} - Mc^{eq} = M^{eq}_{exc}$ is a difference of equivalent power characteristics of external forces (excess moment of external forces).

Differential equation (1) is a non-linear second-order equation with variable coefficients before derivatives and, generally, has no analytical solution.

In the rotative system, the second component of the equation (1) becomes zero, and the equation (1) is simplified to the following:

$$J_{eq}\frac{d\omega}{dt} = Mg^{eq} - Mc^{eq}, \quad (2)$$

from which you can see that velocity vibrations of reduction link motion are entirely defined only by a power function of mismatch in external forces reduced to a reduction link – the differences of moments in equivalent driving forces and useful resistance forces. The equation (2) for most combinations of power characteristics and their linearization has the analytical solution

Thus, if external equivalent forces depend on the motion velocity of a reduction link, i.e. when

$$J_{eq}\frac{d\omega}{dt} = Mg^{eq}(\omega) - Mc^{eq}(\omega),$$

we can obtain an integral expression, having divided variables

$$\int_{t_0}^{t_1} dt = J_{eq}\int_{\omega_0}^{\omega_1}\frac{d\omega}{Mg^{eq}(\omega) - Mc^{eq}(\omega)}$$

and, if difference can be presented as integrable function, that always can be achieved in linearization of components, the solution gives a series of motion time values in a function of velocity, i.e.

$$t_1 = t_0 + J_{eq}\int_{\omega_0}^{\omega_1}\frac{d\omega}{Mg^{eq}(\omega) - Mc^{eq}(\omega)}, \quad (3)$$

this series of values is easily transformed into inverse function $\omega=\omega(t)$.

Similarly, the solution takes place for power characteristics depending on time:

$$J_{eq}\frac{d\omega}{dt} = Mg^{eq}(t) - Mc^{eq}(t)$$

and

$$\int_{\omega_0}^{\omega_1} d\omega = \frac{1}{J_{ep}}\int_{t_0}^{t_1}\left[Mg^{eq}(t) - Mc^{eq}(t)\right]dt$$

or

$$\omega_1 = \omega_0 + \frac{1}{J_{eq}}\int_{t_0}^{t_1}\left[Mg^{eq}(t) - Mc^{eq}(t)\right]dt. \quad (4)$$

In case of constant power characteristics values or their dependence on generalized coordinate φ, selecting Δφ, we can use integral expression:

IOP Conf. Series: Journal of Physics: Conf. Series **1050** (2018) 012008 doi:10.1088/1742-6596/1050/1/012008

$$\frac{1}{2}J_{eq}(\Delta\omega)^2 = \int_{\varphi_0}^{\varphi_1}\left[Mg(\varphi) - Mc(\varphi)\right]d\varphi,$$

hence

$$\Delta\omega = \sqrt{\frac{2\int_{\varphi_0}^{\varphi_1}\left[Mg(\varphi) - Mc(\varphi)\right]d\varphi}{J_{eq}}}$$

and

$$\omega_1 = \omega_0 + \Delta\omega. \quad (5)$$

If the machine has a combination of power characteristics dependent on various arguments, then the solution is possible only with the usage of numerical methods and linearization of power characteristics on intervals of integral computation.

Generalized coordinate φ is divided into small intervals taken as an integration step. On each interval of function for equivalent power moments, Mg_{eq} and Mc_{eq} are assumed as constants, their difference can be written, for example, as

$$Mg^{eq}(\omega) - Mc^{eq}(\varphi) = M(\varphi,\omega),$$

where the initial value $M(\varphi,\omega)$ is determined by initial values $Mg^{eq}(\omega_0)$ and Mc^{eq} (φ_0) with the subsequent step-by-step calculation of change in motion velocity of the reduction link and, as a result, the subsequent determination of level of dynamic loading in links and joints of a mechanical system in a machine drive.

4. Technical realization of the hypotheses

The embedded kinematic chain with a program changeable moment of inertia can be produced on the basis of a fixed flat cam mechanism which curvilinear profile provides a variable distance from a rotation axis of a reduction link for additional movable masses, thus changing their inertial characteristic in a transfer rotary motion. This inertial characteristic combined with the equivalent moment of inertia of the main mechanism potentially provides constancy of the equivalent moment of inertia in the modified system.

Let $J(\varphi)$ be a current variable value of equivalent moment of inertia dependent on generalized coordinate φ in the main general mechanism, $J(\varphi)ad$ be the equivalent inertia moment of additional movable masses of the embedded kinematic chain, then the condition is to be fulfilled

$$J(\varphi) + J(\varphi)ad = Jconst. \quad (6)$$

In each of two components in the equation (6) constant parts determined by rotative properties of the system are removed. This system comprises a motor spindle, a reducer, couplings, a crankshaft of the main mechanism, an arm 3 of rollers 4, minimum inertia moment Jr of rollers 4 in the embedded chain relative to axis 2 (fig. 1). The value Jr can be defined as $Jr = 2\cdot m\cdot r^2_{min}$, where m is a mass of one roller; r_{min} is a minimum size of a radial arrangement of mass center in a roller relative to axis 2.

The profile of cam 1 will determine radius $\rho(\varphi)$, obtained from calculation

$$\rho(\varphi) = r_{\min} + r_r + \sqrt{\frac{J(\varphi)\mathrm{ad}}{2m}}, \quad (7)$$

where r_r is the radius of roller surface 4 sliding on a cam profile.

5. Results and discussion

The hypothesis of possible transformation of a general mechanical system in a rotative one to achieve a constant value of the equivalent inertial characteristic resulting in the elimination of parametric excitation in the system according to mathematical simulation of such transformation is technically achievable. This is proved by the inventor's certificate [10] and engineering calculation of parameters for embedded kinematic chain.

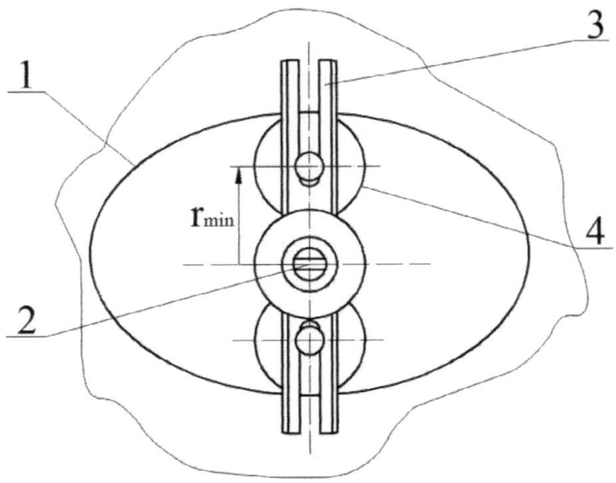

Figure 1. An embedded kinematic chain
1 – a magnetic material cam; 2 – a common rotation axis of a reduction link; 3 – an arm for placement of rollers in embedded kinematic chain; 4 – rollers of the embedded kinematic chain.

The embedded kinematic chain (fig. 1) contains the arm 3, which is rigidly fixed with the reduction link, that has the rotation axis 2 on the arm 3 in special grooves and in diametrically opposite directions there are mutually balanced pairwise rollers 4, everyone with weight "m". The rollers can move radially in arm grooves, these relative movements occur due to running of rollers on the internal curvilinear surface of the cam 1 made of magnetic material. A certain arrangement of rollers on a curvilinear surface of a cam can be provided in another way, for example, by means of resilient elements creating necessary force-closure of an upper couple.

The device works as follows. When the reduction link rotates around the axis 2, the arm 3 rotates too, carrying rollers 4 that run on the cam 1 which profile is executed according to the equation (7). Then the change of equivalent inertia moment of mechanical drive in a general system is compensated by the alteration $J(\varphi)ad$ due to the displacement of rollers 4 to the periphery of arm 3 or to its center thus, $J(\varphi)+J(\varphi)ad = Jconst$, and derivative $\dfrac{dJconst}{d\varphi}=0$.

Transformation of the general drive in the rotative one is technically realizable, it eliminates parametric power inertial excitation of the drive, the level of dynamic loading of links and joints in the drive of general mechanical system thereby decreases that leads to reliability of all elements of the drive and system resource improvement.

6. Conclusions

1) Transformation technique of general mechanical system in a rotative one by introducing an additional kinematic chain with a software-control of equivalent inertial characteristic to the system has been shown to be universal and technically implementable.
2) The constancy of inertial characteristic and equality to zero of its derivative by the generalized coordinate of modified drive means the elimination of parametric inertial power excitation in a power drive of any machines that reduces the level of dynamic loading in links and joints of the drive.
3) Calculation of program change of the inertial characteristic in the embedded kinematic chain is based on the knowledge of a variable part of equivalent moment of inertia in the main mechanism that

IOP Conf. Series: Journal of Physics: Conf. Series **1050** (2018) 012008 doi:10.1088/1742-6596/1050/1/012008

is defined by a known engineering calculation for energy criterion of mass reduction to a reduction link.

References

[1] Landau L D ang Livshits E M 1958 *Theoretical physics. Mechanics* (Moscow: GIFML) p 206
[2] Feyman R et al. 1998 *Lectures on physics. (Perseus Publishing)*
[3] Lurie A I 1961 *Analytical mechanics* (Moscow: Fizmatgiz) p 824
[4] Kolovsky M Z 1989 *Dynamics of machines* (Leningrad: Mechanical engineering) p 262
[5] Shteynvolf L I 1961 *Dynamic calculations of machines and mechanisms* (Moscow: Mashgiz) p 340
[6] Vulfson I I 1976 *Dynamic calculations of cyclic mechanisms* (Leningrad: Mechanical engineering) p 326
[7] Kreynin G V 1988 *Dynamics and control of machines. Reference book* (Moscow: Mashinostroenie) p 239
[8] Ponovko Ya G and Gubanova I I 1964 *Stability and vibrations of resilient systems* (Moscow: Nauka) p 336
[9] Zinoviev V A and Bessonov A P 1964 *Bases of machine units dynamics* (Moscow: Mashinostroenie) p 227
[10] Balakin P D and Troyan O M 1990 Refrigeration and gas machine *Certificate of Authorship* 1550295, MKU 4 F 25 B 9/00 (USSR)

IOP Conf. Series: Journal of Physics: Conf. Series **1050** (2018) 012009 doi:10.1088/1742-6596/1050/1/012009

Calculation of parameters for a resilient element of a flat-belt Continuously Variable Transmission

P D Balakin, I P Zgonnik, O S Dyundik

Omsk State Technical University, 11, Mira ave., Omsk, 644050, Russia

E-mail: tmm@omgtu.ru

Abstract. A design approach of mechanical systems with adaptation property to actual parameters is developed. In particular, a design of a flat-belt Continuously Variable Transmission (CVT) with a variable transfer function depending on the level of a transformed power flow is proposed. This is achieved by means of a control chain, built in directly in auto CVT's pulley. It is shown that if the control chain is built in both CVT's pulleys, then in a broad range of external loading change, the chain can be executed on the basis of resilient elements with a linear characteristic. It is shown that if the control chain is built in both CVT's pulleys, then in a broad range of external loading change, the chain can be executed on the basis of resilient elements with a linear characteristic.

1. Introduction

General problems of dynamics calculation in mechanical drives and, CVTs, in particular, are described in detail in fundamental literature [1-6].

Application of mechanical transmissions in machine drives based on CVT designs with a purpose implementation of a progressive method providing transmissions with adaptation property [1] to variable conditions of external loading on a design stage, gives original, easy implementable technical solutions for drives with new properties.

There are technical solutions of CVT drive designs which kinematic link sizes change depending on the level of transformed power flow, providing a stationary, perfect power performance of engine. As a rule, such systems use an additional power source; electrical, hydraulic devices are a part of controls, which significantly complicate the design of drives. The problem can easy be solved technically by using only those structural elements which functioning is based on laws of mechanics. A resilient element of the control chain built in the structure plays a special role. The resilient element deformation defines the controlled motion leading to auto-change of a transfer function of the drive depending on the level of external loading.

2. Problem statement

There are many variants of specifications to design a mechanical transmission, but we shall focus on a calculation of parameters for a resilient element of a flat-belt automatic CVT with automatic change of kinematic sizes in main links depending on the level of transferred power flow. The example is an auto-CVT according to designs [7, 8]. As a mathematical simulation showed, such design can provide stationary functioning of the engine under variable external loading conditions.

3. Theory

IOP Conf. Series: Journal of Physics: Conf. Series **1050** (2018) 012009 doi:10.1088/1742-6596/1050/1/012009

The text of your paper should be formatted as follows: An automatic change of a transfer function in CVT velocity is achieved by means of the control chain built in a structure of auto-CVT's driven pulley. The control chain provides an additional motion of links to the main motion. A schematic diagram of the proposed pulley option is given in fig. 1 [7].

The transfer function of velocity Tr^{ω} will be defined as $Tr\omega = \frac{\omega_1}{\omega_2}$ where ω_1 is a value of input motion velocity of a driving link, and ω_2 is the velocity of output link. Thus the relation of velocities in a first approximation is determined by the relation of kinematic radiuses in driven and driving links $Tr^{\omega} = \frac{\omega_1}{\omega_2} = \frac{R_2}{R_1}$.

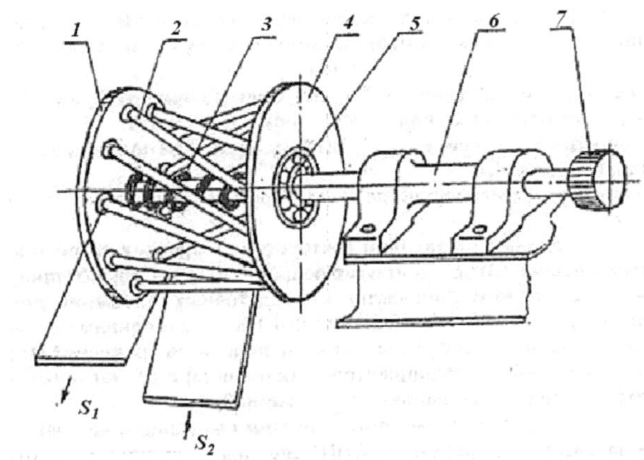

Figure 1. A variable geometry hyperboloid pulley: 1 – a dead sheave – a driven shaft of a belt transmission; 2 – spherically arranged bearing elements; 3 – a tension spring; 4 –a two-degree-of-freedom dead sheave; 5 – a sliding support of two-degree-of-freedom dead sheave; 6 – a support assembly of a driven transmission shaft; 7 – loading; S1 – a pulling transmission side; S2 – a return side.

Flanges 1 and 4 in a start position are located in a minimum distance from each other, and one-sheet hyperboloid of revolution formed by bearing rectilinear rods 2 has a minimum size of diameter in a throat area.

When a loading of driven shaft 6 increases, the flange 4 turns and displaces relative to the driven shaft 6, stretching the spring 3, thereby the diameter size of hyperboloid in the throat area becomes bigger that leads to the automatic increase in auto-CVT gear ratio and, potentially ensures a single-mode permanent engine functioning in variable external loading conditions. Under extreme maximum calculated loading a kinematic surface of a pulley degenerates in a circular cylinder with a diameter, equal to the diameter of placement of bearing rods 2 on flanges 1 and 4.

The evolution of hyperboloid pulley is possible because of the two-degree-of-freedom support 5 in the dead sheave 4.The dead sheave affected by a variable value of power flow determined by the difference of tension in driving and driven sides (S1-S2), can make additional angular and axial motions leading to a spatial change in orientation of bearing elements 2 representing a structure of rectilinear generators of a hyperboloid kinematic surface as a warped surface. Herewith, the ends of elements 2 are arranged movably on dead sheaves on spherical bearing sliding surfaces.

The evolution of hyperboloid driven link, results in the change of radius in its throat area from the initial minimum value r=r0 to the finite maximum r=R corresponding to the location radius of bearing elements on dead sheaves 1 and 4.

Direct and reverse evolution of the hyperboloid is provided in case of power equilibrium interaction of the transferred variable power flow with the resilient element 3 built in the structure.

The resilient element characteristic determines the functioning mode of auto-CVT therefore the stiffness selection of the spring included in a control chain is in general an important stage of auto-CVT design.

Let a kinematic size of a pulley be a variable size of a throat area in a hyperboloid of revolution.

We give a calculation algorithm of some force relations of a pulley with variable geometry [9,10]:

1. Let us calculate the initial and finite value of axial force P_{ax} dependent on the variable moment of resistance forces Mc, initial and finite radius of throat area of the hyperboloid and angle β:

$$P_{ax}{init} = \frac{M_{\text{Cinit}}}{r_0 \cdot tg\beta_0}$$

or

$$P_{ax}{fin} = \frac{M_{\text{Cfin}}}{R \cdot tg\beta_{fin}}$$

where β is an angle of crossing of bearing rods with the hyperboloid axis.

2. Let us define a constant value of stiffness in a linear resilient element $k = \frac{\Delta P_{ax}}{\Delta z_{max}}$,

where $\Delta P_{ax} = \Delta P_{ax}{fin} - \Delta P_{ax}{init}$ is a change of axial force in a resilient element for initial and finite value of the moment; Δz_{max} is an axial deformation of the hyperboloid.

3. Let us determine the required stiffness of resilient element by a given rule of changing in a gear ratio dependant on a variable external loading in a certain range.

The diagram of value changing of the required stiffness in the resilient element depending on a variable external loading is presented in figure 2:

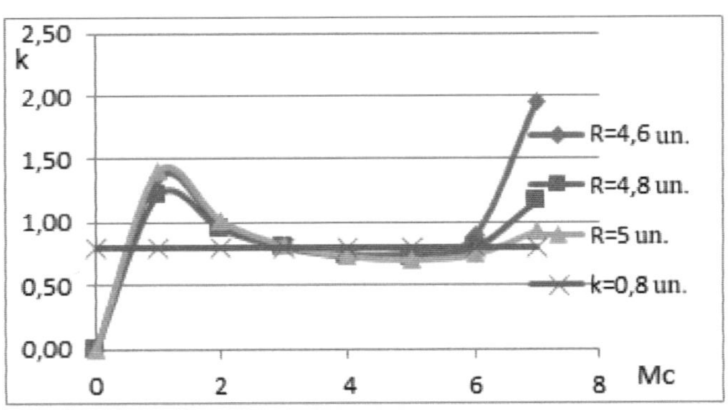

Figure 2. Changing of stiffness in a resilient element depending on the variable external loading Ì ñ under a variable location radius of bearing elements.

4. Results and discussion

Calculations showed that:

1. The required gear ratio autocontrol of auto-CVT in a certain range of external loading change M_c = (2…6) can be provided with a resilient element with a steady stiffness value that is shown in figure 2. Nonlinearity of stiffness value can be technically compensated with a spring without interturn gaps (technologically «pre-stressed» before).

2. In the evolution of hyperboloid from the initial position, an axial shift of two-degree-of-freedom flange is considerable at the beginning, and if the angle β reduces this shift is decelerated. The axial

force P_{ax} taken by the resilient element increases in the process of evolution that results in the overestimated stiffness value of the resilient element. In the process of hyperboloid evolution we should potentially assume that the stiffness k of the resilient element should be a variable one, and in the course of hyperboloid evolution, the stiffness is to increase, thus it is technically implementable. To reduce the influence of a non-linear section of axial deformation, the general deformation of a pulley should be restricted to prevent small values of the angle β.

3. When the external loading increases, a kinematic radius of a driven link grows too, i.e. when the kinematic radius increases by the same times, and ideally, peripheral and axial forces remain invariable, the only factor changing the axial force will be the variable angle β. For adequate changing of auto-CVT transfer function there is a prospective technical solution including two pulleys with variable geometry.

4. If a left part of the pulley is fixed in the axial direction, then the throat area will displace to the right by $\frac{\Delta z}{2}$ in each step, which can be compensated by using of auto-CVT of driving and driven pulleys with variable geometry.

5. Results and discussion

Thus, auto-CVT of the proposed design comprising a variable geometry pulley and a built-in resilient element can provide stationary engine functioning in a certain range of change in external power loading.

Note, however, that in general, the change of auto-CVT transfer function occurs with a certain deviation from the rules of external loading. The auto-CVT with two pulleys of variable geometry will have a great potential in this regard.

In this case, the velocity transfer function, by a linear law of external loading change, will alter, as it is shown in figure 3, i.e. almost linearly.

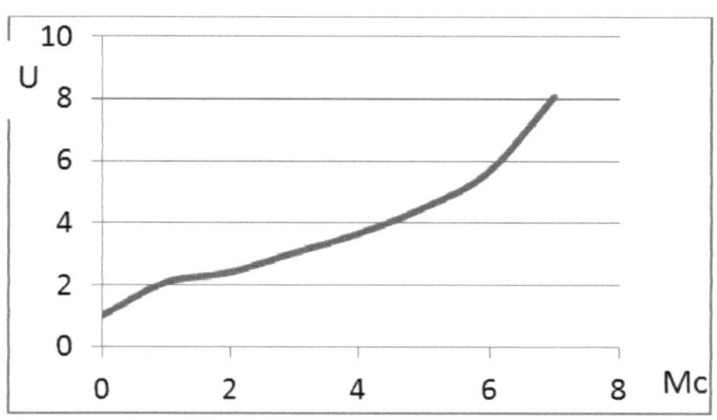

Figure 3. A gear ratio change depending on a variable external loading with a simultaneous operation of two pulleys with variable geometry.

Thereby the simultaneous operation of two pulleys with variable geometry can provide an adequate gear ratio change depending on the variable external loading in a broad range of its change that will ensure the stationary functioning of the engine under variable external loading conditions.

References

[1] Balakin P D 2016 *Theory elements of actual mechanical systems* (Omsk: Publishing house OmSTU) p 272

[2] Yildiz A, Bottiglione F, Carbone G 2015 An experimental study on the shifting dynamics of the chain CVT *Proc. of The Third Intl. Conf. on Advances in Mechanical and Robotics Engineering – AMRE 2015*

[3] Brown AW, van Rooij J and Frank A A 2004 The Design of an Inline GCI Chain CVT For Large Vehicles *International Continuously Variable and Hybrid Transmission Congress* (UC Davis) Paper № 54

[4] Srivastava N and Haque I 2009 A review on belt and chain continuously variable transmissions (CVT) dynamics and control *Mechanism and Machine Theory* **44(1)** pp 19–41, (2009)

[5] Bottiglione F, De Pinto S, Mantriota G and Sorniotti A 2014 Energy Consumption of a Battery Electric Vehicle with Infinitely Variable Transmission *Energies* **7(12)** pp 8317–833

[6] Mangialardi L and Mantriota G 1992 The Advantages of USING Continuously *Variable Transmission in Wind Power Systems, Renewable Energy* **2(3)** pp 201–209

[7] Patent 73425 Russian Federation IPC F 16 N 55/52 Pulley/ Balakin P D Zgonnik I P *Bul.* № 14 2008

[8] Patent 73426 Russian Federation IPC F 16 N 559/52 Pulley Balakin P D Zgonnik I P *Bul.* № 14 2008

[9] Balakin P D Zgonnik I P 2009 Geometrical-kinematic and power relations in a mechanical flat-belt continuously variable transmission *Omsk scientific bulletin* **2(80)** pp 75–78

[10] Balakin P D Zgonnik I P 2010 Calculation algorithm of stiffness for a resilient element as a part of control chain of a mechanical continuously variable transmission *Omsk scientific bulletin* **2(90)** pp 15–18

Method of non-destructive control of the solvent diffusion coefficient in products made from anisotropic porous materials

V P Belyaev, S V Mishchenko and P S Belyaev

Technological Institute, Tambov State Technical University, 106 Sovetskaya St, Tambov, 392000, Russia

Abstract. The development of methods and means of non-destructive products testing in various industries is an actual task. The subject of this study is non-destructive testing of the solvents mass transfer processes in products made from anisotropic porous materials (PM) and the diffusion coefficient determination. The known methods for determining the diffusion coefficient in anisotropic porous media are developed for canonical forms. This leads to the necessity of destroying products for making experimental samples from them of a given configuration and size. The purpose of this study is to develop a dynamic method that allows us quickly determine the diffusion coefficient in products from anisotropic porous materials without destroying them. The method is designed to investigate the diffusion coefficient of solvents, for example, in reinforced building elements from materials having a porous structure. The task of constructing the method mathematical model is solved. The influence of the design and operating measuring device parameters on the accuracy of the determining diffusion coefficient is studied. A theoretical estimate of the error in the measurement result was made. The results of the method application are presented.

1. Introduction

At present, products in the form of blocks or panels made of fibrous composite materials are becoming increasingly widespread in the construction industry [1, 2]. They can have a significant properties anisotropy due to the orientation of the fibers during their production [1-4]. Measurement of the diffusion coefficient in building materials is associated with considerable time expenditure [5, 6]. To implement the known methods, it is often necessary to use expensive and cumbersome equipment that provides the measurement of spatial solvents distributions within the samples [7-9]. Most known methods use samples of special shapes and sizes [5-9]. This makes it difficult to control the diffusion coefficient in finished products from anisotropic porous materials [10]. The point pulse method of nondestructive testing presented by us in [11] is not suitable for the study of anisotropic materials. To calculate the technological processes for the production of anisotropic porous materials, data about the diffusion coefficients in the transverse direction to the fibers location are needed first of all, because diffusion in this direction is the limiting stage of the process [12, 13].

2. Formulation of the problem

The task was to develop a pulse method for measuring the diffusion coefficient of solvents distributed in the anisotropic porous materials (PM) solid phase. The method should provide an opportunity to study products without destroying them and significantly reduce the time spent on monitoring. To achieve the goal it is necessary to solve the following tasks:

– to develop a physical and mathematical model of the method that ensures a fast flow mass transfer process in the controlled product and obtain the necessary measurement information for calculating the diffusion coefficient without destroying the product;

– to investigate the influence of measuring device design parameters and the regime variables of the measurement process on the accuracy and give recommendations about their rational values;

– to develop a mathematical model of measurement errors;

– to test experimentally the operability of the developed method on products from anisotropic porous materials.

3. Theory

The research results are based on the analytical theory of mass transfer, mathematical physics, mathematical modeling, the classical theory of metrology and mathematical statistics. The main idea of the developed method is to use a linear impulse action on anisotropic sample in the form of a solvent dose and to analyze the response in the solvent concentration change at a fixed distance from the source. When we use the proposed method, the conditions for diffusion of the solvent are specially created in the controlled samples, which from the point of the mathematical description view are analogous to the description of radial mass transfer in an unbounded medium when we apply a pulsed action from a linear source of mass. For this purpose, the samples must have a flat section of specified dimensions. This allows one-sided access to the product for placing the measuring device. In such case non-destructive control of the solvents diffusion coefficients is possible. At the beginning of the experiment, a solvent dose with power W along a line of length L and width s is applied to the flat sample surface. After the impulse is applied, the flat product surface in the vicinity of the pulse line is waterproofed. With the help of the electrochemical converter electrodes [14], made in the form of rectilinear segments with length l, which located on both sides of the mass pulse application line at a distance x_0 from it, the change in the electromotive force $E(r_0,\tau)$ in time is measured.

The process of solvent distribution in an anisotropic product made of PM is described by the following equation for mass transfer boundary-value problem in an unbounded medium, when we apply a pulsed action from a linear source:

$$\frac{\partial U(r,\tau)}{\partial \tau} = D\frac{1}{r}\frac{\partial}{\partial r}\left[r\frac{\partial U(r,\tau)}{\partial r}\right] + \frac{W}{\rho_0}\delta(r,\tau),$$

$$\tau > 0,\ 0 \le r < \infty,\ U(r,0) = U_0;;\ \frac{\partial U(0,\tau)}{\partial r} = 0;\ U(\infty,\tau) = U_0,,$$

where $U(r,\tau)$ is the concentration of the solvent in the product at a distance r from the linear source of the mass pulse at time τ; D – the diffusion coefficient; $\delta(r,\tau)$ – Dirac delta function; ρ_0 – the density of the absolutely dry material; W – the power of the "instantaneous" solvent source acting at coordinate $r = 0$, calculated as the ratio of the solvent amount to the impulse line with length L multiplied by the thickness h; U_0 – initial concentration of solvent in the material at the time $\tau = 0$.

The diffusion coefficient can be found from [15]:

$$D = r_0^2/(4\tau_{max}), \tag{1}$$

where τ_{max} is the time corresponding to the maximum on the $U(r_0,\tau)$ curve of the concentration change at a distance r_0 from the source.

Due to the monotonicity of the electrochemical converter static characteristic [14], the moments of reaching the maximum values of the concentration $U(r_0,\tau)$ and the EMF of the converter $E(r_0, \tau)$ coincide. This makes it possible to implement the measurement method without preliminary calibration of the electrochemical converter (ECC). The experience [16, 17] shows that during the research it is expedient to use the range of the static characteristic near the value $0.5U_p$, characterized by high sensitivity and stable signal. Here U_p - the concentration of the diffusing substance in the PM solid phase at equilibrium with saturated solvent vapors at a given temperature. Therefore, it is expedient to obtain the value of the maximum concentration $U_{\max}$ in the vicinity of the given value during the experiment:

$$U_{max} \approx 0.5U_p \tag{2}$$

The values of the ECC location coordinate r_0 and the solvent dose W, supplied at the beginning of the experiment active part exert a significant influence on the value $U_{\max}$. In this case, the change in the concentration of the solvent in the PM in the area of the source action is described by the function:

$$U(r,\tau) - U_0 = W/(4\pi D\rho_0 \tau \exp[r^2/4D\tau]). \tag{3}$$

From (3), taking (1) into account, assuming for simplicity $U_0 = 0$, it is possible to obtain the dependence of the attained maximum $U_{\max}$ at $\tau = \tau_{\max}$ from the solvent dose W:

$$U_{\max}(r_0, \tau_{\max}) = W/(4\pi D\rho_0 e \tau_{\max}). \tag{4}$$

Let us consider the influence of the measuring device design parameters and technological variables of the measurement process on the diffusion coefficient accuracy. The main design parameters for devices implementing the proposed method are: the minimum permissible dimensions of the flat area and the product thickness in the place of the linear pulse source; permissible dimensions of length L and width s of the substance dose line; coordinate r_0 of the ECC electrodes location relative to the location of the pulse line. The main technological variables are the parameters of the applied mass pulse: the duration t and the power W. The basis for finding the design parameters of the device is the coordinate r_0 determination by measuring the EMF change after the application of the pulse.

The mathematical model of the root-mean-square estimate of the relative error δD in determining the sought diffusion coefficient from (1) has the form [16, 18]:

$$\delta D = \sqrt{4(\delta r_0)^2 + (\delta\tau_{\max})^2 + (\delta_{\mathrm{m}})^2}, \tag{5}$$

where $\delta r_0 = \Delta r_0 / r_0$ and $\delta\tau_{max} = \Delta\tau_{max} / \tau_{max}$ are, respectively, the relative error in determining the coordinate of the calculated cross-section and the time to reach the maximum on the curve of the concentration change in the calculated section; δ_{m} is the total value of the methodical error due to the incomplete correspondence of the mathematical model used to real physical processes.

As an object of study, the cement-fiber board sample with a thickness of 50 mm and a density of 1320 kg / m3 in dry state with oriented cotton threads 0.4 mm in diameter is considered. The sample was prepared as follows. Cotton yarns were wound on a mandrel providing their orientation in a given direction, then the mandrel with the wound filaments was placed in a rectangular tray and filled with cement milk. After solidification, the process of setting the sample strength in an autoclave proceeded. Then the sample was dried to constant weight and the diffusion coefficient was investigated. It was shown in [14] that when ECC electrodes are placed on the equipotential product surfaces, value $\pm a/r_0$ can be used as an estimate of δr_0, where a - the radius of the electrodes of the ECC. As for the value $\delta\tau_{\max}$ in (5), it essentially depends on the nature of the change in the concentration $U(r_0,\tau)$ and the error of its determination [17]. Therefore, the choice of distance r_0 from (5), when δ_{m} – const, is achieved by a compromise solution based on the error δr_0 and duration of the experiment. As a result of this analysis, it was decided to place the main electrodes of the electrochemical converter at a distance r_0 = 4 mm from the pulse-action line and additional electrodes at a distance of 3 and 5 mm. Removing the transducer from the source over a distance than 5 mm causes an increase in the experiment duration, in the dimensions of the flat products area and in the overall dimensions of the measuring device with a slight increase in the accuracy. Using (4), and taking into account the calculated constants, it is possible to calculate in advance the required source power of the solvent dose W to ensure the maximum concentration in the calculated cross section r_0:

$$W \approx 8.54 \rho_0 U_{max} r_0^2 \ . \tag{6}$$

The remaining design parameters and regime variables of the diffusion coefficient measurement process was obtained by analyzing the methodical error δ_m. They were chosen from the condition of reaching negligibly small values δ_m. in comparison with the remaining components of the resulting error according to [17, 18].The length L and width s of the impulse line, the dimensions of the measuring device and the minimum permissible dimensions of the product itself were chosen from the condition of ensuring mass transfer under linear impulse action in an unbounded medium [18]. The diffusion coefficient was measured by device pressed against a flat rectangular shape surface 100 × 120 mm. The measuring device had a rectilinear groove 90 mm long to accommodate a linear source of solvent mass. The product should have a flat surface area at least 120 × 120 mm. The permissible product thickness in the control zone is not less than 40 mm. The pulse durability t of the mass must not exceed 0.05 × τ_{max}.This allowed us to take an upper bound estimate of 5% for the total methodical error δ_m. Taking this into account, the root-mean-square estimate of the relative error δD for distances up to a linear source 3, 4 and 5 mm, respectively, is 13; 12 and 12% when δE = 0.5 % and doses for each r_0 optimally calculated by (6).

4. Experimental results

For carrying out experimental studies, we used the information-measuring system developed by the technology of "National Instruments" [19], with a modernized measuring device. Before carrying out the active experiment, we determined first the required pulse mass by formula (6). In Table 1, as an example, the ethanol diffusion coefficient in a cement-fiber board with a thickness of 50 mm and a density of 1320 kg / m3 in the dry state for a transverse direction to the reinforcing yarns is presented. The pulse application line of the solvent mass was oriented in the direction of the reinforcing filaments in the sample. The total number of studies for each PM-solvent system was at least 20 trials. The calculated value of the ethanol pulse for this example is 4.5 × 10-4 kg, r_0=4 mm. Experimental studies have shown that the mathematical expectation of the ethanol diffusion coefficient in a cement-fiber board with a confidence probability α = 0.95 is 2.9 × 10^{-9} м^2/c. The random error of determining the diffusion coefficient is less than 8%. The duration of the experiment does not exceed 30 minutes.

Table 1. Diffusion coefficient of ethanol in cement-fiber plate

Parameter	Parameter Value for τ_{max} (s)				
	1520.9	1716.7	1194.0	1133.1	1219.5
$D_i \cdot 10^9$ (m^2/s)	2.63	2.33	3.35	3.53	3.28
$\Delta D_i = (D_i - \overline{D}) \cdot 10^9$ (m^2/s)	-0.29	-0.59	0.43	0.61	0.36
$\Delta D_i^2 \cdot 10^{18}$ (m^4/s^2)	0.0841	0.3481	0.1849	0.3721	0.1296
Expected value $\overline{D}$ = 2.9 · 10^{-9} (m^2/s); error $\delta D \approx$ 8 (%).					

5. Discussion of the results

IOP Conf. Series: Journal of Physics: Conf. Series **1050** (2018) 012010 doi:10.1088/1742-6596/1050/1/012010

Presented method allows us to measure operatively the diffusion coefficient of the polar solvents in the products of anisotropic porous materials with a thickness more than 40 mm, having at least one flat surface with a size of 120 × 120 mm, without their damage. A significant increase in control performance ensured by:

– eliminating the need for preparing samples for study;

– using a more dynamic process of mass transfer under the pulsed action

– eliminating the need for preliminary calibration of applied electrochemical transducer for each new material and solvent.

The efficiency of this method was checked by comparison our results with the results of water and ethanol diffusion coefficients in samples of solid porous materials, by zonal [20] and adsorption [21] methods. The discrepancy in the numerical values of the desired coefficient does not exceed 10 ... 13 %, what can be considered a satisfactory result in the study of solvents diffusion in the PM.

6. Conclusion

The developed dynamic method allows significantly increase efficiency of the diffusion coefficient control for polar solvents in products made from anisotropic porous materials without their destruction. The method can be used in portable devices for operational diffusion coefficient control in the products. The result of these studies identified the minimum allowable dimensions of the flat surface of the product, on which we can place measuring device. We determined the rational value of the coordinate electrode arrangement for electrochemical transducer relative to the line of pulse impact application from the condition of achieving acceptable accuracy and duration for the measurement process. We also received the resulting equation for calculation the required power line solvent source for pulse exposure. Statistical processing of the obtained experimental data and their comparison with the results obtained from known methods demonstrate the efficiency of the proposed non-destructive testing method.

References

[1] Remond R and Almeida G 2011 *Wood Mater. Sci. Eng.* **1** 23

[2] Nizovtsev M I, Stankus S V, Sterlyagov A N, Terekhov V I and Khairulin R A 2008 *Int. J. Heat Mass Transf.* **51** 4161

[3] Roels S and Carmeliet J 2006 *Int. J. Heat Mass Transf.* **49** 4762

[4] Koptyug I V and Sagdeev R Z 2002 *Russian Chem. Rev.* **71** 789

[5] García-Gutiérrez M J L C, Cormenzana J L, Missana T, Mingarro M and Molinero J 2006 *J. Iberian Geol.* **32** 37

[6] Mishchenko S V, Belyaev P S, Gladkikh V A, and Safronova E N 1999 *Drying Technol.* **17** 2151

[7] Carslaw H S and Jaeger J C 1959 *Conduction of Heat in Solids* (Oxford: Clarendon Press) p 487

[8] Belyaev V P, Belyaev M P, Mishchenko S V and Belyaev P S 2014 *Meas. Tech.* **56** 1190

[9] Belyaev V P, Mishchenko S V and Belyaev P S 2015 *Meas. Tech.* **58** 574

[10] Gurov A V, Sosedov G A, Rodina A E and Ponomarev S V 2013 *Meas. Tech.* **55** 1187

[11] Shashkov A G, Volokhov G M, Abramenko T N and Kozlov V P 1973 *Methods to determine thermal conductivity and thermal diffusivity [in Russian]* (Moscow: Énergiya) p 336

[12] Belyaev M P, Belyaev V P and Divin A G 2010 *Vestn. Tambovsk. Gos. Tekh. Univ.* **16** 797

[13] Kartashov É M and Rudobashta S P 2010 *Diffusion in chemical-technology processes [in Russian]* (Moscow: KolosS) p 478

[14] Scheffler G A and Plagge R 2011 *Proc. 5th Int. Autoclaved Aerated Concrete Conf.* September 2011, Bydgoszcz (Poland) 337

IOP Conf. Series: Journal of Physics: Conf. Series **1050** (2018) 012011 doi:10.1088/1742-6596/1050/1/012011

The research of compensation for changes in air flow rate by increasing the oxygen concentration of oxidant in a fuel cell

P V Belyaev, D A Podberezkin, R A Em

Omsk State Technical University, 11, Mira ave, Omsk, 644050, Russia

e-mail: polly110@yandex.ru

Abstract. In the article an analytical study of a proton exchange membrane fuel cell (PEMFC) is carried out. The mathematical model describing the electrical properties of the fuel element is given. A distinctive feature of a PEMFC is its increased sensitivity to changes in fuel parameters and a lower one – to changes from outside of the element-oxidant. During operation of the system with a fuel cell a number of accidents, for example caused by a decrease in the rate of the supplied oxidant, is possible. For the simulation of the system operation modes a simulation model in the software environment Matlab/Simulink is developed. Using the model, the possible emergency modes of operation are identified, the possibility of compensation for reducing the rate of the oxidant consumption by increasing the concentration of oxygen in it is investigated. Recommendations for individual and mutual regulation of oxidant parameters and ranges of their concentration are developed.
Keywords – fuel cell, PEMFC, simulation

1. Introduction
In the modern world, a growing trend to switch to cleaner sources of energy is observed. This is resulted from the use of traditional resources which is too dangerous to the planet ecosystem. Besides, traditional energy resources are non-renewable, and, according to forecasts, will run out in the nearest future. Therefore, more attention is paid to the development of alternative and renewable energy and to increasing the generation from these sources. One of the promising alternative energy areas are hydrogen fuel cells. Fuel cells (FC) are environmentally friendly and inexhaustible primary resources, as hydrogen is one of the most common elements on the Earth. The study of the FC modes is important, because the method of generating energy using FC has been only recently applied [1 - 2] and is not investigated in full [3 – 4].

Development of mathematical model and simulation are among the main methods of research, along with the full-scale experiment [5 – 7].

2. Problem Statement
Improvement of electrical properties of alternative sources of energy PEMFC requires a detailed study and analysis of its operation modes. With this purpose it is necessary to determine the influence of PEMFC input parameters on its output characteristics, in particular, to identify the adjustment ranges of the fuel element parameters to determine the sustainable and emergency operation.

3. Theory
A fuel cell is an energy source that produces it in the course of a chemical reaction. The process of chemical reactions occurs on porous electrodes. The atoms of the fuel (hydrogen or hydrogen-

containing substance) supplied to the anode are divided into protons and electrons. Electrons move through the external circuit to the cathode, creating direct electric current. Protons move to the cathode through electrolyte. At the cathode reduction reaction occurs, protons and electrons are combined with the oxidant (oxygen or air) to form water. The products of this reaction are heat and water vapor.

The figure shows a schematic diagram of the FC proton exchange membrane [3].

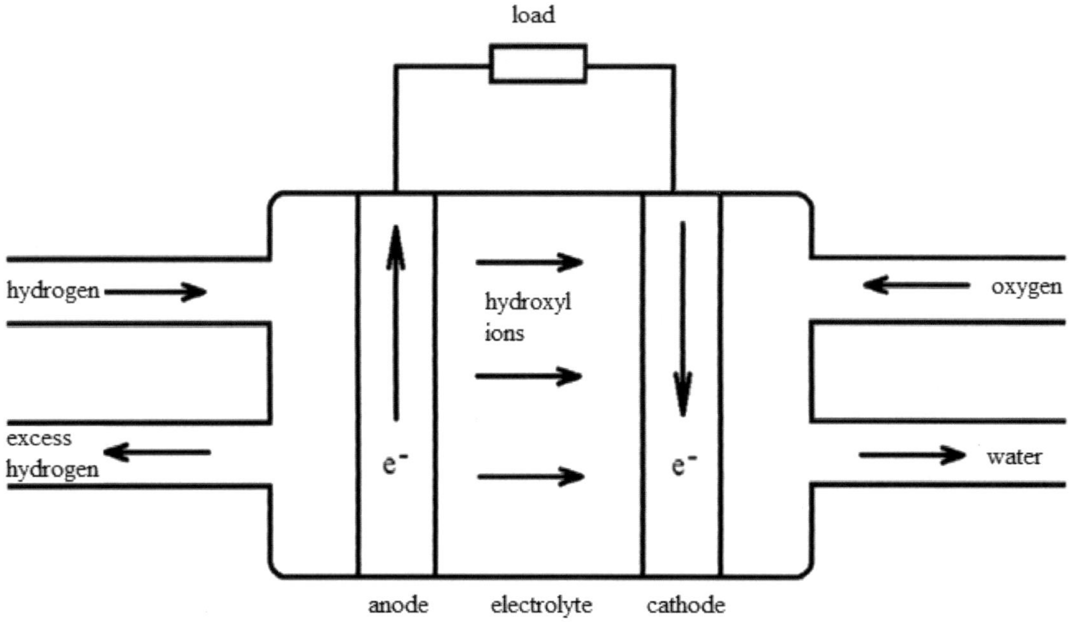

Figure 1. Schematic diagram of fuel cell proton exchange membrane

Reactions occurring in the cell can be described by the following expressions:

On the anode:

$$H_2 \rightarrow 2H^+ + 2e^-; \tag{1}$$

On the cathode:

$$\tfrac{1}{2}O_2 + 2H^+ + 2e^- \rightarrow H_2O\ ; \tag{2}$$

The overall reaction in PEMFC:

$$H_2 + \tfrac{1}{2}O_2 \rightarrow H_2O. \tag{3}$$

4. FC mathematical model

Using a detailed model of a fuel cell, it becomes possible to study changes of its input parameters, such as pressure, temperature, concentration and flow rates of fuel and oxidant. Changes in these parameters affect the open-circuit voltage (E_{OC}), the exchange current (i_0) and the Tafel slope (A). E_{OC}, i_0 and A are modified as follows:

$$E_{OC} = K_c E_n \tag{4}$$

$$i_0 = \frac{zFk(P_{H2}+P_{O2})}{Rh} e^{\frac{-\Delta G}{RT}} \tag{5}$$

$$A = \frac{RT}{z\alpha F} \tag{6}$$

where R – is equal to 8.3145 j/(mol K); F – is 96485 AV/mol, z – is the number of moving electrons, En – is the Nernst voltage, which is the thermodynamics voltage of the cells and depends on the temperature and partial pressure of reactants and products inside the stack (V); α – is the coefficient of charge transfer, which depends on the type of electrodes and catalysts; P_{H2} – is the partial pressure of hydrogen inside the stack (ATM); P_{O2} – is the partial pressure of oxygen inside the stack (ATM); K – is the Boltzmann constant is equal to 1.38×10^{-23}J/ K; h – is the Planck constant – 6.626×10^{-34} J s; ΔG – is the size of the activation barrier which depends on the type of electrode and the catalyst used, T – is the operation temperature (K); KS – is a constant voltage at the nominal operating conditions.

The rate of conversion (use) of hydrogen (U_{fH2}) and oxygen (U_{fO2}) are determined in the following way:

$$\begin{cases} U_{fH2} = \frac{n_{H2}^{r}}{n_{H2}^{in}} = \frac{60000RTNi_{fc}}{zFP_{fuel}V_{lpm(fuel)}x\%} & (7) \\ U_{fO2} = \frac{n_{O2}^{r}}{n_{O2}^{in}} = \frac{60000RTNi_{fc}}{2zFP_{air}V_{lpm(air)}y\%} & (8) \end{cases}$$

where P_{fuel} – is the absolute fuel supply pressure (ATM); P_{air} – is the absolute air supply pressure (ATM); $V_{lpm(fuel)}$ – is the fuel consumption (l/min); V_{lpm}(air) – is the air flow (l/min); x – is the percentage of hydrogen in the fuel (%); Y – is the percentage of oxygen in the oxidant (%); N – is the number of cells.

Constant 60000 comes from the conversion from flow speed l / min used in the model, to the dimensionality of m^3/s (1 liter / min = 1/60000 m^3/sec).

Partial pressure and the Nernst voltage are determined as follows:

$$\begin{cases} P_{H2} = (1 - U_{fH2})x\%P_{fuel} & (9) \\ P_{H2O} = (w + 2y\%U_{fO2})P_{air} & (10) \\ P_{O2} = (1 - U_{fO2})y\%P_{air} & (11) \end{cases}$$

$$E_{nernst} = \begin{cases} 1.229 + (T - 298)\frac{-44.43}{zF} + \frac{RT}{zF} ln(P_{H2}P_{O2}^{1/2}) & \text{при } T \le 100^{\circ}C \\ 1.229 + (T - 298)\frac{-44.43}{zF} + \frac{RT}{zF} ln\left(\frac{P_{H2}P_{O2}^{1/2}}{P_{H2O}}\right) & \text{при } T > 100^{\circ}C \end{cases} \quad (12)$$

where P_{H2O} – is the partial pressure of water vapor inside the stack (ATM); W – is the percentage of water vapor in the oxidant (%).

From the partial gases pressures and the Nernst voltage the new values of open circuit voltage (E_{OC}), the exchange current (i_0) and the Tafel slope (A) can be calculated.

The parameters α, ΔG and Kc are calculated based on the polarization curve at nominal conditions of operation along with some additional parameters, such as the low efficiency of the calorific value of the stack, composition of fuel and air, gas pressure and temperature. They can be easily taken from the manufacturer specifications.

Nominal levels of gases conversion are calculated as follows:

$$U_{f_{H2}} = \frac{\eta_{nom}\Delta h^0(H_2O(gas))N}{zFV_{nom}} \quad (13)$$

$$U_{f_{O2}} = \frac{60000RT_{nom}NI_{nom}}{2zFP_{air_{nom}}V_{lpm(air)_{nom}} \cdot 0.21} \quad (14)$$

where η_{nom} – is the nominal stack efficiency of STC (%), Δh^0 (H2O (gas)) – 241.83 × 103 j/mol; V_{nom} – is the nominal voltage (V); I_{nom} – is the nominal current (A); $V_{lpm(air)nom}$ – is the nominal air flow (l/min); P_{airnom} – is the rated absolute supply pressure of air (PA); T_{nom} – is the nominal operating temperature (K).

The nominal partial pressure of gases and the Nernst voltage can be obtained from the conversion rate. If E_{OC}, i_0 and A are specified and if it can be assumed that the stack operates at a constant rate of conversion or is used under nominal conditions, then α, ΔG and KC can be determined.

5. Simulation

For simulation of fuel cell at constant load and study its performance response to key parameters changes, a simulation quantitative model in software environment MATLAB/Simulink was developed.

The assembled model is shown in figure

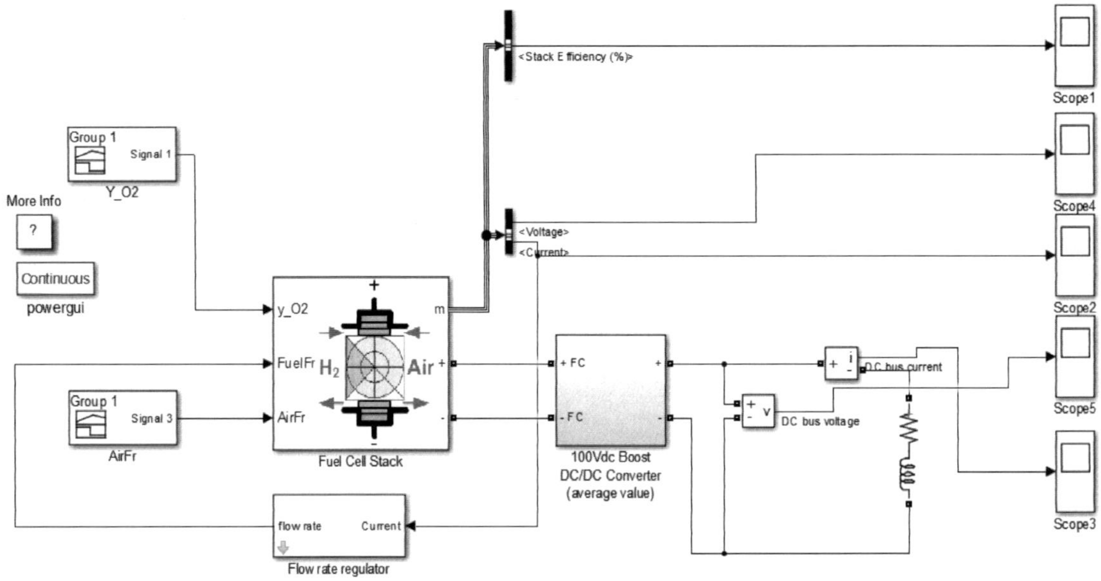

Figure 2. Simulation model of the fuel cell

The nominal voltage of the fuel cell 45 V dc is applied to the voltage converter. The transformer increases the voltage to 100 V dc, then the voltage is applied to a constant load, with interval time of 1 sec. The nominal power of the unit is 6 kW.

6. Experimental results

A distinctive feature of proton exchange membrane fuel cells is their increased sensitivity to changes in fuel parameters and a lower one – to changes from outside of the element-oxidant. Although the change of the oxidant parameters is less critical, it can have a significant impact on the FC functioning and its output characteristics. For this simulation model the nominal values of the oxidant are the following: the oxygen content in the oxidant is at the level of 21 % (normal air), the air flow rate is 300 l/min.

During operation of the system with fuel cell as a power supply, some emergency situations can occur. The reason for operating emergency conditions can be a decrease in the rate of the supplied oxidant. For proper functioning of the fuel cell and the normal flow of chemical reactions, a certain amount of oxidant is necessary.

To avoid FC emergency operation mode, the effect of reducing the air flow rate must be compensated by changes in another parameter. This parameter can be the percentage of oxygen in the oxidant. To study the feasibility of this type of compensation two introductory signals modified at the same time were specified. The first signal (figure 3) is responsible for reducing the air flow rate from

300 l/min to 200 l/min. The second input signal (figure 4) is responsible for the change of the oxygen concentration in the oxidant.

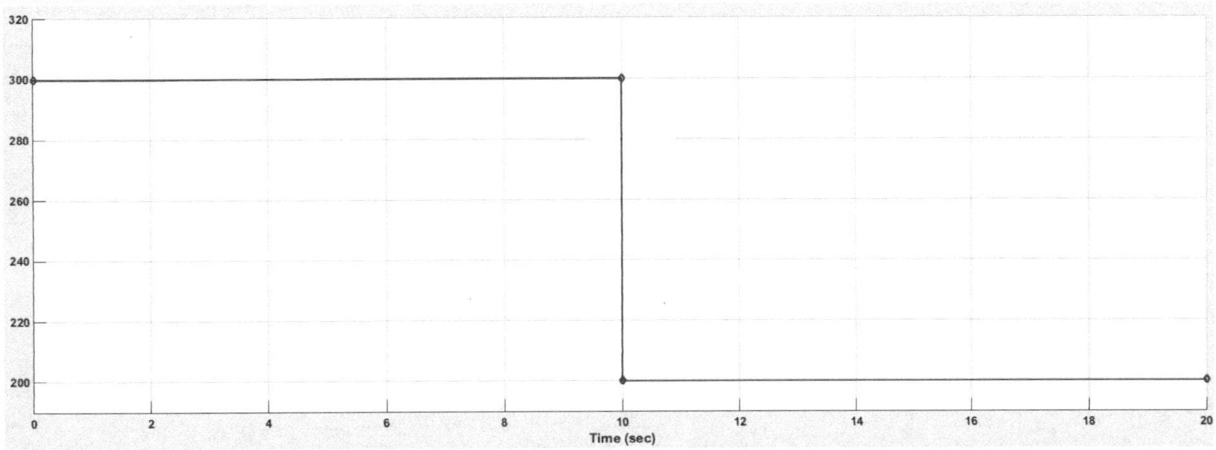

Figure 3. Input responsible for reducing the air flow rate

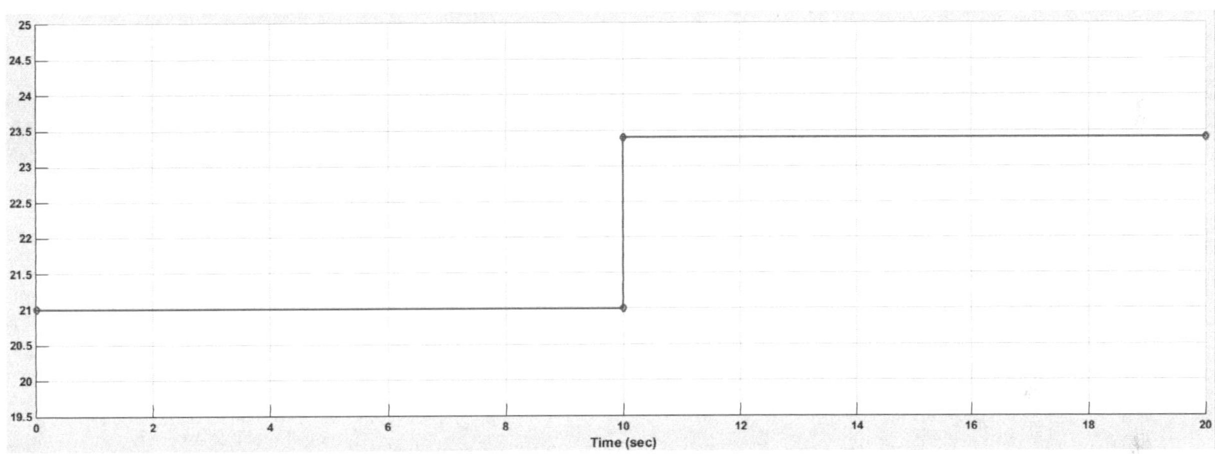

Figure 4. Input responsible for the increase in the oxygen concentration in the oxidant

Up to the tenth second there is the start in the fuel cell, and all the input parameters take the nominal values. Starting with second 10, occurs the ramp change of: the air flow rate is reduced from 300 l/min to 200 l/min, oxygen concentration in the oxidant increases from 21% to 2.5 %. The value of 2.5 % is the minimum increase in the oxidant concentration in such a significant reduction in the rate of its consumption. Changes in the concentration of oxygen in a smaller range prevents normal operation of the fuel cell by reducing the air flow rate.

During the simulation of the air flow rate reduced compensation by increasing the concentration of oxidant by 2.5%, graphs of the fuel cell output current and voltage are shown in figures 5 – 6.

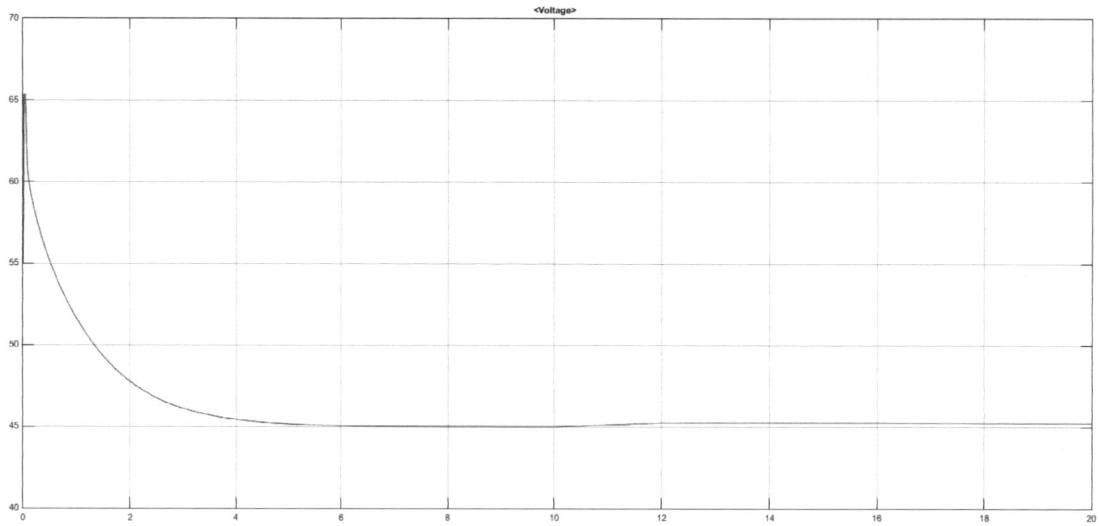

Figure 5. Graph of the FC output voltage from the time of the oxidant adjusting parameters

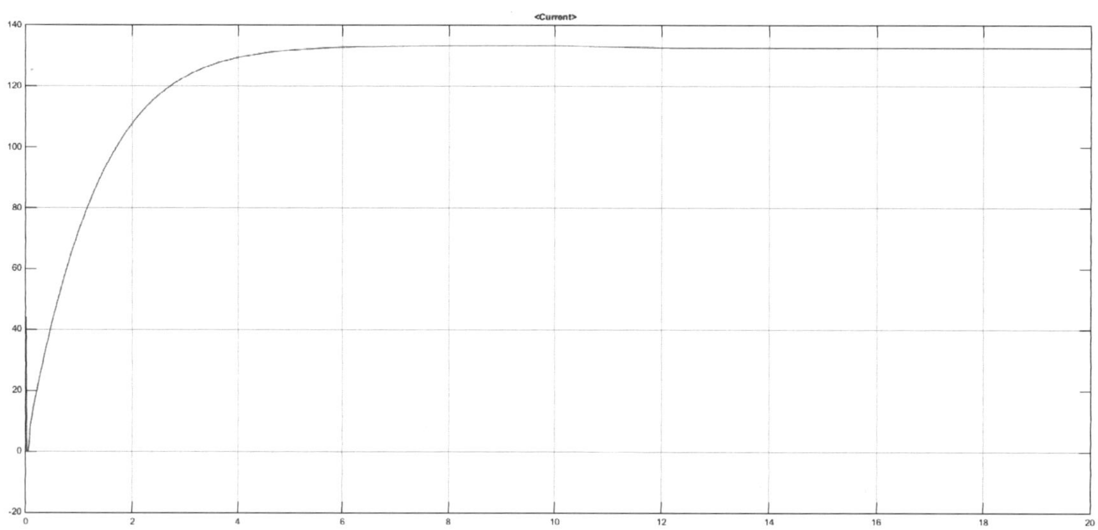

Figure 6. Graph of current output TE from the time when adjusting the parameters of the oxidant.

7. Results and discussion

For the normal operations of the fuel cell, nominal values for all input parameters should be kept including the air flow rate, since at a low value of this parameter the oxidant will not be sufficient for the reaction. As it was found during the simulation, in the studied simulation model the air flow rate reduction up to 200 l/min can be compensated by increasing the oxygen percentage of the oxidant by only 2.5 %.

We can safely adjust these two parameters in two ways:

- simultaneously;
- first, to increase the concentration of oxygen in the oxidant, then, to reduce the consumption of oxidant.

Initially, it is impossible to safely reduce the consumption of the oxidant with a subsequent increase in the oxygen concentration. This will lead to an emergency operation mode of the fuel cell and the current increase up to several kilo- ampers.

IOP Conf. Series: Journal of Physics: Conf. Series **1050** (2018) 012011 doi:10.1088/1742-6596/1050/1/012011

8. Conclusion

- The FC simulation model is developed.
- The possibility of compensation for reducing the rate of oxidant consumption, by increasing the oxygen concentration in it is studied.
- It is revealed that the decrease in the rate of air flow by 100 l/min can be compensated by an increase in the oxygen concentration in the oxidant by 2.5% rather than to exclude the FC emergency mode. Regulation recommendations by using this method are developed.

References

[1] Belmokhtar K et al., 2013 Modeling and fuel flow dynamic control of proton exchange membrane fuel cell in: *Power Engineering, Energy and Electrical Drives* (POWERENG). Fourth International Conference IEEE pp 415–420

[2] Hähnel C, Aul V, and Horn J 2015 Power efficient operation of a PEM fuel cell system using cathode pressure and excess ratio by nonlinear model predictive control in: *Control Conference* (ECC), European. IEEE pp 3340–3345

[3] Belyaev P V, Podberezkin D A 2016 Application of fuel cells proton exchange membrane in: *Journal of science and education*. **5** p 17

[4] Belyaev P et al. 2016 Simulation modeling of proton exchange membrane fuel cells in: *Dynamics of Systems, Mechanisms and Machines (Dynamics), IEEE* pp 1–5

[5] Titov V G, Abramov S V 2007 Simulator of fuel elements characteristics on the basis of semiconductor converters in: *Proceedings of the Tomsk Polytechnic University*. T 311. **4**

[6] Najafizadegan H, Zarabadipour H 2012 Control of voltage in proton exchange membrane fuel cell using model reference control approach in: *Int. J. Electrochem. Sci.* T **7** pp 6752–6761

[7] Nowdeh S A, Ghahnavieh A R, Shojaei H 2012 Reliable Designing of Stand-alone PV/FC Hybrid System in: *Majlesi Journal of Electrical Engineering*. T 7 **2** pp 41–47

Mechanical Science and Technology Update (MSTU-2018) IOP Publishing
IOP Conf. Series: Journal of Physics: Conf. Series **1050** (2018) 012012 doi:10.1088/1742-6596/1050/1/012012

An application of the exponential spline for the approximation of a function and its derivatives in the presence of a boundary layer

I A Blatov[1], **A I Zadorin**[2], **E V Kitaeva**[3]

[1] Volga Region State University of Telecommunications and Informatics, ul. L'va Tolstogo,23, 443010, Samara, Russia
[2] Sobolev Institute of Mathematics, pr. Koptyuga,4, 630090, Novosibirsk, Russia
[3] Korolev Samara State University, Moskovskoe sh., 34, 443086, Samara, Russia

E-mail: zadorin@ofim.oscsbras.ru, blatow@mail.ru

Abstract. The problem of exponential spline interpolation of functions having large gradients in the exponential boundary layer is considered. The spline of the class $C^2[0,1]$ is constructed as the sum of a polynomial of the second degree and a boundary-layer function on each grid interval. Estimates of the error in the approximation of a function and its derivatives are obtained. These estimates are uniform in small parameter. The limiting behavior of the exponential spline is investigated, when the perturbing parameter tends to infinity or to zero. In the first case, the spline becomes cubic, and in the second case it becomes parabolic. The results of numerical experiments are presented.

1. Introduction

Various convection-diffusion processes with predominant convection are modeled on the basis of singularly perturbed problems. The solution of singular perturbed problem has large gradients in the boundary layer. The problem of the interpolation of functions with large gradients in the boundary layer is of interest. It is shown [1] that the interpolation of functions with large gradients in the boundary layer with using of cubic spline on a uniform grid is inefficient. It is proved that the interpolation error can grow without limit when the small parameter ε tends to zero. In [1] it is proved that in the case of Shishkin mesh [2] the interpolation error can grow if ε decreases. In [1] the modification of the cubic spline on Shishkin mesh [2] is made, as result the error of the cubic spline becomes uniform with respect to ε.

In [3] for interpolated functions with large gradients in the exponential boundary layer the exponential spline is applied. In the case of a uniform grid error estimates are obtained, which are uniform with respect to the parameter ε. These results are summarized in [4]. Note that generalized splines have been studied in a number of works, in particular in [5], [6], [7]. However, important questions of the convergence uniform in the parameter ε of interpolation processes for such splines are very poorly investigated.

An interesting question is the approximation of not only the function, but also its derivatives. In this paper we estimate the error in the approximation of the derivatives of functions with large gradients in the boundary layer on the basis of the exponential spline. We assume that the function is defined at the nodes of the uniform grid. We also study the limiting properties of exponential splines.

Introduce the notations. Let $\varepsilon \in (0,1]$, $\Omega = \{x_n : x_n = nh, n = 0,1,\ldots,N,\ x_0 = 0, x_N = 1\}$ be a uniform grid of the interval $[0,1]$ with the step $h = 1/N$. By C and C_j we mean positive constants that do not depend from the parameter ε and the step h. Suppose that $C[a,b]$ is the space of functions continuous on $[a,b]$ with the norm $\|\cdot\|_{C[a,b]}$. We denote by $C_h[0,1]$ the linear normed space of functions from $C[0,1]$ having one-sided second-order derivatives at the points $x = 0$ and $x = 1$, with the norm $\| u \|_{C,h} = \| u \|_{C[0,1]} + h^2(|u''(0)| + |u''(1)|)$.

2. Formulation of the problem and main results

Let us a function $u(x)$ be decomposed in the form of the sum of regular and singular components:

$$u(x) = q(x) + \gamma\Phi(x), \tag{1}$$

where

$$|q^{(j)}(x)| \le C_1,\ 0 \le j \le 4, \Phi(x) = e^{-\alpha x/\varepsilon}, \alpha > 0,\ x \in [0,1].$$

Assume that the regular component $q(x)$ and the constant γ are not given in the representation (1), the function $\Phi(x)$ is known, its derivatives grow unboundedly at the boundary $x = 0$ if $\varepsilon \to 0$. The component $\gamma\Phi(x)$ is responsible for the large gradients of the function $u(x)$ in the boundary layer. Such a representation is valid for the solution of a singularly perturbed problem [2].

We assume that the function $u(x)$ is given at the nodes of the grid Ω, $u_n = u(x_n), n = 0,1,\cdots,N$.

Define the space of L-splines taking into account that the interpolated function $u(x)$ has the representation (1). Let

$$SL(\Omega,3,1) = \{S(x) \in C^2[0,1] : S(x) = a_n + b_n x + c_n x^2 + d_n e^{-\alpha x/\varepsilon},$$

$$x \in [x_n, x_{n+1}], 0 \le n \le N-1\}.$$

We define L-spline $S(x;u) \in SL(\Omega,3,1)$ for the function $u(x)$ from the interpolation conditions

$$S(x_n;u) = u(x_n),\ 0 \le n \le N,\ S''(0;u) = u''(0),\ S''(1;u) = u''(1). \tag{2}$$

Theorem 1 *For a function $u(x)$ having decomposition (1), there exists a unique interpolating spline $S(x;u) \in SL(\Omega,3,1)$ satisfying the conditions (2), and for some constant C and $j = 0,1,2$ the following error estimates are valid*

$$\| S^{(j)}(x;u) - u^{(j)}(x) \|_{C[0,1]} \le C \min\left\{h^{3-j}, \frac{h^{4-j}}{\varepsilon}\right\},\ \varepsilon \in (0,1]. \tag{3}$$

From (3) there follows the uniform in $\varepsilon \in (0,1]$ convergence of the third order accuracy of the interpolation process (2) for functions of the form (1), as well as uniform convergence of the second and first orders for the first and second derivatives, respectively.

Let $S_3(x;u) \in C^2[0,1]$ be a cubic spline with interpolation conditions (2). Similarly $S_2(x;u) \in C^1[0,1]$ is a parabolic spline with interpolation conditions:

$$S_2(x_n;u) = u(x_n),\ 0 \le n \le N,\ S_2''(1-0;u) = u''(1).$$

Theorem 2 *For any function $u(x) \in C_h[0,1]$ the interpolation splines $S_2(x;u)$ and $S_3(x;u)$ exist, they are unique, and for any fixed h the following formulas hold:*

$$\lim_{\varepsilon \to +\infty} \| S(x;u) - S_3(x;u) \|_{C[0,1]} = 0, \tag{4}$$

$$\lim_{\varepsilon \to 0+0} \| S(x;u) - S_2(x;u) \|_{C[0,1]} = 0. \tag{5}$$

As a corollary of Theorem 2 we obtain the following result. If $u(x) \in C^4[0,1]$, then for some constant M the following estimates hold:

$$\| S_3(x;u) - u(x) \|_{C[0,1]} \le Mh^4, \tag{6}$$

$$\| S_2(x;u) - u(x) \|_{C[0,1]} \le Mh^3. \tag{7}$$

Note that in (4)– (7) the form of the function $u(x)$ is not related to the representation (1). The estimate (6) is well known, but for the parabolic spline the estimate (7) is an interesting, because in this case the nodes of the spline with interpolation nodes coincide.

For the convergence of the spline, a more accurate result is also true. Let us $u(x) \in C_h[0,1]$. Then

$$\lim_{\varepsilon/h \to +\infty} \| S(x;u) - S_3(x;u) \|_{C[0,1]} = 0.$$

3. Proof of Theorems

The estimate (3) for $j = 0$ is given in [4] and is proved in [3]. It remains to prove the estimates of the error in the approximation of the derivatives.

We represent the spline $S(x,u)$ in terms of basic splines:

$$S(x;u) = \sum_{n=-3}^{N-1} \alpha_n N_{n,3}(x), \tag{8}$$

where the exponential basis splines $N_{n,3}(x)$ are constructed recurrently.

First we set $N_{n,1}(x)$:

$$N_{n,1}(x) = \begin{cases} \frac{1-e^{\alpha(x_n-x)/\varepsilon}}{1-e^{-\alpha h/\varepsilon}}, \ x \in [x_n, x_{n+1}), \\ \frac{e^{\alpha(x_{n+1}-x)/\varepsilon} - e^{-\alpha h/\varepsilon}}{1-e^{-\alpha h/\varepsilon}}, \ x \in [x_{n+1}, x_{n+2}), \\ 0, \ x \notin [x_n, x_{n+2}). \end{cases} \tag{9}$$

Exponential B-splines of higher orders are defined recurrently:

$$N_{n,k+1}(x) = \frac{1}{h} \int_{x_n}^{x} (N_{n,k}(s) - N_{n+1,k}(s))ds, \quad k = 1,2. \tag{10}$$

In accordance with [3], the following lemma holds.

Lemma 1 *Let the function $u(x)$ be representable in the form (1). Then there exists a function $\tilde{S}(x;u) \in SL(\Omega,3,1)$ and the constant $C > 0$ such that the following estimates are valid*

$$\| \tilde{S}^{(j)}(x;u) - u^{(j)}(x) \|_{C[0,1]} \le C \min\{h^{3-j}, \frac{h^{4-j}}{\varepsilon}\}, \varepsilon \in (0,1], \ 0 \le j \le 2. \tag{11}$$

We introduce the function $err(x) = S(x;u) - \tilde{S}(x;u)$ and represent it in the form

$$err(x) = \sum_{n=-3}^{N-1} \beta_n N_{n,3}(x), \tag{12}$$

where $N_{n,k}(x)$ are normalized exponential B-splines corresponding to (9), (10), generators for each $k = 1, 2, 3$ partition of the unit [3]. We estimate the coefficients β_n, taking into account that $u(x)$ corresponds to (1). Let us consider two cases: 1) $h/\varepsilon \leq C$ and 2) $h/\varepsilon > C$, where C is some constant.

In the first case the following estimate

$$\max_{-3\leq n\leq N-1} |\beta_n| \leq C \max\{h^4/\varepsilon, h^4\} \tag{13}$$

is proved [3].

In the second case according to [3]

$$\max_{-3\leq n\leq N-1} |\beta_n| \leq Ch^3. \tag{14}$$

By virtue of (10), $N'_{i,k}(x) = \frac{1}{h}(N_{i,k-1}(x) - N_{i+1,k-1}(x))$. Therefore, the following relations hold:

$$err'(x) = \frac{1}{h} \sum_{n=-3}^{N-1} \beta_n (N_{n,2}(x) - N_{n+1,2}(x)), \tag{15}$$

$$err''(x) = \frac{1}{h^2} \sum_{n=-3}^{N-1} \beta_n (N_{n+2,1}(x) - 2N_{n+1,1}(x) + N_{n,1}(x)). \tag{16}$$

Since by the non-negativity of exponential B-splines and the fact that they form a partition of unity, for any i will be $0 \leq N_{i,k}(x) \leq 1$. Then from (11), (13), (15), (16) we obtain the estimate (3) for $j = 1, 2$ in the first case. The estimate (3) for $j = 1, 2$ in the second case is obtained similarly with account of (14). This proves Theorem 1.

Now we dwell on the proof of Theorem 2. We represent the cubic spline $S_3(x;u)$ in terms of basic splines:

$$S_3(x;u) = \sum_{n=-3}^{N-1} g_n \tilde{N}_{n,3}(x). \tag{17}$$

First we set $\tilde{N}_{n,1}(x)$:

$$\tilde{N}_{n,1}(x) = \begin{cases} \frac{x-x_n}{x_{n+1}-x_n}, x \in [x_n, x_{n+1}] \\ \frac{x_{n+2}-x}{x_{n+2}-x_{n+1}}, x \in [x_{n+1}, x_{n+2}], \ -1 \leq n \leq N-1 \\ 0, x \notin [x_n, x_{n+2}) \end{cases} \tag{18}$$

Polynomial B-splines of higher orders are defined recurrently:

$$\tilde{N}_{n,k+1}(x) = \frac{1}{h} \int_{x_n}^{x} (\tilde{N}_{n,k}(s) - \tilde{N}_{n+1,k}(s)) ds, \quad k = 1, 2. \tag{19}$$

From (9), (18) it follows that for any n

$$\lim_{\varepsilon\to 0+0} \int_0^1 |N_{n,1}(x) - \tilde{N}_{n,0}(x)| dx = 0, \lim_{\varepsilon\to+\infty} \| N_{n,1}(x) - \tilde{N}_{n,1}(x) \|_{C[0,1]} = 0.$$

Then, taking into account (10), (19) we obtain

$$\lim_{\varepsilon\to 0+0} \| N_{i,k}(x) - \tilde{N}_{i,k-1}(x) \|_{C[0,1]} = 0, \quad \lim_{\varepsilon\to +\infty} \| N_{i,k}(x) - \tilde{N}_{i,k}(x) \|_{C[0,1]} = 0.$$

So,

$$\lim_{\varepsilon\to 0+0} N_{i,k}(x) = \tilde{N}_{i,k-1}(x), \quad \lim_{\varepsilon\to +\infty} N_{i,k}(x) = \tilde{N}_{i,k}(x), \ x \in [0,1]. \tag{20}$$

Taking into account (8), (17), (20) we obtain that in the limit the coefficients α_n and g_n satisfy the same system of equations if $\varepsilon \to +\infty$. This proves that $\lim\limits_{\varepsilon\to+\infty} \alpha_n = g_n$. This proves (4).

The case $\varepsilon \to 0$ is treated similarly. This proves Theorem 2.

4. Results of numerical experiments

We define the function of the form (1)

$$u(x) = \cos\frac{\pi x}{2} + e^{-\frac{x}{\varepsilon}}, \quad x \in [0,1]. \tag{21}$$

Table 1. Errors of the cubic spline on the uniform mesh

ε	N					
	2^4	2^5	2^6	2^7	2^8	2^9
1	$2.82\cdot10^{-7}$	$1.76\cdot10^{-8}$	$1.16\cdot10^{-9}$	$1.02\cdot10^{-10}$	$4.30\cdot10^{-12}$	$2.68\cdot10^{-13}$
10^{-1}	$3.43\cdot10^{-4}$	$2.33\cdot10^{-5}$	$1.51\cdot10^{-6}$	$9.58\cdot10^{-8}$	$6.03\cdot10^{-9}$	$4.11\cdot10^{-10}$
10^{-2}	0.43	$8.38\cdot10^{-2}$	$9.72\cdot10^{-3}$	$8.00\cdot10^{-4}$	$5.59\cdot10^{-5}$	$3.65\cdot10^{-6}$
10^{-3}	9.88	4.58	1.93	0.66	0.15	$2.03\cdot10^{-2}$
10^{-4}	$1.05\cdot10^{2}$	$5.23\cdot10^{1}$	$2.58\cdot10^{1}$	$1.25\cdot10^{1}$	5.90	2.59
10^{-5}	$1.06\cdot10^{3}$	$5.30\cdot10^{2}$	$2.64\cdot10^{2}$	$1.32\cdot10^{2}$	$6.56\cdot10^{1}$	$3.24\cdot10^{1}$
10^{-6}	$1.06\cdot10^{4}$	$5.30\cdot10^{3}$	$2.65\cdot10^{3}$	$1.33\cdot10^{3}$	$6.62\cdot10^{2}$	$3.30\cdot10^{2}$
10^{-7}	$1.06\cdot10^{5}$	$5.30\cdot10^{4}$	$2.65\cdot10^{4}$	$1.33\cdot10^{4}$	$6.63\cdot10^{3}$	$3.30\cdot10^{3}$
10^{-8}	$1.06\cdot10^{6}$	$5.30\cdot10^{5}$	$2.65\cdot10^{5}$	$1.33\cdot10^{5}$	$6.63\cdot10^{4}$	$3.31\cdot10^{4}$

The results of the calculations are summarized in four tables. The tables show the maximum errors in computing the function and its derivatives on the basis of splines, calculated at the nodes of the thickened grid obtained from the given computational grid by dividing each of its grid intervals into 10 equal parts.

In Table. 1 are given the errors of cubic spline in the case of the function (21). The error increases without limit as the parameter ε decreases. This shows that the use of a cubic spline on a uniform grid is not effective in the presence of a boundary layer.

Next, we investigate the accuracy of the approximation of a function and its derivatives on the basis of the exponential spline. Errors and calculated orders of accuracy of the exponential spline for the function (1) are given in the Table 2. In Table. 3 are given the errors and calculated accuracy orders for the first derivative of the function (21), and in Table. 4 the corresponding results for second derivative. The numerical results from Tables 2,3 and 4 are in agreement with the estimates of Theorem 1.

IOP Conf. Series: Journal of Physics: Conf. Series **1050** (2018) 012012 doi:10.1088/1742-6596/1050/1/012012

Table 2. Errors and calculated accuracy orders for the exponential spline

ε	N					
	2^2	2^3	2^4	2^5	2^6	2^7
1	$1.56 \cdot 10^{-4}$	$9.75 \cdot 10^{-6}$ 4.01	$6.05 \cdot 10^{-7}$ 4.01	$3.77 \cdot 10^{-8}$ 4.00	$2.36 \cdot 10^{-9}$ 4.00	$1.47 \cdot 10^{-10}$ 4.00
10^{-1}	$1.25 \cdot 10^{-3}$	$7.18 \cdot 10^{-5}$ 4.12	$4.17 \cdot 10^{-6}$ 4.11	$2.50 \cdot 10^{-7}$ 4.06	$1.53 \cdot 10^{-8}$ 4.03	$9.46 \cdot 10^{-10}$ 4.02
10^{-2}	$3.73 \cdot 10^{-3}$	$4.33 \cdot 10^{-4}$ 3.10	$4.22 \cdot 10^{-5}$ 3.36	$3.02 \cdot 10^{-6}$ 3.81	$1.78 \cdot 10^{-7}$ 4.08	$1.03 \cdot 10^{-8}$ 4.11
10^{-3}	$3.83 \cdot 10^{-3}$	$4.83 \cdot 10^{-4}$ 2.99	$6.02 \cdot 10^{-5}$ 3.00	$7.43 \cdot 10^{-6}$ 3.01	$8.62 \cdot 10^{-7}$ 3.08	$9.21 \cdot 10^{-8}$ 3.25
10^{-4}	$3.95 \cdot 10^{-3}$	$4.83 \cdot 10^{-4}$ 3.03	$6.05 \cdot 10^{-5}$ 2.99	$7.57 \cdot 10^{-6}$ 3.00	$9.46 \cdot 10^{-7}$ 3.00	$1.18 \cdot 10^{-7}$ 3.00
10^{-5}	$3.96 \cdot 10^{-3}$	$4.86 \cdot 10^{-4}$ 3.03	$6.06 \cdot 10^{-5}$ 3.00	$7.57 \cdot 10^{-6}$ 3.00	$9.46 \cdot 10^{-7}$ 3.00	$1.18 \cdot 10^{-7}$ 3.00
10^{-6}	$3.97 \cdot 10^{-3}$	$4.86 \cdot 10^{-4}$ 3.03	$6.06 \cdot 10^{-5}$ 3.00	$7.57 \cdot 10^{-6}$ 3.00	$9.46 \cdot 10^{-7}$ 3.00	$1.18 \cdot 10^{-7}$ 3.00
10^{-7}	$3.97 \cdot 10^{-3}$	$4.86 \cdot 10^{-4}$ 3.03	$6.06 \cdot 10^{-5}$ 3.00	$7.57 \cdot 10^{-6}$ 3.00	$9.46 \cdot 10^{-7}$ 3.00	$1.18 \cdot 10^{-7}$ 3.00
10^{-8}	$3.96 \cdot 10^{-3}$	$4.86 \cdot 10^{-4}$ 3.03	$6.06 \cdot 10^{-5}$ 3.00	$7.57 \cdot 10^{-6}$ 3.00	$9.46 \cdot 10^{-7}$ 3.00	$1.18 \cdot 10^{-7}$ 3.00

Table 3. Errors and calculated accuracy orders for the first derivative of the function (21)

ε	N					
	2^2	2^3	2^4	2^5	2^6	2^7
1	$2.36 \cdot 10^{-4}$	$2.91 \cdot 10^{-4}$ 3.02	$3.61 \cdot 10^{-5}$ 3.01	$4.49 \cdot 10^{-6}$ 3.01	$5.60 \cdot 10^{-7}$ 3.00	$7.00 \cdot 10^{-8}$ 3.00
10^{-1}	$1.53 \cdot 10^{-2}$	$1.86 \cdot 10^{-3}$ 3.04	$2.23 \cdot 10^{-4}$ 3.06	$2.71 \cdot 10^{-5}$ 3.04	$3.34 \cdot 10^{-6}$ 3.02	$4.14 \cdot 10^{-7}$ 3.01
10^{-2}	$6.29 \cdot 10^{-2}$	$1.27 \cdot 10^{-2}$ 2.31	$2.21 \cdot 10^{-3}$ 2.52	$2.89 \cdot 10^{-4}$ 2.93	$3.64 \cdot 10^{-5}$ 2.99	$4.38 \cdot 10^{-6}$ 3.05
10^{-3}	$7.76 \cdot 10^{-2}$	$1.9183 \cdot 10^{-2}$ 2.02	$4.58 \cdot 10^{-3}$ 2.07	$1.04 \cdot 10^{-3}$ 2.14	$2.14 \cdot 10^{-4}$ 2.28	$4.02 \cdot 10^{-5}$ 2.41
10^{-4}	$7.32 \cdot 10^{-2}$	$2.00 \cdot 10^{-2}$ 1.98	$5.00 \cdot 10^{-3}$ 2.00	$1.24 \cdot 10^{-3}$ 2.01	$3.04 \cdot 10^{-4}$ 2.03	$7.30 \cdot 10^{-5}$ 2.05
10^{-5}	$7.95 \cdot 10^{-2}$	$2.01 \cdot 10^{-2}$ 1.98	$5.03 \cdot 10^{-3}$ 2.00	$1.26 \cdot 10^{-3}$ 2.00	$3.14 \cdot 10^{-4}$ 2.00	$7.88 \cdot 10^{-5}$ 2.00
10^{-6}	$7.95 \cdot 10^{-2}$	$2.01 \cdot 10^{-2}$ 1.98	$5.04 \cdot 10^{-3}$ 2.00	$1.26 \cdot 10^{-3}$ 2.00	$3.15 \cdot 10^{-4}$ 2.00	$7.87 \cdot 10^{-5}$ 2.00
10^{-7}	$7.95 \cdot 10^{-2}$	$2.01 \cdot 10^{-2}$ 1.98	$5.04 \cdot 10^{-3}$ 1.99	$1.26 \cdot 10^{-3}$ 2.00	$3.15 \cdot 10^{-4}$ 2.00	$7.88 \cdot 10^{-5}$ 2.00
10^{-8}	$7.95 \cdot 10^{-2}$	$2.01 \cdot 10^{-2}$ 1.98	$5.04 \cdot 10^{-3}$ 1.99	$1.26 \cdot 10^{-3}$ 2.00	$3.15 \cdot 10^{-4}$ 2.00	$7.88 \cdot 10^{-5}$ 2.00

Table 4. Errors and calculated accuracy orders for the second derivative of the function (21).

ε	N					
	2^2	2^3	2^4	2^5	2^6	2^7
1	$4.55 \cdot 10^{-2}$	$1.08 \cdot 10^{-2}$ 2.04	$2.62 \cdot 10^{-3}$ 2.05	$6.42 \cdot 10^{-4}$ 2.03	$1.58 \cdot 10^{-4}$ 2.02	$3.95 \cdot 10^{-5}$ 2.01
10^{-1}	$2.60 \cdot 10^{-1}$	$6.72 \cdot 10^{-2}$ 1.95	$1.66 \cdot 10^{-2}$ 2.02	$4.08 \cdot 10^{-3}$ 2.02	$1.01 \cdot 10^{-3}$ 2.01	$2.51 \cdot 10^{-4}$ 2.00
10^{-2}	$7.87 \cdot 10^{-1}$	$3.53 \cdot 10^{-1}$ 1.15	$1.33 \cdot 10^{-1}$ 1.40	$4.03 \cdot 10^{-2}$ 1.73	$1.04 \cdot 10^{-2}$ 1.95	$2.57 \cdot 10^{-3}$ 2.02
10^{-3}	$9.05 \cdot 10^{-1}$	$4.63 \cdot 10^{-1}$ 0.97	$2.26 \cdot 10^{-1}$ 1.03	$1.07 \cdot 10^{-1}$ 1.08	$4.74 \cdot 10^{-2}$ 1.17	$1.87 \cdot 10^{-2}$ 1.34
10^{-4}	$9.18 \cdot 10^{-1}$	$4.76 \cdot 10^{-1}$ 0.95	$2.40 \cdot 10^{-1}$ 0.99	$1.19 \cdot 10^{-1}$ 1.01	$5.90 \cdot 10^{-2}$ 1.02	$2.88 \cdot 10^{-2}$ 1.03
10^{-5}	$9.20 \cdot 10^{-1}$	$4.78 \cdot 10^{-1}$ 0.94	$2.41 \cdot 10^{-1}$ 0.97	$1.21 \cdot 10^{-1}$ 1.00	$6.03 \cdot 10^{-2}$ 1.00	$3.01 \cdot 10^{-2}$ 1.00
10^{-6}	$9.20 \cdot 10^{-1}$	$4.78 \cdot 10^{-1}$ 0.94	$2.41 \cdot 10^{-1}$ 0.97	$1.21 \cdot 10^{-1}$ 1.00	$6.05 \cdot 10^{-2}$ 1.00	$3.02 \cdot 10^{-2}$ 1.00
10^{-7}	$9.20 \cdot 10^{-1}$	$4.78 \cdot 10^{-1}$ 0.94	$2.41 \cdot 10^{-1}$ 0.97	$1.21 \cdot 10^{-1}$ 1.00	$6.05 \cdot 10^{-2}$ 1.00	$3.02 \cdot 10^{-2}$ 1.00

5. Conclusion

The problem of the interpolation of functions with large gradients in an exponential boundary layer is investigated. On the uniform grid the exponential spline which is exact on the boundary layer component of the interpolated function is constructed. For the constructed spline error estimates are obtained that are uniform with respect to a small parameter. This advantageously distinguishes the constructed spline from the cubic spline, whose error increases with decreasing of perturbation parameter. It is proved that the exponential spline turns into cubic if the perturbing parameter tends to infinity, and parabolic if this parameter tends to zero.

Acknowledgments

The work was supported by the program of fundamental scientific researches of the SB RAS 1.1.3., project 0314-2016-0009.

References

[1] Blatov I A, Zadorin A I and Kitaeva E V 2017 Cubic Spline Interpolation of Functions with High Gradients in Boundary Layers *Comput. Math. Math. Phys.* **57** 9-28

[2] Miller J J H, O'Riordan E and Shishkin G I 2012 *Fitted Numerical Methods for Singular Perturbation Problems: Error Estimates in the Maximum Norm for Linear Problems in One and Two Dimensions* (Singapore: World Scientific)

[3] Blatov I A, Zadorin A I and Kitaeva E V 2018 On the uniform in the parameter convergence of the exponential spline-interpolation with the presence of a boundary layer *Comput. Math. Math. Phys.* **58** 348-363

[4] Blatov I A, Zadorin A I, Kitaeva E V 2017 Proc. Int. Conf. on Computational and Applied Mathematics (Novosibirsk: ICMMG SB RAS) 100-105

[5] Ahlberg J H, Nilson E N and Walsh J L 1967 *The theory of splines and their applications* (New York: Academic Press)

[6] Strelkova E V and Shevaldin V T 2017 On uniform Lebesgue constants of local exponential splines with equidistant knots *Proc. Steklov Inst. Math.* **296** 206-217

[7] Mohammadi R 2015 Exponential B-spline collocation method for numerical solution of the generalized regularized long wave equation *Chinese Phys.* B **24** 050206 https://doi.org/10.1088/1674-1056/24/5/050206

Experimental studies of correction propulsion system elements for small space vehicles manufactured due to additive method

V N Blinov[1], V V Shalay[1], V I Kuznetsov[1], A B Yakovlev[1], P S Yachmenev[1], A I Luyanchik[1], V V Kositsin[2]

[1]Omsk State Technical University, Omsk, Russian Federation
[2] Federal State Unitary Enterprise "Federal Scientific and Production Centre "Progress", Omsk, Russian Federation

Abstract. Producing ammonia correction propulsion system (CPS) elements for maneuvering satellite platforms (MSP) of small space vehicles (SSV) is a relevant problem. The investigation is devoted to the solution of the named problem with the use of direct metal laser sintering (DMLS) method. The research objective is to confirm the feasibility of manufacturing ETMT and CPS evaporator with autonomous heating elements (AHE) by DMLS method, based on the prototypes experimental testing. During the research the following tasks were solved: creating 3D models for ETMT and double-threaded evaporator and producing experimental prototypes by DMSL method. 3D models of ETMT and evaporator casings were developed following the prototypes produced by the conventional methods of turning and milling. 3D models of ETMT and evaporator casings represent complex integral parts with multiple passages for working medium flow. Experimental studies of ETMT and the evaporator were performed with nitrogen as a working medium. ETMT and evaporator temperature characteristics were determined during the experiments. The investigation was made of ETMT with nominal thrust of 30 mN and power consumption in the range of 5-60 W with and without heat insulation. AHE with embedded thermocouples, having the diameter of 6 mm and power consumption of 60 W, was used. AHE temperature was limited by 973 K. A double-threaded evaporator was investigated for power consumption of 5-30 W, the evaporator casing temperature limited by 393 K. The maximal increase in the gas temperature equaled no more than 8.6 % at the nozzle exit in the power consumption range of 10-60 W for ETMT with heat insulation. At ETMT power consumption of 5–50 W, the build-up time for ETMT was 400–600 s. While at power consumption of 50–60 W, it was 200–400 s. At power consumption of 10–30 W, the evaporator casing temperature reached 393 K in 100–340 s, AHE temperature being 400–460 K and the gas temperature at the evaporator throttle exit being no more than 290 K. At power consumption of 5 W, the maximum evaporator casing temperature of 375 K was reached in 1200 s, AHE temperature being 370 K and the gas temperature at the evaporator throttle exit being no more than 302 K.

1. Introduction

Currently the significance of developing MSP with ammonia CPS and ETMT is becoming increasingly important, since the modern SSV call for completing a wide range of orbital maneuvering challenges: error correction of injection by launch vehicles, maintaining orbital parameters, interorbital maneuvering, construction of SSV orbit groups, injection of SSV into utilization orbit, inspection of other SSV and orbital objects, near-space surveillance and etc. [1–3].

Applying ammonia CPS in SSV is possible due to high specific burst of power for ETMT being achievable at minimal energy consumption. It results in decreased fuel weight necessary to realize target characteristic speed of CPS in SSV during orbital maneuvering, as well as the weight of CPS itself and its means of adaptation in SSV.
Recently constructed ammonia CPS with ETMT were a part of SSV "Ugatusat" (Russia, 2008) with 35 kg weight and SSV "VH-2" (China, 2016) with 47 kg weight. Ammonia CPS were created in Ukraine for SSV "SICH - 2M" (2011–2012) and SSV "Microsat".
In such CPS the ammonia becomes gaseous due to its preliminary gasification in the evaporator and final heating in ETMT.
The developed ammonia ETMT with 30 mN thrust are characterized by the cost of thrust up to 2 W/mN and are used as components in SSV with weight of 30–400 kg. The specific burst of power for such ETMT at energy consumption of 60 W is not higher than 250 s. Due to their structure, they can be classified as ETMT with tubular heating element (THE) and ETMT with AHE (figures 1,2) [1–3,10].

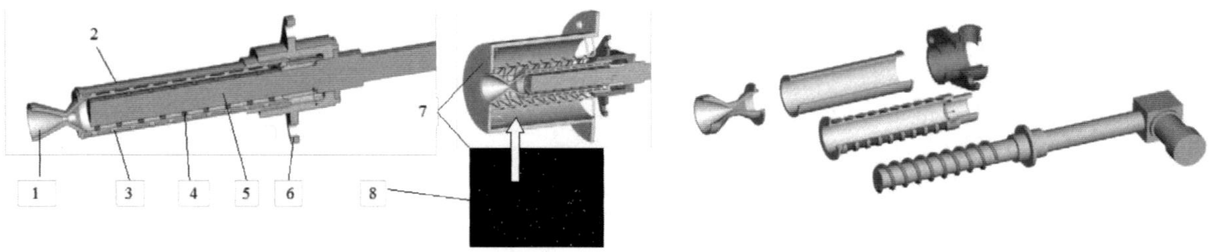

1 - nozzle; 2 - outer casing; 3 - inner casing with gas ducts in double-start thread form; 4 - gas flow conditioner; 5 - AHE; 6 - microthruster attachment point; 7 - heat-protective cover; 8 -heat insulation

Figure 1. Schematics of ETMT with AHE.

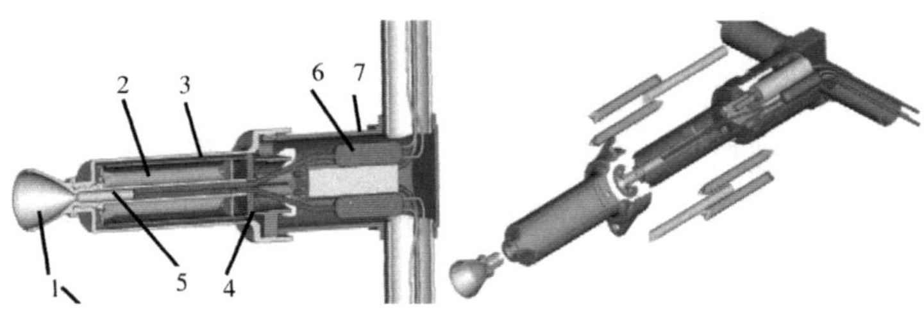

1 - bell nozzle; 2 - THE (main and additional); 3 - casing; 4 - shouldered cup; 5 - gas duct; 6 - thermocouple; 7 - terminal casing

Figure 2. Schematics of ETMT with THE.

The known ETMT represent complex constructions with inner cavities and gas ducts, for example, as shown in figure 3 [1–9].

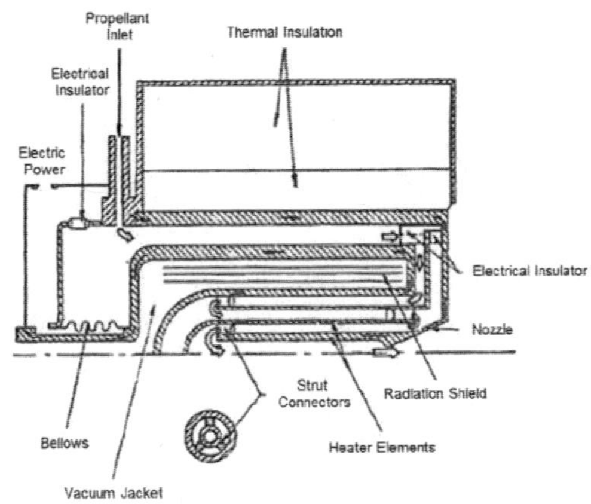

Figure 3. J3 concentric tubular resistojet schematics.

In terms of design ETMT with AHE can be divided into AHE and ETMT casing, that are manufactured separately (figures 1, 4, 5).

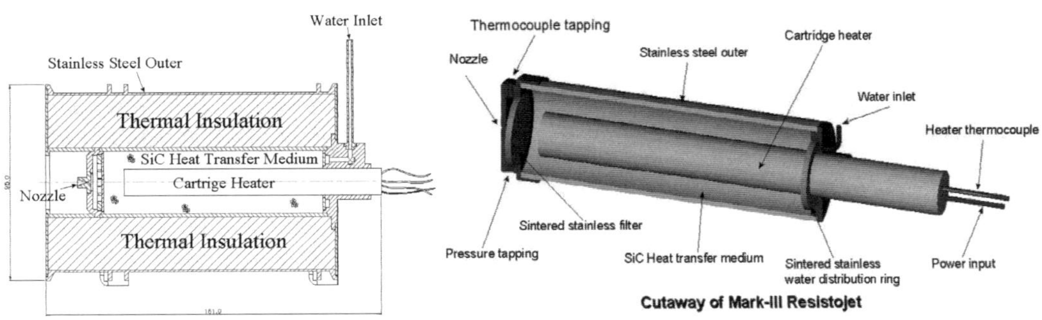

Figure 4. Mark-III, Mark-IY microthrusters schematics.

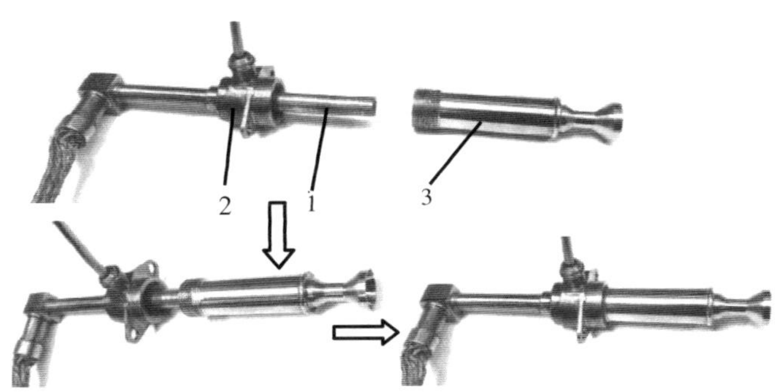

1 - AHE; 2 - fuel inlet and attachment node; 3 - casing with nozzle and gas flow conditioner (located inside).

Figure 5. ETMT with AHE, manufactured by turning and milling.

The conventional method of manufacturing ETMT casing consists of milling and turning its individual parts and forming various cavities and gas ducts. The components are sealed by welding after the casing assembly. Stainless steel 12X18H10T is used as a constructional material. All this results in additional flange joints appearance and ETMT casing walls thickening due to technological characteristics and incorporated equipment limitations. Moreover, technological constraints preclude one from manufacturing inner gaseous cavities and flanges necessary for ETMT optimal performance.

A liquefied ammonia evaporator used in ammonia CPS production also includes AHE and a separate casing (figure 6).

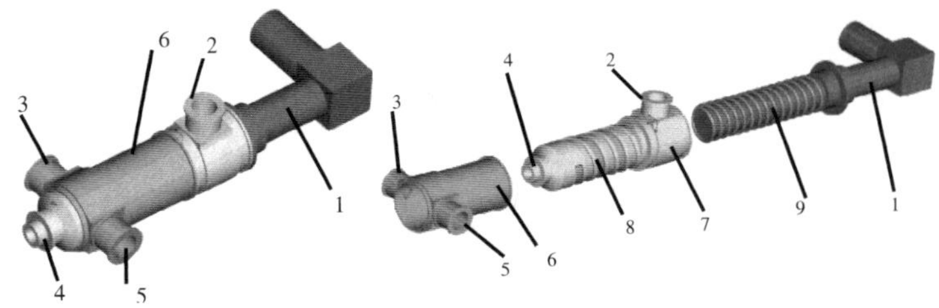

1 - AHE; 2-5 - ammonia inlet/outlet nozzles; 6,7 - inner and outer casing; 8 - ammonia feeding line in double-start thread form; 9 - swirler

Figure 6. Double-threaded liquefied ammonia evaporator.

The example setups for ETMT and the evaporator with AHE as a part of CPS automatic equipment are given in figure 7.

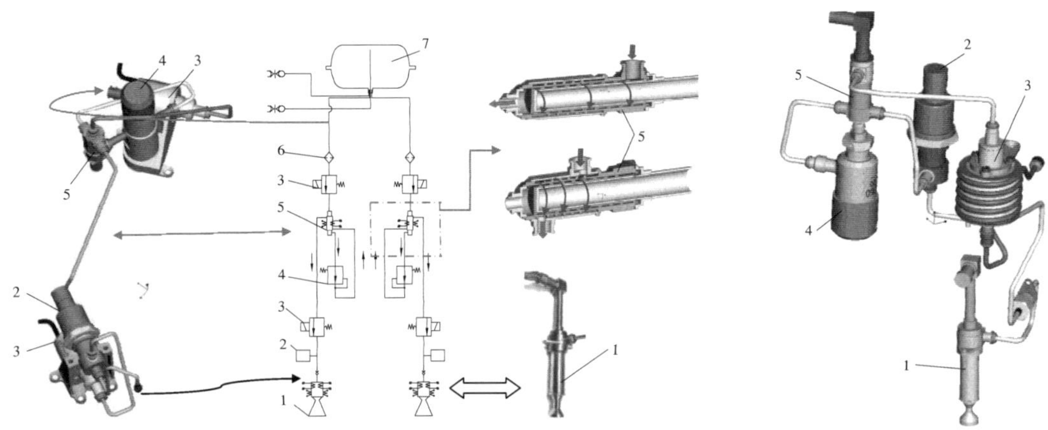

a b

1 - ETMT with AHE; 2 - pressure transducer; 3 - electro-pneumatic valve; 4 - pressure controller; 5 - double-threaded ammonia evaporator with AHE; 6 - filter; 7 - tank

Figure 7. ETMT and liquefied ammonia evaporator with AHE as a part of CPS reserved automatic equipment (a) and unreserved one (b).

Manufacturing AHE and ETMT casing separately allows modern additive methods to be applied to ETMT casing production, this casing being a complex construction with gas ducts and inner cavities.

2. Problem statement

The research objective is to determine whether it is possible to produce ETMT with AHE and the evaporator with AHE, both being a part of the ammonia CPS, following the additive technique based on direct metal laser sintering (DMLS) method.

To solve the task at hand:

- 3D models of ETMT and the evaporator were developed;
- ETMT and the evaporator were produced by DMLS method;
- experimental tests on ETMT and the evaporator with AHE were performed in a vacuum chamber during operation on nitrogen.

Electrical testing of ETMT in the power consumption range of 5-60 W was carried out to study:

IOP Conf. Series: Journal of Physics: Conf. Series **1050** (2018) 012013 doi:10.1088/1742-6596/1050/1/012013

- ETMT heating dynamics at working medium feeding in the vacuum chamber (cold starting method) without heat insulation;
- ETMT heating dynamics at working medium feeding in the vacuum chamber (cold starting method) with heat insulation.

A cylindrical electric heater with two thermocouples was used as AHE, its specifications are:

- nominal power capacity, 60 W;
- working voltage, 12 W;
- diameter, 6.0 mm;
- length, 80 mm;
- weight, 0.03 kg.

To lower heat loss in ETMT, multi-layer heat insulation was used. It was placed into a cylindrical jacket made of composite material, consisting of fabrics HVK-0.05, foil, alloy BD3D, 8 μm.

Evaporator electrical testing was conducted to study the temperatures of AHE, the evaporator casing and the gas at the exit for power consumption of 5–30 W. The maximum temperature of the evaporator casing was maintained at 393 K. This constraint was introduced following the temperature limitations of the pressure controller.

3. Theory

When producing ETMT casing and the evaporator, direct metal laser sintering (DMLS) method was used, this method being widespread in 3D printing of metal products. DMLS is an additive process allowing ETMT casing and the evaporator to be built up gradually layer by layer.

Usually DMLS calls for production of supporting elements which are deleted after the printing. These elements provide the precision of ETMT and evaporator casing and no rigid component parts, considering the thermal processes during melting. All surfaces with the ground plane angle less than 45° are supported. Moreover the ratio of the walls height to their thickness should be maintained within 8:1.

To avoid unnecessary supports, smooth transitions are introduced in 3D model of ETMT and the evaporator. Sharp edges were also rounded at the joints of the walls and inner elements.

3D model of ETMT is constructed in such a way that a single casing includes nozzle, outer casing, inner casing with gas ducts in the double-start thread form, gas flow conditioner and microthruster attachment point (figures 8, 9). 3D model of the evaporator casing (figure 10) is constructed in a similar way.

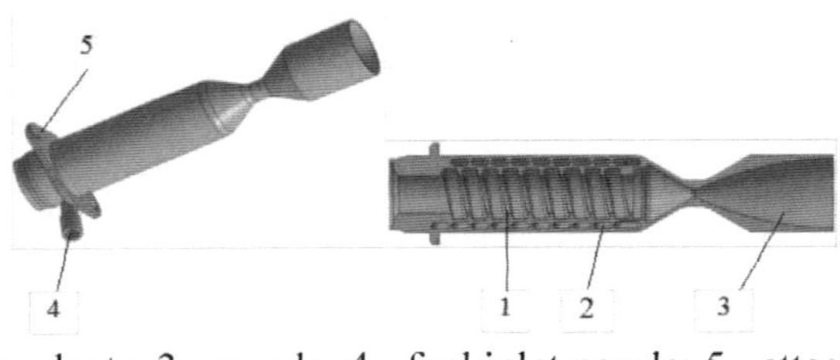

1 - swirler; 2 - gas ducts; 3 - nozzle; 4 - fuel inlet nozzle; 5 - attachment point.

Figure 8. 3D model of ETMT casing.

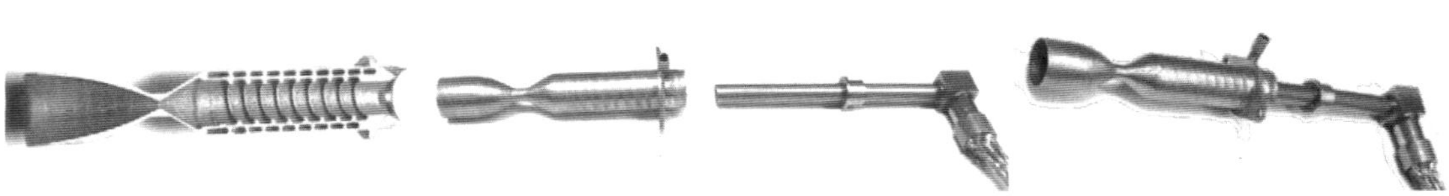

a b c

a - ETMT casing profile; b - ETMT and AHE casing; c - ETMT casing assembled with AHE

Figure 9. Production prototype of ETMT with AHE.

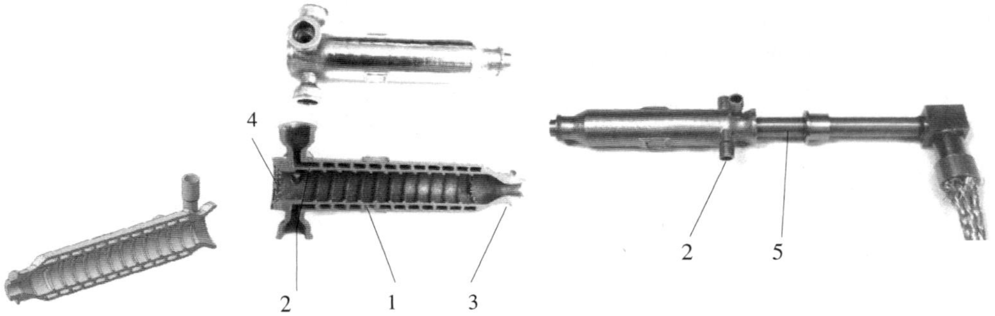

1 - gas ducts; 2 - crosscut working medium inlets/outlets (two variants are provided); 3,4 - axial working medium inlets/outlets; 5 - AHE

Figure 10. 3D model and production prototype of evaporator.

Production prototype of the evaporator assembled with the pressure controller is shown in figure 11.

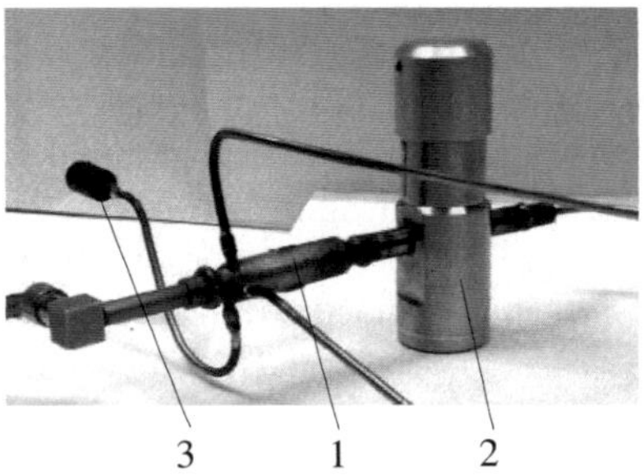

1 - evaporator; 2 - pressure controller; 3 - throttle at evaporator gas outlet

Figure 11. Production prototype of evaporator with pressure controller.

Technological characteristics of manufacturing ETMT casing by the additive technique compared to manufacturing ETMT casing by turning and milling component parts are given in table 1. The evaporator has similar technological characteristics.

Table 1. Technological characteristics of manufacturing ETMT casing.

IOP Conf. Series: Journal of Physics: Conf. Series **1050** (2018) 012013 doi:10.1088/1742-6596/1050/1/012013

№	Characteristics	Manufacturing technique for ETMT casing	
		Turning, milling	Additive, DMLS
1	The number of parts in ETMT casing	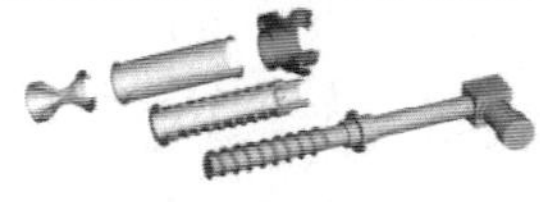 5	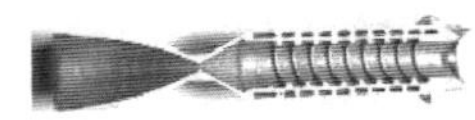 1
2	Gas ducts, fittings and joints sealing provision	Gas migration is possible among cavities	Is provided
3	The possibility of manufacturing any inner cavities, flanges	Is not provided	Is provided
4	The possibility of manufacturing thin-walled shells	Technological constraints are available	Technological constraints are minimal
6	The possibility of machining inner cavities	Is available	Is not available, except nozzle part
7	The achieved ETMT casing weight	25	22
8	The possibility of lowering ETMT casing weight	Technological constraints are available	Technological constraints are minimal

ETMT experimental studies were performed in the vacuum chamber (figure 11). To document the outflowing gas temperature, a thermocouple was installed at ETMT nozzle exit.

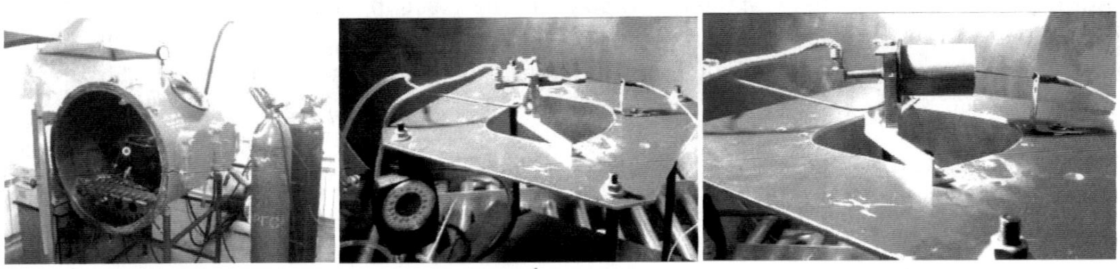

a b c

Figure 11. Installing ETMT in the vacuum chamber (a) without heat-protective cover (b) and with it (c).

Pneumatic-hydraulic diagram of the experimental installation with the vacuum chamber in ETMT testing is given as shown in figure 12.

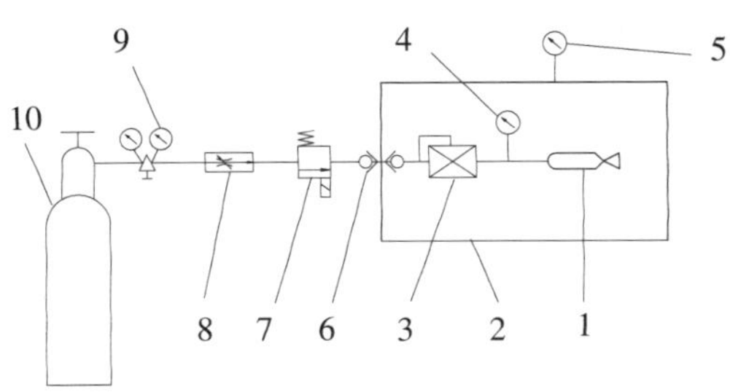

1 - ETMT; 2 - vacuum chamber; 3 - pressure controller; 4 - pressure gauge; 5 - vacuum gauge; 6 - electrofluid connector; 7 - electrovalve; 8 - filter; 9 - pressure gauge; 10 - nitrogen filled vessel.

Figure 12. Pneumatic-hydraulic diagram of the experimental installation.

ETMT electrical tests were performed with heat insulation as well as without one. To lower heat loss in ETMT, multi-layer heat insulation was used, placed in a heat-protective cover from composite material, consisting of fabrics HVK-0.05 TU 6-48-05-786904-151-95, foil, alloy BD3D, 8 μm TU 48-21-151-84.

IOP Conf. Series: Journal of Physics: Conf. Series **1050** (2018) 012013 doi:10.1088/1742-6596/1050/1/012013

Cold starting method was used at ETMT (and evaporator) testing, which means that voltage supply and working medium feeding to ETMT (and evaporator) were performed simultaneously.
When testing the evaporator, pneumatic-hydraulic diagram similar to the one in figure 12 was used. The evaporator was connected to the pressure controller, and the throttle imitating ETMT was installed instead of ETMT. The thermocouples measured the temperatures of the gas at the throttle exit and of the evaporator casing. AHE temperature was measured by the embedded thermocouples.

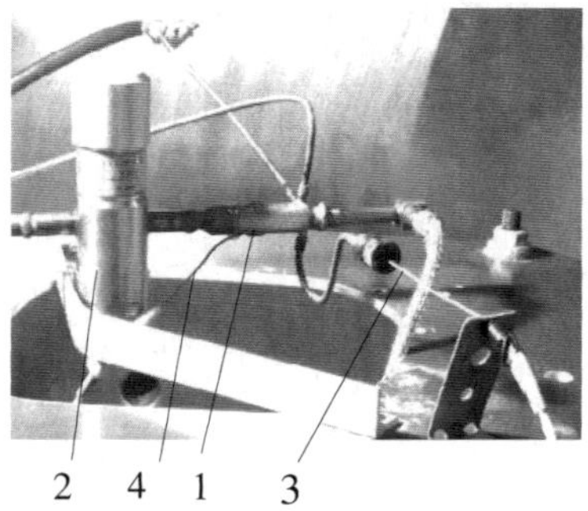

1 - evaporator;
2 - pressure controller;
3,4 - thermocouples

Figure 13. Installing the evaporator in the vacuum chamber.

4. Experimental results

The study subjects were ETMT and the evaporator, produced by DMLS method. More than 12 evaporator and ETMT casings were manufactured. All resulting casing samples were characterized by geometric shapes stability, the nozzle included. All minor shape roughness was removed by machining (figure 14).

Figure 14. Production prototypes of the evaporator and ETMT casings.

The inner nozzle surface can be refined as well, if necessary. All ETMT and evaporator casings were inspected for integrity and gas permeability. Deviations in integrity and throats were not found. When welding was used during ETMT and the evaporator assembly, the material resulting from DMLS method possessed good weld ability.
The experimental testing results for the temperatures of AHE and the gas at ETMT nozzle exit during cold starting method (without preheating) are given in figures 15–18 and tables 2–5.

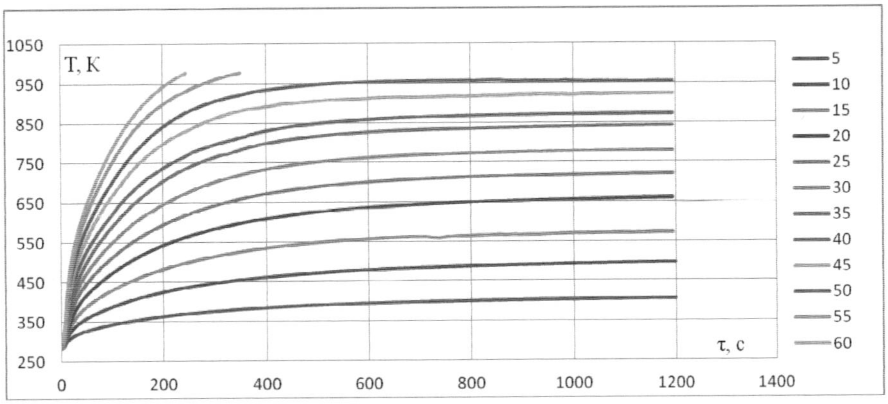

Figure 15. The dependence of AHE temperature on the heating time for ETMT without a heat-protective cover at various values of supplied power with AHE heating temperature limited by 973 K.

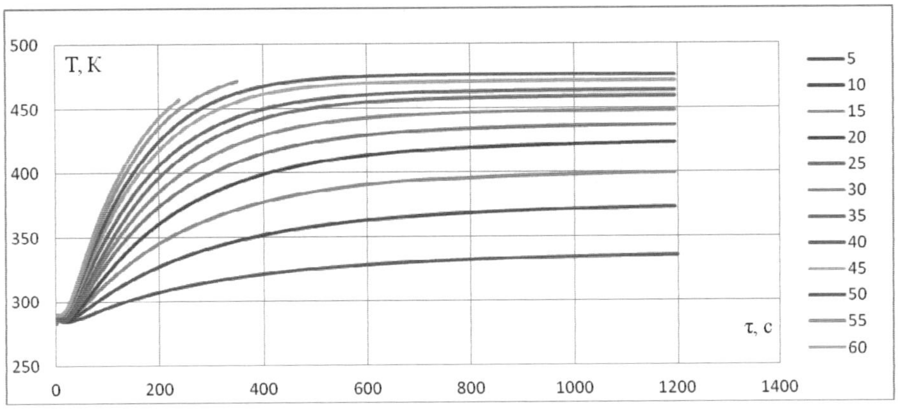

Figure 16. The dependence of the gas temperature at the nozzle exit on the heating time for ETMT without a heat-protective cover at various values of supplied power with AHE heating temperature limited by 973 K.

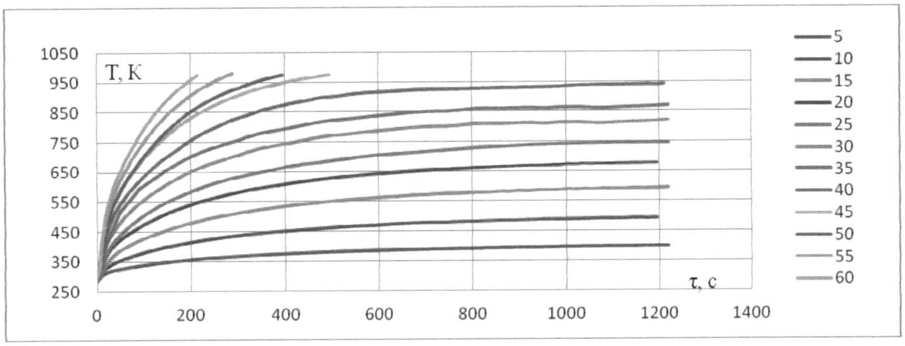

Figure 17. The dependence of AHE temperature on the heating time for ETMT with a heat-protective cover at various values of supplied power with AHE heating temperature limited by 973 K.

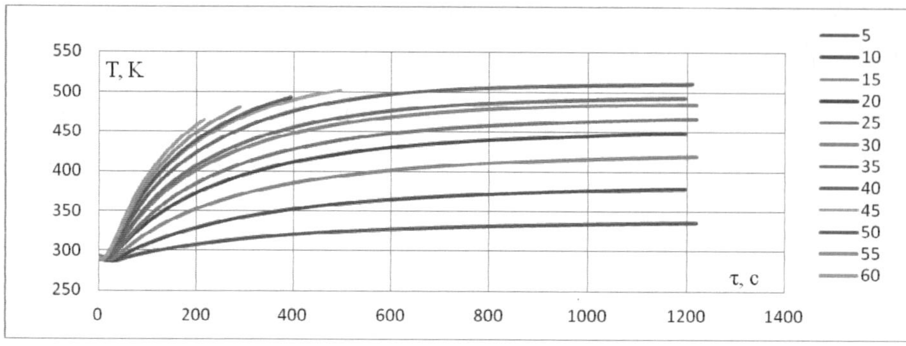

Figure 18. The dependence of the gas temperature at the nozzle exit on the heating time for ETMT with a heat-protective cover at various values of supplied power with AHE heating temperature limited by 973 K.

Table 2. The temperature of AHE and the gas at the nozzle exit for ETMT without heat insulation with AHE heating temperature limited by 973 K.

τ, s	The temperature of AHE (1) and the gas at the nozzle exit (2), K for various values of power, W													
	1	2	1	2	1	2	1	2	1	2	1	2	1	2
	W =5 W		W =10 W		W =20 W		W =30 W		W =40 W		W =50 W		W =60 W	
0	291	287	290	287	289	286	284	282	291	288	297	289	298	289
100	343	295	388	306	475	323	551	337	623	352	704	366	787	379
200	363	307	424	327	544	361	646	385	738	406	841	424	941	442
300	375	315	446	341	584	384	702	413	797	435	905	454		
400	384	321	461	351	610	399	734	429	831	449	933	467		
500	390	325	471	358	626	408	752	438	850	457	946	472		
600	394	328	478	362	637	413	763	442	859	460	952	474		
700	397	330	483	366	644	417	769	445	865	462	955	475		
800	399	332	487	368	649	419	773	446	868	463	956	475		
900	401	333	490	369	653	420	776	447	870	463	955	475		
1000	403	334	493	371	656	421	778	447	872	463	954	475		
1100	404	335	494	372	658	422	779	448	873	464	953	475		
1200	405	335	496	373	659	423	780	449	874	465	954	476		

IOP Conf. Series: Journal of Physics: Conf. Series **1050** (2018) 012013 doi:10.1088/1742-6596/1050/1/012013

Table 3. The temperature of AHE and the gas at the nozzle exit for ETMT with heat insulation with AHE heating temperature limited by 973 K.

τ, s	The temperature of AHE (1) and the gas at the nozzle exit (2), K for various values of power, W													
	1	2	1	2	1	2	1	2	1	2	1	2	1	2
	W =5 W		W =10 W		W =20 W		W =30 W		W =40 W		W =50 W		W =60 W	
0	297	291	296	292	300	291	295	292	299	291	285	289	300	291
100	339	297	380	308	474	334	556	353	634	369	703	379	793	394
200	356	307	413	329	540	372	650	401	756	423	852	438	956	458
300	368	315	435	343	581	395	705	429	830	455	930	472		
400	375	320	450	352	608	41	743	447	874	475				
500	381	324	461	359	627	421	771	460	901	488				
600	385	327	469	365	642	429	786	468	916	496				
700	389	329	476	369	652	435	799	474	923	501				
800	391	331	481	372	660	439	807	478	927	505				
900	394	332	484	374	666	442	811	481	929	507				
1000	395	334	487	376	673	445	812	483	934	508				
1100	397	334	491	378	675	447	811	483	939	509				
1200	400	335	495	379	675	448	817	484	940	509				

Table 4. The temperature of AHE for ETMT without heat insulation and with it with AHE heating temperature limited by 973 K.

τ, s	The temperature of AHE for ETMT without heat insulation (1) and with it (2), K													
	1	2	1	2	1	2	1	2	1	2	1	2	1	2
	W =5 W		W =10 W		W =20 W		W =30 W		W =40 W		W =50 W		W =60 W	
0	291	297	290	296	289	300	284	295	291	299	297	285	298	300
100	343	339	388	380	475	474	551	556	623	634	704	703	787	793
200	363	356	424	413	544	540	646	650	738	756	841	852	941	956
300	375	368	446	435	584	581	702	705	797	830	905	930		
400	384	375	461	450	610	608	734	743	831	874				
500	390	381	471	461	626	627	752	771	850	901				
600	394	385	478	469	637	642	763	786	859	916				
700	397	389	483	476	644	652	769	799	865	923				
800	399	391	487	481	649	660	773	807	868	927				
900	401	394	490	484	653	666	776	811	870	929				
1000	403	395	493	487	656	673	778	812	872	934				
1100	404	397	494	491	658	675	779	811	873	939				
1200	405	400	496	495	659	675	780	817	874	940				

Table 5. The temperature of the gas at the nozzle exit for ETMT without and with heat insulation with AHE heating temperature limited by 973 K.

τ, s	The temperature of the gas at the nozzle exit for ETMT without heat insulation (1) and with it (2), K													
	1	2	1	2	1	2	1	2	1	2	1	2	1	2
	W =5 W		W =10 W		W =20 W		W =30 W		W =40 W		W =50 W		W =60 W	
0	287	291	287	292	286	291	282	292	288	291	289	289	289	291
100	295	297	306	308	323	334	337	353	352	369	366	379	379	394
200	307	307	327	329	361	372	385	401	406	423	424	438	442	458
300	315	315	341	343	384	395	413	429	435	455	454	472		
400	321	320	351	352	399	41	429	447	449	475				
500	325	324	358	359	408	421	438	460	457	488				
600	328	327	362	365	413	429	442	468	460	496				
700	330	329	366	369	417	435	445	474	462	501				
800	332	331	368	372	419	439	446	478	463	505				
900	333	332	369	374	420	442	447	481	463	507				
1000	334	334	371	376	421	445	447	483	463	508				
1100	335	334	372	378	422	447	448	483	464	509				
1200	335	335	373	379	423	448	449	484	465	509				

Experimental testing results for the evaporator are given in figures 19–21.

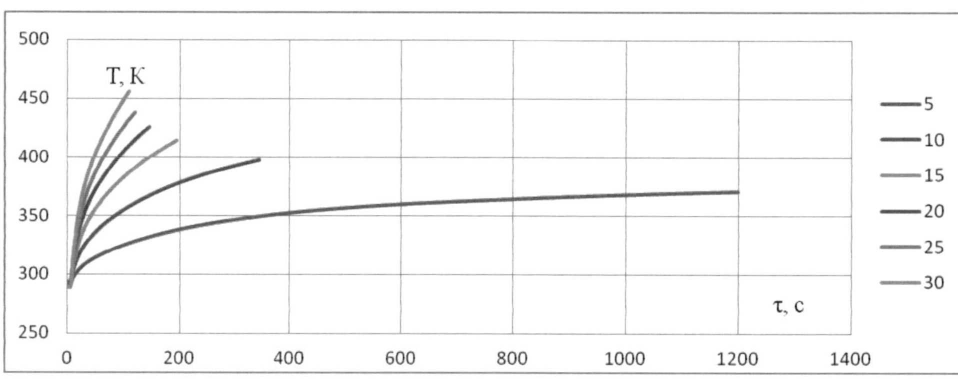

Figure 19. The dependence of AHE temperature on the heating time for various values of supplied power with the casing heating temperature limited by 393 K.

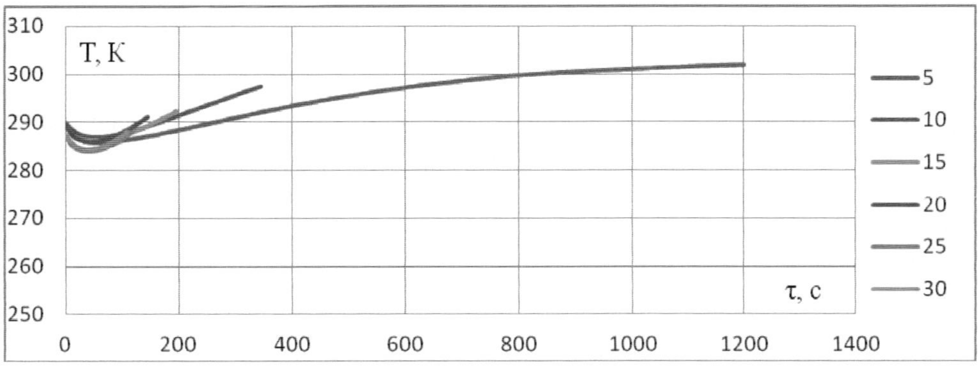

Figure 20. The dependence of the gas temperature at the evaporator throttle exit on the heating time for various values of supplied power with the casing heating temperature limited by 393 K.

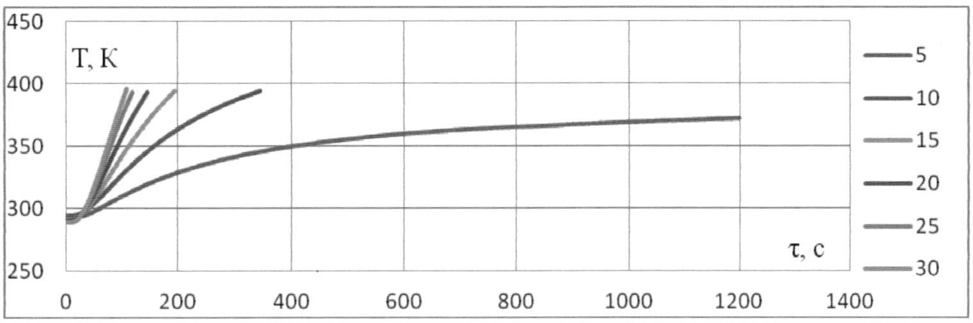

Figure 21. The dependence of the evaporator casing temperature on the heating time at various values of supplied power with AHE heating temperature limited by 973 K.

5. Results and discussion

Applying DMLS method to ETMT manufacturing allowed the following components to be incorporated in one part: outer casing, inner casing with gas ducts in double-start thread form, gas flow conditioner and microthruster attachment point. Therefore, the number of the parts manufactured for ETMT was reduced from 5 to 1, and for the evaporator from 3 to 1. In addition, it is possible to manufacture any inner cavities and flanges.

The analysis of the diagrams in figures 15–18 as well as tables 2–5 showed that at the limited heating temperature for ETMT AHE of 973 K:

- the effect of heat insulation on AHE temperature increase starts at ETMT power consumption of 20 W;
- the maximal increase in AHE temperature in power consumption range of 20–60 W for ETMT with heat insulation is no more than 7 %;
- the effect of heat insulation on the gas temperature increase at the nozzle exit starts at ETMT power consumption of 10 W;
- the maximal increase in the gas temperature equaled no more than 8.6 % at the nozzle exit in the power consumption range of 10–60 W for ETMT with heat insulation;
- the dynamics of ETMT entering the stationary mode largely depends on ETMT power consumption;
- at ETMT power consumption of 5–50 W, the build-up time for ETMT was 400–600 s;
- at power consumption of 50–60 W, the build-up time for ETMT was 200–400s.

The analysis of the diagrams in figures 19–21 showed that at the evaporator casing heating temperature limited by 393 K:

- at power consumption of 10–30 W the evaporator casing temperature of 393 K is reached in 100-340 s, in addition AHE temperature is 400–460 K and the gas temperature at the evaporator throttle exit is no more than 290 K;
- at power consumption of 5 W, the maximum evaporator casing temperature of 375 K is reached in 1200 s, AHE temperature being 370 K and the gas temperature at the evaporator throttle exit being no more than 302 K.

The comparison of the resulting temperature characteristics for ETMT and the evaporator manufactured due to the additive technique with the similar characteristics for ETMT and the evaporator manufactured by the conventional method detected no key differences.

6. Conclusion

1. The possibility of producing ETMT and evaporator casings for SSV CPS by direct metal laser sintering (DMLS) method was experimentally proved, moreover:
   - ETMT and evaporator casings are manufactured as an individual part with gas ducts inside;
   - all manufactured casing prototypes feature stable geometric shapes;
   - the casing material is characterized by integrity and weldability.
2. The results of the undertaken experimental testing point out full operational capability of ETMT and the evaporator. ETMT and evaporator temperature characteristics correspond to the prototypes manufactured by the conventional method (turning and milling).
3. Direct metal laser sintering (DMLS) method allows one to create new designs for ETMT and the evaporator with optimal parameters of gas ducts in terms of quantity and geometry, including the passages to measure pressure (temperature), these being hard or impossible to make with conventional manufacturing methods (turning and milling) (figure 22).

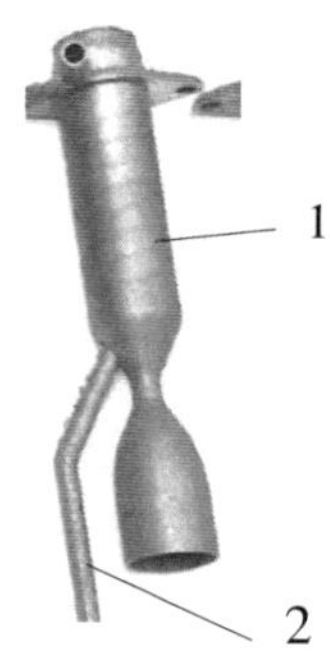

1 - ETMT casing;
2 - passage to measure pressure (temperature)

Figure 22. Resulting experimental ETMT with a passage to measure pressure (temperature) at the entry of the nozzle throat.

References

[1] Blinov V N, Shalay V V, et al. 2014 *The Researches of Correcting Power Units Electro-Thermal Micromotors of Maneuverable Small Space Vehicles* (Omsk: OmSTU Press) p 264

[2] Blinov V N, Vavilov I S, Kositsin V V, Ruban V I, Khodoreva E V and Shalay V V 2015 *Indian J Sci Technol* **8(27)** DOI:10.17485/ijst/2015/v8i27/82937

[3] Blinov V N, Vavilov I S, Kositsin V V, Ruban V I, Khodoreva E V and Shalay V V 2015 *Mod Appl Sci* **9(5)** 337 DOI: 10.5539/mas.v9n5p337

[4] Coxhill I, Gibbon D and Drube M 2008 *Proc. 5th Int. Spacecraft Propulsion Conf.* (Heraklion) pp 1-9

[5] Romei F, Grubišić A and Gibbon D 2017 *35th Int. Electric Propulsion Conf.* (Atlanta) (Atlanta: Georgia Institute of Technology) p 463

[6] Cifali G, Gregucci S, Andreussi T and Andrenucci M 35th *Int. Electric Propulsion Conf.* (Atlanta) (Atlanta: Georgia Institute of Technology) p 371
[7] Passaro A and Bulit A 2013 *33rd Int. Electric Propulsion Conf.* (Washington) (Washington: The George Washington University)
[8] Makled A and Othman M 2016 *Proc. of the 17th Int. AMME Conf.* (Cairo)
[9] Romei F, Grubišić A, Lasagna D and Gibbon D 2017 *7th European Conference for Aeronautics and Space Sciences (EUCASS)* DOI: 10.13009/EUCASS2017-378
[10] Blinov V N, Shalay V V and Khodoreva E V 2012 *Omsk Scientific Bulletin* **1** p 62-67

IOP Conf. Series: Journal of Physics: Conf. Series **1050** (2018) 012014 doi:10.1088/1742-6596/1050/1/012014

DEM Calibration Approach: Implementing Contact Model

A V Boikov, R V Savelev and V A Payor

Saint-Petersburg mining university, 21 line V.O, Saint-Petersburg, 2, 199106, Russia

e-mail: boikov_av@mail.ru

Abstract. The problem of DEM contact models choice is described in the article. An analysis of the input data calibration, including a choosing the contact model, is performed. The main theoretical positions of the most popular models are considered, their comparison is made on the basis of a specially designed stand. Description and justification of the designed stand are considered. The analysis of the obtained results is carried out. Recommendations on the choice of the DEM contact model are given. The results show that it is time-efficient to use non-linear contact model for future model calibration.

1. Introduction

Discrete element method (DEM) is a generalization of the finite element method (FEM) [1] and allows modeling the behavior of a large number of particles of bulk material, which makes it a popular tool for modeling and approbation while designing industrial equipment [2][3]. Usually, bulk material in DEM software characterized by the following granular material parameters:

- Density
- Particle size (distribution)
- Particle shape
- Young's modulus
- Poisson’s ratio
- Particle-particle coefficient of static friction (SFp-p) and particle-boundary coefficient of static friction (SFp-b)
- Particle-particle coefficient of dynamic friction (DFp-p) and particle-boundary coefficient of dynamic friction (DFp-b)
- Particle-particle coefficient of restitution (CoRp-p) and particle-boundary coefficient of restitution (CoRp-b)
- Rolling Resistance (when modelling spheres particles)

The full implementation of software packages based on DEM is slowed down due to the presence of a significant problem: some properties could not be measured directly. Some parameters values of real bulk material can be directly measured and implemented in the model (density, size, shape, Young's modulus, Poisson's ratio), but some parameters values as coefficients of friction requires it's calibration [4][5]. Systematic calibration approach is needed to achieve adequate behavior of bulk material [6]. Nevertheless, there is no single time-efficient method that would allow unambiguously calibrate the inputs for arbitrary real bulk material providing all of the required parameters for its modeling [5][7].

2. Contact models description

In addition to the parameters presented above, it is equally important to choose a contact model that determines the representation of the normal and tangential component of the force acting upon collision of particles. The most popular models are linear and nonlinear Hertzain [7].

In this research linear model is elastic-plastic but allows simulation of the plastic energy dissipation on a contact without introducing the overhead of long simulation times. Since no viscous damping term is used, the energy dissipation is not dependent on the relative velocities of neighboring particles, making the energy dissipation insensitive to other contacts. An additional advantage of this model is that compressible materials can be accurately modeled due to the fact that the contact forces can be almost zero even at residual overlaps [8].

The linear model can be described by the following set of equations:

$$F_n^t = \min\left(F_n^{t-dt} + K_{nu} \bullet ds_n, K_{nl} \bullet s_n\right) \text{ if } ds_n \geq 0 \tag{1}$$

$$F_n^t = \max\left(F_n^{t-dt} + K_{nu} \bullet ds_n, 0.001 \bullet K_{nl} \bullet s_n\right) \text{ if } ds_n < 0 \tag{2}$$

$$ds_n = s_n^t - s_n^{t-dt} \tag{3}$$

where:

- F_n^t is the normal contact force at the current time, t
- F_n^{t-dt} is the force at the previous time, t – dt, where dt is the timestep
- ds_n is the change in the contact normal overlap during the current time
- s_n^t and $s_n^{(t-dt)}$ are the normal overlap values at the previous and the current time
- K_{nl} and K_{nu} are the values of loading and unloading contact stiffnesses respectively

The loading and unloading stiffnesses are defined by the particle size, bulk Young's Modulus, and by the restitution coefficient of contacting materials, the last two of which are the user inputs. The coefficient of restitution ε is a measure of energy dissipation for the contacting pair of materials. The particle's loading stiffness is calculated in the following manner:

$$K_{npl} = E_p \bullet L \tag{4}$$

$$K_{nbl} = E_b \bullet L \tag{5}$$

where:

- E_p is the particle material bulk Young's Modulus or Elastic Modulus (user input)
- E_b is the boundary material Young's Modulus
- L is the particle size

As a matter of fact different DEM codes implement their own tangential contact model but in this research the tangential force is elastic-frictional. It is described by the following equations:

$$F_t^t = \min\left(F_t^{t-dt} + K_{nl} \bullet ds_t, \mu_s \bullet F_n^t\right) \tag{6}$$

if no sliding is taking place for the contact, and

$$F_t^t = \min\left(F_t^{t-dt} + K_{nl} \bullet ds_t, \mu_s \bullet F_n^t\right) \tag{7}$$

if sliding is taking place for the contact

where:

- F_t^t is the tangential contact force at the current time
- F_t^{t-dt} is the same force at the previous time
- ds_t is the tangential relative particles' displacement during the current time

- μ_s and μ_d are the static and dynamic friction coefficients

The sliding is considered to be taking place on the contact the first time the tangential force exceeds the limit of $\mu_s * F_n^t$. Once the tangential force falls below the value of $\mu_d * F_n^t$, the contact is considered non-sliding again.
The Hertz-Mindlin model is a non-linear contact model that can describe the collisions between particle-boundary and particle-particle. The theory is based on the Hertz's Law and can be describes as:

$$F_n = k_n^{'} \varepsilon_n^{3/2} \tag{8}$$

where:

- F_n is a normal contact force
- $k_n^{'}$ is normal stiffness
- ε_n is total normal overlap

The normal stiffness is described by the following equation:

$$k_n^{'} = \frac{4}{3} R_*^{1/2} E_* \tag{9}$$

where:

- R_* is adduced particles radius
- E_* is modulus of rupture (defined by Young's modulus and Poisson ratio as user's input)

Tangential component is described by:

$$F_{t,n+1} = F_{t,n} + k_t \Delta\varepsilon_t \tag{10}$$

where:

- $F_{t,n+1}$ is the tangential contact force at the current time
- $F_{t,n}$ is the same force at the previous time
- k_t is shear stiffness
- $\Delta\varepsilon_t$ is increment of ε_t

3. Experiments handling

At the same time, there is no certain decision about the most optimal model while calibrating user's inputs [9]. In general, it is considered that common models are adequate and accurate for most nonspecific tasks [10]. It is necessary to rely on 2 factors when choosing a contact model: the completeness of the process's physics representation, as well as the calculation time of the model.
Since the calibrated coefficients in this research correspond only to the kinematic and static characteristics of the bulk material [11], it was decided to compare two contact models basing on following important macroscopic characteristics - natural angle of repose and angle of rupture. The angle of repose is an important parameter of the soil strength [12]. Its definition is derived from the Mora-Coulomb destruction criterion, and often used to describe the soil shear resistance along with the normal effective stress [13]. Usually, the maximum permissible repose angles of ledges and sides of quarries, embankments, dumps and piles determination is based on these two angles.
Specialized stand was developed (figure 1) to determine these two parameters. After material's back-pile the damper 1 quickly extends, where one part of the material forms the angle of rupture on the bar. The remaining part is poured into the next compartment, and then after the damper 2 is extended, it flows through the pyramidal funnel, forming the angle of repose on the lower plane.
The developed construction is designed for particles with a diameter from 2 to 10 mm. The funnel is made in the form of a pyramid to ensure the hydraulic flow of material, as well as the absence of possible sticking of the material [14].

An experiment comparing the two contact models was carried out with the modeling parameters for bulk material presented in table 1. The walls stand material is plexiglass, that why values for the Poisson's ratio and Young's modulus were chosen accordingly.

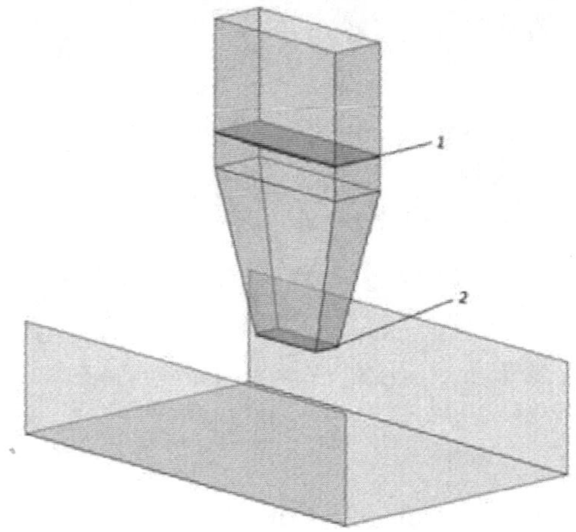

Figure 1. 3D CAD model of the stand

Table 1. Experiment's inputs for each contact model

	1	2	3	4	5
Shape	sphere	sphere	sphere	sphere	sphere
Density	1600 kg/m^3	1600 kg/m^3	1600 kg/m^3	1600 kg/m^3	1600 kg/m^3
Diameter	8 mm	8 mm	8 mm	8 mm	8 mm
Young's modulus	$49*10^9$ Pa	$49*10^9$ Pa	$49*10^9$ Pa	$49*10^9$ Pa	$49*10^9$ Pa
SFp-p	0.5	0.9	0.9	0.5	0.9
SFp-b	0.5	0.5	0.5	0.5	0.5
DFp-p	0.5	0.5	0.5	0.9	0.9
DFp-b	0.5	0.5	0.5	0.5	0.5
CoRp-p	0.5	0.5	0.5	0.5	0.9
CoRp-b	0.5	0.5	0.5	0.5	0.9
RR	0.9	0.5	0.9	0.5	0.9

Thus, 10 experiments were conducted on 5 for each contact model (figure 2), which were then processed using a special algorithm with the application of technical vision. The algorithm is based on image processing and recognition of the contour of the slide. Further, the resulting contour is interpreted into the angles (taking into account the vertex smoothing) and then the average is taken between the two angles. The algorithm is implemented using programming language Python [15].

Figure 2. Final results of modelling examples: left – linear model, right – non-linear model

Results and conclusion

The following results were obtained after analyzing of 10 experiments:

Table 2. Experiment's results for each contact model

№ of experiment	Angle of repose				Angle of rupture			
	Linear	Non-linear	Absolute error	Ratio error	Linear	Non-linear	Absolute error	Ratio error
	degrees			%	degrees			%
1	27.9	27.1	0.8	1.9	59.2	58.5	0.7	1.2
2	23.1	22.3	0.8	2.3	53.0	51.0	2.0	3.8
3	26.2	25.9	0.2	0.5	60.4	59.6	0.8	1.3
4	23.2	23.7	0.5	1.4	54.0	53.1	0.9	1.7
5	7.5	8.5	1.0	8.5	48.5	46.8	1.3	2.7
Average				2.9				2.2

The results showed that the obtained natural angle of repose and angle of rupture did not differ significantly while using different contact models for materials with other equal inputs. Although modeling with the use of a nonlinear model, piles with rounded vertices were formed (authentically displaying the behavior of real bulk materials), almost the same values for the angles of repose and rupture indicate the applicability of both contact models for the modeling of bulk materials. At the same time, it is necessary to take into account the simulation time using different models, since the calibration of the input parameters could be time consuming [16]. Thus, the simulation time of a non-linear model is 2 times less than linear on the average, therefore it is recommended to use non-linear contact model for future model calibration.

References

[1] Cundall P A and Strack O D L 1979 A discrete numerical model for granular assemblies Géotechnique vol 29 (1) pp 47–65. DOI: 10.1680/geot.1979.29.1.47

[2] Miguel Florian-Algarin, Rafael Méndez (2015) Effect of material properties and design parameters on the final blend uniformity using experimental and simulation results *Powder Technology* vol 276 pp 156–165. DOI i: 10.1016/j.powtec.2015.02.024

[3] Feoktistiov A J et al (2015) The application of discete element method to mining and metallurgy process modeling *Journal of Mining Institute* vol 192 pp 145–149

[4] Yan Z et al (2015) Discrete element modelling (DEM) input parameters: understanding their impact on model predictions using statistical analysis *Computational Particle Mechanics* vol 2(3) pp 283–299. DOI: 10.1007/s40571-015-0056-5

[5] Coetzee C J (2016) Calibration of the discrete element method and the effect of particle shape *Powder Technology* vol 297(April) pp 50–70. DOI: 10.1016/j.powtec.2016.04.003

[6] Stratton R E, Wensrich C M (2010) Modelling of multiple intra-time step collisions in the hard-sphere discrete element method *Powder Technology* vol. 199(2) pp 120–130 DOI: 10.1016/j.powtec.2009.12.008

[7] Coetzee C J (2017) Review: Calibration of the Discrete Element Method *Powder Technology* vol 310 pp 104–142. DOI: 10.1016/j.powtec.2017.01.015

[8] *Rocky DEM Software* Rocky Discrete Element Method Package link: http://rocky-dem.com/images/pdf/Rocky_DEMP.pdf

[9] Navarro H A, de Souza Braun M P (2013) Linear and nonlinear Hertzain contact models for materials in multibody dynamics *22nd International Congress of Mechanical Engineering* (COBEM 2013) pp 159–170

[10] Evans T M (2018) Effects of curvature-related DEM contact model on the macro- and micro-mechanical behaviours of granular soils *Géotechnique* vol. January

[11] Beatriz C. Silvério, et al (2014) Effect of the Friction, Elastic, and Restitution Coefficients on the Fluid Dynamics Behavior of a Rotary Dryer Operating with Fertilizer *Industrial & Engineering Chemistry Research* vol 53 (21) pp 8920–8926. DOI: 10.1021/ie404220h

[12] Dai B, Zhou C, Yang J (2017) Micromechanical origin of angle of repose in granular materials *Granular Matter*. DOI: 10.1007/s10035-017-0709-6

[13] Geotechdata.info, “Soil friction angle”, Publication date: 14.12.2013, link: http://www.geotechdata.info/parameter/angle-of-friction.html.

[14] Labra C, Ooi J Y, Luding S T. Weinhart (2016) Influence of coarse-graining parameters on the analysis of DEM simulations of silo flow *Powder Technology* vol 293 pp 138–148 DOI: 10.1016/j.powtec.2015.11.052

[15] Official Python documentation, link: https://docs.python.org/3

[16] McGuire A D, Rosenboom J, Lee K F, Dosta M, Mosbach S (2017) Compartmental residence time estimation in batch granulators using a colourimetric image analysis algorithm and Discrete Element Modelling *Advanced Powder Technology* vol 28(9) pp 2239–2255. DOI: 10.1016/j.apt.2017.06.005

IOP Conf. Series: Journal of Physics: Conf. Series **1050** (2018) 012015 doi:10.1088/1742-6596/1050/1/012015

Solid lubricant ceramic coatings on aluminium and its alloys

A N Bolotov, G B Burdo and V V Novikov

Tver State Technical University, A. Nikitin Emb. 22, Tver, 170026, Russia

Abstract. Intensive development of space technology, chemical and electronic industries, modern machine-building industries requires creation of antifriction materials with significantly higher tribotechnical properties. Ceramic materials based on aluminium oxide and formed in a microarc discharge plasma on the surfaces of aluminium parts are very promising. However they can not work in friction nodes without lubrication and intensive cooling. Their antifriction properties might be improved by including micro- and nanoscale particles of an antifriction filler in a ceramic matrix to obtain ceramic coatings with solid lubricant properties. The authors consider the basics of the original microarc technology that makes it possible to obtain new composite solid lubricating coatings, which consist of various aluminium oxides, as well as solid lubricant dispersed inclusions, such as graphite, molybdenum disulphide and magnetite. They also analyse the effect of inclusion of highly dispersed filler particles from a colloidal electrolyte on the coating formation process. In order to improve the quality of coatings, it is suggested to use filler particles coated with solvated shells. An electrochemical installation for obtaining a composite coating has been improved to create an increased concentration of dispersed phase particles in the coated sample surface area. The paper investigates tribotechnical properties of the obtained wear-resistant oxide coatings on an aluminium alloy. It is shown that the inclusion of special dispersed particles in its composition can ensure a stable and continuous operation of a friction unit, reduction of a friction coefficient and linear wear under conditions of lubricant deficit. A ceramic coating with a molybdenum disulphide has shown the highest antifriction characteristics.

1. Introduction

Nowadays, the problem of ensuring the effective operation of the bearing units of special equipment without lubricating them both with liquid and a lubricating grease has become important again. Several decades ago, the problem of "dry" friction units was solved by using a polymer composition, for example, based on tetrafluoroethylene or polyamide, to produce one of the friction pair elements. People have achieved significant progress in developing antifriction polymers. These materials have generally satisfied the requirements in terms of a friction coefficient, wear resistance, strength, etc. However, recently, as space and military equipment, chemical and electronic industries, as well as textile and weaving factory has been developing intensively, the need in antifriction materials with significantly higher properties has appeared. For example, the proposed composite coatings are extremely important for "dry" friction units installed in modern and advanced bearings that work with gas lubrication, since they make it possible to increase the reliability and durability of supports by increasing the number of "start-stop" cycles and more long-duration preservation of the permissible sizes of working microgaps. We should note that gas-lubricated spin-axis bearing system contribute to a significant increase in the quality of such technically advanced products as gyroscopes, super precision gyroinstruments, super centrifuges, laser measuring instruments, turbomachines for nuclear industry, precision metal processing equipment, etc.

Perspective constructional materials for technically more perfect dry-friction tribounits are ceramic materials based on aluminium oxide, which were obtained in a microarc plasma discharge (microarc oxidation method) [1]. Coatings can be applied to internal and external surfaces of complex parts and do not require preliminary preparation of the original surface. Such coatings have a thickness of up to 200 μm, a complex of high mechanical properties, good adhesion to a base coat. They are heat-resistant, but they do not work in friction nodes without lubrication and intensive cooling. Now there is a need in expanding the range of physical, mechanical and antifriction properties of coatings by obtaining composite coatings according to the abovementioned technology [2]. For this purpose, it is proposed to include antifriction filler micro- and nanoscale particles in a ceramic matrix to obtain solid-lubricating ceramic coatings. This approach allowed creating unique diamond-containing ceramic coatings for an abrasive tool [3, 4].

2. Purpose of the study

The purpose of this work is to develop the basics of the technology for obtaining solid-lubricating composite ceramic coatings on an aluminium alloy with antifriction fillers based on graphite, molybdenum disulphide and magnetite, by microarc oxidation.

3. Description of the conducted research

3.1. Methodical issues of the experimental research

As the basic research materials, the authors have chosen aluminium-based alloys (D1 GOST 4784-74) that are widely used in technology.

Composite coatings with dispersed powders were formed in electrolytes containing disperse materials as filler. These materials modified the structure of the hardened surface layer during oxidation. A nanodisperse phase of a ceramic coating was a magnetite, which is able to increase mechanical properties of coatings and should help reduce friction as a result of reducing adhesion bonds between contacting surfaces.

High-dispersion magnetite was obtained by chemical condensation proposed by W.S. Elmore in 1938. It is based on the following reaction:

$$2\,FeCl_3 + FeCl_2 + 8\,NH_4OH \rightarrow Fe_3O_4\downarrow + 8NH_4Cl + 4\,H_2O.$$

The method of chemical precipitation of highly dispersed magnetite assumes neutralization of salts of bivalent and ferric iron by an excess of an aqueous ammonia solution while stirring constantly. The reaction leads to formation of an ammonium chloride. It is separated from a precipitate by repeated washing with distilled water, which prevents coagulation of magnetite particles and ensures adsorption of SAS stabilizer molecules with the surface of magnetite particles. The precipitate of the chemical reaction contains magnetite particles of size 2–20 nm, most of them are about 7 nm. The surface of magnetite particles has a good adsorption capacity, which is necessary for their colloidal stabilization. In order to protect highly dispersed magnetite particles from oxidation and prevent their coagulation, a peptization process was carried out in a sodium silicate solution to create an electric charge on the surface of particles. Aggregative stability of the colloid was achieved by ionic mechanism.

To obtain a dispersed phase of antifriction coatings, except for magnetite, we used finely divided graphite powders (particle size 2–10 μm) (GOST 18191-78) and molybdenum disulphide (MVCh TsMTU 06-01-68), which traditionally possess high antifriction properties. Glycerine and triethanolamine were surfactants for modifying the surface of particles.

During oxidation, additional dispersed components in the electrolyte are transported to the surface of the processed alloy by an inhomogeneous electric field. When falling into a spark discharge combustion zone, they are included in the coating composition (in fact, this is an arc electrophoresis).

There is an improved laboratory electrochemical installation for producing a composite coating. It includes an electrolyte active mixing system in order to create an increased concentration of dispersed phase particles in a surface area of a coated sample.

IOP Conf. Series: Journal of Physics: Conf. Series **1050** (2018) 012015 doi:10.1088/1742-6596/1050/1/012015

In order to test the obtained antifriction composite coatings, we used countersamples from tool hardened alloy steel KhVG (GOST 5950-73) with a hardness of 45–50 HRC.
Tribotechnical tests of composite coatings with a ceramic matrix and various fillers in the medium pressure range were carried out using a friction machine MTP [5], a disk-finger friction scheme. Table 1 shows operating parameters of the installation.

Table 1. Technical characteristics of the MTP installation.

Friction parameters	Varying range
Sliding speed, m/s	0.15—0.95
Diameter of samples, mm	6
Contact pressure, MPa	0.5—15
Relative error of wear measurement	up to 2%
The relative error in friction moment measurement	up to 0.4%

The method for measuring microhardness of thin ceramic coating was performed according to GOST 15150-80 applying small loads on a pyramidal tip with an angle at the vertex of 136° in order to avoid the influence of mechanical properties of the substrate material half-space on the results.
The microstructure of the obtained coatings was studied on an optical module that included a metallographic microscope of the MIM-8 brand and a digital camera (magnification × 2000) connected to a computer via a video card having a frame capture function. This module makes it possible to quickly and accurately determine the relative proportion of pores and dispersed inclusions on the surface of a ceramic sample.
The developed digital methodology for the quantitative evaluation of foreign inclusions in a ceramic composite coating material is a much more accurate compared with standard metallographic methods due to levelling the human factor influence.
A topographic image of a thin section surface in an intense frictional interaction zone at the micro- and nanoscale was obtained using a scanning probe microscope Solver P47H.

3.2. Technological basis for obtaining coatings

A basis for the technology of composite coatings was the existing technology of microarc synthesis of oxide coating on an aluminium alloy. The application of the coating under consideration does not require thorough preliminary cleaning of the surface and complex laboratory equipment. The coating obtained by the known technology has sufficient thickness, high adhesion to a substrate material (Table 2) and high hardness. The main disadvantage of a base coating is a high dry friction coefficient, which is likely to cause jamming of a friction pair and fast fatigue failure of contacting surfaces. Considering this disadvantage, the coating was modified in the direction of reducing the friction coefficient without using a lubricant.

Table 2. The properties of the anodic coating on the aluminium alloy D1, the formation conditions.

Coating thickness, mm	Dry friction coefficient	Linear wear intensity	Microhardness, GPa	Electrolyte composition
0.1–0.3	0.28–0.35	$\approx 13 \cdot 10^{-9}$	11–12	NaOH alkali, sodium silicate, distilled water

This problem was solved by electrophoretic introduction of solid lubricant particles into a ceramic matrix of a coating. The electric current inclusion leads to the growth of a barrier oxide layer on the surface of the coated aluminium sample, which forms a refractory ceramic matrix. At the same time, fine-dispersed solid lubricant particles are deposited and retained from the electrolyte solution by electromagnetic forces on the surface of the sample. They are gradually incorporated into the matrix. Incorporation of fine-grained magnetite from the colloidal electrolyte into the coating of particles did not cause any significant changes in the formation of the coating.
The inclusion of graphite particles in the coating is complicated by its good electrically conductive properties. When electric current is connected, graphite particles settle on the surface of a part to be

coated and cause corrosion. There is a violation of a surface passive layer and, therefore, nondischarge current flow locations arise.

There are two approaches to coating formation that might solve this problem. According to a first method, the electrochemical coating process is divided into two steps. At the first step, a thin (5–10 μm) technological protective coating is formed on the surface of the sample to be coated in an electrolyte that does not contain graphite particles. At the second step, the ceramic coated sample is placed in an electrolyte containing an elevated concentration of sodium silicate and a finely dispersed (2–10 μm) graphite powder. With the given modes, the process continues until a coating of a given thickness forms. The ceramic coating obtained by this method has a concentration gradient of a dispersed graphite filler. It is completely absent in the layer bordering with aluminium.

The second method for forming the coating uses glycerine as a dielectric film adsorbed on the surface of graphite particles and leading to a decrease in its electrical conductivity. This method gives coatings with a uniform concentration of graphite particles throughout a coating thickness. Non-ionic properties of glycerine determine its choice. Figure 1 shows the microstructure of the coating.

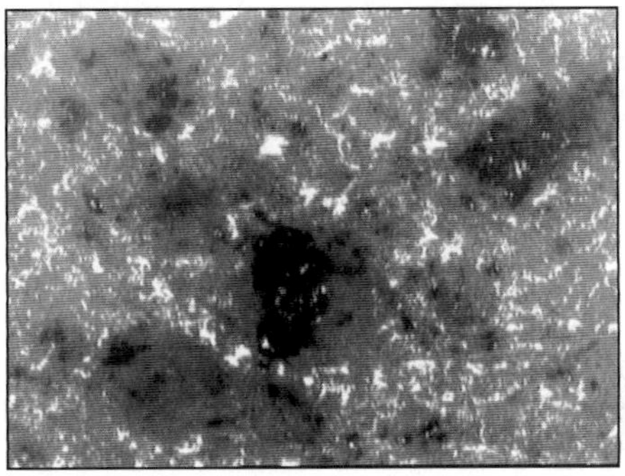
a

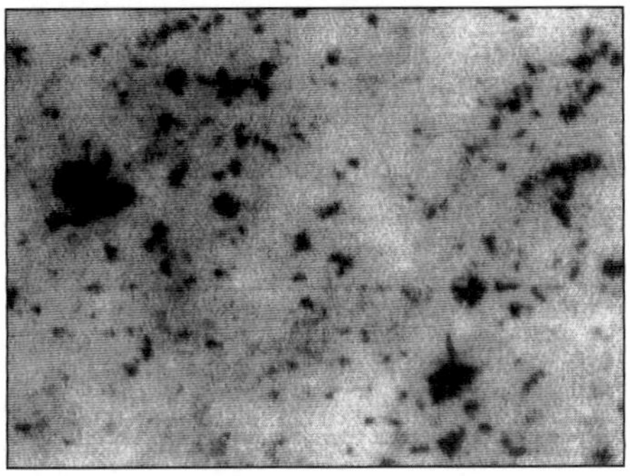
b

Figure 1. A microstructure of a ceramic composite coating surface with a graphite filler on an aluminium alloy D1: a is in ordinary light; b is in polarized light (×500).

The second material of the dispersed phase is molybdenum disulphide, which has good lubricating properties and is a dielectric, in contrast to graphite. A significant disadvantage of molybdenum disulfide is its relatively high chemical activity. More specifically, during a coating formation it can turn into MoO_2 dioxide and MoO_3 trioxide under the influence of high temperature and an oxidizing environment. They substantially reduce antifriction properties of the resulting coating. To prevent molybdenum disulfide oxidation, glycerine protective properties have also been used.

Table 3 shows the optimal composition of electrolytes worked out as the experimental result. The optimal composition includes dispersed phases of various fillers.

Table 3. The composition of electrolytes for obtaining ceramic coatings with antifriction fillers.

Electrolyte constituent elements	Elements concentration in an electrolyte, wt.%		
	Coating with graphite particles	Coating with molybdenum disulphide particles	Coating with magnetite particles
NaOH alkali	0.1	0.1	0.1
Liquid glass	1.5	1.1	1.5
Dispersed particles	3	4.5	4÷6
Surfactant - glycerine	2	3	-
The rest is distilled water			

3.3. Basic tribotechnical properties of coatings

Tribotechnical studies have confirmed the effectiveness of dispersed solid lubricants in the synthesis of ceramic coatings that are operated in dry friction units. They have also shown the advisability of using surfactants (glycerine) to obtain coatings.

It was shown that the friction coefficient for a ceramic coating with molybdenum disulphide, which was not treated in glycerine, is in the range 0.25–0.5 typical for coatings without filler. This confirms the glycerine protective function and is rationale for using it (or similar in surfactant properties) in obtaining a self-lubricating coating with molybdenum disulphide particles as a dispersed phase.
The test results of the obtained coatings are shown in Table 4. It can be seen from the table that the inclusion of magnetite nanodispersed particles in the coating resulted in an increase in coating microhardness. Apparently, this is due to the fact that the coating became less loose as porosity became half as high. The frictional properties of coatings with magnetite during friction without lubrication have slightly improved, however their overall level is lower than for coatings with solid lubricants.
Figure 2 shows photographs of friction paths on the coating with MoS_2 particles. A countersample is KhVG steel. Dark areas on the friction surface of a specimen with a ceramic antifriction coating and a steel countersample indicate forming of a MoS_2 protective lubricant film on them.

Table 4. The properties of the anodic coating on the aluminium alloy D1, the formation conditions.

Filler material	Coating thickness, mm	Friction coefficient	Linear wear intensity	Microhardness, GPa
MoS_2	0.1 – 0.50	0.07 – 0.11	$5 \cdot 10^{-9}$	10 – 12
Graphite	0.1 – 0.45	0.10 – 0.16	$7 \cdot 10^{-9}$	12 – 14
Fe_3O_4	0.1 – 0.30	0.15 – 0.26	$9 \cdot 10^{-9}$	14 – 17
No filler	0.1 – 0.40	0.11 – 0.25	$15 \cdot 10^{-9}$	12 – 14

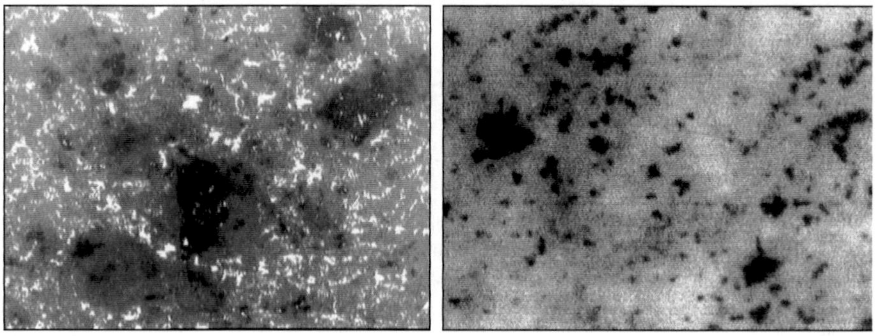

a b

Figure 2. Friction surfaces of the composite coating with MoS2 particles: a is a coated sample; b is a counterbody (KhVG steel).

The influence of contact pressure on linear wear intensity of coatings with graphite and MoS2 particles was investigated. In the investigated working pressure range, a ceramic coating with MoS2 filler have shown the highest antifriction characteristics.
A high concentration of molybdenum disulphide particles in the coating material provides friction under conditions of solid lubrication, and therefore a decrease of the friction coefficient and linear wear. As for a composite coating with graphite particles, its linear wear intensity is 1.5–2 higher than the molybdenum disulfide-modified coating has. It is explained by initially lower antifriction properties of graphite and its lower concentration in the coating material.
During the research, the intensity of counterbody linear wear was monitored. There was a slight excess of the counterbody wear rate comparing with this characteristic for a ceramic coating. This is an undoubted advantage of the created composite coatings in comparison with traditional coatings on aluminum, which significantly wear out the counterbody under dry friction conditions.

4. Conclusion

The paper proposes a new approach to the production of self-lubricating composite coatings on aluminium and its alloys in the anode-spark mode. This approach differs from the known one in that a finely dispersed powder of a molybdenum disulphide, magnetite or graphite is added to the electrolyte by a surfactant. The authors have determined tribotechnical properties of self-lubricating composite coatings on aluminium alloys synthesized by the anodic-spark method. Unlike traditional ocidic ceramic materials obtained by microarc oxidation, they have higher antifriction and antiwear properties, which allows them to be used in extreme conditions without additional lubrication with liquid or consistent materials.

Acknowledgments
The work has been financially supported by RFBR (grant 17-48-690826).

5. References

[1] Suminov I V, Epelfeld A V, Lyudin V B, Krit B L and Borisov A M 2005 *Mikrodugovoe oksidirovanie (teoriya, tekhnologiya, oborudovanie)* [Microarc oxidation (theory, technology, equipment)] (Moscow: EKOMET) 368
[2] Borisov A M, Krit B L, Lyudin V B, Morozova N V, Suminov I V and Epelfeld A V 2016 Microarc oxidation in electrolyte suspensions (review) *Elektronnaya obrabotka materialov* **1** pp 50–77
[3] Bolotov A N, Novikov V V and Novikova O O 2006 The use of microarc oxidation for the production of ceramic diamond-containing material *Uprochnyayushchie tekhnologii i pokrytiya* [Strengthening technologies and coatings] **3** pp 13–16
[4] Bolotov A N, Novikov V V and Novikova O O 2005 Analysis of the operation of tribocouples based on ceramic diamond-containing materials *Friction and Wear* vol 26 **3** pp 279–84
[5] Bolotov A N, Novikova O O and Novikov V V 2017 Studying tribotechnical properties of nanostructured lubricating oils with various dispersive media *Friction and Wear* vol 38 (Springer) **2** pp 121–25

IOP Conf. Series: Journal of Physics: Conf. Series **1050** (2018) 012016 doi:10.1088/1742-6596/1050/1/012016

Multifunctional vibrational source of seismic waves

Yu A Burian[1], V N Sorokin[1] and V V Lizunov[2]

[1] Omsk State Technical University, 11, Mira ave., Omsk, 644050, Russia
[2] Omsk State University, 55A, Mira ave., Omsk, 644077, Russia

Abstract. The paper deals with a powerful low-frequency surface source of seismic waves with an electrohydraulic drive and a force closure in the "inertial mass-soil system". A mathematical model of a hydromechanical vibromodule is developed, and according to the computer simulation, an estimate of the seismic wave generation efficiency is given as well as the operating modes are determined for signals reproduction over a wide range of frequencies and shapes.

1. Introduction

The creation of a powerful, low-frequency and transportable source of seismic waves, designed to solve a wide range of problems, is a relevant problem. Among the priority problems are: effective ground sounding, increasing seismic exploration efficiency (non-explosive seismic exploration), active environment monitoring and determining the strain-stress state of media (earthquake prediction), vibroseis effects on oil-gas formations (oil formation productive capacity increase), etc. [1, 2, 3, 4, 5].
A powerful low-frequency surface source of seismic waves with an electrohydraulic drive and a force closure in the inertial mass-soil system capable of radiating signals in a wide range of frequencies and shapes is considered. A mathematical model of a hydromechanical vibromodule is developed. Based on the computer simulation results, the seismic wave generation efficiency is estimated and the operating modes are determined for playing a "sweep" signal and a complex signal.
In [6] the source efficiency is shown whose construction diagram corresponds to the force closure principle in the "inertial mass-soil" system when operating in the of a monochromatic signal generation mode. It is of interest to consider the possibility of obtaining an impulse seismogram analogue during monochromatic signal emission, as well as the possibility of reproducing a linear-frequency modulated signal ("sweep" signal) by such a source, which will significantly expand its area of application.

2. Problem statement

In case of a force closure in the inertial mass-soil system, the inertial mass, as shown in [6], should be hung on spring elements with high load capacity and low stiffness. As such spring elements, rubber-cord shells (RCS) can be used.
The analytical model of such a source is shown in figure 1.
To generate the multimass system motion differential equations, Z_1, Z_2, Z_3 are taken as generalized coordinates, and it is assumed that the line of $F(t)$ force application by a hydraulic cylinder passes through the centre of mass while spring and dissipation forces application lines are at the same distance ℓ from the mass centre point C.
It is also assumed that at $F(t) = 0$, the inertial mass m_1 is hung on the RCS, and the masses m_2 and m_3 are also in equilibrium positions, while the generalized coordinates values are measured from the equilibrium positions.

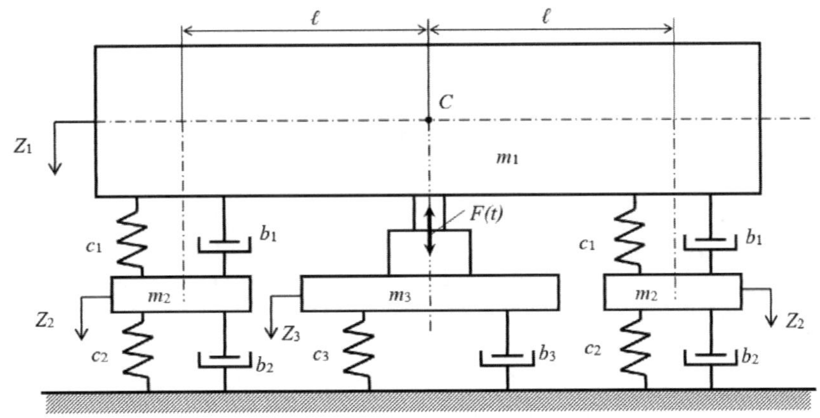

Figure 1. Analytical model: m_1 - inertial mass; m_2 - the mass of the supporting plates together with the apparent soil mass; m_3 – the mass of the base plate, the hydraulic cylinder and the apparent soil mass; c_1, b_1 - stiffness and damping coefficients for the inertial mass suspension spring elements; c_2, c_3, b_2, b_3 - stiffness and damping coefficients characterizing the soil properties under the supporting and base plates; $F(t)$ - the force created by the hydraulic cylinder; Z_1, Z_2, Z_3 – the displacement of the corresponding inertial mass, the supporting plates and the base plate.

Differential equations based on the above-given assumptions and following [6] can be represented in the form of

$$\begin{cases} \ddot{Z}_1 + \dfrac{2}{m_1} f_1(Z) + \dfrac{2b_1}{m_1}\dot{Z}_1 - \dfrac{2b_1}{m_1}\dot{Z}_2 = -\dfrac{F(t)}{m_1}; \\ \ddot{Z}_2 + \dfrac{c_2}{m_2} Z_2 - \dfrac{1}{m_2} f_1(Z) - \dfrac{b_1}{m_2}\dot{Z}_1 + \dfrac{b_1+b_2}{m_2}\dot{Z}_2 = 0; \\ \ddot{Z}_3 + \dfrac{c_3}{m_3} Z_3 + \dfrac{b_3}{m_3}\dot{Z}_3 = \dfrac{F(t)}{m_3}; \\ F(t) = S \cdot P(t), \end{cases} \tag{1}$$

where $f(Z)$ is the RCS nonlinear load characteristic; $Z = Z_1 - Z_2$; S is the hydraulic cylinder piston area; $P(t)$ is the pressure in the hydraulic cylinder cavities; b_1 is the viscous friction coefficient in the RCS; c_2, c_3 are soil stiffness coefficients; m_2, m_3 is the apparent soil mass together with the plates mass; b_2, b_3 are soil damping coefficients.

To ensure the base plate displacements within the soil elastic deformations, under the force of 10^6 N, it is necessary to take the value of S_{pl} of the base plate as equal to 12 m^2 [6], while its mass is $5 \cdot 10^3$ kg.

For low frequencies, the apparent soil mass m_{ap} can be determined from the expression [7,8]

$$m_{ap} = \frac{8(1-\gamma^2)(2-\gamma^2)\rho \cdot r^3}{10 - 2\cdot\gamma^2}, \tag{2}$$

where $\gamma = \dfrac{V_s}{V_p}$; V_s and V_p are the velocities of the transverse and longitudinal waves, respectively; ρ is the soil density; r is the plate conditional radius.

For the values $\gamma = 0.25$ and $\rho = 2000$ kg/m^3, the value $m_{ap} = 2.3\cdot10^4$ kg, and the values m_2 and m_3 can be assumed as equal, i.e. $m_2 = m_3 = 2.8\cdot10^4$ kg.

IOP Conf. Series: Journal of Physics: Conf. Series **1050** (2018) 012016 doi:10.1088/1742-6596/1050/1/012016

The soil stiffness and damping coefficients can be calculated from the expressions:

$$\left. \begin{aligned} c &= 2 \cdot \frac{E \cdot r}{1-\nu^2} \\ b &= b\left(1-\gamma^2\right)V_s \cdot \rho \cdot r^2 \end{aligned} \right\} \tag{3}$$

where $E = 1.67 \cdot 10^8$ Pa is elasticity modulus; $\nu = 0.49$ is Poisson's ratio.

For the adopted values of E, ν, and r, these values are $c = 4.4 \cdot 10^8$ N/m, $b = 1.2 \cdot 10^7$ N·s/m.

If we assume $m_1 = 10^5$ kg, then N-578 type RCS with a quasilinear vertical and lateral rigidity of 739.2 kN/m and 116.7 kN/m, respectively, can be used to hang the inertial mass.

Equations system (1) must be supplemented with the hydraulic drive equations. We assume that force $F(t)$ (figure 1) is created by a double-acting hydraulic motor with a two-stage electrohydraulic distributor [4, 7, 8]. Schematic diagram of the hydraulic system is shown in figure 2.

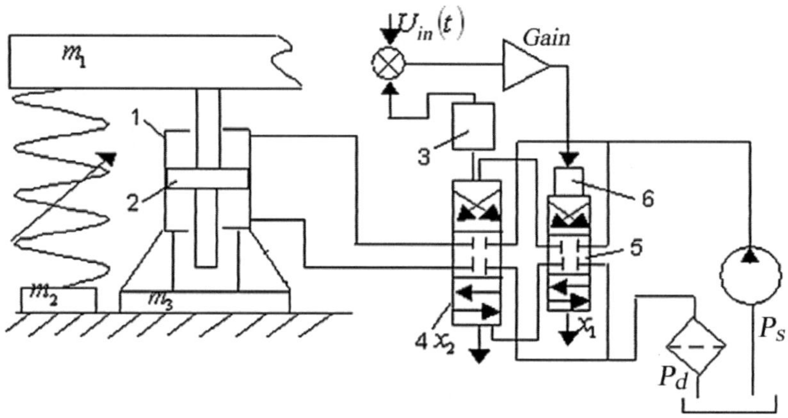

Figure 2. Schematic diagram of the hydraulic drive: 1 – the hydraulic cylinder; 2 – the piston; 3 – the main slide valve position feedback sensor; 4 – the main (second) cascade slide valve; 5 – the control (first) cascade slide valve; 6 – electromagnetic drive (EMD) slide valve 5; P_s, P_d – the supplied working fluid and drainage pressure; $U_{in}(t)$ - the master frequency generator voltage.

A linearized equations system describing the hydraulic system behaviour has the following form:

$$\left. \begin{aligned} &\frac{V_0+V_l}{2B_l}\frac{dp}{dt} + K_{QP} \cdot P = S + x_2 K_{QX} \\ &T\frac{m}{c}\frac{d^3x_1}{dt^3} + \left(\frac{m}{c} + T\frac{b}{c}\right)\frac{d^2x_1}{dt^2} + \left(T + \frac{b}{c}\right)\frac{dx_1}{dt} + x_1 = K_1 u_1 \\ &u_1 = K_g\left(U_{in} - K_{fb}x_2\right) \\ &\frac{dx_2}{dt} = K_{hg} \cdot x_1 \end{aligned} \right\} \tag{4}$$

where V_0 is the hydraulic cylinder cavity space; V_l is the hydraulic line space; B_l is liquid bulk modulus; K_{QP}, K_{QX} are transmission coefficients; x_1 is the first cascade slide valve displacement; x_2 is main cascade slide valve displacement; K_1 is EMD transmission coefficient; T is EMD time constant; m is EMD moving parts mass; h is viscous resistance coefficient in EMD; c is EMD anchor suspension stiffness; K_g is amplifier gain coefficient; K_{fb} is feedback coefficient; K_{hg} - hydraulic gain coefficient according to slide valves 4 and 5 displacements.

Equations systems (1) and (4) together with the tabulated values of the nonlinear function *f*(*z*) with the above-given assumptions describe the hydromechanical oscillatory system dynamics in figures 1 and 2.

3. Theory

Equations systems (1) and (4) are solved numerically using the Matlab application package with the "Simulink" extension (the license was purchased for educational and scientific purposes by OmSTU). The simulation circuit in "Simulink" is shown in figure 3.

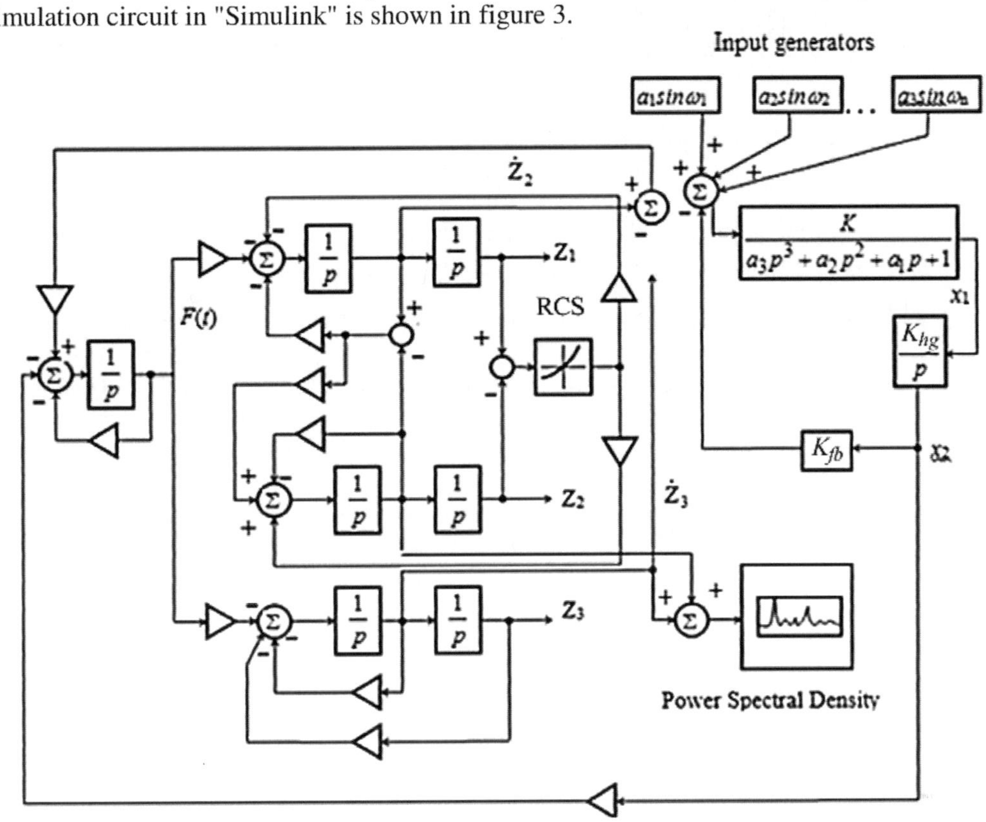

Figure 3. The simulation circuit in the "Matlab" environment with the "Simulink" extension.

It is of interest to evaluate the seismic signals source operation in the following modes:
1 – embodiment of a long monochromatic signal with discrete frequency tuning;
2 – embodiment of a linear-frequency modulated signal (LFM or "sweep" signal);
3 – expansion of the frequency range up to 100–150 Hz;
4 – generation of complex signals.

The long monochromatic signal generation mode is studied in [6]. Moreover, the generation efficiency at a low frequency in a joint operation of two closely located die blocks is determined by the interference interaction of their wave fields if the die blocks interference with their movement is neglected [11, 12]. The radiation efficiency for a system of two such die blocks can be estimated using the coefficient μ.

$$\mu = 1 + \frac{|\dot{Z}_3|}{|\dot{Z}_2|} \cdot \cos|\Delta\varphi|, \tag{5}$$

where $\Delta\varphi$ is the phase shift between $\dot{Z}_3$ and $\dot{Z}_2$.

Masses m_1 and m_2 oscillations analysis showed that the oscillations occur almost in phase in the frequency range (2–5 Hz).

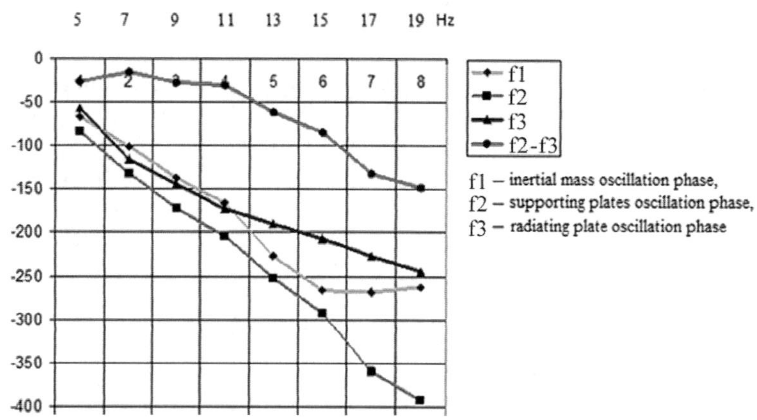

Figure 4. Phase-frequency characteristics of the masses oscillations for a seismic source SV 100/20.

The values of the seismic waves generation efficiency coefficient in the range of 2–20 Hz are given in figure 5 for different values of the relative damping coefficient $\xi = \frac{1}{2} b_1 \sqrt{\frac{m_1}{c_1}}$.

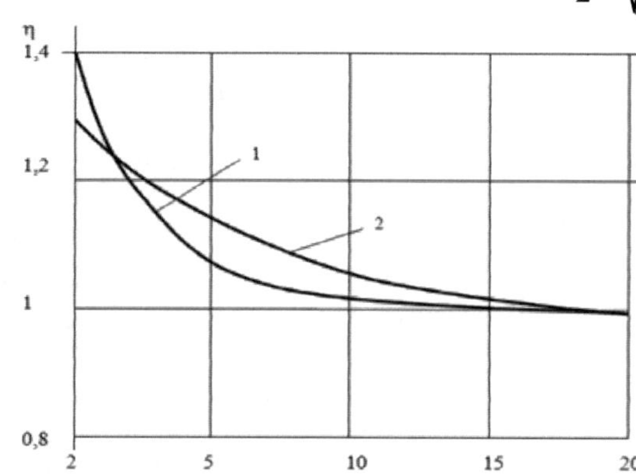

Figure 5. Dependence of the coefficient η on the frequency: 1 - $\xi = 0.1$; 2 - $\xi = 0.3$.

The emission efficiency of the source with a force closure in the "inertial mass - soil" system is preserved in the entire frequency range of 2–20 Hz. It is done at low frequencies by means of almost in-phase oscillations of the supporting plates and the base plate, and at high frequencies it is carried out thanks to the difference in the oscillations amplitude.

It is of interest to consider the possibility of obtaining an impulse seismogram analogue when a discrete series of monochromatic frequencies is generated by such a source.

It is known [5] that in the traditional "vibroseis" method the impulse seismogram $x_i(t)$ is formed by calculating the mutual correlation function between the seismogram x_m measured at the receiving point and the probing (generated) signal from the vibrator $x_v(t)$.

$$x_i(t) = \int_{-T}^{T} x_{\text{m}}(t-\tau)x(\tau)d\tau, \tag{6}$$

where T is the duration of the "sweep" signal.

It is known that for a fixed T the wider the frequency range at sending the linear-frequency modulated signal ("sweep" signal), the closer the transformed signal to the impulse seismogram. For instance, for seismic exploration purposes, the frequency range of two octaves is sufficient, while the signal/noise ratio increase is achieved by looping the "sweep" signal transmission and summing the result. An impulse seismogram can also be obtained by generating a discrete frequency series of monochromatic signals with duration *T*, summing them and calculating the convolution integral of the resulting sum during computer processing, i.e. calculating the following integral

$$x_i(t) = \int_{-T}^{T} \sum_{j=1}^{\mathrm{n}} x_j^{\mathrm{m}}(t-\tau) \cdot \sum_{j=1}^{\mathrm{n}} x_j^{\mathrm{b}}(\tau) d\tau, \qquad (7)$$

where n is the number of discrete frequency changes.
The computer simulation result for expression (7) at different values of T and n and the frequency range shows a good approximation of the calculations result to the impulse seismogram.
For instance, the simulation result with relative values of x_i at a frequency range of 2–8 Hz, a discrete of 0.2 Hz and $T = 10$ s is shown in figure 6.

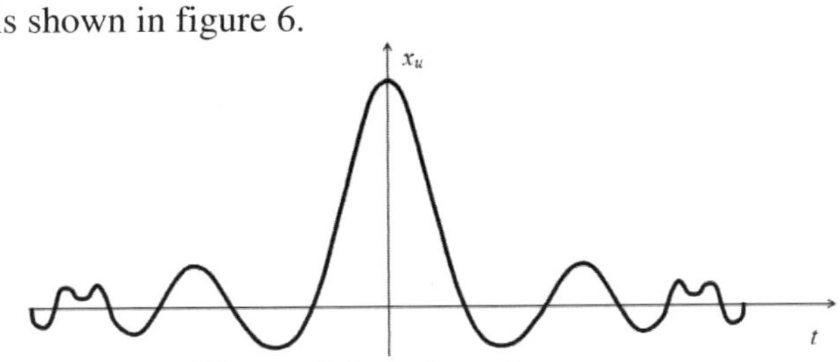

Figure 6. Impulse seismogram.

The results of the outlined above approach show that by means of powerful low-frequency vibrators in a monochromatic emissions mode one can obtain analogues of impulse seismograms with the required degree of approximation to a unipolar impulse. Moreover, in case of an unfavorable signal/noise ratio, the signal summation is possible for multiple emission of frequencies range.
When the vibrator operates in the monochromatic frequencies series generation mode, it is possible to obtain amplitude-frequency and phase-frequency characteristics for active seismic monitoring of the stressed medium state.
It should be noted that the monochromatic mode was realized [9] in a vibrating module with a "force closure to the soil" SV 100/20 (figure 7), which was designed and manufactured in OmSTU for vibroseismic influence on oil-gas formations and it successfully passed the pilot industrial testing in oil-and-gas production department *Buzulukneft*. The main technical characteristics of mobile vibromodule SV 100/20 are as follows:

- vibro-pulling force - 1 MN (100 tnf);
- frequency range - 5–20 Hz;
- relative frequency instability - 10^{-5}.

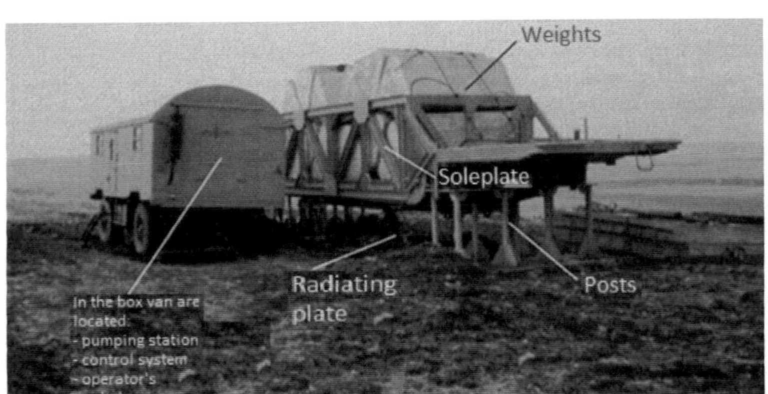

Figure 7. Seismic vibrator SV 100/20.

IOP Conf. Series: Journal of Physics: Conf. Series **1050** (2018) 012016 doi:10.1088/1742-6596/1050/1/012016

The seismic vibrator dynamics in the "sweep" signal mode was estimated from the computer simulation results in accordance with the scheme in figure 3.

The speed graphs $\dot{Z}_2$ and $\dot{Z}_3$ at frequency increase and decrease in the "sweep" for the frequency range of 2–26 Hz and duration τ = 6 c are shown in figures 8 and 9.

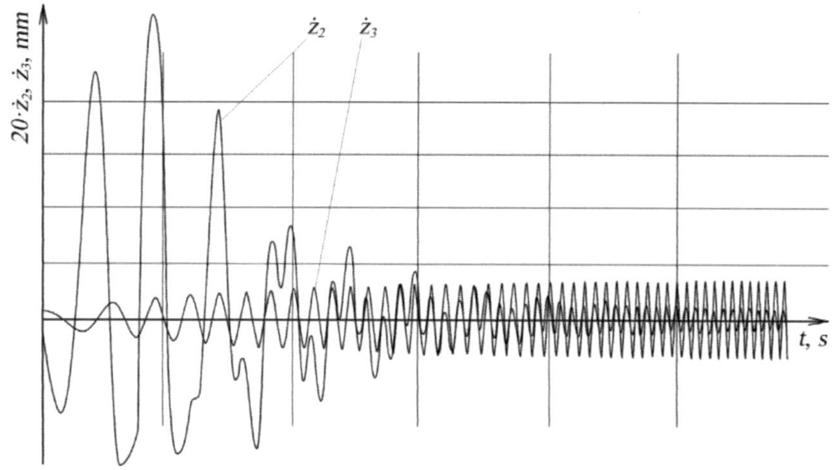

Figure 8. Graph $\dot{Z}_2$ and $\dot{Z}_3$ in the "sweep" mode with frequency increase $20 \cdot \dot{Z}_2, \dot{Z}_3$ (mm).

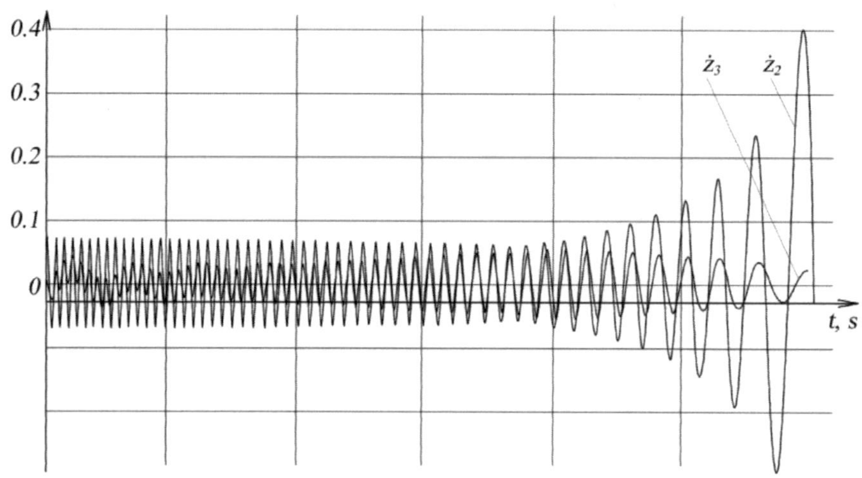

Figure 9. Graph $\dot{Z}_2$ and $\dot{Z}_3$ in the "sweep" mode with frequency decrease.

The motion dynamics analysis for the supporting and base plates in figures 8 and 9 shows that it is necessary to create a LFM from the maximum to the minimum operating frequency for a vibration module with a force closure on the soil.

The vibromodule dynamics modelling for a long "sweep" (τ >> 6 s) revealed the same pattern.

The phase shift |Δφ| between the speed of the base and supporting plates was analysed during the emission in the "sweep" signal mode for a duration of τ = 6 s and a range of 16–2 Hz, and the result is shown in figure 10.

If account is made of the graph in figure 8, expression (5) and the fact that the $\frac{|\dot{Z}_3|}{|\dot{Z}_2|}$ ratio is large at high frequencies, one may conclude that the vibrator under consideration will emit a seismic signal in the "sweep" signal mode efficiently.

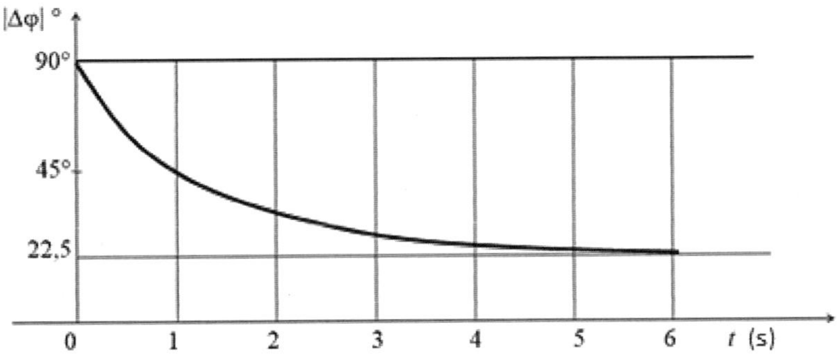

Figure 10. Changing |Δφ| according to time.

The behavior of such seismic waves source is of considerable interest if the frequency range is expanded up to 100–150 Hz, corresponding to the frequency range of seismic sources.
Also, the studies were conducted on a mathematical model.
The modelling results for the source masses oscillations in the frequency range of 5–100 Hz are presented in table 1.

Table 1. Modelling results for the source masses oscillations in the extended frequency range

f (Hz)	5	10	15	20	25	30	35	40	45	50	70	100
$2a_{z3}$ (mm)	6.4	6.2	5.4	3.5	2	1.4	0.8	0.6	0.4	0.35	0.1	0.038
Δz3 (mm)	3.2	3.2	3.2	3.2	3.2	3.2	3.2	3.2	3.2	3.2	3.2	3.2
P_{min}–P_{max} (MPa)	0.2-20	0.2-20	1-19	2.5-17.5	4-16	5-15	6-14.5	6.5-14	6.7-13.7	7-13	7.8-12.8	8.5-11.7

The notations used in the table are:
a_{z3} is base plate oscillations amplitude;
Δz3 is the position of the base plate oscillation center in relation to the day surface;
P_{min}–P_{max} are minimum and maximum values of the pressure oscillations in the hydraulic cylinder.
The analysis of the seismic source elements behavior in a wide frequency range of forced oscillations shows that the working frequency increase results in the force interaction amplitude decrease.
To keep the vibro-pulling force in the high-frequency range area, it is necessary to increase the discharge characteristics of the pump station and the electro-hydraulic amplifier.
At the same time, when expanding the frequency range of a seismic source with a force closure in the "inertial mass-soil" system to the high-frequency area, one should remember that all its elements represent oscillatory systems having their own frequencies.
It is known that when the natural object frequencies coincide with the forcing one, a resonance effect occurs and the oscillation amplitude of these objects can acquire values that may lead to their destruction.
The base plate and the supporting posts of the seismic source are characterized by high stiffness and are likely have relatively large values of the oscillation frequencies. At high frequencies the oscillation amplitudes caused by the exciting force are not large. In addition, they rely on a soil foundation with dissipative properties that smooth out the resonance peak significantly and, therefore, the danger of their destruction is not high.
The inertial mass of the seismic source is a welded volumetric structure assembled from I-shapped rollings and channels into cells with reinforced concrete blocks. These blocks have a triangular cross-section and, during the source operation, they are pulled together with the steelwork by steel ropes, which leads to their interlocking and the provision of a monolithic structure.

Interlocking of the reinforced concrete blocks, however, prevents their oscillations and indirect source emissions.

The inertial mass metal structure oscillations of various configurations, but with strictly defined geometric dimensions, were analysed by means of the finite element method and data processing with the Ansys software package.

Studies have shown that the load-bearing metal structure in the process of loading by the periodic force generated by the hydraulic cylinder produces not only bending oscillations, but also shear and torsional ones. Moreover, the bending oscillations natural frequency lies within the range of 22–26 Hz, the shear oscillations are within 30–36 Hz, and the torsional oscillations are within 14–16 and 30–36 Hz.

Consequently, when expanding the frequency range of a seismic source with a force closure in the "inertial mass-soil" system to the high-frequency part, it is necessary to take into account the values of the elements natural frequencies and to provide the necessary resistance during the design process.

It should be noted that at the discharge characteristics increase for the pumping station and the electro-hydraulic amplifier a hydromechanical source of the seismic waves with a force closure in the "inertial mass-soil" system can operate effectively not only in the low-frequency region, but also in a rather wide frequency range.

In recent years the oil industry has widely used vibroacoustic reservoir treatment to increase their productive capacity. Moreover, a large effect is achieved when vibration is produced at the layer natural frequency (the dominant frequency) and quasiresonant phenomena take place in it. As a rule, oil fields have several fluid deposit layers (formations) and each has its own dominant frequency [12, 13, 14].

To carry out the vibration treatment of two layers simultaneously, it is necessary to use two seismic sources, each operating at the dominant frequency of a specific layer. This method is very expensive and with the current shortage of powerful low-frequency seismic sources it is extremely difficult to implement.

Another way to perform simultaneous treatment of several oil formations is as follows. The seismic source emits a complex signal, being the sum of monochromatic signals, whose frequencies correspond to the formations dominant frequencies [10].

This complex signal puts each of the formations into resonance oscillations. Such treatment promotes fracture porosity development and internal potential energy release, which contributes to the intensification of the jets-clusters formation process. Thus, the production rates as well as the efficiency of its vibroacoustic treatment are increased.

Sources with a force closure in the "inertial mass-soil" system with an electrohydraulic exciter of impulse-type vibrations can radiate complex seismic signals if their control system is supplemented with the working fluid pressure feedback in the power hydraulic cylinder.

A block scheme of the control system in such a source, intended to operate at several dominant frequencies, has the following form (figure 11).

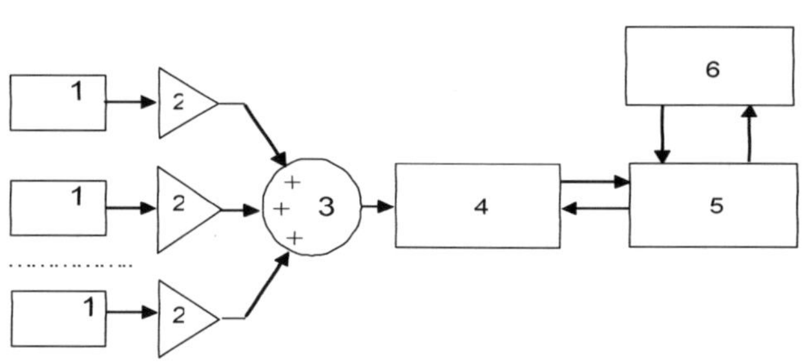

Figure 11. Block scheme of a seismic source for operation at several dominant frequencies: 1 – sin-dominant frequencies generators; 2 – amplifier; 3 – summation unit; 4 – vibrations exciter servo system; 5 – the vibrations generator; 6 – pumping station

Each of the sin generators at the output produces a signal with a frequency corresponding to a dominant one. After amplification and summation these signals are transmitted to the input of the seismic source servo system. The vibrations generator, powered by the pump station and controlled by the servo system, processes a complex signal from the summation unit output.
When producing a complex combined control signal, according to the formation properties, the values of the amplitudes and the combined signals phases may vary greatly. A sufficiently flexible control system for the electrohydraulic vibrations exciter allows a seismic source with such characteristics to operate in a group of similar sources.
Realizing the source simultaneous operation at several dominant frequencies of the productive formation and thus putting blocks of various sizes into the resonance, one should expect a significant increase in productive capacity (with other identical parameters of vibroacoustic reservoir treatment).

4. Conclusion

Thus, the conducted research shows that a low-frequency hydromechanical vibromodule with a force closure in the "inertial mass-soil" system is a universal and convenient tool to solve the problems of active environment seismic monitoring, seismology and regional seismic exploration, and the manufactured and tested prototype SV 100/20 can become a prototype of commercial low-frequency mobile sources of seismic waves.

References

[1] Li X, Li H, Shi Y, Zhou J and Yu H 2010 *Methods and models for rapidly identifying and finely evaluating the ultra-low permeability oil layer* J. Geophys. Eng. **8(1)** pp 13-21
[2] Chuanliang Y, Jingen D, Baohua Y, Hailong L, Fucheng D, Zijian C, Lianbo H, Haiyan Z and Qin H 2014 *Wellbore stability analysis and its application in the Fergana basin, central Asia* J. Geophys. Eng. **11** 015001
[3] Babeshko V A and Chichinin I S 1977 *O dinamicheskih zadachah kolebaniya uprugogo poluprostranstva / Problemy vibracionnogo prosvechivaniya Zemli* (Moscow: Nauka) p 240 (in Russian)
[4] Shaginyan A S 1977 *Sozdanie vibracionnyh istochnikov bol'shoj moshchnosti dlya glubinnogo zondirovaniya Zemli / Problemy vibracionnogo prosvechivaniya Zemli* (Moscow: Nauka) p 240 (in Russian)
[5] Shneerson M B and Majorov V V 1980 *Nazemnaya sejsmorazvedka s nevzryvnymi istochnikami kolebanij* (Moscow: Nedra) (in Russian)
[6] Buryan Yu A and Sorokin V N 2002 *Fiziko-tekhnicheskie problemy razrabotki poleznyh iskopaemyh* (Sibirskoe otdelenie RAN, institut gornogo dela SO RAN) **3 (81-88)** (in Russian)
[7] Baranov V N and Zaharov Yu E 1966 *Elektrogidravlicheskie vibracionnye mekhanizmy* (Moscow: Mashinostroenie) p 180 (in Russian)
[8] Hohlov A V et al. 1971 Elektrogidravlicheskie sledyashchie sistemy (Moscow: Mashinostroenie) p 205 (in Russian)
[9] Buryan Yu A, Sorokin V N and Kuzik V L Patent 2184985 RF Vibroistochnik (in Russian)
[10] Buryan Yu A, Sorokin V N, Serdyukov S V, and Cherednikov E N Patent 2242807 RF Sposob razrabotki neftegazovogo mestorozhdeniya (in Russian)
[11] Babeshko V A, Glushkov E V and Zinchenko Zh F 1989 *Dinamika neodnorodnyh linejno-uprugih sred* (Moscow: Nauka) p 344 (in Russian)
[12] Gurvich I I and Nomokonova V P 1981 *Sejsmorazvedka. Spravochnik geofizika* (Moscow: Nedra) p 464 (in Russian)
[13] Gurvich I I and Boganik G N 1980 *Sejsmicheskaya razvedka* (Moscow: Nedra) p 551 (in Russian)
[14] Aleshin A S, Kuznetsov V V and Tsimmerman V V 1986 *O vzaimodejstvii Sejsmicheskih vibratorov / Razrabotka i issledovanie istochnikov sejsmicheskih signalov i metodov nevzryvnoj sejsmorazvedki* (Moscow: VNIIOEHNG) (in Russian)

Performance characteristics describe the structure of descriptors

I V Chervenchuk, V I Potapov, V V Makarov, O P Shafeeva, M S Doroshenko and A S Gritsay

Omsk State Technical University, 11, Mira ave. , Omsk, 644050, Russia

Abstract. Monitoring systems, as systems for observing the state of objects, usually operate with data of some common nature, which can be successfully described further proposed hierarchical model. The initial data of observations, that is, the data of a certain subject area, which in the future presuppose the conduct of statistical studies over them, as a rule, are homogeneous in nature and, in most cases, are described in terms of descriptors and objects. On this basis, it is possible to build a fairly universal hierarchical model of data storage, focused on their subsequent statistical processing, which, in turn, can serve as an information core for building a specialized information system. The paper proposes a method of evaluating the effectiveness of descriptions of the structures having a natural hierarchy presents a study of the proposed features on the subject of production optimization problems.

1. Introduction
In the description of complex multi-level structures, which has a natural hierarchy, such as classification [3, 4], the communicative structure [8], administrative structures, and the like, the most acceptable is the use of graph structures such as a tree. With such a description excludes unnecessary repetition and redundancy.

2. Formulation of the problem
The aim is to find characteristics, quantitatively determining the winning of the hierarchical approach compared to conventional linear organization in the description of the structures, has a natural hierarchy to explore the data characteristics on the subject of production optimization problems.

3. Theory
We will assume that we are given a certain complex object with a hierarchical structure, therefore, a tree of objects is also specified, which must be described by the user-specialist through the proposed interface. The task is to construct for the given object the most optimal tree of indices or to evaluate any structure of descriptors, where it is proposed to take the characteristics of compactness and high-speed performance of descriptors [2, 5] as criteria for optimality; we have a two-criteria optimization problem.

In this case, by ***compactness*** we mean the characteristic $\lambda(T_x) = \frac{l(T_x)+n}{m+n}$, (1)

where $n = |X|$ - number of indices (of the hanging vertices of the metric tree T_x), m - the number of internal vertices of the tree T_x, $l(T_x)$ - total length of the tree description T_x, $k = m + n$.

Compactness of a tree of indices shows how many times the computer memory costs are reduced when describing the semantics of the system indicators due to the hierarchical structuring of this description.

The value $\tau = \frac{1+n}{2}\frac{n}{l(T_x)}$, (2)

we will call further ***prior high-speed performance of search of an index.*** This characteristic shows the gain from hierarchical structuring when searching for an object [2].

These criteria it is necessary to maximize, that is, to construct such a tree of indices (find the structure of descriptors) for which these characteristics would be as much as possible.

However, the criteria of compactness and high-speed performance of search of an index are obviously contradictory, since the compactness of (λ) increases with increasing $l(T_x)$, and the performance (τ) decreases.

4. Experimental results

Consider the set $\Theta(\kappa)$ of all admissible indices structures (that is, the set of all root trees) with a given total number of vertices $k = m + n$.

Consider the case when the total number of vertices (κ) and the number of hanging vertices (n), that is, the number of descriptors are rigidly set.

In this case according to (1) and (2) if we are by means of some transformation moving from a structure T_x^0 to a structure T_x' to increase the value by one criterion (by changing the value of the lengths of a tree of indices), we will necessarily lose by another criterion, and losing by one criterion we will necessarily win in another way It is possible to give examples of different structures having the same numerical values of compactness and high-speed performance for constant k, it is necessary and sufficient for this that the structures have the same l and n.

Consider the set of all effective structures $P(T_x, k)$, where by effective we will understand the Pareto optimality by the criteria of high-speed performance and compactness. Clearly in this case any structure is effective, i.e. $\{\Theta(\kappa)|\mathrm{n} = n_0 = const\} = P(T_x, k, n)$.

Thus, we have obtained that, in the case of given κ and n, any structure of the exponents is effective, i.e., there is no such structure T_x', that would be inferior to another structure T_x^0 in terms of the compactness and high-speed performance characteristics simultaneously; as well as there is no such structure T_x', that would outperform some structure T_x^0 at the same time in terms of compactness and high-speed performance.

Consider the case where only the fixedly set number of common vertices k, and the number of dangling vertex n (such as, respectively, and internal), freely variable parameter. This is possible when the classification of descriptors is not definitively defined and work is in progress to determine it, or when comparing a number of different indices structures based on performance and compactness.

In this case $\Theta(k) \supseteq P(T_x, k)$, with $k > 2$, inclusion will be strict. It will be shown below, how can one verify the affiliation of some structure to a set of effective structures $P(T_x, k)$.

For the following discussion, we introduce a certain class of trees (sub-class of root trees), which we call TR-Tree– trees with a trunk. TR-Tree have a trunk - not a closed sequence of vertices $< d_1, d_2, \dots, d_r >$, bound ribs; from each such vertex d_i only edges can come out to the hanging vertices or to the vertex d_{i+1}. A case is possible when $r = 1$. The set of hanging vertices issuing from the vertex d_i forms a tier TR-Tree (Here, for convenience, we have moved a little from the traditional concept of "tier", we consider that the root has a 0-tier, and not the first.). Examples of TR-Tree are shown on Figures.1.

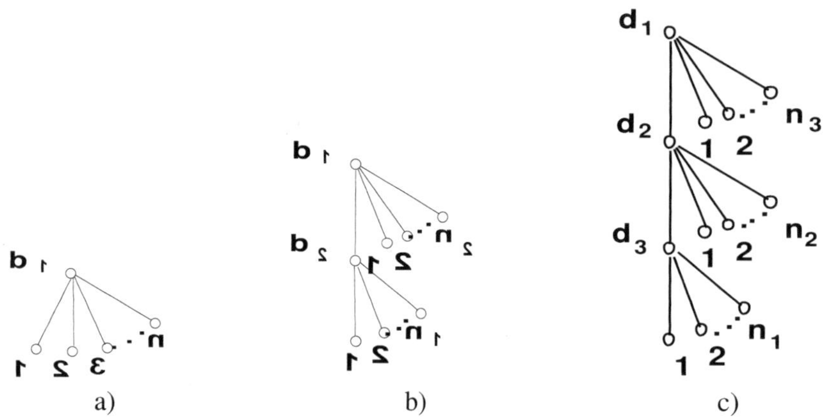

Figure 1. Examples of TR-Tree: a) one-level [n], b) two-level [n1,n2], c) three-level [n1,n2,n3].

For convenience of the further presentation, contrary to tradition, we call a leaf as a hanging vertex in combination with an incident arc. We will assume that the lower tier has n_1 leaves, the next (up) n_2 leaves and the uppermost n_r of the leaves. It is obvious that TR-Tree can be represented by arrays of dimension r of the form $[n_1, n_2, \dots, n_r]$, where r is the number of tiers. The following expressions are valid for TR-Tree

$$n = k - r = \sum_{i=1}^{r} n_i , \quad l(T_x) = \sum_{i-1}^{r} i \cdot n_{r-i+1} \qquad (3)$$

It is possible to prove the following expressions (due to limited volume of the article scope of the article, the evidence is not given)

Statement 1. For each structure T_x^0, that is not TR-Tree, there is a TR-Tree T_x^s, that overlaps T_x^0 by the criteria for speed and compactness.

Evidence. According to (1) and (2) in order for some structure T_x' overlapped some structure T_x^0 by the speed and compactness at the same time , it is sufficient that the number of pendant vertices of the structure T_x' be larger than the number of pendant vertices of the structure T_x^0 at a constant length, i.e. the following conditions would be fulfilled:

$$n(T_x') > n(T_x^0), l(T_x') = l(T_x^0). \qquad (4)$$

Obviously, such structures exist.

We will take arbitrary root tree T_x^0 that is not a CT-Tree.

We will transform this tree in corresponding to it TR-Tree T_x^s. In T_x^0 we will select a branch of the maximum length, beginning from root peak (if there are several such branches, one can take any), the whole branch except for the hanging vertex and the arc (leaf) incident to it will be assigned to the trunk of the future tree T_x^s. In this case the original tree T_x^0 we can consider as some a generalized TR-Tree from the stem vertices d_i of which not only leaves originate, but also complex subtrees (having more than one inner vertex.

We carry out the operation of disassembling of such subtrees R, saving $l(T_x) = l$, which consists in the following:

To tear off a leaf from some difficult subtree with the root d_i,

a) if at the same time peak – the father of this leaf does not become trailing, then to transfer this leaf to a trunk with saving level;

b) if the peak – the father of this leaf becomes trailing, then to transfer this leaf to the first tier.

Figures 2b, c show two successive transformations of R to the structure of Figure 2a, which transform this structure into a TR-Tree.

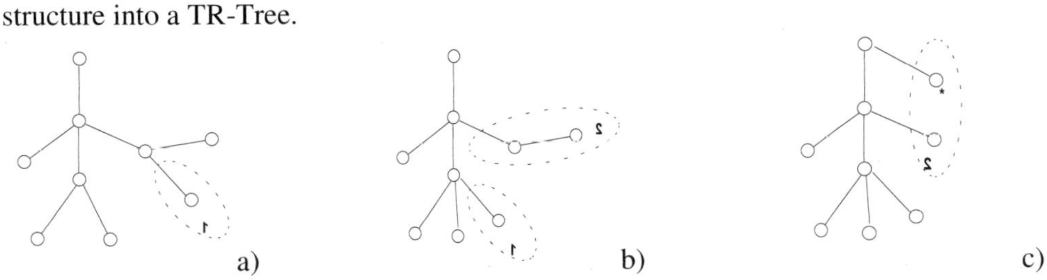

Figure 2. Convert R – disassemble complex subtrees: a) original tree, b) the first step in disassembly, c) the second step of the disassembly and conversion to TR-tree.

As for a trunk the branch of the maximum length is selected, similar disassembling is always possible. It is obvious that, applying to leaves of difficult subtrees, we will surely transform the initial tree to TR-Tree for the finite number of steps depending on complexity of the subtrees which are a part of the initial tree. At the same time, as we subject to disassembling the difficult subtrees (having more than one internal peak), R -conversion for modification b, the increasing number of trailing peaks will be surely applied. As R -conversion does not change total length of a tree, i.e. (4) is executed, in case of analysis of each difficult subtree the value of criteria of high-speed performance and compactness will increase.

The statement is proved.

According to the proved statement we have that the set of TR-Trees with k peaks includes a set of effective structures $P(T_x, k)$, that allows to narrow considerably a set of search effective on τ and λ structures and to formulate sufficient sign of belonging of some structure to a set $P(T_x, k)$. However the set of ST trees, though much less set of all root trees nevertheless is rather great and its power sharply increases $(2^{k-2}, k > 1)$ with growth of k. The following statement gives the conditions allowing to narrow as much as possible a set of search of effective solutions and to formulate sign of belonging of some structure to a set$P(T_x, k)$.

Statement 2. In order that some structure T_x^0 belonged to a set of effective sturture by the criteria of compactness and high-speed performance$P(T_x, k)$, is necessary also enough, that this structure T_x^0 was TR-Treeand the following inequalities were executed:

$$r = 1 \text{, or } \sum_{i=1}^{r-2}(r-i)\cdot n_{ri} > (r-2)\cdot k - r\cdot(r-3),\ r < \frac{k}{2}, \qquad (5)$$

where r –numbers of tiers in this TR-Tree.

Evidence. We will consider structures of TR-Trees with different quantity of tiers r, upon transition to TR-Tree with a large number of tiers, the number of trailing peaks will decrease by unit.

1) We will consider structure of fan type – TR-Tree with $r = 1$ – Figure 1 a), this structure has maximum possible n (i.e. $n_1 = k - 1$) and minimum possible $l(T_x) = k - 1$ – i.e. has the maximum on high-speed performance (τ) also belongs $P(T_x, k)$. At the same time, as we have the two-criteria task, the value of compactness $\lambda(T_{x1}^1)$ will be minimum among all structures $T_x \in P(T_x, k)$ (for example, [1]).

2) We will consider structures of two-tier TR-Tree with $r = 2$ – Figure. 1 b). These structures will have a little smaller values (τ) because of smaller value n *(here* $n_2 = k - 2 = n_1 - 1$). For this purpose, that structures of two-tier TR-Tree T_{xj}^2 (where j – number of structure in case of lexicographic streamlining) exceeded structure T_{x1}^1 on compactness according to (1) is necessary also

enough (since n is strictly set), that inequality was observed $l(T_{xj}^2) > l(T_{x1}^1) + 1$, where unit in the right part compensates reduction of value n therefore

$n_{22} + 2n_{21} > k - 1$,

where the first index at n shows number of tiers in TR-Tree r, the second – number of trailing peaks on a i tier, $i = 1 \ldots r$. Taking into account that $n_2 = k - 2 = n_{21} + n_{22}$ we have

$$n_{21} > 2. \qquad (6)$$

that is in order that two-level ST tree superimposed fan structure on a compactness index it is necessary that the number of peaks on the lower tier (here will be only trailing) was strict more than two. As $n_2 > n_1$, that any structure from T_{xj}^2 will not exceed T_{x1}^1 on compactness. The structure T_{xj}^2 with the with the maximum total length will have total length $l_{max}(T_{xj}^2) = 2 \cdot (k - 2)$, corresponds to structure with $n_{22} = 0$.

3) We will consider structures of TR-Tree with $r = 3$ – Figure 1 c). These structures will have smaller values (τ) because of smaller value of n (here $n_3 = k - 3 = n_2 - 1$). For this purpose, that structures of TR-Tree T_{xj}^3 exceeded all structures T_{xj}^2 on compactness according to (1) it is necessary, that inequality $l(T_{xj}^3) > l(T_{x1}^2) + 1$, where unit in the right part compensates reduction of value n, for this case was observed

$n_{33} + 2n_{32} + 3n_{31} > k - 1$, и т. к. $n_3 = k - 3 = n_{31} + n_{32} + n_{33}$ we have

$$n_{32} + 2n_{31} > k, \qquad (7)$$

and $l_{max}(T_{xj}^3) = 2 \cdot (k - 3)$, corresponds to structure with $n_{32} = n_{33} = 0$.

So, structures T_{xj}^3, fitting (7) will superimpose structures T_{xj}^2, and furthermore T_{x1}^1 (in view of transitivity ">")by criterion of compactness. At the same time with building of total length of value of criterion of compactness will decrease (since the denominator (2) grows), it is obvious as according to (7) of structures T_{xj}^3 we will take only those structures which have big total length, than structures $T_{xj'}^2$, and the number of trailing peaks T_{xj}^3 are less than $T_{xj'}^2$, at what such structure from T_{xj}^2 will superimpose what structure from $T_{xj'}^3$ by criterion of compactness. Thus, all structures $T_{xj'}^2$ fitting (6) and T_{xj}^3 fitting (7) are effective.

Similarly arguing for TR-Trees with any number of tiers we will come to the first part of inequalities (5). However growth of an index of compactness with growth of number of tiers of TR-Trees is limited, adding the greatest possible values of total length of trees for each r, we will find a limit of this growth.

$\lambda_r > \lambda_{r-1}$,

$r \cdot (k - r) + k - r > (r - 1)(k - r + 1) + k - r + 1$,

$r < \frac{k}{2}$.

The structure with the maximum number of tiers, r_{max} satisfying to the last inequality at which all trailing peaks are located at the bottom (first) level will have the maximum total length among effective structures, and respectively, at most on compactness. Such structure reminding a broom is given in a Figure 3 a).

Thus, the sufficiency is proved.

For the proof of need we will assume that there is some structure $T_{xj0}^{r0} \in P(T_x, k)$, and at the same time inequalities (5) are not executed. If $r0 = 1$, that (5) can be executed; if $r0 > 1$, that in case of not execution (5) among TR-Trees with $r = r0 - 1$, there will be a tree with the same total length (or, at least, with total length smaller on unit), but at the same time as such tree will have quantity of

IOP Conf. Series: Journal of Physics: Conf. Series **1050** (2018) 012017 doi:10.1088/1742-6596/1050/1/012017

trailing peaks, bigger on unit, than at T_{xj0}^{r0}, it will superimpose T_{xj0}^{r0} by compactness (or, at least, to be the equivalent) and to superimpose on high-speed performance. Thus, T_{xj0}^{r0} cannot be effective structure.

The statement is proved.

Remark. If to replace signs of a strict inequality (5) with signs of mild inequality we will receive necessary and sufficient conditions for a set of poorly effective structures (optimum according to Slater) by criteria of compactness and high-speed performance.

Evidence. Really, if in (5) to replace signs of a strict inequality with signs of mild inequality, then structures meeting such condition with number of tiers of r will include also structure T_{xj1}^{r} (or a several the equivalent by criteria) the equivalent on compactness with some structure T_{xj0}^{r-1}, having the maximum index of compactness for structures with number of tiers of $r-1$. At the same time among structures T_{xj}^{r-1}, obviously, there will be no structures which are strictly superimposing (i.e. on ">") T_{xj1}^{r} by criteria of high-speed performance and compactness at the same time.

For the final proof of the remark we will prove that among poorly effective estimates there are no structures, the equivalent on high-speed performance and the having different values of criterion of compactness. Let there are two structures T_{x1}^{r1} with number of trailing peaks of n_1 and total length of a tree of l_1, and T_{x2}^{r2} with n_2, l_2 respectively. According to (2) in order that these structures were equivalent on high-speed performance and had different values of criterion of compactness, it is necessary that $n_2 - n_1 > 0, \quad l_2 - l_1 > 0$ or $n_2 - n_1 < 0, \quad l_2 - l_1 < 0$. Let's say that T_{x1}^{r1} are effective, then if $n_2 - n_1 > 0$, then $l_2 - l_1 > 0$ therefore, $\lambda_2 > \lambda_1$ what contradicts efficiencyT_{x1}^{r1}. If $n_2 - n_1 < 0, \quad l_2 - l_1 < 0$, therefore, $r2 > r1$, then among structures with number of tiers of $r1$ we will take structure T_{x3}^{r1} with total length $l(T_{x3}^{r1}) = l_2$, such structure exists as minimum possible value of length of a measure description is less for structures with smaller number of tiers (with $r2 = 1$ (5) it is automatically executed). According to (1) and (2) structure T_{x3}^{r1} will superimpose structure T_{x2}^{r2} on high-speed performance and compactness, i.e. T_{x2}^{r2} cannot be poorly effective.

5. The discussion of the results

For convenience of execution of the procedure of check of some structure regarding belonging to effective boundary $P(T_x, k)$ by the accepted criteria of inequality (5) it is possible to expand. From inequalities (5) it is possible to pass to the system of inequalities which in a general view looks as

$$\begin{cases} jn_{r1} + (j-1)n_{r2} + \cdots + n_{rj} > (j-1)k - (j-2)r; \quad j = 1 \dots r; \\ r < \frac{k}{2}. \end{cases} \tag{8}$$

The system of inequalities (8) allows to realize quickly check regarding belonging of some structure to a set of effective solutions $P(T_x, k)$. If necessary, using these inequalities, it is possible to write easily the procedure allowing to find all effective structures in case of the given k (at the same time also you should not forget that all effective structures belong to a class TR-Trees). In figures 3 a-g the set of $P(T_x, 8)$ is given.

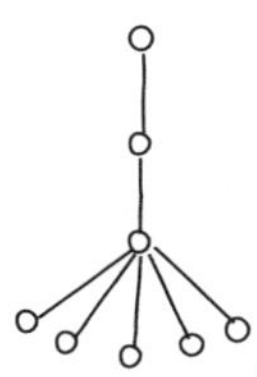
a)

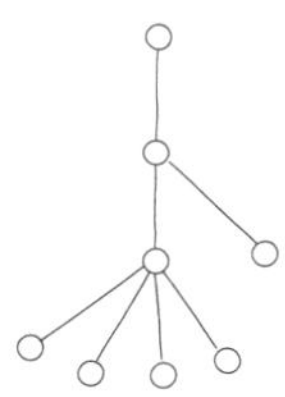
b)

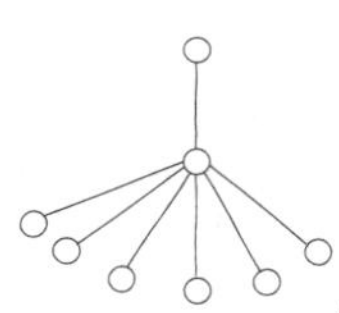
c)

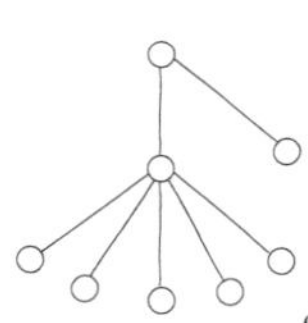
d)

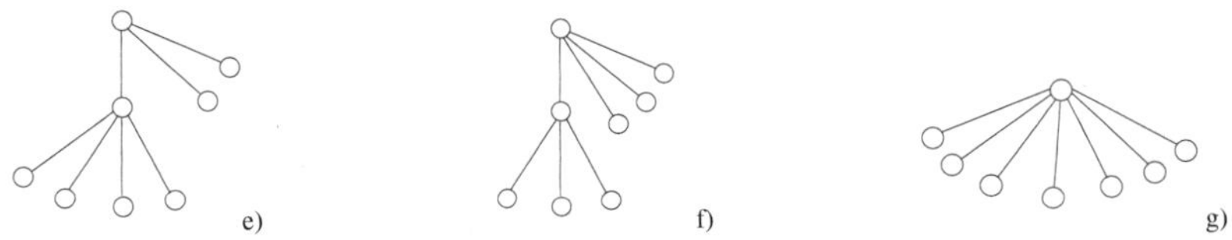

Figure 3. Effective structures for $k = 8$: a) configuration 1, b) configuration 2, c) configuration 3, d) configuration 4, e) configuration 5, f) configuration 6, g) configuration 7.

6. Conclusion

he use of hierarchical structures in the description of objects that have a natural hierarchy is the most appropriate and effective, the characteristics proposed in this article confirm this. In addition, it is possible to identify a number of structures for which such a hierarchical description gives the greatest benefit, to formulate and solve a number of optimization problems.

In view of feature of a structure, use of effective structures in practice is limited, however they, being theoretical result, represent some limit structures in an optimization sense by criteria of compactness and high-speed performance, and can serve as a starting point (a standard for comparing) in case of assessment of arbitrary structures of indices by criteria of compactness and high-speed performance.

Such a hierarchical model descriptions have large scope of application, for example, can be used to describe historical data of electricity load [6, 7].

References

[1] Chao Niu, Yongshun Zhang and Jianrong Guo 2018 Pareto optimal layout of multistatic radar, *Signal Processing* **142** pp 152–156 https://doi.org/10.1016/j.sigpro.2017.07.017.

[2] Michael James Winter, Masayuki Hyodo, Yang Wu, Norimasa Yoshimoto, Muzamir Bin Hasan and Kanta Matsui 2017 Influences of particle characteristic and compaction degree on the shear response of clinker ash *Engineering Geology* **230** 2017 pp 32–45 https://doi.org/10.1016/j.enggeo.2017.09.019.

[3] Cervenchuk I 2015 Model of statistical monitoring data of tap water *Natural and intellectual resources of Siberia* **1** pp 187–191

[4] Haoliang Yuan, Xuecong Li, Fangyuan Xu, Yifei Wang, Loi Lei Lai and Yuan Yan Tang, 2018 A collaborative-competitive representation based classifier model *Neurocomputing* **275** pp 627-635, https://doi.org/10.1016/j.neucom.2017.09.022.

[5] Khamitov R N, Gritsay A S, Tyunkov D A, Sinitsin G E 2017 Day-Ahead Short-Term Forecasting Electricity Load via Approximation *IOP Conf. Ser.: Mater. Sci. Eng.* **189** https://doi.org/10.1088/1757-899X/189/1/012005

[6] Potapov V I, Gritsay A S, Tyunkov D A and Sinitsin G E 2016 Using neural network for building short term forecast of electricity load of LLC "Omsk energy retail company" *Bulletin of the Tomsk Polytechnic University Geo Assets Engineering* **327(8)** pp 44–51

[7] Gritsay A S, Khamitov R N, Sinitsyn G E, Cervenchuk I V 2016 The study of methods of approximation for solving the problem of short-term forecasting of the daily electricity consumption *Vestnik Samgtu. Ser. Tech. science* **4(52)** pp 91–97

[8] Shpynov S Pozdnichenko N and Gumenuk A 2015 Application of Formal Order Analysis (FOA) to Higher Order Grouping of Bacteria in the Genera Rickettsia and Orientia *Microbes and Infection* **17** pp 839–844

IOP Conf. Series: Journal of Physics: Conf. Series **1050** (2018) 012018 doi:10.1088/1742-6596/1050/1/012018

The Kramers problem in the energy diffusion regime: transient times

M V Chushnyakova[1,2], I I Gontchar[1,3], A I Blesman[1]

[1] Physics Department, Omsk State Technical University, Omsk 644050, Russia
[2] Physics Department, Akdeniz University, Antalya 07058, Turkey
[3] Physics and Chemistry Department, Omsk State Transport University, Omsk 644046, Russia

Abstract. Brownian motion driven by thermal fluctuations play an important role in the work of nanomachines and in single-molecule pulling experiments. Some problems in this field are reduced to the so-called Kramers problem concerning the decay rate of a quasistationary state. Provided all Brownian particles start their motion from the potential minimum, the rate increases in time reaching a quasistationary value. We model the process of decay using the Langevin equations. This modeling makes it possible identifying the transient time required for the rate to reach its quasistationary value. We compare the numerical transient times with those known from the literature. It turns out that the difference reaches up to 70% in the case of weak friction. Increasing friction results in decreasing this difference. In order to make our results useful for a larger audience, we present them in the dimensionless form.

1. Introduction

There are many modern examples of the Kramers problem [1] in technology and natural sciences [2-4]. Let us mention electric current through Josephson junction [5], single molecular pulling experiments [6, 7], and atomic nuclear fission [8-10]. Many studies are devoted to the quasistationary stage of the process (see e.g. [9-11]). Yet the transient stage has been investigated significantly poorer [12, 13]. In the present study we aim to obtain the dependence of the transient time upon the friction strength solving numerically stochastic differential equations (SDEs) for the energy diffusion regime.

2. Model

Following [1, 14, 15] we perform modeling for the one dimensional case. The motion of a Brownian particle is described by the dimensionless coordinate q and conjugate momentum p. In discrete form the SDEs read:

$$p^{(n+1)} = p^{(n)}(1 - \eta m^{-1}\tau) + K\tau + g b^{(n)}\sqrt{\tau}, \tag{1}$$

$$q^{(n+1)} = q^{(n)} + \left(p^{(n)} + p^{(n+1)}\right)\tau/(2m). \tag{2}$$

Here the superscripts (n) and $(n+1)$ correspond to two successive time moments separated by the time step of modeling τ. The random numbers b entering the random force possess a normal distribution with zero average and variance 2. In equation (1) η and m are the friction and inertia parameters, respectively; $K = -dU/dq$ represents the conservative force; $g = (\theta\eta)^{1/2}$ is the amplitude of the random force; θ is the mean thermal energy. The potential energy $U(q)$ is parametrized by two parabolas of the same stiffnesses C. The parabolas are smoothly joined at $q_m = 1.3$. All trajectories start from the bottom of the well at $q_c = 1.0$ with zero momentum.

The modeling results in a sequence of N_{tot} trajectories terminated not later than particular time moment t_D. Some of the trajectories reach the absorptive border ($q_f = 2.0$) at $t < t_D$. The algorithms for finding the time dependent decay rate $R_{ft}(t)$ and its quasistationary value R_D are described in detail in [9].

3. Results

In figure 1 we present the dimensionless decay rates (R/R_D) versus the dimensionless time ($t\omega$) for several values of the dimensionless parameter, reflecting the value of friction

$$\varphi = \frac{\eta}{\sqrt{mC}}. \tag{3}$$

Here $\omega = (C/m)^{1/2}$. These calculations have been performed for $U_b\theta^{-1} = 3.00$ (U_b is the barrier height). One clearly sees that the dynamical rates $R_{ft}(t)$ resulting from the SDEs (broken lines with symbols) are subjects of fluctuations. The quasistationary rates R_D are displayed by the horizontal lines. The statistical errors of R_D are about 1-2%. Finally, the smooth lines without symbols $R_m(t)$ denote the approximation of $R_{ft}(t)$; these approximations are found as explained below.

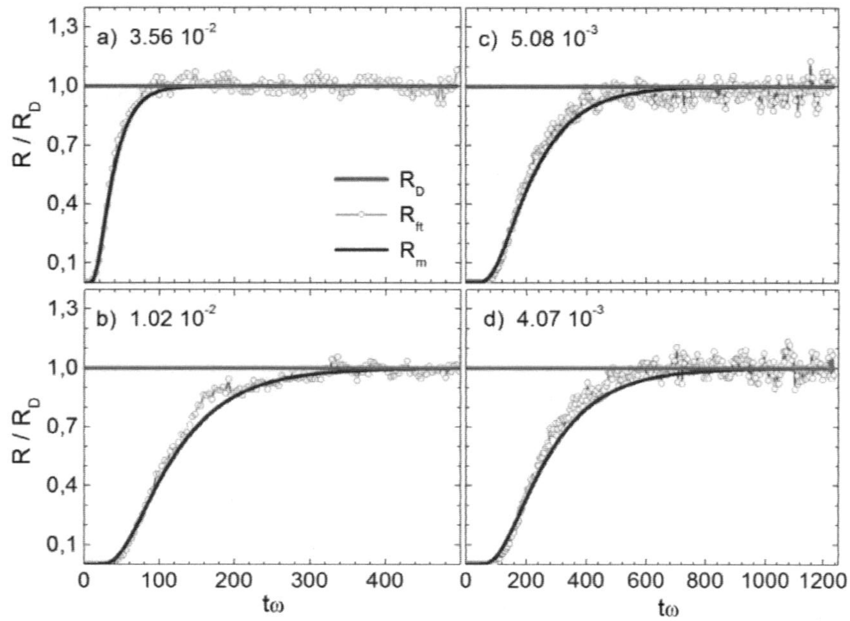

Figure 1. Time dependence of the decay rates for four values of parameter φ indicated in the panels.

In Ref. [12] an approximate analytical expression for the dynamical rate was obtained. Its time dependence basically reads

$$R_W(t) = R_0 \exp\left\{-\frac{U_b}{\theta h_W(t)}\right\}, \tag{4}$$

where

$$h_W(t) = 1 - \exp\left(-\frac{\eta t}{m}\right). \tag{5}$$

One sees that $R_W(t)$ approaches $R_0\exp\{-U_b/\theta\}$ as time increases. It is natural to identify R_0 with $R_D\exp\{U_b/\theta\}$. However, in order to describe the transient behavior of the rate we are forced to modify equation (4) introducing an extra fitting parameter s:

$$R_s(t) = R_D \exp\left\{\frac{U_b}{\theta} - \frac{U_b}{\theta h_s(t)}\right\}, \tag{6}$$

where

$$h_s(t) = 1 - \exp\left(-\frac{s\eta t}{m}\right). \tag{7}$$

The value of s is chosen minimizing the deviation of $R_s(t)$ from $R_{ft}(t)$, i.e. minimizing the following value

$$d_s = \frac{1}{n}\sum_i \frac{|R_s(t_i) - R_{ft}(t_i)|}{R_{ft}(t_i)}. \quad (8)$$

Here n is the number of terms in the sum. In order to avoid the division by zero, the summing is performed starting from such time moment t_t that $R_{ft}(t_t) > 0.02R_D$ for the first time. Typical dependences $d_s(s)$ are presented in figure 2. We see that each dependence indeed possesses a well-pronounced minimum. The approximating curves $R_m(t)$ in figure 1 are obtained with the value of s corresponding to this minimum.

In figure 3 we present the dependences of the optimum values s_m and d_m upon φ. It turns out that s_m decreases monotonically with φ and reaches unity as $\varphi \approx 0.1$, i. e. at this value of φ, equations (5) and (7) coincide and our results agree with Ref. [12].

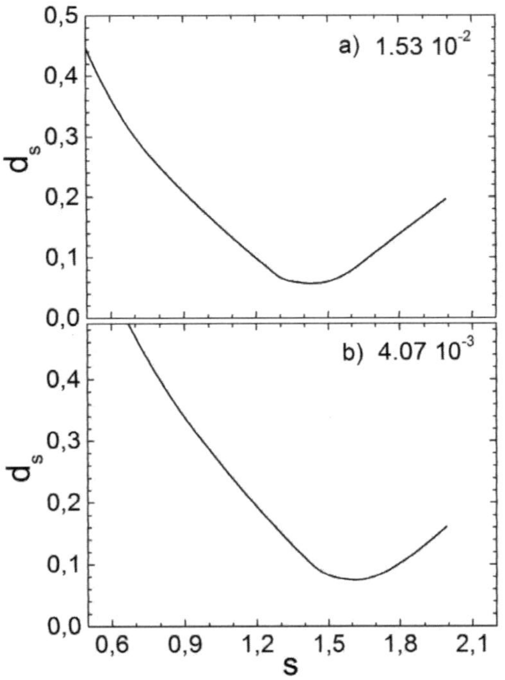

Figure 2. Dependence of the parameter d_s upon s for two values of φ indicated in thepanels.

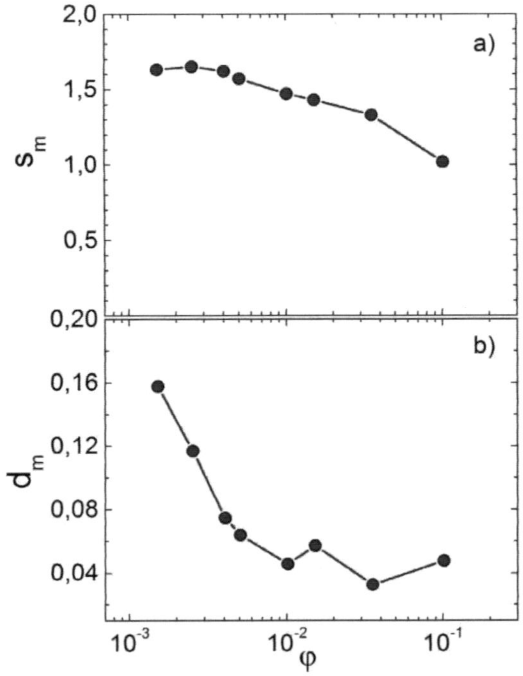

Figure 3. Dependence of the optimum values d_m and s_m upon the parameter φ.

Finally, we show in figure 4a the transient time τ_m (the curve with circles) corresponding to the time moment when $R_{ft}(t = \tau_m) = 0.90 \cdot R_D$. For comparison, we also present here the transient time τ_W (the curve with triangles down) corresponding to equations (4), (5), i.e. to Ref. [12]. As expected, in the energy diffusion regime, the transient time decreases as friction increases, i.e. the decay process accelerates. Note, that in the opposite limiting case corresponding to the overdamping regime, the increase of friction significantly slows down the decay of a metastable state.

In figure 4b the ratio τ_W/τ_m is presented as a function of φ (the latter is proportional to the friction strength). One sees that the ratio, in fact, repeats the $s_m(\varphi)$ – dependence presented in figure 3a. At small values of φ, the transient time τ_W exceeds τ_m by almost 70%. As the value of friction increases the difference becomes significantly smaller.

4. Conclusions

In the present work we have performed numerical modeling of the thermal decay of a metastable state at small values of the friction parameter (the so-called energy diffusion regime). For this aim, the stochastic differential equations (the Langevin equations) have been used. The time dependent rates resulting from the modeling enable us extracting the transient times τ_m. It is shown that at extremely weak friction these transient times are by 50% smaller than those known from the literature (τ_W, see Ref. [12]). As the friction constant increases, our transient times approach those of [12].

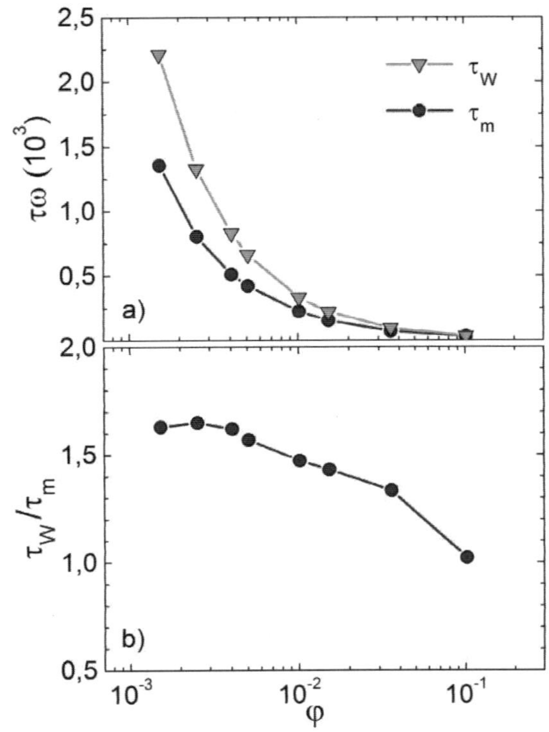

Figure 4. a) Dependence of the relaxation times calculated using the analytical formulas from [12] (τ_W, triangles) and extracted from the numerical simulation (τ_m, circles) upon the parameter φ. b) The ratio of these times as a function of φ.

Acknowledgments
This work was partly supported by the Scientific and Technological Research Council of Turkey (TUBITAK) within the framework of Fellowship Program 2221– for Visiting Scientists and Scientists on Sabbatical Leave.

References
[1] Kramers H A 1940 *Physica* **7** 284-304
[2] Martínez I A et al 2016 *Nature Physics* **12** 67
[3] Höfling F and Franosch T 2013 *Rep. Prog. Phys.* **76** 046602
[4] Hänggi P and Marchesoni F 2009 *Rev. Mod. Phys.* **81** 387
[5] Marchenko I G, Marchenko I I and Zhiglo A V 2018 *Phys. Rev. E* **97** 012121
[6] Abkenar M, Gray T H and Zaccone A 2017 *Phys. Rev.* E **95** 042413
[7] Dudko O K, Hummer G and Szabo A 2006 *Phys. Rev. Lett.* **96** 108101
[8] Nadtochy P N, Kelić A and Schmidt K-H 2007 *Phys. Rev.* C **75**, 064614
[9] Gontchar I I and Chushnyakova M V 2017 *Pramana – J. Phys.* **88** 90
[10] Gontchar I I et al 2010 *Phys. Rev.* C **82** 064606
[11] Pavlova E G, Aktaev N E and Gontchar I I 2012 *Physica* A **391** 6084-100
[12] Grange P, Jun-Qing L and Weidenmüller H A 1983 *Phys. Rev.* C 1983 **27** 2063-77
[13] Gontchar I I and Aktaev N E 2009 *Phys. Rev.* C **80** 044601
[14] Büttiker M, Harris E P and Landauer R1983 *Phys. Rev.* B **28** 1268
[15] Eccles C, Roy S, Gray T H and Zaccone A 2017 *Phys. Rev.* C **96** 054611

Plasmatic hardening application to increase operational durability of rolling mills bearing units

R R Dema[1], O B Kalugina[1*], B A Avdeyev[2] and E P Masyutkin[2]

[1]Federal state Budgetary Institution of Higher Education "Nosov Magnitogorsk State Technical University" (FSBIHE "NMSTU"), Lenin Avenue, 38, Magnitogorsk city, 455000, Russia.
[2]Kerch State Maritime Technological University (KSMTU), Kerch, Russia
[*]Corresponding author: kalugina.olga@bk.ru

Abstract. In the article, ways of increasing wear resistance of rolling mills bearing units under conditions of sliding friction are studied. The working surfaces of these details need the surface hardening. Since the details of the bearing units are traditionally made of 45 grade steel, we have studied the features of the formation of the structure and properties of the steel of this grade after plasma hardening without surface reflow. Variation of microstructure depending on hardening regimes was investigated. The microstructure of the steel in the hardening zone was studied by X-ray diffraction and metallographic methods. The possibility of increasing the operational properties of rolling mills bearing units using of plasma hardening of rubbing surfaces was shown.

Keywords: Plasma hardening, bearing node, rolling mill, wear, microstructure.

1. Introduction

In the rolling production the wear of the most loaded rolling stands parts affects the output characteristics of products and defines the parametric stability of the rolling mill. For example, the wear of rolls and their bearings at rolling mills causes lower quality products or total rejection [1].

Therefore, improving of constructive reliability of rolling mills bearing units at the stage of manufacture at the machine-building enterprises is one of the currently central tasks.

The materials with high wear resistance in the case of sliding friction is the reserve to increase reliability of bearing units. As the use of expensive composite materials is impossible due to economic and technical reasons, the only solution is to use modern high-performance heat treatment methods of traditionally used materials to obtain their exact properties [2].

Distance pieces and rings, protective flanges, distance rings, covers, etc. are the bearing elements most directly involved in sliding friction. Working surfaces of these products demanding of surface hardening is a cylinder or plane (Fig. 1 and Fig. 2).

During the operation bearing unit parts are wearing as a result of abrasion effect and impact fatigue spalling (Figure 3) at particular parts of surface. It indicates at mixed or edge mode of sliding friction and insufficient sealing of the bearing unit leading to the injection of abrasive particles.

Elastic plastic deformation of surface layers of rubbing parts of bearing units are the main factor that determines the nature and depth of the physical-chemical processes into the surface layer of the bearing units. Elastic plastic deformations lead to metal surface layers softening and hardening, fatigue wear processes development and ultimately to wear [3].

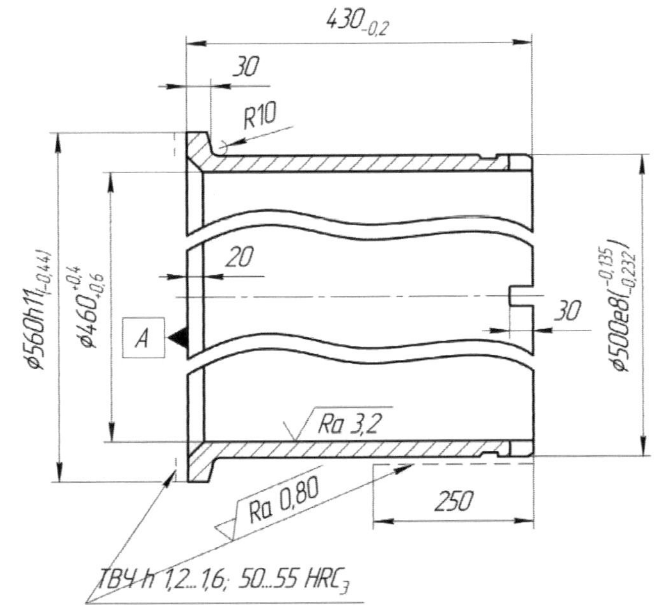

Figure 1. Distance sleeve of rolling mill stand HSM-2000 JSC "MMK"

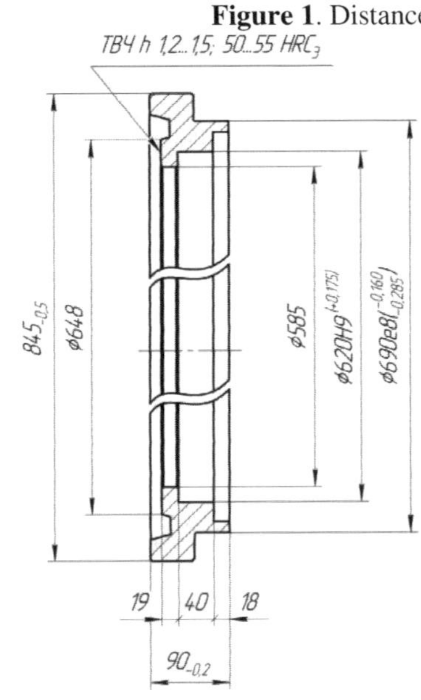

Figure 2. Drawing cover bearing unit of plate mill №4 JSC "MMK"

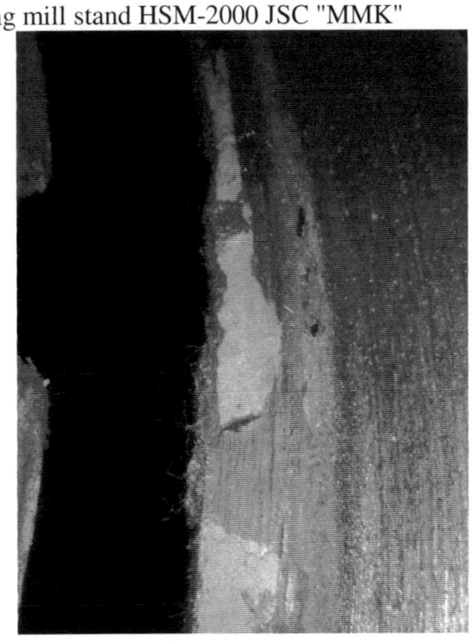

Figure 3. Typical cover wear affected by abrasive particles and fatigue flaking

Thus, to improve wear resistance of bearing units, is necessary to increase their fatigue strength, i.e. the tendency for multiple deformation of the same regions of friction surfaces of actual contact. This can be achieved only by increasing their surface hardness up to the maximum permitted level.

As details of the bearing units are traditionally made out of steel grade 45, heat treatment in batch-type furnace and water chilling cannot provide high operational characteristics of bearing units surface. However, the surface of these details is hardly loaded of contact details, and bearing bushings operate in conditions of dry friction. It predetermines their untimely wear. Application of plasmatic hardening allows us to increase the hardness of the surface up to 62 HRC [4, 5]. It is unattainable by traditional

ways of heat treatment, without resulting in warping of parts[4]. Therefore the plasmatic hardening is the best processing method for hardening of bearing units arrangements.
The purpose of this work is to improve the operational characteristics of the bearing units of rolling mills by the analysis of the structure and properties of 45 steel quenching by plasma at different modes.

2. Materials and methods

Figure 4 shows a panoramic image of the microstructure of hardened strip steel 45 plasma hardened without surface melting. As can be seen from the picture, the effective hardening depth at plasma heating was more than 1.5 mm. It is the layer where the metal has gained austenite structure, and martensite transformation occurs when a plasmotron subsequent gradual moves in it.

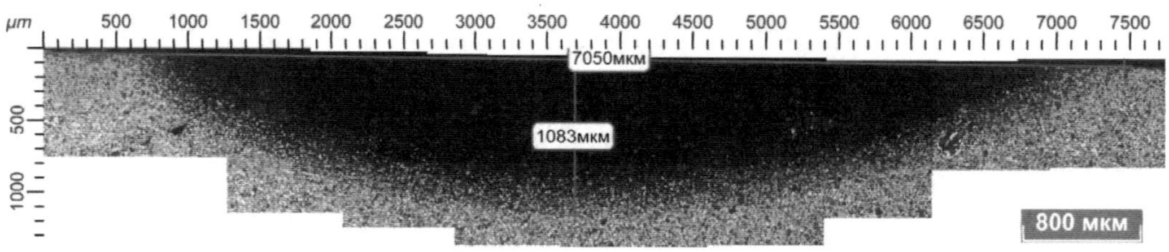

Figure 4. Panoramic image of the microstructure of 45 steel after plasma hardening without surface melting × 50

Deeper layers of metal were insignificantly heatad to tempering temperature or did not undergo significant heating. This zone is no different from the base metal framework in metallographic context, but has a significant effect on the utility characteristics of the reinforced layer, playing a part of a damper perceiving and relaxing external loads acting on the hardened surface [7].
Hardening was carried out of the winding line in a continuously – consistent way. To prevent deformation of thin-walled parts hardening was carried out when plasmotron works in indirect mode.
Plasma hardening was carried out in modes presented in table 1. Plasma jet power and hardened layers blending scheme, i.e. plasma quenching thermal cycle were used as a control variables.

Table 1. Sample processing modes

№	Jet power (of arc), kW	Power weight ratio, kW/cm^2	The overlap tracks
2	2,88	10,08	0
3	2,88	10,08	0,75
4	2,88	10,08	0,5

During hardening without surface melting hardering mode has a great effect on a structure and properties of the hardened zone.

3. Results and discussion

At plasma quenching of 45 steel on mode 1 plasma jet power weight ratio was 10.08 kW/cm^2, and hardened tracks overlapped a butt. The technological scheme of overlay hardening strips may emerge zones where martensite transformation not occur. These zones are located between the tracks and caused by factors such as: current supply gaps, irregularity of supplying of orifice gas, plasma jet feed axis deviation from a motion paths due to external vibrations and vibration, etc. [8].
The presence of unhardened regions or incomplete surface hardening regions confirms data diffraction pattern (Figure 5), taken from a surface of 45 steel, hardened by mode 2 with back to back tracks overlaid. Here martensite and austenite as well as carbide phase and some amount of α-phase with the lattice parameter of 2.861 Å are present, which agrees well with the literary data for ferrite [9].

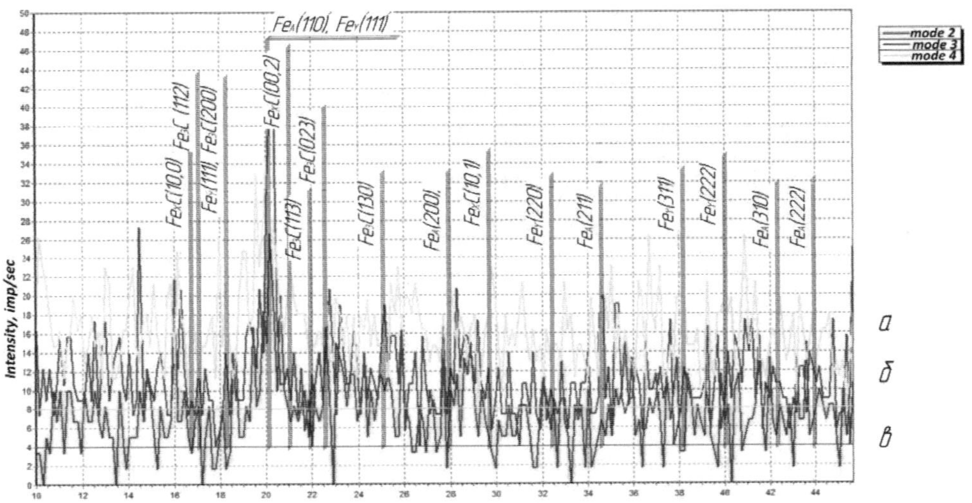

Figure 5. Diffraction pattern of 45steel after plasma quenching by different modes:
a – mode № 2; b – mode № 3; c in mode № 4

Microstructure of 45 steel reinforced by 2 mode is shown in figure 6a. In the zone of hardening martensitic -austenite structure with microhardness 7600 - 7700 Mpa is formed. The average carbon content of martensite is 1.19%, which significantly exceeds the average content of it in steel. The degree of tetragonality of martensite is 1.054.

At plasma quenching on mode 3 (see table 1) with overlaid strips on hardened ¼ the absence of unhardened zones is guaranteed. However, in this case a partial parental tempering of the previously hardened strip takes place when overlapping the next one. Quenching in this mode causes the formation on the surface of the sample in hardened layer regions where single hardening, double hardening followed by overlaying zone of incomplete hardening or tempering occurs. This results in a significant difference in the micro hardness values across the hardened strips from 7300 to 8000 MPA. Metallographic identify and recognizing of these zones is not possible, however, the values of micro hardness, obtained in the various sections of hardened strip indicate the course of processes of partial tempering and repeated hardening on overflow of nearby bands. Hardening zone microstructure is shown in Figure 6b. The x-ray analysis results confirm the presence of martensite and residual austenite and carbide phases as well as type α-cementite in the hardening zone. The average carbon content of martensite is 1.11%, far exceeding the average contents of it in steel, but slightly lower than plasma quenching on mode 2 with butt strips overlaid. The degree of martensite tetragonality is equal to 1.051, which also confirms the assumption of partial martensite tempering processes.

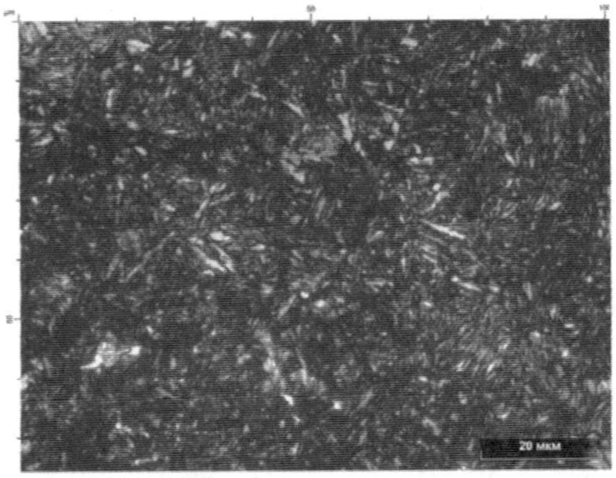

a

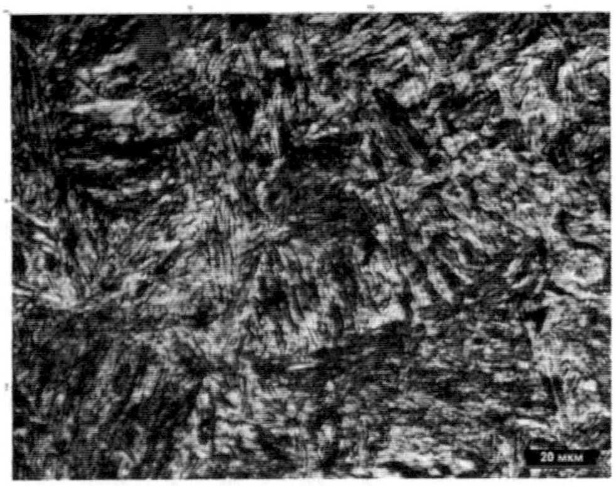

b

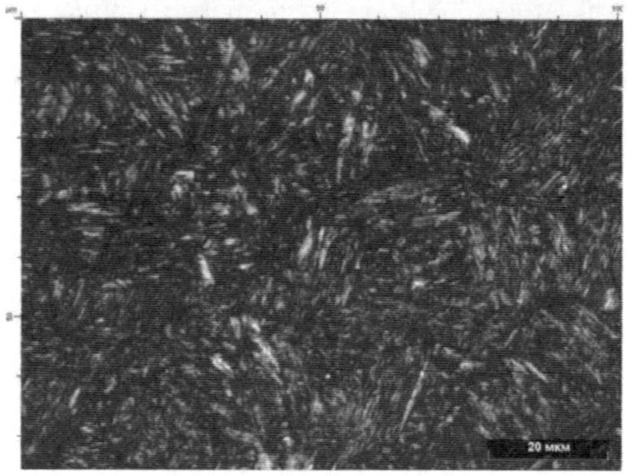

c

Figure 6. Microstructure hardening zone of 45steel hardened by different modes: a – mode № 2; b – mode № 3; c – mode № 4, × 1000

Plasma quenching by mode 4 with tempering strips overlapping on ½ width, leads to the formation of high micro hardness quenching layer on the surface and its even distribution across the hardened layers. There are sections with double plasma hardening and double quenching and incomplete hardening zone imposed on them. This hardening scheme results in the highest surface hardness of all these schemes and the surface hardness is equal to 62 HRC. Hardening zone microstructure is shown in Figure 6b. In the zone of hardening austenito-martensitic structure with 7900–8100 MPa micro hardness is identified by metallographic examination. The average carbon content in the martensite of the hardened zone is 1.04%, its degree of tetragonality is 1.047.

With increasing degree of overlap strips the amount of carbon in the martensite decreases(Figure 7), which can be explained by a more complete burn of the tempering processes [10], and the reinforced layer wear resistance increases, because of lightweight martensite transformation of retained austenite under abrasive wear.

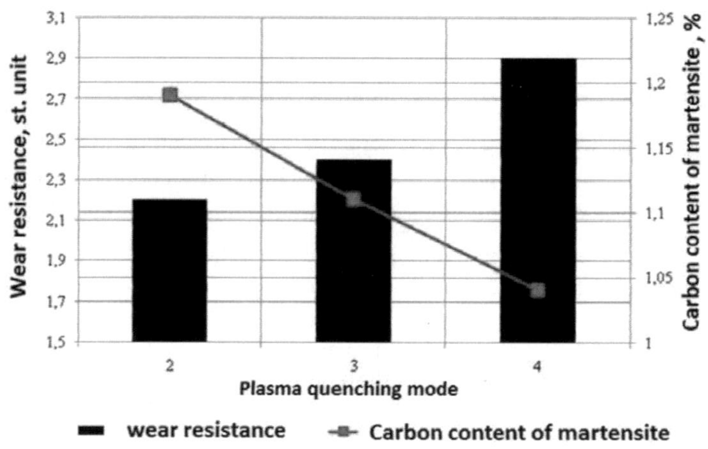

Figure 7. Carbon content of martensite and wear resistance dependence on plasma quenching mode

4. Conclusions

1. Hardening mode, that provides the highest wear resistance is the mode when plasma jet power density is equal to10.08 kW/cm^2 and tempered strips overlapping is ½ of the width.
2. It is shown that the increase of wear resistance is linked to destabilization of retained austenite to deformation martensitic transformation by means of reducing martensite tetragonality while reducing content of carbon in it due to the allocation of ε-cementite and as a result, lowering the activation energy of martensite transformation of the retained austenite.

Acknowledgement

The authors are thankfully acknowledge the financial support by the Ministry of Education and Science of the Russian Federation for the project № 11.2054.2017(Number for publication: 11.2054.2017 / 4.6)/Project task within the framework of the state task for 2017-2019 to carry out this investigation.

References

[1] Kalugina O B 2013 *Optimization of the roll pass shape by the deformation efficiency criterion to energy saving* Dissertation of Cand. of Tech. Sci (Magnitogorsk: NMSTU)

[2] Bandyopadhyay P K, Choudhary R B, Mandal Ch, Prasad R R and Sengupta P P 2017 Application of Oil Condition Monitoring Techniques for Improving Critical Equipment Availability in Steel Plant *Tribology Online* **12** 2 pp 37-41.

[3] Marchenko E A 1979 *About the nature of the fracture surface of metals through friction* (Moscow: Nauka) 118 p

[4] Nefediev S P, Dema R R and Kotenko D A 2015 Possibility of plasma arc hardening application to cutting disk hardening *Bulletin of the South Ural State University, Series: Metallurgy* 1 **15** 70-73

[5] Nefedyev S P, Dema R R, Nefedyeva S A and Yaroslavtcev AV 2015 Microstructure of cast iron after plasma bleaching *Journal of Chemical Technology and Metallurgy* **50** 213-216

[6] Goryany V and Radinsky V 2002 Thermochemical treatment of reinforcing steel *Journal of Mining and Metallurgy* 3-4 **38** 171–177

[7] Balanovsky A.E. 2006 *Plasmatic surface hardening of metals* (Irkutsk: Irkursk National Research Technical University) 180

[8] Korotkov V A, Ananiev S P and Shekurov A V 2012 Influence of cooling rate on the quality of surface layer during plasma hardening investigation *Welding* **3** 23-27

[9] Lahtina Ed Ym and Rahshtadta L G 1980 *Thermal processing in engineering* (Moskow: Mashinostroenie) 783

[10] Korotkov, V.A. 2011 Wear resistanse of plasma quenching materials *Friction and Wear (Trenie i Iznos)* 1 **32** 23-29

[11] Nefed'ev S.P., Dema R.R., Kharchenko M.V., Pelymskaya I.S., Romanenko D.N.and Zhuravlev G.M. 2017 Experience in restoring hydraulic cylinder rods by plasma powder sufacing *Chemical and Petroleum Engineering* 11-12 **52** 785-789

[12] Alexey N. Shapovalov, Olga.B. Kalugina, Roman. R. Dema 2017 Investigation of the structure of wear resistant coatings obtained by plasma powder surfacing *MATEC Web of Conferences* **129**, 02044 (2017) (https://doi.org/10.1051/matecconf/201712902044

[13] Romanenko D N, Artemenko Yu A, Emelyushin A N, Dema R R, Nefed'ev S P and Zhuravlev G M 2017 Physical modeling of mechanism of modification with wear resistant sufacing *Chemical and Petroleum Engineering* **52** 769-773

[14] Shapovalov A N, Nefedieva S A, Nefediev S P and Dema R R 2017 X-ray study of the structure of surface layers of steel 45 after plasma quenching *Factory laboratory Diagnostics of materials* 5 **83** 39-41

[15] Nefediev SP, Dema RR, Tyuyaterikov N.Sh., Morozov AN, Kharchenkov MV, Belotserkovskii MA, Levantsevich M.A 2016 Structure of the fusion zone of a chromate-manganese coating obtained by plasma powder deposition *Problems of ferrous metallurgy and materials science* **1** 68-71

IOP Conf. Series: Journal of Physics: Conf. Series **1050** (2018) 012020 doi:10.1088/1742-6596/1050/1/012020

Control systems design: the technology of stochastic perturbations simulation

L A Denisova[1] **and V A Meshcheryakov**[2]

[1]Omsk State Technical University, 11, Mira ave., Omsk, 644050, Russia
[2]Siberian State Automobile and Highway University, 5, Mira ave., Omsk, 644080, Russia

e-mail: denisova@asoiu.com, meshcheryakov_va@sibadi.org

Abstract. An approach to the random perturbations simulation is considered for performing in the design of control systems for thermal power facilities. In order to take into account the stochastic nature of the technological processes, we proposed the simulation technique that allows us to superimpose random fluctuations on the deterministic model of the control system. To simulate random changes in process parameters, we synthesized the shaping filter that generates a random signal with the required autocovariance function. The developed technology allows us to adequately model the technological processes characterized by the action of random perturbations. This is especially important if the requirements for control quality are high. We have experimentally confirmed the suitability of the developed technology by performing control system simulation using MATLAB / Simulink.

1. Introduction

The simulation of automatic control systems (ACS) processes is one of the main issues that are researched at the initial stage of system design. Development of technological processes models includes two stages of research: parameterization of models and evaluation of model parameters using experimental data [1–6].

At the stage of parametrization, the model is being developed on the basis of a priori data on the technological process, and its structure and number of parameters are being determined. It is taken into consideration that dynamic models of technological processes (for example, channels for regulating temperature, pressure, liquid level in tanks, etc.) are approximately represented by first- or second-degree polynomials, which gives a satisfactory result for practice [1, 4, 6].

At the stage of the model parameters evaluation, information on the dynamic characteristics of the systems is being processed, and, in addition, statistical analysis of the experimental data obtained from the control objects is carried out [4, 5].

Because the data on the technological processes flow in the operating ACS is usually registered and archived by information and computing systems, it is possible to use it as input data for the development of mathematical models of control objects. That is, on the basis of archival information, data on the dynamic characteristics of systems was obtained as well as data used to build the models of random perturbations. In addition, the dynamic stochastic mathematical models of control objects were developed, taking into account the inherent nonlinearities [1–6].

To take into account the stochastic character of the change in the parameters of heat-and-power processes, it is necessary to envisage the possibility of imposing random input effects on the deterministic mathematical model.

The article considers the technology of stochastic disturbances simulation by the example of the automatic control system of a typical automation object: the steam pressure control channel in the deaerator of the turbine section of the nuclear power plant power unit. The dynamic stochastic model of pressure ACS is created by the MATLAB/Simulink tools (which are intended for performing scientific and engineering calculations) [7, 8].

2. The problem statement

The technological requirements to the steam pressure ACS in the deaerator of the Kol'skaya NPP power unit are as follows. The pressure controller prevents the reduction of pressure in the deaerator and maintains the set value of pressure with the required accuracy. The controller forms the control actions on the electric actuating mechanism (AM) of the control valve (CV). The controller perform pulse control on the proportional-integral (PI) control law together with the constant speed actuator.
ACS pressure in the deaerator should ensure stable operation and absence of oscillations in the control loop. The requirement for limiting the frequency of AM actuations (not more than six per minute) must be fulfilled as well [2, 3] by a constant load of the power unit. This requirement is a general one for controllers of turbine unit technological parameters [2–4].
ACS of the deaerator pressure should not allow sudden changes in pressure leading to ebullition on the suction side of the feed pumps and their steaming.
Based on the analysis of the initial data, it is necessary to estimate the dynamic characteristics of the technological object. In addition, on the basis of archival information, it is necessary to obtain data on the statistical characteristics of the operating random perturbations, because pressure changes in the deaerator, i.e., the pulsations of the measured pressure signal, are random processes.
In order to take into account random effects, we must envisage the possibility of superimposing a perturbation with a band-pass frequency spectrum to the object model output. Thus, it is necessary to create a dynamic stochastic model of a control system suitable for simulation and optimization.

3. Theory. Development of dynamic stochastic model of ACS

3.1. The mathematical model description. Preparation of initial data

The mathematical model of the ACS of pressure in the deaerator has been developed with MATLAB/Simulink (see Figure 1). It includes the following submodels: controller, control valve (with AM) and control object. The error signal at the controller's input is formed as the difference between the setpoint signal and the current pressure value.
The controller implements the traditional control method used in control systems of thermal energy parameters with constant-speed actuators [2, 3]. In this case, the controller outputs are the pulses that are sent to the actuator to open or close the control valve. The on-off time ratio of these output pulses depends on the error signal change (the dependence is the proportional-differential (PD)). The controller jointly with the AM approximately realizes PI-control law in the main, so-called sliding mode of operation. In this mode, the AM is being actuated several times in a same direction, until the deviation goes down to the controller's dead zone.
In addition, the model incorporates inertial sections for input signals damping. The control valve with the AM is represented by the gain block modeling the transmission factor (opening $100\%/T_{am}$) taking into account the opening time of the valve T_{am}, and the integration section with a limit of 0 to 100% according to the CV position indicator. The control object, i.e., the channel "AM position – pressure in the deaerator" is represented by a first-order inertial section.

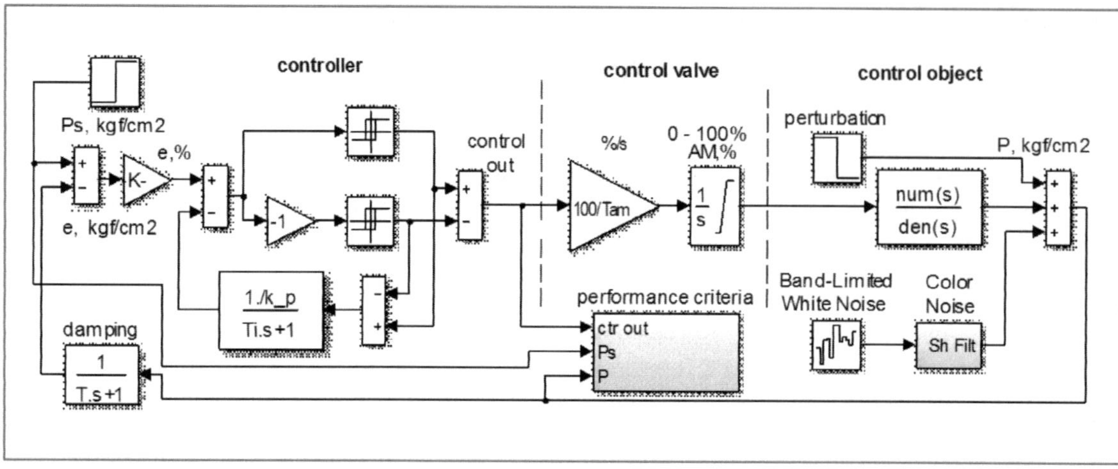

Figure 1. Diagram of the mathematical model of ACS in MATLAB / Simulink.

The fluctuations of random disturbances (pulsations) of the pressure are realized by the shaping filter using the experimentally obtained autocovariance functions of signal fluctuations. The disturbances are further summed up with the output of the inertial section. The block Band-Limited White Noise generates the normally distributed white noise supplied to the input of the shaping filter.

Besides, the mathematical model calculates the ACS performance integral criteria. During the period T (adopted not smaller than the system response time), following quadratic integral criteria are calculated. The criterion for characterizing the magnitude and duration of the existence of control error $e(t)$ (the control parameter deviation from the setpoint): $F_1(X)=\int_0^T e^2(t)dt$. The integral quadratic criterion is calculated as well taking into account the cost of control (u denotes the controller output): $F_2(X)=\int_0^T u^2(t)dt$. In addition, the third criterion is number of pulses formed by the regulator (regulating door actuations): $F_3(X)=n$.

These quality criteria were used to optimize the control system, that is, to determine the best values of the parameters $X=(k_p, T_i)$ (here k_p is the gain and T_i is the integration time of the controller). Minimization of the adopted criteria $F_i(X)$, $i=\overline{1,3}$, ensures the speed of control, the absence of overshoot and reduction of the AM actuations [5].

At the stage of determining the model parameters, the data on the dynamic characteristics of the system provided by the Kol'skaya NPP were processed. In addition, in order to estimate the unknown parameters of the model, statistical processing of the experimental data was carried out.

As a result of the initial data analysis, the dynamic characteristics of technological parameters are estimated. **Table 1** shows the parameters of the control object, the pressure control channel in the deaerator, used for the subsequent calculation of the ACS parameters. In addition, statistical data on the characteristics of the operating random perturbations were obtained on the basis of archival information.

As we have already said, the pressure changes in the deaerator, that is, signal pulsations, are random processes. Figure 2a shows the graph of the pressure change in the deaerator, according to the archival information of Kol'skaya NPP.

From the available data arrays, we have chosen almost stationary realizations in the sense of mathematical expectation constancy and have estimated the statistical characteristics of fluctuations by them. Figure 2b demonstrates the derived estimate of the normalized autocovariance function of these random process for the sample of size n = 400.

Table 1. The control object parameters.

Parameter name	Designation (dimension)	Value
Pressure setpoint	P_S (kgf cm^{-2})	6
Opening time of the control valve	T_{am} (sec)	25
Object transfer factor	K_o (kgf cm^{-2}(% AM position)$^{-1}$)	0.1
Time constant of the object	T_o (sec)	60
Coefficient characterizing the attenuation of the autocovariance function of pressure signal pulsations	α	0.03
Coefficient characterizing the frequency of the periodic component of the signal pressure pulsations	β	0.12
Standard deviation of pressure signal pulsations	σ (kgf cm^{-2})	0.005

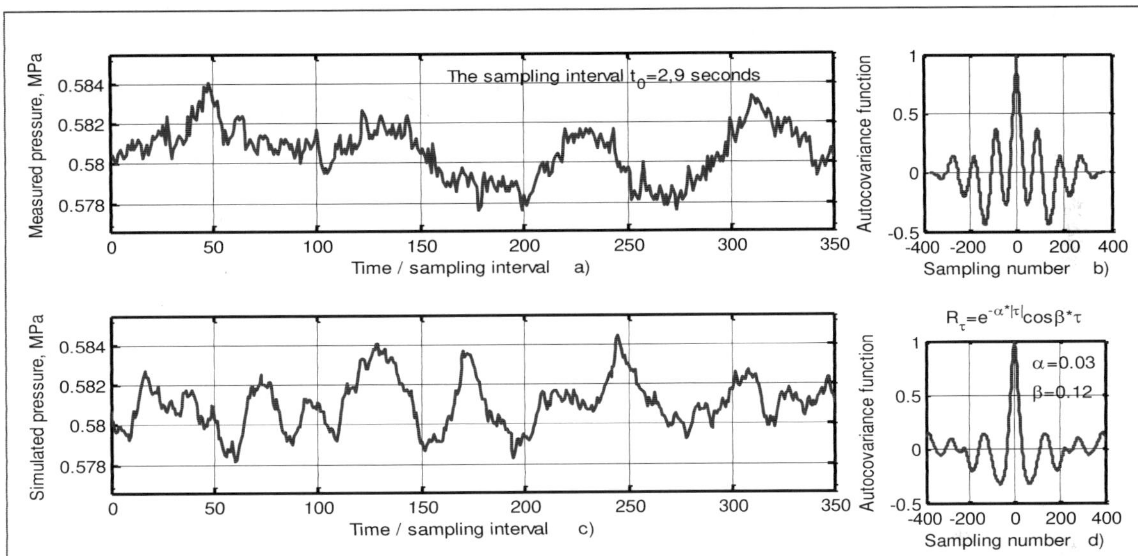

Figure 2. Realizations of random processes and their corresponding autocovariance functions: a), b) – pressure change; c), d) – changing the simulated pressure when using the shaping filter.

3.2. *Synthesis of the forming filter*

To model the random changes of pressure in the deaerator using the obtained initial data, the synthesized shaping filter was used to generate a random signal with the required autocovariance function. The transfer function of the forming discrete filter was obtained on the basis of the spectral decomposition method [1, 6].

The input signal of the shaping filter is the normally distributed discrete white noise $\xi(n)$, i.e., an uncorrelated sequence of random numbers with the mean M_ξ =0 and the variance $\sigma_\xi^2 = 1$.

The experimentally obtained autocovariance functions of signal fluctuations for a continuous process are approximated by the expression

$$R(\tau) = \sigma^2 e^{-\tilde{\alpha}|\tau|} \cos\tilde{\beta}\tau , \quad (1)$$

where the parameters $\tilde{\alpha}$ and $\tilde{\beta}$ characterize the decay of the autocovariance function and the frequency of the periodic component of the process, respectively; σ designates the standard deviation of the signal; and finally, τ represents a time lag.

IOP Conf. Series: Journal of Physics: Conf. Series **1050** (2018) 012020 doi:10.1088/1742-6596/1050/1/012020

A stationary random signal with bandpass frequency spectrum realizes the signal pulsations with the above-mentioned autocovariance function. This allows forming pulsations in the required frequency range through varying the parameters $\tilde{\alpha}$ and $\tilde{\beta}$.

The normalized autocovariance function of the corresponding discrete process is $R(n) = e^{-\alpha n} \cos \beta n$, where $\alpha = \tilde{\alpha} t_0$; $\beta = \tilde{\beta} t_0$, and t_0 stands for time sampling interval.

The discrete transfer function of the shaping filter for realization of a random process with the autocovariance function (1) acquires the form

$$K(z^{-1}) = \frac{a_0 + a_1 z^{-1}}{b_0 + b_1 z^{-1} + b_2 z^{-2}},$$

where z^{-1} is the operator of the inverse shift in time; $a_0 = -\sigma v_1 \sqrt{\rho}$; $a_1 = \sigma \sqrt{\rho}$; $b_1 = -2\gamma \cos\beta$; $b_2 = \gamma^2$; $\gamma = e^{-\alpha}$; $v_1 = v_0 + \sqrt{v_0^2 - 1}$; $v_0 = \dfrac{1+\gamma}{2\gamma\cos\beta}$; $\rho = \dfrac{(1-\gamma^2)\gamma\cos\beta}{v_1}$.

To implement the random signal model, define the shaping filter by the recurrent relationship

$$\tilde{\xi}(n) = a_0 \xi(n) + a_1 \xi(n-1) - b_1 \tilde{\xi}(n-1) - b_2 \tilde{\xi}(n-2), \quad (2)$$

where ξ, $\tilde{\xi}$ are the input and output signals of the shaping filter, respectively; a_0, a_1, b_1, b_2 denote the shaping filter parameters dependent on α and β.

Figure 2d represents a graph approximating the autocovariance function, and Figure 2c shows a corresponding realization of a random process, obtained by the modeling of pulsating pressure signal in the deaerator using the shaping filter (2).

4. Results of experimental studies

Experimental studies were carried out using m-files – MATLAB scripts multiple invoking developed ACS Simulink-model. We used the developed model to determine the optimum parameters of the automatic control system. Preliminary studies and previous experience of the ACS design [3, 5] have shown that the minimal values of the quality criteria are achieved by the different tuning parameters of the system. This suggests the need to reach a compromise between them in solving the optimization problem [9].

The ACS optimal parameters search was carried out by the procedure *gamultiobj* of the multi-objective genetic algorithm (GA) implemented on the basis of the well-known genetic algorithm *NSGA*-II (*Non-Dominated Sorting Genetic Algorithm*) [10, 11]. This algorithm constructs the optimal solutions set (Pareto solutions set) for the case of several criteria.

In comparison with traditional optimization methods, the GA is remarkable for the following substantial advantage. It carries out the solution search using a set of points (a population), but not a single point [10–12]. Using GA, the optimal parameters of the system were found and the corresponding transient processes were obtained. To use the genetic algorithm, the vector of parameters $X=(k_p, T_i)$ is taken as the "individual".

Let us illustrate the use of the developed ACS dynamic stochastic model. Figure 3 shows the graphs of transient characteristics of ACS (perturbation of the control parameter) while minimizing the criteria $F_1(X) – F_3(X)$ in Figures 3a-3c, respectively. In the graphs the dashed curve corresponds to the controlled parameter (pressure), the dash-dotted one is the position of the AM CV, and the solid line represents the pulse output of the controller.

For clarity in Figure 3d the transient graphs are summarized in enlarged scale for the adjustable parameters of the three cases (curves 1, 2, 3 correspond F_1=min, F_2=min, F_3=min, respectively; curve 4 represents setpoint pressure).

As can be seen in Figure 3a, the transient process with the minimum value of the criterion $F_1(X)$ is characterized by the presence of self-oscillations. The transient process with the minimum value of F_2(X) is not only characterized by a significant number of AM actuations (figure 3b), but also have overshooting and is lengthy (Figure 3d, curve 2).

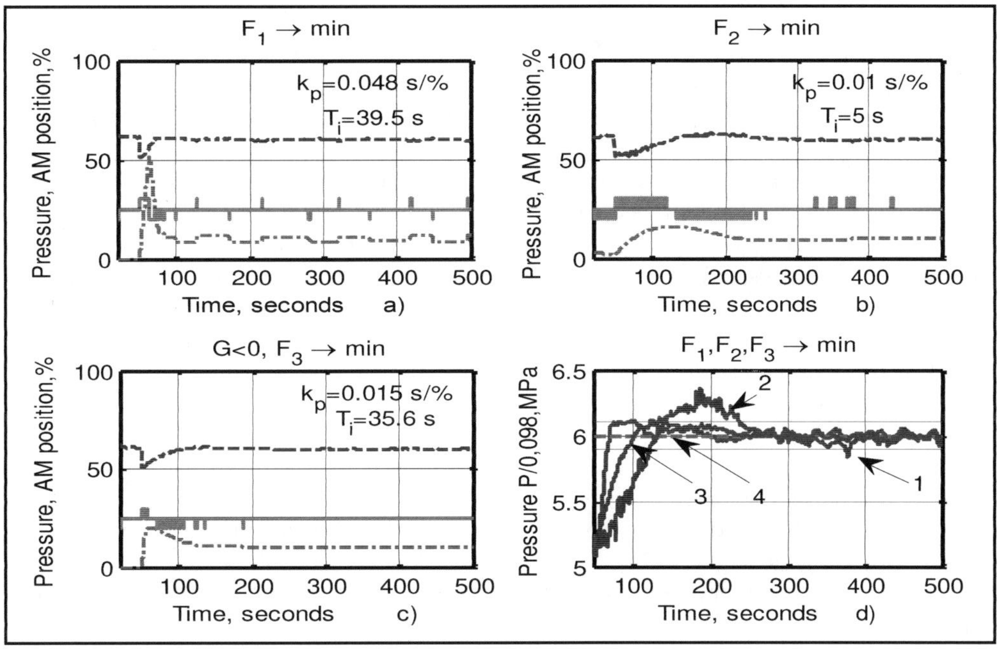

Figure 3. Transient characteristics of the ACS while minimizing the criteria $F_1(X) - F_3(X)$

Figure 3c shows the ACS transient process when the minimum of the criterion $F_3(X)$ is found among the Pareto set of optimal solutions for the criteria $F_1(X)$ and $F_2(X)$. In addition, in this case the constraint $G(X) < 0$ is taken into account (it restricts the parameters to avoid self-oscillations in the system). It can be seen that the transition process is characterized not only by the absence of overshoot and high speed (figure 3d, curve 3), but also by small number of AM actuations (see figure 3c).
It should be noted that using the developed model allowed to take into account the random pulsations of the controlled parameter and to estimate how often the controller actuates the AM under disturbances. For the study, several implementations were carried out. The performed optimization of the automatic control system allowed obtaining high quality of the control process. The values of the ACS parameters chosen from the accepted ranges $k_p \in [0{,}01; 1]$, $T_i \in [1; 70]$.

5. The discussion of the results

Based on archival information on the technological processes flow provided by the Kol'skaya NPP, the estimation of the autocovariance function of random pressure changes in the deaerator was obtained. To simulate random pressure changes, a shaping filter was synthesized, which makes it possible to generate a random signal with the required autocovariance function.
Let's compare the realizations of random processes for changing the pressure in the deaerator (see Figure 2a) and for changing the simulated pressure when using the shaping filter (see Figure 2c). It can be seen that the curve of random changes in the pressure of a real process (in contrast to a simulated process) also contains more high-frequency components.
This indicates that the estimation of the autocovariance function accepted for modeling (see Figure 2d) in accordance with expression (1) only approximates the autocovariance function of the real process (Figure 2b). However, it seems inexpedient to complicate the expression of the autocovariance function by introducing additional harmonic components.
As shown by the performed studies of the model and existing operational experience, the type of autocovariance function we used for synthesis of the shaping filter (used in the ACS model) allows us to perform calculations with sufficient accuracy for practice.

6. Conclusion

As a result of the issue, it was concluded that the proposed technology for modeling stochastic parameter changes is effective in the design of automatic control systems. To simulate parameter fluctuations, the forming filter is synthesized that generates a random signal with the required autocovariance function.

Based on the processed archive data on the flow of technological processes and using the proposed technology, the dynamic stochastic model of the system for controlling the pressure in the deaerator was created. Using the developed model, we performed the optimization of the parameters of the automatic control system, designed to control the pressure in the deaerators at the Kol'skaya NPP power units.

The developed technology allows to adequately simulating the technological processes that are characterized by the action of random disturbances on the control object, which is especially important if high requirements are imposed on the control quality.

References

[1] Dorf R C and. Bishop R H 2011 *Modern control systems* (Prentice Hall)

[2] Denisova L A and Meshcheryakov V A 2015 Automatic parametric synthesis of a control system using the genetic algorithm *Automation and Remote Control* vol **76** no 1 pp 149–156

[3] Denisova L and Meshcheryakov V 2016 Synthesis of a control system using the genetic algorithms *IFAC-PapersOnLine* vol **49** issue 12 pp 156-161

[4] Denisova L A 2016 Automatic feed control of steam generator in the power unit of a nuclear power plant: modeling and optimization *Automation and Remote Control* vol **77** no 6 pp 1084–1092

[5] Meshcheryakov V and Denisova L 2016 Computer-aided design of the fuzzy control system using the genetic algorithm *Dynamics of Systems, Mechanisms and Machines, Dynamics 2016* p. 7819000

[6] Denisova L 2013 Event-driven simulation of control systems with variable parameters *Proceedings of the 7 th IFAC Conference MIM'13 IFAC-PapersOnLine* vol **7** part 1 pp 2179–2184

[7] Xue D and Chen Y Q 2013 *System simulation techniques with MATLAB and Simulink* (John Wiley & Sons)

[8] Purohit G N, Sherry A M and Saraswat M 2013 Optimization of function by using a new MATLAB based genetic algorithm procedure *International journal of computer applications* vol **61**(**15**) pp 1–5

[9] Podinovskiy V V and Noghin V D 2007 *Pareto-optimalnie resheniya mnogokriterialnih zadach (Pareto-optimal solutions of multiobjective problems)* (Moscow: FIZMATLIT)

[10] Deb K 2001 *Multi-objective optimization using evolutionary algorithms* (Chichester: UK, John Wiley & Sons)

[11] Deb K, Pratap A, Agarwal S and Meyarivan T 2002 A Fast and Elitist Multi-objective Genetic Algorithm: NSGA-II *IEEE transactions on Evolutionary Computation* 6 (2) pp 182–197

[12] Holland J H 1994 *Adaptation in natural and artificial systems. An introductory analysis with application to biology, control and artificial intelligence* (London: Bradford book edition)

IOP Conf. Series: Journal of Physics: Conf. Series **1050** (2018) 012021 doi:10.1088/1742-6596/1050/1/012021

Investigation of free oscillations for reasoning constructive decisions of mirror segments of parabolic antenna

S V Doronin and E M Reyzmunt

Institute of Computational Technologies of the Siberian Branch of the Russian Academy of Sciences, 6 Academician M.A. Lavrentiev ave., Novosibirsk, 630090, Russia

E-mail: sdoronin@ict.nsc.ru

Abstract. The paper deals with studying qualitative and quantitative relationships between constructive decisions of mirror segments of parabolic antenna and dynamic characteristics for both individual segments and ready-assembled reflector. It is shown the influence stressed state induced by thermal deformation to natural modes and frequencies of free oscillations of mirror segments. There are justified constructive decisions of the mirror to ensure low sensitivity of dynamic characteristics to service temperature conditions.

1. Introduction

Dynamic characteristics (natural modes and frequencies of free oscillations) of large-size mirror antennas for ground stations of satellite communication have a great effect on antennas productivity. Such is the case of just dynamic characteristics generate structural response to effect of wind flow, acoustic noise and sinusoidal vibrations. Regardless of the fact that the dynamics of antenna system is predominantly determined by configuration of bearing skeleton, the great practical interest is to analyze dynamic behavior of mirror segments. This analysis allows to predict and to prevent development of functional failure of antennas essentially because of vibration of mirror segments despite high dynamic characteristics of skeleton. One of the intelligent design mission is to search constructive decisions to provide overshot of lower natural frequencies of free oscillations for segments of the same for bearing skeleton. The feature of this mission is close connection of constructive decisions for bearing skeleton and mirror because of necessity of guaranteeing their assemblability and associativity as parts of mechanical system. This fact brings to the impossibility of permanent modification of structural parameters for both skeleton and mirror segments. On the contrary, one has to choice of short amount of possible alternative variants of mechanical system consisting of agreed constructive decisions for mirror segments and skeleton.

This paper deals with qualitative and quantitative relationships between constructive decisions of mirror segments of parabolic antenna and their natural modes and frequencies of free oscillations. Deep understanding of these relationships is the basis for intelligent design of mirror segments of parabolic antenna taking into account restraints on its dynamic characteristics.

2. Problem definition

The objects of research are Q/K_a-band parabolic reflector antennas for ground stations of satellite communication. Reflector diameter is in the range 9 to 12 m, and ratio of focal length to diameter is between 0.3 to 0.5. Two alternative variants of mirror organization are considered. The first variant

IOP Conf. Series: Journal of Physics: Conf. Series **1050** (2018) 012021 doi:10.1088/1742-6596/1050/1/012021

consists of six partitions, each of them contains one segment of form factor 1 (segment 1), two segments of form factor 2 (segment 2), and two segments of form factor 3 (segment 3) (figure 1, a). Each segment rests on skeleton by means of five adjusting bays (red spots in fig. 1). The second variant of mirror organization consists of fifteen partitions, each of them contains one segment of form factor 4 (segment 4), and two segments of form factor 5 (segment 5) (figure 1, b). Segments 4 rest on skeleton by means of seven adjusting bays, while segments 5 are connected to eight adjusting bays. Segments 4 and segments 5 are reinforced by radially extended 100 mm high stiffening plates made of 1.4 mm thickness carbon fiber composite. Segments of all five form factors have the same cross-section structure: external layers of 0.6 mm thickness carbon fiber composite, internal layer of 40 mm thickness plastic foam. The fundamentally important fact is that each segment, regardless of others, rests on skeleton by means of adjusting bays; there are no mechanical connections between neighboring segments.

In accordance with regulations the antennas functioning is permitted on condition that ambient air temperature is within the range of minus 50 to plus 55 Celsius degree. Reflector thermal state is developed under the impact of integrated density of Solar flux Q_{sol} in conditions of radiation-convective heat exchange between the mirror and its surrounding.

The research task involves determination and comparative analysis of natural modes and frequencies of free oscillations for mirror segments of five form factors, both taking and without taking their stressed states induced by thermal deformation into account. Result of such comparative analysis is one of the factors for selection preferable variant of parabolic antenna mirror organization.

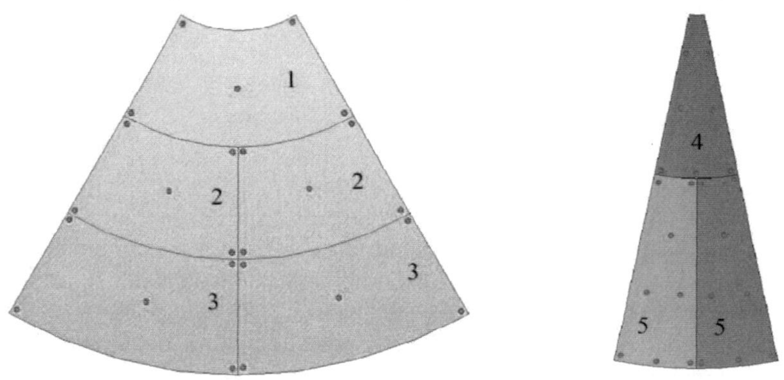

Figure 1. Partition configuration and connections (red spots) between segments and bearing skeleton.

Problem definitions similar by applied-physics implication are also known. The problem definitions involve numerical [1-6] and experimental [1] studies of dynamic behavior of ground-based [1] and space deployable [2-6] reflector antennas. In certain cases the dynamic analysis was performed over wide temperature range: from minus 200 to 180 °C [3], from minus 70 to 120 °C [5]. Natural modes and frequencies were analyzed in connection with prestressed state [2] and some structural features [4] of reflectors.

3. Mathematical formulation of the problem

To execute analysis of free oscillations, it is necessary to develop discrete (finite-element) models of structures and solve following constitutive equations. Natural frequencies ω_i for i-th natural mode can be determined by numerical solving equation

$$\left(\mathbf{K} - \omega_i^2\mathbf{M}\right)\frac{\mathbf{u}}{\cos\omega_i t} = 0,$$

where $\mathbf{M}$ – mass matrix; $\mathbf{K}$ – stiffness matrix; $\mathbf{u}$ – displacement vector; t – time.

The i-th natural mode is determined by eigenvector

$$\varphi_i = \frac{\mathbf{u}}{\cos \omega_i t}.$$

To fulfill modal analysis of prestressed structures the stiffness matrix **K** must be superseded by matrix $\mathbf{K}_c = \mathbf{K} + \mathbf{K}_g$, where $\mathbf{K}_g$ – tangent stiffness matrix. Calculation of $\mathbf{K}_g$ is based on prestress tensor jointly with nonlinear part of strain tensor.

4. Numerical modeling and results

There are realized three series of computing experiments. The first one is devoted to analysis of natural modes and frequencies of mirror segments 1, 2, and 3 (figure 1, a). The second series is subsidiary and aimed to modal analysis of segments 4, 5 (figure 1, b) without stiffening plates. Natural modes and frequencies of segments 4, 5 with radially extended stiffening plates are investigated during the third experiment series.

Let us consider the main tendencies founded in the first experiment series by the example of the first natural mode of free oscillations (table 1).

Table 1. The lowest frequency in the first mode, Hz.

Segment	Without prestressing	The prestressed state at ambient temperature, °C							
		– 50	– 35	– 20	– 5	10	25	40	55
1	44.0	46.4	46.2	46.0	45.7	44.5	42.4	39.3	34.1
2	45.7	49.3	48.9	48.5	47.8	46.8	44.0	37.8	16.0
3	32.6	37.3	36.6	35.8	34.7	33.1	30.2	21.8	35.2

A comparison between the lowest frequencies of the mirror segments (table 1) and bearing skeleton (in the order of 37 Hz) makes it clear that the natural frequency of mirror segment 3 is less than the same for the skeleton, that won't pass.

One can see tendencies of the lowest frequency toward lowering while increasing ambient air temperature. At the same time temperature dependencies for segments 2 and 3 involve some features. On case of increasing ambient air temperature from 40 to 55 Celsius degree the lowest natural frequency of the segment 2 decreases drastically (greater than two), but the same of the segment 3, on the contrary, increases. The last fact breaks general tendency towards decreasing frequency. This is because of qualitative changing nature of the first mode in this temperature range (figure 2).

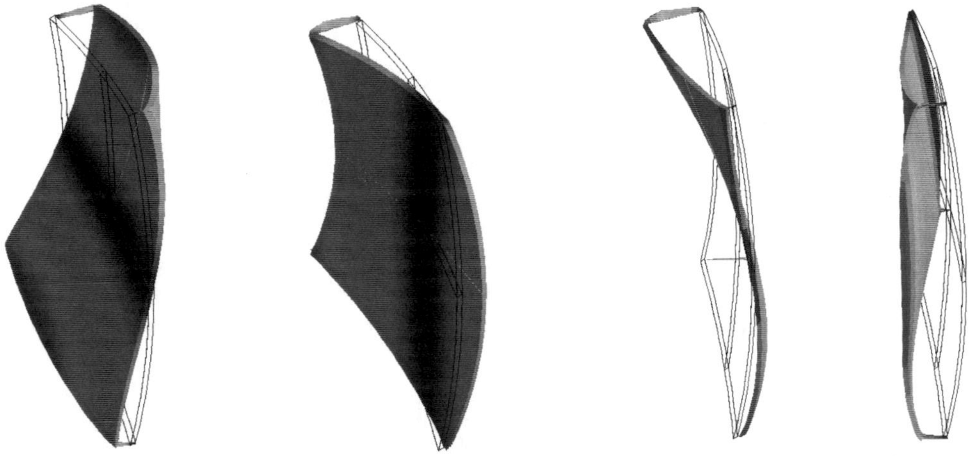

a b c d

Figure 2. The first natural mode of free oscillation of segment2 (a, b) and segment 3 (c, d) at the temperature 40 (a, c) and 55 (b, d) Celsius degree.

Similar patterns may be observed at the least for six lower frequencies of free oscillations. One can see sufficiently large scattering of the first frequency at examined temperature interval. Another feature is big difference in natural frequencies for six lower modes (44.0…144.0 Hz, 45.7…160.2 Hz, 32.6…98.1 Hz for segments 1, 2, and 3 without prestressing accordingly).

At the initial stage of the second series of computing experiments it was detected the availability of numerous natural forms of free oscillations having their origin in oscillations of single mirror segments or groups of segments in various combinations (figure 3). The most obvious reason of the obtained result is following fact. Radial and circumferential dimensions of segments 1, 2, and 3 are great and similar. This circumstance makes possible positive effects of high stiffness of double curved shell. Segments 4 and 5 may be characterized by increasing radial dimensions and decreasing circumferential dimensions. This results in drastic increasing double curvature and stiffness of segments.

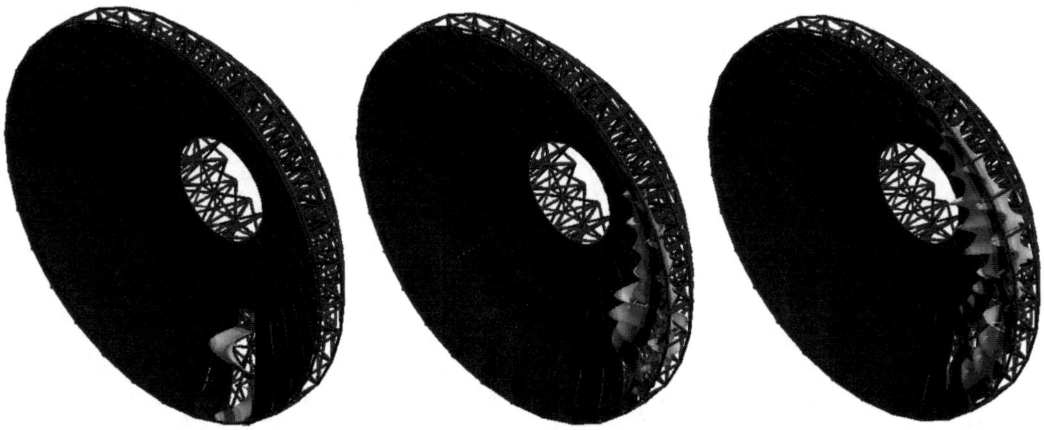

Figure 3. The three lowest natural modes of dree oscillations of ready-assembled reflector.

The next step of investigation lies in insertion of structural connections between neighboring mirror segments. Hypothetically, such connections must activate double curvature effect and make possible co-operative motion of segments during free oscillations.

To proof effectiveness of this decision there were formulated and solved some model problems. The object of research was group of three neighboring mirror segments connected with each other with various numbers of connections.

Model problem 1 – segments are not connected in circumferential direction. In this case each segment oscillates transversely to its surface with frequency about 17 Hz, independently of other segments.

Model problem 2 – segments in pairs are rigidly connected along all the adjacent sides. In this case connected segments form very rigid body with only oscillation of unsupported sides with frequency about 23 Hz. As can be seen, rigid connection between segments excludes their independent oscillations. But rigid connection is not technologically feasible, only limited number of connections in the form of mechanical locks may be realized.

Model problem 3 – segments in pairs are connected by means of single mechanical lock in the middle of common side. In this case one can see both collective and independent modes of segments oscillation with frequencies 17.8, 18.0, 18.4 Hz.

Segments with common sides connected by means of three, seven and fifteen mechanical locks are analyzed in model problems 4, 5 and 6 (figure 4). Obtained results show gradual rise of lower frequencies (table 2) and gradual convergence (as the number of locks rises) of natural modes of oscillations towards the same for the variant with rigidly connected segments. But all the variants failed to escape from oscillations of unsupported sides of mirror segments.

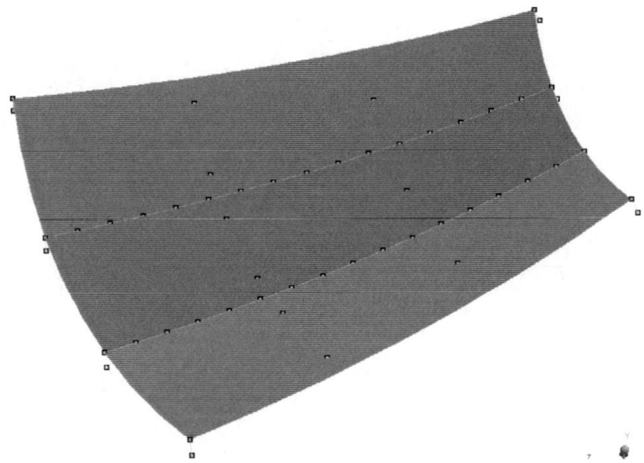

a

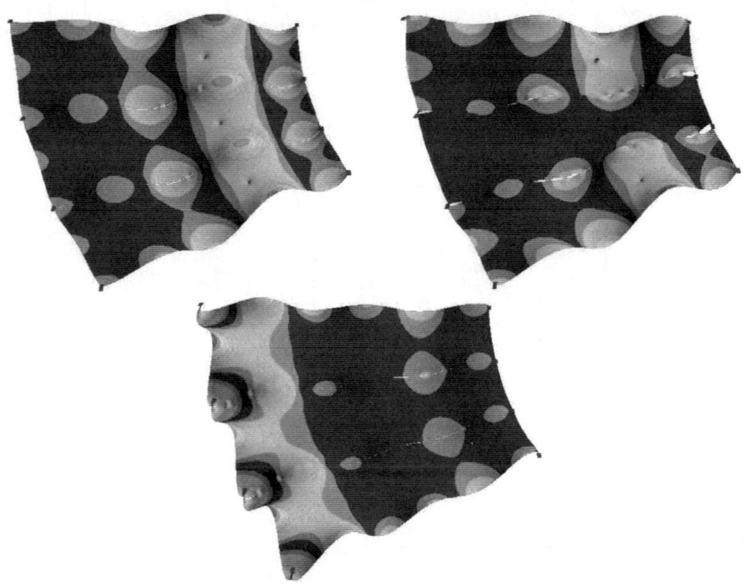

b

Figure 4. Connected segments (a) and three lowest natural modes of oscillation for three segments united with fifteen connections between each other (b).

Table 2. The lower frequencies of oscillations of a group of three segments with different number of connections between them.

Connections between segments	No	One connection	Three connections	Seven connections	Fifteen connections	Rigid along the entire length
1 frequency	16.97	17.83	17.87	18.01	18.23	22.81
2 frequency	17.98	18.05	18.44	18.89	19.97	22.97
3 frequency	17.02	18.38	19.76	21.37	21.92	23.44

On the basis of obtained results, it was recommended to reinforce segments with stiffening plates in radial direction, and to enlarge number of connections between segments and bearing skeleton.

IOP Conf. Series: Journal of Physics: Conf. Series **1050** (2018) 012021 doi:10.1088/1742-6596/1050/1/012021

The third series of computing experiments is devoted to studying segments 4 and 5 rested on skeleton which lowest natural frequency of free oscillation is about 47 Hz. Calculated frequencies of segments free oscillation for six lower natural modes (table 3) at all interval of ambient air temperature and stress state are much more than the lowest frequency of free oscillations for bearing skeleton.

Table 3. The natural frequencies of the first six modes of oscillation, Hz.

Mode	Without prestressing	The prestressed state at ambient temperature, °C							
		– 50	– 35	– 20	– 5	10	25	40	55
Segment 4									
1	133.2	133.3	133.3	133.3	133.3	133.2	133.2	133.1	132.9
2	133.7	133.8	133.8	133.8	133.7	133.7	133.7	133.7	133.7
3	134.8	134.7	134.8	134.8	134.8	134.8	134.8	134.7	134.7
4	135.0	135.0	135.0	135.0	135.0	135.0	135.0	135.0	135.0
5	135.3	135.3	135.3	135.3	135.3	135.3	135.3	135.4	135.4
6	137.4	137.3	137.4	137.4	137.4	137.4	137.3	137.2	137.1
Segment 5									
1	125.1	126.2	126.0	125.7	125.5	125.3	125.0	124.7	124.5
2	137.1	137.2	137.1	137.1	137.1	137.1	137.0	137.0	136.9
3	138.9	139.5	139.4	139.4	139.2	139.0	38.8	138.5	138.1
4	140.7	140.4	140.5	140.6	140.7	140.7	140.7	140.7	140.6
5	141.1	141.7	141.7	141.6	141.4	141.2	141.1	141.0	141.0
6	141.7	142.2	142.0	141.9	141.9	141.8	141.8	141.7	141.5

It is essential to make a note of high constancy of frequencies values: they are almost independent of ambient air temperature and prestressed state. This is because of very high stiffness of three-layered composite shell reinforced by stiffening plates. Finally exactly the segments 4 and 5 are recommended for manufacturing.

5. Conclusion

On the basis of multivariant numerical analysis of mirror segments of parabolic antenna there are determined qualitative and quantitative relationships between structural forms of mirror segments and their dynamic characteristics (natural modes and frequencies of free oscillations) in wide range of temperature conditions. These results are used for substantiation for efficient constructive decisions of the mirror and appropriate skeleton configuration.

It is desirable that dynamic analysis of structures under development involves their natural modes and frequencies of free oscillations in connection with stress state because of thermal deforming in exploitation range of temperatures. Structural variants with low sensitivity of dynamic characteristics to temperature conditions should be considered as preferable.

Acknowledgements. This work was done during the complex project and was financially supported by the Russian Federation Government (Ministry of Education and Science of the Russian Federation). Contract № 02.G25.31.0147

References

[1] Byrnes P W G and Lacy G 2016 *Proc. of SPIE* **9906** pp 99063P-1–99063P-20

[2] Gulyaev V I, Gaidaichuk V V, Chernyavskii A G and Scialino L 2003 *International Applied Mechanics* **39(3)** pp 1084–1088
[3] You B, Zhang H, Wang P and Zhang J 2015 *Aircraft Engineering and Aerospace technology: An International Journal* **87(3)** pp 274–283
[4] Morterolle S, Maurin B, Dube J-F, Averseng J and Quirant J 2015 *Aerosp. Sci. Technol.* **42** pp 74–79
[5] Sharma A, Kumar R, Vaish R and Chauhan V S 2015 *Composite Structures* **128** pp 291–304
[6] Susheel C K, Kumar R and Chauhan V S 2016 *Smart Mater. Struct.* **25** pp 1–17

IOP Conf. Series: Journal of Physics: Conf. Series **1050** (2018) 012022 doi:10.1088/1742-6596/1050/1/012022

Decidability of knapsack problem for Baumslag-Solitar group

Fedor Dudkin[1], **Alexander Treier**[2]

[1] Sobolev Institute of Mathematics Siberian Branch of the Russian Academy of Sciences Omsk, Russia

[2] Omsk State Technical University Omsk, Russia

E-mail: alexander.treyer@gmail.com

Abstract. In this work we investigate a decidability problem of group version of the knapsack problem for Baumslag-Solitar group $BS(p, q)$. We proved, that the knapsack problem is decidable for group $BS(p, q)$ for coprime integers $p > 1$, $q > 1$. In the case when $p = 1$, $q \in \mathbb{N}$, we proved that knapsack problem is decidable for group $BS(1, q)$ with some restriction on the input of the problem. However, the problem of the decidability of the knapsack problem for group $BS(1, q)$ on the whole set of inputs remains open.

1. Introduction

The Baumslag–Solitar groups are particular class of two-generator one-relator groups which have played a surprisingly useful role in combinatorial and geometric group theory. In a number of situations they have provided examples which mark boundaries between different classes of groups and they often provide a testbed for theories and techniques.

These groups have a deceptively simple definition. For each pair p and q of non-zero integers, there is a corresponding group defined by the presentation

$$BS(p, q) = \langle a, t | t^{-1} \cdot a^p \cdot t = a^q \rangle,$$

where, as usual, this notation means that the group is the quotient of the free group on the two generators by the normal closure of the single element $t^{-1} \cdot a^p \cdot t \cdot a^{-q}$. When $|p| = |q| = 1$, $BS(p, q)$ is the fundamental group of the torus or Klein bottle, both of which have been long familiar and well understood — and can therefore be regarded as standing somewhat apart from the remaining groups.

The Baumslag–Solitar groups were introduced in [1] to provide some simple examples of so-called non-Hopfian groups. The main result concerning finiteness conditions and the Baumslag–Solitar groups are following: $BS(p, q)$ is residually finite (i.e. the intersection of all its subgroups of finite index is trivial) if and only if $|p| = |q|$ or $|p| = 1$ or $|q| = 1$. $BS(p, q)$ is Hopfian if and only if it is residually finite or $\pi(p) = \pi(q)$, where $\pi(p)$ denotes the set of prime divisors of p.

Although the above theorem divides the Baumslag–Solitar groups into three classes: those that are residually finite, those that are Hopfian but not residually finite and those that are non-Hopfian, the most marked contrast is between those Baumslag–Solitar groups where $|p| = 1$ or $|q| = 1$ and those for which $|p|, |q| \neq 1$. For a group $BS(1, q)$ there is an obvious homomorphism

IOP Conf. Series: Journal of Physics: Conf. Series **1050** (2018) 012022 doi:10.1088/1742-6596/1050/1/012022

onto the infinite cyclic group, obtained by setting $a = 1$, and standard techniques show that the kernel is isomorphic to the additive group of q-adic rational numbers. Thus, such groups are metabelian and have strong structural properties; in particular, they do not contain a free subgroup of rank two. Moreover, these groups have a particularly simple normal form for their elements.

We give formulation of the knapsack problem for groups, following [2]:

Knapsack Problem. Let the elements $g_1, \ldots, g_k, g$ of the group G are given. Are there numbers $\varepsilon_1, \ldots, \varepsilon_k \in \mathbb{N} \cup \{0\}$ such that

$$g_1^{\varepsilon_1} \cdot \ldots \cdot g_k^{\varepsilon_k} = g. \tag{1}$$

The group version of the knapsack problem generalizes the classical knapsack problem for integers (more precisely, one of its formulations). Taking the additive group of integers as the group G, we obtain the well-known classical version of the knapsack problem. We call attention of the reader that the group G can be either commutative or non-commutative, and the main interest of researchers here is related to the knapsack problem for non-commutative groups. In this direction, naturally arises the question of decidability of the knapsack problem for a group or class of groups and the question of its algorithmic complexity.

We briefly describe some results related to the knapsack problem for groups. In the work [2], in addition to setting the problem of knapsack for groups, the first results in this direction were obtained. In particular, it is shown that the knapsack problem is solvable polynomially for classes of abelian and hyperbolic groups. In the article [4] it is proved that the knapsack problem for partially commutative groups is algorithmically solvable and belongs to the class NP. In articles [6, 5] independently of each other it was proved that there exist nilpotent groups for which the knapsack problem is algorithmically unsolvable. Also, in [6] it is shown that if a two-step nilpotent torsion-free group has a commutator rank greater than 316, then in this group the knapsack problem is algorithmically unsolvable.

In this paper we concentrate on the decidability of the Knapsack problem for Baumslag-Solitaire groups of the form $BS(1,p)$ (section 2) and $BS(p,q)$, where $p, q > 1$ are co-prime natural numbers (section 3). In the case of group $BS(1,p)$, we have proved the decidability of the knapsack problem for a sufficiently large set of inputs; in the case of coprime $p, q > 1$, $BS(p,q)$ is solvable on the whole set inputs. The approaches of proofs for the cases $BS(1,p)$ and $BS(p,q)$ are very different. In the first case, model-theoretic results on the decidability of elementary theories of the systems of the form $\langle \mathbb{N}, +, p^x \rangle$, are used, where p is an arbitrary natural number (for more details, see [3]). The proof of the second case is based on the study of the normal form of elements in HNN — extensions and the number-theoretic approach.

2. Case $BS(1,p)$

2.1. *Normal form*

Let p be a natural number. Denote

$$B = BS(1,p) = \langle a, t | t^{-1} \cdot a \cdot t = a^p \rangle. \tag{2}$$

It is not hard to show, that the group B is metabelian. The group B is semidirect product of the following form:

$$B \cong \mathbb{Z}[\frac{1}{\mathrm{p}}] \rtimes \mathbb{Z}, \tag{3}$$

where $\mathbb{Z}[\frac{1}{\mathrm{p}}] = \{\frac{a}{p^m} | a \in \mathbb{Z}, m \in \mathbb{Z}\}$.

The product (3) defines convenient presentation of elements of group B as a pair (k, i), where $k \in \mathbb{Z}[\frac{1}{\mathrm{p}}]$ and $i \in \mathbb{Z}$. Let a and t are generators from (2) of group B, then $a = (1, 0)$, $t = (0, 1)$ and

$(k,i)\cdot(1,0) = (k+1,i)$, $(1,0)\cdot(k,i) = (k+p^i,i)$, $(k,i)\cdot(0,1) = (pk,i+1)$, $(0,1)\cdot(k,i) = (k,i+1)$. Actually, if (k,i), $(l,j) \in B$, then

$$(k,i)\cdot(l,j) = (kp^j + l, i+j) \tag{4}$$

and inverse element is may be written in the following way:

$$(k,i)^{-1} = (-kp^{-i}, -i). \tag{5}$$

In the next lemma we give a formula for calculating the power of elements of group B:

Lemma 1. *Let $(s,t) \in B$, $\lambda \in \mathbb{N}$, then*

$$(s,t)^\lambda = \begin{cases} (s\dfrac{p^{\lambda t}-1}{p^t-1}, \lambda t) & \textit{if } t \neq 0, \\ (\lambda s, 0) & \textit{if } t = 0. \end{cases}$$

Proof. Case when $t = 0$ follows from 4. Lets prove lemma for the case when $t \neq 0$. Actually,

$$\begin{aligned}(s,t)^\lambda &= (s + p^t s, 2t)\cdot(s,t)^{\lambda-2} = \ldots = \\ &= (s + p^t s + p^{2t}s + \ldots + p^{(\lambda-1)t}s, \lambda t) = \\ &= (s\frac{p^{\lambda t}-1}{p^t-1}, \lambda t).\end{aligned}$$

The lemma is proved. ■

Lemma 2. *Let $g_1, \ldots, g_k, g \in B$ and $\varepsilon_1, \ldots, \varepsilon_k \in \mathbb{N}\cup\{0\}$ is that $g_1^{\varepsilon_1}\cdot\ldots\cdot g_k^{\varepsilon_k} = g$ in the group B. Also $g_i = (s_i, t_i), i = 1, \ldots, k$, $g = (s,t)$. Then for any $m \in \mathbb{Z}[\frac{1}{p}]$, $g'^{\varepsilon_1}_1\cdot\ldots\cdot g'^{\varepsilon_k}_k = g'$ is holds, where $g'_i = (ms_i, t_i)$, $i = 1, \ldots, k, g' = (ms, t)$. And vice versa, if $g'^{\varepsilon_1}_1\cdot\ldots\cdot g'^{\varepsilon_k}_k = g'$, then $g_1^{\varepsilon_1}\cdot\ldots\cdot g_k^{\varepsilon_k} = g$ in the group B.*

Proof. The lemma is follows from (4) and (5).

2.2. Decidability of knapsack problem for $BS(1,p)$

The mail goal for this section is giving a proof of the following theorem:

Theorem 1. *Let p be a natural number and $I = \{(g_1, \ldots, g_k, g \in BS(1,p))\}$ be the set of all inputs of knapsack problem such that, for all $i = 1, \ldots k$, $g_i = (s_i, t_i)$ and $t_i \neq 0$. Then knapsack problem is decidable for group $BS(1,p)$ for any input from I.*

Proof. Let $(g_1, \ldots, g_k, g) \in I$. We will reduce exponential group equation in the group $BS(1,p)$:

$$g_1^{\varepsilon_1}\cdot\ldots\cdot g_k^{\varepsilon_k} = g, \tag{6}$$

where $\varepsilon_i \in \mathbb{N}\cup\{0\}$, $i = 1, \ldots, k$, to existential formula of first order logic for algebraic structure $\mathbb{N}' = \langle\mathbb{N}; +, p^x\rangle$. Decidability of elementary theory of $\mathbb{N}'$ is proved by A. Semenov in [3]. Without loss of generality we may assume that $\mathbb{N}$ contains 0. Note that having binary operation of addition of naturals not hard to define relation of order for natural numbers. Using the order relation we can define constants $0 \in \mathbb{N}$ and $1 = p^0 \in \mathbb{N}$.

Denote by $C_i = \frac{s_i}{p^{t_i}-1}$, $i = 1, \ldots, k$. Let $t'_j = sign(t_j)t_j$, $i = 1, \ldots, k$. Then coefficient C_i is defined by t'_j in the following way:

$$C_i = \frac{s_i}{p^{t'_i}-1},$$

if $t_i > 0$ and

$$C_i = \frac{s_i p^{t'_i}}{1 - p^{t'_i}},$$

if $t_i < 0$.

Divisor of C_i is integer number, since belongs to $\mathbb{Z}[\frac{1}{p}]$. Using *lemma 2*, we will multiply elements $g, g_i,\ i = 1, \ldots, k$ by natural d, such that $u_i = dC_i,\ 1 \le i \le k$ and $u = ds$ is integer. After substitution expression K will be:

We omit the full description of reduction of expression (6) and provide the final system of equations over integers. In notations above the resulting group equation is equivalent to the following system:

$$\begin{cases} \sum\limits_{i=1}^{i \le k} u_i p^{\sum\limits_{j<i, t_j<0} \varepsilon_j t'_j} p^{\sum\limits_{j>i, t_j>0} \varepsilon_j t'_j} sign(t_i)(p^{\varepsilon_i t'_i} - 1) = up^{\sum\limits_{i=1, t_j<0}^{i \le k} \varepsilon_j t'_j}, \\ \sum\limits_{i=1}^{i \le k} \varepsilon_i sign(t_i) t'_i = t. \end{cases} \tag{7}$$

In the system (7) all coefficients except for some u_i and $sign(t_i)$ is non-negative integers. We will send negative coefficients to the right hand side of equation and vice versa. Finally we have the finite system of equation which is equivalent to existential formula of first order logic of decidable theory of algebraic structure $\langle \mathbb{N}, +, p^x \rangle$. The theorem is proved.

Why does not the proof is not true for the whole set of inputs? If we remove the restrictions $t_i \ne 0$ for the input $g_1 = (s_1, t_1), \ldots, g_k = (s_k, t_k), g$, then terms of the form $x \cdot p^y$ appear in the system (7), where x, y are natural variables. According to Yu.G. Penzin [7], the elementary theory of the system $P = \langle \mathbb{N}, +, x \cdot p^y, p^x \rangle$ is undecidable, although it is not yet known whether the existential theory of the system P is decidable or not.

3. Co-prime $p, q > 1$

Let's give several necessary definitions from [8].

Given subgroups A and B of the group H and $\phi\colon A \to B$ is an isomorphism. Then the group

$$H^* = \langle H, t \mid t^{-1}at = \phi(a),\ a \in A \rangle$$

is called the *HNN extension*.

A sequence $g_0,\ t^{\varepsilon_1},\ g_1, \ldots,\ t^{\varepsilon_n},\ g_n$, where $g_i \in H, \varepsilon_i = \pm 1$ is called *reduced* if it contains no consecutive t^{-1}, g_i, t, where $g_i \in\ A$, and t, g_j, t^{-1}, where $g_j \in B$. Substitutions $t^{-1}at = \phi(a)$ and $tbt^{-1} = \phi^{-1}(b)$ are called *t-reductions*.

A sequence $g_0,\ t^{\varepsilon_1},\ g_1, \ldots,\ t^{\varepsilon_n},\ g_n$ is called *cyclically reduced* if every cyclic permutation of the sequence is reduced.

Denote by T_A (T_B) the right transversal of A (B resp.) in H. Then *left normal form* is the sequence $g_0,\ t^{\varepsilon_1},\ g_1, \ldots,\ t^{\varepsilon_n},\ g_n$ such that

(i) g_0 is an element from H,

(ii) if $\varepsilon_i = -1$, then $g_i \in T_A, g_i \notin A$,

(iii) if $\varepsilon_i = +1$, then $g_i \in T_B, g_i \notin B$.

The right normal form defines similarly.

Given $w = g_0 t^{\varepsilon_1} g_1 \ldots g_{n-1} t^{\varepsilon_n} g_n \in H^*$ such that the sequence $g_0, t^{\varepsilon_1}, g_1, \ldots, g_{n-1}, t^{\varepsilon_n}, g_n$ is reduced. The number n of occurrences of $t^{\pm 1}$ in w is called *length* of w and denoted by $|w|$.

For all nonzero integers p, q group $BS(p, q)$ is the HNN extension with base group $\mathbb{Z} = \langle a \rangle$, stable letter t and associated subgroups $\langle a^p \rangle$ and $\langle a^q \rangle$.

The following lemma is useful in the combinatorial approach to the knapsack problem for Baumslag-Solitar groups.

Lemma 3. *Given* $q \in \mathbb{Z}_{>1}, p \in \mathbb{Z}, p \perp q, |p| \notin \{1, 0\}$ *and let* $\{\alpha_i\}_{i=0}^{\infty}$ *be a sequence of integers. Suppose that there exists* $\beta \in \mathbb{Z}$ *and* $\tau \in \mathbb{Z}_{>0}$ *such that:*

$$\alpha_{i+1} = \left(\frac{p}{q}\right)^{\tau} (\alpha_i + \beta), i = 0, 1 \ldots, L,$$

for some $L > \frac{\log_q |\beta \cdot p^{\tau} - \alpha_0 (q^{\tau} - p^{\tau})|}{\tau}$. *Then* $\alpha_0 = k \cdot p^{\tau}, \beta = k \cdot (q^{\tau} - p^{\tau})$ *for some* $k \in \mathbb{Z}$ *and* $\alpha_i = \alpha_0 = k \cdot p^{\tau}$, *for* $i \in \mathbb{Z}_{>0}$.

We will apply the next lemma to describe a solution of some equations in Baumslag-Solitar groups.

Lemma 4. [9] *Given* $g, h \in BS(p, q), p \perp q, p, q \notin \{0, 1, -1\}$. *If* $g^{-1} h^k g = h^l, h \neq 1, k \neq l$, *then*
1) $h = u^{-1} a^s u$, *for some* $u \in BS(p, q), s \in \mathbb{Z}, s \neq 0$,
2) $l = \left(\frac{p}{q}\right)^m \cdot k$, *for some* $m \in \mathbb{Z}$.

Using the following lemma it is possible to bound some variables in knapsack equation.

Lemma 5. *Given* $q \in \mathbb{Z}_{>1}, p \in \mathbb{Z}, p \perp q$ *and* $|p| \notin \{1, 0\}$. *Suppose that* $g_1, \ldots, g_n, g \in BS(p, q), x_1, \ldots, x_n \in \mathbb{Z}_{>0}, g_i = u_i^{-1} w_i u_i, w_i$ *is cyclically reduced and* $|w_i| \geqslant 1$. *There exists computable numbers* M *and* C *depending on* $g_1, \ldots, g_n, g$ *and not depending on* $x_1, \ldots, x_n$ *such that if for some* $1 \leqslant i \leqslant n-1$ *the length of cancellation in word* $g_i^{x_i} \cdot g_{i+1}^{x_{i+1}}$ *greater than* M, *then*

$$g_i^{x_i} \cdot g_{i+1}^{x_{i+1}} = r \cdot a^y \cdot w^z \cdot u$$

for some $z, y \in \mathbb{Z}$, $u, r, w \in BS(p, q)$ *and* $|u|, |w|, |r| \leqslant C$.

Our goal to prove, that knapsack problem is solvable for Baumslag-Soltar groups $BS(p, q)$ with co-prime $p, q > 1$.

Theorem 2. *Given* $q \in \mathbb{Z}_{>1}, p \in \mathbb{Z}, p \perp q, |p| \notin \{1, 0\}$ *and* $g_1, \ldots, g_n, g \in BS(p, q)$. *There is an algorithm to find a solution of equation* $g_1^{x_1} \cdot g_2^{x_2} \cdot \cdots \cdot g_n^{x_n} = g$ *(or to prove that it is unsolvable) in a finite number of steps depending on* $g_1, \ldots, g_n, g$.

4. Conclusion

Groups defined by generators and relations arise from topological and geometric contexts. It is perhaps therefore not altogether surprising that the Baumslag–Solitar groups play a role in questions concerning groups defined by particular topological or geometric conditions. In present paper we tried to solve particular combinatorial problem for Baumslag-Solitar group and we may see that relation played crucial role in approaches and methods of proofs. We hope that our work helps to prove decidability of the knapsack problem for HNN extension of groups.

Acknowledgments

The results of the section 2 was supported by Russian Science Foundation, grant 17-11-01117

References

[1] Baumslag G and Solitar D 1962 Some two-generator one-relator non-Hopfian groups *Bull. Amer. Math. Soc.* **68(3)** pp 199–201
[2] Myasnikov A, Nikolaev A and Ushakov A 2015 Knapsack Problems in Groups *Mathematics of Computation* **84(292)** pp 987–1016

[3] Semënov A L LOGICAL THEORIES OF ONE-PLACE FUNCTIONS ON THE SET OF NATURAL NUMBERS *Mathematics of the USSR-Izvestiya* **22(3)** pp 623–658
[4] Lohrey M and Zetzsche G 2015 Knapsack in graph groups, HNN-extensions and amalgamated products *arXiv:1509.05957*
[5] Koenig D, Lohrey M and Zetzsche G Knapsack and subset sum problems in nilpotent, polycyclic, and co-context-free groups *arXiv:1507.05145*
[6] Mishchenko A and Treier A 2017 Knapsack problem for nilpotent groups. *Groups, Complexity and Cryptology* **9(1)** pp 87–98
[7] Penzin Yu G 1973 Solvability of the theory of integers with addition, order, and multiplication by an arbitrary number *Mathematical notes of the Academy of Sciences of the USSR* **13(5)** pp 401–405
[8] Lyndon R C and Schupp P E 1977 Combinatorial Group Theory *Springer-Verlag, New York*
[9] Dudkin F A 2009 Subgroups of Baumslag–Solitar groups *Algebra and Logic* **48(1)** pp 1–19

Computer-aided design the spherical helical platform intended for the internal surface maintenance of spherical tanks

N V Dyakonov, A G Yanishevskaya

Omsk State Technical University, 11, Mira ave., Omsk, 644050, Russia

Abstract. This paper presents the new construction - spherical helical platform and the method for obtaining spherical helical surface used in the desiging of new platforms. This platforms intended to serve an inner surface of the spherical tank. The mathematical models for external and internal contour of helical platform are presented in this paper. The outer and inner contours of the helical platform are obtained by intersecting the curve of the helicoid with the curves of the sphere and the ellipsoid.

1. Introduction

The spherical tanks are used to store pressurized fuel gases and easily evaporating substances. Park spherical tanks is shown on the Figure. 1. Depending on the function spherical tanks differ in capacity and operating pressure.

Figure 1. The park of spherical tanks

IOP Conf. Series: Journal of Physics: Conf. Series **1050** (2018) 012023 doi:10.1088/1742-6596/1050/1/012023

All spherical (ball) tanks which are in operation have to undergo comprehensive technical examination [1]. A main objective of comprehensive technical examination is definition of a possibility of further safe operation, terms and completeness of the subsequent inspections, need of repair or an exception of operation of the spherical tank at the set technological parameters. A preparation of the spherical tank for full technical inspection is made by forces of the enterprise owner and carried out in such sequence:

- release of the spherical tank from a product
- installation of caps
- removal of the remains of a product
- cleaning of an internal surface of the spherical tank
- lighting installation
- preparation of welded seams and the main metal for carrying out nondestructive control methods of a kachestvaa of metalgraphic researches.

The purpose of internal survey is determination of defects on the internal surface of the spherical tank [1]. At internal visual survey are subject to obligatory check:

- condition of the main metal of a cover
- local deformations, dents are also made buldge out
- condition of welded connections of designs of spherical tanks according to requirements of projects, Construction Norms and Regulations, standards on the corresponding types of welding and types of welded seams.

In the modern petrochemical and oil and gas industry for survey and service of an internal surface of the spherical tank use a scaffolding. Two types of a scaffolding were widely adopted:

- in the form of an internal rotary ladder (Figure 2) which is executed under the copyright certificate [2];
- in the form of the internal multilevel woods from circular platforms with transitions between levels (Figure 3).

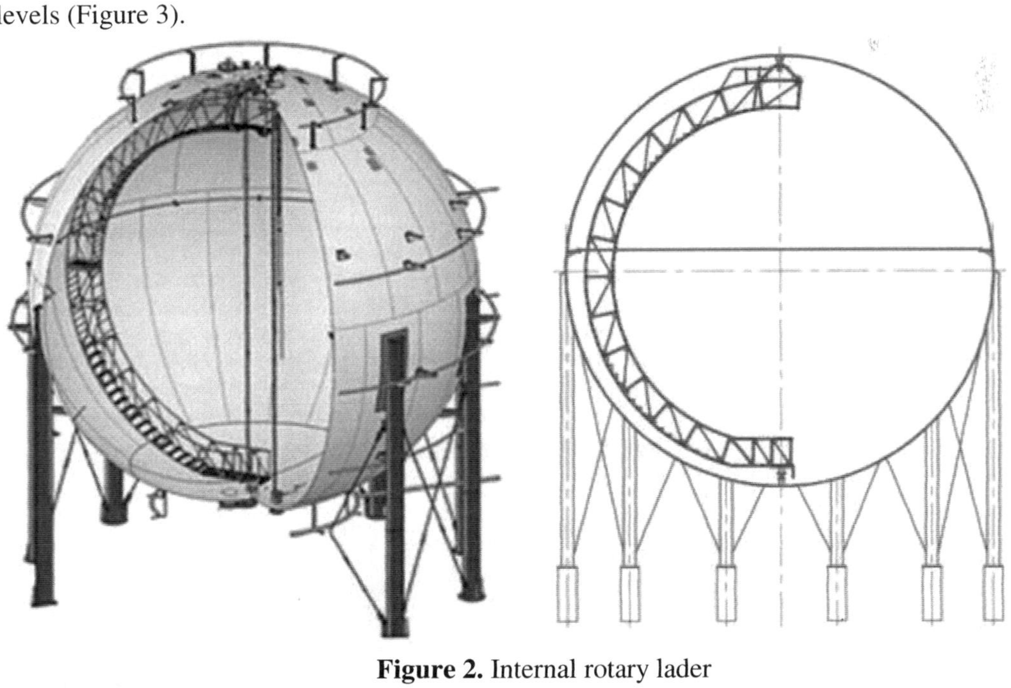

Figure 2. Internal rotary lader

Shortcomings of the first design are:

- need to manually move a ladder on a necessary position;
- possibility of application only on spherical tanks up to 2000 m^3, because of significant increase in mass of a product when providing conditions of durability that leads to impossibility of manual movement without use of additional resources [2];
- complexity of carrying out paint and varnish and other works.

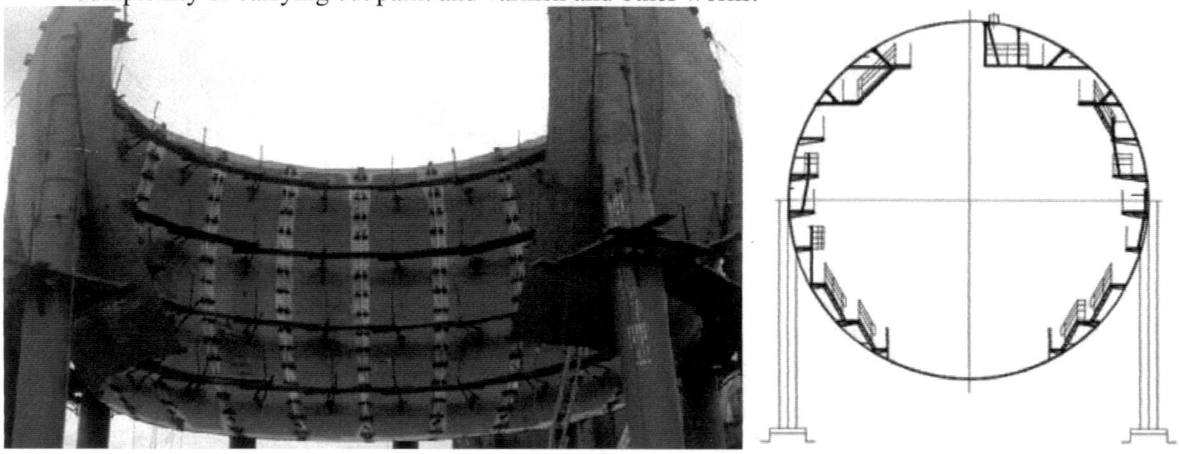

Figure 3. Multilevel construction woods

Shortcomings of the second design:

- Big mass of a product;
- Complexity of evacuation of personnel because of existence of ladder marches and step-ladders as the main way of movement between levels.

History of design and construction of the spherical tanks of various capacity built in the USSR, the USA, Germany, Japan, Great Britain Germany and other countries since the beginning of the 20th century and till present [3, 4] shows that for service of an internal surface of tanks with a capacity up to 2000 m^3 widely use an internal rotary ladder, (look at Figure 2). For tanks with a capacity of 5000 m^3 and more use the multilevel woods (look at YouTube: Inside of a LPG storage tank Erection part 7, URL: https://www.youtube.com/watch?v=o-Hf8wXDaig), or "industrial mountaineering" [3].

The largest producers of vessels for transportation of liquefied natural gas: (Daewoos Shipbuilding & Marine Engineering, Hyundai Heavy Industries, Samsung Heavy Industries, Mitsubishi Heavy Industries) establish for survey and service of an internal surface of tanks the platform with a rotary arrow [7], as shown in Figure 4.

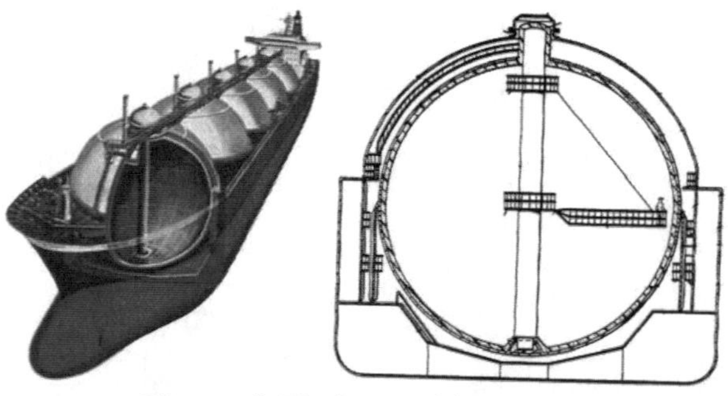

Figure 4. Platform with a rotary arrow

Development of a new scaffolding for service of an internal surface of spherical tanks which is executed in the form of the screw spherical platform (Figure 5) is presented in this article. It is possible to apply a structure of the spherical screw platform in the wide range of the sizes of spherical tanks (volume more than 600 m^3). The offered constructive scheme provides convenience of service of an internal surface of the spherical tank and safety of movement of working personnel. Decrease in weight is reached due to lack of ladder marches and step-ladders for transition between levels.

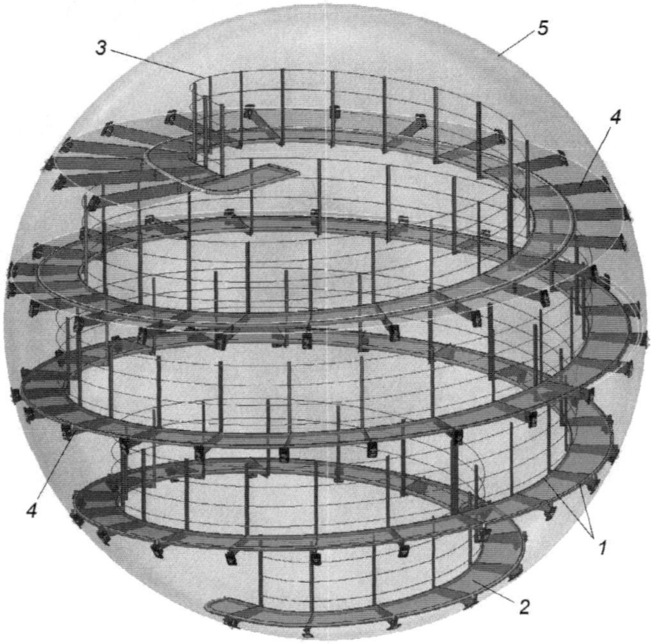

Figure 5. Three-dimensional model of the spherical screw platform

1 – screw beams of the platform;
2 – platform flooring;
3 – protection;
4 – platform consoles;
5 – internal surface of the spherical tank.

The design of a scaffolding offered in Figure 5 has't analogs in the world. The next prototype is a scaffolding in the form of the internal multilevel woods (look at Figure 3). In this regard on this work the patent application for an invention "A screw scaffolding" is submitted.

2. Mathematical model of spherical screw lines of the platform

To questions of a research of display orthogonal projection of a surface on the plane devoted works: [5 - 6], etc. In them, some differential characteristics of a two-dimensional surface or algebraic surface of bigger dimension generally are defined.

The spherical screw line of an external beam of the platform decides by crossing of a surface of the sphere (Figure 6) on a surface of a helicoid (Figure 7).

Sphere equation:

$$f(a,b)=\begin{pmatrix} x(a,b) \\ y(a,b) \\ z(a,b) \end{pmatrix}.$$

Where
$x(a,b)=R\sin(a)\cos(b)$;
$y(a,b)=R\sin(a)\sin(b)$; $z(a,b)=R\cos(a)$;
$a=0...2\pi$; $b=0...\pi$; R – the sphere radius.

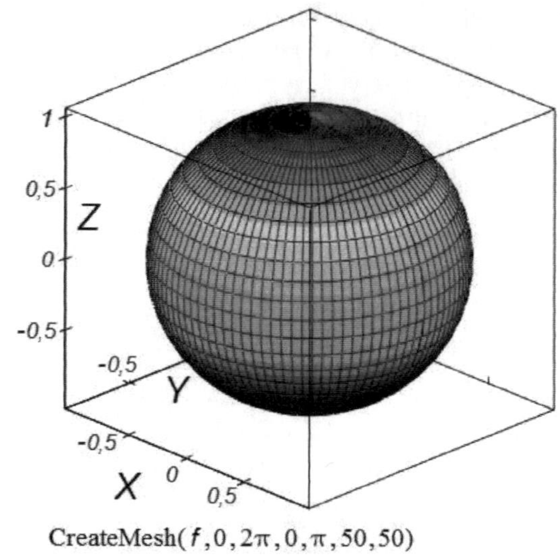

Figure 6. The schedule of the sphere in the Mathcad software product

Single-blade helicoid equation:

$$h(a,b)=\begin{pmatrix} X(a,b) \\ Y(a,b) \\ Z(a,b) \end{pmatrix},$$

Where
$X(a,b)=|a|\cos(b)$; $Y(a,b)=|a|\sin(b)$;
$Z(a,b)=\Delta b$; Δ – step of the screw line.

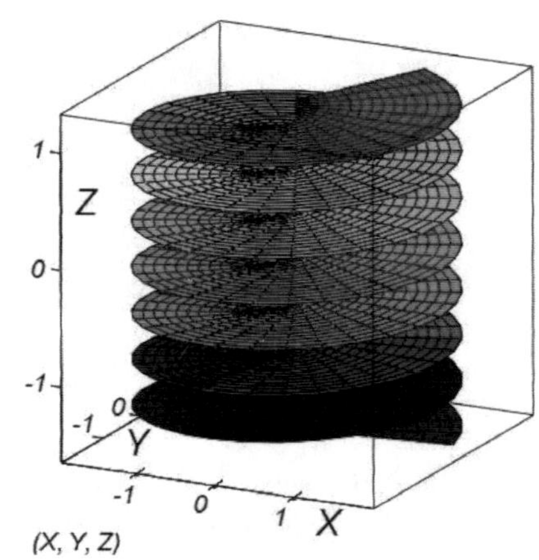

Figure 7. The schedule of the helicoid in the Mathcad software product

The equation of a curve of crossing of surfaces of the sphere and helicoid:

$$f(a,b)-h(a,b)=0\rightarrow\begin{pmatrix} x(a,b) \\ y(a,b) \\ z(a,b) \end{pmatrix}-\begin{pmatrix} X(a,b) \\ Y(a,b) \\ Z(a,b) \end{pmatrix}=0;$$

$0{,}95R<Z(a,b)<0{,}8R.$

The curve of crossing of surfaces is presented in Figure 7.

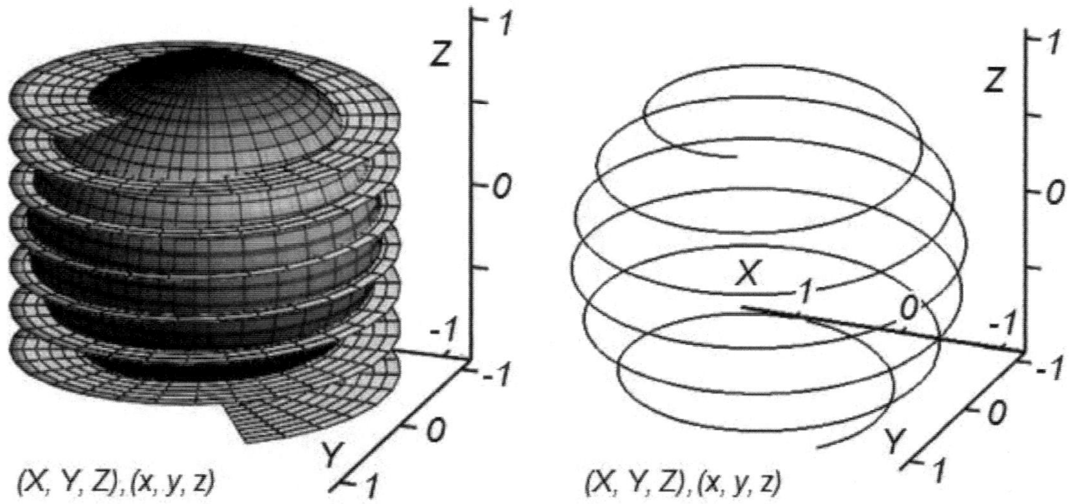

Figure 8. The schedule of a curve of crossing of surfaces of the sphere and helicoid in the Mathcad software product

The spherical screw line of an internal beam of the platform is by crossing of a surface of a ellipsoid with a helicoid surface (see Figure 7), as shown on the Figure 9.

Ellipsoid equation:

$$ff(a,b) = \begin{pmatrix} xx(a,b) \\ yy(a,b) \\ zz(a,b) \end{pmatrix}.$$

Where $xx(a,b) = R_1 \sin(a)\cos(b)$;
$yy(a,b) = R_1 \sin(a)\sin(b)$; $zz(a,b) = R_2 \cos(a)$;
$a = 0...2\pi$; $b = 0...\pi$; R1 and R2 - the ellipsoid radiuses.

The equation of a curve of crossing of surfaces of an ellipsoid and helicoid:

$$ff(a,b) - h(a,b) = 0 \rightarrow \begin{pmatrix} xx(a,b) \\ yy(a,b) \\ zz(a,b) \end{pmatrix} - \begin{pmatrix} X(a,b) \\ Y(a,b) \\ Z(a,b) \end{pmatrix} = 0;$$

$0,95R < Z(a,b) < 0,8R.$

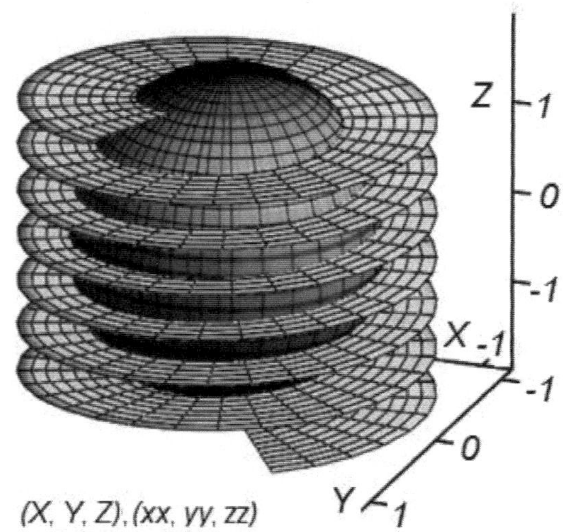

Figure 9. The schedule of a curve of crossing of surfaces of an ellipsoid and helicoid in the Mathcad software product

The spherical screw lines of external and internal contours of the platform received by crossing of a surface of a single-blade helicoid with surfaces of a spheroid and the sphere are presented on the Figure 10.

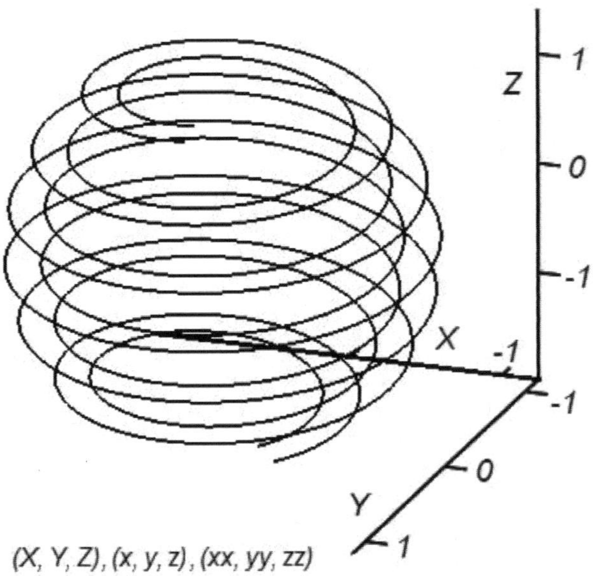

Figure 10. Spherical screw lines of external and internal contours of the platform

The result of plotting of crossings of a surface of a helicoid with the surfaces of the sphere and a spheroid by means of a CAD a "KOMPAS-3D " is presented on the Figure 11.

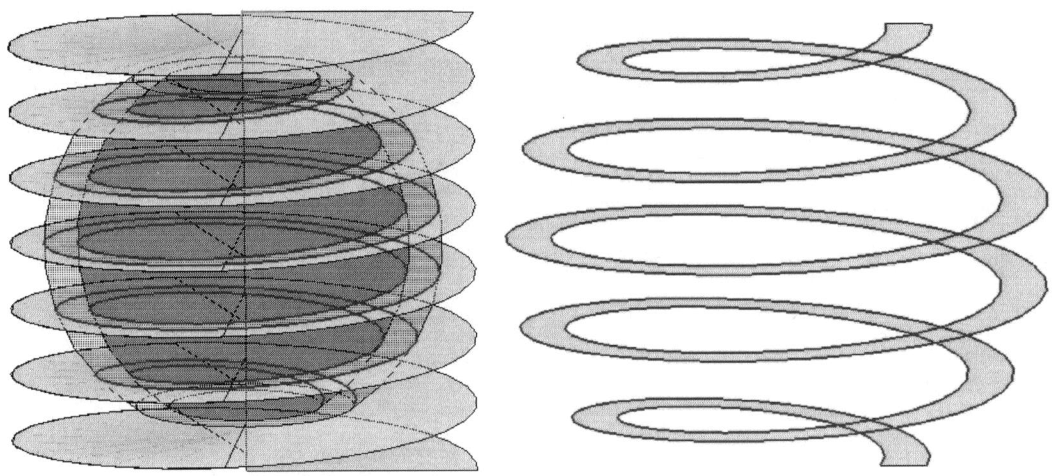

Figure 11. Three-dimensional model of a spherical screw surface

3. Conclusion

The presented mathematical way of receiving curves of crossing of a surface of a helicoid with the surface of the sphere and a spheroid allows to determine key parameters of a new product (the spherical screw platform) prior to design of a product: quantity of rounds, curve length, etc. Also it can serve as a part of uniform system at automation of design of this product.

The presented spherical screw surface can be used at a stage of predesign study of the spherical tank for assessment of quantity of rounds of the internal platform, its weight and cost.

Now in the modern oil and gas and petrochemical industry the design of the internal woods in the form of the spherical screw platform did not meet in world practice. However its application will allow to improve operational qualities of spherical tanks.

4. References

[1] RD 03-380-00 *Instruction by survey ball tanks and gas tanks for store liquefied gases under pressure*
[2] Rvantsev A D, Amelin V S *Scaffold for the production of work inside tanks* Author's certificate of the USSR number 969861 cl E04G3 / 10 1982
[3] Dikun V N, Sheynblit J A 1987 *Construction of spherical tanks* (Moskva: Nedra) p 192
[4] American National Standard 2015 *BPV Code Section VIII Division 2 Design and Fabrication of Pressure Vessels* (New York: Two Park Avenue) NY 10016-5990
[5] Arnol'd V I 1968 *Singularinies of smooth mappings* Russian Math. Surveys vol 23:1 p 1–43
[6] Bruce J W, Giblin P J 1992 *Curves and singularities: a geometrical introduction to singularity theory* 2nd edition (Cambridge University Press) p 321
[7] R Kvamsdal, R Tobiassen *Ships for transport of liquefied gases* Patent US 3968764 A 1976

Choosing modes for thermal treatment of deposited coatings based on chromium steel with carbide-boride-nitride doping

E N Eremin, A S Losev, S A Borodikhin, I A Ponomarev, A E Matalasova

Omsk State Technical University, 11, Mira ave., Omsk, 644050, Russia

E-mail: weld_techn@mail.ru

Abstract. This work explores the regimes of thermal treatment of corrosion-resistant coatings obtained by surfacing with a flux-cored wire with carbide-boride-nitride doping. It was found that tempering at a temperature of 800 degrees Celsius reduces the hardness of the metal to acceptable values for machining. To increase the hardness of the metal after tempering, it was suggested to perform quenching at a temperature of 1020 degrees Celsius. Carrying out such heat treatment ensures the hardness of the metal practically coinciding with the hardness of the metal after surfacing.

1. Introduction
Surface hardening of machine parts and mechanisms operating in difficult operating conditions is one of the important tasks of the machine-building industry. A significant reduction in the price of such products can be ensured by manufacturing them from relatively cheap steels with subsequent surface hardening with the help of wear-resistant flux-cored wires.

Therefore, a large range of details of machine-building production is surfaced with chromium-bearing flux-cored wires creating coatings with high corrosion resistance [1–3].

2. Statement of a problem
However, the operational period of details surfaced with such wires under abrasive wear is low. This is due to the small number of strengthening phases in the structure of the deposited metal. Increasing the wear resistance of such a deposited metal is possible by alloying it with boron [4–8]. A number of studies have established the effectiveness of using the boride compounds in flux-cored wires, which helps to receive coatings with increased wear resistance [9–12]. At the same time, the high hardness of such coatings makes their machining difficult. Hardness regulation of such a depositeded metal is provided by its heat treatment. However, the thermal treatment regimes for such coatings had not been established.

Thus, this work explores the heat treatment regimes for coatings deposited by high-chromium flux-cored wire alloyed with a complex of carbide-boride-nitride compounds.

3. Theory
As the research object we chose the metal of coatings surfaced by high-chromium flux-cored wire, alloyed with a complex of boride compounds, its composition is: 15% Cr + 0.5% B_4C + 0.5% BN + 2.5% TiB_2 + 1.0% ZrB_2. We investigated the effect of heat treatment regimes on the microhardness and microstructure of metal coatings.

Surfacing was carried out on plates of St3 steel with a size of 200×50×10 mm with an experimental flux-cored wire of 2.4 mm diameter in argon in four layers. Surfacing mode: current strength 230 A; voltage 24 V; the rate of surfacing is 20 m / hr.

Metallographic studies of the deposited metal were carried out on an optical microscope AXIO Observer A1m (Carl Zeiss). The microstructure was detected by chemical etching in a reagent of the following composition: $CuSO_4$ – 4 g; HCl – 20 ml; H_2O – 20 ml.

Durometric studies were carried out on metal samples after surfacing and thermal treatment with a hardness tester TK-2 by the method of Rockwell and microhardness tester Shimadzu HMV-2 by the Vickers method. The microhardness was measured over the cross section of the deposited coating, starting with the base metal in 0.2 mm increments.

The metal was studied in states after surfacing, after tempering and after quenching.

4. Results of the experiments and discussion

It was established, that the deposited metal of the coating with borides after surfacing has a complex composite structure. It is characterized by a martensitic matrix and a large amount of eutectic and particles of strengthening phases. As a consequence, such a metal has a high hardness reaching values of 58 HRC.

The results of dyurometric research of the coating metal after surfacing are given in Tabl. 1.

Table 1. Distribution of $HV_{0.2}$ microhardness in the cross-section of the complexly alloyed coating after surfacing

Increment, mm	0	0.4	0.8	1.2	1.6	2.0	2.4	2.8	3.2	3.6	4.0	4.4	4.8
HV	615	674	680	693	773	654	705	680	663	740	725	654	753
Increment, mm	5.2	5.6	6.0	6.4	6.8	7.2	7.8	8.2	8.6	9.0	9.4	9.8	10.2
HV	649	712	894	650	693	674	750	741	789	742	970	800	840

It is noticeable that in the metal of such a coating alloyed with a complex of boride compounds, after surfacing, the microhardness varies within 615–970 HV over the cross section.

The microhardness of the structural components of such a deposited metal is shown in Tabl. 2.

Table 2. Microhardness $HV_{0.01}^{*}$ and $HV_{0.05}$ of structural components of the metal with borids after surfacing

Puncture №	1	2	3	4	5	6	7*	8	9	10	11*	12
HV	978	587	540	552	575	546	1342	521	593	874	1262	829

The microhardness of the structural constituents of the metal after surfacing is 521–593 HV for the matrix, 829–978 HV for the eutectic and 1262–1342 HV for the strengthening phases.

In order to carry out the mechanical treatment of the deposited metal it is necessary to reduce its hardness. For this purpose, tempering was carried out at the regimes recommended for this class of steels at temperatures of 600, 700 and 800 °C for 2 hours [13]. The results of measuring the microhardness over the cross-section of the metal alloyed with the B_4C+BN+TiB_2+ZrB_2 complex after tempering in selected modes are given in Tabl. 3.

Table 3. Distribution of $HV_{0.2}$ microhardness in the cross-section of the complexly alloyed coating after tempering

Microhardness of the coating metal after tempering at 600 ° C													
Increment, mm	0	0.4	0.8	1.2	1.6	2.0	2.4	2.8	3.2	3.6	4.0	4.4	4.8
HV	590	514	585	615	623	580	550	580	565	555	745	598	530
Increment, mm	5.2	5.6	6.0	6.4	6.8	7.2	7.8	8.2	8.6	9.0	9.4	9.8	10.2
HV	545	590	570	627	594	674	851	652	680	715	685	674	692
Microhardness of the coating metal after tempering at 700 ° C													
Increment,	0	0.4	0.8	1.2	1.6	2.0	2.4	2.8	3.2	3.6	4.0	4.4	4.8

mm													
HV	475	490	469	700	470	483	503	530	541	805	550	534	530
Increment, mm	5.2	5.6	6.0	6.4	6.8	7.2	7.8	8.2	8.6	9.0	9.4	9.8	10.2
HV	793	555	530	574	517	642	583	796	513	564	501	553	541
Microhardness of the coating metal after tempering at 800 ° C													
Increment, mm	0	0.4	0.8	1.2	1.6	2.0	2.4	2.8	3.2	3.6	4.0	4.4	4.8
HV	340	375	390	406	419	440	753	400	365	395	353	380	405
Increment, mm	5.2	5.6	6.0	6.4	6.8	7.2	7.8	8.2	8.6	9.0	9.4	9.8	10.2
HV	366	450	380	652	415	418	386	697	428	383	402	419	395

It can be seen that after tempering both at 600 °C and at 700 °C the microhardness has sufficiently high values in the range of 470–850 HV. Tempering at 800 ° C significantly reduces the microhardness to 340–450 HV, although there are structural components with high microhardness values of up to 753 HV.

There is no significant decay of the structural components in the microstructure of the coating metal with borides after tempering at 800 °C. At the same time, the amount of boride eutectic and strengthening phases has decreased.

The microhardness of the structural components of the deposited metal has changed significantly after tempering of the coatings under investigation (Tabl. 4).

Table 4. Microhardness $HV_{0.01}^{*}$ and $HV_{0.05}$ of structural components of the metal with borids after quenching at 800 °C for 2 hours

Puncture №	1	2	3	4	5*	6*	7	8	9*	10*	11	12
HV	358	496	548	420	754	1144	486	358	1071	458	520	387

It can be seen that the microhardness of the structural constituents of the metal with borides after tempering has significantly decreased in comparison with that after surfacing (see Tabl. 2). Microhardness of the matrix has decreased from 521–593 HV to 358–458 HV, for eutectics – from 829–987 HV to 548–754 HV, and for the strengthening phases – from 1262–1342 HV to 1071–1144 HV. The hardness of the metal of such a coating, measured by the Rockwell method, is in the range of 32–37.5 HRC. Such hardness of the metal makes it possible to carry out its machining without any difficulties.

Therefore, we recommend tempering at 800 °C to reduce the hardness of the deposited metal of the examined coatings.

In order to provide high wear resistance of tempered metal after machining it is necessary to increase its hardness. For this purpose, after tempering the samples, we carried out their quenching according to the regimes recommended for this class of steels: at the temperatures of 950 °C, 1020 °C and 1100 °C [13]. The results of measuring the hardness over the cross-section of the coating metal are given in Tabl. 5.

Table 5. Hardness distribution over the cross-section of complexly alloyed coatings after quenching

Quenching temperature, °C	Hardness distribution over deposited metal layers, HRC			
	1	2	3	4
950	52.5	52.5	52	54
1020	54	53	55	58.5
1100	53	54	55	55.5

Analyzing the obtained results, we can note that the hardness over the layers of the coating surfaced with the flux-cored wire with boride compounds after quenching at 950 °C is in the range of 52–54 HRC, after quenching at 1020 °C – 53–58.5 HRC, and after quenching at 1100 °C – 53–55.5 HRC. The best results are obtained by quenching at a temperature of 1020 °C.

The deposited coating metal with borides after such quenching has a complex composite structure with a martensitic matrix, a large amount of eutectic and strengthening phases. The hardness of the metal reaches a maximum value of 58.5 HRC.

The analysis of the microhardness structure of the metal after quenching at a temperature of 1200 °C is given in Tabl. 6.

Table 6. Microhardness $HV_{0.01}^{*}$ and $HV_{0.05}$ of structural components of the metal with borids after quenching

Puncture №	1*	2	3	4	5	6*	7	8	9*	10	11	12
HV	1073	688	971	711	814	1052	773	723	1105	697	784	771

It is noticeable that the microhardness of the matrix is high and lies in the range of 688–784 HV, for eutectic it is 814–971 HV, for particles – 1052–1105 HV.

Thus, tempering at a temperature of 800 °C for 2 hours and subsequent quenching at 1020 °C are rational modes of thermal treatment of the deposited metal of the investigated coatings.

The results of the microhardness distribution of the metal of the coating under investigation after heat treatment performed at optimal conditions are shown in Fig. 1.

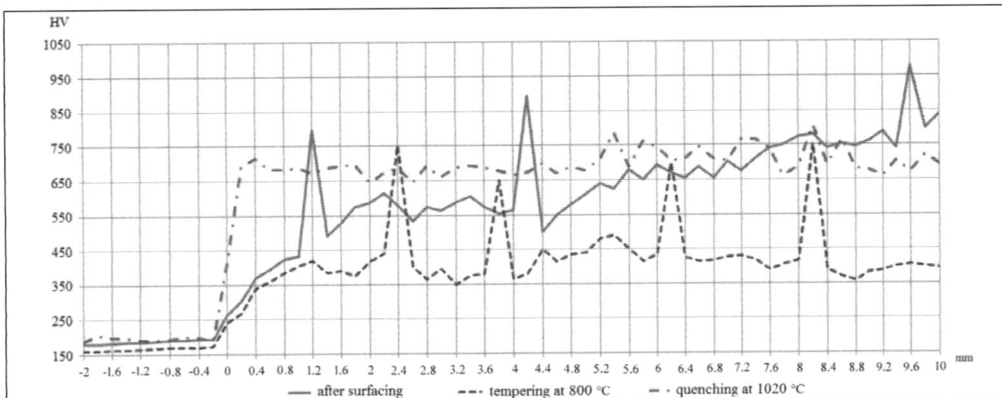

Figure 1. Distribution of microhardness along the cross-section of the complex-doped coating after optimum heat treatment

As it can be seen, such heat treatment leads to the stabilization of microhardness values at a high level, even slightly exceeding the microhardness level of the coating metal after surfacing.

5. Conclusion

The rational thermal treatment of the deposited coatings on the basis of chromium steel with carbide-boride-nitride doping is tempering at 800 °C for 2 hours, providing hardness up to 32–37.5 HRC acceptable for machining, and subsequent quenching at a temperature of 1020 °C, increasing the hardness to 58 HRC.

Acknowledgments

This research was made possible through the Russian Science Foundation grant (project № 17-19-01224).

References

[1] Ryabtsev I A 2004 *Surfacing of machine parts and mechanisms* (Kiev: Ecotechnology) p 159

[2] Frumin I I 1979 *The surfacing materials of the CMEA member countries: catalogue* (Kiev–Moscow: VINITI) p 620

[3] Sokolov G N 2002 *Welding and cladding methods for metals* (Volgograd: VSTU) p 80

[4] Lyakishev N P, Pliper Y L, Lappo S I 1986 *Boron-containing steels and alloys* (Moscow: Mashinostroyeniye) p 192

[5] Artemiev A A, Sokolov G N, Dubtsov Y N and Lysak V I 2011 J. *Izvestiya vysshikh uchebnikh zavedeniy. Poroshkovaya metallurgiya i funktsionalnyye pokrytiya* **2** pp 44–48

[6] Zhong L, Xiang C, Yan-xiang L and Kai-hua H 2009 *J. of Iron and Steel Research, International* **16** 3 pp 37–42
[7] Raghavan V 2003 *Journal of Phase Equlibria* **24** 5 pp 459–460
[8] Liu Y., Li B., Li J. et al. 2010 J. *Materials Letters*. **64** 11 pp 1299–1301
[9] Eremin A E, Eremin E N, Philippov Yu O, Matalasova A E and Katz V S 2014 J. *Omsk Scientific Bulletin* **1** pp 55–58
[10] Eremin E N and Losev A S 2015 J. *Procedia Engineering* **113** pp 435–440
[11] Eremin E N, Losev A S, Borodikhin S A and Ivlev K Ye 2017 J. *Oil and Gas Engineering (OGE-2017) AIP Conf. Proc.* **1876** pp 020071-1–020071-6
[12] Losev A S, Eremin E N and Philippov Yu O 2010 J. *Omsk Scientific Bulletin* **2** pp 131–134
[13] Goldstein M I, Grachev S V and Veksler Yu G 1999 *Special steels* (Moscow: MISIS) p 408

IOP Conf. Series: Journal of Physics: Conf. Series **1050** (2018) 012025 doi:10.1088/1742-6596/1050/1/012025

A Quantitative Criterion for Quality Mixing Assessment for the Effective Unit of Mixed Products

A V Evseev, M S Paramonova, V V Preis

Tula State University, Lenin Avenue, 92, Tula, Russia

Abstract. The paper proposes the new quality assessment criteria of resulting mixture for effectively used unit of mixed products enabling to consider the deviation of all the mix components taking into account the share of each component in the mixture. Quantitative requirements for stability of the dosing devices of mixing apparatus are determined for the specified allowable value of dispersion.

Key words. mixture production, determined formation of homogeneity, mixture quality assessment criteria, mixing devices, the quality of mixed products.

1. Introduction

The general technology approaches to homogeneity formation of loose materials mixtures, which have exclusively probabilistic character [1-8], are known. Such scientific approach was reasonable in the middle of the last century and was not undergone any basic changes up to our time. The most complete generalization of these methods and their characteristics was formulated in the work [1], and as for the last materials in the works [9-11]. All these authors pointed out the complexity of the description and data modeling of mixing processes, and there was not always the predictable use of ready mixas a result.

Currently the production of homogeneous common lots of high-volume mixture products (15 ... 20 tons and more) is often achieved by means of small lots mixing which differ significantly in their characteristics. In this case, the homogeneity of the total lot (the mixing quality) is controlled by the radius of dispersion obtained r_v from the full-scale tests according to formula:

$$r_v = 0,6745\sqrt{\frac{\sum_{i=1}^{n}(V_i - V_{cp})^2}{n-1}}, \tag{1}$$

where V_i – is the main characteristic of the trial test; V_{cp} – is the average value of the main characteristic for a series of tests.

If the dispersion of the basic characteristic exceeds the allowable value r_v, the total lot is subjected to re-mixing.

This quality control of the mixing process is, firstly, expensive, and secondly biased, since the deviation of the main characteristic is caused not only by poor-quality mixing, but also by the dispersion characteristics of small lots.

IOP Conf. Series: Journal of Physics: Conf. Series **1050** (2018) 012025 doi:10.1088/1742-6596/1050/1/012025

The quality of various components mixing in the chemical, medical, construction and other industries, as the analysis shows [1-8], is evaluated by various criteria, however, in the post-Soviet space, the homogeneity coefficient got the most distribution:

$$V = \frac{100}{C_0}\sqrt{\frac{\sum_{i=1}^{n}(C_i - C_0)^2 \cdot n_i}{n-1}}, \% \qquad (2)$$

$$V = \frac{100}{x}\sqrt{\frac{1}{n-1}\sum_{i=1}^{n}(x_i - x)^2}, \% \qquad (3)$$

where C_i- is the concentration value of one of the components in the samples; C_0- is the concentration of the same component with an ideal distribution; n_i- is the number of samples with concentration C_i; $n-$ is the total number of samples; x_i- is the number of particles of a component with an ideal concentration C_0; $x = \frac{\sum_{i=1}^{n} x_i}{n}-$ is the selective average number of particles (tubes, granules, etc.).

In this case, these values were determined only for one of the mixture components, most often for the component of the smallest volume. It is the key component as it is more sensitive in ratio rating violation. Such an assessment is characterized by simplicity, but assesses the process objectively only when two components are mixed. When mixing three components, for example, the diffusion of the smallest component may prove to be insignificant and the coefficient of variation will be close to none, and the actual and required number of particles of the other two components will differ substantially from each other. This will result in considerable heterogeneity of the mixture. It is also difficult to evaluate the mixing quality of several components which have the same volume.

2. Problem formulation

The first scientific justification of refusal of probable homogeneity formation and its technological implementation was proposed in the work [12-14]. The innovative component mixing technique with an ordered component arrange mentrelative to each other was proposed. This provided higher quality of the mixture obtained and the determined formation of their homogeneity taking into account technical parameters of mixing and portioning devices [15, 16]. The scientific novelty of the work is defined by the new approach to the quality forming of future products, and the deductive design of mixing and packing complexes when accounting the ready mixparameters and the equipment parameters .

To solve the technological problem this work proposes to introduce the concept of the effective unit of mixture use which is formed from small lots of homogeneous components with different physical and chemical properties, but providing the uniform, equal effect of its use.

It is expedient to establish the criterion that would allow, firstly, to assess the quality of the mixing process, and secondly, to perform non-destructive testing during the mixing and, thirdly, to correct the process in order to increase product homogeneity.

When developing the criterion for quality mixing assessment of small lots, the number of which is from two to six, often with ratios close to unity, firstly, it is necessary to strive for the criterion to take into account the deviations of all components in the mixture, and secondly, when establishing such an integral indicator it is necessary to take into account that the same absolute ratio deviation of the j-th small lot causes a different mixing quality, depending on the share of this lot in the mixture. Therefore, obviously, it is necessary to take not absolute, but relative deviations of the ratio. Thirdly, such a generalized criterion should take into account not only the mixing quality, but also reflect the quality

IOP Conf. Series: Journal of Physics: Conf. Series **1050** (2018) 012025 doi:10.1088/1742-6596/1050/1/012025

of the mixed product as a whole, i.e. to take into account the degree of influence of the ratio deviation of small lots to the deviations of the output characteristics of the mixture.

3. Theoretical part

When mixing products with a total weight G, the weight ratio $G_1;G_2;...;G_J$ of J small lots is usually given. Then the weight fraction j-th of the small lot is:

$$P_{jb} = \frac{G_j}{G}. \quad (4)$$

For the effective mixture unit, the weight W of the weight j-th of the small lot w_j can be represented as:

$$w_j = w \cdot P_{jb} = w_{0j} \cdot m_j, \quad (5)$$

where w_{0j} – is the arithmetic mean of the mass of one mixture element j-th of the smallest lot; m_j – is the expected number of mixture elements j-th of the smallest lot in the effective unit of the mixture. Total number of elements in the effective unit of the mixture:

$$m = \sum_{j=1}^{J} m_j. \quad (6)$$

Due to value smallness of the one element mass dispersion of j-th of the smallest lot Δw_j (it is estimated that the product of numbers $\Delta w_j \cdot m_j < 1$) next will be assumed to be w_{0j} constant.

Then in the i-th sample of weight w which has m unit mixed elements, the absolute difference between the actual and the required fraction of the j-th small lot is represented as:

$$\Delta P_{\text{абс}i} = \left| \left(\frac{w_{0j} \cdot m_{j\text{д}i}}{w} - \frac{w_{0j} \cdot m_j}{w} \right) \right| = \left| \left(P_{j\text{д}i} - P_j \right) \right|. \quad (7)$$

Graphically, for example, when mixing three lots, this can be represented as a dynamic equilibrium triangle (Fig. 1).

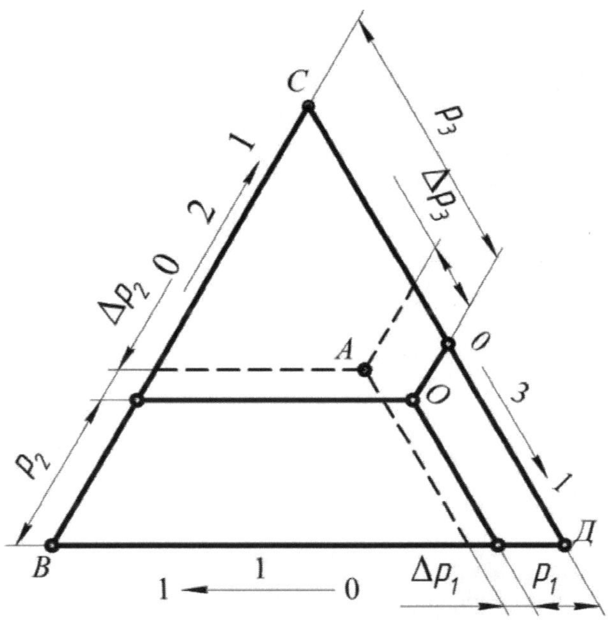

Figure 1. Dynamic equilibrium triangle of three constituent lots.

When dividing this value by the required fraction of the j-th small lot, we obtain the measure of the deviation of the ratio of the j-th smallest lot in the i-th sample:

$$\Delta P_{ji} = \frac{\Delta P_{абсji}}{P_j} = \left| \left(\frac{m_{jдi} - m_j}{m_j} \right) \right| \tag{8}$$

Then, obviously, for i-th sample, the most general indicator of the mixing quality is the measure of the deviation of the total lot, which is the sum of the deviation measures of the constituent lots:

$$\Delta P_i = \sum_{j=1}^{J} \left| \left(\frac{m_{jдi} - m_j}{m_j} \right) \right| \tag{9}$$

The average measure of the deviation between the total lot according to the data of the samples is determined from the formula:

$$\Delta \bar{P} = \frac{\sum_{i=1}^{n} \Delta P_i}{n} = \frac{\sum_{i=1}^{n} \sum_{j=1}^{J} \left| \left(\frac{m_{jдi} - m_j}{m_j} \right) \right|}{n}, \tag{10}$$

where n - is the number of samples.
Selective variance of the discrepancy measure of the common batch:

$$D(\Delta P) = \frac{1}{n-1} \sum_{i=1}^{n} \left(\Delta P_i - \Delta \bar{P} \right)^2 \tag{11}$$

Selective standard deviation:

$$\sigma(\Delta P) = \sqrt{\frac{1}{n-1} \sum_{i=1}^{n} \left(\Delta P - \Delta \bar{P} \right)^2} \tag{12}$$

With continuous mixing of small lots [1,5,12,14], consisting in the single-piece feeding of tubes with metering devices, it is possible to continuously monitor the mixing quality during the operation by determining the actual value of the divergence measure of the common lot $\Delta P_д$ and to control the process, for example, small computing modules and power supplies, ensuring the following:

$$\Delta P_д \le [\Delta P] \tag{13}$$

We can determine the allowable value of the discrepancy measure of the total lot $[\Delta P]$ and establish quantitative requirements for the stability of the metering mixing device for a given permissible value $[r_V]$. It is possible because the main characteristic spread is influenced not only by mixing quality but also the characteristic dispersion of small lots.
We determine the maximum possible relative amplitude of the metering device's output and the relative amplitude δ_{m_0} of the supply of the components δ_{m_j} of the mixture for the given allowable value of the divergence measure of the total lot $[\Delta P]$.
When mixing J small lots on a mixing machine having W_j metering devices for feeding a j-th small lot,

$$\Delta P = \frac{\Delta m_1}{m_1} + \frac{\Delta m_2}{m_2} + \ldots + \frac{\Delta m_J}{m_J} \le [\Delta P], \tag{14}$$

where Δm_j – is the unit amplitude of the supply of the smallest lot.

In the case of the same productivity m_0 of all the metering devices of the mixing machine:

$$m_1 = m_0 \cdot W_1;\ m_2 = m_0 \cdot W_2;\ \ldots\ m_J = m_0 \cdot W_J \tag{15}$$

The experiments proved that [12]

$$\Delta m_j = \Delta m_0 \cdot \sqrt{W_j}, \tag{16}$$

where Δm_0 – is the amplitude of the performance of a single metering device.

Then:

$$\Delta P = \frac{\Delta m_0}{m_0}\left(\frac{\sqrt{W_1}}{W_1} + \frac{\sqrt{W_2}}{W_2} + \ldots + \frac{\sqrt{W_J}}{W_J}\right) = \delta_{m0} \cdot \sum_{j=1}^{J} \frac{1}{\sqrt{W_j}} \leq [\Delta P] \tag{17}$$

Location:

$$\delta_{m_0} \leq \frac{[\Delta P]}{\sum_{j=1}^{J} \frac{1}{\sqrt{W_j}}} \tag{18}$$

Respectively:

$$\delta_{m_j} = \sqrt{W_j} \cdot \delta_{m_0} \leq \frac{[\Delta P] \cdot \sqrt{W_j}}{\sum_{j=1}^{J} \frac{1}{\sqrt{W_j}}} \tag{19}$$

It is convenient to carry out the relative amplitudes determination δ_{m_0} and δ_{m_j} using the nomogram (Fig. 2) or on the computational module.

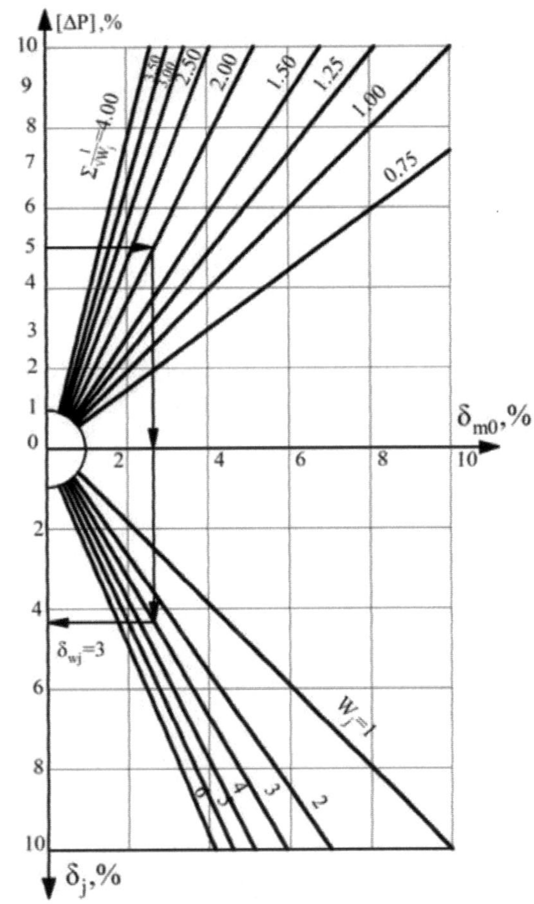

Figure 2. Nomogram of the relative amplitudes determination of the metering device performance δ_{m_0} and the feed unit δ_{m_j}

4. Conclusion

The adapted quantitative criterion for quality mixing assessment is developed. This criterion allows to control the mixing quality without appeal to expensive destructive tests. It can provide high uniformity of composite products due to assignment of admissions on operation stability of metering devices of mixing device. High performance of process was determined using theoretical and pilot studies. In particular, the degree of mixture uniformity was raised in 2-2,5 times, the power expenses due to the exception of long operation of components was decreased in 2,5 -3 times, the working area was decreased in 3-5 times, and labor productivity was increased in 2-2,5 times because of the technology chain reduction [13,17,18].

References

[1] Makarov Yu I Apparatus for loose materials mixing 1973 *Mechanical Engineering*, Moscow. 216 p

[2] Lastovtsev M Efficiency Criterion of the Solids Mixing Process 1950 *Proceedings of NTK-MIHM,* Moscow **7** pp 53–56

[3] Fan L Tand and Shin S H Stochastic diffusion model 1979 *Chem. Eng. Sci.* **34** № 6 p. 811–820

[4] Khan ZS, Van Bussel F, Schaber M, Seemann R, Scheel M and Michiel M Di 2011 High-speed measurement of axial grain transportin a rotating drum *New Journal of Physics* **13** 105005. Doi:10.1088/1367-2630/13/10/105005

[5] Podgornyj Yu I, Martynova T G, Skeeba V Yu, Kosilov A S, Chernysheva A A and Skeeba P Yu 2017 Experimental determination of useful resistance value during pasta dough kneading *IOP Conf. Series: Earth and Environmental Science* **87** 082039 doi: 10.1088/1755-1315/87/8/082039

[6] Chuvpilo A V New in the theory and technology of powder batching 1964 *VNIIEM*, Moscow.

[7] Sokolchik P Yu, Stashkov S I and Malimon M V Forecast and Quality Management of Heterogeneous Bulk Mixes 2013 *Bulletin of the Perm National Research Polytechnic University. Chemical technology and biotechnology* Perm pp 64–83

[8] Aleksandrovsky A A and Ahmadiev F G The Current state and of mathematical modeling problems of mixing bulk materials process 1986 *Abstracts of the reports to the All-Union Conference* **2** *Belgorod* pp 83–84

[9] Barantseva E A, Ponomarev D A, Mizonov V E and Berthiaux N 2003 Nonlinear models of processes of continuous mixing of loose materials. *Proceedings of the XVI International Conference "Mathematical Methods in Engineering and Technology MMTT-16",* St. Petersburg **10** pp 116–117

[10] Weinekotter Rand and Gericke H Mixing of solids 2000 *Kluwer academic publishers*

[11] Arratia R E, Duong Nhat-hang, Muzzio F J, Godbole Pand and Reynolds S 2006 A study of the mixing and segregation mechanisms in the Bohle Tote blender via DEM simulations *Powder Technology* **164** pp 50-57

[12] Lukash A N, Evseev A V and Chuvpilo AV 2000 Development of technologies and equipment for the preparation· of mixtures of loose materials *Newsof Tula State University. Series. Mechanical Engineering* **5** pp 218–224

[13] Evseev A V and Lobanov A V 2014 An estimation of quality of a mix of powder materials for manufacturing of the diamond tool on the basis of a physical model *News of the Tula State University. Technical Science* **3** pp 88–94

[14] Evseev A V and Paramonova M S 2015 Statement of the problem of mathematical modeling of the process of deterministic formation of the homogeneity of a mixture of bulk materials *News of Tula State University. Technical Science* **8–2** pp 203-208

[15] Patent of the Russian Federation No. 2271243. Published. 10.03.2006. Bul. № 7

[16] Patent of the Russian Federation No.2129911. Published 10.05.1999. Bul. № 13

IOP Conf. Series: Journal of Physics: Conf. Series **1050** (2018) 012025 doi:10.1088/1742-6596/1050/1/012025

[17] Evseev AV Classification of Non-Mixing Mixtures and Devices 2014 *News of Tula State University. Technical Science* **3** pp 82–88
[18] Evseev AV The new criterion for the quality mixing assessment of bulk materials 2015 *News of Tula State University. Technical Science* **11–1** pp 139–147

IOP Conf. Series: Journal of Physics: Conf. Series **1050** (2018) 012026 doi:10.1088/1742-6596/1050/1/012026

Designing of special concretes for machine building

R S Fediuk[1], A P Svintsov[2], V S Lesovik[3], A A Pak[1], R A Timokhin[1]

[1]Far Eastern Federal University, 8, Sukhanova str., Vladivostok, 690950, Russia
[2]Peoples' Friendship University of Russia, 6, Miklukho-Maklaya str, Moscow, 117198, Russia
[3]Belgorod State Technological University named after V.G. Shoukhov, 46, Kostyukova str, Belgorod, 308012, Russia

E-mail: roman44@yandex.ru

Abstract. Based on transdisciplinarity research, using the provisions of the geonics, modern composite materials have been developed that can effectively replace steel parts in mechanical engineering. The composite material was prepared by the joint grinding of cement (55%), ash of acid composition (40%) and limestone (5%) to a specific surface area of 550 kg / m^2, with an activity of 70.2 MPa. It was found that the combined effect of mechanical and chemical activation promotes an increase in the pozzolanic activity of acid ash, has a catalytic effect on the reaction activity of the fly ash the Ordinary Portland cement and the limestone crushed waste surfaces during machining in a vario-planetary mill. Thus, in the process of operation of equipment with concrete and reinforced concrete elements, their significant production advantages, determined by the improvement of the precision, strength and other parameters of the machines, due to the use of new materials, will be revealed.

1. Introduction

The task of mechanical engineering is to improve the quality, accuracy and high-speed performance of machinery and equipment, while reducing their cost, labor intensity of manufacturing and increasing the unit capacity. The use of polycrystalline composite materials opens up new possibilities for improving and regulating the characteristics of structures in operation, in comparison with single-crystal materials, whose capabilities are in some cases exhausted. In the US mechanical engineering industry, the share of such materials is approximately 15-20%, in Japan - about 50%, in Germany - up to 30%. In Russia, it is only about 2%, and the trend of its growth largely depends on the absence of a theoretically grounded methodology.

The current level of development of construction technology and structures allows us to successfully solve the problem of improving engineering products through the use of special concretes and reinforced concrete, both conventional and uniaxial, biaxially or volumetrically prestressed. The purpose of this study is to develop a special concrete that can maintain performance in the specific operating conditions of machinery and equipment.

2. Results and Discussion

To achieve the purpose, composite binder, obtained by the joint grinding of Ordinary Portland cement (OPC), superplasticizer (SP), fly ash (FA) and limestone crushed waste, was developed. The OPC of this study was type I (Spassky cement CEM I 42.5N). The physical properties of Spassky Portland

IOP Conf. Series: Journal of Physics: Conf. Series **1050** (2018) 012026 doi:10.1088/1742-6596/1050/1/012026

cement are presented in Table 1.

Table 1. The physical properties of Spassky Portland cement

Compressive strength [MPa]			Setting time [min]		Fineness, passed through sieve 008 [%]	Surface area [m^2/kg]	Normal density [%]
24 hours	3 days	28 days	Start	Finish			
30.2-32.6	40.0-42.2	50.2-52.7	112	256	91.8-93.0	290	27

The The fly ash obtained from Vladivostok thermal power plants was used as components of composite binders (Figure 1).

Figure 1. Typical particle shape of fly ash

In the course of thermal studies it was revealed that in the low temperature interval physically bound water is removed from the FA sample [1–3]. The exothermic effect with a maximum of about 400°C indicates the burnout of organic substances, and the endothermic effect at 712°C is the dissociation of calcite into CaO and CO_2, which was confirmed by X-ray fluorescence analysis (Figure 2).

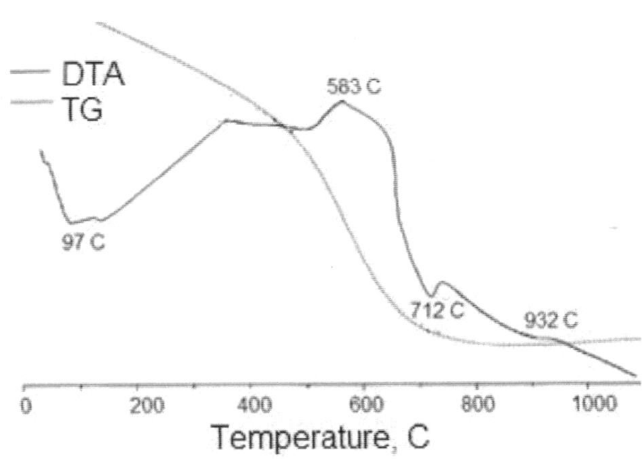

Figure 2. Results of DTA and TG of fly ash of Vladivostok TPP

The limestone crushing waste (LCW) was obtained through pre-drying of carbonate rocks to about 1% moisture content of the raw material and then crushing to 300 microns in the centrifugal mill "Tribokinetika".

PANTARHIT PC160 superplasticizer (Ha-Be Betonchemie, Germany) was used in the mixtures in order to enhance the fresh properties.

The raw materials were dry mixed in various compositions [4]. The mixture was ground by a vario-planetary mill as is more energy-efficient than by either conventional ball mills or vibratory mills. In addition, due to the joint action of shock, centrifugal shock and abrasive forces, it becomes possible to achieve a higher powder dispersion using vario-planetary mill [5]. The movement and the trajectory of the grinding balls can be influenced by varying the gear ratio, so that the balls strike horizontally on the inner wall of the grinding jar (high impact energy), approach tangentially (high friction) or simply roll over the inner wall of the grinding jar (centrifugal mills) [6-7]. The rotational speed of the grinding jars and the support disc was set independently to optimally use high impact energy and high friction, respectively. All intermediate stages and combinations between pressure friction and impact can be freely set (Figure 3). The vario-planetary mill "Pulverisette 4" is controlled by integral software, in which up to 9 programmes can be saved and then loaded quickly and easily via the mill display.

.

Figure 3. Schematic operating principle of a vario-planetary mill

The raw components were milled and mixed in various proportions (OPC 30-100%, FA 0-50%, LCW 0-20%, superplasticizer 0.3%) in a vario-planetary mill for one hours. The materials mixing was carried out using cyclic and continuous flow dispensers with automatic control with a maximum weighing cycle duration of 45-90 s, with a weighing accuracy of 1-3% []8-10. All raw materials were dosed by mass, with the exception of water and liquid additives (if any) that dosed by volume. In this research, there were no liquid additives (the super plasticizer is added dry). Water was added at the last stage, during the mortar preparation.

The control samples were made from a composite paste without addition of sand and fiber. Number of specimens was fabricated to determine the optimum composition of the binder.

3. Results and discussion

To reduce the water requirement of fine-grained cement-sand concrete mixture, while maintaining its required flowability, it is expedient to use a superplasticizer. In the paper, a choice was made of six

IOP Conf. Series: Journal of Physics: Conf. Series **1050** (2018) 012026 doi:10.1088/1742-6596/1050/1/012026

powdered superplasticizers, the most common in the Russian Far East building materials market [11-12]. For the formation of the raw mix of binder used CEM I 42.5N at a water-cement ratio of 0.3. The dosage of the superplasticizer into the binder mixture was 0.3%. The time for the beginning of the measurement of the slump flow was fixed after the end of the mixing of the cement paste. The mixing time of the cement paste was 6 minutes. Compared to other samples of superplasticizers, the attainment of the highest values of the cone spread was experimentally detected on a binder raw mix using a dry polycarboxylic superplasticizer PANTARHIT PC160.

Table 2. Spreading of cement paste with various superplasticizers

Time of measurement start, min	Melflux 1641 F (Germany)	Melflux 5581 F (Germany)	PANTARHIT PC160 (Germany)	FOX^{TM}-8H (Russia)	PC-1030 (China)	JK-04 PPM (China)
	spreading, mm					
0	290	350	370	250	240	130
5	380	390	400	260	280	120
30	390	350	390	240	190	98

Seven binder composites were developed for further research. The "binder: sand" ratio is 1 to 3. To determine the optimal dosage of components in the "cement-limestone-ash" system it was grinded them to specific surface of 600 m^2/kg at various ratios (Table 3).

Table 3. Compositions and properties of composite binders

Mix No.	OPC, wt. %	FA. wt %		LCW, wt. %	Compressive strength, MPa		
		Vladivostok TPP	Artem TPP		3 days	7 days	28 days
1	100 (no milling)	–	–	–	17	32.5	47.5
2	30	–	50	20	30.2	40.1	50.4
3	35	45	–	20	34.2	43.1	53.2
4	40	–	45	15	36.6	48.2	56.6
5	45	45	–	10	39.2	50.1	59.2
6	50	–	40	10	45.1	54.9	65.8
7	55	40	–	5	47.2	54.1	70.2
8	100 (milling)	–	–	–	60.3	81	103.2

Note: reference specimen (no milling); specimens No. 2-8 milled to S_{sp}=600 m^2/кg

In order to determine the optimum particle size the Portland cement (55% wt.), the superplasticizer (less 1% wt.), the fly ash (40% wt.) and the limestone (5% wt.) were ground to different specific surface area (Ss_p): 500, 550, 600, 700, 800, 900 m^2/kg.

This fineness of grinding was chosen based on the results of previous studies, which showed that further reduction of the particle size is economically unprofitable. Positive dynamics of strength growth of the composite binder under the joint influence of the ash fine constituents, limestone crushing screenings and superplasticizer with maximum increase in the activity of the binder at 62% was found. Active mineral components of the composite binder contribute to the binding of Ca(OH)2 produced during cement hydration which results in formation of additional amount of hydrosilicate neoplasms. At the same time optimizing the process of structure formation is achieved by composite polydispersed components. Highly dispersed spherical ash particles act as crystallization centers, and used as filler at the nano- and micro- levels. In conjunction with the larger particles of the mineral component denser filling of intergranular spaces is noted in concrete cement structure with reduction in number of pores and microcracks. This is confirmed by micrographs of composite cement paste derived by joint grinding of clinker and industrial wastes of the Russian Far Eastern region. Cement stone structure is very dense packing of small grains in the crystalline neoplasms total mass (Fig. 6). The additional amount of hydrated crystalline phases contribute to filling of the voids at the micro level in the crystalline matrix of calcium hydrosilicates at the boundary of the contact area, increasing degree of adhesion binder with filler.

4. Conclusion

1. The composite material was prepared by the joint grinding of cement (55%), ash of acid composition (40%) and limestone (5%) to a specific surface area of 550 kg / m^2, with an activity of 70.2 MPa.
2. It was found that the combined effect of mechanical and chemical activation promotes an increase in the pozzolanic activity of acid ash, has a catalytic effect on the reaction activity of the FA, the OPC and the LCW surfaces during machining in a vario-planetary mill.
3. It has been revealed that the addition of fly ash and limestone crushed waste to the composite binder at all dosages increases the strength characteristics of special concretes. Thus, a clear relationship between the properties of concrete and the features of the structure of cement stone is revealed - an increase in the number of hydrosilicate neoplasms, with a complex decrease in gel- and capillary porosity, especially at the molecular and submicroscopic levels, which predetermines the growth of strength and increase in the performance characteristics of concrete.
4. Thus, in the process of operation of equipment with concrete and reinforced concrete elements, their significant production advantages, determined by the improvement of the precision, strength and other parameters of the machines, due to the use of new materials, will be revealed.

References

[1] Ranjbar N, Mehrali M, Maheri M R and Mehrali M 2017 *Cement and Concrete Research* **100** 14-22
[2] Lesovik V S, Urkhanova L A, Gridchin A M and Lkhasaranov S A 2014 *Research Journal of Applied Sciences* **9(12)** 1016-1020
[3] Cai J, Ju Y, Hu X and Yu B 2017 *RSC Detection Science* **9** 58-73
[4] Stelescu D M, Ficai A and Trusca 2017 *International Multidisciplinary Scientific GeoConference Surveying Geology and Mining Ecology Management, SGEM* **17(61)** 445-452
[5] Volodchenko A N, Lesovik V S, Volodchenko A A, Glagolev E S and Bogusevich G G 2016 *International Journal of Pharmacy and Technology* **8(4)** 22673-22686
[6] Lesovik V S, Zagorodnjuk L H, Volodchenko A A, Glagolev E S, Sumskoy D A and Kaneva E V 2016 *International Journal of Pharmacy and Technology* **8(4)** 24868-24879
[7] Bazhenov Y M, Zagorodnjuk L H, Lesovik V S, Yerofeyeva I V, Chernysheva N V and Sumskoy D.A. 2016 *International Journal of Pharmacy and Technology* **8(4)** 22649-22661
[8] Glagolev E, Suleimanova L and Lesovik V 2016 *International Journal of Environmental and Science Education* **11(18)** 12383-12389
[9] Fediuk R S, Smoliakov A K, Timokhin R A, Stoyushko N Y and Gladkova N A 2017 *IOP Conference Series: Earth and Environmental Science* **66(1)** 012026
[10] Fediuk R, Smoliakov A and Stoyushko N 2016 *IOP Conference Series: Materials Science and Engineering* **156(1)** 012042
[11] Ranjbar N, Kuenzel C 2016 *Journal of American Ceramic Cociety* DOI: 10.1111/jace.14848
[12] Ibragimov R, 2016 *ZKG International* **69 (6)** 34-39

IOP Conf. Series: Journal of Physics: Conf. Series **1050** (2018) 012027 doi:10.1088/1742-6596/1050/1/012027

Information technologies usage models during agile systems functioning

A S Geyda, I V Lysenko
St.-Petersburg Institute for Informatics and Automation of the Russian Academy of Sciences, 14 line, 39, St.-Petersburg, Russia

Abstract. The article outlines conceptual and derivative formal models that provide means for estimation of operational properties of agile systems operation with regard to the use of information technologies. The estimation is fulfilled analytically through plotting the dependences of predicted values of operational properties against variables and options of solved tasks. To develop this type of models, the use of information technologies during agile systems functioning is analyzed through a technological agile system. General concepts and principles of modeling of information technology use during operation of agile systems are defined. An exemplary modeling of effects of technological informative and related technological non-informative operations of technological agile systems operation is provided. Based on conceptual models of operation of agile systems with regard to information technologies use, set-theoretical models followed by functional models of agile systems operation using information technologies are introduced. The examples of calculated factors of agile system operational properties and information technologies use are given. The obtained results allow to evaluate predicted values of agile systems operational properties and information technologies use; they help to analytically estimate the efficiency of implementation of new information technologies depending on variables and options in solved tasks.

1. Introduction

The article introduces conceptual and formal models of agile systems functioning with regard to the use of information technologies (hereinafter referred to as “IT”). Their first distinctive feature consists in their revealing of relations between environment impact (that may lead to imperative improvement of the system and its operation) and informative and non-informative actions required for this system operation as a reaction to impact of the environment. Second, the developed conceptual models facilitate transition from geometric graph and set-theoretical to functional models that describe the connections revealed at conceptual modeling using analytic dependences between: Agile system operational properties (from here referred to as “OP”) with regard to IT application, impact of system environment, and informative and non-informative effects of agile system functioning

These models are required to solve research tasks as mathematical problems of operations research and mathematical programming. To develop such models analytic relations between informative and non-informative effects and impact of the environment are revealed in such a way that: First, the models would be based on patterns of operation conditioned by changing environment. Second, the models would register potential demand to implement processes of improvement of systems and their operation (as a reaction to changing environment). Third, they would register subsequent transient behavior of changes in the system and its functioning.

Following these models and given the use of IT, analytic dependences of OP predicted values in an agile system are plotted against variables and options in solved research tasks, namely, in estimation problems, during evaluation of agile systems OP with IT use, in design problems of agile system functioning (with the account of transition processes in operation) on the basis of OP factors of agile system functioning at IT use. To describe the relations between informative and non-informative effects at agile systems functioning (hereinafter referred to as ASF), concepts and principles (conceptual model) of IT application during ASF are suggested. By applying these concepts and principles, general patterns of IT application effects in the context of ASF are revealed. The suggested conceptual model provided for transition first to graph-theoretical, set-theoretical and then to functional model of IT use in the context of ASF that is based on patterns of non-informative effects development with the use of informative effects. Based upon practical application, solution of these problems is relevant when researching: Enterprise architectures, use of IT, implementation of governmental projects and programs involving the use of IT. Research is now becoming ever more relevant due to the fact that analytical research of agile systems OP with regard to IT use supports scientifically based improvements of parameters of informative actions, their relations with other actions depending on changing requirements of the environment and with the account of ASF specifics.

2. Information technologies use during agile systems functioning: general concepts and principles

IT application in the context of ASF is exemplified with IT application in technological agile systems (hereinafter referred to as "TAS") (including the related transition processes and IT application). Suppose, their operation is technological if defined by execution of technological operations, specifically by definition of technological operation execution in technological documentation in TAS. These include, for instance, systems that function to enforce manufacturing of unique products (e.g. in aerospace industry) and systems for implementation of governmental projects and targeted programs. General concepts required for development of models of IT application in the context of TAS include: IT, IT application, information, information use, system, system operation, purposeful changes in system operation, target, outline of changes in system operation, benefit, technological informative operation, technological non-informative operation, system operation effects, effects of transition processes during functioning. Concepts are linked together in a complex on a scheme of purposeful changes to ASF with the application of IT (figure 1). IT effects [1] are manifested at ASF conditioned by changes in operation (for example, by transition processes from reaching one target to reaching another). This change in operation becomes apparent in changes in non-informative actions (their composition, properties and sequence). The changes in non-informative actions are caused by results of the informative actions. Implementation of informative actions is governed by necessary consideration of environment impact at ASF. As a result of the series of changes, the personnel using TAS acquire the effects different from those that would appear, should there have been no changes, that is, not considering environment impact or the new ASF, conditioned by this impact. The operation implementation with new chosen parameters is explained by technological informative operations implemented to consider environment impact. These technological informative operations provide for selection of next technological operations with better parameters (in effected conditions) depending on the changes in the states of TAS and its environment. Best operational effects are achieved through consideration of these changes at execution of technological information operations. The use of different types of technological operations (from now on referred to as "TIOp"), e.g. informative, non-informative, at ASF depending on verified TAS states and its environment is illustrated on figure 1.

IOP Conf. Series: Journal of Physics: Conf. Series **1050** (2018) 012027 doi:10.1088/1742-6596/1050/1/012027

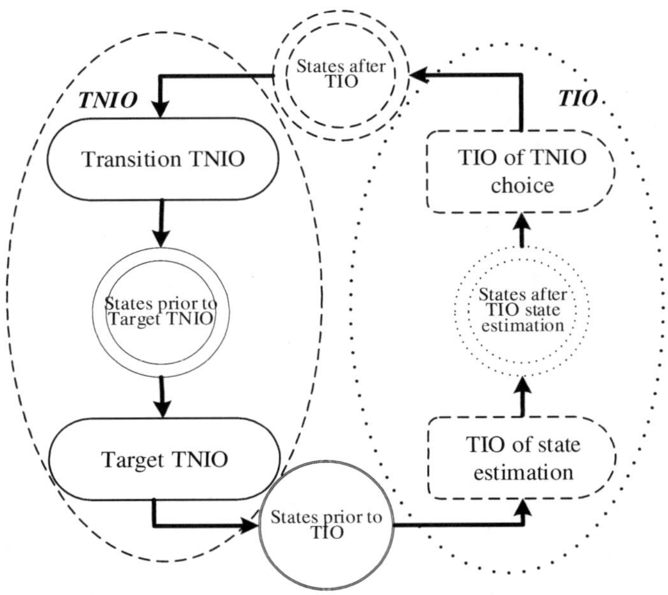

Figure 1. Use of different types of TIOp at ASF

When TIOp sequences are implemented, first technological informative operations (hereinafter referred to as "TIO") are executed. These operations estimate changed states of the environment and system elements with regard to environment impact. Further, TIO liable for changed TIOp are executed (if necessary). Their target result is to obtain information about the TAS state and its environment and what should be changed with this regard. Then technological non-informative operations (hereinafter referred to as "TNIO") connected with informative operations by cause-effect relations are executed through practical implementation. The notions of information and IT, benefits of IT, benefits of information, informative and non-informative actions, TIOp, TIO, TNIO and other related notions were specified in [1]. Principles of agile system research and a number of related notions were introduced in [2]. General OP characteristics were defined in [3]. Let us specify the notions that are used further in the context of functional modeling of ASF.

Technological informative operation (TIO) is an informative action to be taken according to the technological documentation (e.g. manuals, descriptions). Technological non-informative operation (TNIO) is a non-informative action to be taken according to the technological documentation. Technological informative operations are executed according to a certain information technology. TIO (or, as a rule, a number of TIO) aims at obtaining (creation) and transforming the information into such a form, where it could be used by a person or technical equipment to solve a task of choosing (for instance, choosing a mode of TNIO).

During implementation of TIO and TNIO sequences, depending on the occurred events and states of TAS elements and environment, which were revealed as a result, different TIO are executed. Then TIO are used for choosing various TNIO resulting in occurrence of various events and states of TAS. In this regard, the states of TAS and environment do not recur during operation in reality, and sequences of TIOp, events and states (a loop in figure 1) should be expanded into structured sequences of events and states (outcome tree). As a result, numerous possible state sequences are obtained. They are connected by branches (events) depending on states of TAS, environment and implemented sequences of TIOp (TIO and TNIO), and the events, which are revealed during TIOp execution. TAS operation outcome is a sequence of conditioned states of TAS and branches (events) between them caused by TIOp (TIO and TNIO) and actions of TAS environment. During operation, the actual TAS operation outcome is reviewed, being a sequence of implemented TAS states and branches between them caused by TIOp. During planning, the possible operation outcomes are reviewed, being a sequence of possible states and branches between them caused by TIOp (TIO and TNIO). Composition and characteristics of TIOp, which lead to possible operation outcomes, change

IOP Conf. Series: Journal of Physics: Conf. Series **1050** (2018) 012027 doi:10.1088/1742-6596/1050/1/012027

as a result of TIO. TIO lead to various sequences of random events and states revealed as a result of changes in environment states. These events and states form possible outcomes. Each possible outcome, except for various possibility measures of its implementation (depending on states of TAS and environment, and implemented TIOp) complies with different effects (results with specified requirements) of operation and different operation efficiency. Operational properties of TAS and IT, namely TAS potential OP (with regard to IT application), describe system parameters associated with its operational efficiency in changing conditions. This property should be estimated based on the modeling of all possible operation outcomes. TAS potential is a property that indicates if an agile system is suitable to reach changing targets (actual and possible). It would be rational to use the difference between TAS with applied “new” and “old” IT as an indicator of operational properties of the “new” IT compared with previously use one. This indicator should be estimated on the basis of analytic models developed through description of laws and manifestation patterns of effects, as a result of execution of TIO and TNIO sequences of various characteristics at different TAS operation outcomes.

3. Modeling of information technologies use: general concepts and principles

Concepts applied during development of ASF models with regard to transition actions of TAS improvement, and principles applied during conceptual and formal modeling of agile systems were defined in [2]. Let us consider general concepts, which require interpretation due to suggested concept of IT application in the context of ASF. Simplex of TIOp (simplex) is a sequence of initial TIO (TIO required to initiate TNIO), TNIO and final TIO (TIO required to terminate TNIO). Reduced simplex (hereinafter referred to as “RS”) is a simplex containing zero TNIO. There are several types of RS depending on the type of state evaluation task they solve: If RS solves a task of general TAS state evaluation at the moment (to the moment) then it is type one RS. If RS solves a problem of state evaluation of several sites at the moment (to the moment) then it is type two RS. Depending on their specifics, different RS should be executed to evaluate the states of TAS and environment as a result of execution of simplexes.

This rule is fixed by a principle of simplex linking through RS implementation. These RS are implemented differently depending on the results of execution of prior TIOp and environment states. RS target result consists in chosen composition and course of further actions. This result should be used in consequent non-reduced simplexes to achieve target results of TNIO (e.g. treated blanks). While different sequences of simplexes and RS should be executed differently (depending on various recorded states of TAS and environment), different states are implemented as a result. Afterwards these states could lead to implementation of various simplexes and TAS transition into next states, as a result.

Definition of these sequences is given as a principle of functional dependency of TAS operation outcome from simplexes and states of TAS and its environment.

Nodes of an outcome tree are possible states achieved as a result of TIOp (TIO and TNIO by selected means), and tree edges stemming from the parent node are possible outcomes (transitions between states) resulting in TIOp implementation. Tree branching line at TAS operation complies with one of the (possible) functional events providing it became actual. If a TAS state during operation is calculated on the basis of the state of several sites and respective RS of type two, the subtrees complying with possible states of sites and their combinations are connected into the branch. The outcome tree corresponds with all possible TAS operation outcomes. Composition and characteristics of outcomes and the outcome tree depend on the TIOp composition and characteristics, and, as a result, on the used IT. In particular, possibility measure of implementation of every possible outcome (possibility measure of reality of outcome) depends on the composition and characteristics of TIOp (TIO and TNIO) and state of environment at operation. Operation effects achieved as a consequence of certain outcome implementation depend on the composition and characteristics of TIOp (and the IT) and on the states of environment at operation. Knowing the possible outcome and characteristics of the effects, providing this outcome is real, one could calculate TAS potential.

4. An example of functional models of use of information technologies during agile systems functioning

The functional models are developed through systematic re-expression of geometric graph and set-theoretical models. Let us set:

$m_{b.d}$ – d – event method a_b, of a set $M_b = \{m_{b.d}; d = \overline{1,D}\}$, of methods (described as possible implementations and predicted outcomes of events in technological documentation);

C_i – i – TAS operation outcome, possible sequence of outcomes of events (outcome tree branch);

$e_{i.k}(m_{b.d}) =< s_{i.k}, s_{i.k+1}(m_{b.d}) >$ – event consisting in transition of elements to sites involved in the event a_b from state $s_{i.k}$ to state $s_{i.k+1}$ as a result of event execution a_b through $m_{b.d}$ – (i.e. event outcome a_b at site by $m_{b.d}$ – method);

$C_i =< (s_{i.0}, e_{i.0}(m_{0.h})), ..., (s_{i.k}, e_{i.k}(m_{g.f})), ..., (s_{i.K}, e_{i.K}(m_{s.v})) >$:

$s_{i.k+1}(m_{g.f}) =< \tilde{y}_{i.k+1.n}(m_{g.f}) >$ – $k+1$ – state being terminal in event outcome a_f by $m_{g.f}$ – method initiated from $s_{i.k}$, functional event element C_i. It is represented as a vector of effects as a result of k – outcome of $e_{i.k}(m_{g.f})$ event a_f in operation C_i outcome;

$\{m_{0.h}, ..., m_{g.f}, ..., m_{s.v}\}$ – event methods $a_h, ..., a_f, ... a_v$ revealing an outcome of event set of TAS functioning C_i (i.e. outcomes of events $a_h, ..., a_f, ... a_v$ if implemented in such a way that the respective outcomes are included in outcome C_i);

$s_{i.K} =< \tilde{y}_n(C_i, t_\kappa) >$ – state at outcome end of TAS functioning event set C_i, namely, vector of effects $\tilde{y}_n(t_\kappa)$ of functioning, fulfilled to the moment t_K of operation end according to event set outcome C_i, where κ – is a number of the terminal event outcome in C_i (i.e. number of end TIO outcome in C_i);

a_b may be specified by event associated with it $e_{i.k}$ at implementation a_b by methods $m_{b.d}$: $a_b = \{e_{i.k} : e_{i.k} \sim m_{b.d}\}$. $T_c(C_u^\partial, h) = \{C_i, i = \overline{1,I}\}$ – c – tree of possible outcomes of TAS functioning constructed for a preset sequence C_u^∂ of directive states formed at the boundary of TAS and its environment (environment operation outcome at TAS boundary) and for a preset IT h from a set H of possible IT.

Outcome C_u^∂ of environment operation at TAS boundary is set by possible environment impact on TIOp elements at ASF. The outcome in the example is characterized by set, constant effects required only for C_i termination.

Required informative and non-informative effects [2] are understood to be primary results of an action, both target and providing (resources, time). The tree is constructed at the stage of planning depending on the following: outcomes of environment operation C_u^∂ (of which there are several, in general), possible TIO set in relation to the applied h – IT and boundary states; relations between TIO and TNIO depending on the used technology; TAS changing states as a result of environment impact.

$p_{i.k} = Poss(\tilde{E}_{i.k})$, where $Poss(\tilde{E}_{i.k})$ – is a possibility measure of a random event $\tilde{E}_{i.k}$;

The possibility measure has a broad definition here [4] as one the measures (probability, indistinct measure, other types of measures and their combinations).

$\tilde{E}_{i.k}$ – a random event that involves that the k – outcome (transition $e_{i.k}$ between states) at event implementation a_b is accomplished accordingly.

As a result of each environment operation outcome C_u^∂ at TAS boundary and each TAS operation outcome C_i with application of one IT or the other, a different alignment is formed (at boundary of TAS and its environment). This alignment should be checked for the best choice of IT. Assume:

$\tilde{Y}_{j.i}$ – value j – of operation effect to the end of C_i; $\tilde{Y}_{j.i} = f(C_i)$. For instance, labor content of service, power costs.

$\tilde{Y}_{j.i}^\partial$ – requirements from the environment to j – operation effect to the end of C_i; $\tilde{Y}_{j.i}^\partial = g(C_i, C_u^\partial)$.

$p_i(C_i, C_u^o) = \prod_{k: e_{i.k} \in C_i(C_u^o)} p_{i.k}(C_i)$ – possibility measure of implementation of TAS functioning outcome C_i providing the outcome at boundary between environment and TAS is C_u^o; Then:

$W_i(C_i, C_u^o) = \prod_{j=1,J} Poss(\tilde{Y}_{j.i}(C_i) \lesssim \tilde{Y}^o_{j.i}(C_i, C_u^o))$ – possibility measure of meeting the requirements at TAS functioning outcome C_i and environment operation outcome C_u^o (TAS performance measurement at preset outcomes of TAS and its environment operation). Results of TIOp package should be estimated with the account of all possible functional events at different environment impact and during different TIOp (TIO implemented according to preset IT and TNIO, which are carried out with regard to preset non-informative technology). Scalar characteristic of TAS potential meets this requirement, potential function [2] when h IT is applied, calculated across all environment operation outcomes C_u^o, with regard to possibility measures p_u of their implementation:

$$\psi(T_c, h) = \sum_{C_u \in T_c} [\sum_{C_i \in T_{c.u}(C_u^o, h)} W_i(C_i, C_u^o) \cdot p_i(C_i, C_u^o)] \cdot p_u .$$

Assume there are two IT, h ("new") and f ("old"). For them $c-$ (for $h-$IT) and $p-$ (for $f-$IT) trees of outcome for TAS operation were constructed $T_{c.u}(h)$ and $T_{p.u}(f)$ for every possible outcome C_u^o of environment operation at boundary with TAS from the tree of environment operation outcome T_c (at fixed outcome C_u^o TAS functioning outcomes form trees $T_{c.u}(h)$, $T_{p.u}(f)$ of TAS operation outcomes when $h-$and $f-$IT are applied). Let us define the difference between TAS potential function values when old and new IT are applied. $\Delta\psi(h, f, T_c) = \psi(T_c, h) - \psi(T_c, f)$.

This difference defines OP of the "new" IT application as compared to the "old" one given the execution of the same outcomes C_u^o of environment operation at the TAS boundary with the same measures of possibility p_u, according to the preset T_c. The difference defines the target effect of new IT application. The target effect reflects how suitable the IT is to reach changing targets (according to the outcomes of environment operation at boundary with TAS). The described target effect is to be compared with effects that ensure the target effect is reached (at IT application). This should be done using efficiency factors of the new IT application.

Efficiency factor of the new IT application could be introduced traditionally [3] as a possibility measure evaluating if the characteristics of the effects meet the requirements. The factor should be estimated in the context of possible outcomes of environment operation C_u^o at TAS boundary with the account of value of potential $\psi(T_c, h)$ function for h IT. If it measures positively meaning the new IT should be implemented then the result should be used as a performance indicator of h IT as compared with the previously used IT. For further calculation of efficiency factor of h IT implementation, the agility of h IT implementation and resource intensity of h IT implementation should be evaluated as well.

5. References

[1] A. Geida, I. Lysenko, and R. Yusupov, *Main concepts and principles for information technologies operational properties research*, *Tr. SPIIRAN,* **42,** 5–36 (2015)

[2] A. Geida and I. Lysenko, *Operational properties of agile systems and their functioning investigation problems: conceptual aspects*, J. of Appl. Informatics, **12,** 93-106 (2017)

[3] A. Geyda and I. Lysenko, *Schemas for the analytical estimation of the operational properties of agile systems, SHS Web Conf.* **35**, 01058 (2017)

[4] E. Zio and N. Pedroni, *Literature review of methods for representing uncertainty* (Toulouse, France: FONCSI), 61 (2013)

Acknowledgments
Performed under support of the RFBR grant No. 16-08-00953

Methodical uncertainty in the two-point measurement of parts prismatic elements linear dimensions

V I Glukhov, V V Shalay

Omsk State Technical University, 11, Mira ave., Omsk, 644050, Russia

Abstract. The article is devoted to the topical problem of the reliability of measurements of geometrical specifications of technical products. The purpose of this article is to improve the quality control of the linear dimensions of the parts by the two-point measurement method. The goal of the article is the study of advanced methodical uncertainties of measurement of linear dimensions of plane-parallel prismatic elements. The method of research is geometric modeling of deviations of the element surfaces form and position in the rectangular coordinate system. The research was carried out for various executive elements with regard to their information content corresponding to the classes of kinematic pairs in the theory of mechanics and the number of constrained degrees of freedom in function of the basic elements. Prismatic elements with informative content 3, 2, 1 and θ (zero) were studied. Uncertainty assessment of two-point measurements were made by comparing the measurements results of linear dimensions and functional dimensions of the element maximum and minimum material constrained by the limits of maximum and minimum. As a result of studies of the two-point measurements accuracy, it is found that the methodical uncertainty is formed when measuring the average size of elements for all types of form deviations. The two-point measurement method cannot account for deviations of the location dimension of the element and therefore its use for items with less than the maximum informativeness creates unacceptably large methodical uncertainties of measurement the maximum, minimum and medium linear sizes. The same methodical uncertainties occur under the arbitration control of the parts linear dimensions by the limit two-point gauges.

Keywords – Technical measurements, linear sizes, elements informativeness, methodical uncertainty.

1. Introduction

The International Organization for Standardization technical Committee TC213 “Dimensional and geometrical product specifications and their control” updated all terminology to linear and angular values at the level of international standards [1-5] in the first decade of the 21st century without convincing justification. Millions of engineers – users of standards in all processes of the product life cycle – designers, technologists, metrologists, control masters, university and college teachers found themselves in a difficult position. New terms and definitions did not specify or improve old terms that served for 25-50 years. The new terms abolished the old engineering ones and replaced them by more complex terms [6].

Still, between the new and old terminology there must be continuity in the development and improvement of terms, including on the following unresolved problems:

– there is no classification of geometric specifications for a functional purpose, coordinating dimensions of the elements of parts are not introduced; geometric and kinematic specifications of parts connections are not analyzed;
– coordinate systems for measuring the geometrical specifications of parts, parts elements, and assembled products are not applied; despite the fact that the design of electronic products in the computer-aided design is carried out in a three-dimensional rectangular 3d coordinate systems, the production of components of products provided by multi-axis CNC machines and the control - by coordinate measuring machines;
– the engineering principles for ensuring the quality of products in terms of the geometric characteristics accuracy are not used: the principle of inversion, the principle of the unity of design, technological and measuring datums, the principle of linear-angular two-dimensionality of geometric specifications, and the principle of geometric elements informativeness [7];
– systematic functional deviations of the component surfaces shape parts are considered as random variables and are estimated by statistical characteristics.

2. Statement of the problem

The possibility of reliable measurements of the size and shape of the parts is essential for the precise products manufacturers [8]. Know-how – to know how to measure – is a basis of knowledge about the measurements required by a designer, a technologist and a metrologist [9]. The evaluation of the uncertainty in measuring geometric characteristics is a prerequisite for products quality control [10].
Technical measurements make it possible to evaluate the part quality by the accuracy of geometric specifications in the production process. The quality evaluation is carried out by controlling the dimensional and geometric accuracy of the component parts to meet the requirements of the design project. The transition of international GPS standards to the evaluaion of the dimensional accuracy of parts based on the average size of elements using the two-point method requires estimation of the measurements uncertainty, which is the aim of this work.
The specific problem of evaluating the influence of form and an joints deviations of cylindrical dimensional elements surfaces of various informative content is stated and solved on the expanded uncertainty of measurements of average, maximum and minimum linear sizes of elements by the two-point method.

3. The theory of elements functional linear sizes

Any part of a technical product consists of geometric elements, which are parts of the material of the part, constrained by one or several geometric surfaces, primarily flat and cylindrical. Every part element performs a specific function. The set of main design datums performs the functions of positioning the part in the product and identifying its generalized coordinate system[7]. sliding joints, the set of main datums performs the third function - it provides the kinematics of rotation and (or) the translational movement of the product part with the fit gap. The set of auxiliary design datums performs the functions of positioning the attached part to the considered one and identifying the auxiliary coordinate system for the attached part. The actuating elements perform the working functions of the part. Dimensional elements determine the overall dimensions of the part, and free elements with dimensions without tolerances perform the functions of connecting elements that unite all the geometric elements into a single solid body of the part.
The geometric element datum may constrain the detail one to six degrees of freedom (up to three linear "t" and three angular "r") that in total is element informativeness specification "c" and coincides with the class of kinematic pairs in theoretical mechanics. Maximum informativeness of the flat datum of the prismatic element is equal to three (3c) and consists of a single linear constraint (1t) normal datum-wise and two angular (2r) – around the two perpendicular directions in the plane of the datum (3c=1t+2r).
Datums are combined into sets of datums to constrain the part of the six degrees of freedom (6c = 3t + 3r) and form a generalized or auxiliary coordinate system in which the dimensions and coordinates of the elements should be measured. The parts are joined on sets of datums in order to combine the

generalized coordinate system of the attached part with the auxiliary coordinate system of the base part.

The surfaces of all geometric elements have shape deviations in the form of alternating smooth harmonic protrusions and cavities. In addition, elements with informative content less than the maximum have protrusions and cavities of surfaces in the form of location deviations. Surface protrusions form the elements datums. Fits in the joints of the shafts and holes are formed by protrusions of the elements real surfaces. C Consequently, the dimensions of the elements along the projections of real surfaces are the functional maximum material dimensions of the elements.

The cavities of the elements real surfaces constrain the shape and location deviations, and together with the micro-roughness, they are the containers for the joint lubrication. Therefore, the size of the element along the cavities of the real surface is the functional minimum material size.

Thus, each real dimension of the geometric element has two boundary functional sizes: maximum material size and minimum material size. According to the principle of two-dimensionality of linear-angular values [7], these are the sizes of two regular (ideal) elements of the same type as the dimensional element, nominally located relative to the coordinate system of the datums set, in which the dimensional element fulfills its executive function, and covering the real surfaces of the dimension element with a tangent: the first - from the material element - forms the maximum material size; the second - from the cavities of the element material - forms the minimum material size.

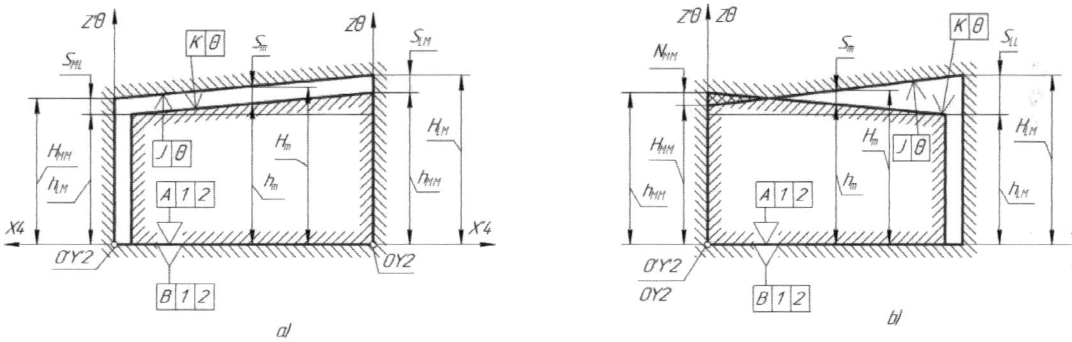

Figure 1. The model of the interference formation (b) along the height maximum material sizes in the fit with a gap (a) along the average sizes of the prismatic hole and shaft with informativeness 3

The sizes of the maximum and minimum material are constrained by the upper and lower limits of the element size. Their reliable measurement guarantees the quality of the connections in the assembly by ensuring the required fits.

The average element size should be calculated as half the sum of the sizes of the maximum and minimum material. However, the connection with the gap between the average sizes of the height of the hole and shaft Hm and shaft hm (Fig.1,a) can only be provided by positive differences of the hole minimum material size HLM and the shaft maximum material size hMM that can take place in a single phase distortions of the executive surfaces J and K opposite to the elements datums A and B, but leads to the undesirable mismatch of the coordinate systems of the hole and the shaft. When there are anti-phase distortions of the executive surfaces (Fig.1,b), in the connection interference NMM can be formed between the dimensions of maximum material of holes height NMM and shaft hMM that will preclude forward movement of the prismatic shaft.

Hereinafter, note that the maximum material size of the element is the largest size of the element (shaft) external dimension and the minimum size of the internal dimension element (holes). The minimum material size of the element is the smallest size of the external dimension element (shaft)

and the maximum dimension size of the internal element (holes). It is for these functional sizes a system of tolerances and fits is developed [2].
According to the new terminology [4], the element maximum material size is the minimum described size of the outer element and the maximum inscribed size of the inner element. The minimum material size is the maximum inscribed size of the outer element and the minimum described size of the inner element. Fitting in the compound is formed by the maximum inscribed size of the internal element (hole) and minimum described size of the outer element (shaft).

4. Prismatic elements dimensions measurement uncertainty

Dimensional prismatic elements constrain a part of the material by two nominally parallel planes, with informativeness 3, 2, 1 and θ (zero), and form, respectively, the dimensions of height, width, length and overall size of the element.
The outer prismatic element with the maximum informative value 3 (3s=2t+1r) has a flat datum D3 (Fig.2) and the executive plane Jθ with zero informativeness, the distance between which form the size of the height h. Deviations of the element plane form are the following: a V-shape (a), a dual convexity (b) a dual concavity (c) and an incurvation (d) form two boundary height sizes – maximum height size hMM and the size of the minimum height hLM. These dimensions are respectively equal to the sizes of the two right prismatic elements nominally positioned relative to the datum D3 of the dimension element, and covering with tangent the actual surface of the element. The average element size hm is equal to the half-sum of the height maximum h MM and minimum hLM sizes.
Measurements of maximum hMM and minimum hLM height of the prismatic element of the two-point method is carried out without methodical uncertainties in the deviation of the shape of flat surfaces such as in-curvature, dual convexity and the dual concavity. However, the estimatoion of the average size of the height hm can differ from the maximum material size and minimum material size and will therefore have methodical uncertainties ULM and UMM, numerically equal to half the deviation of the shape of each real flat surface.

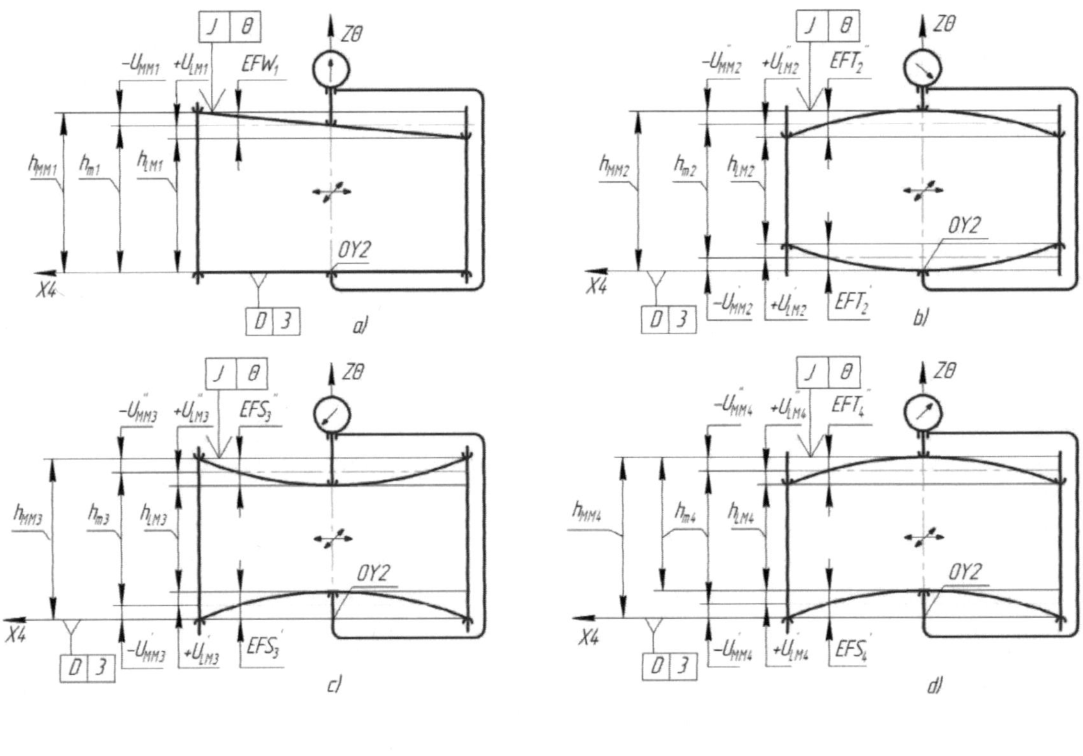

Figure 2. The model of the measurements uncertainties when applying the two-point method of the average height outer prismatic element with informativeness 3

If the prismatic element is curve (Fig.2,d), D3 base surface has concavity EFS while the executive surface - convexity EFT. The two-point measurement method in this case allows to measure only the average size hm of the element height with uncertainties UMM and ULM, which are numerically equal to the sum of deviations of the two flat surfaces form . The sizes of the maximum and minimum material height are in this case measured by using two-point measurement method, thus, the evaluation of boundary sizes on the average size, their uncertainty is equal to the uncertainty of the average element size.

Prismatic elements with informative content 2 (Fig.3,a) have the dimensions of maximum and minimum wMM and wLM of width between the flat datum B2 with informative content 2 and executive nominally parallel plane Dθ. Datum B2 and executive surface Dθ have deviations of perpendicularity relative to datum A3. Datum B2 and executive surface Dθ have form deviations of flatness type with deviations from perpendicularity EPE form 2 width sizes - the size of the maximum material of width wMM and minimum material size of width wLM as the sizes of the two correct prismatic elements nominally positioned relative to the datum A3 and covering with tangent the actual surfaces of the dimensional element: the first – out of the element material, the second from the element material . The average size of the element widthwwm is the half-sum of the maximum and minimum material sizes of the element width.

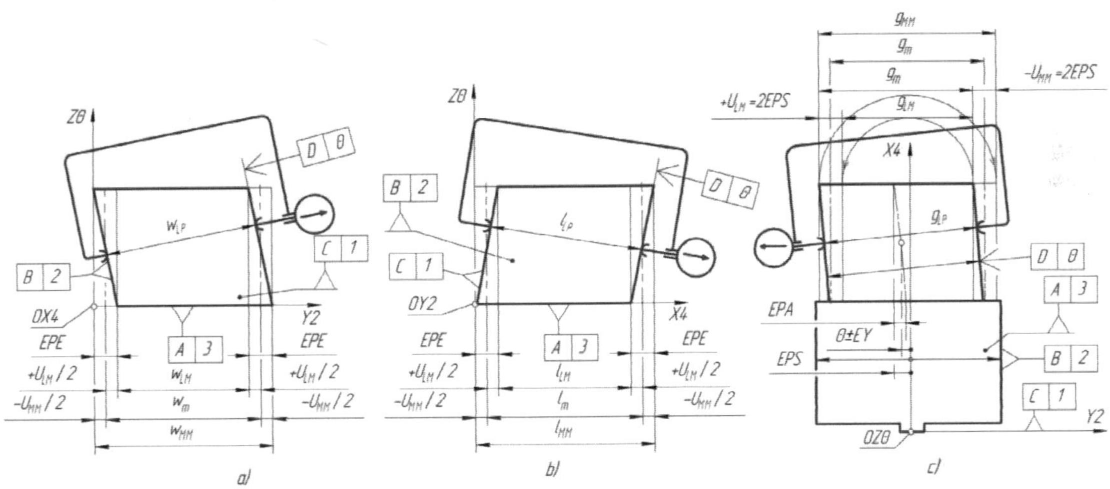

Figure 3. Uncertainty measurement model of the two-point method of average widths (a) and length (c) of prismatic elements with informativeness 2, 1 and θ (zero)

The two-point method can only measure the average size of width wm with the methodical uncertainty equal to the half-sum of the deviations from perpendicularity EPE of the elements flat surfaces taking into account the form deviations of the surfaces (Fig. 2). The sizes of the maximum and minimum widths cannot be measured by the two-point method. Attributing a medium size to them gives the boundary widths the uncertainties of the average size U LM and U MM.

A similar approach to the dimensions measurements uncertainties of length l from a flat datum with informativeness 1 is shown in Fig.3,b.

Prismatic elements with zero informativeness (θ) perform an executive function of the executive or dimension elements, which form fittings with a gap in the connections of details or dimensions of parts (Fig.3 c).

Prismatic elements with the informativeness θ do not participate in the formation of coordinate system OX4Y2Zθ of the datums set of part A3B2C1. The position of the element in the coordinate system is

IOP Conf. Series: Journal of Physics: Conf. Series **1050** (2018) 012028 doi:10.1088/1742-6596/1050/1/012028

defined, generally, by using the center of the element symmetry plane by the linear coordinate θ±EY and two angular coordinates θ ° ± AEZ and 90 °±AEX symmetry plane.
The primary position deviation symmetry EPS and parallelism EPA in relation to the datum B2 and perpendicularity relative to the datum A3 (not shown) of the prismatic element with informativeness θ influence the functional size of the element – increase the maximum material size gMM and reduce the minimum material size g LM.
These dimensions ensure the required clearance fit when assembling parts. However, the two-point measurement method cannot account for deviations of the location of the dimension element in the part coordinate system.

5. Results and discussion
The results of the modeling uncertainties were verified through experimental measurements of the precision products parts prismatic elements linear dimensions: hydraulic motors, bearings, gear and vane pumps, and compressors. As a result of studies of the accuracy of two-point measurements of the linear dimensions it is established that methodical uncertainties do not exist only for the maximum and minimum sizes at maximum informativeness 3 of prismatic elements having shape deviations of the wedge-shape type, double convexity or concavity. Methodological uncertainty is formed while measuring the boundary functional linear dimensions of the prismatic elements with maximum informativeness at the deviation of the shape of the incurvation type. The methodical uncertainty is formed when measuring the average size of elements for all types of geometric tolerances. The two-point measurement method cannot account for deviations of the location dimension of the element and therefore its use for items with less than the maximum informativeness creates unacceptably large methodical uncertainty of measuring the maximum, minimum and medium linear sizes. The same methodical uncertainties occur under the arbitration control of the parts linear dimensions by the limit two-point gauges. The results of the research cover the outer and inner prismatic elements.

6. Conclusion
1. Each dimensional geometric element, constrain ed by a real surface, has two dimensions: two permissible size covers for the normalization of the accuracy class and two functional sizes for the operation of the part in the product: the maximum material size and the minimum material size.
2. According to the principle of sizes two-dimensionality, the size of the material maximum and the size of the material minimum of the dimensional element are the dimensions of two regular (ideal) elements of the same type as the dimension element, nominally located relative to the coordinate system in which the dimensional element fulfills the executive purpose, with the tangentially covering the real surface of the dimensional element: the first one - outside the element material - materializes the size of the material maximum, the second - from the element material - materializes the size of the material minimum.
To implement these definitions, when measuring two sizes of an element, we need a coordinate system in which the dimensional element operates, and knowing the base informativeness of an element, i.e. the number of degrees of freedom constrained by the element datum.
4. Measurement of local linear sizes according to ISO standards by the two-point method is carried out by the shortest distance between opposite points of a dimensional element real surface in a set of pairs of points. By the processing of the measurement results the maximum size, the minimum size and the average size, which is preferred as an ordered one and having the biggest methodical uncertainty, are found.
5. Scientific novelty of the work is the proof of the existence of two functional actual linear dimensions of each of the prismatic elements of the shafts and the holes forming the fitting – the maximum material size and minimum material size. The character of fitting in the connection is determined by the dimensions of the mating elements maximum material. The average sizes are only statistical specifications of prismatic elements real dimensions.

6. The greatest influence on the uncertainty of two-point measurements of prismatic elements linear sizes with maximum informativeness 3 has the elements form deviation of an incurvature (straightness) type.
7. The uncertainty in the two-point measurements of the linear dimensions of the prismatic elements having the informative value of the maximum deviation impact location of the type of parallelism, perpendicularity and symmetry bases of the measured elements, which leads to unacceptably large values of uncertainties and output sizes for the intervals of tolerances.
8. Only one way for solving this problem is offered: the transition of the ISO standards to the functional approach to geometrical specifications of the products based on the principle of two-dimensional size classification of elements in terms of information, with the introduction of coordinate systems to reference specifications, and with the inclusion of these applications in the strategic plan of the standards updating [5].

References

[1] ISO 8015:2011, Geometrical product specifications (GPS)– Fundamentals–Concepts, principles and rules
[2] ISO 286-1:2010, Geometrical product specifications (GPS) – ISO code system for tolerances on linear sizes – Part 1: Basis of tolerances, deviations and fits
[3] ISO 17450-1:2011,Geometrical product specifications (GPS) – General concepts- Part 1: Model for geometrical specifications and verification
[4] ISO 14405-1:2010, Geometrical product specifications (GPS) – Dimensional tolerancing. Part 1: Linear sizes
[5] Nielsen H S 2013 *Recent developments in International Organization for Standardization geometrical product specification standards and strategic plans for future work* J. Eng. Manufact. **227** pp 643-649
[6] Wiesner H ISO 199:2014 and ISO 492:2014 standards – complexity versus unambiguity J. SKF Evolution l pp 27-30
[7] Glukhov V I *Geometrical product specifications: Alternative standardization principles, coordinate systems, models, classification and verification* 2014 Dynamics of Systems, Mechanisms and Machines (Dynamics) (IEEE Conf. Pub.) pp 1-9
[8] Saunders P, Gludice S and Swart J 2014 *Identifying measurement knowledge and its relationship to engineering design* Int. J. Metrol. Qual. Eng. **5** pp 203-212
[9] Maropoulos P, Ceglarek D, 2010 *Design verification and validation in product lifecycle* CIRP Annal. Manufact. Technol. **59** pp 740-759
[10] Vetturi D, Lancini M, Bodini I and Pasinetti S 2013 *Relationship between measurement uncertainty and verifiability of geometric specifications: the case study of drilled hole axis orthogonality* Int. J.Metrol. Qual. Eng. **4** pp 35-39

IOP Conf. Series: Journal of Physics: Conf. Series **1050** (2018) 012029 doi:10.1088/1742-6596/1050/1/012029

Applicability of Kramers rate formulas in the energy diffusion regime

I I Gontchar[1,2], M V Chushnyakova[2,3], A I Blesman[2]

[1] Physics and Chemistry Department, Omsk State Transport University, Omsk 644046, Russia
[2] Physics Department, Omsk State Technical University, Omsk 644050, Russia
[3] Physics Department, Akdeniz University, Antalya 07058, Turkey

Abstract. The work of nanomachines, which is a contemporary subject of the numerous investigations, is permanently affected by thermal fluctuations. Some aspects of this phenomenon might be reduced to the so-called Kramers problem concerning the decay rate of a metastable state. In the present work we compare the result of computer modelling, performed by solving the Langevin equations, with the analytical formulas for the decay rate for the case of energy diffusion regime (relatively small values of the friction strength). Qualitatively the agreement between these two approaches is rather good. However, the quantitative analysis reveals a difference of 10-20% which depends upon the thermal energy of the system.

1. Introduction

Nowadays the work of nanomachines is under extensive experimental study [1-3]. For these objects, the thermal fluctuations driving Brownian motion play an important role. Some challenges in this field are related to the so-called Kramers problem [4-6] concerning the quasistationary thermal decay rate (QDR) of a metastable state. The QDR depends significantly upon the friction strength η: at larger values of η (the spatial diffusion regime) the rate decreases with η whereas at smaller values of friction (the energy diffusion regime) opposite is predicted. The approximate formulas for the QDR obtained in [4] and modified in [7, 8] were checked thoroughly by comparing with the numerical modeling for the spatial diffusion regime [9-11]. Yet for the energy diffusion regime no quantitative tests of the approximate formulas are known. This is the goal of the present study to perform these tests.

2. Model

For this aim we model the decay process using the Langevin equations. Since in [4, 7] only the one-dimensional (1D) analytical formula was derived, we perform our modeling for 1D too. The Brownian motion is described by a dimensionless coordinate q and the conjugate momentum p. In the discrete form the equations used for modeling read:

$$p^{(n+1)} = p^{(n)}(1 - \eta m^{-1}\tau) + K\tau + gb^{(n)}\sqrt{\tau}, \quad (1)$$

$$q^{(n+1)} = q^{(n)} + \left(p^{(n)} + p^{(n+1)}\right)\tau/(2m). \quad (2)$$

The superscripts represent two moments of time separated by the time step of numerical modeling τ. The random numbers b entering the random forces have a Gaussian distribution with zero average and variance equal to 2. In equation (1) m and η are the coordinate independent inertia and friction parameters, respectively; $K = -dU/dq$ is the driving force; $g = (\theta\eta)^{1/2}$ is the amplitude of the random force; θ stands for the average thermal energy. The potential $U(q)$ is represented by two parabolas of the same stiffness C smoothly jointed at q_m.

The modeling results in a sequence of N_{tot} trajectories terminated not later than at the time moment t_D. Some of the trajectories reach the absorptive border $q_a = 2$ before t_D. The algorithm for calculating the QDR resulting from the dynamical modeling, R_D, is described in Ref. [9].

Let us now discuss the approximate formula for the QDR when there is almost no dissipation in one bounce. This formula was obtained in [4]. It reads

$$R_{KL} = \gamma \frac{\omega}{2\pi} \exp\left(-\frac{U_b}{\theta}\right). \quad (3)$$

Here ω is the frequency of oscillations near the parabolic bottom of the potential well; U_b is the barrier height.

$$\gamma = \frac{I_b \eta}{\theta m}; \quad (4)$$

I_b denotes the classical non-dissipative action at the collective energy equal to U_b:

$$I_b = 2\int_{q_l}^{q_b} \sqrt{2m[U_b - U(q)]} dq. \quad (5)$$

It is evaluated from the left turning point (q_l) up to the location of the top of the barrier (q_b). In our case $I_b = 2\pi \cdot 1.07 \cdot U_b/\omega$. Equation (3) is supposed to be valid provided $\gamma < 1$.

In [7] a modification of equation (3) was proposed allowing a smooth transition from the energy diffusion regime to the spatial diffusion regime. Denoting this modified rate as R_{KLB} we write it as

$$R_{KLB} = \frac{\delta - 1}{\delta + 1} R_{KL}. \quad (6)$$

Here

$$\delta = \left(1 + \frac{4\alpha}{\gamma}\right)^{1/2}; \quad (7)$$

α is a dimensionless adjustable parameter of the order of unity.

3. Results

In figure 1 we compare the approximate rate R_{KLB} with the numerical one R_D in the wide range of the friction strength in a qualitative manner. For convenience of a wider audience, the rates and damping coefficient are given in units of ω ($\varphi = \eta(m\omega)^{-1}$). These calculations are performed for three values of the controlling parameter $\varepsilon = U_b\theta^{-1}$. The values of R_{KLB} are calculated using $\alpha = 1.3$. In figure 1 one sees good qualitative agreement of the approximate rates with the numerical ones for all values of ε and φ. Note, that the numerical rate R_D is a subject of statistical errors which are between 1% and 2% in our calculations.

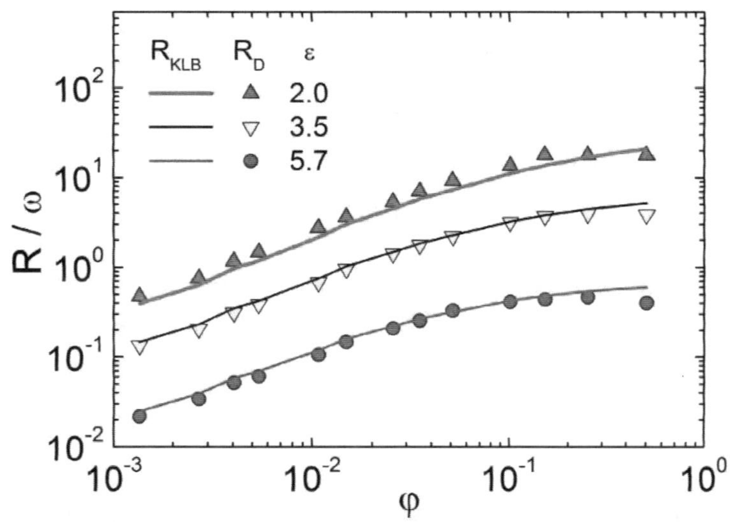

Figure 1. The decay rates R_{KLB} and R_D versus the dimensionless friction strength φ for three values of controlling parameter ε indicated in the figure.

In order to consider the problem in a more quantitative manner, in figure 2 we present the fractional difference

$$\xi = \frac{R_{KLB}}{R_D} - 1. \tag{8}$$

for the same values of the controlling parameter ε as in figure 1. Analyzing the values of ξ is more convenient because the rates themselves cover several orders of magnitude whereas the deviation of the approximate rate from the dynamical one is at most several tens percent.

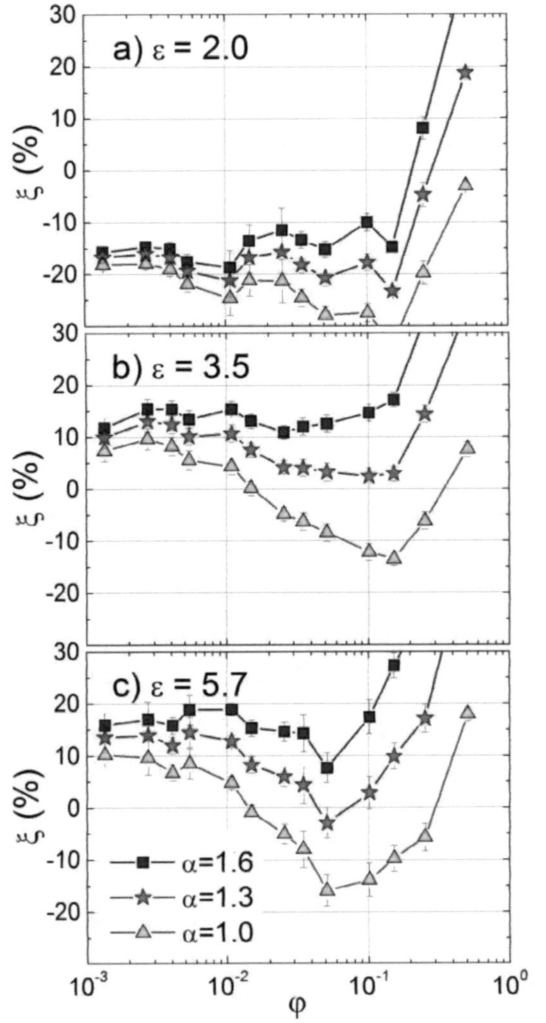

Figure 2. The fractional differences between R_{KLB} and R_D as functions of the dimensionless friction strength φ for three values of controlling parameter ε and adjustable parameter α (both are indicated in the figure).

First of all, we note in in figure 2 that no value of fitting parameter α provides an ideal agreement of R_{KLB} with R_D, i.e. the value of ξ never stays around zero within the error bars. In each panel the values of ξ corresponding to different values of α converge as φ decreases. This is expected from equations (6), (7). Another interesting feature observed in each panel is that the fractional difference ξ saturate at extremely weak friction. The saturated values are different depending upon ε. Generally speaking, the Kramers approach is expected to become poorer as ε decreases (see details in [4, 10, 12]): the larger ε the larger excess of R_D over the Kramers rate, i.e. the smaller ξ. Results presented in figure 2c seem corresponding to this consideration. However, the value of $\xi = 10\%$ at $\varepsilon > 3$ in figures 2a and 2b is unexpected. It can be related to the circumstance that the Kramers formula (3) is obtained using the "Fokker-Planck" equation for the action variable and is approximate by itself whereas we model the decay process using the Langevin equations which are equivalent to the Fokker-Plank equation for the phase-space density.

4. Conclusions

In the present work we have compared an approximate analytical formula for the thermal decay rate in the energy diffusion regime, i. e. at small values of friction, with the exact quasistationary numerical rate R_D. The latter was obtained modeling the decay process by solving the Langevin equations. Our calculations demonstrate good qualitative agreement between the approximate rate R_{KLB} and R_D.

The quantitative comparison between these two rates shows that there is no way to reach an ideal agreement of R_{KLB} with R_D within the statistical errors of the latter which do not exceed 2%. Moreover, at extremely weak friction an approximate approach of [4] overestimates the numerical (exact) rate by 10% when all conditions of applicability of this approach are fulfilled (including the controlling parameter $\varepsilon \gg 1$). As ε becomes comparable to unity, the numerical rate exceeds the approximate one. This observation agrees with the general expectation.

Acknowledgments

This work was partly supported by the Scientific and Technological Research Council of Turkey (TUBITAK) within the framework of Fellowship Program 2221– for Visiting Scientists and Scientists on Sabbatical Leave.

References

[1] Seifert U 2012 Stochastic thermodynamics, fluctuation theorems and molecular machines *Rep. Prog. Phys.* **75** 126001
[2] Howard J 2001 *Mechanics of Motor Proteins and the Cytoskeleton* (Sunderland, MA: Sinauer Associates) 119-263.
[3] Blickle V and Bechinger C 2012 Realization of a micrometre-sized stochastic heat engine *Nature Phys.* **8** 143-6
[4] Kramers H A 1940 Brownian motion in a field of force and the diffusion model of chemical reactions *Physica* **7** 284-304
[5] Talkner P and Hanggi P 2012 *New Trends in Kramers' Reaction Rate Theory* (Berlin: Springer Science & Business Media).
[6] Dudko O K, Hummer G and Szabo A 2006 *Phys. Rev. Lett.* **96** 108101
[7] Büttiker M, Harris E P and Landauer R 1983 Thermal activation in extremely underdamped Josephson-junction circuits *Phys. Rev.* B **28** 1268
[8] Jing-Shang Z and Weidenmüller H A 1983 Generalization of Kramers's formula: Fission over a multidimensional potential barrier *Phys. Rev.* C **28** 2190
[9] Gontchar I I and Chushnyakova M V 2017 Thermal decay rate of a metastable state with two degrees of freedom: Dynamical modelling versus approximate analytical formula *Pramana – J. Phys.* **88** 90
[10] Gontchar I I et al 2010 Disentangling effects of potential shape in the fission rate of heated nuclei *Phys. Rev.* C **82** 064606
[11] Gontchar I I and Kosenko G I 1991 Is the Kramers formula applicable for describing the decay of highly excited nuclear systems? *Yad. Fiz.* **53** 133-42
[12] Abkenar M, Gray T H and Zaccone A 2017 Dissociation rates from single-molecule pulling experiments under large thermal fluctuations or large applied force *Phys. Rev.* E **95** 042413

Software for energy efficient control system over process parameters of MIMO objects on a set of functioning states

A N Gribkov[1], D Yu Muromtsev[1], V N Shamkin[1], O A Belousov[1], M P Belyaev[2]

[1]Tambov State Technical University, 106, Sovetskaya str., Tambov, 392000, Russia
[2] The Zhukovsky-Gagarin Air Force Academy, 133, Marshala Nedelina ul, Voronezh, 394052, Russia

E-mail: mdjur@mail.ru

Abstract. Theoretical and practical aspects of building software architecture for energy efficient control of MIMO systems on a set of functioning states are considered. The control systems are classified depending on the nature of changes in the functioning state variable and the possibility of its identification on the control interval. The specifics of software implementation for the systems of various classes are shown. The software is structured with the allocation of functional subsystems and modules. The generalized software architecture for energy efficient control systems on a set of functioning states is proposed.

1. Introduction

Currently, energy efficient control systems (EECS) are used in virtually all industries. The use of the EECSs not only to significantly reduces energy costs and increases the productivity of process units, but also ensures the required quality of products. In this regard, issues related to the development of such systems are very relevant and attract considerable attention of researchers.

From the perspective of automation, most modern industrial installations are complex MIMO systems that have a lot of interrelated input and output variables [1, 2]. Such systems include multipurpose vehicles in which phase composition changes, structural transitions, etc. occur. One of the most widespread and energy-consuming MIMO systems are heat engineering equipment, for example, drying plants, industrial furnaces, boiler plants, etc.

At present, a great deal of papers and monographs on the problems of designing automation and control systems have been published. The works of many scientists working in this subject area are known (e.g. L.S. Pontryagin, V.G. Boltyansky, RV Gamkrelidze, E.F. Mishchenko, R. Bellman, A.M. Letov, A.A. Krassovsky, A.M. Lyapunov, M. Atans, P. Falb., A. Hurwitz, H. Nyquist, A.V. Mikhailov, V.A. Trapeznikov, N.N. Krasovsky, V.A. Besekersky, A.G. Aleksandrov, V.V. Solodovnikov, A.A. Voronov, K.A. Pupkov, N.D. Egupov, V.I. Zubov, V.S. Pugachev, B.T. Polyak, I.P. Norenkov, G.K. Goodwin, S.F. Grebe, M.E. Salgado, J. Lanning, R. Betting, D. Newton, L. Gould, D. Kaiser, A.A. Feldbaum, Yu.L. Muromtsev, A.M. Tsykunov, E.A. Parsheva, E.L. Eremin, A.Z. Asanov, etc.)

Many Russian and foreign scientists are working on the problem of building intelligent control systems. This is due to a number of reasons, the most important of which are: the continuous complication of control objects; the emergence of high-performance computer facilities; a significant increase in the requirements for the accuracy, reliability, and efficiency of control algorithms under uncertainty; the need for synthesis of optimal control actions in real time; the complexity of obtaining an exact mathematical model of the object, etc.

2. Statement of the problem

The process of developing the ECCS for MIMO systems is quite laborious and knowledge-intensive both at the stage of creating the algorithmic support of the system and at the stage of software implementation of the created algorithms [3]. This is because the development of the algorithmic support of the system requires a rather complex mathematical apparatus based on the methods of optimal control theory, decision-making under uncertainty, artificial intelligence, etc. In turn, the algorithmic complexity of the system leads to a complex software architecture of the EECS, which must ensure the possibility for a quickly control over the process system in real time.

When designing the software for the EECS, it is necessary to take into account that in the course of its further real operation, changes in the parameters of the technological process, environment, intensity of disturbing effects, etc., might occur. These changes can have a significant impact on the accomplishment of the goal of control, so the system must take them into account and react quickly to them.

In this connection, the task of designing the software architecture of energy efficient control for MIMO systems, taking into account the possible changes in the functioning states in the process of real operation, is quite relevant.

3. Theoretical grounds

The effectiveness of any system is determined by its properties (reliability, accuracy, speed, etc.), working conditions, the influence of external disturbances and other factors that can be either deterministic or random. For the complex accounting of the main factors affecting the operation of the EECS, the concept of a set of functioning states (SFS) is introduced [4].

When considering the EECS on the SFS, a variable h is used to indicate the current state of functioning. In this case $h \in \boldsymbol{H}$, where $\boldsymbol{H}$ is a SFS, i.e. a set of possible values of the variable h. The change in the value of h can occur at certain or random times. The factors influencing the change in the value of the variable h may include the changes in the ambient temperature, atmospheric pressure, humidity, process parameters, and other factors that may affect the system efficiency and the accomplishment of the control goal. In general, the SFS is determined by the design features of the EECS for a particular process facility. Depending on the nature of changes in the variable h and the possibility of its identification on the control interval, four main classes of systems in the SFS are distinguished [4].

The Class 1 EECSs on the SFS are characterized by the fact that the value of the state variable h is known before the start of the control and it remains constant on the control interval. An example of the first class system can be a reliable control system for simple batch process plants, in which changes in the functioning state occur outside the control interval, while the interval itself is usually short.

The Class 2 EECSs on the SFS are characterized by the fact that the value of the state variable h before the start of control is unknown and does not change over the control interval. In this case, the value h is also unknown on the control interval, but probabilities of possible values of the variable h may be known. The examples of systems of the second class are stochastic systems and systems with differential inclusions.

The Class 3 EECSs on the SFS are characterized by the fact that the value of the state variable h before the start of control is known and it can change over control interval, with the value of h at each time instant being identifiable with high accuracy. Dynamic modes of the third class systems are conveniently described with the help of "multistage" models in the form of differential controls systems with a discontinuous right-hand side.

The Class 4 EECSs on the SFS are characterized by the fact that the value of the state variable h before the start of control can be either known or unknown and it can change over control interval, but it cannot be accurately identified. When designing systems of the fourth class, their algorithms can include models that make it possible to simulate possible trajectories of changing the state variable h in the course of actual operation.

IOP Conf. Series: Journal of Physics: Conf. Series **1050** (2018) 012030 doi:10.1088/1742-6596/1050/1/012030

An important condition for the effective functioning of Classes 1-4 systems is the fulfillment of inclusion conditions. It should also be noted that the above classification of the EECS on the SFS does not show all possible options for the systems that are used in practice. In addition to the four classes identified, various intermediate options are also possible, depending on conditions of actual operation. For example, one and the same system can belong to different classes depending on the mode of operation.

4. Experimental

One of the most important stages in the design of the EECS software architecture is its structuring, i.e. allocation of subsystems and modules. Obviously, the composition of the subsystems and modules, as well as the complexity of their software implementation, will largely be determined by the class to which the designed system belongs to on the SFS. From the above classification of systems on the SFS, it can be seen that some classes are special cases of others, while the EECSs belonging to third and fourth classes are the most complicated in terms of software implementation.

For MIMO systems, the generalized software architecture is presented in Figure 1. The generalized architecture of the system includes a knowledge control subsystem, an interface subsystem and six program modules (in Figure 1, to the left of the software module name, the classes of systems that can include this module are listed).

The knowledge control subsystem includes:

– the knowledge base, in which procedural knowledge that implements algorithmic support of the EECS is stored (the knowledge base can be implemented using an object-oriented approach based on the frame and product knowledge models [5]);

– the output machine, which is a software module that manages knowledge;

– the relational database containing arrays of source data and the outcomes of the EECS operation;

– the database management system (DBMS), providing the necessary interface to work with the database.

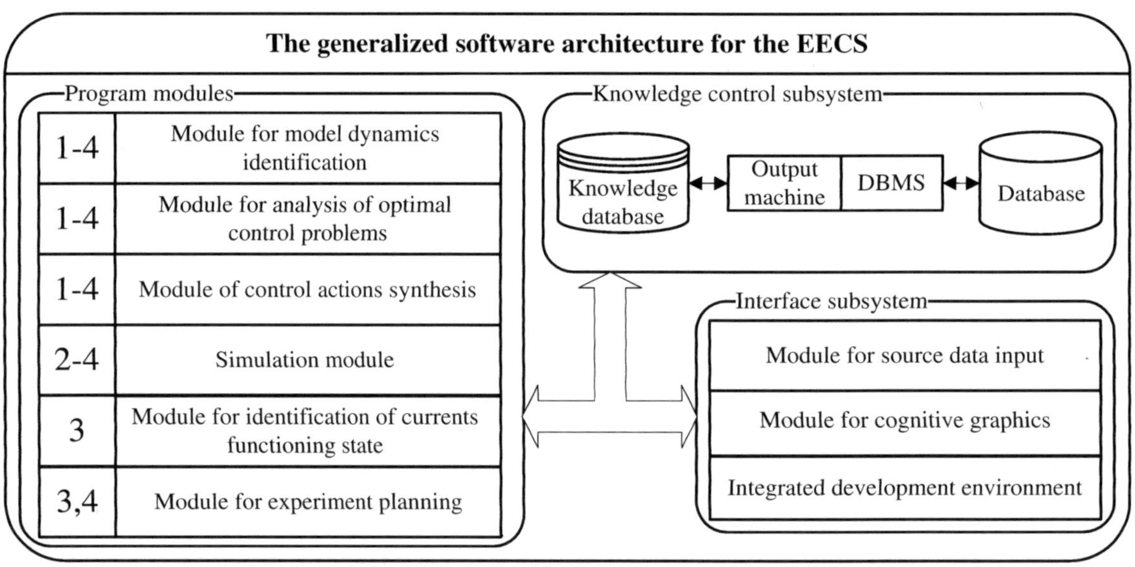

Figure 1. The software architecture for the EECS of Classes 1-4 on the SFS

The interface subsystem includes:

– the module for source data input;

– the module for cognitive graphics that provides a visual interface for displaying the results of the EECS functioning (cognitive models for solving energy efficient control problems [6], mimic diagrams of the control object, etc.);

– the integrated development environment that allows making changes to the knowledge base and configuring the software modules of the EECS (the need for making changes arises when the process equipment is changed or its operating modes are changed, the product range is changed, etc.).

The software modules included in the generalized architecture of the EECS have the following functional purpose: [7]

– the module for model dynamics identification solves the problem of parametric identification of model dynamics of the MIMO system [8];

– the module for analysis of optimal control problems is used to quickly obtain the results of a complete analysis of the optimal control problem, i.e. to study the domain of existence of the solution of the problem, determine the types of optimal control functions and relationships for calculating their parameters [9, 10];

– the module for control actions synthesis provides the choice of the optimal control algorithm for the system, while the results obtained during the operation of other modules of the system are used.

– the simulation module is designed to compare the effectiveness of control algorithms with the possible changes in the functioning state variable of the control object (the results can later be used to select the most effective control algorithm in real operation);

– the module for the functioning state identification is used in Class 3 systems and is designed to determine the current functioning state of the control object, using the data received through the measurement channel (different types of sensors and measuring devices, for example, temperature, humidity sensors, etc. can be used to obtain data;);

– the module for experiment planning provides an experimental plan and statistical data for a given number of experiments (usually applied in conjunction with the simulation module in Class 3 and Class 4 systems);

The software modules of the EECS exchange data with each other using a database in which the results of operation of all the system modules are stored.

5. Results and discussion

The proposed modular architecture of the control system allows taking into account the design features of the EECS on the SFS. It is obvious that the EECSs of different classes will differ from each other by the presence or absence of a number of software modules. For example, for the functioning of EECSs of Class 1, it is sufficient to have the three upper program modules shown in Fig. 1, while the systems of the class 2 must be complemented with a simulation module, etc.

At the same time, the proposed structuring of the system into subsystems and modules is sufficiently universal and invariant for various MIMO systems. At the same time, in spite of the general functional purpose of the program modules considered for systems of different classes, their immediate software implementation can vary significantly depending on the complexity of the particular control object for which the EECS is designed.

In the practical development of the system software modules, it is possible to use both specialized instrumental software, for example, SCADA systems, and the application development tools based on CASE and RAD technologies. The first approach is very convenient for the development of the EECS on the basis of the existing industrial management system and makes it possible, without creating a system from scratch, to add new modules or expand the functionality of the modules already available in the system.

6. Conclusion

The features of designing the software architecture of the EECS for MIMO systems have been considered. The problem of structuring the software of the EECS with the allocation of functional subsystems and modules has been solved. A generalized software architecture for the EECS on the SFS, which allows developing control systems of various classes, has been proposed.

This approach has been applied in the development of software architecture of energy efficient control systems for the dynamic modes of roller-drum and drum-type dryer installations. The developed control systems have made it possible to reduce energy costs in dynamic operating modes by 5-10% and provide the required values of the product quality indicators [11, 12].

Acknowledgements
The work was supported by the Russian Foundation for Basic Research, Grants No. 17-08-00457-a, No. 18-08-00555-a.

References

[1] Goodwin G C, Graebe S F, and Salgado ME 2000 *Control system design* (New Jersey: Prentice Hall)

[2] Khalil H K 1996 Adaptive output feedtach control of nonlinear systems represented by input-output models *IEEE Trans. Automat. Control* 41(2) 177-188

[3] Muromtsev D Y, Gribkov A N, Shamkin V N, Divin A G, and Savenkov A P 2017 Functional modeling of business processes for development of control and monitoring systems *Proceedings of the 2017 International Conference "Quality Management, Transport and Information Security, Information Technologies" (IT&QM&IS)* pp 440-442

[4] Muromtsev Yu L, Ljapin L N and Popova O V 1992 Modelirovanie i optimizacija slozhnyh sistem pri izmenenijah sostojanija funkcionirova-nija (Voronezh : Izd-vo VGU)

[5] Muromtsev D Y, Gribkov A N and Kurkin I A 2015 Baza znanij intellektual'noj informacionno-upravljajushhej sistemy mnogomer-nym ob'ektom *Informacionno-upravljajushhie sistemy* 5 60-64

[6] Muromtsev D Y and Pogonin V A 2015 Specificity of analysis and synthesis of energy efficient control *Jour. of Eng. and App. Sciences* 10(3) 37-39

[7] Gribkov A N and Kurkin I A 2013 Struktura programmnogo obespechenija informacionno-upravljajushhih sistem mnogomernymi ob'ektami na mnozhestve sostojanij funkcionirovanija *Novye informacionnye tehnologii v nauchnyh issledovanijah: mat-ly XVIII Vserossijskoj nauch.-tehn. konf. studentov, molodyh uche-nyh i specialistov.* (Rjazan': Rjazanskij gosudarstvennyj radiotehnicheskij universitet) pp 244-245

[8] Muromtsev D Y, Gribkov A N and Kurkin I A 2014 Certificate of computer programs of state registration 2014611295 of 30.01.2014 "Programmnyj modul' parametricheskoj identifikacii matematicheskih modelej dinamiki mnogosvjaznyh ob'ektov na mnozhestve sostojanij funkcionirovanija"

[9] Pchelintsev A N and Pogonin V A 2015 A method of determining the control parameters in the energy-saving control problem *Jour. of Eng. and Appl.Sciences* 10(2) 33-36

[10] Gribkov A N and Kurkin I A 2011 Analitiko-graficheskij metod issledovanija oblasti sushhestvovanija reshenija zadachi optimal'nogo upravlenija mnogomernym ob'ektom *Informatika i sistemy upravlenija.* 3(29) 141-152

[11] Gribkov A N 2009 Informacionno-upravljajushhaja sistema dinamicheskimi rezhimami v mnogosekcionnyh sushil'nyh ustanovkah *Informatika i sistemy upravlenija* 2(20) 123-129

[12] Gribkov A N, Kurkin I A and Gavrilov A O 2016 Programmnoe obespechenie informacionno-upravljajushhej sistemy dinamicheskimi rezhimami processov sushki poslespirtovoj bardy *Aktual'nye problemy jenergosberezhenija i jenergojeffektivnosti v tehnicheskih sistemah: tezisy dokladov 3-oj Mezhdunar. konf. s jelementami nauchnoj shk* (Tambov: Izd-vo Pershina R.V.) pp 61-62

IOP Conf. Series: Journal of Physics: Conf. Series **1050** (2018) 012031 doi:10.1088/1742-6596/1050/1/012031

Method of the cast-in-place friction pile well walls local soil compaction

A V Gruzin[1] and V V Gruzin[2]

[1]Omsk State Technical University, 11, Mira ave., Omsk, 644050, Russia
[2]Saken Seifullin Kazakh Agrotechnical University, 62 Zhenis avenue, Astana, 01001, Kazakhstan

Abstract. For improving the bearing capacity of the foundations on cast-in-place friction piles, the well walls local soil compaction technology is proposed to be used. The idea of the local soil compaction technology is to form the well having a cross-section different from the traditional round one. The application prospects theoretical justification of the shortened hypocycloid as a cast-in-place friction pile cross-section shape was carried out. The possibility of the cast-in-place friction pile bearing capacity considerable improvement is shown. The pile bearing capacity as compared to the traditional circle section without additional costs can theoretically be increased by the value of up to 80%. The basic arrangement of the attached working body for forming the well having a cross section in the shortened hypocycloid shape is presented. The proposed technical solution application when constructing foundations only in housing construction will make it possible to save annually on the piles material over 400 million rubles.

Key words: cast-in-place pile, bearing capacity, shortened hypocycloid

1. Introduction

The modern stage of the building technologies development is characterized by searching for the material resources and technical equipment rational use technologies [1-4]. The manufactured products competitiveness is currently possible including by reducing the capital costs during the construction process of the industrial facilities and technological infrastructure ones. The materials efficient use implies not only the energy and resource private saving, but also the search for the technologies comprehensively solving the problem of simultaneous providing the constructed foundation necessary lifting properties and costs reducing.

Currently the cast-in-place friction piles (CFP) supported foundations have become widespread. The main advantage of this pile as compared to the driven friction piles (DFP) is the possibility of using the lower quality and strength materials aimed only for operating loads. Another advantage is CFP installation practically without wastes. At the same time, CFP main disadvantage is low (reduction to 40 %) bearing capacity in comparison with the similar in size DFP. FCP bearing capacity improving problem solution is seen in the comprehensive problem solution. The bearing capacity improvement of the pile should not be followed by the materials consumption increase for producing, should not be followed by the increased environmental impact, should not result in the increased complexity of the developed technology and equipment used for its implementation. The local compaction technologies have a certain potential to deal with such complex problems. At present, various technologies are used for the formation and compaction of CFP well walls. One of such technologies is punching the wells with compacted walls by means of gravitational, hydraulic or vibrating punches (figure 1). Such technologies advantage is the possibility of the cross-section of different shapes formation including

IOP Conf. Series: Journal of Physics: Conf. Series **1050** (2018) 012031 doi:10.1088/1742-6596/1050/1/012031

non-circular section, as well as the formation of the side surface inclination to the CFP axis which significantly improves its bearing capacity. However, a significant disadvantage of such technologies is the specialized energy-intensive equipment requirement for sinking into the soil and subsequent pulling the profiled punch out of it. Another promising direction of the well walls soil compaction is rolling with the help of the specialized working body. The advantage of this technology is the low energy consumption, only rotary motion using when developing the well body which essentially simplifies the base machine choice (figure 2). The technology main drawback is the impossibility to form the well cross-section different from the round one.

Figure 1. The attached equipment UKS-7 for sinking the punch into the soil (A) and the hydraulic power unit for the subsequent pulling out (B).

The development of the equipment that would simultaneously allow, by using operating base machines, to form the wells with the compacted soil into the walls and cross-section, different from the round one, would make it possible to solve the material costs reducing problem comprehensively and, as the ultimate objective, the capital costs reducing problem at the stage of the foundation works.

Figure 2. The attached equipment for the wells rolling WR-250A (A) and the rolling working body (B).

The use of the wells rolling attachable equipment technical idea (figure 2) seems to be the most reasonable one as the CFP well walls local soil compaction working body prototype due to the low power-to-weight ratio and easy technical implementation.

2. Problem statement

The CFP well walls local soil compaction problem solution, hence the bearing capacity improving one seems to be in the search of the closed smooth curve to be used as the working body sealing elements motion trajectory and further development of the mechanism design for the chosen soil moving elements motion trajectory execution.

3. Theory

According to the current normative documents, the CFP bearing capacity is defined by the following dependence [5]:

$$F_d = \gamma_c \left(\gamma_{cR} RA + \gamma_{cf} u \sum_{i=1}^{n} f_i h_i \right) \qquad (1)$$

where γ_c is the pile operating conditions coefficient; γ_{cR} is the soil operating conditions coefficient under the pile tip; R is the soil calculated resistance under the pile tip, kPa; A is the pile bearing area, m^2; γ_{cf} is the soil operating conditions coefficient of the pile skin; u is the cross-sectional perimeter of the pile shaft, m; f_i is the calculated resistance of the soil i-layer of the pile shaft side surface, kPa; h_i is the thickness of the soil i-layer touching the pile skin, m.

The dependence (1) analysis makes it possible to define the promising directions of CFP bearing capacity improving. The pile skin area increase without its cross-section area one may be one of these directions. In the formula (1) the parameter u, the pile shaft cross-section perimeter, directly impacts on the CFP pile skin area. Thus, the perimeter length increase without increasing the area limited by the perimeter will result in the pile bearing capacity improvement. In such a case, the design of the working body required to form the well having the cross-section complex profile in the ground, should meet several requirements:

- the working body moving elements should trace out closed smooth trajectory;
- the working body kinematic scheme should be as simple as possible.

The flat curve, hypocycloid, kinematically defined as the trajectory of the point on a circle rolling along another inner circle without or with sliding (figure 3), satisfies the mentioned requirements.

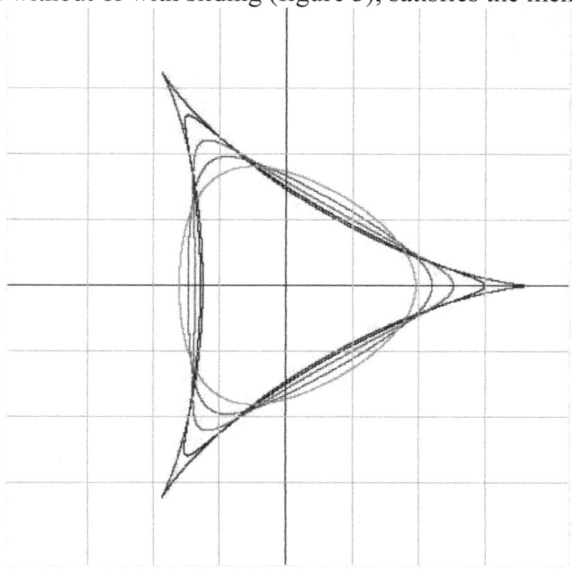

Figure 3. The examples of hypocycloids.

The hypocycloid equation in the parametric form is as follows [6]:

$$\begin{cases} x = (R-r)\cos\varphi + \lambda r \cos\left(\dfrac{R-r}{R}\varphi\right) \\ y = (R-r)\sin\varphi + \lambda r \sin\left(\dfrac{R-r}{R}\varphi\right) \end{cases} \quad (2)$$

where R is the director circle radius; r is the generating circle radius; φ is the parameter changing in the following range $[0; 2\pi]$; λ is the sliding coefficient. The radii of the director R and generating r circles have the dependence:

$$n = \frac{R}{r} \quad (3)$$

where n is the integer number equal to 3, 4, 5, etc. The integer value of the variable n is required to form non-self-intersecting closed curve.

4. Results discussion

Calculations show that at the same limited area, three-rayed hypocycloid (deltoid) has more length than four-rayed astroid, therefore the hypocycloid with a relation of the director circle radius R to the generating circle one r equal to n=3 is suggested to be chosen for further research. For CFP bearing capacity improvement possibility evaluation the necessary theoretical calculations were performed. CFP wells cross-section areas values were used as the initial data of the comparative analysis. The required areas values were calculated in advance for CFP installation typical wells diameters: 0.1, 0.15, 0.2, 0.25, 0.3, 0.4 and 0.5 m. Additionally, the geometric data of the sections in the form of a square and an equilateral triangle were used in the comparative analysis. For the ease of comparison, the obtained theoretical data are presented in the graphical form (figure 4).

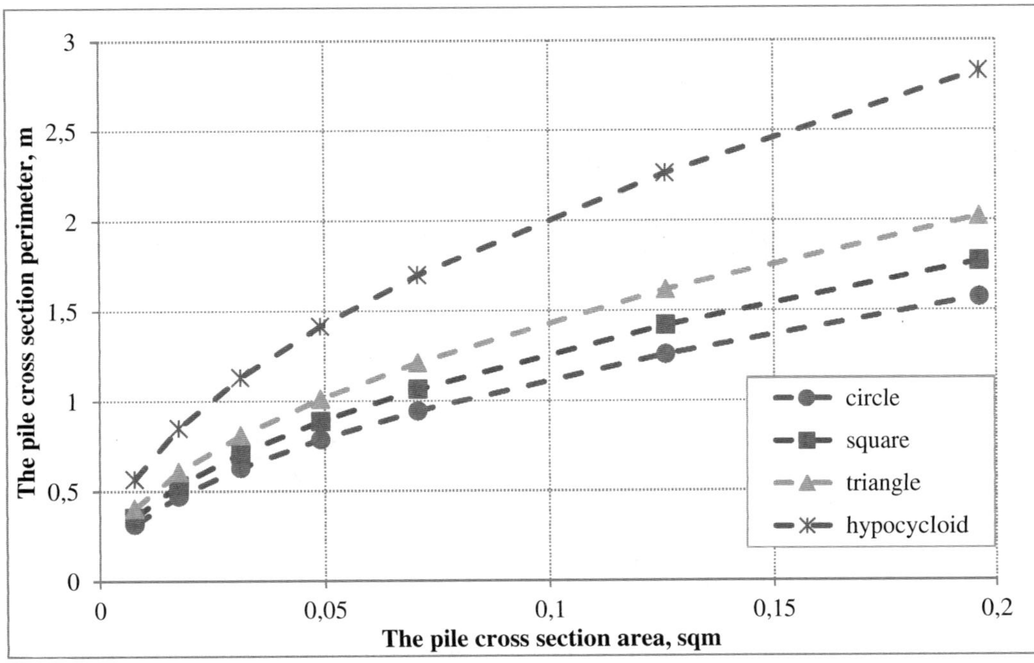

Figure 4. The impact of CFP cross-section area on the perimeter length depending on the cross section shape.

The presented graphs show that the pile having the cross-section in the shape of hypocycloid has the perimeter maximum value. The given curve using as a working body moving elements motion trajectory can theoretically improve CFP bearing capacity up to 80% compared to the traditional CFP having the circular cross-section. It is evident from the graphical data that the square cross-section applying provides CFP bearing capacity improvement due to the pile skin increase of 12.8% as compared with a circular cross-section. The triangular section using improves the bearing capacity in comparison with the circular section by 28.6%.

The design implementation is proposed in the form of the combined working body mounted as the attached equipment on the hydraulic excavators of II and III dimensional groups. The attached equipment consists of the guide support 2 attached to the base machine handle 1 (figure 5). In the guide support upper part 2 there is the hydraulic drive 3 which provides the rotational movement of the working body consisting of three consecutively assembled interconnected units: cone auger 6, roller 5 and hypocycloidal crossing profile shaper 4.

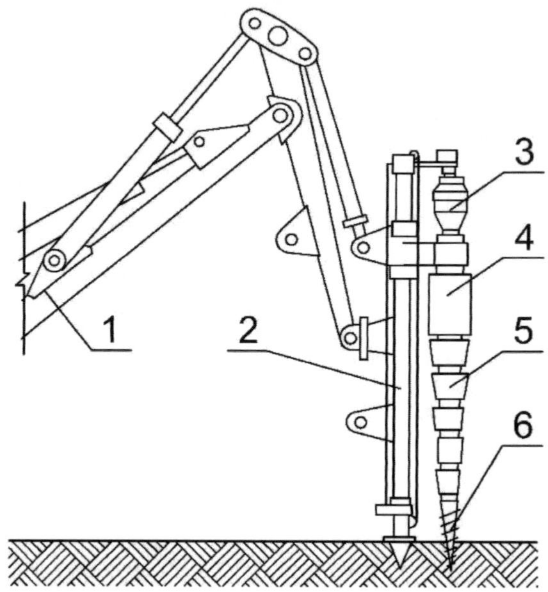

Figure 5. The working body mounted on the base machine.

The shaper design provides for the coaxial arrangement with the roller and leader auger and drive rotary motion sharing for the following screwing, well rolling and final forming the well cross-section in the form of hypocycloid.

For evaluating the economic efficiency of CFP bearing capacity improvement proposed way, the comparative analysis was carried out. The expenses calculation both for DFP producing, necessary for the commissioned housing pile foundations construction and for CFP manufacturing for the same purposes was conducted. According to the Federal State Statistics Service (Rosstat), the total area of the residential buildings put into operation in the last years constituted: in 2015-85.3 mln. m^2, in 2016 – 80.2 mln. m^2, during the period of January-September 2017-61 mln. m^2 [7] (figure 6).

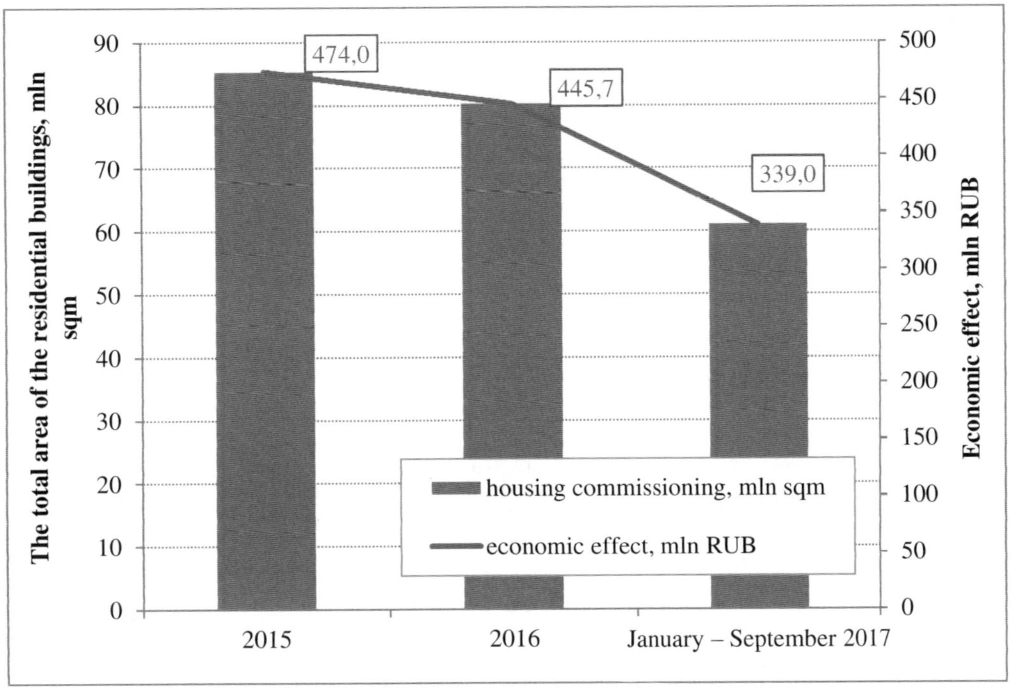

Figure 6. The total area of the commissioned residential buildings and economic effect from the applying of CFP produced by the local compaction technology.

For the comparative analysis conducting, the following assumptions were accepted:

- for the foundations construction, DFP of grade S120.35 GOST 19804-2012 and 12 meters long CFP having the hypocycloidal cross-section equal to DFP area of 0.1225 m^2 are used;
- 1 m^2 of housing is equivalent to 12.5 kN of pile load;
- pulverescent clay soil with the flow index of $I_L = 0.3$ was chosen as a foundation soil;
- DFP are made of concrete of class B22.5 in the compressive strength, CFP are made of concrete of class B15 in the compressive strength;
- calculations do not take into account the «cluster» effect of the friction piles combined action.

The calculations show that considering the assumptions made, the bearing capacity of DFP of grade S120.35 GOST 19804-2012 equals to 1105.65 kN, while the bearing capacity of 12 meters long CFP having the hypocycloidal cross-section equal to DFP area is 1182.13 kN. Thus, taking into account the used concrete cost differential in manufacturing only DFP, the total costs in 2015 would be 4 bln 253 mln RUB, and when using only CFP - 3 bln 779 mln RUB. The economic effect CFP using in 2015 would be 474 mln RUB, in 2016 - 445.7 mln RUB and for the first nine months of 2017 - 339 mln RUB (figure 6).

5. Conclusions

The choice of the hypocycloid as a working body moving elements motion trajectory for CFP well walls soil local compaction will theoretically make possible to improve CFP bearing capacity up to 80% without additional material costs. In this case, the trajectory is kinematically formed only by rotation, which simplifies the design implementation of CFP well walls soil local compaction working body mechanism. Using the cast-in-place friction pile well walls soil local compaction technology will make possible to save over 400 mln RUB annually just while constructing the residential buildings.

IOP Conf. Series: Journal of Physics: Conf. Series **1050** (2018) 012031 doi:10.1088/1742-6596/1050/1/012031

6. References

[1] Abramenkov D E, Gruzin A V and Gruzin V V 2012 *Mechanical means and construction technology* (Saarbrucken: Palmarium Academic Publishing)

[2] Zavyalov M A, Zavyalov A M, Gruzin A V and Kucherenko M V 2013 Thermal stability ensuring of artificial constructions *Oil Industry* **8** 105-7

[3] Gruzin A V, Tokarev V V, Shalai V V and Logunova Yu V 2015 *Proc. Engineering Oil and Gas Engineering* vol 113, ed A V Myshlyavtsev, V A Likholobov and V L Yusha (Amsterdam: Elsevier) pp 158-68

[4] Gruzin A V, Gruzin V V and V V Shalay 2016 *Proc. Engineering Oil and Gas Engineering* vol 152, ed A V Myshlyavtsev, V A Likholobov and V L Yusha (Amsterdam: Elsevier) pp 182-9

[5] Set of Rules 2011 SP 24.13330.2011 Pile foundations Revised Edition SNiP 2.02.03-85 (Moscow: Minregion Russia)

[6] Bronshtejn I N and Semendjaev K A 1980 *A handbook on mathematics for engineers and students of VTUZ* (Moscow: Nauka)

[7] Construction Federal State Statistics Service (Rosstat), available at http://www.gks.ru/wps/wcm/connect/rosstat_main/rosstat/ru/statistics/enterprise/building/

The description of the Industry 4.0 cyber and physical production division based on Petri net models

A V Gurjanov[1], D A Zakoldaev [2], A V Shukalov[2], I O Zharinov[2]

[1] Director, Stock Company «Experimental Design Bureau «Electroavtomatika» Name After P A Yefimov, 40, Marshala Govorova St., Saint Petersburg, 198095, Russia
[2] Department of Computer Design of On-Board Computer Equipment, Saint Petersburg National Research University of Information Technologies, Mechanics and Optics, 49, Kronverksky Av., Saint Petersburg, 197101, Russia

E-mail: mpbva@mail.ru

Abstract. The Industry 4.0 production division base is cyber and physical systems. They function automatically using some additive technologies, the Internet of things technology, sensors technologies and etc. To project the cyber and physical production divisions they can use some mathematical models based on Petri nets. The Petri nets let describe and research discrete events processes which are a part of cyber and physical production division. A Petri net structure is given to describe production processes in the Industry 4.0 devices designing company. The Petri net model is done in the environment VisualPetri 1.2.

1. Introduction

The modern industrial production develops rapidly in the direction of new information technologies and equipment (cyber and physical systems) implementation in the devices designing components manufacturing technological processes. The actual point of company industrial capabilities transferring to working with the new technological equipment is the exchange of the automatized lines of component production for the robotized automatic lines. Such a transfer exists today as practical implementation of the ideas of «the fourth industrial revolution» which is widely known in the science literature under common definition Industry 4.0.

One of approaches how to create and research the digital twin of the Industry 4.0 devices designing mechanical assembly production is the approach when they describe the devices designing component (parts) using the oriented graphs model. The graph model of the adequate discrete events dynamics of the devices designing manufacturing process can be described using a general theory mathematical apparatus known as nets of Petri.

The Petri nets theory [1-3] and available means how to model discrete processes (events) can be used to describe a variety of discrete states of the Industry 4.0 [4-6] production division of the components manufacturing technological routes which have the following properties:

- manufacturing operations completion parallelism in different mechanical assembly production sections technological lines;
- manufacturing operations completion asynchronism in different mechanical assembly production sections technological lines;
- the hierarchy of component structure and its assembly algorithm which is based on the

technological difficulty increase of some assembly units.

The parallelism, asynchronism and hierarchy of the devices designing components manufacturing technological routes can be explained with some specific features of the Industry 4.0 devices designing company mechanical assembly production organization — systems of discrete events which includes:

- the incoming flow of materials and assembly units in separate production lines (component transferring inside the workshop with transport robotized system);
- state transformation of materials, components and parts in separate production lines and sections which is based on technological operation completion (for example, component unit automatic assembly, lacquering and anti-moisture covering application on a printed circuit board with installed components, component electrotyping which was obtained with some additive technologies application and etc.);
- floor plan and nomenclature of the engaged technological equipment to manufacture the component in the devices designing company mechanical assembly production sections.

So the task of the mathematical model of creation and research of the Industry 4.0 devices designing company mechanical assembly production based on the Petri nets theory is actual one.

2. Petri nets component to describe mechanical assembly production

The primary components of Petri nets graphical descriptions are:

- place (in several literatures they use some equivalents «position», «top»);
- transfer (sometimes they use equivalents «barrier»);
- connection (sometimes they use equivalents «arc»);
- marker (sometimes they use equivalents «chip»).

The figure 1 shows the typical components of Petri net which are used to describe devices designing company manufacturing processes.

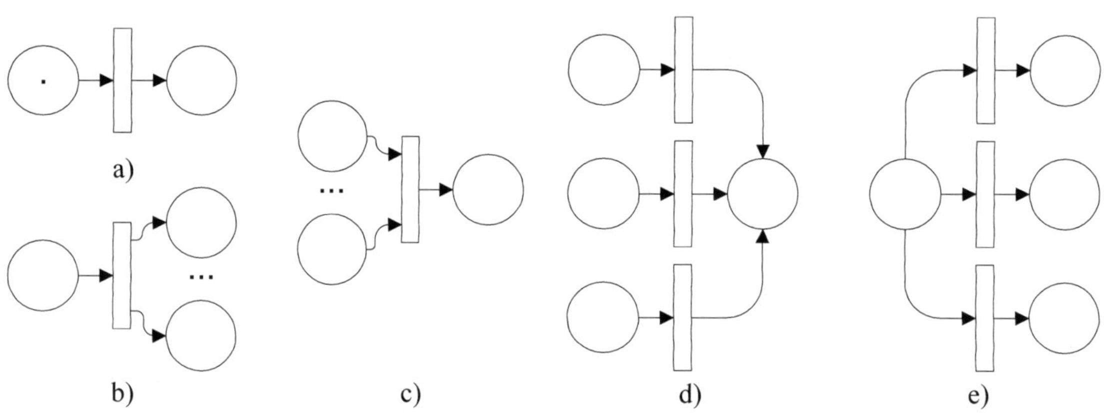

Figure 1. Typical components of a classical Petri net:
a) a net of a single transfer with one entering and one exiting places;
b) a net of a single transfer with one entering and several exiting places;
c) a net of a single transfer with several entering and one exiting places;
d) a net of three transfers with three entering and one exiting places;
e) a net of three transfers with one entering and three exiting places.

When they describe discrete events of mechanical assembly production processes the place of Petri nets is each production position in the workshop section and condition of manufacturing operation completion in this position. The graph shows the place of Petri net as a circumference.

When they describe discrete events of mechanical assembly production processes the transfer of Petri nets is the completion result of separate (or enlarged — which depends on the technological process research scale) technological operation. The graph shows the transfer of Petri net as a vertical rectangle («barrier»).

When they describe discrete events of mechanical assembly production processes the connection of Petri nets is component transfer (part, assembly unit, material, component and etc.) in the technological lines and production sections during component manufacturing. The graph shows the connection of Petri net as a graph arc — line (one ordinary or several repeated) with an arrowhead which connects several net components (place, transfer). The arrowhead direction is the component transfer direction (its assembly units) in the mechanical assembly production during its manufacturing in the devices designing companies.

When they describe discrete events of mechanical assembly production processes the marker of Petri nets is an object (component, part, assembly unit and etc.) on which the manufacturing operation is being done in a particular manufacturing position. The graph shows the marker of Petri net as a point («chip») which is located inside the circumference — place. The marker position in the place circumference is the initial state — the initial mesh of Petri net. In a single circumference there could be several marker points. If there are more than seven points than according to the rules of Petri net they show only the resulting number (the number of markers) inside the circumference.

The marker transferring from one circumference to another which means transfer completion is being directed according to the direction of the connection arrowhead (mingled states) which shows the component transfer in the production and which is the result of condition completion — a separate manufacturing operation in the devices designing component manufacturing technological route.

So the marker transferring (the marker transferring from entering place to the exiting place for corresponded transfer) in Petri net describes the dynamics of mechanical assembly production sections is a result of technological operation completion (a flow of events) of the general technological route component manufacturing and characterize the dynamics of manufacturing system behavior in the space of devices designing company states.

According to the general rule when completion is done the marker from its entering place transfers to its exiting place. The number of transferring markers is equal to the number of connections. The marker transferring can be done only if the number of markers in its entering place is equal or higher than the times of the corresponded connections «place-transfer». So two varieties provide the functionality of Petri net:

- a variety of step by step transfer completions which shows the completion of separate technological operations in the production;
- a variety of available Petri net meshes which number can characterize the transfer of materials, parts and components in the mechanical assembly production technological lines.

It is clear that after the completion of each transfer Petri net becomes an information similarity of discrete events manufacturing process for the Industry 4.0 devices designing company.

3. A Petri net model for the Industry 4.0 cyber and physical system

The Industry 4.0 devices designing company mechanical assembly production organization is based on the robotized cyber and physical systems automatic technological lines implementation. The industrial purpose of cyber and physical system according to its functionality manner belongs to the automatic control digital system class with a random delay. The property of system «digitalizing» can be explained with some digital components which work in discrete time with the principles of program control.

Cyber and physical system is realized [7] with a system of detectors (sensors technology), gears, controller and a feedback channel to provide control and quality of technological operations completion inside the system work chamber. Cyber and physical system random delay can be explained with the system support of the user remote access technology using the Internet of Things (IoT – Internet of Things) using protocols of guaranteed pings provision and remote storage technology to organize the control of system functionality [8-10].

It is clear that IoT technology application to organize the Industry 4.0 devices designing company production section is more profitable than the technology M2M (Machine-to-Machine) which is applied today in the Industry 3.0 companies.

The analysis of the IoT technology properties which is the base of cyber and physical system functionality shows that not only the physical channel of the manufacturing technological operations completion exists which is the purpose of that system but there is also parallel virtual flow where the technological operation completion is the parameter calculation of component digital model. The virtual flow is done on the level of IoT remote storage servers.

So each Petri net transfer on the technological process completion physical level is equal to the corresponded transfer on the cyber level. The figure 2 shows Petri net which corresponds to the cyber and physical system and includes the flow parallel structure of the Industry 4.0 devices designing company manufacturing process.

The transfers that shows the technological operations completion on the physical level in the figure 2 are shown as «barriers with fillings». The transfers that shows the technological operations completion on the cyber level in the figure 2 are shown as «barriers without fillings». The dot and line arrows show the correspondence of barrier and technological operation (on physical level) or barrier and component digital model (on cyber level) in figure 2.

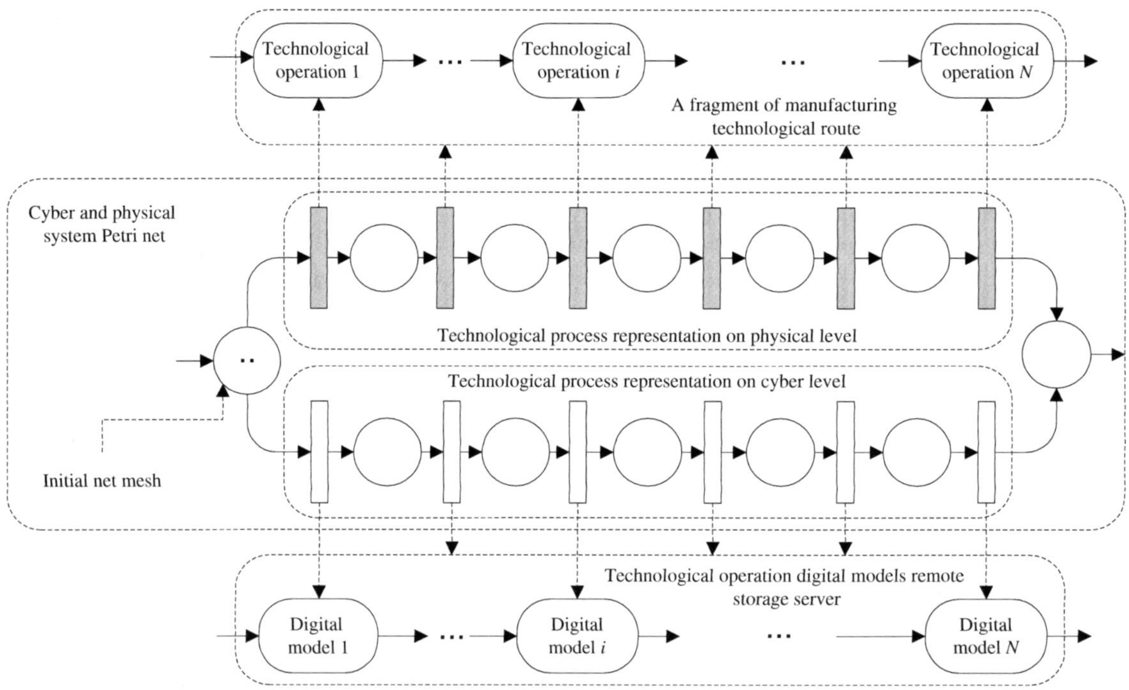

Figure 2. Petri net that shows technological process of the component manufacturing (part) with cyber and physical application in the Industry 4.0 production company.

Additionally to the process description of the Petri net cyber and physical systems shown in figure 2 where places on physical level are corresponded to:

- cyber and physical systems controller;
- the IoT wireless net controller to organize information exchange between technological equipment in a production section (workshop);
- the control command digital to analog converter which the controller sends to the work chamber gear where component manufacturing technological operation is being done (part);
- the cyber and physical system work chamber gear;
- devices designing component technological process manufacturing work chamber (assembly unit);

- the technological process completion state detectors and the technological equipment state detectors;
- an industrial data analog to digital converter which are received from the cyber and physical system sensors sub-system.

The marker location in the Petri net initial place circumference is according to the rule of the initial net mesh. The number of located markers depends on the initial functionality condition of a particular cyber and physical system and can vary. For example, for a robotized automatic component installer which is a part of the SMD (SMD – Surface Mounted Devices) section the number of the initial marker is equal to the number of the installing on PCB (PCB – Printed Circuit Board) radio devices components (resistance, microchips, capacitors and etc.) and one more (the PCB itself). For a 3D-printer (D – dimension) a single marker inside the place circumference is equal to the desired volume of metal powder which will be baked with a laser to manufacture a component which is a part of target device and etc. The figure 3 shows the generalized Petri net structure for the Industry 4.0 devices designing company mechanical assembly production description.

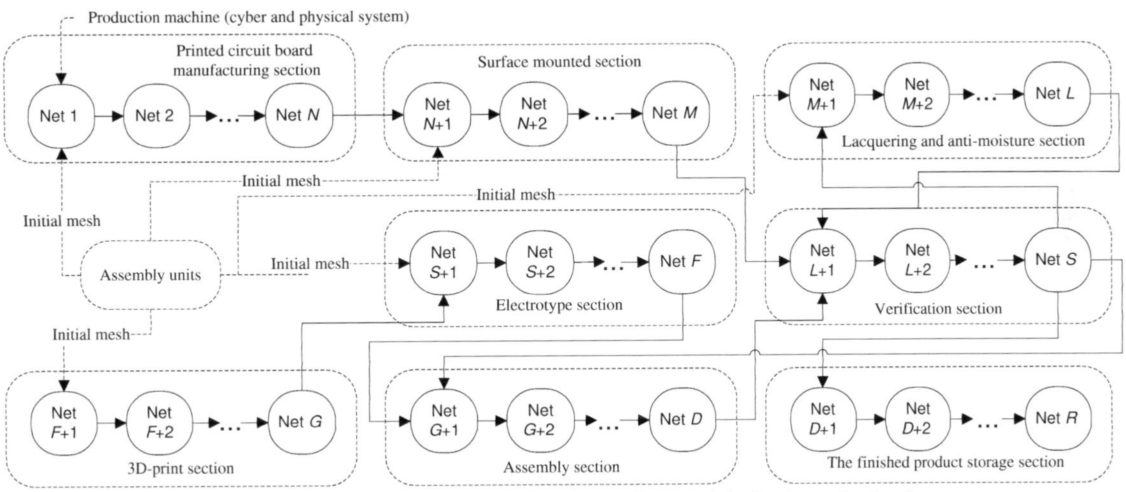

Figure 3. Petri net structure for the Industry 4.0 devices designing company mechanical assembly production.

4. Petri net modeling which describes the Industry 4.0 mechanical assembly production

For the available instrument research of the proposed Petri net and properties of the corresponded devices designing company mechanical assembly production on the level of discrete events there can be used several program environments and modeling instruments. The most popular instruments according to the practice are:

- Petri NetWork 2.0 has an advantage of easily usable graphical editor to synthesize a Petri net;
- The Petri Net Kernel 2.0 has an advantage of easily for user to apply some modeling instruments when they create a Petri net;
- HPSim has an advantage of included modeling time support in the modeling environment;
- VisualPetri 1.2 has an advantage to model several technological and business processes which are part of automatized (automatic) productions.

The instrument VisualPetri 1.2 is a program instrument with an open source-resource with a free code freely distributed in the internet with the license General Public License which explains its application to model Petri nets which describe the Industry 4.0 devices designing company mechanical assembly production in the initial project stages of the future digital company production sections. The figure 4 shows an example of the VisualPetri 1.2 modeling program work window of the discrete events system with a Petri net fragment to describe a cyber and physical system.

The Petri net discrete events analysis which structure is shown in figure 3 in the modeling environment VisualPetri 1.2 shows that the proposed Petri net structure option has the property of attainability which shows the possibility of practical tasks solutions which has the Industry 4.0 devices designing company mechanical assembly production.

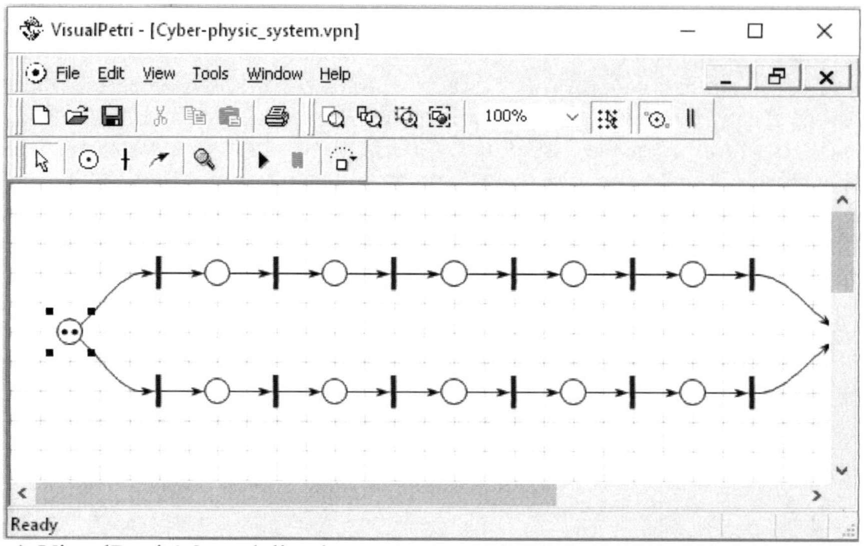

Figure 4. VisualPetri 1.2 modeling instruments work window of the discrete events system with a Petri net fragment to describe a cyber and physical system.

5. Conclusion

To research the Industry 4.0 devices designing company industrial processes on the stage of sketch and technical project it is necessary to create a digital twin of the mechanical assembly production. A mechanical assembly production digital twin can be a model which made graphically as a Petri net and with program support of matrix algebra mathematical apparatus and formal logic theory.

Petri nets modeling in the specialized imitation environments gives a chance to research process character of the discrete events states space which are part of the production and install possibility and practical significance of the digital industrial company practical realization on the base of the Petri nets approved basic properties (attainability, limitation, security, viability.

References

[1] Huang J, Zhu Y, Cheng B, Lin Ch and Chen J 2016 *Sensors* **382** 1-10
[2] Chen T, Sanchez-Aarnoutse J C, Buford J 2011 *IEEE Transactions on smart grid* **2** 741-749
[3] Qian Zh, Yu Y, Fan G 2015 *Applied mathematics & information sciences* **9** 1981-92
[4] Theorin A, Bengtsson K, Provost J, Lieder M, Johnsson Ch and Lundholm Th 2017 *International journal of production research* **55** 1297-11
[5] Liao Y, Deschamps S, Loures E F and Ramos L F 2017 *International journal of production research* **55** 3609-29
[6] Jung K, Choi S S, Kulvatunyou B, Cho H and Morris K S 2017 *Production planning & control* **28** 108-122
[7] Choi S S, Jung K, Kulvatunyou B and Morris K S 2016 *Journal of research of the national institute of standards and technology* **121** 422-433
[8] Hwang G, Lee J, Park J and Chang T-W 2017 *International journal of production research* **55** 2590-02
[9] Qu T, Thurer M, Wang J, Wang Z, Fu H and Li C 2017 *International journal of production research* **55** 2622-49
[10] Zuehlke D 2010 *Annual reviews in control* **34** 129-138

Mathematical modeling of the using of the innovative intermediate products at the stage of production of gross regional product

O D Kazakov, S P Novikov and N A Afanasyeva

Federal State-Funding Educational Institution of Higher Education "Bryansk State University of Engineering and Technology", 3, Stanke Dimitrov Av., Bryansk, 241037, Russia

Abstract. This article presents a mathematical model describing interindustry industrial interrelation. Special attention is paid to the analysis of the production structure in the context of innovative products. The imitation model of using of the innovative intermediate product at the stage of production of gross regional product is developed on the basis of the research. This model allows to predict the volume of output of innovative intermediate product taking into account the need for production of new goods in the first phase of their life cycle.

1. Introduction

The enormous number of the mathematical balance models characterizing interindustry production relationships is described in modern scientific literature. For example, the issues of improving the structure of production based on the best results of the balancing production using the method Moodie Young and Comsoal are explored in the works of Ikhsan S., Ulina T. Sipangkar A. Prasetio A. [1]. Vasin A.A., Morozov V.V., Morgenschtern O., Mulen A., Petrosyan L.A., Zenkevich N.A., Ouan G.,Karaseva A.G., Orlova T.A., Pecherskiy S.L., Belyaeva A.A., Blakuall D., Grischik M.A., Chebotarev V.I., Zolotuhin A.Y., Gubko M.V., Novikov D.A., Dixit A., Skit S. explore the issues of development of the economic-mathematical models and the decision-making methods on the basis of the game theory, linear programming, dynamic programming, queueing theory. Siregar I, Nasution A., Prasetio A., Fadillah K. Analysis research the issue of optimization of the structure of production in the context of the concept "Lean manufacturing" [2]. However the idea of mathematical modeling of the using of the innovative intermediate products in the open literature is unexplored.

The proof of striving to balance production of the intermediate product and the production of the product in the structure of gross regional product with the help of the mathematical methods is one of the main tasks of this study. The study is also analyzed in the context of the production of innovative products.

2. Model description

Let's present the model of using innovative intermediate product at the stage of production of gross regional product (Figure 1).

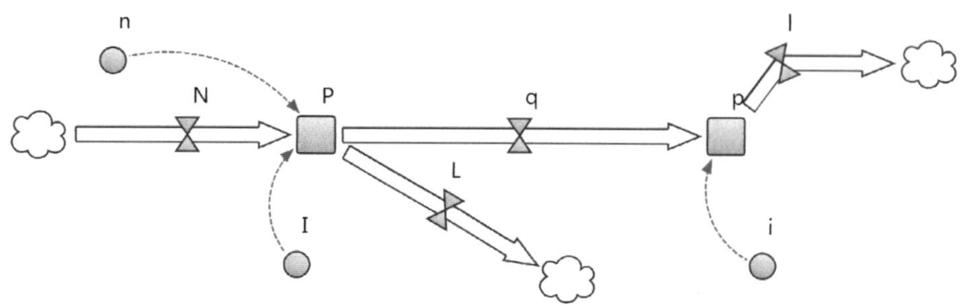

Figure 1. The model of using innovative intermediate product at the stage of production of gross regional product

If P and p – current production volumes of the intermediate and the gross regional product

N – the amount of new products in the first phase of their life cycle.

n – the amount of innovative products in the structure of new products in the first phase of their life cycle.

L and l – the amount of output of production of goods in the last phase of their life cycle taking into account the intermediate and the gross regional product.

q — the volume of production of intermediate on the basis of the production need of the new products in the first phase of their life cycle.

I and i — the amount of innovative products in the production structure of the intermediate and the gross regional product.

μ — the ratio of innovative products in the gross regional product.

Then the mathematical model of the dynamics of production volumes of the intermediate and the gross regional product has the form:

$$\left.\begin{aligned} \frac{dP}{dt} &= N - L - q, \\ \frac{dp}{dt} &= q - l, \\ \frac{dI}{dt} &= n \times N - \mu q, \\ \frac{di}{dt} &= \mu q. \end{aligned}\right\} \tag{1}$$

The intensity of output from production of goods in the last phase of their life cycle taking into account the gross regional product L is a function of P. The intensity of output the goods from production in the last phase of their life cycle taking into account intermediate l is a function of p. The intensity of the manufacture intermediate products on the basis of need in the manufacture of new products in the first phase of their life cycle q is a function of P and p. Let's suppose that the function $L(P)$, $l(p)$ and $q(P,p)$ have next features:

$$L(0) = l(0), \frac{dL}{dP} > 0, \frac{dl}{dp} > 0, \frac{dq}{dP} > 0, \frac{dq}{dp} \leq 0. \tag{2}$$

We can talk that all the variables of the system (P, p, I, i) are greater than zero so the phase space of the system (1) is R_+^4 (nonnegative orthant of the four-dimensional space). In the current system (1) the first equations (for P and p) are closed system with phase space R_+^2 (the first quadrant of the plane):

$$\frac{dP}{dt} = N - L - q, \frac{dp}{dt} = q - l. \tag{3}$$

3. The proof of the balance of production of innovative intermediate product and the production of an innovative product in the structure of gross regional product

The states of balance $(P = P', p = p')$ of the system (3) are determined by the solving the system of equations:

$$\begin{cases} L(P) + q(P, p) = N, \\ l(p) + q(P, p) = 0. \end{cases} \tag{4}$$

Let's replace the second equation by the sum of both equations. In the result we can get:

$$\begin{cases} L(P) + q(P, p) = N, \\ L(P) + l(p) = N. \end{cases} \tag{5}$$

The system (5) shows us that there is some implicit function $p = f_1(P)$, which can be determined with the help of the first equation. Also it can be concluded that there is some implicit function $p = f_2(P)$, which can be determined with the help of the second equation of this system.

Differentiating the second equation (5) according (P), we get:

$$\frac{dL}{dP} + \frac{dl}{dp}\frac{dp}{dP} = 0. \tag{6}$$

Earlier we had determined, that formula. Than from (6) follows, that for the function $p = f_1(P)$:

$$\frac{dp}{dP} < 0. \tag{7}$$

So the function $p = f_1(P)$ is decreasing. Differentiating the first equation (5) according (P), we get:

$$\frac{dL}{dP} + \frac{dq}{dP} + \frac{dq}{dp}\frac{dp}{dP} = 0. \tag{8}$$

So from (2) we see that for the function $p = f_2(P)$:

$$\frac{dp}{dP} > 0. \tag{9}$$

So the function $p = f_2(P)$ is increasing. On the basis of (6) and (7) we can say that graphs of functions $p = f_1(P)$ and $p = f_2(P)$ can intersect no more than in one point. This means that there is the balance of production of the intermediate product and the production of the gross regional product.

We will try to prove that we can speak about balance between the production of the intermediate product and the production of the gross regional product when we have any starting conditions R_+^2 in the system (1). This can be verified by approximation of function values. Let's change variables: $\xi = P - P', \eta = p - p'$. Than $\dot{\xi} = \dot{P} \dot{\eta} = \dot{p}$ and $\dot{\eta} = \dot{p}$, and the balance variables (ξ, η) has coordinate system $(0,0)$. Let's decompose the right parts of the system of getting variables (2.5) in a Taylor series in the variable ξ and η near the point $(0,0)$:

$$\left.\begin{aligned}\dot{\xi} &= \left(-\frac{\partial q}{\partial P}-\frac{dL}{dP}\right)\xi-\frac{\partial q}{\partial p}\eta,\\ \dot{\eta} &= \frac{\partial q}{\partial P}\xi+\left(\frac{\partial q}{\partial p}-\frac{dl}{dp}\right)\eta\end{aligned}\right\}, \tag{10}$$

where derivatives are calculated at the point $(P = P', p = p')$.

The achieving of the balance of the production of the intermediate product and the production of the gross regional product can be if in the characteristic equation $2\lambda + A\lambda + B = 0$ the coefficients A and B will be more than 0.

So (2), we get:

$$\left.\begin{aligned}A &= \frac{\partial q}{\partial P}+\frac{dL}{dP}-\frac{\partial q}{\partial p}+\frac{dl}{dp}>0,\\ B &= \frac{\partial q}{\partial p}\frac{dL}{dP}+\frac{\partial q}{\partial P}\frac{dl}{dp}+\frac{dL}{dP}\frac{dl}{dp}>0.\end{aligned}\right\} \tag{11}$$

Our first statement is proved. Next let's discuss the model of the balance of production of innovative intermediate product and production innovation of the gross regional product (the subsystem of equations for I and i in the system (1) if $P = P'$ and $p = p'$.

The ratio of innovative products in the total volume of the intermediate product $\mu' = \frac{I'}{P'}$ satisfies the equation:

$$\mu' q(P', p') - nN = 0. \tag{12}$$

We can find the ratio of innovative products in the total volume of the intermediate product:

$$\mu' = \frac{nN}{q(P', p')} \tag{13}$$

The meaning $\mu = \mu'$ is balanced index as the equation for $\frac{dI}{dt}$ in (1) the value $\frac{dI}{dt}$ will be affirmative if $\mu < \mu'$ and negative if $\mu > \mu'$

Taking into account that balanced volumes P' and p' can be found as the solving of the system (5), so according (13) we can get:

$$\mu' = \frac{nN}{q(P', p')} = n\frac{q'+L'}{q'} = n\left(L + \frac{L'}{q'}\right) = n\left(L + \frac{L'}{l'}\right). \tag{14}$$

4. The realization of mathematical models in the system AnyLogic

A mathematical model of the using of innovative intermediate product at the stage of production of gross regional product was implemented in the system AnyLogic (Figure 2).

There are some oscillograms of changes of the volumes of innovative products in the structure of production of intermediate and gross regional product in the animation window.

The model parameter (the amount of innovative products in the structure of new products in the first phase of their life cycle; the amount of output of production of the intermediate product in the last phase of its life cycle; the amount of output of production of gross regional product in the last phase of its life cycle) can be changed with the help of scrollbars.

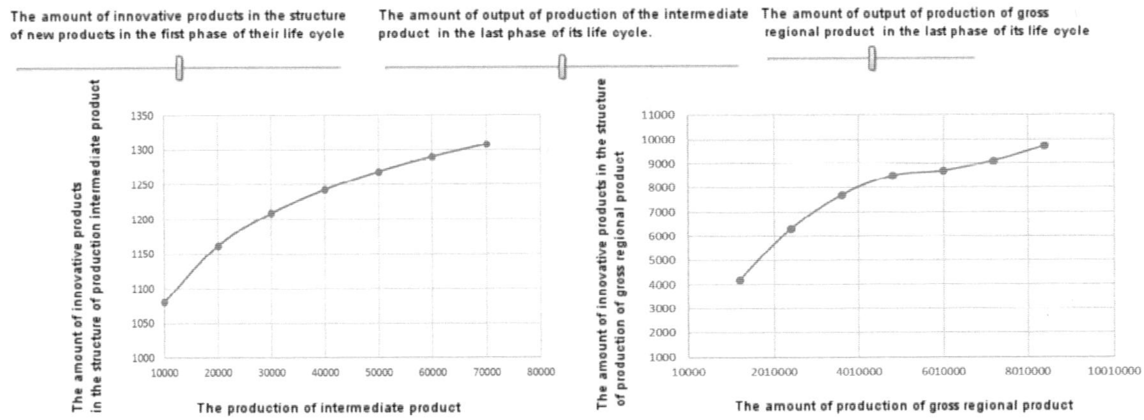

Figure 2. A mathematical model of the using of innovative intermediate product at the stage of production of gross regional product

This simulation model allows to predict the volume of production of innovative intermediate product according the need in the manufacture of new products in the first phase of their life cycle.

5. Conclusions

So it was proven that there is the opportunity to achieve a balance between the production of intermediate product and the production of the gross regional product in any initial conditions R_+^2.

When we modeled the balance of production of innovative intermediate product and the production innovation of the gross regional product we determined the method of calculation of a balanced ratio of innovative products in the amount of intermediate product.

The applied significance of our scientific research is to develop software product based on the simulation model of using the innovative intermediate product at the stage of production of gross regional product.

This software product will improve the quality of forecasting the amount of production of innovative intermediate product on the basis of need in the manufacture of new products in the first phase of their life cycle.

6. References

[1] Ikhsan S, Ulina T, Sipangkar A, Prasetio A 2017. Improvement of production layout based on optimum production balancing scale results by using Moodie Young and Comsoal method. 1757-899X 237 012003

[2] Siregar I, Nasution A, Prasetio A, Fadillah K. Analysis of production flow process with lean manufacturing approach. 2017 1757-899X 237 012002

[3] Bandyopadhyay M, *Chattopadhyay* J 2005. Ratio-dependent predator–prey model: effect of environmental fluctuation and stability. Nonlinearity 18 913

[4] Assaf M, Mobilia M 2010. Large fluctuations and fixation in evolutionary games. *J. Stat. Mech.* 2010 P09009

[5] De Martino A, Marsili M 2006 Statistical mechanics of socio-economic systems with heterogeneous agents. *J. Phys. A: Math.* Gen. 39 R465

[6] Samanidou E, Zschischang E, Stauffer D and Lux T 2007. Agent-based models of financial markets. *Rep. Prog. Phys.* 70 409

[7] Shaleh W, Rasim and Wahyudin 2018. The System of Inventory Forecasting in PT. XYZ by using the Method of Holt Winter Multiplicative

Injection of hydrogen sulfide into a porous medium partially saturated with water

M K Khasanov

Sterlitamak Branch of Bashkir State University, pr. Lenina 49, Sterlitamak 453103, Russia

E-mail: hasanovmk@mail.ru

Abstract. In the basis of equations of the mechanics of multiphase media a mathematical model of injection of hydrogen sulfide in liquid state into a porous medium saturated with water and oil is constructed. The dependence of the coordinate and temperature of the front of formation of hydrogen sulfide hydrate on the flow rate of hydrogen sulfide injection was studied. It is established that at rather high intensity of injection of hydrogen sulfide temperature on border of phase transitions can rise above the equilibrium temperature of dissociation of hydrogen sulfide hydrate.

1. Introduction

One of methods of decrease in emission in the atmosphere of the hydrogen sulfide produced by the oil processing enterprises is its burial in the exhausted oil and gas collectors [1, 2]. For the purpose of decrease in risk of leak of recyclable gases on a surface the possibility of their transfer in a gas hydrate state which allows to store identical volumes of gas with smaller pressure [3-6] is considered now.

The mathematical models of a heat and mass transfer in the porous medium considering hydrate formation processes are necessary for creation of technologies of underground utilization of hydrogen sulfide in a hydrate state. Mathematical models of formation of gas hydrates when injection gas on the porous medium which is partially saturated with water are formulated, in particular, in works [7-9]. However in these works it is injected, the same gas which saturates the porous medium in an initial state. Mathematical model of injection carbon dioxide are provided to the layer saturated with water and methane in work [10]. In the real work the mathematical model of formation of H_2S hydrate at injection of hydrogen sulfide in liquid state on the porous medium saturated with water and oil is provided.

2. Mathematical model

Let a pore of natural layer in initial time be saturated with water with an initial water saturation of S_{w0} and oil, pressure of p_0 and T_0 temperature in which initial state formations of H_2S hydrate meet conditions. Let's believe that through the well which has opened layer on all thickness hydrogen sulfide in liquid state is injected.

Conditions of formation of H_2S hydrate are given in the phase diagram (fig. 1) [5]. On this chart curve *gh* corresponds to three-phase balance between gaseous hydrogen sulfide, H_2S gas hydrate and water, curve *lh* – to balance between liquid hydrogen sulfide, H_2S gas hydrate and water, and curve *lg* – to two-phase balance between gaseous and liquid hydrogen sulfide. Respectively H_2S gas hydrate

IOP Conf. Series: Journal of Physics: Conf. Series **1050** (2018) 012034 doi:10.1088/1742-6596/1050/1/012034

stable exists in the field of more to the left curve *lh* and *gh*.

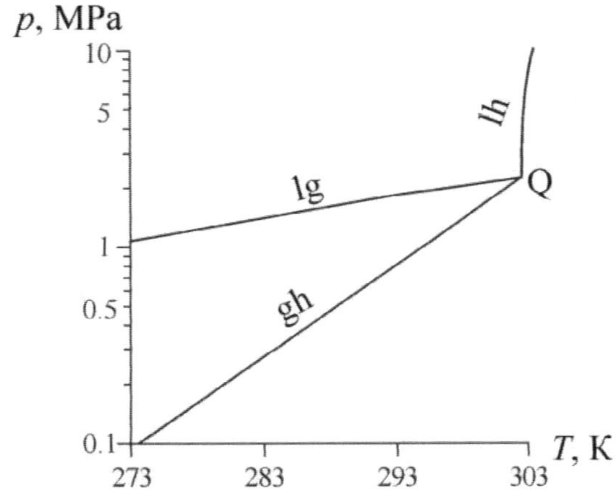

Figure 1. Phase diagram of the H_2S-H_2O system.

In this work we will neglect water filtration (since the case when initial value the water saturation of the porous medium does not exceed 0.2 is considered). Also we will neglect kinetics of process of hydrate formation and to consider that oil displacement by liquid hydrogen sulfide happens in the piston mode. Then it is possible to consider that H_2S gas hydrate is formed on frontal border of displacement of oil by hydrogen sulfide. In this case in the porous medium two zones are formed. In the first (near) zone of a pore are saturated with hydrogen sulfide and H_2S gas hydrate, and in the second (distant) zone of a pore are saturated with water and oil.

At the mathematical description of processes of a heat and mass transfer at injection of liquid hydrogen sulfide on porous medium we will believe: water and H_2S gas hydrate are not mobile and are not squeezed, oil and hydrogen sulfide are elastic liquids, and the porosity is constant.

The system of the main equations describing processes of heat transfer and filtrations of oil and water in natural layer consists equations of continuity, inflow of heat and state, and also Darcy's law [8]:

$$\frac{\partial}{\partial t}\left(\rho_i m S_i\right)+\frac{1}{r}\frac{\partial}{\partial r}\left(r\rho_i m S_i \upsilon_i\right)=0,$$

$$\rho\tilde{n}\frac{\partial T}{\partial t}+\rho_i c_i m S_i \upsilon_i \frac{\partial T}{\partial r}=\frac{1}{r}\frac{\partial}{\partial r}\left(r\lambda\frac{\partial T}{\partial r}\right),$$

$$m S_i \upsilon_i=-\frac{k_i}{\mu_i}\frac{\partial p}{\partial r}, \qquad (1)$$

$$\rho_s=\rho_{0s}\exp\left(\beta_s(p-p_0)\right),$$

$$\rho_l=\rho_{0l}\left(1+\beta_l(p-p_0)\right).$$

Here r – radial coordinate; t – time; m – porosity; T – temperature; p – pressure; the lower indexes $i = l, s$ belong respectively to parameters of oil and liquid hydrogen sulfide; k_i – phase permeability; μ_i – dynamic viscosity; υ_i – the actual average speed; ρ_i – density; c_i – specific heat; ρc and λ – effective values of heat capacity per unit volume and coefficient of heat conductivity of saturated stratum. As in values ρc and λ the main contribution is made by parameters of oil stratum, will consider further them constants.

We will set dependence of coefficient of phase permeability of k_i on absolute permeability of k_0 and saturation of S_i as follows:

$$k_i = k_0 S_i^3 \ (i = s,\ l).$$

On frontal border between zones the ratios following from conditions of balance of heat and mass are carried out:

$$\frac{k_s}{\mu_s}\frac{\partial p_{(1)}}{\partial r} = m\left(\frac{\rho_h}{\rho_s} S_h G + S_s\right) V_n,$$

$$-\frac{k_l}{\mu_l}\frac{\partial p_{(2)}}{\partial r} = m S_l V_n,$$

$$m S_h \rho_h (1-G) V_n = m S_{w0} \rho_w V_n, \qquad (2)$$

$$\lambda \frac{\partial T_{(1)}}{\partial r} - \lambda \frac{\partial T_{(2)}}{\partial r} = m \rho_h L_h S_h V_n,$$

where ρ_h, S_h and L_h – density, a saturation and specific mass heat of formation of H_2S hydrate, S_{w0} and ρ_w – initial water saturation of the porous medium and density of water, V_n – the speed of a frontal surface of hydrate formation. The lower indexes $j = 1, 2$ correspond to the parameters of the near and distant zones, respectively. The lower index n belongs to parameters on a frontal surface between zones.

Let's consider that hydrogen sulfide in liquid state is injected on porous Medium with the fixed mass flow rate of Q (on well height unit) and the continuous temperature of T_e. Let's consider rather great values of time after the beginning of injection of hydrogen sulfide. Then the radius of the first $r_{(n)}$ area will exceed considerably the well radius of r_e and the size of the well will practically not exert impact on features of course of process. In this case conditions on well border taking into account Darcy's law have an appearance:

$$r = r_e \to 0, \quad t > 0: \quad -\frac{2k_s\pi}{\mu_s}\left(r\rho_s \frac{\partial p}{\partial r}\right) = Q, \quad T = T_e.$$

3. Self-similar solution

Let's enter a self-similar variable:

$$\xi = r\big/\sqrt{\chi^{(T)} t},$$

where $\chi^{(T)} = \lambda / \rho c$.

Then from (1) we will receive the following decisions for temperature and pressure in each of layer zones:

$$P_{(1)} = P_{(n)} + \frac{Q\mu_s\beta_s}{2\pi\rho_{0s}k_s}\int_{\xi}^{\xi_{(n)}} \frac{1}{\xi}\exp\left(-\frac{\xi^2}{4\eta_{(1)}}\right) d\xi,$$

$$T_{(1)} = T_{(n)} + \frac{\left(T_e - T_{(n)}\right)\int_{\xi}^{\xi_{(n)}} \frac{1}{\xi}\exp\left(-\frac{\xi^2}{4} - \frac{\rho_{0s}c_s k_s}{\lambda\mu_s\beta_s} P_{(1)}\right) d\xi}{\int_{0}^{\xi_{(n)}} \frac{1}{\xi}\exp\left(-\frac{\xi^2}{4} - \frac{\rho_{0s}c_s k_s}{\lambda\mu_s\beta_s} P_{(1)}\right) d\xi}, \qquad 0 < \xi < \xi_{(n)}, \qquad (3)$$

$$p_{(2)} = p_0 + \left(p_{(n)} - p_0\right)\frac{\int\limits_{\xi}^{\infty}\frac{1}{\xi}\exp\left(-\frac{\xi^2}{4\eta_{(2)}}\right)d\xi}{\int\limits_{\xi_{(n)}}^{\infty}\frac{1}{\xi}\exp\left(-\frac{\xi^2}{4\eta_{(2)}}\right)d\xi},$$

$$T_{(2)} = T_{(0} + \frac{\left(T_{(n)} - T_0\right)\int\limits_{\xi}^{\infty}\frac{1}{\xi}\exp\left(-\frac{\xi^2}{4} - \frac{\rho_{0l}c_l k_l}{\lambda\mu_l\beta_l}p_{(2)}\right)d\xi}{\int\limits_{\xi_{(n)}}^{\infty}\frac{1}{\xi}\exp\left(-\frac{\xi^2}{4} - \frac{\rho_{0l}c_l k_l}{\lambda\mu_l\beta_l}p_{(2)}\right)d\xi}, \qquad \xi_{(n)} < \xi < \infty, \qquad (4)$$

where $P_{(1)} = \exp\left(\beta_s\left(p_{(1)} - p_0\right)\right)$, $\eta_{(1)} = k_s / \mu_s m S_s \beta_s \chi^{(T)}$, $\eta_{(2)} = k_l / \mu_l m S_l \beta_l \chi^{(T)}$.

On the basis of relations (2) - (4) we obtain equations for finding the coordinate of the formation boundary of hydrogen sulfide hydrate $\xi_{(n)}$ and the values of the parameters $p_{(n)}$ and $T_{(n)}$ on it:

$$\frac{Q\mu_s\beta_s}{\pi\rho_{0s}k_s}\exp\left(-\frac{\xi_{(n)}^2}{4\eta_{(1)}}\right) = A_1 P_{(n)}\xi_{(n)}^2,$$

$$\frac{\left(p_{(n)} - p_0\right)\exp\left(-\frac{\xi_{(n)}^2}{4\eta_2}\right)}{\int\limits_{\xi_{(n)}}^{\infty}\frac{1}{\xi}\exp\left(-\frac{\xi^2}{4\eta_{(2)}}\right)d\xi} = A_2\xi_{(n)}^2,$$

$$\frac{\left(T_{(n)} - T_e\right)\exp\left(-\frac{\xi_{(n)}^2}{4} - X_1 P_{(n)}\right)}{\int\limits_0^{\xi_{(n)}}\frac{1}{\xi}\exp\left(-\frac{\xi^2}{4} - \frac{\rho_{0s}c_s k_s}{\lambda\mu_s\beta_s}P_{(1)}\right)d\xi} - \frac{\left(T_0 - T_{(n)}\right)\exp\left(-\frac{\xi_{(n)}^2}{4} - X_2 p_{(n)}\right)}{\int\limits_{\xi_{(n)}}^{\infty}\frac{1}{\xi}\exp\left(-\frac{\xi^2}{4} - \frac{\rho_{0l}c_l k_l}{\lambda\mu_l\beta_l}p_{(2)}\right)d\xi} = B\xi_{(n)}^2,$$

where $A_1 = \frac{m\mu_s\beta_s\chi^{(T)}}{k_s}\left(\frac{\rho_h G S_h}{\rho_{0s}} + 1 - S_h\right)$, $A_2 = \frac{m\mu_l\chi^{(T)}}{k_l}\left(1 - S_{w0}\right)$, $B = \frac{m\rho_h L_h S_h}{2\rho\tilde{n}}$.

4. Calculations results

Dependences of coordinate and temperature of the front of phase transitions (full curve) and the equilibrium temperature of dissociation of H_2S hydrate (dotted curve) from a mass flow rate of injection hydrogen sulfide are given in fig. 2. For system parameters adopted values: $S_{w0} = 0.12$, $m = 0.3$, $p_0 = 6$ MPa, T_e=292 K, $T_0 = 294$ K, $k_0 = 2\cdot10^{-13}$ m^2, $G = 0.24$, $\lambda = 2$ W/(m·K), $\rho c = 2\cdot10^6$ J/(K·kg), $L_h = 4.1\cdot10^5$ J/kg, $\beta_s = 3\cdot10^{-9}$Pa^{-1}, $\beta_l = 1\cdot10^{-9}$ Pa^{-1}, $\rho_h = 1003$ kg/m^3, ρ_{0s} =890 kg/m^3, ρ_{0l} =900 kg/m^3, $\mu_s = 2\cdot10^{-4}$ Pa·s, $\mu_l = 2\cdot10^{-3}$ Pa·s, $c_s = 1800$ J/(K·kg), $c_l = 1900$ J/(K·kg).

According to fig. 2 with increase in a mass flow rate of injection hydrogen sulfide the speed of the front of hydrate formation increases. It is caused by the fact that in this task the movement of a phase boundary is defined by the oil displacement speed by liquid hydrogen sulfide which increases with increase in intensity of injection.

From fig. 2 it is visible that temperature at the front of formation of H_2S hydrate increases with in a mass flow rate of injection hydrogen sulfide. From Fig. 2 that the temperature at the formation front of the H2S hydrate increases with the mass flow of injected hydrogen sulphide. It is caused by the fact that with increase in a mass flow rate of injection hydrogen sulfide the speed of the front of hydrate formation increases and respectively intensity of allocation of the latent heat of phase transitions increases. As temperature on the well is lower than the temperature at the formation front of H_2S

hydrate, the part of heat which is marked out on it is taken away through well walls. On the other hand with increase in speed of border of phase transitions intensity of heat extraction through walls of the well decreases that leads to increase in temperature H_2S gas hydrate formations at the front. As a result at rather high values of a mass flow rate and according to the high speed of the movement of border of phase transitions temperature on this border can exceed the equilibrium temperature of dissociation of H_2S hydrate. This case corresponds to incomplete transition of water to a condition of gas hydrate.

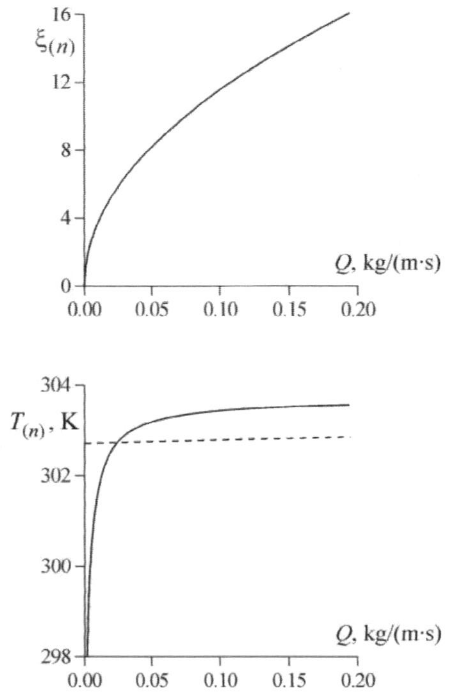

Figure 2. Dependence of coordinate and temperature of the front of phase transitions on a mass flow rate of injection H_2S.

5. Conclusions

Specific features of formation of H_2S hydrate due to injection of a liquid hydrogen sulfide into a natural layer initially filled with oil and water are considered. It is found that, depending on the mass flow rate of hydrogen sulfide injection, the H_2S hydrate forms either at a frontal surface or in an extended zone.

Acknowledgement

This work was supported by the Russian Foundation for Basic Research and the Republic of Bashkortostan (project no. 17-48-020123).

References

[1] Machel H G 2005 *Oil and Gas Science and Technology* **60** 51–65
[2] Xu T, Apps J A, Pruess K and Yamamoto H 2007. *Chemical Geology* **24 (3)** 319–346
[3] Chuvilin E M and Guryeva O M 2009 *Earth's Cryosphere* **13 (3)** 70–79
[4] Duchkov A D, Sokolova L S, Ayunov D E and Permyakov M E 2009 *Earth's Cryosphere* **13 (4)** 62–68
[5] Byk S Sh, Makogon Y F and Fomina V I 1980 *Gas Hydrates* (Moscow: Khimiya)
[6] Makogon Y F 1997 *Hydrates of Hydrocarbons* (Tulsa: PennWell Publishing Company)
[7] Shagapov V Sh, Rafikova G R, Khasanov M K 2016 *High Temperature* **54 (6)** 858–866

[8] Gimaltdinov I K, Khasanov M K 2016 *Journal of Applied Mathematics and Mechanics* **80 (1)** 57–64
[9] Shagapov V Sh, Chiglintseva A S, Belova S V 2017 *Journal of Engineering Physics and Thermophysics* **90 (5)** 1147–1161
[10] Tsypkin G G 2016 *Fluid Dynamics* **51 (5)** 672–679

Numerical study of the process of injection carbon dioxide into a porous formation saturated with methane and its hydrate

M K Khasanov[1], M V Stolpovsky[1,2] and I K Gimaltdinov[2]

[1]Sterlitamak branch of Bashkir state University, Lenin Avenue 49, Sterlitamak, Russia
[2]Ufa State Petroleum Technological University, Kosmonavtov street 1, Ufa, Russia
E-mail: hasanovmk@mail.ru

Abstract. In this article, in a planar one-dimensional approximation, we examine the problem of injecting gaseous carbon dioxide into a porous formation saturated in its initial state with methane and its hydrate. On the basis of the equations of multiphase systems mechanics, a mathematical model of the formation of CO_2 hydrate from methane gas hydrate is constructed, acknowledging the real properties of gases. It is shown that the formation of carbon dioxide hydrate can occur in two regimes: 1) with the formation of an area saturated with methane and water, or 2) without water at all. It has been established that the main parameters determining the hydrate formation regimes are: the formation permeability, the parameters of the injected gas, and the initial parameters of the system. Here, we construct and analyze critical diagrams separating these two modes.

1. Introduction

Gas hydrates are solid crystalline compounds formed from gas and water (ice) at certain pressures and temperatures. The huge reserves of natural gas, concentrated in hydrates, make them an almost inexhaustible source of energy, the fuel of the future. Another aspect of their research is associated with the sequestration of greenhouse gases into the hydrate phase. A number of researchers have suggested an idea regarding the injection of CO_2 into gas hydrate formations in order to extract natural gas from them and the carbon dioxide compound into the hydrate form [1].

This is possible due to the fact that carbon dioxide hydrate, other conditions being equal, is more stable than methane hydrate. Laboratory experiments on the effect of CO_2 (in the liquid or gaseous phase) on methane hydrate are examined, in particular, in [2-4]. It is shown that the proportion of methane substituted in this way from the hydrate composition can make up to 70% [2, 3]. Theoretical models of injecting CO_2 into methane hydrate formations are examined in [5, 6]. In this paper, we present a mathematical model of the substitution of CH_4 for CO_2 in the methane hydrate formation during the carbon dioxide injection, taking into account the finite extent of the formation, the real gas properties, the Joule-Thomson effect, and the adiabatic expansion of the gas.

2. Mathematical model

We examine a homogeneous porous medium of length L, the roof and sole of which are impermeable and thermally insulated. We assume that at the initial moment the reservoir is completely saturated with methane and its hydrate, the pressure p_0 and temperature T_0 of which correspond to the conditions for their stable existence. Let us suppose that carbon dioxide is pumped through the left boundary of the formation ($x = 0$), the pressure p_e and the temperature T_e correspond to the conditions for the stable existence of a mixture of CO_2 and its hydrate. As a result of this injection of CO_2 in the formation, two areas are formed. Thus, the pores of the near area adjacent to the left border of the formation contain carbon dioxide and its hydrate; the pores of the far field are filled with methane and its hydrate. In this case, a stable mobile boundary arises that moves into the depth of the reservoir over time and separates the above-mentioned regions. Its stability comes from the fact that flows in porous media are laminar, and also greater, in comparison with methane, viscosity of carbon dioxide. In describing the processes of heat and mass transfer in porous media, we accept the following assumptions: the porosity is constant; the skeleton of the porous medium, hydrate and water are incompressible and immobile; the temperature of the porous medium and the saturating substance is the same. In this case, we assume that methane and carbon dioxide hydrates are two-component systems with mass concentration of gas G_m and G_c, .

The system of basic equations describing the processes of filtration and heat transfer in a porous medium and representing the laws of conservation of masses and energy, the Darcy law and the equation of state, in the rectilinear parallel case with the assumptions noted above in each of the regions has the form:

$$\frac{\partial}{\partial t}\left(\rho_g \phi S_g\right) + \frac{\partial}{\partial x}\left(\rho_g \phi S_g \upsilon_g\right) = 0,$$

$$\rho C \frac{\partial T}{\partial t} + \rho_g C_g \phi S_g \upsilon_g \left(\frac{\partial T}{\partial x} + \varepsilon_g \frac{\partial p}{\partial x}\right) - \frac{\partial}{\partial x}\left(\lambda \frac{\partial T}{\partial x}\right) - \rho_g C_g \phi S_g \eta_g \frac{\partial p}{\partial t} = 0, \quad (1)$$

$$\phi S_g \upsilon_g = -\frac{k_g}{\mu_g}\frac{\partial p}{\partial x}, \quad \rho_g = \frac{p}{z_g R_g T}, \quad \eta_g = \frac{1}{\rho_g C_g} - \varepsilon_g,$$

$$\rho C = (1-\phi)\rho_{sk} C_{sk} + \phi \sum_{j=g,h,l} S_j \rho_j C_j, \quad \lambda = (1-\phi)\lambda_{sk} + \phi \sum_{j=g,h,l} S_j \lambda_j$$

Here, ϕ – porosity; p – pressure; T – temperature; S_j, ρ_j, C_j and λ_j ($j = g, h, l$) – pore density, density, specific mass heat capacity and thermal conductivity of the j phase (g – gas, h – gas hydrate, l – water); υ_g, k_g, R_g, μ_g, ε_g, η_g ($g = c, m$) – speed, permeability, gas constant, dynamic viscosity, throttling factor and adiabatic coefficient of the g-th gas phase (c – carbon dioxide, m – methane); ρ_{sk}, λ_{sk} and C_{sk} – density, thermal conductivity and specific heat of the skeleton of the porous medium; ρC and λ – specific volume heat capacity and coefficient of thermal conductivity of the «porous medium-saturating fluid» system.

The dependence of the phase permeability coefficient for the gas k_g on the gas saturation S_g and the absolute permeability k_0 is given in the following way [7]:

$$k_g = k_0 S_g^3 \ (g = c, m).$$

The supercompressibility coefficient for methane will be found on the basis of the Latonov-Gurevich equation [8]:

$$z_m = \left[0.17376 \cdot \ln\left(T/T_c\right) + 0.73\right]^{p/p_c} + 0.1 \cdot p/p_c,$$

here T_c and p_c are empirical critical parameters for the gas.

For the coefficient of supercompressibility of carbon dioxide, we will use a linear approximation [9]:

$$z_c = -0.006p/10^6 + 0.998$$

From the system (1) it is possible to obtain the equations of piezoconductivity and thermal diffusivity:

$$\frac{1}{z_{(i)}T_{(i)}}\frac{\partial p_{(i)}}{\partial t}=\chi_{(i)}^{(p)}\frac{\partial}{\partial x}\left(\frac{p_{(i)}}{z_{(i)}T_{(i)}}\frac{\partial p_{(i)}}{\partial x}\right), \tag{2}$$

$$\frac{\partial T_{(i)}}{\partial t}=\chi^{(T)}\mathrm{X}_{(i)}\frac{p_{(i)}}{z_{(i)}T_{(i)}}\frac{\partial p_{(i)}}{\partial x}\left(\frac{\partial T_{(i)}}{\partial x}+\varepsilon_g\frac{\partial p_{(i)}}{\partial x}\right)+\chi^{(T)}\frac{\partial}{\partial x}\left(\frac{\partial T_{(i)}}{\partial x}\right)+\mathrm{Y}_{(i)}\frac{p_{(i)}}{z_{(i)}T_{(i)}}\frac{\partial p_{(i)}}{\partial t}, \tag{3}$$

in which $\chi_{(1)}^{(p)}=k_c\left(\mu_c\phi(1-S_{hc})\right)^{-1}$, $\chi_{(2)}^{(p)}=k_m\left(\mu_m\phi(1-\nu)\right)^{-1}$ $\mathrm{X}_{(1)}=C_ck_c\left(\lambda R_c\mu_c\right)^{-1}$, $\mathrm{X}_{(2)}=C_mk_m\left(\lambda R_m\mu_m\right)^{-1}$, $\mathrm{Y}_{(1)}=\phi C_cS_c\eta_c\left(R_c\rho C\right)^{-1}$, $\mathrm{Y}_{(2)}=\phi C_mS_m\eta_m\left(R_m\rho C\right)^{-1}$, $\chi^{(T)}=\lambda\cdot\rho C^{-1}$. Here and below, the subscripts in parentheses i = 1, 2 refer to the parameters of the first and second regions accordingly.

As noted above, when injecting CO_2 into the methane hydrate formation, two areas and the mobile boundary of the phase transition $x = x\ (n)$ separating them appear. Therefore, the conditions for the conservation of masses of gases and water on it have the form [5, 6]:

$$\phi\rho_cS_c(\upsilon_c-\dot{x}_{(n)})=\phi\rho_{hc}S_{hc}G_c\dot{x}_{(n)},\ \phi\rho_mS_m(\upsilon_m-\dot{x}_{(n)})=\phi\rho_{hm}S_{hm}G_m\dot{x}_{(n)},$$

$$\phi\rho_c\left(1-G_c\right)S_{hc}\dot{x}_{(n)}=\phi\rho_m\left(1-G_m\right)S_{hm}\dot{x}_{(n)} \tag{4}$$

Here S_{hc} and S_{hm} - pore saturation with hydrate of carbon dioxide and methane, accordingly; $\dot{x}_{(n)}$ – the velocity of the boundary of the phase transition.

Since the volume saturation of the methane hydrate equals ν ($S_{hm}=\nu$), from the last equation of the system (4) for volume saturation with hydrate CO_2, we get:

$$S_{hc}=\rho_{hm}\left(1-G_m\right)\left(\rho_{hc}\left(1-G_c\right)\right)^{-1}\nu.$$

Applying Darcy's law to the first two equations of system (4), the conditions for the mass balance of gas and heat on the mobile boundary of the phase transition $x = x_{(n)}$ can be represented as:

$$-\frac{k_c}{\mu_c}\frac{\partial p_c}{\partial x}=\phi\dot{x}_{(n)}\left(S_c+\frac{\rho_{hc}G_cS_{hc}}{\rho_c}\right),$$

$$-\frac{k_m}{\mu_m}\frac{\partial p_m}{\partial x}=\phi\dot{x}_{(n)}\left(S_m+\frac{\rho_{hm}G_m\nu}{\rho_m}\right), \tag{5}$$

$$\lambda\frac{\partial T_c}{\partial x}-\lambda\frac{\partial T_m}{\partial x}=\phi\left(\rho_{hc}S_{hc}L_{hc}-\rho_{hm}\nu L_{hm}\right)\dot{x}_{(n)}.$$

Here L_{hc} and L_{hm} –accordingly , are the specific heats of formation of carbon dioxide and methane hydrates.

The initial conditions, as well as the conditions at the boundaries of the formation, can be represented as:

$$\begin{aligned} t=0:&\quad p=p_0,\quad T=T_0 &&(0\le x\le L);\\ x=0:&\quad p=p_e,\quad T=T_e &&(t>0);\\ x=L:&\quad p=p_0,\quad \frac{\partial T}{\partial x}=0 &&(t>0). \end{aligned} \tag{6}$$

To solve the problem (2), (3) with initial and boundary conditions (6) and conditions (5) on the moving boundary of the phase transition, the original equations are represented in an implicit finite-difference form. To solve such a system, the method of catching the front in a spatial mesh node was used [5, 6], in combination with the sweep algorithm and the iteration method.

3. Calculations results

In Figure 1, for the time t = 200 days, the temperature and pressure distributions in the formation are shown when carbon dioxide is injected with temperature $T_e = 280$ K. For other parameters characterizing the system, the following values are assumed: $L = 100$ m, $\phi = 0{,}1$, $\nu = 0{,}2$,

$\lambda = 2$ W / (m·K), $k_0 = 10^{-15}$ m^2, $p_0 = 3$ MPa, $T_0 = 274$ K, $\rho C = 2{,}5\cdot10^6$ J/(K·m^3), $G_c = 0{,}28$, $G_m = 0{,}13$, $R_c = 189$ J / (kg·K), $R_m = 520$ J/(kg·K), $\mu_c = 1{,}3\cdot10^{-5}$ Pa·s, $\mu_m = 10^{-5}$ Pa·s, $\rho_{hd} = 1100$ kg/m^3, $\rho_{hm} = 900$ kg/m^3, $L_{hc} = 4{,}1\cdot10^5$ J/kg, $L_{hm} = 4{,}5\cdot10^5$ J/kg, $C_c = 800$ J/(kg·K), $C_m = 1560$ J/(kg·K). For the Joule-Thomson coefficients of gases, the values $\varepsilon_c = 13{,}9$ K/MPa , $\varepsilon_m = 4{,}5$ K/MPa.

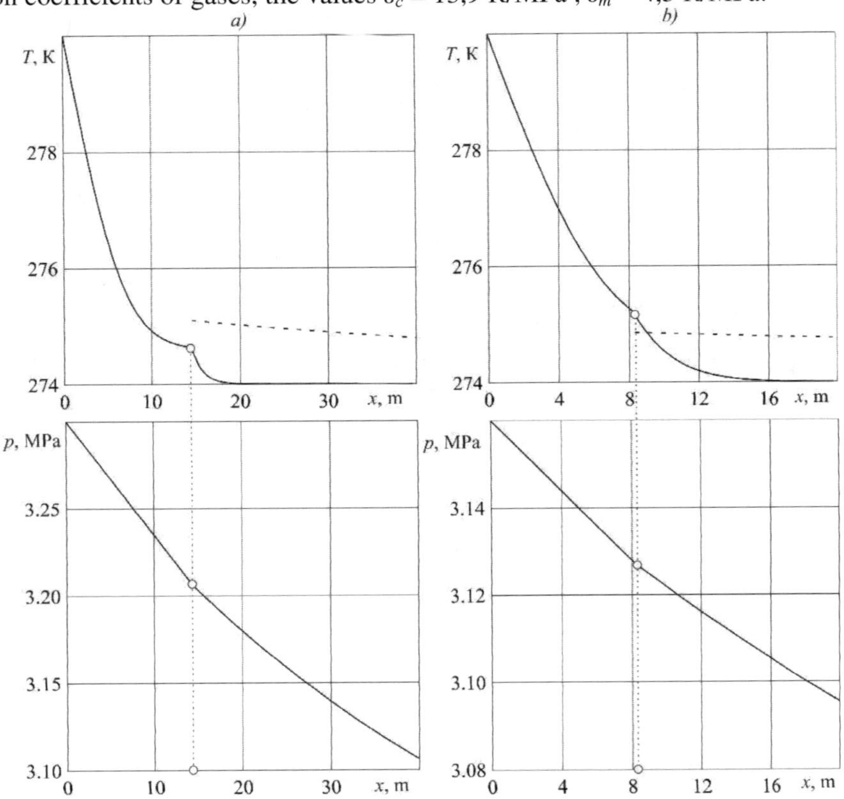

Figure 1. Distribution of temperature and pressure in the reservoir at time t = 200 days. at $p_e = 3.3$ MPa – *a*; $p_e = 3{,}16$ MPa –*b*.
The dashed line is the equilibrium temperature of methane hydrate.

In fig. 1 it is stated that for various parameters of the injected gas, the formation of a carbon dioxide hydrate can occur both with the dissociation of methane hydrate into gas and water, and without it. In Fig. 1a the temperature is shown to be in the area saturated with methane and its hydrate is below the equilibrium temperature, which is well described by a curve of the form [10]:

$$T_{(s)} = T_0 + T_* \ln\left(p/p_{s0}\right).$$

Here T_* and p_{s0} – empirical parameters, depending on the type of gas hydrate. In particular, for methane hydrate $T_* = 10$ K, $p_{s0} = 2{,}87$ MPa. Consequently, in this case, the decomposition of methane into gas and water does not occur. At lower values of the pressure of the injected gas, as followed from fig. 1 b, the temperature in the region of methane and its hydrate rises above the equilibrium temperature. In this case, it is necessary to enter the area containing methane and water in the free state; the formation of the hydrate of CO_2 now occurs from gas and water.

In fig. 2 for the instant $t = 200$ days, the temperature dependences $T_{(n)}$ at the boundary of the phase transition on the injection gas pressure p_e, the absolute permeability k_0 of the formation, and the initial pressure p_0 of the system are shown. From fig. 2 follows the fact that an increase in both the initial pressure of the system and a decrease in the permeability of the formation and the pressure of the injected gas in comparison with the parameters corresponding to fig. 1 lead to the fact that the velocity of the front of the phase transition, which is limited by the intensity of the supply of CO_2 to it, decreases. In this case, the effect of a higher temperature on the well, which results in the temperature in the area containing methane and hydrate rising above the equilibrium temperature, is indeed significant.

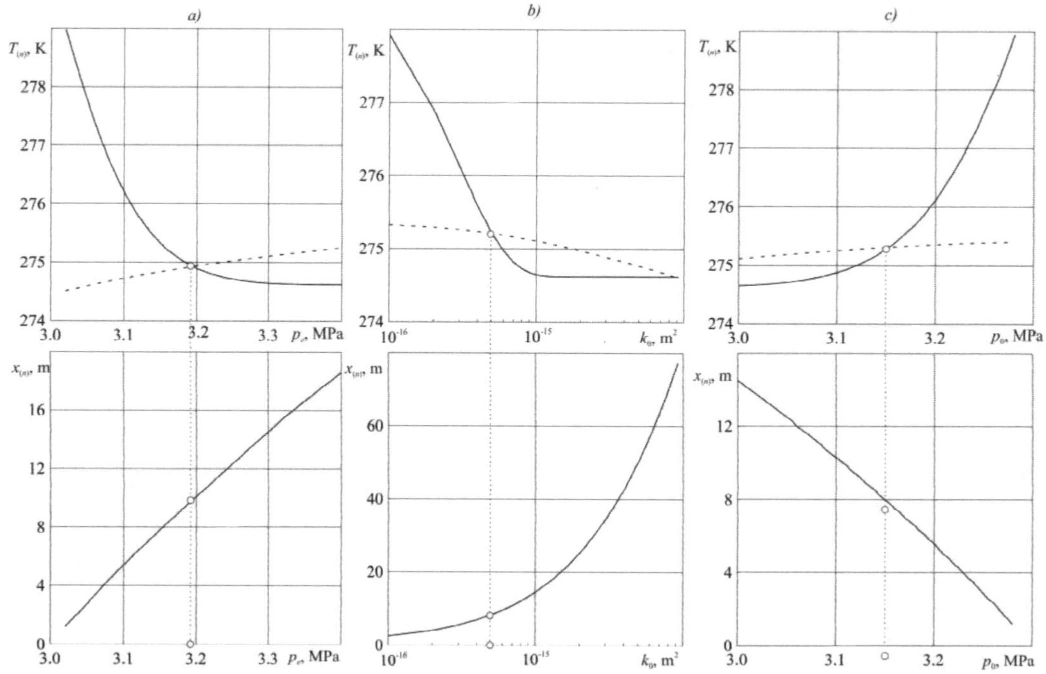

Figure 2. Dependences of the temperature at the boundary of the phase transition and its coordinates as a function of the pressure of the injected gas – *a*;
absolute permeability of the formation – *b*;
the initial pressure in the system is *c*.
The dashed line is the equilibrium temperature of methane hydrate.

In fig. 3 for the time t = 20 days, the dependence of the critical temperature is shown, above which the regime with methane hydrate decomposition into gas and water is realized, from the initial pressure p_0 and the pressure of the injected gas, p_e.

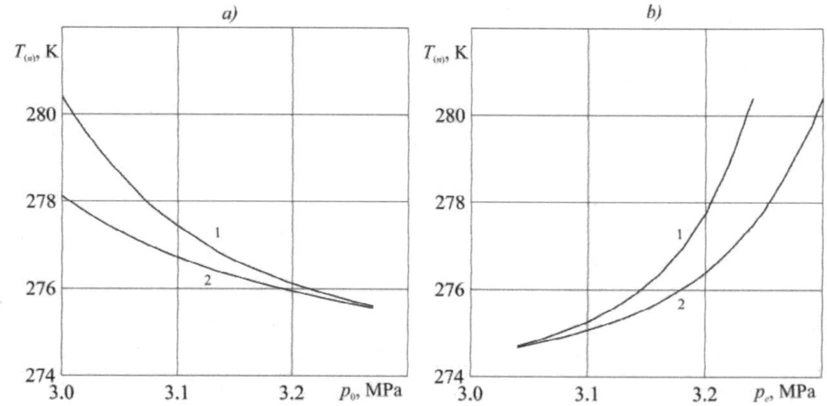

Figure 3. The dependence of the critical temperature on

a – the initial pressure of the system: $k_0 = 10^{-15}$ m^2 – 1, $k_0 = 7{\cdot}10^{-16}$ m^2 – 2;
b – discharge gas pressure: ν = 0,2 – 1; ν = 0,25 – 2

As follows from the fig. 3, the value of the critical temperature drops, as the initial pressure increases and the permeability of the formation decreases. This is due to the fact that the velocity of the boundary of the phase transition is limited by the rate of supply of CO_2, which, in turn, depends on the pressure drop in the formation. Similarly, the increase in the critical temperature is observed with an increase in the pressure of the injected gas. At the same time, with a decrease in the initial hydration of the reservoir at relatively high values of the pressure of the injected gas, the value of the critical temperature increases. This is true due to the fact that with decreasing hydratonsaturations, gas saturation increases, therefore increasing the intensity of carbon dioxide intake to the boundary of the phase transition.

Acknowledgement

This work was financially supported by the grant of Russian Science Foundation (project 17-79-20001).

References

[1] Ohgaki K, Takano K, Sangawa H, Matsubara T, Nakano S. 1996 *J Chem Eng Jpn* **29** 478–483.
[2] Ota M, Morohashi K, Abe J et al 2005 *Energy Conversion and Management* **46** 1680–1691.
[3] Qing Y., Chang-Y. S., Bei L. et al. 2013 *Energy Conversion and Management* **67** 257–264.
[4] Espinoza D N, Santamarina J C 2011 *Int. J. Greenhouse Gas Control* **5** 1032–1038.
[5] Shagapov V Sh, Khasanov M K, Musakaev N G, Duong Ngoc Hai 2017 *International Journal of Heat and Mass Transfer* **107** 347–357.
[6] Gimaltdinov I K, Stolpovsky M V, Dodova M I *Bulletin of the Tomsk Polytechnic University. Geo Assets Engineering* 2017 **328 (6)** 91–98
[7] Barenblatt G I, Entov V M, and Ryzhik V M 1982 *Motion of Liquids and Gases in Natural Strata* (Moscow: Nedra).
[8] Musakaev N G; Khasanov M K, Borodin S L 2018 *International Journal of Heat and Mass Transfer* **118** 455–461.
[9] Grobe M, Pashin J C, Dodge R 2009 *AAPG Studies in Geology* **59** 521–543.
[10] Byk S Sh, Makogon Y F and Fomina V I 1980 *Gas Hydrates* (Moscow: Khimiya).

IOP Conf. Series: Journal of Physics: Conf. Series **1050** (2018) 012036 doi:10.1088/1742-6596/1050/1/012036

Experimental study of heat transfer on the rubber-cord cased pneumatic element surfaces under the natural convection

V S Korneyev, I A Pen'kov, I N Kvasov and S A Korneyev

Omsk State Technical University, 11, Mira ave., Omsk 644050, Russia

Abstract. One of the ways of improving the calculations accuracy of the stationary objects and vehicles protection systems from vibrations and impacts is taking into account the influence of heat exchange processes with the environment occurring in the rubber-cord cased pneumatic elements and obeying Fourier's law of heat conduction and Newton's heat-transfer law. In the conducted study, the heat transfer coefficient values on the pneumatic element walls in the natural convection conditions are defined experimentally. The test bench, testing technique and used measuring instruments which are the heat flux sensors with the internal thermocouple are described. The heat transfer coefficient on the rubber-cord casing curved surface and on the metal flanges flat ones is found out to assume constant values being sufficient for practical application in the operating temperature range based on the process data. When applying the heat exchangers technical theory methods for the pneumatic elements calculation, it is acceptable to consider as a first approximation that the heat transfer coefficient has the identical values on all outer surfaces of the pneumatic element.

Key-words: pneumatic element, heat transfer coefficient, test bench

1. Introduction

The pneumatic elements (PE) with rubber-cord casing (RCC) became widely used in vibration protection and vibration isolation systems of stationary objects and vehicles, they are used both as air springs and air damping shock absorbers [1-6]. When air in PE is compressed, gas temperature rises, resulting in the heat release into the environment. On the contrary, when the gas expands, the temperature decreases and related heat input from the environment takes place. Since under the sprung mass stationary variations the expansion and compression processes succeed one another, heat will periodically take turns to be supplied and released.

Traditionally [1-5] PE gas (air) expansion and compression processes are described by the polytropic equation with non-defined (assigned) index n: at low frequencies n is recommended to be equal to 1 (isothermal process), while at high frequencies it is n=1.4 (adiabatic process). In some cases the mathematical model using the polytropic processes engineering thermodynamics specified method leads to the unacceptable discrepancy between the calculated and experimental data, especially when describing the air damping shock absorbers operation [6]. This is due to the fact that the polytropic mathematical model assumes the heat exchange with the environment processes reversibility. To overcome the given and other drawbacks, the extended mathematical model based on Fourier's law of heat conduction and Newton's heat transfer law is proposed in paper [6]. It is performed following the example of the reciprocating compressors technical theory by using the heat exchangers technical theory methods. The given direction main difficulty is defining the coefficient of heat transfer α between solid walls and gas included into the empirical dependence

$$q = \alpha(T - T_0). \quad (1)$$

where q is the heat flux [W/m^2]; T , T_0 is the temperature of the wall and gas correspondingly [K or °C]. According to [7] the indicative values of the gases heat transfer coefficient under the natural convection is α = 0.5…1000 W/(m^2·K). Such values range objectively requires experimental studies to specify the heat transfer coefficient numerical values under the natural convection on the RCC and PE flanges surfaces. The experimental data obtained in this case is of great practical importance during the calculation of PE applied in vibration protection and vibration isolation systems of stationary objects and vehicles pneumatic suspensions.

2. Problem statement and test bench description

The test bench scheme and its general view are represented in figure 1 and figure 2 correspondingly. The cylinder-stored type and N-50 model RCC manufactured by FSUP «FRPC «Progress» was used as a test object. The test bench consists of two fixed grips of Instron testing machine (position 1 in figure 1 and figure 2). The grips hold the RCC in the middle position (position 5 in figure 1, position 2 in figure 2) with two flanges. The electric heater of the clamp type (position 4 in figure 1) and the THA 9608-00 brand temperature sensor (position 6 in figure 1) designed for temperature monitoring in the vicinity of the heater are installed inside the RCC. The DTP 0924-R-P-90-25-1 brand heat flux sensor with the internal thermocouple (position 7 in figure 1, position 3 in figure 2) is glued on the RCC end surface. Six heat flux sensors of the DTP 0924-E-D-27-1 brand with the internal thermocouple (position 8 in figure 1) are glued on the metal flanges covering the RCC. The sensor readings are taken and digitized by the multichannel temperature meter IT-6-16 (position 4 in figure 2) and the data are subsequently transmitted to the PC. The measurements results are displayed on the computer screen in the form of tables and graphs. The heater is supplied with power from the household wiring by the laboratory autotransformer of the regulating LT-1M (position 5 in figure 2). The power supplied to the heater is measured by using the M890 series digital multimeters (position 6 and position 7 in figure 2).

The thermoelectric transducer THA 9608-00 is designed for liquid and gaseous media and solids temperature measuring in hard-to-reach places due to the possibility of the attachment section bending when installed on the monitoring object. The measured temperatures range from 40° to 800°C; the thermal inertia index is 0.5 ... 2.5 sec; the error is not more than 2.5°C. The filling compound of the DTP 0924-R-P-90-25-1 brand heat flux sensor installed onto the RCC is rubber, while the filling compound of the DTP 0924-E-D-27-1 brand heat flux sensor installed on metal flanges is epoxy. The sensors of both types have an internal thermocouple, the heat flux measured densities range is from 10 to 1000 W/m^2, the operating temperature range is from 10° to 100°C, the error is less than 8%. I, II, III heat flux sensors centres at the top flange are apart from the RCC axis at the distance of 83 mm, 55 mm, 33 mm and IV, V, VI sensors centres at the bottom flange are apart from the RCC axis at the distance of 56 mm, 31 mm, 83 mm respectively (figure 3).

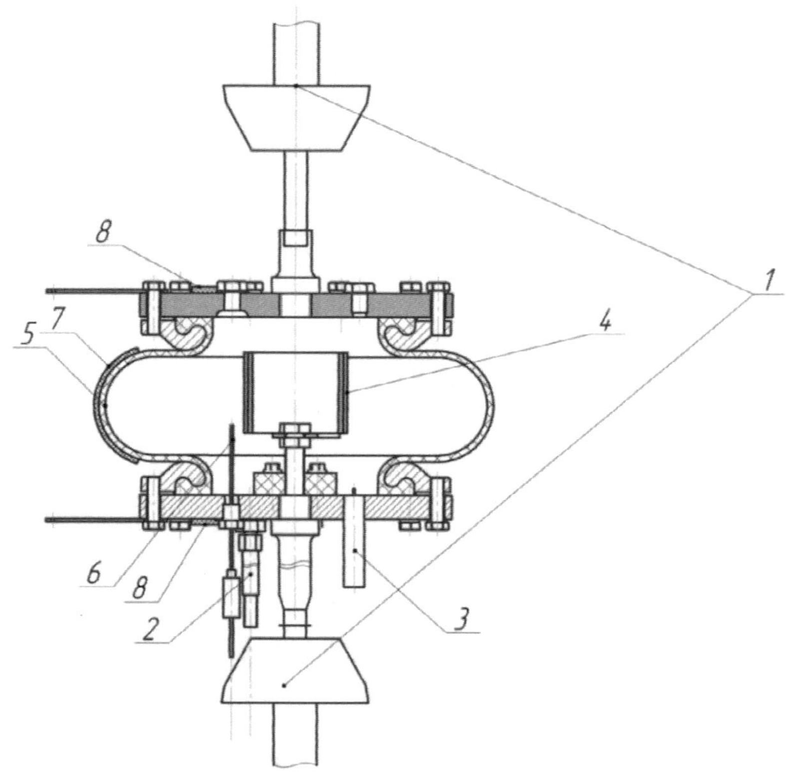

Figure 1. The test bench scheme.

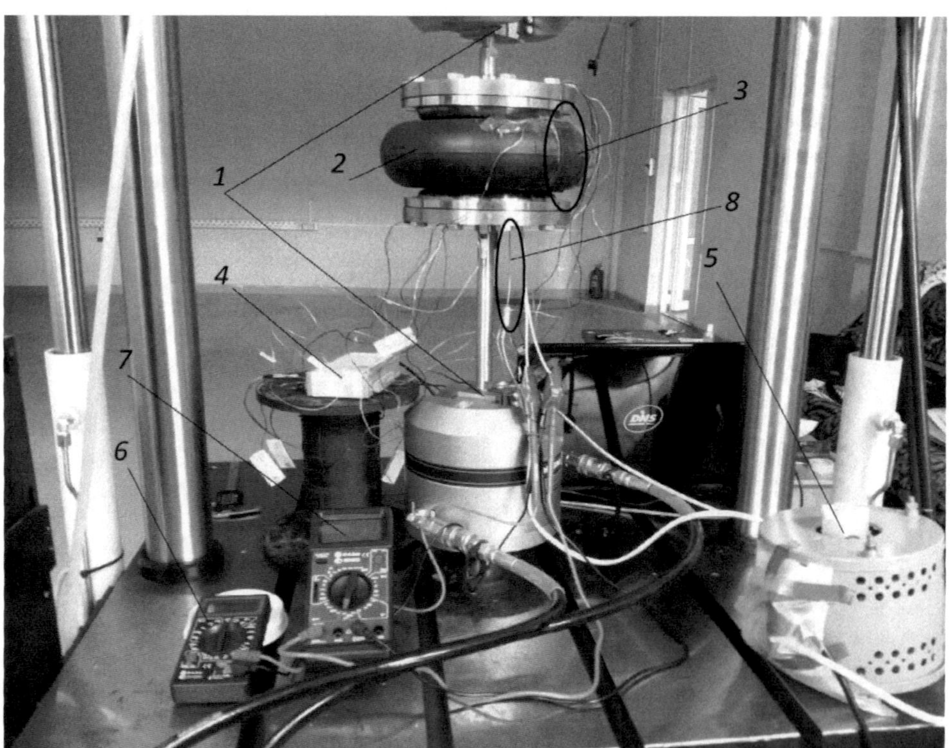

Figure 2. The general view of the test bench.

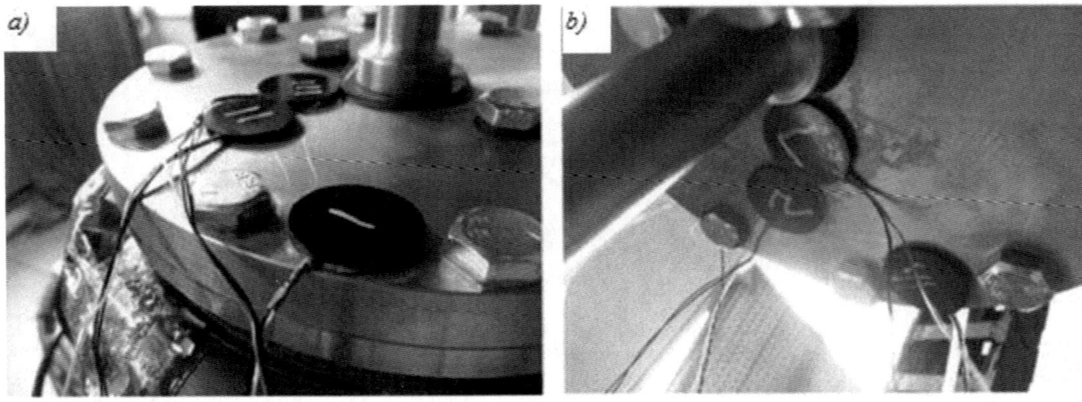

Figure 3. The heat flux sensors position at the top flange *(a)* and at the bottom flange *(b)*.

3. Theory with test procedure description

During testing, the air inside the rubber-cord casing hermetically sealed by two flanges is heated according to four stages followed by cooling to ambient temperature (the fifth stage) in the conditions of the slow natural process. In this case the voltage supplied, the current magnitude and the power consumed by the heater at every stage are presented in table 1.

Table 1. The conducted tests parameters.

Stage	1	2	3	4	5
Voltage, V	58	78	94	110	0
Current, A	0.18	0.25	0.31	0.37	0
Power, W	10.3	19.5	29.6	40.7	0
Duration, minute	100	100	85	90	–

4. Experimental results

The temperature change in the vicinity of the heater during the tests, as well as the ambient temperature value are shown in figure 4. The readings of the heat flux sensors with the internal thermocouple at the top flange (sensors I, II, III), at the bottom flange (sensors IV, V, VI) and on the casing surface (RCC sensor) are represented in figure 5, figure 6, figure 7 correspondingly.

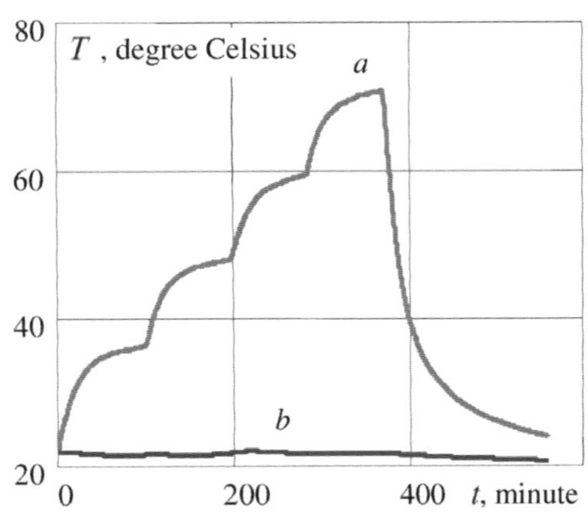

Figure 4. The readings of the temperature sensor in the vicinity of the heater (a) and ambient temperature sensor (b).

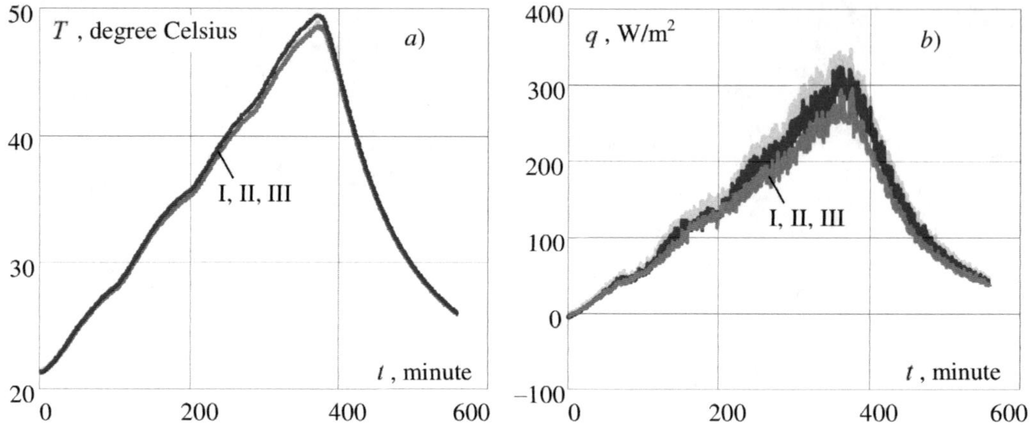

Figure 5. The readings of the heat flux sensors with the internal thermocouple at the top flange (sensors I, II, III): *a* is the temperature *T*; *b* is the heat flux *q*.

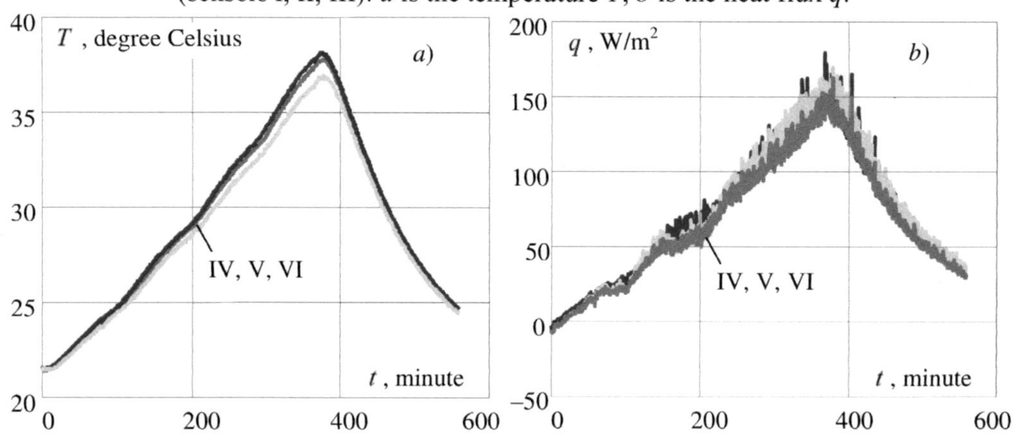

Figure 6. The readings of the heat flux sensors with the internal thermocouple at the bottom flange (sensors IV, V, VI): *a* is the temperature *T*; *b* is the heat flux *q*.

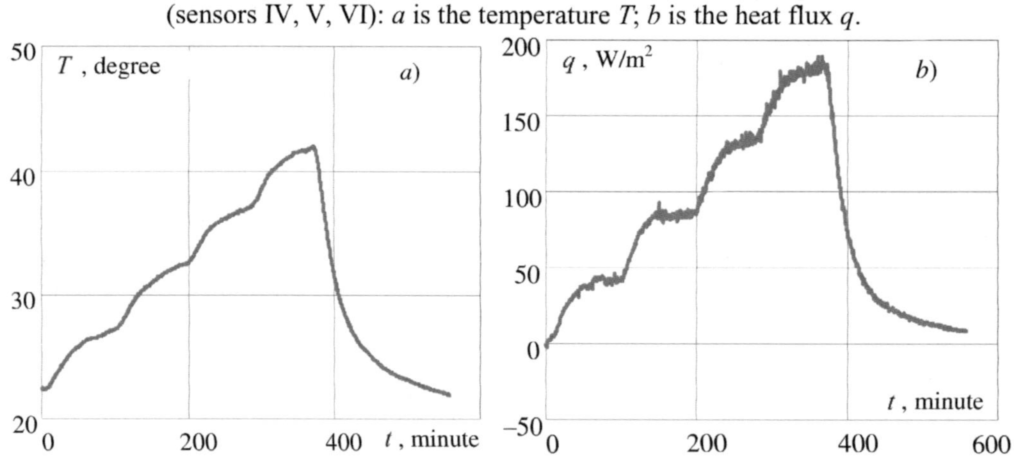

Figure 7. The readings of the heat flux sensor with the internal thermocouple outside on the casing surface (RCC sensor): *a* is the temperature *T*; *b* is the heat flux *q*.

The ambient temperature sensor readings were processed first. The average value was T_0 = 21.53°C with a standard deviation of the mean σ_{T_0} = 8.87·10^{-3}°C. The warming area during which the sensor temperature increased by 1°C from the initial value was separated and thrown away from the heat flux sensors experimental data. The heat transfer coefficient α included in Newton's empirical law (1) and supposed to be the constant value was defined by minimizing the discrepancy

$$\Omega_q = \sum_{k=1}^{N} \left[q_k - \alpha(T_k - T_0)\right]^2$$

calculated according to the readings T_k, q_k ($k = 1..N$) of each sensor separately (N is the total number of test points). The results obtained by the standard least squares method are compiled in table 2. The heat flux calculated value standard deviation (1) from the experimental data was also presented

$$\sigma_q = \sqrt{\frac{1}{(N-1)N} \sum_{k==1}^{N} \left[q_k - \alpha(T_k - T_0)\right]^2}$$

which is the generally accepted error estimation [8, 9].

Table 2. Heat transfer coefficient and its error estimation.

Sensor	I	II	III	IV	V	VI	PKO
α, W/(m^2·K)	9.51	10.28	11.48	8.59	8.76	9.90	8.76
σ_q, W/m^2	0.23	0.30	0.22	0.11	0.14	0.16	0.17

5. Results discussion

Visually the experimental data description accuracy degree by formula (1) with the obtained values of the heat transfer coefficient (table 2) can be qualitatively assessed according to figure 8 for every sensor separately. As can be seen, the agreement is quite acceptable. However, the small values of the heat flux mean values standard deviation σ_q (table 2) are mainly caused by the fact that the total number of test points N is sufficiently high. The given reason mitigates the presence of the experimental data high frequency pulsation on the heat flux in each particular measurement (figure 5-figure 7, *b*). The mentioned pulsations are not observed in the experimental temperature data (figure 5 figure 7, *a*). This is resulted from the heat flux sensor high sensitivity and its appropriate response even to the ambient air weak flows caused by the accidental factors. The internal temperature sensor with a higher response time does not have such a reaction.

IOP Conf. Series: Journal of Physics: Conf. Series **1050** (2018) 012036 doi:10.1088/1742-6596/1050/1/012036

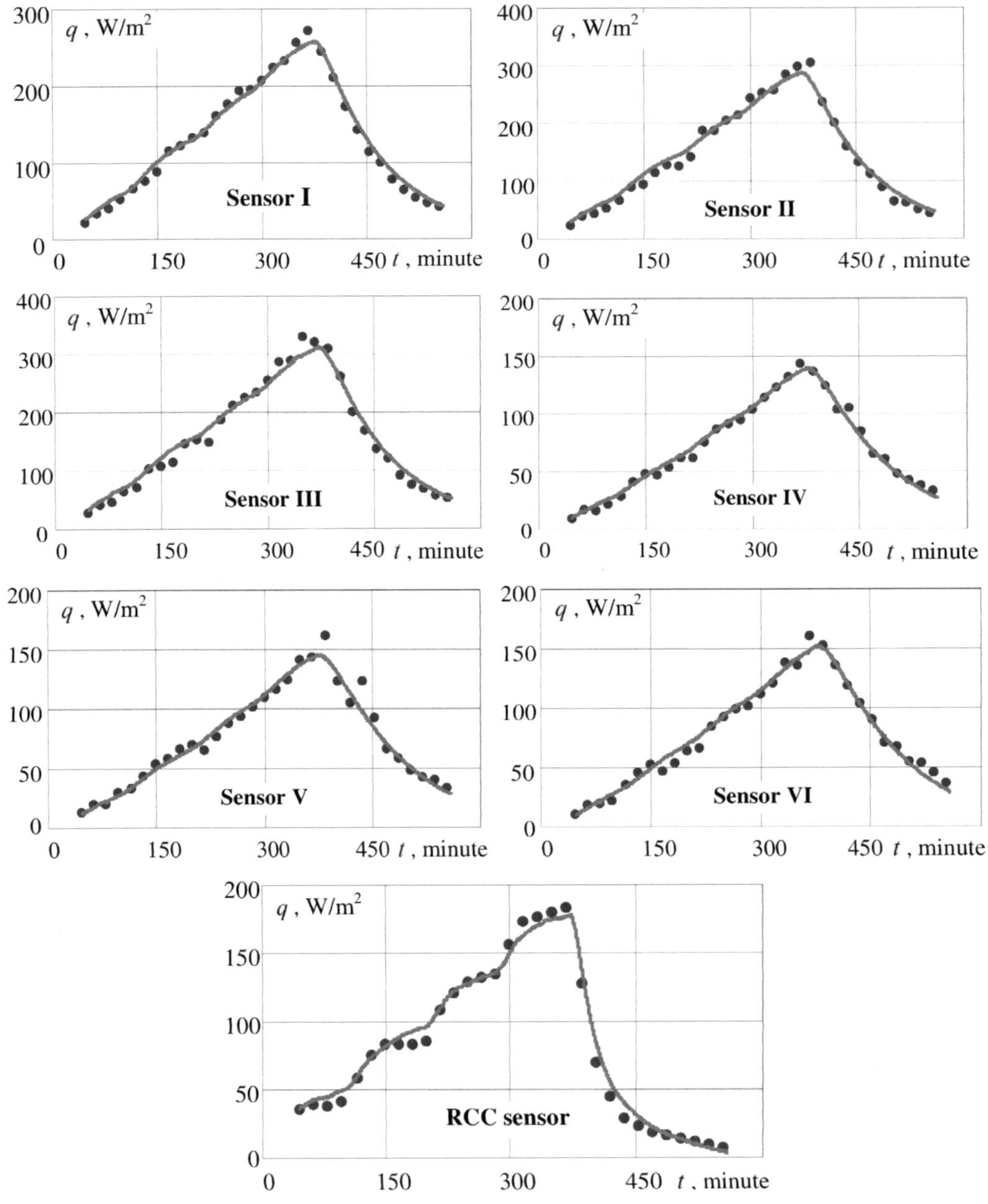

Figure 8. The heat flux calculated according to the formula (1) (the solid line) and the experimental data on the heat flux sensors (dotted line).

6. Conclusions

The data of table 2 lead to the conclusion that when carrying out the calculations by using the heat exchangers technical theory methods it is acceptable to consider in the first approximation that the heat transfer coefficient is constant and identical in magnitude on all outer surfaces limiting the pneumatic element.

7. References

[1] Burian Yu A and Silkov M V 2017 J. Phys.: Conf. Ser. 858 012007 URL: https://doi.org/10.1088/1742-6596/858/1/012007 (Date of the application 14.01.18)

[2] Tesfay A H and Goel V K 2015 *IOP Conf. Ser.: Mater. Sci. Eng. 100 012020* URL: https://doi.org/10.1088/1757-899X/100/1/012020 (Date of the application 14.01.18)

[3] Gavriloski V and Jovanova J 2010 *Dynamic behaviour of an air spring element* (Machines & industrial design engineering vol 4-5) pp 24–7

[4] Presthus M 2002 *Derivation of Air Spring Model Parameters for Train Simulation* (Master Thesis: Lulea University of Technology) p 74

[5] Pevzner Ya M, Gorelik A M 1963 *Pneumatic and hydropneumatic suspension* (Moscow: GNTIML) p 319

[6] Korneyev S A, Korneev V S, Zubarev A V, Klimentyev E V 2016 *Fundamentals of the technical theory of pneumatic shock absorbers* (Omsk: OmGTU) p. 148

[7] *The Engineering ToolBox*, available at URL: https:// www.engineeringtoolbox.com/ convective-heat-transfer-d_430.html (Date of the application 14.01.18)

[8] Squires G L 1968 *Practical Physics* (London: McGraw-Hill) p 246

[9] Hudson D J 1964 *Statistics. Lectures on Elementary Statistics and Probability* (Geneva: CERN) p 296

IOP Conf. Series: Journal of Physics: Conf. Series **1050** (2018) 012037 doi:10.1088/1742-6596/1050/1/012037

Formation of wear resistant nanostructural topocomposite coatings on metal materials by ionic-plasma processing

D N Korotaev[1], E E Tarasov[2], K N Poleschenko[3], E N Eremin[4], E V Ivanova[5]

[1]Siberian State Automobile and Highway University, 5, Mira ave., Omsk, 644080, Russia
[2]FGUP Federal Research and Development Centre "Progress", 4, 5th Kordnaya street, Omsk, 644018, Russia
[3]Omsk Research Institute of Process and Engine Manufacture (RIPEM), 283, street of B. Khmelnytskyi, Omsk, 644021, Russia
[4]Omsk State Technical University, 11, Mira ave., Omsk, 644050, Russia
[5]Omsk Tank-Automotive Engineering Institute, 119, 14th military town, Omsk, 644098, Russia

E-mail: korotaevd99@mail.ru

Abstract. The article regards formation processes of activation and dissipative energy flows under ion-plasma sputtering of coatings. It proposes an advanced system of ionic-plasma processing with the activation of the processed surface by high voltage pulses of electric potential. The authors carry out the experimental evaluation of tribological properties for the samples with coatings and study the features of their surface damage. It is established that the topokompozitny coverings, containing high-disperse nanodimensional clusters reduce probability of brittle destruction during an initial stage of operation of tribocouplings. They show that the use of high-frequency pulse oscillations under ion-plasma sputtering decreases friction ratio and increases wear resistance of coatings.

1. Introduction

Reliability of current technology and the equipment of the machine-building complex depends first on service properties of the materials the details of friction units are made of. The increase in working capacity of surfaces of the metal products, operated under high loadings and speeds, is achieved by coating process providing the formation of optimum meso - micro - nanostructures of the materials. Consequently, the creation of new materials and wear resistant functional coatings, achieved by means of concentrated energy flows, is a promising direction of material science and tribology.

Vacuum ionic-plasma processing (VIPP) possesses wide technological capabilities of composite coatings formation for various service conditions of constructional parts for tribocouplings made of metallic. VIPP with nanostructural topocomposite coatings (NSTCC) application allows modelling the modified structure corresponding to the positive gradient code of mechanical properties in depth. It favorably influences tribotechnical characteristics of friction units [1, 2].

2. Problem statement

There are stress concentrators of a various scale which cause crack nucleation and fatigue failure initiation under ion-plasma sputtering at the interface «a hardened layer (coating) - a base material» [3]. Firstly, this is because when the coating is enlarged for more than 4-5 μm thick, significant

IOP Conf. Series: Journal of Physics: Conf. Series **1050** (2018) 012037 doi:10.1088/1742-6596/1050/1/012037

locked-up stresses can be formed in it. They lead to fast fracture under operating conditions of the product. Secondly, low coating adhesion with the base material can lead to stripping at the initial stage of the product operation. It sharply reduces its service life. Therefore, the development of minimization methods for sudden failures of products in the process of their operation is a major materials science problem.

On the contrary, the creation of nanostructural surface layers with the properties, changing in depth, provides the development of relaxational and adaptive processes under the tribocontact due to the size reduction of deformable volumes, stress decrease in their shearing stability and, consequently, level decrease in peak stresses under operating conditions of modified constructional parts [4].

Considering multifactoring of the phenomena proceeding in surface coatings of the materials under VIPP, it is appropriate their synthesis and analysis from the perspective of the thermodynamic approach [5]. It enables to analyze the influence of structural and technological parameters on adhesion strength of constructional parts with topocomposite coatings considering their adaptive behaviour under operating conditions [6].

The purpose of the given work is the creation and research of new nanostructural topocomposite coatings under structural modification of metal surfaces by the method of vacuum ion-plasma sputtering.

3. Materials and methods

Formation and comparative effectiveness research of nanostructural topocomposite coatings is carried out on the samples of different sorts of structural alloy steels with the use of two VIPP variants. In the work the results of the research for the structural alloy steel 15HGN2TA are given. In the first variant, the three-cathodic system [7] is used. In the second variant, the reflector of the ions is applied, besides the cathodic system. It enables to carry out ionic-plasma processing in the conditions of ionic assisting (fig. 1).

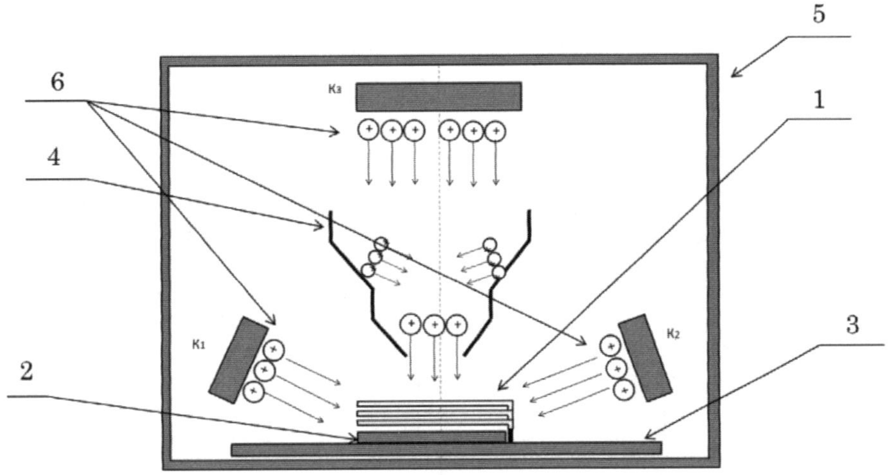

Figure 1. The scheme of the cathodic system with ionic assisting. Designations: K1, K2, K3 – cathodes; 1 – a template; 2 – a sample; 3 – a holder of samples; 4 – a reflector; 5 – a vacuum chamber; 6 – flows of ions.

In the first variant (mode I) coating formation is carried out from the flows of gas and metal plasma without impulse voltage input. In the activation mode the voltage across the underlayer changes in the range of 80-600V at current values of 100A. Coating deposition is carried out at the pressure of 900V with the current strength of 180A.

In the second variant (mode II) under activation across the underlayer there are high-voltage pulses of electric potential with the range of about 10kV, with the period of 10-20 μs and with the frequency of

10-15 kHz under constant negative voltage of 1000V. In the course of the coating deposition on the underlayer the voltage impulse amplitude is about 5kV, the impulse period is about 50-60 μs, the frequency is of 15-25 kHz under constant negative voltage of 1000V. Besides, the reflector of the special design (fig. 1) and the grid pattern are used to receive multicharged ions in the stream.
In both cases, the titanic alloy BT5 is used as the material for the cathodes. Nitrogen is used as a working substance.
Researches of changes of the surface morphology after various modes of ionic-plasma activation and coating deposition are carried out on the scanning probe microscope NTEGRA Prima in the mode of the contact atomic force microscopy (c-AFM). The research of wear resistance of the samples is carried out with the friction factor on the vibration shaker according to the scheme "a shaft- a shoe" under the following conditions: the pressure P = 4 MPa; the traversing speed V = 1 m/s; the testing time t = 60 mines; the vibration frequency f = 10 Hz; the displacement amplitude Ap = 0.2 mm.

4. Results and discussion

There are interacting processes of energy flows, ions or diffusing elements with the process material under ionic-plasma impact on the material surface.
It is possible to present the process of formation NSTCC in the form of the scheme given in fig. 2 that is based on the three-cathodic system [7, 8], which is used earlier.
There is a cathode sputtering with ions, metal and gas mixture production under the voltage generation between the cathode and the anode and in the process of the delivery of the working substance (gas) to the working chamber. It leads to primary surface sputtering or to ions deposition according to the energy of ions and the location of cathodes. There is a jump of the temperature spurring trouble processes of its crystal structure, mass transfer, defect and metastable structure (MS1, MS2) formation because of the energy deposition on the surface layer of the process material. The response of two processes simultaneously (both activation and dissipation) on VIPP leads to the disbalance of energy flows. The prevalence of the dissipative process leading to the ionic flows deposition on the surface with the formation of nanophase germs and their subsequent growth, promotes the nanostructural coating layers [8]. Thermodynamic processes of energy and mass transfer are developed at all stages of ionic-plasma impact.

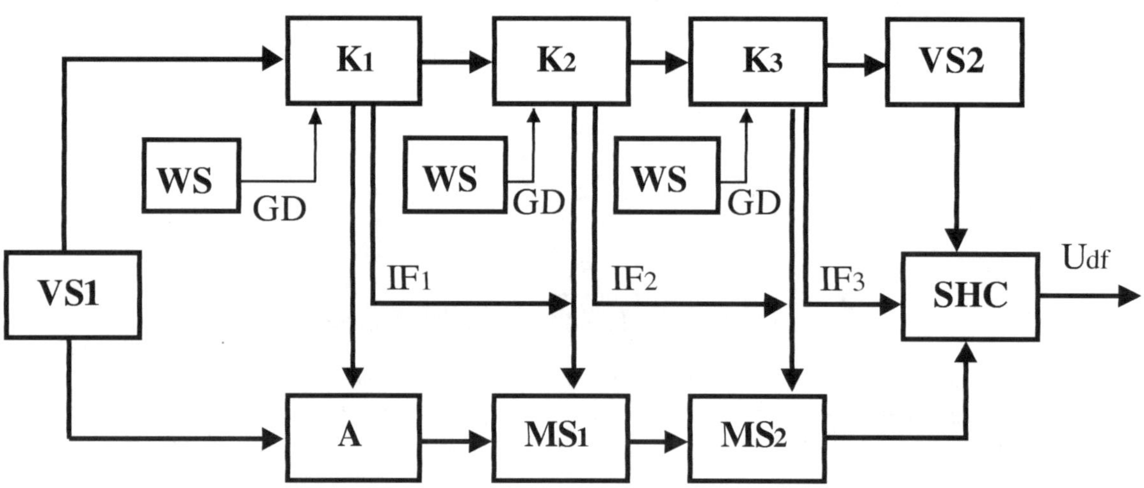

Figure 2. Structural diagram for producing nanostructural topocomposite coatings by ionic-plasma processing. Designations: VS1 – a voltage supply; VS2 – a high-voltage impulse supply; WS – working substance (gas); GD – a glow discharge; A – an anode (a process sample); K1, K2, K3 – cathodes; MS1, MS2 – metastable structures; SNC – a stabilized nanostructural surface layer

IOP Conf. Series: Journal of Physics: Conf. Series **1050** (2018) 012037 doi:10.1088/1742-6596/1050/1/012037

condition; IF1, IF2, IF3 – ion flows generated by corresponding cathodes; Udf – dissipative thermal flows.

High density of the discharge power under VIPP and short duration of its active interaction period with the material keep from generating effective thermal dissipative channels. In this connection, there is a disbalance share of energy deposition, which prevents from relaxation and dissipation of discharge thermal energy.

In fig. 3 there are hypothetical dependences of power density of the energy flow under the activation process (curve 1), the thermal dissipative process (curve 2) and the current density of power consumption in the course of disbalance (zone 3 and curve 4). The charge power density reaches a maximum level at the end of the initial activation period t_a, and then reduces to zero after the time t_n.

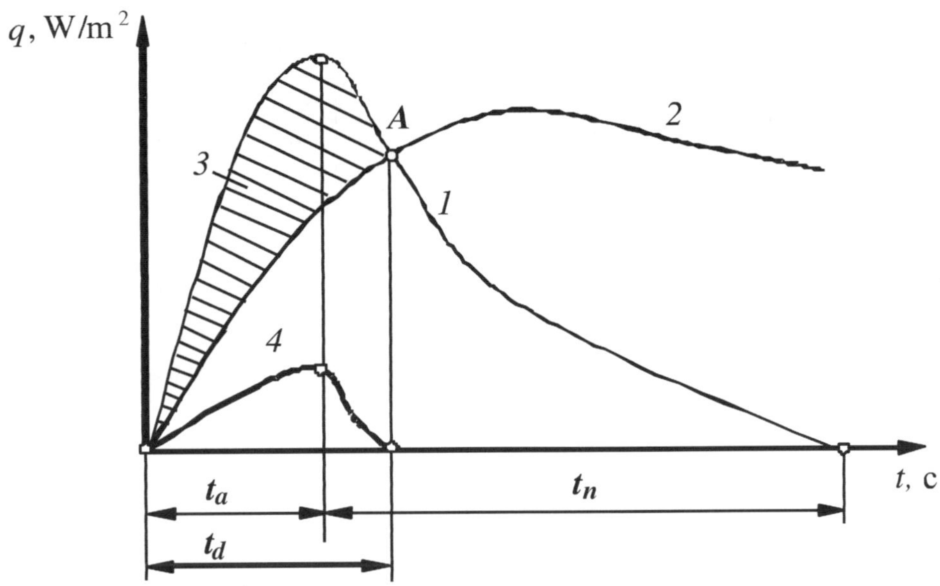

Figure 3. Energetic characteristics of the ionic flow under VIPP: 1 – Power density of the ionic flow for the activation process; 2 – Power density of the flow for the dissipative thermal process; 3 and 4 – disbalance of energy flow

Potentialities of the dissipative thermal channel grow up according to the temperature field generation, however, during the initial period of the ionic-plasma impact its intensity is considerably behind the intensity of the activation process. The maximum activity of the dissipation thermal process is achieved when the peak of energy brought by the ion charge is passed. The balance between the energy of the activation process and the energy taken away by the thermal process is achieved at point A. Hence, the formation of metastable structures takes place in the period from zero to t_d when a dissipative thermal process cannot balance the input charge energy. The disbalance of energy flows under VIPP is defined by the area of the shaded zone 3, and the time dependence for power density of power consumption for the disbalance neutralisation can be presented by curve 4, as the difference between the ordinates of curve 1 and curve 2 in the time span from 0 to t_d.

Thus, in the system two energy flows are set and operate simultaneously: 1 – an activation flow of the input energy, 2 – a dissipative flow, which has a thermal nature and is accompanied by energy and mass transfer processes. The simultaneous development of the two energy flows leads the system to the stabilized nanostructural condition (SNC).

According to the results of the thermodynamic analysis of activation and dissipative flows, developing under VIPP, it is required to provide the stabilization of the nanostructural condition for the coating. For this purpose, it is necessary to increase the values of power density of energy deposition at the

expense of voltage and current increase, because the achievement of their maximum values is limited by technological opportunities of the unit. For the solution of the problem, a cathodic system is developed. The scheme of this system is presented in fig. 1.
The analysis of the given electron microscope images (fig. 4) demonstrates that the surface activation before the coating process in the mode of ionic assisting leads to the development of high-dispersity nanosized formations (fig. 4, a) and to the increase in their density in comparison with the activation without the accelerating voltage supply(fig. 4, b).

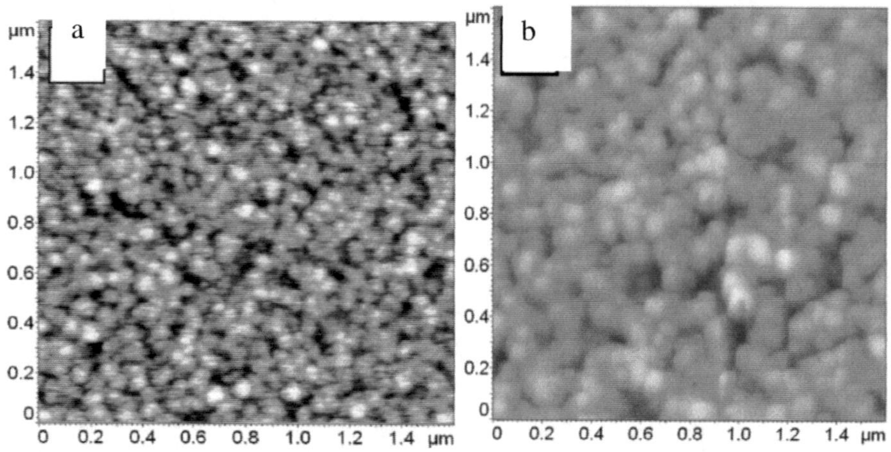

Figure 4. Surface images of the nanostructural topocomposite coating after ionic-plasma processing: a) - voltage with the amplitude of 10 kV, the duration of 10-20 µs, the frequency of 10-15 kHz under constant negative voltage of 1000V; b) – the voltage across the underlayer of 180V, the current strength of 100A.

Fig. 5 presents the given changes of the friction ratio *f* in the course of the surface wear process.

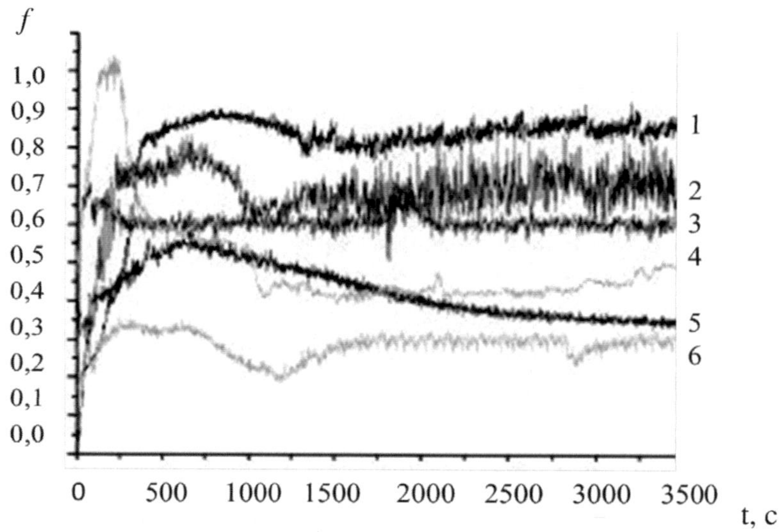

Figure 5. Friction ratio dependence on the testing time of steel samples 15HGN2TA with nanostructural topocomposite coatings: 1 – the voltage across the underlayer of 80V, the current strength of 100A; 2 – the voltage across the underlayer of 100V, the current strength of 120A; 3 – the voltage across the underlayer of 120V, the current strength of 130A; 4 – the voltage across the underlayer of 150V, the current strength of 150A; 5 – the voltage with the amplitude of 5 kV, the duration of 10-20 µs, the frequency of 10-15 kHz, the negative voltage of 1000V; 6 – the voltage with

the amplitude of 10kV, the duration of 10-20 μs, the frequency of 10-15 kHz, the negative voltage of 1000V.

The analysis of the quoted results suggests that the greatest decrease in the friction ratio is achieved involving the cathodic system, given in fig. 1 under the surface activation by means of high-voltage impulses supply across the underlayer. Moreover, the increase in the amplitude of impulses from 5 to 10 kV has a great effect on the decrease in the friction ratio (dependences 5 and 6 in fig. 5).

5. Conclusion

1. The development of the structural diagram for nanostructural topocomposite coatings is proposed based on the analysis of activation and dissipative processes under forming functional coatings by ionic-plasma processing.
2. The system of ionic-plasma processing including a cathodic complex with ionic assisting and the reflector of the special design are developed for the efficiency upgrading of the ion-plasma sputtering method and for the generation of the stabilized condition of nanostructural coatings. It enables to receive topocomposite surface layers with the structure, containing high-dispersity nanosized formations, reducing the probability of fracture failure of coatings during the initial operating period for modified structural components of tribocouplings.
3. The result of the tribological tests is that the greatest decrease in the friction ratio is achieved involving the advanced cathodic system under the surface activation in the course of the formation of nanostructural topocomposite coatings by means of high-voltage impulses supply across the underlayer.

References

[1] Upadhyay R, Brossard N and Chen J 2016 Mechanisms underlying astringency: introduction to an oral tribology approach *J. Phys. D: Appl. Phys.* **49** pp 104003

[2] Matthews A, Franklin S and Holmberg K 2007 Tribological coatings: contact mechanisms and selection *J. Phys. D: Appl. Phys.* **40** pp 5463

[3] Panin V E, Moiseenko D D, Panin S V, Maksimov P V, Goryacheva I G and Cheng C H 2014 Mechanisms of elastic energy dissipation in the transition layer between a coating and a substrate under contact interaction *Journal of Applied Mechanics and Technical Physics* **55(2)** pp 318326

[4] Poleschenko K N and Khudyakov O D 2016 Fretting-resistance of nanostructured topo composites on the titanic basis *Strengthening technologies and coatings* **11(143)** pp 44-48

[5] Prigogin I and Kondepudi D 2002 *Modern thermodynamics* (Moscow: World) p 461

[6] Pogrebnyak A D, Bagdasaryan A A, Pshik A V and Dyadyura K A 2017 Multicomponent nanocomposite coatings with adaptive behaviour in surfacel engineering *Successes of physical sciences* **6(187)** pp 629-652

[7] Poleschenko K N, Korotaev D N and Tarasov E E 2016 Structural and morphological features of nanostructural topo composite coatings for tribotechnical purpose *Vestnik SibADI* **4(50)** pp 126-132

[8] Grinberg P B, Poleschenko K N, Goryunov V N and Tarasov E E 2012 Method of nanostructural topo composites producing for load-bearing improvement of constructional parts of power equipment *Vestnik of Omsk University* **2** pp 253-258

IOP Conf. Series: Journal of Physics: Conf. Series **1050** (2018) 012038 doi:10.1088/1742-6596/1050/1/012038

Comparative analysis of methods of cargo vibration damping moved by overhead crane

M S Korytov[1], V S Shcherbakov[1] and V V Titenko[2]

[1]Siberian automobile and highway university, 5, Mira ave, Omsk, Russia
[2]Omsk State Technical University, 11, Mira ave., Omsk, 644050, Russia

Abstract. A comparative analysis of three different methods of damping cargo plane oscillations at the approach of the lifting body of the overhead crane to the point of delivery: using a proportional-integral-derivative controller for speed load, using proportional-integral-differential regulator for moving cargo and the braking point of suspension of cargo with constant acceleration. We consider the advantages and disadvantages of each method, the recommended range of applications.

Keywords: overhead crane, PID, velocity, displacement, load, vibration damping, relay drive

1. Introduction

It is necessary to fulfil complete damping of residual pendulum oscillation of cargo when cargo lifting part overhead crane (OC) approaches (come nearer to the point of load delivery [1 – 6]. It is used to reduce time of cycle of load displacement by means of OC. Let us consider the methods of solving this task without using additional mechanical arrangement and without devices of oscillation damping, i.e. using only OC drives and cargo trolley which OC has.

We examine the process of load oscillation in plane, i.e. the separate co-ordinate of pendulum system OC. Spatial load cargo oscillations can be presented as superposition of cargo oscillations in two mutually perpendicular planes[1]. These oscillations can be represented in the sense of (by) slight amplitude.

The following parameters are taken in the capacity of initial data of task (figure 1):

– length of OC cargo rope from the point of cargo suspension to the cargolifting part;
– mass of cargo m_{car};
– damping coefficient anglewise b_r;
– initial velocity of translational displacement of point of suspension (speed of bridge movement and cargo trolley OC) v_{px};
– angle of initial rope displacement from gravity vertical R_x;
– initial velocity of angle changing of rope slope vR_x;
– finite value of the horizontal coordinate of cargo movement $X_{car\,fin}$.

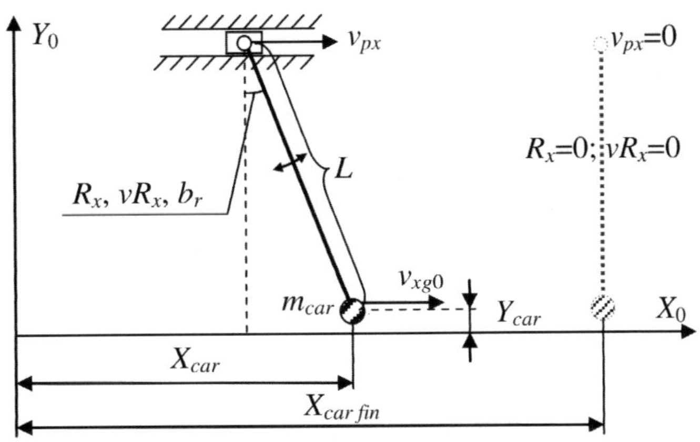

Figure 1. Computational scheme of .task of damping cargo oscillations when lifting body of the bridge crane is approaching to the point of delivery

The task can be formulated as the damping of the random load oscillations which are at the initial time moment. They are on the flexible lifting bar with simultaneous full stopping of the suspension point in a given position. Prior to the start time, the targeted management of suspension point with the purpose of damping oscillations of the load were not performed.

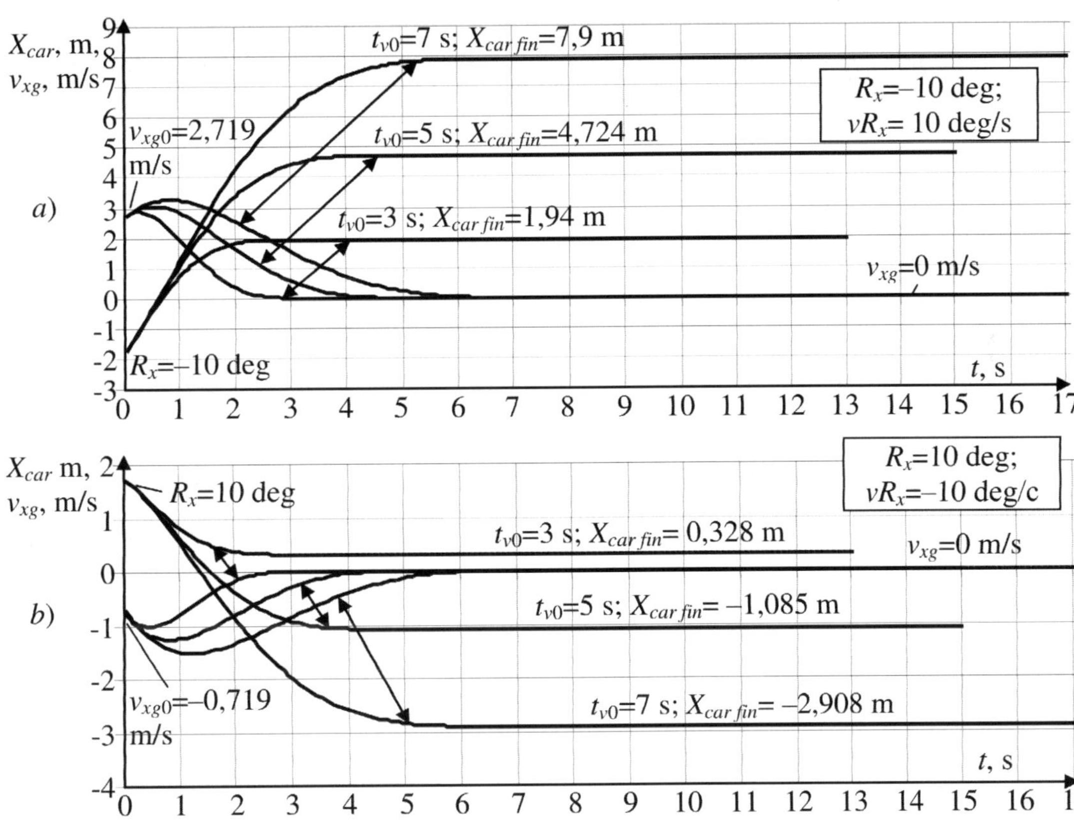

Figure 2. Set (required, reference) time dependences of the displacement $X_{car}(t)$ and velocities of the cargo $v_{xg}(t)$ for full-time damping t_{v0}=3, 5 and 7 s: *a*) when the initial conditions R_x=–10 deg; vR_x=10 deg/s; *b*) with initial conditions R_x=10 deg; vR_x=–10 deg/sec (example for L=10 m, v_{px}=1 m/s)

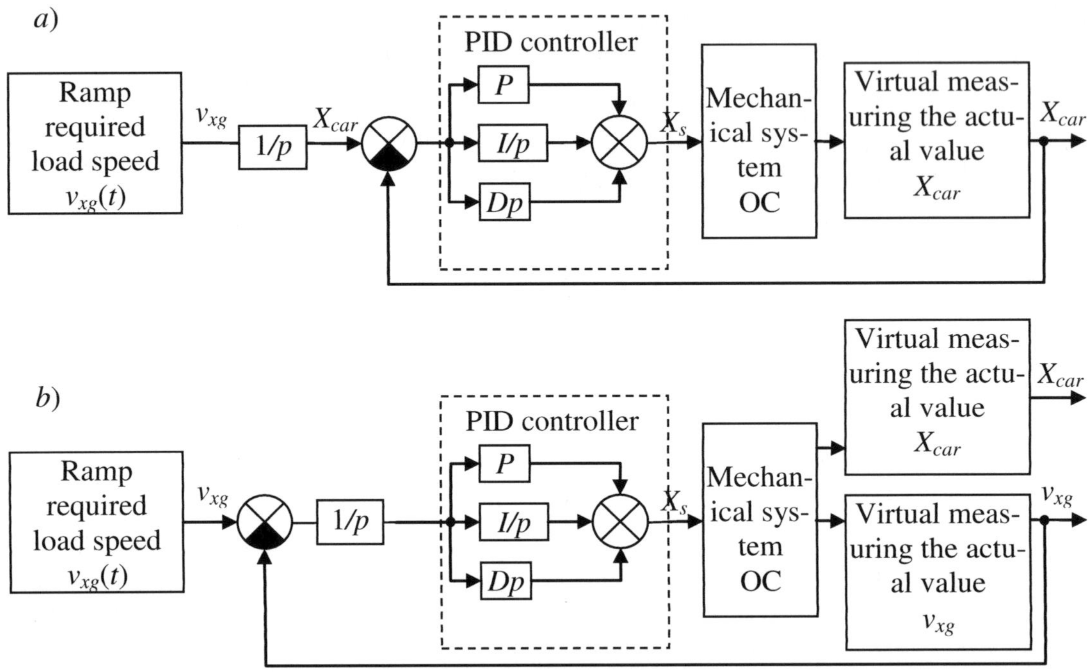

Figure 3. Schematic relations of the two variants of the model of the mechanical subsystem of OC with PID controller: *a*) method № 1, the development of the movement; *b*) method № 2, velocity development

For OC with flexible cable suspension lifting body there are known ways to ensure that freight movement on the required trajectory using proportional-integral-differential (PID) control of each of the two axes of the load (bridge and trolley) [7 – 16].

For their application we have developed simulation model OC with PID controller [8, 15, 16] in the MATLAB/Simulink system.

For their application we use simulation model OC with PID controller [8, 15, 16] developed in the MATLAB/Simulink system.

The final value of the horizontal coordinates of the movement of cargo $X_{car\,fin}$ is functionally related to the time of complete stop of cargo t_{v0}. Taking into account that in a simulation Simulink-model input stimulus are given in the form of time functions, as well as the need to ensure the boundary conditions of equality to the specified values of the first two derivatives of displacement of the load along the axis X_0 (v_{xg0} speed and acceleration a_{xg0}), in the case studies as parameters that affect the coordinate of the end position of the cargo, were: the value of time of full cargo stop t_{v0}, the initial velocity of the load along the axis X_0 v_{xg0} (further – initial velocity) and an initial acceleration of the load along the axis X_0 a_{xg0} (further – initial acceleration). In the final moment of time we have the following corresponding boundary conditions, based on the wording of the tasks – v_{xg}=0; a_{xg}=0.

In figure 2 as the examples we have given with the use of splines (in B-form 6-degree) time dependence of the displacements $X_{car}(t)$ and velocities of the cargo $v_{xg}(t)$ for full-time damping t_{v0}=3, 5 and 7 s for two variants of initial conditions (R_x=–10 deg; vR_x=10 deg/s and R_x=10 deg; vR_x=–10 deg/sec). The length of the hoist rope took the value L=10 m, the initial velocity of the suspension point a v_{px}=1 m/s.

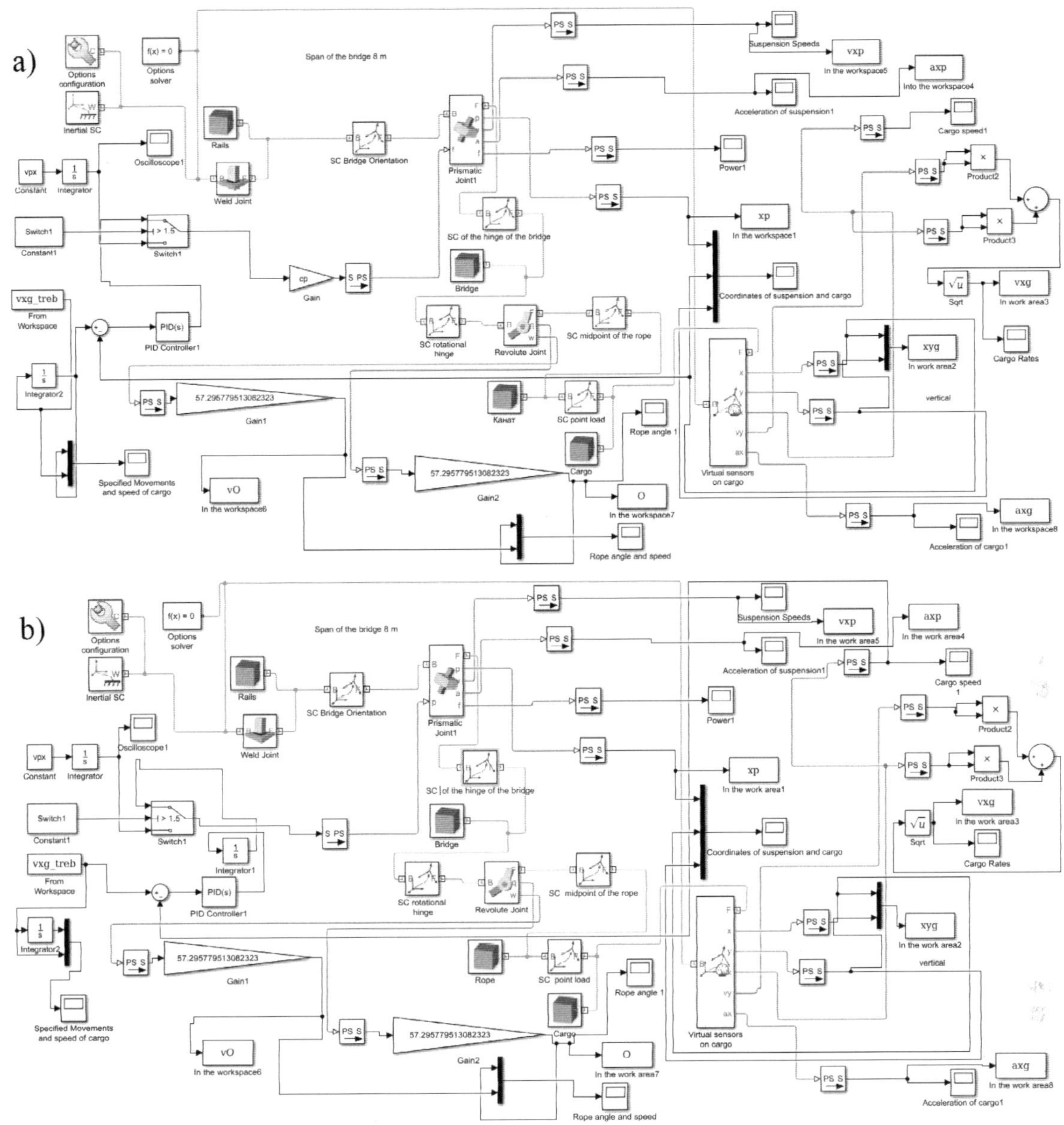

Figure 4. Simulated model of OC with PID controller in symbols of SimMechanics Second Generation и Simulink: *a*) method № 1; *b*) method № 2

Let us consider, on a simulation of mathematical model, feasibility of using for the task of PID controller for horizontal coordinate movement of cargo X_{car}. This regulator may be installed in at least two places: as practicing the desired movement of the load $X_{car}(t)$ (method № 1, figure 3, *a*) and as attribute of maintaining the required load velocity $v_{xg}(t)$ (step 2, figure 3, *b*). Given displacement of the suspension point in a simulation model is in to form of rigidly prescribed displacement to the input of block hinge with translational degree of freedom (Prismatic Joint1, figure 4).

Method № 1. When developing the PID controller moving on the 6 trajectories shown as an example in figure 2, using the simulation model, shown in figure 4, and constructed according to the scheme in figure 3, we have obtained values of the coordinates of the suspension and of the cargo and velocities of the point of suspension which are shown in figure 5, 6.

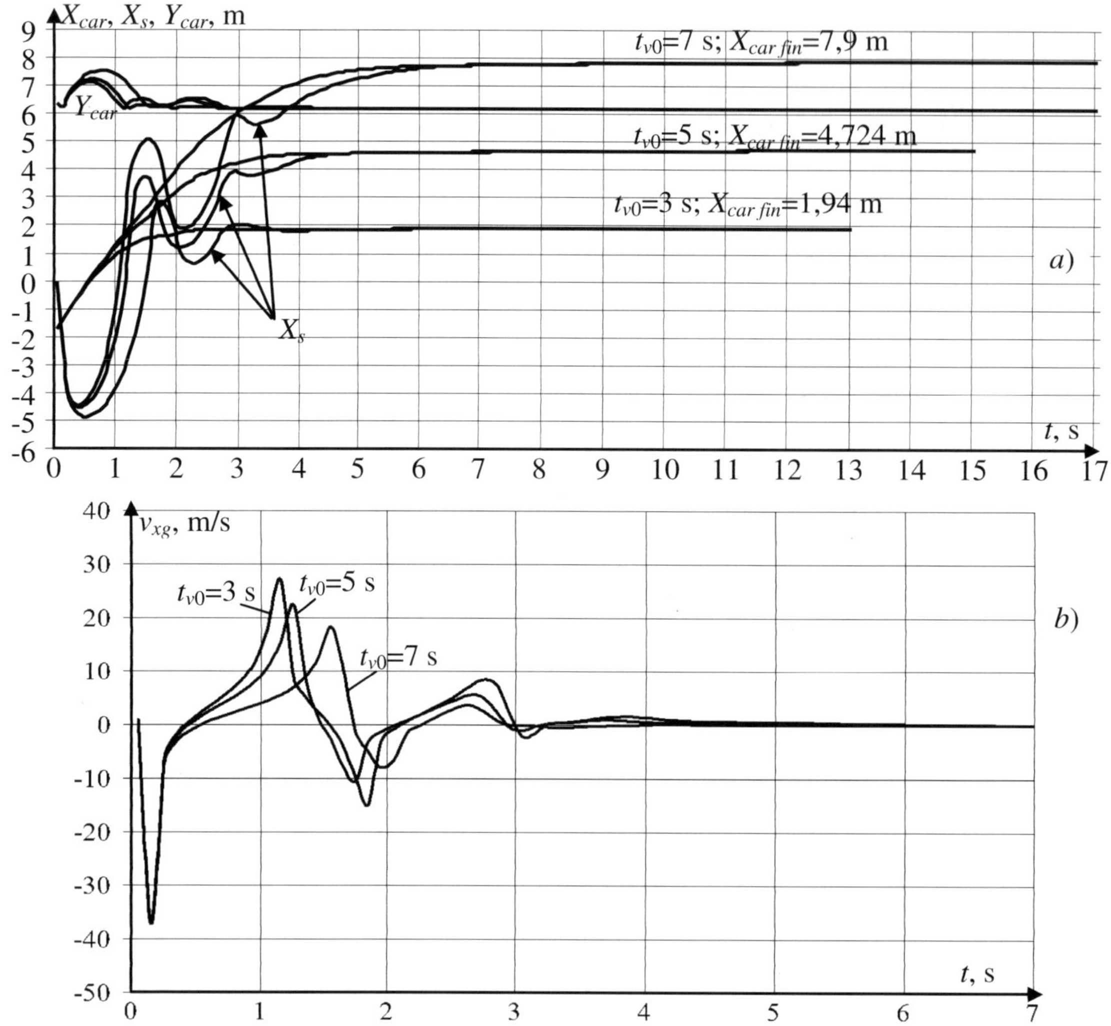

Figure 5. Graphs of time histories of the coordinates of the lifting bar and cargo (*a*) as well as velocities of suspension points (*b*) for method № 1, with initial conditions R_x=–10 deg; vR_x=10 deg/s for different values t_{v0}=3, 5, 7 s

Method № 2. When PID controller has development of cargo velocity on 6 trajectories, shown as an example in fig 2 and using simulation model, shown in figure4, based upon the scheme in figure 3, *b*, we have obtained the values of value of coordinates of lifting bar and cargo as well as velocities of suspension point, shown in figure 7, 8.

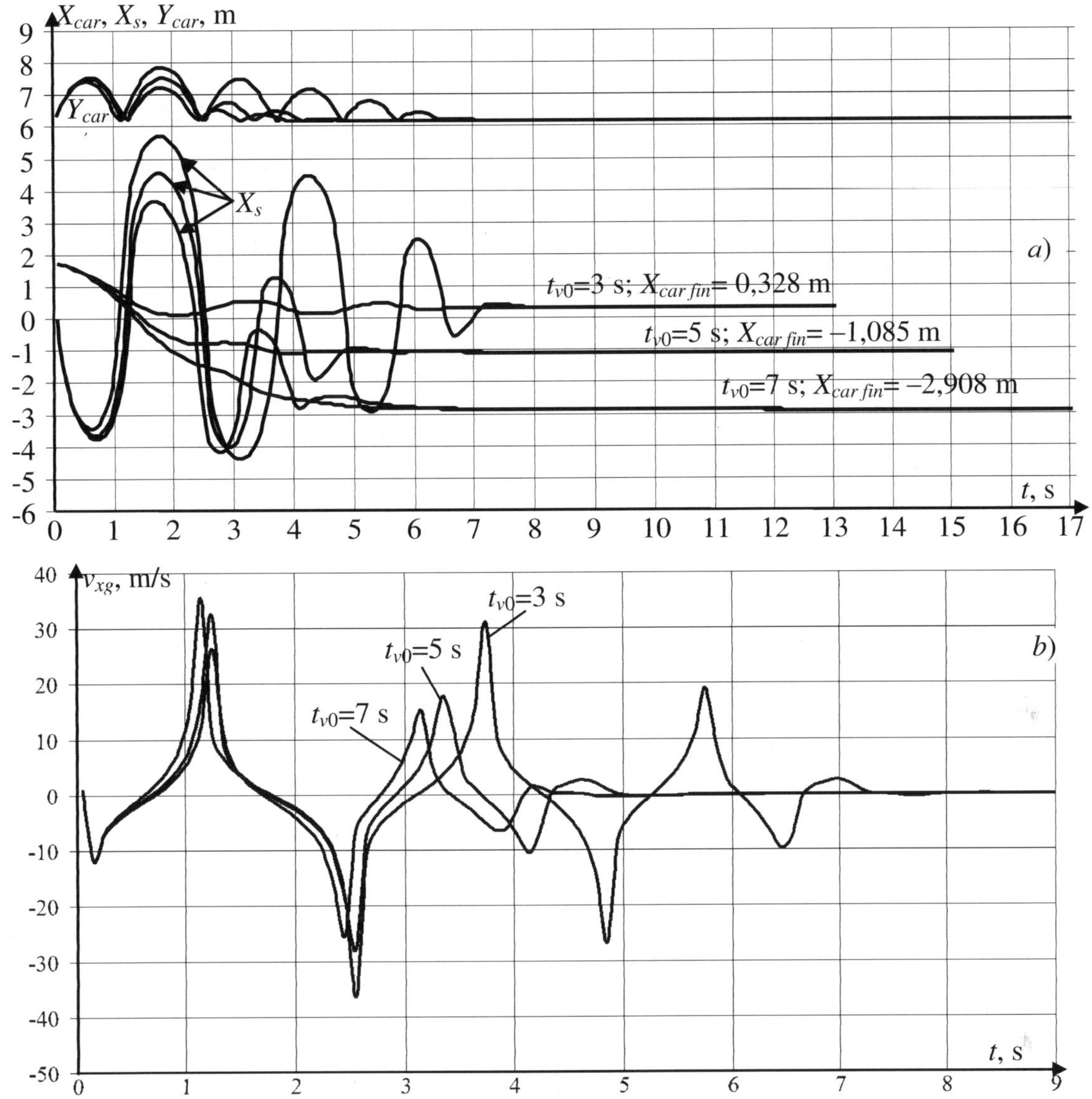

Figure 6. Graphs of time histories of the coordinates of the lifting bar and cargo (*a*) as well as velocities of suspension point (*b*) for method № 1, with initial conditions R_x=10 deg; vR_x=–10 deg/s for different values t_{v0}=3, 5, 7 s

The constant data of this math model take the value: L=10 м; b_r=100 N·m/(rad/s); m_{car}=100 kg; v_{px}=1 m/s. Varied data take the value R_x=±10 deg; vR_x=±10 deg/s; t_{v0}=3, 5, 7 s.

Analysis of testing using a PID controller of the same trajectories movement of cargo, compared with the actual displacement of the load (see figure 5, 6) and the speed of movement of the load (see figure 7, 8) have shown that method № 2 testing the speed of movement of the load gives an order of magnitude smaller velocity and acceleration of suspension point with the same final result. That is, allows the use of less powerful and fast-speed drive of suspension point of the cargo.

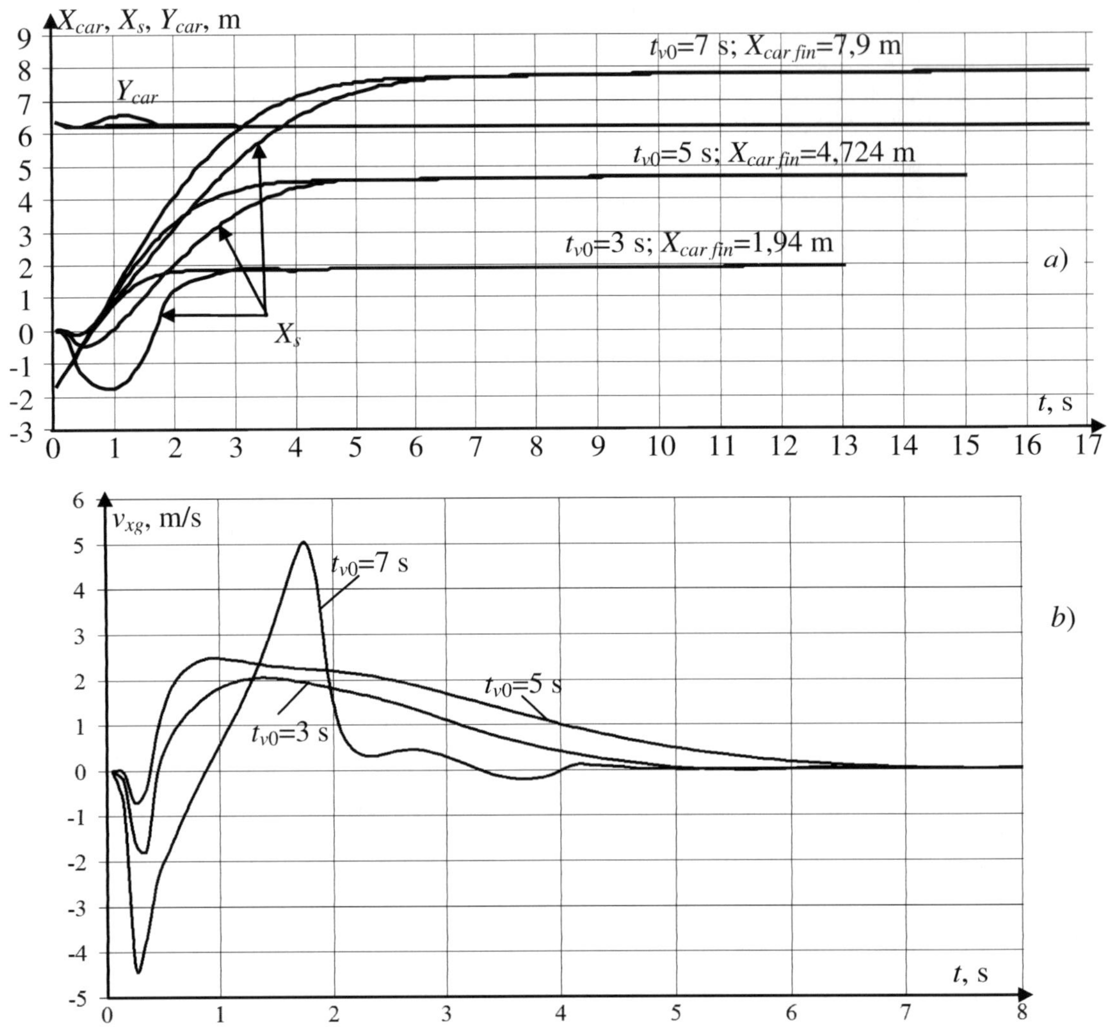

Figure 7. Graphs of time histories of the coordinates of the lifting bar and cargo (*a*) as well as velocities of suspension point (*b*) for method № 2, with initial conditions R_x=–10 deg; vR_x=10 deg/s for different values t_{v0}=3, 5, 7 s

The velocities and accelerations synthesized by the method № 1, are actually not realizable using existing drives based on electric engines. For their implementation it is possible to use hydraulic drive, which has its limitations, in particular, on the maximum displacement of the cylinder rod.

Method № 3. When using in present on the majority of the working OC holding brake we can make the assumption that braking of OC and its cargo trolley occur with constant acceleration [17].

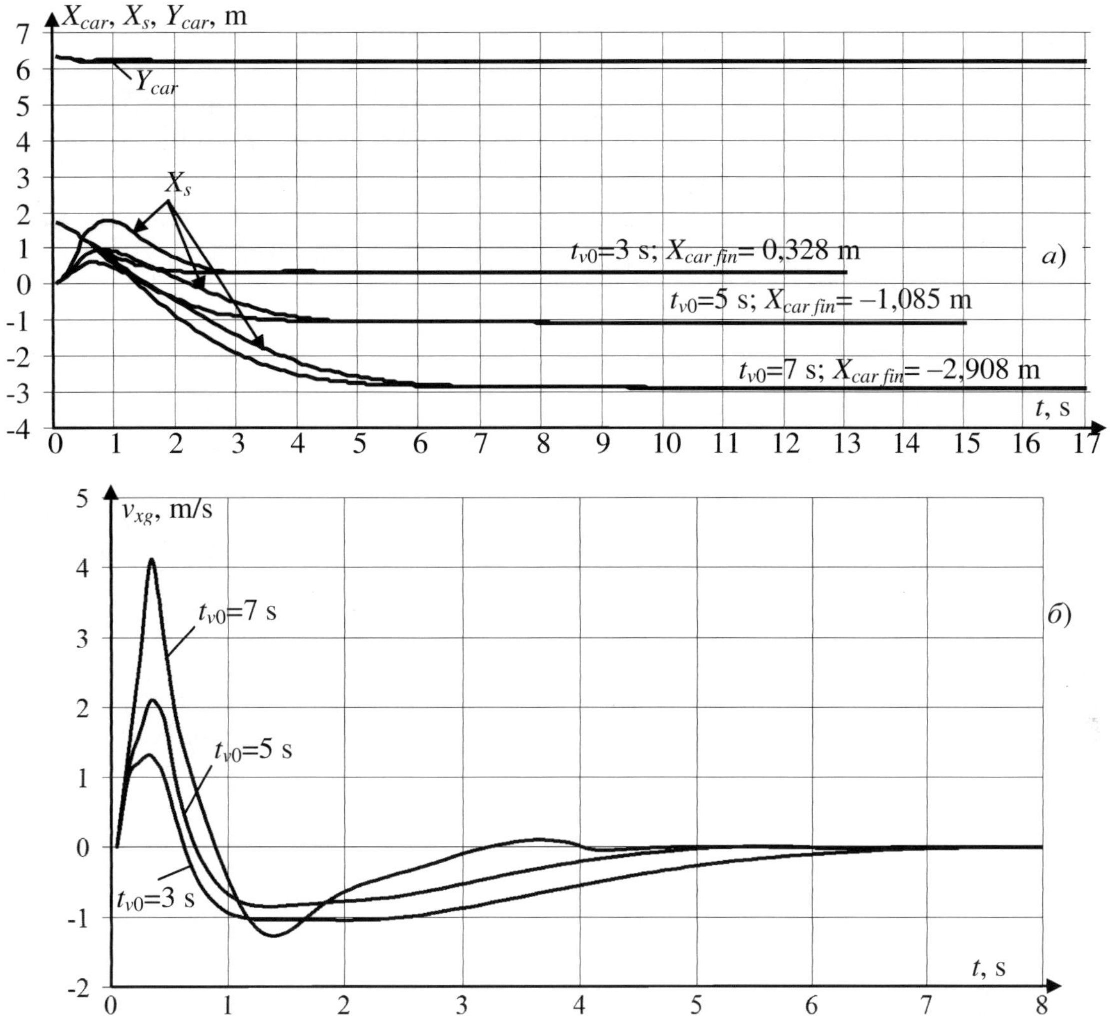

Figure 8. Graphs of time histories of the coordinates of the lifting bar and cargo (*a*) as well as velocities of suspension point (*b*) for method № 2, with initial conditions R_x=10 deg; vR_x=–10 deg/s for different values t_{v0}=3, 5, 7 s

A constant value of braking acceleration limits the possibility of damping the load close to the zero values of the residual rate of change of the angle of inclination of the hoist rope and the residual velocity of the suspension point can only be attained when all the values of the system parameters OC (v_{px} R_x vR_x) at time t=0, after which it is permissible and desirable braking. In addition, the movement of the load along the axis X_0 in the process of braking and the time interval until complete OS installations of cargo and suspension (options $X_{car\,fin}$ and t_{v0}, see figure 1) depend on combination of the initial values of the parameters v_{px} R_x vR_x. When using method № 3 we can observe that the braking of suspention point with constant acceleration a_{xg}=const, in general case, does not coincide with time t=0 (starts later), after which it is permissible and desirable braking. Time of actual deceleration with constant acceleration $\Delta Term$, based on the conditions of tasks, is functionally related to the initial constant velocity of the suspension point v_{px}:

$$\Delta Term = v_{px}/a_{xg}. \quad (1)$$

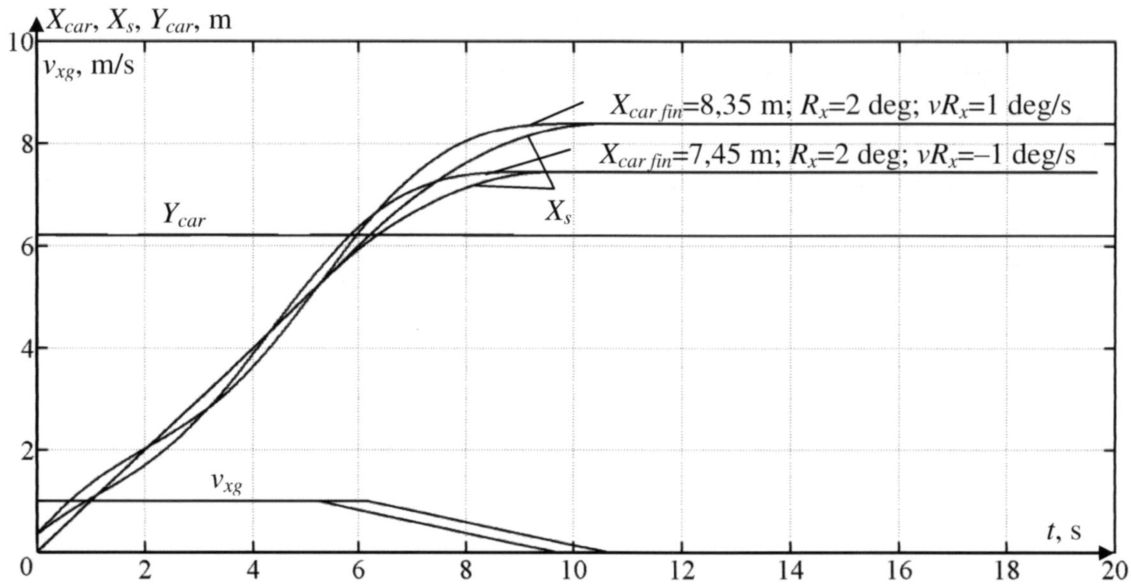

Figure 9. The obtained optimal time dependence of the coordinates of the lifting bar and cargo

In figure 9 as an example, temporal dependencies of co-ordinates of lifting bar and cargo (*a*) and velocity of the suspension point (*b*) for the method № 3 for several combinations of initial conditions (v_{px}=1 m/s; R_x=2 deg; vR_x=1 deg/s), (v_{px}=1 m/s; R_x=2 deg; vR_x=–1 deg/s) are given. Constant data of the mathematical model took the values given above. The optimal values a_{xg} and $\Delta Term$ were obtained by Simplex-method taking into account the condition of minimization of the function

$$y=\dot{\theta}_{res}+v_{res}, \tag{2}$$

where $\dot{\theta}_{res}$ is the maximum residual rate of change of the deflection angle of the hoist rope from the vertical after the end of braking; v_{res} – maximum residual velocity of the suspension point of the cargo after a moment the termination of braking. Given in (2) the indicators were determined using the signal processing virtual meter in the simulating model OC.

Dependencies illustrate a significant difference in obtained values $X_{car\,fin}$ and t_{v0}.

Conclusions. To implement the first two considered methods it is necessary to use continuously variable ratio drives which are able not only to brake, but to accelerate to different speeds in a wide range in the damping of the residual oscillations of the load, which can be considered a significant limitation. The advantages of the first two considered methods with using of PID controllers might be the ability to stop cargo at the point with exactly the given coordinates.

Method № 3 does not allow to stop the load at a point with precisely defined coordinates and at a specified time. The damping of vibrations of deposits is limited. At the same time, it is the only one of the three considered methods allows to use already available on most construction-structures OC brake lever. It does not require the complication of a design of drive and the brake drive and does not require to use continuously variable ratio drives . That is, from a practical point of view, method № 3 is the least expensive and easily implemented.

At the same time, method № 2 is perspective – development of braking on velocity of load displacement only when OC has continuously variable ratio drives.

References

[1] Shchedrinov A V, Serikov S A and Kalmykov V V 2007 Automated system for quieting the fluctuations of the load for the overhead crane *Devices and systems. Management, control, diagnostics* **8** 13–7

[2] Blackburn D, Singhose W, Kitchen J, Patrangenaru V and Lawrence J 2010 Command Shaping for Nonlinear Crane Dynamics *Journal of Vibration and Control* **16** 477–501

[3] Tolochko O I and Bazhutin D V 2010 Comparative analysis of methods of damping oscillations of the load, to the mechanism of the translational motion of overhead crane *Electrical machinery and electrical equipment* **75** 22–8

[4] Omar H M 2003 *Control of gantry and tower cranes : PhD Dissertation* (Blacksburg, Virginia: Virginia Polytechnic Institute and State University) p 100

[5] Abdel-Rahman E M, Nayfeh A H and Masoud Z N 2003 Dynamics and control of cranes: a review *Journal of Vibration and Control* **9** 863–908

[6] Fang Y, Dixon W E, Dawson D M and Zergeroglu E 2003 Nonlinear coupling control laws for an underactuated overhead crane system *IEEE/ASME Trans. Mechatronics, Vol. 8* **3** 418–23

[7] Shcherbakov V S, Korytov M S and Wolf E O 2014 Method of increasing the accuracy of the trajectory of the moving object loading crane by compensating its unmanaged spatial oscillations *Mechanization of construction* **2** 21–5

[8] Shcherbakov V S, Korytov M S, Sukharev R U and Volf E O 2015 Mathematical modeling of process of moving cargo by overhead crane *Applied Mechanics and Materials* **701-702** 715–20

[9] Denisenko V V 2007 Varieties of PID controllers *Automation in the industry* **6** 45–50

[10] Korytov M S, Glushets V A and Zyryanova S A 2004 Simulation of working movements of the crane using SimMechanics and Virtual Reality Toolbox *Exponenta Pro. Mathematics in applications* **3** 94–102

[11] Korytov M S and Zyryanova S A 2004 Modeling a dynamic system of the crane with assistance package block "SIMMECHANICS" of the system MATLAB *Omsk scientific Bulletin* **4** 88–90

[12] Shcherbakov V S, Korytov M S and Kotkin S V 2012 *Automation System simulation boom cargo cranes: a monograph* (Omsk: SibADI) p 143

[13] Shcherbakov V S, Korytov M S and Wolf E O 2014 Increase of accuracy and speed of cargo moving along the required trajectory by a bridge type crane *Systems. Methods. Technology* **4** 52–7

[14] Shcherbakov V S, Korytov M S and Wolf E O 2014 The system of extinction of spatial oscillations of cargo, moved by a bridge crane *Vestnik SibADI* **6** 56–61

[15] Korytov M S, Shcherbakov V S and Volf E O 2015 Impact sigmoidal cargo movement paths on the efficiency of bridge cranes *International Journal of Mechanics and Control* vol 16 **2** 3–8

[16] Shcherbakov V S, Korytov M S and Volf E O 2015 Reduction of the acceleration drive of bridge cranes through judicious choice of the trajectory of a moving load *Journal of the Serbian Society for Computational Mechanics* Vol 9 **1** 57–64

[17] Ivanchenko F K 1978 *Calculations of hoisting and conveying machines* (Kiev: Vishha shkola) p 576

IOP Conf. Series: Journal of Physics: Conf. Series **1050** (2018) 012039 doi:10.1088/1742-6596/1050/1/012039

Modeling of a Temperature Distribution in Optomechanical Thermal Microsystems

A G Kozlov

Omsk State Technical University, 11, Mira ave., Omsk, 64450, Russia
kozlov1407@gmail.com

Abstract: An approach for modeling the temperature distribution in optomechanical thermal microsystems is presented. Three types of the microsystems are considered: microsystem with two supporting microbeams in adjacent corners of the microplate; microsystem with two supporting microbeams in opposite corners of the microplate; microsystem with four supporting microbeams. In the structure of each type of the microsystems, the modeling domains are marked out, which are divided into zones with homogeneous equivalent parameters. Analytical expressions for the temperature distribution in the zones of optomechanical thermal microsystems are obtained. The method of modeling was used to determine the temperature distribution in bi-material sections in cantilever microbeams in the optomechanical thermal microsystems under consideration. The dependence of the temperature difference on the boundaries of the bi-material section on the thickness of the aluminum layer in this section was determined.
Keywords: optomechanical thermal microsystem, temperature distribution, method of eigenfunctions, microplate, cantilever microbeam, bi-material

1. Introduction

The modern stage of the mechanical engineering development is characterized by the extensive use of microsystems. Optomechanical thermal microsystems represent a new type of microsystems, in which the advantages of the three physical effects are combined: optical, mechanical and thermal. The development and investigation of these microsystems began about twenty years ago and are intensively continuing at the present time. The feature of these microsystems is that in them the electrical part is excluded from the transformation processes and, accordingly, the influence of electromagnetic effects on the characteristics of the microsystem is not significant.

The operation of the optomechanical thermal microsystems is based on the following principles:
– formation of an elevated temperature field in the microsystem structure by converting a measured or control action into a thermal signal;
– usage of a bimetallic effect in the cantilever microbeams of the structure to move the microplate;
– usage of optical reading to determine the amount of a microplate movement;
– usage of structures made by MEMS technology to increase a temperature sensitivity of bimetallic elements.

For the first time, optomechanical thermal microsystems in the form of bimetallic cantilever microbeams were used as chemical sensors for measuring the amount of heat released in a chemical thermocatalytic reaction by the substance being determined [1]. At present, an important and promising application of optomechanical thermal microsystems is their use as sensors for measuring temperature [2], uncooled IR detectors [3-6] and elements of scanning IR images [7-12].

The characteristics of optomechanical thermal microsystems depend on three subsystems: optical, mechanical and thermal. Among these subsystems, the main role is played by the thermal subsystem, which to a large extent affects the dynamic characteristics of the optomechanical thermal microsystem [13, 14] and its sensitivity. To take into account this influence of the thermal subsystem, it is necessary to know the temperature distribution in the structure of the optomechanical thermal microsystem. Currently, this issue is being intensively developed.

2. Formation of the Problem

Optomechanical thermal microsystems are manufactured using surface micromachining. They consist of the following elements:

1) the microplate with a functional layer in which the energy of input action is converted into thermal energy;
2) the cantilever microbeams which on the one side are fixed on the base and on the other side support the microplate over the base;
3) bi-material section that are part of the microbeams and induce the deflection of the microplate a result of the bimetallic effect.

The basic structures of optomechanical thermal microsystems are shown in Fig. 1. In these microsystems, an important role is assigned to bi-material sections that are part of cantilever microbeams. From the structure and geometric dimensions of these sections and the distribution of temperature in them depends the displacement of the microplate. In its turn, the temperature distribution in the bi-material sections is determined by the structure of an optomechanical thermal microsystem. Thus, the determination of the temperature distribution in the structure of the optomechanical thermal microsystem is an important task, the solution of which makes it possible to determine the displacement of the microplate and, accordingly, the output signal of the optical subsystem.

In this connection, the aim of this paper is to simulate the temperature distribution in the structure of an optomechanical thermal microsystem to determine the temperature distribution in the bi-materials sections of, and to assess the influence of the design parameters of the microsystem on this distribution.

3. Theory. Method of Modeling

To find the temperature distribution in the structure of the optomechanical thermal microsystem, an analytical method, proposed specifically for these purposes and taking into account the features of thermal microsystems [15], will be used.

The algorithm of the method has the following sequence of steps.

1. Taking into account the symmetry of the structure of the optomechanical thermal microsystem (a microplate with a functional layer and supporting cantilever microbeams), the domain of modeling is highlighted in it (Fig. 1). The conditions for highlighting this domain of modeling are, firstly, the formation of the entire structure from the domain of modeling by means of the corresponding symmetry transformation, and, secondly, accounting the peculiarities of the temperature modeling method presented in [15]. This method allows us to reduce the size of the domain of modeling due to symmetric boundary conditions. As can be seen from the data presented in Fig. 1, the domains of modeling for the microsystems under consideration have the same form.

2. The domain of modeling is divided into a series of rectangular zones (Fig. 2). The dimensions of these zones are determined by the composition of the layers. Each zone is replaced by an equivalent zone with homogeneous parameters: equivalent thickness, d_j, and an equivalent thermal conductivity, λ_j, whose values are determined as follows

$$d_j = \sum_{i=1}^{n} k_i^{(j)} d_i^{(j)} ; \tag{1}$$

IOP Conf. Series: Journal of Physics: Conf. Series **1050** (2018) 012039 doi:10.1088/1742-6596/1050/1/012039

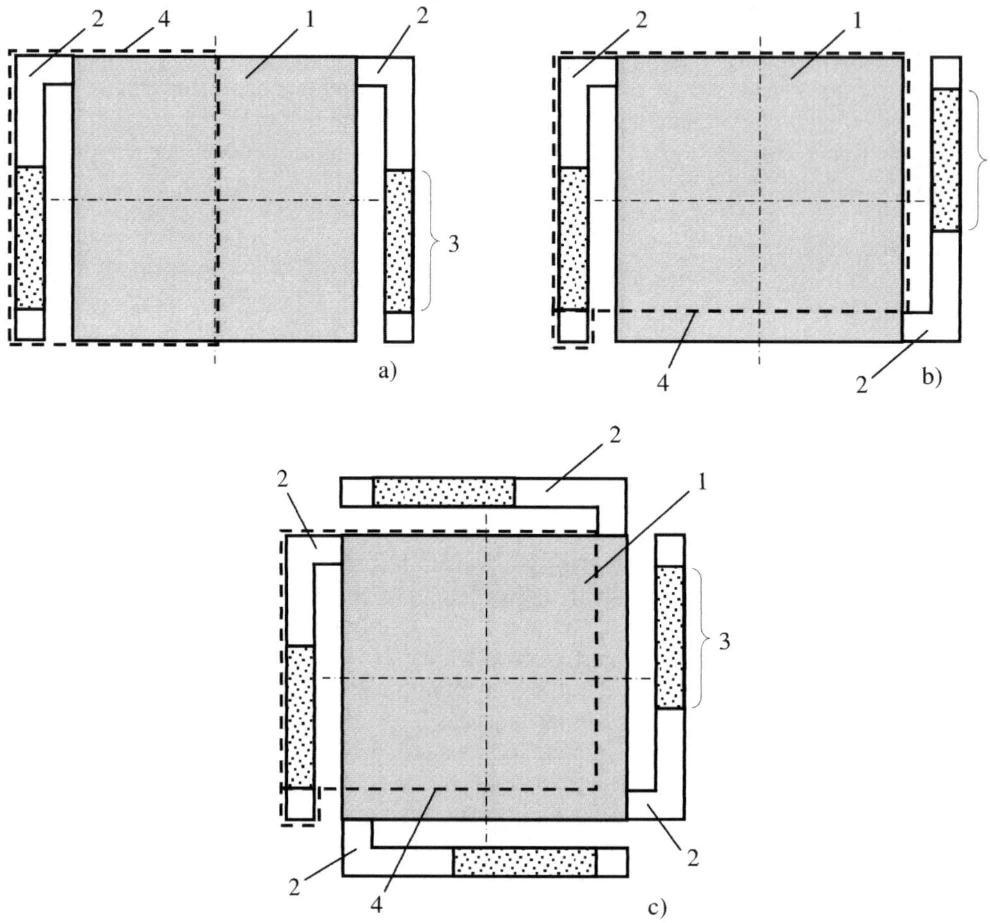

Figure 1. Optomechanical thermal microsystems on the basis of rectangular microplate supported by (a) two microbeams in adjacent corners, (b) two microbeams in opposite corners, (c) four microbeams: 1 – microplate with a functional layer; 2 – supporting microbeam; 3 – bi-material section; 4 – domain of modeling.

$$\lambda_j = \left(\sum_{i=1}^{n} k_i^{(j)} d_i^{(j)} \lambda_i^{(j)} \right) \Big/ d_j , \tag{2}$$

where $d_i^{(j)}$ is the thickness of layer i in zone j; n is the number of layers in zone j; $k_i^{(j)}$ is the coefficient equal to the relation of the area of layer i in zone j to the total area of this zone; $\lambda_i^{(j)}$ is the thermal conductivity of layer i in zone j. The division of the domains of modeling into zones for the microsystems under consideration is different due to the features of their structures.

3. For each zone of the domain of modeling, the conditions of its heat exchange with the environment are determined. Through the boundary with the neighboring zones and the substrate, heat transfer is carried out by thermal conductance. From the lower and upper surfaces of the zone heat exchange is carried out by means of convective and conductive heat transfer through the gaseous medium and by radiative heat exchange. The total heat loss from the upper and lower surface of the zone j, normalized per unit area ($N_t^{(j)}$), can be represented as follows

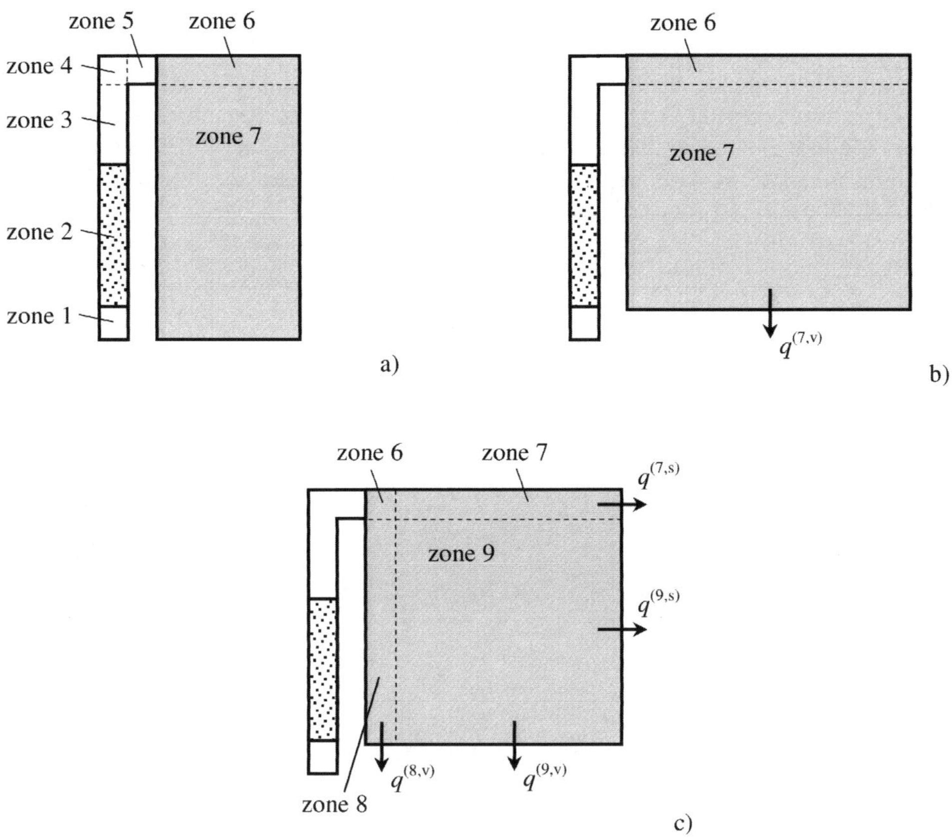

Figure 2. Division of the domain of modeling into zones for microsystems based on rectangular microplate supported by (a) two microbeams in adjacent corners, (b) two microbeams in opposite corners, (c) four microbeams; the division of the microbeam into zones in the variants (b) and (c) is analogous to the variant (a).

$$N_{\mathrm{t}}^{(j)} = A_j\left(T_j - T_{\mathrm{en}}\right), \tag{3}$$

where

$$A_j = h_{\mathrm{c}} + \frac{\lambda_{\mathrm{a}}}{d_{\mathrm{l}}^{(j)}} + 4\sigma\left(\varepsilon_{\mathrm{l}}^{(j)} + \varepsilon_{\mathrm{u}}^{(j)}\right)T_{\mathrm{en}}^3; \tag{4}$$

T_j and T_{en} are the temperature of zone j and the environment temperature, respectively; A_j is the total coefficient of the heat transfer from the lower and upper surfaces of zone j; h_{c} is the convective coefficient of the heat transfer for the given structure of the microsystem; λ_{a} is the thermal conductivity of the air medium around the microsystem; $d_{\mathrm{l}}^{(j)}$ is the distance between the lower surface of zone j and the substrate surface; σ is the Stefan-Boltzmann constant; $\varepsilon_{\mathrm{l}}^{(j)}$ and $\varepsilon_{\mathrm{u}}^{(j)}$ are the emissivities of the lower and upper surfaces of zone j, respectively.

In the equation (4), the total heat transfer coefficient takes into account the following ways of heat transfer from the lower and upper surfaces of the zone. The first term determines the heat transfer from the upper surface of the zone by convection. The second term characterizes the heat removal from the lower surface of the zone by means of thermal conductance through the air gap between this surface

IOP Conf. Series: Journal of Physics: Conf. Series **1050** (2018) 012039 doi:10.1088/1742-6596/1050/1/012039

and the substrate surface. The third term determines the dissipation of heat from the upper and lower surfaces of the zone by radiation heat exchange.

4. For each zone, a stationary differential heat equation and boundary conditions are determined. The solution of the heat equation is carried out by the method of eigenfunctions and is written in an analytical form. The heat fluxes densities between the zones included in the solution are represented as a sum of orthogonal functions with weight coefficients whose values are unknown. The stationary differential heat equation for all zones can be represented in the following general form

$$\frac{\partial^2 T_j'}{\partial x_j^2}+\frac{\partial^2 T_j'}{\partial y_j^2}-p_j^2 T_j'=\varphi_j, \tag{5}$$

where

$$T_j' = T_j - T_{\text{en}};$$

$$p_j=\sqrt{\frac{A_j}{\lambda_j d_j}};$$

$$\varphi_j=-\frac{q_j}{\lambda_j d_j};$$

q_j is the heat power density in zone j.

The solution of the differential heat equation (5) for all zones has the following form

$$\begin{aligned} T_j &= D_{0,0}^{(j)}+2\sum_{k=1}^{\infty} D_{k,0}^{(j)}\cos(k\pi x_j/l_j)+2\sum_{m=1}^{\infty} D_{0,m}^{(j)}\cos(m\pi y_j/b_j)+ \\ &+\frac{4}{l_j b_j \lambda_j}\sum_{k=1}^{\infty}\sum_{m=1}^{\infty} D_{k,m}^{(j)}\cos(k\pi x_j/l_j)\cos(m\pi y_j/b_j), \end{aligned} \tag{6}$$

where

$$D_{k,m}^{(j)}=\frac{-(-1)^k\delta_m^{(j,s)}-(-1)^m\delta_k^{(j,t)}+\delta_m^{(j,u)}+\delta_k^{(j,v)}}{l_j b_j \lambda_j[(k\pi/l_j)^2+(m\pi/b_j)^2]}; \tag{7}$$

l_j, b_j are the length and the width of zone j, respectively; $\delta_m^{(j,s)}$, $\delta_k^{(j,t)}$, $\delta_m^{(j,u)}$, and $\delta_k^{(j,v)}$ are the weight coefficients determining heat flux densities on boundaries of the zone j and the zones s, t, u, and v, respectively; k, m are the summation indexes on x and y coordinates, respectively.

The zones s, t, u, and v are localized, respectively, on the right, above, left, below the zone j. The expressions for the temperature distribution in the microsystem zones contain the weight coefficients $\delta_m^{(j,s)}$, $\delta_k^{(j,t)}$, $\delta_m^{(j,u)}$, and $\delta_k^{(j,v)}$, which characterize the heat flux densities on the boundaries of the zones and whose values are unknown.

5. To determine the unknown values of the weight coefficients entering into equation (6), we use the equations for the temperature equality from the conjugate boundary conditions between neighboring zones. In the general case, the problem of determining the values of the weight coefficients reduces to solving a system of linear differential equations. After finding the values of the weight coefficients, the temperature distributions in all zones of the optomechanical thermal microsystem are determined.

Knowing the temperature distribution in the zone with bi-material (zone 2), one can find the deflection of microbeams and, accordingly, the deviation of the microplate. For this, the differential equation of the bending of a cantilever beam with a bi-material is used [3]

$$\frac{d^2 z}{dy^2} = 6(\alpha_2 - \alpha_1)\frac{d_1^{(2)} + d_2^{(2)}}{\left[d_2^{(2)}\right]^2 K}\left[T_2(y_2) - T_2(y_2)\big|_{y_2=0}\right], \tag{8}$$

where α_1 и α_2 are the thermal expansion coefficients of layers 1 and 2 in zone 2, respectively;

$$K = 4 + 6\left(\frac{d_1^{(2)}}{d_2^{(2)}}\right) + 4\left(\frac{d_1^{(2)}}{d_2^{(2)}}\right)^2 + \frac{E_1}{E_2}\left(\frac{d_1^{(2)}}{d_2^{(2)}}\right)^3 + \frac{E_2}{E_1}\left(\frac{d_2^{(2)}}{d_1^{(2)}}\right);$$

E_1 и E_2 are the Young's modulus of materials of layers 1 and 2 in zone 2, respectively. The solution of the differential equation (8) allows one to find the microplate deflection.

4. Results of Modeling

The proposed model was applied to determine the temperature distribution in the bi-material sections in three types of optomechanical thermal microsystems with an absorbing layer (Fig. 1). In the simulation, the following values of parameters were used:

– size of microplate: length – 50 μm; width – 50 μm; thickness – 0,5 μm;

– the dimensions of the supporting L-shaped microbeams: the length of the long section is 50 μm; the length of the short section attached to the microplate is 4 μm; width – 4 μm; thickness – 0,5 μm;

– material of the microplate and the microbeams – silicon dioxide with the following properties: thermal conductivity – 1.2 W/(m·K); emissivity – 0.5; coefficient of thermal expansion – $6.1\cdot10^{-6}$ K^{-1}; Young's modulus is 0.65 GPa;

– material of the absorbing layer - platinum with the following properties: thermal conductivity – 72 W/(m·K); emissivity – 0.9;

– the thickness of the absorbing layer is 0.4 μm;

– dimensions of the section with bi-material in supporting microbeams: length – 38 μm; width – 4 μm;

– the thickness of the layer 2 of the section with bi-material is 0.05 ... 0.5 μm;

– material of layer 2 of the section with bi-material - aluminum with the following properties: thermal conductivity – 236 W/(m·K); emissivity – 0.55; the coefficient of thermal expansion – $23.3\cdot10^{-6}$ K^{-1}; Young's modulus is 70 GPa.

The absorbing layer occupies the entire area of the microplate on which it is located. The microsystem is located in a housing whose dimensions are larger than the dimensions of the microsystem itself. The distance between the lower surface of the microplate and the substrate was assumed to be 20 μm. The values of the other parameters were chosen as follows: the thermal conductivity of the ambient air was 0.026 W/(m·K); coefficient of convective heat transfer - 5.6 W/(m^2·K); ambient temperature - 300 K; the power density of the input radiation is 10,000 W/m^2.

In Fig. 3, the overheating temperature distribution on the surface of the the bi-material section of the cantilever microbeam in the optomechanical thermal microsystems of the types under consideration is shown for a thickness of layer 2 (aluminum) of the bi-material section equal to 0.1 μm. The overheating temperature distribution along the bi-material section of the cantilever microbeam is shown in Fig. 4. The dependence of the temperature difference on the boundaries of the bi-material section of the cantilever microbeam in the optomechanical thermal microsystems under consideration on the thickness of layer 2 (aluminum) of the bi-material section is shown in Fig. 5.

5. Results Discussion

The analysis of the overheating temperature distribution on the surface of the bi-material section of the cantilever microbeam in the optomechanical thermal microsystems presented in Fig. 3, shows that for identical designs of the microplates and the microbeams and the same heating conditions in microsystem designs with two supporting microbeams, the temperature distribution on the surface of the bi-material section is practically identical. The same is confirmed by the data for the overheating temperature distribution along the bi-material section of the cantilever microbeam (Fig. 4). In comparison with the microsystems in which the microplate is supported by two microbeams, in the microsystem with four microbeams, the temperature of the bi-material section of the cantilever microbeam has the less value (Fig. 3).

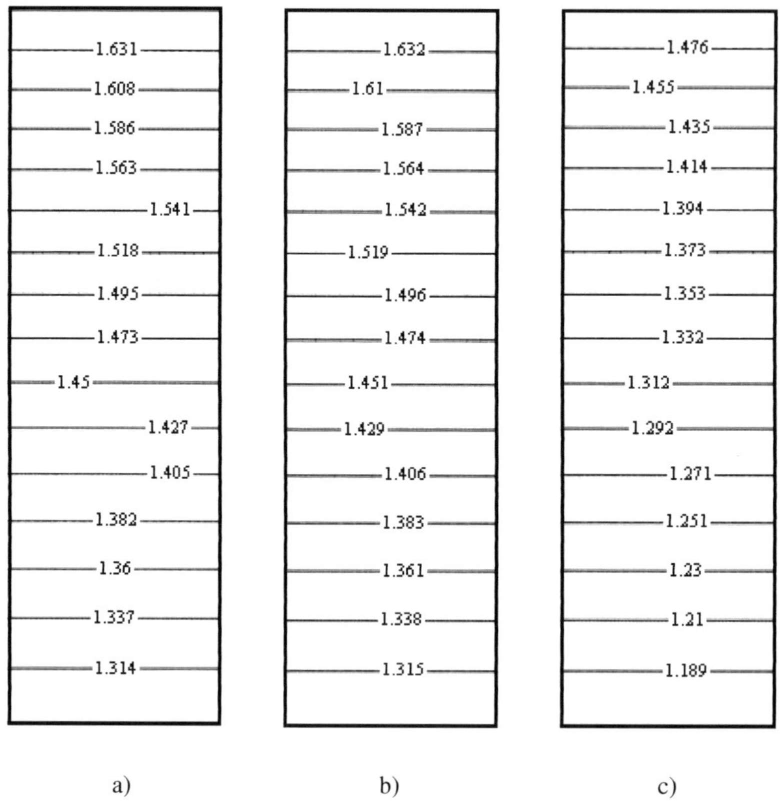

Figure 3. The overheating temperature distribution in the bi-material section of the cantilever microbeam (zone 2) in the optomechanical thermal microsystems on the basis of rectangular plate supported by (a) two microbeams in adjacent corners, (b) two microbeams at opposite corners, (c) four microbeams.

In all types of microsystems, the overheating temperatures along the bi-material section of the cantilever microbeam in a direction from the attachment point of the microbeam to the place of its attachment to the microplate increase linearly (Fig. 4). The temperature difference at the ends of the bi-material section of the cantilever microbeam for the thickness of the aluminum layer of 0.1 μm is: a) for microsystems with two support microbeams (regardless of the attachment point) – 0.36 K; b) for microsystem with four supporting microbeams – 0.33 K.

As can be seen from the presented data, the replacement of four supporting microbeams with two support microbeams in the considered structures of the optomechanical thermal microsystems makes it possible to increase the difference in temperature at the ends of the bi-material section of the cantilever microbeam by no more than 10 percent.

The importance in designing optomechanical thermal microsystems has the choice of the thickness of layer 2 of the bi-material section. This layer is applied to the microbeam and together with the microbeam material forms a bi-material section. Its thickness, mechanical and thermophysical properties determine the displacement of the microbeam end and influence on the sensitivity of the microsystem. From the data shown in Fig. 5, it is seen that with increasing the thickness of the aluminum layer (layer 2 of the bi-material section in the microsystems under consideration), the temperature difference at the ends of the bi-material section decreases nonlinearly. At small layer thicknesses (0.05 ... 0.2 μm), a significant decrease in the temperature difference (by 3.5 times) is observed, and for large layer thicknesses (0.2 ... 0.5 μm), the temperature difference decreases by 2.4 times.

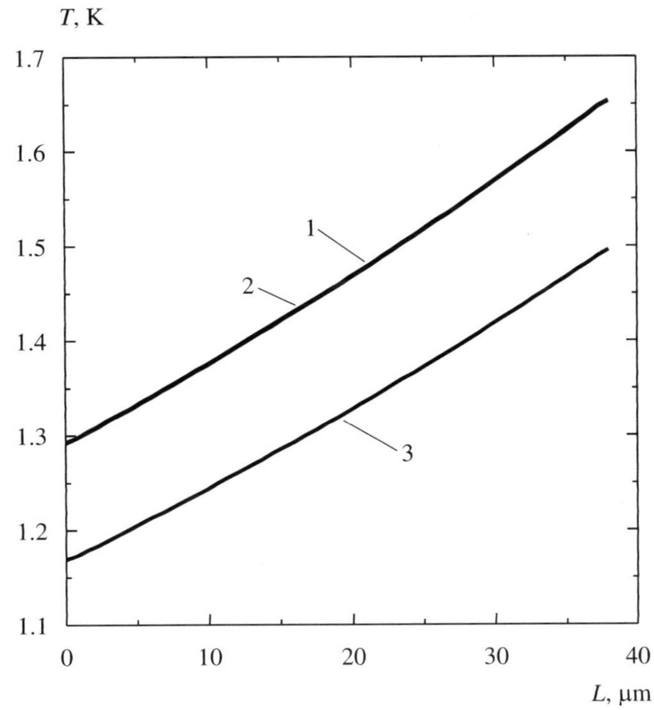

Figure 4. The distribution of the overheating temperature along the bi-material section of the cantilever microbeam in optomechanical thermal microsystems on the basis of rectangular plate supported by (1) two microbeams in adjacent corners, (2) two microbeal beams at opposite corners, (3) four microbeams.

6. Conclusion

The approach for modeling the stationary temperature distribution in the structure of the opto-mechanical thermal microsystems consisting of microplate with the functional layer and the supporting cantilever microbeams with the bi-material section is proposed. Based on this approach, the temperature distribution in the structure of microsystems and, in particular, in the bi-material sections of the cantilever microbeams is determined. It is established that the distinction in the temperature difference at the ends of the bi-material section of the cantilever microbeam in the microsystems with two and four supporting microbeams does not exceed 10 percent. With increasing thickness of layer 2 of the bi-material section, the temperature difference decreases according to a nonlinear law.

The proposed approach for modeling the stationary temperature distribution in the structure of optomechanical thermal microsystems can be used to develop the design of these microsystems, the choice of materials and the design dimensions of the microsystem elements. Using the proposed approach, it is possible to establish the influence of the properties of the used materials and the geometric dimensions of the elements on the important characteristic of microsystems - the temperature difference at the ends of the bi-material section of the cantilever microbeams.

References

[1] Gimzewski J K, Gerber Ch, Meyer E and Schlittler R R 1994 *Chemical Physics Letters* **217** 589

[2] Sohi A N and Nieva P M 2014 *J. Micromech. Microeng.* **24** 115004

[3] Barnes J R, Stephenson R J, Woodburn C N, O'Shea S J, Welland M E, Rayment T, Gimzewski J K and Gerber Ch 1994 *Rev. Sci. Instrum.* **65** 3793

[4] Barnes J R, Stephenson R J, Welland M E, Gerber Ch and Gimzewski J K 1994 *Nature* **372** 79

[5] Varesi J, Lai J, Perazzo T, Shi Z and Majumdar A 1997 *Appl. Phys. Lett.* **71** 306

[6] Liuqiang Z, Daiwei D and Qiang W 2006 *Journal of Physics: Conference Series* **34** 990

[7] Zhang L and Yang G 2001 *J. Micromech. Microeng.* **11** 85

[8] Miao Z, Zhang Q, Chen D, Guo Z, Dong F, Xiong Z, Wu X, Li C and Jiao B 2007 *Ultramicroscopy* **107** 610

[9] Lin I-K, Zhang Y and Zhang X 2008 *J. Micromech. Microeng.* **18** 075012

[10] Gao J, Zhang Q, Jiao B and Chen D 2009 *J. Micromech. Microeng.* **19** 095018

[11] Fu J, Shang H, Shi H, Li Z, Ou Y, Chen D and Zhang Q 2015 *J. Micromech. Microeng.* **25** 065012

[12] Fu J, Shang H, Shi H, Li Z, Ou Y, Chen D and Zhang Q 2016 *J. Micromech. Microeng.* **26** 025001

[13] Kozlov A G 2014 *Proc. 15th Int. Conf. on Thermal, Mechanical and Multi-Physics Simulation and Experiments in Microelectronics and Microsystems, "EuroSimE 2014".* (Ghent: Belgium) p 174

[14] Kozlov A G 2016 *Problems of Advanced Micro- and Nanoelectronics Systems Development* part IV (Moscow: IPPM RAS) p 112

[15] Kozlov A G 2002 *Sensors Actuators A. Physical.* **101** 283

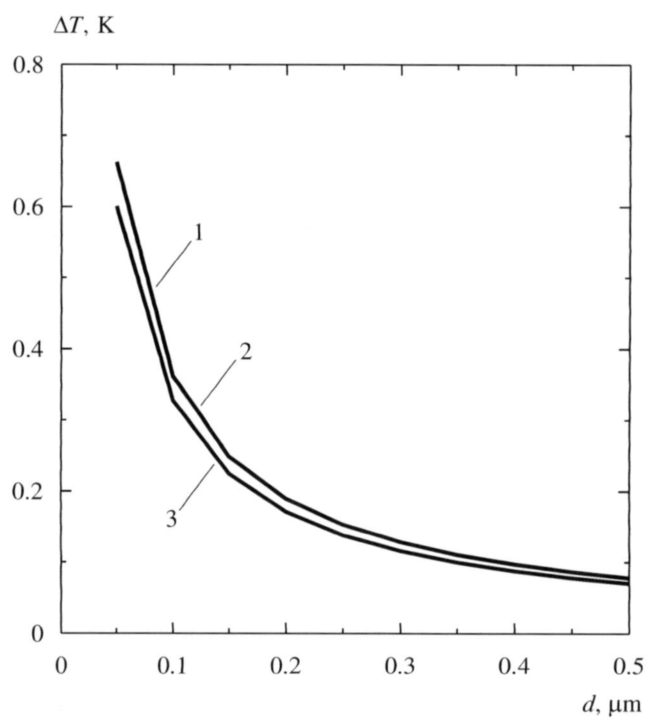

Figure 5. Dependence of the temperature difference on the boundaries of the bi-material section of the cantilever microbeam in optomechanical thermal microsystems on the basis of rectangular plate supported by (1) two microbeams in adjacent corners, (2) two microbeams at opposite corners, (3) four microbeams on the thickness of the aluminum layer.

IOP Conf. Series: Journal of Physics: Conf. Series **1050** (2018) 012040 doi:10.1088/1742-6596/1050/1/012040

Studies on the dynamics of impact interaction of the mechanoreological model under elastic plastic transformation of its mechanical system

V L Lapshin[1], **A V Eliseev**[2]

[1]Irkutsk National Research Technical University, 83, Lermontov St., Irkutsk, 664074, Russia
[2]Irkutsk State Transport University, 15, Chernyshevsky St., Irkutsk, 664074, Russia

Abstract. Dynamic interaction of machine and equipment parts and elements is a widespread process. Therefore, simulation and analysis of these processes are relevant tasks. To study impact interaction, a mechanorheological viscoelastic plastic model was developed. The force in the beginning of plastic deformations is an important model parameter. However, previous researches were carried out under impact loading of the model. They assumed that the plastic deformation occurs along with the elastic one. However, impact processes are more sophisticated. At first, only elastic deformations occur. When stresses achieve a yield point, plastic deformations occur. Therefore, the influence of the force on the model impact interaction dynamics is of great interest. The results let us draw the following conclusions. At the loading stage, when a viscoelastic model transforms into a viscoelastic plastic model, plastic deformations decrease the impact force and rebound height. Due to plastic deformations, the surface becomes more flexible, the impact duration increases body braking acceleration and inertia decrease. It leads to the decrease in the dynamic load and rebound height.The research results can be used for further development of methods for calculating model parameters, enhancing impact process simulation accuracy and reliability.

1. Introduction

Dynamic interaction of machine and equipment parts and elements is a widespread process. Impact processes accompanied by high dynamic loading play a crucial role. Therefore, simulation and analysis of these processes are relevant tasks. Dynamic processes are studied by many researchers [1-5].

To study impact interaction processes, a more general mechanorheological viscoelastic plastic model was developed [6]. It includes a shift element which takes into account energy losses due to plastic deformations under impact interaction. To adapt the model to real dynamic processes, several researches were carried out [7-11].

The scheme of the viscoelastic plastic mechanorheological model describing impact interaction of a spherical body with a surface is presented in Figure 1. The model consists of two consecutive units: the viscoelastic unit $K_1 - C$ and elastic plastic unit $K_2 - f_2$. The resistant force of the viscoelastic deformation is calculated as $N_1 = F_{DIS} + F_{EL1}$; $F_{DIS} = C(\dot{y}_1 - \dot{y}_2)^{a1}(y_1 - y_2)^{a2}$; $F_{EL1} = K_1(y_1 - y_2)^{n1}$, where $y_1, y_2, \dot{y}_1, \dot{y}_2$ is the shift and mass speed m_1 and m_2; K_1 is the ratio of stiffness of an element of the viscoelastic unit; C is the ratio of viscosity of an element of the viscoelastic unit.

The resistant force of the elastic plastic deformation is calculated as $N_2 = F_{PL} + F_{EL2}$; $F_{EL2} = K_2\, y_2^{\,n2}$; $F_{PL} = f_2\, y_2^{\,n3} + F_{ST}$, where F_{ST} is the force corresponding to the occurrence of plastic deformation; K_2 is the ratio of stiffness of an elastic element of the elastic plastic unit; f_2 is the ratio of shift of the elastic plastic unit or the ratio of ductility.

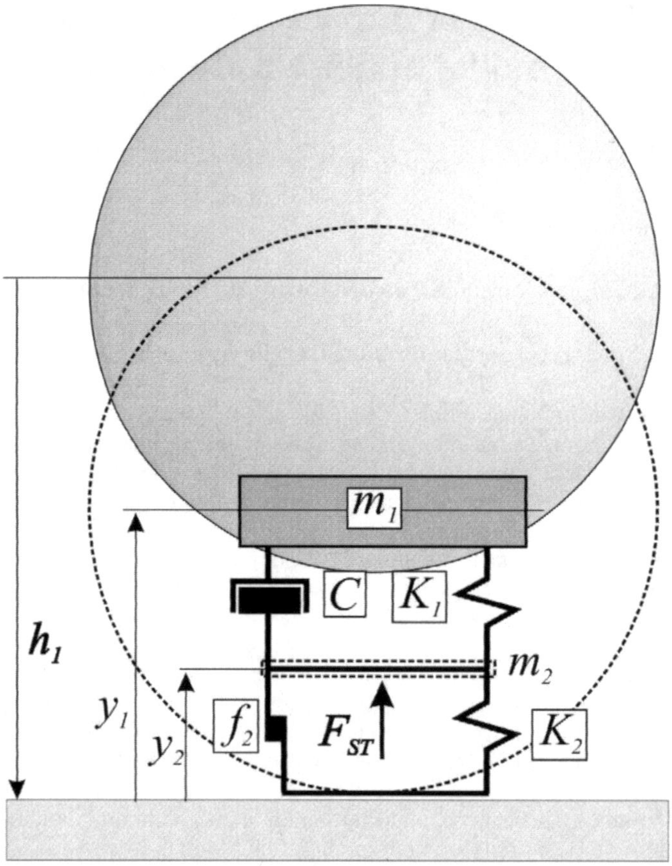

Figure 1. The scheme of the viscoelastic plastic model

The total body mass is concentrated in inertia element m_1, the mass of element m_2 is negligibly small ($m_2 \to 0$) and cannot influence the dynamics of mechanical system movement. It is used for mathematical description of the system with two second order differential equations.

The viscoelastic plastic model can work in the following way: at the initial stage of impact interaction, only elastic deformations occur, so the viscoelastic unit $K_1 - C$ is deformed. The unit describes elastic deformations and energy losses. The elastic plastic unit comes into operation when the dynamic force achieves the target value of force F_{ST} corresponding to intensive development of plastic deformations. Varying the value of force $F_{ST,}$ one can ensure viscoelastic or viscoelastic plastic interaction of mechanical systems when one or two units of the model come into operation.

When a dynamic force value is maximum N_{MAX}, the model starts unloading. At this stage, only the viscoelastic unit describing disappearance of elastic deformations comes into operation. The elastic plastic unit remains deformed as it characterizes plastic (residual) deformations.

All model elements have linear characteristics. The degree index is set in general terms (a_1, a_2, n_1, n_2, n_3). The approach expands model abilities and can be used in researches aimed to study the effects of non-linear characteristics of model elements on the dynamics of its impact interaction [8]. As for specific values of power indices, under impact interaction of the spherical body, the elastic component will be $n_1 = n_2 = 3/2$ [2-4], the plastic component can be proportional to the acting force ($n_3 = 1$) [2,3]. Unloading of materials follows the linear elasticity law [2].

2. Purpose statement

When using theoretical models, it is important to adapt the simulation model to real dynamic (impact) processes. Model adequacy is assessed based on the coincidence of main parameters of the dynamic process determined using a model with experimental data. As for the impact process under study, these parameters are the impact interaction force variation regularity, impact time, body rebound height, and a deformation value. To calculate model parameters K_1, K_2, f_2, C by experimental impact parameters, a special method was developed [12,13]. However, previous researches [7-14] were carried out for the case when $F_{ST}=0$, i.e. under the impact loading. They assumed that plastic deformations occur in the beginning of the impact process along with the elastic ones.

However, the impact process can be more sophisticated. At first, only elastic deformation occurs. When stresses achieve a yield point (when $N=F_{ST}$), plastic deformations described by the model unit $K_2 - f_2$ occurs. Therefore, the effects of a force value F_{ST} on the model dynamics of impact interaction are of great interest. To study the process, researches were carried out.

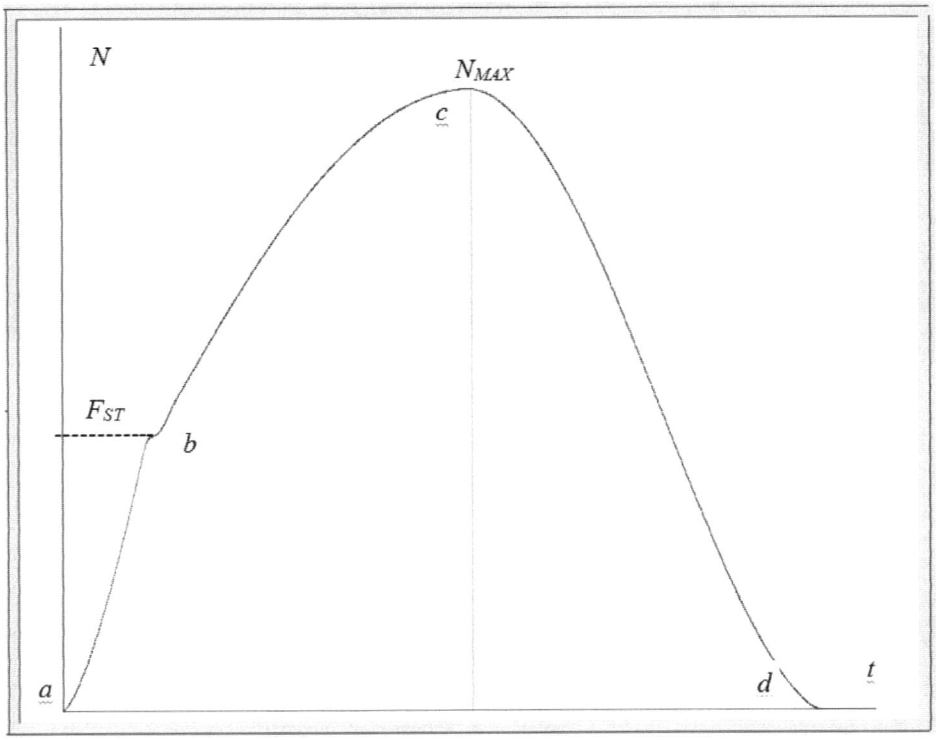

Figure 2. Diagram of changes in force N

3. Theory

Differential model movement equations are as follows:

$$m_1 \ddot{y}_1 + C_1(\dot{y}_1 - \dot{y}_2)^{a1}(y_1 - y_2)^{a2} + K_1(y_1 - y_2)^{n1} = -m_1 g;$$

$$m_2 \ddot{y}_2 + K_2 y_2^{n2} + f_2 y_2^{n3} + C_1(\dot{y}_2 - \dot{y}_1)^{a1}(y_2 - y_1)^{a2} + K_1(y_2 - y_1)^{n1} = -m_2 g + F_{ST}.$$

The mathematical model can be used to study the effects of viscoelastic and plastic properties of materials on the dynamics of impact interaction of a spherical body and a surface. The model served a basis for a special research program designed to calculate dynamic interaction parameters: impact time, impact interaction force, deformation value, speed and rebound height.

Figure 2 presents a diagram of changes in the impact interaction force (normal response force N) of a steel spherical body. Calculations are carried out using the program developed. The following parameters were used: $a_1 = 1, a_2 = 1, n_1 = 3/2$, $n_2 = 3/2, n_3 = 1, K_2 = 0{,}1K_1$. At the initial stage of contact interaction (segment $a-b$), elastic deformations occur. When normal response force N

achieves a target value of force F_{ST}, the elastic plastic unit comes into operation (segment $b-c$). When elastic deformations disappear, segment $c-d$ corresponds to the unloading stage (the viscoelastic unit operates). At point *d*, the body leaves the surface ($N=0$).

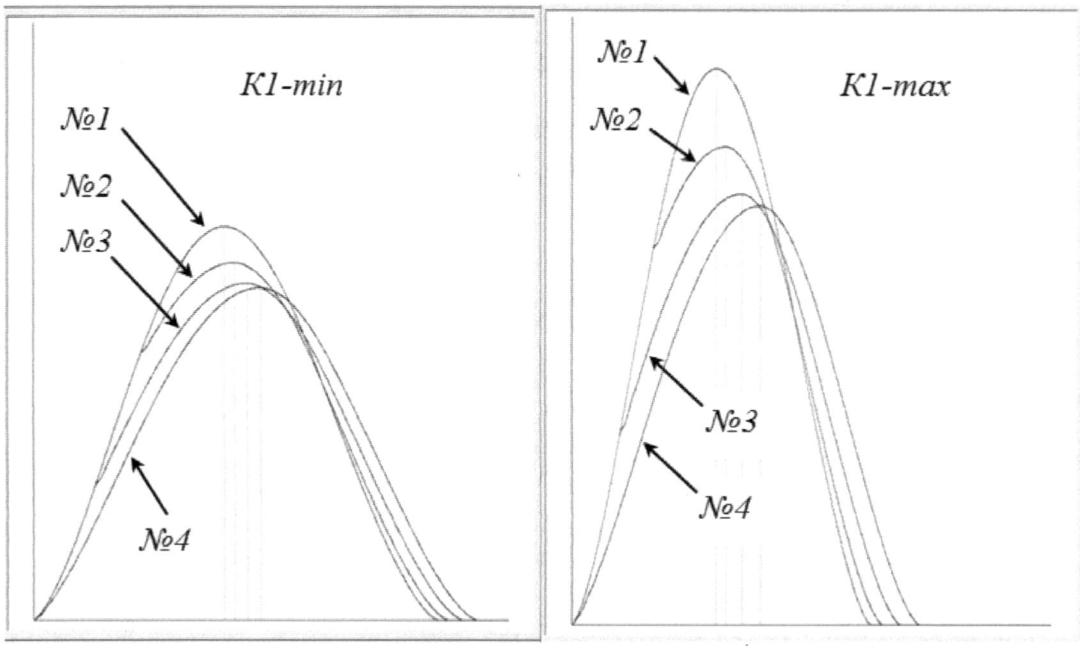

Figure 3. Diagram of changes in force *N*

4. Analysis of research results

Two calculation variants were used (Figure 3). Variant 1: $F_{ST}=N_{max}$ (only the elastic deformation occurs). Variant 2: $F_{ST}=2N_{max}/3$. Variant 3: $F_{ST}=N_{max}/3$. Variant 4: $F_{ST}=0$ (the plastic deformation occurs since the beginning of the impact process). The following parameters of the impact interaction were analyzed: impact force *N* and relative body rebound height $dh=h2/h1$ ($h1$ is the drop height, $h2$ is the rebound height). Model parameters K_1, K_2, f_2, C varied within the prescribed limits.

Figure 3 presents diagrams of the normal response force *N* under different values of F_{ST} and K_1. The research results are presented in Figures 4 and 5. The following conclusions can be drawn from the results: at the loading stage, when a viscoelastic model transforms into a viscoelastic plastic model ($F_{ST}=N_{max}\rightarrow F_{ST}=0$), impact force N_{max} and rebound height *dh* decrease at different model parameters K_1, K_2, f_2, C. These changes are significant. The phenomenon occurs due to the fact that the surface becomes more ductile due to plastic deformations, the impact force increases, and the body braking acceleration and inertia decrease. It decreases the dynamic load (impact interaction force). Due to plastic deformations, the impact energy dissipates and rebound height decreases.

5. Conclusions

Thus, the research results for dynamic interaction of the mechanorheological model under the elastic plastic transformation of its mechanical system speak for significant changes in its behavior dynamics. Therefore, the model parameter F_{ST} along with other parameters (K_1, K_2, f_2, C) are important control factors influencing the model behavior.

The practical application of the model can enhance accuracy and reliability of impact process simulation, and can be used for further development of methods for calculating physical and mechanical properties of materials. Knowledge of mechanical properties of materials is used for solving different research tasks by means of mathematical vibration and impact process simulation. The key task is to adapt the simulation model to real impact processes by parameter F_{ST}. To this end, appropriate methods should be developed. These tasks can be solved using the research results.

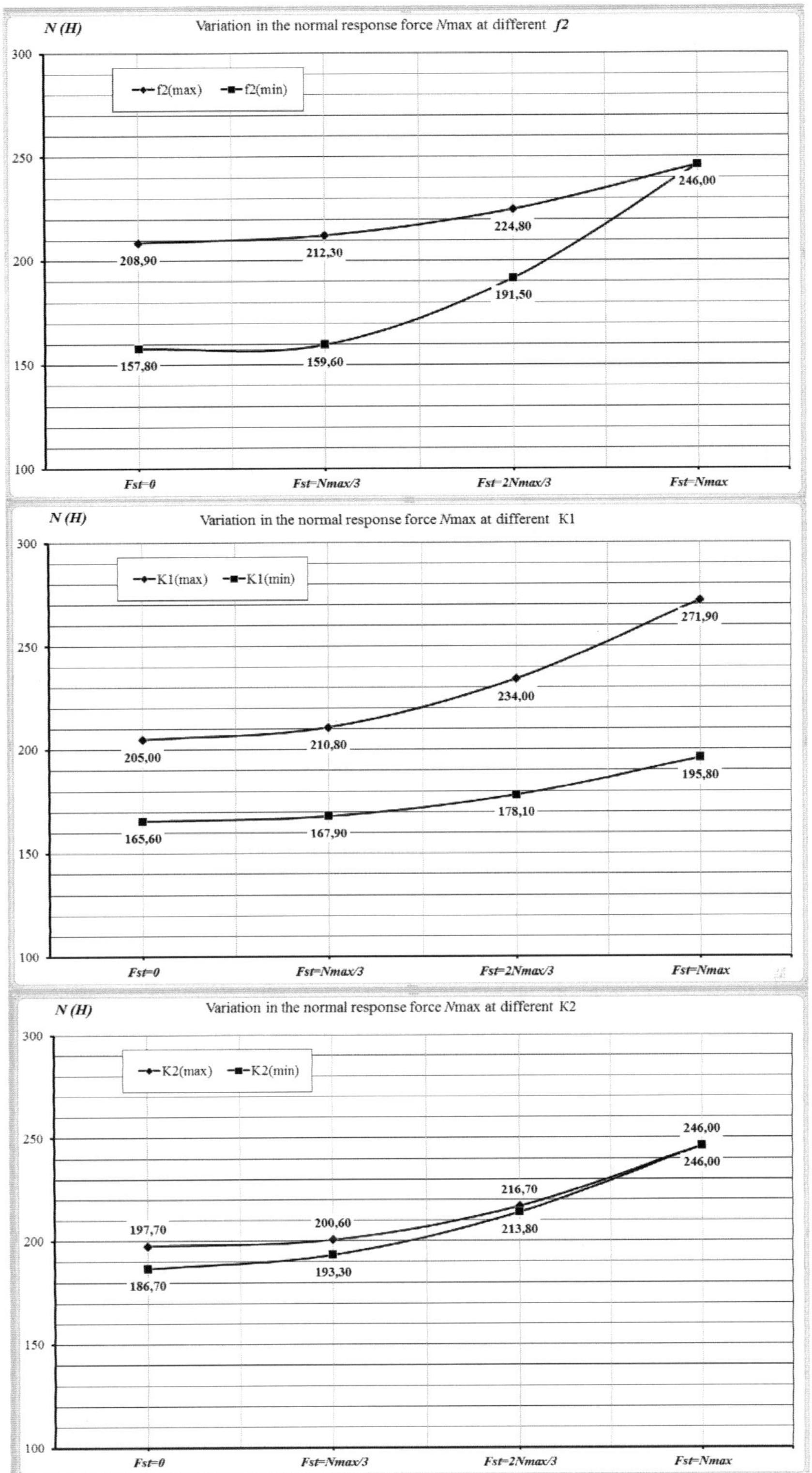

Figure 4. Changes in the normal response force

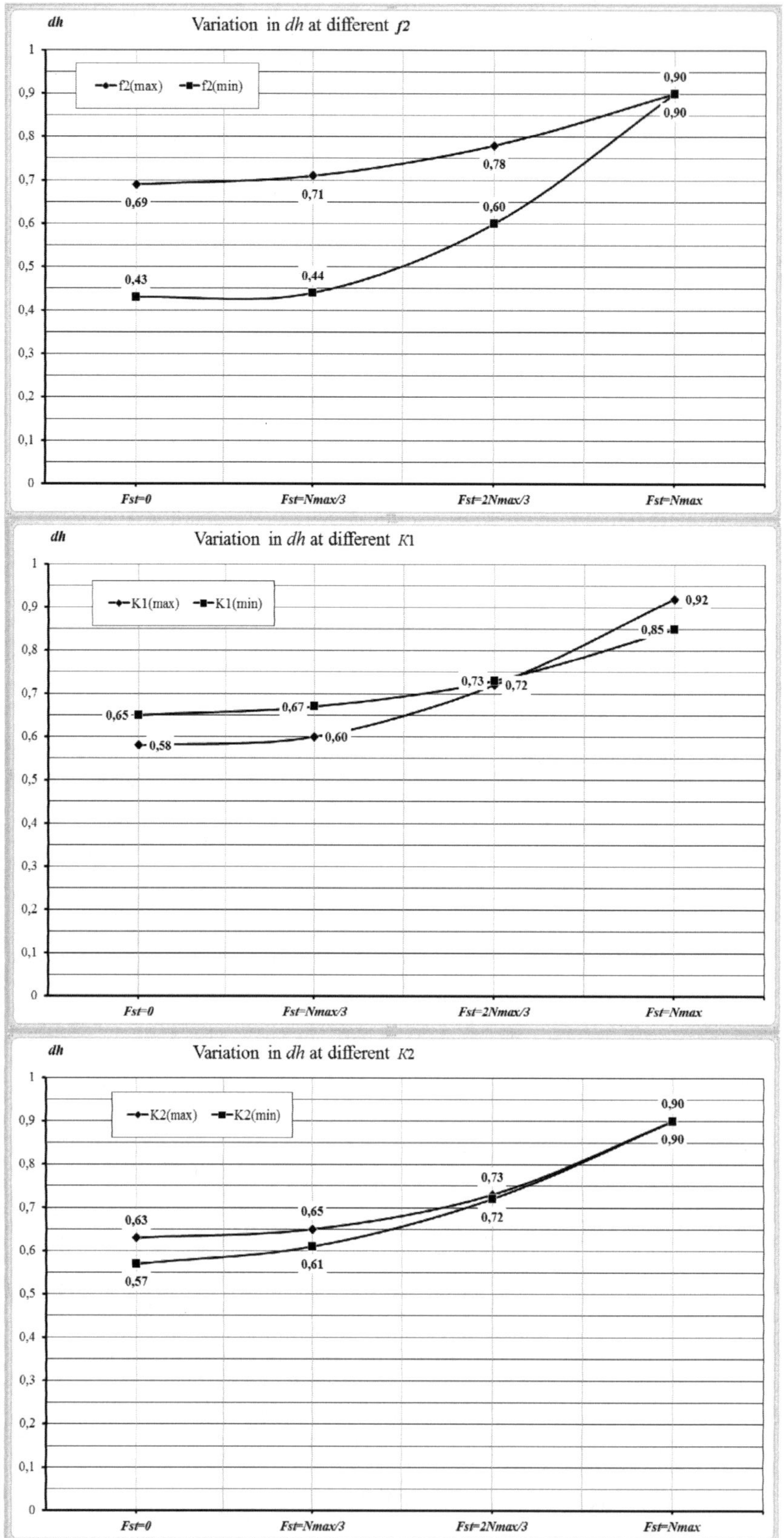

Figure 5. Changes in the rebound height

6. References

[1] Panovko Y G 1990 *Fundamentals of the fluctuation and impact theory* (Leningrad: Politechnik) p 272

[2] Batuev G S, Golubkov Y V, Efremov A K and Fedosov AA 1977 *Engineering impact process research methods* (Moscow: Mashinostroenie) p 240

[3] Kilchevskiy N A 1976 *Dynamic contact compression of solid bodies. Impact* (Kiev: Naukova dumka) p 319

[4] Lapshin V L, Yaschenko V P and Rudykh A V V 2006 *Bulletin of Irkutsk State Technical University* **2(26)** pp 110-115

[5] Lapshin V L, Rudykh A V and Glukhov A V 2017 2017 *Int. Conf. on Industrial Eng., Applications and Manufacturing* (St. Petersburg Institute of Electrical and Electronics Engineers IEE Explore) pp 1–4

[6] Lapshin V L, Demakov E I 2007 *All-Russian Sc. Eng. Conf. Mechanics in the 21st century* (Bratsk: BSU) pp 67–71

[7] Lapshin V L, Glukhov A V 2011 *Systems. Methods. Technology* **3 (15)** pp 14-19

[8] Lapshin V L, Rudykh A V and Glukhov A V 2012 *Systems. Methods. Technology* **3 (15)** pp 21-25

[9] Lapshin VL, Glukhov AV 2011 *Proc. International Sc. Conf. Modern equipment and technology: studies and development* (Lipetsk: Gravis) pp 20-27

[10] Lapshin V L, Glukhov A V 2011 *Bulletin of Irkutsk State Technical University* **10 (57)** pp 44-49

[11] Lapshin V L, Glukhov A V 2011 *Modern technology, system-based analysis, modelling* (Irkutsk: ISRU) pp 39-45

[12] Lapshin V L, Rudykh A V and Glukhov A V 2014 Patent No 2526233, Russia, IPC G 01 N 3/48. Bulletin 23.

[13] Lapshin V L, Rudykh A V and Glukhov A V 2015 *Systems. Methods. Technologies* **2 (26)** pp 37-43

[14] Lapshin V L, Glukhov A V 2011 *Proc. of the 65th Scientific Conf. Fundamental and applied studies – a basis for modernization and innovation development of architectural and construction and road transport systems of Russia* (Omsk: SibADI) vol 1 pp 54-59

Attitude motion of space debris during its removal by ion beam taking into account atmospheric disturbance

A S Ledkov[1], V S Aslanov[2]

Samara National Research University, 34, Moscovskoe shosse, Samara, 443086, Russia

[1]E-mail: ledkov@inbox.ru, www.ledkov.com
[2]E-mail: aslanov_vs@mail.ru, aslanov.ssau.ru

Abstract. The paper is devoted to the issue of active large space debris removal. The contactless method of influencing space debris through the ion flow is considered. The main advantage of this method is absence of a dangerous and complex stage of space debris capturing or docking. In this study, the project "Ion Beam Shepherd" was taken as a basis. The aim of this paper is to study the effect of the atmosphere in LEO on the space debris behaviour during its removal. The equations of plane motion of space debris under the action of the ion beam, gravitational and aerodynamic forces and torques were obtained using Lagrange formalism. For calculation ion beam force and torque the self-similar model of ion engine plume exploration and fully diffused reflection model of ions interaction with the surface were used. Simulation of deorbiting the spent upper stage of the Cosmos 3M rocket was carried out. Phase portraits of the rocket stage attitude motion for various heights were constructed and analysed. It was shown that the motion of the stage around its centre of mass has a noticeable effect on the average ion beam force and descent time.

1. Introduction

The space debris issues are widely discussed in the scientific literature in recent decades. Much attention is paid to removal of large space debris, which includes upper rockets stages and broken satellites [1]. Collisions of large space objects are very dangerous, since they lead to the formation of small debris clouds [2] and can cause the Kessler syndrome [3]. The scientific community discusses various ways of large space debris removal. Detailed reviews of diverse technical solutions and projects are given in the works [4, 5]. This study focuses on the contactless method of deorbiting space debris. The main advantage of contactless methods is the absence of a complex and dangerous stage of space debris capture by an active spacecraft-cleaner. To date, a few methods of contactless impact on a space object for the purpose of changing the parameters of its motion are considered: using Coulomb interaction [6, 7], by means of the ion beam [8, 9], due to laser action [10, 11].

In this study, the project "Ion Beam Shepherd" was taken as a basis [8, 12, 13]. A space debris object is removed from orbit by an active spacecraft (shepherd), which is equipped with two ion engines. The spacecraft is approaching the object at a distance of about ten meters. Thereafter, one of the engines is directed toward the object. The ion beam created by this engine blows the space object and thus momentum transfer occurs. The second engine is needed to compensate for the thrust of the first one and to hold the spacecraft near the space debris. Modern ion thrusters allow exerting a force of the order of a dozen mN on a space object [14]. Calculations show that Ion Beam Shepherd can deorbit space debris with a mass of several tons from an orbit with a radius of one thousand kilometers per few months [8]. The attitude motion of space debris has a significant effect on its descent time. This factor was considered in [15]. Another important factor that influences the motion of a body at a given altitude is the effect of the atmosphere. A rough estimation shows that for a body on a circular orbit of 250 km altitude the magnitude of the drag force is comparable to the force created by the ion flow. The purpose of this work is to study the effect of the atmosphere on the space debris behavior when it is removed from the orbit by an Ion Beam Shepherd.

2. Mathematical model

The plane motion of space debris under the influence of gravitational, aerodynamic and ion beam forces and torques is considered. The spacecraft-shepherd is considered as a material point (point A on figure 1). It is held at a constant distance from the space debris by its control system. In the orbital reference frame associated with the space debris Bx_Oy_O, the radius vector of the shepherd has coordinates $\boldsymbol{\rho}_A = [0, \rho_A]^T$. The task of the spacecraft control is beyond the scope of this study. It is supposed that the space debris is a rocket stage which is modeled by a rigid body of cylindrical shape with the center of mass at point B. The inertial frame is denoted as Ox_py_p. The axis x_p passes through the pericenter of initial space debris orbit. Axis x_O of the orbital frame lies along the radius vector $\mathbf{r}$ of the space debris center of mass, and the axis y_O is directed towards the orbital flight. The body frame Bx_by_b is fixed relative to the space debris. Its axes coincide with the principal axes. The ion beam reference frame Ax_ay_a is connected with shepherd. The axis x_a is directed along ion beam axis to flight direction, and the axis y_a lies in the plane of the orbit and completes the right-handed set.

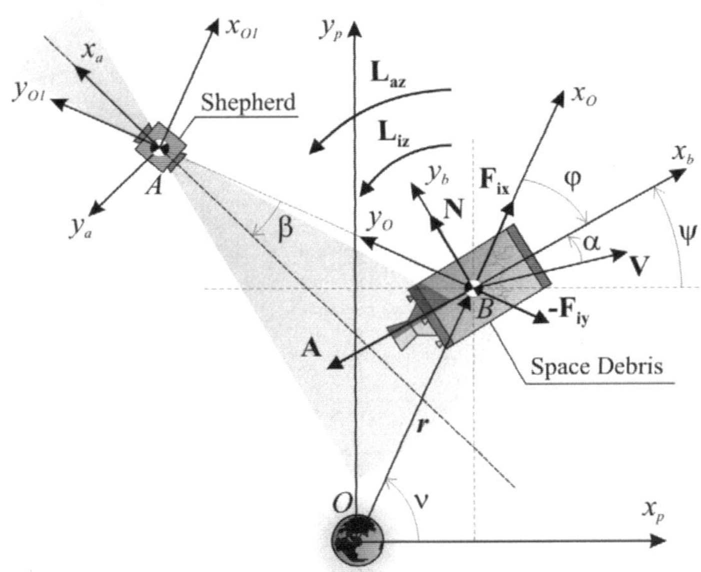

Figure 1. Space debris motion under the influence of aerodynamic, and ion beam forces and torques.

Equations of the space debris can be obtained with by the means of Lagrange equations of the second kind.

$$\frac{d}{dt}\frac{\partial L}{\partial \dot{q}_j} - \frac{\partial L}{\partial q_j} = Q_j \tag{1}$$

where L is the Lagrangian, $q_j = \{\nu, r, \varphi\}$ is the generalized coordinates, ν is the true anomaly angle, r is the distance between the center of Earth and the space debris center of mass, φ is the deflection angle of the space debris longitudinal axis from its local vertical OB, Q_j is the non-potential generalized forces. The Lagrangian of the space debris can be written in the form

$$L = \frac{m(r^2\dot{\nu}^2 + \dot{r}^2)}{2} + \frac{J_z(\dot{\nu} - \dot{\varphi})}{2} + \frac{\mu m}{r} + \frac{\mu(J_x + J_y + J_z)}{2r^3} - \frac{3\mu(J_x \cos^2\varphi + J_y \sin^2\varphi + J_z)}{2r^3} \quad (2)$$

Here μ is the gravitational constant of the Earth, m is the mass of the space debris, J_x is the longitudinal moment of inertia of the space debris, J_y and J_z are the transversal moments of inertia.

Generalized forces can be calculated through the virtual work

$$\delta W = (\mathbf{F}_i + \mathbf{F}_a)\cdot\delta\mathbf{r} + (\mathbf{L}_i + \mathbf{L}_a)\cdot\delta\boldsymbol{\psi} = \sum_j Q_j \delta q_j \quad (3)$$

where $\mathbf{F}_i = [F_{ix}, F_{iy}]^T$ is the ion beam force, which is specified in the orbital reference frame, $\mathbf{F}_a = [-A, N]^T$ is the aerodynamic force in body frame, A is the axial aerodynamic force, N is the normal aerodynamic force, $\mathbf{L}_i$ is the ion beam torque, $\mathbf{L}_a$ is the pitch moment, $\psi = \nu - \varphi$.

$$Q_r = -A\cos\varphi + N\sin\varphi + F_{ix},$$
$$Q_\nu = Ar\sin\varphi + Nr\cos\varphi + F_{iy}r + L_a + L_{iz},$$
$$Q_\varphi = -L_a - L_{iz} \quad (4)$$

Aerodynamic force and torque can be calculated as

$$A = \frac{\rho V^2}{2} SC_A(\alpha), \quad N = \frac{\rho V^2}{2} SC_N(\alpha), \quad L_{iz} = \frac{\rho V^2}{2} SlC_{mz}(\alpha) \quad (5)$$

where ρ is the density of the atmosphere, V is the velocity of the space debris, S is the space debris cross-section area, l is the characteristic length of the space debris, $C_A(\alpha)$, $C_N(\alpha)$, and $C_{mz}(\alpha)$ are dimensionless aerodynamic coefficients of the axial force, normal force and pitch moment, α is the angle of attack.

$$\alpha = \mathrm{sign}(V_x \sin\psi - V_y \cos\psi)\arccos\left(\frac{V_x \cos\psi + V_y \sin\psi}{V}\right) \quad (6)$$

Here V_x and V_y are projections of the velocity vector on the axis of the inertial reference frame $Ox_p y_p$.

For calculation ion beam force and torque the self-similar model of ion engine plume exploration and fully diffused reflection model of ions interaction with the surface can be used. The surface of the space debris can be divided into triangles and force acting on each *j*-th triangle can be calculated as [14, 15]

$$\mathbf{F}_{ij} = -\frac{m_0 n_0 R_0^2 S_j}{x_{aj}^2 \tan^2\alpha_0} \exp\left(-\frac{3(y_{aj}^2 + z_{aj}^2)}{x_{aj}^2 \tan^2\alpha_0}\right)(\mathbf{V}_j \cdot \mathbf{N}_j)\mathbf{V}_j \quad (7)$$

where m_0 is the ion mass, n_0 is the plasma density at the beginning of the far region, R_0 is the radius of the beam at the beginning of the far region, α_0 is the divergence angle of the ion beam, $\mathbf{V}_j = \left[-u_0, -u_0 y_{aj} x_{aj}^{-1}, -u_0 z_{aj} x_{aj}^{-1}\right]^T$ is the velocity of the ion flux at the barycenter P_j of *j*-th triangle, given by its coordinates in $Ax_a y_a z_a$ frame (figure 1); u_0 is the axial component of the ion flus velocity, the radius vector of the barycenter point P_j in $Ax_a y_a z_a$ frame has coordinates $\mathbf{d}_j = \left[-x_{aj}, y_{aj}, z_{aj}\right]^T$, S_j is the area of *j*-th triangle. The total ion beam force and torque vector, relative to the space debris center of mass, can be calculated as

$$\mathbf{F}_i = \sum_{j \in J} \mathbf{F}_j, \quad \mathbf{L}_i = \sum_{j \in J} \mathbf{d}_j \times \mathbf{F}_j \tag{8}$$

where J is an index set including the subset of triangles inside the ion beam.

Equations (1) after substitution of (2) and expression of second derivatives take the form

$$\ddot{r} = r\dot{v}^2 - \frac{\mu}{r^2} + \frac{3\mu(3J_x \cos^2\varphi + 3J_y \sin^2\varphi - J_x - J_y + 2J_z)}{2mr^4} + \frac{Q_r}{m}, \tag{9}$$

$$\ddot{v} = -\frac{2\dot{r}\dot{v}}{r} + \frac{3\mu(J_x - J_y)\cos\varphi \sin\varphi}{mr^5} + \frac{Q_\varphi + Q_v}{mr^2}, \tag{10}$$

$$\ddot{\varphi} = -\frac{2\dot{r}\dot{v}}{r} + \frac{3\mu(J_x - J_y)\cos\varphi \sin\varphi(J_z + mr^2)}{J_z r^5 m} + \frac{Q_\varphi}{J_z} + \frac{Q_\varphi + Q_v}{mr^2} \tag{11}$$

3. Results and Discussion

As an example, the removal of Cosmos 3M rocket stage from a circular orbit of 500 km altitude by ion beam is considered. The stage has a mass $m = 1400$ kg, its length is L=6.5m, and its radius is 1.2 m. The moments of inertia are $J_x = 1300\,\mathrm{kg \cdot m^2}$, $J_y = J_z = 6800\,\mathrm{kg \cdot m^2}$. The shepherd-spacecraft is held at a constant distance $\rho_A = 15\mathrm{m}$ from the stage. It creates ion beam with the following parameters: the mass of particle (xenon) is $m_0 = 2.18 \cdot 10^{-25}\,\mathrm{kg}$, the plasma density is $n_0 = 2.6 \cdot 10^{16}\,\mathrm{m^{-3}}$, the radius of the beam at the beginning of the far region is $R_0 = 0.1\mathrm{m}$, the axial component of the ion flus velocity $u_0 = 38000\,\mathrm{m/s}$, the divergence angle of the beam is $\alpha_0 = 15°$. Ion beam axis passes through the geometric center of the stage ($\beta = 0$ on figure 1), by which the point lying at the intersection of the axis and the plane of symmetry is meant. The Cosmos 3M rocket was modeled by a cylinder with flat butt ends, which were divided into 54 thousand triangles. The ion beam force and torque were calculated by equations (8), and they are shown in figures 2 and 3. The geometric center of the stage was chosen as the pivot point of the ion beam torque. The dimensionless aerodynamic coefficients were calculated using the Newton impact theory [16]. They are shown in figure 4. Torque coefficient $C_{mz}(\alpha)$ was calculated with respect to point O_1. This point is shown in figure 5.

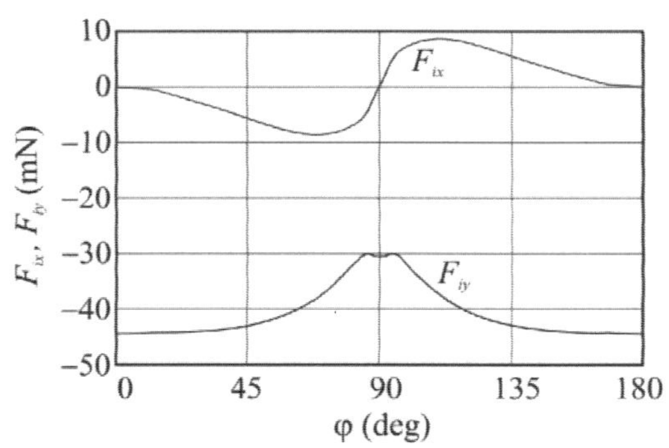

Figure 2. Dependence of ion beam force projections on deflection angle.

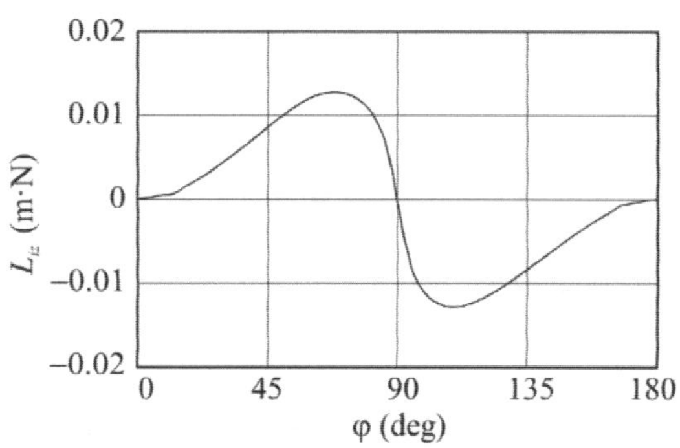

Figure 3. Dependence of ion beam torque projection L_{iz} on deflection angle.

The simulation of the Cosmos 3M stage descent from the orbit was conducted by the means of equations (7)-(8). Figure 6 shows dependence of the stage radius vector r on time for the cases that take into account the influence of the atmosphere (solid line) and do not take it into account (dashed line). In the first case descent to an altitude of 100 km takes about 55 days, in the second case the process lasts more than 85 days. Figure 6 demonstrates that the atmosphere begins to have an effect comparable to the ion beam influence at a height of 300 km.

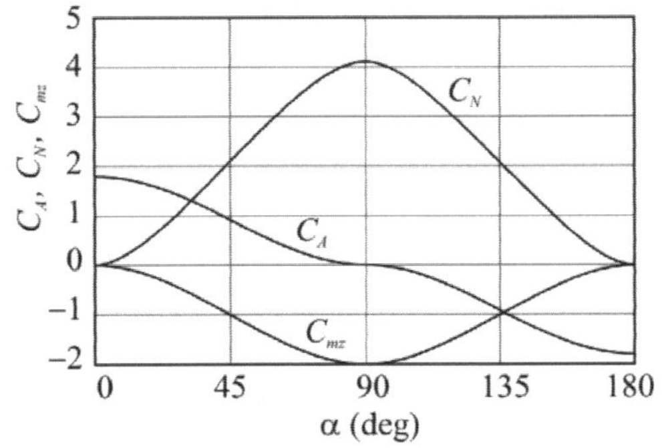

Figure 4. Dimensionless aerodynamic coefficients of the cylindrical stage.

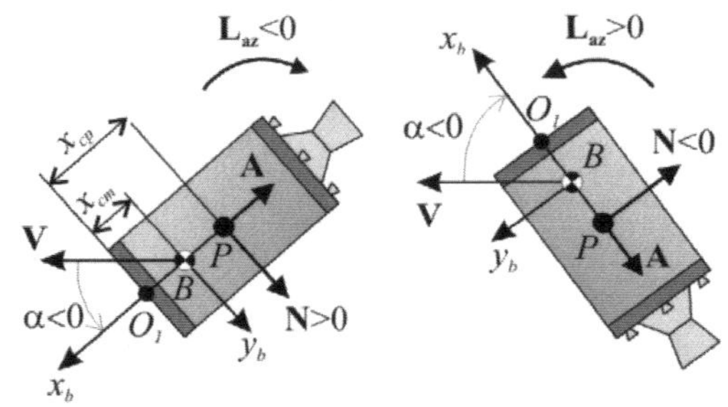

Figure 5. Aaerodynamic forces and torque.

In the real situation, the center of mass of the stage does not coincide with its geometric center. Let us study the attitude motion of the rocket stage when its center of mass is at a distance of $x_{cm} = L/4$ from point O_1 (figure 5). In this case, aerodynamic and ion beam torques should be recalculated relative the center of mass. We construct a phase portraits of equation (9) for fixed values of r and $\dot{v}$. Figure 7 demonstrates the phase portrait for the rocket stage altitude of 500 km. At this altitude, the aerodynamic forces and moments are negligibly small. Five areas can be distinguished in the phase portrait: three oscillation areas A_1, A_2, A_3; and two rotation areas A_4, A_5. These areas are separated from each other by separatrices that connect the saddle points s_1, s_2, and s_3. The separatrices are shown in figure 7 in bold lines. The instability of position s_2 and the existence of two oscillation areas A_1, A_2 in its vicinity is caused by the influence of the gravity gradient torque on the rocket stage. It should be noted that the rocket stage can oscillate with a large amplitude in the area A_3 or with a rather small amplitude in the areas A_1, A_2.

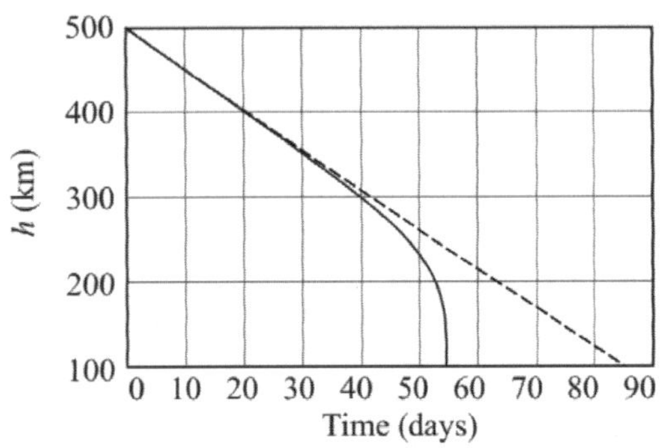

Figure 6. Change in the Cosmos 3M stage height over time.

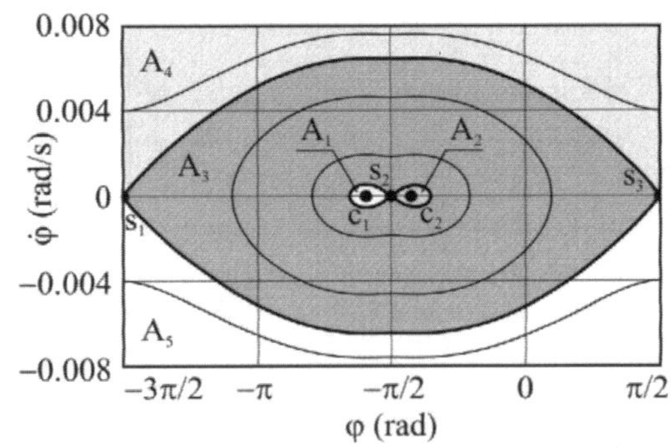

Figure 7. The phase portrait of equation (9) for $r = 6871\text{km}$.

When the altitude of the orbit decreases, the density of the atmosphere increases, and, accordingly, the values of the aerodynamic forces and moments acting on the rocket stage increase too. Figure 8 shows the phase portraits that were constructed for an altitude of 250 km taking into account the influence of the atmosphere (solid line) and without taking this influence into account (dotted line). In both cases the phase portraits not differ qualitatively from that shown in Figure 7. The increase of the atmosphere influence leads, on the one hand, to a decrease in the areas of the oscillation regions A_1, A_2. The centers c_1 and c_2 approach the saddle point s_2. On the other hand, when we move away from the saddle point s_2, the phase trajectories become more elongated vertically (Figure 8). In Figure 9, the areas A_1, A_2 are shown enlarged. Phase portraits were obtained for a height of 150 km. It can be seen that the areas A_1, A_2 are even more compressed, and the oscillations inside the region A_3 can occur with a rather small amplitude around the inner separatrix.

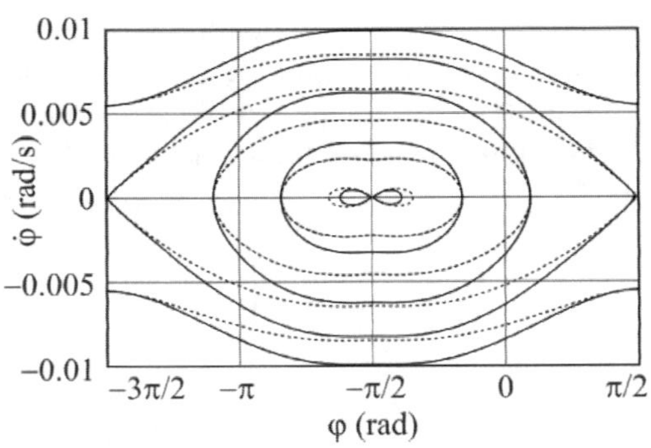

Figure 8. Comparison of phase portraits for 250km when the atmosphere is taken into account (solid line), and when it is not taken into account (dotted line).

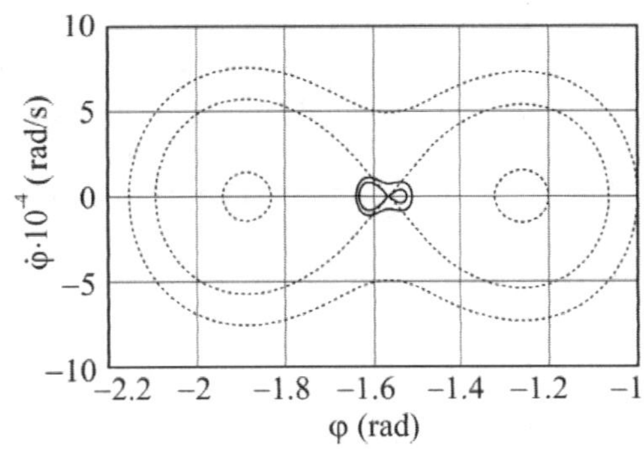

Figure 9. Comparison of phase portraits for 150km when the atmosphere is taken into account (solid line), and when it is not taken into account (dotted line)..

The amplitude of the oscillations has a noticeable effect on the time of descent, in order to demonstrate this, we calculate the average for the period of oscillations force F_{iy} for various initial positions. The results of the calculations are shown in Table 1. The calculation was carried out for a height of 150km, taking into account the influence of the atmosphere ($\nu_0 = \pi / 2$). Analysis of the table shows that the most favorable situation is when the stage at the initial instant of time is oriented perpendicular to the ion beam axis ($\varphi_0 = 0$). Deorbiting in oscillation mode near s_2 or inside A_1, A_2 areas (when $\varphi_0 \approx -\pi / 2$) is not effective.

Table 1. The influence of the space debris amplitude of oscillations on the ion beam average force.

φ_0 (rad)	$\dot{\varphi}_0$ (rad/s)	Area	F_{iy} (mN)	F_{ix} (N)
-1.64	0	A_3 near inner separatrix	-30.4	$-2.56 \cdot 10^{-5}$
-1.58	0	A_1	-30.4	$-2.16 \cdot 10^{-4}$
-1.56		A_2	-30.5	$1.39 \cdot 10^{-4}$
-0.785	0	A_3 inside	-38.8	$-4.44 \cdot 10^{-6}$
0	**0**	**A_3 inside**	**-41.9**	**$-4.97 \cdot 10^{-6}$**
1.40	0	A_3 near outer separatrix	-37.9	$2.96 \cdot 10^{-5}$
1.57	0.002	A_4 near separatrix	-35.7	$-1.81 \cdot 10^{-5}$

IOP Conf. Series: Journal of Physics: Conf. Series **1050** (2018) 012041 doi:10.1088/1742-6596/1050/1/012041

4. Conclusion

The problem of large space debris removal, by an active spacecraft with two ion engines was considered in this article. The influence of the atmosphere on the space debris attitude motion under the action of the ion beam created by the active spacecraft engine was studied. The mathematical model, which describes a plane motion of space debris under the action of the ion beam, gravitational and aerodynamic forces and torques, was constructed using Lagrange formalism. For calculation ion beam force and torque the self-similar model of ion engine plume exploration and fully diffused reflection model of ions interaction with the surface were used. As an example, the removal of Cosmos 3M rocket stage from a circular orbit of 500 km altitude by ion beam was considered. The dimensionless aerodynamic coefficients were calculated using the Newton impact theory. It was shown that the presence of an atmosphere significantly reduces the time of descent from orbit. The analysis of phase portraits describing the deviation of space debris from the local vertical was carried out. It was shown that in the case of a cylindrical space debris, which center of mass lies on the axis of symmetry, taking into account the influence of the atmosphere leads to deformation of the phase portrait, but does not lead to its qualitative changes. It was found that the amplitude of oscillations of space debris has a noticeable effect on the ion beam force, averaged over the period of oscillation, and therefore it also affects total descent time. The most favorable situation is when the rocket stage at the initial instant of time is oriented perpendicular to the ion beam axis.

Acknowledgments

This work was supported by the Russian Foundation for Basic Research (Projects No. 18-01-00215 A, No. 17-41-630274 R_A).

References

[1] Bonnal C, Ruault J M and Desjean M C 2013 Active debris removal: Recent progress and current trends *Acta Astronautica* **85** 51–60

[2] Brito T P, Celestino C C and Moraes R V 2013 A brief scenario about the" space pollution" around the Earth *Journal of Physics: Conference Series* **465** 012020

[3] Kessler D J and Cour-Palais B G 1978 Collision frequency of artificial satellites: The creation of a debris belt *Journal of Geophysical Research: Space Physics* **83** (A6) 2637–2646

[4] Hakima H and Emami M R 2018 Assessment of active methods for removal of LEO debris. *Acta Astronautica* **144** 225–243

[5] Shan M, Guo J and Gill E 2016 Review and comparison of active space debris capturing and removal methods. *Progress in Aerospace Sciences* **80** 18–32

[6] Schaub H, Parker G G and King L B 2004 Challenges and prospects of Coulomb spacecraft formation control. *Journal of Astronautical Sciences* **52** (1) 169–193

[7] Aslanov V S 2017 Exact solutions and adiabatic invariants for equations of satellite attitude motion under Coulomb torque. *Nonlinear Dynamics* **90** (4) 2545–2556

[8] Bombardelli C and Pelaez J 2011 Ion Beam Shepherd for Contactless Space Debris Removal *Journal of Guidance, Control, and Dynamics* **34** (3) 916–920

[9] Kitamura S, Hayakawa Y and Kawamoto S A 2014 Reorbiter for large GEO debris objects using ion beam irradiation *Acta Astronautica* **94** (2) 725–735

[10] Bondarenko S, Lyagushin S and Shifrin G 1997 Prospects of using lasers and military space technology for space debris removal *Second European Conference on Space Debris* **393** 703–706

[11] Khalifa N S 2015 Dynamical study of low Earth orbit debris collision avoidance using ground based laser *NRIAG Journal of Astronomy and Geophysics* **4** (1) 25–31

[12] Bombardelli C, Merino-Martínez M, Galilea E A, Pelaez J, Urrutxua H, Herrera-Montojo J and Iturri-Torrea A 2011 *Ariadna Call for Ideas: Active Removal of Space Debris* Final Report. (Ariadna ID 10-6411, Contract Number: 4000101447/10/NL/CBi)

[13] Bombardelli C, Urrutxua H, Merino M, Pelaez J and Ahedo E 2013 The ion beam shepherd: A new concept for asteroid deflection *Acta Astronautica* **90**(1) 98–102

[14] Alpatov A P, Fokov A A, Khoroshylov S V and Savchuk A P 2016 Error Analysis of Method for Calculation of Non-Contact Impact on Space Debris from Ion Thruster *Mechanics, Materials Science & Engineering Journal*

[15] Aslanov V S and Ledkov A S 2017 Attitude Motion of Cylindrical Space Debris during Its Removal by Ion Beam *Mathematical Problems in Engineering* (Article ID 1986374)

[16] Jaslow H 1968 Aerodynamic Relationships Inherent in Newtonian Impact Theory. *AIAA Journal* **6** (4) 608–612

IOP Conf. Series: Journal of Physics: Conf. Series **1050** (2018) 012042 doi:10.1088/1742-6596/1050/1/012042

Estimation of the failure-free operation for deployment of transformable space structures

A M Lepikhin, A E Burov and Yu P Pokhabov

Institute of Computational Technologies, Siberian Branch of Russian Academy of Science (ICT SB RAS), Krasnoyarsk, 660049, Russia

E-mail: aml@ict.nsc.ru

Abstract. High reliability is required for the deployment of the transformable structures as this process is crucial for successful spacecraft mission. Computational evaluation of the reliability parameters for the deployment is of a great importance as it is not possible to fully reproduce the real conditions of outer space during ground-based tests. The presented approach to estimate the probability of failure-free deployment is based on the condition of exceeding the driving torque over the resisting moment at each of the angular positions of the structure. The approach was used to compute the reliability functions for deployment of a space reflector. The calculations were carried out based on kinematic and dynamic analysis of the structure.

1. Introduction

The development of transformable space structures demands the solution of a number of technical and mechanical problems conditioned by the uniqueness of objects and the combination of conflicting requirements to increase overall dimensions and provide a high rigidity under strict weight limitations on the power frame [1-3]. As a rule, such structures are delivered to space orbits within a space vehicle in a transportation state being folded in compact volume. Bringing them to the operating position is associated with the implementation of the deployment process. High requirements for the probability of failure-free operation are imposed on the transformable structures as the deployment process belongs to dynamic operations that directly affect the successful execution of a spacecraft's mission. Ground testing is mandatory for the deployment process. However, the ground-based tests are limited as it is not possible to fully reproduce the real conditions of the deployment in outer space. Therefore, evaluation of the reliability parameters for the deployment of mechanical systems is of a great importance [4, 5].

Reliability of the deployment of transformable structures from the transportation to operating position is determined by the functioning of mechanisms when turning to the latching position. The failure-free operation is satisfied through the fulfillment of the equilibrium condition in the case of rotational motion [3, 6]:

$$M_{\mathrm{d}}(\varphi)-M_{\mathrm{r}}(\varphi)=0 \quad \forall\varphi\in[\varphi_{\mathrm{i}},\varphi_{\mathrm{f}}], \tag{1}$$

where M_d – is the actual value of the drive torque; M_r – is the actual values of the moment from resistance forces on the motion path of the angle bar; φ_i, φ_f – are the initial and final angle, respectively.

The drive torque is produced by the deployment actuators such as tape-springs, pneumatic piston, electromechanical devices etc. The resisting forces are generated by friction, loads from gravity etc. In the general case, the deployment is possible provided that the inequality of moments is observed (the sign of the functional dependence on the rotation angle is omitted hereinafter):

$$M_{\mathrm{d}} > M_{\mathrm{r}} \tag{2}$$

The probability of failure-free deployment P_{d} in this case is defined as the probability of meeting the operating conditions (2) at each of the angular positions of the structure:

$$P_{\mathrm{d}} = P(M_{\mathrm{d}} > M_{\mathrm{r}}) \tag{3}$$

In the domestic practice of design for single-used mechanical devices of rotary type, the condition of exceeding the drive torque over the moment from resistance forces on the turning path of the structure is generally defined in the following form [5, 6]:

$$M_{\mathrm{d}} > k \cdot M_{\mathrm{r}} \tag{4}$$

where k is the margin of driving torque.

The margin expresses the idea of necessity to separate the average values of the driving torque and the resistance moment in order to increase the reliability. Torque or force margins are similar to safety factors used in structural design and relate to the ratio between the driving torque (force) and resistance moments (loads) [7]. By established practice, the margins assigned for driving torques (or forces) in movable mechanical units (hinges, friction pairs, etc.) are no less than 100% (the ratio 2:1) to the worst-case value of the moments from resistance forces. This rule was applied in the Russian space engineering industry before the implementation of joint projects with European partners aimed to development of communication satellites, the first of which was Sesat (Siberian-European Satellite). Under the influence of the requirements from European partners, the margins of driving torques began to be assigned at least 200% (ration 3:1), although the ratio 2:1 was used in some cases, depending on the dynamics of deployment processes and available (confirmed) experience. Choice criteria and values of margins for driving torques (forces) have not yet been standardized neither by industrial nor state standards in Russia.

It should be noted that in the United States there is a somewhat different approach. The first standard MIL-A-83577 issued in 1975 provided for the presence of static margins for driving moments and forces in the movable mechanical units of spacecraft. The required margins in the specified standard were recommended to be assigned in the ranges from 100% to 200% to the worst case of resistance moments or forces. In cases of weight restrictions and available practice, it was allowed using the margins less than 100%. Kinetic torque and force margins are defined as follows:

$$\text{Kinetic Torque Margin} = \left[\frac{\text{Drive Torque} - \text{Resisting Torque}}{\text{Torque Required for Acceleration}} - 1\right] \times 100 \tag{5}$$

$$\text{Kinetic Force Margin} = \left[\frac{\text{Drive Force} - \text{Resisting Force}}{\text{Force Required for Acceleration}} - 1\right] \times 100 \tag{6}$$

At present, standards AIAA S-114-2005 (issued by American Institute of Aeronautics and Astronautics, AIAA) and NASA-STD-5017 (National Aeronautics and Space Administration, NASA) substitute MIL-A-83577B and specify general requirements for the testing of movable mechanical units. In AIAA S-114-2005, the kinetic margins specified in (5) and (6) are named as dynamic margins.

The maximum value of the moment from the resistance forces $M_{\mathrm{r}}^{\max}$ is unknown in advance, since it is determined by the conditions and modes of the structure operation in outer space. However, the actual values of the resistance moment have a pronounced dependence on the ambient temperature. Coefficients of sliding friction in friction pairs and the stiffness of transit electric cables, as a rule, increase with changing ambient temperatures from $T_{\mathrm{warm}}^{\max}$ to $T_{\mathrm{cold}}^{\max}$. The maximum actual value of the resistance moment $M_{\mathrm{r.actual}}^{\max}$ can be determined as a result of ground tests of the rotary structure with transit electric cables at an abnormally low temperature $T_{cold}^{\max}$ corresponding to the operating conditions. If the maximum value of the moment from resistance forces is taken as the maximum value at the abnormally low temperature:

$$M_{\mathrm{r}} = M_{\mathrm{r.actual}}^{\max}\left(T_{\mathrm{cold}}^{\max}\right) \tag{7}$$

then the maximum value of the resistance forces moment in the formula (4) can be determined by the relation:

$$M_{\mathrm{r}}^{\max} = k \cdot M_{\mathrm{r.actual}}^{\max}\left(T_{\mathrm{cold}}^{\max}\right) \tag{8}$$

The influence of various technological and operational factors, including temperature, can lead to fluctuation of the driving and resisting moment values. Assume that this fluctuation is defined by the Gauss probability functions with given expected values and coefficients of variation (standard deviations). Taking into account the foregoing, the expression (3) can be written in the following form:

$$P_{\mathrm{d}} = \Phi\left\{\frac{\mu_{\mathrm{r}} - \mu_{\mathrm{d}}}{\sqrt{\sigma_{\mathrm{d}}^2 - \sigma_{\mathrm{r}}^2}}\right\} = \Phi\left\{\frac{k-1}{\sqrt{V_{\mathrm{r}}^2 k^2 - V_{\mathrm{d}}^2}}\right\} \tag{9}$$

where μ_{d}, μ_{r} – are the expected values of the driving moment and the moment of resistance; σ_{d} and σ_{r}, V_{d} and V_{c} are the corresponding standard deviations and coefficients of variation; k is the torque margin.

The above approach was used to compute the reliability functions for deployment of a space reflector. The calculations were carried out, based on kinematic and dynamic analysis of the structure. The results of the analysis included the forces and moments in kinematic pairs, velocities, accelerations, trajectories, driving and resisting torques. The calculation scheme is presented on the figure 1. The displacement rate of the input link of 4.23 mm/s was taken to ensure an average angular velocity of the radial ribs of about 0.03 rad/s, which corresponded to 30 seconds of deployment time.

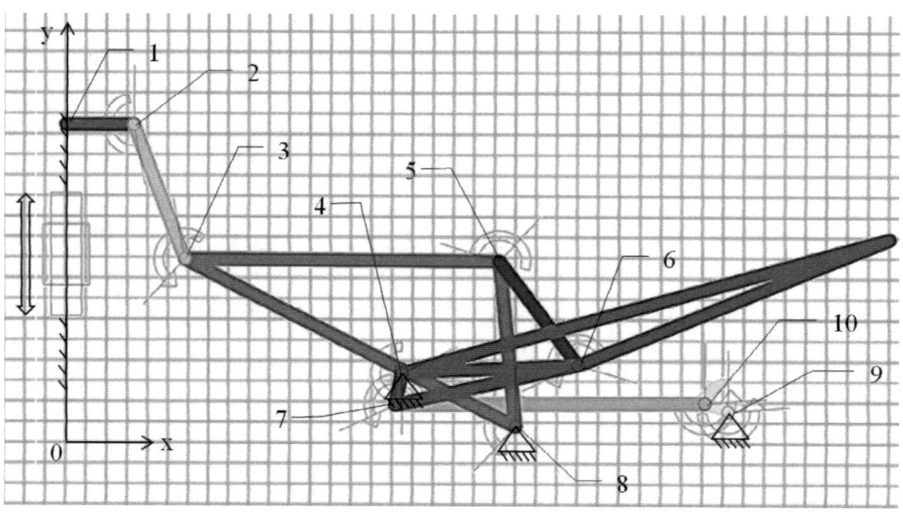

Figure 1. Kinematic scheme of the reflector deployment mechanism in the operation position: 1 – connection of the input link to the base; 2-10 – pivot points

The reliability functions calculated for the range of V_r from 0.08 to 0.12 and V_d = 0.10÷0.15 are presented on figure 2. As seen in figure 2, the deployment reliability does not fall below 0.99 even for maximum coefficients of variation. The results of ground tests on deployment of reflectors and solar arrays [3, 6] revealed that the coefficient of variation V_d does not exceed 0.12, while V_c does not exceed 0.08.

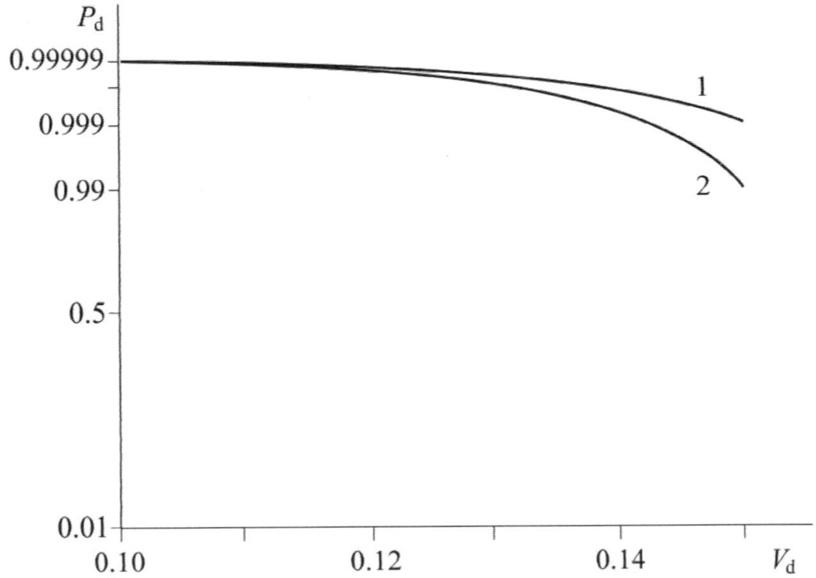

Figure 2. Projected reliability functions: 1 - $V_r = 0.08$; 1 - $V_r = 0.12$

Assuming that the torque margin equals 3.0, according to (9) one gets the probability of failure-free deployment above 0.999999. At k = 2.0, the probability of failure-free deployment still remains at a level above 0.99999. However, if we examine the available statistics on failure of deployment of the transformable mechanical systems on orbit, the above estimates are overly optimistic.

The history of space launches shows, the failures of deployment and fixing mechanisms are not such rare events. In the review [8] the exploitation of 449 European and American geostationary satellites for the period from January 1995 to October 2008 was analyzed. The given data demonstrate that the combined share of failures of mechanisms and structures (including systems of deploying) amounted to 2.4% of the total number of failures.

Statistical information on the implementation of 236 missions on the orbital stations Salyut-4, -6, -7, the Mir orbital complex and the ISS for the period 1974-2012 was presented in [9]. It was noted that during 130 spacewalk performed, about 4% of the total number of operations were caused by the need to act upon the undeployed structures. Summarizing these data yields the average probability of failure Q_{av} of $1.698 \cdot 10^{-2}$ per spacecraft. The average probability of failure at deployment of transformable space structures in 2009-2016 amounts to $1.072 \cdot 10^{-2}$ per spacecraft.

Taking into account these statistical estimates and the results of calculations based on the above scheme, the average probability of failure-free deployment comes to 0.9892. Since the analysis did not include all the failures, the estimate obtained can be considered as the lower probability level for failure-free deployment of space transformable structures.

Thus, the proposed model considering the features of kinematic and dynamic analysis can be used to estimate the reliability functions for the transformable space structures.

Acknowledgments
This work was done within a complex project and financially supported by the Russian Federation Government (Ministry of Education and Science of the Russian Federation), contract No. 02.G25.31.0147.

IOP Conf. Series: Journal of Physics: Conf. Series **1050** (2018) 012042 doi:10.1088/1742-6596/1050/1/012042

References

[1] Nikolsky V V 2003 *Design of Space Vehicles* (St Petersburg: Voenmeh) p 131
[2] Romanov A V and Testoedov N A 2015 *Fundamentals of Information and Control and Mechanical Systems of Spacecraft* (St Petersburg: NGO Professional) p 240
[3] Zimin V N, Krylov A V, Meshkovskii V E, Sdobnikov A N, Fayzullin F R and Churilin S A 2014 *Sci. Educ.: Scient. Publ.* **10** 179–91
[4] Chodimella S P, Moore J, Otto J and Fang H 2006 *Proc. 47th AIAA/ASME/ASCE/AHS/ASC Structures, Structural Dynamics, and Materials Conf. (Newport, Rhode Island)* AIAA-2006-1603
[5] Saleh J H and Caster J-F 2011 *Reliability and Multi-State Failures: A Statistical Approach* (Chichester: John Wiley & Sons) p 206
[6] Shatrov A K, Nazarova L P and Mashukov A V 2009 *Fundamentals of Designing of Mechanical Devices of Spacecraft. Constructive Decisions, Dynamic Characteristics* (Krasnoyarsk: SibGAU) p 144
[7] Postma R W 2004 *Proc. 37th Aerospace Mechanism Symp.* (Galveston, Texas) edited E A Boesinger (NASA Center for AeroSpace Information, Hanover, MD) pp. 107–18
[8] Gorbenko A V, Zasukha S O, Ruban V I, Tarasyuk O M and Kharchenko V S 2011 *Aeros. Tech. Techn.* **1** 9–20
[9] URL: http://ecoruspace.me

IOP Conf. Series: Journal of Physics: Conf. Series **1050** (2018) 012043 doi:10.1088/1742-6596/1050/1/012043

The sapphire tips of the active control devices of detail's dimensions of produced with hollows and ridges, with the possibility of determining the lateral approximation of ridges

E V Leun[1], V I Leun[2], A E Shahanov[1]

[1]Lavochkin Association, 24 street Leningradskaya, Khimki, 141402, Russia
[2]Omsk State Technical University, 11, Mira ave., Omsk, 644050, Russia

Abstract. The high speed rotation of ≥10 turn/s details with intermittent surfaces when machining on grinding machines and small time tabs detail past the tip with a duration of less than 1-5 ms necessitates synchronization of the measurements in the active control devices (ACD) depending on the current position of the ridge and it lateral approach. The implementation of optical methods to control the shape of details and high-precision measurement of movements are also in demand in the ACD. The use of sapphire tips allows to solve three problems simultaneously: to create an asymmetric lateral optical flow to determine the lateral approach of the ridge of the detail, to use the tip as a reflector to measure the movements of the l_x, and to form a cut-off border to control the shape of the product.

The article presents two optical ACD schemes implementing these capabilities with amplitude conversion of the optical signal and phase synchronization of the output signal frequency with the rotation of the detail. The conditions for ensuring a small roughness of the surface of the sapphire tip R_z under active control of the detail dimensions and the peculiarities of the formation of refractive and reflected optical flows with the help of it are considered. It is shown that the small roughness of the surface of the sapphire tip $R_z \leq 0{,}1$ μm, ensuring the coherence of the reflected optical flux, is achievable under the condition of a small effort of its clamping to the product ≤3H, which is similar to the processing of sapphire in quasi-plasticity to obtain a faultless surface. The article discusses the optical-mechanical, dynamic and metrological characteristics of the ACD using a laser interferometer movements.

1. Introduction

The tasks of increasing labor productivity, set by the leadership of Russia, and successfully solved by the active control devices (ACD) of the sizes of the details present new requirements for them in the current state of the art and technology. Since the 60s of the XX century, leading in the field of active control of the dimensions of details in the process of machining on metal cutting machines are the scientific teams of MGTU "STANKIN", OmSTU, MSTU name N.E. Bauman, JSC "NIIizmereniya" [1-6]. Recently, a sufficient number of publications have appeared in Russia concerning various aspects of the development and application of methods and means of active detail control and developing this direction [7-11].

Judging by the publications, the issues of the development and effective use of the ACD are relevant for representatives of scientific teams from different countries [12-16], but the Italian firm Marposs [17] is the recognized leader in this field.

IOP Conf. Series: Journal of Physics: Conf. Series **1050** (2018) 012043 doi:10.1088/1742-6596/1050/1/012043

It should be noted in common, which unites all the existing designs ACD is the application for the contact measurement of the dimensions of the measuring tips (in next text - tip) of the high strength, but non-transparent materials. This is the simplest and most reliable, but does not allow the fullest use of optical methods and controls, limiting the prospects for the implementation of high accuracy and extensive functionality. However, attempts of active use optical directly ACD with contactless measurement methods also face significant technological constraints due to the presence of flow, vapour, mist from the cutting fluid (coolant), complex conditions, etc.
In this regard, the most promising are fundamentally new hybrid ACD with the use of tips made of high-strength, wear-resistant, but optical transparent materials, such as diamonds and especially widely available artificial corundums, such as sapphire, ruby, which allow to implement contact and non-contact measurement methods. Such opportunities are most demanded for active control of the most difficult products in the form of products of tool production having intermittent surfaces with hollows and ledges. So in Russia since 2013 in this area appeared and began to be created laser ACD with sapphire tips [18-22]. Also, new contactless [19, 22, 23] and hydraulic jet [24-26] ACD were created, which expand the possibilities of detail size control. However, the reserve of improvement in the direction of enhancing the functionality of such ACD is far from exhausted and this article is devoted to the consideration of one of these possibilities.

2. Formulation of the problem

Now the process of measuring the dimensions of details with a intermittent surface made with hollows and ridges, implemented in ACD with high – strength and optically transparent sapphire tip, based on mechanical contact, being essentially tactile, and does not allow to determine the lateral approximation (next-the approximation) of moving ridges of products. Since for different conditions of control contact time can be no more than 1-5 ms, for quick measurements need to advance information about the approach of the ridge. At the same time, fixing the moment of touching between the ridge of the detail and the tip can be quite late, potentially leading at best to a dynamic component of the error for cyclic measuring instruments, and at worst – to the defective operation of the entire ACD.
In addition, the control of the frictional drive, to ensure the ceasing of the tip in the troughs between the ridges and are practically binding site of the modern ACD also implies the availability of information about the current position and the approximation of the ridge of the detail.
This task is complicated by the possible changes in the speed of rotation of the machine and the presence of a large number of details, such as tool production with short and/or asymmetric ridges (**Figure 1**), the movement of ridges which are difficult to accurately determine in advance by calculation.

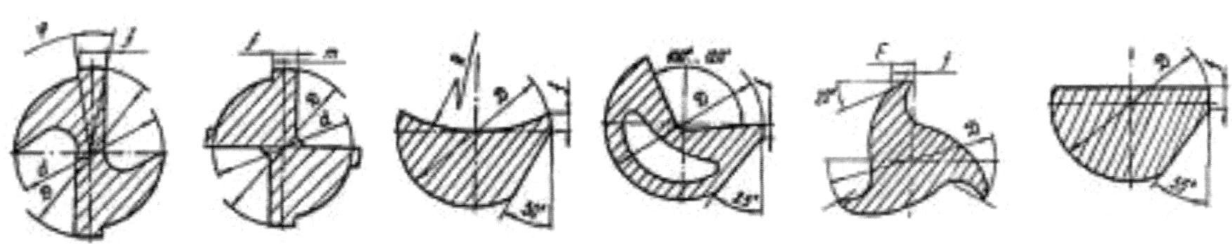

a) b) c) d) e) f)

Figure 1. Cross section with short and asymmetrical ridges of the drills (a-d): spiral, four-belt with straight fluted, for gun, for rifle and for core drill (d,e): for light alloy and combined.

The solution of the problem of determining the lateral approximation of moving ridges of details in laser ACD, simultaneously intersects with the problem of optical control of the detail shape [22], solved by ACD, and the desire to improve the accuracy of measurements of the movements of the sapphire tip when using it as a reflector for the laser interferometer movements [18, 19]. The latter is

possible, first of all, while maintaining the coherence of the optical flow passing through the sapphire tip.
In this regard, the research task is to develop and study a sapphire tip ACD with the ability to determine the lateral approximation of the ridges of the detail implemented in conjunction with the optical control of the detail shape and the use of a sapphire tip as a reflector for the laser interferometer movements.

3. Theory
In order to solve this problem, two designs of ACD with asymmetric optical flow were developed: with amplitude registration of the reflected signal (**Figure 2**) and phase synchronization of rotation of the detail (**Figure 3**). Their composition, principle of operation and calculation of the main parameters are considered further on the example of one-contact measurements at unidirectional motion (rotation) on a cylindrical grinding machine with the implementation of a black and white method for controlling the shape of the detail and high-precision measurement of the movement of the sapphire tip laser interferometer.

3.1. The composition and principle of operation of amplitude ACD
The developed optical scheme ACD (**Figure 2**) includes a tip 1 consisting of a sapphire fiber 2 and a protective coating 3, optical elements 4 and 5, a laser interferometer of movements 6, working with the input 7 and reflected 8 optical streams, a illuminator 9, a lens 10, a camera 11, a comparator 12 and a recorder 13, which can be created on the basis of a video camera and a thermal imager operating in different spectral wavelength ranges, as described in [21]. Processing of the detail is carried out 14 grinding wheel 15.
The operation of the device in the processing of the detail and the active control of its size can be divided into 3 stages depending on the position of the ridge relative to the tip:
1) is substantially remote;
2) approaching;
3) passing by.

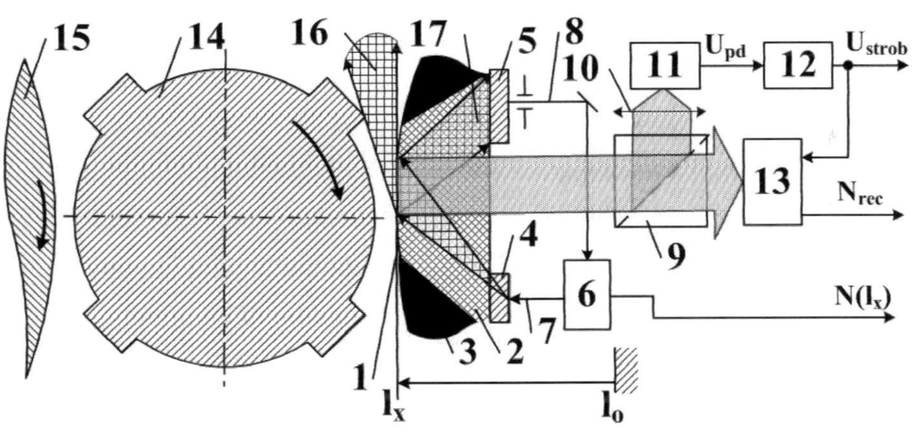

Figure 2. ACD' optical scheme with a sapphire tip and a single-ended output of the optical flux by using the amplitude method of forming a Gating signal.

During the first stage, the ridge of the detail is essentially removed from the sapphire fiber 2 of the tip 1. Laser interferometer 6 generates an input optical flux illuminating optical element 4, made in the form of a connection of an optical wedge and a scattering lens based on the elements of integral and/or Fresnel optics. Its output is formed a divergent optical flow, the next through the sapphire fiber 2 and illuminating the inside surface in the range of angles: α_{inc1}- α_{inc2}, provided condition $\alpha_{inc1}<\alpha_{lim\ inc}<\alpha_{inc2}$, where $\alpha_{lim\ inc}$ – limit (critical) angle of incidence, which defines the condition of total internal reflection. The area of the illuminated spot is smaller than the area of the sapphire fiber 2 and this creates a cut-off line used to control the shape of the detail 14.

Part the next through the sapphire fiber 2 divergent optical flow and then exits out outside the tip 1, forming asymmetric side stream 16 composed of refracted rays within a specified range of angles: from $\alpha_{refl}=60°$ to $\alpha_{ref2}=90°$. This flux of refracted rays follows in the direction of the expected appearance of the ridge of the detail 14.

The other part of the optical flow following the sapphire fiber 2 is mirrored from the contact surface of the tip 1 to the optical element 5, passes through the diaphragm, the mirror and returns as a reflected stream 8 to the laser interferometer of movements 6. Thus, a tip 1 with a similar triangular stroke of light for the laser interferometer 6 acts as a prism, the movements of lx which are associated with the size of the detail are measured with high accuracy.

In the second stage, when approaching the ridge of the detail 14 illuminated by an asymmetric optical flow 16 formed from refractive rays, a multidirectional diffusely scattered radiation is formed. Part of it should be in the opposite direction, passes through a sapphire fiber 2, the beam splitter 9 and then to the entrance of the photodetector 11, the output signal U_{fd} which follows to the input of the comparator 12 forming the signal U_{strob}. The output of the comparator 12 is connected to the input of the recorder 13.

As the ridge of the detail approaches 14, the level of the reflected optical signal and U_{fd} signal increases and after reaching a certain threshold level corresponding to the specified position of the protrusion of the detail before the mechanical touch with the tip 1 comparator 12 forms a positive pulse Delta of the U_{strob} signal corresponding to the beginning of the third stage and allowing the operation of the recorder 13 for video fixation of the image of the end surface of the protrusion passing through the cut-off border, , thermal imaging pattern for the correction of measurement results with the formation of a digital signal N_{reg}.

After passing the ridge past the tip, the level of the reflected signal, the signals of the U_{fd} sharply decreases forcing the comparator 12 to form a negative pulse drop, which eventually blocks the operation of the recorder 13 until the next protrusion approaches, after which the cycle repeats.

In accordance with the algorithm of operation, as well as, judging from the diagram of the refractive flow direction and the route of the reflected flow from the detail, it is possible to form an equivalent optical scheme that is close to the flow path in the fiber-optic reflectometric displacement sensors with the corresponding conversion function (**Figure 3b**) previously well studied in [27].

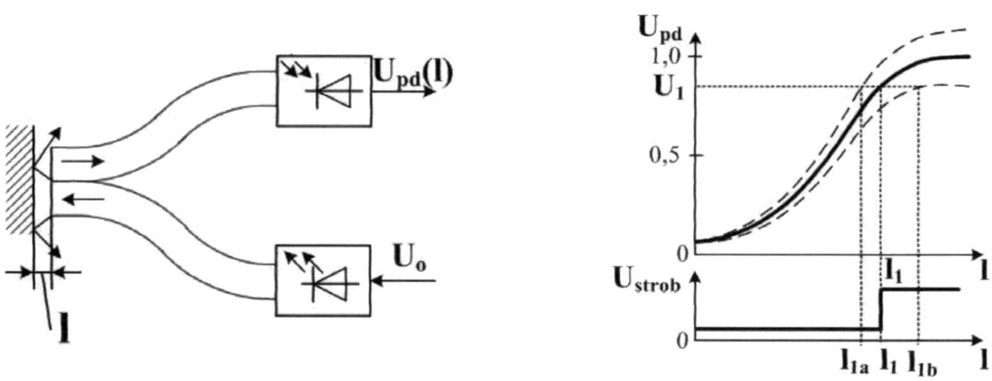

a) b)

Figure 3. The equivalent optical scheme of the widely used fiber optic time domain reflection displacement sensors (a) and the transformation function (b)

The formation of the threshold level U_1 allows in the end to create the gate signal U_{strob} for a specific current position of the projection of the detail to control operations in the ACD. Despite its simplicity, this technical solution has the following disadvantages:

- The amplitude of the $U_{fd}(l)$ from a variety of external flashes, different "wildness" in turbulent or dispersions the coolant flow, the reflection properties of the moving ridges, which ultimately leads to errors in the formation of the gating signal U_{strob}: $\Delta l = l_{1b} - l_{1a}$ **(Figure 3a);**
- It is impossible to realize the synchronized rotation of the detail for its full turn with frequency signal and to implement the coordinate–temporal reference of the external surface of the object to check its surface at the given coordinates.

3.2. The composition and principle of action ACD with phase synchronization of the rotation of details
Further expansion of the functionality associated with the introduction of phase synchronization of the rotation of the ridges of the detail with a frequency signal by using the phase automatic frequency control (PAFC), shown in **Figure 4**. This device consist of a phase detector 18, a low-pass filter 19, a generator controlled by a voltage of 20 and a divider 21.

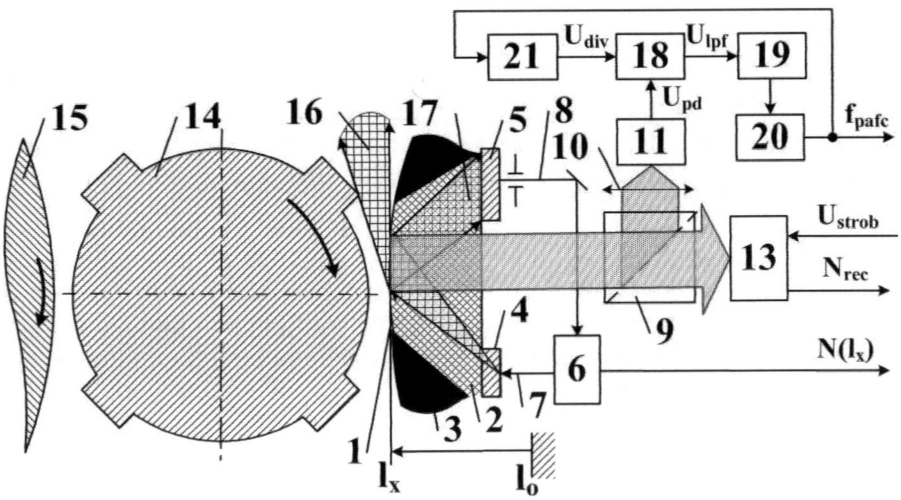

Figure 4. ACD's optical scheme with phase synchronization of the output signal frequency and rotation of the detail

The use of the PAFC system is quite common, and its mode of operation is well studied. A feature of this PAFC system is the formation of the output signal with frequency f_{pafc}, which after divider 21 corresponds to the frequency of rotation of the ridges of the detail. For the balanced mode of the PAFC, the frequency of its output signal is equal to the frequency of pulses generated by the photodetector from the running ridges of the detail $f_{pafc} = f_{rid}$, from which the equation is derived

$$\frac{f_{pafc}}{k_{div}} = N n_{rid}\,, \tag{1}$$

where k_{div} - the division coefficient, N - the rotational speed in turn/s, n_{rid} - the number of ridges of the detail.
Using the formula (1) to determine the value of f_{pafc} frequency of the signal used for gridding, for example, with a step l_{step}=1 μm on the outer surface of the cutter

$$f_{pafc} = \frac{v}{l_{step}} = \frac{2\pi RN}{l_{step}} \tag{2}$$

For a five-cogs cutter (n_{rid}=5) with a diameter of 20 mm and a rotational speed of 10 turn/c, the frequency of f_{pafc}=600 kHz, and the division coefficient determined from formula (1) as will be 12000.

3.3. Optical and mechanical characteristics
Earlier in the works [19, 28] the results of theoretical and experimental studies of the loading capacity of sapphire tips under shock loads in the process of active control of the sizes of details with an intermittent surface and in particular with a five-tooth cutter rotating at an angular velocity of 600 turns/min =10 turns/c were presented. it was shown that the surface roughness R_z was 0.15 μm. This

was the first experimental test of high strength parameters of sapphire, which confirmed the possibility of its use as a material for ACD tips. However, the obtained R_z value corresponds to a significant, up to 30-70 times the permissible load, which, as shown in [6], should not exceed 3 N. And this gives reason to assume that such a mode of interaction of sapphire tip with the processed detail is close to mirogrinding in the quasi-plasticity mode to obtain a faultless surface, well studied in [29, 30] and the possibility of reducing the R_z, at least 2-3 times, to values of 0.05-0.08 μm.

These calculated values of R_z make it possible to ignore the parasitic diffuse scattering of light in the wavelength range of 0.6-0.76 μm, characteristic of most of the measuring laser and read the reflected light from the surface of the sapphire tip of the mirror, and the image captured by the recorder 13 allowing to reach the highest possible resolution.

To calculate the parameters of the refracted stream 16 can be used a known expression $\frac{\sin\alpha_{inc}}{\sin\alpha_{ref}} = \frac{n_{cool}}{n_{spf}}$, where α_{inc} and α_{ref} – the angles of incidence and refraction of light to the surface of the sapphire fiber, n_{cool} and n_{spf} - the refractive indices of sapphire and the coolant.

The boundaries of the optical flow formed by the optical element 4 can be calculated using the parameters of the deflected flow. The limiting angle of incidence of $\alpha_{lim\ inc}$, at which a refractive stream is formed, directed along the surface of the tip 1, vertically upwards according to **Figure 2**. The refractive index of sapphire n_{spf}=1,76, and for coolant, for example, based on glycerol, n_{cool}=1,47 and in accordance with this $\alpha_{lim\ inc} = \arcsin\frac{n_{cool}}{n_{spf}} = \arcsin\frac{1.47}{1.76} \approx 56.6°$.

Due to the fact that part of the optical flow covering the inside surface of the sapphire tip must mirror effect, it is taking its width is equal to ≈9,4°, we obtain the value of the upper boundary of the illuminating optical flow α_{inc1} equal to 66°.

The lower value of the angle of incidence of light α_{inc2} falling from the inside to the surface of the sapphire tip is associated with the maximum value of the refraction angle for the exiting refractive rays. Taking the width of the formed asymmetric flow equal to 30° with the corresponding refractive angle α_{ref}=60°, it is possible to calculate the value of the angle of incidence α_{inc2} – $\alpha_{inc2} = \arcsin\frac{n_{cool}\sin\alpha_{lim\ inc}}{n_{spf}} = \arcsin\frac{1.47\cdot 0.86}{1.76} \approx 46°$.

Thus, for the designed optical circuit ACD with the development of emerging optical flow with a width of 30°, consisting of refracted rays, and creating mirrored beams for use sapphire tip as a reflector in cooperation with a laser interferometer it is necessary to form the optical element of the optical flow with a width of 20°. This thread should cover the surface of the sapphire tip from the inside and within the angular sector from α_{inc2}≈46° to α_{inc2}≈66°.

According to the geometric rules and the flow of optical rays in the contact area of the detail 14 and sapphire tip 1 for isosceles triangle ABC (**Figure 5**) the ratio of the triangle side to the sine of the opposite angle is equal to two radii of the circumference described near this triangle

$$\frac{|AC|}{\sin\alpha_{inc}} = 2R \quad (3)$$

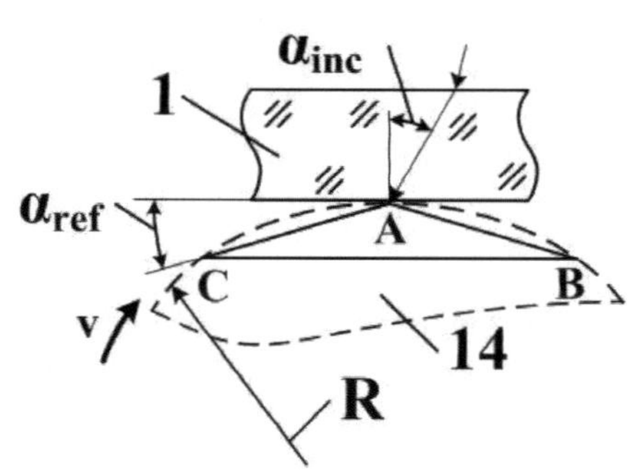

Figure 5. Formation of an asymmetric optical flow in the form of a flux of refracted rays within the angle of ABC from incident rays at an angle of α_{inc2} in the contact zone of the detail 14 and

sapphire tip 1

Based on this formula, the rotation time t of the surface of the detail from point C to point A can be approximately determined as follows:

$$t \approx \frac{|AC|}{v} = \frac{2R\sin\alpha_{ref}}{2\pi RNn_{rid}} = \frac{\sin\alpha_{ref}}{\pi Nn_{rid}} = \frac{\sin\alpha_{ref}}{\pi Nn_{rid}n_{cool}} \qquad (3)$$

As you can see, formula (3) demonstrates the one of the main relationship for these ACD between the size of the detail, its rotational speed and the optical parameters of the tip and the environment. And so allows to calculate their values for optimal scheme ACD.

3.4. Dynamic characteristic

In this section will be the estimation of the dynamic characteristics for two modes of operation ACD:
1) determining a lateral proximity of the ridges of the detail;
2) high-precision measurement of sapphire tip movements by laser interferometer.

3.4.1. Dynamic characteristics in determining the lateral approximation of the ridges of the detail. Modern photodetectors working in the fiber-optic communication lines allow recording frequencies of the optical signals until 1 GHz, having a delay time of <1 ns. Taking the comparator 12 delay time of with the reserve equal to ≈1 μs $=10^{-6}$ c and assuming that during this time the ridge moves at a distance of 3 mm, the displacement speed will be $0.003/10^{-6}=3\bullet10^{3}$ m/s. The obtained value is comparable to the first cosmic velocity and has never been realized on metal-cutting machines, which confirms a significant margin of speed and the overall feasibility of this possibility.

3.4.2. Dynamic characteristics of high-precision measurement of sapphire tip movements by laser interferometer. In the process of active control of detail dimensions, the tip carries out various movements. Thus, during movement to (or from) detail the speed usually does not exceed 50 mm/s, and at different stages of processing the detail [23] is the following values:

- For black grinding 200-500 μm/s,
- For fine grinding 10-50 μm/s,
- For fine grinding 10-50 μm/s.

The rate of exit of the tip of the $v_н$ to the ledge from the cavity of the detail during the so-called return movements in the process of active control can be determined for the worst conditions, when the fall between the ledges is ≈ 50 μm, and the exit from it to the ledge occurs during 1 ms. In these data, the value $v_н$ is equal to 50 mm/s.

As you can see, the maximum speed of the tip movements at different stages of the active control does not exceed 50 mm/s and when the laser interferometer with a wavelength λ=0,63 μm receives the maximum value of the Doppler frequency range taking into account the double stroke of the beam $f_{dop}= 2v_{вс}/\lambda\approx170$ kHz. For acousto-optic (AO) laser interferometers with a modulation frequency of AO f_{mod}=8 MHz, this is only ≈ 2% and is not a critical value, confirming the possibility of their use.

3.5. Metrological characteristics

As follows from the author's studies [31] the resolution of AO laser interferometers of movements with phase-digital transformation can be reduced to λ/3000, making ≈0.2 nm. It follows from this that the total measurement error will largely consist of substantially larger components, such as temperature error and error of the mechanical system ACD: tip shape, straight guides, etc.

The possibility of thermal imaging measurements carried out using the recorder 13, allows you to enter the correction of the measurement results depending on the temperature of the detail and the tip, significantly reducing its share. More detailed metrological analysis is beyond the scope of this article and will be considered in the following publications.

4. Experimental result

IOP Conf. Series: Journal of Physics: Conf. Series **1050** (2018) 012043 doi:10.1088/1742-6596/1050/1/012043

Currently Omsk state technical University conducted testing of a prototype of the ACD laser with a sapphire tip. The mechanical scheme providing repeatability of movements of a tip at the level of 1 μm is created.

Experimentally confirmed the presence of a defect-free surface (**Figure 6a**) as a result of sapphire surface treatment with micro-tool quasi-plasticity's mode [29, 30]. This mode allowed to obtain the surface roughness from R_a=1.95 nm, as follows from the profilogram in **Figure 6b**, according to the established ratio of R_z=5R_a allows us to estimate R_z≈10 nm=0.01 μm. The value comparable in order with the obtained value can be obtained by reducing the clamping force of the protrusion to the tip to 3N, i.e. up to 30-70 times relative to the force specified in [28]. This is confirmed by numerous experimental studies of doctor of technical Sciences, prof. Leun VI for almost during about 20-year period in the 70-90 years of the XX century to obtain the surface of the opaque tip almost polished mirror surface.

a) b)

Figure 6. The treated surface of the sapphire cylinder (diameter 31 mm) and profilogram the surface of the sapphire element R_a=1,946 nm after mirogrinding by quasi-plasticity's mode

The possibility of measuring the shape of the surface of the detail is clearly visible when registering the image (**Figure 7**) obtained by processing the cutting tool [32].

Figure 7. The image of cutting tools using a cut-off line

5. Discussion of results

1. The article analyzes two designs of the ACD with the possibility of determining the lateral approximation of the protrusion of the detail: with the amplitude transformation of the optical flow and phase synchronization of the output frequency signal and the rotation of the detail. It is shown that the first variant is simpler, and the second one is more accurate and allows to carry out coordinate – time binding of the external surface of the detail with an accuracy comparable to ≈ 1 μm to register its surface at the given coordinates.

2. The equations linking time and optical parameters of the considered ACD structures are obtained, which allow to optimize them for different processing conditions, sizes and parameters of details, characteristics of measuring devices.
3. Discusses the issues of ensuring minimum roughness of the surfaces of sapphire tips in the measurement process based on the execution conditions of quasi-plasticity's mode. The results of experimental work confirming the achievement of such.
4. Discusses the issue of implementation of the cut-off method to control the shape of the detail with the help of sapphire tips with the application of the Gating pulse emerging from the lateral approximation of the projection of the detail.

6. Conclusion

1. Determining a lateral proximity of the projection of the detail allows you to synchronize the measurement process of the laser ACD and to provide control of the sizes of details having a discontinuous surface with short ridges or asymmetrical arrangement. Implementation of this functionality can be carried by amplitude signals with the formation of a Gating pulse delta or phase synchronization of the frequency output signal with the rotation of the detail using the PAFC system. The first option is simpler, but the error of forming a Gating pulse delta depends significantly on the optical conditions of the ACD. The second option is more noise-resistant with the additional possibility of coordinate–temporal reference of the external surface of the detail with a resolution ≈1 μm for the registration of the surface at the given coordinates.
2. As a result of experiments it was shown that the sapphire can be successfully used as a bits ACD. The clamping force of the lugs ACD details having a discontinuous surface, for example, five-cogs cutters that are significantly greater than 3 N for 30-50 and up to 70 times, i.e. equal to ≈ 90210 premiere. this N leads to wear of the sapphire tip with the formation of surface roughness Rz=0.15 μm. The resulting surface roughness for such large loads does not allow it to be used without parasitic scattering for visible wavelengths.
At the same time, it can be confidently assumed that the reduction of 30-70 times to the level of ≈ 90-210 H efforts of pressing the ACD tips to details with an intermittent surface will allow to obtain the surface roughness of R_z, at least 2-3 times and bringing the value of R_z to the values of 0.05-0.07 μm, i.e. to the level of the so – called "mirror" surface for red wavelengths of 0.6-0.76 μm. This has the result successfully achieved in terms of mirogrinding sapphire details in the quasi-plasticity's mode obtaining a defect-free surface have been well studied in numerous works.
3. Improving the surface quality of sapphire tips, allowing the use of a visible part of the wavelength spectrum, simplifies the optical control of the surface shape of the details, effectively separating it from thermal measurements of temperature using thermal radiation in the near infrared range.
The developed design of ACD extends the functionality of laser ACD with sapphire tips increasing the Arsenal of promising methods and tools for controlling the size of details.

7. References

[1] Volosov S 1970 *Appliances for automatic control in mechanical engineering* (Moscow: Mechanical Engineering) p 310
[2] Vysotsky A 1979 *Active control in metalworking* (Moscow: Publishing House of standards) p 175
[3] Kondashevsky V and Lotze V 1976 *Active control of part dimensions on machine tools* (Omsk: Omsk State Technical University) p 431
[4] Sobolev M 2005 *Automatic dimensional inspection on machine tools* (Smolensk:Oikumena) p 300
[5] Leun V and Tignibidin A 2010 *New principles of construction of active control devices for products of tool production and engineering*“ (Omsk: Omsk State Technical University) №2(90) pp. 165-169
[6] The website of JSC Niizmereniya 2017 http://www.micron.ru/production/active_instruments/bv-4304/ (accessed: 05.02.2018)

[7] Belolapotkov D, Dobrovinsky I and Bear J 2007 *Improving the accuracy of control of dimensions of parts in the manufacturing process* (Moscow: The World measurements) №7 pp. 43-46.

[8] Bobkov D, Pudovkin A and Moskvitin 2006 *Active control of wall thickness of the bearing shells* (Design and technology of electronic means) №4. pp. 20-23.

[9] Pudovkin A, Chernyshov V and Kolmakov A 2004 *Active control of the geometric dimensions of bearings* (Measurement techniques) No. 9. pp. 32-35.

[10] Karpeeva E, Yakovlev A and Gorbunov V 2003 *The Prerequisites for creation of the device of the active control a new generation of grinding machine* (Saratov: Autom. and control in mechanical engine. and instrum. making: mezhvuz. science. Sat) pp. 122-125.

[11] Matyushkov V, Nazarenko V and Shumyacher V 2001 *Control of the process of machining grinding wheels, with the help of an active control system* (Shlifabraziv: Processes abrasive machining, abrasive tools and materials: Sat. articles of the Intern.scien.-techn. conf.) pp. 272-273.

[12] Industrial Metrology: Modern Measuring Systems for Monitoring Quality in Production Proc./ URL: https://www.jenoptik.com (accessed-05.02.2010).

[13] Keferstein C, Honegger D , Thurnherr H and Gschwend B 2008 *Process monitoring in non-circular grinding with optical sensor* (CIRP Annals - Manufacturing Technology) 57 (2008), pp.533-536, URL: https://www.kellenberger.com/newswriter_files/CIRP_Tagungsband_S_533-536_Keferstein.pdf (accessed-05.02.2018)

[14] Lee M. H., Bae J. Il, Yoon K. S. and Harashima F 2001 *Real time and an in-process measuring system for the grinding process cylindrical workpieces using Kalman filtering* (IEEE Transactions on Ind. Electr.) 47(6): pp. 1326 - 1333. URL: https://www.researchgate.net/publication/3217758_Real_time_and_an_in-process_measuring_system_for_the_grinding_process_cylindrical_workpieces_using_Kalman_filtering (accessed -05.02.2018)

[15] Gao Y ,Huang X and Zhang Y *Development of an in-process form error measurement system for surface grinding* // URL: http://ets.ifmo.ru/tomasov/konferenc/AutoPlay/Docs/Volume%202/4_22.pdf (accessed -05.02.2018)

[16] Yuan L, Järvenpää V, Virtanen S and Shiravani H Tampere *Numerical and experimental approach for roll grinding process* //Finland, URL: https://surveillance7.sciencesconf.org/conference/surveillance7/46_numerical_and_experimental_approach_for_roll_grinding_process.pdf (accessed -05.02.2018)

[17] The website of the company Marposs. URL: http://www.marposs.com (accessed: 05.02.2018).

[18] Leun E and Leun A 2015 *Method of active control of linear dimensions in the processing of the detail and the device for its implementation* " (Moscow:Patent of the RF 2557381, IPC 24 B 49/00 G 01 B 7/12, publ. 20.07., bul.№20)

[19] Leun E 2016 *Development of active control devices dimensional parameters of products by using sapphire measuring tips*(Omsk: Omsk State Technical University) №4(148), pp. 123-127

[20] Leun E 2016 *Features of construction of perspective laser methods of active control of the sizes of details* (Omsk: Mat-ly X all-Russian nauchn. Konf. memory of the chief designer AT "Flight" by A Klinyshkova) pp. 68-74

[21] Leun E ,Leun V ,Sysoev V ,Zanin K ,Shulepov A,Vyatlev P 2018 *The active control devices of the size of products based on sapphire measuring tips with three degrees of freedom* // Journal of Physics: Conference Series, Vol., conf. 1, URL: http://iopscience.iop.org/article/10.1088/1742-6596/944/1/012073 (accessed 04.02.2018)

[22] Leun E Leun V 2018 *Questions of construction of multifunctional devices for active control of linear and angular dimensions of products and their surface shape* (Omsk: Omsk Scien. Herald) №1 (157) pp. 89-95

[23] Leun E 2016 *Method of measuring the linear dimensions of products made with depressions and ridges on the surface*" (Moscow:Patent of the RF 2603516, IPC B24B 49/04, 49/12 B, publ. 27.11., bul.№33)

[24] Leun E 2017 *Intelligent lathe cutter with devices for active control of cutting zone temperature, product size and surface shape parameters* (Omsk: Omsk Scientific Herald) №4(154) pp. 87-93
[25] Leun E 2016 *Inkjet control method of the linear dimensions of products* (Moscow:Patent of the RF 2612349, IPC G01B 11/02, publ. 07.03, bul.№7)
[26] Leun E 2016 *Hydro-interferometric method control dimensions of details*"Proc. Int. Conf. IEEE X Dynamics of systems, mechanisms and machines, №1, vol. 2, (Omsk:Omsk State Technical University), pp. 101-109
[27] Zak E 1989 *Fiber-optic converters with external modulation* (Moscow: Energoatomizdat) p 128
[28] Leun E and Shulepov A 2017 *Research and development of sapphire crystal measuring tips for active control devices dimensional parameters of products*"(Omsk: Omsk State Technical University) № 3(153) pp. 91-95
[29] Konshin A, Teplova T and Solov'ev V 2010 *Surface treatment of solid minerals and crystals for microelectronics* (Moscow: Scient. Herald of the Moscow state mining University) №8, pp. 41-50.
[30] Teplova T 2010 *Quasiplastic removal of the surface layer of hard brittle materials to obtain nanometer surface relief* (Moscow: Scient. Herald of the Moscow state mining University) №8, pp. 73-88
[31] Leun E 2002 *Features of circuitry of acousto-optic laser systems for measuring movements with phase-digital conversion* (Moscow: Technology of mechanical engineering) № 5, pp. 33-39
[32] Optical vs Electron Microscope URL:https://scienceofsharp.wordpress.com/2016/11/03/optical-vs-electron-microscope (accessed 04.02.2018)

IOP Conf. Series: Journal of Physics: Conf. Series **1050** (2018) 012044 doi:10.1088/1742-6596/1050/1/012044

Simulated Annealing for Competitive p-Median Facility Location Problem

T V Levanova, A Y Gnusarev

Sobolev Institute of Mathematics SB RAS, Omsk Branch, Omsk, Russia

E-mail: levanova@ofim.oscsbras.ru, alexander.gnussarev@gmail.com

Abstract. There is a number of models that take into account the market competition in the location theory. Aboulian R. et al. formulated the location and design problem (CFLDP), in which the share of the market being served elastically varies depending on the location. The models with elastic demand are considered to be rather complicated and advantageous. We developed this problem and proposed the new one which combines CFLDP and classical p-median problem. In the given formulation a number of new facilities is fixed and is equal to p, so competitive p-median facility location and design problem (CPFLDP) has been derived. Therefore, the task can be considered as a clustering problem. In this paper we develop local search algorithms for CPFLDP. A computational experiment is carried out, the results are discussed.

1. Introduction

Facility location problems take a special place in the location theory. During last few years the competitive models, which describe the modern economy processes most properly, have gained the ground. There are different types of the state above problems, depending on the rival parts and customers behavior, possible facility locations [1]. Aboolian R., Berman O., Krass D. have formulated the location and design problem, where the rival parts struggle for the demand share [2]. The share changes elastically depending on the customers choice of the facility in order to satisfy their demands. The authors [2] describe this type of demand through a nonlinear function, traditionally used in the special interaction models in marketing. The developed mathematical model with a nonlinear objective function complicates the optimum search. Most commonly the known commercial software is not able to find any possible solution [3, 4]. The approximate approach becomes actual in the given situation. It should be mentioned that the number of algorithms to solve the elastic demand problems is quite limited. Probabilistic weighted greedy heuristics was offered to solve the location and design problem in [2]. We developed and researched a number of local search algorithms for the given problem [3, 4]. In this work we continue investigations in the given direction for the new competitive p-median facility location problem.

2. Competitive p-Median Facility Location Problem

One of the most known facility location problems is the p-median problem, which has a wide employment range and is often formulated as a clustering problem. However, the competitive location models attract the main interest at this time. This work researches the offered

Competitive p-Median Facility Location and Design Problem (CPFLDP) which unites the statement of the elastic demand problem from [2] and the classical p-median problem. Let us introduce the problem formulation and its mathematical model.

A new company is to locate a limited number of its facilities according to the given volume of money. It can state both the locations and the types of the facilities, for example: a supermarket, a hypermarket, etc. It has to rival for the demand with the existing companies, taking into account the fact that the customers choose the facilities depending on the distance, attractiveness and other factors. That is why the share of the serviced demand changes elastically depending on the companys and customers decisions. The aim of the company is to define the facility locations and types in order to attract the biggest share of the total demand.

Let us introduce the following notation with the consideration of [2]: $N = \{1, 2, \ldots, n\}$ is a discrete set of the demand points, so that each point is able to locate a facility; $C \subset N$ are the points of the demand occupied by competitor; $S = N \setminus C$ are the possible facility location points for the new company; R is a number of facility types; w_i is the demand weight in the point $i \in N$; c_{jr} is the facility opening cost if the variant $r \in R$ is used in the point $j \in S$; p is the facility number to be opened; β is the distance sensitivity parameter; λ is the demand elasticity coefficient.

Variables $x_{jr} = 1$, if the facility of the type $r \in R$ is located in the point $j \in S$, $x_{jr} = 0$ otherwise.

The usefulness u_{ij} of the company's facility $j \in S$ for a customer $i \in N$ is calculated the following way: $u_{ij} = \sum_{r \in R} k_{ijr} x_{jr}$, where $k_{ijr} = a_{jr}(d_{ij} + 1)^{-\beta}$ are the special coefficients which take into consideration the distance d_{ij} between the points i and j, the customers sensitivity to it β and the attractiveness a_{ir} of the facility type $r \in R$, $i, j \in N$. The total facility usefulness $U_i(S)$ of the point $i \in S$ is estimated through $U_i(C) = \sum_{j \in C} u_{ij} = \sum_{r \in R} k_{ijr} x_{jr}$. The notation for the competitors is similar.

In this model the demand function has an exponential form: $g(U_i) = 1 - \exp(-\lambda \cdot U_i)$, where U_i is the total utility for a customer at $i \in N$ from all open facilities:

$$U_i = \sum_{j \in S} \sum_{r=1}^{R} k_{ijr} x_{jr} + \sum_{j \in C} \sum_{r=1}^{R} k_{ijr} x_{jr}.$$

The share of the new facilities in the total volume of customer $i \in N$ servicing is equal to:

$$MS_i = \frac{\sum_{j \in S} \sum_{r=1}^{R} k_{ijr} x_{jr}}{\sum_{j \in S} \sum_{r=1}^{R} k_{ijr} x_{jr} + \sum_{j \in C} \sum_{r=1}^{R} k_{ijr} x_{jr}}.$$

Based on above notation, the mathematical model looks as follows:

$$\max \sum_{i \in N} w_i \cdot \left(1 - \exp\left(-\lambda \left(\sum_{j \in S} \sum_{r=1}^{R} k_{ijr} x_{jr} + \sum_{j \in C} \sum_{r=1}^{R} k_{ijr} x_{jr}\right)\right)\right) \cdot \tag{1}$$

$$\cdot \left(\frac{\sum_{j \in S} \sum_{r=1}^{R} k_{ijr} x_{jr}}{\sum_{j \in S} \sum_{r=1}^{R} k_{ijr} x_{jr} + \sum_{j \in C} \sum_{r=1}^{R} k_{ijr} x_{jr}} \right),$$

$$\sum_{j \in S} \sum_{r \in R} c_{jr} x_{jr} \leq B, \tag{2}$$

$$\sum_{r \in R} x_{jr} \leq 1, j \in S, \tag{3}$$

$$\sum_{j \in S} \sum_{r \in R} x_{jr} = p, \tag{4}$$

$$x_{jr} \in \{0, 1\}, \quad r \in R, j \in S. \tag{5}$$

The objective function (1) reflects the company's goal to maximize its demand share. Inequation (2) allows locate the facilities taking in account the budget amounts available. Conditions (3) show that there is a possibility to locate the facilities of only one type. Equation (4) sets the condition for the number of facilities to be opened.

Theoretically the given problem is NP-hard. For the implementation of experimental studies a set of 192 test examples have been used, which were constructed earlier on the basis of a real applied problem [3]. The set consists of two series, where the distances between the points are defined with the equally uniform distribution (Series 1) and satisfy the triangle inequality (Series 2). A set of 16 examples of the dimension $|N| = 60, 80, 100, 150, 200, 300$ with three possible projects and budget limits of 3, 5, 7 and 9 units have been formed.

3. Development of Simulated Annealing Algorithm

The research of the problem is frequently started from the attempt to use the existing software. The usage of the commercial software GAMS (solver CoinBonmin) [5] for the CPFLDP does not always allow find a permissible solution. Furthermore, the computational experiments [6] showed that CPFLDP is more complicated for the solver than the Competitive Facility Location and Design Problem (CFLDP), which is set by the conditions (1)-(3), (5). The data about average CPU time for Series 1 is given in the Table 1.

Table 1. Average CPU time of CoinBonmin, sec.

	60	80	100	150	200	300
CPFLDP	181	329	482	2351	2650	11831
CFLDP	106	161	295	1000	1325	4561

Taking into account the given above information the local search algorithms were chosen as the solution methods. The Variable Neighborhood Search (VNS) algorithm was offered in order to solve the CPFLDP in [6]. In this work we offer the realization of the simulated annealing (SA) algorithm. This algorithm is currently known for its successful usage for a wide range of optimization problems, for example [4, 7].

The simulated annealing algorithm starts from some initial solution and initial temperature parameter. At each value of temperature a certain number of iterations is carried out. On each iteration the algorithm selects from neighbourhood of the current solution a new solution randomly. This solution is accepted as a new current one according to some probabilistic law. Then the temperature decreases. The process continues until the system reaches the frozen state or until other stopping criteria are satisfied (e.g. maximal number of iterations, a maximal number of steps without improvements etc.).

The algorithm has following controlling parameters: τ^0 is the temperature of the frozen state; L is the length of a temperature interval; $\varphi(\tau)$ is the low of reduction of temperature.

Consider the scheme of the SA algorithm.

(1) Set the initial solutions S and temperature τ.

(2) While $l < L$ do

  (a) choose solutions S' from the neighbourhood $N(S)$ of solutions S;
  (b) $\Delta := F(S') - F(S)$;
  (c) if $\Delta \leq 0$ then $S := S'$;
  (d) if $\Delta > 0$ then $S := S'$ with probability $p = \exp\{-\Delta/\tau\}$.

(3) Reduce temperature $\tau := \varphi(\tau)$.

(4) If time to terminate then return the best found solution S^* else go to step 2.

On the basis of the competitive p-median facility location and design problem specificity the special kinds of neighborhoods have been built. The Boolean matrix X can be associated an integer vector $z = (z_j)$, so that $z_j = r$, if $x_{jr} = 1$. Let us call the neighborhood N_p of a solution z a set of vectors received as a result of the following steps:

(1) choose a facility opened in the point $j = s$ with the project variant z_s;

(2) choose a point $j = q$, where a facility is not opened yet;

(3) open a facility in the point q with the project variant z_s and close the facility in the point s, i.e. execute the following transformation: $z_q := z_s$, $z_s := 0$.

It should be noted that there is a problem of selecting parameter values, so that the algorithm can give good results for most instances. After the series of preliminary experiments the following parameter values for the simulated annealing algorithm have been found: the threshold value equal to 5; for the simulated annealing algorithm the temperature interval length $L = 10$, the initial temperature $\tau = 150$, the cooling (minimal) temperature value $\tau^0 = 5$, the low of reduction of temperature is $\varphi(\tau) = 0.99 \cdot \tau$.

Table 2. Average deviations from upper bounds in case of uniform distribution distances.

	60	80	100	150	200	300
min	0.000	0.000	0.000	0.000	0.000	0.000
aver	0.483	0.101	0.450	0.263	0.190	0.384
max	3.326	0.825	2.421	2.212	3.011	3.162

Table 2 shows the minimum (min), average (aver) and maximum (max) values of deviations from the upper bounds (UB) for test instances with uniform distribution of distances in a single run of the algorithms. The average deviation for the uniform distribution of distances instances of 300 locations is 0.384%, and the maximal deviation not exceeds 3.162%.

The test instances with Euclidean distances proved to be difficult for all considered algorithms (see Table 3). The average deviations in this case for the same dimension is 15.775%, the maximal deviation is 18.912%. Previously it was shown that the upper bound used is rough for the Series 2 for the CFLDP problem [8]. The large deviations in table 3 may be due to the same reason.

Table 3. Average deviations from upper bounds in case of the Euclidean distances.

	60	80	100	150	200	300
min	19.274	18.802	10.027	18.349	10.178	9.330
aver	29.695	26.495	21.577	25.666	16.123	15.775
max	49.883	33.252	33.029	32.189	22.535	18.912

The described algorithm were tested on a computer with Intel Xeon X5675 @ 3.07 GHz processor, 32 GB RAM. In Table 4 the minimum, average and maximum running time for Series 1 are done. For example, the average CPU time of 300 locations is 166.465 seconds, compared to 11831 seconds of solver CoinBonmin of system GAMS (see Table 1). This advantage can be explained by the relatively simple type of neighborhood used. This fact once again confirms that the simulation annealing algorithm for should be applied to problems in which it is possible to construct a neighborhood that allows one to simply move from one solution to another. However, due to the presence of a large number of parameters in the algorithm, a wide computational experiment is required to configure them. For Series 2, similar results were obtained.

Table 4. Average CPU time for case of uniform distribution distances, sec.

	60	80	100	150	200	300
min	5.238	9.126	12.972	28.346	43.980	94.818
aver	12.174	18.855	25.287	51.234	75.961	166.465
max	24.052	33.777	48.124	106.724	146.520	312.111

4. Conclusions

In this paper we have suggested simulated annealing algorithm for new Competitive p-Median Facility Location Problem. This algorithm allow to find solutions close to the upper bounds on the special test instances in case of uniform distribution of distances. On another series in case of the Euclidean distances the deviations are greater. This confirms that the upper bounds used are rough for these examples. It was necessary to continue research using other rules for the upper bound construction. It should be noted that the proposed algorithm works on average faster than the GAMS up to 70 times. It is interesting to continue these studies for other threshold algorithms.

Acknowledgments

Section 3 studies has been supported by the Russian Foundation for Basic Research grant 18-07-00599.

References

[1] Karakitsiou A 2015 *Modeling Discrete Competitive Facility Location* (Springer, Heidelberg)
[2] Aboolian R, Berman O, Krass D 2007 Competitive facility location and design problem *Eur. J. Oper. Res.* **182(1)** pp 40–62
[3] Levanova T, Gnusarev A 2016 Variable neighborhood search approach for the location and design problem *DOOR-2016 LNCS* ed Kochetov Y et al (Springer, Heidelberg) vol 9869 pp 570–577
[4] Levanova T, Gnusarev A 2018 Development of threshold algorithms for a location problem with elastic demand *Lirkov I, Margenov S (eds) Large-Scale Scientific Computing LSSC 2017 LNCS* (Springer, Cham) vol 10665 pp 382–389
[5] The General Algebraic Modeling System (GAMS) URL: http://www.gams.com
[6] Levanova T, Gnusarev A 2017 Variable neighborhood search approach for competitive p-median facility location problem *Abs. of the 17th Baikal int. school-seminar Methods of Optimization and Their Applications* (Irkutsk: ESI SBRAS) p 112
[7] *Local Searh in Combinatorial Optimization* 1997 ed E Aarts and J K Lenstra (J Wiley and Sons Ltd)
[8] Levanova T V , Belan S E 2017 Application of upper bounds for analysis of approximate algorithms for a competitive location problem with elastic demand *Vestnik Omskogo universiteta = Herald of Omsk University* **4(86)** pp 4–10. DOI: 10.25513/1812-3996.2017.4.4-10. (in Russ.)

IOP Conf. Series: Journal of Physics: Conf. Series **1050** (2018) 012045 doi:10.1088/1742-6596/1050/1/012045

Numerical study of printing ink structuring

S N Litunov, E N Gusak, Yu D Toshhakova

Omsk State Technical University,11, Mira ave., Omsk, 644050, Russia

Abstract. The computational experiment showing the structuring of the printing ink, being a highly-filled suspension and possessing the property of thixotropy. The areas of structuring are shown. It is proved that the center of rotation of the structured area, and consequently their radius is little dependent on the ink viscosity and the rotational speed of the ink fountain roller. The volume of the structured area rotating without stirring is calculated, which makes 20% of the volume of the ink can filled with ink by 70%.

1. Introduction

The quality of the printout in the offset printing depends, among other things, on the ink device operation. The supply of ink into an inker depends on the correct operation of the feeding group. The supply of ink into the inker feeding group is significantly influenced by its structuring, caused by the thixotropic character. It is difficult to visually estimate the influence of ink structuring in the ink can on its supply into the printing area. To estimate the impact of ink structuring, it is necessary to develop a mathematical model. Paulo R. de Souza Mendes developed a model of mechanical behavior of thixotropic fluids using equations for voltage and structural parameter [1]. It predicts voltage overshoot and viscosity bifurcation. A number of works are devoted to the study of structuring of printing ink in gravure printing. In particular, the transfer of the printing ink from the idealized cell of the form [2], the influence of the constricting shift on the ink transfer from the gravure print cell [3], the influence of the viscoelastic properties of ink [4, 5] is investigated.

Figure 1. Ink in an offset printing machine ink can when working

During a flat offset printing machine operation, the circulating movement of ink is formed in an ink can (Fig 1). The figure shows that the ink roller rotates without ink stirring. This leads to a deterioration of the ink flow supply to the ink-distribution system.

In machines of relief offset printing there is also a problem of insufficient ink stirring which in technical literature is usually considered as a problem of ink non-rotation. If there is ink non-rotation, the gap is formed between the ink fountain roller and ink body, and the ink supply into the ink-distribution system stops. The phenomenon of the ink non-rotation precedes the formation of the area, which moves without stirring. Due to the fact that this area has a tendency to grow, gradually increasing, it covers all the ink in the ink can.

2. Problem Statement

The ink movement in a can is rotational. The ink velocity at the flow boundary decreases from the periphery to its rotation axis. Tangent stresses are also reduced from the periphery to the axis of rotation. In the area close to the axis of rotation, the velocity of the ink flow and tangent voltages are equal to zero, that is, the ink is at rest (is static). In this case, due to the ink thixotropy character, an internal structure obtaining mechanical strength begins to form. The dimensions of this structure increase as long as the tangent stresses in the flowing ink do not exceed the shear stress limit. This leads to the formation of the area that rotates without stirring. The dimensions of this area depend on the ink properties. For example in UV dry ink, applied in relief offset printing, this area extends on all the ink volume. The task of the paper was the conducting of computational experiment, which shows the formation of the area rotating without stirring in the ink can of the flat offset printing machine.

3. Theoretical rationale

The dimensions of the area rotating without stirring depend on the of the ink rheological properties (viscosity, shear stress limit) and machine parameters (geometric parameters of the ink can, rotational velocity of the ink fountain roller). Since the transverse dimensions of the circulating ink flow are much smaller than its longitudinal dimensions (Fig. 1), the flow pattern in two parallel planes, located perpendicular to the axis of rotation, will slightly differ from each other. This allows to consider the flow in one plane perpendicular to the axis of rotation and located in the middle of the length of the ink roller.

To carry out the calculations, the following assumptions were made: ink flow is laminar; viscosity and temperature do not change over time; ink is incompressible and isotropic. Tangent stresses of thixotropic fluid are described by the following equation:

$$\tau = \tau_0 + \mu' \cdot \varepsilon', \qquad (1)$$

(at $\tau > \tau_0$), where τ is the tangent component of voltage tensor (shear stress of moving ink); τ_0 is the limit shear stress required to destroy the structure of the thixotropic fluid; μ'' is the dynamic coefficient of structural viscosity (viscosity of fluid with completely destroyed structure); ε' are tangent voltages.

In the case when the relative velocity of the neighboring ink layers is so small that the tangent voltages are less than the limit shear stress, i. e. $\tau < \tau_0$, the structure of the thixotropic fluid will begin to recover, which will lead to the formation of the area rotating without stirring.

The generalized Law of Newton for incompressible viscous fluid in the analytical form for a rectangular coordinates system has the following form:

$$\begin{aligned}\tau_{11} &= -\tau + 2\mu'\frac{\partial u}{\partial x},\\ \tau_{22} &= -\tau + 2\mu'\frac{\partial v}{\partial y},\\ \tau_{33} &= -\tau + 2\mu'\frac{\partial w}{\partial z}\end{aligned} \tag{2}$$

$$\begin{aligned}\tau_{12} = \tau_{21} &= \mu'\left[\frac{\partial u}{\partial y} + \frac{\partial v}{\partial x}\right],\\ \tau_{23} = \tau_{32} &= \mu'\left[\frac{\partial v}{\partial z} + \frac{\partial w}{\partial y}\right],\\ \tau_{13} = \tau_{31} &= \mu'\left[\frac{\partial w}{\partial x} + \frac{\partial u}{\partial z}\right]\end{aligned} \tag{3}$$

where τ_{11}, τ_{22}, τ_{33} are the normal voltages; $\tau_{12} = \tau_{21}$, $\tau_{23} = \tau_{32}$, $\tau_{13} = \tau_{31}$ are the tangent voltages; μ'' is the dynamic coefficient of structural viscosity (viscosity of liquid with completely destroyed structure); *u, v, w* is the projection of the rate vector on the x, y, z coordinate axis.

As the flat circulating ink flow around the axis of the perpendicularly to the considered plane is analyzed, then: $\tau_{11} = \tau_{22} = \tau_{33} = \tau_{12} = \tau_{13} = \tau_{23} = 0$. Then, the flow tangent voltages are recorded as follows:

$$\tau_{12} = \tau_{21} = \mu'\left[\frac{\partial u}{\partial y} + \frac{\partial v}{\partial x}\right], \tag{4}$$

Thus, in order to find the tangent stresses, it is necessary to locate the projection of the rate vector on the coordinate axis U and V, and then the u and v change in the y and x direction respectively. For this purpose it is necessary to carry out numerical differentiation, defining partial derivatives from the following expressions

$$\frac{\partial v}{\partial x} \approx \frac{v_{i+1} - v_i}{x_{i+1} - x_i}, \quad \frac{\partial u}{\partial y} \approx \frac{u_{i+1} - u_i}{y_{i+1} - y_i},$$

where u_i, u_{i+1}, v_i, v_{i+1} are the rate projection values in the two adjacent grid cells in the y and x direction respectively. Further, using the viscosity values of the ink μ'' the values of tangent stresses τ_{12} were obtained. Knowing the values of tangent stresses, the sectors in which $\tau < \tau_0$ where the areas where the paint rotates without stirring are formed, were found within the flow.

Thus, the computational experiment consists of the following stages.

1. Finding the velocity of the paint flow in a plane perpendicular to the axis of rotation of the ink fountain roller.
2. Finding tangent stresses in the flow.
3. Defining the areas that are rotating without stirring.

4. Conducting the experiment

To find the flow rate a specialized program FlowVision, which implemented a numerical solution to the Navier-Stokes equation and the equation of continuity were used. The calculation was carried out for the ink can of light-duty offset printing machine Gronhi 1800 YK. Fig. 2 shows the 3D model of the flow area.

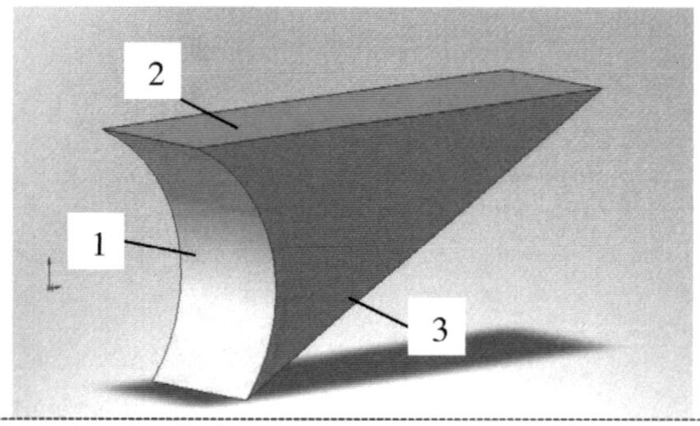

Figure 2. Calculation area of the ink can: 1 – ink fountain roller; 2 – free surface; 3 – Ink knife

Figure 3 shows a calculation example for ink with viscosity of 30 Pa s, at the rotation velocity of the ink fountain roller equal to 4 rpm.

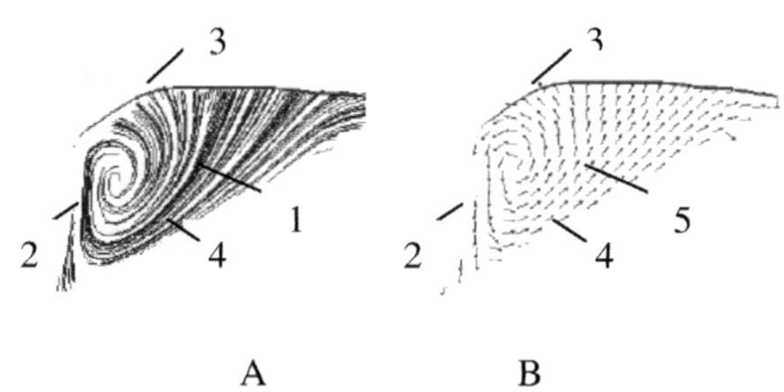

Figure 3. Current lines (a) and velocity fields with vectors of the same length (b) 1 – flow lines; 2 – ink fountain roller; 3 – free surface; 4 – ink knife; 5-velocity field

To find the tangent stresses and the area in which the condition $\tau < \tau_0$ is met, a custom designed program was used. The result is presented as a numerical matrix (Fig. 4), in which the cells where the condition $\tau < \tau_0$ is met do not have numerical values, the cells in which this condition is not met have the value of tangent stresses n/m^2, while the cells that go outside the flow area have zeros.

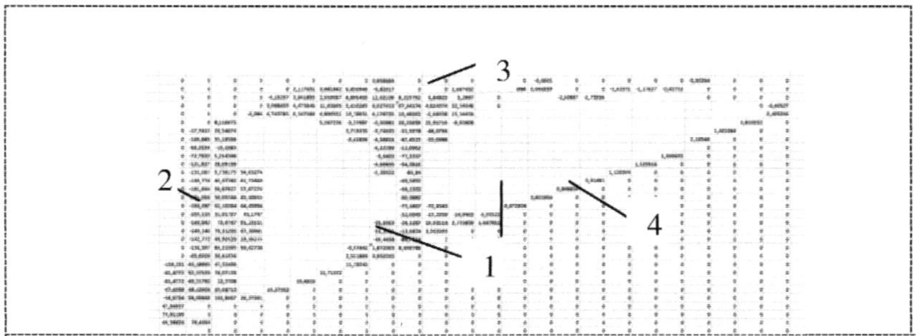

Figure 4. Tangent stress values: 1 – the area of structuring; 2 – Ink fountain roller; 3 – Free surface; 4 – Ink knife

5. Results and discussion

According to the calculations results, the coordinate of the ink rotation center depending on the viscosity and rotation velocity of the ink roller was found. The difference in the coordinate values of the of the structured area rotation center was less than 4% for the same amount of ink. The calculated structured areas for the different viscosity values of ink and the rotation velocity of the ink fountain roller are shown in Fig. 5 and 6.

The figures show that with the increase in the ink viscosity and the decrease in the ink fountain roller velocity, the dimensions of the structured area increase, which, however, does not affect the size of the body rotating without stirring.

Although the area where $t<t_0$ has an irregular shape, the structured area is a circle (a cylinder in three-dimensional space) when moving. The radius of this area is equal to the minimum distance from the center of its rotation to the ink fountain roller. As the coordinates of the rotation center for different parameters are little different from each other, the radii of the structured area are not very different from each other. Thus, for inks with viscosity of 40 Pa s and 50 Pa s at the velocity of rotation of the ink fountain roller 1-4 rps the radius of the structured area is 8.43 mm, and for ink with viscosity 30 Pa – 8.11 mm. Thus, the volume of the body that rotates without stirring, is about 20% of the total amount of ink in the ink can. In this case, the body volume increases as the amount of ink decreases. So, fig. 1 shows a structured area with a small amount of ink, which is about 90% of the total amount of ink.

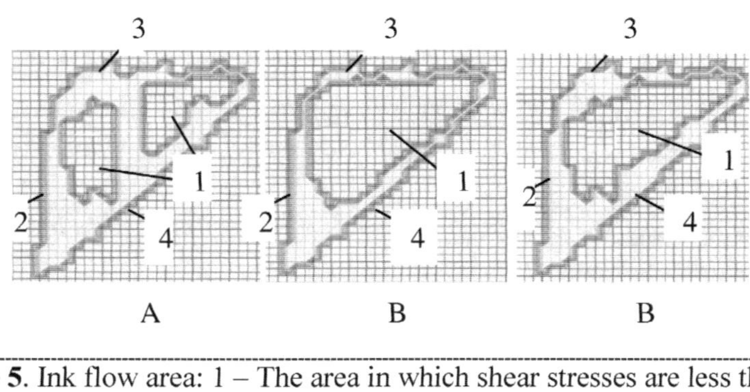

Figure 5. Ink flow area: 1 – The area in which shear stresses are less than the shear stress limit; 2 – ink fountain roller; 3 – Free surface; 4 – ink knife.
The rotating frequency of the ink fountain roller is 1 rps.
Ink viscosity (Pa s) A – 30, B – 40, V – 50

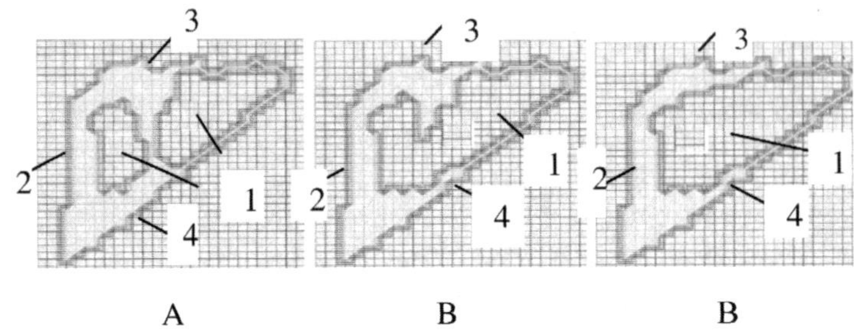

Figure 6. Ink flow area: 1 – The area in which shear stresses are less than the shear stress limit; 2 – ink fountain roller; 3 – Free surface; 4 – ink knife.
The rotating frequency of the ink fountain roller is 4 rps.
Ink viscosity (Pa s) A – 30, B – 40, V – 50;

6. Conclusion

1. The coordinates of the rotation center of the structured area do not depend on the rotation velocity of the ink fountain roller and ink viscosity.
2. The dimensions of the structured area are directly proportional to the ink viscosity and inversely proportional to the ink fountain roller rotation velocity.
3. The dimensions of the structured area make up 20% of the total amount of ink in the ink can, when it is 70% full.

References

[1] Paulo R. de Souza 2009 Mendes Modeling the thixotropic behavior of structured fluids / Paulo R. de Souza Mendes *Journal of Non-Newtonian Fluid Mechanics* **7(5)** pp 66-75 DOI: 10.1016/J. Jnnfm. 2009.08.005

[2] Khandavalli S and Rothstein J P 2017 Ink transfer of non-Newtonian fluids from an idealized gravure cell: The effect of shear and extensional deformation *Journal of Non-Newtonian Fluid Mechanics* **243** pp 16-26 https://doi.org/10.1016/j.jnnfm.2017.02.005

[3] Sunilkumar Khandavallia J. Alex Lee bMatteo Pasqualibcd Jonathan P. Rothsteina The effect of shear-thickening on liquid transfer from an idealized gravure cell *Journal of Non-Newtonian Fluid Mechanics* 221, July 2015, pp 55-65 https://doi.org/10.1016/j.jnnfm.2015.03.007

[4] J. Alex Lee Jonathan P. Rothsteinb Matteo 2013 Pasqualiac Computational study of viscoelastic effects on liquid transfer during gravure printing *Journal of Non-Newtonian Fluid Mechanics* Volume **199**, September, Pages 1-11 https://doi.org/10.1016/j.jnnfm.2013.05.005.

[5] Ashwin K. Sankaran Jonathan P. Rothstein 2012 Effect of viscoelasticity on liquid transfer during gravure printing *Journal of Non-Newtonian Fluid Mechanics* **175–176** pp 64-75https://doi.org/10.1016/j.jnnfm.2012.03.011

IOP Conf. Series: Journal of Physics: Conf. Series **1050** (2018) 012046 doi:10.1088/1742-6596/1050/1/012046

Perspectives of subjects' psychophysiological state identification using dynamic biometric features

P S Lozhnikov[1], **A E Sulavko**[1], **R V Borisov**[2], **S S Zhumazhanova**[1]

[1] Omsk State Technical University, pr. Mira, 11, 644050, Omsk, Russia

[2] Siberian State Automobile and Highway University, pr. Mira, 5, 644080, Omsk, Russia

E-mail: `lozhnikov@gmail.com`

Abstract. The hypothesis on drivers psychophysiological state (PPS) identification before and during driving by dynamic biometric features is analyzed. An experimental evaluation of the reliability of three psychophysiological states identification (normal, drowsiness, alcohol intoxication) in the space of handwritten-signature and voice features was provided.

1. Introduction

Today, with a high level of control processes automation, there is an increase in errors committed by humans. The problem of the "human factor" is especially evident in the field of transport management. According to the data provided by the WHO (World Health Organization [1]), more than 1.25 million people die each year as a result of an accident, between 20 and 50 million people are injured. Among the causes of accidents are excess speed, driving in a state of intoxication or under the influence of psychoactive drugs, distraction, vehicle technical defects. The WHO also reports that almost 90% of accidents on the roads of Europe can refer to the error under influence of "human factor". According to the statistics of the US Department of Transportation, in the United States every day almost 29 people die in alcohol-impaired vehicle crashes. The financial damage in this case is estimated at 44 billion dollars per year [2]. The share of alcohol-related fatalitis varies between 5% and 35% for different countries, with a weighted average of 21.8% [3]. The above statistics show that effective methods are needed to eliminate this problem. The present work is devoted to the analysis of perspectives of development and use of the driver's psychophysiological state recognition system while driving a vehicle using dynamic biometric features.

2. Briefly on shortcomings of existing solutions on driver's state monitoring

To date, many governments have used different strategies to solve the problem of a road accident before they occur: mass media campaigns, specialized school programs, organization of sobriety checkpoints, and the introduction of drunk driving laws the level of permissible blood alcohol concentration (BAC).

Concerning the last one, it can be noted that in 47% of countries the BAC level is less or equal to 0.05 g/dl, for a number of countries (19%) this level is below 0.02 g/dl, and such restrictions are introduced currently only in 134 countries. There is a considerable amount of works and even industrial model that to some extent solve the problem of alcohol intoxication identification. For example, the ignition interlocks installed in cars measure alcohol on the drivers breath. It keeps the car from starting if the driver has a BAC above a certain level, usually 0.02. However, their use is limited and applied for people convicted of drunk driving and are highly effective at preventing repeat offenses. The most famous solutions include the Attention Assist system, installed on Mercedes-Benz vehicles from 2011. On the signals of the sensors of steering wheel rotation, engine control, interaction with the braking system, traffic stability, etc. an individual driver's model of the trip (template) is formed. If the computer notices significant deviations of the model parameters during the next driver's trip on the vehicle, a decision is made to interrupt the further movement. In [4], the drawbacks of the described system are indicated: complexity, which excludes its use in most cars produced, cases of making false decisions are possible.

However, not only alcohol intoxication is the main cause of an accident. In 10-20% of cases, an accident is associated with driver fatigue (with "fatigue" the term "drowsiness" is used). A person driving a car for more than 17 hours awake is also at risk of crashing, as being at 0.05 BAC. Such a risk arises from a combination of several factors: biological, lifestyle, work, etc. [5]. The systems for determining fatigue, depending on the method of obtaining the identifying features/parameters, are divided into two categories: invasive and non-invasive. The first ones include systems based on the analysis of ECG [6] and EEG [7]. Non-invasive systems acquire information about drivers state and vehicle without contact with driver, so the monitoring process will not interfere with the driver. Received data can better reflect real driver behavior and real-time status. Among the identifying parameters in such systems could be determined the following: 1) the frequency of blinking (in particular, the period/ percentage of time when the eyelid is raised/closed) [8]; 2) the characteristics of the steering [9] (angles of rotation); 3) the movement of mouth [10] and other parts of the face for the analysis of facial expressions, on the basis of which the conclusion is made, including drivers fatigue level [11]. Unfortunately, most research in this direction is carried out in the driving simulator, and their effectiveness under real conditions remains unproven. For example, on a real road, a stochastic displacement due to uneven road coverage will impose additional noise on driving characteristics and increase state detection error. When using video cameras to control the eyelid-moving activity or drivers facial expressions, the quality of the images obtained is sufficiently sensitive to lighting and other external factors. When using EEG features (also EDR or ECG), there is a need to use expensive specialized contact equipment, making it difficult for driver to control the vehicle.

There are no encouraging studies and results confirming the applicability, for example, of the alcohol intoxication system for assessing fatigue and same on the contrary. In addition, there are other features (dynamic biometric features) that have proved their applicability in problems of recognition the subjects and their state, exceeding reliability of mentioned above, including by combining several groups of features obtained from different, but economically and technically accessible equipment. These characteristics include voice and handwriting dynamics features.

The specified characteristics of the subject refer to subconscious movements of human [12]. The authors of [13] found that considering the PPS when creating a users template and in the process of identifying the user by handwriting dynamics, the number of identification errors is reduced by an average of 28%, which indicates the existence of information on PPS hidden in the features of subconscious movements. Later in work [14], it was shown the principal possibility of identifying the PPS using signature parameters that were used earlier to recognize the signers. In [15] the authors investigated the possibility of identifying alcohol intoxication through a handwritten signature and reached a maximum probability of correct decisions of 95% after 35 min from alcohol consumption. It is also known that voice characteristics reflect human emotions [16]. Studies show that the percentage of errors in recognizing some emotions by voice reaches 5%. Attempts are made to determine the stress state of a person by voice [17]. The presented data point to the possibility of creating a system for recognizing an inadequate human condition based on dynamic biometric features.

3. Biometric features

The basis for this work is a hypothesis: dynamic biometric features (handwriting dynamics and voice features) characterizing the subject also characterize his/her PPS. This hypothesis is confirmed in [14, 18]. The issue of drivers identification is not considered in this paper. Handwritten signature consists of functions of the pen tip position (coordinates) on the tablet $x(t)$, $y(t)$ and pen tip pressure on the tablet$p(t)$, where t is the time in discrete form. As the biometric features in the present study, the following parameters characterizing signers and proposed in [19] were tested:

(1) Distances between some signature points, normalized by length of the signature, in three-dimensional space (the third dimension is the pressure $p(t)$). Points are selected evenly with a certain step.

(2) Some characteristics of the static image of the signature: the ratio of signature length to its width, the center of the signature, the angle of signature inclination, the angle of inclination between the centers of the signatures halves.

(3) The energy-normalized amplitudes of the first 16 (most low-frequency) harmonics of the velocity function of the pen on the tablet $v_{xy}(t)$ and the pen pressure function on the tablet $p(t)$. The first 16 harmonics contain more than 95% of the energy of the signals $v_{xy}(t)$ and $p(t)$, which is typical for all subjects.

(4) Correlation coefficients between all pairs of signature functions $x(t)$, $y(t)$, $p(t)$ and their derivatives $x'(t)$, $y'(t)$, $p'(t)$.

(5) Some values of the functions $x(t)$, $y(t)$ and $p(t)$, as well as the speed of the pen on the tablet $v_{xy}(t)$. Points are selected evenly with a certain step.

The process of voices features extraction is based on the approach from [20]. For digital recording of the speech signal, the sampling parameters were 8 Hz and the number of quantization levels was 2^{16}. The use of the parameters makes it possible to obtain sound of good quality, but which does not provide unnecessary information. After receiving the signal, it is processed, after that an array of statistical characteristics of this signal is provided, for which histograms of relative frequencies are built, the column values of which are used as identifying features. Such features include:

(1) Characteristics of the integral frequency of speech signal fragments (zero formant).

(2) Characteristics of the integral frequency of the transitions of the speech signal through the energy envelope coefficients.

(3) Frequency characteristics of positive and negative pressure level.

(4) Correlation coefficients of the characteristics of the signal integral frequency and the transitions of the speech signal through the energy envelope.

(5) Correlation coefficients of the frequency characteristics of the positive and negative pressure levels.

The resulting discrete signal Y is divided into intervals Y_i to 0.025 sec. With the shift τ_{PF} = 0.0125 sec. The value is 0.025 sec. is explained by the frequency of the pitch (PF) (min (PF) = 80 Hz) due to the vibration frequency of the vocal cords. The frequency - ν_{PF} is the smallest significant frequency of the spectrum of the voice signal. The length of the interval is chosen to be twice as long as the period of the smallest frequency min (PF), so that a minimum interval of 1 PF period is placed in a separate interval. The length of the shift is equal to the length of the period min(PF) for more accurate localization in time of changes in the processed signal. For each interval i, the number of transitions of the signal Yi through the zero is calculated.

A correction should be made for the case of the appearance of an active outgoing air pressure deflecting the microphone membrane from a quiet state and subtracting from their sample Y_i their mean - $Y_i^{mean} = \sum_N^{j=0} Y_i$ value, where $N = \tau_{PF}\nu_D$ if Y_i^{mean} is greater than zero, or add Y_i^{mean} is less than zero. Thus, we get a centered signal - Y_i^0. Each interval Yi is associated with the number of transitions of the signal Y_i^0 through zero - $T0(Y_i)$, calculated by the formula:

$$T0(Y_i) = \sum_{x=0}^{N-1} (y(x) = \begin{cases} 1, Y_i^0(x) \cdot Y_i^0(x+1) < 0 \\ 0, Y_i^0(x) \cdot Y_i^0(x+1) \geq 0 \end{cases}) \tag{1}$$

where $N = \tau_{RFD}$. The intervals Y_i are grouped into arrays of intervals belonging to the interval Z_i in z sec., in accordance with which a histogram of the relative frequencies $G(T0(Y_i))$ of the values $T0(Y_i)$ of the intervals Y_i belonging to Z_i is plotted. The length of the interval $Z_i - |Z_i| = z$ sec. is chosen due to the number of intervals Y_i, the increment of the number of which to $Y_{(i+1)}$ gives a change in the histogram $G(T0(Y_i))$ less than 0.03. The set of histogram column values $G(T0(Y_i))$ and will be the signs of the j-th realization of the k-th voiceprint - $S(G(T0(Y_i)))$.

Within a few days, a full-scale experiment was conducted to collect a database of the features, involving 60 subjects being in following states (confirmation of the "transition" to the corresponding PPS was done on Holter's monitor "Cardiotechnics-04", since it is the cardiovascular system that is one of the main markers for assessing the current psychophysiological state or changing this state):

(1) Adequate (or normal) state, in which the subject was not exposed to any influences. The experiment was conducted at the beginning of the working day.
(2) Drowsiness. To simulate this state, participants took natural herbal sedative remedies, which include motherwort, mint, valerian, and listened to soothing music.
(3) Alcohol intoxication. The subject took alcohol, the dosage was calculated according to the Widmark formula. The weight of the drink corresponded to the amount of alcohol in the blood, for which the concentration was more than 0.5 . This level exceeds the permissible level of alcohol concentration for drivers of vehicles in Russia and other countries and leads to statistically significant changes in the parameters of the cardiovascular system.

4. Evaluation of the effectiveness of the driver's PPS identification based on the Bayesian hypothesis formula

Recognizing the PPS, it is more convenient to use the Bayesian strategy, since in this case, it is not necessary to expose a number of parameters characteristic, for example, for the neural network approach: the number of inputs of neurons, the Hamming distance between the generated code and the correct one. The optimal values of these parameters should be determined separately for each subject based on a computational experiment, which in practice is problematic to implement. Integral estimates of the proximity of the sample presented to the PPS standards during the identification stage can be obtained by repeatedly applying the Bayesian rule [21] in a number of steps equal to the number of features. Each template of subjects PPS is associated with hypothesis. At each step, the a posteriori probabilities of the hypotheses are calculated using formula (2), considering the value of one of the feature, while the a priori probability of the hypothesis assumes its a posteriori probability calculated at the previous step. In the first step, all hypotheses are equally probable$P(H_i|A_0) = n - 1$, where n is the number of hypotheses. At the last step, the preference is given to a hypothesis with a maximum a posteriori probability.

$$P(H_i \mid Aj) = \frac{P(H_i \mid A_{j-1})P(A_j \mid H_i)}{\sum_{i=1}^{n} P(H_i \mid A_{j-1})P(A_j \mid H_i)} \tag{2}$$

where $P(H_i|A_j)$ is the posterior probability of the i-th hypothesis, calculated at the j-th step at entrance of the j feature, $P(A_j|H_i)$ is the conditional probability of the i-th hypothesis at the j-th step (equal to the probability density attribute A_j, obtained from the parameters of the i-th template).

A computational experiment was performed in which biometric data of 60 subjects received in the normal (adequate) state, as well as in drowsiness (after using sleeping pills) and light alcohol intoxication (blood alcohol concentration from 0.5 to 1) were fed to the entrance method of consistent application of the Bayesian hypothesis formula. Identification of PPS was carried out separately for each subject according to a vector of features values, obtained from one sample of biometric data, which means: 1 signature, 8-12 seconds of continuous speech of arbitrary content. For the training of the network, 20 samples of the signature and the voice of each person participating in the experiment were used. For each examinee, the number of errors was calculated. The error was the situation when the wrong hypothesis had the maximum a posteriori probability at the last step in making Bayes strategy decisions. The error probability for each subject was calculated as the ratio of the number of errors in the last step of the Bayesian strategy to the total number of experiments. Next, the mean M_x of error probabilities for all subjects was determined, as well as the root-mean-square deviationS_x of these probabilities. The results are shown in Figure 1.

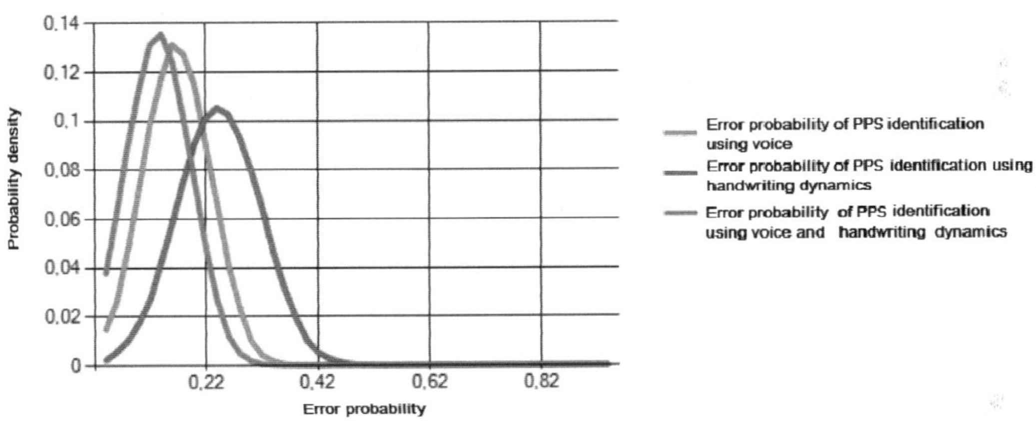

Figure 1. Results on drivers' PPS identification.

5. Conclusion

In the present study, the hypothesis about the possibility of recognition of PPS by dynamic biometric features was tested . The results on identification of 3 psychophysiological states (normal, intoxicated, drowsiness) of vehicle drivers on the basis of real biometric data of 60 subjects with a confidence of 0.99 were obtained. The number of PPS identification errors, depending on the characteristics used for different subjects, was: from 0.5% to 29.7%, the average error was 14.5%. Thus, this hypothesis is confirmed, but it is not performed for all subjects. It is necessary to carry out additional studies to determine the dependence of the error probability on such "personal factors" as the type of human temperament, gender, etc. Another important point that limits the use of this method in practice: the need to create a template for each recognizable state of a person. In order to avoid this limitation, using these biometric features, it is required to develop a model that allows to go from the template of the normal state of subject to the template of the changed state. This model should be created considering "personal factors". The question of the possibility of developing such a model is still open and waiting for special studies.

References

[1] *Road traffic injuries.* World Health Organization Media Centre. Available URL: http://www.who.int/mediacentre/factsheets/fs358/en/ (accessed on 30.01.2018).

[2] *Drunk Driving: Overview.* National Highway Traffic Safety Administration. Available URL: https://www.nhtsa.gov/risky-driving/drunk-driving (accessed 30.01.2018).

[3] Vissers L, Houwing S and Wegman F 2017 Alcohol-Related Road Casualties in Official Crash Statistics, Paris, *Int. Transport Forum ITF*, p 55, Ref .; International Traffic Safety Data and Analysis Group IRTAD Research Report.

[4] Pat. 2413633 RF: IPC B60K 28/06: Device to prevent sleep driver at the wheel / V.E. Pilkin; V.E. Pilkin. - No. 2010101263/11; claimed. 19.01.2010; publ. 10.03.2011, Bul. no 7.

[5] *Fatigue.* European Commission. Available URL: https://ec.europa.eu/transport/roadl_safety/specialist/knowledge/fatiquel_en (accessed 30.01.2018).

[6] Picot A, Charbonnier S, Caplier A 2012 On-Line Detection of Drowsiness Using Brain and Visual Information *IEEE Trans. Syst. Man Cybern.* **42** pp 764-775. doi: 10.1109 / TSMCA.2011.2164242.

[7] Chai R, Naik G, Nguyen T N, Ling S, Tran Y and Craig A 2016 Driver Fatigue Classification with Independent Component by Entropy Rate Bound Minimization Analysis in an EEG-Based System *IEEE J. Biomed. Health Inform.* **21** pp 715-724. doi: 10.1109 / JBHI.2016.2532354.

[8] Mandal B, Li L., Wang G S, Lin J 2016 Towards detection of bus driver fatigue based on robust visual analysis of eye state. IEEE Trans. Intell. Transp. Syst. **18** pp 545-557. doi: 10.1109/TITS.2016.2582900.

[9] Qu X, Cheng B, Lin Q and Li S 2013 Drowsy driving detection *Autom. Eng.* **35** pp 288-291.

[10] Wang R, Guo L, Tong B and Jin L 2004 Monitoring Gesture for Driver Fatigue or Distraction with One Camera *In: The 7th Int. IEEE Conf. on Intelligent Transportation Systems* pp 314-319.

[11] Assari M A, Rahmati M 2011 Driver Drowsiness Detection Using Face Expression Recognition *In Proceedings of the IEEE Int.lConf. on Signal and Image Processing Applications* (Kuala Lumpur, Malaysia, 16-18 Nov. 2011) pp 337-341.

[12] Eremenko A V, Lozhnikov P S, Sulavko A E 2017 Generatsiya klyuchevykh posledovatel'nostey na osnove parametrov podsoznatel'nykh dvizheniy [Generation of key sequences based on parameters of subconscious movements] (in Rusian) *Informatsionnyye sistemy i tekhnologii* OGU: Orel no 1 pp 99-109.

[13] Lozhnikov P S, Sulavko A E, Samotuga A E 2015 Personal Identification and the Assessment of the Psychophysiological State While Writing a Signature *Information* no 6 pp 454-466.

[14] Epifantsev B N, Lozhnikov P S, Sulavko A E, Zhumazhanova S S 2016 Identification Potential of Online Handwritten Signature Verification *Optoelectronics, Instrumentation and Data Processing* no 3(52) pp 238-244. DOI: 10.3103/S8756699016030043.

[15] Jungpil Shin, Toshiki Okuyama 2014 Detection of alcohol intoxication via online handwritten signature verification *Pattern Recognition Letters* vol 35(1) pp 101-104.

[16] Ayadi M El, Kamel M S, Karray F 2011 Survey on speech emotion recognition: Features, classification schemes, and databases *Pattern Recognition* vol 44(3) pp 572-587.

[17] Hansen J, Patil S 2007 Speech under stress: analysis, modeling and recognition *Speaker classification I: Fundamentals, features, and methods* Berlin: Springer pp 108137.

[18] Epifantsev B N 2013 Skrytaya identifikatsiya psikhofiziologicheskogo sostoyaniya cheloveka-operatora v protsesse professional'noy deyatel'nosti: monografiya [The hidden identification of operators psychophysiological state in the process of professional activity: monograph] (in Russian) SibADI: Omsk 198 p.

[19] Lozhnikov P S, Sulavko A E, Eremenko A V, Volkov D A 2016 Eksperimental'naya otsenka nadezhnosti verifikatsii podpisi setyami kvadratichnykh form, nechetkimi ekstraktorami i perseptronami [Experimental evaluation of reliability of signature verification by quadratic forms networks, fuzzy extractors and perceptrons] (in Russian) *Informatsionno-upravlyayushchiye sistemy* GUAP, Sankt-Peterburg no 5 pp 73-85.

[20] Sulavko A E, Eremenko A V, Tolkacheva E V, Borisov R V 2017 Kompleksirovaniye nezavisimykh biometricheskikh priznakov pri raspoznavanii sub"yektov na osnove setey kvadratichnykh form [Integration of independent biometric features in subjects recognition on the basis of networks of quadratic forms, perceptrons and measures of the CHI-module] (in Russian) *Informatsionno-upravlyayushchiye sistemy* no 1 (86) pp 50-62. doi: 10.15217 / ISSN 1684-8853.2017.1.50.

[21] Vasiliev V I, Lozhnikov P S, Sulavko A E, Zhumzhanova S S 2016 Otsenka identifikatsionnykh vozmozhnostey biometricheskikh priznakov ot standartnogo periferiynogo oborudovaniya [Evaluation of the identification

capabilities of biometric features from standard peripheral equipment] (in Russian) *Voprosy zashchity informatsii* FGUP VIMI: Moscow no 1 pp 12-20.

Acknowledgments
The work was performed with the financial support of the Ministry of Education and Science of the Russian Federation within the scope of the base part of a State Assignment within the sphere of scientific activity (Project No. 2.9314.2017).

IOP Conf. Series: Journal of Physics: Conf. Series **1050** (2018) 012047 doi:10.1088/1742-6596/1050/1/012047

Generation of a biometrically activated digital signature based on hybrid neural network algorithms

P S Lozhnikov, A E Sulavko

Omsk State Technical University, 11, Mira ave., Omsk, 644050, Russia

Abstract. This paper suggests a model of a hybrid wide neural network based on perceptrons, quadratic form networks and multidimensional difference and hyperbolic Bayes functionals. It is experimentally proved that this model is highly efficient when used for biometric authentication and generation of a digital signature activated biometrically. The paper suggests methods of generating keys of a digital signature and personal authentication by handwritten patterns, a key stroke manner and facial parameters. Comparatively high rates of reliability for taken solutions were achieved that were estimated taking into account the variability of dynamic biometric patterns over time.

1. Introduction

In the 21st century a majority of documents is created using a special software: text editors, accounting systems, etc. Next, legally valid documents may be distributed in two ways: in analog (a paper document) or in digital (an electronic document). In the first case the document is printed and verified with a handwritten signature and a company's stamp. In the second case the document stays in a digital form, and a digital signature ensures its authenticity and integrity. Nowadays we see the whole segments of document flow are migrated into the digital environment: public services, banking, e-procurement.

Consequently, a number of court proceedings that deal with unauthorized e-document creation increases. Users allow strangers to access secret keys of their digital signatures, give these keys to unauthorized people, allow a presence of malicious software. These violations mark the main security issue of the digital environment: in comparison with a handwritten signature a digital signature is detachable. If an intruder takes possession of a secret key of a non-owned digital signature and signed a document using it, this document will be legally valid according to the legislation. Today a majority of judicial acts is decided against the owners of such discredited digital signatures.

In an analog environment a paper document becomes legally valid with a handwritten signature, and the issue of its detachment does not exist. Another issue exists instead: an intruder's simulation of a signature. In order to prove the fabrication in a court large time expenditure is required for a graphologic examination. In these circumstances, the owner of the simulated signature always has chances to prove a fact of forgery.

Thus, two issues connected with the usage of signatures for attesting documents in analog and digital environments are stated. The term "signature" has different natures in different environments: in an analog environment it is a dynamic biometric image fixed on a paper as a picture of a signature, in a digital environment it is a result of a cryptographic modification of a secret (private) key and a hash-function of a document. Therefore, a hybrid document flow management is suggested to move to that implies the mutual exclusion of issues stated above. A hybrid document may be stored in a digital form or as a paper, and contain a picture of a signature protected with a digital signature and private or

IOP Conf. Series: Journal of Physics: Conf. Series **1050** (2018) 012047 doi:10.1088/1742-6596/1050/1/012047

open biometric patterns, that allow fast timely checking its integrity and authenticity irrespective of a data carrier type. A key feature of a hybrid document is a digital signature that is formed using a personal biometric pattern. In this case algorithms of translating biometric features into a secret key of a digital key are used.

In this paper we suggest to use a so called biometric handwritten signature (an autograph or a handwritten password) or a dynamic facial image and parameters of key stroking while typing a password phrase as biometric patterns.

Russia's interest in the development of a market of biometric systems reflects in the structure of the National Technological Initiative (NTI), In a draft of the NTI there is a market called SafeNet with a segment "Applied systems for security" that is almost completely devoted to solving a task of biometrics: deployment the first ever national biometric authentication platform, a biometrically activated digital signature and so on.

2. Using "wide" hybrid neural network to generate a secret key of a digital signature

Currently scientists work on the development and research of the methods of building and training large neural networks with a small number of layers (wide neural networks) for biometric authentication [1, 2] and other [3]. Fast algorithms of training and testing wide neural networks appeared in Russia several years ago [4] and provided the basis of the series of the Russian State Standard GOST P 52633 standards that deal with the development of systems of highly-reliable biometric authentication. These algorithms are not unique and give rise to a class of potentially more efficient algorithms. This approach is being developed by Russian and Kazakhstani scientists mainly.

As opposed to deep artificial neural networks, wide networks have several benefits (in comparison with other approaches of image recognition, wide neural networks, in particular):

1. High performance speed that allows using these algorithms with low-end devices.
2. A possibility to use absolutely stable training procedures for a small number of instances (15-25) of an image to be recognized regardless the complexity of the artificial neural networks (they learn in layers with every neuron being trained independently from other network neurons).
3. Fast procedures of evaluating possibilities of image recognition errors (using comparatively small test sets on the analogy with the method described in the GOST P52633.3-2011 standard) may be applied.
4. Unsupervised learning.
5. High potential of increasing the reliability of decisions made due to different functionals used by neurons.

Wide neural networks may be configured to generate a fixed binary sequence when a pattern belonging to a specific class inputs and a random equally distributed bit sequence (white noise) when a pattern belonging to an unknown class inputs. Thus, network data may be used to generate encryption keys or a digital signature activated biometrically.

By modification of the functionality we mean that a basis of a neuron may form not only the functionals of weighted sum (1) that is a standard for a classic perceptron, but any other function of several arguments as well. Any proximity measure may be selected as such functional. Arguments of the functional are values of the features. Input data of a neuron is enriched and a calculated value of a functional enters an activation function. In a simple case the activation function is a threshold function. This case is considered in this paper. In theory, a neuron may have several functionals and several activation functions. Every activation function produces one neuron output. In this paper every neuron have one activation function and one output, consequently.

This class of neural network algorithms under consideration is flexible with regard to a feature space. Different segments of a wide network process features with a specific level of mutual correlative dependence. Every artificial neuron has inputs associated with values of features that are the most relevant to its functional.

The back propagation principle is not used for training wide neural networks. Every neuron is trained independent on others, its parameters are configured based on the law of distribution of the processed features. With regards to the biometric authentication the probability density function may be defined in accordance with a known law of distribution as a rule (normal in most cases, or close to it). A wide neural network is trained layer by layer, that means every next layer is trained on the base of output values of neurons form the previous layer taking them are feature values. It may be said that every layer of a wide network consists of several subnets, the process of neuron configuration is specific for every subnet.

In the simplest case, only one layer of neurons can be configured. This is often enough for most image recognition tasks. The increase of the layers number requires justification. A reasonable explanation in this case may be the use of another principle of enriching the data entering the next layer of the network, to take into account additional information about the image.

There are functionals for processing conditionally independent and strongly dependent data. In this work we use a one-layer network of neurons based on classic functionals of perceptrons, quadratic forms and multidimensional Bayesian functionals (Fig.1). At the output of the network there is a neuron correcting erroneous bits of a binary code generated using the network. The neuron may be configured to correct a certain number of bits, changing the number it is possible to achieve the best proportion of probabilities (percentage) for False Rejection Rates (FRR) and False Acceptance Rates (FAR). This neuron is built using error correcting coding suggested in the work [5] and developed specially for biometrics. Intrinsically a task of recognition of a biometric pattern differs form a task of generating a digital signature key activated biometrically by the fact that in the first case Hamming measure acts as a neuron or other functional (Hamming weighted measure, for instance, the Bayesian formula [6] or others).

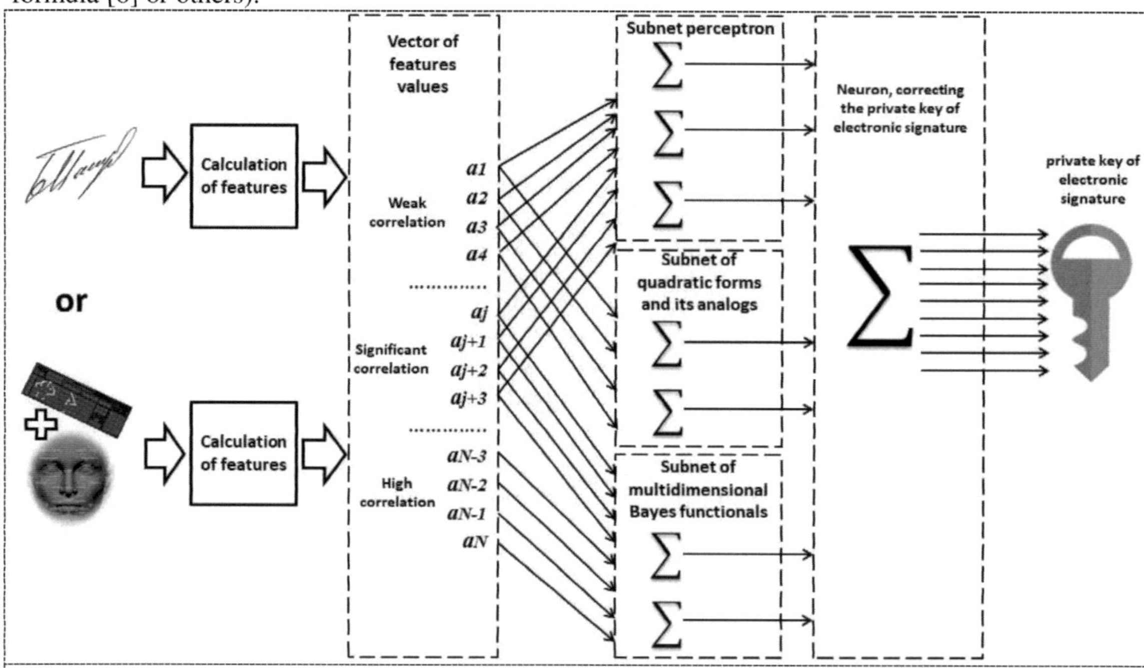

Figure 1. Work of the biometrics-code translator based on hybrid networks

Within such architecture of the neuron network an increase in a number of neurons does not lead to the complication of the learning process and to a growth in a size of a learning sampling. It leads to an almost linear increase of time required for learning and generating keys. A possibility of comparatively fast learning of wide neuron networks consisting of 20 samples of human biometric patterns is a significant feature and defines the viability of its application for biometric authentication and generating digital signatures with biometric activation.

When a biometrics-code translator is applied it is important to protect the biometric pattern and the key from being compromised. A network segment being a perceptron may be treated as rather protected from this threat [2]. It is impossible to extract the data of the learning sampling from the neuron weights and recreate a template of the biometric pattern in a timely manner (it is a computationally complex and poorly formalized task). However, quadratic forms and the Bayesian functional deal with parameters of feature distribution laws directly, that leads to a necessity of storing parameters data. If a server keeping a table of neural network functionals is not trusted, there is a threat of restoring key fragments and a template of the user's biometric pattern using the data from the table of neural network functionals. In this case a mechanism of a protected neural network container [1] may be applied to protect the biometric template at the stage of storage. This mechanism means the parameters of a neuron from a hybrid network are encrypted at the outputs of the perceptron's neurons. If the perceptron's neurons output the correct fragments of the key, the parameters of other neurons will be decrypted. Otherwise the network will generate noise as the decrypted values of the weight coefficients will not match the personal template.

Correcting codes from the work [5] allow secure storing error syndromes and do not allow restoring a key without presenting a biometric pattern that is rather close to the authentic one. In combination with the mechanism of a protected neural network container [1] these error correcting codes [5] solve the task of secure storage of the table of neural network functionals. However, the influence of the mechanism of protected neural network container on the probability of erroneous decisions made by the wide neural network is still open.

2.1. A subnet based on a perceptron

A perceptron, an artificial neuron of a subnet, is based on a functional (1) and a threshold activation function. Modules of weights of the neurons from the first layer are calculated using the formula (2). The Russian standard GOST P 52633.5-2011 and works [1, 2] describe the operations of network configuration in details.

$$y = \sum_{j=1}^{N} \mu_j a_j, \tag{1}$$

$$\mu_j = |m_ч(a_j) - m_c(a_j)| \,/\, \sigma_ч(a_j)\cdot\sigma_c(a_j), \tag{2}$$

where $m_c(a_j)$ are mathematical expectations of values of the j-th feature of an authentic pattern ("Own"), $\sigma_c(a_j)$ is a root-mean-square deviation of values of the j-th feature of an pattern ("Own"), $m_ч(a_j)$ and $\sigma_ч(a_j)$ are analogical figures for an unknown pattern ("Non-Own"). If a neuron is configured to output "one" when the "Own" pattern inputs, then a sign of the weighting coefficient is selected following the rule: "+" if $m_ч(a_j)<m_c(a_j)$, otherwise "-". If the neuron is configured to zero, the signs are inverted.

A single-layer perceptron with a threshold activation function is characterized by a decrease in efficiency if the neuron outputs are correlated. The stronger the relation between the features, the higher the correlation of the outputs. Researches show that the quality of perceptrons is accessible if the module of correlation coefficients between features is less that 0.7 ($\phi< 0.7$)

2.2. Subnet of quadratic forms

If the feature space is "plane" then the shortest path from the image to the center of the template can be found using the Euclidean measure. If the ranges of values of particular features vary significantly in scale, the space ceases to be "ideally plane", but it can be normalized by calculating distances with the help of the Pearson measure (3) [7]. Chi-module measure provides similar results while recognizing images (4). In the tasks of biometric authentication (identification) correlation between the features "bends" the space in the neighborhood of images templates. In order to take into account the heterogeneity of the curvature/bending of the feature space in the theory a quadratic form of (5), also called the Mahalonobis measure, can be used.

$$\Pi = \sqrt{\sum_{j=1}^{N} \frac{(m_j - a_j)^2}{\sigma_j^2}}, \tag{3}$$

$$\chi = \sum_{j=1}^{N} \frac{|m_j - a_j|}{\sigma_j}, \tag{4}$$

$$y = (\overline{m} - \overline{a})^T \cdot [R]^{-1} \cdot (\overline{m} - \overline{a}), \tag{5}$$

where a_j – is the value of the j-th feature, N is the number of features processed by the functional (the dimension of the functional), m_j and σ_j are the mathematical expectation and the standard deviation of the j-th feature values calculated from the training sample data, $\bar{a}$ is the vector of features values in the normalized coordinate system, $[R]$ is the matrix of pair correlation coefficients between the features. The use of this quadratic form is complicated by the fact that it is required to perform the inversion of the correlation matrix $[R]$, which is not possible in practice. The point is that this operation is ill-conditioned and performed with significant errors [8]. An attempt of such treatment leads to a problem called the "dimensional curse" [9], at least in the problems of dynamic biometric images recognition on training samples in dozens of examples.

The work [10] suggests an adaptable algorithm of learning a perceptron based on the GOST methodology R 52633.5-2011 to configure a quadratic form (5). However, this algorithm does not solve the task of processing strongly dependent features.

For these reasons, proximity measures, which are quadratic forms, are focused on processing only independent and weakly dependent data.

2.3. Subnet of Bayesian functional

The results of recent studies indicate that there are functionals (proximity measures) that can take fewer wrong decisions if the correlation between the features is high. The first group of such proximity measures was called multidimensional Bayesian functionals. They evaluate not only the proximity of the image to the standard, but also the dependence of a certain feature of the image on several other features. There are multidimensional correlation [11], difference (6) [8], Bayesian hyperbolic functionals (7) and (8) [12]).

$$d_t = \sum_{j=1}^{N} \left| \frac{|m_t - a_t|}{\sigma_t} - \frac{|m_j - a_j|}{\sigma_j} \right|, j \neq t \tag{6}$$

$$g_{t-} = \sum_{j=1}^{N} \left(\frac{(m_t - a_t)^2}{\sigma_t^2} - \frac{(m_j - a_j)^2}{\sigma_j^2} \right), j \neq t, \tag{7}$$

$$g_{t+} = \sum_{j=1}^{N} \left(\frac{(m_t - a_t)^2}{\sigma_t^2} + \frac{(m_j - a_j)^2}{\sigma_j^2} \right), j \neq t, \tag{8}$$

The greater value of N and higher the coefficient of equal correlation of features, the more effectively the Bayesian functional works. Equal correlation in this case is meant that the coefficient of pair correlation between the t-th and j-th features is within certain limits (the interval of the ±0.1 order is admissible one) [1]. The work [1] suggested a methodology of building a network of multidimensional Bayesian functionals possessing a moderate redundancy and high efficiency.

3. Dynamic biometric patterns

We suggests 2 variants of using a technology of generating secret keys of a digital signature based on biometric images:

1. The first variant presupposes a usage of a handwritten signature or a handwritten password as a biometric pattern. In this case the generation of a digital signature activated biometrically will not significantly influence current business-processes in an organization as a signature is a traditional method of verifying the authenticity of paper documents.
2. The second suggested method of generating a digital signature is the usage of data obtained from standard equipment like a keyboard and a web-camera to create a secret key. In this case facial parameters and a key stroke manner are used. We suggest an "advanced" technology of protecting electronic samples of documents based on this method that is as follows. Having generated a hybrid documents, its owner can restrict an access of other people to its selected parts,

IOP Conf. Series: Journal of Physics: Conf. Series **1050** (2018) 012047 doi:10.1088/1742-6596/1050/1/012047

and restrict certain actions (printing, editing, etc.). At this time the content of any of these parts of the document will be encrypted using an open key of a person who is granted an access. If more than one person have access to any part of the document, several copies of the document are created and each one is encrypted using the corresponding open key. While a user is working with an electronic hybrid document, a real-time continuous monitoring of facial parameters and a key stroke manner is performed. These parameters are used for generating a private key used further for decrypting the corresponding parts of the document (Fig. 2). While fixing changes of personal biometric characteristics recording in a process of work, the document temporarily "changes" or "hides" its content completely or totally, block some functions of editing it. A user cannot "bypass" the program as all confidential information is encrypted by corresponding cryptographic keys. Public information is not encrypted. Such technology of an active protection is called a technology of "an enlivened document". In fact, documents will contain information about who created and edited it and what was the psychophysiological state of this person.

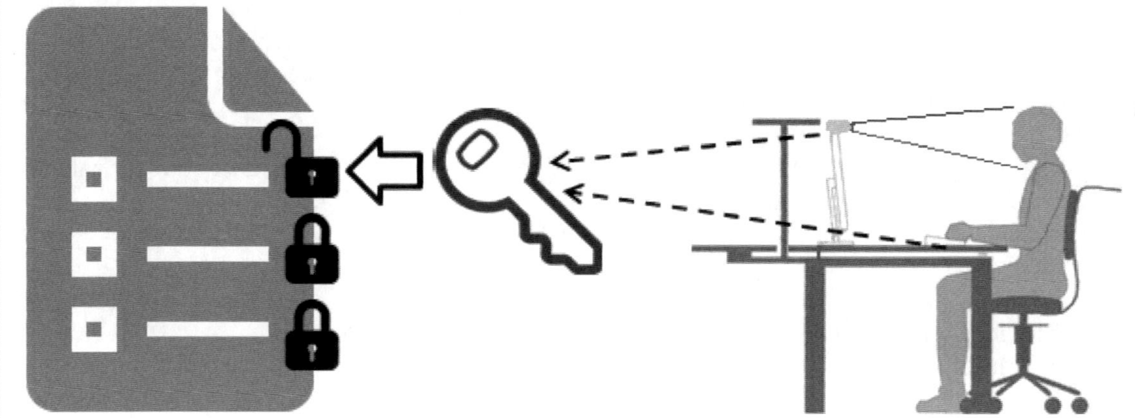

Figure 2. A picture of a process of accessing a fragment of an electronic hybrid document.

To increase the reliability of the process of key generation, a multi-factor method that combines both suggested variants is used.

3.1. Parameters of a handwritten password or a signature

In a computer representation a signature can consist of functions of a pen position on a tablet $x(t)$, $y(t)$ and pen pressure on a tablet $p(t)$, where t is a time in a discrete form. Every handwritten pattern passes through spectral and correlative analyses to compute a fixed number of informative features. This vector includes both values characterizing the outer appearance (distances between certain points of a signature image, parameters of its slope, width, length) and the dynamics of its reproduction (amplitudes of harmonics for functions $x(t)$, $y(t)$, $p(t)$, that correspond a frequency of vibration of a signer's hand (approximately 1-10 Hz), coefficients of correlation between these and derived functions, coefficients of the Daubechies D6 wavelet transform). The process of computing these features in detail is described in the work [13].

3.2. Parameters of a face and a key stroke manner

Several characteristics from the work [14] were used as facial parameters, in particular:

- Distances between eyes, a center of a face, a tip of a nose (in pixels, values are normalized by a diagonal of a face in a frame).
- Area of eyes, a nose, a mouth (values are normalized by a face area).
- Coefficients of correlation of intensity and color parts of pixels (according to RGB) between all pairs of the following area of a face: eyes, a nose, a mouth. These features describe facial asymmetry.
- Parameters describing a color of eyes and skin.

To isolate the face (skin), eyes, nose, mouth and the subsequent analysis of these areas, the Viola-Jones method was used. This detector has a low probability of false detection of the face and well recognizes facial features at an angle (up to 30 degrees) [15]. To determine the color components of the eye, the area of the iris and the pupil are initially separated using the algorithm for detecting circles on the basis of the Hough transformation.
Time of key retention and pauses between key strokes were used as features of a key stroke dynamics [16].

4. Reliability scoring of generating keys of a digital signature based on hybrid wide networks

An experiment to estimate the reliability of generating keys by a hybrid neural network was carried out (Fig. 1). The experiment involved 90 people, each of them entered the biometrical data during one month. The first day the person entered 40-45 samples of a certain handwritten pattern, and tried to type a password phrase with a keyboard ("система защиты должна постоянно совершенствоваться", that means "the protection system must be constantly improving"). Next, once a week a person repeated entering biometric data (20-25 attempts). After each procedure of typing a phrase with a keyboard a test person looked in the direction of a web-camera to take a photo that fixed facial parameters. To collect data we used Wacom tablet, a standard mechanical keyboard and a web-camera with a resolution 640x480.
As a result every person gave 120-145 biometric images of each type (a face, a key stroke manner, a handwritten pattern). A total number of samples of each type exceeds 10000.
For each person three hybrid neural networks were formed to generate a personal key based on the following biometric patterns:

- Handwritten passwords/signatures (Fig. 3);
- Patterns of a face and a key stroke manner (Fig. 4);
- Patterns of a face, a key stroke manner and handwritten passwords/signatures (Fig. 5).

The authentication (or generation of cryptographic keys) by facial data only or a key stroke manner is not secure. A face is an open biometric pattern that may be forged (using a photo or other "replicas"). An open pattern may be used for identification purposes only or as additional information for personal verification. A key stroke manner recording with a standard keyboard is too unstable (more unstable than handwriting dynamics as multiple researches show). The reliability of biometric authentication via a key stroke manner is comparatively slow. However, this pattern may be secret therefore it may be compensated with a facial image.
There is no single method of configuring hybrid wide neural network in any specific case. However, to define limit conditions the following arguments may be used.
First of all, a wide network of different functionals must provide a large range of input information processing methods. As long as neurons make different mistakes, their number may be increased adding new computing elements into the network structure. Solutions made by these new elements do not repeat the solutions of the existing neurons. An estimation of the correlation between the work of all elements of a neural network is unreasonable as it is an extremely resource-intensive process that is impossible to complete using a small learning sampling. For this reason it is required to select a number of different neurons and their synapses on a basis of a principle of work of any used functional, much relying on an intuitive approach and empirics.
It is self-explanatory that a new neuron must be efficient, that means it makes errors as rare as possible. To provide non "zero" contribution of a new neuron, it is required to build it based on another functional or that it processes other (unique) combinations of features. In this paper we take three types of functionals and three subnets consisting of neurons based on corresponding functionals (Fig. 1). A number of neurons of every subnet may be controlled independently. Gradually increasing a number of neurons in the network we can come to a situation when they all will process a significantly intersecting set of features and therefore a larger number of neurons will make similar decisions. It is necessary to find a moment when a further increase in a number of neurons of a certain type does not influence a decrease in error probability of pattern recognition.

The work [1] state that to process the considered biometric features with ϕ>0.5 it is necessary to use difference (6) and two type of hyperbolic (7) and (8) Bayes functionals (using features with ϕ<0,5 does not lead to a significant gain in recognition reliability). Researches showed [1, 8, 12] that the networks based on above mentioned Bayes functionals can be configured in the same way and produce the similar results on average, but at the same time their particular solutions differ. Thus, functional (6), (7) and (8) complement each other. A procedure of defining a configuration of Bayes-Hamming network is suggested in [1]. This procedure may be applied one-time to build three Bayes-Hamming subnets at once (three segments inside one network) with the same number of neurons and the same synapses that differ only in a functional in a basis of a neuron. This approach was used in this experiment.

To process features with ϕ<0.7 a single-layer perceptron configured according to the GOST R 52633.5-2011 was used. To define a minimal required number of neurons it is possible to rely on a principle of minimal complexity of a neural network [17], the maximum limits of the network size may be found based on results of the previous experiment on estimation of the reliability of solutions made by perceptrons while recognizing signature, face and key stroke patterns [1, 10]. A number of neurons and inputs was selected empirically taking into account these data.

To process features with ϕ<0.5 a subnet of quadratic forms consisting of 3 segments based on different functionals was used: Pearson measure (3), a Chi-module (4) and Mahalonobis (5). A number of neurons and their inputs is selected empirically but focusing on the results of the works [1, 10]. The usage of a wider range of features than when ϕ<0.3 is explained by the fact that significant errors in calculating correlative coefficients are met when the sampling is small [8].

Thus, this simulation experiment consisted of several series of tests, in each series parameters of neural networks being generated change in a certain range. Each network was trained using 20 examples of corresponding biometric patterns of a certain person ("Own") and 64 examples of other persons ("Non-Own"). The process of learning was based on biometric samples obtained in the first day of the experiment. Other patterns were used to evaluate the reliability of person's recognition (more than 8000 of every type). Error probabilities were evaluated for two cases:

1. Immediately after learning. To test the neural networks, biometric patterns obtained in the first day of the experiment were used (an ideal variant).
2. During a month after learning. To test neural networks biometric patterns were used obtained during the next period of time of the experiment. In this case an aspect of variability of a personal biometric pattern in a course of time is taken into account. This aspect was not considered in previous experiments [1, 7, 10, 13]. This test scenario meets real conditions.

The results of the experiment is shown in Figures 3-5.

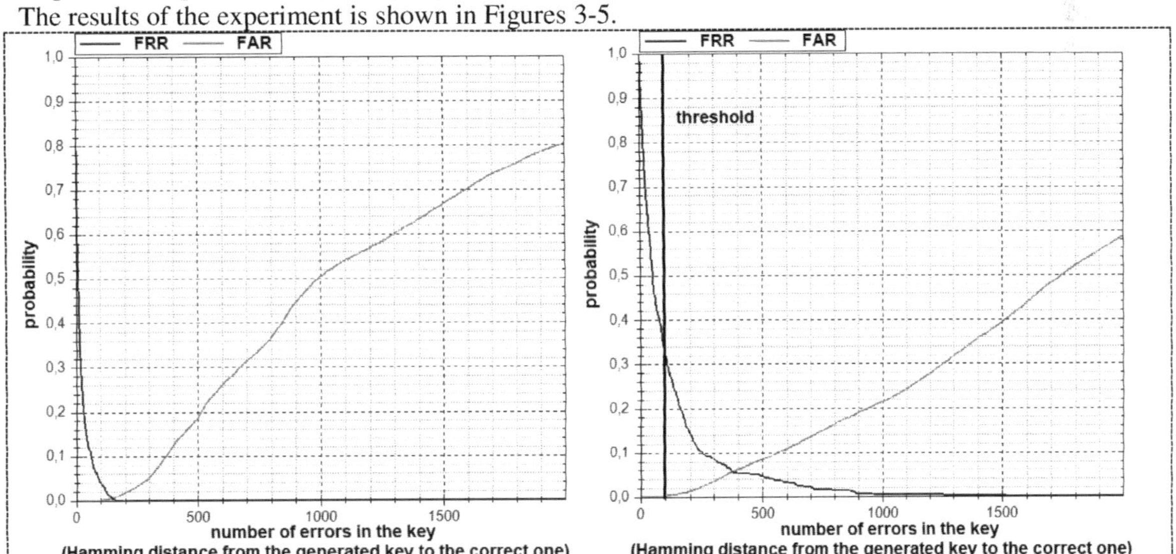

Figure 3. Estimation of errors of generating keys from signatures (depending on a number of

corrected bits): with regard to (on the right) and without regard to variability of features over time (on the left)

While testing hybrid networks every person was considered "Non-Own" in relation to other persons. Thus, FRR and FAR were computed based on data from different experiments, in the first case keys were generated from the "Own" patterns, in the second case they were generated from the "Non-Own" patterns correspondingly.

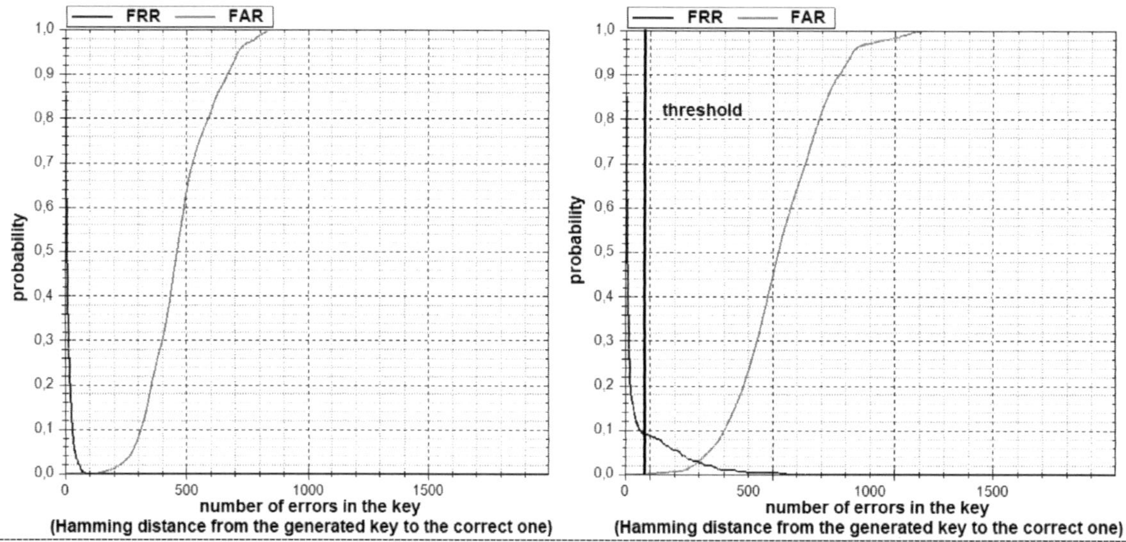

Figure 4. Estimations of errors of generating keys by a two-factor system based on face and key stroke patterns (depending on a number of corrected bits): with regard to (on the right) and without regard to the variability of features over time (on the left).

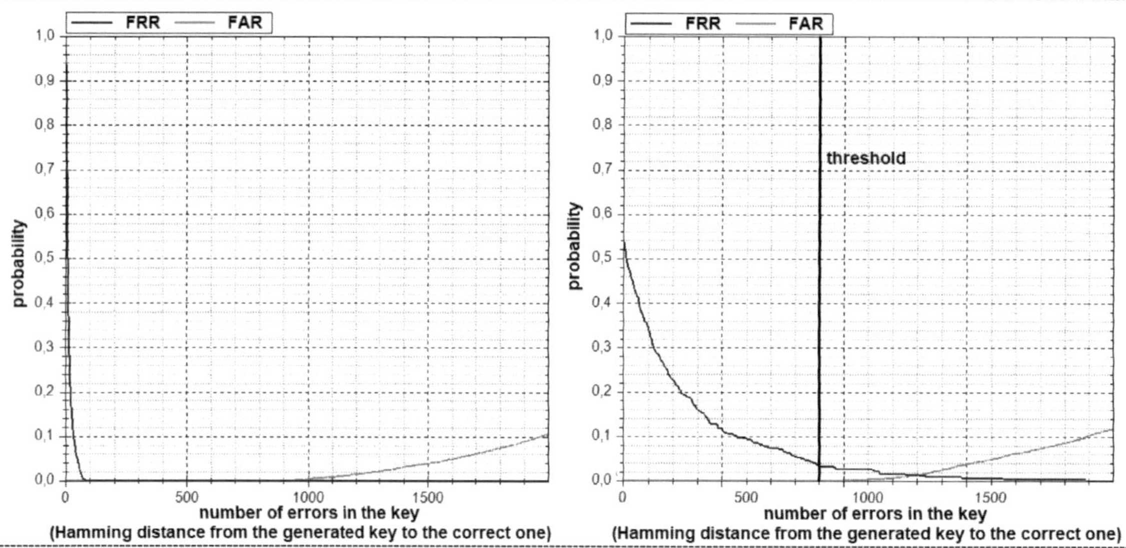

Figure 5. Estimations of errors of generating keys by a three-factor system based on signatures, a face and a key stroke manner (depending on a number of corrected bits): with regard to (on the right) and without regard to the variability of features over time (on the left).

5. Discussion of results and its comparison with previous ones

As shown in Figures 3-5 the network configurations used allow generating a key with a length up to 2048 bits (each neuron generates one bit) in all cases. This is not a limiting ability of a wide network, however according to the calculations a further increase in a number of neurons does not lead to a decrease in error probabilities, at least if functionals described in this work are used. For comparison, we give graphs of signatures verification errors by the Bayes-Hamming networks (Figure 6). They show that the use of a network of one type of functional gives more errors and a slightly different picture of the characteristic curves, which indicates that the measures (6), (7) and (8) are not completely correlated.

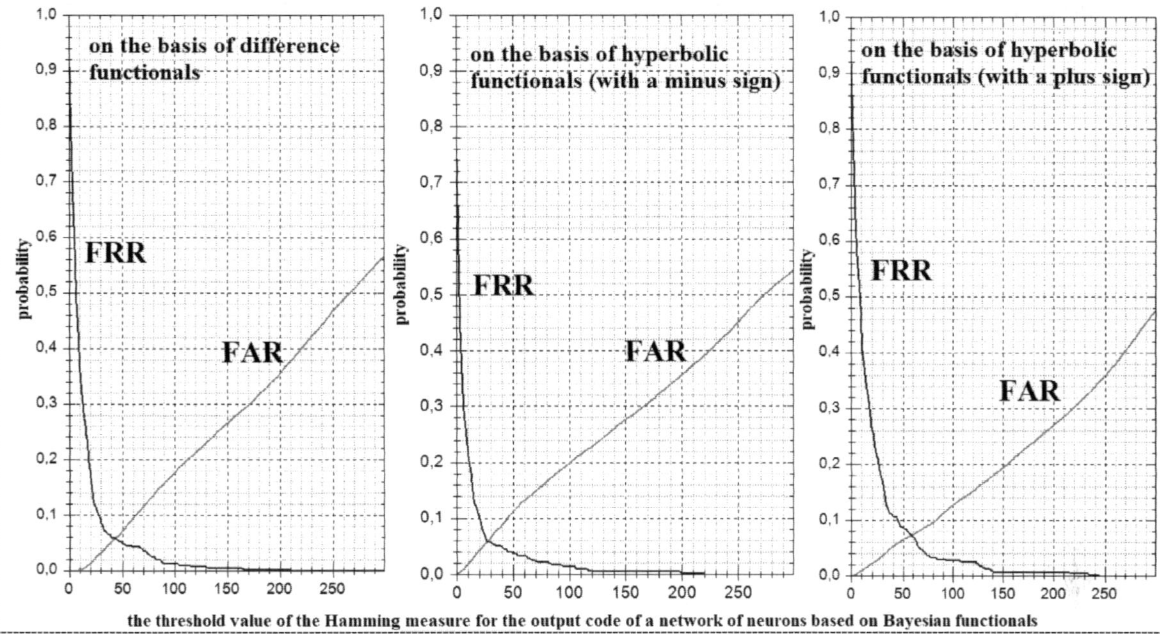

Figure 6. The results of verification of signatures by the Bayes-Hamming networks with regard to the variability of features over time.

If a biometric system is tested immediately after its learning (or using testing samplings) the hybrid wide networks divide users' patterns almost error-free. An equal error rate for such ideal case is:

– EER=0.5% when a user is recognized/a key is generated using a handwritten pattern;
– EER≈0.02% when a user is recognized/a key is generated using a key stroke manner and a face;
– EER<0.02% when a user is recognized/a key is generated using a handwritten pattern, a face and a keystroke manner.

These results significantly exceed the previous ones [1, 7, 10, 13] that proves high efficiency of wide hybrid neural networks.

As you see, due to the variability of patterns of key strokes, signatures and photography conditions in an extended period, the reliability of personal recognition decreases. Researches show that time variations of dynamic biometric features are spontaneous and cannot be forecasted. However even considering this, the system may be configured to the following reliability indicators:

– When a person is recognized/a key is generated using a handwritten pattern: FRR=31.5% in case of FAR<0.01% (EER=5.1%, Fig. 3);
– When a person is recognized/a key is generated using a key stroke manner and a face: FRR=9.2% in case of FAR<0.01% (EER=2.28%, Fig. 4);
– When a person is recognized/a key is generated using a handwritten pattern, a face and a keystroke manner: FRR=3.5% in case of FAR<0.01% (EER=1.15%, Fig. 5).

These estimations are generalized for all users. In practice a trigger threshold of a protection system must be configured for every person individually. A basic threshold value (Fig. 3-5) must be corrected on a basis of a neuron response to patterns of a learning sampling, that enters a hybrid network immediately after learning. The custom configuration allows decreasing the obtained estimations by 10-20%.

6. Conclusion

This work suggests a model of a wide hybrid neural network based on perceptrons, quadratic form networks and multidimensional difference and hyperbolic Bayes functionals. It was experimentally proved that this model is highly efficient for biometric authentication and generating a digital signature activated biometrically. The work suggests methods of generating keys of a digital signature, and authenticating persons by handwritten patterns, a key stroke manner and facial parameters. While attempting to generate a digital signature activated biometrically based on a handwritten pattern immediately after system learning, the erroneous solution rates were about 0.5%, and for a multi-factor system EER≤0,02%. Taking into account changes in dynamic biometric features over time (a time period of about a month is considered) the erroneous solution rates increase many times. However even taking into account the variability of patterns over time, with fine-tuning of the biometric system according to the peculiarities of every user's patterns the reliability rates were:

- When a person is recognized/a key is generated using handwritten patterns: FRR=24.5% in case of FAR<0.01% (EER=3.9%, Fig. 3);
- When a person is recognized/a key is generated using a key stroke manner and a face: FRR=8.3% in case of FAR<0.01% (EER=2%, Fig. 4);
- When a person is recognized/a key is generated using handwritten patterns, a face and a keystroke manner: FRR=3% in case of FAR<0.01% (EER=0.9%, Fig. 5).

These indicators are enough for authentication systems.
The result of this work may be used in practice. However it is recommended to start learning the recognition system over time again, in this case erroneous solution probabilities will be lower. Researches show that for different persons the recommended period of time differs, and it ranges from two weeks to several months.

7. References

[1] Ivanov A I, Lozhnikov P S, Sulavko A E 2017 Evaluation of signature verification reliability based on artificial neural networks, Bayesian multivariate functional and quadratic forms *Computer Optics* vol 5 p 765-774

[2] Malygin A, Seilova N, Boskebeev K, Alimseitova Zh 2017 Application of artificial neural networks for handwritten biometric images recognition *COMPUTER MODELLING & NEW TECHNOLOGIES* vol 21(1) p 31-38

[3] Ivanov A I, Kulagin V P, Kuznetsov Y M, Chulkova G M, Ivannikov A D 2016 High-dimensional neural-network artificial intelligence capable of quick learning to recognize a new smell, and gradually expanding the database *Digital Information Processing, Data Mining, and Wireless Communications (DIPDMWC), Third International Conference* (Moscow)

[4] Volchikhin V.I., Ivanov A.I., Funtikov V.A. 2005 Fast learning algorithms of neural network mechanisms of biometric-cryptographic information protection (Penza: Penza State University Press) 273 p.

[5] Bezev A V, Ivanov A I, Funtikova U V 2014 Optimization of the structure self-correcting bio-code, storing syndromes error as fragments hash-functions *UrFR Newsletter. Information Security* vol 3(13) p 4–13

[6] Epifantsev B N, Lozhnikov P S, Sulavko A E, Zhumazhanova S S 2016 Identification Potential of Online Handwritten Signature Verification *Optoelectronics, Instrumentation and Data Processing* vol 3(52) p 238-244

[7] Sulavko A E, Fedotov A A, Eremenko A V 2017 Users' identification through keystroke dynamics based on vibration parameters and keyboard pressure *XI International IEEE Scientific and Technical Conference "Dynamics of Systems, Mechanisms and Machines (Dynamics)* (Omsk: Omsk State Technical University) p 1-7

[8] Ivanov A I, Lozhnikov P S, Serikova Yu I 2016 Reducing the Size of a Sample Sufficient for Learning Due to the Symmetrization of Correlation Relationships Between Biometric Data *Cybernetics and Systems Analysis* vol. 52(3) p 379–385

[9] Akhmetov B, Doszhanova A, Ivanov A, Kartbayev T, Malygin A 2013 Biometric Technology in Securing the Internet Using Large Neural Network Technology *World Academy of Science,*

Engineering and Technology (Singapore)

[10] Lozhnikov P S and Sulavko A E 2017 Usage of quadratic form networks for users' recognition by dynamic biometric images *XI International IEEE Scientific and Technical Conference "Dynamics of Systems, Mechanisms and Machines" (Dynamics)* (Omsk: Omsk State Technical University) p 1-6

[11] Ivanov A I, Kachajkin E I, Lozhnikov P S 2016 A Complete Statistical Model of a Handwritten Signature as an Object of Biometric Identification *Control and Communications (SIBCON)* (Moscow) p 1–5

[12] Ivanov A I, Lozhnikov P S, Vyatchanin S E Comparable Estimation of Network Power for Chisquared Pearson Functional Networks and Bayes Hyperbolic Functional Networks while Processing Biometric Data *Control and Communications (SIBCON)* (Astana) p 1-3

[13] Lozhnikov P S, Sulavko A E, Eremenko A V, Volkov D A 2016 Methods of Generating Key Sequences based on Parameters of Handwritten Passwords and Signatures *Information* vol *7*(4)

[14] Vasilyev V I, Sulavko A E, Eremenko A V, Zhumazhanova S S 2016 Identification potential capacity of typical hardware for the purpose of hidden recognition of computer network users *X International IEEE Scientific and Technical Conference "Dynamics of Systems, Mechanisms and Machines" (Dynamics)* (Omsk: Omsk State Technical University) p 1-5

[15] Cho H and Hwang SY J 2015 High-performance on-road vehicle detection with non-biased cascade classifier by weight-balanced training *EURASIP Journal on Image and Video Processing*

[16] Lozhnikov P S, Sulavko A E, Eremenko A V, Buraya E V 2016 Methods of generating key sequences based on keystroke dynamics *X International IEEE Scientific and Technical Conference "Dynamics of Systems, Mechanisms and Machines" (Dynamics)* (Omsk: Omsk State Technical University) p 1-5

[17] Vasilyev Vladimir 2016 Structural design of shallow neural networks on the basis of minimal complexity principle *Control and Automation (MED) 24th Mediterranean Conference.*

Acknowledgments
The research was supported by the Russian Scientific Foundation (project №17-71-10094).

IOP Conf. Series: Journal of Physics: Conf. Series **1050** (2018) 012048 doi:10.1088/1742-6596/1050/1/012048

Geometrical and computer modeling of the technical workpieces main objects shaping

A A Lyashkov

Omsk State Technical University, Mira ave, 11, Omsk, 644050, Russia
E-mail: 3dogibmod@mail.ru

Abstract. Geometrical and computer modeling of the main objects of technical workpieces shaping is considered: envelope, real envelope and removable layers based on the common methodology. The carried out research of two-dimensional surfaces and three-dimensional hyper-surfaces discriminants allowed to offer a definition of the envelope and real envelope families of curvatures and surfaces from a single viewpoint of analytical and numerical methods, respectively. This methodology is developed to perform 3D-modeling by means of CAD tools of both an envelope curvature and removable layers. The obtained surface models are used for analyzing the influence on the envelope shape of the of the of shape parameters profile family of the shaping curvature, as well as the parameter of setting a product relatively to the tool.

1. Introduction

The shaping of a detail surfaces is an important engineering and technological challenge in the production of various workpieces in several mechanical engineering branches [1–4]. In this task, a significant place is given to issues related to the interaction of workpiece and a tool [5]. The surface obtained after the processing by the tool will consist of the surfaces family envelope [6] and, rather often, of the transition curves. Therefore, along with the development of a mathematical model of the envelope, in the process of shaping the issues of defining a model of a real surface – envelope, are considered [7]. For obtaining models of the envelope and real envelope families of curvatures and surfaces a lot of methods based on different approaches to obtaining these shaping objects are developed. Most of them apply the methods of differential geometry or the kinematic method. Recently, the possibilities of modern computer technologies are successfully used for simulation of the surfaces shaping modeling [8–16].

In addition to the envelope and real envelope, in practical applications the removed layers models are often required [17]. For the study of removable layers the methods different from those used for modeling the envelope and the real envelope surface are used. However, as a rule, in known publications, the study of only one of the objects of shaping by different methods is discussed.

Thus, the main shaping objects are an envelope , real envelope and removable layers. During their modeling, in many tasks there are a number of unsolved or not properly solved problems. This is resulted from the lack of a common methodology that would allow, if necessary, to determine the discriminant and real envelope families of lines and surfaces, as well as removable layers using the capabilities of modern geometric, polygon, and solid computer modeling in their full.

2. Problem Statement

The studies of a family of lines or surfaces envelope are carried out in many papers. Mostly, they are based on the methods of differential geometry, or on the kinematic method. In these methods, obtaining an equation that establishes the relationship of the parameters of the line or surface and their family is carried out. As a rule, in applications this equation is transcendental, which is associated with significant difficulties in both its solution and defining the envelope as a whole.

Another area of defining the envelope of a family of lines or surfaces is based on the study of the surface or hyper-surface discriminants. These surfaces and hyper-surface are formed by displaying the collections of lines and surfaces obtained in the process of shaping, into a space of higher dimensionality. Its dimensionality one greater than the dimensionality of the space where the lines and surfaces are [18–22]. In these works, basically, a number of differential characteristics of the discriminants of two-dimensional surfaces or algebraic surfaces of higher dimensionality is determined.

Publications [20, 21] and some others contain the study of peculiarities of displaying the orthogonal projection of different dimensions algebraic surfaces. For example, in [20] the examples of obtaining hyper-surface discriminant defined by a polynomial are given. Displaying the orthogonal projection of some two-dimensional and three-dimensional surfaces defined by the equation in implicit or parametric equations for a plane and a hyper-plane is considered in [23–25].

The analysis of these works shows that the study of peculiarities of displaying the surfaces and hyper-surfaces orthogonal projection on the plane and the hyper-plane allows to solve the problem of determining the envelope of a family of lines or surfaces, by both analytical and numerical methods that in some cases seems to be most efficient. Following are the research results of the displaying by orthogonal projection of two-dimensional surface and three-dimensional hyper-surface, in various forms of their setting, on the coordinate plane and hyper-plane analytical methods and using the capabilities of CAD systems, which in many cases allows us to obtain the best possible results.

3. Theory

The envelope of the family of lines and surfaces as the discriminant of a surface or hyper-surface.

Let the investigated two-dimensional surface is given by the equation in implicit form

$$F(x, y, z) = 0 \tag{1}$$

and it is displayed by the orthogonal projection on the coordinate *XY-plane*. A characteristic feature of this display is a discriminant or a contour of the surface. The corresponding line on the surface is a criminant or a contour line. At criminant points tangent plane to the surface is parallel to the coordinate axis *Z*, denoted by equation in the following form

$$F_z(x, y, z) = 0 \tag{2}$$

It is suggested to consider equation (1) as the equation of a new surface. Then, equations (1) and (2) determine their line of intersection, which is the contour line of the surface (1). The studies show that on the curves obtained as the intersection of the surface with planes parallel to the coordinate planes *XZ* and *YZ*, there are extreme points relative to the coordinate plane *XY*. Cutting planes are parallel to the coordinate axes that defines the direction of projection that is Z axis (fig.1). Such points may include the point of folds of the assemblage point and the singular points of the surface and they define the contour line of the surface.

On this basis, discriminant *D* of surface (1) is the union of sets of extreme points of view

$$D = \sum_{i=1}^{n} \min f(x, z) \vee \max f(x, z)\Big|_{x=a_i},$$

if equation (1) is written down explicitly $y = f(x, z)$, and the variable is the z-coordinate in its region of definition.

The result of analyzing the differential characteristics of surface criminant allows us to propose the methodology for calculating the coordinates of the considered points using numerical methods to define the conditional extremum of one of the coordinates, e.g. y, with constraining to the other coordinate – x. The independent variable in this case is z-coordinate.

The illustration of the obtained result can be a model of the workpiece with a helical groove and tool surface with a cylindrical surface (fig. 2). The surfaces of these workpieces are conjugated. Fig. 2 shows that a cross section of a screw surface of the workpiece by the coordinate plane *YZ* contains the

IOP Conf. Series: Journal of Physics: Conf. Series **1050** (2018) 012048 doi:10.1088/1742-6596/1050/1/012048

extreme point A with respect to coordinate plane XY. This point defines one of the points of the profile of the tool cylindrical surface mating with a given workpiece screw surface. The coordinates of the point can be determined by both analytical and numerical methods.

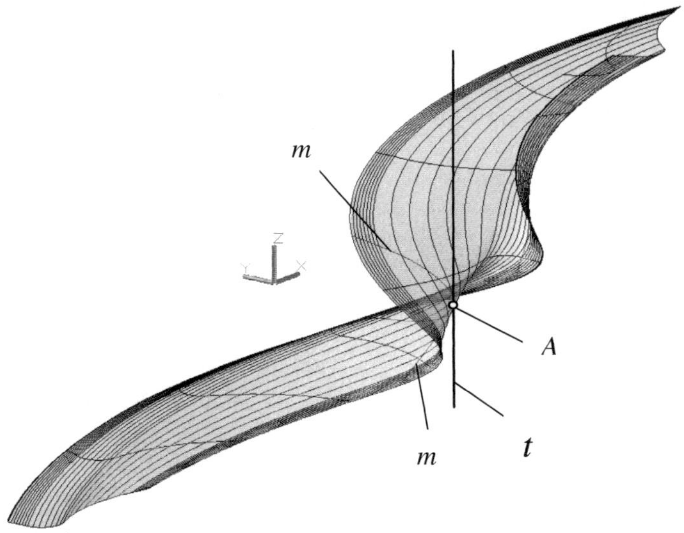

Figure 1. Quasi-screw surface, its cross sections m and n by planes parallel to the coordinate ZY and ZX, respectively, and tangent t to these cross sections at point A

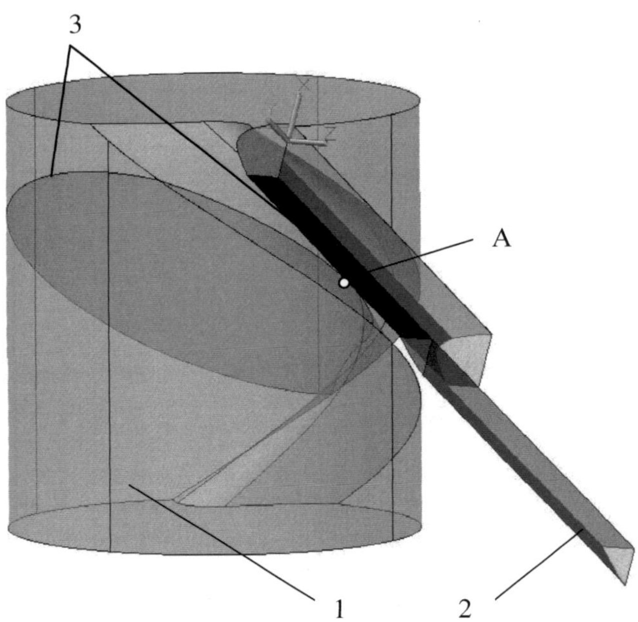

Figure 2. Solid models of the compartments of the workpiece 1 and tool 2, their sections 3 by the coordinate plane YZ and tangency point A to these sections

Thus, if equation (1) is considered as the equation of the surface obtained by displaying congruent plane curves in space R^3, the envelope of this family can be commonly identified by both analytical and numerical methods. By analytical methods - by solving equation (1) and (2), while numerically - by imposing the conditions of one of coordinates and calculating the extreme value of another coordinate. The third coordinate takes the discrete values.

Let us now consider the mapping of the orthogonal projection of a two-dimensional surface defined by parametric equations:

$$x = f_1(u,v),\ y = f_2(u,v),\ z = f_3(u,v). \tag{3}$$

The condition under which the plane tangential to surface (3) is parallel to axis Z – to the direction of projection is written as:

$$F(u,v) = f_{1u} \cdot f_{2v} - f_{2u} \cdot f_{1v} = 0 \tag{4}$$

This equation establishes a link between the parameters u, v. It, together with equations (3) determine the discriminant of the surface. In addition, this equation can be regarded as the equation of some curve μ in the Cartesian coordinate system U and V. Criminant μ' of surface (3) is obtained by mapping curve μ on the given surface. This mapping can be written in the form: $(\mu \subset F) \subset R^2 \xrightarrow{F} (\mu' \subset \Phi) \subset R^3$, where F is given by equations (4).

The conducted research of curves μ and μ' allowed to establish the location of surface (3) criminant points relative to the respective coordinate planes. This result is similar to the one obtained above for the surface specified in implicit form.

In this case, discriminant D of surface (3) is the union of the set of extreme points, namely:

$$D = \sum_{i=1}^{n} \min f_1(u,v) \vee \max f_1(u,v)\Big|_{f_2(u,v)=a_i},\ z = f_3(u,v).$$

The variable in this relationship is one of the parameters of the studied surface, and the surface itself is dissected by a family of planes parallel to the coordinate plane ZY.

In a number of applied tasks there is a need to define the envelope of the two-parameter family of surfaces. So, if a family of surfaces is defined by two independent parameters, the study of the orthogonal mapping to the hyper-plane is a subject to a four-dimensional hyper-surface Σ_1, having the following form:

$$F(x,y,z,u,v) = 0, \tag{5}$$

where u and v are the independent variable parameters.

In this case, the hyper-surface display is performed in the directions of axes u and v on the corresponding coordinate hyper-plane. For such a hyper-surface, the equation of the tangent to the hyper-plane is written in the form:

$$F_x \cdot (x - x_0) + F_y \cdot (y - y_0) + F_z \cdot (z - z_0) + F_u \cdot (u - u_0) + F_v \cdot (v - v_0) = 0, \tag{6}$$

where x_0, y_0, z_0, u_0, v_0 are the coordinates of point N of the surface.

At the points of hyper-surface, where the tangent points to this hyper-surface of a hyper-plane are parallel to axis $0U$, the following condition is met:

$$F_u(x,y,z,u,v) = 0. \tag{7}$$

If equation (7) is considered as the equation of additional four-dimensional hyper-surface Σ_1^1, the intersection of hyper-surfaces (5) and (7) defines a three-dimensional hyper-surface Σ_2. It is a criminant of hyper-surface Σ_1 at its orthogonal mapping along axis u.

The equation of the tangent hyper-plane to hyper-surface (7) has the form

$$F_{ux} \cdot (x - x_0) + F_{uy} \cdot (y - y_0) + F_{uz} \cdot (z - z_0) + F_{uu} \cdot (u - u_0) + F_{uv} \cdot (v - v_0) = 0. \tag{8}$$

where x_0, y_0, z_0, u_0, v_0 are the coordinates of a point K of this hyper-surface.

The obtained hyper-planes (6) and (8) intersect in three-dimensional hyper-planes tangent to hyper-surface Σ_2.

At hyper-surface (5) criminant points the following condition is met

$$F_v(x, y, z, u, v) = 0. \tag{9}$$

As a result, equation (9) defines one more additional four-dimensional hyper-surface Σ_1^2. Then hyper-surface Σ_1 criminant when it is mapped orthogonally along axis $0V$ is determined by the intersection of four-dimensional hyper-surfaces (5) and (9) and is a three-dimensional hyper-surface Σ_3. The equation of a hyper-plane relating to hyper-surface (9), is written in the form:

$$F_{ux}\cdot(x-x_0)+F_{uy}\cdot(y-y_0)+F_{uz}\cdot(z-z_0)+F_{uu}\cdot(u-u_0)+F_{uv}\cdot(v-v_0)=0. \tag{10}$$

where x_0, y_0, z_0, u_0, v_0 are the coordinates of a point L of the surface.
The intersection of three-dimensional hyper-surfaces Σ_2 and Σ_3 sets a two-dimensional surface Σ_4, which is a hyper-surface (5) criminant when it is orthogonally mapped along axes u and v on hyper-plane *XYZ*.
Since the points *N, K* and *L* belong not only to the hyper-surfaces, but also to a two-dimensional surface Σ_4, the tangent plane to the surface Σ_4 is determined by the intersection of hyper-planes (6), (8), and (10). From these equations we get:

$$u-u_0=\frac{-A\cdot F_{uv}+A_1\cdot F_v}{\Delta},\ v-v_0=\frac{A_1\cdot F_u+A\cdot F_{uu}}{\Delta},$$

where: $A=F_x\cdot(x-x_0)+F_y\cdot(y-y_0)+F_z\cdot(z-z_0)$, $A_1=F_{ux}\cdot(x-x_0)+F_{uy}\cdot(y-y_0)+F_{uz}\cdot(z-z_0)$, $\Delta=F_u\cdot F_{uv}+F_v\cdot F_{uu}$,
and the equation of the plane tangent to the surface Σ_4, will have the following form

$$(x-x_0)\cdot\begin{vmatrix}F_x & F_u & F_v\\ F_{ux} & F_{uu} & F_{vu}\\ F_{vx} & F_{uv} & F_{vv}\end{vmatrix}+(y-y_0)\cdot\begin{vmatrix}F_y & F_u & F_v\\ F_{uy} & F_{uu} & F_{vu}\\ F_{vy} & F_{uv} & F_{vv}\end{vmatrix}+(z-z_0)\cdot\begin{vmatrix}F_z & F_u & F_v\\ F_{uz} & F_{uu} & F_{vu}\\ F_{vz} & F_{uv} & F_{vv}\end{vmatrix}=0. \tag{11}$$

As a result, the envelope of the two-parameter family of surfaces (5) is determined by the system of equations (5), (7) and (9), under conditions:

$$|F_x|+|F_y|+|F_z|\neq 0 \text{ and } \begin{vmatrix}F_{uu} & F_{vu}\\ F_{uv} & F_{vv}\end{vmatrix}\neq 0.$$

Thus, the obtained results allow to define the envelope and a real envelope for families of lines and surfaces by both analytical methods, by obtaining the constraint equations of line or surface parameters, and the parameter of their family (which is often time-consuming), and numerical methods not requiring such equations. These results are used for computing polygonal and solid modeling not only of the envelope and real envelope, but also of the third object of shaping that are removable volumes. The accuracy and efficiency of the above stated is confirmed by the experiments.

4. Experimental results

4.1. Polygon modeling of congruent curves families

In applied problems, solving a design study of a cutting tool, there is a need to define the envelope of the family of congruent curves associated with the centroid of the tool rolls without slipping on the centroid of the workpiece. When mapping families of congruent curves in space R^3 some auxiliary surfaces are obtained [22]. The analysis of the geometry of these surfaces by means of CAD-systems allows to identify the effect of the tool settings relative to the workpiece, as well as the shape of the curve family on the envelope shape. As an example, consider the family of curves associated with centroid – a circle rolling on another one, the centroid of the circle. Let the radii of these centroid be R_1 and R_2, respectively, and the curve of the family is defined by the parametric equations:

$$x = x(t),$$
$$y = y(t).$$

After the mapping of this family into space R^3 we get a surface which equation has the following form

$$x_s = x(t)\cdot\cos k\varphi - y(t)\cdot\sin k\varphi - A\cdot\sin\varphi,$$
$$y_s = x(t)\cdot\sin k\varphi + y(t)\cdot\cos k\varphi + A\cdot\cos\varphi, \quad (12)$$
$$z_s = p\cdot\varphi,$$

where φ is the parameter of family of curves $k = \frac{R_1 + R_2}{R_1}$, and p is a large zero constant.

Comparing the system of equations (12) with the equation of cylindrical screw surface, it can be stated that surface (12) is obtained by affine transformation of the cylindrical screw surface. Consequently, the resulting surface is quasi-winding. A computer model of one of these surfaces is shown in fig.3a.

To obtain the envelope of the considered family of lines, mapping the orthogonal projection of the surface onto the coordinate plane (fig. 3b) is made. For the model presented in fig. 3a, the horizontal outline of the surface is the envelope of family of curves related to the circle (fig. 3b). This model is used in the visualization mode to study the effect of setting the parameters of the workpiece relative to the tool on the envelope shape.

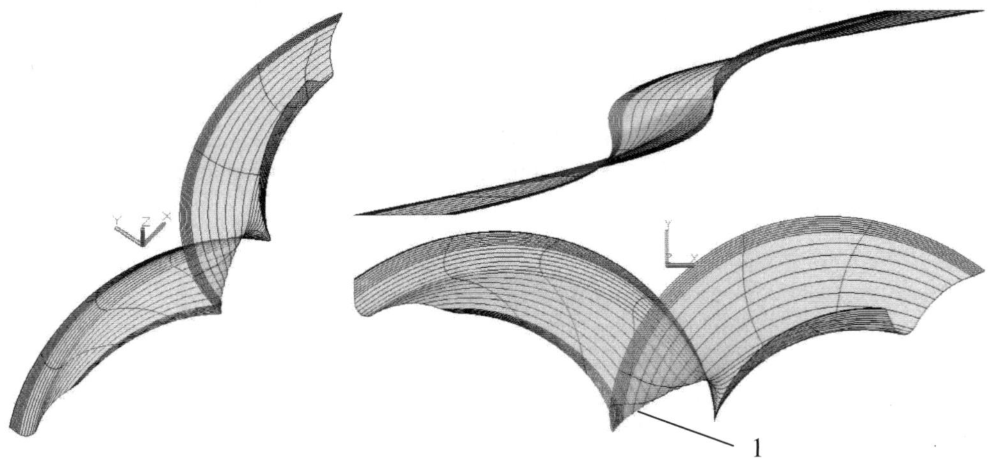

a) b)

Figure 3. Model of a quasi-screw surface (a) and its orthogonal projection (b); 1 – the discriminant of the surface (envelope of a family of profiles)

Another example of computer modeling is determining the envelope of the family of curves (tool profiles) associated with the direct rolling without slipping on a circle. The mapping of curves into space is not applied here. Fig. 4 shows two families of curves for two positions of the tool relative to the workpiece. In fig.4a, the tool has a positive displacement (convergence) relative to the workpiece, while fig. 4b – negative (receding). As a result, in the first case, the tooth profile is outlined not only by the involute, but also by a significant value of the transition curve. In the second case (fig. 4b) there is no transition curve, the tooth profile increases, which increases the strength of the tooth when bending; still, its vertex is pointed.

The carried out experiments on computer modeling of families of lines on the plane or their mapping into space with subsequent projection onto a plane, allow to obtain a qualitative picture of the shape of the family of curves envelope depending not only on the shape of the source profile, but on the relative positions of a workpiece and a tool.

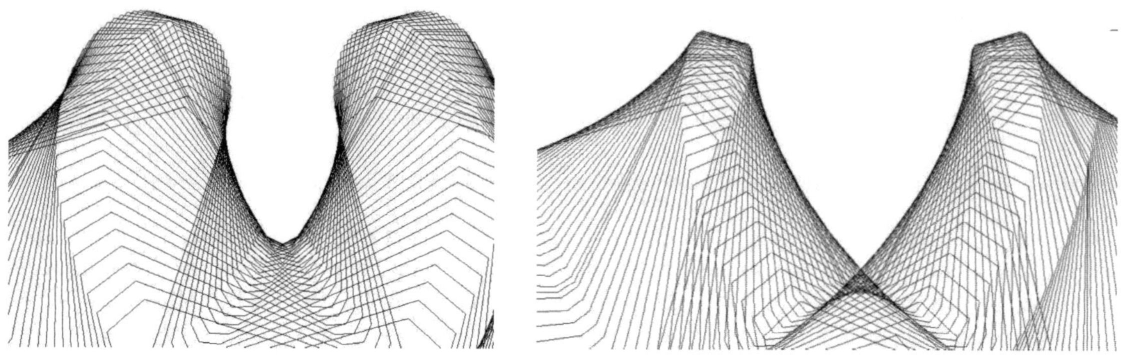

a) b)

Figure 4. Collection of tool profiles related to direct rolling circumference: (a) if positive and (b) if negative displacement of the tool relative to a workpiece

4.2. Computer solid modeling of shaping

Another direction of studying the main objects of shaping is solid-state computer modeling of the interaction models of the tool and the workpiece in accordance with the selected kinematic scheme. In the process of designing a cutting tool along with a profiling of the forming part, an important role is played by the study of cutting process.

The developed computer programs in the CAD environment perform the shaping of the workpiece surface by the tool on the basis of solid modeling in an automated mode. During modeling, the interaction of solid models of the tool and workpiece is performed using Boolean operations. Fig. 5 shows the simulation of the shaping by a helical groove disc mill on a cylindrical workpiece. The modeling result is real envelope surface of this groove.

In addition to the real envelope surface formed by the workpiece of the modeled process to create that surface, and, therefore, set of possible transitions, the configuration of the removed layer and the load on the cutting edges of the tool.

Some of the possibilities of computer modeling of the spatial schema of the gear are illustrated by fig. 6. It shows a model of a parts cylindrical billet and a model of layered, removed from it by an end mill. Based on the solid models, it is possible to determine the qualitative characteristics of the form, removable layers, and quantitative ones – of their volumes. So, from fig. 6 it follows that for the case under consideration, the configuration and volumes of the removable layers do not change during processing. If, after shaping, to remove the models of the removable layers from the model of the workpiece, it is possible to investigate the existence of possible transition curves on the real envelope surface, obtained by lateral surface of the cutter and its face plane.

Shaping solid modeling algorithms and programs in the tasks of flat gear scheme along with getting paired profile, and conjugated spatial pattern allow us to observe the process of sequential cutting of the hollows between the wheel teeth. The opportunity to obtain quantitative parameters is also provided: the volume of the layers, removed in one double stroke of the tool; the volume of the layer, removed by the side cutting edges of the tool and its peripheral edges. These parameters are used to establish the relationship of the removed volume from the value of the cutting parameter and their analysis allows to assign the optimal parameter values of cut, number of passes and depth of cut for each pass.

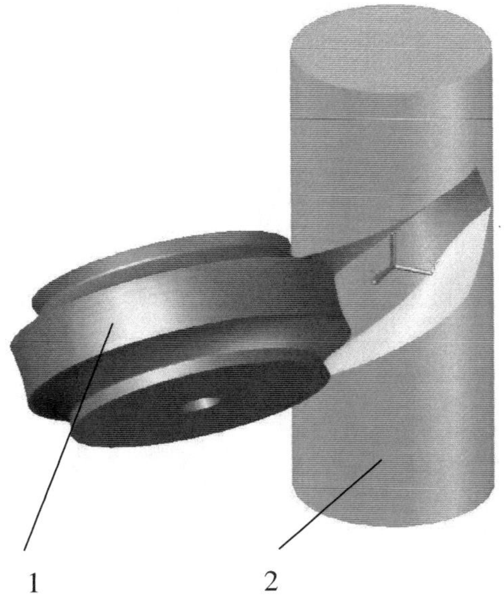

Figure 5. The model of the disk cutter 1 in the process of shaping screw grooves in a cylindrical workpiece 2

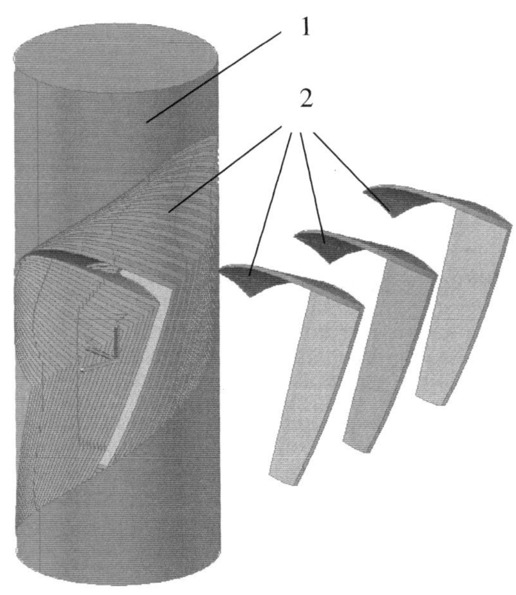

Figure 6. The model of the workpiece 1 after its shaping, and the 2 removable layers

The described possibilities are illustrated by fig.7. It shows the model of the tool – shaping cutter and a gear workpiece in the process of shaping, and the models of the layers removed by the tool.
Thus, the computer solid modeling allows to model automatically the two main objects of shaping– real envelope surface and removable layers. In some tasks this approach may be a key for solving the described problem or it can add, if necessary, to geometric modeling.

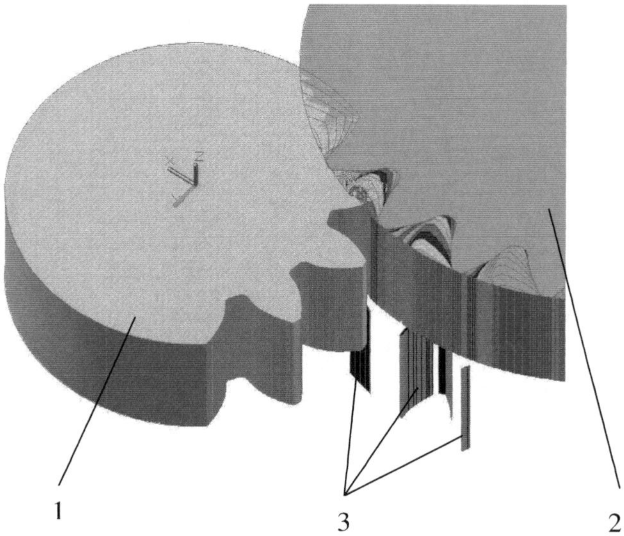

Figure 7. Models of the tool 1, the workpiece 2, in the process of its shaping, and removable layers 3

5. Results and discussion

The experiments showed that the proposed methodology allows to commonly determine the discriminant and real envelope of the family of lines and surfaces by analytical and numerical methods, as well as removable layers successfully using the possibilities of modern geometric, polygon, and solid computer modeling. Further improvement of this methodology will allow to model the shaping of workpieces with non-circular centroid, too, which is a complex task even by using the known methods.

6. Conclusion

The proposed methodology of modeling the shaping of technical products surfaces allows to solve the following tasks:

- to develop mathematical models of surfaces and hyper-surfaces, obtained from the display of families of lines and surfaces into the space of dimensionality one greater than the dimensionality of the space, where these families are;
- to perform the mapping of the obtained surfaces and hyper-surfaces by orthogonal projection on the corresponding plane and hyper-plane; as a result, the envelope and envelopes of lines or surfaces families can be determined by both analytical and numerical methods, without obtaining complex rationalized equations of line or surface forms parameters and the parameter of the family;
- to obtain a model of new surfaces formed by the family of plane curves associated with the centroid of the tool, rolling without slipping on the centroid of the product; the computer visualization of such surfaces allows to track the change in the envelope of the profiles family depending on the profile shape and the relative positions of a workpiece and a tool;
- to create solid models of the removed volumes, based on the analysis which provides the ability to assign the optimal supply parameter value and the number of passes while shaping.

Thus, the proposed methodology allows, based on a common theoretical framework, to model the envelope, real envelope and removable layers in the process of shaping parts with a tool by both analytical and numerical methods and also with the use of modern computer technology, which provides the necessary quality of designed products.

References

[1] Lashnev S I and Yulikov M I 1975 *Calculation and design of cutting tools using a computer.* (Moscow: Machinostrienie / Mechanical engineering) p 392 (in Russian)
[2] Lyukshin V S 1968 *Theory of helical surfaces in the planning of cutting tools.* (Moscow/ Mashinostroenie) p 371 (in Russian)
[3] Litvin F L and Fuentes A 2004 *Gear Geometry and Applied Theory*, 2nd Edition (Cambridge University Press: Cambridge)
[4] Chemborisov N A and Devzheeva T G 2010 Review of methods for a worm mill for gear wreaths profiling of, in: Metalworking 4 pp 2-6 (in Russian)
[5] Radzevich S P 2001 *Formoobrazovanie poverkhnostei detalei: osnovy teorii (Surface Shaping of Machine Parts: Theoretical Principles)* (Kiev: Rastan) p 592
[6] Zalgaller V A 1975 *Theory of Envelopes* (Moscow Nauka) p 104 (in Russian)
[7] Sheveleva G I 1999 *The theory of shaping and contact of moving bodies* (Moscow: Mosstankin) p 494 (in Russian)
[8] John Argyris and Litvin Faydor L and Qiming Lian and Lagutin Sergey A 1999 *Determination of envelope to family of planar parametric curves and envelope singularities. Computer Methods in Applied Mechanics and Engineering.* Vol 75 Issue 1-2 pp 175-187
[9] Lopatin B A and Khaustov S A 2008 *Automated modeling and analysis of ways of forming the gear wheels teeth* / B. A. Lopatin, in: Herald YurSu. Ser. Engineering Vol 11 No 10 pp 72-77 (in Russian)
[10] Lyashkov A A [et al.] 2010 *Simulation of the shaping of complex surfaces* in: Metalworking 4 pp 36-42 (in Russian)
[11] Lyashkov A. A. 2011 *Modeling of screw surfaces by lath and worm mill* in: Metalworking 1(61) pp 2-7 (in Russian)
[12] Dimitriou Vasilis, Vidakis Nectarios, Antoniadis Aristomenis 2006 *Advanced Computer Aided Design Simulation of Gear Hobbing by Means of Three-Dimensional Kinematics Modeling.* Journal of Manufacturing Science and Engineering Vol 129 Issue 5 pp 911 – 918
[13] Dimitriou V and Antoniadis A 2012 *CAD-Based Calculation of Cutting Force Components in Gear Hobbing* Journal of Manufacturing Science and Engineering Vol 134 p 89
[14] Lyashkov A A 2012 *Shaping of harts with a helical surface by means of a disk mill* in: Russian Engineering Research Allerton Press Inc Vol 32 4 pp 404-406.
[15] Lyashkov A A and Panchuk K L 2015 *Computer modeling of the pump screw and disc tool cross shaping process* Procedia Engineering Vol 113 pp 174-180
[16] Lyashkov A A and Vasil'ev E V and Popov A Y 2017 *Development of 3D modeling technology for manufacturing finned ribbons from heat-resistant steelss* in: IOP Conf. Series: Earth and Environmental Science IPDME 2017 Vol 23
[17] Shishkov V A 1951 *The formation of surfaces by cutting according to the form-generating method* (Moscow: Mashgiz) (in Russian)
[18] Thom R 1962 *Sur la theorie des envelopes* in: J. de math. pur et apple Vol 41 № 2 pp 177-192
[19] Arnold V 2000 *"Singularity theory"*, Development of mathematics 1950–2000 (Birkhäuser, Basel) pp 63–95
[20] Bruce and Giblin P J 1992 *Curves and Singularities* (Cambridge University Press, Cambridge)
[21] Platonova O A 1984 *Characteristics of smooth surfaces projections* in: Uspekhi (advances) of Mat. sciences. 39 vol 1 pp 149-150 (in Russian).
[22] Lyashkov A A and Kaneva Yu A 2011 *Auxiliary surfaces in the modeling of details shaping by means of computer graphics* in: Bulletin of the Kuzbass State Technical University 5(87) pp 75-80 (in Russian)

[23] Lyashkov A A and Volkov V Ya 2012 *Displaying the orthogonal projection of hypersurface at the hyperplane* in: Bulletin of Irkutsk State Technical University 3(62) pp 18-22 (in Russian)

[24] Lyashkov A A 2012 *Displaying orthogonal projection of the surface given by parametric equations* in: Omsk scientific Bulletin Ser. Appliances, machinery and technology 2 (110) pp 9-13 (in Russian)

[25] Lyashkov A A 2015 Mathematical and Computer modeling of component surface shaping *IOP Conference Series: Materials Science and Engineering. International Conference on Mechanical Engineering, Automation and Control Systems* MEACS 2015

Automated Geometric and Computer-aided Non-Circular Gear Formation Modeling

A A Lyashkov, K L Panchuk, I A Khasanova

Omsk State Technical University, 11, Mira ave., Omsk, 644050, Russia

E-mail: 3dogibmod@mail.ru

Abstract. In non-circular gears of hydromachines used for transit of oil, fuel-oil residue and water, difference takes place not only between gear teeth profiles, but also between lateral profiles of the same tooth. Profiling of such gears requires application of complex mathematical apparatus. The solution to the inverse task of formation is even more complex in this case. This paper proposes a geometrical model of solution to the task of profiling a non-circular gear, the centroid of which consists of interconnected arcs. It is effective in case all the conditions of formation are met. The proposed automated solid computer-aided modeling of the direct and the inverse tasks of formation realized on virtual imitation level allows us not only to acquire the envelope of the profile, but also to observe the possible formation of transition curves and undercuts. It reveals constructive and technological conditions under which they appear, presents an opportunity to conduct the respective research in order to introduce the required corrections into the kinematic scheme of formation, which results in achieving a high-grade solution to the task of profiling. Automated solid modeling of the inverse task of formation allows us to validate the solution to the allocated task and, if necessary, to refine the initial data and therefore exclude the necessity to conduct the expensive full-scale experiments. The results of the study can be used in non-circular gear design in the field of hydromachines used for transit of oil, fuel-oil residue and water.

1. Introduction

Items comprising non-circular gears are applied in numerous fields of industry. They have found the widest application in looms, measuring devices, flow meters and various other mechanisms and machines [1-5]. The application of non-circular gears appears promising in planetary rotary hydromachines used in pumps that transit oil, fuel-oil residue and water, especially contaminated i.e. on drill sites, as well as in dosing pumps for various liquids [6,7].

The two main problems that restrain the spread of non-circular gear drives are the complexity of non-circular gear teeth machining and the complexity of tooth profile calculation. At the present time the first problem is effectively solved by the use of CNC machines, i.e. EDM machines. The solution to the second problem is acquired by the use of various approaches. One of such approaches is based on the classical theory of envelopes [1]. This approach is characterized by substantial complexity in case the centroid line is represented by an ellipse. In this case, difference takes place not only between all of the teeth of the same gear, but also between lateral profiles of the same tooth. The calculations require complex mathematical apparatus or the use of mathematical and design CAD [8-12]. In case the precision of tooth profile is not subject to strict requirements, the elliptic centroid line is approximated by arcs [6, 13-15] and the possibility of application of classical profiling methods emerges. However, in this case there are complications emerging from the fact that the centroid

IOP Conf. Series: Journal of Physics: Conf. Series **1050** (2018) 012049 doi:10.1088/1742-6596/1050/1/012049

consists of a number of interconnected sections. At the present time the field of workpiece profiling based on computer-aided modeling of actual process of workpiece formation by an instrument [9-11], [16-18] with further integration into CAM system is successfully being developed. Since the teeth profiles acquired as a result of formation modeling do not always consist of the envelope of the corresponding family of profiles, but rather have undercuts and transition curves, it is required to verify the result. In this case, it is required to solve the inverse task of formation. As shown by the analysis of the existing scientific literature, such approach to non-circular gear formation has not yet been considered.

2. Problem Definition

The aim of this paper is the automation of the process of geometric and computer-aided non-circular gear teeth formation modeling. The objectives of the study are: acquiring a mathematical model of gear tooth profile as an envelope of a family of instrument profiles; development of an algorithmic kinematic formation model that would provide the solution to the direct and the inverse tasks of non-circular gear profiling as well as the possibility of acquiring their actual profile that would include the specific elements such as transition curves and undercuts; realization of said model in automated mode on the basis of computer-aided solid modeling methods in order to acquire the digital model of a workpiece adapted to CAM systems; establishment of automated solid modeling technology of solving the inverse task of formation, that would include the creation of removed stock model and allow us to exclude the necessity to conduct the expensive full-scale experiments.

3. Theory

1. Formation of initial data for modeling.

A number of applications [6, 7] provide the ability to approximate the elliptic centroid with elementary geometric primitives – circles and lines. In this case the kinematic scheme of formation of a workpiece with an instrument is reduced to rolling motion of one centroid along the other. On the first stage the two tasks are to be solved: 1) to construct out of primitives a centroid that would sufficiently closely approximate the initial centroid; 2) to select the parameters of the primitives so that the total length of the centroid would be divisible by the pitch of teeth of the generating gear.

Consider a centroid of a non-circular gear outlined by four coupled arcs of radiuses R_1 and R_2 (fig.1). In order to acquire the points A of coupling of circles of radiuses R_1 and R_2 let us put down the equation of a circle with center in point O_2 and radius equal to R_1 - R_2 (fig.2):

$$\left(x-\frac{d}{2}\right)^2+y^2=(R_1-R_2)^2. \quad (1)$$

The intersection of the circle (1) with the axis $0Y$ defines the point B, which is the center of the circle R_1 with coordinates $X=0$, $y=-\sqrt{(R_1-R_2)^2-\frac{d^2}{4}}$. Then angle β can be acquired, for example, from the following correlation:

$$\beta=arctg\left(\frac{d}{2\cdot\sqrt{(R_1-R_2)^2-\frac{d^2}{4}}}\right). \quad (2)$$

This angle is further used to define boundaries of rolling motion of the centroid of the circular gear along the non-circular gear. Furthermore, the point of A coupling of two arcs is defined. Its coordinates are acquired from the following correlations:

$$x_A=-\frac{d}{2}-R_1\cdot\sin\beta,\ y_A=R_1\cdot\cos\beta. \quad (3)$$

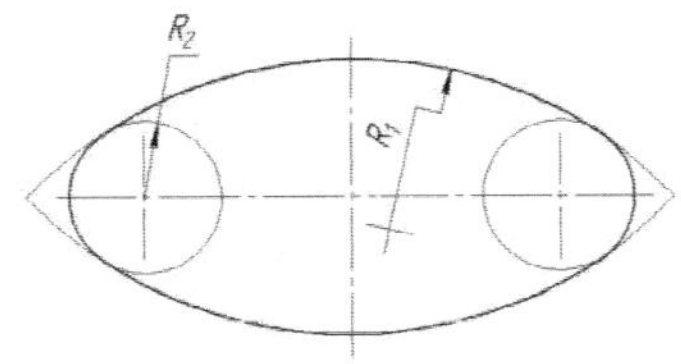

Figure 1. Centroid of a gear combined from four conjugate arcs of circles

2. Geometric modeling of formation.

Geometric modeling constitutes realization of the algorithm of kinematic formation scheme, in which the centroid of the generating gear is rolling along the centroid of a workpiece without slipping. One of such algorithms is performed in the following sequence: first, the generating gear turns around the axis of the modeled gear on the angle φ_2, then around its own axis on the angle φ_1. The angles φ_1 and φ_2 are linked with a the following corellation: $\varphi_1 = \frac{D_2}{D_1} \cdot \varphi_2$. This geometric formation modeling algorithm is described by formulas of transition from the moving coordinate system of the generating gear to the coordinate system of the workpiece. As a result of the mentioned turns, the family of instrument profiles appears. Its envelope represents the sought profile of the workpiece. The analytical solution of gear profiling comes to determination of correlation between the parameters of the initial profile and its family, which for the case under consideration is of the form of equation [19]

$$k \cdot x(t) \cdot \frac{\partial x(t)}{\partial t} + k \cdot y(t) \cdot \frac{\partial y(t)}{\partial t} + A \cdot \frac{\partial x(t)}{\partial t} \cdot \sin(\frac{R_2}{R_1} \cdot \varphi_1) + A \cdot \frac{\partial y(t)}{\partial t} \cdot \cos(\frac{R_2}{R_1} \cdot \varphi_1) = 0, \tag{4}$$

where the initial profile is defined by parametric equations of the form $x = x(t), y = y(t)$; $\frac{\partial x(t)}{\partial t}, \frac{\partial y(t)}{\partial t}$ represents partial derivatives with respect to parameter t; $k = \frac{R_1 + R_2}{R_1}$; R_1 represents the centroid radius of the instrument; R_2 represents the centroid radius of the workpiece; φ_1 represents the family parameter.

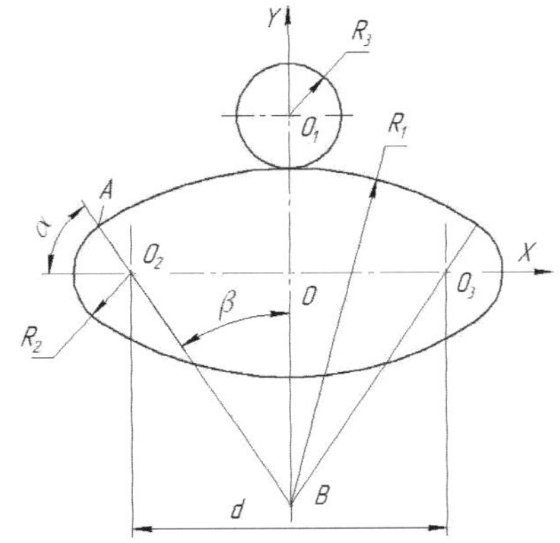

Figure 2. Centroids of circular and non-circular gears

The acquired equation together with the equations of the initial profiles define the workpiece profile. The disadvantages of geometric modeling are the following: certain difficulties in solving the equation (4); the possible transition curves on the sought profile are not taken into account, which requires additional research.

3. Computer-aided modeling of formation.

In the present paper, following from the essence of solid modeling, the technical sequence of solution of the formation task under consideration is developed and realized in automated mode in the

AutoCAD environment in AutoLISP programming language. The algorithm of this technical sequence includes the following stages:
1) Initialization of the solid models of the generator instrument and the workpiece.
2) Execution of the direct task of formation in accordance with the developed modeling programs, which constitutes acquiring non-circular gear profile using the initial model of a cylindrical gear with intermittent teeth.
3) Solution of the inverse task of formation, which constitutes acquiring the generating gear model given the acquired non-circular gear model; this task is solved in case it is necessary to confirm the significance of the result of direct task solution.
4) In case of necessity, execution of removed stock modeling in order to appoint optimal technological parameters of formation.
Contents of the suggested automated computer-aided modeling are demonstrated by the conducted experiments.

4. Results of Experiments.
In a computer experiment the parameters of the generating gear and the non-circular gear centroid have been selected according to fig. 3. The diameter of the centroid of the gear with intermittent teeth has been calculated so that the length of non-circular gear centroid is divisible by the pitch of teeth of the generating gear. In order to achieve this, the points of coupling of arcs of non-circular gear centroid have been calculated beforehand using the correlations (1) – (3), and their lengths have been acquired subsequently. Solid models of the generating gear and the workpiece with non-circular centroid have been created with said parameters. The solid models constitute the initial data for the automated computer-aided modeling of the direct task of formation.

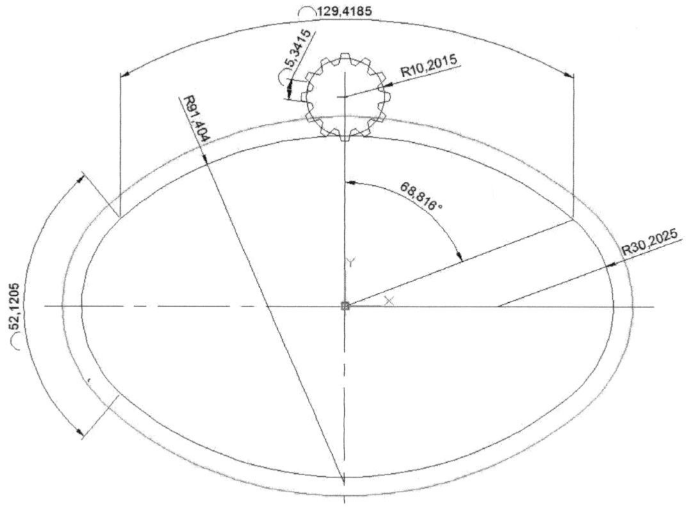

Figure 3. Centroids of circular and non-circular gears and their parameters

On the subsequent stage of task solution, in accordance with developed programs, formation modeling of the workpiece upon rolling motion of the generating gear centroid along the non-circular gear is performed on the basis of boolean operations. First, the rolling motion occurs along the arc of radius R_1, then along the arc of radius R_2.
Since in the considered example the length of the non-circular gear centroid is divisible by four lengths of instrument centroid, the modeling is performed only for one quarter of its length. In order to compare the acquired teeth profile of various sections of non-circular gear, fig. 4 depicts its fragment, while fig. 5 depicts teeth profiles scaled up. As follows from fig. 5, pitch of teeth along the centroid is

equal on all sections, which confirms that the calculations were performed correctly. However, tooth top on section 2 is insignificantly sharpened. Undercut of dedendum is present on both sections.

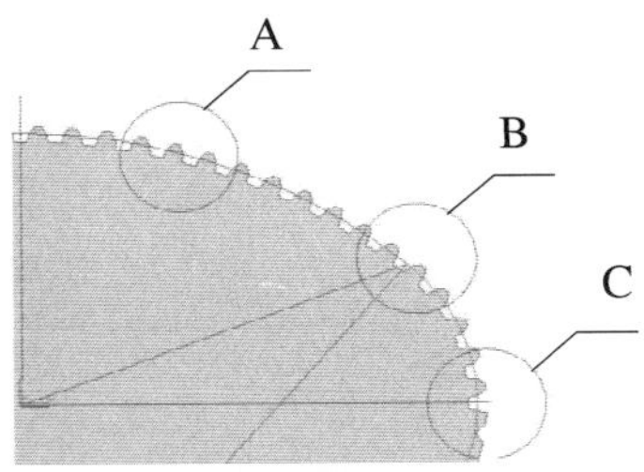

Figure 4. Non-circular gear workpiece fragment after its formation

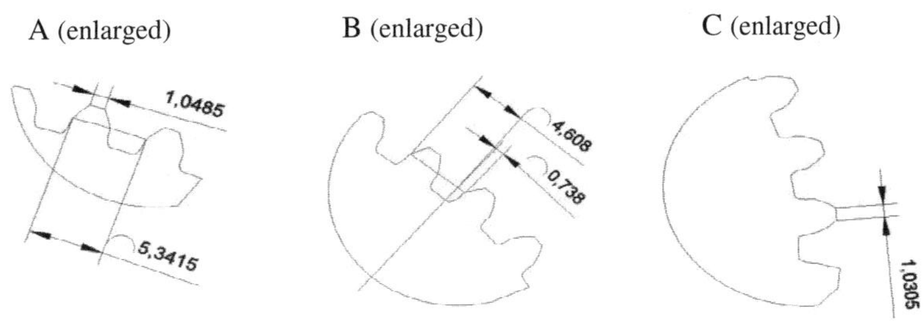

Figure 5. Teeth profile fragments of various sections of workpiece centroid

The result of solid modeling of a non-circular gear by means of an instrument with intermittent teeth is rendered on fig. 6. The acquired digital model of workpiece can be used on a CNC-operated machine without further preparation in order to produce non-circular gears of required quality.

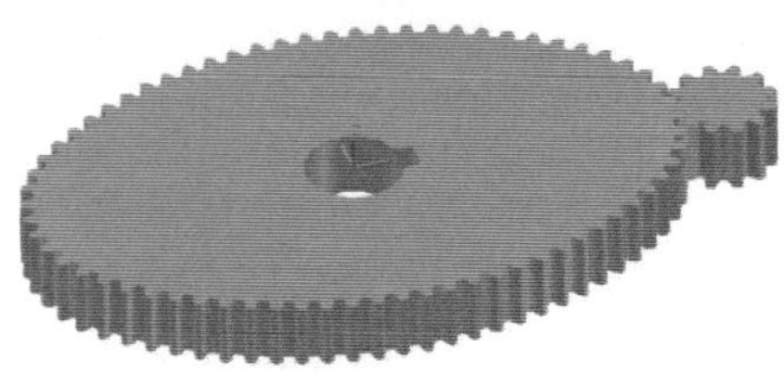

Figure 6. Computer solid models of the generating and the non-circular gears after formation

The developed software allows us to solve the inverse task of formation, i.e. acquiring the instrument tooth model given the acquired non-circular gear model. Fig. 7 depicts a fragment of modeling of inverse formation task. It validates the acquired results and drawn conclusions.
The developed software executes removed stock modeling in the process of gear teeth formation modeling, which allows us to assign optimal tool advance and number of passes as well as detect cutting edge workload on the basis of its attribute-based and quantitative parameters.

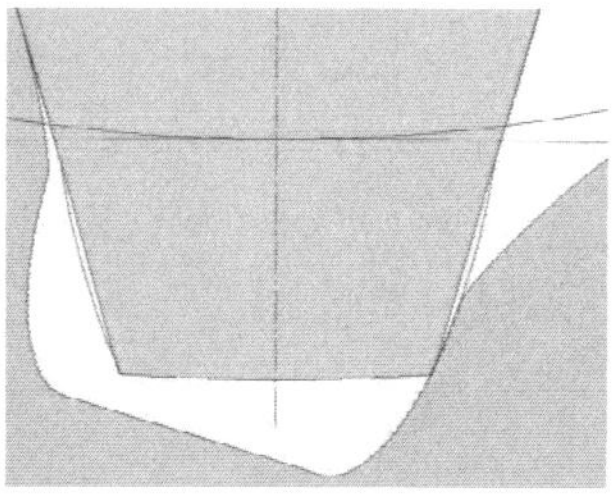

Figure 7. Computer-aided modeling of inverse task of formation

5. Consideration of the Results

The more complex the form of non-circular gear centroid, the more complex the geometry of teeth profile formation of such gear. In general, difference takes place not only between all the teeth of the non-circular gear, but also between lateral profiles of the same tooth. The solution to such task by means of analytic methods requires complex mathematical apparatus and presents significant difficulties. The reverse task of formation is even more complex in this case. Computer-aided solid modeling realized on a virtual imitation level in automated mode allows us to observe the formation of transition curves and undercut. It presents an opportunity to envision design-manufacturing conditions under which they appear, allows us to conduct the respective research and introduce the required corrections into the kinematic scheme of formation. Since complex form of non-circular gear centroids corresponds to complex kinematic formation scheme, it is essential to refine and develop new approaches and algorithms of solid computer-aided formation.

6. Conclusions

1. A geometrical model of solution to a task of profiling of non-circular gear, the centroid of which comprises interconnected arcs, is proposed. A calculation of generating wheel with intermittent teeth parameters as well as workpiece centroid parameters is performed. Such calculation is effective only in case the sought teeth profile consists exclusively of envelopes of families of circular gear profile.
2. The automated solution to the task of non-circular gear profiling is acquired on the basis of solid modeling. The profiling is executed on virtual imitation level on the basis of algorithms, which model the formation of a workpiece with intermittent teeth by means of a gear cutter. Such solution presents an opportunity not only to acquire the sought profile as an envelope of a family of curves, but also to model the undercut and transition curves, which allows us to achieve a high-grade solution to the task of profiling.
3. The proposed solid modeling also executes the solution to the inverse task of formation in automated mode. The inverse task of solution constitutes acquiring a profile of a gear with intermittent teeth given the previously acquired model of a non-circular gear. This modeling presents an opportunity to validate the solution of the assigned task, to adjust the initial data if needed and to exclude the necessity to conduct the expensive full-scale experiments.
4. The developed software presents an opportunity to create solid models of removed stock, analyzing which it is possible to solve technological tasks including the assignment of optimal tool advance and number of passes.

References

[1] Litvin F L and Fuentes A 2004 *Gear Geometry and Applied Theory* 2nd Edition (Cambridge: Cambridge University Press) p 800

[2] Mundo D and Danneli G A 2004 *Use of Non-Circular Gears in Pressing Machine Driving Systems* [Internet source] URL:http://www.wseas.us/e-library/conferences/udine2004/papers/483-172.

[3] Han J Li, D Z Gao T and Xia L 2015 *Research on Obtaining of Tooth Profile of Non-circular Gear Based on Virtual Slotting* The 14th IFToMM World Congress (Taipei: Taiwan) October 25-30

[4] Kiryeev S O, Ershov Yu V and Padalko N A 2010 *Calculation of teeth parameters for oval gear manufacture* (University News of North-Caucasian Region: Technical Sciences Series) №1 pp 76-78 (in Russia)

[5] Xia L Li D and Han J. 2014 *Research on Mathematical Modeling and Kinematic Simulation of Elliptic Family Gears* Key Engineering Materials pp 300-304

[6] I-Kan An Ilin A S and Lazurkevich A V 2016 *Aspects of geometric calculation of the planetary gear train with intermediate rollers* Part 1 [Electronic resource] IOP Conference Series: Materials Science and Engineering Vol 124: Mechanical Engineering, Automation and Control Systems (MEACS2015).

[7] Volkov G Yu and Kurasov D A 2014 *Planetarno-rotornaya gidromashina* (Past-Ural Science Journal) №2 (6) pp 135-136 (in Russia)

[8] Litvin F L Gonzalez-Perez I Yukishima K et al 2007 *Generation of planar and helical elliptical gears by application of rack-cutter, hob, and shaper* (Comput. Methods Appl. Mech. Engrg.) 196(41-44) pp 4321–4336

[9] Litvin F L, Gonzalez-Perez I, Fuentes A et al 2008 *Design and investigation of gear drives with non-circular gears applied for speed variation and generation of functions. Computer methods in applied mechanics and engineering* 197(45): pp 3783-3802

[10] Han J, Li D Z, Gao T and Xia L 2015 *Research on Obtaining of Tooth Profile of Non-circular Gear Based on Virtual Slotting* The 14th IFToMM World Congress (Taipei: Taiwan) October 25-30 p 5 URL:http://www.iftomm2015.tw/IFToMM2015CD/PDF/PS6-001.pdf

[11] Marius Vasie and Laurenţia Andrei 2012 *Analysis of noncircular gears meshing* (Mechanical Testing and Diagnosis) Vol. 4 pp 70-78

[12] Li J G et al 2007 *Numerical computing method of noncircular gear tooth profiles generated by shaper cutters* (The international journal of advanced manufacturing technology) Vol. 33 pp 1098-1105

[13] Chang S L and Tsay C B 1998 *Computerized Tooth Profile Generation and Undercut Analysis of Noncircular Gears Manufactured With Shaper Cutters* (Journal of Mechanical Design, ASME Transactions) Vol. 120 pp 92-99

[14] Bair B W 2002 *Computerized Tooth Profile Generation of Elliptical Gears Manufactured by Shaper Cutters* (Journal of Materials Processing Technology) Vol. 122 pp 139-147 URL:http://www.scielo.org.co/scielo.php?script=sci_arttext&pid=S0012-73532016000400027

[15] Padalko A P and Padalko N A 2013 *Zubchataya peredacha s nekruglym kolesom* (Theory of Mechanisms and Machines) №2 Vol.11 pp 89-96 (in Russia)

[16] Lyashkov A. A. 2012 *Shaping of harts with a helical surface by means of a disk mill* (Russian Engineering Research, Allerton Press, Inc) Vol. 32 № 4 pp 404-406

[17] Lyashkov A A and Panchuk K L 2015 Computer modeling of a pump screw and disc tool cross shaping process (Procedia Engineering 113) pp 174-180

[18] Lyashkov A A, Vasil'ev E. V and Popov A Y 2017 *Development of 3D modeling technology for manufacturing finned ribbons from heat-resistant steelss* (IOP Conf. Series: Earth and Environmental Science. IPDME) Vol. 23

[19] Lyashkov A A and Kaneva Yu A 2011 *Auxillary surfaces used in workpiece formation modeling by means of computer graphics* (Bulletin of Kuzbass State Technical University) №5(87) pp 75-80 (in Russia)

IOP Conf. Series: Journal of Physics: Conf. Series **1050** (2018) 012050 doi:10.1088/1742-6596/1050/1/012050

Architecture of the automated information system for monitoring and resources accounting

Andrey G Malyutin, Andrey A Lavrukhin, Andrey S Okishev

Omsk State Transport University, Omsk, Russian Federation

Abstract. We solve the problem of minimization of material and time costs on scaling, deployment and expansion of functionality of the distributed automated information system for monitoring and resources accounting (hereinafter–the system), that have been created for experimental research of complicated transition processes in high-voltage circuits of power supply networks (to 25 kV). The solution of this task became possible by developing the new system architecture (software, technological and information support) and functional modules, considering the geographical distribution and heterogeneity of data channels and transmission technologies used to information exchange between the elements of the system.

We explain the features of the new original architecture for systems of this class, considering the logical division into functional levels: measuring, intermediate storage and processing, data representation for users. We have developed mobile information and measuring complexes that allow making a research of continuous dynamic processes in high-voltage circuits using wireless technologies and independent power supplies. A set of solutions is proposed to provide modularity of the architecture for the rapid addition of new devices and functions using intermediate software environment that is actual for a changing infrastructure and variety of interfaces between the parts of the system. The architectural solutions offered in the paper provide a comprehensive analysis and evaluation of power losses due to suboptimal operation of electrical substations and electric rolling stock (high and low voltage, power flows between traction substations, losses in step-down transformers, voltage surges due to recuperation, etc.), as well as the formulating of recommendations for improving the energy efficiency of power supply systems. The system can be adapted to various enterprises needing the account of any production resources and technological parameters that can be synchronously measured in real time for research and operational decision-making.

1. Introduction

The monitoring task is considered and accounting parameters and dynamic characteristics of the process as a continuous process for observation and recording of object parameters. The monitoring result is a plurality of measured values of the parameters obtained in the continuously adjoining time intervals, for which parameter values do not vary significantly. Accordingly, the system for monitoring and recording system is hierarchical multilevel geographically distributed computerized system and makes it possible to control and analyze a variety of technological resources consumption modes, decisions regarding their rational use and aggregated enterprise resource data. The main objective is to obtain reliable information on the embedding amount produced, transmitted, distributed and consumed resources. The need to improve such systems due to the constant rise of a variety of resources indexability of their application and accounting, intensification of modern information technologies in the industry. The system put into operation at the enterprise of the Sverdlovsk Railway, a branch of Russian Railways, is described in [1][2][3]. Consider the basic approaches and architectures used in the modernization of this system.

IOP Conf. Series: Journal of Physics: Conf. Series **1050** (2018) 012050 doi:10.1088/1742-6596/1050/1/012050

2. Analysis of requirements for information systems

The task of designing an automated corporate info-system for monitoring and accounting resources is associated with the need to select and develop a wide range of technical solutions considering the specificity of the automation object. First, it is necessary to carry out research or study of the object, features and modes of its operation, and the dynamics of its processes, to clarify the technical task requirements.

It is necessary to identify the functional tasks associated with the main purpose of the operation of the particular service of the system or the part of it. Wherein, functional tasks can be conditionally subdivided into internal service tasks and user tasks focused on servicing the system user functions. There are often two main levels in corporate informational systems (CIS), that respectively implement the system and application functions (for example, management). The system component of the CIS is a software and hardware base for the application layer operation and it solves the problems of forming a single universal information environment for the transmission, processing and storage of information. The application layer, implemented exclusively in the form of software applications, uses these three main services of the system level.

When developing the automated system architecture, it is necessary to ensure that the requirements of reliability, economy, scalability and configurability are simultaneously fulfilled. The system building solutions must meet the information security requirements and have a margin for expanding the system functionality and its future development.

In most cases, automated systems can be represented as a hierarchical structure, at the top of that are automated workstations, operator stations, various servers and services, whose tasks are integrated throughout the system, centralized collection and comprehensive data analysis, changing the system operating modes of the system as a whole thing. At the same time, the top-level subsystem can be both centralized and distributed, using all modern information technologies, providing different access levels for various categories of personnel working with the system. To transfer information, global or corporate networks can be involved, that can reserve each other.

The greatest complexity, and therefore, greatest interest are distributed systems, in which the automation object consists of a set of local subsystems located at large distances from each other. An example is the power supply system of the railway, which relates to the task of monitoring and technical metering of electricity at the traction substations inputs and railway contact network feeders in order to detect current and power imbalance and to carry out measures to improve the electricity supply system energy efficiency. System peculiarity needs accurate synchronous measurements at objects remotely located from each other.

In distributed automated systems, a typical solution is to use an intermediate level—process controllers or industrial computers that collect data and real-time control on local sites of the automation object and perform a time synchronization of all devices in the system. The lower system level consists of single measuring devices and actuators.

Now, there are two main ways that should be used when building the architecture of an automated system: providing full determinism and providing the possibility of system self-organization. In most cases when working with technological objects, self-organization is justified only at the lower system level, while the structure and mode of operation of the upper two levels remains stationary. For example, there is a flexible automated robotic factory or warehouse, on the lower level of which, on the scale of workshops or rooms, individual operations are performed by robots not according to a predetermined algorithm, but in accordance with the constructed object-oriented or event model. Such a system is characterized by the mobility of lower level devices, the use of wireless data transmission technologies, dynamic changes in the data network configuration and parameters.

When inspecting and performing diagnostic activities at automated facilities, it may be necessary to install and dismantle measuring and other low-level technological devices quickly. For example, it is necessary for monitoring and examining the operating modes of devices in a distributed power supply system. In this case, the application of wireless data transmission technologies with the quick self-tuning of the information network possibility will be optimal.

IOP Conf. Series: Journal of Physics: Conf. Series **1050** (2018) 012050 doi:10.1088/1742-6596/1050/1/012050

The following general requirements for the information system for monitoring and accounting of resources can be signed out:

- the possibility of expanding the functionality, flexibility and infrastructure controllability, scalability (without stopping the work of the already configured system);
- simple development and configuration; reliability and fault tolerance of the system, which are achieved due to backups and independent redundancy;
- real-time operation of the entire transport and server infrastructure;
- guaranteed transparent access to the information 24 hours a day, 7 days a week, 365 days a year;
- a high degree of the system software components integration. For example, this is achieved using a unified configuration database and the same type of data transfer protocols;
- low response time, adequate bandwidth, no bottlenecks and traffic isolation without additional delays;
- accurate time synchronization on a system-wide scale.

3. General structure of the information system and its features

From the point of view of the main functional tasks, the system for monitoring and accounting of resources should ensure the collection of information from data sources (meters), intermediate information processing and storage, and further centralized storage and provision of information to users of the system. In accordance with this functional tasks set, the structure of the system consists of four main levels (Figure 1).

The lower system level—Data measurement systems (DMS)—provides the measurement and primary transformation of information parameters, their short-term or medium-term storage in memory devices, their necessary processing (averaging, filtering, spectral analysis, event detection) and communication to the next level. In the control systems, this level includes actuators and local control systems. The article [5] describes algorithms that are applied at this level.

The second level is the Local data computing systems (LDCS) and the corresponding data acquisition and transmission devices (DATD). The main device of this level is an industrial controller (which performs the functions of local and coordinated control), or an industrial computer (which collects and stores medium-term or long-term data within a single site). The device can also handle events related to processes generally occurring on the site.

The third level—Data computing system (DCS)—provides centralized data collection from LDCS, long-term storage and data transmission to the next level.

The fourth level (external)—the main users of information—is represented by a variety of clients (automated workplaces) and service servers. The peculiarity of the interaction of this level with the DCS level is the use of standard Internet/Intranet-oriented technologies, tools and protocols for data exchange using a universal web client—browser, which allows the client part to be implemented on various software and hardware platforms. This architecture has a high degree of scalability, stability, availability, universality and unification. The latter characteristics are expanded using a service-oriented architecture (SOA), based on the use of services with standardized interfaces, for example, based on the XML language.

The connection between the first two layers of the system can be based on any available technologies: wired (including fiber-optic) and wireless (Wi-Fi, ZigBee, GSM, GSM-R, UMTS, LTE, etc.). The use of wireless technologies is associated either with the principle of mobility of devices within the local site, or with the need for a rapid deployment of the system. In case the DMS's electricity meters are installed on the railway rolling stock, they can transmit information, either via Wi-Fi or ZigBee technology, to a data concentrator located in locomotive depots (C1, C2, C3 or B1, B2, B3), or using GSM mobile operators' networks (D1, D2, D3) as shown in figure 1. In the latter case, an additional mobile data concentrator installed on the rolling stock may be required, which transmits information to the stationary concentrator located in the depot.

IOP Conf. Series: Journal of Physics: Conf. Series **1050** (2018) 012050 doi:10.1088/1742-6596/1050/1/012050

The interaction of LDCS and DCS is carried out via a wired datanet or wireless networks (GSM, GSM-R, UMTS, LTE) in the coverage area. Similar technologies and networks can also be used to implement the interaction between the DCS and the fourth level.

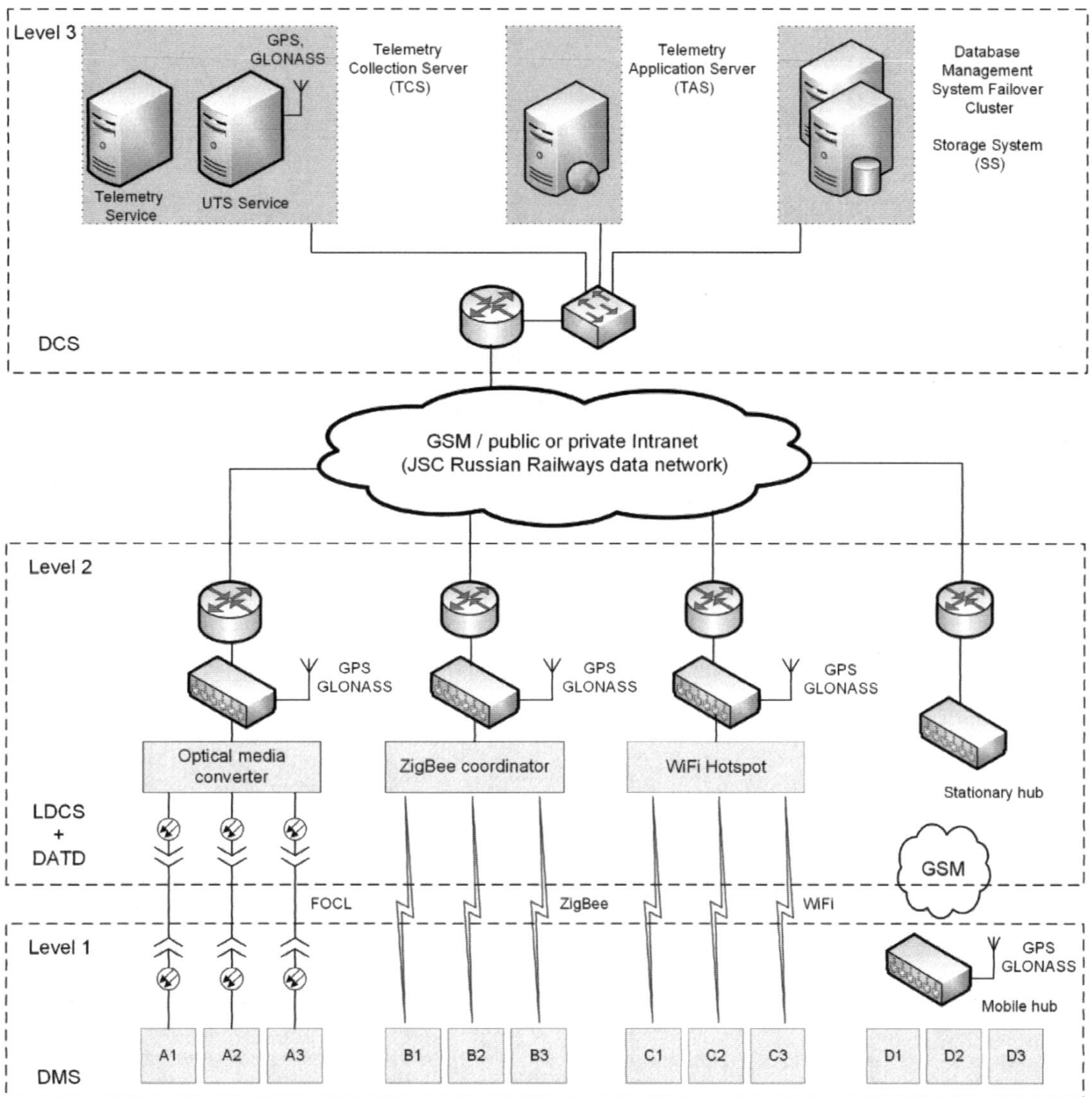

Figure 1. The distributed corporate information system structure.

4. Application the concept of intermediate software environment

The system implements the intermediate software environment (middleware) concept, which primarily provides a unified and independent mechanism for intercomponent messaging and functional tasks execution. In addition, tasks are solved such as component authentication and authorization, protection of data transferred between components, ensuring data integrity, load balancing and detection of remote components. The architecture that implements the intermediate software environment concept is shown in figure 2.

The system uses several intermediate media, that leads to the formation of a heterogeneous medium. This makes it possible to simplify the level interaction implementation considering the

IOP Conf. Series: Journal of Physics: Conf. Series **1050** (2018) 012050 doi:10.1088/1742-6596/1050/1/012050

transport level specifics and the geographical distance of the components of different levels of the system relative to each other.

Figure 3 shows a typical system architecture with an intermediate software environment, which is implemented at each system level.

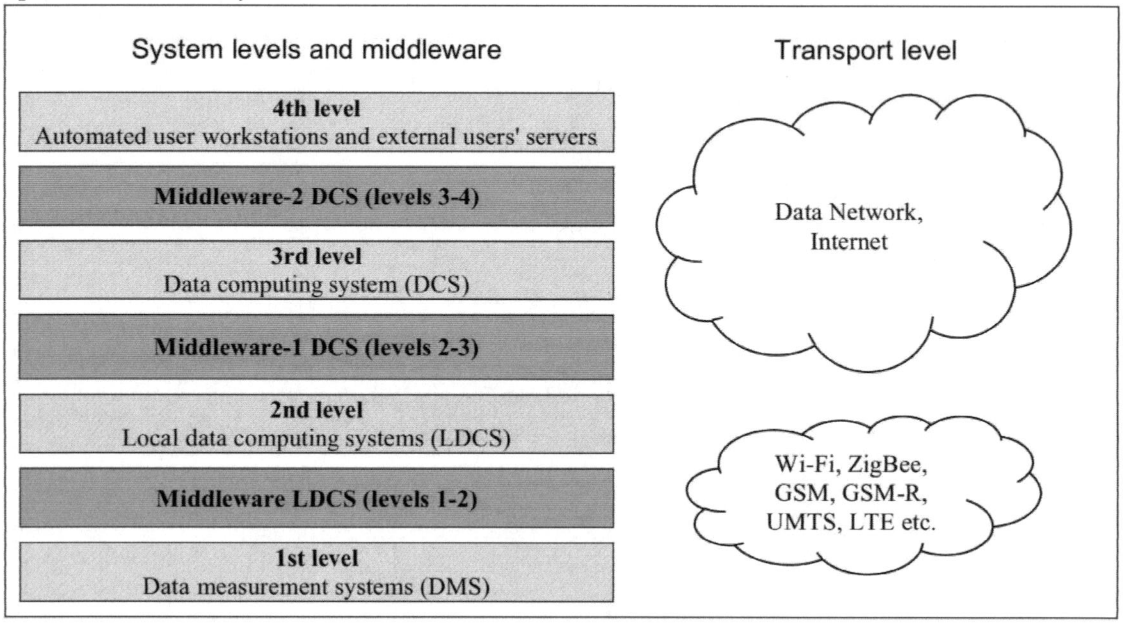

Figure 2. The architecture of a heterogeneous distributed system with intermediate software environments.

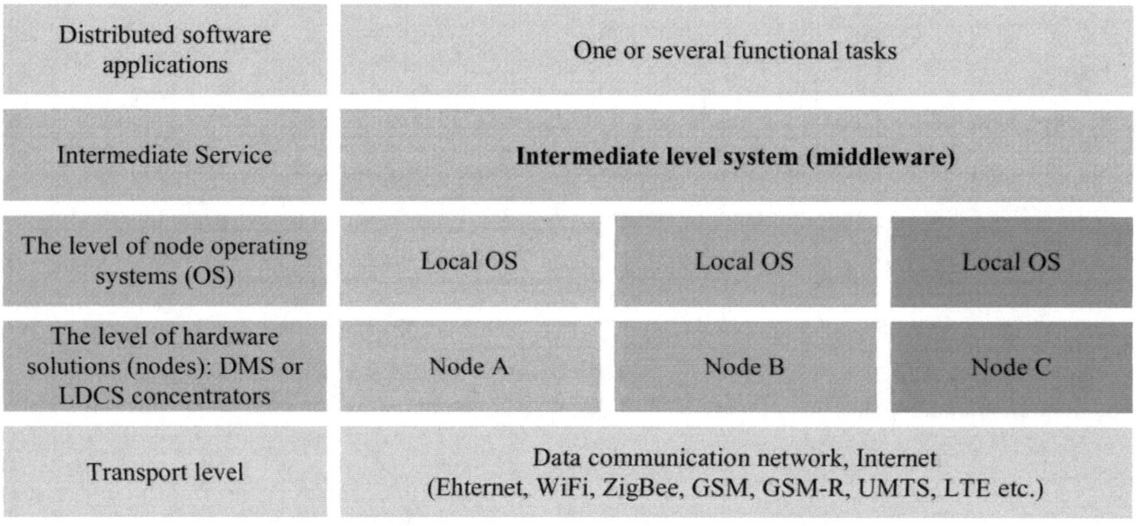

Figure 3. The architecture of a system with an intermediate software environment.

5. Technological features of the implementation the information system

During the technological architecture of the system and the DATD development, the following general principles for the data transmission channels organization were taken into account:

- transport subnets of data transmission are logically divided at the LDCS level;

- the wireless network and the core package network may have different addressing schemes. Note that the functions of the wireless network or the core package network can be physically implemented in one equipment;
- control of the availability and routing of wireless network nodes is entirely entrusted to the wireless network itself;
- the functional separation of the wireless network interfaces must have several possible options;
- interfaces should be based on the logical model of the block controlled by this interface;
- one physical network element can contain several logical blocks.

The technological architecture provides the following basic transport functions:

- transmission of information and diagnostic messages to a higher level;
- receiving command messages (control information) from a higher level;
- detection of errors in messages;
- discarding corrupted messages;
- retransmission of messages;
- information protection, including management of encryption keys;
- quality of service management.

The following specific transport functions are assigned to the LDCS level:

- sending command messages (control information) to a lower level;
- transit of command messages (control information) from a higher to a lower level;
- re-connection at the DMS-LDCS, LDCS-DCS levels;
- installation, configuration, adjustment of the wireless network.

The specific transport functions of the DCS level are:

- sending command messages (control information) to a lower level;
- obtaining control information from external administrative systems.

The basis for building transport communications of remote system elements is the OSI / ISO open systems interaction model and the TCP / IP protocol stack, which is a standard, cross-platform, low-level data exchange service. At the same time, all elements of the system have a network component of operating systems based on the socket interface, which provide low-level primitives for the byte stream direct exchange between software processes. Of the two most popular concepts for the software components interaction (messaging and procedure calling or the remote component object methods), the first one has been applied. At that, the asynchronous variant of interaction with the message queue is used. In this case, the mediator—the message queue manager is used. The program component sends a message to one of the manager queues, after which it can continue its work. Later, the receiving part will extract the message from the manager queue and begin its processing.

6. Organization of data storage

According to the domain approach [6] to describing the information systems architecture, the system under consideration belongs to the class of information management systems (IMS), which implements the process of collecting data coming from various sources, in particular from meters. The data is processed, and a variety of reports intended for different categories of users are generated on their basis. Reports can be considered when deciding on the effectiveness of the automation object.

IOP Conf. Series: Journal of Physics: Conf. Series **1050** (2018) 012050 doi:10.1088/1742-6596/1050/1/012050

Figure 4 presents the architecture of distributed storage, processing and presentation of information in the context of IMS, which contains the following main components:

- data sources (DS)—DMS meters;
- a database for storing intermediate results (DBSIR) for processing time-synchronized measurement data from the DS ("raw data");
- a data processing subsystem (PS), which processes the measurement results and prepares them for storing and report generation;
- the main database (MDB), which provides storage of monitoring and accounting data in the form suitable for report generation;
- subsystem of report generation (RG)—provides information representation in a user-friendly form;
- a set of software man-machine interfaces—provides access to reports on a regular basis, or in case of exceptional situations. In general, interfaces are used to implement operational, tactical and strategic management.

Data sources are implemented at the first level of the system, DBSIR and PS—partially on the second and third levels. The main database, the report generation subsystem and a set of software man-machine interfaces are implemented on the third and fourth levels of the system.

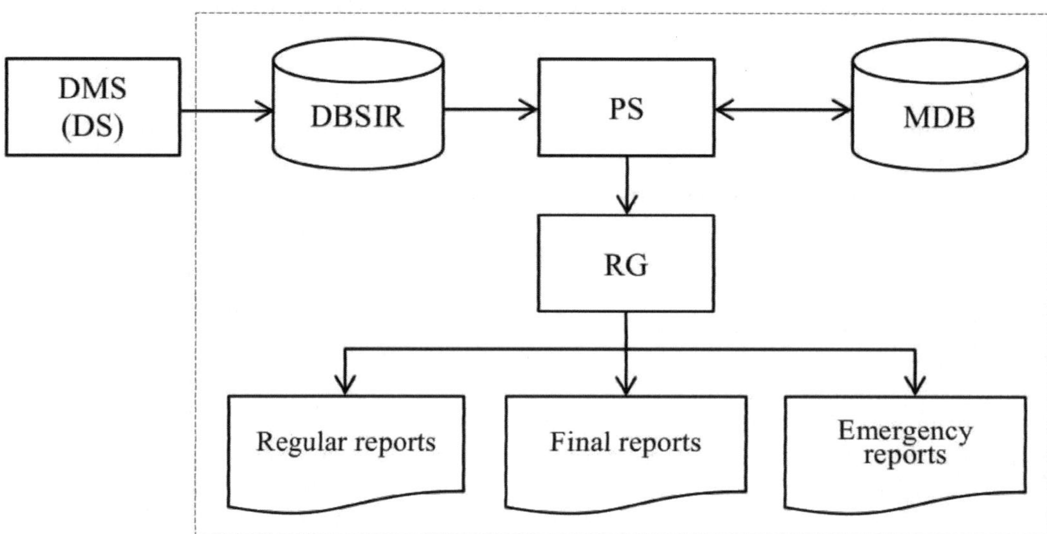

Figure 4. The architecture of distributed storage, processing and presentation of information.

DBSIR and MDB are the elements of a data architecture that includes logical and physical data stores and data management tools. The software architecture is implemented in the data processing subsystem, the report generation subsystem and the software man-machine interfaces. The technological architecture, which characterizes hardware and system software, is implemented in server equipment and DATD.

7. Decomposition of functional tasks of the data computing level

Let's consider the architectural solutions implemented at the third level of the system. Figure 5 presents the three-tier architecture of the information and computing complex, in accordance to which the functions of the DCS are divided into three layers (levels): presentation layer, business logic, data access layer. These functions are implemented in full or in part on different elements of the DCS. So a set of software man-machine interfaces is fully implemented on the clients of the DCS, and data storage is located on database management system (DBMS) servers and in the data storage system.

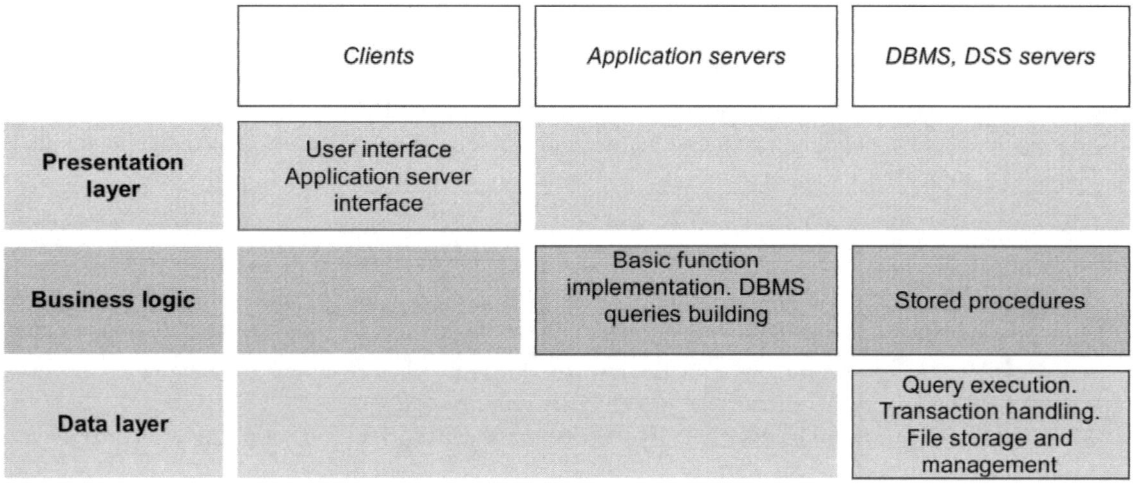

Figure 5. Three-tier architecture of the DCS.

The DCS consists of the following main elements:

- telemetry collection server (TCS);
- telemetry application server (TAS);
- database management system failover cluster and data storage system;
- access router and switch.

The TCS and TAS servers are related to application servers. They implement the core functionality of the DCS for the collection and processing of information and the report generation. The TCS (telemetry service) directly interacts with the LDCS concentrator and receives DMS telemetry data from it. It also performs time synchronization with the use of precise time signals from GPS / GLONASS satellites. The TAS manages external system users to provide access to telemetry data. For this, it includes an FTP-service and a web-service that provide access to information via FTP and HTTP protocols. The proxy service provides regulated access to services, monitoring and balancing of application traffic.

The DCS provides the solution for the following tasks:

- automatic and / or on-demand collection and storage of measurement results from all LDCS included in the system;
- logging of events from all LDCS;
- databases backup and copying of their archive to an external storage;
- automatic time correction in the DCS and their time synchronization with a common calendar time;
- automatic transmission of measurement results and status of measuring instruments to the DCS and other interested entities, provision of regulated access to visual, printed and electronic data to users and operating personnel;
- mechanical and software protection against unauthorized access to the DCS elements.

In the DCS, the following functional subsystems can be distinguished, which are of an auxiliary nature:

1) the network subsystem, designed to organize the interaction between the DCS components by the TCP / IP family protocols. The subsystem consists of switching and traffic routing equipment. The equipment is intended for routing and switching traffic inside the DCS and between the system elements;

2) the information security subsystem (IS), is designed to reduce the probability of confidential information leakage, unauthorized modification or loss of information (destruction, irreversible distortion with loss of meaning, failure to get the information to the addressee or block the access to it). The information security subsystem includes a set of the following tools:

- firewall means—necessary for package filtering of network traffic inside the DCS and between the system elements;
- means to detect and counter attacks—designed to detect and exclude network attack traffic from the data stream;
- means of access control of end devices to the network—designed to prevent access to a system of devices that do not comply with the security policy;
- means of centralized access control—designed for centralized control of administrator access to the system equipment;
- management tools of the IS subsystem—designed for IS resources centralized management.

3) management and monitoring subsystem—designed for system elements secure management, including centralized administration, system messages collection, statistics collection and software update.

8. Conclusion

Thus, we considered the main architectures used in the corporate information system for resources monitoring and accounting. The choice of certain architecture depends on the position in hierarchy and functions of the system element. We reviewed the current trends in the information systems development and those concepts of building that provide the requirements established for systems of similar class and purpose.

9. References

[1] Cheremisin V T, Ushakov S U, Kashtanov A L 2013 The uniform automated system of electricity metering in electric rolling stock (EASA EPS) DC *Proceedings of the Trans-Siberian Railway* Vol 3 p 108–113

[2] Cheremisin V T, Pashkov D V, Ushakov S U 2014 Automated monitoring of energy efficiency operation of the electric rolling stock of JSC "Russian Railways" *Proceedings of the Trans-Siberian Railway* Vol 3 p 87–91

[3] Chizhma S N, Lavrukhin A A, Okishev A S, Malyutin A G 2015 Principles of construction of information system of monitoring and metering on the feeders of the contact network *Proceedings of the Trans-Siberian Railway* Vol 3 p 94–104

[4] Lavrukhin A A, Malyutin A G, Okishev A S, Kildibekov A B 2016 Algorithms and structural solutions for construction the system of monitoring and electricity metering on the feeders of the contact network *Proceedings of the Trans-Siberian Railway* Vol 4 p 124–134

[5] Kuznetsov A A, Lavrukhin A A, Kuznetsova M A 2017 Schemes and problems of SCADA system for electrical energy control on railway rolling stock *Proc. Int. Conf. on Industrial Engineering, Applications and Manufacturing (ICIEAM)* (St. Petersburg, Russia)

[6] Duffy D 2004 Domain Architectures: Models and Architectures for UML Applications *Datasim Education BV, Amsterdam, Netherlands*

IOP Conf. Series: Journal of Physics: Conf. Series **1050** (2018) 012051 doi:10.1088/1742-6596/1050/1/012051

Improvement of aluminothermic welding on the basis of the experimentally-theoretical research of welding seam cooling process

A L Manakov, A D Abramov, A S Ilinykh, M S Galay [a], Je S Sidorov

Siberian Transport University, 191 Dusi Koval'chuk Street, Novosibirsk, 630049, Russia

E-mail: [a] galayms@mail.ru

Abstract. The distinctive features of welding rails in the aluminothermic way are considered. Defects appearing in the process of operation of welded rail seams are indicated. A computer simulation of the aluminothermic welded rail connection cooling process using the LVMFlow software package was carried out. Thermograms were obtained showing the temperature distribution over the cross-section of the rail during cooling. To confirm the results of computer simulation, we conducted field experiments, based on the results of which time-temperature dependences of cooling of the rolling surface of an aluminothermic welded seam were constructed. The time of the surface cooling of the welded seam after welding has been determined to reach a temperature of 600 ° C, which is necessary for grinding.

1. Introduction

The railway track today has more than 10 million welded seams [1]. The safety of movement depends on the quality of the welded seams. Welded seams are obtained by contact and aluminothermic welding methods. Contact welding is the most common on the railway network, because this type of welding is carried out by special rail welding machines. In contrast to the contact method for welding rails, aluminothermic welding is carried out manually and the quality of the welded seams in most cases depends on the level of specialist's training. However, the need for aluminothermic welding is growing, since this welding method is mobile and less expensive [1].

In recent years, there has been a trend towards the increase in the number of defective welded seams, as well as the number of seized highly defective rails (HDR) for welding (Fig. 1) [2]. The most common defects that arise during the operation of welded seams obtained by the aluminothermic welding method are fractures of welded seam and crushing of the rail head rolling surface in the weld.

Previous studies [3] have shown that the defect of the rail head crushing in the field of the weld during operation is associated, apparently, with a lower value of the weld hardness as compared to the hardness of the rail base metal. According to the measurements, the weld hardness is 220-245 HB, and the rail hardness is 340-360 HB. The cause of breaks in aluminothermic welded seams is an increased value of hardness at the border of the cast metal alloying and rail metal of about 380 HB. The increased value of hardness can be explained by the fact that at the fusion boundary of the cast metal and the rail metal, an increase in temperature occurs during welding, whereby the metal of the rail at the fusion interface heats up, the austenite grain grows and this coarse granulation persists, which leads to the increase of hardness and embrittlement.

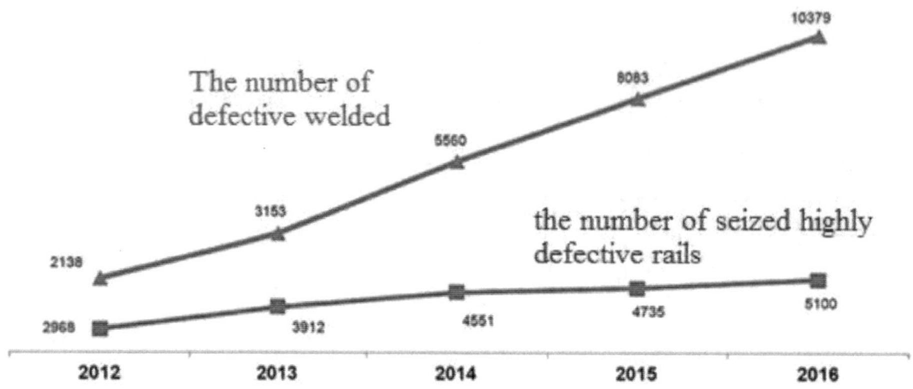

Figure 1. The number of defective and highly defective rails (HDR) for welding in the period from 2012 to 2016

The literature review showed that the problem of the defect appearance and the disclosure of their occurrence causes in welded aluminothermic seams is topical. The paper [4] noted that the cause of surface defect appearance in aluminothermic welded seams is a violation of the casting mold installation technology. Investigations aimed at improving the mechanical properties of welded seams due to the doping of a thermite powder are reported in [5-7]. The articles contain the results of heat treatment effect study [8] and surface plastic deformation [9] on the structure and properties of welded rail seams. The results of simulating the loading process of aluminothermic welded seams of rails with imperfections and porosity defects are presented in [10, 11].

2. Statement of the problem

It is possible to influence the mechanical properties of the rolling surface of the aluminothermic welded rail connection either before welding by changing the chemical composition of the foundry termite powder or after welding by post-welding machining operations. Earlier, we carried out a complex of studies on the effect of grinding performed at different residual post-weld temperatures of welded seams on hardness and roughness [12]. It has been established that the grinding of welded seams immediately after welding at welding temperatures of 850, 800, 700 and 600 °C leads both to a decrease in hardness before and to an increase in hardness. It should be noted that for today in the normative documentation there is no information on the optimum temperature of welded rail connections grinding. In practice, grinding is carried out immediately after removing the grate from the rolling surface of the rail head. According to the results of the studies presented in [12], to ensure the hardness of the welded seam close to the hardness of the base metal 340-360 HB, grinding should be carried out at a residual post-weld temperature of a welded seam of the order of 600 C. The task of this work is to determine the aluminothermic welded rail connection to a temperature of 600 ° C after welding.

3. Simulation results

Experimental studies and computer modeling were carried out to determine the holding time of the welded seam. Computer simulation was carried out using the software product LVMFlow, in which, based on the finite element method, it is possible to determine the temperature value at any point of the model and construct a temperature field in the selected section of the calculated region. In accordance with the existing technological process for the production of aluminothermic welded rail seam [13], after the completion of the thermite reaction and filling the mold with liquid metal, the mold is first held for 8-10 minutes, and then removed and started to perform mechanical processing. Therefore, for modeling using the program "KOMPAS-3D" two three-dimensional models were created. Model No. 1 consisting of two rails, a welded seam, a mold, gates and a profitable part is designed to simulate the process of cooling the weld metal from the metal pouring into the mold until the mold and excess metal are removed (Figure 1). Model No. 2 consisting of two rails and a welded seam allows measuring the change in the temperature of the metal without a mold (Fig. 2).

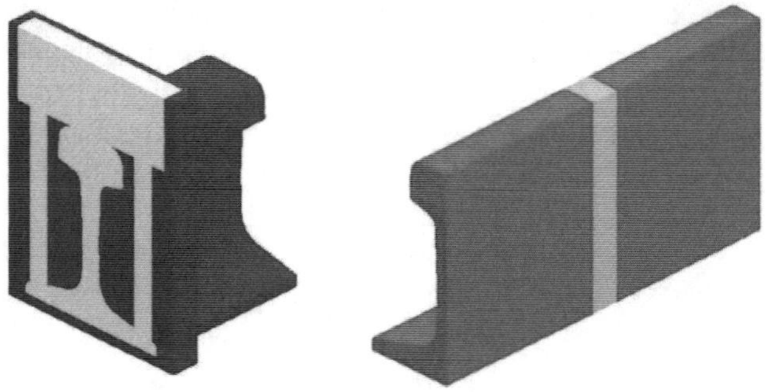

№1 №2

Figure 2. Three-dimensional models of LVMFlow

High accuracy of computations during modeling was provided due to the small size of the cells of the difference grid equal to 1 mm. Initial installations for modeling the cooling process of a welded seam with a mold (Model No. 1): steel grade - E76; the material of the form is quartz sand; the temperature of the filled metal is 2500 °C; The temperature of the rails and mold is 850 °C. Initial data for modeling the cooling of the rail seam without a mold (Model No. 2): steel grade - E76; the temperature of the weld metal is 1500 °C; the temperature of the connected rails is 1200 °C. The temperature of the weld metal was determined at certain points that were installed on the models in the cross section of the rail and along its length. The simulation results are shown in Fig. 3.

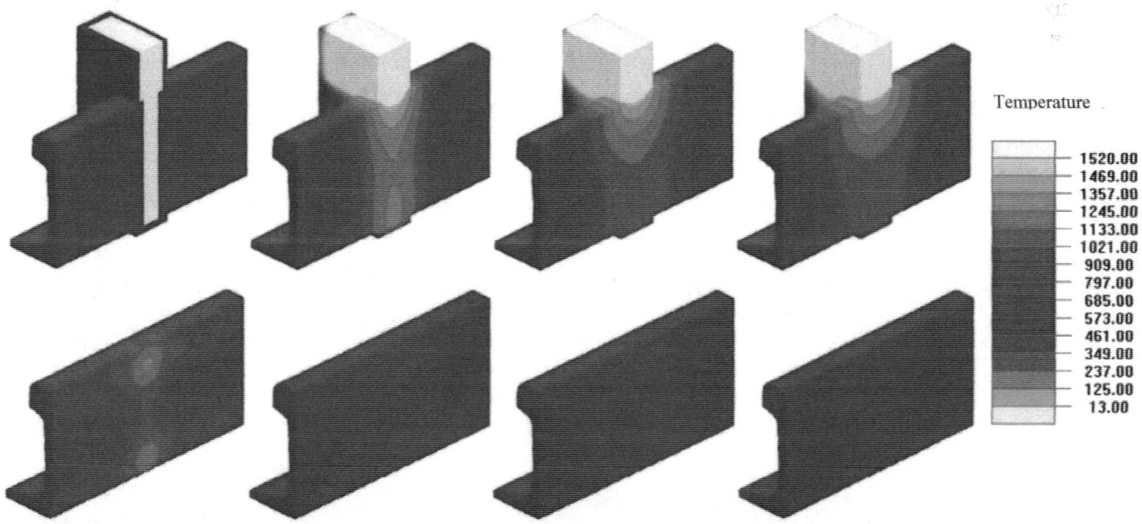

Figure 3. Results of simulating the temperature distribution of the aluminothermic welded rail connection in the process of cooling after welding

At the first stage of the simulation, the change in the temperature of the welded seam metal was estimated from the moment when molten metal was poured into the mold before the operations for removing the shape and excess metal (Fig. 3, a). The second stage consisted in determining the temperature of the welded seam metal during the post-welding treatment of the welded seam rail head (Figure 3, b).

4. Results of field experiments

After the simulation, a series of full-scale experiments was carried out to measure the surface temperature of the rolling aluminothermic welded rail seam and determine the holding time to reach a temperature of 600 ° C during cooling after welding the rails. The experimental conditions (steel grade, mold material, cast metal temperature, rails and mold temperature) are similar to those in the simulation. Measurements the surface temperature of the rolling aluminothermic welded seam were carried out with a pyrometer. In contrast to the simulation, the measuring the surface temperature of the rolling rail head was carried out only after the mold was removed, because the casting mold does not have access to the rolling surface. The results of the experiments are presented in the table.

5. Discussion of results

The results obtained in the course of the experiments and modeling are presented in the form of temperature-time dependences (Fig. 4). It should be noted that the results obtained during the experiment are consistent with the results of the simulation. The greatest drop in temperature is noted in 2 cases: when the metal is poured into the welding mold (the beginning of the crystallization of the metal) and when the welding mold is removed in the weld region.

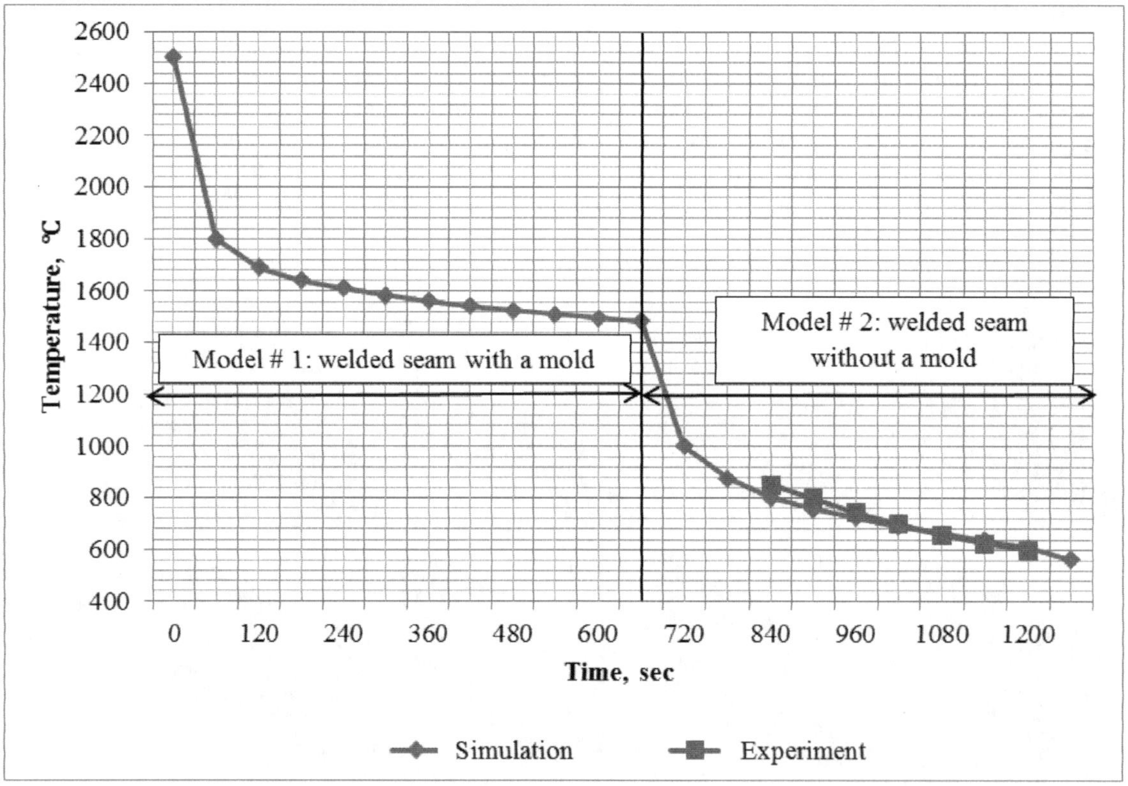

Figure 4. Temperature-time dependence upon cooling of the rolling surface of an aluminothermite welded rail connection after welding

The decrease in temperature at the end of the pouring metal into the welding mold from 2500 to 1800 °C is explained by the intense heat removal to the welded rails, which is confirmed by the thermograms in Fig. 3. When removing the casting mold, there is also a significant drop in the temperature of the metal from 1600 to 900 °C, which is associated with a decrease in the volume of the casting.

As noted earlier, grinding to increase the hardness of the rolling surface of aluminothermic welded seam should be carried out at a residual post weld temperature of welded seam of about 600 ° C. It can be seen from the graph that this temperature is reached after 1200-1300 s after pouring the metal into the welding mold.

6. Conclusions and inclusion

Experimental and theoretical studies of the cooling process of the aluminothermite welded rail seam after welding have been carried out. The results of the investigation of the temperature field distribution arising during the welding of rails in an aluminothermic way, carried out using the computer simulation of the molding processes of LVMFlow, are presented. As a result of the simulation, thermograms were obtained showing the distribution of heat in the region of the rail seam in the process of its cooling. The results of computer simulation are confirmed by the results of field experiments. The time for holding the weld seam in the mold after the welding of the rails for grinding at the metal temperature of the weld head of the welded seam 600 ° C is determined. The results of experimental and theoretical investigation of the process of cooling the welded seam in aluminum thermite welding make it possible to develop a number of measures for perfecting the organization of aluminothermic welding technology.

References

[1] Evseev S. 2017 Seam block *Gudok* **202** **(26341)** URL http://www.gudok.ru/newspaper/?ID=1393150 (accessed: 16.01.2018).

[2] Orlov S E 2016 The aluminothermic rail welding on domestic railways *Put' i putevoe hozjajstvo* **10** 26-28.

[3] Kargin V A, Tikhomirova L B , Galay M S and Kuznetsova Y S 2015 Improving service properties of welded joints produced by aluminothermic welding *Welding International* **29(2)** 155-157

[4] Yamamoto R , Terashita Y, Tatsumi M, Itoh H and Umenai K 2016 Investigation of Cause and Prevention Measures against Surface Defects on Thermit Welds *Quarterly Report of RTRI* **2** **(57)** 112-117

[5] Costea D M, Gåman M N and Dumitru G 2014 The weld aluminothermic optimization of rail track by microalloyed *Bulletin of the Transilvania University of Braşov CIBv* **7** **(56)** 285 – 288.

[6] Brânzei M and Coman T 2012 Structure Improvement of Aluminothermic Welding Joints by Using Modifiers *International Journal of Chemical, Molecular, Nuclear, Materials and Metallurgical Engineering* **6** **(8)** 745-748.

[7] Saarna M and Laansoo A 2004 Rail and rail weld testing *Proc. 4th International DAAAM Conference "Industrial engineering – innovation as competitive edge for SME"* (Tallinn, Estonia, 29-30 April 2004) pp. 217-219

[8] Tikhomirova L B, Ilinykh A S, Galay M S and Sidorov Je S 2016 The effect of normalization on the structure and mechanical properties of rails aluminotermitic welded joints *Obrabotka metallov (tekhnologiya, oborudovanie, instrumenty) – Metal Working and Material Science* **1(70)** 60-66 DOI: 10.17212/1994-6309-2016-1-60-66

[9] Kargin V A, Tikhomirova L B, Abramov A D and Galai M S 2014 Effect of vibroimpact treatment on the physical and mechanical properties on the surface layer of welded joints in rails

Welding International **28 (3)** 245-247
[10] Mouallif Z, Radi B and Mouallif I 2015 The Thermomechanical Modeling of Aluminothermic Welds Affected by Different Defects *International Journal of Engineering Research and Development* **11** 44-48
[11] Sidki F, Mouallif I, Mouallif Z and Benali A 2012 Finite element modeling of the aluminothermic welding with internal defects and experimental analysis, *Proc. International Conference on Structural Nonlinear Dynamics and Diagnosis CSNDD'2012 (MATEC Web of Conferences1)* (Marrakech, Maroc, 2012) DOI: 10.1051/matecconf/2012010000
[12] Ilinykh A S, Galay M S and Sidorov Je S 2017 Enhancing the effectiveness of the diamond metal bond instrument when grinding high-strength materials *Obrabotka metallov (tekhnologiya, oborudovanie, instrumenty) – Metal Working and Material Science* **3 (76)** 28-34 DOI: 10.17212/1994-6309-2017-3-28-34
[13] Specifications 0921-127-01124323-2005 2005 The aluminothermic rail welding methods-house of the intermediate casting Moscow VNIIZhT Publ. p 27

IOP Conf. Series: Journal of Physics: Conf. Series **1050** (2018) 012052 doi:10.1088/1742-6596/1050/1/012052

Calculation of linear technological dimensions based on the results of workpiece control

V B Masyagin[1], R I Bazhenov[2], B A Avdeyev[3], R L Artykh[1], D D Primak[1] and I A Volkov[1]

[1]Omsk State Technical University, 11, Mira ave., Omsk, 644050, Russia
[2]Sholom-Aleichem Priamursky State University, Shirokaya 70a, Birobidzhan 679015, Russia
[3]Kerch State Maritime Technological University, Ordjonikidze 82, Kerch 298309, Russia

Abstract. It is necessary to provide optimal processing conditions for each particular initial workpiece, based on reliable information about the allowances at each transition, during processing on software machines. However, according to the existing methods, the required dimensions of the initial workpiece are determined as a result of the calculation, while the same dimensions can already be known as a result of the control of the workpieces already prepared, i.e., known in advance and should be taken as initial in the calculation. The article presents a method of calculation that solves the given problem, including three main points: calculation of relatively constant dimensional accuracy parameters – technological tolerances, calculation of conditionally variable size-accuracy parameters – allowances and calculation of technological dimensions, ensuring the removal of previously calculated allowances.
Keywords: linear technological dimensions, workpiece, general allowance, operational allowance.

1. Introduction

Technological dimensional calculations are an indispensable stage in the engineering process and determine the quality and economy of production [1, 2, 3, 4, 5]. At present, it is necessary to provide optimal processing conditions for each particular initial workpiece, based on reliable information about the allowances at each transition, during processing on software machines. However, according to the existing methods, the required dimensions of the initial workpiece are determined as a result of the calculation, while the same dimensions can already be known as a result of the control of the workpiece already prepared, i.e., known in advance and should be taken as initial in the calculation.

A method of calculation of linear technological dimensions was proposed, in which the dimensions of the workpiece were taken as initial data in determining the technological dimensions [6]. However, according to this method, the dimensions of the workpiece are determined by the standards for the design of workpiece, i.e. it is assumed that the dimensions of the workpiece must have significant tolerances in comparison with the tolerances of the dimensions of the workpiece as a result of their control.

Below we propose the upgraded method based on the method described in the source [6], which takes into account the requirement for preliminary calculation of the technological dimensions for each particular workpiece with known high precision dimensions based on the results of their control

IOP Conf. Series: Journal of Physics: Conf. Series **1050** (2018) 012052 doi:10.1088/1742-6596/1050/1/012052

2. Upgraded method of calculation of linear technological dimensions

We will consider the linear dimensions, these dimensions for the finished part - design, are known - 'Figure 1': $S_1 = 75_{-0,4}$ mm, $S_2 = 25_{-0,35}$ mm, $S_3 = 26,5^{+0,1}$ mm, $S_4 = 12^{+0,43}$ mm.
The method includes the following stages.
1. Control of dimensional and precision parameters of the workpiece.
All dimensions of the workpiece are controlled according to the drawing of the workpiece 'Figure 2'. In the drawing, the sizes with the departures to which the prepared workpieces should correspond are indicated. However, each particular workpiece has its dimensions with the departures, which can be determined by high precision measuring. Let the measurement results be the following (see 'Figure 2'): $A_{0.1}$ = 19,6±0,1 mm, $A_{0.2}$ = 32,4±0,1 mm, $A_{0.3}$ =50±0,1 mm.

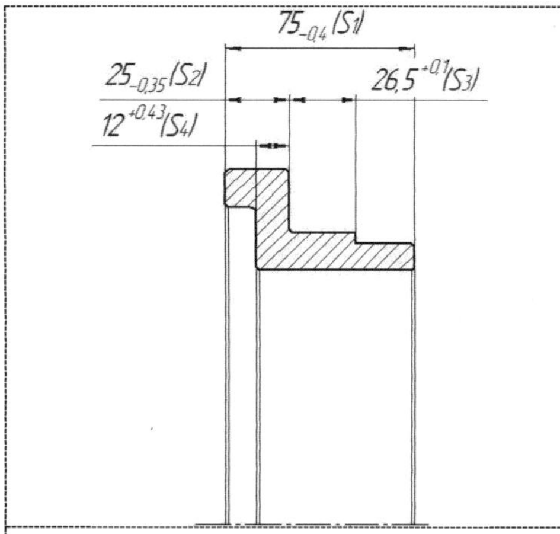

Figure 1. Detail drawing "Bushing"

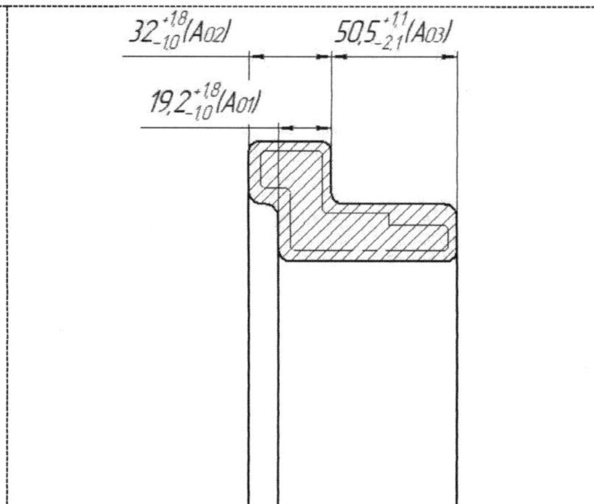

Figure 2. Workpiece drawing of the part "Bushing"

2. Determination of parameters of general processing allowances.
According to the known method [1], a processing scheme 'Figure 3' and a graph are constructed according to the technological process 'Figure 4'.
On the graph 'Figure 4' new edges are introduced - general allowances for processing Z_3, Z_7, Z_8, Z_{13}, which are shown by thickened wavy lines. The indices of the general allowances are taken from the boundary number of the finished part, on which the general allowances are located. Then, dimensional chains are detected along the graph and equations of size chains are formed, which include general allowances Z_i, the dimensions of the workpiece $A_{0.i}$, and the design dimensions of S_i.
We will calculate using the average dimension, then we compile two systems of equations. The first system of equations will include the average values of the dimensions - the general allowances Z_{im}, the dimensions of the workpiece $A_{0.im}$, and the design dimensions of S_{im}. In the second system of equations - the errors of the general allowances δZ_i, the tolerances for the dimensions of the workpiece $\delta A_{0.i}$ and the tolerances for the design dimensions δS_i.
For each general allowance, it is possible to write down three equations, into which there are two unknowns - general allowances, only 12 equations.
By eliminating repetitive pairs of general allowances, and by expressing some general allowances through other equations, these can be reduced to three equations, each of which includes the same general allowance, for example, Z_{8m}

$$
\begin{aligned}
&Z_{7m} - S_{4m} + S_{2m} + Z_{8m} - A_{0.1m} = 0, \\
&Z_{3m} + S_{2m} + Z_{8m} - A_{0.2m} = 0, \\
&Z_{13m} - A_{0.3m} - Z_{8m} - S_{2m} + S_{1m} = 0.
\end{aligned}
\tag{1}
$$

IOP Conf. Series: Journal of Physics: Conf. Series **1050** (2018) 012052 doi:10.1088/1742-6596/1050/1/012052

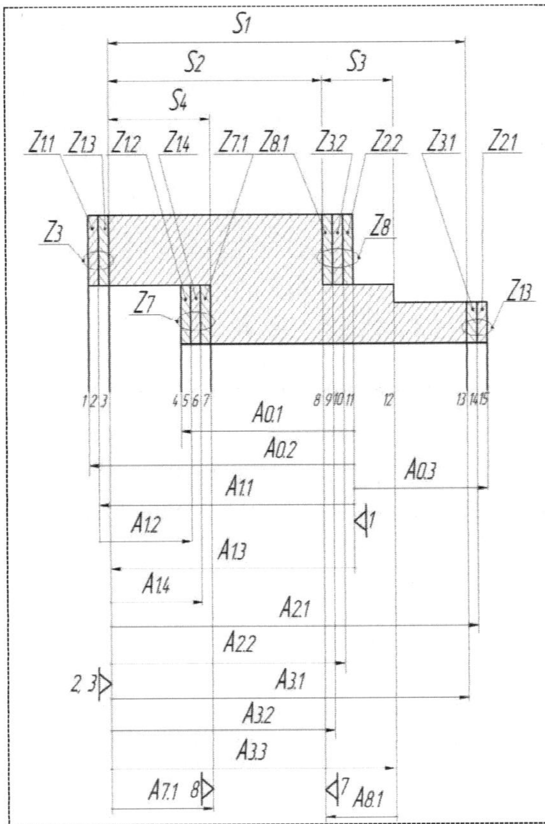

Figure 3. Processing scheme

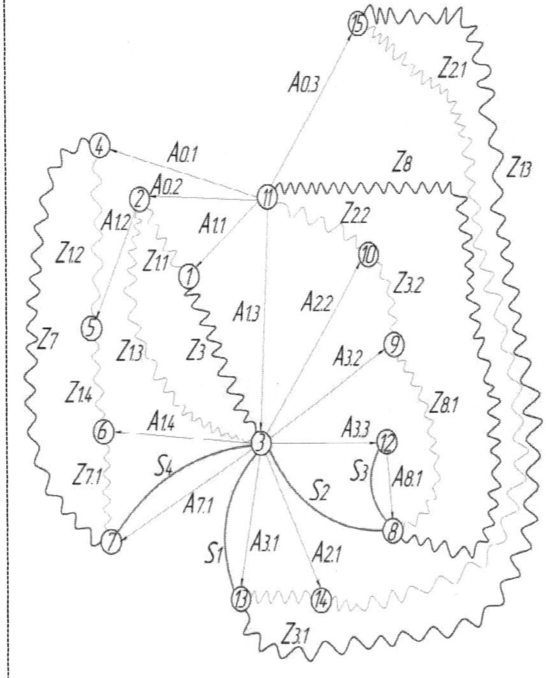

Figure 4. Graph of technological dimensional chains

From this we obtain the error equations for the general allowances δZ_i.

$$\delta Z_7 = \delta S_4 + \delta S_2 + \delta Z_8 + \delta A_{0.1},$$
$$\delta Z_3 = \delta S_2 + \delta Z_8 + \delta A_{0.2}, \quad (2)$$
$$\delta Z_{13} = \delta A_{0.3} + \delta Z_8 + \delta S_2 + \delta S_1.$$

However, the number of unknowns - four in the received systems (Z_{7m}, Z_{3m}, Z_{13m}, Z_{8m} for the first system and δZ_7, δZ_3, δZ_{13}, δZ_8 - for the second) is more than the number of equations - three for both systems, therefore for each system there is an infinite number of solutions. In order to obtain a unique solution, it is necessary to someway reduce the number of unknowns or increase the number of equations. However, the number of equations can not be increased, since their number is determined by the structure of the dimensional constraints, so it remains possible only to reduce the number of unknowns. Let's pay attention that in systems of the equations parameters of one of the general allowances (Z_{8m}, δZ_8) are present in all corresponding equations. If someway to determine these parameters, the number of unknowns will be reduced by one, and the system of equations will have unique solutions.

The general allowance Z_8 is located on the initial base (the base on the first operation), it consists of three operational allowances $Z_{2.2}$, $Z_{3.2}$ and $Z_{8.1}$. Knowing the parameters of operational allowances, you can determine the parameters of the general allowance. For a general allowance, it is possible to determine its minimum required value equal to the sum of the minimum required operational allowances, and an error equal to the sums of the expected errors in the components of the operational allowances, therefore, its average dimension and error can be determined.

Assigned to the minimum necessary values of the operating allowances on the initial database, perform their summation and determine the minimum required general allowance on the initial base (Z_{min8}) and similar values of the general allowances on the other workpiece surfaces (Z_{min7}, Z_{min3}, Z_{1min13}).
The minimally necessary values of the general allowances are equal to the sums of the minimum required operational allowances ($Z_{mini.j}$), assigned according to the reference data, depending on the number of machining of the workpiece surface (for example, at the first machining - 0,49 mm, at the second - 0,2 mm, third - 0,05 mm).

$$Z_{min8} = Z_{min2.3} + Z_{min3.2} + Z_{min8.1} = 0{,}49 + 0{,}2 + 0{,}05 = 0{,}74 \text{ mm.}$$

$$Z_{min7} = Z_{min1.2} + Z_{min1.4} + Z_{min7.1} = 0{,}49 + 0{,}2 + 0{,}05 = 0{,}74 \text{ mm.}$$

$$Z_{min3} = Z_{min1.1} + Z_{min1.3} = 0{,}49 + 0{,}2 = 0{,}69 \text{ mm.}$$

$$Z_{min13} = Z_{min2.1} + Z_{min3.1} = 0{,}49 + 0{,}2 = 0{,}69 \text{ mm.}$$

Assigning to the reference data the technological tolerances for dimensions ($\delta A_{1.3}$= 0,584, $\delta A_{2.2}$=0,210, $\delta A_{3.2}$=0,084, $\delta A_{3.3}$=0,300, $\delta A_{8.1}$=0,033), which form on the graph 'Figure 4' contours with the operational tolerances of the initial database ($Z_{2.2}$, $Z_{3.2}$, $Z_{8.1}$), after which we determine the expected errors in the operating allowances on the source database.

$$\delta Z_{2.2} = \delta A_{1.3} + \delta A_{2.2} = 0{,}584 + 0{,}210 = 0{,}794 \text{ mm.}$$

$$\delta Z_{3.2} = \delta A_{2.2} + \delta A_{3.2} = 0{,}210 + 0{,}084 = 0{,}294 \text{ mm.}$$

$$\delta Z_{8.1} = \delta A_{3.2} + \delta A_{3.3} + \delta A_{8.1} = 0{,}084 + 0{,}300 + 0{,}033 = 0{,}417 \text{ mm.}$$

Implementing their summation, we determine the error of the general allowance δZ_8 on the initial base.

$$\delta Z_8 = \delta Z_{2.2} + \delta Z_{3.2} + \delta Z_{8.1} = 0{,}794 + 0{,}294 + 0{,}417 = 1{,}505 \text{ mm.}$$

After that, we determine 'by the equations (2)' the errors of the other general allowances (δZ_7, $\delta Z3$, δZ_{13}). To calculate the 'equations (2)', you must first find the values of design tolerances and dimensions tolerances of the workpiece. These values are determined according to the dimensions indicated above: : δS_1 = 0,4 mm, δS_2 = 0,35 mm, δS_4 = 0,43 mm, $\delta A_{0.2}$ =0,2 mm, $\delta A_{0.2}$ =0,2 mm, $\delta A_{0.3}$ =0,2 mm.

$$\delta Z_7 = \delta S_4 + \delta S_2 + \delta Z_8 + \delta A_{0.1} = 0{,}43 + 0{,}35 + 1{,}505 + 0{,}2 = 2{,}485 \text{ mm.}$$

$$\delta Z_3 = \delta S_2 + \delta Z_8 + \delta A_{0.2} = 0{,}35 + 1{,}505 + 0{,}2 = 2{,}055 \text{ mm.}$$

$$\delta Z_{13} = \delta A_{0.3} + \delta Z_8 + \delta S_2 + \delta S_1 = 0{,}2 + 1{,}505 + 0.35 + 0{,}4 = 2{,}455 \text{ mm.}$$

We determine the average general allowance Z_{8m} on the initial base for the minimally necessary general allowance Z_{min8} and the tolerance of the general allowance δZ_8.

$$Z_{8m} = Z_{min8} + 0{,}5 \cdot \delta Z_8 = 0{,}74 + 0{,}5 \cdot 1{,}505 = 1{,}4925 \text{ mm.}$$

Then, 'using equations (1)', we determine the average values of the remaining general allowances. The average dimensions of S_{im} and $A_{0.im}$ entering 'into equations (1)' are calculated from the given drawing dimensions: S_{1m} = 74,8 mm, S_{2m} = 24,825 mm, S_{4m} = 12,215 mm, $A_{0.1m}$ = 19,6 mm, $A_{0.2m}$ = 32,4 mm, $A_{0.3m}$ = 50 mm.

$$Z_{7m} = S_{4m} - S_{2m} - Z_{8m} + A_{0.1m} = 12{,}215 - 24{,}825 - 1{,}4925 + 19{,}6 = 5{,}4975 \text{ mm.}$$

$$Z_{3m} = - S_{2m} - Z_{8m} + A_{0.2m} = - 24{,}825 - 1{,}4925 + 32{,}4 = 6{,}0825 \text{ mm.}$$

$$Z_{13m} = A_{0.3m} + Z_{8m} + S_{2m} - S_{1m} = 50 + 1{,}4925 + 24{,}825 - 74{,}8 = 1{,}5175 \text{ mm.}$$

We determine the actual values of the minimum general allowances and perform the verification of the received real minimum values of the general allowances for compliance with the requirement $Z_{dmin\ i} \geq$

$Z_{min\ i}$. If this requirement is not met, then it is necessary to increase the actual value of the minimum general allowance on this surface to the required minimum required value by correcting the actual minimum general allowance on the initial base.

$$Z_{dmin7} = Z_{m7} - 0{,}5 \cdot \delta Z_7 = 5{,}4975 - 0{,}5 \cdot 2{,}485 = 4{,}255 \text{ mm} \geq Z_{min7} = 0{,}74 \text{ mm}.$$

$$Z_{dmin3} = Z_{m3} - 0{,}5 \cdot \delta Z_3 = 6{,}0825 - 0{,}5 \cdot 2{,}055 = 5{,}055 \text{ mm} \geq Z_{min3} = 0{,}69 \text{ mm}.$$

$$Z_{dmin13} = Z_{m13} - 0{,}5 \cdot \delta Z_{13} = 1{,}5175 - 0{,}5 \cdot 2{,}455 = 0{,}29 \text{ mm} \leq Z_{min13} = 0{,}69 \text{ mm}.$$

The resulting negative value of the allowance Z_{dmin13} does not satisfy the requirement $Z_{dmin\ i} \geq Z_{min\ i}$. To ensure the minimum required positive value of this allowance Z_{min13} = 0,69 mm, it is required to increase the obtained value Z_{dmin13} = 0,29 mm by $\Delta Z = Z_{min13} - Z_{dmin13}$ = 0,69 – 0,29 = 0,4 mm due to the increase of the minimum general allowance on surface - the initial base to the value of Z_{dmin8} = 0,74 + 0,4 = 1,14 mm. Then the values of Z_{8m}, Z_{13m} and Z_{dmin13} increase, and the values of Z_{3m}, Z_{dmin3} and Z_{7m}, Z_{dmin7} decrease by 0.4 mm.

$$Z_{8m} = 1{,}4925 + 0{,}4 = 1{,}8925 \text{ mm}.$$

$$Z_{dmin8} = 1{,}14 \text{ mm} \geq Z_{min8} = 0{,}74 \text{ mm}.$$

$$Z_{13m} = 1{,}5175 + 0{,}4 = 1{,}9175 \text{ mm}.$$

$$Z_{dmin13} = 0{,}29 + 0{,}4 = 0{,}69 \text{ mm} \geq Z_{min13} = 0{,}69 \text{ mm}.$$

$$Z_{3m} = 6{,}0825 - 0{,}4 = 5{,}6825 \text{ mm}.$$

$$Z_{dmin3} = 5{,}055 - 0{,}4 = 4{,}655 \text{ mm} \geq Z_{min3} = 0{,}69 \text{ mm}.$$

$$Z_{7m} = 5{,}495 - 0{,}4 = 5{,}0975 \text{ mm}.$$

$$Z_{dmin7} = 4{,}255 - 0{,}4 = 3{,}855 \text{ mm} \geq Z_{min7} = 0{,}74 \text{ mm}.$$

The obtained values are satisfy the requirement $Z_{dmin\ i} \geq Z_{min\ i}$.

Based on the obtained actual values of the minimum general allowances, we calculate the actual values of the minimum operating allowances that provide the values of the actual minimum general allowances.

This calculation is carried out by distributing the above actual values of the minimum general allowances for components - the actual minimum operating allowances in accordance with the coefficient K_i, which takes into account the dependence of the amount of the overshot allowance on the side of the number of processing (table 1).

Table 1. Determination of actual minimum operating allowances.

Actual minimum general allowance			Actual minimum operating allowance			
Index	Number of processing	Value, mm	Index	Processing number	Coefficient K_i	Value, mm
Z_{dmin3}	2	4.655	$Z_{dmin1.1}$	1	5/7	3.325
			$Z_{dmin1.3}$	2	2/7	1.33
Z_{dmin7}			$Z_{dmin1.2}$	1	5/8	2.409
	3	3.855	$Z_{dmin1.4}$	2	2/8	0.964
			$Z_{dmin7.1}$	3	1/8	0.482
Z_{dmin8}			$Z_{dmin2.2}$	1	5/8	0.712
	3	1.14	$Z_{dmin3.2}$	2	2/8	0.285
			$Z_{dmin8.1}$	3	1/8	0.142
Z_{dmin13}	2	0.69	$Z_{dmin2.1}$	1	–	0.49
			$Z_{dmin3.1}$	2	–	0.2

The received actual minimum values of operational allowances are the initial data for the program for calculating the linear technological dimensions "DIAMOND" [7].

3. Determination of operating tolerances that ensure the values of errors in common allowances.

Previously, according to the reference data, technological tolerances for all technological dimensions are excluded, excluding the previously assigned tolerances for the dimensions of the workpiece ($\delta A_{0.1}$, $\delta A_{0.2}$, $\delta A_{0.3}$) and the dimensions associated with the operating allowances on the initial base ($\delta A_{1.3}$, $\delta A_{2.2}$, $\delta A_{3.2}$, $\delta A_{3.3}$, $\delta A_{8.1}$): $\delta A_{1.1}$ = 0,365 mm, $\delta A_{1.2}$ = 0,149 mm, $\delta A_{1.4}$ =0,07 mm, $\delta A_{2.1}$ = 0,30 mm, $\delta A_{3.1}$ = 0,12 mm, $\delta A_{7.1}$ = 0,027 mm.

Using by the computer program "DIAMOND" [7], the expected errors in the design dimensions ρS_i and the general allowances ρZ_i are determined with the initial process tolerances assigned by the program.

For design tolerances and errors in general allowances, which are less than the expected errors at the initial designed values of technological tolerances, the corresponding technological tolerances are provided compressed. For design tolerances and errors in general allowances, which are greater than the expected errors, i.e. allow the expansion of technological tolerances, the corresponding technological tolerances are provided expand. We get the final tolerances:

$$\delta A_{1.1} = 0{,}710\text{-}0{,}075\text{=}0{,}635 \text{ mm},$$

$$\delta A_{1.2} = 0{,}18\text{-}0{,}031\text{=}0{,}149 \text{ mm},$$

$$\delta A_{1.4} = 0{,}07\text{-}0{,}018\text{=}0{,}052 \text{ mm},$$

$$\delta A_{2.1} = 0{,}300\text{+}0{,}3355\text{=}0{,}6355 \text{ mm},$$

$$\delta A_{3.1} = 0{,}120\text{+}0{,}280\text{=}0{,}4 \text{ mm}.$$

4. Calculation of technological dimensions, including the dimensions of the workpiece.

It is carried out according to the known method by using the program "DIAMOND". The dimensions of the workpiece($A_{0.1}$ = 19,7±0,1 MM, $A_{0.2}$ = 32,5±0,1 MM, $A_{0.3}$ =50±0,1 MM) obtained as a result of calculation (table 2) can differ from the previously specified dimensions ($A_{0.1}$ = 19.6 ± 0.1 mm, $A_{0.2}$ = 32.4 ± 0.1 mm, $A_{0.3}$ = 50 ± 0.1 mm) due to the adjustment of the technological dimensions during the calculation . The final dimensions of the workpiece are determined according to the standard, together with the operational technological dimensions calculated according to the program.

Table 2. Linear technological dimensions.

Sign	Nominal value	Upper deviation	Lower deviation
$A_{0.1}$	19.700	0.100	-0.100
$A_{0.2}$	32.500	0.100	-0.100
$A_{0.3}$	50.000	0.100	-0.100
$A_{1.1}$	29.050	0.000	-0.635
$A_{1.2}$	11.900	0.149	0.000
$A_{1.3}$	27.050	0.000	-0.584
$A_{1.4}$	11.650	0.052	0.000
$A_{2.1}$	75.850	0.000	-0.635
$A_{2.2}$	25.750	0.000	-0.210
$A_{3.1}$	75.000	0.000	-0.400
$A_{3.2}$	25.250	0.000	-0.084
$A_{3.3}$	51.540	0.000	-0.300
$A_{7.1}$	12.200	0.027	0.000
$A_{8.1}$	26.550	0.033	0.000

5. Verification of ensuring minimum operating allowances.

IOP Conf. Series: Journal of Physics: Conf. Series **1050** (2018) 012052 doi:10.1088/1742-6596/1050/1/012052

It is performed by a well-known method by using the program "DIAMOND-B", designed to solve the inverse problem of dimensional calculations [8]. For the case under consideration, all actual minimum operating allowances are provided. If the actual minimum operating allowances are not provided, it is checked whether the corresponding minimum required operational allowances are ensured. If they are ensured, the calculation is completed, otherwise the actual minimum general allowances and actual minimum operating allowances are redistributed to increase the unrealized actual minimum operating allowances, and the calculation is repeated in accordance with stages 3 and 4 until the minimum required operational allowances are provided.

3. Conclusion

The presented example of calculation of linear technological dimensions includes three main points: calculation of relatively constant dimensional accuracy parameters - technological tolerances, calculation of conditionally variable size-accuracy parameters - allowances and calculation of technological dimensions, ensuring the removal of previously calculated allowances. It is possible to take into account the results of control of the initial workpiece when calculating the technological dimensions.

4. References

[1] Mordvinov B S and Ogurtsov E S 1975 Calculation of technological dimensions in the design of technological processes of mechanical processing (Omsk) p 160

[2] Britton G A and Thimm G 2002 A Matrix Method for Calculating Working Dimensions and Offsets for Tolerance Charting *International Journal of Computer Applications in Technology* (London: Springer) pp 448–453

[3] Kai W Ning H and Wenhe L 2005 Selection of cutting parameters in high speed milling of thin-walled structure components based on the machining errors control *Mechanical Science and Technology* (Guangdong: Hindawi) 24(7) pp 788–791

[4] Legoff O and Hascoet J - S 2010 Technological form defects identification using discrete cosine transform method *International Journal of Advanced Manufacturing Technology* (Malvern: Springer) pp 52–58

[5] Thimm G Britton G A and Fok S C 2004 A graph theoretic approach linking design dimensioning and process planning Part 1 Designing to process planning *International Journal of Advanced Manufacturing Technology* (London: Springer) pp 261–271

[6] Bartholomew V A and Masyagin V B 2011 Technique of calculation of linear technological dimensions on the basis of dimensional and precision data on workpiece *Omsk Scientific Bulletin. Series Devices, machines and technologies* **1(97)** pp 41–45

[7] Masyagin V B, Bushkov I A and Bartholomew V A 2010 Design Calculation of linear technological dimensions in the development of technological processes of machining *Problems in the development, manufacture and operation of rocket and space and aircraft: mater V all-Russia sci. conf., dedicated to the memory of the chief designer of PO "Polet" A S Klinyshkov* (Omsk: OmSTU) pp 169–173

[8] Masyagin V B, Bartolomey V A and Bushkov I A Verification calculation of linear technological dimensions in the design and analysis of technological processes of mechanical processing *Problems in the development, manufacture and operation of rocket and space and aircraft: mater V all-Russia sci. conf., dedicated to the memory of the chief designer of PO "Polet" AS Klinyshkov* (Omsk: OmSTU) pp 173–177

IOP Conf. Series: Journal of Physics: Conf. Series **1050** (2018) 012053 doi:10.1088/1742-6596/1050/1/012053

Comparative erosion tests results of titanium alloy Ti-6Al-4V samples obtained by using 3D - printing technology and manufactured by the traditional technological method

A F Mednikov, A B Tkhabisimov, A A Makeeva and M R Dasaev

National Research University «MPEI», Moscow, 111250, Russia

E-mail: MednikovAlF@mpei.ru

Abstract. Rotating blades of steam turbines and compressors operate in conditions of various damaging factors, including shock-dynamic impacts of liquid droplets contained in steam or in air. One of the most promising ways to reduce damage from water droplet erosion can be the application of the 3D - printing technology that will allow to replace damaged parts of structural elements, as well as cut repair costs and prolonging the service life of the equipment.

The aim of present work was to determine the possibility of using the 3D - printing technology of the turbine blades for use in power engineering. The main task of the work was to check samples of the material obtained by this method for water droplet erosion resistance, which is one of the regulated tests type at the turbine blades design stage.

This paper presented the results of water droplet erosion process experimental studies of the titanium alloy Ti-6Al-4V samples prepared by the method 3D - printing and by the traditional technological process at a collision velocity Cimp = 300 m/s with a monodisperse flow of liquid droplets with a diameter dd = 1000 μm. Erosion tests were carried out by using a set of research equipment URI «Hydroshock rig «Erosion-M» NRU «MPEI». It has been established that the erosion resistance for the incubation period duration of the Ti-6Al-4V titanium alloy samples manufactured by the 3D - printing method is 1.4 times less than for the samples manufactured by the traditional technological process. The data obtained can be used in the future in the development of complex passive methods for protection steam turbines and compressors last stages rotating blades: creation of replaceable erosion-resistant inserts by 3D - printing with a protective coating.

1. Introduction

In present, according to available data, foreign aircraft manufacturers widely use titanium alloys, such as Ti-6Al-4V [1] to create structural elements of flying devices, in particular, compressor blades of gas turbine engines [2]. These elements are subject to water droplet erosion due to the use of primary air cooling technology supplied to the compressor inlet. This is accomplished by dispensing liquid droplets, several microns in size, in order to lower the temperature of the primary air and increase the efficiency of the installation as a whole. Thus, the problem of water droplet erosion is relevant not only for power engineering (wear of turbines wet steam stages rotating blades), but also for civil and military aviation.

One of the promising ways to reduce the cost of replacing damaged parts of power equipment, such as blades of power turbines and gas turbine engines, is the use of materials obtained with the help of additive technologies [3], the use of which [1] is gaining popularity now [4]. An advantage of additive technologies is the reduction in the time of production: according to various estimates [5], the time from design to manufacturing is reduced from 2-3 years to 5-6 months [6].

Nowadays, rapidly developing additive technologies for metal products printing have wide perspectives of use not only as prototyping of machine elements, but also their subsequent replacement. Implementation of the transition to partial or complete reconstruction of the equipment critical elements with 3D - printing technology use, which have new properties compared to the traditional methods manufactured, is impossible without full research and testing of metals created by using additive technologies.

Today, the overwhelming number of works is aimed to study the strength characteristics of materials used for the manufacture of turbomachines structural elements by the methods of additive technologies, such as: tensile-compressive strength [7], bending 8], torsion and rupture under the action of corrosive media [9]. Another part of the research is devoted to the study of the sample parameters effect manufactured by the methods of additive technology [10], as well as the subsequent surface treatment on their strength characteristics.

Despite the large amount of data available, at present there is no information on the verification of such materials for water droplet erosion resistance. According to the guidance documents [11], this type of test is mandatory for any new manufacturing and/or hardening technology.

Proceeding from this, the problem arises of determining the fundamental patterns of erosive destruction of materials obtained by 3D - printing, promising both for the full-scale production of powerful steam turbines last stages blades and the first stages of gas turbine engines compressors, and for creating passive methods for their protection by using hardening of the surface and the application of "erosion-resistant" inserts on the periphery of the blades. (see figure 1b), acting as a promising replacement for stellite plates installed at the entrance edge (see figure 1a) in the zone most strongly affected by the impact of water droplet erosion.

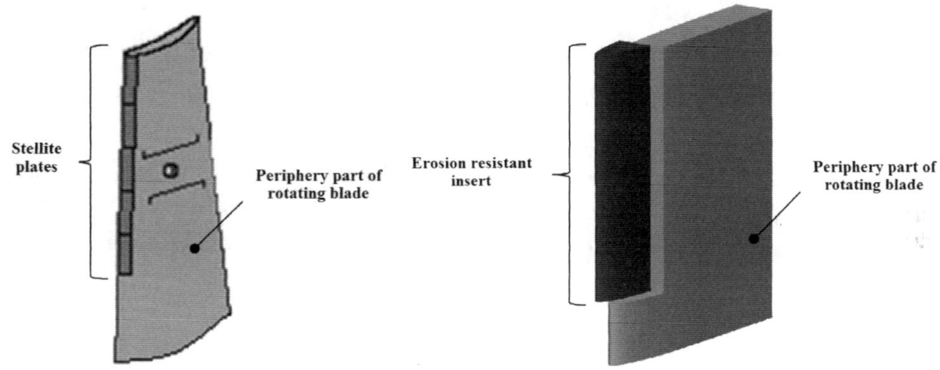

a) b)

Figure 1. Scheme of stellite plates application (a) and special erosion-resistant insert (b), manufactured by 3D - printing technology.

2. Experimental procedures

The manufacturer was provided with a previously developed sketch for production the erosion testing samples produced by a traditional technological method (Figure 2a). Company-manufacturer was provided model created at the CAD environment by the SolidWorks software package for production samples by using 3D - printing technology (Figure 2b).

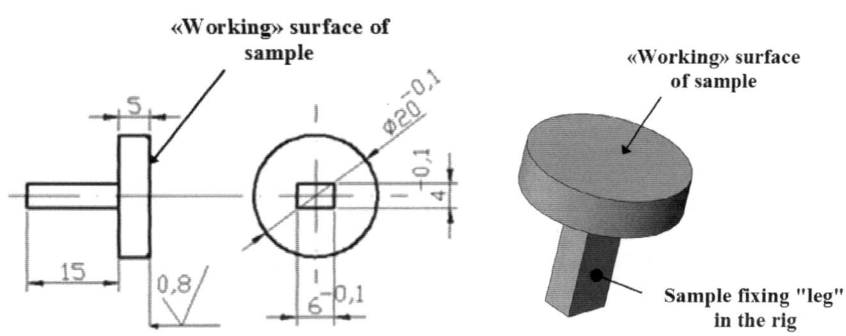

a) b)

Figure 2. Sketch of the experimental samples (a) and it's 3D - model (b).

For experimental researches were produced the samples of titanium alloy Ti-6Al-4V (this alloy contains up to 90% of the mass titanium fraction, 6.75% aluminum and 4.5% vanadium and other trace elements) using «Concept Laser Mlab_cusing» printer by means of selective laser sintering technology. The form of the «grown» samples at special printer platform is shown in Figure 3a. After manufacturing, the samples were removed from the tooling and the working surface was grinded to remove the excess material remaining after printing (see figure 3b).

a) b)

Figure 3. Titanium alloy Ti-6Al-4V samples, manufactured by the 3D - printing method.

Samples form produced by traditional technological method of the hot rolled titanium alloy Ti-6Al-4V twig (the chemical composition is presented in Table 1) by means of turning, milling and grinding is shown in Figure. 4.

Table 1. The chemical composition of the titanium alloy Ti-6Al-4V.

Element	Ti	Al	Mo	V	Zr	Fe	N	C	H	O	Si	Admixtures
Composition, %	Base	5.9	-	4.1	0.07	0.2	0.03	0.07	0.01	0.1	0.04	< 0.3

Tests of both types titanium alloy Ti-6Al-4V samples were carried out with research equipment of the URI «Hydroshock rig «Erosion-M» NRU «MPEI». Hydroshock rig – is rotary test rig that works as follows: two test samples are affixed to the ends of a titanium alloy rod rotating in a vacuum chamber and intersect a vertical liquid droplets flow emerging from a special drops generator. The samples were tested in pairs, i.e. one titanium alloy Ti-6Al-4V sample obtained by the 3D - printing method and manufactured by a conventional technological method were simultaneously tested in one experiment.

Figure 4. Titanium alloy Ti-6Al-4V samples, manufactured in the traditional way.

Experimental studies were carried out at a collision rate C_{imp} = 300 m/s of sample with droplets d_d = 1000 μm. The time for samples testing varied. Six different exposure times at least were selected to construct an erosion curve. To perform a comparative analysis of the erosion tests results, the sample mass loss Δm, (g) as a function exposure time at the test rig t, (s) was made.

The sample mass loss was calculated of according to formula:

$$\Delta m_i = m_0 - m_i$$

where, m_0 is the initial samples weight; m_i is the sample mass after experiment; i is the experiment number.

The results of the tests are presented as a function $\Delta m = f(t)$, with the display of experimental points and the relative error to determining Δm, is δm = 5%.

3. Results and discussion

After the manufacture both types of sample were made metallographic sections to reveal differences in the specimen microstructure (Figure 5).

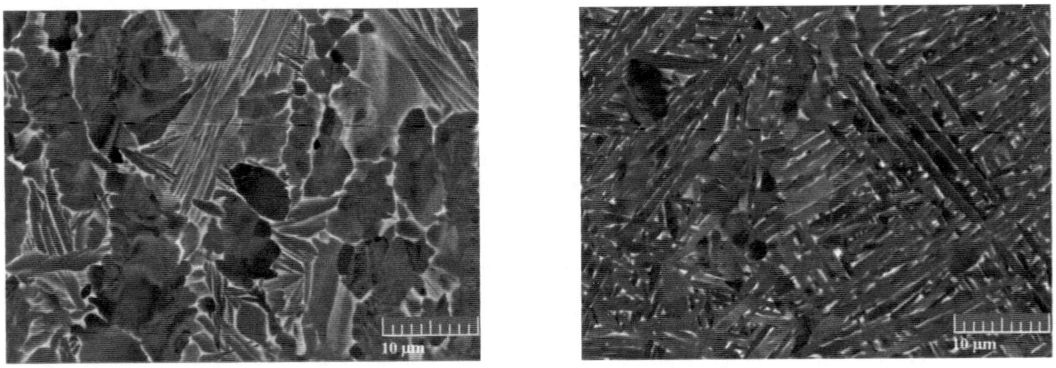

a) b)

Figure 5. Image of the titanium alloy Ti-6Al-4V sample microstructure manufactured in a conventional process (a) and with additive technology (b).

The alloy obtained by the traditional method (Figure 5a) is represented by a two-phase structure ($\alpha + \beta$). In this case, the structure is characterized by the presence of heterogeneity in the shape and size of the grains-among the crystallites of the elongated shape, it is possible to observe α - phase grains close to equiaxial. It can be assumed that such heterogeneity arose as a result of the action of the thermal deformation cycle during the manufacture or subsequent thermal treatment of the alloy. An alloy made by using additive technologies (Figure 5b) has a characteristic for this combination of material and technology disperse needle-shaped martensitic structure.

Kinetic erosion curves at $\Delta m = f(t)$ coordinates (Figure 6) were obtained as a result of the tests series of titanium alloy Ti-6Al-4V samples produced by 3D - printing and the traditional technological method at the impact samples velocity $C_{imp} = 300$ m/s with liquid droplets with the diameter $d_d = 1000$ μm.

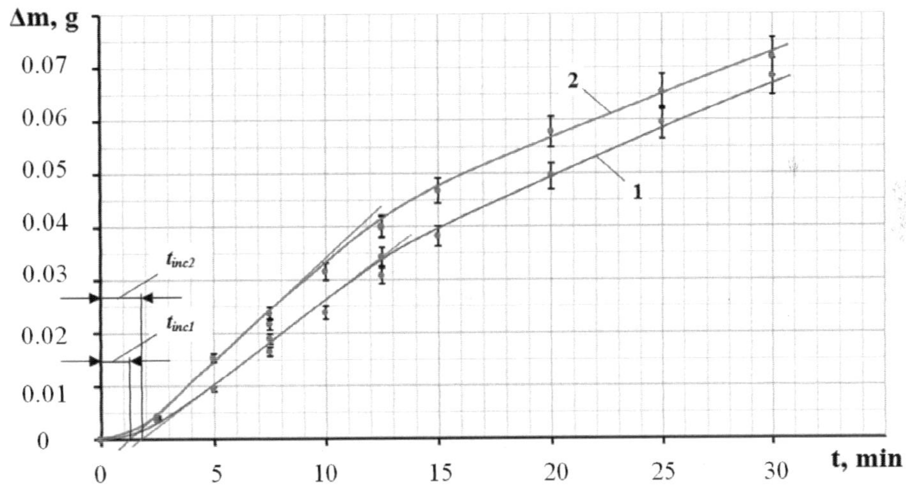

Figure 6. The erosion wear curves of titanium alloy Ti-6Al-4V samples obtained by the 3D - printing method (1) and manufactured by the traditional technological method (2) at the impact velocity $C_{imp} = 300$ m/s and a droplet diameter $d_d = 1000$ μm.

The sample surface topography change dynamics after water droplet erosion tests over time is shown in Table 2.

IOP Conf. Series: Journal of Physics: Conf. Series **1050** (2018) 012053 doi:10.1088/1742-6596/1050/1/012053

Table 2. Changes in the experimental samples surface topography after erosion tests.

Material mark	Test time, min									
	0	2.5	5	7.5	10	12.5	15	20	25	30
Ti-6Al-4V (Traditional method)	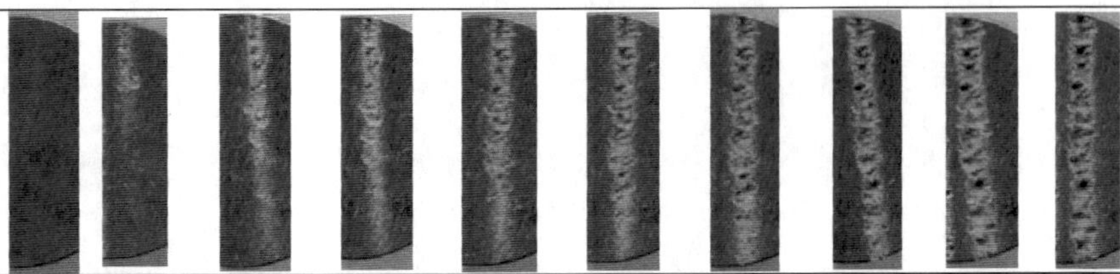									
Ti-6Al-4V (3D - printing)	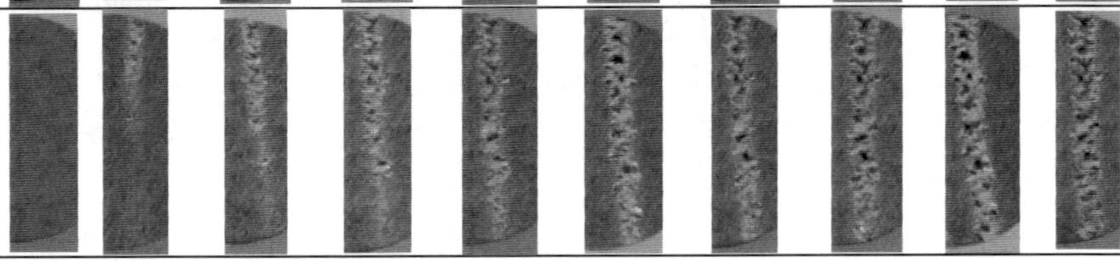									

As a result of the carried out experimental studies and subsequent processing of the obtained data, it was found that under the given experimental conditions:

– the duration of the incubation period for Ti-6Al-4V titanium alloy samples manufactured by the traditional technological method is 2 minutes;

– the duration of the incubation period for Ti-6Al-4V titanium alloy samples made by 3D - printing is 1.5 minutes.

Comparing the relative erosion resistance of titanium alloy Ti-6Al-4V samples obtained by 3D printing and manufactured by the traditional technological method was carried out according to the incubation period duration characterizing the time before the surface destruction began (see figure 7).

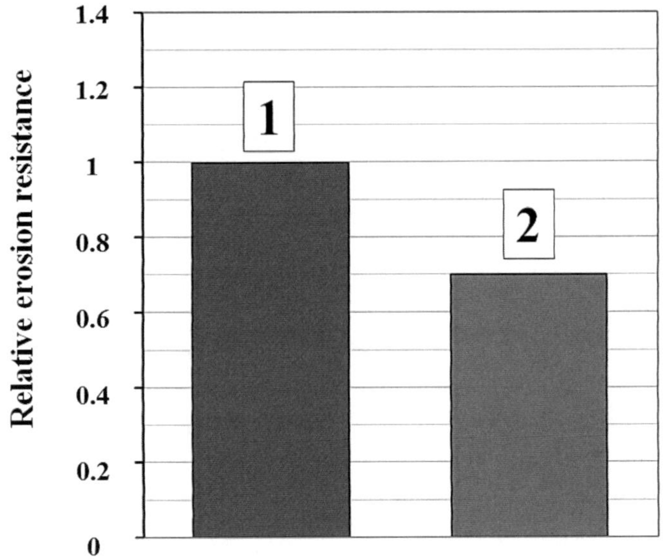

Figure 7. A bar of the relative erosion resistance over the incubation period duration of the titanium alloy Ti-6Al-4V samples manufactured by a traditional technological method (1) and produced with 3D - printing (2).

Obtained results analysis showed that the relative erosion resistance of the titanium alloy samples produced by the traditional method in the incubation period duration is higher than for titanium alloy Ti-6Al-4V samples manufactured by 3D - printing at least than in 1.4 times.

4. Conclusions

Conducted researches illustrates that it is possible to make titanium alloy Ti-6Al-4V samples of the required geometry and chemical composition by using 3D - printing technology.

The relative erosion resistance over the incubation period duration of the titanium alloy Ti-6Al-4V samples manufactured in a traditional technological way is higher than for Ti-6Al-4V titanium alloy samples manufactured using 3D - printing at least than in 1.4 times.

Determination the erosive destruction dependencies of the metal specimens created with 3D - printing technology, with both another collision speeds and droplet diameter values is a solutions for creating technological possibilities for increasing the efficiency of using various passive anti-erosion methods of protection: creating easily replaceable inserts with protective coating for the peripheral part of blades. It is possible to use inserts from a titanium alloy created by using 3D - printing technology to replace stellite plates, but additional tests are necessary.

Further carrying out erosion tests of titanium alloy Ti-6Al-4V samples manufactured by 3D - printing with various hardening options will allow to determine the optimal composition of the protective coating for this material, increasing it's resistance to the water droplet erosion and allow to realize the possibility of complex protection using.

Gratitude

The study results have been achieved thanks to financial support of the Russian Science Foundation pursuant to the Agreement No. 17-79-10462 of 01.08.2017.

References

[1] Navrotsky V, Graichen A and Brodin H Industrialization of 3D printing (additive manufacturing) for gas turbine components repair and manufacturing // *VGB PowerTech* №12, 48-52 (2015).

[2] https://www.siemens.com/innovation/en/home/pictures-of-the-future/industry-and-automation/additive manufacturing-3d-printed-gas-turbine-blades.html Breakthrough with 3D printed Gas Turbine Blades

[3] Wycisk E, Solbach A, Siddique S, Herzog D, Walther F and Emmelmanna C Effects of Defects in Laser Additive Manufactured Ti-6Al-4V on Fatigue Properties // *Physics Procedia* 56, 371-78 (2014).

[4] Zopp C, Blümer S, Schubert F and Kroll L Processing of a metastable titanium alloy (Ti-5553) by selective laser melting Ain Shams // *Engineering Journal* 8 475–79 (2017).

[5] Sufiiarova V S, Popovicha A A, Borisova E V, Polozova I A, Masayloa D V and Orlov A V The effect of layer thickness at selective laser melting // *Procedia Engineering* 174, 126 - 34 (2017).

[6] Dovbysh V M, Zabednov P V and Zlenko M A Additivnye tekhnologii i izdeliya iz metalla [Additive technologies and metal products]. *FGUP NAMI, FGUP Vneshtechnika, SPbGPU.* — Retrieved from http://nami.ru/uploads/docs/centr_technology_docs/55a62fc89524bAT_metall.pdf [in Russian].

[7] Babentsova L P and Antsiferova I V Quality and ecology of selective laser sintering technology // *MASTER`S JOURNAL* № 1, 87-92 (2017).

[8] Edwards P and Ramulu M Fatigue performance evaluation of selective laser melted Ti–6Al–4V // *Materials Science & Engineering* A 598, 327-37 (2014).

[9] Konečná R, Kunz L, Bača A and Nicoletto G Long fatigue crack growth in Ti6Al4V produced by direct metal laser sintering // *Procedia Engineering* 160, 69 - 76 (2016).

[10] Chandramohan P, Shepherd B, Obadele B A and Olubambi P A Laser additive manufactured Ti–6Al–4V alloy: tribology and corrosion studies // *Int J Adv Manuf Technol* 92, 3051-61 (2017).

[11] RD 153-34.1-17.462-00. Methodical instructions on the procedure for evaluating the performance of steam turbines working blades at the manufacture process, operation and repair.

[12] Mednikov A F, Ryzhenkov V A, Seleznev L I and Lebedeva A I Studying the variation of parameters characterizing the material surface during the droplet erosion incubation period //*Thermal Engineering*. 2012. V.**59**. № 5. P. 414-20.

IOP Conf. Series: Journal of Physics: Conf. Series **1050** (2018) 012054 doi:10.1088/1742-6596/1050/1/012054

Universal input of knapsack problem for groups

Alexei Mishchenko[1], **Alexander Treier**[1]

[1] Omsk State Technical University, Omsk, Russia

E-mail: alexei.mishenko@gmail.com, alexander.treyer@gmail.com

Abstract. In this work we investigate the group version of the well known knapsack problem (**KP**) in class of nilpotent groups. We show that there exists the universal input for this problem. In other word it mean that there is the two step nilpotent group G_U and parametric input I_n, where n is natural parameter, such that for any torsion free nilpotent group G and input I for **KP** there exists a natural number m such that **KP** for group G_U on input I_m is equivalent to **KP** for group G on input I.

1. Introduction

In the paper [1], A. Myasnikov, A. Nikolaev, and A. Ushakov stated a group version of the well known knapsack problem. The motivation for our research and initial results in this direction may be found in that paper. Other results in this direction may be found in [2, 3, 4].

We give a definition of knapsack problem for groups following [1]. Let G be an arbitrary group with a presentation $G = \langle X | R \rangle$. Let $g_1, \ldots, g_k, g$ be finite words in the alphabet $X \cup X^{-1}$. Then the knapsack problem for the group G is stated in the following way.

The Knapsack Problem (KP): *Given input words $g_1, \ldots, g_k, g$, decide whether there exist non-negative integers $\varepsilon_1, \ldots, \varepsilon_k$ such that the equality*

$$g_1^{\varepsilon_1} \ldots g_k^{\varepsilon_k} = g \tag{1}$$

holds in the group G.

There is also variation of this problem, termed integer knapsack problem (**IKP**), when the coefficients ε_i are arbitrary integers. In [1] shown that **IKP** is polynomial time reducible to **KP** for any group G. It is not hard to show that **KP** stated above and **KP** over positive integers are easily reduce to each other for any group G.

Note that a group G should has solvable word problem, otherwise **KP** will be undecidable in the group G.

There are several notable questions related to **KP**. One such question is that of decidability of **KP** for a specific class of groups $\mathcal{K}$. In the case when **KP** is decidable for a class $\mathcal{K}$, another natural question is how computationally hard **KP** for a class $\mathcal{K}$ is. In this regard, it is known that **KP** is decidable in polynomial time for abelian and hyperbolic groups.

In [5] shown that knapsack problem is undecidable for any torsion free group of nilpotency class two if the rank of the derived subgroup is at least 316. This result together with the fact that if **KP** is undecidable for a subgroup then it undecidable for the whole group allows us extend this result to certain classes of polycyclic groups, linear groups and nilpotent groups of higher nilpotency class (≥ 2). In [4] shown that **KP** is undecidable for a direct product of sufficiently many copies of the discrete Heisenberg group $H_3(\mathbb{Z})$. This implies that **KP** is generally undecidable for nilpotent groups.

2. Nilpotent Groups

Recall the definition and basic properties of nilpotent groups. A group G is called a nilpotent group of class c if it has a lower central series of length c:

$$G = G_1 \trianglerighteq G_2 \trianglerighteq \ldots \trianglerighteq G_c \trianglerighteq G_{c+1} = \{1\},$$

where $G_{k+1} = [G_k, G]$, $k = 1, \ldots, c$ and $G_1 = G$.

Let $X = \{x_1, \ldots, x_n\}$ be a set of letters, and $G = \langle X \rangle$ be a free nilpotent group of class 2. By definition, the following identity holds for the group G:

$$\forall x, y, z \in G \; [x, [y, z]] = 1. \tag{2}$$

Using identity (2), the collection process in the group G is organized via the transformation

$$yx = xy[x, y]^{-1}, \tag{3}$$

where x, y are any elements of G. Using the equality (3), we can reduce any word g in the alphabet $X \cup X^{-1}$ to the normal form for elements of the group G:

$$g = x_1^{\alpha_1} \ldots x_n^{\alpha_n} \prod_{i<j} [x_i, x_j]^{\beta_{ij}}, \tag{4}$$

where $\alpha_i, \beta_{ij} \in \mathbb{Z}, i, j = 1, \ldots, n, i < j$ and $[x_i, x_j] = x_i^{-1} x_j^{-1} x_i x_j$.

Using (2), it is not hard to show that for any two elements a, b of the group G and $\alpha, \beta \in \mathbb{Z}$ we have the following equality:

$$[a^\alpha, b^\beta] = [a, b]^{\alpha\beta}. \tag{5}$$

3. Knapsack Problem

We stated the knapsack problem (**KP**) for groups in Introduction. Recall that the **KP** is called decidable for the class of groups $\mathcal{K}$ if there is an algorithm that for any group $G \in \mathcal{K}$ and given any input $g_1, \ldots, g_k, g$ answers the question whether or not the exponential group equation (1) has a solution in the group G. We can restrict the notion of decidability of **KP** and explore **KP** for single group or for some type of inputs of **KP**.

Let G be a free nilpotent group of class 2 and let $g_1, \ldots, g_k, g$ be presented in the form (4). Using (3) and (5), we can reduce the expression $g_1^{\varepsilon_1} \ldots g_k^{\varepsilon_k}$ to the form (4). The following proposition shown in [5]:

Proposition 3.1 *Let G be a free two-step nilpotent n-generated group. Then* **KP** *stated above for the group G is equivalent to a system of Diophantine equations with unknowns $\varepsilon_1, \ldots, \varepsilon_k$ of degree 2. Moreover, the number of linear equations in the system is not greater than n and the number of quadratic equations is not greater than $\frac{n(n-1)}{2}$.*

4. Universal Knapsack Problem

Proposition 3.1 shows that **KP** for nilpotent groups is closely related to Diophantine equations. A polynomial equation $D(x_1, \ldots, x_n) = 0$ with integer coefficients is called Diophantine. In 60-70th of the previous century, M. Davis, J. Robinson, H. Putnam, and Yu. Matyasevich proved that there is no algorithm to decide whether an arbitrary Diophantine equation has a solution

IOP Conf. Series: Journal of Physics: Conf. Series **1050** (2018) 012054 doi:10.1088/1742-6596/1050/1/012054

in integers or not. For more details we refer the reader to the book of Yu. Matiyasevich [6], which, in addition to the solution of the problem, provides a historical survey and describes a number of applications of the negative solution of Hilbert's Tenth Problem. In some cases of Diophantine equations, there exists an algorithm to decide whether the equation has a solution. In [7], C. Siegel gives an algorithm for a single Diophantine equation of degree ≤ 2.

From papers of Julia Robinson, Martin Davis, Hilary Putnam [8] and Yu. Matiyasevich [9], every recursively enumerable set W can be presented in Diophantine form:

$$x \in W \iff \exists x_1, \ldots, x_n \ P(x, x_1, \ldots, x_n) = 0, \tag{6}$$

where the variables $x_1, \ldots, x_n$ are positive integers and $P(x, x_1, \ldots, x_n)$ is a Diophantine polynomial. Since there exist recursively enumerable but non recursive sets, then there is no algorithm to decide for arbitrary Diophantine equation whether it has a solution. Moreover, if $W_1, W_2, \ldots$ is a list all recursively enumerable sets, then there is a polynomial U such that for any $k \in \mathbb{N}$

$$x \in W_k \iff \exists x_1, \ldots, x_n \ U(x, k, x_1, \ldots, x_n) = 0. \tag{7}$$

The polynomial $U(x, k, x_1, \ldots, x_n)$ has fixed degree and fixed number of variables. Such polynomial U is called a universal polynomial. J.P. Jones in [10] constructed the universal system of equations (D1)–(D37) that can be reduced to the universal single diophantine equation of degree 4 with 58 variables.

We are not aware of any published work that provides an explicit version of the universal system of equations of degree 2, so we give this system (8)–(57) in the present paper. To obtain universal system of diophantine equation (8)–(57) we take system (D1)–(D37) from [10] which has 53 variables and apply to them transformations and substitutions which are described by Jones. First of all we eliminate 17 variables by simple substitution, viz. A, b, C, D_0, M_1, Q, V, W, N_1, N_2, N_3, S_1, S_2, S_3, T_1, T_2, T_3. Then we introduce 22 new variable $\Gamma_1, \ldots, \Gamma_{22}$ to define b^2, MU, $2AB - B^2 - 1$, AC_1, c^2, c^4, $<$, Q^2, λB, GH, F^2, AE, C^2, AC, N^2, YK, $<$, PK, $\Box$, Q^3, Q^4 and c^4Q^3 correspondingly. Then we add 4 new variables $\Gamma_{23}, \ldots, \Gamma_{26}$ and 4 new equations (54)–(57) to reduce number of bindings of the same variables (see next sections for details). After applying transformations to Jones system, we obtain the following universal system of degree 2:

$$\Gamma_1 = \Gamma_{26}^2, \tag{8}$$
$$\Gamma_2 = MU, \tag{9}$$
$$\Gamma_3 = B(2\Gamma_{23} - B) - 1, \tag{10}$$
$$\Gamma_4 = \Gamma_{23}C_1, \tag{11}$$
$$\Gamma_5 = c^2, \tag{12}$$
$$\Gamma_6 = \Gamma_5^2, \tag{13}$$
$$\Gamma_8 = \Gamma_{24}^2, \tag{14}$$
$$\Gamma_9 = \lambda B, \tag{15}$$
$$\Gamma_{10} = GH, \tag{16}$$
$$\Gamma_{11} = F^2, \tag{17}$$
$$\Gamma_{12} = \Gamma_{23}E, \tag{18}$$
$$\Gamma_{13} = \Gamma_{25}^2, \tag{19}$$
$$\Gamma_{14} = \Gamma_{23}\Gamma_{25}, \tag{20}$$
$$\Gamma_{15} = N^2, \tag{21}$$

$$\Gamma_{16} = YK, \tag{22}$$
$$\Gamma_{18} = PK, \tag{23}$$
$$\Gamma_{20} = \Gamma_8\Gamma_{24}, \tag{24}$$
$$\Gamma_{21} = \Gamma_8^2, \tag{25}$$
$$\Gamma_{22} = \Gamma_6\Gamma_{20}, \tag{26}$$
$$B = 2\Gamma_1^2(2\blacksquare_z)^{5^{59}+1}, \tag{27}$$
$$D_1 = \Gamma_{24} + C_1(\Gamma_{23} - B) + \alpha\Gamma_3, \tag{28}$$
$$(\Gamma_4 - C_1)(\Gamma_4 + C_1) + 1 = D_1^2, \tag{29}$$
$$C_1 = 5^{59} + \Delta(\Gamma_{23} - 1), \tag{30}$$
$$c = 1 + (\Gamma_{26} - \varepsilon)B + g, \tag{31}$$
$$e + 2\blacksquare_z\Gamma_{26}l + 2\blacksquare_z B\Gamma_6 + \Gamma_7 = 2\blacksquare_z\Gamma_{24}, \tag{32}$$
$$l = \blacksquare_u + t(B - 2\blacksquare_z), \tag{33}$$
$$e = \blacksquare_y + m(B - 2\blacksquare_z), \tag{34}$$
$$S = g - 4\blacksquare_z^2\Gamma_{22} + l\Gamma_{24} + e(\Gamma_8 + 4\blacksquare_z\Gamma_{22}) + 2\blacksquare_z\Gamma_9(-2\blacksquare_z\Gamma_{22} + \Gamma_{20} + \Gamma_{21}), \tag{35}$$
$$T = \Gamma_{24} - 1 - (\Gamma_{26} - 1)l + (\Gamma_9 - 2\lambda\blacksquare_z)(\Gamma_{24} + \Gamma_8) + 2\blacksquare_z(B - 2)\Gamma_{21}, \tag{36}$$
$$N = 16\blacksquare_z\Gamma_{20}\Gamma_8, \tag{37}$$
$$R = S(\Gamma_{15} - N) + (T + 1)(\Gamma_{15} - 1), \tag{38}$$
$$P = 2M\Gamma_2, \tag{39}$$
$$(K - \Gamma_{18})(K + \Gamma_{18}) + \Gamma_{19}^2 = 1, \tag{40}$$
$$(2\Gamma_{25} - 2\Gamma_{16} - K)(2\Gamma_{25} - 2\Gamma_{16} + K) + \Gamma_{17} = 0, \tag{41}$$
$$K = R + 1 + h(P - 1), \tag{42}$$
$$M = RY, \tag{43}$$
$$U = \Gamma_{15}w, \tag{44}$$
$$Y = \Gamma_{15}s, \tag{45}$$
$$D = -2\Gamma_{25} - 5\gamma + \Gamma_{26}w + \Gamma_{23}(\Gamma_{25} + 4\gamma), \tag{46}$$
$$I = D + oF, \tag{47}$$
$$(D - \Gamma_{14})(D + \Gamma_{14}) + \Gamma_{13} = 1, \tag{48}$$
$$E = i\Gamma_{13} + 1, \tag{49}$$
$$(\Gamma_{12} - E)(\Gamma_{12} + E) - \Gamma_{11} + 1 = 0, \tag{50}$$
$$G = \Gamma_{23} + \Gamma_{11}(\Gamma_{11} - \Gamma_{23}), \tag{51}$$
$$H = 2R + 1 + j\Gamma_{25}, \tag{52}$$
$$I^2 + H(H - \Gamma_{10}) = 1, \tag{53}$$
$$\Gamma_{23} = \Gamma_2 + M, \tag{54}$$
$$\Gamma_{24} = 1 + \Gamma_9 - \lambda, \tag{55}$$
$$\Gamma_{25} = 2R + 1 + C_1 + \varphi, \tag{56}$$
$$\Gamma_{26} = \varepsilon + x. \tag{57}$$

Any letter symbols in system (8)–(57) are variables except $x, \blacksquare_z, \blacksquare_y, \blacksquare_u$, which are positive integer parameters of U. The constants $\blacksquare_z, \blacksquare_y, \blacksquare_u$ encode a recursively enumerable set which determines the universal system. So if we put $\blacksquare_z, \blacksquare_y, \blacksquare_u$ that encode a non-recursive set W, then there is no algorithm for any $x \in W$ to answer the question whether the equation has a solution.

The main result of this work is the construction of a universal input of **KP** based on system (8)–(57). This result is formulated as the following theorem:

Theorem 4.1 *There exists two step nilpotent free group* G_U *which has* 26 *generators and parametric input* $u_1, \ldots, u_{334}, u_D$ *of* **KP**, *where* D *is natural parameter, such that for any nilpotent torsion free group* G *and any input* $I = \{g_1^{\varepsilon_1} \ldots g_k^{\varepsilon_k}, g\}$ *there exists a natural* D_I *such that* **KP** *for group* G_U *on input* $u_1, \ldots, u_{334}, u_{D_I}$ *has solution iff* **KP** *has solution for input* I *and group* G.

Acknowledgments

The work was supported by Russian Science Foundation, grant 17-11-01117.

References

[1] Myasnikov A, Nikolaev A and Ushakov A 2015 Knapsack Problems in Groups *Mathematics of Computation* **84(292)** pp 987–1016.
[2] Frenkel E, Nikolaev A and Ushakov A 2016 Knapsack problems in products of groups *Journal of Symbolic Computation* **74** pp 96–108
[3] Lohrey M and Zetzsche G 2015 Knapsack in graph groups, HNN-extensions and amalgamated products *arXiv:1509.05957*
[4] Knig D, Lohrey M and Zetzsche G Knapsack and subset sum problems in nilpotent, polycyclic, and co-context-free groups *arXiv:1507.05145*
[5] Mishchenko A and Treier A 2017 Knapsack problem for nilpotent groups *Groups Complexity Cryptology,* **9(1)** pp 87–98
[6] Matiyasevich Yu. 1993 Hilberts Tenth Problem *MIT Press, Cambridge, Massachusetts* ISBN 0-262-13295-8
[7] Siegel C L 1972 Zur Theorie der quadratischen Formen *Nachr. Akad. Wiss. Gttingen Math.-Phys. KL II* pp 21–46
[8] Davis M, Putnam H and Robinson J 1961 The decision problem for exponential diophantine equations *Ann. of Math.* **74(3)** pp 425–436
[9] Matijasevic Ju V 1970 Enumerable sets are Diophantine *Dokl. Akad. Nauk SSSR* **191** pp 279–282 English transi.: 1970 *Soviet Math. Doklady* **11** pp 354–358
[10] Jones J P 1982 Universal Diophantine equations *Journal of symbolic logic* **47(3)** pp 549–571

IOP Conf. Series: Journal of Physics: Conf. Series **1050** (2018) 012055 doi:10.1088/1742-6596/1050/1/012055

On NP-completeness of subset sum problem for Lamplighter group

Alexei Mishchenko[1], **Alexander Treier**[1]
[1] Sobolev Institute of Mathematics Siberian Branch of the Russian Academy of Sciences Omsk, Russia
E-mail: alexei.mishenko@gmail.com, alexander.treyer@gmail.com

Abstract. In the paper "Knapsack problem for groups" A. Myasnikov, A. Nikolaev, and A. Ushakov stated a group version of well known Subset sum and Knapsack problems and many others. Actually they created a new direction of research as intersection of group theory and discrete optimization. They called the new direction as non-commututative optimization. Our work belongs to this new direction. In present paper we show that Subset sum problem is NP-complete for lamplighter group. Our result was obtained by polynomial reduction of well known Exact set cover problem which is strongly NP-complete to Subset sum problem for lamplighter group. As we expected, this result provides a convenient approach to prove NP-completeness of Subset sum problem for a wide class of groups.

1. Introduction

In present paper we investigate one of non-commutative discrete optimization problem. The main aim of the non-commutative discrete optimization is to study complexity of the classical optimization problems in non-commutative groups. For example, several classical optimization problems concerning integers transform to problems where the group of additive integers $\mathbb{Z}$ is replaced by an arbitrary group G. In the paper [1] A. Myasnikov, A. Nikolaev, and A. Ushakov stated a group version of the well known Subset sum problem, Knapsack problem and many others. The motivation for our research and initial results in this direction may be found in [1], and further results in [2, 3, 4, 5]. For reader convenience let us give some excerpt of introduction of [1]: "This direction of research is gaining in popularity in recent years. It is caused by several reasons. One of them is algorithmic reason: these are algorithmic problems which are very interesting from the computational algebra viewpoint. They unify various techniques in group theory which seem to be far apart now. On the practical level, non-commutative discrete optimization problems occur in many everyday computations in algebra, so it is important to study their computational complexity and improve the algorithms." Let us briefly describe our previous results in this direction that may be interesting to reader. In the papers [3, 5] authors indepenedtly shown that group version of Knapsack problem is undecidable for nilpotent groups and it was surprisingly since the similar by definition Subset sum problem has polynomial complexity for nilpotent groups. As stated above, we investigate the subset sum problem (SSP, see section 3 for definition, recent results and motivation to study SSP for lamplighter group). More precisely we research complexity of subset sum problem for well known lamplighter group (see section 2 for definition of lamplighter group). We show in *theorem 2* that subset sum problem is NP-complete for lamplighter group.

2. Lamplighter Group

Lamplighter group L is the wreath product $\mathbb{Z}_2 \wr \mathbb{Z}$, where $\mathbb{Z}$ is additive group of integer numbers, and $\mathbb{Z}_2$ is group isomorphic to quotient group $\mathbb{Z}/2\mathbb{Z}$. The lamplighter group L is metabelian and has standard presentation:

$$\langle a, t | a^2, [t^m a t^{-m}, t^n a t^{-n}], m, n \in \mathbb{Z} \rangle. \tag{1}$$

It is convenient for our purposes to look at the group L as semidirect product of two groups B and Z, where B is direct sum of countable many copies of $\mathbb{Z}_2$. The structure of group L gives very useful presentation for group which clarifies the word "lamplighter" in the name of group L. Let us briefly describe this presentation. To any element g of L we put the pair (b, z) where b is infinite string of ones and zeroes with finite number of ones and marked starting position, and z is the integer which denotes current position of the caret in string b. The string b is called lamp line where symbol "1" means that lamp is switched on and "0" means that lamp is switched off. The integer z is called position of lamplighter. To multiply two elements (b, z) and (c, y) of group L we need to make a right shift of the lamp line c to y positions and then take a sum of two strings b and shifted c with respect to addition in $\mathbb{Z}_2$. In other words $(b, z) \cdot (c, y) = (b + c(\to y), z + y)$.

Example. Let g, h are elements of group L and

$$g = (\ldots 01\mathbf{1}010 \ldots, 1),$$

$$h = (\ldots 10\mathbf{1}110 \ldots, 2).$$

Then

$$\begin{aligned} g \cdot h = (\ldots 01\mathbf{1}010 \ldots + \ldots 10\mathbf{1}110(\to 2) \ldots, 1 + 2) = \\ = (\ldots 01\mathbf{1}01000 \ldots + 00\mathbf{1}01110 \ldots, 3) = \\ = (\ldots 01\mathbf{0}00110 \ldots, 3). \end{aligned}$$

3. The Subset sum problem

In this paper we focus on Subset sum problem for a given group G which has solvable word problem. We formulate the subset sum problem as follows:

The subset sum problem (SSP): Given elements $g, g_1, \ldots, g_n$ of group G, decide if

$$g = g_1^{e_1} \ldots g_n^{e_n},$$

for some $e_1, \ldots, e_n \in \{0, 1\}$ in group G. Let us briefly describe previous results for Subset sum problem in groups. First of all we note that if equality problem is decidable in group G for some generating set X then SSP is decidable for group G for words presented in alphabet X. This implies that SSP is decidable for lamplighter group given in presentation (1). In [1] is shown that complexity of subset sum problem for additive group of integer numbers $\mathbb{Z}$ depends whether the set X of generators $\mathbb{Z}$ is finite or infinite. If group $\mathbb{Z}$ generated by the set $X = \{1\}$ then SSP for $\mathbb{Z}$ is in polynomial time. However, if group $\mathbb{Z}$ is generated by the set $X = 2^n | n \in \mathbb{N}$ then SSP for $\mathbb{Z}$ is NP-complete. Moreover, in [1] is shown that SSP is NP-complete in metabelian groups of finite rank $r \geq 2$, the wreath product $\mathbb{Z} wr \mathbb{Z}$, each of the Baumslag-Solitar metabelian groups $B(1, p)$, $p \geq 2$. On the other hand, in [2] is shown that SSP is in polynomial time for every finitely generated nilpotent group and for any hyperbolic group.

4. Exact Set Cover

In this section we give definition of classical discrete optimization problem which is named "exact set cover". Let $X = \{x_1, \ldots, x_n\}$ be a finite set and S be the set of subsets of X. We call the set $S* \subseteq S$ exact cover of X if the following two conditions hold:

(i) The union of all sets from S^* is equals to X;

(ii) Any two sets from S^* are disjoint.

The decision search of exact cover problem for finite set S belongs to famous R. Carp list of NP-complete problems [6]. Two sets $X = \{x1, \ldots, x_n\}$ and $S = \{A_1, \ldots, A_m\}$ are input for exact set cover problem and size of input defined as sum of numbers $n = |X|$ and $k = |S|$.

The Exact Set Cover (ESC) problem belongs to the class of strongly NP-complete problems which means that ESC is NP-complete even in the case of numeric value of the input. We note that classical Subset sum problem for integers is not strongly NP-complete and there is the pseudo-polynomial algorithm for SSP over integers. For any input X and S of exact set cover problem we define the input for subset sum problem for lamplighter group L, such that searching decision of ESC is equivalent to searching decision of SSP for group L.

5. Reduction

In this section we describe the polynomial reduction of classical Exact set cover problem to Subset sum problem over Lamplighter group. Let $X = \{1, \ldots, n\}$ and $S = \{A_1, \ldots, A_k\}$ be an input of Exact set cover problem. Denote by BS the correspondence matrix with k rows and n columns. There is 1 on the i-th row and j-th column if $j \in A_i$ and 0 otherwise. On the next step we define the extended correspondence matrix BES. To do this we mark all elements of A_i by additional index i and denote by A'_i the set of all marked elements from A_i. Denote by $S' = \{A'_1, \ldots, A'_k\}$. Finally, we put $BES = BS'$.

Example. Let $X = \{1, 2, 3, 4\}$. $A_1 = \{1, 2\}$, $A_2 = \{2, 3\}$, $A_3 = \{1, 2, 3\}$, $A_4 = \{4\}$.

$$B_S = \begin{array}{c} \\ A_1 \\ A_2 \\ A_3 \\ A_4 \end{array} \begin{array}{c} \begin{array}{cccc} 1 & 2 & 3 & 4 \end{array} \\ \begin{pmatrix} 1 & 1 & 0 & 0 \\ 0 & 1 & 1 & 0 \\ 1 & 1 & 1 & 0 \\ 0 & 0 & 0 & 1 \end{pmatrix} \end{array}, BE_S = \begin{array}{c} \\ A_1 \\ A_2 \\ A_3 \\ A_4 \end{array} \begin{array}{c} \begin{array}{cccccccc} 1_1 & 1_3 & 2_1 & 2_2 & 2_3 & 3_2 & 3_3 & 4_4 \end{array} \\ \begin{pmatrix} 1 & 0 & 1 & 0 & 0 & 0 & 0 & 0 \\ 0 & 0 & 0 & 1 & 0 & 1 & 0 & 0 \\ 0 & 1 & 0 & 0 & 1 & 0 & 1 & 0 \\ 0 & 0 & 0 & 0 & 0 & 0 & 0 & 1 \end{pmatrix} \end{array}.$$

Denote by C_i the set of columns in BE_S corresponded to element $i \in X$. Let BBE_S be the matrix which we getting from BE_S by inserting $n \cdot l$ columns filled by zeroes between C_i and C_{i+1}. Denote by $\overline{\mathbb{O}}$ zero valued vector of length $n \cdot l$.

Example. In notations of examples above, we have the following matrix BBE_S :

$$BBE_S = \begin{array}{c} \\ A_1 \\ A_2 \\ A_3 \\ A_4 \end{array} \begin{array}{c} \begin{array}{ccccccccccccc} 1_1 & 1_3 & & 2_1 & 2_2 & 2_3 & & 3_2 & 3_3 & & 4_4 \end{array} \\ \begin{pmatrix} 1 & 0 & \overline{\mathbb{O}} & 1 & 0 & 0 & \overline{\mathbb{O}} & 0 & 0 & \overline{\mathbb{O}} & 0 \\ 0 & 0 & \overline{\mathbb{O}} & 0 & 1 & 0 & \overline{\mathbb{O}} & 1 & 0 & \overline{\mathbb{O}} & 0 \\ 0 & 1 & \overline{\mathbb{O}} & 0 & 0 & 1 & \overline{\mathbb{O}} & 0 & 1 & \overline{\mathbb{O}} & 0 \\ 0 & 0 & \overline{\mathbb{O}} & 0 & 0 & 0 & \overline{\mathbb{O}} & 0 & 0 & \overline{\mathbb{O}} & 1 \end{pmatrix} \end{array},$$

where $\overline{\mathbb{O}}$ — is zero valued vector of length 32.

We continue to build an input of SSP. An input of SSP will contain two type of elements. The elements of first type $h_1, \ldots, h_k \in L$ will be corresponded with $S = \{A_1, \ldots, A_k\}$. We put $h_i = (a_i, z_i)$, where $a_i \in \oplus_{-\infty}^{+\infty} Z_2$ (lamp string) and $z_i \in Z$. Let $z_i = 0$ and a_i is i-th row of matrix BBE_S, and first element of row a_i of matrix BBE_S has position 1 as element of group L. We will assume that 1 denotes a switched on lamp, and 0 — is switched off lamp.

Example. For examples above we have the the following elements $h_1, \ldots, h_4 \in L$:

$$
\begin{array}{lcccccccccccccc}
 & & 1 & 2 & \ldots & 35 & 36 & 37 & \ldots & 70 & 71 & \ldots & 104 & \\
h_1 & = & (1 & 0 & \overline{\mathbb{O}} & 1 & 0 & 0 & \overline{\mathbb{O}} & 0 & 0 & \overline{\mathbb{O}} & 0, & \mathbf{0}), \\
h_2 & = & (0 & 0 & \overline{\mathbb{O}} & 0 & 1 & 0 & \overline{\mathbb{O}} & 1 & 0 & \overline{\mathbb{O}} & 0, & \mathbf{0}), \\
h_3 & = & (0 & 1 & \overline{\mathbb{O}} & 0 & 0 & 1 & \overline{\mathbb{O}} & 0 & 1 & \overline{\mathbb{O}} & 0, & \mathbf{0}), \\
h_4 & = & (0 & 0 & \overline{\mathbb{O}} & 0 & 0 & 0 & \overline{\mathbb{O}} & 0 & 0 & \overline{\mathbb{O}} & 1, & \mathbf{0}).
\end{array}
$$

Denote by $g_1, \ldots, g_e$ the second type of elements for input of SSP, where e — is the number of non zero columns of matrix BBES. In other words, $e = \sum_{i=1}^{k} |A_i|$ — the sum of cardinals of for all sets from S. Let $\alpha(i)$ be the element number of the of the set X which corresponds to i-th non-zero column of matrix BBES. Let $\beta(i)$ be the number of column of matrix BBES which is corresponded to i-th nonzero column of matrix BBES. The values $\alpha(i)$ and $\beta(i)$ are connected by the following formula: $\beta(i) = l \cdot n \cdot (\alpha(i) - 1) + i$.

Example. In notations above, $\alpha(1) = \alpha(2) = 1, \alpha(3) = \alpha(4) = \alpha(5) = 2, \alpha(6) = \alpha(7) = 3, \alpha(8) = 4$. $\beta(1) = 1, \beta(2) = 2, \beta(3) = 2 + 32 + 1 = 35, \beta(4) = 36, \beta(5) = 37, \beta(6) = 70, \beta(7) = 71, \beta(8) = 104$.

Define $g_i = (..010.., n)$, where position of unique 1 in the lamp line is equals to $\beta(i) - n * (\alpha(i) - 1)$. Finally we put $g = (0, n^2)$.

Example. Write down the whole input for Subset sum problem for group L which is equivalent to the posted Exact set cover problem:

$$
\begin{array}{llllccccccccccccc}
 & & & & 1 & & & \beta(3) & & & & \beta(6) & & & \beta(8) & \\
h_1 & = & (& & 1 & 0 & \overline{\mathbb{O}} & 1 & 0 & 0 & \overline{\mathbb{O}} & 0 & 0 & \overline{\mathbb{O}} & 0, & \mathbf{0}), \\
h_2 & = & (& & 0 & 0 & \overline{\mathbb{O}} & 0 & 1 & 0 & \overline{\mathbb{O}} & 1 & 0 & \overline{\mathbb{O}} & 0, & \mathbf{0}), \\
h_3 & = & (& & 0 & 1 & \overline{\mathbb{O}} & 0 & 0 & 1 & \overline{\mathbb{O}} & 0 & 1 & \overline{\mathbb{O}} & 0, & \mathbf{0}), \\
h_4 & = & (& & 0 & 0 & \overline{\mathbb{O}} & 0 & 0 & 0 & \overline{\mathbb{O}} & 0 & 0 & \overline{\mathbb{O}} & 1, & \mathbf{0}), \\
 & & & & & & & & & & & & & & & \\
g_1 & = & (& & 1 & 0 & \overline{\mathbb{O}} & 0 & 0 & 0 & \overline{\mathbb{O}} & 0 & 0 & \overline{\mathbb{O}} & 0, & \mathbf{4}), \\
g_2 & = & (& & 0 & 1 & \overline{\mathbb{O}} & 0 & 0 & 0 & \overline{\mathbb{O}} & 0 & 0 & \overline{\mathbb{O}} & 0, & \mathbf{4}), \\
g_3 & = & (& (\leftarrow 4) & 0 & 0 & \overline{\mathbb{O}} & 1 & 0 & 0 & \overline{\mathbb{O}} & 0 & 0 & \overline{\mathbb{O}} & 0, & \mathbf{4}), \\
g_4 & = & (& (\leftarrow 4) & 0 & 0 & \overline{\mathbb{O}} & 0 & 1 & 0 & \overline{\mathbb{O}} & 0 & 0 & \overline{\mathbb{O}} & 0, & \mathbf{4}), \\
g_5 & = & (& (\leftarrow 4) & 0 & 0 & \overline{\mathbb{O}} & 0 & 0 & 1 & \overline{\mathbb{O}} & 0 & 0 & \overline{\mathbb{O}} & 0, & \mathbf{4}), \\
g_6 & = & (& (\leftarrow 8) & 0 & 0 & \overline{\mathbb{O}} & 0 & 0 & 0 & \overline{\mathbb{O}} & 1 & 0 & \overline{\mathbb{O}} & 0, & \mathbf{4}), \\
g_7 & = & (& (\leftarrow 8) & 0 & 0 & \overline{\mathbb{O}} & 0 & 0 & 0 & \overline{\mathbb{O}} & 0 & 1 & \overline{\mathbb{O}} & 0, & \mathbf{4}), \\
g_8 & = & (& (\leftarrow 12) & 0 & 0 & \overline{\mathbb{O}} & 0 & 0 & 0 & \overline{\mathbb{O}} & 0 & 0 & \overline{\mathbb{O}} & 1, & \mathbf{4}), \\
 & & & & & & & & & & & & & & & \\
g & = & (& & 0 & 0 & \overline{\mathbb{O}} & 0 & 0 & 0 & \overline{\mathbb{O}} & 0 & 0 & \overline{\mathbb{O}} & 0, & \mathbf{16}),
\end{array}
$$

The notation "$\leftarrow s$" means that positions of all lamps in the lamp line is less that position of corresponded lamps for elements h_j by the number s. In other words the lamp line for element h_j is shifted by s positions to the left.

Finally we need to describe a map between decisions of Subset Sum Problem and Exact Set Cover problem. Let $E = \{\varepsilon_1, \ldots, \varepsilon_r\}$ be the decision of SSP. Then the corresponded decision of Exact Set Cover Problem will be the set $S' = \{A_{s_1}, \ldots, A_{s_t}\}$, such that $A_i \in S'$ if and only if $\varepsilon_i = 1$ in decision E of SSP for group L.

Example. Decision of ESP from examples above are sets $A3 = \{1, 2, 3\}$ and $A4 = \{4\}$. Decision of equivalent SSP are elements $h3, h4, g2, g5, g7$ and $g8$.

6. Main results

Let us formulate main results of present paper.

Theorem 1. *There is the polynomial reduction of the exact set cover problem to subset sum problem for Lamplighter group.*

Theorem 2. *Let L be the Lamplighter group. Then the subset sum problem for group L is strongly NP-complete.*

7. Conclusion

In conclusion we want to say some words about some unsuccessful naive attempts of solving our problem and some words on the class of groups that our approach can cover with NP-completnes of Subset sum problem. It is not hard to show that Subset sum problem is polynomial hard for some subgroups of L which is "simpler" than group L and thus NP-completeness of SSP for lamplighter group can't be proven by reduction to SSP for this subgroups. To demonstrate it we note that for $B < L$ (subgroup B is defined above) finding decision of SSP is equivalent to solving a finite system of linear equations over F_2 (the finite field of order 2) and it can be solved using Gauss method in polynomial time. The subgroup of integers $\mathbb{Z}$ of L is generated by $\{1\}$ and hence SSP is polynomial hard for this subgroup. On the other hand, lamplighter group is a subgroup for a wide class of solvable groups and wreath products, so our results give an answer to the question about complexity class for Subset Sum Problem for this class of groups too.

Acknowledgments

The work was supported by Russian Science Foundation, grant 14-11-00085

References

[1] Myasnikov A, Nikolaev A and Ushakov A 2015 Knapsack Problems in Groups *Mathematics of Computation*, **84(292)** pp 987–1016
[2] Frenkel E, Nikolaev A and Ushakov A 2016 Knapsack problems in products of groups *Journal of Symbolic Computation* **74** pp 96–108
[3] Lohrey M and Zetzsche G 2015 Knapsack in graph groups, HNN-extensions and amalgamated products *arXiv:1509.05957* Accepted in STACS16
[4] Konig D, Lohrey M and Zetzsche G Knapsack and subset sum problems in nilpotent, polycyclic, and co-context-free groups *arXiv:1507.05145*
[5] Mishchenko A and Treier A 2017 Knapsack problem for nilpotent groups *Groups, Complexity and Cryptology* **9 (1)** pp 87–98
[6] Karp R 1972 Reducibility Among Combinatorial Problems *Complexity of Computer Computations* (New York: Plenum) pp 85–103

Experimental studies on the mechanical properties of a woven composite material for space antenna reflector

E V Moskvichev[1] and A Yu Larichkin[2]

[1]Institute of Computational Technologies SB RAS, PO box 25515, Krasnoyarsk, 660049, Russia
[2]Lavrentyev Institute of Hydrodynamics SB RAS, 15 Acad. Lavrentyev Ave., Novosibirsk, 630090, Russia

E-mail: jugr@icm.krasn.ru

Abstract. In this paper the results of mechanical tests of a biaxial woven composite are presented. This material was used for a new design of deployable precision space antenna reflector of large aperture. The experiments included standard tensile tests at low, normal and high temperatures, compression and shear tests. An additional nonstandard bending test was performed to determine the smallest radius of curvature that a specimen could withstand without fracture. The experiments showed the perspectives for the use of the studied composite in the reflector structure. The determined mechanical properties can be used for numerical modeling of the reflector behavior.

1. Introduction

To meet the growing demands on signal quality of space antennas it is necessary to use large precision reflectors with high surface accuracy. In traditional design of deployable space antennas, the reflector is represented by a metallized mesh which requires additional tension to ensure the surface accuracy. To eliminate disadvantages of mesh reflectors, a new concept of the large 4 m diameter reflector was considered by JSC Information Satellite Systems Reshetnev [1]. According to this concept shown in figure 1 the reflector consists of rigid spokes and a thick flexible composite shell of corresponding parabolic shape. For placement in a launch vehicle, the reflector is folded into the transportation position by turning the spokes closer together and bending the composite shell in a compact form. Thus, the composite shell material needs to have an appropriate stiffness and strength to ensure the shape stability in deployed position after large bending deformations.

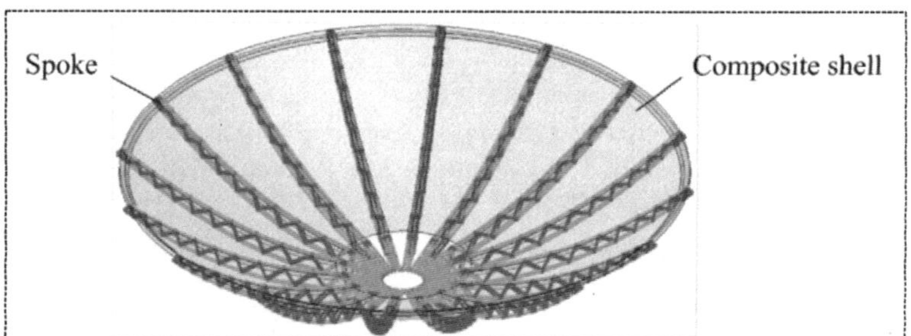

Figure 1. The space antenna reflector.

2. Composite shell material and testing methods

A biaxial carbon fiber woven composite was considered as a material having the desired properties. To fabricate the reflector shell from this material the out-of-autoclave method was used. The resulting composite shell shown in figure 2 consisted of few layers of carbon-epoxy fabric with plain weave style and the orientation angle equal to 0°. The average thickness of the composite shell was 0.5 mm.

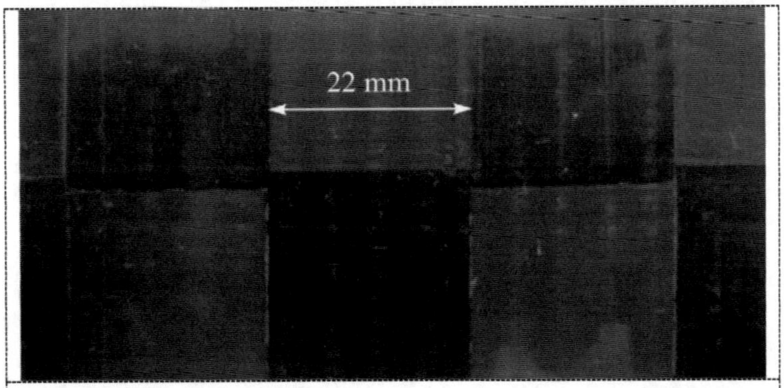

Figure 2. The specimen of composite shell material.

According to this structure, the mechanical properties of the composite shell material are equal in longitudinal and transversal directions. Due to the small thickness, the shell is in a state of plane stress. In this case the elastic properties of the reflector material can be characterized by Young's modulus, Poisson's ratio and shear modulus.

To determine the elastic and strength properties, a series of experiments were carried out on specimens of the composite shell. The experiments included standard tensile tests, compression and shear testing. The tensile tests were carried out at the temperature range from -150 to +125 °C which corresponded to the operation temperature of the reflector. The testing procedures were mostly based on commonly used national standards. At least 5 specimens were tested for each type of experiment.

To determine the smallest radius of curvature that the composite shell material can withstand in process of folding the reflector an additional nonstandard bending tests were performed. The standard three-point and four-point bending tests were not applicable because the thickness of the composite shell is too small and it can be bended to very small radius [2,3]. To perform a test the specimen was taped to the edges of two flat plates mounted on a testing machine as shown in figure 3. The plates were moved closer together making the specimen to bend between them. When the plates were sufficiently close, the specimen failed reaching the failure curvature. The radius of curvature was calculated as half of the distance between two plates.

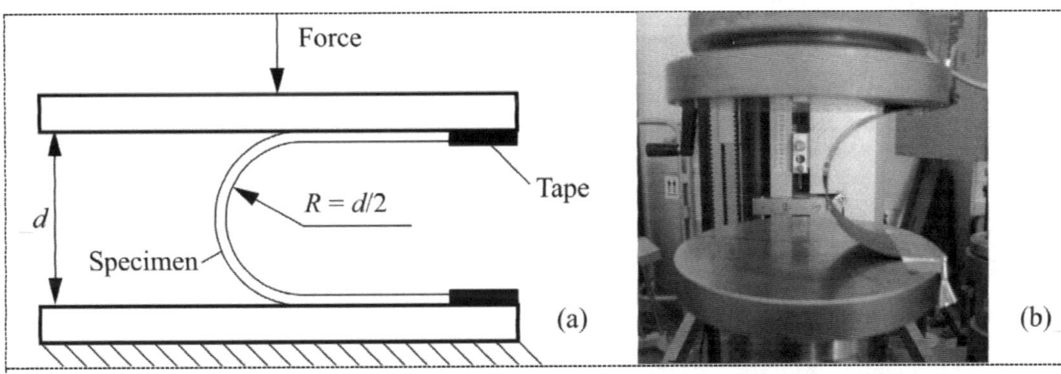

Figure 3. The bending loading scheme (a) and bending test (b).

To carry out the entire test program, three universal testing machines were used. The tension at normal and high temperatures, compression and bending tests were carried out using the Zwick/Roell Z100 electromechanical testing machine equipped with a contact extensometer and a thermal chamber. The tensile tests at low temperatures were carried out on the modified United Smart 1 SFM-300KN testing machine with an environmental chamber. The measurement of Poisson's ration and shear tests were performed on the Tinius Olsen 100ST equipped with a video extensometer and a three-rail shear test fixture.

3. Results and discussion

During the tests, the loading diagrams were recorded and processed to determine the corresponding mechanical properties. The minimum radius of curvature R_{min} was determined as the intersection point of two linear parts of the bending diagram shown in figure 4. This point corresponds to the transition of the specimen from elastic deformation to fracture.

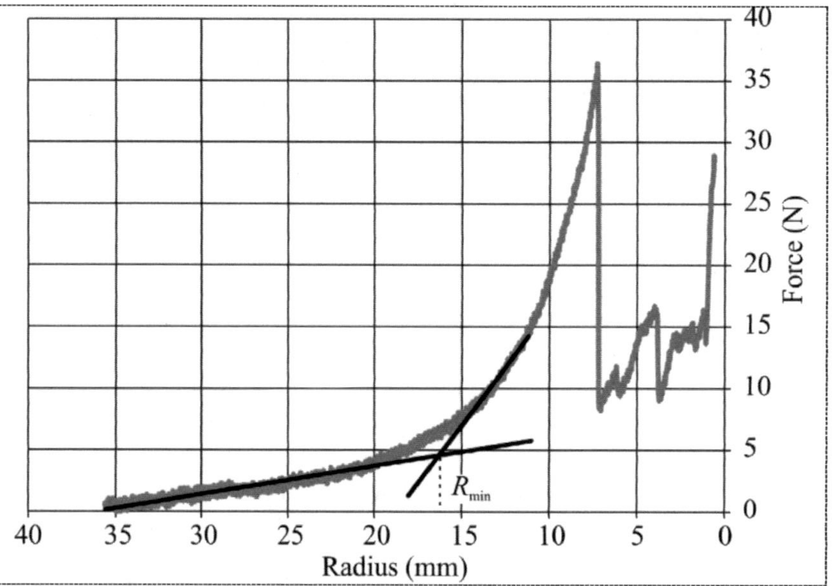

Figure 4. The loading diagram of the bending test.

The experimental results are presented in table 1. As the results show, Young's modulus and tensile strength do not have pronounced temperature dependence. Thus, it can be assumed that the reflector fabricated from the studied material is capable of stable operation under considered temperature conditions. The minimum radius of curvature limits the reflector shape in the folded position that should be taken into account both in the modeling and full-scale experiment of folding the reflector. The obtained mechanical properties can be used for refined numerical calculations of the surface accuracy of the reflector under mechanical and thermal loading [4].

Table 1. The mechanical testing results of the composite shell material.

Testing temperature (°C)	Young's modulus (MPa)	Tensile strength (MPa)	Poisson's ration	Compressive strength (MPa)	Shear modulus (MPa)	Shear strength (MPa)	Minimum radius of curvature (mm)
-150	66773	809	–	–	–	–	–
+23	64452	853	0.08	171	6151	257	16.58
+125	64951	–	–	–	–	–	–

Acknowledgements
This work was done during the complex project and was financially supported by the Russian Federation Government (Ministry of Education and Science of the Russian Federation). Contract No. 02.G25.31.0147.

References
[1] Goldobina Y L and Shalkov V V 2015 *Proc. 19th Inter. Sci. Conf. "Reshetnev readings"* vol **1** (Krasnoyarsk: Reshetnev Siberian State Aerospace University) pp 99–101
[2] Sanford G, Biskner A and Murphey T 2010 *Proc. 51st AIAA/ASME/ASCE/AHS/ASC Structures, Structural Dynamics, and Materials Conf.* (Orlando: AIAA)
[3] Kueh A, Soykasap O and Pellegrino S 2005 *Proc. Eur. Conf. on Spacecraft Structures, Materials and Mechanical Testing* (Noordwijk: ESA Publications Division, ESTEC)
[4] Moskvichev E V and Khakhlenkova A A 2017 *AIP Conf. Proc.* **1915** 040006

IOP Conf. Series: Journal of Physics: Conf. Series **1050** (2018) 012057 doi:10.1088/1742-6596/1050/1/012057

Vibration protection of a human-operator based on the application of disturbance-stimulated control mechanism

V A Nehaev[1], V A Nikolaev[1], N V Zakernichnaya[2] and S V Baglaychuk[2]

[1]Omsk State University of Railways, Omsk, Russia
[2]Military Academy of logistics behalf of the army General A. V. Khrulyov, Saint-Petersburg, Russia

Abstract. The article presents a short design algorithm for an effective vibration protection system of the human-operator of a special-purpose vehicle based on the principle of external disturbances compensation. The authors regard the results of its tests with random, deterministic and impact actions, which show that in the low-frequency range the maximum efficiency is 26 dB in acceleration and up to 46,69 dB on impact. The data are obtained which will allow to make timely changes in the design of the seat suspension of a special-purpose vehicle at the stage of its design.

1. Introduction

The critical need to ensure the functional readiness of the human operator for a variety of transport and technological vehicles (tractors, armored vehicles, helicopters, locomotives, trams, rock handlers, etc.) and conservation of his health in today's challenging real conditions of their operation, poses the urgent task of ensuring his effective protection from vibration.

As shown in works [1, 2], vibration can weaken the perception of information (e.g., visual one), which reduces the efficiency of vehicle control. Intense long-term vibration in an undesirable manner affects the spine, causing degenerative changes in the vertebrae and discs, and leads to the change of physiological functions of the human operator's body.

2. Vibration protection of a human-operator based on the application of disturbance-stimulated control mechanism

It is known that the optimal stiffness of the vibration isolation system of any object under the kinematic perturbation action is its smallest value. In the case of the suspending system implementation that has a zero dynamic stiffness, the independence of the output coordinate of the object from the kinematic perturbations induced by the motion of the vehicle over bumps, can be ensured. It is absolute invariance to external disturbance. Having some small positive resulting stiffness we call such a system as a quasiinvariant one with the accuracy to the size of ε, providing the vibration reduction, acting on the operator, to levels regulated by the requirements and by other normative-technical standard documents.

As noted in work [3], it is not always possible to achieve high up-to-date requirements to the quality of vibration protection by means of standard passive vibration isolation systems, based on the use of elastic elements and vibration absorbers. These systems cannot always provide a desired kind of the transition process, the stabilization of the object in a wide range of frequencies and so on. Recently there is a great interest to the development and research of the vibration protection system, based on the disturbance stimulated control mechanism, i.e., the compensation principle of external disturbances, ensuring the invariant character of the output coordinate to disturbance.

A necessary (but not sufficient) indicator of the feasibility of a completely invariant system is when the system has at least two channels of transmission of disturbances between the point of its application and the point of coordinate measuring, for which the invariance is achieved. The second sufficient indicator of the invariance is the insurance of physical realizability requirement. It implies the satisfaction, firstly, to the requirements of functional, dimensions and structural limitations for the values of the structural parameters of the vibration protection system, and, secondly, the equality to zero of the transfer and the weight functions of the system under t<0 (the physical realizability in the sense of Wiener) [4].

The creation of such systems is becoming possible with the introduction of high-speed computers and microprocessor technology. This way increases opportunities for development of invariant systems implemented in the form of active vibration isolation systems. The creation of such systems requires the complex of precision vibration-measuring instruments (sensors, preamplifiers, bandpass filters, etc.), analog-to-digital converters, an observer, a regulator with a fast controlling computer and executive (electrohydraulic, electromechanical, electropneumatic, etc.) devices for forming a compensating action.

A great number of Russian and foreign scholars [5 – 7] dedicate their works to the development of such systems focused on the latest achievements in the creation of microprocessor technology and computers. The creation of the quasiinvariant system for the vibration protection of the object can also be achieved without the use of external energy sources. It can be done through the introduction of the mechanical elastic system into the structural diagram of standard passive mechanical vibration isolation [8]. It has the greatest reserve of potential energy in the position of the static equilibrium and compensates the work of the dynamic response of the main elastic element with the kinematic effect on the protected object.

The idea of using such elastic systems to improve the efficiency of vibration protection of dynamic objects was first proposed by Professor P. M. Alabuzhev [9].

A block diagram of the vibration isolation system of the human operator presented in the form of a biomechanical two-mass model (one part of the body weight is calculated with the mass of the seat and the other one - with the mass of the head [10, 11]) are presented in 'Figure 1'. Here $z_м$ is input disturbance on the system of vibration isolation from the frame of the vehicle; c_p – stiffness of the main springing medium of the seat. – $c_к(\delta)$ – negative stiffness of the compensating device depending on the deflection of the suspension δ; β_p – an equivalent coefficient of viscous friction, which is also dependent on the deflection of the suspension; m_{pr} – an equivalent mass of the seat, c_h and β_h - elastic and dissipative characteristics of the human body.

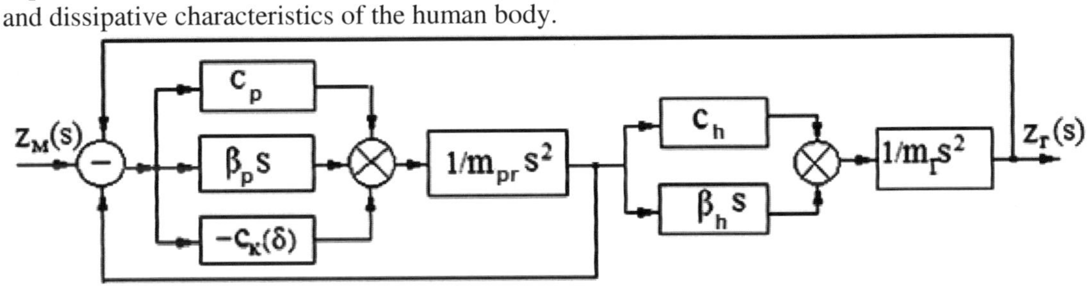

Figure 1. Structural diagram of the two-channel vibration protection system for the human operator

When designing a compensating device to ensure absolute invariance of the protected object under small deflections of the vibration isolation system it is necessary to implement the following condition [8]:

$$P_p + F_k(\delta) = 0, \tag{1}$$

where $P_p = c_p\delta$ and $F_\kappa(\delta) = -c_\kappa\delta$, accordingly, the forces formed by the basic elastic element of a suspension bracket and the compensating device: c_p and δ - rigidity and a deflection of a suspension bracket. We assume that the condition (1) is to be performed in the mean-square approximation; in this case, the criterion of finding the optimal values of the design parameters of the compensating device can be written as:

$$J = \int_{-D}^{D} [P(\delta) + F_\kappa(\delta)]^2 d\delta \rightarrow \min \tag{2}$$

where D is the set value of the deflection of the vibration isolation system, within which its maximum efficiency must be ensured under functional, dimensions and structural limitations on the values of these parameters.

As a result of the joint action of the primary and follow-up compensation device for the elastic elements of the system of vibration isolation the total power characteristic is implemented, which is approximated from the equipoise of the system by a cubic parabola in the form

$$F_\Sigma = \alpha\delta + \beta\delta^3 \tag{3}$$

Such a power characteristic is favorite for the soft limiting of dynamic deflection of the suspension and the achievement of the required quality for the protection of the human operator from vibration impacts. We created several models of the vibration isolation system for the human operator. One of the options is installed on the platform of the electrohydraulic vibration table "Karl Shenk", equipped with devices to set and control mechanical vibrations. It is part of the assembly aggregate for measuring means of vibration. The registration of acceleration is carried out with the precision vibration measuring equipment "Brüel and Kjær". The test results are presented in the form of spectrograms in 1/3-octave RMS amplitude spectrum.

At the baseline, free oscillation monitoring of the system "man–machine" is executed. For the implementation of this process, you can assess the level of dissipation in the vibration isolation system, to determine its main parameter – the frequency of free oscillations and to identify the nature of nonlinearity. The option of vibration isolation system with parameters, providing its maximum protective move, was selected for testing. Implementation of recording free oscillations for the vibroprotective seat with a man of 78 kg weight, sitting on it, is shown in 'Figure 2'. The initial displacement in the position of statistical equilibrium is 0,12 m.

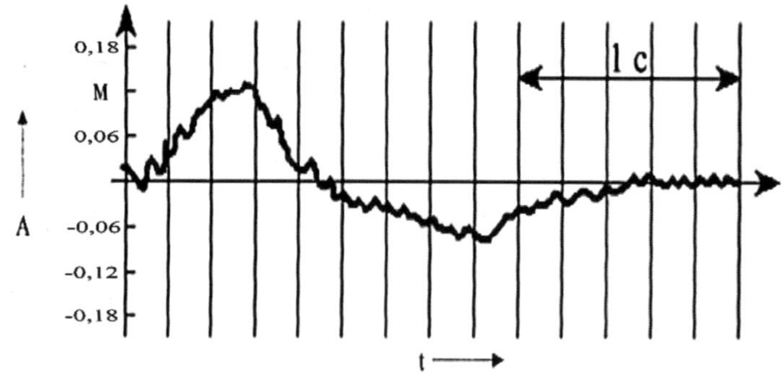

Figure 2. Free oscillations of the system "man – machine"

IOP Conf. Series: Journal of Physics: Conf. Series **1050** (2018) 012057 doi:10.1088/1742-6596/1050/1/012057

The analysis of this implementation enables to deduce that the total dynamic stiffness of the elastic suspension for the vibration isolation system is weakly nonlinear. With a decrease in the amplitude of the vibration, the natural frequency of the system vibration is reduced; the medium frequency of free oscillations for the vibration isolation system when you change the deflection from 0,12 m to zero is 0,42 Hz. To obtain such a value of the natural frequency of the lincar system, the static deflection of the elastic element must be the value of 0,7 m, which is, of course, impossible to realize because of the high dimensions restrictions defined by the ergonomic requirements. In addition, the damping of the oscillations of the object takes place practically during one oscillation. It shows that the damping is close to optimal. At the second stage of the full-scale experiment a random signal in the frequency range of 1,6 – 63,0 Hz (curve 1, 'Figure 3'), which is the implementation of one of the adverse impacts on the human operator, is applied as the disturbance input of the system "man–machine". The analysis of these results enables to conclude the following:

1. The values of the RMS accelerations in the output from the system at all frequencies do not exceed the comfort limit regulated by the standard vibration with the eight-hour working day. Therefore, a system, based on the principle of compensation of external disturbances, has large vibration protection and can operate effectively on a variety of transport and technological machines with high levels of disturbance action.
2. At the frequency of 6,3 Hz efficiency of vibration protection is 26 dB, i.e., the acceleration at the seat cushion compared to the same indicator at the input to the system is reduced by a factor of 20.

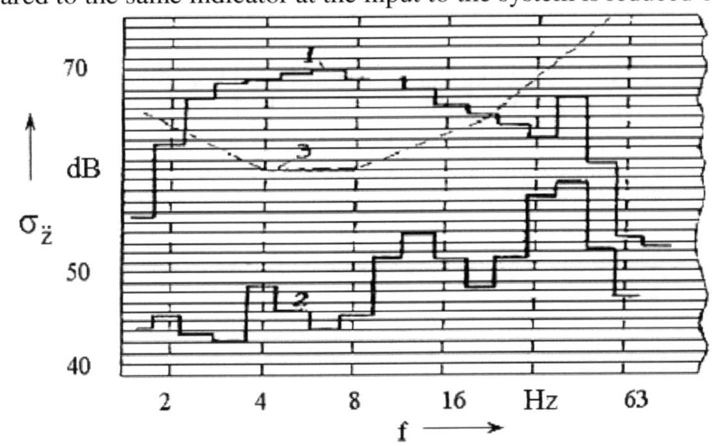

Figure 3. Spectrograms of RMS vibration acceleration at the input (1) or output (2); 3 – survivability border

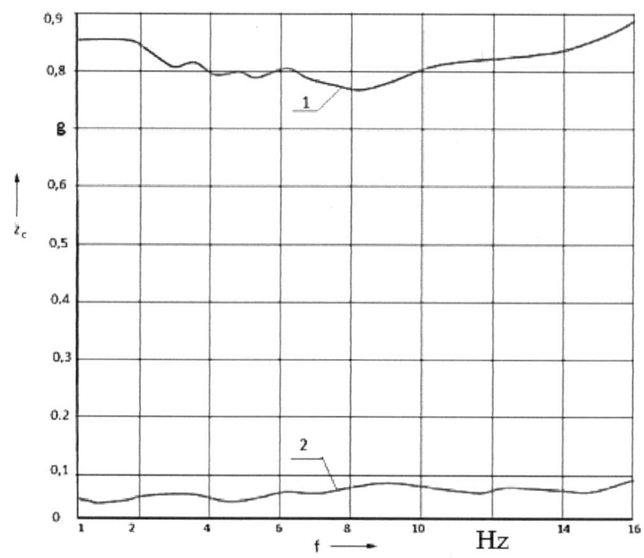

Figure 4. Accelerations in the input of vibration isolation (1) and in its output (2)

Another example of a physical model of the similar system for vibration protection of the human operator was tested on the harmonic effects at high amplitude levels of the vibration acceleration disturbance, typical of a certain class of vehicles. The test results ('Figure 4') show that the vibration isolation system effectively provides the protection of the human operator: in its output on different disturbing frequencies, the level of vibration acceleration remains minute. On the perturbation frequency of 2 Hz, accelerations reduce by a factor of 10,87. When measuring we use piezoelectric accelerometers VC 111 (the frequency range is from 0,5 to 10000 Hz, the amplitude range is ± 510 g, the axial sensitivity is 103,4 mV/g) and the spectrum analyzer ZET 017.
The test results of the system of vibration protection on the shock impact at different levels of step action of the base are given in table 1.

Table 1. Dynamic parameters of the vibration protection system under shock impact

	15 мм	30 мм
$\ddot{z}_{input}$	49.719 g (123.91 dB)	49.746 g (124.27 dB)
$\ddot{z}_{ouput}$	0.00753 g (47.82 dB)	0. 2219 g (77.22 dB)
Efficiency	76.75 dB	46.69 dB

3. Conclusion

Thus, the introduction of the second, compensating, channel in the block diagram of the vibration protection system of the human operator gives it a high dynamic quality. In this case, the compensating device, calculated by the presented algorithm, is a controller (or an actuator), effectively absorbing the disturbance. There is no need for sensors, amplifiers and electro-hydraulic executive devices, which complex electro-hydraulic and electro-pneumatic active systems consist of.
The introduction of this system in practice enables to reduce fatigue, and, consequently, improve the functional state of the human operator and the efficiency of technological processes. In addition, the decrease in vibration load of the operator's position brings significant economic benefits, as it reduces the likelihood of vertebrogenic pathology occurrence for the human operator's spine. And it cuts expenditures for sick-pay.

References

[1] Kalashnikov B A 2006 Nonlinear oscillations of mechanical systems (Omsk: Omsk State Technical University) p 206

[2] Vasiliev V V, Poklad M P and Seryakov O A 2013 Design of multi-purpose tracked vehicles. Theory and Motion and Dynamics of Multipurpose Track Machines (Omsk) p 436

[3] Galiev I I, Nehaev V A and Nikolaev V A 2010 Metody i sredstva vibrozaschity zheleznodorozhnyh ekipazhey (monografiya) (Moscow: Izd-vo Uchebno-metodicheskiy tsentr po obrazovaniyu na zh.-d. transporte) p 340

[4] Tarasov V N 2006 Theory in construction and engineering (Moscow: Publishing house of construction high schools) p 336

[5] Vasilchenkov V F 1998 Military caterpillar vehicles. Construction and calculation (Ryazan) p 448

[6] Rybak L A, Sinev A V and Pashkov A I 1997 Sintez aktivnyh sistem vibroizolyatsii na kosmicheskih ob'ektah (Moscow: Yanus-K) p 159

[7] Tibilov T A and Tsisovski T 2001 Proektirovanie regulyatorov, ispol'zuyuschih vozmuscheniya, s tsel'yu uluchsheniya dinamicheskih kachestv zheleznodorozhnogo ekipazha *IV nauchno-prakt. konf. «Resursosberegayuschie tehnologii na zheleznodorozhnom transporte»* (Moscow: MIIT) pp 49–50

[8] Nekchaev V A and Nikolaev V A 1993 Synthesis of invariant vibroprotection system (theory and practice) *Proceedings of Twelve International Scientific Conference «Nonlinear vibration problems»* (Warszawa: DWH-Polish Sientific Publishers) pp 224-236

[9] Vibratsii v tehnike 1981 **6** (Moscow: Mashinostroenie) p 456

[10] Wittenburg J 2002 (Springer) p 230

[11] Furunzhev R I and Ostanin A I 2004 Management of combinations of multi-bearing machines (Moscow: Mechanical Engineering) p 206

Criterion of equationally Notherian property for posets

A Yu Nikitin[1] and I D Kudyk[2]

[1] Complex Information Defense Department, Omsk State Technical University, Omsk pr. Mira 11, Russian Federation
[2] Institute of Mathematics and Informational Technologies, Omsk State University n.a. F.M. Dostoevskiy, Omsk pr. Mira 55a, Russian Federation

E-mail: [1]nikitinlexey@gmail.com, [2]volume_8091@mail.ru

Abstract. Many modern models of informational defence represents by graphs and partial orders (posets). It is very important to resolve such algorithmic problems as searching of elements that satisfies some conditions (or, another words, to resolve recognition problems) in this models. Algebraic geometry is resolving such problems and searches elements that is solution of system of equations. But systems could be infinite. In this article we're formulating the condition for poset that every infinite system under such posets can be reduce to finite system and still be equivalent to origin system.

1. Introduction

Decision of equations and system of equations with various coefficients is one of the oldest problem in mathematics. Classical algebraic geometry studies solutions of system of equations (algebraic sets) under the real and complex fields. Classical algebraic geometry got great development in XIX century. But since the middle of XX century big development was received by universal algebraic geometry. Universal algebraic geometry studies algebraic sets under the arbitrary algebraic structures whether graph, group or universal algebra. For the present, a series of contributions by E.Yu. Daniyarova, A.G. Myasnikov and V.N. Remeslennikov [1], [2], [3], [4] and [5] is some kind of development's result of universal algebraic geometry. In this article studies algebraic geometry under partial order. Inspite of theory of partial orders gained great development [6], many problems of algebraic geometry under partial orders doesn't solve yet. In the works [7], [8], [9] some kind of partial orders such as lattices, semilattices and boolean algebras was studied from algebraic-geometric point of view. In [7] was observed coordinate algebras structure, algebraic sets, equationally Notherian property and compactness property for partial orders was studied . Equationally Notherian property of algebraic structure is very important property that allows to use variety of algebraic geometry results. In this article we're looking at equationally Notherian property for arbitrary partial order and, as result, we're formulate the criterion of it.

2. Preliminaries

In this section we reminds basic definitions from theory of partial orders and universal algebraic geometry under partial orders.

Partial ordered set (partial order or poset) is algebraic system $\mathcal{P} = \langle P| \leqslant^{(2)}, A\rangle$, where $\leqslant$ – binary order relation and A – constant symbols, and the structure satisfies 3 axioms:

(i) $\forall p \in P\ p \leqslant p$.
(ii) $\forall p_1, p_2 \in P\ p_1 \leqslant p_2 \wedge p_2 \leqslant p_1 \rightarrow p_1 = p_2$.
(iii) $\forall p_1, p_2, p_3 \in P\ p_1 \leqslant p_2 \wedge p_2 \leqslant p_3 \rightarrow p_1 \leqslant p_3$.

Language (signature) of posets with constants we're note as L_A.

Elements x and y of poset $\mathcal{P}$ are *comparable* if $x \leqslant y$ or $y \leqslant x$. Otherwise, elements are *incomparable*.

Define for every subset of elements A of poset $\mathcal{P}$ sets $A^{\uparrow} = \{x \in \mathcal{P} \mid \forall a \in A\ a \leqslant x\}$ and $A^{\downarrow} = \{a \in \mathcal{P} \mid \forall a \in A\ x \leqslant a\}$. These sets calls upper set (upset) and lower set (down set) of A. For one-element set $A = \{a\}$ notations are $a^{\uparrow}$ and $a^{\downarrow}$ correspondence.

The set of *generators* of up(down) set $A^{\uparrow}(A^{\downarrow})$ is subset B of poset's elements $\mathcal{P}$ that $B^{\uparrow} = A^{\uparrow}$ ($B^{\downarrow} = A^{\downarrow}$). Up(down) set $A^{\uparrow}$ ($A^{\downarrow}$) is *finitely generated* if there is exist finite set B of generators of up(down) set.

Let's separate some subset A of elements of $\mathcal{P}$. *Upper A-fan* is pair $(A, A^{\uparrow})$, where A is *base* and $A^{\uparrow}$ is upset of A. Lower A-fan defines in a similar way. Upper A-fan is *finitely generated* if there is exist finite subset $B \subseteq A$ that $B^{\uparrow} = A^{\uparrow}$.

We're formulate important technical lemma.

Lemma 1 *For every poset $\mathcal{P} = \langle P \mid \leqslant, A\rangle$ there is true two properties: (a) $\forall A \subseteq P\ (A^{\downarrow})^{\uparrow} \supseteq A$ and (b) $\forall A \subseteq P\ (A^{\uparrow})^{\downarrow} \supseteq A$.*

Proof. $(A^{\downarrow})^{\uparrow} = \{x \mid x \geqslant b\ \forall b \in A^{\downarrow}\}$ by the definition. But if $\forall a \in A\ \forall b \in A^{\downarrow}\ a \geqslant b$ then $A \subseteq (A^{\downarrow})^{\uparrow}$. Property (b) is proving the same way. ■

Next, instead of $(A^{\downarrow})^{\uparrow}$ we are writing simple $A^{\downarrow\uparrow}$.

Let note $X = \{x_1, \ldots, x_n\}$ as a set of variables. *Term* of poset $\mathcal{P}$ in language L_A from variables X is any constant from A or any variable from X. *Atomic formula* of poset $\mathcal{P}$ in language L_A from variables X defines through next two statements:

(i) $t_1 = t_2$ is atomic formula, where t_1, t_2 is a terms;
(ii) $t_1 \leqslant t_2$ is atomic formula, where t_1, t_2 is a terms.

Equation under poset $\mathcal{P}$ in language L_A from variables X is any atomic formula under $\mathcal{P}$. It is easy to see all types of equations under $\mathcal{P}$:

(i) $a_i = a_j$, where $a_i, a_j \in A$,
(ii) $a_i \leqslant a_j$, where $a_i, a_j \in A$,
(iii) $x_i = a_j$, where $x_i \in X,\ a_j \in A$,
(iv) $x_i = x_j$, where $x_i, x_j \in X$,
(v) $a_i \leqslant x_j$, where $x_j \in X,\ a_i \in A$,
(vi) $x_i \leqslant a_j$, where $x_i \in X,\ a_j \in A$,
(vii) $x_i \leqslant x_j$, where $x_i, x_j \in X$.

System of equations $S(X)$ from variables $X = \{x_1, \ldots, x_n\}$ is any set of equations from variables X. Point $p = (p_1, \ldots, p_n) \in P^n$ is a *solution* of system $S(X)$ if substitution of point's coordinates in system $S(X)$ instead of correspondence variables is providing a true statement under $\mathcal{P}$: $\mathcal{P} \vDash \varphi(p_1, \ldots, p_n)$. *Set of solution* is defining in a natural way. Set system's $S(X)$ solutions denotes as $V(S)$. Two systems S_1 and S_2 is *equivalent* if their sets of solutions are the same (denotes as $S_1 \sim S_2$).

IOP Conf. Series: Journal of Physics: Conf. Series **1050** (2018) 012058 doi:10.1088/1742-6596/1050/1/012058

Lets reach agreement about notation of equation's sets. Equations $a_i = a_j$, where $a_i, a_j \in A$ in system $S(X_n)$ denote as $S_{a=a}$. Set of the rest equations are similar: $S_{a\leqslant a}, S_{x=a}, S_{x=x}, S_{a\leqslant x}, S_{x\leqslant a}$ and $S_{x\leqslant x}$ for system $S(X_n)$.

It says that poset $\mathcal{P}$ has *equationally Notherian property* is every system of equations $S(X_n)$ from n variables under $\mathcal{P}$ has equivalent finite subsystem $S'(X_n) \subseteq S(X_n)$. And if for every system $S(X_n)$ under $\mathcal{P}$ there is exist equivalent finite system $S'(X_n)$ then $\mathcal{P}$ has *weak equationally Notherian property.*

Class of posets $\mathcal{K}$ has (weak) equationally Notherian property if every poset $\mathcal{P} \in \mathcal{K}$ has (weak) equationally Notherian property.

3. Main result

In this section we're formulate the main result.

Theorem 1 *Poset $\mathcal{P}$ has equationally Notherian property iff for every subset A of $\mathcal{P}$ upper and lower A-fans are finitely generated.*

Proof. Firstly, we assume for $\mathcal{P}$ every A-fans is finitely generated. There is system of equations $S(X_n)$ under $\mathcal{P}$ in language L_A from n variables. Also assume that the system contains every 7 types of equations and equations of every type are infinite.

Without loss generality we take an agreement that every separate constants in $S(X_n)$ has separate interpretation in $\mathcal{P}$. This condition isn't necessary, but it makes further proof more easier and it hasn't essential influence to proof.

It has to prove the system $S(X_n)$ has finite subsystem $S'(X_n)$ that equivalent to $S(X_n)$. Note that system $S(X_n)$ is divides on 7 noncrossing subsystems $S_{a=a}, S_{x=a}$, etc. It can define solution set of system $S(X_n)$ as $V(S) = V(S_{a=a}) \cap V(S_{a\leqslant a}) \cap V(S_{x=a}) \cap V(S_{x=x}) \cap V(S_{x\leqslant a}) \cap V(S_{a\leqslant x}) \cap V(S_{x\leqslant x})$. Next we show that it could chose finite subsystem from every type of equations that equivalent to origin. For instance, for system $S_{x=x} \subset S(X_n)$ there is exist finite subset $S'_{x=x} \sim S_{x=x}$.

Firstly, lets observe subsystem $S_{a=a} \subset S(X_n)$. If this system are compatible under $\mathcal{P}$ then it hasn't influence on the rest system and subsystem $S'_{a=a} = \emptyset$. But if $\mathcal{P} \not\models S_{a=a}$, then there is exist equation $a_i = a_j \in S_{a=a}$, that doesn't compatible under $\mathcal{P}$ and this equation is equivalent to $S_{a=a}$. Subset $S_{a\leqslant a} \subset S(X_n)$ observes the similar way.

There is exist finite subsystems $S'_{x=x}$ and $S'_{x\leqslant x}$ that equivalent to $S_{x=x}$ and $S_{x\leqslant x}$, because of finite power of variables. Note that subsystems $S_{x=x}$ and $S_{x\leqslant x}$ can't be incompatible.

If subset $S_{x=a} \subset S(X_n)$ is compatible then there is exist finite subsystem $S'_{x=a} \subset S_{x=a}$ that equivalent to origin, because of finite power of variables. Otherwise, $S_{x=a}$ contains next pair of equations $x_i = a_j, x_i = a_k,\ a_j \neq a_k$. This pair of equations is equivalent to $S_{x=a}$.

Note that we still didn't use condition from criterion. The rest subsystems $S_{x\leqslant a} \subset S(X_n)$ and $S_{a\leqslant x} \subset S(X_n)$ use this condition for chose finite subsystems. Firstly, we observe $S_{x\leqslant a}$.

Lets select subset $S_{x_i\leqslant a} \subset S_{x\leqslant a}$ every equation of it depends from only one variable x_i. This set could be represent as $S_{x_i\leqslant a} = \{x_i \leqslant a_j \mid j \in J, |J| = \infty\}$. It's easy to see from noted representation that $V(S_{x_i\leqslant a}) = A_i^{\downarrow}$, where $A_i = \{a_j \mid j \in J\}$. Using property of finite generation of every A-fan from poset $\mathcal{P}$, chose subset $J' \subset J$, that $A'_i = \{a_j \mid j \in J'\}$ and $(A'_i)^{\downarrow} = A_i^{\downarrow}$. It means that there is exist finite subsystem $S'_{x_i\leqslant a} \sim S_{x_i\leqslant a}$.

Using this procedure for every variables from $S_{x\leqslant a}$, we chose finite subsystem $S'_{x\leqslant a} \sim S_{x\leqslant a}$. Existing of finite subsystem $S'_{a\leqslant x} \sim S_{a\leqslant x}$ is proving the same way.

We showed the way to construct subsystems $S'_{a=a}, S'_{a\leqslant a}, S'_{x=a}, S'_{x=x}, S'_{a\leqslant x}, S'_{x\leqslant a}$ and $S'_{x\leqslant x}$, equivalent to $S_{a=a}, S_{a\leqslant a}, S_{x=a}, S_{x=x}, S_{a\leqslant x}, S_{x\leqslant a}$ and $S_{x\leqslant x}$ from system $S(X_n)$. It means that $V(S) = V(S'_{a=a}) \cap V(S'_{a\leqslant a}) \cap V(S'_{x=a}) \cap V(S'_{x=x}) \cap V(S'_{x\leqslant a}) \cap V(S'_{a\leqslant x}) \cap V(S'_{x\leqslant x})$.

Now, lets assume that poset $\mathcal{P}$ has equationally Notherian property. Select infinite set A of $\mathcal{P}$ and observe $A^{\downarrow}$. Next, we build system of equations $S(x) = \{x \leqslant a_i \mid a_i \in A\}$ under $\mathcal{P}$. There is exist finite subsystem $S'(x) = \{x \leqslant a_i \mid a_i \in A', |A'| < \infty\} \sim S(x)$ because of $\mathcal{P}$ is equationally Notherian. It means that $A^{\downarrow} = A'^{\downarrow}$, and all lower A-fans is finitely generated. Finite generation of upper A-fans is proving the same way.

■

Note that if A-fan isn't finitely generated it doesn't mean $A^{\uparrow}$ or $A^{\downarrow}$ isn't finitely generated. Let's observe the next example: poset $\mathcal{P} = \mathbb{Z} \cup \{q\}$, where $\mathbb{Z}$ is linear order of integers and q is single element that is not comparable with all elements of $\mathbb{Z}$. Scheme of this poset is shown on fig. 1.

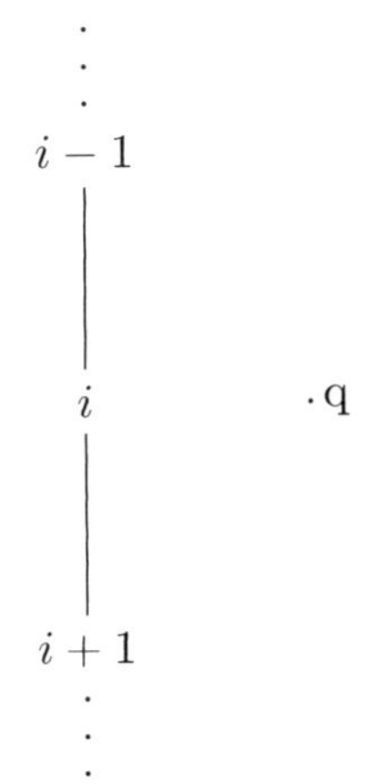

Figure 1. Poset $\mathcal{P} = \mathbb{Z} \cup \{q\}$

Observe $\mathbb{Z}$-fan. It's easy to see that $\mathbb{Z}^{\downarrow} = \emptyset$. Any finite subset of $\mathbb{Z}$ generates nonempty lower set and it means that lower $\mathbb{Z}$-fan isn't finitely generated. But if we take subset $B = \{0, q\}$, then $B^{\downarrow} = \emptyset = \mathbb{Z}^{\downarrow}$. And it means that $\mathbb{Z}^{\downarrow}$ is finitely generated.

For this reason and with formulated equationally Notherian property for posets it could be formulate next criterion.

Consequence 1 *Poset $\mathcal{P}$ has weak equationally Notherian property iff for every subset A of $\mathcal{P}$ $A^{\uparrow}$ and $A^{\downarrow}$ is finitely generated.*

It's easy to produce criterion of equationally Notherian property for the class of posets from criterion for single poset. Class of posets has an equationally Notherian property iff every poset from this class has equationally Notherian property.

Acknowledgments
The research was supported by Russian Science Foundation (project No. 17-11-01117)

References

[1] Daniyarova E Y, Myasnikov A G and Remeslennikov V N 2012 *Journal of Mathematical Sciences* **185** 389–416
[2] Daniyarova E Y, Myasnikov A G and Remeslennikov V N 2011 *Southeast Asian Bulletin of Mathematics* **35** 35–68
[3] Daniyarova E Y, Myasnikov A G and Remeslennikov V N 2011 *Algebra and Logic* **49** 483–508
[4] Daniyarova E Y, Myasnikov A G and Remeslennikov V N 2012 *Algebra and Logic* **51** 28–40
[5] Daniyarova E Y, Myasnikov A G and Remeslennikov V N 2012 *Algebra and Logic* **56** 281–94
[6] Gratzer G 1978 *General Lattice Theory* (Basel: Birkhäuser)
[7] Shevlyakov A N 2015 *Algebra and Logic* **54** 258–71

[8] Harju T, Karhumaki J and Petrich M 1997 *Structures in Logic and Computer Science* **LNSC 1261 Springer** 268–80
[9] Rudeanu S 2001 *Lattice Functions and Equations* (London: Springer)

IOP Conf. Series: Journal of Physics: Conf. Series **1050** (2018) 012059 doi:10.1088/1742-6596/1050/1/012059

Efficient topological generation in compact Lie groups

Gennady A. Noskov

The Sobolev Institute of Mathematics, Pevtsova 13, Omsk 644099, Russia

E-mail: g.noskov@gmail.com

Abstract. We prove that for any compact simple Lie group G and any nonidentity element g of G the subset of $h \in G$, for which g, h topologically generate G, is nonempty and Zariski open in G. A connected compact Lie group G satisfies 1.5-generation property iff it is simple or abelian. Any compact simple Lie group G has a conjugacy class C such that for every nontrivial elements g_1,g_2 of G there exists $y \in C$ so that $\overline{\langle g_1, y\rangle} = \overline{\langle g_2, y\rangle} = G$. In particular, the generating graph of G has diameter 2.

1. Introduction

Let G be a topological group. For $g_1, \ldots, g_n \in G$ denote by $\langle g_1, \ldots, g_n\rangle$ the subgroup of G generated by these elements and by $\overline{\langle g_1, \ldots g_n\rangle}$ the closure in G of the group $\langle g_1, \ldots g_n\rangle$. If $\overline{\langle g_1, \ldots, g_n\rangle} = G$, we say $g_1, \ldots, g_n$ are topological generators of G.

The generation problem, widely speaking, is to describe possible generating sets for groups in a given class. The generation problem for topological groups has a short history but a long past. The first result in this context is the theorem of L.Kronecker (1884) [Kr], which asserts, in modern terms, that for every natural number n and every point $(t_1, \ldots, t_n) \in \mathbb{R}^n$ the image of this point in the torus $\mathbb{T}^n = (\mathbb{R}/\mathbb{Z})^n$ is a topological generator of $\mathbb{T}^n$ if and only if the numbers $1, t_1, \ldots, t_n$ are linearly independent over $\mathbb{Q}$. H.Auerbach proved in 1934 that a compact connected Lie group G could be generated (topologically) by two elements and that the set of generating pairs has full measure in $G \times G$ [Aue34]. Subsequently, in 1935, J.Schreier & S.Ulam proved that for compact connected metrizable topological groups G the set of pairs generating G is dense in $G \times G$ [SU35] . We refer to the article by K.H.Hoffman and S.A.Morris [HM90] for more recent results on the minimal number of generators needed for locally compact groups.

In 1949, M.Kuranishi showed that connected semisimple Lie groups could be generated by two elements [Kur49]. Later, in 1951, Kuranishi showed that if G is a perfect (i.e. coincides with the closure of its commutator subgroup) Lie group, then the set of pairs generating G is open and dense in $G \times G$ [Kur51].

Let G be a compact connected semisimple Lie group. It follows from the result of M.Goto [Got49] that G is perfect and so by Kuranishi the set of pairs generating a compact connected semisimple Lie group is open and dense. The proof of Kuranishi's theorem uses the Hausdorff topology on compact subsets of G. In particular, the proof gives little insight into the structure of the pairs which do not generate G. This problem had been investigated by M.Field, who proved in [Fie99] that the subset of $G \times G$ consisting of pairs (g, h) which topologically generate G is nonempty and Zariski open. Later Field and Parry used openness results of this type to prove results on the stable ergodicity of Holder continuous skew-extensions by compact Lie groups. The paper [Fie99] is concluded with a conjecture suggested by Peter Rowley: If G is a compact simple Lie group and g is a nonidentity element of G, then the subset of $h \in G$ for which $\overline{\langle g, h\rangle} = G$ is non-empty and Zariski open. (Recall that any compact Lie grouo G admits a continuous real linear representation and it is real algebraic relative to any such representation, in particular admits Zariski topology whose closed subsets by definition are the zero sets of the families of polynomials in matrix coefficients x_{ij} and inverse determinant $\det(x_{ij}^{-1})$). In this paper we prove this "1.5-generation conjecture" by Field-Rowley in a stronger form.

Theorem 1 (1.5-generation) *Let G be a compact simple Lie group and let g be a nonidentity element in G. (1) For every regular t there is h in the conjugacy class t^G such that g, h topologically generate G, i.e. $\overline{\langle g, h\rangle} = G$. (2) The subset of $h \in G$ for which $\overline{\langle g, h\rangle} = G$ is nonempty and Zariski open.*

The result was announced without proofs in [No10]. A close result and for noncompact groups is obtained by H.Abels and E.B.Vinberg [AV11]. However, note that in our theorem 1 the "complementary" element h can be found out in a conjugacy class of an arbitrary regular element whereas in [AV11] it is taken in the conjugacy class of g.

Our next main result gives description of connected compact Lie groups satisfying 1.5-generation property.

Theorem 2 (characterization of compact groups with 1.5-generation) *A connected compact Lie group G satisfies 1.5-generation property iff it is simple or abelian.*

The generating graph $\Gamma(G)$ of a topological group G is the graph defined on the elements of G with an edge connecting two distinct vertices if and only if they topologically generate G. The following is the analog of Breuer-Guralnick-Kantor result [BGK] on finite simple groups.

Theorem 3 (diameter of generation graph) *Any compact simple Lie group G has a conjugacy class C such that for every nontrivial elements g_1, g_2 of G there exists $y \in C$ so that $\overline{\langle g_1, y\rangle} = \overline{\langle g_2, y\rangle} = G$. In particular, the generating graph of G has diameter 2.*

2. Preliminaries on compact Lie groups

We refer to the books of J.Adams and N.Bourbaki [Bou07]. A compact Lie group G is a compact differentiable manifold G together with a smooth multiplication map $G \times G \to G$ which gives G the structure of a group. The inverse map is automatically smooth [[Br07], p. 22]. When the multiplication is clear from context, we usually just say "G is a Lie group". For instance, G might be the multiplicative group of unit complex numbers, the multiplicative group of unit quaternions, or the group SO_n of rotations in Euclidean n-space. A component of the identity G^0 of a Lie group G, that contains the identity, is a normal Lie subgroup. The remaining components are the cosets of G^0. Every open subgroup of the group G contains G^0. There are only finitely many connected components in a compact Lie group.

A compact Lie group is said to be simple if it has no nontrivial closed normal subgroups. Any connected compact Lie group with a trivial center is isomorphic to a product of simple compact Lie groups. Call a connected compact Lie group almost simple if it has a finite center. Any connected compact Lie group G has a finite cover which is isomorphic to a product $T \times G_0$, where T is a torus and G_0 is a product of almost simple groups. In particular, any simply connected compact Lie group is isomorphic to a product of almost simple groups and any connected compact Lie group with a trivial center is isomorphic to a direct product of simple compact Lie groups [DW98].

A torus is a compact Lie group which is isomorphic to $\mathbb{R}^n/\mathbb{Z}^n = (\mathbb{R}/\mathbb{Z})^n$ for some $n > 0$. The number n is called the rank of the torus. It is clear that the rank of a torus is equal to its dimension as a manifold. In fact any connected compact abelian Lie group is isomorphic to a

IOP Conf. Series: Journal of Physics: Conf. Series **1050** (2018) 012059 doi:10.1088/1742-6596/1050/1/012059

torus. From now on G is a compact connected Lie group. A maximal torus $T \leq G$ is a subgroup which is a torus, such that if $T \leq U \leq G$ and U is a torus then $T = U$. Any subtorus of G is contained in a maximal torus. An element $g \in G$ is regular if it is contained in just one maximal torus. If g is contained in more than one maximal torus, it is called singular. An element $g \in G$ is regular iff $\dim Z(g) = \mathrm{rank}(G)$, where the rank of G is defined to be the rank of any maximal torus of G. An element $g \in G$ singular iff $\dim Z(g) > \mathrm{rank}(G)$. Every regular element is a topological generator of the corresponding maximal torus. Conversely, the topological generator of a maximal torus T is a regular element.

2.1. *Finiteness for closed subgroups of maximal rank*

We will rely on finiteness of the number of closed connected subgroups containing the maximal torus.

Let G be a compact Lie group and let T be its maximal torus. A multiplicative character of T is a continuous homomorphism of T into the topological group $\mathbb{S} = \{z \in \mathbb{C} : |z| = 1\}$. By $X(T)$ we denote a (commutative) group of multiplicative characters of T. Usually we consider $X(T)$ as an additive group. Let $T \curvearrowright \mathfrak{g}$ be the adjoint representation of T on the Lie algebra $\mathfrak{g}$ of G. For any $a \in X(T)$ define the root subspace in $\mathfrak{g}^{\mathbb{C}} = \mathbb{C} \otimes_{\mathbb{R}} \mathfrak{g}$

$$\mathfrak{g}_a^{\mathbb{C}} = \left\{ v \in \mathfrak{g}^{\mathbb{C}} : gv = a(g)\, v, \forall g \in T \right\}.$$

Since T is commutative the space $\mathfrak{g}^{\mathbb{C}}$ is a direct sum of the subspaces $\mathfrak{g}_a^{\mathbb{C}}$ for $a \in X(T)$. The roots of representation $T \curvearrowright \mathfrak{g}$ are by definition those characters a for which the root subspaces $\mathfrak{g}_a^{\mathbb{C}}$ are nonzero. The set of all roots is denoted by $R(G, T)$.

The basic result we need is

Theorem 4 (basic tool) *([Bou07],§4,Prop.12.7) Let G be a connected compact Lie group and T its maximal compact torus. Let $\mathscr{H}$ be the set of all connected closed subgroups of G, containing T and ordered by inclusion. Then the map $H \mapsto R(H, T)$ is an order preserving bijection from $\mathscr{H}$ onto the set of all symmetric subsets of $R(G, T)$, ordered by inclusion. In particular the set $\mathscr{H}$ is finite.*

Corollary 5 (finiteness) *Let G be a connected compact simple Lie group and T its maximal torus. Then there are only finitely many closed subgroups of G that contain T.*

Proof. Indeed, we know from theorem 4 that the set $\mathscr{H}$ is finite. If $H \in \mathscr{H}$ is a closed subgroup of G then its component H^0 belongs to $\mathscr{H}$ and thus H is a subgroup in the normalizer $N_G(H_0)$ of H^0 and it is known that the quotient $N_G(H_0)/H_0$ is finite ([Bou07] ,§2, n°4, proposition 4 and remark 2). Hence there are only finitely many possibilities for H with fixed H^0 and there are only finitely many possibilities for H^0 in view of theorem 4. It follows that there are finitely many closed subgroups of G that contain T. ■

3. Proofs of the main results

Throughout this section G denotes is a connected compact simple Lie group and T denotes its maximal torus.

Lemma 6 (proper subgroup is almost abnormal) *For every non-identical $g \in G$ and every proper closed subgroup S of G there is a nonempty Zariski open subset X of G such that $x^{-1}gx \notin S$ for $x \in X$.*

Proof. Note that S, being a compact Lie group is Zariski closed in G. Hence the subset

$$Y = \left\{ y \in G : y^{-1} g y \in S \right\}$$

is also Zariski closed in G. We assert that Y is proper in G. We argue by contradiction. If the equality $Y = G$ were true, then this would mean that the conjugacy class g^G would contained in S. But then the closed subgroup generated by this class would be proper and normal in G contradicting the simplicity of G. We conclude that $X = G - Y$ is the set in question. ■

The following lemma is a generalization of lemma 6 for several proper subgroups.

Lemma 7 (finitely many proper subgroups are almost abnormal) *For any finite family $\mathscr{S}$ of proper closed subgroups of G and any nonidentical $g \in G$ there is a nonempty Zariski open subset $X \subseteq G$ such that $x^{-1}gx \notin \cup_{S\in\mathscr{S}} S$ for $x \in X$.*

Proof. First note that $\cup_{S\in\mathscr{S}} S$ is a proper subset since its dimension is strictly less than that of G. Next the set $Y_S = \{y \in G : y^{-1}gy \in S\}$ is Zariski closed and proper in G by previous lemma. Hence by dimension argument $Y = \cup_{S\in\mathscr{S}} Y_S$ is Zariski closed and proper subset of G. The complement $X = G - Y$ is the set in question. ■

A subset $S \leq G$ of G is called abnormal if $\cap_{x\in G} S^x = 1$. For instance every proper subgroup of G is abnormal by simplicity of G. On the other hand any nontrivial conjugacy class is "normal", i.e. not abnormal in G.

Lemma 8 (abnormality) *Let $\mathscr{S}_T$ be the set of all proper connected closed subgroups of G containing T and let S_T be the union of all subgroups from $\mathscr{S}_T$. Then the set S_T is abnormal.*

Proof. We argue by contradiction. By corollary 5 the set $\mathscr{S}_T$ is finite. Suppose that the intersection $\cap_{x\in G} S_T^x$ contains nonidentical element g. Then $g^x \in S_T$ for all $x \in G$. On the other hand by lemma 7 $g^x \notin S_T$ for x from nonempty Zariski open subset of G, so we have reached a contradiction. ■

3.1. Completing the proof of theorem 1.

Let $g \in G$ be an arbitrary non-identical element and let t be any regular element. Fix a maximal compact torus $T < G$, containing t. By abnormality lemma there is $s \in G$ such that $g \notin s^{-1}S_T s$. We assert that the elements $g, s^{-1}ts$ topologically generate G, i.e. that the group $H = \overline{\langle g, s^{-1}ts\rangle}$ coincides with G. We argue by contradiction and suppose that H is properly contained in G. The maximal torus $s^{-1}Ts$ is contained in H hence $H \in \mathscr{S}_{s^{-1}Ts} = s^{-1}\mathscr{S}_T s$. Thus $g \in H \subseteq s^{-1}S_T s$ - contradiction with $g \notin s^{-1}S_T s$. Thus we have proved the nonemptiness assertion. It remains to infer that the subset of $h \in G$ for which $\overline{\langle g, h\rangle} = G$ is Zariski open or dually that the "non-generation" subset N_g consisting of all $h \in G$, for which $\overline{\langle g, h\rangle} < G$, is Zariski closed. We know from the Field's theorem that the set N_G of pairs $\{g, h\}$ for which $\overline{\langle g, h\rangle} < G$ is Zariski closed in $G \times G$. Since $(g, N_g) = N_G \cap (g, G)$ we conclude that (g, N_g) is Zariski closed in $G \times G$ and it it clear that N_g is Zariski closed in G.

3.2. Proof of theorem 2

If G is abelian then it is a torus and thus by Kronecker it is 1-generated. If G is nonabelian and has a nontrivial center then every central nonidentity element can not generate G since otherwise G would be abelian. Thus in this case G is not 1.5-generated. Finally, suppose that G is nonabelian and with trivial center. Then it is a product of simple compact Lie groups. If G has at least two nontrivial factors then any nonidentity element g lying in one factor can not generate the whole group together with any other element and so G is not 1.5-generated. In the remaining case G is simple and so it is 1.5-generated by theorem 1.

3.3. Proof of theorem 3

Let $g_1, g_2 \in G$ be arbitrary non-identical elements. Fix a maximal compact torus $T < G$ and its topological generator $t \in T$. By lemma 7 there are nonempty Zariski open subsets X_1, X_2 in G such that $g_1 \notin x^{-1}S_T x$ for $x \in X_1$ and $g_2 \notin x^{-1}S_T x$ for $x \in X_2$. In view of Zariski connectivity of G the intersection $X_1 \cap X_2$ is nonempty. Fix $s \in X_1 \cap X_2$. By the argument of the previous section G is topologically generated by each of the pairs $\{g_1, s^{-1}ts\}, \{g_2, s^{-1}ts\}$. This proves the theorem with $C = t^G$ and $y = s^{-1}ts$.

3.4. Problems and questions

- Define and prove "x.5-generation" for compact Lie groups for suitable natural x.
- Let G be a separable compact connected group. Is the set of pairs in which topologically generate G necessarily of full measure in $G \times G$ [GlM17] ?

Acknowledgments

The work was supported by the program of fundamental scientific research of the SB RAS I.1.1., project 0314-2016-0004.

I also thank Herbert Abels, Grisha Margulis, Grisha Soifer and Ernest Borisovich Vinberg for helpful discussions.

References

[AV11] H. Abels, E. B. Vinberg. *Generating semisimple groups by tori.* J. Algebra 328. (2011), 114–121.

[Ad] J.F.Adams. *Lectures on Lie Groups.* University of Chicago Press, 1969.

[Aue34] H. Auerbach. *Sur les groupes linéaires bornés (III)'.* Studia Mat., V:43–49, 1934.

[Bou07] N. Bourbaki. *Groupes et algebres de Lie. Chapitre IX.* Springer, 2007.

[BGK] T. Breuer, R.M. Guralnick and W.M. Kantor. *Probabilistic generation of finite simple groups, II,* J. Algebra Vol. 320. 2, (2008), 443–494.

[Br07] R. Bryant. *Introduction to Lie Groups and Symplectic Geometry,* in Geometry and quantum theory, IRS/Park City Math, vol.I., 2007.

[DW98] W.G. Dwyer and C.W. Wilkerson. *The elementary geometric structure of compact Lie groups.* Bull. London Math. Soc., 30:337–364, 1998.

[Fie99] Michael Field. *Generating Sets for Compact Semisimple Lie Groups.* Proceedings of the American Mathematical Society, 127(11):3361–3365, 1999.

[GlM17] Tsachik Gelander and François Le Maître, *Infinitesimal topological generators and quasi non-archimedean topological groups,* Topology and its Applications, Volume 218, 1 March 2017, Pages 97–113.

[Got49] M. Goto. *A theorem on compact semi-simple groups.* J. Math. Soc. Japan, 1:270–272, 1949.

[HM90] K. H. Hofmann and S. A. Morris. *Weight and* $\mathfrak{c}$. J. of Pure and Applied Algebra, 68:181–194, 1990.

[Kr] L. Kronecker. *Näherungsweise ganzzahlige Auflösung linearer Gleichungen.* Monatsber. Koniglich. Preuss. Akad. Wiss. Berlin (1884), 1179-1193, 1271-1299..

[Kur49] M. Kuranishi. *Two element generations on semi-simple Lie groups.* Kodai math. Sem. Report, pages 9–10, 1949.

[Kur51] M. Kuranishi. *On everywhere dense imbedding of free groups in Lie groups.* Nagoya Math. J., 2:63–71, 1951.

[No10] G. A. Noskov, *1.5-generation of compact Lie groups and conjecture by Field and Rowley,* "Stochastice models in biology and limit algebras", OMSK, 2010.

[SU35] J. Schreier and S. Ulam. *Sur le nombre des générateurs d'un groupe topologique compact et connexe.* Fund. Math., 24:302–304, 1935.

IOP Conf. Series: Journal of Physics: Conf. Series **1050** (2018) 012060 doi:10.1088/1742-6596/1050/1/012060

Peculiarities of mathematical modeling of contact interaction of massive bodies and shells

S P Novikov[1], **O D Kazakov**[1] **and A V Iakovlev**[1]

[1] Federal State Budgetary Educational Institution of Higher Education "Bryansk State Engineering Technological University", 3, Stanke Dimitrov Av., Bryansk, 241037, Russia

Abstract. The limits of using a mathematical model of contact interaction of elastic bodies, based on the Hertz's solution in engineering practice are defined. Numerical investigation of the contact of a massive body and a shell has been conducted when the parameters of the latter vary. Three-dimensional finite-element models of contacting bodies have been developed in the case of different curvatures and thicknesses of the shell walls. A solution of the normal contact problem is given by the force method for various configuration models of the shell. Ratios of the geometric parameters of models the finite-element schemes of the contacting bodies are ere determined where the problem is described with sufficient accuracy by the Hertz's solution.

1. Introduction

A solution of the contact problem is very often required in engineering calculations [1]. In some cases, the contacting bodies have complex geometry and are thin-walled shells. Axisymmetric shells are widespread in engineering. Their median surfaces are rotational surfaces. Boilers, tanks, pipes, and cisterns can serve as examples. As shown in paper [2], where the contact problem for the roller and the cylindrical guidance has been solved, the stress-strain state in the roller support has a rather complex nature, which is due to the high compliance of the shell and local deformation of its wall. For these reasons, the use of analytical methods of solution is difficult, and they are used only for an approximate evaluation.

Most of the analytical approaches to static contact problem solving are based on Hertz's solution. Hertz's work [3] gives a solution of the normal contact problem for solid elastic entire bodies. The normal contact stresses in the contact spot area are distributed according to the equation of the ellipsoid, and according to Hertz's assumptions about the smallness of the sizes of the contact patch, the contacting bodies can be represented by half-spaces and the solution of the contact problem can be obtained on the basis of the Boussinesq's solution. However, if one of the contacting bodies is a shell, the latter cannot be represented in the form of a half-space, which imposes substantial limitations on the Hertz's solution.

This work is aimed at research peculiarities of mathematical modelling contact interaction of elastic bodies and revealing deviations of the parameters of the contact from the Hertz's ones in the case when one of the contacting bodies is a shell. In addition, this article studies the nature of these deviations when the magnitude of a load of contacting bodies, the geometric shapes of the shell and the thickness of its wall vary. Based on the research, the ratios of the geometric sizes of bodies for which the problem is fairly accurately described by Hertz's solution are established.

IOP Conf. Series: Journal of Physics: Conf. Series **1050** (2018) 012060 doi:10.1088/1742-6596/1050/1/012060

2. Peculiarities of modeling the finite-element schemes of the contacting bodies

Due to the fact that the application of analytical methods for solving the contact problem in this situation is problematic, the most common numerical approaches are used. One of such approaches is the application of the finite element method (FEM), according to which finite element calculation schemes are built at the first stage [4]. In our case, it is necessary to study the change in the basic parameters of the contact when the shell geometry and the wall thickness are varied. A series of experiments with various geometric parameters of the contacting bodies was conducted with the help of computer modeling.

To conduct any multivariate study and to reveal common patterns, it is necessary to provide the identity of the initial conditions prior to the calculation and, first of all, the identity of the mathematical models of the studied bodies. When using FEM [5], all finite element models should have the same structure and the same boundary conditions, and in the process of solving the contact problem using the force method described in [6], in refining the calculation results, it is necessary to ensure that all allocated and then refined local areas, containing the contact spot have the same size.

Thus, when composing finite-element computational schemes of contacting bodies, it is necessary to satisfy three conditions of identity:

- to provide the same size and the same structure of rough finite element calculation schemes of contacting bodies with superimposed identical boundary conditions;
- for contact finite elements to provide the identity of the ratio of their linear sizes;
- to ensure their equal size and structure in the process of calculation by the force method when allocating local areas.

If these conditions are met, errors that always give finite element schemes will have the same influence on the calculation results of the contacting pairs of bodies with different geometric parameters. Thus, the errors in the FEM will not affect the common patterns of a particular phenomenon, thereby ensuring the purity of model experiments.

Because of the fact that contact spot size is very small in comparison with the sizes of the contacting bodies, it is necessary to solve the problem of sizes of the rough finite element calculation schemes that are used in the first stage of solving the contact problem by the force method. Considering a large number of the proposed calculations and the rigid requirements applied by the FEM to computer resources, the minimum size of the schemes and a small number of finite elements should be ensured. To reveal the acceptable size of finite element schemes, the following experiment was conducted. We considered the contact of a massive sphere with a radius of 1 m, which was presented taking into account the symmetry, by the calculation scheme $100 \times 100 \times 100$ mm, and a shell with a radius of 1 m and a thickness of 20 mm.

The elasticity modulus and the Poisson's ratio were chosen to be the same for both bodies and equal, respectively $E = 2 \times 10^{11}$ and $\mu = 0.25$. Three computational schemes of contacting bodies, differing in the size of the finite element calculation scheme of the shell, were prepared (Figure 1).

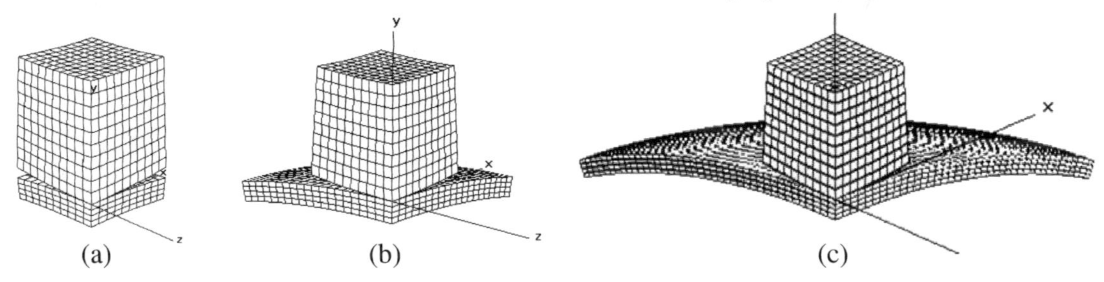

Figure 1. Calculation schemes of contacting bodies with varying sizes of the shell scheme: (a) - $100 \times 100 \times 20$ mm; (b) - $200 \times 200 \times 20$ mm; (c) - $400 \times 400 \times 20$ mm

Taking into account that finite element models are axisymmetric, this figure represents ¼ of the design, which was used for the subsequent calculation.

The links were superimposed accordingly in the direction of the symmetry axes and along the computational domain selection boundaries.

IOP Conf. Series: Journal of Physics: Conf. Series **1050** (2018) 012060 doi:10.1088/1742-6596/1050/1/012060

The results of calculations [7] when using the indicated schemes by the force method are presented in the table 1.

Table 1. The results of solving the contact problem for various shell models

The type of the shell scheme (Figure 1)	Number of elements in the scheme	Load $P/4$, (кN)	Maximum pressure in the contact p_0, (MPa)	Contact spot radius a, (mm)	$\sigma^M_{IV\max}$, (MPa)	$\sigma^S_{IV\max}$, (MPa)
1a	400	24.954	980.3	4.21	662.32	596.9
1b	1000	24.956	977.8	4.27	660.83	577.6
1c	6400	24.968	976.4	4.30	660.64	574.9

Here $\sigma^M_{IV\max}$ and $\sigma^S_{IV\max}$ are the maximum equivalent stresses calculated using the IV strength theory for a massive body and a shell, respectively.

Based on these results and taking into account that the difference between the two schemes (Figure 1a and 1c) differing fourfold in size, by the maximum pressure p_0 is 0,4%, by the size of the contact spot is 2%, and by the maximum equivalent stresses are 3,6%, quite small, then to achieve acceptable accuracy for further calculations the scheme 1a may be used. To carry out planned researches five configurations of shells with different radii of curvature were considered.

It is these dimensions of the chosen scheme that were used to construct finite element models of all subsequent calculation schemes with the variable geometry.

To carry out the planned researches the five configurations of shells with different radii of curvature R_1^S and R_2^S were considered (Table 2).

Table 2. Configurational shell models

Designation of the shell configuration	Shell type	R_1^S, (mm)	R_2^S, (mm)	Wall thickness of the shell δ, (mm)
$S1$	Spherical, convex	1000	1000	Varies from 10 to 100 with a pitch of 10
$S2$	Spherical, convex	2000	2000	
$C1$	Cylindrical	∞	1000	
$P1$	Plate	∞	∞	
$\overline{S2}$	Spherical, concave	-2000	-2000	

Thus, up to 50 calculation schemes were prepared, for which the wall thickness and shell geometry varied. For each calculation scheme, a normal contact problem was solved for various implementations in order to select a given load, which amounted to several hundred calculations by the force method. The choice of a correct small finite element model of contacting bodies and the use of an algorithm for improving the calculation results by isolating the local calculation scheme in the region of the contact spot and grinding of the latter described in [3] allowed to perform such a large number of calculations. All calculations were conducted using a software package FEMS developed by the author [4].

3. Investigation of the contact of the sphere and a spherical shell

One of the simple analytical solutions that can be obtained according to Hertz is the contact of two elastic spheres. Let us consider how the parameters of the contact will change if one of the spheres is a spherical shell, the wall thickness of which will vary. We take the radius of the sphere equal to 1 m. We choose the elastic modulus and Poisson's ratio to be the same for the both bodies and equal, respectively, $E = 2\times10^{11}$ Pa and $\mu = 0.25$. The calculated finite element scheme of contacting bodies is shown in Figure 2.

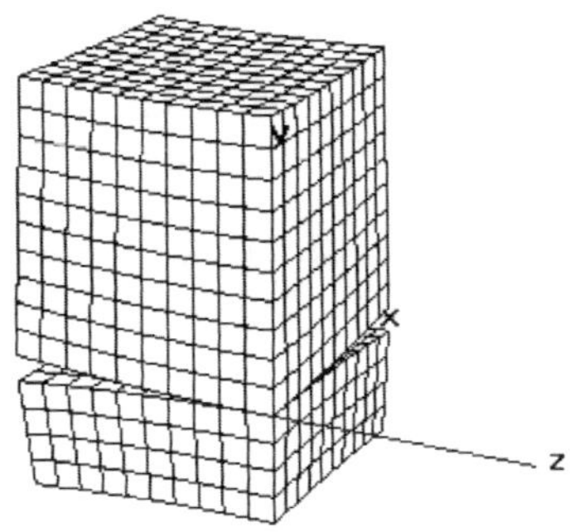

Figure 2. The calculated scheme of contacting bodies

The shell with a wall thickness of 40 mm is shown here. All other calculation schemes will be obtained by changing the wall thickness from 10 to 100 mm with a pitch of 10 mm, while the calculation scheme of the sphere and the finite element mesh on the contact surface will remain unchanged. According to the force method, it is necessary to specify the penetration of one body into another, according to which the load can be defined after solving the contact problem. Let us select such a penetration for each calculation scheme so that the load is $P =$ 100 kN. Taking into account that $1/4$ part of the contacting body is taken as the calculation scheme, the load attributable to the finite element model will be $\overline{P} = P/4 = 25$ kN. The results of the calculation by the force method with a variation of the shell wall thickness are accumulated in Table 3.

Table 3. The results of solving the contact problem for two spherical bodies with a variation in the wall thickness of the shell

Sizes of the shell model, (mm)	The number of nodes in the contact	Load $P/4$, (kN)	Contact spot radius a, (mm)	Maximum pressure in the contact P_0, (MPa)	$\sigma^M_{IV\max}$ (MPa)	$\sigma^O_{IV\max}$ (MPa)
According to Hertz	-	25.00	7.06	959.8	-	-
$100\times100\times100$	52	24.98	7.15	962.1	643.5	642.3
$100\times100\times60$	52	24.97	7.25	960.1	643.3	637.7
$100\times100\times40$	52	24.97	7.32	958.6	642.5	625.4
$100\times100\times20$	60	24.99	7.64	950.0	637.4	565.6
$100\times100\times10$	62	24.98	8.25	780.9	524.8	746.2

According to the presented data, the contact parameters smoothly change with a decrease of the shell thickness, and apparent deviations from the Hertz's values are observed only with a wall thickness of less than 20 mm. With a shell wall thickness of 10 mm, the deviation of the maximum pressures from the values obtained by the Hertz's solution is 18,6%, while the sizes of the contact patch increase by 13,9%. With a shell thickness of less than 10 mm, a dip in the central point of the ellipsoid of the distribution of contact pressures (Figures 3b and 3c) is observed, in comparison with the pressure distribution at a shell wall thickness of 100 mm (Figure 3a).

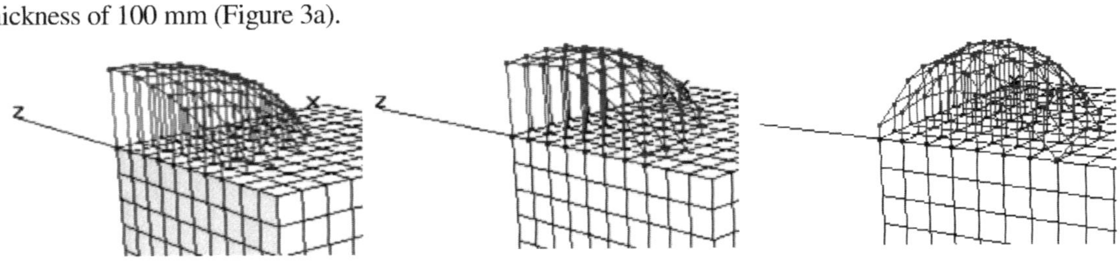

(a) (b) (c)

Figure 3. Distribution of contact pressures when shell wall thickness is: (a) – 100 mm; (b) – 10 mm; (c) – 5 mm.

Thus, the nature of distribution of contact pressures in the area of the contact spot varies significantly for thin-walled shells. This affects the distribution of equivalent stresses in the vicinity of the contact spot. When there is a dip of the central point of the pressure diagram, the arrangement of the points corresponding to the maximum equivalent stresses calculated according to the IV strength theory changes. They start to shift from the location below the central point of the pressure ellipsoid and are located under the corresponding points of maximum contact pressures. Moreover, when a thickness of the shell wall is small, they come out on the inner surface of the shell, which indicates large tension stresses in this area.

The behavior of another parameter of the contact - the approach of contacting bodies α -will be considered when the wall thickness of the shell varies. Let us proceed to non-dimensional parameters and introduce the notation:

$$\overline{\delta} = \delta K \ , \ \overline{\alpha} = \frac{\alpha(a+b)}{2ab} ,$$
$$\overline{p_0} = \frac{p_0(k1+k2)(a+b)}{2abK} , \tag{1}$$

where δ – is the shell wall thickness;

a, b – are dimensions of the semiaxes of the contact spot;

p_0 – is the maximum contact pressure.

The remaining values included in the presented expressions are determined by Hertz's solution [2]: $K = A + B = (\frac{1}{R_1^M} + \frac{1}{R_2^M} + \frac{1}{R_1^S} + \frac{1}{R_2^S})$, $k_1 = \frac{1-\mu_1^2}{\pi E_1}$; $k_2 = \frac{1-\mu_2^2}{\pi E_2}$, where R_1^M, R_2^M, R_1^S and R_2^S - are the radii of curvature of the massive body and shell, E_1, E_2, μ_1 and μ_2 - are elastic characteristics of the contacting bodies, respectively.

If the contacting bodies have a spherical surface of the same radius $a = b$; $R = R_1^M = R_2^M = R_1^S = R_2^S$; $k = k_1 = k_2$; $\mu = \mu_1 = \mu_2$ and $K = A + B = \frac{1}{4R}$, the entered parameters, in this case, can be represented as follows:

$$\overline{\delta} = \frac{\delta}{4R} , \ \overline{\alpha} = \frac{\alpha}{a} , \ \overline{p_0} = \frac{2p_0 k}{aK} . \tag{2}$$

The dependence of the $\overline{\alpha}$ parameter on $\overline{\delta}$ in the case of constant curvature of the contacting bodies according to the relations (2) is presented in Figure 4a.

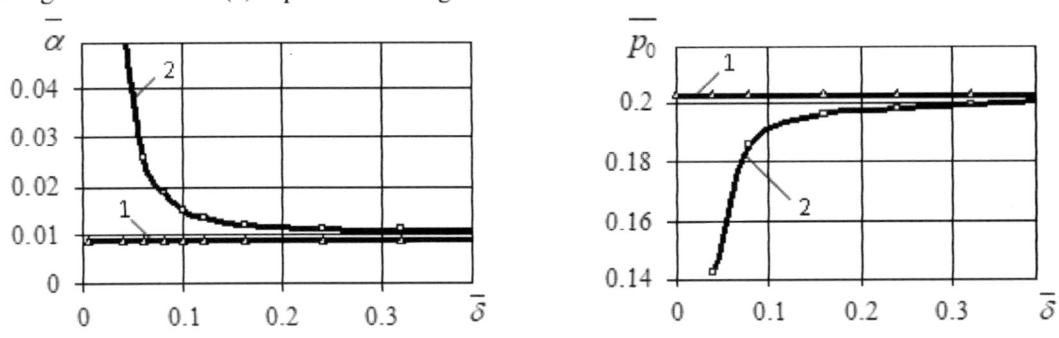

Figure 4. Dependences of $\overline{\alpha}$ and $\overline{p_0}$ on the thickness of the shell wall at constant curvature and load $P = 10000$ kg in case of the contact of two spherical bodies: 1 – theoretical curve (according to Hertz); 2 – calculated curve

According to this dependence, the fracture point of the calculated curve corresponds to the value $\bar{\delta}$, equal to 0.008, and consequently, the shell wall thickness $\delta \approx 20$ mm, characterizes the sharp change in the contact parameters for the load $P = 100$ kN. Consequently, when the shell thickness decreases from 20 mm or less, the deviation of the value $\bar{\alpha}$ from the corresponding value according to the Hertz's solution sharply rises (from 40% or more). In addition, the approach of the contacting bodies increases quite dramatically, which can lead to a loss of stability of the shell, and therefore a verification calculation of the stability for a thin-walled shell is required.

Another dependence (Figure 4b) characterizes the behavior of the maximum pressure in the contact with a decrease in the shell thickness. Here, the fracture of the calculated curve 2 is observed practically in the same parameter range $\bar{\delta}$, as for the previous dependence. However, at this point, corresponding to $\bar{\delta} = 0.08$, the deviation of the $\overline{p_0}$ parameter from the theoretical value is 8.5 %. And only in the case of $\bar{\delta} = 0.04$ the deviation reaches 30 %, which corresponds to the shell wall thickness of 10 mm.

Thus, for a given load $P = 100$ kN, when two spherical bodies of the same curvature are contacting, the most significant deviations of the contact parameters from the theoretical ones are observed when a shell wall thickness is 10 mm or less. It is the thickness at which a sharp change in the distribution of contact pressures is observed (Figure 3). In this case, the application of the Hertz's solution gives sufficiently significant errors and it is necessary to use numerical methods of calculation.

To establish similar limitations of the Hertz's solution for other loads, we will perform a series of calculations for identical spherical bodies in the case of various penetrations. We introduce the notation $\bar{P} = \dfrac{P(k1 + k2)}{ab}$, where $P-$ is an external load on the contacting bodies. In this case, the results of calculations can be represented by the dependences, shown in Figure 5.

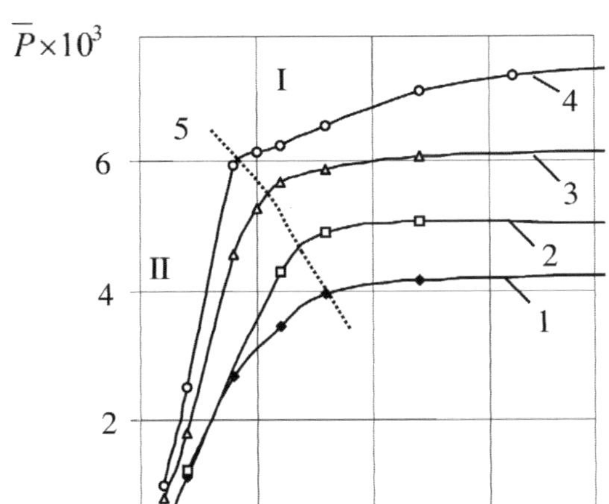

Figure 5. Dependencies that determine the nature of load change from the thickness of the pipe with constant curvature for the penetrations:
1– 0.05 mm; 2 – 0.07 mm;
3 – 0.1 mm; 4 – 0.14 mm;
5 – boundary curve

The points of fracture of the given curves indicate a sharp change in contact parameters in the case of varying the thickness of the shell. Then curve 5 (Figure 5) constructed from these points will reflect the nature of the change in the contact parameters for varying the given approach of contacting bodies according to the force method and known values $\bar{P}$ and $\bar{\delta}$. This curve can approximately be considered as the boundary of two sections (Figure 5): I - characterizes the behavior of contacting bodies, as massive, where the assumptions are valid and the contact parameters obtained by Hertz's solution can be used with sufficient accuracy;

II - characterizes the influence of shell properties on the contact parameters, accordingly, the increasing deviation from the Hertz's solution. However, section II can be considered as a transitional from thick-walled to thin-walled shells. Significant deviations of the contact parameters and the form of the distribution of contact pressures from the Hertz values are observed when the thickness is sufficiently small, from 10 mm or less. In this case, the application of Hertz's solution for an approximate evaluation is impossible.

4. The study of the effect of shell configuration on contact parameters

Let us consider the effect of shell configuration on contact parameters taking into account the fact that most contacting bodies in real engineering problems have complex geometry. One of the contacting bodies we will represent as a sphere with a radius of 1 m, and the other in the form of a shell, the geometry of which will vary. We will choose five types of shells (Table 2). For all types of shells, we will select the value of penetration of one body into another to obtain a load $P = 40$ kN. Considering that significant deviations from the Hertz's solution are observed when the shell thickness is 10 mm or less, we will choose the shell thickness $\delta = 10$ мм in all calculation schemes.

The calculation results using the force method when the shell curvature varies are summarized in Table 4. The presented data indicate that the parameters of the contact change substantially when the shell curvature changes. The deviation of the parameters values from the Hertz's solution increases when curvature decreases.

Table 4. The results of solving the contact problem for varying the shell curvature and the constant thickness of its wall $\delta = 10$ mm.

Type of the shell (Table 2)	Number of nodes in the contact	Load $P/4$, (kN)	Sizes of contact spot semi-axes a/b, (mm)	Maximum pressure in the contact P_0, (MPa)	Deviation of P_0 from the Hertz's solution, (%)
$S1$	31	10.007	5.2/5.2	614.22	13.1
$S2$	33	10.002	5.6/5.6	494.95	15.1
$C1$	36	10.002	7.1/4.2	476.12	15.9
$P1$	43	10.009	6.2/6.2	364.94	17.9
$\overline{S2}$	65	10.006	8.1/8.1	216.97	22.6

In addition, a study of the deviation of parameters values determined by the relation (1) from the Hertz's ones when the shell curvature changes was conducted for all types of shells.

The results of this study are shown in Figure 6.

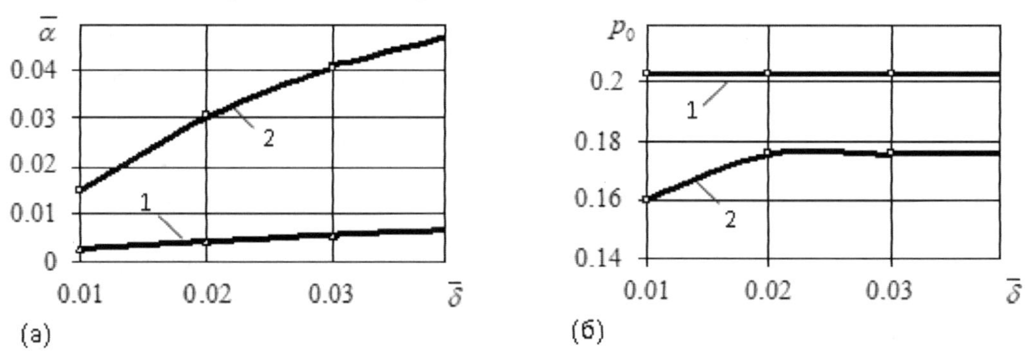

Figure 6. Dependences $\bar{\alpha}$ and $\bar{p}_0$ on the shell curvature when the thickness of its wall $\delta = 10$ mm and the load $P = 40$ кН: 1 – theoretical curve (according to Hertz); 2 – calculated curve.

On the given diagram (Figure 6a) for the shell wall thickness $\delta = 10$ мм, when its curvature decreases, the deviation $\bar{\alpha}$ from the corresponding value according to the Hertz's solution decreases and deviation $\overline{P_0}$ increases.

Thus, according to the obtained results, a change in the shell curvature with a small thickness of its wall also has an effect on the values of main contact characteristics and their deviation from the Hertz's solution. In the case of a four-fold increase in the curvature for the shell from 0.01 to 0.04 when the parameter $\overline{\delta}$ changes, the deviation of the $\overline{\alpha}$ value from the corresponding value according to the Hertz's solution increases by 3.5 % and the deviation of the value $\overline{p_0}$ decreases by 8.1%.

5. Conclusions

The presented work considers the behavior of contact parameters and quite completely analyzes the deviations of the latter from the corresponding values according to the Hertz's solution when the shell characteristics such as curvature and wall thickness change. Thus, the boundaries of the Hertz's solution are determined in the case when one of the contacting bodies is a shell. The presented dependences study the behavior of such contact characteristics as the approach of contacting bodies α and the maximum contact pressure p_0 when the geometry of the shell and the thickness of its wall change. The dependence (Figure 4a) allows one to control α when the thickness of the shell wall changes and, accordingly, to carry out verification calculations for stability. The dependence (Figure 5) shows the possibility of applying Hertz's solution for an approximate evaluation of the contact characteristics for known values of the load, curvature, and shell wall thickness.

Thus, the presented study of the case of contact between a massive body and a shell demonstrates when it is sufficient to use the Hertz's solution to solve the contact problem and when it is necessary to use numerical methods. However, any model studies need an appropriate experimental verification.

6. References

[1] Johnson K L 1989 *Contact Mechanics* (Moscow: Mir) 509
[2] Novikov S P, Ivchenko V N, Podoprigora Yu A, Davydov S V 2000 Analysis of the stressed state of a tubular guide and supports of a belt conveyor *Dynamics and strength of transport vehicles* (*Bryansk*) pp 79-86.
[3] Hertz H. 1895 *Gesammelte Werke.* Bd.1. (Leipzig) P.155-96
[4] Segerlind L 1979 Applied finite element analysis (Moscow: Mir) 392
[5] Strang G, Fix J 1977 Theory of the finite element method (Moscow: Mir).
[6] Novikov S P, Ivchenko V N, Podoprigora Yu A, Sakalo V I 2000 Contact of a massive body with a cylindrical shell *Dynamics and strength of transport vehicles* (*Bryansk*) pp 72-78.
[7] Zenkevich O 1975 Finite Element Method in Engineering (*Translation from English*) (Moscow: Mir) 544

IOP Conf. Series: Journal of Physics: Conf. Series **1050** (2018) 012061 doi:10.1088/1742-6596/1050/1/012061

Research of Electro-Thermal Aging Process of Cross-Linked Polyethylene

D A Polyakov[1], K I Nikitin[2], A N Novozhilov[3], M Ya Kletsel[3]

[1]Omsk State Technical University, 11, Mira ave., Omsk, 644050, Russia
[2]Tyumen Industrial University (Tobolsk Industrial Institute), 5 VUZ Zone, Tobolsk, Tyumen Region 626158 Russia
[3]S. Toraighyrov Pavlodar State University, 64 Lomov st. Pavlodar 140006 Kazakhstan

Corresponding author's e-mail address: polyakowdmitry@yandex.ru

Abstract. Cable XLPE-insulation operation capability and reliability highly depends on its insulation condition. Statistical analysis showed that more than 70% of cable technological disturbances are insulation breakdowns, caused by insulation aging or poor quality of cable jointing. This paper is devoted to the cable XLPE-insulation aging under the influence of electro-thermal stress. The goal of the research is to develop the technique of cable insulation residual life calculation. Theoretical and experimental research was carried out for this goal. During theoretical research Kuchinsky insulation aging model was considered and statistical research of cable insulation breakdown was performed. Experimental research of cable insulation accelerated electro-thermal aging was carried out. XLPE-insulated cable samples were exposed to the influence of temperature and voltage above rated. The results of theoretical and experimental research showed that partial discharge dependence on electric field has variable coefficients and requires an additional research, taking all possible influencing factors into account.

1. Introduction

XLPE-insulated cables are one of the most widespread elements of electric power systems. It is known that cable insulation is the most destructible element of the cable system. The most frequent cause of technological disturbances on the cable power transmission lines is insulation breakdown. Statistical analysis showed that the cable insulation breakdown is usually 70% or more of all technological disturbances of cable power transmission lines. Cable insulation breakdown is usually caused by insulation aging or by poor quality of cable jointing. Taking this problem into account, a wide spectrum of the cable insulation tests are developed and being applied:

- High voltage VLF test [1];
- Impulse voltage test [2];
- Partial discharge (PD) measurement and analysis [3], [4];
- Insulation resistance measurement [5];
- Tan δ measurement and dielectric response test [6];
- Condition assessment using polarization/depolarization current method [7];
- Reflectometry method [8];
- Cable insulation aging assessment using terahertz waves [9].

The most significant disadvantage of the listed tests is inability of the insulation remaining life prediction. Therefore, insulation materials aging and remaining life determination research continues to be an urgent task.

There are three common types of cable XLPE-insulation aging mechanisms:

- Thermal aging [10], [11];
- Partial discharges [12], [13];
- Electrical and water treeing [14], [15].

Thermal aging has an influence on the cable insulation condition and decreases its electrical strength [16]. However, the recent research [10] showed that thermal aging has to be considered only with the electrical aging effects. Moreover, the other research [17] showed the dominant effect of thermal aging in comparison of the DC stress. Electrical and water treeing significantly effect on the insulation material characteristics, e. g. insulation material conductivity increases [17]. Partial discharges (PD) occur in the insulation defects and lead to their growth and insulation breakdown. Usually these defects are electrical treeing [18], insulation impurities [2] and cavities in the cable joints [19]. All listed mechanisms deteriorate the insulation and lead to the insulation electrical strength decrease. This paper is devoted to the problem of insulation aging and insulation electric strength decrease during operation.

2. Insulation aging simulation

2.1. Mathematic model

Cable insulation remaining life is to be calculated using the data of insulation degree of degradation and statistical data of insulation breakdowns. Cable insulation degradation degree determination is possible using the insulation aging mathematic models [20] and monitoring of the insulation destruction factors [21]. In spite of insulation aging models diversity, in the literary sources wasn't found the model taking all factors into account. Not all of them were considered on their trustworthiness. So in this paper mathematic insulation aging model, proposed by Kuchinsky, is considered [22]:

$$\tau_o = AE^{-n}\exp\left(\frac{W_a}{kT}\right) \tag{1}$$

where W_a – Arrhenius activation energy, k – Boltzmann constant, T – absolute temperature, E – electric field, A – constant.

The main idea of Kuchinsky model is multiplication of partial discharge influence and thermal effect on the insulation. So the formula (1) is based on two formulas: partial discharge power dependency on the electric field:

$$P_q = BE^{-n} \tag{2}$$

and Arrhenius equation, calculating the chemical reactions rate constant:

$$K = K_0\exp\left(\frac{W_a}{kT}\right) \tag{3}$$

But the proposed model cannot be used for the insulation remaining life calculation without improvement and transformation. The basic parameter determining the insulation remaining operation life is insulation destruction degree. It is to be determined using the formula of no destructed molecules number of the insulation material:

$$N(t) = N_0 e^{-C\cdot B\cdot E^n\cdot K_0\cdot\exp\left(-\frac{W_a}{kT}\right)\cdot t} \tag{4}$$

where C – parameter, depending on cable size, t = τ_0 if N(t) has critical value and insulation operation is not recommended.

But electric field and temperature in the cable insulation depend on time. Therefore, insulation destruction rate is to be calculated:

$$v(t) = -\frac{dN(t)}{dt} = c \cdot B \cdot E^n \cdot K_0 \cdot \exp\left(-\frac{W_a}{kT}\right) \cdot N_0 \cdot e^{-c \cdot B \cdot E^n \cdot K_0 \cdot \exp\left(-\frac{W_a}{kT}\right) \cdot t} \quad (5)$$

In this case the no destructed molecules number of the insulation material is to be computed using the formula:

$$N(t) = N_0 - \int_0^t v(t)dt \quad (6)$$

This equation allows calculating the no destructed molecules number on the basis of the integrated influences on the insulation. The obtained dependence on time is to be extrapolated using least squares method. It allows determining the N(t) empiric function and the time to the insulation critical destruction and insulation breakdown probability increase. The cable XLPE-insulation critical degree of destruction is to be determined using the insulation breakdown statistical analysis and simulation.

2.2. Statistical analysis and simulation of the insulation breakdown
Statistical simulation of the cable insulation breakdown was carried out using the computer program developed for this goal. The main idea of the simulation was to represent the insulation as a three-dimensional massive of cells (molecules) with proportions of cable insulation size. It was assumed that the insulation breakdown occurs when 80% of the molecules in one line between cable core and screen are destructed. Every moment of time one random molecule is destructed. The result of breakdown simulation allowed determining the insulation breakdown probability dependence on the insulation degradation degree (figure 1).

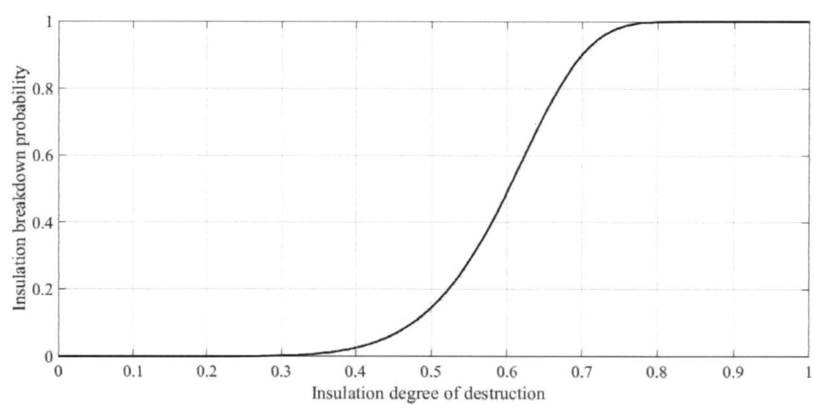

Figure 1. Insulation breakdown probability dependence on insulation degree of destruction.

Obtained characteristic shows that insulation breakdown probability is less than 1% if the insulation is less than 35% destructed. Cable operation until 35% of the insulation destruction is mostly safe. But operation with more than 35% material destruction tends to periodical insulation breakdowns because of the insulation defects occurrence. Therefore, it is recommended to operate cables until their 35% insulation degradation.

3. Experimental research

Experimental research of electro-thermal cable XLPE-insulation accelerated aging was carried out. To exclude the effects of insulation water absorption the insulation samples were aged in a dry air. Experimental setup was developed for this goal (figure 2). It consists of asbestos-cement pipe containing cable samples, tubular electric heating elements for 1 kW in summary and basaltic heat insulation. Insulation samples were 6 kV jointed cables. During experimental research insulation samples were exposed to the heating and AC 50 Hz voltage. Temperature inside the experimental setup was maintained at the level of 100°C using electronic scheme with a symistor and temperature sensor. This temperature was chosen because of the cable sheath melting with a temperature about 105-110°C. AC voltage was obtained using testing transformer up to 100 kV. Experimental samples voltage was periodically increased from 10.8 kV to 50 kV. The experiment time was 2500 h. Partial discharge power were periodically calculated on the basis of measurement.

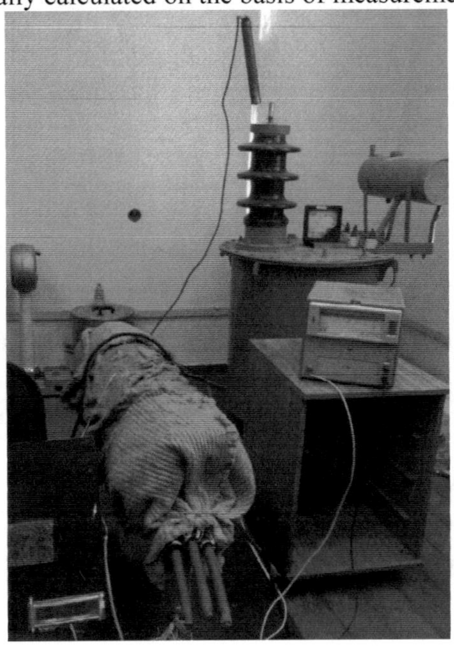

Figure 2. Experimental setup.

The results of experiment showed sufficient difference of PD power for different measurements. Therefore, this research allows only determining the range of PD power dependences on electric field (figure 3), and the parameters B and n (2) are variables on different factors, e. g. insulation moisture, temperature etc.

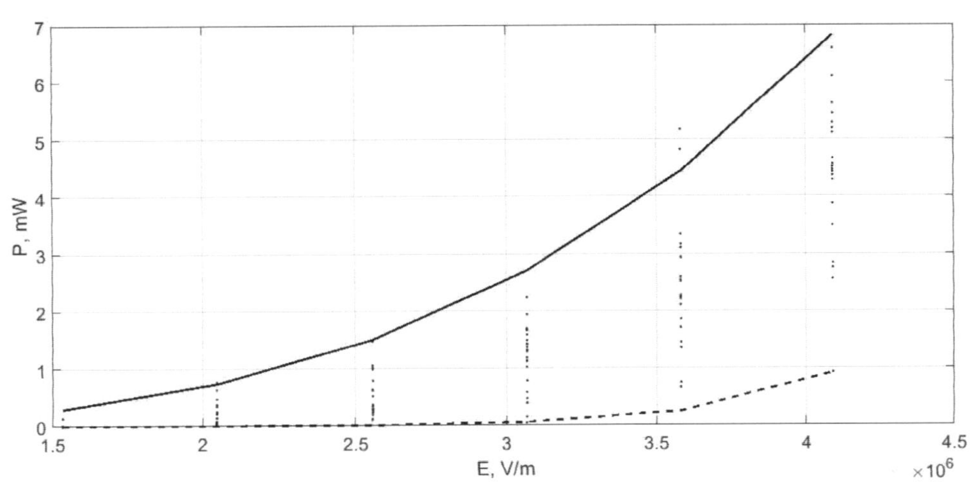

Figure 3. PD power dependence on electric field.

4. Results and discussion

Theoretical research of the insulation aging process consisted of Kuchinsky mathematic model considering and statistical simulation of the insulation breakdown. Kuchinsky insulation aging model was transformed and improved to calculate the insulation residual life. Statistical simulation of the insulation breakdown allowed determining the critical insulation destruction rate. 35% of the insulation destruction or more leads to periodical insulation breakdowns. It is recommended to carry out the predictive maintenance to reduce the number of short circuits caused by insulation breakdown. Experimental research showed a significant difference of PD power for different measurements. This result shows that PD power dependence on electric field has variable coefficients. Their determination requires a research, taking all possible factors into account.

5. Conclusion

Cable XLPE-insulation remaining life calculation model based on Kuchinsky insulation aging model is proposed. Insulation voltage and temperature monitoring during operation allows calculating its residual life. The obtained experimental results showed that PD power function has variable coefficients. Their determination requires a research, taking all possible influencing factors into account. The experimental data and experience are to be used to improve the proposed mathematic model for its possibility of residual life calculation of any type and voltage class of XLPE-insulated cables.

References

[1] Eager G S, Katz C, Fryszczyn B, Densley and Bernstein B S High voltage VLF testing of power cables 1997 *IEEE Trans. Power Deliv* vol 12 no 2 pp 565–570

[2] Woschitz R, Muhr M and Sumereder C. Quality control of XLPE cables by means of impulse voltage test *Proceedings of 2005 International Symposium on Electrical Insulating Materials, 2005. (ISEIM 2005)* 2005 vol 2 pp 526–528

[3] Ahmed Z, Hussain G A, Lehtonen M, Varacka L and Kudelcik J Analysis of partial discharge signals in medium voltage XLPE cables *Proceedings - 2016 17th International Scientific Conference on Electric Power Engineering, EPE 2016.*

[4] Chen X, Xu Y and Cao X Nonlinear time series analysis of partial discharges in electrical trees of XLPE cable insulation samples *Dielectr. Electr. Insul. IEEE Trans* vol 21, no 4 pp 1455–1461 2014

[5] Qureshi M I, Malik N H, Al-Arainy A A, Saati M N, Al-Nather O A and Anam S Investigation of resistance toward water tree growth in XLPE insulated MV cables produced in the Arabian Gulf region *2008 International Conference on Condition Monitoring and Diagnosis* 2008 pp 1195–1198

[6] J. Liu, D. Zhang, C. Bin, L. Huang, Y. Gao, and X. Li, “A New method of aging assessment for XLPE cable insulation based on dielectric response,” in *2015 IEEE 11th International Conference on the Properties and Applications of Dielectric Materials (ICPADM)* 2015 pp 560–563

[7] G. Ye, H. Li, F. Lin, J. Tong, X. Wu, and Z. Huang, “Condition assessment of XLPE insulated cables based on polarization/depolarization current method,” *IEEE Trans. Dielectr. Electr. Insul.*, vol. 23, no. 2, pp. 721–729, 2016.

[8] J. Wang, P. E. C. Stone, D. Coats, Y.-J. Shin, and R. A. Dougal, “Health Monitoring of Power Cable via Joint Time-Frequency Domain Reflectometry,” *IEEE Trans. Instrum. Meas.*, vol. 60, no. 3, pp. 1047–1053, 2011.

[9] Z. Yan *et al.*, “Investigation of aging effects in cross-linked polyethylene insulated cable using terahertz waves,” *Mater. Res. Express*, vol. 4, no. 1, p. 15304, Jan. 2017.

[10] M. Talreja, J. D. Douglas, T. Person, P. Caronia, J. Overberg, and P. Flood, “The effect of thermal aging on the electrical breakdown strength of crosslinked polyethylene,” in *2014 IEEE Electrical Insulation Conference (EIC)*, 2014, pp. 162–166.

[11] G. C. Montanari and A. Motori, “Thermal endurance evaluation of XLPE insulated cables,” *J.*

IOP Conf. Series: Journal of Physics: Conf. Series **1050** (2018) 012061 doi:10.1088/1742-6596/1050/1/012061

Phys. D. Appl. Phys., vol. 24, no. 7, pp. 1172–1181, Jul. 1991.

[12] Y. Tian, P. L. Lewin, A. E. Davies, S. J. Sutton, and S. G. Swingler, "Partial discharge detection in cables using VHF capacitive couplers," *IEEE Trans. Dielectr. Electr. Insul.*, vol. 10, no. 2, pp. 343–353, 2003.

[13] B. Sheng, W. Zhou, J. Yu, S. Meng, C. Zhou, and D. M. Hepburn, "On-line PD detection and localization in cross-bonded HV cable systems," *IEEE Trans. Dielectr. Electr. Insul.*, vol. 21, no. 5, 2014.

[14] W. Tao *et al.*, "Study on the electric-field characteristics of water tree region on the dry or wet condition in XLPE cables," in *2016 IEEE International Conference on High Voltage Engineering and Application (ICHVE)*, 2016, pp. 1–4.

[15] A. Xie, S. Li, X. Zheng, and G. Chen, "The characteristics of electrical trees in the inner and outer layers of different voltage rating XLPE cable insulation," *J. Phys. D. Appl. Phys.*, vol. 42, no. 12, p. 125106, Jun. 2009.

[16] A. Motori, F. Sandrolini, and G. C. Montanari, "Chemical, physical and microstructural properties of thermally aged XLPE cable models," in *Proceedings., Second International Conference on Properties and Applications of Dielectric Materials*, 1988, pp. 764–768 vol.2.

[17] X. Chen, A. R. Mantsch, L. Hu, S. M. Gubanski, J. Blennow, and C. O. Olsson, "Electrical treeing behavior of DC and thermally aged polyethylenes utilizing wire-plane electrode geometries," *IEEE Trans. Dielectr. Electr. Insul.*, vol. 21, no. 1, pp. 45–52, 2014.

[18] A. Najafi Pour and S. M. Shahrtash, "A comprehensive circuit model for partial discharge activities in cable insulation due to electrical treeing," *Int. J. Electr. Power Energy Syst.*, vol. 44, no. 1, pp. 646–655, 2013.

[19] F. C. Gu, H. C. Chang, F. H. Chen, and C. C. Kuo, "Partial discharge pattern recognition of power cable joints using extension method with fractal feature enhancement," *Expert Syst. Appl.*, vol. 39, no. 3, pp. 2804–2812, 2012.

[20] G. Mazzanti, "The combination of electro-thermal stress, load cycling and thermal transients and its effects on the life of high voltage ac cables," *IEEE Trans. Dielectr. Electr. Insul.*, vol. 16, no. 4, pp. 1168–1179, 2009.

[21] S. Liu, Y. Wang, and F. Tian, "Prognosis of Underground Cable via Online Data-Driven Method With Field Data," *IEEE Trans. Ind. Electron.*, vol. 62, no. 12, pp. 7786–7794, 2015.

[22] G. S. Kuchinsky, *Chastichnyye razryady v vysokovoltnykh konstruktsiyakh*. Leningrad: Energiya, 1979.

IOP Conf. Series: Journal of Physics: Conf. Series **1050** (2018) 012062 doi:10.1088/1742-6596/1050/1/012062

Numerically-analytical solution of problem gaming confrontation hardware-redundant dynamic system with the enemy operating in conditions of incomplete information about the behavior of participants in the game

V I Potapov, O P Shafeeva, M S Doroshenko, I V Chervenchuk and A S Gritsay

Omsk State Technical University, 11, Mira ave., Omsk, 644050, Russia

Abstract. The mathematical model of a game task in which conflicting parties act and make decisions in conditions of incomplete information about the behavior of participants in the game. The attacked side has a hardware-redundant dynamic system, which is not only resource of protection from enemy attacks, but also the resources of the active influence on the probability of finding an attacker in the appropriate States of the attack. It is believed that the behavior of warring parties approximarely Markov process. The solution of the considered differential game is reduced to a multi-step matrix game and its sequential solution at intervals of discretization with constant average probabilities of finding the attacker on these intervals. When solving a problem use numerical and analytical methods.

1. Introduction

Statement of the problem and the solution of game problems of confrontation of different nature systems in conflict situations, many papers [1-3], the closest of which to the contents of this article [10], devoted to the issues of formulating and solving the game problem of a technical system of confrontation with the enemy attacking in force under conditions of incomplete information in the course of the game.

In [10] has set and resolved the task's antagonism attacked, defending against attacks from its own resources, hardware and excessive dynamic system $S_A(n, m, \bar{s})$, with the attacking enemy force in the conditions of incomplete information about the behavior of the enemy attacked during the conflict. The attacking side in the game aims at the expense of resources of the attacks increase the failure rat $\lambda_0(t), \lambda_1(t), \ldots, \lambda_q(t)$ primary and backup units attacked $S_A(n, m, \bar{s})$ - System up to its complete rejection.

The attacked party (player A) by appropriate strategy $W^A = \{\bar{r}, s(\bar{\tau})\}$ redistribution $m(m = s_1 + s_2 + \cdots + s_q)$, broken spare blocks on q corresponding groups between the number of the failed main functional $n(n = n_1 + n_2 + \cdots + n_q)$, blocks hardware redundant system at appropriate times τ during the playing time $[0, t_f]$, aims to maximize its probability of failure-free operation $P(t_f) = \max P(t_f, \bar{s}(\tau_k))$ by the end of the game (confrontation) with the attacking enemy.

The above notation corresponding notation in [10], have the following meanings: $\bar{s} = (s_1 + s_2 + \dots + s_q)$ - integral vector redundancy; $\lambda_0(t)$ - the failure rate of reserve units; $\lambda_1(t), \lambda_2(t), \dots, \lambda_q(t)$ - the failure rate corresponding basic unit; $\bar{\tau} = \{\tau_0, \tau_1, \dots, \tau_L\}$ - adjustment vector (redistribution spare blocks between the failed basic) attacked $S_A(n, m, \bar{s})$ - systems whose elements correspond torque redistributing spare blocks, wherein $\tau_0 = 0,\ \tau_L < t_f$, where t_f - the end of the game; $\bar{s}(\tau_k)$ - vector distribution of reserve units at a time $\tau_k (0 \le k \le L)$.

In [10] It is also believed that the attacker system $S_A(n, m, \bar{s})$, opponent (player B). During the game it can be located in one of the states $B_1, B_2, \dots, B_N$, characterized by the corresponding result of the attack resource attacks on hardware-redundant dynamic system as a vector of the system failure rates $\bar{\lambda}(t) = \{\bar{\lambda}^i(t)\}$, each element of which represents the totality of intensities in failure q groups major functional blocks and not included in job spare blocks targeted system $\bar{\lambda}^i(t) = \{\lambda_0^i(t), \lambda_1^i(t), \dots, \lambda_q^i(t)\}$, $i = 1, 2, \dots, N$. In this case, if for each state B_i attacking enemy given the probability of its presence in these states $\bar{Q}(t) = \{Q_i(t)\}$, the set of policies attacking the opponent (Player B) is defined as $W^B = (\bar{Q}(t), \bar{\lambda}(t)\}$. Thus, the player's actions in the process of the game is a random selection of N states, which correspond to intensity non-stationary poisson flows bounce primary and backup units $S_A(n, m, \bar{s})$, - system in order to minimize the probability of failure-free operation of the targeted system.

For the numerical solution of the problem of confrontation game players A and B [10] are two algorithms. First - for calculating the vector $\bar{s}(\tau_k)$ aximizing the probability of failure-free operation $P(t_f, \bar{s}(\tau_k))$ targeted system $S_A(n, m, \bar{s})$, by the end of the game, and the second - the actual game problem solving using a first algorithm for finding the vectors $\bar{\tau}$ and $\bar{s}$ unknown player A strategy.

2. Formulation of the problem

In this paper, the development of the above stated game problem is put more complicated game problem, which proposes the development and expansion of the game model discussed above under conditions of incomplete information about the behavior of its members, taking into account the fact that the attacked party A is not only passively protected against enemy attacks, (side B) due to its resource protection - hardware redundancy, but is actively using the resource and the impact on the attacking side in that as a result of exposure unlike [10] changes the probabilities be $Q_i(t)$ in an appropriate state finding B_i among the plurality of states $(i = 0, 1, 2, \dots, N+1)$ not abruptly in a predetermined fixed times and continuously for the whole time interval $[0, t_f]$ games.

The solution to this problem the game with the payoff function $P(t_f) = \max P(t_f, \bar{s}(\tau_k))$ based on the methods and numerical algorithms developed by the author of this article [10].

3. Experimental results

The solution to this problem the game with the payoff function $P(t_f) = \max P(t_f, \bar{s}(\tau_k))$ based on the methods and numerical algorithms developed by the author of this article [10].

We assume that in this game G_2 under conditions of incomplete information from the participants in the game on the behavior of the opposing side are two players A and B .

Player A has a system $S_A^*(n, m, \bar{s})$ structurally and functionally identical hardware and excessive dynamic system, discussed in [10], and differs from it in that it has not only a "passive" protection against attacks of enemy resources (player B), but also the resources active influence on the probability of being in the respective states $B_i (i = 0, 1, \dots, N+1)$ offensive player B . As a probabilistic model of changes in the state of the attacking side (of the enemy) will use homogeneous Markov processes with discrete states and continuous time. We assume that the enemy of the state transitions $B_i (i = 0, 1, \dots, N+1)$ with one resource attack $\bar{\lambda}^i(t) = \{\lambda_0^i(t), \lambda_1^i(t), \dots, \lambda_q^i(t)\}$, $i = 1, 2, \dots, N$ he player A to the state from other resources occurs under the active influence on the part of player A in the form of non-stationary poisson flow of events with intensities $y_j(t)$, the transitions from state B_0 in state $B_j (j = 1, 2, \dots, N)$ and with intensities $\delta_j(t)$ in transitions from B_j to B_{N+1}. It is easy to understand that in the state B_{N+1}. The player has exhausted all its resources to attack the player .

In the course of the game G_2 in the time interval $[0, t_f]$ player A optimally using its resource protection in the form of m standby units, their redistribution in the game at the appropriate times $\tau_k (0 \leq k \leq L)$ between the basic, refuse from the enemy attacks (player B), functional n blocks hardware excessive dynamic system $S_A^*(n, m, \bar{s})$ to maximize the probability of its uptime $P(t_f)$ by the end of the game.

Given the above assumptions the probability $Q_i(t)$ Spent attacking side in the states $B_i(i = 0, 1, \ldots, N + 1)$ describes a system of differential equations with varying coefficients according to the known procedure [11]

$$\frac{dQ_0(t)}{dt} = -\sum_{j=1}^{N} y_j(t) Q_0(t).$$

$$\frac{dQ_i(t)}{dt} = y_i(t) Q_0(t) - \delta_i(t) Q_i(t). \qquad (1)$$

$$i = 1, 2, \ldots, N$$

$$\frac{dQ_{N+1}(t)}{dt} = -\sum_{j=1}^{N} \delta_j(t) Q_j(t).$$

with the initial conditions at the beginning of the game when the attacking side is in a state B_0, equal

$$Q_0(0) = 1, \ Q_1(0) = Q_2(0) = \cdots = Q_{N+1}(0) = 0.$$

The analytical solution of the equations (1) with variable coefficients is not possible, therefore, for an approximate calculation of probabilities $Q_i(t)$. We will use numerical-analytical method of solving the problem.

We use for this sampling technique [7], the essence of which is as follows.

We compute the minimum positive integer r satisfying the following conditions:

$$r \geq 2.$$

$$\max_{1 \leq i \leq N} \ \max_{1 \leq v \leq r} \ \max_{t \in \Delta_v} \ |y_i(t) - y_{iv}| \leq \varepsilon,$$

$$\max_{1 \leq i \leq N} \ \max_{1 \leq v \leq r} \ \max_{t \in \Delta_v} \ |\delta_i(t) - \delta_{iv}| \leq \varepsilon,$$

Where

$$\Delta_v = [t_{v-1}, t_v], \ t_v = v \cdot \Delta t, \ \Delta t = T/r,$$

$$y_{iv} = \frac{1}{2}[y_i(t_{v-1}) + y_i(t_v)],$$

$$\beta_{iv} = \frac{1}{2}[\delta_i(t_{v-1}) + \delta_i(t_v)],$$

ε - the maximum allowable deviation functions $y_i(t)$, and $\delta_i(t)$ from the respective constants y_{iv}, δ_{iv} at sampling intervals Δ_v for all $v(1 \le v \le r)$, $T = t_f$.

Then the system of equations (1) splits into r systems with constant coefficients:

$$\frac{dQ_0^v(t)}{dt} = -\sum_{j=1}^{N} y_j^v\, Q_0^v(t),$$
$$\frac{dQ_i^v(t)}{dt} = y_i^v Q_0^v(t) - \delta_i^v Q_i^v(t), \tag{2}$$
$$i = 1, 2, \dots, N$$
$$\frac{dQ_{N+1}^v(t)}{dt} = \sum_{i=1}^{N} \delta_i^v\, Q_i^v(t),$$

with the initial conditions

$$Q_i^v(t_{v-1}) = \begin{cases} Q_i(0), & \text{если } v = 1, \\ Q_i^{v-1}(t_{v-1}), & \text{если } 2 \le v \le r. \end{cases}$$

Integrating the system (2) by means of Laplace transform we obtain the expression for the aggressor state probabilities for v-th sampling interval:

$$Q_0^v(t) = Q_0^{v-1}(t_{v-1}) \exp\left(-\sum_{j=1}^{N} y_j\, t\right),$$

$$Q_i^v(t) = Q_0^{v-1}(t_{v-1}) y_i \frac{\exp\left(\sum_{j=1}^{N} y_j t\right) - \exp(-\delta_i t)}{\delta_i - \sum_{j=1}^{N} y_j} + Q_i^{v-1}(t_{v-1}) \exp(-\delta_i t),$$

$$i = 1, 2, \dots, N$$

$$Q_{N+1}^v(t) = \sum_{i=0}^{N+1} Q_i^{v-1}(t_{v-1}) - \sum_{i=0}^{N} Q_i^v(t).$$

In the interval $[\tau_k, \tau_{k+1}]$ determine the probability of finding the mean values of the attacking side in the states B_i:

$$\tilde{Q}_i = \frac{1}{\tau_{k+1} - \tau_k} \int_{\tau_k}^{\tau_{k+1}} Q_i(t)\, dt = \frac{\varphi_i(\tau_{k+1}) - \varphi_i(\tau_k)}{\tau_{k+1} - \tau_k}, \tag{3}$$

Where

$$\varphi_i(\tau) = \int_0^\tau Q_i(t)dt.$$

Based on the above arguments, if $T = \tau$ we can write the approximate equality

$$\varphi_i(\tau) \approx \sum_{v=1}^{\tau} \int_{t_{v-1}}^{t_v} Q_i^v(t)dt \approx \sum_{v=1}^{\tau} \psi_i^v. \quad (4)$$

expression for ψ_i^v are as follows:

$$\psi_0^v = \frac{Q_0^{v-1}(t_{v-1})}{C}[\exp(-ct_{v-1}) - \exp(-ct_v)], \quad (5)$$

$$\psi_i^v = Q_0^{v-1}(t_{v-1})y_i\left[\frac{\exp(-ct_{v-1}) - \exp(-ct_v)}{c(\delta_i - c)} + \frac{\exp(-\delta_i t_{v-1}}{\delta_i(c - \delta_i}\right] +$$

$$+\frac{Q_i^{v-1}(t_{v-1})}{\delta_i}[\exp(-\delta_i t_{v-1}) - \exp(-\delta_i t_v)] \quad (6)$$

$$i = 1, 2, \dots, N$$

$$\psi_{N+1}^v = (t_v - t_{v-1})\sum_{i=0}^{N+1} Q_i^{v-1}(t_{v-1}) - \sum_{i=0}^{N} \psi_i^v, \quad (7)$$

Where

$$c = -\sum_{j=1}^{N} y_i.$$

The solution of the game G_2 multiple-matrix reduces to the game and the successive solution in accordance with the procedure and numerical algorithms presented in [10], $(L+1)$ games slots $[\tau_k, \tau_{k+1}]$ $(0 \le k < L;\ \tau_{L+1} = t_f)$ with constant average probability of finding the attacker in the states $B_i(i = 0, 1, \dots, N+1)$. Calculated by the formula (3-7). set of solutions $(L+1)$ games, for which the probability of failure-free operation $P(t_f)$ the targeted hardware redundant dynamical system has the highest value, yields the desired settings vector $\bar{\tau} = \{\tau_0, \tau_1, \tau_2, \dots, \tau_L\}$, $\tau_0 = 0$, $\tau_L < t_f$ system $S_A^*(n, m, \bar{s})$, whose elements correspond to the moments of reserve elements redistribution attacked system between the failed basic in q-groups and corresponding to these settings backup vectors $\bar{s}(\tau_k) = \{s_1(\tau_k), s_2(\tau_k), \dots, s_q(\tau_k)\}$ dynamic system, which are the solution G_2 games.

4. Conclusion

In conclusion, it should be noted that the considered numerical and analytical model of the game problem of confrontation of two dynamical systems under incomplete information from players of the behavior of participants in the game in the process of conflict is applicable not only to study and optimize the behavior of conflicting technical systems but also for systems of different physical nature with appropriate identification.

It will be appreciated that the analytic solution of the problem is possible only in the simplest cases. In other cases it is necessary to use numerical methods and modeling with the help of personal computers.

References

[1] Bingzhen Sun, Weimin Ma and Haiyan Zhao, 2016 Rough set-based conflict analysis model and method over two universes *Information Sciences* **372** pp 111-125 https://doi.org/10.1016/j.ins.2016.08.030.

[2] Dejan Verčič, Ralph Tench and Ana Tkalac Verčič 2018 Collaboration and conflict between agencies and clients, *Public Relations Review* **44**(1) pp 156-164 https://doi.org/101016/j.pubrev.2017.11.001.

[3] Iain Phillips and Irek Ulidowski 2015 Reversibility and asymmetric conflict in event structures *Journal of Logical and Algebraic Methods in Programming* **84**(6) pp 781-805 https://doi.org/10.1016/j.jlamp.2015.07.004

[4] Bingzhen Sun, Weimin Ma and Haiyan Zhao 2016 Rough set-based conflict analysis model and method over two universes *Information Sciences* **372** pp 111-125 https://doi.org/10.1016/j.ins.2016.08.030.Nartov B.K. Conflict of Moving Systems. AMSE Press, France.

[5] Tamer Başar, Jun Moon 2017 Riccati Equations in Nash and Stackelberg Differential and Dynamic Games *IFAC-PapersOnLine* **50**(1) pp 9547-9554 https://doi.org/10.1016/j.ifacol.2017.08.1625.

[6] Potapov V I 2015 Tasks and numerical algorithms to optimize the reliability of hardware and redundant technical system in a conflict situation with different strategies to protect against enemy attacks *Mechatronics, Automation, Control* **16** pp 617-624

[7] Potapov V I Formulation and solution of the problem of confrontation gaming hardware-redundant dynamic system with the attacking enemy force in the conditions of incomplete information in the course of the conflict *Mechatronics, Automation, Control* **(9)** pp 18- 24

IOP Conf. Series: Journal of Physics: Conf. Series **1050** (2018) 012063 doi:10.1088/1742-6596/1050/1/012063

Algorithms and software for obtaining dissimilar order and high order of symbolic sequences

N N Pozdnichenko, M S Stupin, A S Gumenyuk, M S Doroshenko and O P Shafeeva

Omsk State Technical University, 11, Mira ave., Omsk, 644050, Russia

Abstract. Generally accepted methods for processing symbolic sequences existing nowadays do not take into account the arrangement of components in those sequences or do it in an inefficient way. In this paper, we propose the development of previously presented means of formal order analysis (FOA). In particular, two new methods of order transformation are proposed here: the secondary order (the high order or the order derivative) and the dissimilar order. First method takes into account not only the intervals between the individual elements, but also the arrangement of the intervals themselves. And the second method takes into account the order of elements occurrences. Algorithms and software for implementing the above transformations are presented. Software approbation is carried out, and adequacy for study of the arrangement of components in symbolic sequences (on the example of nucleotide sequences) is shown.

1. Introduction

In a number of works by other authors [1, 2, 3, 4], the intervals between the closest identical elements (in form of inter-nucleotide distance) were used to study nucleotide sequences. However, after extracting the intervals, the authors used the statistical apparatus and compared the truncated rank distribution of intervals. In some works even intervals of congeneric sequences were described. Nevertheless, the concept of the order of elements as a separate entity has not been defined, and no integral characteristic that allows describing separate nucleotide sequence by a single number has been defined. All the above makes the application of such approaches not very convenient in practice.

In works [5, 6, 7, 8, 9] methods of formal order analysis were presented, they provide means for the analysis of symbolic sequences of different nature, in particular nucleotide sequences. Previously various approaches, methods and software tools for processing symbolic sequences were developed, in particular: functions of order (sliding window local characteristics)[9, 10], matrixes of accordance and binary characteristics [7], rank distributions of congeneric sequences characteristics [7, 8], statistical characteristics analogs, characteristics of order analogous to the moments of random variables.

In a number of works, the mathematical properties of the order were considered [5, 6, 11]. However, all considered approaches use only intervals between identical components. This paper proposes the approach using the arrangement of intervals for obtaining the secondary and higher orders, and the approach taking into account the intervals between the n-th (first, second, etc.) occurrences of different elements to obtain an order on the basis of dissimilar sequences, hereinafter referred to as a dissimilar order.

2. The Formulation of the problem

The subject of this work is developmentof algorithms and software implementing extraction of a high order and a dissimilar order as well as their numeric characteristics with the subsequent approbation of their adequacy and effectiveness for the study of nucleotide sequences.

To obtain a secondary order or a higher order from the existing one, it is necessary to construct a sequence where each element is replaced by an interval either to the next same element or from the previous one, depending on the binding (link), and then to extract the order of such a sequence in the standard way [8].

To obtain a dissimilar order it is necessary to replace all n-th occurrences of the sequence elements with the current numbers of these occurrences, and then to extract the order of such a sequence in the standard way [8].

3. Algorithms and software development

3.1. Dissimilar order extraction algorithm

Two new methods of order transformation were developed in this work. First method extracts dissimilar order from symbolic sequence. To obtain dissimilar order one needs to replace each element with number of its occurrence in congeneric sequence.

In pseudocode form the procedure of dissimilar order extraction looks as follows:

```
Chain function dissimilar (Chain source)
{
  result = new Chain(source.length);
  entries = new int[source.Alphabet.Length];
  for (j = 0; j < source.Length; j++)
  {
    index = source.Alphabet.IndexOf(source[j]);
    entry = ++entries[index];
    result[j] = entry;
  }
  return (result);
}
```

Where «source» is a representation of symbolic sequence as object, storing an alphabet and an order of this sequence; «entries» is an array storing current number of occurrences for each element from alphabet; «result» is a resulting dissimilar sequence.

A	C	C	T	G	A	C	T	G	C	T	A	T	C	G	G	A	T	T	G	A	T	A	C
1	2	2	3	4	1	2	3	4	2	3	1	3	2	4	4	1	3	3	4	1	3	1	2
1	1	2	1	1	2	3	2	2	4	3	3	4	5	3	4	4	5	6	5	5	7	6	6

Figure 1. The example of initial sequence, its order and its dissimilar order.

In figure 1 on the first line is an example of a nucleotide (symbolic) sequence, on the second line is its order, and on the last line is the resulting dissimilar order sequence. The alphabet of the initial sequence being {A, C, T, G}.

3.2. High order extraction algorithm

In this work we define the original order of symbolic sequences as primary order (or just order); derivative from the primary order as secondary order (or order derivative); derivative from the secondary order as tertiary order; etc.

IOP Conf. Series: Journal of Physics: Conf. Series **1050** (2018) 012063 doi:10.1088/1742-6596/1050/1/012063

Thus, the second order transformation method produces secondary order or, in general, high order. To obtain a secondary order, it is necessary to extract an array of intervals between the nearest identical elements from each congeneric sequence, where one of the following bindings can be used: to the beginning, to the end, the cyclic binding to the beginning and the cyclic binding to the end. Binding defines the reading direction (bindings to the beginning correspond to the reading from the right to the left and bindings to the end correspond to the reading from the left to the right). Further, the elements of the order are replaced by the lengths of the corresponding intervals of congeneric chains.

To obtain a higher order, the same steps are applied cyclically to the results of the previous transformation.

In figure 2 on the first line there is an example of a nucleotide (symbolic) sequence, on the second line is its order, and on the last line is the resulting secondary order sequence extracted from the congeneric sequences using cyclic binding to the end, and the alphabet of the initial sequence being {A, C, T, G}.

In pseudocode form the procedure of high order extraction looks as follows:

```
Chain function high order (Chain source, Link link)
{
  result = new Chain(source. length);
  entries = new int[source. Alphabet .Length];
  intervals = new int[source. Alphabet. Length][];
  for (j = 0; j < source .Alphabet. Length; j++)
  {
    Intervals[i] = Extract Intervals( source . Congeneric Chain[i], link);
  }
  for (j = 0; j < source .Length; j++)
  {
    index = source. Alphabet.IndexOf(source[j]);
    entry = entries[index]++;
    interval – intervals[index][entry];
    result[j] = entry;
  }
  return (result);
}
```

Where «source» is a representation of symbolic sequence as an object, storing an alphabet and an order of this sequence; «link» is a binding of the intervals (reading direction) for high order; «entries» is an array storing current number of occurrences for each element from the alphabet; «intervals» is a two-dimensional array of intervals between the nearest similar elements; «result» is a resulting high order sequence. Intervals array is filled by «Extract Intervals» method, which extracts all the intervals for given congeneric sequence (sequence containing elements of only one type).

A	C	C	T	G	A	C	T	G	C	T	A	T	C	G	G	A	T	T	G	A	T	A	C
1	2	2	3	4	1	2	3	4	2	3	1	3	2	4	4	1	3	3	4	1	3	1	2
5	1	4	4	4	6	3	3	6	4	2	5	5	10	1	4	4	1	3	9	2	6	2	2

Figure 2. The example of initial sequence, its order and its secondary order.

3.3. Software development

A software module was developed to obtain a high order and a dissimilar order. Also this module calculates the characteristics of these sequences. The developed module was integrated into the program complex Libiada [12] In figure 3, 4 there are the examples of the interface of this software module.

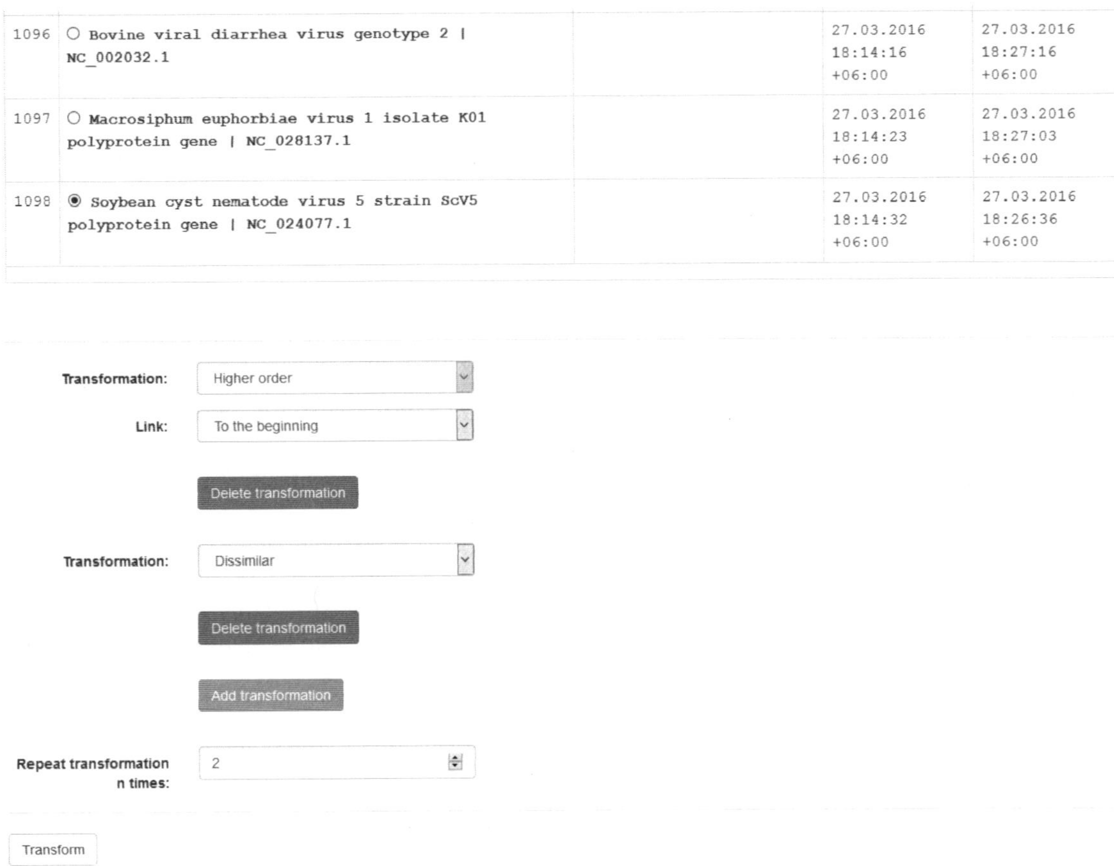

Figure 3. The High order and dissimilar order extraction interface.

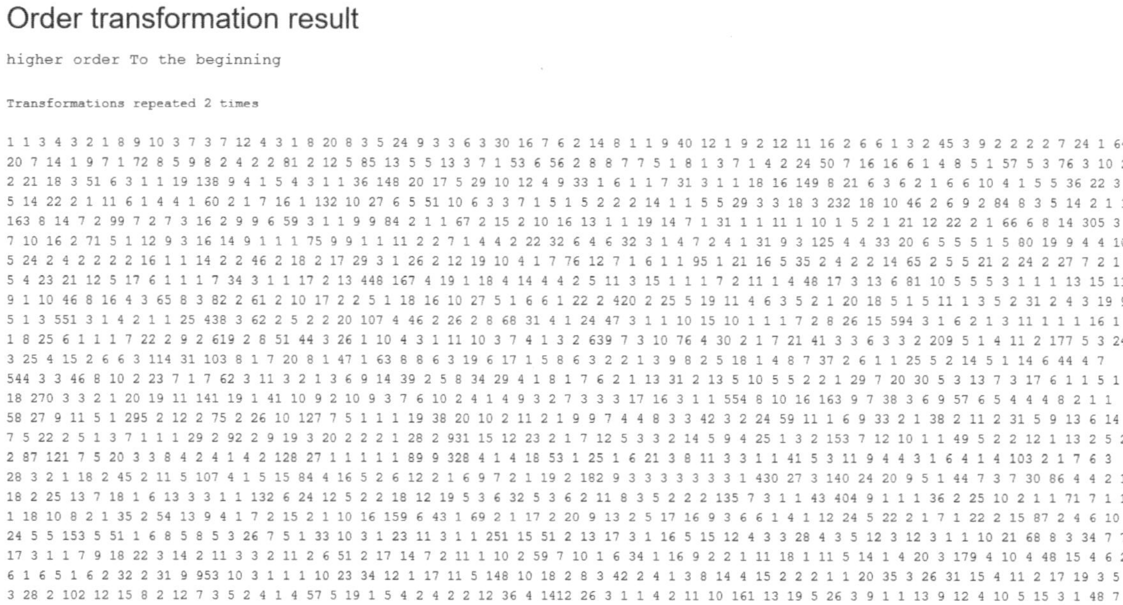

Figure 4. The **t**ertiary order in the application interface.

4. The software approbation

In order to test the adequacy and the informativeness of the characteristics of new representations for the description and study of nucleotide sequences (and other data sets), as well as the approbation of the software complex, a study of complete genomes from the representatives of the family Rickettsiaceae was conducted. As a reference distribution of organisms, the classification obtained in the study [13] was used.

In Table 1 thirty six studied organisms from the family Rickettsiaceae are presented, 34 of which are *Rickettsia* and two are *Orientia*.

Table 1. The secondary order characteristics of complete procariotic genomes of representatives of family rickettsiaceae.

Sequence name	n	g SD	H
R. prowazekii str. Katsinyian | NC_017050.1	88	1.67598359	3.19711214
R. prowazekii str. BuV67-CWPP | NC_017056.1	88	1.67601146	3.19712179
R. prowazekii str. Madrid E | AJ235269.1	88	1.67586287	3.19712541
R. prowazekii str. GvV257 | NC_017048.1	88	1.67594252	3.19712833
R. prowazekii Rp22 | CP001584.1	88	1.67593515	3.19713386
R. prowazekii str. RpGvF24 | NC_017057.1	88	1.67581072	3.19715095
R. prowazekii str. Chernikova | CP003391.1	88	1.67590190	3.19719181
R. prowazekii str. Breinl | NC_020993.1	88	1.67572321	3.19723555
R. typhi str. TH1527 | NC_017066.1	85	1.67253300	3.19796155
R. typhi str. Wilmington | NC_006142.1	85	1.67263713	3.19796214
R. typhi str. B9991CWPP | NC_017062.1	85	1.67246556	3.19796671
R. bellii OSU 85-389 | NC_009883.1	87	1.64546795	3.20362886
R. bellii RML369-C | NC_007940.1	87	1.64473158	3.20374837
R. canadensis str. CA410 | NC_016929.1	90	1.64759932	3.20380032
R. canadensis str. McKiel | NC_009879.1	88	1.64680584	3.20424057
R. felis URRWXCal2 | NC_007109.1	87	1.63000272	3.20732618
R. australis str. Cutlack | NC_017058.1	84	1.62661138	3.20839627

R. japonica YH | NC_016050.1	81	1.62614132	3.20863694
R. parkeri str. Portsmouth | NC_017044.1	84	1.62486505	3.20909549
R. slovaca str. D-CWPP | NC_017065.1	86	1.62410819	3.20913409
R. conorii str. Malish 7 | NC_003103.1	84	1.62468248	3.20914311
R. slovaca 13-B | NC_016639.1	86	1.62407562	3.20915111
R. rickettsii str. Arizona | NC_016909.1	88	1.62408424	3.20952379
R. rickettsii str. Iowa | NZ_CP000766.1	88	1.62393110	3.20952385
R. rickettsii str. Brazil | NC_016913.1	88	1.62407860	3.20953596
R. rickettsii str. Hino | NC_016914.1	88	1.62349341	3.20956609
R. rickettsii str. Hlp#2 | NC_016915.1	86	1.62332926	3.20957332
R. rickettsii str. Colombia | NC_016908.1	88	1.62341171	3.20959700
R. rickettsii str. "Sheila Smith" | NC_009882.1	89	1.62369530	3.20960625
R. massiliae MTU5 | NC_009900.1	84	1.62279389	3.20968071
Candidatus R. amblyommii str. GAT-30V | NC_017028.1	83	1.62417867	3.20976029
R. philipii str. 364D | NC_016930.1	88	1.62311406	3.20977191
R. peacockii str. Rustic | NC_012730.1	82	1.61984729	3.21070805
R. akari str. Hartford | CP000847.1	82	1.61912378	3.21182222
Orientia tsutsugamushi str. Ikeda | AP008981.1	90	1.61471078	3.21378315
Orientia tsutsugamushi str. Boryong | NC_009488.1	90	1.61340758	3.21411961

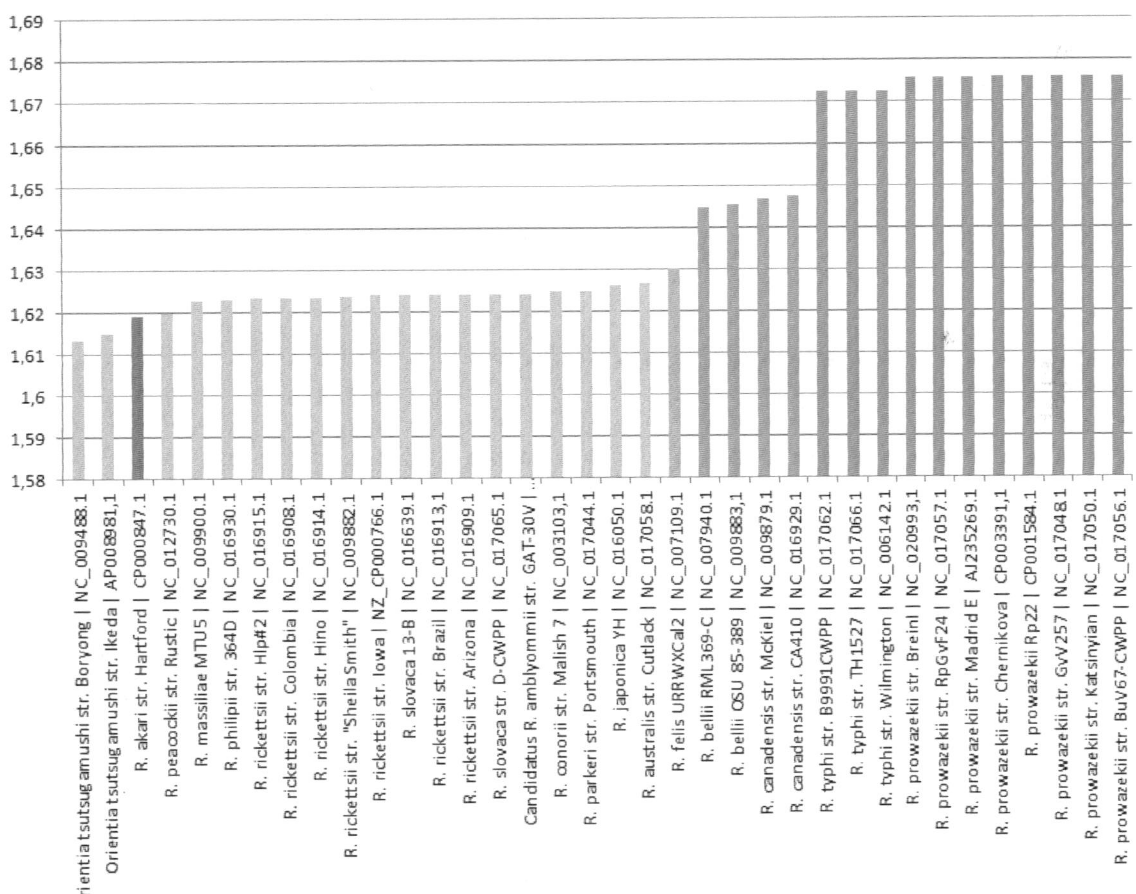

Figure 5. The Rickettsiaceae family complete genomes classification using secondary order average remoteness standard deviation characteristic.

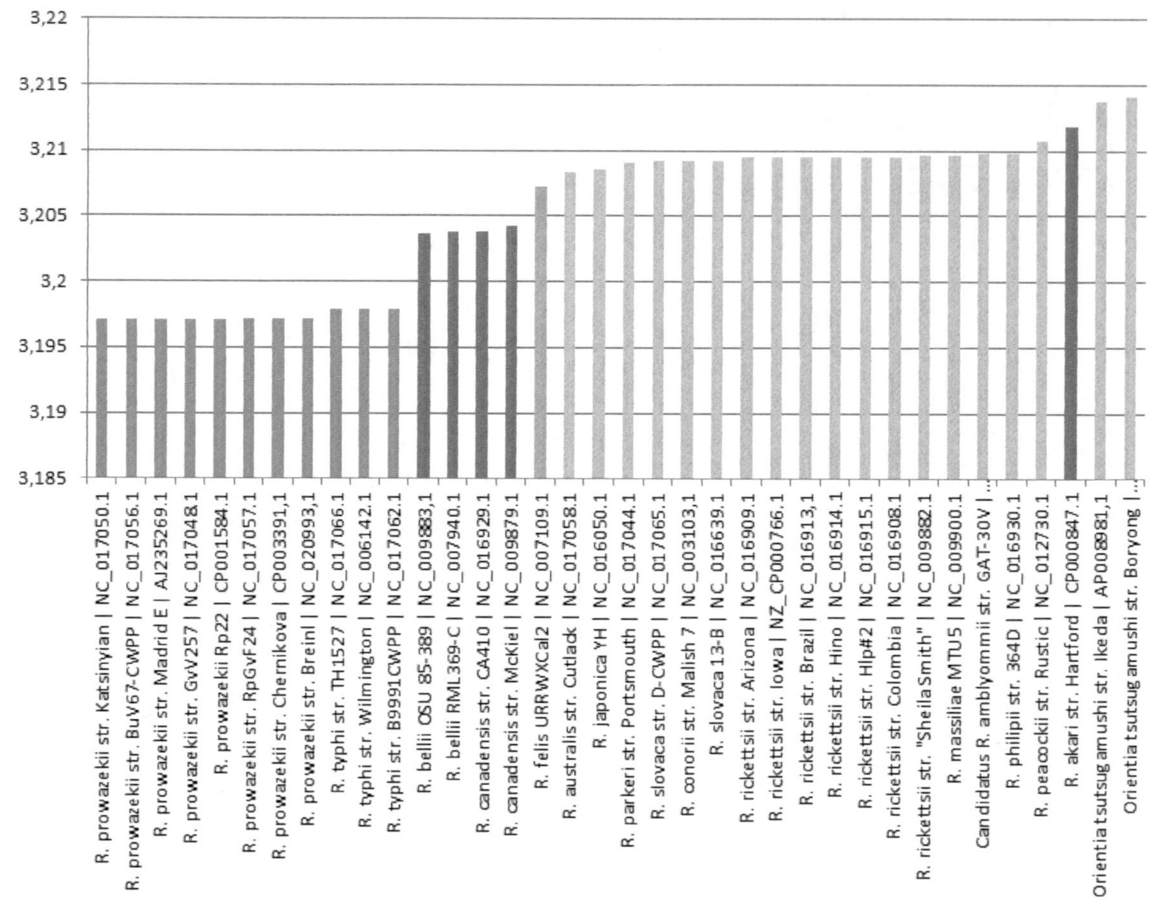

Figure 6. The Rickettsiaceae family complete genomes classification using secondary order entropy characteristic.

5. Discussion

Out of eight computed characteristics of the secondary order of complete genomes, only two characteristics were informative and they fully confirmed the previously constructed classification of rickettsia into three groups. Moreover, the characteristic of the average remoteness standard deviation (*g SD*) made it possible to determine more distinctly the boundaries of the groups (Figure 5). And according to the characteristic of the entropy (H), the genomes were located in exactly the same order as in the original study (Figure 6).

Thus, the developed tools and the obtained characteristics give vast opportunities for both studying real nucleotide sequences and other information chains, and for revealing the mathematical patterns of such transformations and the arrangement of components in sequences as a whole. A test of convergence of a sequence of transformations of different orders to a constant value can be an example of a mathematical study of the obtained transformations.

6. Conclusion

The software extracting a high order and a dissimilar order is designed and implemented in this work. The calculators of order characteristics for obtained sequences are also developed. The approbation of the developed software was carried out. The informativeness of the calculated characteristics was verified by validating the previously conducted research. The perspective of the proposed approaches for further studies, both real sequences, and mathematical regularities was noted.

References

[1] Achuth Sankar S. Nair, T.Mahalakshmi Visualization of genomic data using inter-nucleotide distance signals. *Proceedings of IEEE Genomic Signal Processing*. Romania: Bucharest. 2005.

[2] Vera Afreixo, Carlos A. C. Bastos, Armando J. Pinho, Sara P. Garcia, and Paulo J. S. G. Ferreira. Genome analysis with inter-nucleotide distances. *Bioinformatics*, **25(23)**: pp. 3064–3070, 2009.

[3] Chang G, Wang T. Genome analysis with the conditional multinomial distribution profile. *Journal of Theoretical Biology*, V. **271**, I. 1, 2011, pp. 44–50, 10.1016/j.jtbi.2010.11.034.

[4] Vera Afreixo, Carlos A.C. Bastos, Armando J. Pinho, Sara P. Garcia, Paulo J.S.G. Ferreira. Genome analysis with distance to the nearest dissimilar nucleotide. *Journal of Theoretical Biology*, V. **275**, I. 1, 2011, pp. 52–58, 10.1016/j.jtbi.2011.01.038.

[5] Rodionov I.N., Semenov A.V. On the comparison of the two approaches to the study of genetic texts. *Mathematical Biology and Bioinformatics: III international conf.*, Pushchino: Reports. Edited by Lahno V.D. OOO «MAKS Press», 2008. – pp. 125-126.

[6] Semenov A.V. On one algorithm for generation of all possible orders for sequences with given parameters and a procedure for calculating their number. Information technology and control automation. *Materials of III scientific-practical conference OmSTU (regional).* – Omsk: OmSTU, 2011. – pp. 166 – 168.

[7] Gumeniuk A.S., Klikushin Iu.N., Kobenko V.Iu., Tsiganenko V.N. *Algorithms for the analysis of structure of signals and data*. Omsk. 2010. 272 p.

[8] Gumenuk A.S., Pozdnichenko N.N., Shpynov S. N., Rodionov I.N. Formal analysis of structures of nucleotide chains. *Mathematical Biology and Bioinformatics*. 2013. vol. **8**. № 1. pp. 373–397. URL: http://www.matbio.org/article.php?journ_id=15&id=158

[9] Gumenuk A.S., Pozdnichenko N.N., Shpynov S. N. Formal analysis of order in the local structure of the nucleotide sequences. *Tomsk State University Journal of Control and Computer Science*. 2014. №**4(29)**. pp. 23–30. URL: http://journals.tsu.ru/informatics/en/&journal_page=archive&id=1119&article_id=17725

[10] Tsymbal V.S., Pozdnichenko N.N. The study of mean values and spread of local characteristics of order of the 16S and 18S ribosomal RNA of organisms from the three the realm of life. *Problems of the development, manufacture and operation of rocket and space technology and training of engineering personnel for the aerospace industry: Materials of IX All-Russian scientific conference dedicated to the memory of the chief designer of PA "Poljot" A.S. Klinyshkov* (Omsk, 17 February 2015). – Omsk: OmSTU, 2015. pp. 324-336.

[11] Semenov A.V. On the investigation of combinatorial properties of the order of the symbolic chains. *Modern Problems of Mathematics, Informatics and Bioinformatics. International conference, dedicated to the 100th anniversary of the birth of Alexei Andreevich Lyapunov, corresponding member of the USSR Academy of Sciences: report theses.* – Novosibirsk: Akademgorodok, 2011. – pp. 66.

[12] Skiba A.A., Pozdnichenko N.N. On the development of a software package for computing the characteristics of the structure of ordered data sets. Information technology and mathematical modeling: *Materials of XII All-Russian scientific-practical conference with international participation of A.F. Terpugov* (29-30 Novrmber 2013). – Tomsk: Tomsk University, 2013. – V.**1**. – pp. 50-55.

[13] S. Shpynov, N. Pozdnichenko, A. Gumenuk Application of Formal Order Analysis (FOA) to Higher Order Grouping of Bacteria in the Genera Rickettsia and Orientia // *Microbes and Infection* **17** (2015) pp. 839-844.

IOP Conf. Series: Journal of Physics: Conf. Series **1050** (2018) 012064 doi:10.1088/1742-6596/1050/1/012064

Dynamic Loading of Cogwheels at the Time of Start-Up

N E Proskuriakov and I V Lopa

Tula State University, Lenin Ave., 92, Tula, Russia

Abstract. Loading of cogwheels of the high-loaded transfers to the start-up moment is considered. It is shown that at torque transfer on a surface of a nave of a cogwheel are generated and extend in material in the radial direction of a wave of tension of torsion. The equations for definition of distribution of tension of torsion on the radius of a cogwheel and for definition of a torsion torque on an outer surface of a cogwheel at the time of the beginning of its rotational motion are received. It is shown that thus overloads have more considerable sizes, than used at modern calculations. Results of calculations for specific input data allow us to draw a conclusion that the dynamic torsion torque at start-up is more, than a rated torsion torque more than twice.

1. Introduction

Due to the continuous growth of power capacity of tooth gears, increase in speed ranges of their work there is a problem of more exact accounting of dynamism of loading. At calculations of tooth gears, it is supposed that loadings from a shaft are transferred to a wheel without changes. However, at the time of start-up, loading has dynamic (shock) character [1–3]. As a result, in material of cogwheels, waves of tension are generated and extended, and characteristics tensely - the deformed state in them are heterogeneous and non-stationary. At torque transfer on a surface of a nave of a cogwheel radial waves of torsion in process of which distribution of the forefront there is a reduction of amplitude of tension owing to increase in a surface of the front are generated. At achievement, rotational motion of the last will begin a forefront wave of tension of an outer surface of a cogwheel.

2. Formulation of problem

The intense deformed condition of material of a cogwheel is defined by the components of tensors of tension $\sigma_{R\theta}$ and deformations $\varepsilon_{R\theta}$, which are not depending on coordinates θ and z, the being functions only of the coordinate R and time t. In addition, other components of tensors of tension and deformations are equal to zero. These assumptions reduce process of deformation to distribution in material of radial waves of torsion. The system of equations describing a task includes the movement equation, a condition of continuity and the defining ratio - Hooke's law [4]:

$$\left.\begin{aligned}
&\rho\frac{\partial V(R,t)}{\partial t}=\frac{\partial\sigma_{R\theta}(R,t)}{\partial R}+2\frac{\sigma_{R\theta}(R,t)}{R}\\
&\frac{\partial\varepsilon_{R\theta}(R,t)}{\partial t}=\frac{\partial V(R,t)}{\partial R}-\frac{V(R,t)}{R}\\
&\frac{\partial\varepsilon_{R\theta}(R,t)}{\partial t}=\frac{1}{G}\frac{\partial\sigma_{R\theta}(R,t)}{\partial t}
\end{aligned}\right\}; \qquad (1)$$

where ρ – cogwheel material density; $V(R,t)$ – circumferential speed of material particles; G – module of elasticity of the second sort (rigidity modulus).

3. Theory

Entry conditions correspond to unstressed and undeformed condition of material:

$$\sigma_{R\theta}(R,0) = \varepsilon_{R\theta}(R,0) = V(R,0) = 0. \quad (2)$$

As boundary condition, the dependence considering reduction of tension on a contact area in process of distribution of a front wave deep into of cogwheel material was used:

$$\sigma_{R\theta}(R_0,t) = \sqrt{\left(\frac{R_0}{R(t)}\right)^3}\,\tau_{tor}, \quad (3)$$

where $\tau_{tor} = \frac{T_{tor}}{W_\rho}$ – torsion tension on a shaft surface; T_{tor} – the torsion torque transferred from a shaft to a wheel; W_ρ - polar drag torque of section of a shaft.

The system of equations (1) describing a task represents system of quasi-linear differential equations in partial derivatives of first order of hyperbolic type and therefore, it is reasonable to solve it by method of characteristics [5]. Let's add it with entry conditions (2)-(3) and identical ratios for total differentials of required functions (only 3). As a result, we will receive system of 6 linear algebraic equations with unknown functions:

$$\frac{\partial\sigma_{R\theta}}{\partial z}, \frac{\partial\sigma_{R\theta}}{\partial t}, \frac{\partial\varepsilon_{R\theta}}{\partial z}, \frac{\partial\varepsilon_{R\theta}}{\partial t}, \frac{\partial V}{\partial z}, \frac{\partial V}{\partial t}.$$

Let's demand that it had infinite set of decisions. For this purpose it is necessary that its main determinant equaled to zero. From this condition received three families of the equations of characteristics:

$$dR = 0,\ dR = \sqrt{\frac{G}{\rho}}dt = a\,dt,\ dR = -\sqrt{\frac{G}{\rho}}dt = -a\,dt,$$

where a – a propagation velocity of an elastic wave of tension of torsion deep into cogwheel material.

That the decision along characteristics final zero other determinants of system needed to consider equality. For this purpose, we will replace the last column of the main determinant with the corresponding right parts of the equations and we will consider its equality to zero. Consistently substituting in the received equality of the equation of characteristics, we will construct differential ratios between required functions.

Thus, integration of quasi-linear system of differential equations of hyperbolic type at the set entry and boundary conditions is reduced to integration of ratios along the respective characteristic directions.

The equations of the characteristic directions and ratio along them were received:

– along characteristics $dR = 0$:

$$d\sigma_{R\theta}(R,t) - G\,d\varepsilon_{R\theta}(R,t) = 0; \quad (4)$$

– along characteristics $dR = \sqrt{\frac{G}{\rho}}dt = a\,dt$:

$$d\sigma_{R\theta}(R,t) - a\,\rho\,dV(R,t) + G\,\frac{V(R,t)\,dt}{R} + \frac{2\,a\,\sigma_{R\theta}(R,t)\,dt}{R} = 0; \quad (5)$$

– along characteristics $dR = -\sqrt{\frac{G}{\rho}}dt = -a\,dt$:

$$d\sigma_{R\theta}(R,t) + a\,\rho\,dV(R,t) + G\frac{V(R,t)\,dt}{R} - \frac{2\,a\,\sigma_{R\theta}(R,t)\,dt}{R} = 0\,; \tag{6}$$

The forefront wave of torsion tension extends along the characteristic $dR = \sqrt{\frac{G}{\rho}}dt = a\,dt$, on which the ratio is carried out (5). Besides, on a forefront wave, conditions of a dynamic and kinematic continuity must be satisfied [6]:

$$\sigma_{R\theta}(R,t) = -a\rho V(R,t) \text{ and } V(R,t) = -a\varepsilon_{R\theta}(R,t)\,. \tag{7}$$

Substituting the equations (7) in (5), received the differential equation describing tension variation of torsion on a forefront wave of tension:

$$2d\sigma_{R\theta}(R,t) = -a\frac{\sigma_{R\theta}(R,t)}{R(t)}dt\,. \tag{8}$$

The decision (8) taking into account boundary condition (3) has an appearance:

$$\sigma_{R\theta}(R,t) = \tau_{\kappa p.}\sqrt{\frac{R_0}{R_0 + at}} = \tau_{\kappa p.}\sqrt{\frac{R_0}{R(t)}}\,. \tag{9}$$

Thus, at distribution of a forefront wave of torsion tension of amplitude of tension decreases under the law (9).

4. Results and discussions

It is known that tension of torsion is distributed on radius under the linear law. Therefore, in case of homogeneous materials of a cogwheel and shaft, in each point of material of system "a shaft – a wheel" torsion tension will also change under the linear law. In this case, on the neutral line, it will be equal to zero, and in the point corresponding to arrival to it of a forefront wave of torsion tension – it will be defined from the decision (9). In process of distribution of the forefront deep into of cogwheel material, his material particles will receive rotational motion with line speed $V(R,t)$.

At achievement by a forefront wave of tension of torsion of a ring gear (a dividing circle of a cogwheel) the wheel will begin rotational motion. In Figure 1 distribution of tension of torsion is presented on a forefront wave of tension in an instant of its arrival on a dividing circle of a wheel at the following input data: diameter of a shaft is 60 mm, dividing wheel size of 460 mm, a torsion torque equal, $T = 2000\ Nm$. It is visible that in the considered case tension falls on 2/3 from initial value at the front. In addition, the more diameter of a cogwheel, the more will be tension drop.

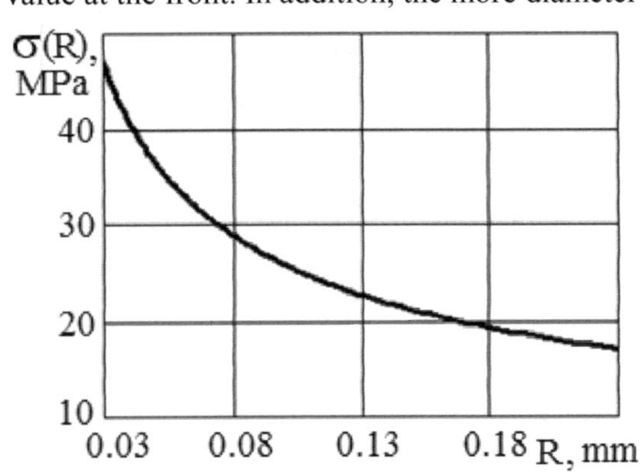

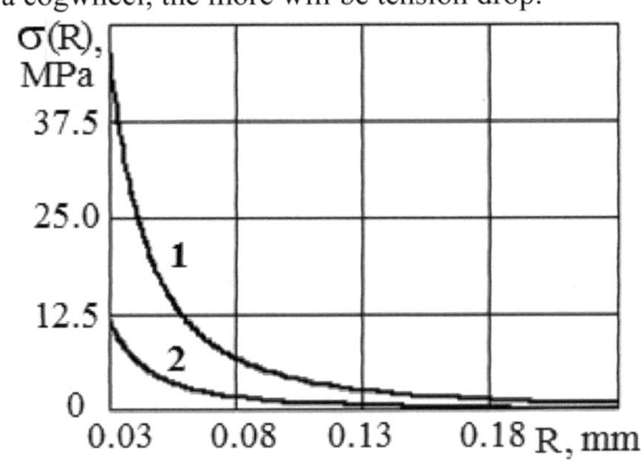

Figure 1. Distribution of tension of torsion on a forefront of wave

Figure 2. Distribution of torsion tension on wheel radius

At the same time, behind the front of the extending wave there will be a tension drop due to its reduction on a boundary surface (a surface of contact of a shaft and a wheel). Numerical solutions of the corresponding wave tasks [4-6] allow to draw a conclusion that tension drop behind the front happens under the law to the similar law of tension drop at the front, that is corresponds to the equation (9) where as a multiplier before a root it is necessary to use boundary condition:

$$\sigma_{R\theta}(R,t)=\sqrt{\left(\frac{R_0}{R(t)}\right)^3}\tau_{\kappa p.}\sqrt{\frac{R_0}{R_0+at}}=\left(\frac{R_0}{R(t)}\right)^2\tau_{\kappa p}\,. \qquad (10)$$

In Figure 2 distribution of tension on wheel radius at the same input data is provided to different instants of time: the curve 1 corresponds to run time a wave from a shaft to an external wheel rim; a curve 2 – time at run by the reflected front wave front from an external wheel rim to a shaft (time is twice more). It is visible that distribution of tension on the radius of a wheel is leveled. It comes as because of partial reflection of the falling wave of tension of torsion from a surface of contact with the second cogwheel entering gearing and because of the subsequent interference of the corresponding waves. The reflected front extends in previously heterogeneous strained and deformed material. Tension at the front of a back wave unloads material in process of distribution of its front to the loaded surface of contact of a shaft and a wheel. Thus, the total state on the forefront of a back wave was presented in the form [7]:

$$X_{II}^{\Sigma}(R,t)=X_{II}(R,t)+X_{I}(R,t),$$

where X_{II}^{Σ} – total value of required sizes at the front a back wave as a result of its interaction with the falling wave; X_{II} – the amplitude value of required size introduced in considered point by the forefront of a back wave; X_I – the corresponding parameter of a state in the considered rod point before arrival of the front of a back wave to it.

Then, the dynamic torsion torque in an instant of the beginning of rotational motion of a wheel will be defined so:

$$T_d=2\pi\int_{R_0}^{R} r^2\sigma_{R\theta}(R,t)dr\,.$$

Taking into account (10) expression for definition of a dynamic torsion torque has an appearance:

$$T(t)=2\pi\int_{R_0}^{R} r^2\left(\frac{R_0}{R_0+at)}\right)^2\tau_{\kappa p.}dr\,. \qquad (11)$$

Results of calculations at the considered earlier input data allow us drawing conclusion: the dynamic torsion torque (at start-up) is significantly more than nominal torque. At run time a wave from a shaft to an external wheel rim (at the time of a start of motion), a torsion torque equal $T=5330\ Nm$, that is 2.66 times more rated, and then at double increase in time a torsion torque equal $T=2667\ Nm$, that is rather Use the "Insert Citation" button to add citations to this document.
close to nominal value of a torsion torque.

5. Conclusions

Thus, the scheme of wave loading of material of cogwheels at the time of start-up is considered. The equations for definition of distribution of tension of torsion on the radius of a cogwheel and for definition of a torsion torque on an outer surface of a cogwheel at the time of the beginning of its

IOP Conf. Series: Journal of Physics: Conf. Series **1050** (2018) 012064 doi:10.1088/1742-6596/1050/1/012064

rotational motion are received. It is shown that thus overloads have more considerable sizes than used at modern calculations, but they have short-term character and quickly fade.

References

[1] Baranov V L, Karpukhin V P, Lopa I V 2003 Features of design of dynamic drives of locks of pipelines (Tula: Tula State University) 202 p

[2] Tobe T, Takatsu N 1989 Dynamic Loads on Spur Gear Teeth caused by Teeth Impact *Bull. JSME* **16** pp 1031-1037

[3] Velex P, Berthe D 1990 A Survey of Dynamic Loading Conditions on Gear Teeth Contacts *NATO ASI Series* (Series E: Applied Sciences, Springer, Dordrecht) **174** pp 631-652

[4] Baranov V L, Lopa I V 1990 Radial waves of pressure in thermo-elastic and visco-plastic plate *Izvestiya VUZov. Mashinostroyeniye* **2** pp 16-19

[5] Baranov V L, Lopa I V 1989 Radial waves of torsion and longitudinal shift in an elastic and viscoplastic thick plate in no isothermal statement *Izvestiya VUZov. Mashinostroyeniye, Moscow* 7 pp 27-35

[6] Baranov V L, Lopa I V 1993 Longitudinal elastic and visco-plastic waves in rods of final length *Izvestiya VUZov. Mashinostroyeniye* **1** pp 54-57

[7] Proskuriakov N E, Lopa I V 2017 The monitoring of pipeline strength in dynamic loading AIP Conference Proceedings **1876** 020095 https://doi.org/10.1063/1.4998915

A method for the measurement of the diameters of large-scaled shafts

H Radev, V Bogev, V Vassilev

Mechanical Engineering Faculty, Technical University of Sofia, Sofia, Bulgaria

Abstract. This article discusses a method for the measurement of the root-mean-square circles of the important cylindrical surfaces of large-scale shafts. The method allows the realization of the measurement directly on the machine by the use of the results of measuring the radial runout of the relevant cross-sections of the shaft and of one attached to the shaft forehead etalon ring which has pre-calibrated both the diameter and as well as its roundness deviation. The main sources of uncertainty have been analysed.

Key words: large shafts, measurement of diameters, uncertainty.

1. Introduction

High requirements are demanded for the geometric accuracy of the surfaces of rotation of the large-scaled shafts in the transport and heavy machinery industry, as well as the power industry and other industries, such as turbine, roller, crankshaft, etc. Tolerances of such diameters are usually within 15-60 μm.

The large sizes and weight of these objects (diameters up to 1500 mm, length up to 20000 mm and weight up to 80-100 t) specify the measurement of such tolerances.

The most widely used measurement methods by the means of mechanical measuring instruments (like micrometres) do not assure the necessary accuracy. The application of coordinate measuring machines (CMM) is highly restricted by shape of the shafts and also their positioning in the machines working space as well as economic reasons make their use impossible [4].

Indirect method of measurement using measurement data by the circle elements or by contact rollers also do not give the required accuracy, which in turn is highly influenced by the measured cross sections form deviations.

Furthermore, these methods give information only for some average value of these diameters [3].

The main goal of this paper is a new method of the measurement of the diameters of large-scaled shafts developed by the authors that assures high measurement accuracy directly on the production machine [2].

2. Description of the method

2.1. Essentials of the measurement

The principle of the operation during the positioning of the shaft on the machine centres is shown in Fig. 1.

IOP Conf. Series: Journal of Physics: Conf. Series **1050** (2018) 012065 doi:10.1088/1742-6596/1050/1/012065

A ring with a calibrated diameter is attached to the one of the shaft faces. The diameter of the ring also its form deviation are preliminary calibrated in marked in one longitudinal and one cross section by the means of height gauge and roundness measuring machine.

The angular rotation of the shaft is measured by a rotary encoder, attached in a device with contact roller, which touches the outer surface of the shaft, and a linear encoder measures the radial runout of the surface. This linear encoder is connected to the production machine's carriage. The both encoders are connected by the means of a display gauge to a PC.

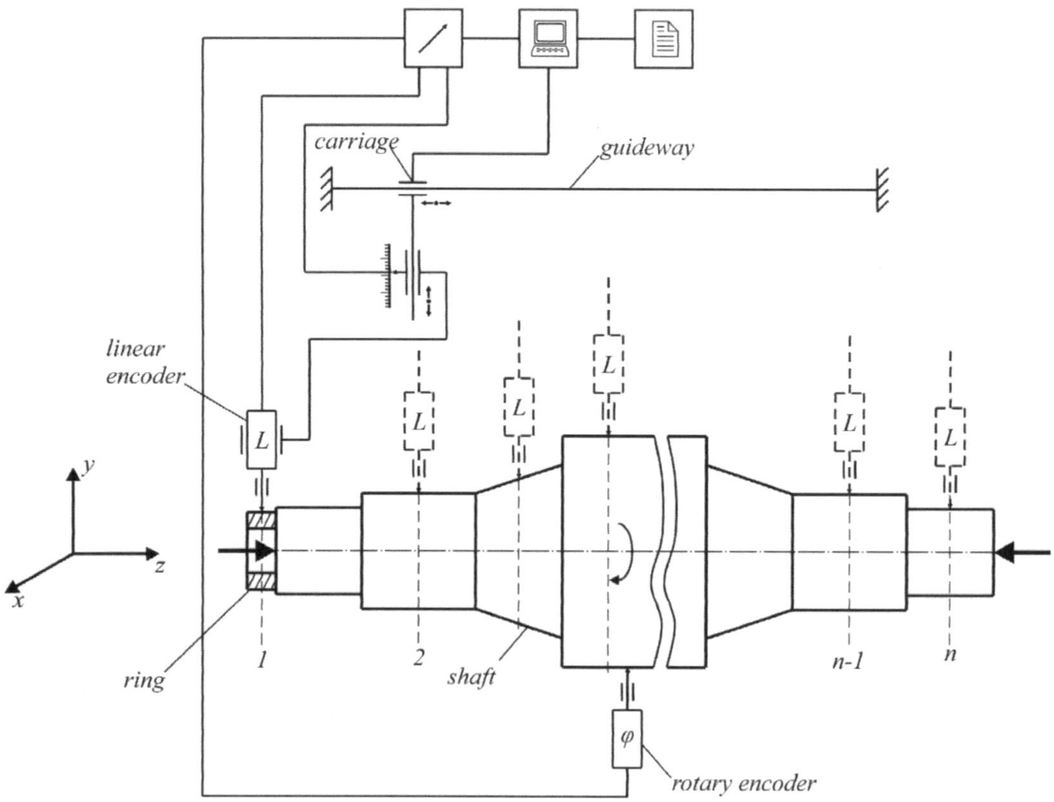

Figure 1. Principle diagram of shaft's diameters measurement when based on the production machine

It is assumed that the carriage travel is straight and parallel to the shaft's axis of rotation.

The measuring devices of the machine determine the longitudinal and transverse displacement of the support.

2.2. Measurement procedure

The symbols and numeration are as follow:

i – number of the current cross section;

$i = 1 \dots n$

$i = 1$ – the cross section of the ring;

$i = n$ – the last measured cross section of the shaft;

j – current measured point of cross section's profile;

$j = 1 \dots m$, (m is even);

$j = 1$ – the marked points of both the shaft and ring.

D_{1i} – local diameter as a distance between two opposite points of the profile;

R_{av_i} – root mean square (RMS) circle's radius;
D_{av_i} – RMS circle's diameter;
e_i – eccentricity;
$e_{x_{1i}}$ – eccentricity projection on the line of measurement;
Δ_{1i} – local roundness deviation of profile's marked point;
$\Delta_{1+\frac{m}{2},i}$ – local roundness deviation of the point opposed to the marked one where m is the number of the profile's measured points;
0 – axis of rotation;
0_{av_i} – centre of RMS circle.

The measurement procedure is as it follows:

The tip of the linear encoder is brought in contact with the ring in the marked point ($j = 1$) of the calibrated cross section ($i = 1$) (Fig. 2).

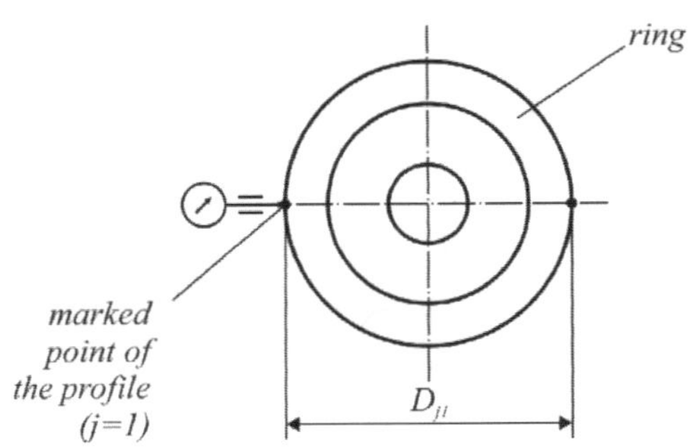

Figure 2. The ring with calibrated diameter D_{11}

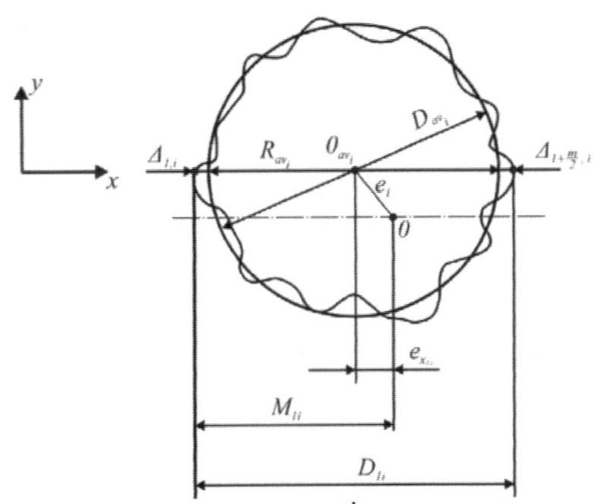

Figure 3. Diagram of the ith cross section

The readings A_{11} of the display gauge and also the reading B_{11} of device measuring the carriage transverse travel are input in the PC. Their sum is calculated:

$$C_{11} = A_{11} + B_{11} \tag{1}$$

The shaft is rotated by one revolution and by the values of the local runout, i.e. C_{j1}, a diagram of the cross section is created (Fig. 3).

The values Δ_{11}, $\Delta_{1+\frac{m}{2},1}$ and $e_{x_{11}}$ are calculated using the diagram.

The value of D_{av_1} is calculated by the equation:

$$D_{av_1} = 2(M_{11} - \Delta_{11} - e_{x_{11}}), \tag{2}$$

But

$$D_{av_1} = 2\left(D_{11} - \Delta_{11} - \Delta_{1+\frac{m}{2},1}\right) \tag{3}$$

The value of D_{11} is measured during the ring's calibration. The following equation is drown from substituting (3) into (2):

$$M_{11} = 0{,}5\left(D_{11} - \Delta_{11} - \Delta_{1+\frac{m}{2},1}\right) + \Delta_{11} + e_{x_{11}} \tag{4}$$

By moving the support the linear encoder is brought in contact with the shaft in the next i^{th} section at a marked point (j_i, where $j = 1$) in the plane of the marked point of the ring.

The readings A_{1i} of the display gauge and B_{1i} of the device measuring the transverse displacement of the carriage are imputed in the PC and their sum C_{1i} is calculated:

$$C_{1i} = A_{1i} + B_{1i} \tag{5}$$

The shaft is rotated at one turn and the results of the radial runout, i.e. by the values of C_{1i} the ith cross section's roundness graph is drawn.

The values of Δ_{1i} and $e_{x_{1i}}$ are evaluated by the roundness graph.

$$R_{av_i} = M_{1i} - \Delta_{1i} - e_{x_{1i}}, \tag{6}$$

$$D_{av_i} = 2\left(M_{1i} - \Delta_{1i} - e_{x_{1i}}\right) \tag{7}$$

$$M_{1i} = M_{11} + \Delta_{C_i} \tag{8}$$

$$\Delta_{C_i} = C_{1i} - C_{11} \tag{9}$$

After substitution of (8) in (7):

$$D_{av_i} = 2\left(M_{11} + \Delta_{C_i} - \Delta_{1_i} - e_{x_{1i}}\right) \tag{10}$$

The local diameter D_{av_i} can be evaluated as the distance between two opposite points:

$$D_{ji} = D_{av_i} + \Delta_{ji} + \Delta_{j+\frac{m}{2},i}$$

In the case of a straightness deviation of the movement of the carriage and a deviation from parallelism of this movement with respect to the axis of rotation, these deviations are determined by testing the geometrical accuracy of the production machine by appropriate methods.

Corresponding corrections of the values of C_{ji} shall be entered to exclude the influence of the both the deviations of straightness and parallelism of the carriage's movement.

3. General sources of uncertainty

In accordance with this method, the model function (the functional relation between the diameter of the RMS circle D_{av_i} in a given section i and the input values) is given by the expression:

$$D_{av_i} = D_{11} - \Delta_{1+\frac{k}{2},1} + \Delta_{11} + 2e_{x_{11}} + 2\Delta A_{1i} + 2\Delta B_{1i} - \Delta_{1i} - e_{x_{1i}} \tag{11}$$

It is assumed that the input quantities are non-correlated. Then:

$$u\left(D_{av_i}\right) = \sqrt{\begin{array}{l} u(D_{11})^2 + u\left(\Delta_{1+\frac{k}{2},1}\right)^2 + u(\Delta_{11})^2 + 2u\left(e_{x_{11}}\right)^2 + \\ +2u(\Delta A_{1i})^2 + 2u(\Delta B_{1i})^2 + u(\Delta_{1i})^2 + u\left(e_{x_{1i}}\right)^2 \end{array}}, \tag{12}$$

but

$$u\left(\Delta_{1+\frac{k}{2},1}\right)^2 = u(\Delta_{11}) = u(\Delta A_{1i}) = u(\Delta_{1i}),$$

IOP Conf. Series: Journal of Physics: Conf. Series **1050** (2018) 012065 doi:10.1088/1742-6596/1050/1/012065

then:

$$u\left(D_{av_i}\right) = \sqrt{u(D_{11})^2 + 4(\Delta_{11})^2 + 2u\left(e_{x_{11}}\right)^2 + u\left(e_{x_{1i}}\right)^2 + 2u(\Delta B_{1i})^2} \tag{13}$$

$u(D_{11})$ is calculated by measuring instruments, which have been used during the calibration of the ring. When measuring with TESA MicroHite-M356, when the diameter of the ring is 200 mm, $u(D_{11}) = 0.7\mu\text{m}$.

$u(\Delta_{11})$ is defined by the accuracy of the measuring instrument. According to the datasheet of the linear encoder, the maximum error is ± 1 μm.

With adopted rectangular distribution law (evaluation type B), $u(\Delta_{11}) = \frac{2}{3{,}46} = 0{,}6$ μm.

$u\left(e_{x_{11}}\right)$ and $u\left(e_{x_{1i}}\right)$ depend on the roundness deviation values in the corresponding cross sections, and also on the errant runout of the axis of rotation. Research shows that when the number of profile's measured points is more than 128, $u\left(e_{x_{11}}\right)$ and $u\left(e_{x_{1i}}\right)$ do not exceed 1% of the roundness value [1].

$u(\Delta B_{1i})$ depends on the accuracy of the transverse movement of the carriage and requires special inspection.

For example, when assessing the accuracy of the positioning of a Hercules circular-grinder at the NKMZ - Kramatorsk, Ukraine, the estimated uncertainty $u(\Delta B_{1i})$ does not exceed 2.2 μm.

4. Conclusion

1. To the important rotating surfaces of large-scaled shafts in the transport and heavy machinery industry, and other industrial sectors, do exist high requirements for the accuracy of the processing, respectively to the accuracy of their diameters.
2. The developed measurement method ensures high measurement accuracy when the shaft is based directly on production machine.
3. The method allows the determination of both the diameters of the RMS circles of the measured cross sections and of the local diameters as the distance between two opposite points of the profiles.
4. The review of major sources of uncertainty indicates that the method and measurement tools used for its implementation provide uncertainty of measurement less than the target-specific uncertainty associated with the tolerance.

5. References

[1] Vasilev V Radev H Bogev V Blagov I 2012 *Izsledvane na postoyanstvoto na virtualna bazova os pri izmervane na otkloneniyata na formata i razpolozhenieto na rotatsionni detayli (Sozopol)* (Sofia: Technical University of Sofia Press) p 161

[2] Radev H Bogev V Vasilev V Panchev N Dzhambazov M Naduev I Voloshin A Samoylov V Ignatova E *Patent Application №111501/08.08.2013 A Method for measuring diameters of large-sized shafts.*

[3] Radev H *et al.* 2010 *Metrology and Measurement Instruments Vol. 2* (Sofia: Softtrade) chapter 6 pp 378-445

[4] Rubinov A 1982 *Kontrol bolshikh razmerov v mashinostroyenii* (Leningrad: Mashinostroyeniye)

IOP Conf. Series: Journal of Physics: Conf. Series **1050** (2018) 012066 doi:10.1088/1742-6596/1050/1/012066

Measurement of the macro geometry of large-scale rotary workpieces type "ring"

H Radev, V Bogev, V Vassilev, I Blagov

Mechanical Engineering Faculty, Technical University of Sofia, Sofia, Bulgaria

Abstract. The measurement of geometric parameters (diameters and deviations of the shape and position) of large-sized details such as "ring" with the intended purpose is explored. The measurement system is built on the basis of a vertical turning machine, universal measuring instruments of linear displacement, specialized measuring instruments and the special software and allows to measure the respective parameters directly on the production machine. The developed measuring procedures are based on the determination of the deviation of the form and the position from the results of the radial and axial runout, with or without exclusion of the errant runout. In the first case, these results are brought to the virtual datum (axis) by introducing the appropriate corrections.

Key words: measurement, large-scale rotary workpieces, macro geometry.

1. Introduction

Large ring-shaped parts are one of the main functional elements of hydroelectric power plant turbines. The typical requirements for their macro geometry are shown in Fig. 1.

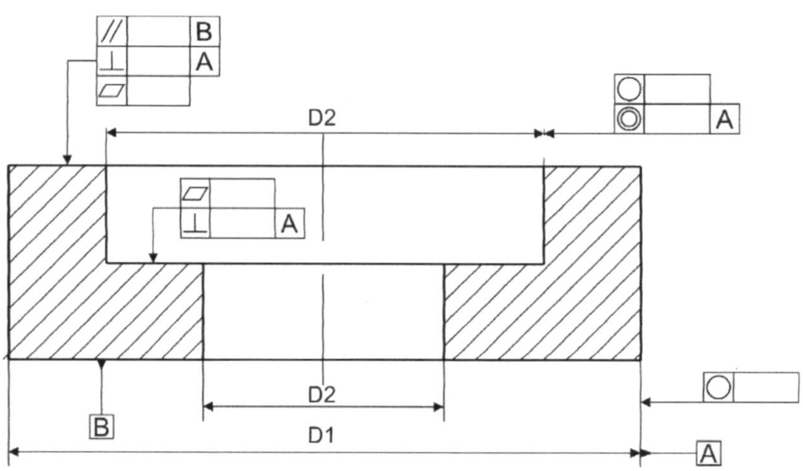

Figure 1. Typical requirements for the macro geometry of large-scaled workpieces type "ring"

For similar sized parts, measurement of these parameters is not a problem. However, for workpieces with sizes up to and over 10 m, very serious metrological problems arise due to the difficulty or inability to use routine (standard) measurement schemes.

IOP Conf. Series: Journal of Physics: Conf. Series **1050** (2018) 012066 doi:10.1088/1742-6596/1050/1/012066

These problems can be solved with the developed measuring system, based on a vertical turning machine, universal linear displacement measuring instruments, specialized measuring equipment and software [3].
Measurements can be carried out directly at the workplace without removing the workpiece from the machine.

2. Principle of operation of the measuring system

The principle of operation of the measuring system is shown in fig.2 [3]. The vertical turning machine is a main component of it.

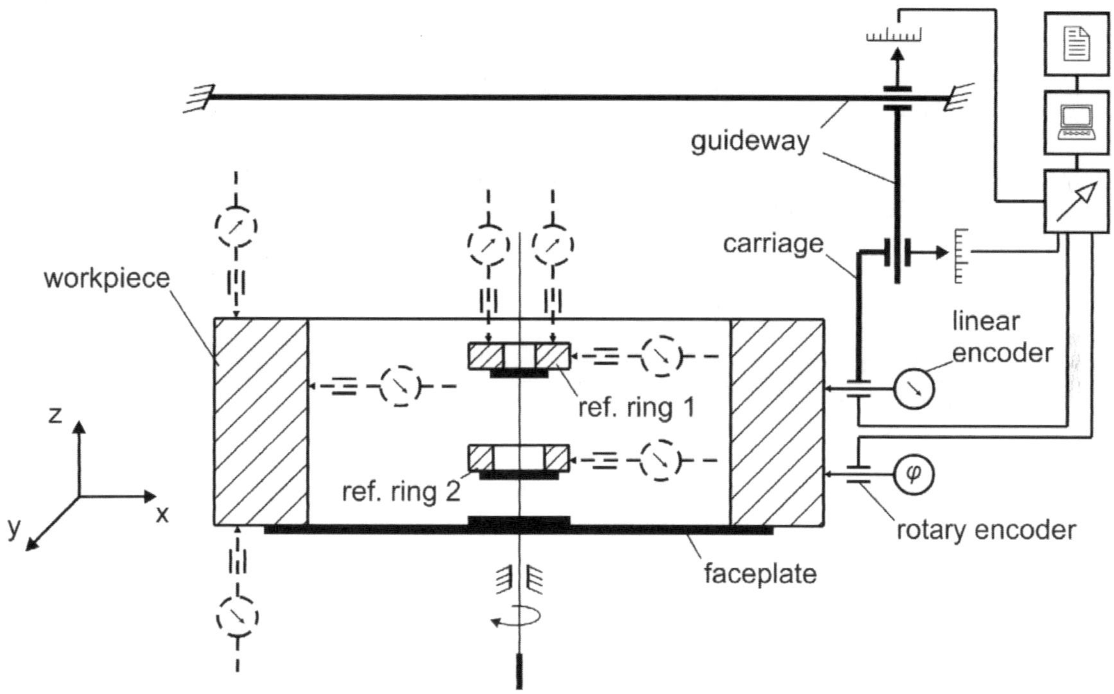

Figure 2. Principle of operation, based on a vertical turning machine

The measured part is located on the faceplate of the lathe and is centred relative to the axis of rotation by means of the jaws of the faceplate.

In the centre, roughly aligned with the axis of rotation, a mandrel is mounted with two reference rings 1 and 2, spaced apart at a distance H of each other. Both rings have calibrated diameters and roundness deviation EFK in marked cross sections, and ring 1 has also calibrated flatness deviation EFE of the front surface.

A linear sensor is installed in place of the tool post, and the rotation of the faceplate is measured by a rotary encoder built into a device, which contacts with the outer rotational surface of the workpiece.

Both the linear and rotary encoders are connected to the reading unit (the computing and reading unit of the system).

Both the horizontal (*X* axis) and vertical (*Z* axis) linear encoders are also connected to the reading unit. The primary measurement information is input for processing and registration to the computing unit (laptop) and the register unit (printer).

The software of the system ensures acquisition of primary measurement information, data processing and presentation of measurement results [1]. The results shall be documented in the form of relevant protocols presenting the information in textual and graphical form.

3. Methods of measurement

3.1. Measuring the deviation from roundness

Determination of the deviation from roundness is made on the results of the measurement of the radial runout in the respective cross sections both without and with excluding the errant runout by special programs [2].

In the first case, an exact datum (axis of rotation) is required, and in the second, this requirement is not necessary.

The decision to use one way or the other is made after evaluating the significance of the errant runout. When the errant runout is excluded, the results are brought to the virtual datum (axis of rotation) by entering appropriate corrections [2].

3.2 Measuring the deviation from flatness

The deviation from flatness is determined by the results of the measurement of the axial runout of one or more circular planar profiles by a given radius *R*. When measuring one profile (one track), the program for calculating the deviation from roundness is used. On the profile diagram, fig. 3, Δ_j represents the flatness deviation from the root mean square (RMS) associated plane at the jth point of the profile ($j = 1 \div k$) and the eccentricity Δ_x and Δ_y determine the position and the magnitude of the projection e' (e_{xy}) of the normal vector to the RMS associated plane in a plane, perpendicular to the axis of rotation, i.e. in the *XOY* plane.

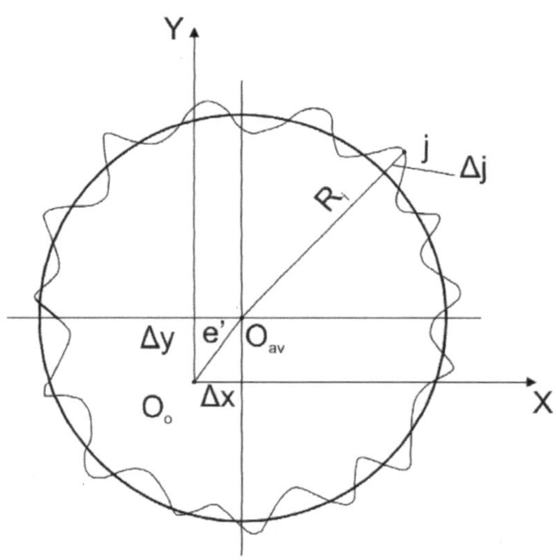

Figure 3. Profile diagram for calculating the deviation from roundness

When determining the deviation from flatness from the measurement results of the axial runout of more than one concentrically arranged round profiles, it is necessary, in addition to exact rotational movement of the workpiece (minimum axial runout) and straight, perpendicular to the axis of rotation, movement of the linear encoder, fig. 4.

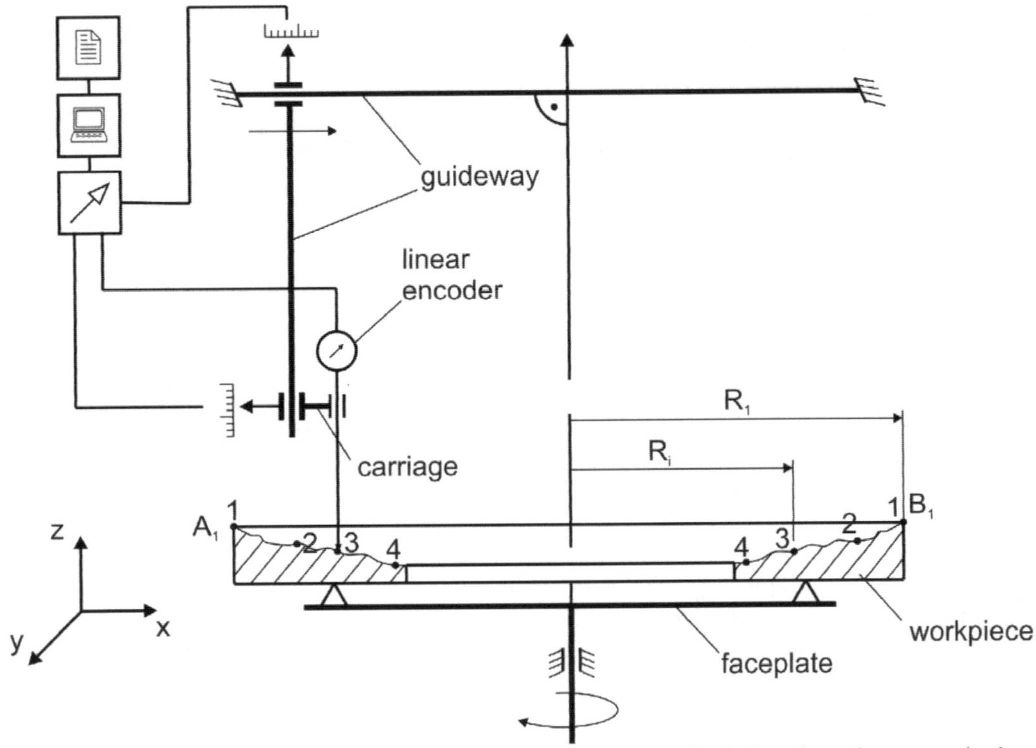

Figure 4. Arrangement for the measurement of the flatness deviation, based on a vertical turning machine

The radius of the current circular profile is denoted by R_i in fig. 4, where i=1÷n. In fig. 4 i=1÷4.

Generally, the face is not perpendicular to the axis of rotation. The axial runout is measured along a radius R_i. There are determined the z_{ij} coordinates of the points of the profile in the corresponding *XYZ* coordinate system. It is assumed that the *Z* axis coincides with the axis of rotation.

By the z_{ij} values of all measured profiles, with the corresponding coordinates x_{ij} and y_{ij}, an RMS associated plane is calculated and the deviations from flatness EFE are determined, as well as the position of the normal vector of this RMS plane relative to the axis of rotation.

3.3 Measurement of perpendicularity deviation

The deviation of perpendicularity of a given axial face relative to the axis of the rotary outer or inner surface is defined as the perpendicularity of its RMS associated plane to the axis of the rotational surface defined by the centers of the associated RMS circles of the profiles in at least two cross sections.

This is made in the following order:

1. Measure the axial runout of the given surface. After excluding the axial errant runout (if it is

IOP Conf. Series: Journal of Physics: Conf. Series **1050** (2018) 012066 doi:10.1088/1742-6596/1050/1/012066

significant), an RMS associated plane is built into the special program and the position of the normal vector is determined. It reflects the angle between the associated plane and the axis of rotation.

2. The radial runout shall be measured in at least two cross sections of the datum rotating surface. After excluding the errant radial runout (if significant), the position of the centers of their RMS associated circles relative to the axis of rotation is determined, i.e. the eccentricity Δx and Δy, and the axis is defined by these two points. It is determined the angle between this axis and the axis of rotation. The difference between the two angles – this one of the normal vector and the axis of the rotational surface relative to the axis of rotation determines the perpendicularity deviation between the RMS plane of the face and the axis of the rotational surface.

When measuring in more than two cross sections of the rotary cylindrical surface, the axis is defined as the RMS straight line build by the centers of the RMS circles of the measured profiles.

3.4 Measurement of coaxiality deviation

Similarly, to 3.1, the radial runout is measured in at least two cross sections of the rotation surfaces under consideration. After corrections for radial errant runout (if significant), the eccentricities Δx and Δ_y of the centers of the RMS circles of the measured profiles with respect to the axis of rotation are determined, i.e. their position in the adopted coordinate system *XYZ* (the Z axis coincides with the axis of rotation). A corresponding axis is calculated by the coordinates of the centers of the RMS circles (such as the straight line connecting the centers of the two end sections, or as the RMS straight line, when the cross sections are more than two). The deviation from coaxiality is defined as the distance from the endpoints of the axis to the reference axis normal to it, or as a distance from the endpoints of the two axes to their common reference axis. All calculations are carried out by a special program, part of the measurement system software [1].

3.5 Measurement of the deviation from parallelism

The parallelism deviation of the two faces of rotational workpieces is determined by the angle between the normal vectors of the RMS planes of these faces obtained by measuring the axial runout (Figure 5).

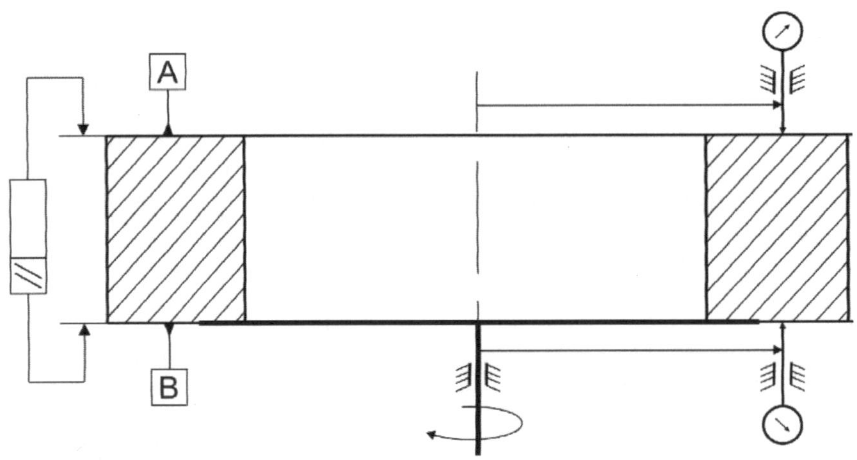

a)

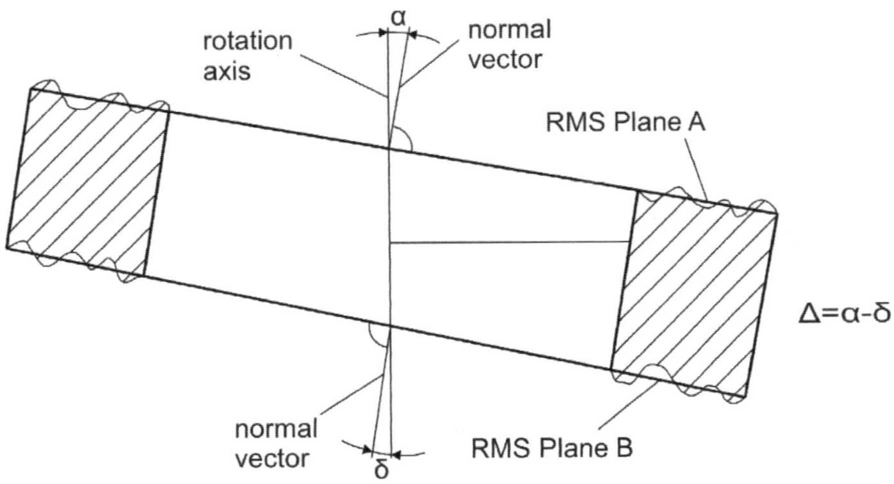

b)

Figure 5. Measuring the parallelism of the face surfaces of the ring-shaped rotary parts.
(a) a measurement scheme
b) construction of normal vectors

The measurement is carried out in the following sequence.

1. Measure the axial runout of the two faces along a path of radius R.

After excluding the errant runout (if it is significant) by the results of the axial runout measurement, according to the special program, the RMS planes are built against the measured circular profiles, as well as the position of the normal vectors of these RMS planes relative to the axis of rotation.

2. The differences between the angles α and β of the normal vectors of the two RMS planes relative to the axis of rotation determine the parallelism deviation in angular or linear quantities.

When measuring the axial runout in more than one circular profile of the face surfaces, RMS planes shall be constructed according to the procedure described in paragraph 3.2 and the position of the normal vectors of these planes relative to the axis of rotation shall be determined.

3.6. Measurement of the diameters of outer and inner rotating surfaces

3.6.1. Measurement of internal diameters

It is done in the following order:

– The radial runout of the internal rotational surfaces of the workpiece and the reference ring is measured in the respective cross sections.

– After corrections for the errant axial runout (if significant), profile diagrams of these two profiles are built. The roundness deviations shall be calculated by these graphs at the jth and the $(j+\frac{k}{2})$th points of the profiles, as well as the eccentricity of the centers of their RMS associated circles relative the axis of rotation, fig. 6.

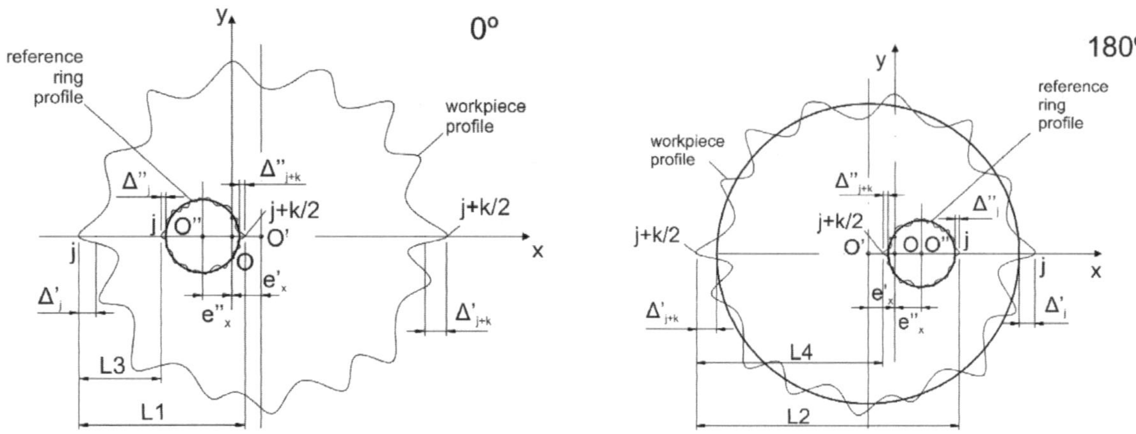

a) b)

Figure 6. Measurement of internal diameters
a) before the turn of 180°
b) after turning 180°

It is measured the distance L_1 between the jth point of the profile of the of the workpiece and the $(j + \frac{k}{2})^{th}$ point of the reference ring, as well as the distance L_2 between the $(j + \frac{k}{2})^{th}$ point of the section profile of the workpiece and the jth point of the reference ring profile after rotating the face plate at 180°. When determining L_1 and L_2, both the linear encoders travel and the horizontal displacement of the carriage are taken into account.

On the basis of this primary measurement information, both the local diameters of the workpiece and the diameters of the RMS circles of the measured profile can be determined by the formulas:

$$D_j = L_1 + L_2 - r_{av} - \Delta_j'' - r_{av} - \Delta_{j+k/2}'' \tag{1}$$

$$D_{av} = L_1 + L_2 - r_{av} - \Delta_j'' - r_{av} - \Delta_{j+\frac{k}{2}}'' + \Delta_j' + \Delta_{j+\frac{k}{2}}' \tag{2}$$

Where:

Δ_j' and $\Delta_{j+\frac{k}{2}}'$ are the deviations from roundness of the workpiece section at points *j* and *j+k/2*;

Δ_j'' and $\Delta_{j+\frac{k}{2}}''$ are the deviations from roundness of the reference ring at points *j* and *j+k/2*.

The values of the average diameters of the reference ring are determined when it is calibrated.

3.6.2. Measurement of external diameters

The procedure is analogous to the measurement of internal diameters.

The radial runout of the outer rotational surfaces of the workpiece and the reference ring is measured in the respective sections.

After entering a correction for errant radial runout (if it is significant) profile graphs of the two circular profiles are built. The profile graphs reflect the deviation from roundness at j^{th} and $(j + k/2)^{th}$ points of the profiles, as well as the eccentricity of their RMS circles relative to the axis of rotation. The distances L_3, between the jth point of the profile of the of the workpiece and the $j+j^{th}$ point of the reference ring, as well as the distance L_4 between the $(j + \frac{k}{2})^{th}$ points of the two profiles are determined after rotating the face plate at 180°.

D_j and D_{av} are calculated using the formulas:

$$D_j = L_3 + L_4 + r_{av} + \Delta_j'' + r_{av} + \Delta_{j+k/2}'' \quad (3)$$

$$D_{av} = L_3 + L_4 + r_{av} + \Delta_j'' + r_{av} + \Delta''_{j+\frac{k}{2}} - \Delta_j' - \Delta'_{j+\frac{k}{2}} \quad (4)$$

4. Conclusions

1. Measurement of geometric parameters of responsible large-scale ring-shaped parts is a serious metrological problem that can be solved successfully with the developed measurement system.

2. The system is based on a vertical turning machine, universal linear displacement measuring instruments, specialized measuring equipment and special software, that allow the measurement of the relevant parameters of the workpieces directly on the production machine.

The developed measurement methods are based on the determination of the deviation of the form and the position from the results of the radial and the axial runout measurement results with and without exclusion of the errant runout. In the first case, these results are matched to a virtual datum (axis) by entering appropriate corrections.

3. The methodology developed for measuring the diameters of the external and internal rotational surfaces of the workpieces allows the determination of both the local diameters and the diameters of the RMS circles of the measured profiles.

5. References

[1] Bogev V Vasilev V 2010 *Razrabotvane na softuer za kraglomerni izmervaniya* (Sofia: Technical University of Sofia Press) p 401

[2] Radev H *Patent №65461 V1/27.10.2008 Method for measuring the shape and position deviations of the surfaces and the axis of the rotational parts relative to the virtual datum*

[3] Radev H Bogev V Vasilev V Blagov I 2013 *Novaya tehnologiya izmereniya geometricheskih parametrov krupnogabaritnayh rotatsionn,h detaley* (Metrologiya ta prikladi Vol 2)

[4] Rubinov A 1982 *Kontrol bolshikh razmerov v mashinostroyenii* (Leningrad: Mashinostroyeniye)

[5] *Geometrical product specifications (GPS) - Geometrical tolerancing - Tolerances of form, orientation, location and run-out* (ISO 1101:2017)

[6] Whitehouse D 2004 *Surfaces and Their Measurement* (Kogan Page Science)

Research on the high quality replacement carbide plates operability with Al-Si-N hardening coating

D S Rechenko[1], B P Gritsenko[2], A Yu Popov[1], D G Balova[1]

[1]Omsk State Technical University, Mira ave., Omsk, 644050, Russia
[2]Institute of Strength Physics and Materials Science of Siberian Branch of Russian Academy of Sciences (ISPMS SB RAS), Tomsk, Russia

Abstract. Quality requirements for the manufacturing parts made of hard-to-machine materials with cutting tools are quite high. At the same time, modern carbide cutting tools are not always able to maintain working capacity throughout the entire mechanical operation, providing these requirements. The article considers test results of replaceable carbide plates during finish processing of a heat resistant alloy. Carbide plates were sharpened with ultra high speed grinding, and then hardened with Al-Si-N coating, which made it possible to obtain tool blade sharpness ρ = 3...5 μm and to achieve resistance several times higher compared to a similar tool.

1. Introduction

The achievement of the required accuracy of parts made of hard-to-machine materials used in modern military engineering is often a very difficult task, and sometimes practically impossible without the use of finishing operations. It is known that any of the finishing operations are expensive, low-productive, demand highly skilled workers, special equipment and devices, which significantly increase manufacturing costs. The elimination of finishing operations from the technological process is possible only if the required quality of processing with a cutting tool is ensured. The problem is that a modern metal cutting tool is not able to provide the required quality of processing, due to its insufficient sharpness.

The quality of a processed surface defines the operation efficiency of a metal part and it is characterized by a number of criteria. Modern technological processes of high precision unit making for aviation, automobile, rocket production industries etc. cannot be performed without high quality carbide cutting tool. Often it forms the required surfaces of parts and determines the roughness and accuracy of processing. Therefore, creating a high quality cutting tool is an actual and a complex task.

The quality of the cutting tool is a complex characteristic, including following parameters: blade sharpness, defined by a conditional inscribed rounding radius; lack of microchips and microcracks; roughness of the cutting wedge surfaces; geometric correspondence of the cutting wedge profile; the cutting wedge oxidation level. These parameters determine the operability of the cutting tool. The creation of a carbide tool with high parameters of quality makes it possible to exclude grinding operations from the technological process of manufacturing high precision parts from hard-to-machine materials, which of course reduces production costs.

At present, there are several basic technologies for grinding and obtaining high quality carbide cutting tools. The use of fine-grained diamond wheels, fine-tuning and electro-diamond sharpening makes it possible to obtain blade sharpness no less than 5 microns, but on the cutting wedge there can be formed microcracks, front and back surface backfalls and sometimes microchips. Meanwhile, the surface roughness obtained by grinding can reach Ra = 0.04 μm. The main and essential disadvantage of all these sharpening methods is the oxidized surface layer formation on the treated surfaces and the volume near the cutting wedge blade. This happens as a result of thermal loads, while the use of cutting fluid does not exclude the occurrence of oxygen on the treated surfaces.

2. Problem formulation

The aim of the work is to obtain a high quality carbide tool with ultra high speed grinding followed by a hardening coating application to exclude finishing operations in precision parts manufacturing. In the present study the following tasks were solved: the creation of a high speed grinding machine VZ-326F4; sharpening of a carbide blade using various modes; determination of the sharpened tool quality using raster microscopy; and application of a hardening coating by deposition in an argon-nitrogen gas mixture using a vacuum machine "KVANT".

3. Theory

Some directive technological processes of manufacturing parts for military equipment do not allow grinding, polishing and finishing operations. It is possible to use only blade operations, while the quality requirements for the treated surfaces are quite high. The blade operations requirements for dimensional accuracy can reach 3...5 μm, and for roughness Ra = 0.1...0.2 μm. Of course, to achieve such quality indicators for a blade tool is very problematic, primarily because of the quality of the carbide tool itself, e.g. the sharpness of the blade, the presence of microchips and microcracks and defective layers of the cutting wedge.

High quality cutting edge of the metal cutting tool is currently achieved by various methods - finishing, electro-diamond grinding, grinding with fine-grained wheels, etc., but these methods are low-productive and do not allow to obtain a stable result, namely, sharpness of the blade – ρ = 3...5 μm; absence of microchips and microcracks; the roughness of the cutting wedge surfaces – Ra = 0.04...0.10 μm; geometric correspondence of the cutting wedge profile; the cutting wedge condition.

The method of ultra high speed grinding with the use of traditional grinding and shaping movements combined with a reinforcing nanostructured coating makes it possible to produce a high quality carbide tool.

Classical sharpening methods and modes followed by finishing operations make it possible to obtain a blade with sharpness no less than 5...6 μm. The sharpening of a replaceable carbide plate (RCP), made of hard alloy GC1105, was carried out at a speed V = 30...40 m/s, longitudinal feed F = 1...1.5 m/min and a cutting depth t = 10...20 μm/double stroke. In this case, cobalt burns out from the cutting wedge surface layers, which indicates a low grinding quality.

The blade sharpness was measured by MikroCad Premium, manufacturer GF Messtechnik GmbH, using software for measuring ODSCAD in OOO PROMTEKH (St. Petersburg).

The elemental composition of the surface was determined by X-ray spectral analysis, the condition of the carbide plates cutting edge was examined by means of a scanning electron microscope EVO50 (Carl Zeiss, Germany) and a scanning electron microscope Jeol JCM-5700 (Japan), with an energy-dispersive spectrometer JED-2300 attachment. It was found that the WC content (tungsten carbide) is about 92%, and Co (cobalt) - about 5.5% and O (oxygen) - about 2.3%, which indicates the formation of oxides on the hard alloy surface (atomic percentages are used). Studies have shown, that the percentage of oxygen is in the range between 1.7...6.3%, depending on the grinding modes and the application of the cutting fluid. The change in the chemical composition indicates the occurrence of high contact temperatures during grinding.

Ultra high speed grinding method for grinding modes V = 300...400 m/s, F = 1…1.5 m/min and t = 3...5 μm/double stroke makes it possible to obtain a blade with a sharpness of 2...3 μm. It was found that the WC content (tungsten carbide) is about 94%, and Co (cobalt) - about 6%, which corresponds to the initial composition and condition of the alloy GC1105 (Sandvik Coromant) for CNMG120404 type plates. At the same time, no detectable oxides are observed on the controlled surface, and no cobalt burning from the cutting wedge body is observed.

After superfast grinding and obtaining sharpness of the blade 2...3 μm, a hardening coating was applied to the front surface. Ensuring of the required carbide tool blade firmness during finishing operations is possible due to the use of selection of optimal processing modes and application of hardening coating on the cutting wedge surfaces [1-4]. Currently there are several basic methods for modifying tool surfaces. Of the total amount of surface modification operations, the share of strain hardening is about 10%, thermal hardening is 15%; surface alloying - 25%, coating application - 48% and combined methods - up to 2%.

The analysis of coatings applied to cutting tools revealed two groups that differ in their application technology and use features: chemical vapour deposition (CVD) and physical vapour deposition (PVD) (Table 1). High requirements are imposed on modern coatings: the coefficient of friction μ should be in the range from 0.1 to 0.35; heat resistance from 900 to 1200 °C; chemical stability; hardness 35-50 GPa; adhesion strength 50-150 N, etc. Many studies have been written on the application of these methods [5-9]. There are significant disadvantages that limit the use of the CVD method on the finishing tool. This is due to the rounding of the cutting edge during deposition, as well as to the uneven coating of tools with complex shapes.

The use of PVD coatings allows to coat sharp edges without causing significant dulling due to uniform deposition. With the help of the PVD process, coatings up to 5 microns thick are obtained, which do not require additional processing. The PVD method allows coatings to be applied without increasing the brittleness of the tool, since the application temperature is about 500 °C, in contrast to 900-1000 °C for the CVD method. The PVD method allows the use of coating compositions such as Ti-CN, Al-Si-N, Ti-Al-N and Al-Cr-N, oxides of Al_2O_3, $(Al\text{-}Cr)_2O_3$, Ti-Al-Si-Cr-N. According to the authors of the studies [8; 12], multilayer coatings have great prospects due to the increase in the specific loads on the cutting tool, caused by the wide spread of titanium and high temperature alloys. At the same time, optimal coating thicknesses are 4.0-12.0 μm, at which minimum residual stresses and maximum adhesion forces with the base are achieved in the coating [5; 6].

Table 1. Comparative characteristics of coating application methods.

Characteristic	CVD	PVD
Heating temperature during application, °C	700-1100	200-500
Coating area	External, internal open	Reversed to ion source
Thickness of applied coating, microns	2-10 Increases the rounding radius of the blade	0,1-6
Composition of the applied coating	Ti-N, Ti-C, Ti-C-N, Al_2O_3	Ti-N, Ti-C, Ti-C-N, Al_2O_3, Ti-Al-N, Cr-N, Al-Si-N, Zr-C, Ti-Al-Si-N
Microhardness of coating, GPa	20-24	21-35
Coefficient of friction	0.4-0.7	0.07-0.5
Oxidation temperature, °C	400-600	400-1100
Residual stresses	High tensile	Tensile- compressive
Heat treatment after coating application	Required to remove residual stresses	Not required

According to the data of study, the additional element introduction of silicon Si or boron B makes it possible to increase the hardness, as well as the stability of the coating while operating. In addition, it is also important to improve other properties of the coating, for example, improving heat resistance and oxidation resistance, reducing the coefficient of friction and wear, increasing endurance, etc. Current experiments show that nanocomposite coatings consisting of nanograins exhibit such properties. Such nanocomposites can be formed either by reactive sputtering, when the nanocrystalline phase is formed at a low deposition temperature or by nanocrystallizing from an amorphous material.

The unique properties of coatings are stipulated by their nanostructure. However, nanostructures remain stable only at temperatures below the recrystallization temperature, at which grain growth and the formation of new crystalline phases begin [6].

The thermal stability of most hard coatings lies at temperatures below 1000 °C, this is certainly enough for their use in many areas. However, cutting tools for processing heat-resistant and titanium alloys can experience extreme local temperatures (up to 2000 °C).

Therefore, it is extremely important to use protective coatings that are thermally stable at temperatures T > 1000 °C and at the same time provide good protection against oxidation. These two requirements can only be met with amorphous composite nanostructured coatings, because amorphous materials do not contain grains.

An important characteristic of coatings is the elastic modulus of coating E, which shows the ability of the coating to be elastically deformed when a force is applied. According to the existing studies, the elastic modulus of Ti-N is 590 GPa, Ti-C is 460 GPa, Al-C is 350 GPa, Ti-Al-N is 350 GPa, Si-Ti-N is 350 GPa, Si-Al – 200 GPa, Si-BCN – 210 GPa [3].

Studies [6; 7; 9] show that the best results for resistance to oxidation have Si-Ti-N, Al-Si-N and Si-B-C-N coatings. The hardness of these coatings is in the range 25-30 GPa, and the modulus of elasticity is in the range of 200-350 GPa.

4. Experimental results

In order to clarify the effect of the hardening coating on the cutting properties of the tool, a study was made of the tool life with an Al-Si-N coating, and Mo ions were implanted in coatings of some samples. The coating thickness on the cutting wedge front surface was 2.3 μm, and a coating with a much worse adhesion of 0.26 μm thickness due to dusting was also formed on the rear surface.

Resistance tests of the RCP with different geometry (Figure 1) were carried out when turning the steel XH77TЮР on an Okuma ES-L8II lathe with BlasoCut 2000 cutting fluid. Machining conditions are as follows: cutting speed V = 40 m/min, longitudinal supply F = 0.05 mm/rev and cutting depth per diameter t_d = 15 μm.

a) b)

IOP Conf. Series: Journal of Physics: Conf. Series **1050** (2018) 012067 doi:10.1088/1742-6596/1050/1/012067

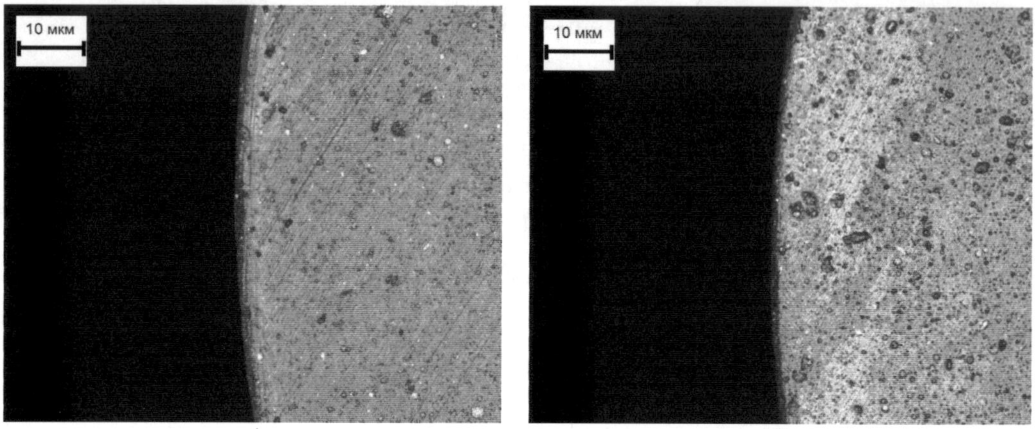

c) d)

Figure 1. Blades of carbide plates razor-sharpened and with chamfer: a) razor-sharp, coating Al-Si-N; b) razor-sharp, coating Al-Si-N + Mo; c) with chamfer, coating Al-Si-N; d) with chamfer, coating Al-Si-N + Mo

The razor-sharp plate (Figure 2, a) has a pronounced cut and a classic wear on the back surface. The razor-sharp plate with a coating and implanted molybdenum Mo (Figure 2, b) has a classic wear on the back surface. The plate with a chamfer and an Al-Si-N coating (Figure 2, c) has residual outgrowth traces (stagnant zone) with coating remained, and the plate with a chamfer and implanted molybdenum Mo (Figure 2, d) has a pronounced wear on the front surface with a peeling coating and residual outgrowth traces. Meanwhile, this plate has chipping on the blade. The best result for resistance and quality of processing was shown by a plate with negative chamfer and coating. According to the test results, the plates had different wear.

a) b)

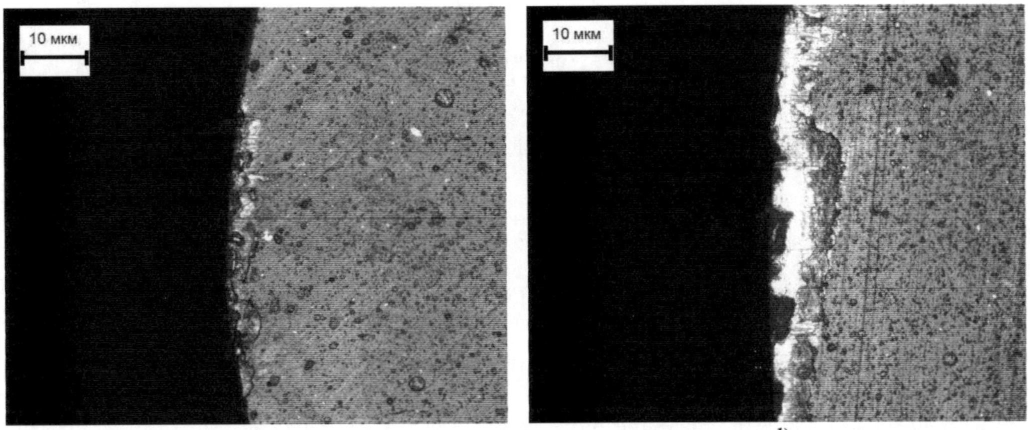

c) d)

Figure 2. Wear of carbide plates razor-sharpened and with chamfer: a) razor-sharp, coating Al-Si-N; b) razor-sharp, coating Al-Si-N + Mo; c) with chamfer, coating Al-Si-N; d) with chamfer, coating Al-Si-N + Mo

Tool resistance in the processing of the high temperature alloy XH77TЮР was evaluated by changing the quality parameters of the workpiece, namely, when the roughness of the machined workpiece surface Ra = 0.2 μm was increased, after which the RCP was replaced.

5. Results discussion

The obtained stability results in comparison with the initial RCP are given in table 2.

Table 2. Stability of carbide plates razor-sharpened and with chamfer

Plate	Resistance T, min	Cutting path L, m	Processed area S, cm^2
Razor-sharp, coating Al-Si-N	42,4	1696	847,8
Razor-sharp, coating Al-Si-N + Mo	59,5	2379	1189,7
With chamfer, coating Al-Si-N	169,6	6782	3391,2
With chamfer, coating Al-Si-N + Mo	43,5	1741	870,4
Initial	31,3	1252	626,1

Obtained results of processing alloy XH77TЮР, presented in Table 2, show that the best result for resistance and quality of processing has a plate with chamfer and Al-Si-N coating. The resistance is 169.6 minutes, which is 3.9 times higher than the standard RCP (initial).

6. Conclusion

The carried out researches have shown, that high speed sharpening provides much higher quality of the RCP blade and of the cutting wedge processed surfaces. Classical grinding modes provide the sharpness of the blade at 5...10 μm, and high speed grinding - at 2...3 μm, meanwhile the treated surfaces do not have traces of oxides. The operation efficiency of the tested RCP, evaluated by the resistance when turning the heat resistant alloy XH77TЮР, is 3-4 times higher than with the standard RCP.

The application of thin Al-Si-N type coatings results in an effective increase in the RCP resistance, without deterioration of the machined parts quality.

Funding source. Acknowledgments
The work was carried out with the financial support of the Fundamental Scientific Research Program of the State Academies of Sciences, Project III.23.1.1 "Physical mesomechanics of nonlinear multilevel hierarchically organized systems in fields of external influences".

Supervisor – Denis Rechenko

References

[1] Sinopalnikov V A and Tereshin M V 1987 Monitoring the state of high-speed drills to limit their temperature *Soviet engineering research* (vol 7 Issue 6) pp 59–62
[2] Filimonov L N 1979 *High-speed grinding* (Leningrad: Machine building) p 248
[3] Wang C, Fang Q, Chen J, Liu Y and Jin T 2016 Subsurface damage in high-speed grinding of brittle materials considering the kinematic characteristics of the grinding process *International Journal of Advanced Manufacturing Technology* (vol 83) pp 937–948
[4] Zhu Y, Lu W, Sun Y and Zuo D 2016 Grinding characteristics in high-speed grinding of boron-diffusion-hardened TC21-DT titanium alloy with vitrified CBN wheel *The International Journal of Advanced Manufacturing Technology* pp 0–9
[5] Gribkov V A, Grigoriev F I, Kalin B A and Yakushin V L 2001 *Perspective radiation-beam technologies of material processing* (Moscow: All the year round) ed Kalina B A p 528
[6] Gribkov V A, Demina E V, Dubrovsky A V, Maslyaev S A, Pimenov V N, Sasinovskaya I P, Miklaszewski R and Scholz M 2009 Application of a plasma accelerator of the dense plasma focus type in simulation of radiation damage and testing of materials for nuclear systems *International Topical Meeting on Nuclear Research Applications and Utilization of Accelerators* p 8
[7] Korotaev A D, Borisov B D, Moshkov Yu V, Ovchinnikov S V, Pinzhin Yu P and Tyumentsev A N 2009 Elastic stress state in superhard multielement coatings *Physical Mesomechanics* (vol 12 (5-6)) pp 269–279
[8] Rechenko D S, Popov A Y, Gritsenko B P, Sungatulin A R, Titov Y V, Sergeev V P, Voronov A V, Deev K A and Pupchin V A 2016 Formation of wear-resistant structures on solid alloy for superfinish processing *AIP Conference Proceedings* (vol 1783 part 020191 (1-4))
[9] Mitterer C, Mayrhofer P H and Musil J 2003 Thermal stability of PVD hard coatings *Vacuum* (vol 71 (1)) pp 279–284

IOP Conf. Series: Journal of Physics: Conf. Series **1050** (2018) 012068 doi:10.1088/1742-6596/1050/1/012068

Limit combinatorics as a method for investigating of mathematical models

Vladimir Remeslennikov

Sobolev Institute of Mathematics, Siberian Branch of Russian Academy of Science, Omsk, Russia

E-mail: remesl@ofim.oscsbras.ru

Abstract. In recent years, due to the applications in theoretical informatics, the researches on finite combinatorics (graphs, directed graphs, partial orders, and so on) and the applications of these studies in the economy, in the social sphere and in statistical physics have been intensively developing. The main objective is to present the latest mathematical studies on formal languages, large networks, finite combinatorics and model theory to a wide range of researchers involved in the development of economic and social models, information security models, the growth of the Internet models, and so on, for whom, by the author's opinion, such the studies would be useful.

1. Introduction

The article is on the popular science side. The main objective is to present the latest mathematical studies on formal languages, large networks, finite combinatorics and model theory to a wide range of researchers involved in the development of economic and social models, information security models, the growth of the Internet models, and so on, for whom, by the author's opinion, such the studies would be useful. For specific practical tasks there are the complicated working models to be built up, while our current task is to show them in the very core of the subject, to reveal their unification mathematical idea. It would be naive to suppose that our recommendations can satisfy the needs of any and all mathematical models developers. This is not so, and we formulate the requirements that are imposed on the mathematical models. They are the following:

(i) the model must be dynamical in discrete time, i. e. must consist of a finite large (or infinite) number of finite combinatorics

$$C_1 \to C_2 \to \ldots \to C_k \to \ldots; \tag{1}$$

(ii) it should be described in formal mathematical language transitions

$$\alpha_i : C_i \to C_{i+1}.$$

Good examples of such models are currently served by different models of Internet growth [1].

For the models above, the notion of a Δ-index (this is a rational number) is defined, as well as the suitable Δ-properties, where Δ is a concrete finite combinatorics. The main mathematical problems here are the following two mathematical problems having good practical applications:

(i) Is the combinatorics sequence (1) convergent, and how does its limit appear C_∞?

(ii) Which Δ-properties are quasi-random, which means their stability with respect to the passage to C_∞? (The exact definition, see, for example, in [12, 13]).

The universality of the mathematical approach to a wide range of such models is achieved, first, by abstracting to arbitrary categories, and secondly, by treating objects of categories as models of a certain language, which allows describing equations, inequalities, and so on in them. Let us explain in more detail what is being discussed.

2. The category of combinatorics

The definition of the category theory can be found in many sources. For readers convenience, we provide the definition from Wikipedia (in Russian): "The category theory is a section of mathematics that studies the properties of relations between mathematical objects that do not depend on the internal structure of objects". More formally, a category $\mathcal{C}$ consists of the following three mathematical entities:

(i) a class of objects C;

(ii) for every pair of objects A and B it is defined a set of morphisms $\mathrm{Hom}(A, B)$;

(iii) for every pair of morphisms $f \in \mathrm{Hom}(A, B)$ and $g \in \mathrm{Hom}(B, C)$ it is defined a composition $f \circ g \in \mathrm{Hom}(A, C)$;

(iv) for each object A it is defined an identity morphism $id_A \in \mathrm{Hom}(A, A)$,

such that the following two axioms hold:

(i) *associativity*: $h \circ (g \circ f) = (h \circ g) \circ f$;

(ii) *identity*: $f \circ id_A = id_B \circ f = f$ for all $f \in \mathrm{Hom}(A, B)$.

Let us list the most important examples of categories in the study of mathematical models.

- The category of sets. Objects here are sets and morphisms are maps between sets.
- The category of simple graphs. Objects here are simple graphs (i. e., undirected graphs in which both multiple edges and loops are disallowed) and morphisms are maps which preserve edges.
- The category of partial orders. The partial order is a set with a binary relation $\leq$ with the following axioms:
  (i) *reflexivity*: $\forall a\ a \leq a$;
  (ii) *transitivity*: $\forall a, b, c\ (a \leq b) \wedge (b \leq c) \rightarrow (a \leq c)$;
  (iii) *antisymmetry*: $\forall a, b\ (a \leq b) \wedge (b \leq a) \rightarrow (a = b)$;
  Morphisms in this category are maps preserving the order relation.

3. The language of equations and inequalities for combinatorics

The concepts of the equation and inequalities are well known for fields, including the field of real and complex numbers. In the monograph of three authors E.Yu. Daniyarova, A.G. Myasnikov and V.N. Remeslennikov "Algebraic geometry over algebraic structures" [2], the concepts of equation and inequality for combinatorics of all types are introduced, if a language is defined for them. Here the language is understood as those relations (predicates) and functions that are defined for the class of combinatorics $\mathcal{C}$. For example, for the category of simple graphs, the language will consist of two predicates — the vertex equality predicate and the neighborhood predicate for two vertices $E(x, y)$. For these predicates, the following natural axioms are fulfilled: $\neg E(x, x)$ (hence, there are no loops in the graph), $E(x, y) \longleftrightarrow E(y, x)$ (in other words, this predicate is symmetric with respect to the pair of vertices). For the category of partial orders, the language will consist of predicates of equality and a predicate of order $\leq$, and the axioms for it are given in the previous section. The monograph [2] explains what is an equation and an inequality in this case, and how to solve such systems of equations and inequalities.

We also note that many properties of simple graphs can be written in this language. For example, the fact that in a finite graph there are 10 different triangles or there are no pentagons and so on.

Let us explain this more formally. Let Δ be a fixed finite simple graph with n vertices and Γ be another finite simple graph. It is not hard to write a formula in the language $\mathcal{L}$ that states "graph Γ has hereditary subgraph Δ". Let $n = 3$ and Δ be the triangle, then the desired formula is of the form: $\exists x_1, x_2, x_3\ (x_1 \neq x_2 \wedge x_1 \neq x_3 \wedge x_2 \neq x_3 \wedge E(x_1, x_2) \wedge E(x_1, x_3) \wedge E(x_2, x_3))$. Hence it is clear that we can write down the formula by the fact that Γ contains exactly k of different triangles. If k is the maximum number contained in Γ triangles, then we say that the Δ-index of the graph Γ is k.

4. Quasi-random combinatorics

In many cases, the finite combinatorics must be digitized either by elements from the field or by elements of some Boolean algebra. The notion of a random graph and a quasi-random graph arise on this path. Let us give these definitions following the classical books on the graph theory. We start with the definition of a quasi-random graph. By definition, a quasi-random graph on n vertices is a graph Γ satisfying the following properties:

(i) Γ is a complete graph;
(ii) to each edge on the vertices x, y the probability of appearance of this edge in the graph Γ is assigned. For example, if this probability is 0, then this means that corresponding edge actually absent in the graph Γ.

A quasi-random graph Γ is called a random graph if the real number $p \in (0, 1]$ is chosen and the probability of occurrence of each edge in the graph Γ is equal to the number p.

5. Large networks and Δ-indexes

In this report, we will only work with so-called dynamic large networks. Let us explain what this means. We have a finite or infinite sequence of finite combinatorics:

$$C_1 \to C_2 \to \ldots \to C_k \to \ldots,$$

where C_k are graphs (their nature can be different: quasi-random graphs, multi-graphs and so on), symbol " $\to$ " means a morphism in this category, and the number of elements in combinatorics C_k increases asymptotically to infinity.

Let us list the main tasks associated with such sequences.

(i) It is necessary to correctly define the notion of convergence with the help of Δ-indexes.
(ii) Determine when the limit combinatorics C_∞ exists.
(iii) If it possible, determine Δ-index for the limit combinatorics C_∞.
(iv) Determine what properties of this sequence are quasi-random.

Over the past 10 years, many mathematicians dealt with these problems. Also there have been published a large number of works in this area, many of which are with a practical trend [3–14].

IOP Conf. Series: Journal of Physics: Conf. Series **1050** (2018) 012068 doi:10.1088/1742-6596/1050/1/012068

6. Limit combinatorics

Let us define one useful metric on graphs, which occurs in many mathematical models. Let H and G be two simple graphs. Assume that the set of vertices $V(H)$ of the graph H have cardinality m, and $|V(G)| = n$, $m \leq n$. Suppose that an embedding $\alpha : V(H) \to V(G)$ is given.

We define the metric $ED(\alpha)$ as the number of errors in the embedding α. We consider an error to be the mapping of incident vertices into nonincident vertices or the mapping of nonincident vertices into incidence vertices. The *edit distance* between graphs H and G is the following value:

$$ED(H, G) = \min_{\alpha} ED(\alpha)/C_n^2.$$

With this distance, the sequence (1) turns into a metric space, and some properties of combinatorics can be extracted with the help of topological arguments.

7. Statistical approach to research

Let us define the central concept of the statistical method for research of combinatorics. This is the concept of the density of one combinatorics relative to the other one. It allows us to enter the correct definition of the convergent sequence (1) with respect to the selected Δ-property. If convergence exists and the Δ-property is satisfied in the limit combinatorics C_∞, then it is quasi-random. Particularly useful are the combinatorics C_∞, which are convergent with respect to all finite combinatorics Δ. This is a fairly rare case. Often for practical applications, it is important that convergence is relative to the most important properties that are defined by the combinators $\Delta_1, \Delta_2, \ldots, \Delta_k$. If we talk about stock exchanges, then there is a finite set of main exchange indices.

Let Δ and G be two simple graphs, $|V(\Delta)| = m$, $|V(G)| = n$, $m \leq n$. Denote by $\rho(\Delta, G)$ the number of induced subgraphs in G, that are isomorphic to Δ. Then denote the density of the combinatorics Δ in G by $\rho^*(\Delta, G) = \frac{\rho(\Delta,G)}{C_n^m}$, where C_n^m is the binomial coefficient. Further, fixing a graph Δ and using the sequence (1) one can construct the following sequence of rational numbers:

$$\rho^*(\Delta, C_1), \rho^*(\Delta, C_2), \ldots, \rho^*(\Delta, C_k), \ldots \tag{2}$$

Definition. *The sequence (1) is called convergent, if for any Δ the sequence (2) is a convergent rational sequence.*

This definition is very rigid, and it is often not fulfilled. Fortunately, the research customer is only interested in the density with respect to a finite number of properties $\Delta = \{\Delta_1, \ldots, \Delta_k\}$, so we give a Δ-variant of this definition.

Definition. *The sequence (1) is called Δ-convergent, if the sequence (2) is a convergent rational sequence for all selected combinatorics $\Delta_1, \ldots, \Delta_k$.*

References

[1] Raygorodskiy A M 2010 Models of random graphs and its applications *Proceedings of MIPT* **4(2)** pp 130–140

[2] Daniyarova E Yu, Myasnikov A G and Remeslennikov V N 2016 Algebraic geometry over algebraic structures (Novosibirsk: SO RAN) 243 p.

[3] Chung F and Graham R 1991 Quasi-random tournaments *Journal of Graph Theory* **15(2)** pp 173–198

[4] Chung F, Graham R and Wilson R 1989 Quasi-random graphs *Combinatorica* **9(4)** pp 345–362

[5] Diaconis P and Janson S 2008 Graph limits and exchangeable random graphs *Rendiconti di Matematica, Serie, VII* **28** pp 33–61

[6] Erdös P 1962 On the number of complete subgraphs contained in certain graphs *Publ. Math. Inst. Hungar. Acad. Sci* **7** pp 459–464

[7] Griffiths S 2013 Quasi-random oriented graphs *Journal of Graph Theory* **74(2)** pp 198–209

[8] Jagger C, St'ovicek P and Thomason A 1996 Multiplicities of subgraphs *Combinatorica* **16(1)** pp 123–141

IOP Conf. Series: Journal of Physics: Conf. Series **1050** (2018) 012068 doi:10.1088/1742-6596/1050/1/012068

[9] Krivelevich M and Sudakov B 2006 Pseudo-random graphs *In: More Sets, Graphs and Numbers. Bolyai Society Mathematical Studies* **15** (Spring-Verlag) pp 199–262
[10] Razborov A 2007 Flag algebras *Journal of Symbolic Logic* **72(4)** pp 1239–1282
[11] Razborov A 2013 On the Caccetta-Haggkvist conjecture with forbidden subgraphs *Journal of Graph Theory* **74(2)** pp 236–248
[12] Razborov A 2013 What is a Flag Algebra? *Notices of the AMS* **60(10)** pp 1324–1327
[13] Coregliano L and Razborov A On the Density of Transitive Tournaments *arXiv:1501.04074*
[14] Lovasz L 2012 Large networks and graph limits *AMS Colloquium publications* **60** pp 473

Numerical analysis of thermal deformation for constructive variants of mirror segments in a parabolic antenna

E M Reyzmunt and S V Doronin

Institute of Computational Technologies of the Siberian Branch of the Russian Academy of Sciences, 6 Academician M.A. Lavrentiev ave., Novosibirsk, 630090, Russia

E-mail: sdoronin@ict.nsc.ru

Abstract. This work is devoted to the research of thermal deformation of segments in a parabolic reflector mirror in connection with the limitations determined by the radio frequency range of antenna operation. The thermal deformation analysis is performed based on the results of a numerical solution of the coupled problem of stationary thermal conductivity under conditions of radiation-convective heat transfer and thermoelasticity. A comparative analysis of five form factors of mirror segments was performed, which allowed choosing its preferred configuration.

1. Introduction

The efficiency of mirror reflectors in terrestrial satellite communication systems is directly determined by their geometric stability - the shape constancy of the reflecting surface of the mirror under the action of complex operational loads and impacts. As the aperture of the reflector increases, it is natural that the absolute deformations of the mirror increase with constant loads and impacts. However, as the frequency of the radio bands increases, the requirements to geometric stability sharply increase. This is based upon the following considerations.

As is well known, the relation between the wavelength λ and its propagation velocity v is expressed by the formula $\lambda = v/f$, where f is the oscillation frequency. For electromagnetic waves in a vacuum, their propagation velocity is equal to the speed of light, and the wavelength expressed in meters is

$$\lambda = \frac{299792458}{f}, \quad (1)$$

where the frequency f is measured in hertz.

In modern communication systems, radio bands with frequencies reaching hundreds of gigahertz are used [1, 2]. As the frequencies increase, the wavelength obviously decreases. For example, for the Y-band frequencies are from 325 to 500 GHz [3], and the wavelength in accordance with (1) is in the range from 0.60 to 0.92 mm. It is known that the maximum linear deviation of the actual shape of the mirror from the theoretical should usually not exceed [4] $(1/10 \ldots 1/16) \cdot \lambda$. According to other sources [5], this value is $1/8 \cdot \lambda$. Otherwise, there is a significant decrease in the efficiency of the antenna. Then, for the Y-band, the tolerance should not exceed 0.06 mm. This value can be considered as the limiting value of absolute deformation.

Thus, the reflectors of parabolic antennas operating in the high-frequency bands belong to the class of precision structures with increased requirements for geometric constancy. To achieve levels of geometric constancy, characterized by limiting deformations that are in millimeters and tenths of a millimeter, with an increase in the aperture of the reflector it turns out to be a nontrivial task. An obligatory condition for its solution is a comprehensive calculation analysis of the nature of the deformation of the reflector in the range of operational loads and impacts.
The present paper takes stock of the experience of the calculation analysis of the thermodynamic state and deformation of the reflector mirror in a wide range of operating temperatures.

2. Problem definition

This study is dedicated to reflector antennas for terrestrial satellite communication systems working in conditions of Q/K_a frequency band. The reflector diameter is in the range 9 to 12 m and the ratio of the focal length to the diameter is in the range 0.3 to 0.5. At the stage of research and development, several alternative constructive variants of the mirror using segments of the reflecting surface are considered. These options are determined by the configuration of the power frame and the limitations of the technological process of manufacturing segments of polymer composite materials.
The common for the alternative constructive solutions of the mirror segments considered below is their cross-section structure: the external layers of 0.6 mm thickness carbon fiber composite, the internal layer of 40 mm thickness plastic foam. Two essentially different constructive variants of the mirror structure are considered (Figure 1). In one case (configuration I), the mirror consists of six parts, each of which includes one segment of form factor 1 and two segments of form factors 2 and 3 (Figure 1, a). In the other case (configuration II), the mirror is formed by fifteen parts consisting of one segment of form factor 4 and two segments of form factor 5 (Figure 1, b). Another difference is that the reflecting segments of configuration II (form factors 4 and 5) are reinforced by radially extended 100 mm high stiffening plates made of 1.4 mm thickness carbon fiber composite.

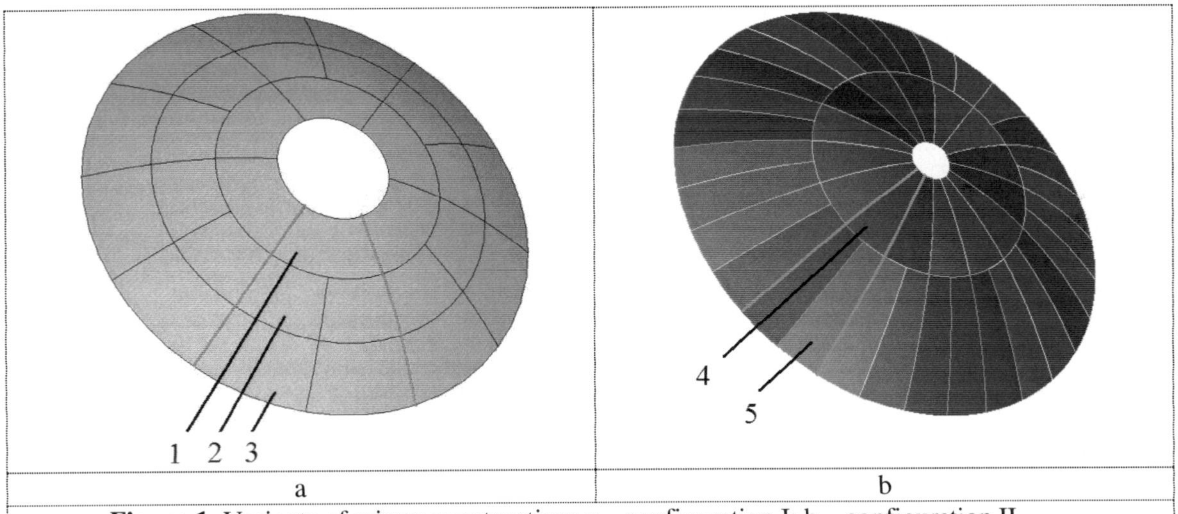

Figure 1. Variants of mirror construction: a – configuration I; b – configuration II.

In accordance with regulatory requirements, the antennas are operated in the ambient air temperature range from minus 50 to plus 55 ° C. The thermal state of the reflector is formed under the influence of the integrated flux density of solar radiation Q_{sol} under conditions of radiation-convective heat exchange with the surrounding medium.
The research task is to determine the fields of temperature, strain and stress for the mirror segments of five form factors, and to perform a comparative analysis of the results and an estimate of the operability of the antenna mirror for the two configurations.

3. Mathematical formulation of the problem

To determine the thermodynamic and stress-strain states of mirror segments, it is necessary to solve the conjugate problems of stationary thermal conductivity and thermoelasticity. At the first stage, the temperature distribution over the volume of the segments is calculated under the boundary conditions of radiation-convective heat transfer (Figure 2).

Under the action of solar radiation on the mirror surface, most of the radiation is reflected (radiation heat exchange). The effective value of the integrated flux density of solar radiation, expended on the heated surface of the segments, is

$$q^{+} = Q_{sol}\mu ,$$

where μ is the reflection coefficient of the surface (albedo).

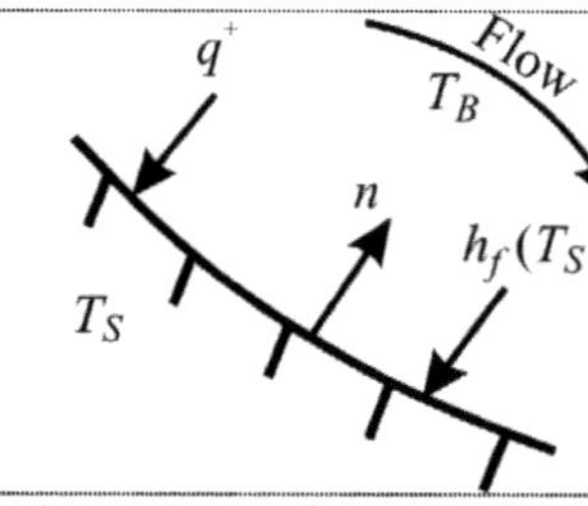

Figure 2. Calculation scheme for radiation-convective heat transfer.

$$q^{-} = \int_{S} q^{c} dS , \tag{2}$$

where S is the outer surface of the segments; q^{c} is the amount of heat removed from a unit surface area due to convective heat transfer.

Under the Newton-Richman law

$$q^{c} = h_f (T_S - T_B), \tag{3}$$

where h_f is the coefficient of convective heat transfer (heat transfer); T_S, T_B – respectively the surface temperatures of the segments and ambient air temperature.

The temperature distribution over the volume of segments in the Cartesian coordinate system *xyz* (the temperature field $T(x,y,z)$) is determined by the equation of stationary thermal conductivity

$$\frac{\partial^2 T}{\partial x^2} + \frac{\partial^2 T}{\partial y^2} + \frac{\partial^2 T}{\partial z^2} = 0 , \tag{4}$$

and the Fourier thermal conductivity law, which establishes a unique relationship between the temperature gradient along the normal to the isothermal surface n (Figure 2) and the heat flux vector

$$\vec{q} = -k\frac{\partial T}{\partial n} , \tag{5}$$

where k is the coefficient of thermal conductivity.

The result of solving equations (2) - (5) is the calculated temperature field $T(x,y,z)$ in the volume of the mirror segments, which makes it possible to determine the temperature difference $\Delta T(x,y,z)$ on the faces of the elementary volume at the point with the coordinates x, y, z. Then for the reflector mirror, which configuration is described by the domain Ω in three-dimensional space, the absolute deformations (displacements) along each axis of the Cartesian coordinate system are determined by summing the deformations along elementary volumes in the region Ω

$$\{\Delta x, \Delta y, \Delta z\} = \alpha \int_{\Omega} \Delta T(x,y,z) d\Omega, \quad (6)$$

where α is the coefficient of linear temperature expansion.
Further, the relative deformations and stresses are determined in accordance with the classical equations of the three-dimensional theory of elasticity.

4. Numerical modeling and results

Equations (2) - (6) were solved for discrete (finite-element) models of mirror segments I and II configurations in a three-dimensional formulation. A regular mesh of finite elements was constructed under the control of the step size of the mesh, which was varied in order to ensure convergence of the results along the finite element mech. Finally, for all five mirror segments, the same degree of sampling is adopted. The kinematic boundary conditions consist of assigning zero displacements along all translational degrees of freedom at the points of segments in which they are mounted on the reflector skeleton: five points for segments 1, 2, 3, seven points for segment 4 and eight points for segment 5.
Figures 3 and 4 show the typical distribution fields of the calculated temperature values and absolute deformations (displacements) for segment 4 at an ambient air temperature of minus 20 °C

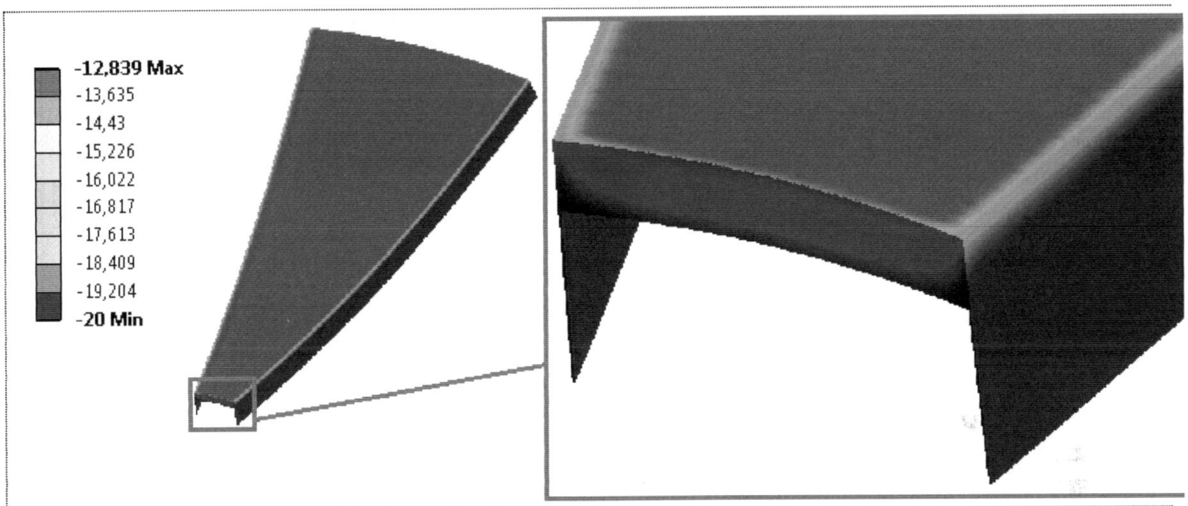

Figure 3. Typical temperature distribution field (°C).

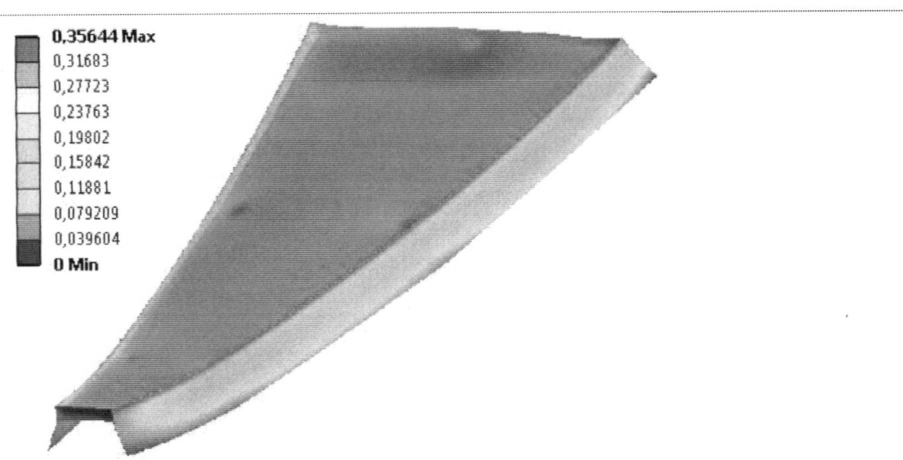

Figure 4. Typical displacements distribution field (mm).

Analysing the obtained results we found that in the range of ambient air temperatures from minus 50 to plus 55 °C with an effective integral flux density of solar radiation q^+ = 112 W/m2 and convective heat transfer coefficient h_f = 1.5·10^{-5} W/(m^{2}·°C), the temperature difference across the segment volume for different configurations is in the range 7.03...7.62 °C. This difference is equal to the excess of the reflecting surface temperature of the mirror segment due to solar heating above the ambient temperature.

These results practically coincide with the following values obtained by other authors. The temperature regime of the radio telescope RT-70 is characterized by an excess of the temperature of the reflecting surface above the ambient temperature by an average of 7.1...9.8 °C [6]. Studies [7] have shown that the maximum temperature changes in the reflector constructions do not exceed 5...8 °C, respectively, reflecting the possibilities of paint and varnish protection. Similar estimates are known for building structures: it is reported that the maximum temperature of roofing membranes with a very high reflectivity above the ambient air temperature is only 8.0...8.5 ºC [8].

Interest in analyzing the deformed state was the maximum displacement of the outer layer of carbon fiber - the reflecting surface proper (Table 1). These are the movements that are compared with the maximum permissible deviations of the surface from the original shape in accordance with the wavelength by (1). For Q/K_a band antenna reflectors in accordance with (1), the wavelength of the electromagnetic wave is in the range of 6.0 to 11.3 mm. We apply the rule of multiplication of intervals [9]

$$[a,b]\times[c,d]=[\min(ac,ad,bc,bd),\max(ac,ad,bc,bd)],$$

where a = 1/16, b = 1/8, c = 6.0, d = 11.3. We obtain the range of the maximum permissible deviations 0.375 ... 1.413 mm. Comparing these values with the data of Table 1, we arrive at the following conclusions.

For segments of form factors 1, 2, 3 of mirror configuration I, the maximum displacements do not exceed the upper limit of the range of permissible deviations (1.413 mm) in the range from minus 35 to plus 55 °C. At a temperature of minus 50 °C, the calculated displacements exceed the permissible values.

For segments of form factors 4, 5 of mirror configuration II, the maximum displacements do not exceed the upper limit of the range of permissible deviations in the full range of operating temperatures.

Table 1. Maximum displacement of the reflection surface of the mirror.

	Ambient air temperature (°C)							
Form factor	– 50	– 35	– 20	– 5	10	25	40	55
1	1.60	1.23	0.86	0.49	0.16	0.26	0.63	1.00
2	1.09	0.85	0.60	0.35	0.14	0.15	0.39	0.64
3	1.82	1.40	0.98	0.56	0.23	0.29	0.71	1.13
4	0.70	0.53	0.36	0.23	0.11	0.17	0.34	0.51
5	0.98	0.76	0.53	0.31	0.11	0.14	0.36	0.58

5. Conclusion

As a result of the formulation and numerical solution of the problem of analyzing the thermodynamic and deformed states of the five form factors of reflector segments, the regions (intervals) of the characteristic values of temperature differences across the volume of segments and absolute deformations of the reflecting surface of the mirror are established. Temperature differences are

characterized by high stability and correlate with similar results of other authors. This makes it possible to use the temperature range of 7...8 °C as the temperature boundary conditions for the preliminary thermoelastic analysis of alternative constructive variants of the mirror without first solving the problem of stationary thermal conductivity under conditions of radiation-convective heat transfer.
The results of the analysis of the deformed state demonstrate a high sensitivity to structural inhomogeneities that ensure the stiffness of the structure. The introduction of radial stiffeners results in both a significant reduction in the overall level of deformations and a decrease in their spread within the operating temperature range.

Acknowledgements. This work was done during the complex project and was financially supported by the Russian Federation Government (Ministry of Education and Science of the Russian Federation). Contract № 02.G25.31.0147

References

[1] Nomenclature ITU-R V. 431-8 (08/2015). Nomenclature of the frequency and wavelength bands used in telecommunication
[2] IEEE Std 521-2002. IEEE Standard for Letter Designation for Radar-Frequency Bands
[3] https://en.wikipedia.org/wiki/Radio_spectrum#Waveguide_frequency_bands
[4] Metal constructions. 1999 *Steel structures, structures from aluminum alloys. Reconstruction, inspection, strengthening and testing of structures of buildings and structures (Handbook of the designer)* **3** (Moscow: Publ. ASV) 528
[5] Kondratenkov G S and Frolov A Yu 2005 Radio broadcasting. Radar systems for remote sensing of the Earth (Moscow: Publ. Radiotechnics) 368
[6] Borovkov A I , Shevchenko D V, Gimmel'man V G, Machuev Yu I and Gaev A V *Finite element modeling and investigation of the temperature fields of the main reflector of the radio telescope RT-70* *http://fea.ru/research/RT70/RT70_research_Compmechlab.pdf*
[7] Polyak V S and Bervalds E Ya 1990 *Precision constructions of mirror radio telescopes: Experience of creation, problems of analysis and synthesis* (Riga: Publ. Zinatne) 526
[8] Roodvoets D L, Miller W A and Desjarlais A O Long term reflective performance of roof membranes / Available at: http://coolroofs.org/documents/LongTermReflectivePerformanceofRoofMembranes_000.pdf
[9] Shokin Yu I 1981 Interval analysis (Novosibirsk: Nauka) 112

IOP Conf. Series: Journal of Physics: Conf. Series **1050** (2018) 012070 doi:10.1088/1742-6596/1050/1/012070

Development of a mathematical model of sound insulation with flexible panels

E A Romanenko, G S Russkikh, Z N Sokolovskiy

Omsk State Technical University, Mira ave., 11, Omsk, 644050, Russia

E-mail: elinaromanenko18@gmail.com, russkikh_gs@mail.ru, ninasok@yandex.ru

Abstract. For sound insulation in the presence of restrictions related to transportation and installation conditions, practically the only possible constructive solution is the use of flexible panels. Also, the flexible panel provides reliable protection in the most difficult area - low frequencies, so modernization of the panel design is an actual topic for research. The object of the study is a mesh-plate panel. Flexibility is ensured by the following constructive solution - fixing the main soundproofing plates on a steel mesh bearing with basalt fabric with a gap. The sound insulation of the panel has been investigated experimentally. The goal is to develop analytical techniques evaluation flexible sound insulation panels based on the fundamental positions of industrial and architectural acoustics, sound insulation, and synthesis of published experimental studies of the effect of gaps (acoustic holes and voids). The calculation results are consistent with tests insulation panels manufactured at a sufficient level. As a result, it was described how the sound insulation of the panel is affected by the size, shape and filling of acoustic holes in the gap between the plates. Suggested ways to reduce their negative impact and optimize the design of a flexible soundproof panel.

1. Introduction

Currently, the development of structures for effective sound insulation of low frequencies gives rise to many technical solutions. The usual sandwich panels are replaced by panels with different perforation configurations [1], [2], acoustic membranes [3], [4] and fibrous structures [5]. Acoustic holes are present in all these cases. The using of adequate theoretical models allows adjusting the design parameters to achieve the desired level of sound insulation.

In this paper we consider a flexible sound insulating barrier in the form of a plate-mesh panels (hereinafter - PMP) protected Russian Federation patents [6, 7].

The PMP has a woven mesh, two layers of square (TxT) metal plates with total thickness *h*, arranged symmetrically relative to the mesh. Four basalt fabric layers are used as an elastic gasket. To provide the desired flexibility of the plate are mounted at a distance *t* from each other, and also have bevels in the region of the side surfaces, thus increasing PMP's flexibility.

2. Statement of the problem

To develop a complete mathematical model, it is necessary to evaluate the sound insulation of the working panel, taking into account main features. The general view and the element of the PMP are shown in Fig. 1.

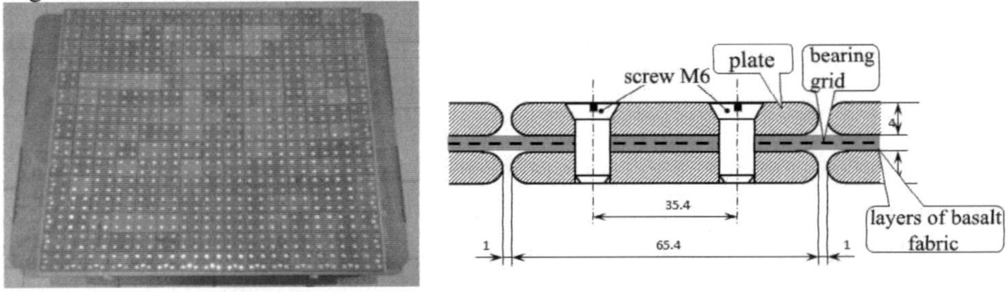

Figure 1. General view and element of the PMP

The ratio of the total area of PMP to size LxL square slots between the plates is:

$$\frac{S_0}{S} = 0.0301 \text{ for plates}$$

The layers of the basalt fabric liner practically do not resist bending, they are not a continuous obstacle to the sound within the width of the gaps *t* and are also characterized by a certain ratio of the area of the gaps and the total area:

$$\frac{S_0}{S} = 0.0258 \text{ for gasket}$$

Depending on the type of fabric and its weaving method, this parameter may be different. Influence on the permeability of the mutual arrangement of layers of fabric during installation is not controlled and not taken into account.

In the general case, the varied design parameters that affect the sound insulation of the PMP are:

- material, thickness *h* and dimensions T_1 and T_2 of the main insulating plates,
- the *t* value (i.e. $\frac{S_0}{S}$ plates) and the shape of the gap,
- material and thickness *h* of the gasket (i.e. $\frac{S_0}{S}$ gaskets),
- the sizes of the PMP L1 and L2 (in the actual PMP T1 = T2, L1 = L2).

The authors do not know purposeful experimental studies of the degree of sound insulation of PMP, depending on the design parameters, now. There is reliable data only on experimental studies of the sound insulation of a working PMP [8], carried out by one of the authors, and the results of the experiments [9] in comparison with sound insulation: a solid plate, a plate having a narrow slit, a circular hole and plates with a 16th round holes uniformly distributed over the area at

$$\frac{S_0}{S} = 0.001$$

in all three experiments..

Theoretical relationships for calculating sound insulation obtained in the basic studies on architectural acoustics, including taking into account acoustic holes and slits. In [9] it is partially used for the assessment of flexible insulation panels, but not taken into account:

- the influence of the gaps between the plates and the sound permeability of basalt fabric,
- features of sound insulation during acoustic tests,
- relatively large amplitude of the oscillations of the PMP with a limited free (protected) volume.

3. Theory

Investigation of the effect on the sound insulation holes slots devoted a number of works by foreign authors. A mathematical model and a series of experiments to determine the sound transmission through the holes and cracks were studied in [10]. Shows three options of the depth of circular openings - 0.036 m, 0.072 m and 0.144 m; three options slit depth - 0.051 m, 0.076 m and 0.152 m.

Diffraction phenomenon in acoustics discussed in [11]. Three different cases are given: the width and depth of the slit are equal to the wavelength, the width of the slit is equal to the wavelength (depth is 10 times smaller), the width of the slit is equal to the wavelength (depth is 5 times larger).

An analytical model of the acoustic resistance of round holes of finite depth is given in [12].

Application of these models to the studied case we are difficult because of differences in the key geometric dimensions and, consequently, the physical processes occurring in the system.

In this paper attempted to closest approach, and corrections of the theoretical curves and noise calculation algorithm applied to the design features the PMP and its operating conditions, including on the basis of mathematical processing of results known experiments. Program and setting of further clarifying the test was determined.

When calculating the sound insulation known PMPs are sequentially used for sound insulation according to the individual layers of the structure with and without taking into account the acoustic holes.

The sound insulation of the continuous layer without taking into account the dimensions and vibrations of the PMP was calculated using the impedance method:

$$R_C = 10\lg\left[\cos^2(hk_2) + \frac{1}{4}\left(\frac{\rho_1 c_1}{\rho_2 c_2} + \frac{\rho_2 c_2}{\rho_1 c_1}\right)\sin^2(hk_2)\right] \quad (1)$$

where is:

- k– wave number,
- ρ_i, c_i – acoustic resistance of media, at the interface of which there is an effect of sound insulation,
- h – layer thickness.

The sound insulation of a layer having cracks - acoustic holes, is calculated by the formula [9]:

$$R = R_1 - 10\lg\left[\frac{1+\varphi\cdot\frac{S_0}{S_1}\cdot 10^{\frac{R_1-R_0}{10}}}{1+\frac{S_0}{S_1}}\right] \quad (2)$$

where is:

- R_0 – sound insulation of acoustic hole,
- $\frac{S_0}{S}$ – calculated (geometric) ratio of the area of the acoustic hole and the area of the layer,
- φ – dimensionless coefficient taking into account an increase in sound transmission diffuse field conditions depends on the hole size and frequency and, in the author's opinion, can be φ =1..10. The value of the coefficient is refined on the basis of mathematical processing of the experiments [9].

The dependence of φ on the ratio of the effective length of the hole and the wavelength of sound represented as:

$$\varphi(l_w, \varphi_w) = \begin{cases} 1, \text{ if } l_w \geq \lambda \\ \varphi_w \cdot (\lambda - l_w) \cdot \frac{1}{\lambda}, \text{ if } l_w < \lambda \\ \varphi \geq 1 \end{cases} \quad (3)$$

where l_w, φ_w are the effective hole size and the effective maximum value φ, respectively.

In the literature the concept of a "large" and "small" acoustic aperture is introduced, determined by the ratio of its dimensions to the wavelength, i.e. the same hole may have the properties of a large acoustic hole in the high-frequency range and a small one in the low-frequency region. It is also obvious the effect on the soundproof ratio of the wavelength of sound and the size of the barrier. In this paper, these positions are specified by introducing a correction function for the equivalent relative size of the acoustic hole

$$\frac{S_0}{S}(l_w, \varphi_w) = \frac{S_0}{S} \cdot \begin{cases} 1, \text{ if } l_w > \lambda \\ \frac{l_w}{\lambda}, \text{ if } l_w < \lambda \end{cases} \quad (4)$$

Formulas (3), (4) in the calculation of sound insulation with allowance for acoustic holes are substituted in (2), which takes the form:

$$R_c^S = R_1 - 10\,lg\left[\frac{1+\varphi\left(l_w,\varphi_w\right)\cdot\frac{S_0}{S_1}\left(l_w,\varphi_w\right)\cdot 10^{\frac{R_1-R_0}{10}}}{1+\frac{S_0}{S_1}\cdot\left(l_w,\varphi_w\right)}\right] \quad (5)$$

The effective values of the parameters l_w, φ_w were determined from the condition of the best agreement (the maximum value of the reliability of the approximation $R^2 \to 1$ due to minimization of the sum of the squares of the experimental differences [9] and the values of sound insulation calculated by (5)) and amounted to:

Table 1. Treatment of experiments [9]

Experiment	Size, m	Effective Size l_w, m	Maximum effective coefficient, φ_w	Reliability of approximation R^2	Average error of approximation $\bar{A}$, %
Narrow gap	0.001*1	0.055	10	0.996	2.69
Round hole	d=0.036	0.087	8.69	0.998	3.80
16 round holes	d=0.009	0.173	12.78	0.924	8.79

The reliability of the approximation calculates by the formula:

$$R^2 = 1 - \frac{\sum_1^n (y-Y)^2}{\sum_1^n Y^2 - \frac{\sum_1^n Y^2}{n}} \quad (6)$$

where y and Y are calculated and experimental values of sound insulation, n - the number of experimental values.

The average error of approximation calculates from the formula:

$$\bar{A} = \frac{1}{n}\sum_n \frac{\left|(y_i - Y_i)\right|}{y_i}\cdot 100\% \quad (7)$$

Taken into account design features and applications PMP:

- reduced stiffness and dependence of the natural oscillation frequency both on the cylindrical rigidity, and on the dimensions of the plates, gaps and stiffness of the carrier grid;
- relatively large amplitude oscillations PMP U_{max} under sound pressure as a consequence of the flexibility of the PMP;
- the commensurability of the volume of air displaced by vibrations with the volumes of the protection zone V_2 and the source zone V_1

$$\Delta V = S \cdot U_{max} \cdot K_u \quad (8)$$

where K_u is coefficient that takes into account the shape and amplitude of the oscillations. For a working PMP having a square shape, when pivoted $K_u \approx 4/\pi^2$

- the additional pressure fluctuations Δp in these zones were calculated according to the Boyle-Mariotte law:

$$(V_i - (-1)^i \cdot \Delta V)\left(p + (-1)^i \cdot \Delta p_i\right) \cdot V_i \Delta p \to \Delta p_i = \frac{\Delta V}{V_i}$$

- the change in the phase of the oscillations K_{uD} upon passing through resonance taking into account the loss coefficient η under oscillations.

To take into account these features, the derivation of the formula for the calculation of sound insulation at low frequencies was modernized and an equation obtained for calculating the sound insulation at frequencies of sound effect ω close to the frequency of the first harmonic of natural oscillations ω_r

$$R_\omega = 10\lg\left[(1+\frac{\eta\cdot m\cdot\omega}{2\rho_0\cdot c_0}\left((\frac{\omega_r}{\omega})^2\right)^2 + (\frac{m\cdot\omega}{2\rho_0\cdot c_0}(1-\left((\frac{\omega_r}{\omega})^2 - \frac{K_{uD}\cdot K_u\cdot p\cdot S\cdot(V_1+V_2)}{V_1\cdot V_2\cdot\omega\cdot 2\rho_0\cdot c_0}\right)^2\right] \quad (9)$$

where $\omega_r = 2\pi(5\ldots 27)$ Hz (when the position of the PMP is changed from vertical to horizontal); the loss factor η = 0.6, corresponding to the normal damping coefficient of mechanical oscillations β≈0.3.

For working PMP obtained $\frac{S_0}{S} = 0.0301$ for plates, $\frac{S_0}{S} = 0.0258$ for gasket. To take into account the uncertainty of the overlapping of holes in the layers of basalt fabric, the coefficient K_{hole} = (1/3 ... 1/2) was introduced.

The algorithm for calculating the sound insulation of the PMP includes the following sequence of calculations:

1. Calculation R_ω according to (9) for a solid plate without taking into account the holes and gaps.
2. Calculation of R_c from (1) for laying layers without taking into account permeability.
3. Calculation R_C^S according to (5) for layers of the gasket taking into account the slits at R_c from point 2 and R_0 (for air).
4. Calculation of R_C^S PMPs according to (5) with $R_c = R_\omega$ from item 1 and $R_0 = R_C^S$ from item 3.

The algorithm provides a lower bound, since it ignores small (due to the small differences impedances) soundproofing gasket between the base plate and the support mesh.

4. Experimental results

The experiments [9] were carried out on an aluminum plate $L = 1\ m$ and $h = 0.003\ m$ at a constant size of the holes in the thickness, in contrast to the variable cross section and a larger slit depth in the working PMP.

Research of influence of variable cross-section and the depth of the effect on the sound insulation gap requires further experimentation and refinement. In the present work, when calculating the sound insulation of the working memory bandwidth, the following are first taken as the closest results of determining the effective parameters for the gap: t = 0.001 m and for basalt tissue: an experiment with 16 uniformly distributed holes

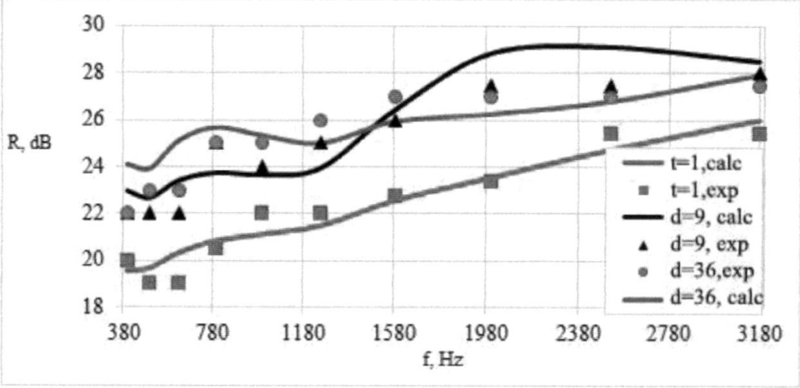

Figure 2. Comparison of the experimental values of sound insulation [6] and calculated by (5)

If we take as a calculated curve approximating a curve, then R^2 = 0.97, and the average approximation error does not exceed 15%.

Figure 3 shows the result of calculating the insulation operating PMP developed algorithm at ω_r = 2·π·13.5 and the set of experimental values f = (15 ... 200) Hz and f = (200 ... 10000) Hz averaged values.

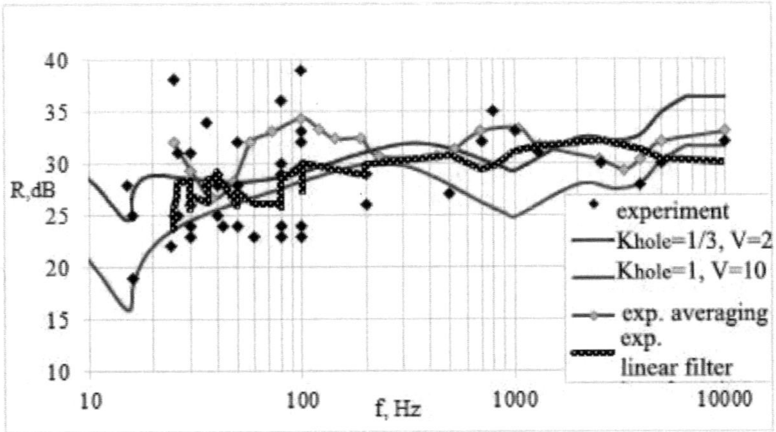

Figure 3. Comparison of the experimental values of soundproofing of the memory bandwidth and calculated according to the algorithm with different parameters V and K_{hole}

The correspondence between experiments and calculations was estimated by the value of the indicators R^2 and $\bar{A}$. If $V_1 = V_2 = (2 \ldots 10)$ m^3, K_{hole} = (1/3…1), then $R^2 = 0.965 \ldots 0.976$, $\bar{A}$ = (13.467 ... 14.377) %.

5. Discussion of results

Comparison of the graphs shows that after f ≥ 200Hz, the calculation results are almost identical, i.e. the effect of the dynamics of the memory bandwidth on vibration isolation is manifested only at low frequencies.

For the working PMP, the results of the sound insulation calculations are in satisfactory agreement with the experimental data (taking into account the lower estimate and spread of the experimental data for the 6 tests for the determination of sound insulation - the line shows a linear filter line at 5 points).

6. Conclusions

The technique of analytical assessment soundproofing flexible panels having structural features as cracks and holes.

To refine the algorithm, the following additional experiments are necessary:

1. To definition of sound insulation for different combinations of T, t and h in a conventional plate and PMP, as well as for different orientations and the number of layers of basalt fabric. The purpose of the tests: refinement of the quantities l_c, φ_c and their adequate geometric calculation.
2. To develop the design of the memory bandwidth with a significantly smaller t, for example, with one-sided flexibility.
3. Development of the methodology for calculating the first form of natural oscillations of the memory bandwidth for a different from the horizontal position in space.

References

[1] Meng H, Galland M A, Ichchou M, Bareille O, Xin F X and Lu T J 2017 *J. Composite struct.* **182** pp 1-11
[2] Simon F 2018 *J. of Sound and Vibr.* **421** pp 1-16
[3] Langfeldt F, Riecken J, Gleine W and Estorff O 2016 *J. of Sound and Vibr.* **373** pp 1-18

[4] Li S, Mao D, Huang S and Wang X 2018 *J. Appl. Acoustics* **130** pp 92-98
[5] Yang Y, Li B, Chen Z, Sui N, Chen Z, Saeed M, Li Y, Fu R, Wu C and Jing Y 2016 *J. Composites B:Engin.* **96** pp 281-286
[6] Zubarev A V, Tribelskiy I A, Adonin V A and Malyutin V I 2007 *Sound Insulation Panel* [http://www1.fips.ru/wps/portal/IPS_Ru#1518716254883] Patent 2340478 Russian Federation
[7] Tribelskiy I A, Adonin V A, Bobrov S P, Denisov V D, Bokhan V V and Gidion V A 2012 *The sound-insulating panel and the method of its manufacturing* [http://www1.fips.ru/wps/portal/IPS_Ru#1518716165441] Patent 2457123 Russian Federation
[8] Tribelskiy I A 2011 *The results of tests, design estimates and justification of the choice of design options for ZP panels and passage nodes through ZP, schemes for fixing ZP panels to the supporting contour* Omsk FSUE "Progress" p 145
[9] Bogolepov I I 2009 *J. Engineering and construction* **1** pp 17-20
[10] Oldham D J and Zhao X 1993 *J. of Sound and Vibr.* **161** pp 119-35
[11] Berezovski A and Engelbrecht J 2016 *J. Wave Motion* **60** pp 35-45
[12] Yang D and Morgans A S 2016 *J. of Sound and Vibr.* **384** pp 294-311

IOP Conf. Series: Journal of Physics: Conf. Series **1050** (2018) 012071 doi:10.1088/1742-6596/1050/1/012071

Analysis of Approaches for Synthesis of Networks-on-chip by Using Circulant Topologies

A Yu Romanov[1,2,3], **A A Amerikanov**[1] **and E V Lezhnev**[1]

[1]National Research University Higher School of Economics, 20 Myasnitskaya Ulitsa, 101000, Moscow, Russia

[2] e-mail: a.romanov@hse.ru
[3]Author to whom any correspondence should be addressed

Abstract. The article gives a review of existing methods of network-on-chip design based on the approach in which mapping of the characteristic tasks graph is performed on a given regular topology. The networks-on-chip synthesis problem is generally characterized. The analysis and comparison of standard topologies (mesh and torus) with circulant topologies are performed. Advantages and disadvantages of mesh and torus topologies usage, and the effect, achieved by their application to various implementations of networks on chip, are analyzed. Extension of the scope of solutions for standard regular network topologies mesh and torus on the circulant topologies with better characteristics is proposed. This will make it possible to take advantage of the deterministic approach, but with the use of more effective NoC topologies optimized for a particular task.

1. Introduction

The architectural solutions within networks-on-chip (NoCs) become more and more widespread in modern embedded systems. Such systems are generally based on hierarchical principle containing a large number of processors and peripherals. There is a trend of constantly increase of the number of processor cores, as uniprocessor systems cannot cope with applications requiring high computing performance and large data flows (exchange processing of video and still images in IP-telephony, etc.) [1]. In this way, the development of effective communication subsystems is necessary. Under such conditions, implementation of new approaches for NoCs becomes of great importance.

2. NoC synthesis problem

Due to a complex structure of NoCs, their synthesis problem is difficult to formulate and has a lot of ways to be solved. The initial data for NoC synthesis problem is a characteristic task graph. Its vertices correspond to the IPs (intellectual properties – separate modules, usually computational nodes), and are characterized by injection of packages in NoC environment. Mathematically, it can be described as follows: $G=G(C,A)$, a directed graph, where C – set of vertices (IPs), and A – set of communication processes between them. In turn, NoC system architecture is characterized by the followings: $T(R,Ch)$ topology (R and Ch – sets of routers and physical links between them); a routing mechanism (P_R); a mapping function of application characterization graph vertices onto NoC routers (Ω (C)).

IOP Conf. Series: Journal of Physics: Conf. Series **1050** (2018) 012071 doi:10.1088/1742-6596/1050/1/012071

The characteristic task graph's directed edges correspond to data transmission process between two IPs. They can be characterized by minimal and maximal communication delay, communication intensity and bandwidth [2].
According to the above definitions, the model can be complemented by communication energy cost dependence formula:

$$E = \sum_{\forall a_{i,j}} \upsilon(a_{i,j}) \times E_{bit}(\Omega(c_i), \Omega(c_j)) \tag{1}$$

where $\upsilon(a_{i,j})$ – capacity of data transmission process between nodes i, j; $E_{bit}(\Omega(c_i), \Omega(c_j))$ – energy spent on 1 bit of data transmission between nodes c_i and c_j.
And the communication energy minimization problem can be expressed by a formula (2). It consists in finding such Ω (C) which arranges for connections of communication process with high capacity to have a low energy consumption to transfer 1 bit of data [3].
Similarly, the formula of the total volume of data, transmitted between nodes in a NoC, is as following:

$$V = \sum_{\forall a_{i,j}} \upsilon(a_{i,j}) \times L_{i,j}(\mathrm{P}_R(r, i, j)) \tag{2}$$

where $L_{i,j}(\mathrm{P}_R(r, i, j)$ – distance between nodes i and j according to the routing algorithm.
According to the above formulas, the application problem can be represented as a set of an individual subtasks graphs whose edges describe the data flow between them. In this case, the characteristic task graph is obtained by mapping of one or several subtasks on the IPs and by data flow distribution between subtasks on the edges of the graph.
In this case, the synthesis problem is to choose a NoC topology and projection of a characteristic task graph onto it in accordance with the demands for the cost of the bandwidth, power, chip area, as well as minimal and maximal delay of the packages transmission. Under this approach, the topology is represented as a numbered graph whose vertices correspond to the routers (every router is connected to the corresponding IP) and edges – to the physical data channels (trunks). In addition to the topology, a synchronization and communication subsystem method, a transmission of data streams technology, a quality of service insurance method, a routing strategy, a structure of routers, and so on are chosen.
Thus, according to [4], the synthesis of NoC design requires the solution of four distinct tasks:

- Selection of an appropriate NoC topology.
- Task graph mapping onto the topology.
- Efficient data streams routing insurance.
- Decision on NoC functioning technology.

The second [5, 6], third [7], and fourth [8] tasks are discussed in the scientific literature, but they aren't the object of this research. As for the problem of the efficient NoC topology synthesis, it is chosen either an application-specific topology, or platform dependent topology (in case when the information about the future computing tasks which will be performed by the finite system is known, or topology is defined by hardware), or a regular topology is used. The approaches, based on application-specific topologies, are observed in works [9, 10], and they aren't a subject of this research either. The approach with regular topologies implementation is analyzed below.

2.1. The common regular topologies: mesh and torus

The definition of a regular topology in various sources is different. Thus, in [11] there is a definition according to which topologies with the same degree of vertices are regular – that is, those having homogeneous routers. According to this definition, torus and hypercube topologies, unlike mesh, are regular. But in practice, all topologies with a scalable homogeneous structure (we will also use this

IOP Conf. Series: Journal of Physics: Conf. Series **1050** (2018) 012071 doi:10.1088/1742-6596/1050/1/012071

definition) belong to regular topologies, therefore, mesh topology can also be referred to regular topologies [3, 8, 12, 13]. Most topologies can be multidimensional [14–16], but for their implementation, it is necessary that NoC components can be located on several physical layers of the crystal; therefore, 2D flat topologies are most often used in NoC design.

The main characteristics of the topology are as follows: number of vertices – routers (N); number of edges – physical connections between the routers (Ed); order of vertex – the number of edges emanating from it (St); diameter of the graph – the maximum among minimum distance between any two nodes (D); average distance between the shortest paths among all nodes of the graph (L_{av}).

The most common among regular topologies is mesh topology which is a network of $N = m{\cdot}n$ nodes of rectangular type, or a network of $N = n{\cdot}n$ nodes of square type. Each node is connected with the four neighboring ones (Figure 1). The peripheral nodes have some unused ports [8, 17].

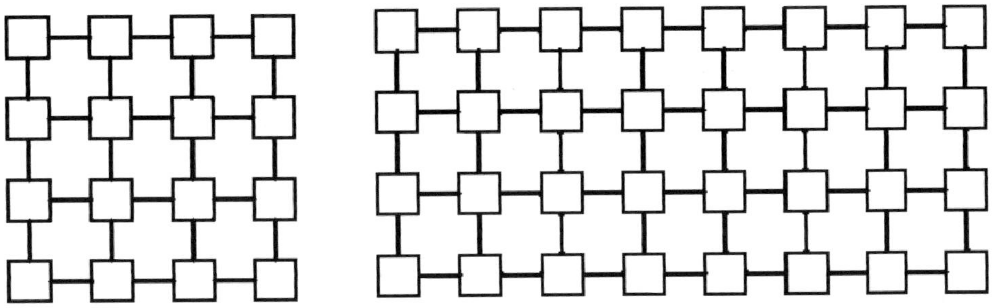

(a) (b)

Figure 1. Mesh topology for 16 (a), and 32 (b) nodes.

Mesh $n \times n$ (Figure 1 (a)) networking characteristics are [8, 17]:

$$Ed = 2\cdot(N-\sqrt{N}),\ \ St = 2\div 4,\ \ D = 2\cdot(\sqrt{N}-1),\ \ L_{av} = \frac{2(N-1)}{3\sqrt{N}},\ \ Ext = 2\sqrt{N}+1. \qquad (3)$$

Characteristics of a rectangular mesh $m\times n$ (Figure 1 (b)) are:

$$Ed = 2mn-m-n,\ \ St = 2\div 4,\ \ D = m+n-2,\ \ L_{av} = \frac{(m+n)(mn-1)}{3mn},\ \ Ext = \min(m,n), \qquad (4)$$

where $\mathrm{mod}(x, y) = x - y\left\lfloor \frac{x}{y} \right\rfloor$.

The main disadvantage of mesh topology is a too large diameter. So, for a network with 3x3 nodes, the distance between routers in the opposite corners is of 4 hops. Moreover, in such networks, the load on the routers is distributed unevenly.

The attempt to eliminate these disadvantages at the expense of the large resource costs is the topology of torus obtained by the combining of extreme nodes of mesh with its opposite ones (Figure 2 (a, b)). In this case, $D = 2\cdot\left\lfloor \sqrt{N}/2 \right\rfloor$, but $Ed = 2\cdot N$, and St=4. A kind of a modification is the folded torus which is to reduce the length of the connecting lines between the extreme nodes (Figure 1 (c)) [8, 11].

IOP Conf. Series: Journal of Physics: Conf. Series **1050** (2018) 012071 doi:10.1088/1742-6596/1050/1/012071

(a) (b) (c)

Figure 2. Torus topology for 9 (a) and 16 (b) nodes and folded torus for 16 (c) nodes.

Characteristics of a square torus $n \times n$: $Ed = 2N$, $St = 4$, $D = 2\left\lceil \sqrt{N}/2 \right\rceil$, $L_{av} = \frac{N\sqrt{N}(\text{mod}((\sqrt{N}+1),2))}{2(N-1)} + \frac{\sqrt{N}(\text{mod}(\sqrt{N},2))}{2}$, $Ext = 2\sqrt{N}+1$. For a torus topology of 3x3 in size, the number of connections is already 18, and the diameter is 2 hops, that is, an increase in the number of connections by 50% provides a reduction in diameter by a factor of 2. In addition, the degree of vertices in this topology is fixed which makes it possible to use homogeneous routers. The disadvantage of this topology is the large length of the end trunks, so, its modification – folded torus (Figure 2 (c)), which allows reducing the length of connecting lines between the end nodes [18], is often used.

All mentioned topologies reside in the restriction in the number of nodes to which they can be applied. For example, mesh and torus topologies are applicable for networks of the square form with the number of the nodes equal to the natural numbers to the power of 2 (4, 9, 16, 25, ...), but to create a network of 11 routers is not possible at all. The use of a rectangular topology (Figure 1 (b)) is possible, but this leads to an increase in the diameter of the graph and a decrease in the width of the division of the graph in half. And for the number of nodes, which is a prime number (5, 7, 11, ...), the task becomes even more complicated, since one must abandon either deterministic routing algorithms (for example, XY [19]), or take a topology for more nodes, leaving one router as an auxiliary and unconnected to the computing node. Thus, there is a need for the new, more flexible approaches able to offer the efficient topologies for the networks with the arbitrary number of nodes and the configured number of connecting lines.

2.2. Circulant networks for NoC development

There are a number of topologies, such as: chord rings, octagon, spidergon, etc. which are variations of circulant networks [20].

Let's define the circulant network. Circular networks (graphs) are Cayley graphs of Abelian groups. Let N – number of vertices of the graph, and $S = \{1 \le s_1 < \ldots < s_k < N\}$ – set of k integers (generatrices). Then, an undirected graph $C(N;\ s_1, \ldots,\ s_k)$ with a set of vertices $V = \{0,\ 1,\ \ldots,\ N-1\}$ and a set of edges $E = \{(v, v \pm mod(s_i, N))\ |\ v \in V, i = \overline{1,k}\}$ is a circulant of dimension k.

Thus, the circulating network $C(N;\ s_1, \ldots,\ s_k)$ can be represented as a ring structure, where each vertex is associated with k successive vertices and k previous vertices in steps of $s_1, \ldots,\ s_k$ (Figure 3).

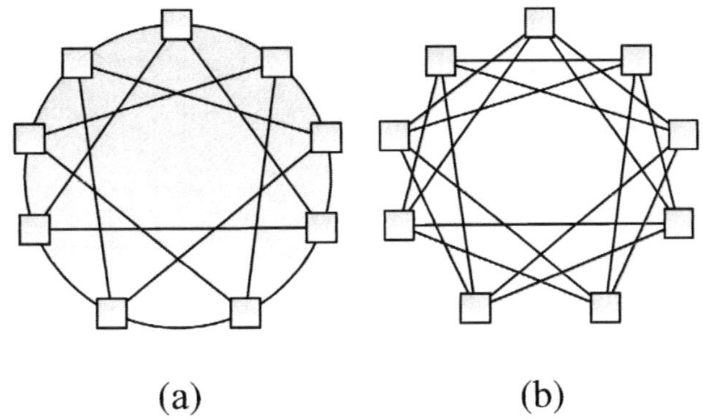

Figure 3. Circulant topology: (a) – $C(9;1,3)$; (b) – $C(9;2,3)$.

For implementation as NoC topologies, circulants of dimension 2 are of primary interest. This class of circulants has a degree of vertices 4 and good topological characteristics which makes it promising for use as a basis for networks-on-chip. Since relatively small routers with 4 inputs / outputs are suitable for their construction, plenty of examples were developed (for example, Netmaker library [4]), and their high efficiency and balance were shown.

The theory of second-order circulants is well developed, and it was shown in [20, 21] that circulants with generatrices, calculated by formula

$$C(n;\ D-1,\ D),\ \text{ where } D=\sqrt{n/2},\ n>2, \tag{5}$$

are optimal ones.

Thus, by synthesizing them, we were able to compare their characteristics with the characteristics of the mesh and torus graphs. Since the diameter and average distance between the nodes are one of the most important characteristics of NoC, we give the obtained characteristics for topologies with number of vertices from $3^2 = 9$ to $23^2 = 529$ (we take topologies only for the number of vertices which are numbers in the second degree; for mesh and torus topologies, the square form is the most optimal). Dependencies of L_{av} and D on the number of vertices are shown in the figures below.

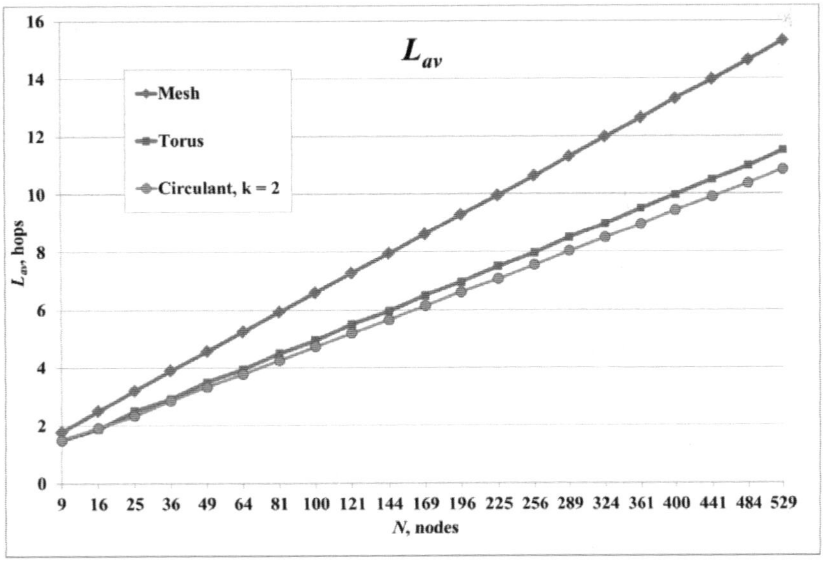

Figure 4. Dependence of average distance between the shortest paths among all the nodes on the number of nodes.

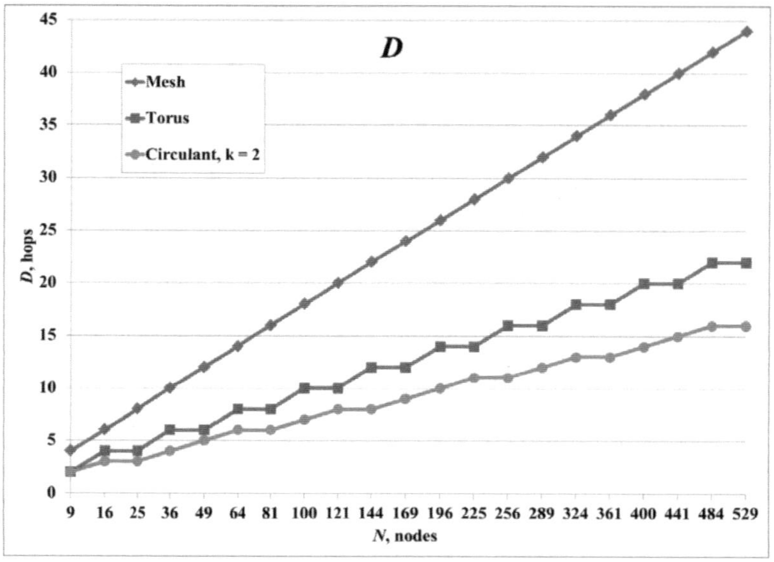

Figure 5. Dependence of the diameter on the number of nodes.

Based on the obtained graphs, it can be concluded that circulants of form $C(n;\ D-1,\ D)$, at equal costs of connective resources as torus topology, make it possible to reduce: the diameter to 20.0–59.4% in comparison with torus, and to 50.0–63.9% in comparison with mesh; the average distance – to 2.1–6.7% in comparison with torus, and to 15.6–29.2% – in comparison with mesh. This is generally obvious, since torus can be represented as a circulant-like graph, but not the optimal one. In addition, in comparison with mesh and torus topologies, it is possible to construct circulant graphs for any number of vertices without reducing their effectiveness. Thus, the use of circulant graphs as a topological basis in NoC development makes it possible to improve NoC characteristics in comparison with mesh and torus topologies without losing the regularity of networks and increasing the dimension of routers. Moreover, like in other regular topologies, circulants interconnects can be reconfigured according to the application task graph by adding and removing some links, as proposed in work [22].

3. The review of methods of NoC synthesis by projecting the characteristic task graph onto a regular topology

A widely used approach is when one of the regular topologies is selected at the beginning of the synthesis and remains unchanged in the course of the further mapping of the characteristic tasks graph onto it. The advantage of this approach is that such well-known regular topologies as torus, mesh, hypercube, star, etc. are sufficiently investigated. They have predictable characteristics and deterministic routing algorithms [12]. In addition, there is a wide experience of their use, as well as circulants [23], in global networks which can be used in NoC design, taking into account the specificity of ASIC and FPGA hardware platforms, of course.

An example of such an approach is the algorithm for the energy-efficient construction of a NoC based on mesh topology, proposed in [3]. The characteristic tasks graph is mapped onto a mesh-like topology consisting of homogeneous cells (a computational core and a router) and using the tree search algorithm for node mapping alternatives: at each step of descending through branches of the alternative tree, the consumed power of the mapped nodes is calculated, and less optimal alternatives are discarded. According to the authors, this makes it possible to reduce the power consumption of the NoC by approximately 60%, in comparison with the arbitrary mapping of the problem, and to reduce the search time by 27 times, in comparison with the full-search algorithm. In [24] the same authors use linear programming in optimizing delays in packet transmission by eliminating data collisions in the transmission path and reducing the weighted communication distance when mapping the characteristic tasks graph onto a mesh topology. This approach gives an increase in capacity by 17%, compared to the previous solution, but leads to an increase in energy costs by 9%. In the next paper [13], the authors attempted to compensate for the shortcomings of mesh topology by realizing the possibility of adding regions that occupy the dimensions of several cells, as well as pre-fixing the placement of certain nodes (for example, those related to the inputs / outputs of the chip). The represented algorithms are of a rather universal nature and can be adapted relatively easily for use with circulant topologies; moreover, a number of problems, associated with uneven load of nodes located on the periphery of the mesh graph, are eliminated.

The method of linear programming, used in [25], is to search for the optimal mapping of the characteristic tasks graph onto a mesh topology with a reduction in the average distance between nodes and a decrease in the maximum load of communication lines. Mapping is performed in two stages: nodes are mapped at the first stage, and connections between them – at the second stage. This allowed accelerating the solution of the problem of linear programming $10 \sim 10^3$ times. The result of simulation of various video applications was a decrease in power consumption and bandwidth requirements of NoCs by 55.5% and 49.2%, respectively, compared to heuristic algorithms. Despite the fact that the results of the proposed method are demonstrated using the mesh topology example, we did not find any obstacles to its extension to circulant topologies.

In [26] NMAP algorithm is proposed for optimal mapping of the characteristic tasks graph onto the mesh topology, taking into account the bandwidth requirements. According to the algorithm, at first, the most connected node is mapped to the center of the network on the router having the largest number of neighbors, and after that, depending on the degree of connectivity with nodes already located on the free routers, other nodes are sequentially mapped in order to minimize the cost of communication between them. The obtained result is improved by pairwise permutations of the nodes in order to achieve a minimum average distance between the nodes. Data flows, to which capacity of the communication lines is not sufficient, are split into several threads, and the routing table is given various alternative shortest paths for them. This algorithm was used in synthesis of NoCs for video applications using XPipes library [27] and, according to the authors, this allowed reducing communication costs by 32% and increasing the throughput by 53%. This approach was developed in [28] by planning the optimal placement of components on a chip by linear programming and introducing the possibility of selecting the characteristic tasks graph for various regular topologies from XPipes library. And in [29], specialized software SUNMAP was proposed to map the characteristic tasks graph on a given topology included in NoC NetChip generation environment [30]. The proposed algorithm can be applied to NoC on the basis of circulant topologies, taking into account the fact that the process of mapping the most loaded node of the tasks characteristic graph can be started from any node of the circulant topology, in view of its peculiarity that it is completely symmetric. This will simplify and speed up the algorithm under consideration.

The MOCA algorithm [31], according to the authors, ensures close results, in comparison with the previous approach, but in less time. This algorithm is used to divide the task graph into two equal parts in the form of an artificial tree which are then easily mapped to a mesh topology. In addition, optimization of the capacity of trunks and data transmission delays is performed. According to the authors of the algorithm, this led to a decrease in energy consumption by 14%, compared to the mapping, obtained by linear programming for processing tasks of video and audio data. This algorithm can also be easily applied to circulant topologies.

In work [32], based on Aethereal NoC, it is proposed a solution that can be applied to arbitrary NoC topology, and distribution of data flows with different quality of service is considered. In this case, the UMARS+ algorithm is used, according to which, when choosing the SCC topology, data streams with a guaranteed minimum delay service are mapped on it. Incorrect turns [33] are detected to avoid deadlocks and, in view of this, flows with guaranteed bandwidth service are mapped when routing tables are created, as well as assigning time slots to the ports of the routers. Using the MPEG decoder as an example, it is shown that the use of information about data flows with various services and their spatial time distribution, when mapping the task graph to the NoC topology, gives a 33% gain in resources and 35% – in energy consumption, and also reduces by four times the maximum delay of passing packets, in comparison with the approach without taking into account information about the quality of service. This approach does not impose any restrictions on the topology.

In [34] it was suggested to generate the variants of mapping the characteristic tasks graph onto a mesh-shaped topology with the help of genetic algorithms. The behavioral model of the obtained mappings is then subjected to discrete-event modeling to estimate the gain in the performance and power consumption of the NoCs and to compare them with the results of other mappings to select the best ones. A two-stage genetic algorithm for mapping the graph of problems onto NoC mesh topology is used in [35]. The algorithm takes into account that the compute nodes are inhomogeneous and are divided into groups with different performance. Each possible mapping of the tasks graph is expressed in the form of a chromosome, where the genes correspond to the vertices of the graph, and their values – to NoC nodes. At the first step of the algorithm, a simplified formula is used to calculate the fitness function of chromosomes, according to which the transmission delay between the vertices of the tasks graph depends directly on the average minimum distance between nodes for mesh topology. The best individuals of the population undergo mutations and rearrangements. The most optimal instances become the initial data for the second step of the genetic algorithm which differs from the first one by a more precise formula for calculating the data transmission delay and depends directly on the minimum distance between nodes and specifies the placement of the vertices of the task graph on the nodes of the corresponding type. Taking into account the fact that, unlike mesh, vertices in the circulants are completely homogeneous, this approach with some improvements can be extended to circulant topologies.

A similar representation of chromosomes is used in [36], where the optimal parameter of the specimen is the operating temperature of the crystal which is evaluated using the HotSpot utility [37]. The energy consumption values of each compute node of the two-dimensional mesh network, required for the HotSpot utility, are estimated using Synopsys Power Compiler synthesis. To generate test sequences and to simulate this approach, exemplified by development of a code decoder with a low density of parity checks, NoCSim simulator is used [38]. The development of this approach is in [16], where the projection is performed on 3D mesh, and also in [39], where, in addition to the optimality parameter for energy consumption, the parameter of minimization of the occupied area of the crystal was added. Attention is also worthy of the work [40], where the questions of energy-efficient mapping of tasks on the already established architecture of the location of NoC nodes are considered. This approach is based on the analysis of temperature distribution on the crystal at the stage after synthesis, therefore, the logical topology, which is the basis of NoC communication system, does not play any role here.

In work [41], it is proposed a congestion-aware applications mapping heuristic algorithm, based on betweenness centrality metric of links, which makes it possible to alleviate congestion from highly loaded NoC links. The proposed algorithm utilizes the volume of communication traversing through NoC links to obtain the operational and dynamic characteristics of the system. Compared with other baseline application mapping algorithms, the proposed approach achieved up to 46% and 12% lower channel load and end-to-end latency. Despite the fact that the work of the algorithm is demonstrated on the example of mesh topology, this approach can be extended to any regular topology including circulant graphs.

In [42] the process of synthesis of an energy-efficient NoC consists of such stages as: mapping the graph of subtasks to the characteristic tasks graph; mapping the processor elements to cells in a 2D mesh-like topology; mapping routing information to connectivity resources; assigning priorities between tasks and setting the operating frequency lines of communication depending on their loading according to the DVS algorithm [43]. The first three steps are performed using genetic algorithms. This approach makes it possible to reduce energy consumption by 39%, in comparison with the algorithm which is based on the use of a random mapping of the task on NoC and XY routing. This approach fully adapts to our task by performing the processor elements mapping on the circulant topology on the second stage instead of mesh and selecting the routing algorithm suitable for circulant topology.

In the latest researches, we observed a two-step approach, when several methods of the optimal mapping of the application task graph onto a regular network topology of the NoC are combined together, and it mutually reinforces the resulting performance gain.

For example, in work [44], it is proposed a dynamic resource balance algorithm to achieve a higher system performance by balancing the utilization of on-chip computing resources and communication resources: at first, a dynamic communication optimization algorithm is applied, and then a multi-rectangle selection algorithm, to choose a corresponding number of resource regions for the constructed mapping scheme to allocate the application, is used. The authors state that their approach improves the system throughput by at most up to 31.6% compared with First Free algorithm [45]; up to 25.2% – compared with Nearest Neighbor algorithm [46]; and up to 9.4% – compared with CoNA-SHiC algorithm [47]. The work contains a rather simplified model of the location of compute nodes and their corresponding routers on the chip area; in reality, they can intersect and overlap, as well as do the communication wires between them. The usage of circulant graphs as topologies for NoCs will require a more accurate calculation of the Manhattan distances between routers which can be obtained after the NoC synthesis stage and its placement on the crystal resources, especially, when it is an FPGA. Nevertheless, this task does not seem impossible.

In other work [48], the problem of energy-efficient contention-aware application mapping and scheduling on NoC is considered. It is presented a model, where voltage scaling techniques for processors are combined with frequency tuning techniques for NoC links to save overall system energy consumption. The two-step approach consists of: 1) application mapping problem formulated as a quadratic binary programming problem to minimize the communication energy and solved by using a relaxation-based iterative rounding algorithm; 2) application scheduling problem which aims to find the optimal voltage level for each application task and optimal frequency level for each communication link to minimize the overall system energy consumption and solved by genetic algorithm to search the solution space for the voltage and frequency assignment that minimizes the overall system energy consumption and meets the application's deadline. As a result, the authors argue that energy saving achieves up to 26.5% in simulations with real application graphs. In this study, the NoC structure and topology are presented at a high level of abstraction with a logical spatial connection of heterogeneous computing nodes to mesh topology, but there are no obstructions to prevent it from being modified to a more optimal circulant topology.

4. Conclusion

Thus, the analysis of various ways of NoC synthesis by projecting the characteristic tasks graph onto a predefined regular topology has shown that all optimization methods consist in finding the arrangement of computational tasks on topology nodes capable to minimize data flows and distances between nodes with the most intensive data exchange. This is achieved through the application of various approaches based on the solution of the problem of linear programming, the use of genetic algorithms and various tools for estimating the distribution of energy consumption, and other performance NoC characteristics.

In most cases, mesh topology is used as the predefined NoC topology, characterized by the simplicity of the organization and use of deterministic routing algorithms, which is its advantage, but, at the same time, determines the drawbacks – the limitations associated with the non-optimal form of topology, especially, when mesh topology of rectangular shape is used. For this reason, the approach to NoC synthesis by projecting the characteristic tasks graph onto a predefined regular topology is mainly used when NoC has a homogeneous structure with homogeneous nodes and a uniform distribution of data flows. Although, even in this case, it is actual to use circulants as NoC topological solutions with better characteristics. In addition, most of the approaches to NoC synthesis are considered to be quite universal and can be applied for NoC synthesis on the basis of circulant topologies which will make it possible to take advantage of the deterministic approach, but with the use of more effective NoC topologies optimized for a particular task.

Acknowledgments
This work is an output of a research project implemented as part of the Basic Research Program at the National Research University Higher School of Economics (HSE).

References

[1] Meenderinck C, Azevedo A, Juurlink B, Messa M A and Ramires A 2009 Parallel Scalability of Video Decoders *Journal of Signal Processing Systems* **57** pp 173–194
[2] Marculescu R and Ogras U 2009 Outstanding Research Problems in NoC Design: System, Microarchitecture, and Circuit Perspectives *IEEE Transactions on Computer Aided Design of Integrated Circuits and Systems* **28** pp 3–21
[3] Hu J and Marculescu R 2003 Energy-aware mapping for tile-based NOC architectures under performance constraints *Proceedings of Asia South Pacific Design Automatic Conference.* pp 233–239
[4] Romanov O and Lysenko O 2012 The Comparative Analysis of the Efficiency of Regular and Pseudo-optimal Topologies of Networks-on-Chip Based on Netmaker *Proceedings of Mediterranean Conference on Embedded Computing* pp 13–16
[5] Naresh K, Vasantha M and Nithin K 2017 System level fault-tolerance core mapping and FPGA-based verification of NoC *Microelectronics Journal* **70** pp 16–26
[6] Marcon C, Webber T and Susin A 2017 Models of computation for NoC mapping: Timing and energy saving awareness *Microelectronics Journal* **60** pp 129–143
[7] Gabis A and Koudil M 2016 NoC routing protocols – objective-based classification *Journal of Systems Architecture* **66–67** pp 14–32
[8] Dally W J and Towles B 2004 *Principles and practices of interconnection networks* (San Francisco, CA: Elseiver) p 550
[9] Huang J, Zhong W, Li Z and Chen S 2017 Lagrangian relaxation-based routing path allocation for application-specific network-on-chips Integration, *the VLSI Journal*. In press.
[10] Cilardo A and Fusella E 2016 Design automation for application-specific on-chip interconnects: A survey *Integration, the VLSI Journal* **52** pp 102–121
[11] Saldana, M, Shannon L and Chow P 2006 The Routability of Multiprocessor Network Topologies in FPGAs *Proceedings of the 2006 international workshop on System-level interconnect prediction (SLIP'06)* (NY: ACM) pp 49–56
[12] Benini L and Micheli G 2006 *Networks on Chips: Technology and Tools* (San Francisco, CA: Morgan Kaufmann) p 408
[13] Hu J and Marculescu R 2005 Energy- and Performance-Aware Mapping for Regular NoC Architectures *IEEE Transactions on Computer-Aided Design of Integrated Circuits and Systems.* **24** 4 pp 551–562
[14] Joseph J M, Blochwitz C, García-Ortiz A and Pionteck T 2017 Area and power savings via asymmetric organization of buffers in 3D-NoCs for heterogeneous 3D-SoCs *Microprocessors and Microsystems* **48** pp 36–47
[15] Pavlidis V F and Friedman E G 2007 3-D Topologies for Networks-on-Chip *IEEE Transactions on Very Large Scale Integration (VLSI) Systems* **15** 10 pp 1081–1090
[16] Addo-Quaye C 2005 Thermal-aware Mapping and Placement for 3-D NoC Designs *IEEE International SOC Conference* pp 25–28
[17] Suboh S, Bachouya M, Gabe J and El-Ghazawi T 2008 An interconnection architecture for network-on-chip systems *Telecommunication Systems* **37** 1–3 pp 137–144
[18] Asaduzzaman A, Chaturvedula S R and Pendse R 2013 A novel folded-torus based network architecture for power-aware multicore systems *Computers & Electrical Engineering* **39** 8 pp 2494–2506

[19] Chawade S D, Gaikwad M A and Patrikar R M 2012 Review of XY Routing Algorithm for Network-on-Chip Architecture *International Journal of Computer Applications* **43** 21 pp 0975–8887

[20] Monahova E A 2011 Strukturnye i kommunikativnye svojstva cirkuljantnyh setej *Prikladnaja diskretnaja matematika* **3**(13) pp 92–115

[21] Beivide R, Herrada E and Balcazar J L 1991 Optimal distance networks of low degree for parallel computers *IEEE Trans. Computers.* **40** 10 pp 1109–1124

[22] Ajwani D, Hackett A, Ali S, Morrison J P and Kirkland S 2016 Co-optimizing application partitioning and network topology for a reconfigurable interconnect *Journal of Parallel and Distributed Computing* **96** pp 12–26

[23] Wang D 2015 Tradeoff study among traffic egression schemes and member allocation optimization for link aggregation groups in integrated switching systems *Computer Networks.* **77** pp 56–72

[24] Chou C and Marculescu R 2008 Contention-aware Application Mapping for Network-on-Chip Communication Architectures *IEEE International Conference on Computer Design (ICCD 2008)* pp 164–169

[25] Rhee C, Jeong H Y and Ha S 2004 Many-to-many Core-switch Mapping in 2-D Mesh NoC Architectures *IEEE International Conference on Computer Design: VLSI in Computers and Processors (ICCD 2004)* pp 438–443

[26] Murali S and De Micheli G 2004 Bandwidth-constrained Mapping of Cores onto NoC Architectures // *Proceedings of the Conference on Design, Automation and Test in Europe (DATE'04)* **2** pp 16–20

[27] Bertozzi D and Benini L 2004 Xpipes: A Network-on-chip Architecture for Gigascale Systems-on-chip *IEEE Circuits and Systems Magazine* **4** 2 pp 18–31

[28] Murali S, Benini L and De Micheli G 2005 Mapping and Physical Planning of Networks-on-chip Architectures with Quality-of-service Guarantees *Proceedings of the Asia and South Pacific Design Automation Conference (ASP-DAC 2005)* **1** pp 27–32

[29] Murali S and De Micheli G 2004 SUNMAP: A Tool for Automatic Topology Selection and Generation for NoCs // *41st Conference on Design Automation (DAC'04)* pp 914–919

[30] Murali S 2007 *Methodologies for Reliable and Efficient Design of Networks on chips* (Stanford University) p 272

[31] Srinivasan K and Chatha K S 2005 A technique for Low Energy Mapping and Routing in Network-on-chip Architectures *Proceedings of the 2005 International Symposium on Low Power Electronics and Design (ISLPED'05)* pp 387–392

[32] Hansson A, Goossens K and Radulescu A 2007 A Unified Approach to Mapping and Routing on a Network-on-chip for Both Best-effort and Guaranteed Service Traffic *VLSI Design* pp 1–16

[33] Starobinski D, Karpovsky M and Zakrevski L A 2003 Application of Network Calculus to General Topologies Using Turn-prohibition *IEEE/ACM Transactions on Networking (TON)* **11** 3 pp 411–421

[34] Ascia G, Catania V and Palesi M 2004 Multi-objective Mapping for Mesh-based NoC Architectures *Proceedings of the 2nd IEEE/ACM/IFIP International Conference on Hardware/software Codesign and System Synthesis (CODES+ISSS'04)* pp 182–187

[35] Lei T and Kumar S 2003 A Two-step Genetic Algorithm for Mapping Task Graphs to a Network on Chip Architecture *Euromicro Symposium on Digital System Design* pp180–187

[36] Hung W, Addo-Quaye C, Theocharides T, Xie Y, Vijakrishan N and Irwin M J 2004Thermal-aware IP Virtualization and Placement for Networks-on-chip Architecture *IEEE International Conference on Computer Design: VLSI in Computers and Processors (ICCD 2004)* pp 430–437

[37] Skadron K, Stan M R and Huang W 2003 Temperature-aware Microarchitecture *Proceedings of the 30th Annual International Symposium on Computer Architecture (ISCA'03)* **31** 2 pp 2–13

[38] Whelihan D 2003 *The NOCsim Simulator Users Guide: Version 2.0* (Pittsburgh: CMU) p 51

[39] Hung W, Xie Y and Vijaykrishnan N 2005 Thermal-aware Floorplanning Using Genetic Algorithms *Sixth International Symposium on Quality of Electronic Design (ISQED 2005)* pp 634–639
[40] Xie Y and Hung W 2006 Temperature-Aware Task Allocation and Scheduling for Embedded Multiprocessor Systems-on-Chip (MPSoC) Design *Journal of VLSI Signal Processing Systems for Signal, Image and Video Technology* **45** 3 pp 177–189
[41] Maqsood T, Bilal K and Madani S A 2016 Congestion-aware core mapping for Network-on-Chip based systems using betweenness centrality *Future Generation Computer Systems* In press
[42] Shin D and Kim J 2004 Power-aware Communication Optimization for Networks-on-chips with Voltage Scalable Links *International Conference on Hardware/Software Codesign and System Synthesis (CODES+ISSS 2004)* pp 170–175
[43] Schmitz M T and Al-Hashimi B M 2001 Considering Power Variations of DVS Processing Elements for Energy Minimization in Distributed Systems *Proceedings of the 14th International Symposium on Systems Synthesis (ISSS'01)* pp 250–255
[44] Wang C, Zhu Y, Jiang J, Qiu M and Wang X 2017 Dynamic application allocation with resource balancing on NoC based many-core embedded systems *Journal of Systems Architecture* **79** pp 59–72
[45] Carvalho E, Calazans N and Moraes F 2007 Congestion-aware task mapping in NoC-based MPSoCs with dynamic workload *VLSI, 2007. ISVLSI '07. IEEE Computer Society Annual Symposium* pp 459–460
[46] Carvalho E, Ipiranga A and Alegre P 2007 Heuristics for dynamic task mapping in NoC-based heterogeneous MPSoCs *Rapid System Prototyping (RSP 2007) 18th IEEE/IFIP International Workshop* pp 34–40
[47] Fattah M, Daneshtalab M, Liljeberg P and Plosila J 2013 Smart hill climbing for agile dynamic mapping in many-core systems *Proceedings of the 50th Annual Design Automation Conference on - DAC '13* p 201
[48] Li D and Wu J 2016 Energy-efficient contention-aware application mapping and scheduling on NoC-based MPSoCs *Journal of Parallel and Distributed Computing* **96** pp 1–11

IOP Conf. Series: Journal of Physics: Conf. Series **1050** (2018) 012072 doi:10.1088/1742-6596/1050/1/012072

Generic NP-incomplete problems

Alexander Rybalov
Omsk State Technical University, prospekt Mira 11, Omsk 644050, Russia.
E-mail: alexander.rybalov@gmail.com

Abstract. Ladner in 1975 proved that if P $\neq$ NP then there exist problems in NP, which are not decidable in polynomial time and not NP-complete. Kapovich, Myasnikov, Schupp and Shpilrain in 2003 developed a theory of generic-case complexity. Generic-case approach considers an algorithmic problem on "most" of inputs instead of all domain and ignores its behaviour on the rest of inputs. In this paper we prove a generic analog of the Ladner's theorem: If P $\neq$ NP and P = BPP, then there exists a set $S \in$ NP, such that S is not strongly generically decidable in polynomial time and not generically NP-complete. Supported by Russian Science Foundation, grant 17-11-01117.

1. Introduction

Polynomial reducibility of algorithmic problems is one of the most important notion in classical computational complexity theory. It enables to compare the complexity of algorithmic problems and to develop rich theory of NP-completeness. Excellent book of Garey and Johnson [1] on NP-complete problems was recognized in 2006 as the most cited reference in computer science literature. It is amazingly, that the majority of algorithmic problems, interesting both from a practical and a theoretical point of view, are classified either as polynomial time decidable problems or as NP-complete problems. Ladner in [4] proved that if P $\neq$ NP then there exist problems in NP, which are not decidable in polynomial time and not NP-complete. Perhaps, famous graph isomorphism problem is the one such NP-intermediate problem.

In [3] a theory of generic-case complexity was developed. Generic-case approach considers an algorithmic problem on "most" of inputs instead of all domain and ignores its behaviour on the rest of inputs. This generic complexity approach has numerous applications, especially in cryptography, where cryptosystems must be based on problems which are hard on almost all (random) instances. Usually such behavior of problems is studied in terms of average-case complexity, but generic-case complexity and generic computability are much more convenient in applications, in particular it allow to study complexity of undecidable problems. Rybalov in [5] defined a notion of generic polynomial reducibility and proved that the Boolean satisfiability problem is generically NP-complete.

In this paper we prove a generic analog of the Ladner's theorem: If P $\neq$ NP and P = BPP, then there exists a set $S \in$ NP, such that S is not strongly generically decidable in polynomial time and not generically NP-complete. Here class BPP consists of problems, decidable in polynomial time by probabilistic Turing machines. The equation P = BPP means that every polynomial probabilistic algorithm can be effectively derandomized to an equivalent polynomial deterministic algorithm. There are significant results toward the proving of equation P = BPP (see [2]), but this hypothesis is still unproven.

IOP Conf. Series: Journal of Physics: Conf. Series **1050** (2018) 012072 doi:10.1088/1742-6596/1050/1/012072

2. Preliminaries

Let I be the set of all inputs and I_n be the set of all inputs of size n (sphere of radius n). For a subset $S \subseteq I$ define the following sequence

$$\rho_n(S) = \frac{|S \cap I_n|}{|I_n|}, \quad n = 1, 2, 3, \ldots$$

The value $\rho_n(S)$ is probability to generate an input from S during random and uniform generation of inputs from sphere I_n. The asymptotic density of S is the following limit (if it exists)

$$\rho(S) = \lim_{n\to\infty} \rho_n(S).$$

S is called *generic* if $\rho(S) = 1$ and *negligible* if $\rho(S) = 0$. Clearly, S is generic if and only if its complement in I is negligible.

Following [3] we call a set S *strongly negligible* if sequence $\rho_n(S)$ exponentially fast converges to 0, i.e. there are constants $0 < \sigma < 1$ and $C > 0$ such that for every n

$$\rho_n(S) < C\sigma^n.$$

Now S is called *strongly generic* if its complement is strongly negligible.

A set $S \subseteq I$ is *generically decidable in polynomial time* if there exists a set $G \subseteq I$ such that

(i) G is generic,

(ii) G is decidable in polynomial time,

(iii) $S \cap G$ is decidable in polynomial time.

If G is strongly generic, then S is called *strongly generically decidable in polynomial time.* A generic algorithm $\mathcal{A}$ for S works on an input $x \in I$ in the following way. At first $\mathcal{A}$ decides whether $x \in G$. If $x \in G$ then $\mathcal{A}$ can decide S on G, else $\mathcal{A}$ says "I don't know". So $\mathcal{A}$ correctly decides S on "almost all" inputs (inputs from generic set).

Let I, J be sets of inputs. A set $A \subseteq I$ is *generically polynomially reducible* to a set $B \subseteq J$ (written $A \leq_{GenP} B$), if there are a polynomial probabilistic algorithm $\mathcal{R} : I \times \mathbb{N} \to P(J) \cup \{?, !\}$ (here $P(J)$ is the set of all subsets of J), a polynomial $p(n)$, a polynomial $q(n)$ of degree greater than 2 and a constant $C > 0$ such that

(i) for all $x \in I$ either $(\forall n\, \mathcal{R}(x, n) = ?)$, or for all $n \geq q(k)$, where $k = \text{size}(x)$, it holds:
   (a) $\forall y \in \mathcal{R}(x, n)\ (y \neq !\ \Rightarrow\ \text{size}(y) = n)$;
   (b) $\mathcal{R}$ outputs all elements of $\mathcal{R}(x, n) \setminus \{!\}$ uniformly;
   (c) the probability of $\big(\mathcal{R}(x, n) = !\big)$ is not greater than 2^{-Ck};
   (d) $\dfrac{|\mathcal{R}(x, n)|}{|J_n|} > \dfrac{1}{(p(n))^k}$;
   (e) $x \in A \Rightarrow \mathcal{R}(x, n) \subseteq B$;
   (f) $x \notin A \Rightarrow \mathcal{R}(x, n) \subseteq J \setminus B$;

(ii) the set $\{x \in I : \forall n\ (\mathcal{R}(x, n) = ?)\}$ is strongly negligible.

Now define a generic analog of class NP. A set $S \subseteq I$ belongs to the class sgNP, if there exists a polynomial strongly generic set $G \subseteq I$ such that $S \cap G \in$ NP. A set $S \in$ sg NP is *generically NP-complete*, if $A \leq_{GenP} S$ for all $A \in$ sg NP.

IOP Conf. Series: Journal of Physics: Conf. Series **1050** (2018) 012072 doi:10.1088/1742-6596/1050/1/012072

3. Representation of Boolean formulas

Classical representation of the Boolean formulas by truth tables is not practical because the size of this representation is exponential in the number of Boolean variables. We will represent Boolean formulas by labeled binary trees. This more compact way of representation is used in many real programs and it is convenient for calculations.

Let φ be a Boolean formula in the basis $\{\vee, \wedge, \neg\}$. Without loss of generality we will assume that all negations in φ are only above variables. One can naturally associate with the formula φ a binary tree T_φ that presents the construction of φ from Boolean variables and their negations by means of disjunctions and conjunctions. The internal vertices of T_φ are labeled by symbols $\vee$ and $\wedge$, and the leafs of T_φ are labeled by Boolean variables or their negations. Conversely, given a binary tree T as above one can uniquely reconstruct a Boolean formula φ_T. This gives a one-to-one representation of the Boolean formulas φ by the binary trees T_φ. If T_φ has n leaves then at most n variables may occur in T_φ, so we may assume from the beginning that all variables in T_φ belong to the set $x_1, \dots, x_n$. The size of formula φ is the number of leafs in T_φ.

From now on we identify Boolean formulas φ with their trees T_φ. Denote by $\mathcal{F}$ the set of all Boolean formulas, and by $\mathcal{F}_n$ the set of all formulas in $\mathcal{F}$ of size n.

Lemma 1. $|\mathcal{F}_n| = 2^{n-1}(2n)^n C_{n-1}$, *where* $C_n = \dfrac{1}{n+1}\dbinom{2n}{n}$ *is the n-th Catalan number.*

Proof. Any formula from $\mathcal{F}_n$ is a binary labeled tree with n leafs and $n-1$ internal vertices. There are (see [?]) C_{n-1} non-labeled binary trees with n leafs. Every internal vertex of such tree can be labeled by $\vee$ or $\wedge$, so there are 2^{n-1} such labelings. Every leaf can be labeled by variables $x_1, \dots, x_n$ or its negations, so there are $(2n)^n$ such labelings. So $|\mathcal{F}_n| = 2^{n-1}(2n)^n C_{n-1}$. □

For every Boolean formula φ define the set

$$Eq(\varphi) = \{\varphi \vee ((x_1 \wedge \neg x_1) \wedge \psi),\ \psi \text{ — arbitrary Boolean formula}\}.$$

It is clear, that φ is satisfiable if and only if every formula from $Eq(\phi)$ is satisfiable.

Lemma 2. *For any Boolean formula* φ *it holds*

$$\frac{|Eq(\varphi) \cap \mathcal{F}_n|}{|\mathcal{F}_n|} > \frac{1}{(16(n-2))^k}$$

for any $n > k+2$, *where* k *is the size of formula* φ.

Proof. Let formula φ has size k. Then for any formula $\varphi \wedge \psi$ from $Eq(\varphi) \cap \mathcal{F}_{n+2}$ formula ψ has size $n-k$. Also ψ can contain any variable from $x_1, \dots, x_n$ As in the proof of Lemma 1 we can count

$$|Eq(\varphi) \cap \mathcal{F}_{n+2}| = 2^{n-k-1}(2n)^{n-k} C_{n-k-1}.$$

Therefore

$$\frac{|Eq(\varphi) \cap \mathcal{F}_{n+2}|}{|\mathcal{F}_{n+2}|} = \frac{2^{n-k-1}(2n)^{n-k}C_{n-k-1}}{2^{n-1}(2n)^n C_{n-1}} = \frac{1}{(4n)^k} \cdot \frac{C_{n-k-1}}{C_{n-1}} = \frac{1}{(4n)^k} \cdot \frac{n}{n-k} \cdot \frac{\binom{2(n-k-1)}{n-k-1}}{\binom{2(n-1)}{n-1}} >$$

$$> \frac{1}{(4n)^k} \cdot \frac{(n-1)!}{(n-k-1)!} \cdot \frac{2(n-k-1)\dots(n-k)}{2(n-1)\dots n} = \frac{1}{(4n)^k} \cdot \frac{((n-1)\dots(n-k))^2}{2(n-1)\dots(2n-2k-1)} >$$

$$> \frac{1}{(4n)^k}\left(\frac{(n-1)\dots(n-k)}{2(n-1)\dots(2n-2k-1)}\right)^2 > \frac{1}{(4n)^k} \cdot \frac{1}{2^{2k}} = \frac{1}{(16n)^k}.$$

IOP Conf. Series: Journal of Physics: Conf. Series **1050** (2018) 012072 doi:10.1088/1742-6596/1050/1/012072

So we can bound

$$\frac{|Eq(\varphi)_{n+2}|}{|\mathcal{F}_{n+2}|} > \frac{1}{(16n)^k},$$

and finally

$$\frac{|Eq(\varphi)_n|}{|\mathcal{F}_n|} > \frac{1}{(16(n-2))^k}.$$

□

4. Generic Ladner's theorem

Theorem 1. *If P ≠ NP and P = BPP, then there exists a set S such that*

(i) $S \in sgNP$,

(ii) S *is not strongly generically decidable in polynomial time,*

(iii) S *is not generically NP-complete.*

Proof. Denote the set of all satisfiable Boolean formulas by SAT. It follows from the proof of original Ladner's theorem (see [4]) that there exists a set $A \subseteq$ SAT such that $A \in$ NP, $A \notin$ P and A is not NP-complete. Consider the following set

$$Eq(A) = \bigcup_{\varphi \in A} Eq(\varphi).$$

We prove that $Eq(A)$ is a set we need to construct. At first note, that $Eq(A) \in$ NP, because $A \in$ NP. Therefore $Eq(A) \in$ sgNP.

Now prove that $Eq(A)$ is not strongly generically decidable in polynomial time. Suppose, to the contrary, that there are a strongly generic polynomial set of formulas G and a polynomial algorithm $\mathcal{A}$, deciding $Eq(A)$ for every formula $\varphi \in G$. We construct a polynomial probabilistic algorithm $\mathcal{B}$, deciding A for every formula φ. Since P = BPP, then there is a deterministic polynomial algorithm deciding A – a contradiction with $A \notin$ P. Algorithm $\mathcal{B}$ works on a formula φ of size n in the following way.

(i) $\mathcal{B}$ checks does φ belong to G. If $\varphi \in G$, then $\mathcal{B}$ determines whether $\varphi \in A$ by algorithm $\mathcal{A}$ and stops. If $\varphi \notin G$, then go to step 2.

(ii) $\mathcal{B}$ generates a random formula $\psi \in Eq(\varphi)$ of size n^2.

(iii) $\mathcal{B}$ checks does $\psi \in G$? If $\psi \in G$ $\mathcal{B}$ determines whether $\varphi \in A$ by algorithm $\mathcal{A}$ and stops.

(iv) If $\psi \notin G$, then $\mathcal{B}$ outputs NO.

Note that polynomial algorithm $\mathcal{B}$ outputs correct answer on steps 1 and 3, but can output incorrect answer on step 4. We need to prove that the probability of step 4 is less than 1/2. The probability that a random formula $\psi \in Eq(\varphi)$ of size n^2 does not belong to G is at most

$$\frac{|(\mathcal{F} \setminus G)_{n^2}|}{|Eq(\varphi)_{n^2}|} = \frac{|(\mathcal{F} \setminus G)_{n^2}|}{|\mathcal{F}_{n^2}|} \times \frac{|\mathcal{F}_{n^2}|}{|Eq(\varphi)_{n^2}|}$$

Since G is a strongly generic set, then there exists a constant $\alpha > 0$ such that

$$\frac{|(\mathcal{F} \setminus G)_{n^2}|}{|\mathcal{F}_{n^2}|} < \frac{1}{2^{\alpha n^2}}$$

for all n. By Lemma 2

$$\frac{|\mathcal{F}_{n^2}|}{|Eq(\varphi)_{n^2}|} < (16(n^2-2))^n.$$

IOP Conf. Series: Journal of Physics: Conf. Series **1050** (2018) 012072 doi:10.1088/1742-6596/1050/1/012072

Therefore the probability we need is less than

$$\frac{(16(n^2-2))^n}{2^{\alpha n^2}} < \frac{(32n^2)^n}{2^{\alpha n^2}} = \frac{2^{5n+2n\log n}}{2^{\alpha n^2}} < \frac{1}{4}$$

for large enough n. So the probability of step 3 is less than 1/2.

Now prove that $Eq(A)$ is not generically NP-complete. Assume, to the contrary, that $Eq(A)$ is generically NP-complete. Then for any $S \in$ NP $S \leq_{GenP} Eq(A)$ and there is a polynomial probabilistic algorithm $\mathcal{R} : I \times \mathbb{N} \to P(\mathcal{F}) \cup \{?, !\}$ such that $x \in S \Rightarrow \mathcal{R}(x,n) \subseteq Eq(A)$ and $x \notin A \Rightarrow \mathcal{R}(x,n) \subseteq \mathcal{F} \setminus Eq(A)$ for large enough n. Since P = BPP, there is a deterministic polynomial algorithm $\mathcal{A}$ such that $x \in S \Leftrightarrow \mathcal{A}(x) \in Eq(A)$, i.e. S is polynomially reducible to $Eq(A)$. But obviously, $Eq(A)$ is polynomially reducible to A, so S is polynomially reducible to A. A contradiction. □

5. Summary

We proved a generic analog of the Ladner's theorem: If P $\neq$ NP and P = BPP, then there exists a set $S \in$ NP, such that S is not strongly generically decidable in polynomial time and not generically NP-complete. This result can be applied in cryptography for generic and average-case complexity analysis of public-key cryptosystems, widely used in Internet security and blockchain technology.

[1] Garey M and Johnson D 1979 *Computers and intractability: a guide to the theory of NP-completeness.* W. H. Freeman & Co. New York, NY, USA

[2] Impagliazzo R and Wigderson A 1997 P=BPP unless E has subexponential circuits: derandomizing the XOR lemma *Proceedings of the 29th STOC* pp 220–229

[3] Kapovich I Myasnikov A Schupp P and Shpilrain V 2003 Generic-case complexity, decision problems in group theory and random walks *J. Algebra* **264(2)** pp 665–694

[4] Ladner R 1975 On the structure of polynomial time reducibility *Journal of the ACM* **22** pp 155–171

[5] Rybalov A 2017 On generic NP-completeness of the Boolean satisfiability problem *Prikladnaya Diskretnaya Matematika* **36** pp 106–112 (in Russian)

IOP Conf. Series: Journal of Physics: Conf. Series **1050** (2018) 012073 doi:10.1088/1742-6596/1050/1/012073

The acoustic waves propagation laws in the force-fit connections for test of the interference fit

A O Ryzhova, S A Bekher and A L Bobrov

Siberian Transport University, Novosibirsk, Russia

Abstract. Calibrated tuning samples of compression joints with an interference of 29, 72, 126 μm were developed and manufactured. The samples were tested by ultrasonic control with an pulse echo technique. The coefficients of reflection of longitudinal waves from the boundary of the parts of different values of interference are experimentally determined under real operating conditions at frequencies of 2.5, 5 and 10 MHz.

1. Introduction

One of the most common ways to install bearings on technical objects and vehicles includes stationary detachable joints created by interference fit. The violation of the manufacturing technology and operating modes, defects of the parts to be connected can reduce the values of interference fit (the difference between the diameter of the shaft and the hole before assembling). The weakening of the interference fit is one of the most common causes of the rotation of the inner ring of the bearing, and subsequently its jamming and destruction. According to the data of car repair companies, the failures of the axle blocks of rolling stock in operation are in the second place by the frequency of freight cars stopping (25%) and their subsequent repairing for 12 months of 2017 on railway transport. Malfunctions of the axle boxes due to the common causes of the occurrence are distributed as follows: heating (77%), defects (17%), axle shift (6%). Thus, the internal defects of the axle boxes (83%) are among the most common reasons for failure in operation, and the problem of checking the tightness of the bearing inner rings on the neck of the shaft is rather vivid. The task of monitoring joints planted with interference has been repeatedly considered in research papers [1-4], but the variety of such compounds requires a correction of the approach to it in each case.

2. Formulation of the problem

Currently, the interference fit bearing inner rings is determined indirectly by the difference in the results of measurements of the inner diameter of the ring and the outer diameter of the axis before mounting in a free state. The reliability of the measurements is influenced by: the difference in the temperature of the parts, the surface condition, the ovality and the conicity. Pressing the rings on the shaft leads to the occurrence of scoring of the neck of the axle, therefore, when repairing the axle boxes of wheel pairs, it is allowed not to remove the bearing inner rings of bearings provided that the interference is checked. Currently, the most common method of control is checking the density of interference fit on the rotation using specialized devices. The drawbacks of this method include the possibility of damaging the raceway of the bearing ring, leading to its rejection, and the low reliability of the method which is associated with the influence of the human factor.

The railways use specialized measuring blocks for checking the tightness of the bearing rings on the neck of the axis: PS-219 (No. 22100-01 in the State Register of Measuring Instruments) [5] and

IOP Conf. Series: Journal of Physics: Conf. Series **1050** (2018) 012073 doi:10.1088/1742-6596/1050/1/012073

UDS1-SYN (No. 21975-09 in the State Register of Measuring Instruments). The principle of the UDS1-SIN device is to measure the amplitude-frequency characteristic, the decay time and the intensity of repeatedly reflected ultrasonic pulses. A special program sets the operating modes of the controller, collects, processes and displays the data of the interference fit [6]. The PS-219 is based on measuring the time interval, during which a pendulum colliding with a ring-mounted ring performs a predetermined number of oscillations. Planting (interference) is considered normal if the measured time interval is greater than the set limit value. These instruments were designed to be used together with with the "hot" ring-to-axle interference fit technology, where the heated ring is freely mounted to the desired position on the axis, and the fixed connection is formed after the temperature of the parts to be connected is adjusted to an interference fit of 30 to 65 μm. At present, a cold press fit technology is applied in the rolling-stock sector, where the ring is mounted on the axle using a press with a difference in the diameters of the axis and rings from 45 to 130 microns. At the same time, the surfaces of the joined parts are rubbed, which, as experience of using devices in production shows, reduces the reliability of the results.

A separate line of research is related to the use of the influence of the landing density on the propagation velocity and amplitude of ultrasonic waves. Several methods are considered, in one of which [7] two types of waves come into action in the bearing ring and the value of the interference fit is judged by the difference in propagation times. Transverse waves propagate from the radiator to the receiver and are reflected from the cylindrical interface of the parts, and the inhomogeneous wave propagates along the outer cylindrical surface of the ring. However, fluctuations in temperature, chemical composition and structural state lead to the fact that the measurement error is comparable to the measured values of stresses arising in the rings, due to elastic deformation when planted on the axis.

The second method was carried in such a way that ultrasonic vibrations are excited by a combined ultrasonic transducer whose acoustic axis is oriented in the direction of the cylindrical aperture under the fastening element, signals reflected from the first and second contact paths along the beam are recorded. The value of the interference is determined by the difference in the amplitudes of the received signals. This method is aimed at controlling the quality of compact joints with interference with an air gap between parts, such as bolted and riveted joints.

In this paper, we investigate the influence of the interference fit of the bearing inner rings on the shaft (axis) after press fit on the pattern of propagation of acoustic waves through the joint boundary to assess the possibility of controlling the detachable fixed cylindrical joints manufactured by the press fit method.

3. Theoretical aspects

The connection with the interference is the interface between the three media (figure 1): the material of the ring - the gap between the ring and the shaft - the material of the shaft. A gap of thickness h between the ring and the shaft is completely or partially filled with lubricant. For acoustic control methods, the thickness of the gap serves as the informative parameter, which determines the acoustic properties of the medium interface and characterizes the tightness of the controlled connection.

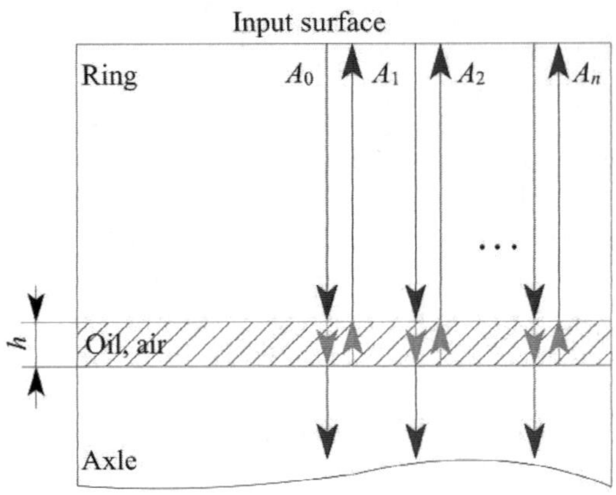

Figure 1. The scheme of interaction of the ultrasonic wave pulse with the ring-shaft boundary and the outer boundary of the ring

In the case of pulsed emission of acoustic waves A_0, a series of acoustic pulses (A_1, A_2, .., A_n in figure 1) which are repeatedly reflected from the interface and the ring insertion surface by the receiving transducer is captured. In the plane model, the amplitude of the nth pulse is proportional to the amplitude of the (n - 1) th pulse and depends on the damping ratio and the reflection ratios from the input surface R_c and the media interface R_g according to the law [9]:

$$A_n = A_0 \cdot \left(R_c \cdot R_g \cdot e^{-\gamma \cdot 2 \cdot l}\right)^n \tag{1}$$

where A_n – is the amplitude of the signal reflected n times from the interface, rel. units; A_0 – is the amplitude of the emitted signal, rel. units; γ – attenuation coefficient, m^{-1}; R_c – is the reflection coefficient from the input-boundary surface – the ring-air material; R_g – is the reflection coefficient from the ring-gap-shaft boundary; l – is the thickness of the ring, m.

In relative logarithmic units, dB, expression (1) takes the form:

$$N = 20 \cdot \lg\left(\frac{A_n}{A_0}\right) = \alpha \cdot n \tag{2}$$

where α – is a ratio of proportionality, which depends on the damping ratio, the path length of the ultrasonic wave in the ring, and the reflection rate from the input surface:

$$\alpha = 20 \cdot \mathrm{Lg}(R_g) + 20 \cdot \mathrm{Lg}\left(R_c \cdot e^{-\delta \cdot 2 \cdot l}\right) \tag{3}$$

Expression (2) relates the coefficient of proportionality of the amplitude and the number of the pulse with the reflection coefficient from the interface of media, which in turn depends on the thickness of the layer h [10]:

$$R_g = \frac{\left(\frac{z_1}{z_2} - \frac{z_2}{z_1}\right)^2}{4\,\mathrm{ctg}\left(\frac{2 \cdot \pi \cdot h}{\lambda_2}\right) + \left(\frac{z_1}{z_2} + \frac{z_2}{z_1}\right)^2} \tag{4}$$

where z_1 – is the acoustic resistance of the material of the ring, equal to steel 46 MPa· s/m; z_2 – the acoustic resistance of the gap medium, taken for an oil equal to 1.3 MPa· s/m, the acoustic resistance of the air – 0.00043 MPa · s/m; δ – width of gap, m; λ_2 – The length of the acoustic wave in the second material, m.

Thus, the experimentally determined proportionality factor 20 · Lg (R_g) of the dependence of the amplitude of the echo pulse N on the momentum number n in expression (2) with the help of expression (4) makes it possible to estimate the thickness of the contact layer of the gap, which, in other words, characterizes mechanical stress in the parts to be connected.

4. Experimental results

The experiments were carried out on specially designed samples with a known interference fit. Samples were made from a shaft and a bearing ring 36-42726E2M in accordance with GOST 18572-

2014. Calibration is performed before the rings are placed on shafts in FGUP SNIIM with an error of no more than 1.5 μm (Certificate of calibration No. 10-1291 dated August 1, 2017). The conicity of shafts and axes does not exceed 2.0 microns, ovality doesn't exceed 0.5 microns. The roughness of the landing surfaces of the bearing and shaft rings did not exceed the standard value R_a=1.25 μm. The rings and shafts were selected in such a way as to provide minimum, maximum and medium tension with a difference in the diameters of the axis and bearing inner rings: 29, 72 and 126 μm.
In the manufacture of tuning samples, they tried to implement the technology used in car repair depots if it was possible. Inside, the rings were preliminarily worn on the shafts by the method of thermal fit for 20% of the length. For this purpose, the EMPI-1 grease was applied to the mating surfaces of the parts. The inner rings were heated to a temperature of 120 ° C, and the shafts were cooled to -15 ° C. At the last stage, the joints (figure 2) were formed on the press with a force of 400, 800, 1500 kN, respectively.

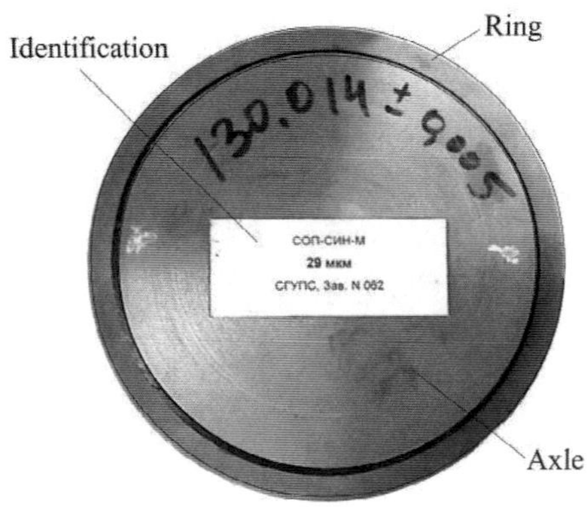

Figure 2. Appearance of adjusting sample of connection with interference fit

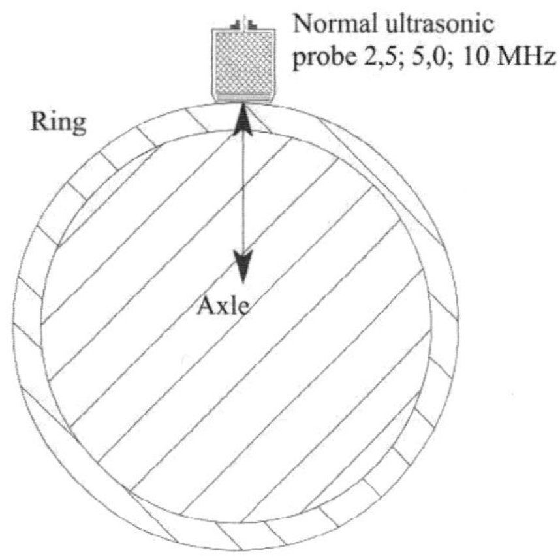

Figure 3. Scheme of test of stationary detachable connection of a ring with an axis by an pulse echo technique

The studies of the density of rings on the neck of the axis were carried out by pulse echo technique. Direct combined converters with an operating frequency of 2.5, 5.0 and 10 MHz were used for the excitation and reception of ultrasonic waves. (figure 3). The converters were connected to the universal echo-pulse flaw detector USD-50 (No. 52657-13 in the State Register). The connection of the ring to the axle was radially observed from the rolling surface of the bearing ring. A series of echo pulses, re-reflected from the inner and outer cylindrical surfaces of the bearing rings, were registered.
The smallest uncertainty in the amplitude of the re-reflected pulses at various points on the ring surface was observed when monitoring direct PES with an operating frequency of 5 MHz. All the results of further measurements are given for a PET operating at a frequency of 5 MHz. Dependences of the amplitudes of echo pulses, expressed in dB, on their ordinal number (figure 4) are satisfactorily described by a linear relationship with a correlation ratio greater than 0.95. The method of least squares is used to calculate the proportionality ratios for different values of the interference (figure 5, item 1). The lowest slope of the dependence is observed for the free bearing ring - 1.9 dB.

5. Discussion

The oscillation damping, the input surface deviation from planeness, the ultrasonic beam divergence are characterized by a polar pattern which has an influence on the slope of the dependence between the echo amplitudes and their serial number for the free ring (0 mm). The difference between the proportionality factor for joints with interference and for a free ring is directly proportionally to the logarithm of the reflectance for the ring–gap boundary R_g:

$$\mathrm{Lg}(R_g) = \frac{\alpha_\Delta - \alpha_0}{20} \quad (5)$$

IOP Conf. Series: Journal of Physics: Conf. Series **1050** (2018) 012073 doi:10.1088/1742-6596/1050/1/012073

where α_Δ and α_0 are proportionality factors (slope) of the dependence between the echo amplitudes and their serial number for the reference blocks with an interference fit Δ and for a free ring, respectively.

The reflectance from the boundaries in the reference blocks is determined by the expression (5):

$$R_g(29\ \mu m) = 0.66,\ R_g(72\ \mu m) = 0.58,\ R_g(126\ \mu m) = 0.62 \quad (6)$$

The average design thickness of the gap, filled with industrial oil, in the reference blocks based on (4) and (5) is:

$$h(29\ \mu m) = 0.27,\ h(72\ \mu m) = 0.19,\ h(126\ \mu m) = 0.23 \quad (7)$$

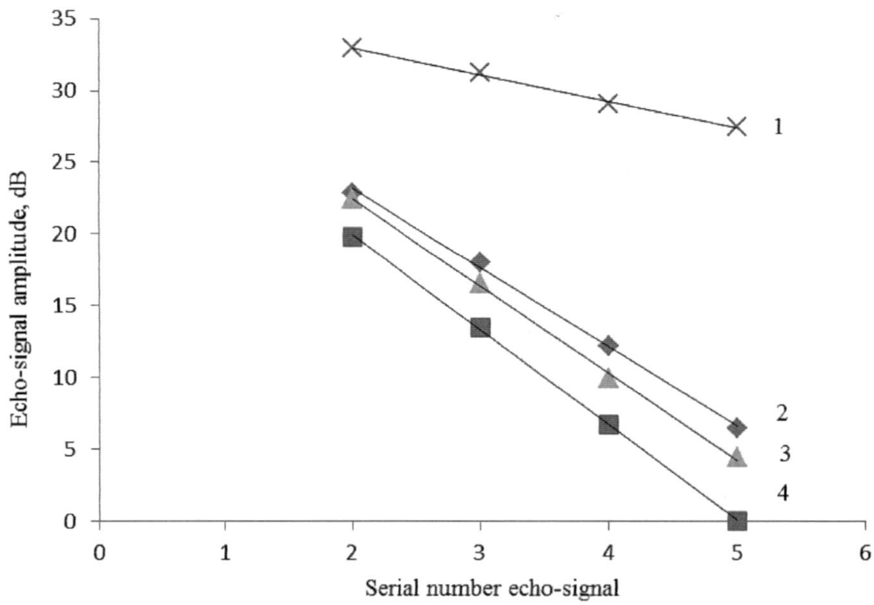

Figure 4. The chart of the dependence between the echo amplitudes and their serial numbers: 1 – free ring, 2 – interference fit is 29 μm, 3 – 126 μm, 4 – 72 μm

The obtained values of the thickness of the gap with the ointment correspond to their roughened surface for the ring which is less than Ra = 1.25 μm, and for the shaft - Ra = 0.8 μm. The estimated thickness of the gap when it is filled with air is less than 10^{-6} μm, which significantly exceeds the surface roughness.

The main change of the proportionality factor (3) and, respectively, of the reflection coefficients (6) occurs in the range of the interference fit range from 0 to 30 μm (see figure 5). The test results of real connections of rings with axles with a weak interference by the ultrasonic method in the car repair depot conditions are shown in figure 5, pos. 2. The part of the connections has a proportionality factor close to the proportionality factor for the free ring, which may be due to the difference between the shape of the axis and the cylindrical: out-of-roundness and taper. An indirect confirmation of this fact is the standard deviation in a series of connections' measurements. This deviation is always higher with a weak interference fit than the standard deviation for the connections with a strong interference fit.

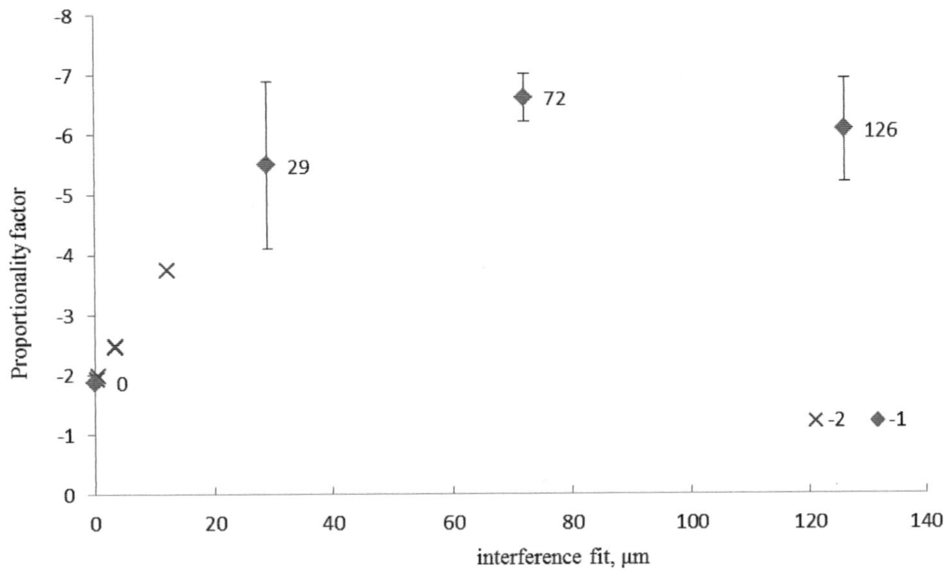

Figure 5. The chart of the dependence between the proportionality factor and the interference fit: 1 – reference blocks, 2 – real objects

6. Conclusions

A technique for measuring the reflectance from the force-fit connection boundary based on a comparison of the proportionality factor between the echo amplitude and the serial number on the controlled connection and the free bearing ring has been proposed and implemented.The reflectance of 0.66, 0.58 and 0.62. on reference blocks with the interference fit of 29, 72 and 126 µm was experimentally established. The calculated oil contact layer thickness of the connection is 0.27, 0.19 and 0.23 mm, which corresponds to 20-30% of the surface roughness. The greatest gradient of the reflection coefficient on the reference blocks corresponds to the interference fit range from 0 to 30 µm. In actual service conditions, the deflection of the axis surface from the cylindrical shape has a significant effect on the connection gap thickness and on the acoustic characteristics of the connection. Consequently, the ultrasonic method can be used to assess the closeness of parts to be fit and detect the out-of-roundness and taper.

References

[1] Senko V I, Chernin I L, Chernin R I and Senko N G 2010 Improvement of the control of the joints with the interference of the wheel bearing *Modern technologies. System analysis. Modeling* **3** pp 18-22.

[2] Glinnikova T P 2011 Determination of the interference of the shut-off valve of the piston compressor *News of higher educational institutions. Mountain magazine* **7** pp 110-113

[3] Muraviev V V, Strizhak V A, Volkova L V and Pryakhin A V 2014 Control of interference fit of the wheel band using the acoustoelasticity method *Locomotive* **5** pp 39-41

[4] Stepanova L N, Tenetilov E S and Bekher S A 2009 The control rings of the bearings of the locomotive by means of acoustic emission *Defectoscopy* **9** pp 49–55

[5] Belonogov A G, Rodionov S A and Shanaurin A M 2004 The IIC-219 instument for testing the tightness of bearing races *Russian Journal of Nondestructive Testing* **7** pp 496-497

[6] Muraviev V V, Glukhov B V, Kozyatnik I I und Kurbatov A N 2000 Investigation of the possibility of determining the value of interference fit by ultrasonic method *Collection of scientific articles: «New technologies – railway transport: training of specialists,*

organization of the transportation process, operation of technical means» (Omsk) pp 268-272.

[7] Ryzhova T B and Tcherban K S 1995 *Pat. 2045059 RF, G01N29/14. The method of ultrasonic quality control of assembling of joints with interference fit*

[8] Ryzhova T B 1996 *Pat. 2063027 RF, G01N29/10. The method of ultrasonic quality control of assembling of joints with interference fit*

[9] Muraviev V V, Muravieva O V and Platunov A V 2016 *Acoustic tensometry and structuroscopy of thin steel wires (Izhevsk: Izhevsk technical university)* p 108

[10] Troitsky V A and Popov V U 2010 *Ultrasonic inspection of welded joints (Kiev*) 2010 p 224

IOP Conf. Series: Journal of Physics: Conf. Series **1050** (2018) 012074 doi:10.1088/1742-6596/1050/1/012074

Experimental registrations of plain sieve operation, making transverse fluctuations

U K Sabiev, E V Demchuk, A Yu Golovin and I V Skysanov

Omsk State Agrarian University named after P.A. Stolypin, Omsk, Russia

Abstract. Possible methods of attachments improvement of grain cleaning machines by means of mobilization factors, influencing direct on qualitative and quantitative separation are represented. Graphical dependences of grain separation completeness from the parameters of sieve vibration for different values of specific loads are also given. Basic relationships intensified the separation process of grain on the sieve, making transverse fluctuations has been detected according to the classical experiment results, as well as the influence of more important factors by the results of sieving experiment on separation completeness as on optimization criteria. The range vibrations of significant factors is also proved.

1. Introduction

Grain production is the basis of stable development of agricultural sector. The most resource-intensive process in all technological production chain is post harvest treatment. The aim of post harvest treatment is to obtain seed, productive and feed grains, corresponding to the requirements provided for standards.

Feed grain can be processed into compound feed, developing own feed production, which enables to reduce costs for purchase of raw material and its transportation. It provides profitable conducting of livestock breeding [1]. In order to provide the population with food it is necessary to get seed and productive grain.

Qualitative and quantitative conservation of grain depends upon providing enterprises of agro-industrial complex with modern equipment for post harvest treatment, its technical level and efficient usage. Foreign grain cleaning machines of different makes are recommended good enough itself at farm commodity producers but not every farm can buy these machines because of the cost level for foreign technique and others connected with its maintenance and at present days it is impossible to buy foreign machines.

Post harvest treatment of grain is carried out by current grain cleaning lines where all grain passes through after combine harvesting. Characteristics of harvested grain lots are greatly influenced on the quality of grain cleaning machines operation. That is why agricultural enterprises of our country need innovative grain cleaning machines for post harvest treatment of grain. It is the main problem in the conditions of the Western Siberia with high moisture of grain during post harvest treatment.

Carried-out review and analyses of literature show that at present in creation of new grain cleaning methods, there are not any cardinal changes to the best in developments of progressive attachment of grain cleaning machines.

The most widespread separate elements for grain cleaning are penetrative sieves with rectangular openings separating the seeds on thickness. These sieves have greater values of specific productiveness than the sieves with round openings.

R. Aguirre [2] studied the problems of increasing efficient separation of grain on plain sieves in the result of the whole complex variation of kinematic parameters for the sieves making harmonious fluctuations. However they have essential disadvantages. Existing sieves and kinematics of motions don't promote the orientation of grain mixture particles with regard to openings. The sieve in the machine operates as vibroconveyor. But the fluctuations of cleaning shoes are accompanied by considerable dynamic loads caused the increasing frame vibration and as a result the reduction of reliability grain cleaning aggregate.

2. Problem definition

Scientific researches on post harvest treatment of grain are carried out in the agroengineering department of the Omsky SAU. It is suggested to intensify the grain separation without increasing outline dimension of attachments by means of mobilization factors direct influencing on qualitative and quantitative operations of separative attachments of the grain cleaning machines. Constructive solutions on sieve separators improvement and researches of rational kinematic regimes of cleaning shoes operations are related to these factors. Especially perspective suggestions permitted to obtain essential effect not only under conditions of new machines creation, but can be used for intensification of cleaning process and grain separation on serial grain cleaning machines. One variant of increasing fractional, plain sieve grain cleaning machine productiveness by usage the sieves with rectangular openings located under an optimal angle towards its motion is suggested in this work [3]. It is established that it allows to increase the possibility of particle pass in opening in 1,7 times under angle of opening edge arrangement 20º. Productive check of sieves operation suggested by construction showed the increasing of productiveness on 23% in comparison with serial analog.

There are another methods influencing on separation intensification process, for example, complication of fluctuations law of plain sieves according to different trajectory [4.5].

Sieves of grain cleaning machines located in such a way that the direction of sieve long cross points match with fluctuations direction and perpendicular to the direction of grain movement are of great interest. Search experiences are carried out according to the classical plan with the aim to define the arrangement range of the main factors, influencing on technological process. Experimental researches of the plain sieve operation results making transverse fluctuations are carried out on modernized laboratory arrangement in two stages.

Classical experiment is done during the first stage. Its aim is to detect the area regimes of the sieve transverse fluctuations intensified the grain separation process and to define the frontier lines and the content of the planned experiment. Usage of classical experiment for these purposes are distinguished from planned search methods (steepest ascent method, gradient method and others) are justified because every value of specific load on the sieve corresponds the definite optimal value of the sieve acceleration. Using transverse fluctuations of the sieve we change in a great range of acceleration and for wide range of specific load values, extremums of the response function it is very difficult to seek for every fixed values of delivery.

According to the results of sieving experiments of plain sieve operation making transverse fluctuations, statistic significants on Student's criteria on the 5% level are the following factors: specific load on the sieve, angle of sieve forward pitch to the level and fluctuations parameters of the sieve (amplitude and frequency).

Variable arranged parameters (factors) according to the sieving experiment are as follows:

- specific load on unit area of the sieve (G) has been changed in limits 0,335...1,34 kg/(m² • c), that corresponds to the regimes of the grain cleaning machines operation;
- angle of sieve forward pitch α has been changed from 8° to 16°. Essential increasing of this parameter in comparison with serial machines is explained by necessity to increase (compensations) of medium rate of material movement on sieve surface because under forward travels of cleaning shoe, fluctuations help this process, taking into account that_there movement corresponds to material movement;
- frequency of fluctuations n has been changed in limits from 300 to 500 min^{-1};
- amplitude equals $7{,}5 \cdot 10^{-3}$ m, that corresponds to fluctuations regime of plain sieves of home machines of initial and secondary cleaning CM-4 and MC-4,5.

Completeness of separation has been calculated according to the standard method. Obtained meanings were statistically treated and systematized.

The dependences of qualitative indicator of sieve operation from variable parameters of experience are built-up according to the experiments results.

3. The results of experiments

The dependences of completeness separation from regimes parameters of sieve transverse fluctuations, particularly from the angle of sieve forward pitch for different value meanings of specific load are obtained in the result of the first stage of classical experiment (figures 1....3).

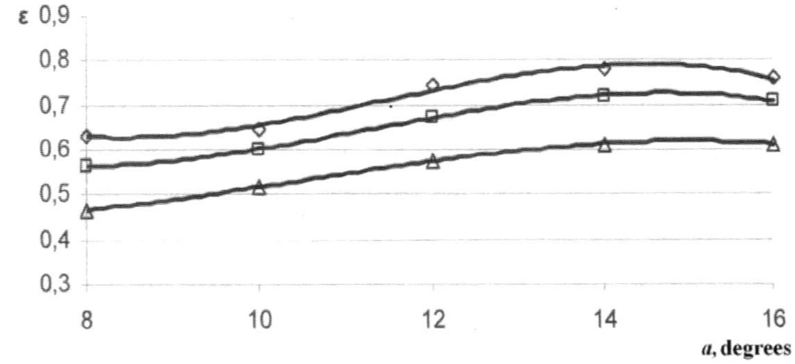

Figure 1. The dependence of completeness separation from the angle of sieve forward pitch for value of specific load (◊ – 0,335; □ – 0,838; Δ – 1,34) kg/(m² • c) if n= 300 min^{-1}.

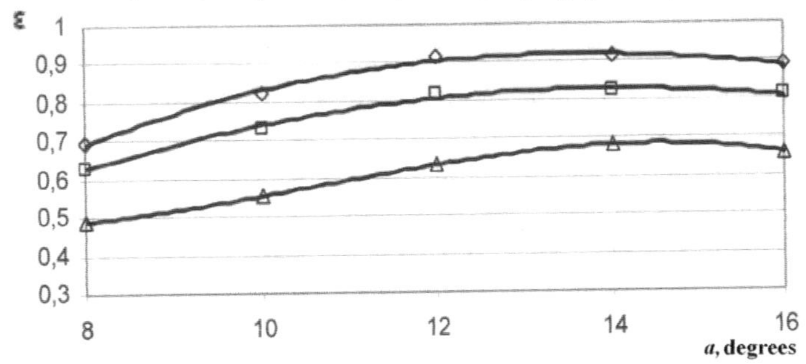

Figure 2. The dependence of completeness separation from the angle of sieve forward pitch for value of specific load (◊– 0,335; □– 0,838; Δ– 1,34) kg/(m² • c) if n= 400 min^{-1}.

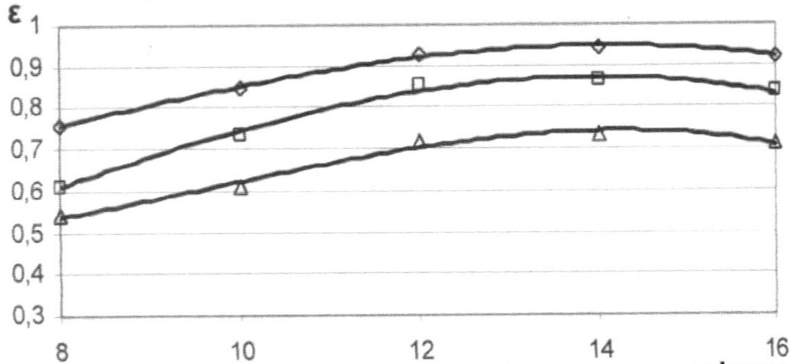

Figure 3. The dependence of completeness separation from the angle of sieve forward pitch for value of specific load (◊– 0,335; □– 0,838; Δ– 1,34) kg/(m² • c) if n= 500 min^{-1}.

From analyses of figures 1-3 we can see:

1. Under increasing of the angle of sieve forward pitch α, value of completeness separation ε is increased for all regimes of specific load on the sieve. And with subsequent increasing α, completeness separation begins to reduce, it can be caused by influence of this angle on the movement rate of grain material on the sieve and by excess of maximum rate, after that cleaning is not so

IOP Conf. Series: Journal of Physics: Conf. Series **1050** (2018) 012074 doi:10.1088/1742-6596/1050/1/012074

effective. Under increasing rate characteristics of grain material the cleaning regimes changes into the stage of movement with separation from sieve surface. At the same time the particles are orientated transverse with regard to the direction of descent of grain flow from the sieve.

2. Increasing frequency n completeness separation is increased and explained that any value of specific load on the sieve corresponds its own rate value. It is clear from told: the more the specific load value on the sieve, the more acceleration value must be. That just achieved by increasing n.

3. The influence of angle sieve forward pitch is reduced with the growth of specific load. According to the figures 1.1…1.3 we can see that completeness separation under specific load G = 1,34 kg/(m² • c) with the change of angle sieve forward pitch is changed less than under specific loads G = 0,335...0,838 kg/(m² • c).It is explained by considerable grain layer on the sieve under heavy load, that increase two-three thickness of wheat grain. Only lower grain layer which is located directly on the sieve surface connects with the sieve and its openings. Thus we draw conclusion that this sieve is used more effectively for leading grain material to the basic conditions

We conducted experiments with variation of radius eccentric value, generating the reciprocating motion of cleaning shoe for investigation of fluctuations amplitudes influence of the sieve fluctuations on completeness separation. At the same time for shortage the number of experiences it was taken into consideration to eliminate following established regimes, more perspective completeness separation value.

As value of specific load G = 1,34 kg/(m² • c) in series of experiments increase completeness separation it was decided not to use it. Frequency of the sieve fluctuations during next series of experiments has been fixed on the level of 500 min¯1. The fluctuations amplitude varies on the following levels (7,5; 8,1; 8,7; 9,3; 9,9) • 10^{-3} m.

The dependences ε = f(R), shown in the figures 4….6 are obtained in the results of experiments.

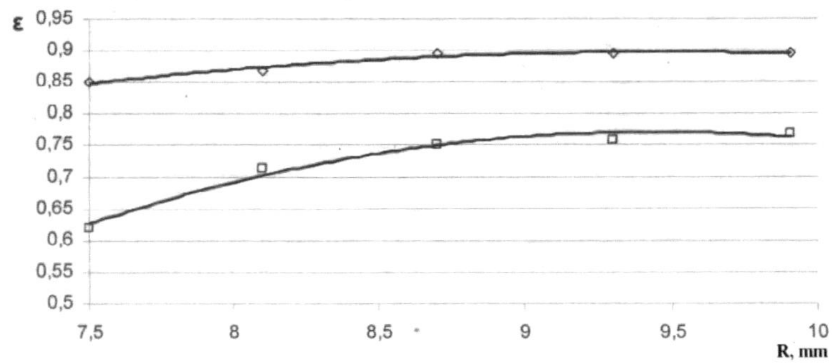

Figure 4. The dependence of completeness separation from the fluctuations amplitude for angle of forward pitch α= 10°, value of specific loads: (◊– 0,335; □– 0,838) kg/(m² • c).

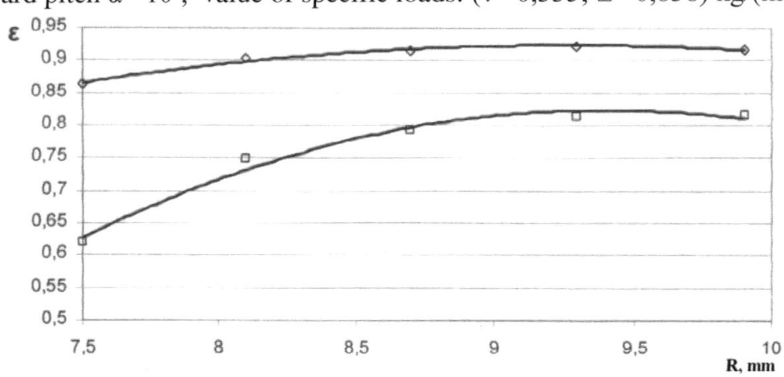

Figure 5. The dependence of completeness separation from the fluctuations amplitude for angle of forward pitch α= 12°, value of specific loads: (◊– 0,335; □– 0,838) kg/(m² • c).

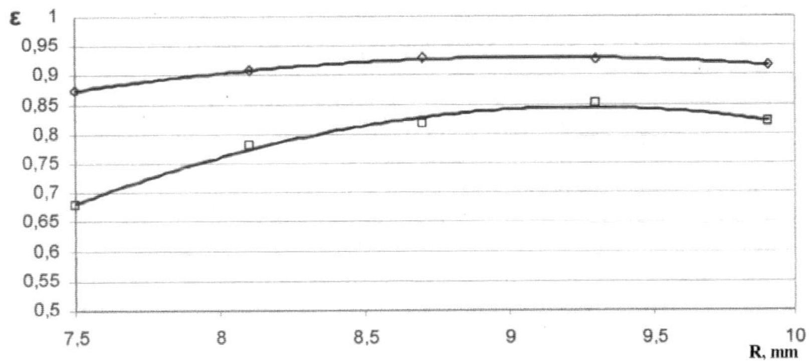

Figure 6. The dependence of completeness separation from the fluctuations amplitude for angle of forward pitch α= 14°, value of specific loads: (◊– 0,335; □– 0,838) kg/(m² • c).

From the analysis of figures 4....6 it is possible to reveal the following

1. At increase of amplitude fluctuations of sieve the completeness of separation increases too, and for the size of specific load G = 0,838 kg/(m² • c) its influence is more expressed. Also as well as in case with change of frequency fluctuations it is possible to explain with the fact that to any value of specific load on the sieve corresponds the value of acceleration. And acceleration of the sieve is in direct ratio to amplitude of its fluctuations. In addition it is possible to note that at the value of specific load G=0,838 kg/(m² • c) the completeness of separation changes in big limits, then at G=0,335 kg/(m² • c). The conditions under which mixing processes in grain mass proceed quicker are created at increase of fluctuations amplitude on this mode, larger impurities rise upward and small impurities fall to a lower layer. It leads to their faster allocation from the grain lots. In a case when specific load on the sieve is 0,335 kg/(m² • c), the thickness of grain layer on it is rather small and increase of fluctuations amplitude leads to slight increase of completeness of separation.

Apparently from figures further increase of fluctuations amplitude of the sieve is inexpedient, as it leads to decrease in completeness of separation and respectively will lead to decline in quality of separation, at the expense of unload of grain material towards fluctuations of cleaning shoe.

2. Increase of forward pitch angle of the sieve increases completeness of separation and it reaches the maximum value at the modes which are shown in the figure 4.6. Increase *α* in our case increases the resultant rate of grain flow on the sieve, from this we can draw the conclusion that with increase of grain flow rate the completeness of separation increases too. And we compensate by increase of rate that increased way of grain material on the sieve which takes place at transverse vibrations. Respectively, the motion rate of grain material on the sieve remains former at increase of separation completeness.

3. At value of fluctuations amplitude (8,75...9,5) • 10^{-3} some stagnation of separation completeness at the certain level is observed. It is obvious that in this case, grain motion rate on the sieve comes nearer to limit at which process of separation stops. Further increase of fluctuations amplitude will negatively affect the value of separation completeness and increase in cleaning shoe imbalance and all grain cleaning machine.

4. Conclusions

1. On the results of classical experiment the main patterns intensifying process of separation of grain on the sieve making transverse vibrations and also influence of the most significant factors by results of the sieving experiment on completeness of separation as on criteria of optimization are revealed.

2. The variation range of following factors: for the value of specific load (G) – 0,335...0,838 kg/(m^2 • c); angle of forward pitch of the sieve (a) – 10...16°; amplitude of sieve fluctuations (R) - (7,25...9,25) • 10^{-3} m; frequency of sieve fluctuations (n) – 450...550 min^{-1} has been proved.

References

[1] Sabiev U K Demchuk E V Myalo V V and Soyunov A S 2017 *Innovative equipment and production method for mixed fodder in the conditions of agricultural enterprises* (United Kingdom: In collection: IOP Conference Series: Materials Science and Engineering 8. Сер. "VIII International Scientific Practical Conference "Innovative Technologies in Engineering"" p 012020.

[2] Aguirre R and Garay A E 1999 *Continuous – flowing portable separator for cleaning and upgrading bean seeds and grains* (Agr. Mech, Afr. and Lat. Amer.) vol 1 pp 59-63.

[3] Sabiev U K and Zilbernagel A V 2011 *Theoretical research of trajectory particle movement influence on its interaction possibility with the edge of oblong sieve opening* (Barnaul: Vestnik of the Altaj SAU) vol 9 pp 78-81

[4] Sabiev U K and Golovin AYu 2012 *Cleaning shoe* (The Russian Federation Patent for useful model) № 117838 Publish in B I. vol 19.

[5] Golovin A Yu Sabiev U K and Soyunov A S 2017 *Trajectory grain movement on the sieve, making round motions* (Omsk: Vestnik of the Omsk state agrarian university) vol 4 pp 204-210

Improving the processing accuracy of the valve seats of internal combustion engines using diagnostic measurements

D T Safarov, A G Kondrashov

Naberezhnochelninsky Institute (branch) of Kazan (Volga Region) Federal University, Naberezhnye Chelny, Russia

safarov-dt@mail.ru

Abstract. Currently, the main tool of quality management at the enterprises of various industries are statistical methods of quality management. In the literature, mostly on examples of the successful application of control charts for dimensions. More sophisticated measures of quality details not given. The article illustrates a practical application of control charts applied to technologically sophisticated measure of the accuracy of the key indicator affecting the operation of the engine - radial runout of valve seat cylinder head. Data of measurements made in accordance with the standard metrological definitions are processed in the software product "Attestator". As a result, identify the potential percentage of possible marriage, revealed the existence of special reasons to change the values of individual indicators, the index of reproducibility and stability of the process, the decision about the certification process, but the factors of the process that need to be addressed to improve the quality of the products is not revealed. A universal methodology consisting of four steps, the implementation of which allows to develop such a scheme of measurement, which significantly improves the search performance of important technological factors. For example, the same technological operations and the accuracy rate shows the application of the methods. The result of the survey revealed two main factors of the process is bending of the boring bar and the error of the satellite-based suppression will significantly improve the quality of manufacturing of parts.
Keywords: improving the quality, accuracy, processing, automotive components, diagnostic components

1. Introduction

Enterprise for the production of components should follow the requirements of standard IATF 16949:2016. According to his requirements the organization shall:

– To monitor and measure product characteristics to verify and compliance with production requirements.

– To determine the process of continuous improvement.

– To take corrective action to eliminate causes of nonconformities to prevent their repeated occurrence.

The technological operations most often for this purpose apply statistical methods of quality management. The main tool is the control chart.

Under statistical control technologist puts the key (most important) quality characteristics. The management plan includes the scope and frequency of taking the sample. Examples of the application of quality management techniques described in the literature, as a rule, do not provide additional identification of the factors of the process [1-7]. Information about the efficient use of control charts for structurally complex in terms of accuracy is almost not given. In the works dedicated to the detection equipment includes elements of identification of the factors of the process, but in turn poorly applied statistical methods of quality control [8-10].

2. Problem definition

Consider how effective the use of control charts as a separate management tool on the example of processing of the important parts of a car engine – head-cylinder. The key characteristics of items affecting the operation of the parts – the value of total radial runout of valve seats. Beating breaks the air tightness of the combustion chamber, reducing engine power. This indicator is formed in a continuous automatic line for boring operations. Runout measurement was performed by special measuring device, equipped with a dial indicator type, according to their metrological determination (the difference between the largest and smallest distances from all points of the real surface within the regulated area to the base axis). The measurements were performed according to the requirements of the current management plan, with the registration of the full magnitude of the radial runout every 15 min.

The measurement data were processed by the software product "Attestator" [11]. First, the constructed histogram and curve of the normal distribution (Fig. 1 a). The main parameters of a distribution the standard distribution and the mean value of the heartbeat, as well as indices of stability and reproducibility of the process. In the histogram the presence of two indicators, the limits of tolerance of the accuracy rate. Indexes of reproducibility have unsatisfactory values of RR=0,84 RRB=0,82. The potential number of nonconforming parts is 1.3%. This may not be certificated (tabl. 1).

Further analysis of Stewhart charts (Fig. 1 b) confirmed the conclusions of the statistical analysis and further gave the following results: unusual change in the sequence of points C1, C6, C7, C9, C10, indicating the presence of special causes in the process of boring seats.

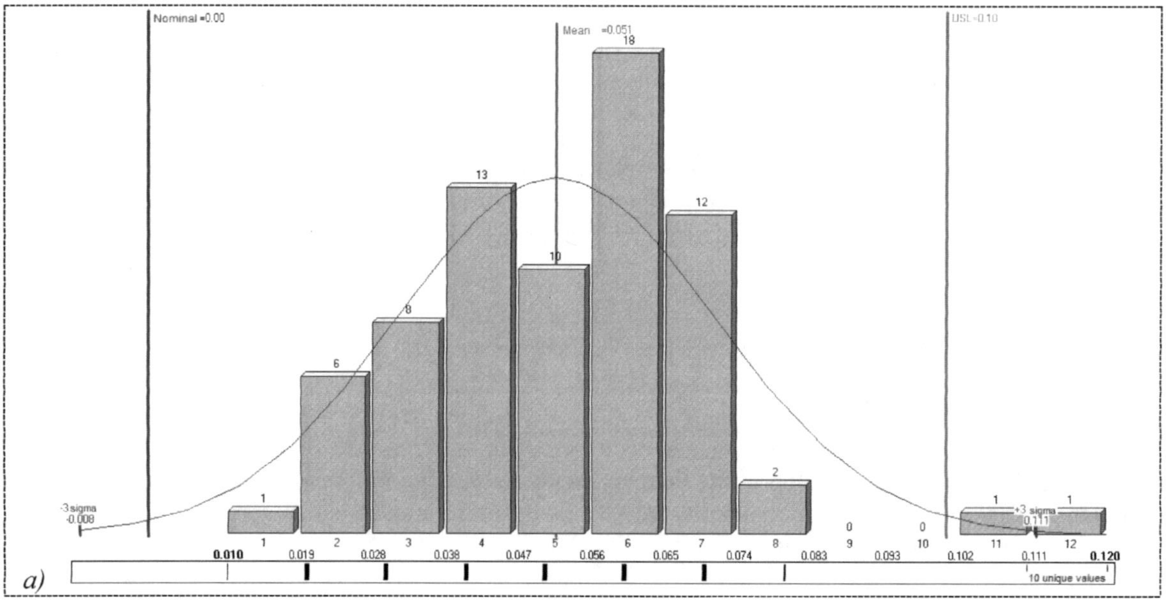

a)

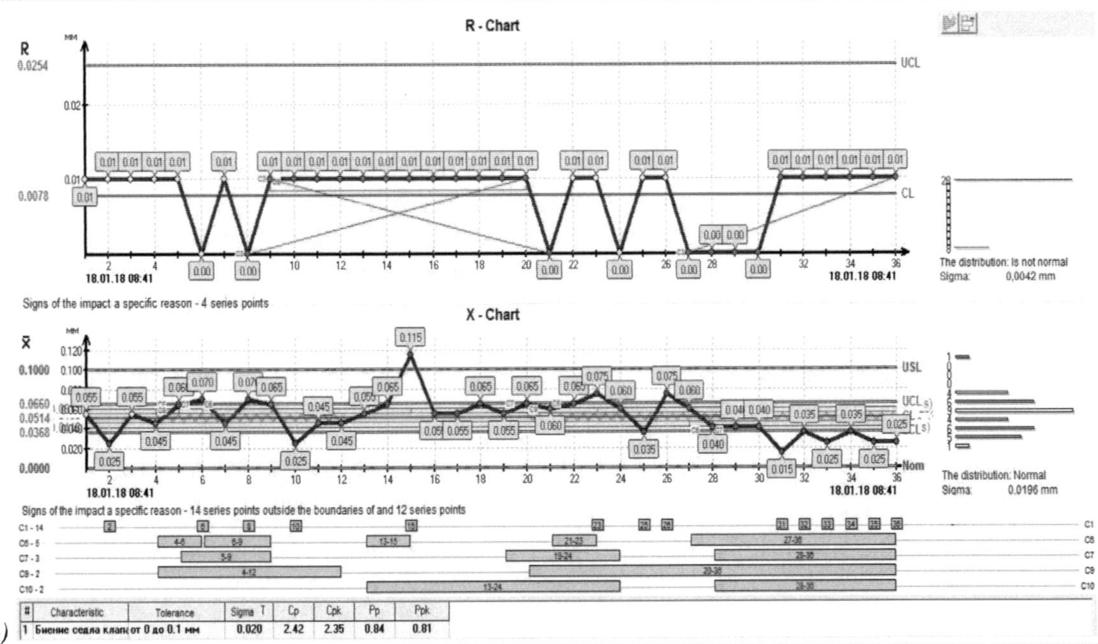

b)

Figure 1. Results for data processing in accordance with the requirements of the standards [1, 2]: a) the histogram curve normal distribution, process indicators, b) the control chart of ranges and average values of the runout of the valve seat in) table of certification process. Setting the runout of the valve seat Pp Ppk 0,84 0,81

Table 1. Process certification

#	Characteristic	Condition	The actual value	Execution
1	The runout of the valve seat	Pp>=1.33	0.835	Not
2	The runout of the valve seat	Pp>=1.33	0.835	Not
3	The runout of the valve seat	Cp>=1.33	2.417	Yes
4	The runout of the valve seat	Cpk>=1.33	2.35	Yes
5	The runout of the valve seat	Stability on X and R chart	X:points outside of the control of borders and a series; R series	Not

Thus the analysis only established only the presence of special causes process instability, but the specific causes of instability and special reasons can not be identified. To find them you need to perform additional research – such as disassembly of the individual modules with the assessment of wear and tear, or the performance of coordinate measuring machined features of a part. Beating is technologically complex figure. Affect the level, to reduce its size and to provide stability through a single characteristic is impossible. When there is insufficient training of engineers and workshop staff are the reason of instability could be found. The risk of issue of marriage will continue.

3. Theory

IOP Conf. Series: Journal of Physics: Conf. Series **1050** (2018) 012075 doi:10.1088/1742-6596/1050/1/012075

Let's consider what information the measurements should give to immediately reveal the limiting factors that act in the process of forming the indicator without carrying out additional engineering studies. At the first stage, we establish the structure and functions of the modules of the technological system in the process of forming the indicator, and form the list of key control characteristics.
At the second stage, we determine the technological structure of the measured quality index and the response plans (corrections) for controlling the value of the significant technological components of the deviation. At the third stage, a measurement scheme is developed, which, based on the results of the measurement of the indicator, allows us to isolate as far as possible all the important key structural components. In the fourth stage, the diagnostic data of the key control characteristics are preliminarily registered, the absolute and relative values of the significant components of the quality index are modeled in the beat measurement field.
Processing of saddles is carried out on a two-spindle boring position using a special combined tool. First, the guide hole of the bushing is unfolded by machine scanning, and then, based on the shank of the sweep, the boring bar moves, chamfering the seat of the valve seat. Analysis of the scheme of processing allowed to identify the process factors and technological components, given in Table 2.

Table 2. The process factors affecting the value of the technological components of the total radial runout

Technological level	**The technological components of the runout**	**Tolerance of the indicator**	**Factors of the process of boring valve seat**
Rotation	Undulation	Not regulated	The uneven size of the allowance
Cutting stroke	Deviation of the profile of the valve seat adjacent the circumference	0,005 mm	The bending axis of the tool Axial runout of spindle
Cutting pass	The radial size of the surrounding circle	25,25±0,25 mm	The wear of the forming tool feature The error of basing the insert
Position	The deviation of the axis of the valve seat from the axis of the guide hole	Not regulated	Error insertion of the guide sleeve
Production task	The deviation of the axis of the valve seat from the axis of the guide hole	Not regulated	Error-based satellite in the position processing Error insertion of the guide sleeve

4. Experimental results

From here the measurement scheme was developed, which allows using standard measuring means, with the registration of the values of 4 technological components. For this purpose, the readings of the instrument needle in the coordinate system of the part were registered in four angular coordinates (Fig. 2).

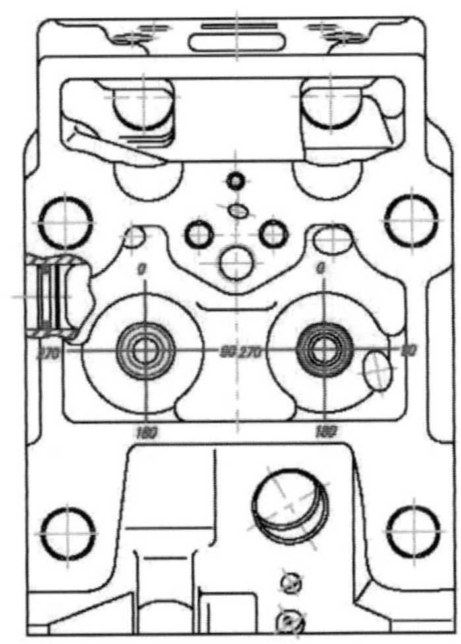

Figure 2. Angular coordinates of the valve seat

The initial adjustment of the device to zero value is performed on the reference part, in which all technological deviations are minimized. Before measuring, the instrument is set to zero, then the whole batch of parts is measured with the setting saved (Table 3). The table shows the scale of the instrument.

Measurements of the details of the wire are strictly in the order of their manufacture with the registration of the adjustment moments for the size or tool change.

Table 3. Fragment of coordinate measurement protocol (μm)

Measurement No	Angular phase, deg.			
	0	90	180	270
1	-15	-12	-10	-15
2	-21	-20	-19	-21
3	-18	-13	-12	-17
4	3	7	6	2
5	-21	-16	-14	-19
6	-5	2	2	-5
7	6	9	11	7
8	-15	-10	-8	-12
9	-36	-31	-29	-29
10	-11	-8	-9	-10

The largest values of the indices are shown in Fig. 3. As seen from the graph, the most significant is the change in the position of the valve seat axis from the boring bar (1) in the batch, then the deviation of the position of this axis from the axis of the base hole (2) for each individual operation, and finally the smallest deviation is the deviation from the roundness of the profile on the diametrical size of the measuring tip (3).

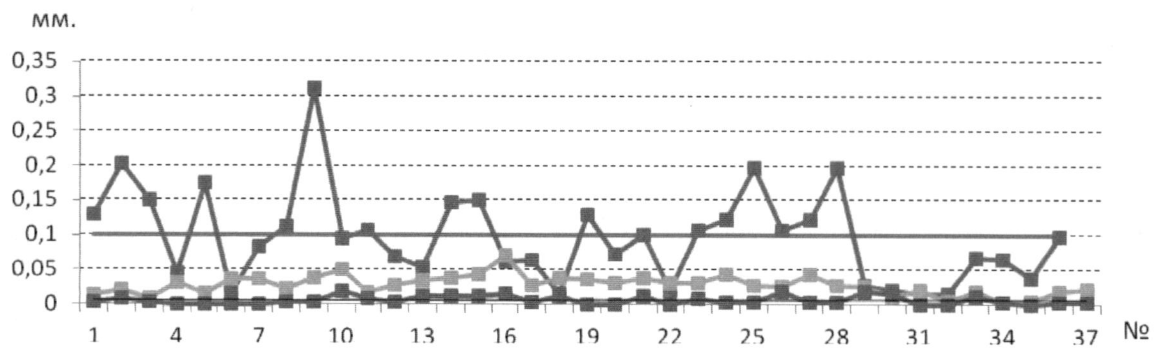

Figure 3. Graph of changes in the values of technological components in a batch of parts (1 – boring bar axis; 2 – axis of the base hole, 3 – deviation from the circularity of the profile)

The coordinates of the valve seat axis are shown in the graph (Fig. 4, 5). Their position from detail to detail is in the first and third quarters of the graph. The reason for this may be the error of positioning the satellite holding and moving the part from one operation to another, together with the bending of the axis with the guide.

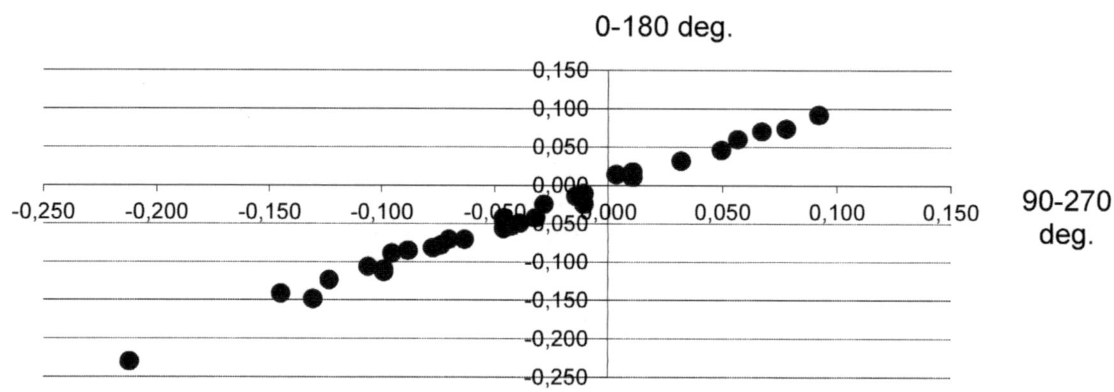

Figure 4. Graph of deviation of the position of the deviation of the position of the boring bar in the batch of parts.

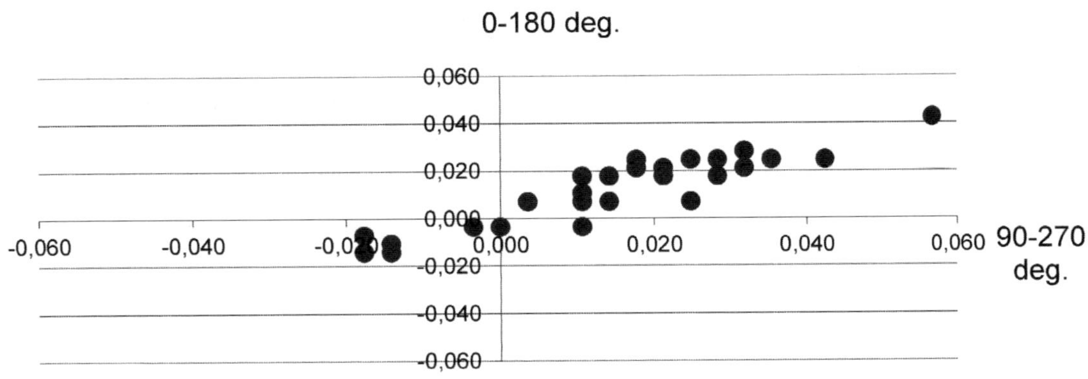

Figure 5. Diagram of deviation of the position of the deviation of the position of the boring bar

5. The discussion of the results

Technologists, and after them mathematicians, systematizing the errors that lead to deviations from the ideal values of the output produce a systematic and random among them. Systematic errors can be controlled, directly affect them, and random errors can not be eliminated, they are the consequence of the variability of the process itself. For example, a change in hardness in a batch of parts is considered random. But if we identify at least the coordinates of the distribution of parts inside the furnace for the tempering of the parts and associate it with the numbers of the measured parts, then in a random manifestation of the hardness value a regularity is revealed that depends on the propagation of the temperature isotherms inside the furnace chamber, which can already be controlled, for example, . The same applies, for example, to the indicator - beating. If only its value is known, but not known, its structure, or processing conditions, then we can not interpret the data for effective control of the process. Once a single measuring coordinate system has been introduced, associated with the technological coordinate system, the structure of the indicator has been singled out, so it becomes possible to effectively identify the specific factors that operate in the process. In this case, the processing of measurement data immediately gives specific suggestions for improvement, bypassing the various kinds of engineering searches by personnel for jobs.

It is due to the increase in the degree of identification of the conditions for the performance of the process that the developed methodology and the completed measurement data proved to be more effective in finding the causes of deviations than standard statistical methods. The result of the application of the methodology is a sharp increase in management efficiency, especially indicators with a complex technological structure, as well as a reduction in the time for corrective and preventive actions.

6. Conclusion

Thus, based on the results of measurements, immediately follow the suggestions for improvement. As a result, management effectiveness has sharply increased, as well as the required time for planning and corrective actions. Effective use of control charts is possible only for the indicator with the simplest internal technological structure, or on condition that the embedded technological components of the indicator are negligible or only one known cause predominates.

In the case where the equipment is worn out and the structure of the technological indicator is developed, the direct application of control charts or other statistical methods for improving the quality of products is rendered ineffective.

References:

[1] Vasiliev V A, Odinokov S A, Serov M M Modern methods for the quality management of high-rate melt solidification *Russian Metallurgy (Metally)*, Vol 2016 Is 13 pp 1321–1324

[2] Vasiliev V A, Odinokov S A, Borisova E V, Letuchev G M Methods of quality management of innovation process *2016 IEEE Conference on Quality Management, Transport and Information Security, Information Technologies, IT and MQ and IS 2016* Nalchik; Russian Federation 4–11 October 2016 pp 239–241

[3] Kasjanov S V, Kalyashina A V The Entry of the Russian Industry in the global market as a source of experience for managers and specialists from other countries *Innovations of Sustainable Production for Green Mobility. 3-d International Chemnitz Manufacturing Colloquium ICMC 2014* – Fraunhofer Institute for Machine Tools and Forming Technology IWU 2014 Part 1 pp 733–741

[4] Khafizov I I, Nurullin I G Product quality as the main factor of increase of competitiveness (on the example of JSC Kazan helicopter plant) *IOP Conference Series: Materials Science and Engineering* 2017 Vol 240 Is 1 article # 012039

[5] Khafizov I I, Galimov A N IT-strategy and major aspects of quality management on the market of goods and services *IOP Conference Series: Materials Science and Engineering* 2017 Vol 240 Is 1 article # 012038

[6] Kol'zov A G, Blokhin D A, Khabarov A V, Redorovich D A The influence of kinematic characteristics of stewart platform for precision moving measuring mechanism *2014 Dynamics of Systems, Mechanisms and Machines, Dynamics 2014 Proceedings* 2014 P 7005669

[7] Elakova A A, Abyzov A P, Stupko V B Quality characteristics study of complex shape parts after vibroabrasive treatment *International Journal of Pharmacy and Technology* 2016 Vol 8 Is 3 pp 14815–14821

[8] Kas'yanov S V, Safarov D T Diagnosis of technical state of equipment and tools according to indices of technological accuracy *Avtomobil'naya Promyshlennost* Is 5 2004 pp 24–28

[9] Kondrashov A G, Kasyanov S V, Safarov D T and Kuznetsova A V Diagnostic measurements of geometrical parameters of spatial and difficult details of autocomponents by one-coordinate altimeter *Control. Diagnostics (Kontrol. Diagnostika)* 2013 Is 8 pp 60–64

[10] Kasjanov S V, Kondrashov A G, Safarov D T Regulation of Geometrical Parameters Deviations of Automotive Components Parts through Diagnostic Measurements Organization *Procedia Engineering* Vol 206 2017 pp 1508–1514

[11] Attestator. The demo version (available at: http://centr-prioritet.ru/ct/txt/184 date of access: 22.01.2018)

IOP Conf. Series: Journal of Physics: Conf. Series **1050** (2018) 012076 doi:10.1088/1742-6596/1050/1/012076

A mathematical model of changes of parameters of handwritten patterns depending on the signers states

A E Samotuga

Omsk State Technical University, 11, Mira ave., Omsk, 644050, Russia

Abstract. Authors of some studies point out the need of taking into account the state of employee to reduce the risk that he will cause damage to organization including information resources. An assessment of psychophysiological state is possible on the basis of the dynamic parameters of a signature. Etalons creation of handwritten signature for each state significantly complicates the formation of an etalons base of subjects. The purpose of the study is to provide a mechanism for the formation of subjects etalons in different states without the need to collect of handwritten signature in each states. To achieve the goal we should solve the problem of finding the regularity of the signature parameters changes while changing the states of signers. We develop a model of changing signature parameters depending on the state based on the discovered regularities. We compared the accuracy of the obtained model by estimating the probabilities of erroneous decisions obtained in the experiment of states recognition using etalons obtained in a natural way and using a model. To carry out the research, we used the apparatus of probability theory and mathematical statistics. This included using the multidimensional Bayes functional and Bayes hypothesis formula. As a result the author describes the model of biometric etalons of handwritten signatures of the user. The model assumes the use of formulas and calculated coefficients for calculating the distribution of signature features characteristics in the changed state from the values of the signature features in the normal state. The introduction of this model will significantly simplify the procedure for obtaining the etalons in changed state The mentioned above model can find applications in systems of dynamic biometric identification.

1. Introduction

At present, to effectively protect the information resources of an organization, it is necessary to take into account the possibility of infliction of harm by its own employees. This statement is confirmed by studies of sources of information security threats [1]. An employee of an organization may cause harm maliciously or because of he being in a non-optimal state, for example, into alcohol intoxication. The solution to this problem is in the application of developments in the field of assessment the psychophysiological state (PPS) of employees, for example, at the time of biometric authentication. The most suitable for this purpose are dynamic biometric images. There are studies on the recognition of the subjects states by thermograms of the face and neck [2] and handwritten signature [3,4].

It is required to create a biometric etalons for each psychophysiological state in order to recognize them later. This significantly complicates the process of creating a etalons base and leads to a multiple increase in the expenditure of time. This approach is shown in detail in [3]. It consists in the fact that at the time of the formation of the etalon the physiological parameter is registered using the experience of the existing technologies of polygraph tests [4]. As a physiological parameter is the heart rate variability (HRV) [5]. The range of values of the parameter is divided into ranges, each of which is associated with the signatures written by the subject in a certain psychophysiological state. Etalon is formed from the signatures assigned to each range. This approach also assumes the registration of the physiological parameter at the stage of recognition of the subject for comparing the values of the biometric features of the presented signature with the etalons of subjects in this state.

2. Proposed Approach

2.1. Research objective

It is evident that the implementation of described etalon forming procedure for each of the states in practice will be difficult. In this paper, we propose solution that will allow the conversion of the etalon of the handwritten signature formed in the "normal" state of the signers to the etalon of the same signer formed in the "changed" state with sufficient accuracy. It is necessary to achieve results, according to which it will be not significant excess of the probability of errors achieved using the approach given in [6] and applying the similar methods.

2.2. Forming the base of handwritten signatures of various psychophysiological states

To study the regularities of changing the parameters of handwritten images, signatures were collected using a Wacom graphic tablet. By analogy with [3] the HRV was chosen as physiological parameter. Information on HRV was recorded on Holter's monitor "Cardiotechnics 04".
An experiment on the formation of a signature database was conducted with the involvement of test subjects that was entered alternately into the following states (the change of state was confirmed by information from the Holter monitor) [6]:

- Adequate (or normal) state – the optimal state in which there are the best results of the subject activity.
- Excitement – in this state for the subject is characterized by psychic activity, increased heart rate, rapid breathing. Test subject The experiment participant took coffee, which increased his heart rate by an average of 10%.
- Fatigue (after physical exertion) – is a state that occurs after an intense physical exertion which volume for each subject was set individually, on the basis of functional tests. It is characterized by a heart rate increase of subject by 10-30%.
- Relaxed state - is characterized by mild drowsiness, low working capacity. To simulate present state, subjects took natural herbal remedies for sedation. Subjects heart rate returned to the values at rest, or there was decreased by 3-5%.
- Alcoholic intoxication –subject took alcohol, which dosage was calculated according to the Widmark formula (1) so that its concentration value was at least 0.5 ‰.

$$c = \frac{A}{m \cdot r}, \tag{1}$$

where c – concentration of alcohol in the blood in ‰, A – weight of drinked alcohol in grams, m – subject body weight in kilograms, r – the distribution coefficient of Widmark (0.70 for men, 0.60 for women).
In the experiment, 60 subjects were involved. Each subject reproduced signature at least 30 times on a Wacom graphic tablet in each of the listed states.
For convenience, we use the term "modified" PPS of subject which images will include samples that are the subjects entered in any state different from normal. For each subject who was in a certain PPS a separate etalon of the handwritten patterns was formed to further evaluate the accuracy of the etalon conversion by means of a model. The etalon consists of mathematical expectations and mean-square deviations of the features values calculated from the corresponding samples.

2.3. Extraction of biometric features

Each time the user entered a signature on the graphic tablet, the signature coordinates x,y and the pen pressure on the tablet p were fixed. Signature realizations were equalized in duration before we calculating the parameters that characterizing the appearance of the signature and the dynamics of the

signature reproduction. In addition at this stage an initial and final values of all the points with zero pressure was discarded.
Each signature representing the functions *x (t), y (t), p (t)* was transformed into a vector of values of a set of biometric parameters (further features) that given in table 1

Table 1. Groups of features are used for identification of subjects PPS.

№	Short Description	Number of Features	Nearest distribution law
1	The first 16 amplitudes (the most low frequency) of function $p(t)$ harmonics normalized by energy	16	normal distribution, rarely lognormal distribution
2	The first 16 amplitudes (the most low frequency) of function $v_{xy}(t)$ harmonics normalized by energy	16	normal distribution
3	Correlation coefficients between pairs of signature $x(t)$, $y(t)$, $p(t)$ functions and their derivatives — $x'(t)$, $y'(t)$, $p'(t)$ functions	15	normal distribution
4	Distances between some signature dots are normalized on the signature length in three-dimensional space (the third dimension is pen pressure on the tablet)	120	normal distribution, rarely lognormal distribution
5	Some characteristics of the static signature image: ratio of the signature length to its width, signature center, signature slope angle, angle of inclination between the halfs centers of the signature.	5	normal distribution
6	Daubechies wavelet transform coefficient D6 of $v_{xy}(t)$ and $p(t)$ functions	70	Laplace distribution, normal distribution

2.4. Model of changes of parameters of handwritten patterns depending on the signers states
During the analysis of handwritten etalons we found that the changes of the most features during the transition from the normal signer state to any other PPS are similar in character and sufficiently stable [7]. These changes can be manifested to varying degrees in different subjects, some of the features vary in different ways depending on the state, a small part of the features varies chaotically.
Bringing the etalon to the desired form is proposed to be carried out according to the formulas (2), discarding the features which changes while changing the PPS are chaotic. Correction coefficients using to convert the etalon in the normal state to the etalon in the changed state were calculated.

$$\mathrm{E}'_c = [m'_{c1} \ \ldots \ m'_{cN}] = \mathrm{E}_n \times \mathrm{K}_m = [m_{n1} \ \ldots \ m_{nN}] \times \begin{bmatrix} \kappa_{m1} \\ \ldots \\ \kappa_{mN} \end{bmatrix} \approx \mathrm{E}_c = [m_{c1} \ \ldots \ m_{cN}],$$
$$\Theta'_c = [\sigma'_{u1} \ \ldots \ \sigma'_{cN}] = \Theta_n \times K_s = [\sigma_{n1} \ \ldots \ \sigma_{nN}] \times \begin{bmatrix} \kappa_{\sigma 1} \\ \ldots \\ \kappa_{\sigma N} \end{bmatrix} \approx \Theta_c = [\sigma_{c1} \ \ldots \ \sigma_{cN}], \quad (2)$$

where *Ec* and *En* – are matrixes of mathematical expectations of the subject features in the changed and normal state respectively, *Θc* and *Θn* – are matrices of mean-square deviations of the subject features in the changed and normal state, respectively, *Km* and *Ks* – are vectors of correction coefficients for mathematical expectations and mean square deviations, respectively, *N* – is a number of features. We call a transformed etalon consisting of matrices *E'c* and *Θ'c* synthetic.

A computational experiment to recognize PPS was carried out. The method of sequential application of the two-dimensional (ordinary) of Bayes hypothesis formula (3) and η-dimensional functional (4) were applied, similarly to the way it was done in [6] to form solutions.

$$P(H_h|A_j) = \frac{P(H_h|A_{j-1})P(A_j|H_h)}{\sum_{h=1}^{n} P(H_h|A_{j-1})P(A_j|H_h)}, \quad (3)$$

where $P(H_h|A_j)$ – is posteriori probability of h hypotheses, depending on j feature, $P(A_j|H_h)$ – is conditional probability h hypotheses, depending on j feature.

$$P(H_h|A_s) = \frac{P(H_h|A_{s-1})\prod_{x=1}^{n} f_h(a_{j(s,x)})}{\sum_{h=1}^{n}(P(H_h|A_{s-1})\prod_{x=1}^{n} f_h(a_{j(s,x)}))}, \quad (4)$$

where $j(s, x)$ – number of feature, which does not coincide with the s step number but depends on it, $f_h(a_{j(s, x)})$ is a probability density of $a_{j(s, x)}$.feature

3. Results and Discussions

We use two hypotheses associated with the etalons of normal and changed states. The probability of the states recognition error using synthetic etalons of the changed PPS exceeded the analogous estimate from [6] and amounted to 0.118 and 0.105.

Despite the fault of probabilities the result can be called satisfactory. We can conclude that the constructed model simplifies describing the changes in the parameters of handwritten patterns that occur when the signer uses alcohol, caffeine or sedatives, and also after physical exertion.

The reliability of recognition of subjects PPS by handwritten patterns can be increased if an apparatus of neural networks will be used. Neurons can be constructed on the basis of one of the functionals mentioned in the present article.

4. Conclusions

The paper presents model of changing the features of a handwritten patterns depending on the signers PPS. The model allows transforming etalons formed in a normal PPS, to the etalons of the changed state of the signers. It was possible to achieve an average error probability of 0.105 while applying the method of recognizing the changed state of the signer using this model and the multidimensional Bayesian functional.

Acknowledgments
The author expresses his gratitude to Pavel Sergeevich Lozhnikov and Alexey Evgenievich Sulavko for the help in writing this article.

References

[1] PricewaterhouseCoopers *The global state of information security. Prospects for 2014* pp 1–24
[2] Sulavko A E, Zhumazhanova S S 2017 Human psychophysiological state recognition based on analysis of thermograms of face and neck regions *Dynamics of Systems, Mechanisms and Machines (Dynamics), 2017* pp 143–51
[3] Lozhnikov P S, Sulavko A E, Samotuga A E 2015 Personal identification and the assessment of the psychophysiological state while writing a signature *Information* **6** pp 454–66
[4] Aldert Vrij 2000 *Detecting Lies and Deceit: The Psychology of Lying and Implications for Professional Practice* (Wiley-Blackwell) p 276

[5] Epifantsev B N, Lozhnikov P S, Samotuga A E, Sulavko A E Method of Personal Identification by Features of Password Writing Dynamics: Russian Federation MPK G06K9/00/; he applicant and the patentee: Omsk State Technical University. № 2543927. Date of filing. 22.04.14; Date of publication. 10.03.2015.

[6] Sulavko A E, Eremenko A V, Levitskaya H A, Samotuga A E 2017 Identification of psychophysiological states of signers by autograph reproduction features *Information-measuring and Control Systems* **3** pp 40–8.

[7] Sulavko A E, Samotuga A E 2017 Impact of psycho-physiological state of signers on biometric parameters of manuscript images and results of their verification *Information and Control Systems* **6(91**) pp 29–42

Mathematical models for estimating production risks

V S Serdyuk, A M Dobrenko, O A Tsorina, E V Bakiko and S V Yanchij

Omsk State Technical University, 11, Mira ave.,Omsk, 644050, Russia

Abstract. When designing and operating machine-building facilities, their operation reliability is significant. The quantitative measure of reliability is production risks. The article proposes a mathematical model to determine probabilities of unfavorable events: the occurrence of a risk factor at a given time interval at some point in time; a worker's exposure at a given workplace in the adverse event area; negative changes in the worker's ability to fulfill his production duties. The technique to define production risks based on the knowledge of likelihood of adverse events occurrence is resulted. Examples of the application method for solving safety management problems of a typical technological operation of assembling an all-in-one connection have been presented.

1. Introduction

Risks are studied from different disciplines points, with discussion of various methods and tools used to optimize technological processes and to reduce adverse events [1]. The improvement of the technological process should be aimed at meeting requirements of international standards ISO 9001:2015 and ISO 14001:2015 and the draft standard ISO 45001, that consider a problem to minimize production risks [2].

The idea of using simulation to estimate the production risk in different engineering areas has been discussed by researchers from different countries. At the same time, it is important to realize the nature of the initial data included into the model, the field of its application and the result obtained [3]. However, the application of mathematical models to estimate occupational risks in the workplace for the management of occupational safety is currently lacking.

The quantification of production risks should be based on expert estimates obtained as a result of experiments or accumulated statistical data [4].

2. Problem statement

The technological process in time interval T is serviced with n working places $w_1,...,w_n$ and to analyze its danger level there are m independently acting hazards $f_1,...,f_m$.

This situation corresponds to a chart of technological process risk levels in time interval T

$$\begin{bmatrix} \mathrm{P}_{11} & \cdots & \mathrm{P}_{1n} \\ \cdots & \cdots & \cdots \\ \mathrm{P}_{m1} & \cdots & \mathrm{P}_{mn} \end{bmatrix}$$

where P_{ij} is probability of a risk factor f_i exposure to a working place w_j ($i = 1,\dots, m; j = 1,\dots, n$).

IOP Conf. Series: Journal of Physics: Conf. Series **1050** (2018) 012077 doi:10.1088/1742-6596/1050/1/012077

According to [5] $P_{ij} = P_i \cdot q_{ij}$, where P_i is a probability estimation of occurrence of the factor f_i, q_{ij} is assessment of the likelihood of worker's exposure in the working place w_j into the zone of its impact if it is. The overall level of risk (the hazards level of the technological process relatively $f_1,...,f_m$) is estimated with the formula

$$\overline{P} = 1 - \prod_{i=1}^{m}\prod_{j=1}^{n}(1 - P_{ij}) \tag{1}$$

3. Theory

When solving problems of managing the production safety, the following questions are very important:

1. How will the overall level of risk $\overline{P}$ decrease if some elements P_{ij} of risk levels charts are managed to be reduced?
2. What elements P_{ij} of risk levels chart and how many of them to be reduced to achieve reduction in the overall level of risk $\overline{P}$ to the given value?

To obtain the dependence of changing the overall level of risk $\overline{P}$ on changes in the elements of the risk level chart P_{ij} the following well-known fact from mathematical analysis is used: for the function $f(x_1,...,x_N)$ in the point $M_0(x_1^{(0)},...,x_N^{(0)})$ there is approximate equality

$$f(x_1^{(0)} + \Delta x_1,...,x_N^{(0)} + \Delta x_N) - f(x_1^{(0)},...,x_N^{(0)}) \approx \left.\frac{\partial f}{\partial x_1}\right|_{M_0} \cdot \Delta x_1 + ... + \left.\frac{\partial f}{\partial x_N}\right|_{M_0} \cdot \Delta x_N$$

(the right-hand side of the equation is a total differential of the function $f(x_1,...,x_N)$ in the point M_0). The order of errors of this equation can be estimated as $\max\left((\Delta x_i)^2\right)$.

One considers $\overline{P}$ as a function $N = m \cdot n$ of variables

$$f(x_{11}, x_{12},...,x_{mn}) = 1 - \prod_{i=1}^{m}\prod_{j=1}^{n}(1 - x_{ij}),$$

And for it

$$\frac{\partial f}{\partial x_{ij}} = \prod_{\substack{k=1\\k\neq i}}^{m}\prod_{\substack{s=1\\s\neq j}}^{n}(1 - x_{ks}).$$

One calculates $\frac{\partial f}{\partial x_{ij}}$ in the point $M_0(P_{11}, P_{12},...,P_{mn})$ using equation (1).

$$\left.\frac{\partial f}{\partial x_{ij}}\right|_{M_0} = \prod_{\substack{k=1\\k\neq i}}^{m}\prod_{\substack{s=1\\s\neq j}}^{n}(1 - P_{ks}) = \frac{1 - \overline{P}}{1 - P_{ij}}$$

There is $\Delta P_{ij} \geq 0$. Then

$$f(P_{11} - \Delta P_{11}, P_{12} - \Delta P_{12},...,P_{mn} - \Delta P_{mn}) - f(P_{11}, P_{12},...,P_{mn}) \approx \sum_{i=1}^{m}\sum_{j=1}^{n}\frac{1 - \overline{P}}{1 - P_{ij}} \cdot (-\Delta P_{ij})$$

Hence, denoting through $\overline{P}_H$ general level of risk when decreasing P_{ij} by ΔP_{ij} (i = 1,..., m; j = = 1,..., n), the formula is obtained

$$\Delta\overline{P} = \overline{P} - \overline{P}_H \approx \sum_{i=1}^{m}\sum_{j=1}^{n}\frac{1 - \overline{P}}{1 - P_{ij}} \cdot (\Delta P_{ij}). \tag{2}$$

The order of errors of the formula is $\max_{ij}\left((\Delta P_{ij})^2\right)$.

This ratio analyzes quantitatively the following situations:

1. To estimate the magnitude of the reduction in the overall level of risk with a reduction in the likelihood of risk factors occurrence (change in technological production parameters) and (or) with a decrease in the likelihood of workers entering the zone of its exposure (organizational changes in the operation of the relevant technological processes).
2. To estimate the value ΔP_{ij} of reducing the likelihood of exposure of any i-th risk factor to j-th working place (if possible) to reduce the overall level of risk by a given amount $\Delta\bar{P}$.

There is formula (2) in the first situation, there is another formula obtained from formula (2) in the second situation:

$$\Delta P_{ij} \approx \frac{1-P_{ij}}{1-\bar{P}} \cdot \Delta\bar{P}. \tag{3}$$

4. Experimental results

These situations are presented on the example of a typical technological operation of assembling an fixed joint (tubular assembly) on a magnetic pulse installation.

Time T, the exposure of risk factors is estimated in 8 hours (working shift), there are 2 workers (an operator, a mechanical technician), number of manufactured products is 200 pcs.

The following dangerous risk factors are identified:

1) f_1 is a dangerous factor: fragments of the inductor formed during its destruction, an expert estimate of the number of possible fractures is 1 per 800 discharges;

2) f_2 is a dangerous factor: a situation when the electrical circuit is closed through the human body - 1 case for 2000 hours of operation.

As expert estimates of time for entering working places (an operator, a mechanical technician) into the zone of exposure to risk factors under these conditions, the following estimates are considered:

1) risk factor f_1: for the operator is 0.01 % working shift time; for the mechanical technician it is 0.03 % working shift time;

2) risk factor f_2: for the operator is 10 % working shift time; for the mechanical technician it is 5 % working shift time

A chart of technological operation risks levels relative to risk factors $f_1, f_2\,(m=2)$ at time interval T = 8 hours for working places w_1, w_2 (an operator, a mechanical technician, n = 2) is calculated.

Using the Poisson distribution, the probability of occurrence of a risk factor f_1 according to the formula $P_1 = 1 - e^{-NP_0}$ is calculated considering initial data, $N = 200$, $P_0 = 1/800$:

$$P_1 \approx 1 - e^{-0.25} \approx 0.221199.$$

Using an exponential distribution, the probability of occurrence of a risk factor f_2 according to the formula $P_2 = 1 - e^{-\lambda T}$ is calculated considering initial data, $\lambda = 1/2000$, T = 8:

$$P_2 \approx 1 - e^{-0.004} \approx 0.003992.$$

Estimates of the probabilities of getting working places w_1, w_2 into the exposure area of risk factors f_1, f_2 if they are, according to the initial data are

$$q_{11} = 0.00001, \quad q_{12} = 0.0003, \quad q_{21} = 0.1, \quad q_{22} = 0.5.$$

A risk levels chart of the technological operation in question during the work shift T ($P_{ij} = P_i q_{ij}$) is:

$$\mathrm{P} = \begin{bmatrix} 0.000022 & 0.000066 \\ 0.000399 & 0.000199 \end{bmatrix}.$$

The overall level of risk according to formula (1) is

$$\overline{\mathrm{P}} = 1 - \prod_{i=1}^{2}\prod_{j=1}^{2}(1 - \mathrm{P}_{ij}) = 0.000686.$$

Coefficients $\frac{1-\overline{\mathrm{P}}}{1-\mathrm{P}_{ij}}$ are calculated according to formula (2):

$$\frac{1-\overline{\mathrm{P}}}{1-\mathrm{P}_{11}} \approx 0.959336, \quad \frac{1-\overline{\mathrm{P}}}{1-\mathrm{P}_{12}} \approx 0.959380, \quad \frac{1-\overline{\mathrm{P}}}{1-\mathrm{P}_{21}} \approx 0.959713, \quad \frac{1-\overline{\mathrm{P}}}{1-\mathrm{P}_{22}} \approx 0.959513.$$

Thus, formula (2) is in the case

$$\Delta\overline{\mathrm{P}} = 0.959336 \cdot \Delta\mathrm{P}_{11} + 0.959380 \cdot \Delta\mathrm{P}_{12} + 0.59719 \cdot \Delta\mathrm{P}_{21} + 0.959513 \cdot \Delta\mathrm{P}_{22}. \tag{4}$$

Example 1. How will the overall level of risk (level of danger) of the said technological operation decrease if the inductor reliability is doubled (an estimate of the possibility of destruction is 1 per 1600 discharges)?

In this case

$$\mathrm{P}_1^{(\mathrm{H})} = 1 - e^{-200\cdot(1/1600)} \approx 1 - e^{-0.25} \approx 0.117503.$$

The new risk level chart is

$$\mathrm{P}_{11}^{(\mathrm{H})} = \mathrm{P}_1^{(\mathrm{H})} \cdot q_{11} = 0.000012, \quad \mathrm{P}_{12}^{(\mathrm{H})} = \mathrm{P}_1^{(\mathrm{H})} \cdot q_{12} = 0.000035,$$

wherein,

$$\Delta\mathrm{P}_{11} = \mathrm{P}_{11} - \mathrm{P}_{11}^{(\mathrm{H})} = 0.00001, \quad \Delta\mathrm{P}_{12} = \mathrm{P}_{12} - \mathrm{P}_{12}^{(\mathrm{H})} = 0.000031, \quad \Delta\mathrm{P}_{21} = \Delta\mathrm{P}_{22} = 0.$$

According to formula (4) one obtains

$$\Delta\overline{\mathrm{P}} = 0.959336 \cdot 0.00001 + 0.959380 \cdot 0.000031 = 0.000039,$$

And it is 5.7 % from the total risk level, i.e. the level of danger will decrease by 5.7 %.

Example 2. How will the overall level of risk decrease if the probability of electric shock for an operator is reduced by 2?

In this case

$$\mathrm{P}_{21}^{(\mathrm{H})} = 0.5 \cdot \mathrm{P}_{21} = 0.0001995, \quad \Delta\mathrm{P}_{21} = \mathrm{P}_{21} - \mathrm{P}_{21}^{(\mathrm{H})} = 0.0001995, \quad \Delta\mathrm{P}_{11} = \Delta\mathrm{P}_{12} = \Delta\mathrm{P}_{22} = 0.$$

According to formula (4) one obtains

$$\Delta\overline{\mathrm{P}} = 0.959713 \cdot 0.0001995 = 0.0001915,$$

And it is 27.9 % from the total risk level, i.e. the level of danger will decrease by 27.9 %.

Example 3. How will the overall level of risk decrease if the probability of electric shock for a mechanical-technician is reduced by 2?

In this case

$$\mathrm{P}_{22}^{(\mathrm{H})} = 0.5 \cdot \mathrm{P}_{22} = 0.0001, \quad \Delta\mathrm{P}_{22} = \mathrm{P}_{22} - \mathrm{P}_{22}^{(\mathrm{H})} = 0.0001, \quad \Delta\mathrm{P}_{11} = \Delta\mathrm{P}_{12} = \Delta\mathrm{P}_{21} = 0.$$

According to formula (4) one obtains

$$\Delta\overline{\mathrm{P}} = 0.959513 \cdot 0.0001 = 0.000096,$$

And it is 14 % from the total risk level, i.e. the level of danger will decrease by 14 %

Example 4. Suppose that it is necessary to reduce the overall level of risk $\overline{\mathrm{P}} = 0.000686$ by 10%, i.e. $\Delta\overline{\mathrm{P}} = 0.000068$.

According to formula (3) one obtains

$$\Delta P_{11} = \frac{1-\overline{P}}{1-P_{11}} \cdot \Delta\overline{P} = 0.00071, \quad \Delta P_{12} = \frac{1-\overline{P}}{1-P_{12}} \cdot \Delta\overline{P} = 0.000071,$$

$$\Delta P_{21} = \frac{1-\overline{P}}{1-P_{21}} \cdot \Delta\overline{P} = 0.000070, \quad \Delta P_{22} = \frac{1-\overline{P}}{1-P_{22}} \cdot \Delta\overline{P} = 0.000070.$$

Comparing ΔP_{11}, ΔP_{12} with P_{11}, P_{12} relatively proves that a decrease in the likelihood of exposure to a risk factor f_1 to working places w_1 or w_2 it is impossible to achieve such a reduction in the overall level of risk.

Comparing ΔP_{21}, ΔP_{22} with P_{21}, P_{22} proves that to do this it is necessary to reduce the likelihood of exposure of a risk factor f_2 to working place w_1 by 17 % or to reduce the likelihood of exposure f_2 to working place w_2 by 35.2 %.

5. Discussion

The mathematical model for calculating operational risks was used by the authors for the quantitative estimation of operational risk in the working places in the engineering operations. The calculation results allow:

1. To implement an automated quantitative risk estimation using software based on the proposed use of mathematical models.
2. To apply mathematical simulating in the process of managing production risks, it is proposed by international standards.
3. To apply the proposed mathematical models to quantify the production risks in organizations with different activities for different processes.
4. To formulate recommendations for the development of effective events to manage the production risks in organizations.
5. To use the proposed methodological approach when training specialists in the field of occupational safety in the production management.

6. Conclusion

The proposed mathematical model for calculating production risks has both a theoretical and an applied nature. It gives a quantitative estimation of production risks in working places in the engineering operations according to the international standards requirements. The use of risk estimation techniques increases their management effectiveness.

The mathematical model offers an unambiguous algorithm for calculating production risks and can therefore be formalized using computational methods of mathematics.

7. References

[1] Wua D D and Olson D L 2013 Computational simulation and risk analysis: An introduction of state of the art research *Mathematical and Computer Modelling* **58**. pp. 1581–87
[2] Karkoszka T 2017 Operational monitoring in the technological process in the aspect of occupational risk *MESIC* (Vigo (Pontevedra) Spain: Procedia Manufacturing) pp. 1463–69
[3] Rosenberg Z and Dekel E 2010 On the deep penetration of deforming long rods *International Journal of Solids and Structures* **47** pp. 238–250
[4] Kudryavtsev S S, Yemelin P V and Yemelina N K 2017 The Development of a Risk Management System in the Field of Industrial Safety in the Republic of Kazakhstan *Safety and Health at Work* pp. 1–12.
[5] Serdyuk V S, Goryaga A V and Dobrenko A M 2005 *Models of Quantitative Estimates of Risk Levels of Production Processes* (Omsk: OmGTU) p18.

IOP Conf. Series: Journal of Physics: Conf. Series **1050** (2018) 012078 doi:10.1088/1742-6596/1050/1/012078

Logical aspects of machine learning

A N Shevlyakov[1] [2]

[1] Sobolev Institute of Mathematics, Pevtsova 13, Omsk, 644099, Russia
Omsk State Technical University, pr. Mira 11, Omsk, 644050, Russia

E-mail: a_shevl@mail.ru

Abstract. Using the basic ideas of universal algebraic geometry, we define a distance function on the set of quantifier-free formulas. We explain the general definition by two examples (partial orders and feature selections) and prove basic facts about the distance in such examples.

1. Introduction

Many machine learning algorithms of prediction and clustering work in metric spaces generated by features of learning objects. Recall that learning objects may have arbitrary types, and there is a need to define a metric between such objects.

One can explain the importance of such metrics by the following example. Suppose we have N forecast statements asserted by N experts. If N is large, we can not effectively process this data. Thus, we have to decrease the number of forecasts removing the "almost identical" statements. If we define a distance function over the set of statements, then a pairs of statements with a small distance can be treated as a pair of "almost identical" statements.

In the current paper we define a distance function on the set of first-order $\mathcal{L}$-formulas in variables X, where the values of X belong to a fixed $\mathcal{L}$-structure $\mathcal{A}$ (Section 3). The idea of our approach is based on the results of universal algebraic geometry [1]. In Sections 4, 5 we consider two applications of the defined metric. Namely, in Section 4 we study statements with partial order, whereas in Section 5 we show the application of the defined metric in feature selection.

Our approach was inspired by the following problem in theoretical machine learning. There are popular algorithms of clustering (e.g. FOREL, MeanShift...) which claim the information about the distance distribution of clustered objects. If we give a wrong information to such algorithm, the results of clustering will be either trivial or not interpretable. Our definition of the metric allows to estimate a priori the average distance between objects before clustering (see Sections 4, 5).

2. Main definitions

Let $\mathcal{L}$ be a language. Below we do not assume that the symbol $=$ belongs to $\mathcal{L}$ on default.

An $\mathcal{L}$-equation is an atomic formula of $\mathcal{L}$. An arbitrary set of $\mathcal{L}$-equations is called an $\mathcal{L}$-*system* (*system*, for shortness). A disjunction of a finite number of $\mathcal{L}$-systems is a $\vee$-*system*.

Let $\mathcal{A}$ be an $\mathcal{L}$-structure. A point $P \in \mathcal{A}^n$ is a *solution* of a system $\mathbf{S}$ if P satisfies all equations in $\mathbf{S}$. A point $P \in \mathcal{A}^n$ is a *solution* of a $\vee$-system $\mathbf{S}_1 \vee \mathbf{S}_2 \vee \ldots \vee \mathbf{S}_r$ if P is a solution of at least one system $\mathbf{S}_i$. The set of all solutions of an ($\vee$-)system $\mathbf{S}$ is denoted by $\mathrm{V}_{\mathcal{A}}(\mathbf{S})$.

[2] Present address: Sobolev Institute of Mathematics, Pevtsova 13, Omsk, 644099, Russia

A set $Y \subseteq \mathcal{A}^n$ is called *($\vee$-)algebraic over* an $\mathcal{L}$-structure $\mathcal{A}$ if there exists an ($\vee$-)system $\mathbf{S}$ in n variables with $\mathrm{V}_{\mathcal{A}}(\mathbf{S}) = Y$. An ($\vee$-)algebraic set Y is *irreducible* if Y is not a proper finite union of other $\vee$-algebraic sets. The *radical* $\mathrm{Rad}_{\mathcal{A}}(Y)$ of an algebraic set Y is the set of all $\mathcal{L}$-equations satisfying all points $P \in Y$.

An algebraic $\mathcal{L}$-structure $\mathcal{A}$ is *equationally Noetherian* if for any infinite system $\mathbf{S}$ in variables $X = \{x_1, x_2, \dots, x_n\}$ there exists a finite subsystem $\mathbf{S}' \subseteq \mathbf{S}$ with the same solution set in $\mathcal{A}$.

Proposition 2.1. [1] *Any algebraic set Y over an equationally Noetherian algebraic $\mathcal{L}$-structure $\mathcal{A}$ is a finite union of irreducible sets*

$$Y = Y_1 \cup Y_2 \cup \ldots \cup Y_m, \quad Y_i \not\subseteq Y_j \textit{ for all } i \neq j,$$

and this decomposition is unique up to a permutation of components.

The subsets Y_i from Proposition 2.1 are called the *irreducible components* of Y.

The following result gives the decomposition of a $\vee$-algebraic set into the union of its irreducible components.

Proposition 2.2. *Let Y be the solution set of a $\vee$-system $\mathbf{S}_1 \vee \mathbf{S}_2 \vee \ldots \vee \mathbf{S}_r$ over an equationally Noetherian $\mathcal{L}$-structure $\mathcal{A}$, and*

$$Y_i = \mathrm{V}_{\mathcal{A}}(\mathbf{S}_i) = \bigcup_j Y_{ij}$$

be the decomposition of Y_i into the union of irreducible components.

Then the sets

$$\{Y_{kl} \mid Y_{kl} \not\subseteq Y_{ij} \textit{ for all } (i,j) \neq (k,l)\}$$

are irreducible components of Y.

3. The distance

In the current section we define the distance function on the class of first-order formulas according to the following principles:

(i) we define a distance $d(\psi_1, \psi_2)$ just for a pair (ψ_1, ψ_2), where each ψ_i is a $\vee$-system;

(ii) we reduce the computation of $d(\psi_1, \psi_2)$ to distances between irreducible components of $\mathrm{V}_{\mathcal{A}}(\psi_i)$ (where $\mathcal{A}$ is an appropriate algebraic $\mathcal{L}$-structure).

Let us explain the assumptions above.

(i) Let us explain why we do not consider firs-order formulas with with quantifiers or negations. Actually, this assumption is not strict. Indeed, for any formula $\psi(X)$ with quantifier one can introduce a new predicate symbol $R_{\psi}(X)$ with the same domain (Morleyization). Similarly, one can eliminate negations in any formula.

(ii) We state that the differences in irreducible components of $\vee$-algebraic sets $Y = \mathrm{V}_{\mathcal{A}}(D)$, $Z = \mathrm{V}_{\mathcal{A}}(E)$ naturally express the distance between the formulas D, E. Let us consider a simple example.

Let $\mathcal{L} = \{\leq\}$, $\mathcal{A}$ be a linearly ordered $\mathcal{L}$-structure. How can we estimate the distance (difference) between:

$$\psi_1\colon (x_1 \leq x_2) \wedge (x_1 \leq x_3), \tag{1}$$

$$\psi_2\colon (x_1 \leq x_3) \wedge (x_2 \leq x_3), \tag{2}$$

The first (second) formula does not assert about the relation between x_2 and x_3 (x_1 and x_2). Thus, ψ_1 admits either $x_2 \leq x_3$ or $x_3 \leq x_2$, and ψ_2 admits either $x_1 \leq x_2$ or $x_2 \leq x_1$. Therefore, ψ_1, ψ_2 are respectively equivalent to

$$(x_1 \leq x_2 \leq x_3) \vee (x_1 \leq x_3 \leq x_2),$$

$$(x_1 \leq x_2 \leq x_3) \vee (x_2 \leq x_1 \leq x_3),$$

and we actually decompose ψ_i into the union of its elementary cases. The difference between ψ_1, ψ_2 is hidden in the differences between the elementary cases of the statements ψ_i, However, the elementary cases of ψ_i are exactly the irreducible components of $\vee$-algebraic sets $\mathrm{V}\mathcal{A}(\psi_i)$.

Let us mention the another advantage of the metric based on irreducible components. For many important (see Sections 4, 5) algebraic structures $\mathcal{A}$ first-order definable sets and $\vee$-algebraic sets have a complicated structure, and it is difficult to directly work with them. Whereas the description of irreducible sets over $\mathcal{A}$ is often simple and there are clear algorithms which decompose any $\vee$-algebraic set into the union if its irreducible component.

To compute the distance $d(\psi_1, \psi_2)$ between $\vee$-systems ψ_1, ψ_2 we propose the following:

(i) fix an $\mathcal{L}$-structure $\mathcal{A}$;

(ii) denote $Y = \mathrm{V}_{\mathcal{A}}(\psi_1)$, $Z = \mathrm{V}_{\mathcal{A}}(\psi_2)$.

(iii) let $Y = \bigcup_{i=1}^{k} Y_i$, $Z = \bigcup_{i=1}^{l} Z_i$ be the irreducible decompositions of Y, Z;

(iv) put $d(\psi_1, \psi_2) = d(Y, Z) = \{\max d(Y_i, Z_j) \mid Y_i \nsubseteq Z, Z_i \nsubseteq Y\}$.

Thus, the distance between Y, Z is reduced to the distance between their irreducible components. The choice of a distance between irreducible sets depends on the universal theory of $\mathcal{A}$. One can naturally define d for the most important algebraic structures $\mathcal{A}$ (see the next two sections).

4. Partial orders

Motivation. There are many models in economics and machine learning, where the order between parameters is significant as well as their exact values. For example, the information about the order between the parameters

$$x = \{\text{US dollar to ruble rate for tomorrow}\},$$

$$y = \{\text{US dollar to ruble rate for the day after tomorrow}\}$$

is sufficient for the optimal tomorrow‘s currency exchange.

Thus, there appears an important problem in the studying of first-order formulas with the relation of partial order. The approach developed in Section 3 allows us to define a distance between such formulas.

Let $\mathcal{L} = \{\leq\}$ be the language with the predicate symbol of a partial order, and we consider $\vee$-systems of $\mathcal{L}$-equations over a linearly ordered $\mathcal{L}$-structure $\mathcal{A}$. One can directly prove that $\mathcal{A}$ is equationally Noetherian, and Proposition 2.1 holds.

Let us define an algorithm which decomposes any $\vee$-algebraic set into a union of its irreducible components. By Proposition 2.2, it is enough to give such algorithm for algebraic sets.

Proposition 4.1. *Let* $\mathbf{S}$ *be a system in variables* $X = \{x_1, x_2, \dots, x_n\}$. *The irreducible components of* $Y = \mathrm{V}_{\mathcal{A}}(\mathbf{S})$ *are all possible linear orderings of* X *consistent with* $\mathbf{S}$. *For example, the irreducible components of* $\mathrm{V}_{\mathcal{A}}(\{x_1 \leq x_2, x_3 \leq x_4\})$ *are defined by the following systems:*

$$\{x_3 \leq x_4 \leq x_1 \leq x_2\}, \{x_3 \leq x_1 \leq x_4 \leq x_2\}, \{x_3 \leq x_1 \leq x_2 \leq x_4\},$$

$$\{x_1 \leq x_3 \leq x_4 \leq x_2\}, \{x_1 \leq x_3 \leq x_2 \leq x_4\}, \{x_1 \leq x_2 \leq x_3 \leq x_4\}.$$

Propositions 2.2,4.1 allows us to decompose any $\vee$-algebraic set over $\mathcal{A}$. Let us define a distance between irreducible algebraic set over $\mathcal{A}$ by the following way:

(i) since any irreducible set Y defines a linear order on the set of variables $X = \{x_1, x_2, \ldots, x_n\}$, one can denote the ordered vector of X by $\overrightarrow{Y}$ (for, example, for $Y = \mathrm{V}_{\mathcal{A}}(\{x_3 \leq x_1 \leq x_4 \leq x_2\})$ we have $\overrightarrow{Y} = (3, 1, 4, 2)$);

(ii) the distance between irreducible algebraic sets Y, Z is the Hamming distance between $\overrightarrow{Y}, \overrightarrow{Z}$.

Example 4.2. Suppose ψ_1, ψ_2 are defined by (1,2) then:

(i) using $\mathcal{A}$ defined above, we have

$$Y = \mathrm{V}_{\mathcal{A}}(\psi_1) = \mathrm{V}_{\mathcal{A}}((x_1 \leq x_2 \leq x_3)) \cup \mathrm{V}_{\mathcal{A}}(x_1 \leq x_3 \leq x_2),$$

$$Z = \mathrm{V}_{\mathcal{A}}(\psi_2) = \mathrm{V}_{\mathcal{A}}(x_1 \leq x_2 \leq x_3) \cup \mathrm{V}_{\mathcal{A}}(x_2 \leq x_1 \leq x_3).$$

(ii) by the definition,

$$d(\psi_1, \psi_2) = d(\mathrm{V}_{\mathcal{A}}(x_1 \leq x_3 \leq x_2), \mathrm{V}_{\mathcal{A}}(x_2 \leq x_1 \leq x_3)) = d((1, 3, 2), (2, 1, 3)) = 3.$$

One should remark in conclusion that the definition of $d(\psi_1, \psi_2)$ allows to obtain upper and lower bound for the average value of $d(\psi_1, \psi_2)$ on the set of $\vee$-systems in n variables.

5. Unary predicates

Motivation. In machine learning and statistics, feature selection is the process of selecting a subset of relevant features (variables, predictors) for further use in model construction. Any algorithm of feature selection divides the set of original features into the groups of relevant and irrelevant features (also, an algorithm may decline to classify some features). Different algorithms of feature selections may give different answers about the relevancy (irrelevancy) of features. Therefore, there arises a question: how to estimate the closeness between the results of feature selection algorithms?

The answer comes from the general scheme defined in Section 3. Let $\mathcal{L} = \{R, R'\}$ be the language of two unary predicates, and $\mathcal{A}$ an $\mathcal{L}$-structure, where the predicates $R(x), R'(x)$ satisfy $\mathcal{A} \models \forall x\ \neg(R(x) \leftrightarrow R'(x))$. The $\mathcal{L}$-structure $\mathcal{A}$ is the set of features, and the predicate $R(x)$ ($R'(x)$) has the interpretation: "a feature x is relevant (irrelevant)".

The result of any feature selection algorithm may be expressed by an appropriate $\vee$-system. For example, the following $\vee$-systems

$$\psi_1\colon\ (R(x_1) \vee R'(x_2)) \wedge (R'(x_1) \vee R(x_2)), \tag{3}$$

$$\psi_1\colon\ (R(x_1) \vee R(x_2)) \wedge (R'(x_1) \vee R'(x_2)) \tag{4}$$

respectively state that

(i) "exactly one of the features x_1, x_2 is relevant";

(ii) "the features x_1, x_2 are relevant or irrelevant simultaneously".

The distance on the set of $\vee$-systems induces a metric on the class of feature selection algorithms. The following statement is an analogue of Proposition 4.1.

Proposition 5.1. *Let* $\mathbf{S}$ *be a system in variables* $X = \{x_1, x_2, \ldots, x_n\}$. *A variable* $x \in X$ *is called undefined if neither* $R(x)$ *nor* $R'(x)$ *belongs to* $\mathbf{S}$. *Then* $Y = \mathrm{V}_{\mathcal{A}}(\mathbf{S})$ *is the proper union*

$$Y = \mathrm{V}_{\mathcal{A}}(\mathbf{S} \cup \{R(x)\}) \cup \mathrm{V}_{\mathcal{A}}(\mathbf{S} \cup \{R'(x)\}). \tag{5}$$

Applying the splitting (5) of Y to all undefined variables, we obtain the irreducible decomposition of the original algebraic set.

One can directly check that $\mathcal{A}$ is equationally Noetherian, and Proposition 2.1 holds. Let us define a distance between irreducible algebraic set over $\mathcal{A}$ by the following way:

(i) by Proposition 5.1, for any irreducible set Y over $\mathcal{A}$ and for any variable $x \in X$ we have either $R(x) \in \mathrm{Rad}_{\mathcal{A}}(Y)$ or $R'(x) \in \mathrm{Rad}_{\mathcal{A}}(Y)$;

(ii) therefore Y is defined by a $(0,1)$-vector $\overrightarrow{Y} = (y_1, y_2, \ldots, y_n)$ such that

$$y_i = \begin{cases} 0, & \text{if } R'(x_i) \in \mathrm{Rad}_{\mathcal{A}}(Y), \\ 1, & \text{if } R(x_i) \in \mathrm{Rad}_{\mathcal{A}}(Y) \end{cases}$$

(iii) the distance between irreducible algebraic sets Y, Z is the Hamming distance between $\overrightarrow{Y}, \overrightarrow{Z}$.

Example 5.2. Let $\vee$-systems ψ_1, ψ_2 be defined by (3,4). Let $\mathcal{A}$ be an $\mathcal{L}$-structure satisfying the formula

$$\forall x \neg(R(x) \leftrightarrow R'(x)).$$

We have that the sets $Y_i = \mathrm{V}_{\mathcal{A}}(\psi_i)$ have the following irreducible decompositions

$$Y_1 = \mathrm{V}_{\mathcal{A}}(\{R(x_1), R'(x_2)\}) \cup \mathrm{V}_{\mathcal{A}}(\{R'(x_1), R(x_2)\}) = Y_{11} \cup Y_{12},$$

$$Y_2 = \mathrm{V}_{\mathcal{A}}(\{R(x_1), R(x_2)\}) \cup \mathrm{V}_{\mathcal{A}}(\{R'(x_1), R'(x_2)\}) = Y_{21} \cup Y_{22}.$$

The irreducible sets Y_{ij} define the following vectors $\overrightarrow{Y_{ij}}$:

$$\overrightarrow{Y_{11}} = (1,0),\ \overrightarrow{Y_{12}} = (0,1),\ \overrightarrow{Y_{21}} = (1,1),\ \overrightarrow{Y_{22}} = (0,0).$$

We have

$$d(Y_{11}, Y_{21}) = d(Y_{11}, Y_{22}) = d(Y_{12}, Y_{21}) = d(Y_{12}, Y_{22}) = 1.$$

Thus, $d(\psi_1, \psi_2) = 1$.

The definition of the distance allows us to estimate the average distance between two irreducible sets. One can directly check that the number of $(0,1)$-vector pairs $(\overrightarrow{Y}, \overrightarrow{Z})$ with the Hamming distance k is equal $2^n \binom{n}{k}$. Therefore, the average distance between two irreducible sets is

$$\frac{1}{2^n \cdot 2^n} \sum_{k=0}^{n} k 2^n \binom{n}{k} = n/2.$$

Acknowledgments

The author was supported by Russian Research Fund (project 17-11-01117).

References

[1] Daniyarova E, Myasnikov A and Remeslennikov V 2017 *Algebraic geometry over algebraic structures* (Novosibirsk: SO RAN)

On equations over Brandt semigroups

A N Shevlyakov[1], **M A Vakhrameev**[2]

[1] Sobolev Institute of Mathematics, Pevtsova 13, Omsk, 644099, Russia
[2] Sobolev Institute of Mathematics, Pevtsova 13, Omsk, 644099, Russia

E-mail: `vahrmih@yandex.ru`

Abstract. In this paper, we study equations in one variable over Brandt semigroup B_n. We show that the average number of solutions of these equations is asymptotically equal to n^2. We prove that the proportion of unsolvable equations is asymptotically equal to $\frac{2}{n^2}$.

1. Introduction

Algebraic set is one of the basic concepts of universal algebraic geometry, which is developed in the works of E. Yu. Daniyarova, A. G. Myasnikov, V. N. Remeslennikov (see paper [1]). Algebraic set over an algebraic system A is defined as the set Y of solutions of a system of equations over this algebraic system A. Among all algebraic sets of fixed dimension, 1-algebraic sets, i.e. sets that are defined by a single equation, are especially important. Generate a random 1-algebraic set is a quite actual problem for many applications. The simplest solution is to choose a pair of terms $t(x), s(x) \in T$ relative to some probability measure μ defined on the set of terms $T(x)$. By using these terms, we can define 1-algebraic set of the equation $s(x) = t(x)$.

In [2], dedicated to equations over free semilattices, it was noticed that the uniform measure μ generates distribution that is far from a uniform distribution on the class of 1-algebraic sets. This is due to the fact that for the 1-algebraic sets the numbers $|Eq(Y)| = |\{s(x) = t(x) \mid V_A(s(x) = t(x)) = Y\}|$ substantially differ for different Y. The similar problem was also considered in [3]

We consider another way to generate a random 1-algebraic set. Our method partially corrects the drawbacks of the first solution. We introduce an equivalence relation on the set of terms over algebraic system A:

$$s \sim t \iff s(x) = t(x) \ for\ all\ x \in A.$$

By choosing one representative from each class of equivalent terms, we obtain a set of terms $\bar{T}(x)$ and generation of a random 1-algebraic set reduces to the choice of a pair of terms $s(x), t(x) \in \bar{T}(x)$.

In this paper, we apply the second method to algebraic sets defined by equations in 1 variable over the Brandt semigroup B_n, where an equation is defined as equality of two termal functions. We prove, that the number of classes of equivalence of termal functions over B_n is equal to $n^4 + 3n^2 + 3$. Using this result, we calculate the numbers of solvable and unsolvable equations over B_n and establish that the average number of solutions of equations over B_n is asymptotically equal to n^2.

2. Basic definitions

The Brandt semigroup B_n is a semigroup $B_n = \{0\} \cup \{(i,j), 1 \leq i,j \leq n\}$, $|B_n| = n^2 + 1$ whose multiplication operation is defined as follows:

1. for all $b \in B_n$: $b \cdot 0 = 0 \cdot b = 0$;
2. for all $(e,k),\ (l,m) \in B_n \setminus \{0\}$:

$$(e,k) \cdot (l,m) = \begin{cases} (e,m), & \text{if } k = l; \\ 0, & \text{else.} \end{cases} \tag{1}$$

The B_n*-term* is the following expression: $w(x) = w_1 \cdot w_2 \dots w_s$ such that for all $1 \leq i \leq s$ either $w_i \in B_n$ or w_i is the variable x. The elements of the semigroup B_n which are contained in a notation of a term we will call as *constants*. Two terms s and t we will call as *equals* if they have the same notation. Let's define the equivalence relation on the set of all terms:

$$s(x) \sim t(x) \iff s(x) = t(x) \text{ for all values of variable } x.$$

The termal function is a function $f : B_n \to B_n$ which is defined by some B_n-term. The set of all different termal function will be denoted by T. We suggest that the termal function f defined by term w is equivalent to term v if $w \sim v$.

The equation over semigroup B_n is the equality of two termal functions $f_1 = f_2,\ f_1, f_2 \in T$. *The solution* of the equation $f_1 = f_2$ is a value of variable x that makes the equality in this equation true. The equation which has at least one solution is called *solvable*. The equation which has no solutions is called *unsolvable*. The set of all solutions of the equation $f_1 = f_2$ will be denoted as $V_{B_n}(f_1 = f_2)$. The set of all equation over semigroup B_n will be denoted as Eq_n.

3. Equivalence of termal functions

Let's define the following sets of terms:

- $C_1 = \{b \in B_n\}$, $|C_1| = n^2 + 1$;
- $C_2 = \{x\}$, $|C_2| = 1$;
- $C_3 = \{x^2\}$, $|C_3| = 1$;
- $C_4 = \{bx,\ b \in B_n \setminus \{0\}\}$, $|C_4| = n^2$;
- $C_5 = \{xb,\ b \in B_n \setminus \{0\}\}$, $|C_5| = n^2$;
- $C_6 = \{bxd,\ b,d \in B_n \setminus 0\}$, $|C_6| = n^4$.

Let $C = C_1 \cup C_2 \cup C_3 \cup C_4 \cup C_5 \cup C_6$. Since sets C_i are mutually disjoint $|C| = |C_1| + |C_2| + |C_3| + |C_4| + |C_5| + |C_6| = n^4 + 3n^2 + 3$.

Statement 3.1. *For any terms* $s, t \in C$ *such that* $s \sim t$ *it follows that* $t = s$.

Proof. In order to verify the correctness of this statement, it suffices to compare the sets of values for all terms from C:

- $t \in C_1,\ t = b,\ b \in B_n$
- $t \in C_2,\ t = x$
- $t \in C_3,\ t = x^2 = \begin{cases} (i,i), & \text{if } x = (i,i); \\ 0, & \text{else;} \end{cases}$
- $t \in C_4,\ t = bx = (b^1, b^2)x = \begin{cases} (b^1, i), & \text{if } x = (b^2, i); \\ 0, & \text{else;} \end{cases}$

- $t \in C_5,\ t = xb = x(b^1, b^2) = \begin{cases} (i, b^2), & \text{if } x = (i, b^1); \\ 0, & \text{else}; \end{cases}$
- $t \in C_6,\ t = bxd = (b^1, b^2)x(d^1, d^2) = \begin{cases} (b^1, d^2), & \text{if } x = (b^2, d^1); \\ 0, & \text{else}. \end{cases}$

Thus, the set C doesn't contain different equivalent terms.

□

Further, any termal function will be identified with the corresponding term from the set C.

Statement 3.2. *For any termal function $f : B_n \to B_n$ there is an equivalent term $t \in C$.*

Proof. Suppose that the termal function f is defined by the term $w = w_1 \cdot w_2 \ldots w_k$. Multiplying every pair of neighboring constants, this term can be reduced to the term $v = v_1 \cdot v_2 \ldots v_s$ which is not containing two consecutive constants in the notation, or to the 0. If $f \sim 0$ then the statement is automatically proved because $0 \in C_1$. So let's consider the term $v = v_1 \cdot v_2 \ldots v_s$.

If $s \leq 2$ then the term v is a term of one of the following types: b, x, x^2, bx, xb. All such terms are contained in the set C, so the statement is proved as well.

It remains to consider the case when $s > 2$. Notice, that

$$x^s = \begin{cases} (i, i), & \text{if } x = (i, i); \\ 0, & \text{else}, \end{cases}$$

so, $x^s \sim x^2$. Thus, if $v = x^s$ then the statement is proved. Therefore suggest that v contains at least one constant. Let's prove that in this case, term v is either identically equal to 0 or not equals to 0 only for one value of variable x. Notice, that the term v necessarily contains one of the following expressions:

1. $(b^1, b^2) \cdot x \cdot x = \begin{cases} (b^1, b^2), & \text{if } x = (b^2, b^2); \\ 0, & \text{else}; \end{cases}$
2. $x \cdot (b^1, b^2) \cdot x = \begin{cases} (b^2, b^1), & \text{if } x = (b^2, b^1); \\ 0, & \text{else}; \end{cases}$
3. $x \cdot x \cdot (b^1, b^2) = \begin{cases} (b^1, b^2), & \text{if } x = (b^1, b^1); \\ 0, & \text{else}; \end{cases}$
4. $(b^1, b^2) \cdot x \cdot (d^1, d^2) = \begin{cases} (b^1, d^2), & \text{if } x = (b^2, d^1); \\ 0, & \text{else}. \end{cases}$

Any of the expressions 1-4 doesn't equal to 0 only for the single value of variable $x \neq 0$. So, the term v is either identically equal to 0, or not equals to 0 only for the single value of$x \neq 0$. Suppose that the term v is not equal to 0 for $x = (a^1, a^2)$ and $v((a^1, a^2)) = (e^1, e^2)$. In this case the term $v \sim (e^1, a^1)x(a^2, e^2) \in C$, and the statement is proved.

□

Consequence 3.3. The number of the different termal functions over B_n is equal to n^4+3n^2+3.

IOP Conf. Series: Journal of Physics: Conf. Series **1050** (2018) 012079 doi:10.1088/1742-6596/1050/1/012079

4. The average number of solutions of the equations over B_n

By definition, an equation over B_n is the equality of two termal functions $f_1, f_2 \in T$: $f_1(x) = f_2(x)$. Because of $|T| = n^4+3n^2+3$, $|Eq_n| = (n^4+3n^2+3)^2 = n^8+6n^6+15n^4+18n^2+9$.

Let's compute the average number of solutions for equations from Eq_n. Each termal function is equivalented to some term from classes $C_1, \ldots, C_6$. It allows us to split all equations into $6 \cdot 6 = 36$ disjoint groups. Consider few examples.

Example 4.1.

$f_1 \sim t_1 \in C_1,\ f_2 \sim t_2 \in C_3:\ b = x^2,\ b \in B_n.$

The number of these equations is equal to $|C_1| \cdot |C_3| = n^2 + 1$. Notice, that

$$x^2 = \begin{cases} (t,t), & \text{if } x = (t,t),\ 1 \le t \le n; \\ 0, & \text{else.} \end{cases}$$

Consider all possible cases:

- if $b \neq 0$ and $b \neq (i,i),\ 1 \le i \le n$, then there is no solutions; The number of such equations is equal to $(n^2+1) - 1 - n = n^2 - n$;
- if $b = (i,i),\ 1 \le i \le n$, there is a single solution – (i,i). The number of such equations is equal to n;
- if $b = 0$ – single equation. Any value of x except $x = (i,i),\ 1 \le i \le n$ is a solution, so there are $(n^2+1) - n$ solutions.

The summary result is presented in the table 1.

Table 1.

Number of equations	Number of solutions
$n^2 - n$	0
n	1
1	$n^2 - n + 1$

Example 4.2.

$f_1 \sim t_1 \in C_1,\ f_2 \sim t_2 \in C_4:\ b_1 = b_2 x,\ b_1, b_2 \in B_n, b_2 \neq 0$

The number of these equations is equal to $|C_1| \cdot |C_4| = (n^2+1)n^2$.

Suppose that $b_1 \neq 0$ and $b_1 = (b_1^1, b_1^2),\ b_2 = (b_2^1, b_2^2)$. We have two possible cases:

- if $b_1^1 = b_2^1$ ($n \cdot n \cdot n = n^3$ equations), then there is only a single solution $x = (b_2^2, b_1^2)$;
- if $b_1^1 \neq b_2^1$ ($n \cdot n \cdot (n-1) \cdot n = n^4 - n^3$ equations), then there is no any solution.

Now suppose that $b_1 = 0,\ b_2 = (b_2^1, b_2^2)$. The number of such equations is equal to $n \cdot n = n^2$. Any value of variable x except $x = (b_2^2, i),\ 1 \le i \le n$ will be a solution. In this way there will be $n^2 + 1 - n$ solutions for each of n^2 equations.

The summary result is presented in the table 2.

IOP Conf. Series: Journal of Physics: Conf. Series **1050** (2018) 012079 doi:10.1088/1742-6596/1050/1/012079

Table 2.

Number of equations	Number of solutions
$n^4 - n^3$	0
n^3	1
n^2	$n^2 - n + 1$

All groups of equations were considered in [4]. It was calculated that the average number of solutions of equations from Eq_n is asymptotically equal to n^2. Also in [4] it was proved that the proportion of unsolvable equations asymptotically equals to $\frac{2}{n^2}$.

Acknowledgments
The work was supported by the program of fundamental scientific researches of the SB RAS I.1.1., project 0314-2016-0004.

References
[1] E Yu Daniyarova, A G Myasnikov and V N Remeslennikov 2016 *Algebraic Geometry over Algebraic Structures* (Novosibirsk: Publishing House SB RAS) p 243

[2] A N Shevlyakov 2016 Equivalent equations in semilattices *Sib. Elektron. Mat. Izv.* **13** 478–90

[3] A N Shevlyakov 2017 Equationally extremal semilattices*Journal of Siberian Federal University. Mathematics and Physics* **10 :3** 298–304

[4] M A Vakhrameev 2017 On equations over Brandt semigroups *Preprint* 1709.08043

Research and software development using AR technology

D A Smuseva, A Y Rolich

National Research University "Higher School of Economics", 34 Tallinskaya str., Moscow, Russia

Abstract. The paper considers programs and devices of augmented reality, examines the general environments and methods of software development and the rationale for their selection. The work describes in detail the operating principle of the software, the pattern recognition algorithm, the UML class diagram, the UML usage diagram, and the architecture of the 3D rendering engine and a description of its operation. An example of practical application of software with pattern recognition is offered. The paper examines the impact of virtual reality on human health, as well as the problem of assimilation of educational material in preschool education. To solve the problem, various algorithms for the program are proposed. Based on the conducted studies, it was decided to create the software for the experiment on the basis of developed algorithms for preschool education. The results of the work can be used for further research in the field of expanded reality, for new developments in this field and improvement of the quality of education.

1. Introduction

The mobile market shows high growth rates [1], with gadget users becoming not only adults but also children. This is the reason for the rapid development of the mobile application market for children's audiences. A separate segment of this market is software for mobile devices using the technology of augmented reality. Augmented Reality (AR) – a technology that allows you to combine the virtual world with the real, integrating virtual objects in real time [2].

Augmented reality can change the perception of the modern world and make it more convenient and interactive. The spheres of application of this technology are endless, it all depends on the person's imagination, this can be medicine, military science, education, the sphere of state. services, etc. The fact is that augmented reality, unlike virtual reality, complements the real world with virtual objects, rather than replacing it. This means that AR requires direct user contact with reality through additional content, which greatly simplifies the solution of certain tasks.

When considering the interaction of children with gadgets, it should be borne in mind that in France in 2013 the Academy of Sciences published the book "The Child and the Screens" (L'enfant et les écrans) [3], which discusses the interaction of children and gadgets with the results of recent work in medicine, neurology and psychology. Here it is described that the child can be allowed to use smartphones with 1.5 years.

According to studies conducted by Common Sense Media across the United States, as early as in 2011, 38% of children from 0 to 8 years used mobile devices for watching video and games [4], but in 2013 their number increased to 72 % [5]. The latest study of 2017 notes that to date, 84% of children of the age under consideration are actively using mobile devices [6].

In 2015, the expert project Hi-Tech.Mail.Ru conducted a survey in Russia among more than 5,000 parents. According to the results of the survey, it became clear that children begin to use gadgets for the first time between the ages of 1 and 3, with 85% of parents giving out "smart" adults to children.

IOP Conf. Series: Journal of Physics: Conf. Series **1050** (2018) 012080 doi:10.1088/1742-6596/1050/1/012080

Also during the survey, it was found that 66% of children use gadgets for entertainment, and 20% – use smart devices for training.

Taking into account the above-described indicators, it can be concluded that today the implementation of software with augmented reality for mobile devices is actual in the educational process. In this case, it is not necessary to replace books familiar to parents and children, on the contrary, this technology will allow to expand the possibilities for studying them. Thus, the methodology of teaching does not change radically, but at the same time, children receive a complete presentation of the material and its assimilation.

2. Formulation of the problem

The problem is that traditional methods of learning do not work effectively on the assimilation of information through oral speech and text. The human brain is more inclined to process visual images than a simple text. On this subject, there are many studies [7], which show that the technologies of augmented reality in the field of education enrich the visual and contextual learning, improving the content of information. If 25% of information is retained while listening to lectures in short-term memory, then this number reaches 80% with visual training. To date, the implementation of various methods for visualization and gaming of education is topical, and the use of augmented reality in the sphere of education will only benefit people, especially children. This technology will not only develop associative thinking, but also increase the effectiveness of mastering the educational material.

Existing mobile applications are similar in content and are not of great interest, since they do not use the technology of augmented reality. In addition, the existing software for pre-school children using the technology of augmented reality is aimed at a certain area of knowledge and requires additional paid printed materials (books, magazines, etc.). There is a need to develop software for preschool education using AR-technology, which allows to increase the level of mastering of educational material in preschool education, which proves the relevance of this work.

In this regard, the purpose of this work is to increase the level of learning material in pre-school education by developing software for interactive visualization of educational material using algorithms of augmented reality. Using this technology, you can deliver the material necessary for learning in a more interesting and accessible form for students, building an activity on the basis of fascinating games and demonstrations. The convenience of using virtual 3D-objects simplifies the process of explaining new material. At the same time, mastering the technology of augmented reality, the level of information literacy of parents and their children.

To achieve this goal, an analysis of the subject area was carried out, which showed that to date there is no comprehensive software with augmented reality for the education of children on the market. In this regard, applications that are available on the Russian market for mobile phones on the iOS platform, in which the augmented reality is used for the education of children, were considered. To work with the programs in question, only a smartphone with a built-in camera is required and no additional equipment, for example glasses, is required. It is important to note that for each software reviewed, printed material is required from which the camera of the device will read the markers and then display its content. However, some of the applications require the purchase of a printed publication, while for others it is sufficient to print materials at home.

In the course of the analysis (Table 1), it turned out that for children there are not so many training applications with augmented reality, despite the fact that demand for them is growing. At the same time, educational programs are divided into two categories: teaching the alphabet and the mathematical account. As for applications with ABC, there are at least 3 programs, but for them you need to purchase additional content. It should also be noted that among the programs found Russian-language content is only - Devar kids and Live ABCs 3d. Programs for studying mathematics in which the technology of augmented reality is used also do not represent such a wide variety, there is not even a single software in Russian, and only AR Flashcards software is used to teach young children under 5 years of age.

Table 1. Comparative analysis of existing software with augmented reality

	AR Flashcards – Animal Alphabet	Devar kids	3D live alphabet	Maths Teach-AR	FETCH! Lunch Rush	AR Flashcards	Funny lessons
Cost of the full version (rub)	-	-	-	379	-	299	-
Built-in purchases	-	+	-	-	-	-	±
Free additional printed material	+	-	-	+	+	+	+
Age (years)	to 5	4+	to 5	12+	6-8	to 5	to 7
Sound Assistant	-	-	+	-	+	-	+
Russian language support	-	+	+	-	-	-	+
Subject	Alphabet			Mathematics			Mathematics, Alphabet etc.

That is why it is important to create a comprehensive application for children in Russian, in which augmented reality will be used both for studying mathematics and for the Russian language, namely, alphabet. The presented comparative analysis (Table 1) showed that the new software will be unique and have the advantage of Russian-speaking users.

3. Methods for solving the problem

To solve the problem, various algorithms for the program are proposed. Based on the conducted studies, it was decided to create the software for the experiment on the basis of the developed algorithms called "Funny lessons" for teaching children, where for today there are two sections: "Mathematics" and "Alphabet", for the study of numbers and letters, respectively.

The software consists of the following components:

1. The main menu that meets the user and provides access to the instructions for working with the software, a page with general information about the product and authors, and also directly to work with a conventional camera and a stereo camera;
2. «Start» – window for selecting a section;
3. «Mathematics» – window of work with the AR-camera for the study of figures;
4. «Alphabet» – window for working with AR-camera for studying letters;
5. The window for working with a stereocamera in the section «Mathematics»;
6. Window for working with a stereocamera in the section «Alphabet»;
7. «How to play» – a window containing instructions for working with the program;
8. «About» – a window of general information about the application and about the authors.

IOP Conf. Series: Journal of Physics: Conf. Series **1050** (2018) 012080 doi:10.1088/1742-6596/1050/1/012080

A more detailed structure of the software is shown in Fig. 1.

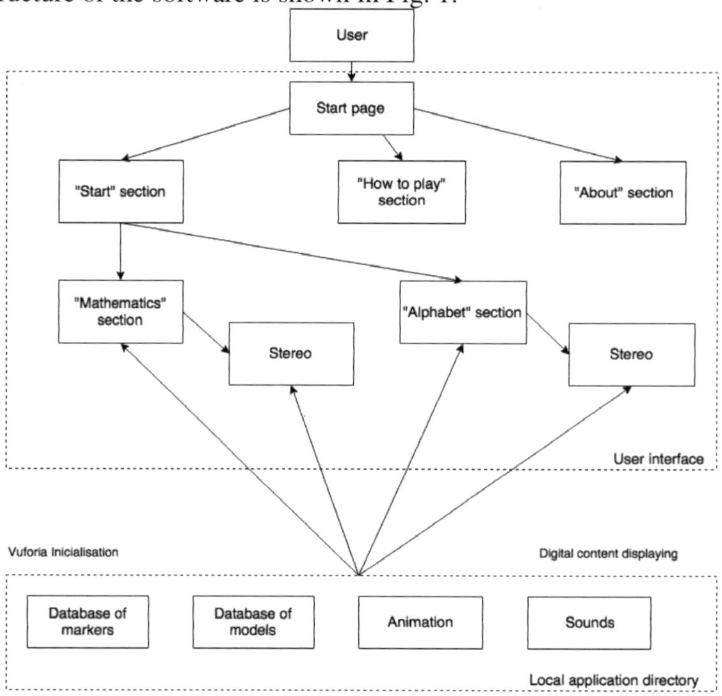

Figure 1. Software structure

The class diagram of the software being developed is shown in Fig. 2. It should be noted that the VuforiaBehavior class is the main class of the software, since it initializes all the main packages, in addition, this class is responsible for tracking and controlling the camera.

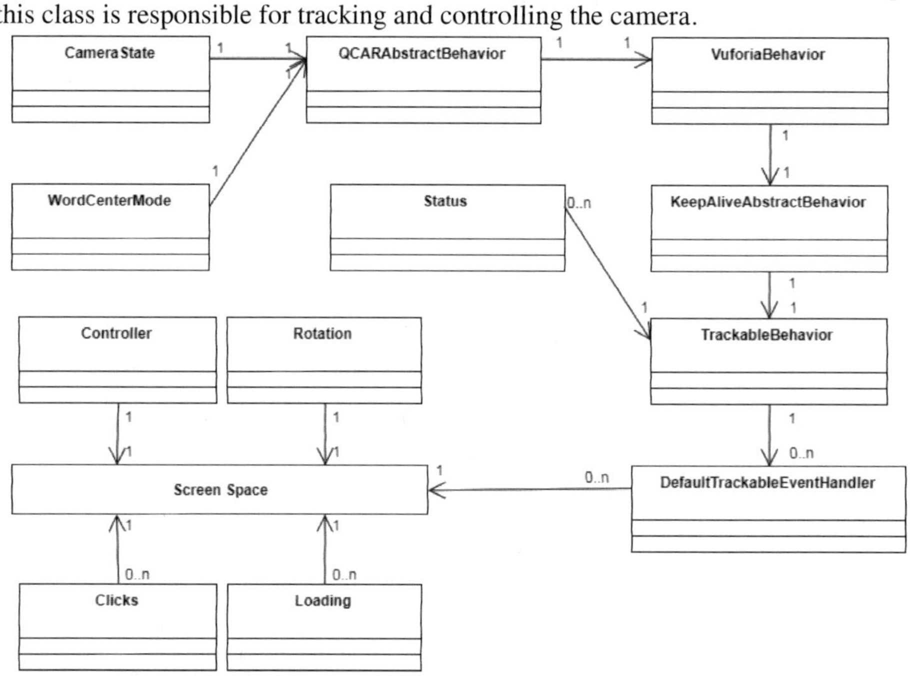

Figure 2. UML class diagram

In Fig. 3 is a UML diagram of use cases.

Figure 3. UML diagram of use cases

The blocks shown in Fig. 4 and Fig. 5, expand the software architecture, providing more detailed information about the 3D rendering module.

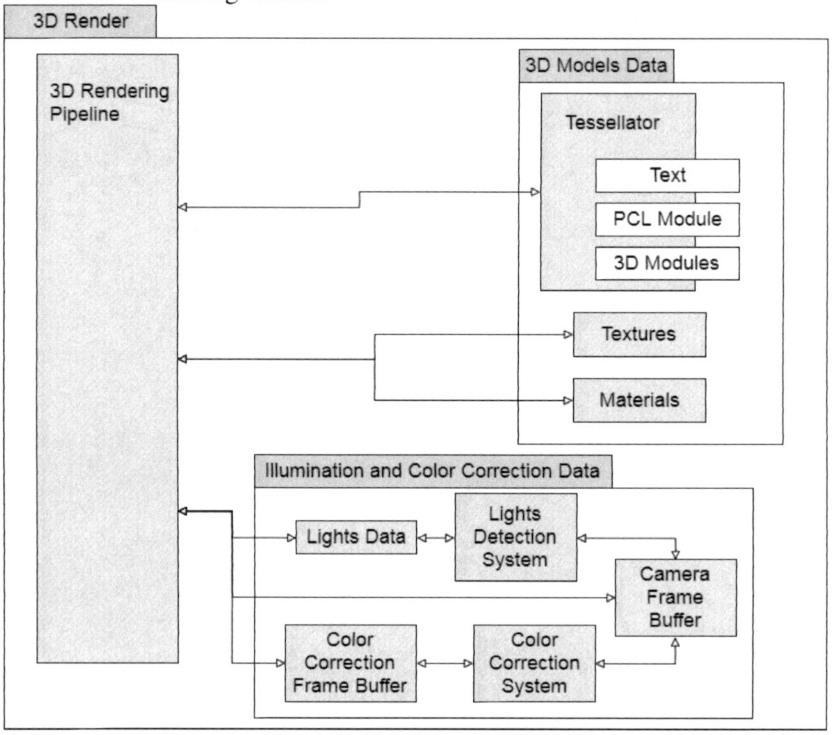

Figure 4. The architecture of the 3D rendering block

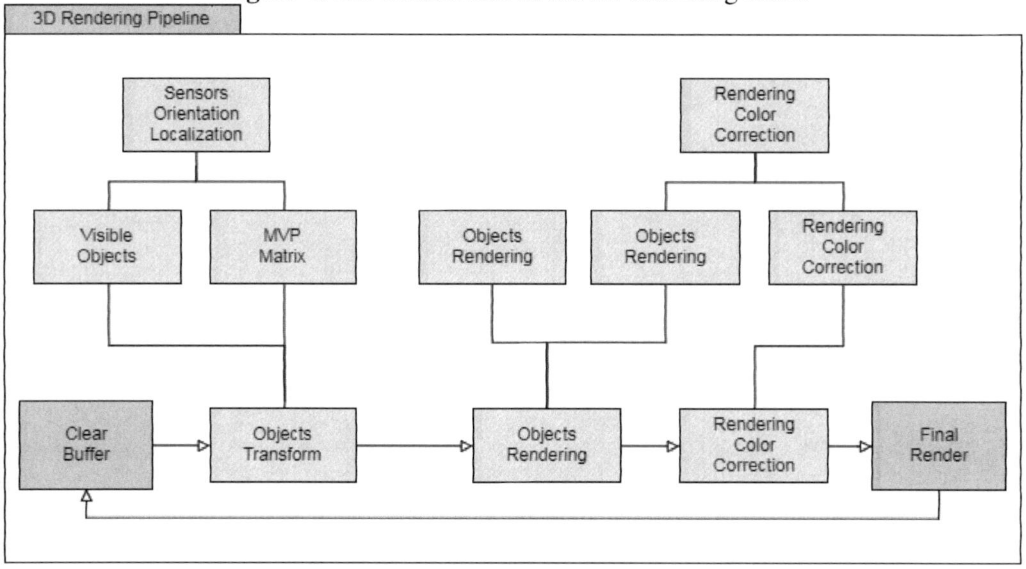

Figure 5. 3D rendering pipeline

The UML state diagram displays the algorithm of the software operation, describing possible sequences of states and transitions (Figure 6).

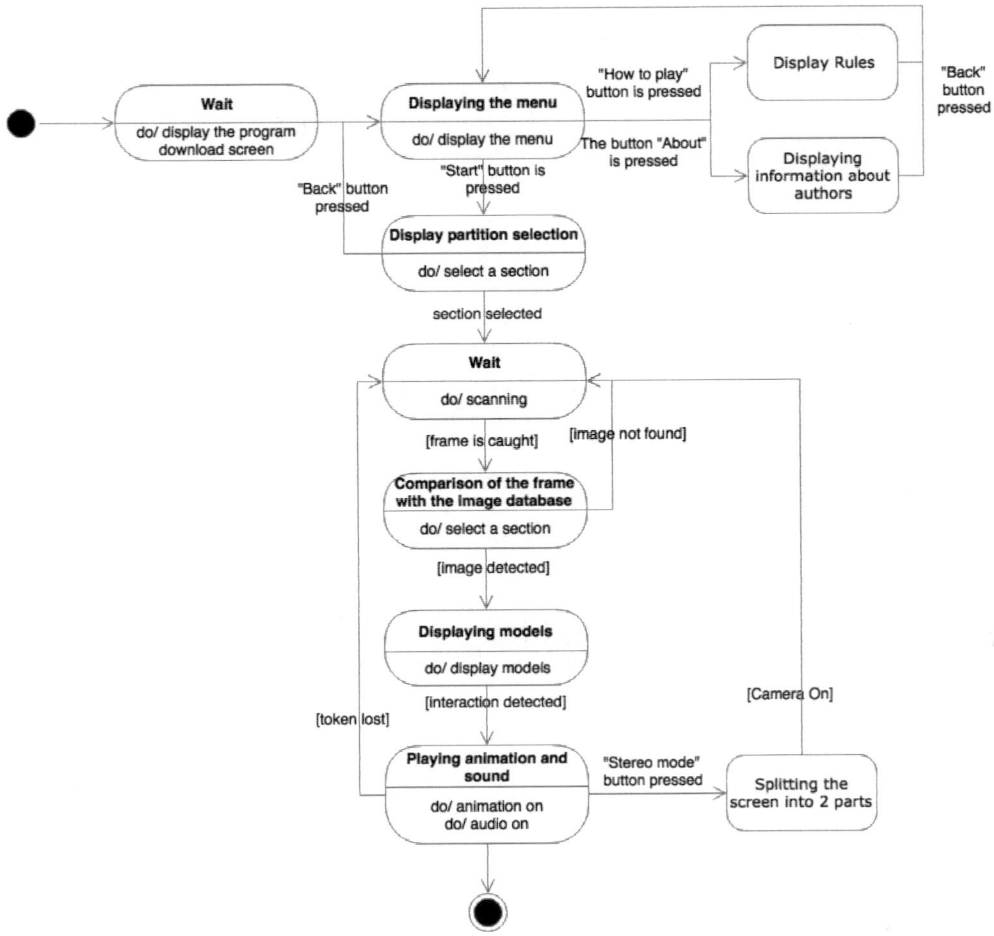

Figure 6. UML state diagram

IOP Conf. Series: Journal of Physics: Conf. Series **1050** (2018) 012080 doi:10.1088/1742-6596/1050/1/012080

A more detailed algorithm for the operation of the software can be found in Fig. 7.

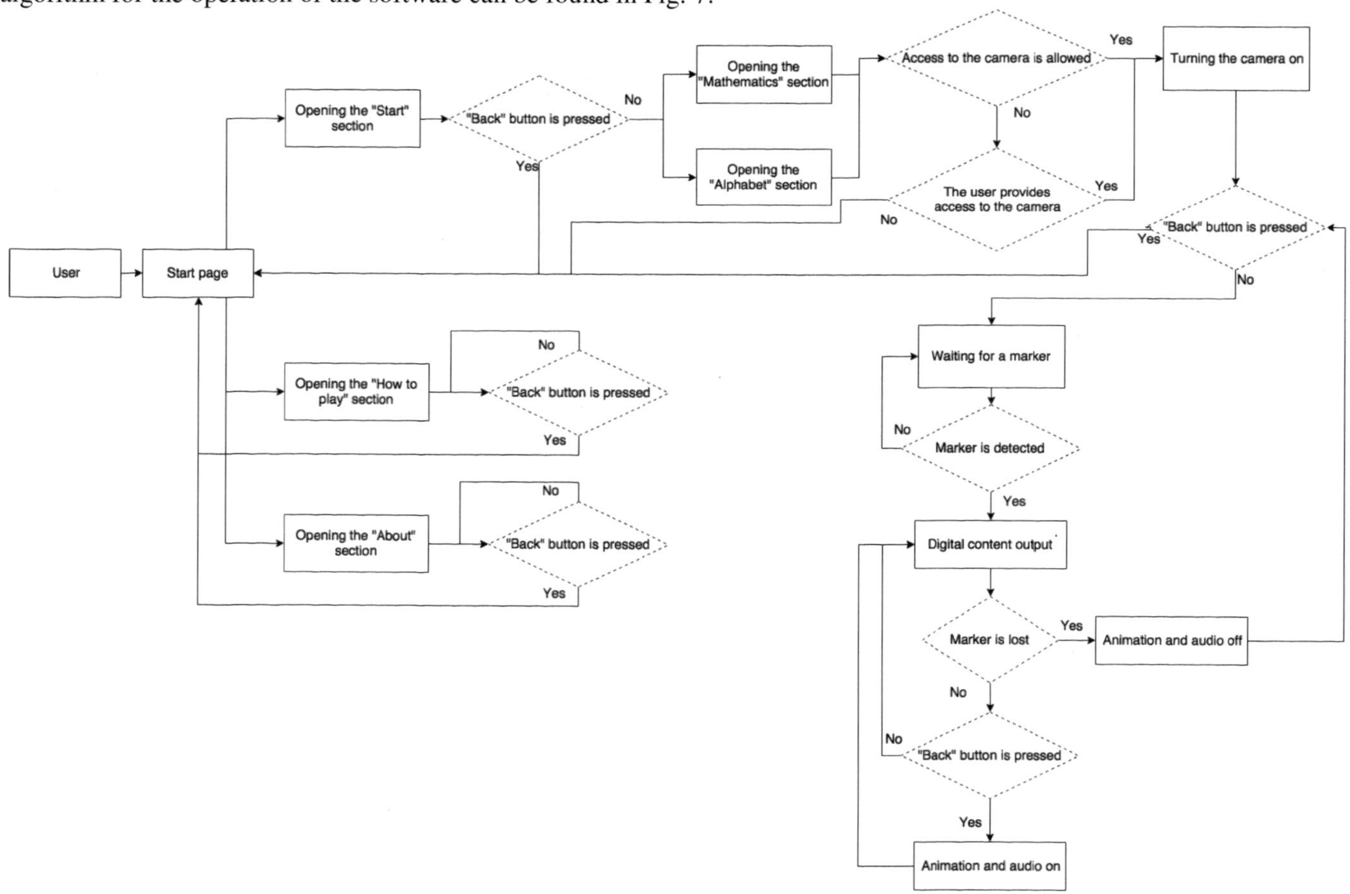

Figure 7. Algorithm of software operation

In the software, augmented reality is realized. For this purpose, the following steps were taken (Fig. 8) for image recognition and implementation of the corresponding digital content:

1. Capturing images from the real world, using the camera, each frame is captured and transferred to a tracker.
2. Next, the image converter that changes the format of the captured image provided by the camera into a format suitable for rendering OpenGL ES and for subsequent binding to the displayed materials.
3. In the tracker, using computer vision algorithms, image recognition occurs.
4. At this time, the software code must regulate all the work and update the application logic with each new input data.
5. Computer vision algorithms compare the resulting image from the camera with an existing marker database.

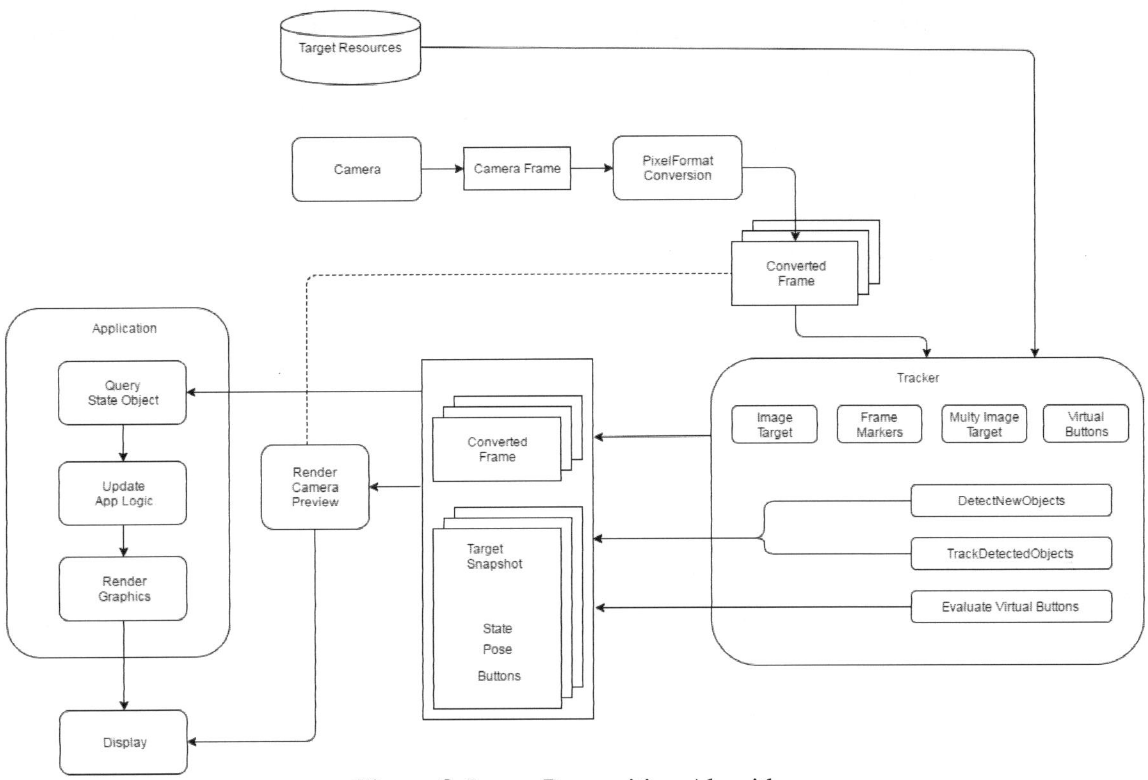

Figure 8. Image Recognition Algorithm

IOP Conf. Series: Journal of Physics: Conf. Series **1050** (2018) 012080 doi:10.1088/1742-6596/1050/1/012080

4. Experiment

Since the software being developed is oriented towards the children's audience, it was decided to make a simple interface and animation of the augmented reality models to attract children to the learning process, which was successfully implemented with the help of Unity and scripts in C #.

Simulation of the animation (Figure 9) took place in the Unity editor using the Animation component.

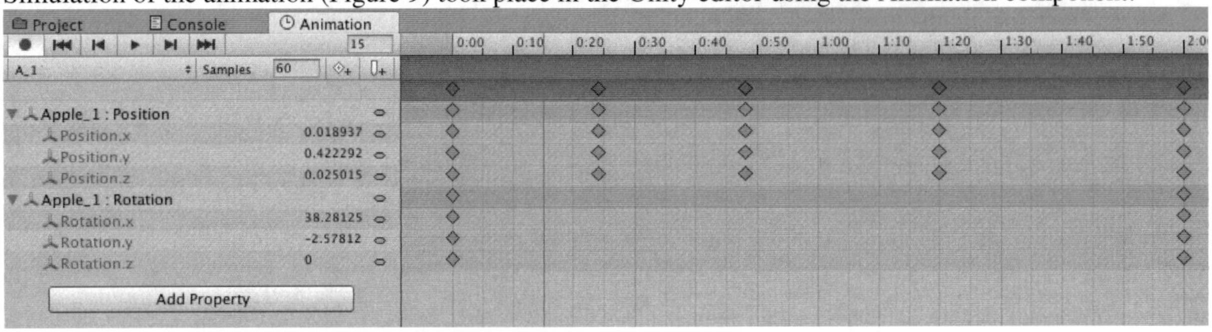

Figure 9. Animation of the object

Then the voice acting of the animation was recorded and reproduced. An example of a function that plays animation and sound when you press a virtual button:

```
public void OnButtonPressed(VirtualButtonAbstractBehaviour vb) {

    Debug.Log ("button Pressed");

    Apple_1.GetComponent<Animation> ().Play ();
    Apple_2.GetComponent<Animation> ().Play ();
    Apple_3.GetComponent<Animation> ().Play ();
    Apple_4.GetComponent<Animation> ().Play ();

    Apple_1.GetComponent<AudioSource> ().Play (34000);

}
```

Function, to stop all audio components:

```
private AudioSource[] allAudioSources;

void StopAllAudio()
{
    allAudioSources = FindObjectsOfType (typeof(AudioSource)) as AudioSource[];
    foreach (AudioSource audioS in allAudioSources) {
        audioS.Stop ();
    }
}
```

Moreover, the interaction of the user with virtual objects was realized. It was also decided to make an assignment to the picture on the "Press" marker, the OnButtonPressed() function. For this it was necessary to make an animation for each scene object, as well as create a virtual button with the following parameters:

```
private GameObject vbButtonObject;

void Start() {

    vbButtonObject.GetComponent<VirtualButtonBehaviour> ().RegisterEventHandler (this);

}
```

To reduce the waiting time for the user to load the camera for augmented reality, a function was implemented that, during the loading of a new level, shows the download window:

IOP Conf. Series: Journal of Physics: Conf. Series **1050** (2018) 012080 doi:10.1088/1742-6596/1050/1/012080

```
using System.Collections;
using System.Collections.Generic;
using UnityEngine;

public class loading : MonoBehaviour {

    public int Level;

    IEnumerator Start()
    {
        AsyncOperation async = Application.LoadLevelAsync (Level);
        yield return async;
        Debug.Log ("Загрузка");
    }
}
```

The experimental testing of the developed algorithms was carried out, which showed the operability of the software. In Fig. 10 shows an example of the work of image recognition and the implementation of digital content. Software using the camera of the mobile device reads the image, compares it with the loaded database and when the markers match, they display the corresponding digital content.

Figure 10. Marker recognition and visualization of digital content (on the left – before, on the right – after)

Also, a functional was implemented that reproduces the animation of objects when pressed (Figure 11). When you click on the "Start" button, the animation is played, which ends when the corresponding image is lost and starts again, if you press the button again. Also when the button is pressed, the sound is played.

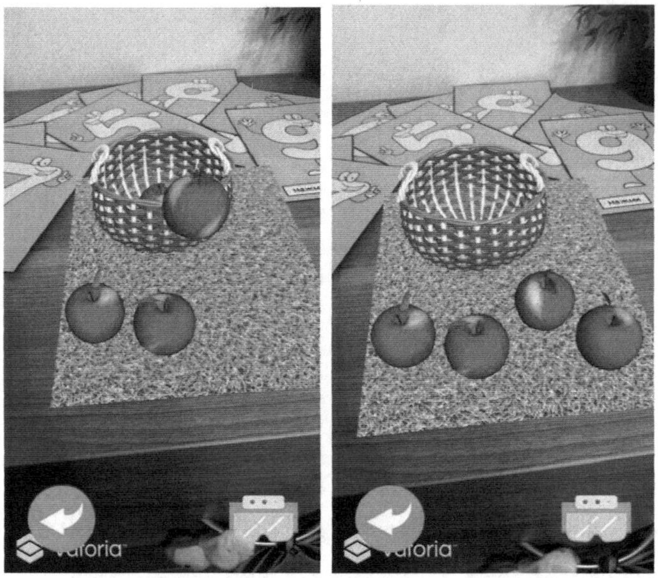

Figure 11. Realization of animation after interactive interaction

The interface realization function in stereopair mode for the virtual reality helmet was tested (Fig. 12). The software functions do not differ from the standard mode, when the button is pressed, animation and sound are played.

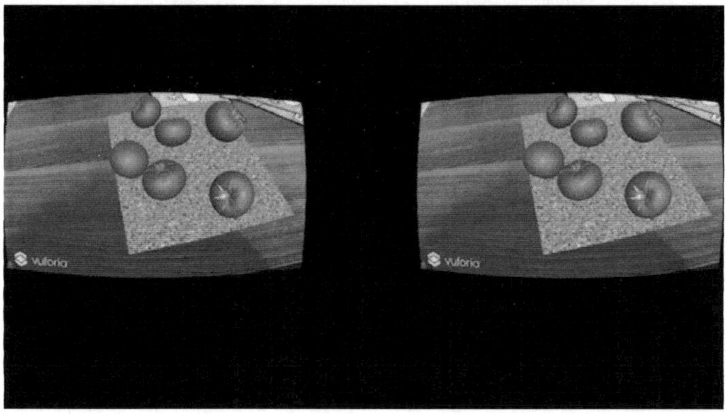

Figure 12. An example of the application operation in stereopair mode for the virtual reality helmet

5. Conclusion

As a result of the work carried out, it was shown that augmented reality is an effective tool for creating software in the field of education. This technology allows filling printed materials with digital three-dimensional content, which simplifies the learning process and increases the interest of children in textbooks and books.

As part of the work, the need to create software using the technology of augmented reality for preschool education is identified. Algorithms for the operation of software, natural human-computer interaction have been developed. The efficiency of the developed algorithms is confirmed by the results of experimental testing.

The offered software is expedient for applying in education of children for development at children of associative thinking and increase of level of mastering of a teaching material. Possible ways to improve the effectiveness of preschool education by adding the opportunity to work in glasses augmented reality, subject to the creation of special glasses for children.

6. References

[1] Agapov V, Pratusevich V, Jakovlev S 2014 *Obzor i ocenka perspektiv razvitija mirovogo i rossijskogo rynkov informacionnyh tehnologij*
[2] Agogi E 2011 *Augmented Reality in Education* (Greece: Proceedings of Science Center to Go Workshopp) pp 1–88
[3] Bourcier S 2010 *L'enfant et les écrans* (Éditions du CHU Sainte-Justine)
[4] Common sense media 2011 *Zero to Eight: Children's Media Use in America*
[5] Common sense media 2013 *Zero to Eight: Children's Media Use in America*
[6] The Common Sense Census 2017 *Media Use by Kids Age Zero to Eight*
[7] Nesloney T 2013 *Augmented Reality Brings New Dimensions to Learning*

IOP Conf. Series: Journal of Physics: Conf. Series **1050** (2018) 012081 doi:10.1088/1742-6596/1050/1/012081

Increasing efficiency of vibration protection system by using pneumatic rubber cord devices

V N Sorokin and N V Zakharenkov

Omsk State Technical University, 11, Mira ave., Omsk, 644050, Russia

Abstract. The article is devoted to investigation of the vibration protection system based on pneumatic rubber-cord devices. The device and principle of operation of the combined vibration protection system with the use of rubber-cord shells as power elements are presented. The analysis of the developed mathematical model is carried out. New methods of construction and algorithms for controlling the active vibration protection system are investigated on the created experimental complex. The results obtained at the experimental complex are presented and analyzed. The study of the developed mathematical model of combined vibration protection system with throttle control of gas pressure in power cells has shown its efficiency in the low-frequency range. The evaluation of the quality of the combined vibration protection system using integral criteria has shown its effectiveness up to 40% in comparison with the passive vibration protection system with harmonic excitation. The obtained test results have a discrepancy in the low-frequency region of the operating range with mathematical modeling data at the level of 10%.
Key words: active suspension, control strategy, vibration protection, rubber cord device

1. Introduction

Modern machinery production is aimed at implementing high-precision equipment and manufacturing processes. The process of achieving high precision of the component dimensions is, however, related to significant challenges due to negative vibration impact of natural and production-induced character [1].

The analysis of natural and production-induced vibration sources indicates their frequency range lying in the region of 0.1 Hz to 10 kHz [8–11].

Modern vibration protection systems, however, have frequency range bottom margin ≈2 Hz with maximal oscillation suppression ratio of 35–45 dB achievable at ≈10 Hz frequency [6, 12].

With due consideration of high-precision equipment cost and the importance of the research and production challenges solved with this equipment [3, 6, 12], it is unmistakable that vibration protection systems development is significant for such equipment.

At present time passive vibration protection systems are wide spread, but they have a number of drawbacks. Vibration protection systems with active elements connected to the external power source are in fact automatic control systems, and they nearly always work together with passive systems.

As passive vibration protection elements one uses pneumatic rubber-cord shells (RCS) having high lifting capacity and reliability [1, 2, 5]. Their usage as active systems actuating elements is, however, challenging, since they may create force only in one direction. Reverse movement at pressure relief is carried out either due to protected equipment gravity or additional spring element action. As the spring element one should use RCS incorporated into the system via lever similar to that of a tensile testing machine for compressing specimen.

2. Problem statement

The present work is concerned with the development of theoretical and practical approaches to produce combined vibration protection support based on RCS including both active and passive components (figure 1).

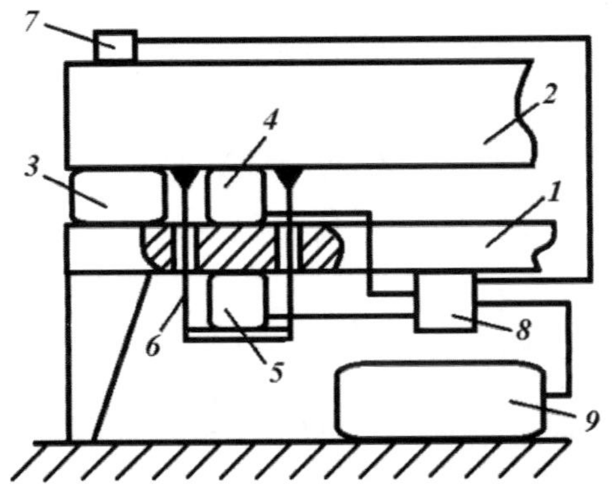

Figure 1. Scheme of combined vibration protection system with RCS: 1 – table; 2 – vibration-proof platform; 3 – RCS of the passive vibration protection system; 4, 5 – upper and lower RCS of active vibration protection system; 6 – lever; 7 – speed sensor; 8 – control unit; 9 – compressor

Vibration protection support bearing element is passive system RCS 3, whose pressure is not controlled during operation. Pressure in active system RCS 4 and 5 is altered by control unit 8 in antiphase to platform 2 oscillations. When it is moving downwards, the pressure is applied to RCS 4, and when it is moving upwards then to RCS 5 hindering this displacement by lever 6. Control unit activates according to the signals of speed sensor 7 installed on platform 2.

3. Theory

Calculation scheme of the combined vibration protection system is given in Figure 2.

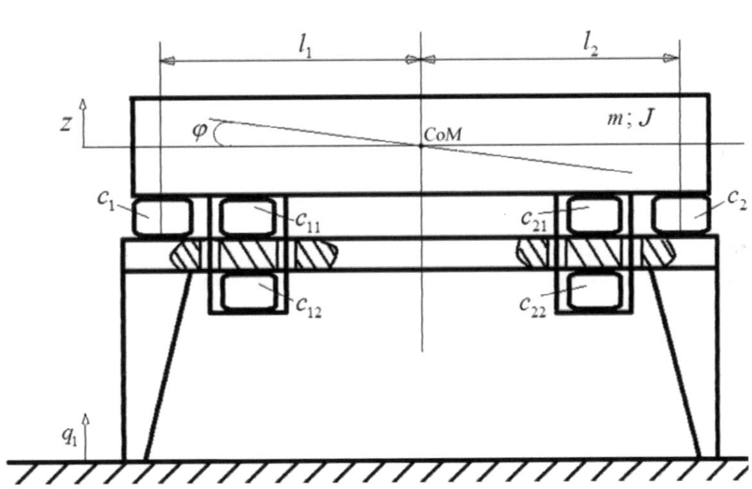

Figure 2. Calculation scheme: m – protected platform mass; J – the moment of inertia of the protected platform relative to the center of mass; l_1, l_2 – distance from the center of mass to the suspension points; Δl – distance from the center of mass to the corresponding axis of the protected platform

Platform oscillations are examined in two generalized coordinates z and φ, as well as subsidiary coordinates q (foundation displacements). The system gas two degrees of freedom, a vibration-proof platform with mass m and inertia moment J.

Also, it is assumed that: $c_{11} = c_{12} = c_{p1}$, $c_{21} = c_{22} = c_{p2}$, $q_1 = q_2 = q$, $l_l = l_2$.

Allowing for the assumptions and the fact that protected platform gravity is balanced by the passive system RCS string forces, and the values of generalized coordinates are measured from the equilibrium states, and neglecting the friction forces in suspension as well as RCS damping properties, the differential equations system takes the form of:

$$\left.\begin{aligned} & m\ddot{z} + 2c_1\left(z - q + l_1\varphi\right) + 2c_2\left(z - q + l_2\varphi\right) \\ & \qquad = 2S_{эф}P_1 + 2S_{эф}P_2 - 2c_{p1}\left(z - q + l_1\varphi\right) - 2c_{p2}\left(z - q + l_2\varphi\right); \\ & J\ddot{\varphi} + 2c_1 l_1\left(z - q + l_1\varphi\right) + 2c_2 l_2\left(z - q + l_2\varphi\right) \\ & \qquad = -2l_1 S_{эф}P_1 + 2l_2 S_{эф}P_2 - 2c_{p1}l_1\left(z - q + l_1\varphi\right) - 2c_{p2}l_2\left(z - q + l_2\varphi\right). \end{aligned}\right\},$$

where z, φ are linear and angular displacements of protected mass; S_{ef} is RCS effective area; $P_{1,2}\left(\dot{\varphi},t\right)$ is pressure in RCS.

When analyzing the system configuration and considering that its protected part mass has significant value and exceeds the supported equipment mass, the center of mass is practically coincides with the elastic centre. This configuration feature lets us assume that vertical protected mass oscillations are independent of longitudinal-angular ones because of the suspension symmetry, and the differential equations system takes the form of:

$$m\ddot{z} + 2c_1\left(z-q\right) + 2c_2\left(z-q\right) = 2S_{ef}P_1 + 2S_{ef}P_2 - 2c_{p1}\left(z-q\right) - 2c_{p2}\left(z-q\right);$$
$$J\ddot{\varphi} + 2\left(c_1 l_1^2 + c_2 l_2^2\right)\varphi = -2l_1 S_{ef}P_1 + 2l_2 S_{ef}P_2 - 2\left(c_{p1}l_1^2 + c_{p2}l_2^2\right)\varphi.$$

Structural diagram of gas pressure regulating system in active system RCS is given in figure 3.
When constructing mathematical model of distributing pneumatic devices (DPD), the equation for gas consumption via cylindrical throttle is of importance.

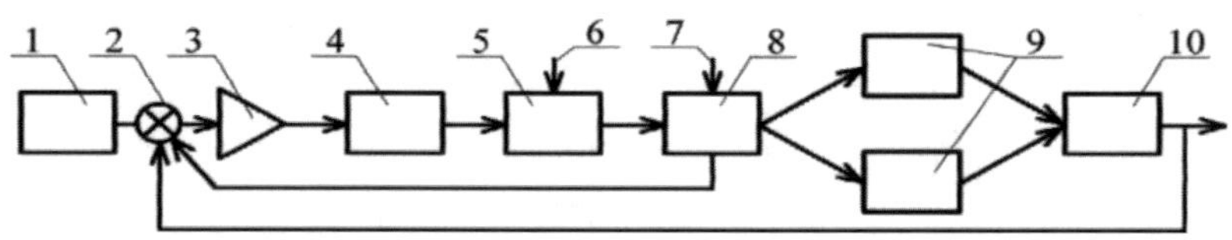

Figure 3. Structural diagram of the system for regulating the gas pressure in the RCS of the active system: 1 – speed sensor; 2 – sum block; 3 – amplifier; 4 – electromechanical converter; 5 – control cascade slide valve; 6 – gas supply line; 7 – fluid supply line; 8 – main slide valve; 9 – rubber-cord shells; 10 – control object

Passive vibration protection system is constructed on the basis of I-09 type RCS. The pressure in these RCS is set according to the load on the protected equipment plate and does not change in operation.
The filling time of I-09 type RCS in active vibration protection system in isothermal process can be determined by the expression:

$$t_н = \frac{\left(P_{\max} - P_{\min}\right)\cdot V_0}{10\alpha\cdot f\cdot P_н\sqrt{T}\cdot N},$$

где $\alpha = \sqrt{gRTN} \approx 20\sqrt{T}$ m/s is elastic wave propagation speed in medium; f is internal cross section area of the main distributor slide valve; T is absolute temperature; N is heat capacity ratio; V_0 is RCS volume in middle position; R is gas constant.
RCS drainage time under the same conditions is determined by the expression:

$$t_0 = -\ln\frac{P_{\min}}{P_{\max}}\cdot\frac{V_0}{0.58\cdot\alpha^2\cdot f}.$$

Allowing for $V_0 = 1 \cdot 10^{-3}$ m^3; $T = 298$ °K; $N = 1.4$; $f = 3.14 \cdot 10^{-4}$ m^2, RCS filling and drainage process control at small deviations is described by the following equations:

$$\Delta Q_2 = K_{Qx}\Delta x_2 - K_{Qp}\Delta p_н;$$

$$\Delta p_н S_{ef} - c_c\Delta z + c_c\Delta z_m = 0;$$

$$m\frac{d^2(\Delta z_m)}{dt^2} + k_{fr}\frac{d(\Delta z_m)}{dt} + (c_c + c_н)\Delta z_m = c_c\Delta z;$$

$$\Delta Q_2 = S_{ef}\frac{d(\Delta z)}{dt} + \frac{V_0}{2E'_ц}\frac{d(\Delta p_н)}{dt};$$

$$\Delta x_2 = K_{xh}\Delta h - K_{oc}\Delta z.$$

where z is RCS displacement; z_m is protected plate displacement; $\Delta p_н$ is pressure drop in RCS; $x_2 = x_{з0} + \Delta x_з$; $z_{m1} = z_{m0} + \Delta z_m$; ΔQ_2 is main cascade slide valve consumption alteration; x_2 is main cascade slide valve displacement; K_{Qp} is transmission gains; m is protected plate mass; K_{fr} is friction coefficient; c_c is contact stiffness; K_{fb} is feedback gain.

Assuming that forces created in active system RCS and acting on the protected platform are controlled by a two-stage electrohydraulic pneumatic distributor with main slide valve position feedback system, the schematic diagram of the electrohydraulic system can be represented as in figure 4.

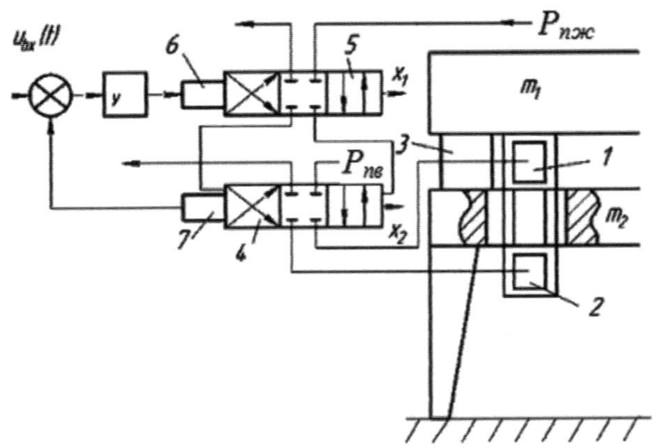

Figure 4. Schematic diagram of the pneumatic drive unit: 1 – upper RCS of active vibration protection system; 2 – lower RCS of active vibration protection system; 3 – RCS of passive vibration protection system; 4 – the distributor of the main (second) cascade; 5 – the distributor of the control (first) cascade; 6 – electromechanical converter (EMC) of the control valve cascade; 7 – feedback sensor of the slide valve position of the main cascade; x_1, x_2 – slide valve movement of the control and main cascade; P_n, P_a – pressure in the power line and atmospheric pressure

Перемещения управляющего золотника (золотника первого каскада) x_1 определяются уравнениями электромагнитного привода (ЭМП).

$$\left.\begin{aligned} &\tau\frac{m}{c}\frac{d^3x_1}{dt^3} + \left(\frac{m}{c} + \tau\frac{h}{c}\right)\frac{d^2x_1}{dt^2} + \left(\tau + \frac{h}{c}\right)\frac{dx_1}{dt} + x_1 = K_1U_1; \\ &U_1 = K_{amp}\left[U_{in}(t) - K_{fb2}x_2\right]. \end{aligned}\right\}$$

Where K_1 is EMD transmission gain; τ is EMD time constant; m is transducer moving members mass; h is viscous resistance coefficient; c is EMD anchor suspension stiffness; U_1 is voltage applied to EMD; K_{amp} is amplifier gain; K_{fb} is feedback gain.

The equation for the main slide valve motion takes the form of: $x_2 = K_{ny} \cdot x_1$,

where K_{ny} is gain ratio by slide valve displacement.

In adiabatic process, at heat capacity ratio $N = 1.4$, $\varepsilon_{кр} = 0.528$. Setting $P_n = 0.17$ MPa, $P_a = 0.1$ MPa, one may assume that maximum working pressure in RCS (P_{max}) should not exceed 0.15 MPa, while minimal one (P_{min}) should not exceed 0.1 MPa.

Passive system RCS stiffness is determined by the expression:

$$c_0 = \frac{TP_0S_{ef}^2}{V} + \frac{\partial S_{ef} P_0}{\partial z}.$$

Assuming in the first approximation that value $\frac{\partial S_{ef}}{\partial z}$ is small and RCS volume V is a linear function of displacement, RCS stiffness ratio can be estimated using the following expression:

$$c_0 = \frac{NP_0S_{ef}^2}{h_0},$$

where h_0 is RCS height in middle position; c_0 is passive system RCS stiffness ratio; P_0 is pressure in passive system RCS.

Therefore, in terms of the previously given assumptions, the above-mentioned equations determine longitudinal movement dynamics of a pneumatic-mechanical system with passive and active vibration protection system.

To analyze protected platform oscillations (acting on one support), let us take $m_1 = 80$ kg. I-09 type RCS was also chosen as active system spring elements.

Mathematical model was studied in two stages, and the investigation was made of:

- vertical oscillations of the platform with active system speed control;
- vertical oscillations of the platform with active system displacement control.

In simulation it was assumed that, with ideal measuring element for protected platform vertical displacements speed, gas pressure application to active system RCS is performed by a two-stage slide valve, it being immediately connected with the platform in displacement control.

Structural diagram of vibration protection system with linear servo system for the first stage (in MATLAB environment with Simulink) is given in figure 5.

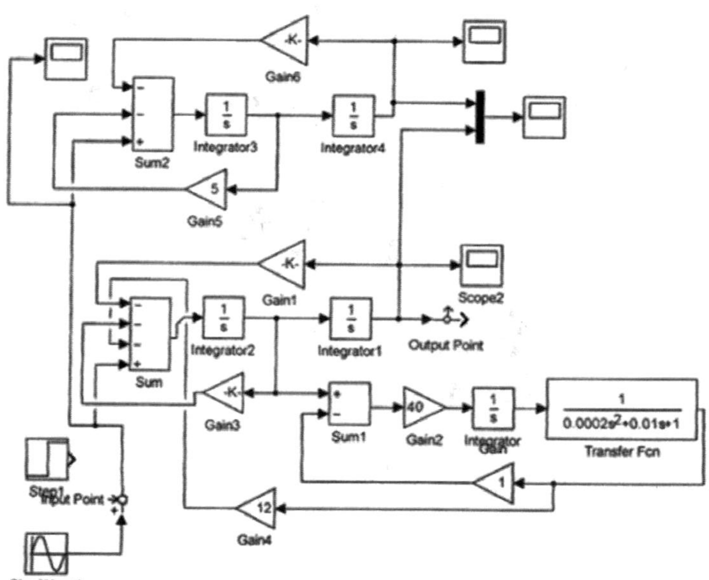

Figure 5. Structural diagram of vibration control system with speed control

Step-function response type at out-of-operation active system of protected platform vibration protection system (line 2) and at putting the active system in operation (line 1) is shown in figure 6.

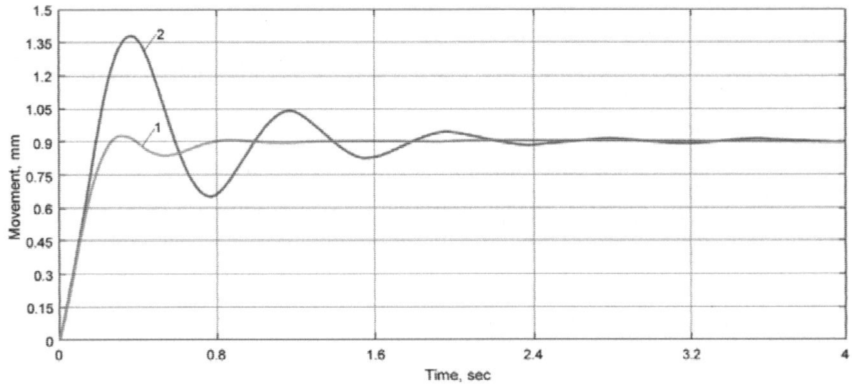

Figure 6. Graphs of step-function response: 1 – for a passive vibration protection system; 2 – when the passive and active vibration protection system works together

The simulation results of protected platform vertical oscillations depending on the disturbing oscillation frequency are presented in figure 7.

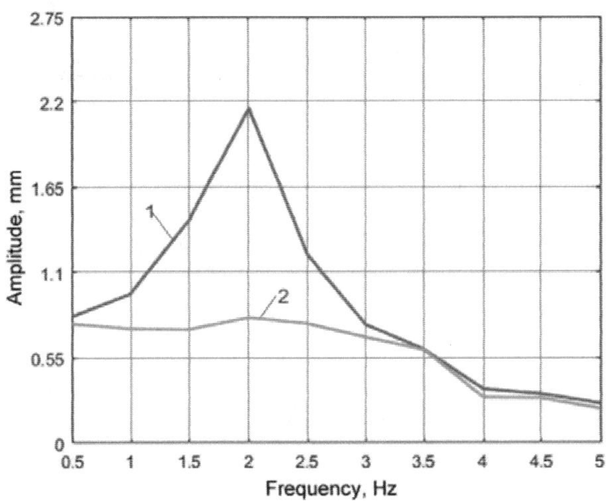

Figure 7. The results of the simulation of the combined vibration protection system: 1 – only the passive vibration protection system works; 2 – combined vibration protection system works

The obtained graphs allow for the following conclusions to be made. First, realization of the combined vibration protection system with displacements speed control decreases the amplitude of the protected platform forced oscillations. Second, and most importantly, this occurs at low frequencies, where standard and modified passive damping systems are inefficient.

Realization of the combined vibration protection system with displacements value control has almost no effect on the forced vertical oscillations amplitude of the protected platform.

4. Experimental results

The experimental results were obtained with the use of the physical model, that is experimental complex for investigating combined vibration protection systems, allowing for the proposed design of the combined vibration protection system to be realized and various control algorithms to be studied.

Field tests were aimed at evaluating operation efficiency of the combined vibration protection system.

When experiments were conducted, the variable was the actuating signal frequency of the vibrations initiator.

Moreover, linear displacements and table and platform displacements speed were measured.

IOP Conf. Series: Journal of Physics: Conf. Series **1050** (2018) 012081 doi:10.1088/1742-6596/1050/1/012081

5. Discussion of the results

The graphs of platform displacements gain frequency characteristic, obtained on the workbench and in the mathematical model, were compared, and the results are given in figure 8. The received results are comparable. The analysis of the oscillations amplitude changes by frequency demonstrates that the graphs type is similar except the region in the range of 3...4 Hz, where in full-scale study one can notice the resonance caused by RCS frequencies overlapping.

Then the operation of combined vibration protection system with working active and passive systems was analyzed.

The diagrams in figure 9 indicate the efficiency of platform oscillations suppression by combined vibration protection system in the whole given frequency range. However, the maximum is observed in the region of 2.15...2.3 Hz just as in case of tests only with the passive system. In addition resonance phenomena are not explicit.

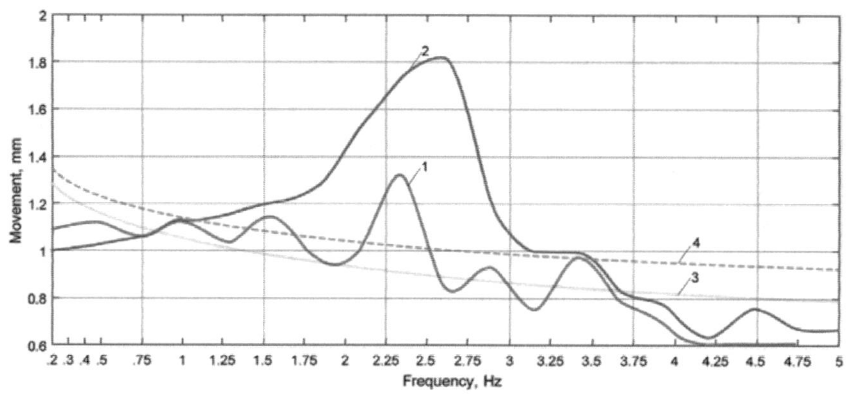

Figure 8. Comparison of gain frequency characteristic of the movements of the upper beam only with the passive system obtained on the experimental complex (curve 1) and gain frequency characteristic of the movements obtained on the mathematical model (curve 2); curve 3 and 4 are trend lines

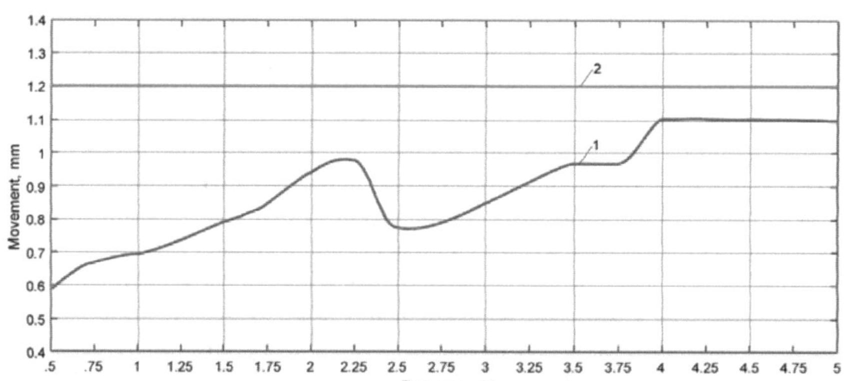

Figure 9. Gain frequency characteristic of movements of the upper beam with a combined vibration protection system (curve 1) and gain frequency characteristic of movements of the lower beam (curve 2)

To evaluate the adequacy of the mathematical model, the results of full-scale experiment and mathematical simulation were used. As the diagrams (figure 10) show, dynamic characteristics of combined vibration protection system and mathematical model correlate satisfactorily.

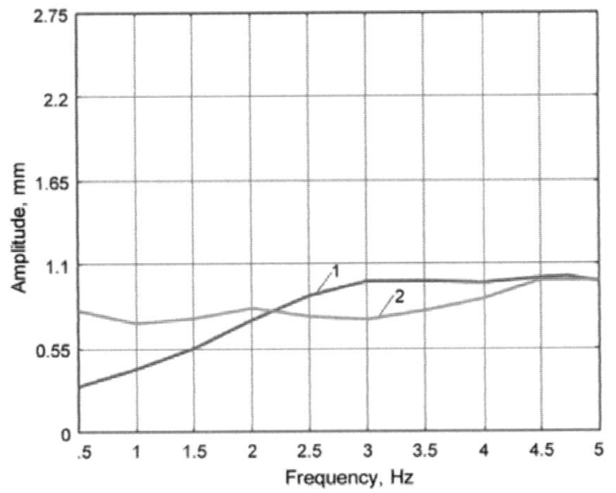

Figure 10. Gain frequency characteristic of oscillations of the upper beam with a combined vibration protection system obtained at the experimental complex (curve 1) and the results of mathematical modeling (curve 2)

By using linear integral estimation it was specified that results have a discrepancy of 7.4%. This proves that the experiment and mathematical simulation results are adequate.

6. Conclusions

The study of the developed mathematical model demonstrated that the combined damping system is efficient in the frequency range of 0.5...2.5 Hz following the effect of the pneumatic system operation speed. Consequently, active system must be switched off in the range higher than 2.5 Hz.

The examinations of the combined vibration protection system conducted on the experimental complex demonstrated that equipment vibration load level decreases by 40%. Amplitudes of the oscillations speed caused by singe impact are reduced by up to 21%. At periodical disturbance the combined vibration protection system efficiency reaches 47% compared to the passive system. The operation efficiency of the combined vibration protection system with pneumatic drive unit control by the plate displacement is no more than 14% in comparison with the passive system.

The obtained results provide us with new directions for investigating new control algorithms of active vibration protection systems and optimizing the existing ones.

References

[1] Adamo F, Attivissimo F and Lanzolla A M L 2014 Assessment of the Uncertainty in Human Exposure to Vibration: An Experimental Study *IEEE Sensors Journal* **14 (2)** pp 474–81

[2] Caterino N, Spizzuoco M and Occhiuzzi A 2016 Skyhook-based monitoring and control of a steel building under seismic action *Environmental, Energy, and Structural Monitoring Systems (EESMS), 2016 IEEE Workshop on* pp 1–6

[3] Gardonio P and Elliott S J 2005 Smart panels with velocity feedback control systems using triangularly shaped strain actuators *The Journal of the Acoustical Society of America* **117 (4)** pp 2046–64

[4] Herfst R, Dekker B, Witvoet G, Crowcombe W, de Lange D and Sadeghian H 2015 A miniaturized, high frequency mechanical scanner for high speed atomic force microscope using suspension on dynamically determined points *Rev. Sci. Intsrum.* **86 (11)** 113703

[5] Palacios-Quiñonero F and Karimi H R 2013 Passive-damping design for vibration control of large structures *Control and Automation (ICCA), 2013 10th IEEE International Conference on* pp 33–8

[6] Kodera N, Yamashita H and Andoa T 2005 Active damping of the scanner for high-speed atomic force microscopy *Review of scientific instruments* **76** 053708

[7] Tsuruta K, Ojiro T, Ushimi N and Shibata H 2013 Vibration Suppression Control for Multiaxis Table Drive System *Mechatronics and Automation (ICMA), IEEE International Conference on* pp 1135–40

[8] Droppa P, Kalna P and Filípek S 2015 Application diagnostics methods for modernization vehicle IFV-2 *Military Technologies (ICMT), 2015 International Conference on* pp 1–5
[9] Dimirovski G M Vuk and Georgi 2013 An adventure into active systems via mechatronics, robotics and manufacturing engineering *Intelligent Systems and Informatics (SISY), 2013 IEEE 11th International Symposium on* pp 11–9
[10] Capatti M C, Carbonari S, Gara F, Roia D and Dezi F 2016 Experimental study on instrumented micropiles *Environmental, Energy, and Structural Monitoring Systems (EESMS), 2016 IEEE Workshop on* pp 1–6
[11] Bian J and Jing X 2014 Biomimetic design of woodpecker for shock and vibration protection *Robotics and Biomimetics (ROBIO), 2014 IEEE International Conference on* pp 2238–43
[12] Paulitsch C, Gardonio P and Elliott S J 2006 Active vibration control using an inertial actuator with internal damping *The Journal of the Acoustical Society of America* **119 (4)** pp 2131–40
[13] Pat. 159456 RF, IPC F 16 F 9/05. Kombinirovannaya vibrozashchitnaya sistema [Combined vibration protection system] / Buriyan Yu. A., Sorokin V. N., Zelov A. F., Kondyurin A. Yu. №2015123195/05; Publ. 10.02.2016, Bul. No 4
[14] Buriyan Yu A, Sorokin V N, Zakharenkov N V and Zelov A F 2016 Stand for the study of combined vibration protection system *Vestnik SibADI* **6 (52)** pp 13–9 (In Russ.)

Biometric pattern recognition using wide networks of gravity proximity measures

A E Sulavko, S S Zhumazhanova

Omsk State Technical University, 11, Mira, ave., 644050, Omsk, Russia
E-mail: sulavich@mail.ru

Abstract. In this article a new interpretation of the pattern recognition problem is proposed, in accordance with the problem the own domain of pattern (image) classes are gravitational fields generated by immovable objects (classes of patterns) that propagate in the feature space. A measure of proximity is proposed for calculating the 'attractive force' of the pattern to existing templates in the feature space, which can work with both strongly dependent features and with independent ones. A network of gravitational proximity measures was configured to recognize a person by the characteristics of voice and handwritten passwords reproducing. Comparatively low error rates are obtained.
Keywords. Quadratic forms, attraction of a pattern, curvature of feature space, correlation dependence of dynamic biometric parameters.

1. Introduction

The problem of confidential information and computer resources protection from unauthorized access becomes more actual than ever. Mechatronic, computer, and robotic systems have to be protected. The classic way to solve the problem is to protect the data with encryption and authentication based on passwords and hardware keys. However, these authenticators are alienated from their owner, the protected information (system) can be compromised in the case it getting to third parties.

This problem is being solved with the help of key and passwords binding to the biometric parameters of a person. In Russia within the national technology initiative (NTI) program, a 'road map' concept is being projected for the development of the SafeNet safe and protected computer technologies market. Key segments of the market, as well as current scientific and technical tasks and pilot projects have already been identified. They include: the introduction for the first time in the world of a national biometric authentication platform, an electronic signature with biometric activation, multimodal biometrics for a personal data management system, etc.

Most of the current commercial products for biometric protection are based on static biometric patterns (three-dimensional models of face and skull, fingerprints, iris patterns, etc.). The fundamental problem lying in the use of open biometric patterns is that they are not secret and can be falsified (by creating a fake finger, using high-quality photography, etc.).

To date, active studies of dynamic biometric features [1] are being conducted to solve similar problems (authentication by handwriting and keystroke dynamics, voice). Dynamic biometric patterns can be secret, however, as many studies have shown, these patterns consist mainly of low informative and interdependent features (biometric parameters, quantity characterizing subjects). This circumstance makes it very difficult to develop procedures for highly reliable authentication in the space of dynamic biometric features. The present study is aimed at finding effective proximity measures for pattern recognition in feature space with different mutual correlation dependencies.

2. Problem statement

The classical statement of the pattern recognition problem (its geometric interpretation) leads to the construction of separating hyperplanes in hyperspace. Each dimension of this space is a range of values of a certain feature. If the pattern classes are considered inseparable in the original feature space, then it is necessary to go to a space of higher dimension for the construction of hyperplanes. In this case, the derived (derivatived) hyperspace is sometimes called rectifying.

In the process of learning the recognition system by learning examples of the pattern, the example's template is being created (we introduce this term, replacing the notion of the pattern class). For the purpose, each presented example of the pattern is transformed into a vector of features, i.e. to a point with coordinates equal to the values of the corresponding features. The template is a cluster of such points. Next, hyperplanes are constructed that separate their own domains (scope) of templates. At the stage of identification (or verification) of the pattern, a vector of values of similar features is assigned to the pattern and it is required to determine which new point belongs to its own domain. The existing variety of methods and approaches to pattern recognition allows to solve this problem in some ways with some error.

If the feature space is 'plane', i.e. n-dimensional Euclidean space, then the shortest path from the pattern to the center of the template can be found using the Euclidean measure (1). If the ranges of values of particular features vary significantly in scale, the space ceases to be 'ideally plane', but it can be normalized by calculating distances with the help of the Pearson's measure (2) [2].

$$e = \sqrt{\sum_{j=1}^{N} (m_j - a_j)^2}, \qquad (1)$$

$$\Pi = \sqrt{\sum_{j=1}^{N} \frac{(m_j - a_j)^2}{\sigma_j^2}}, \qquad (2)$$

where a_j is the value of the *j-th* feature, N is the number of features processed by the functional (the dimension of the functional), m_j and σ_j are the mathematical expectation and the standard deviation of the *j-th* feature values calculated from the training sample data. In the problems of biometric authentication (identification), such an idealized case is rare one, because most of the features are correlated differently. Correlation 'bends' the features space in the neighborhood of patterns templates. The force of 'bending' in each spatial dimension can be different, as in the template neighborhood (Figure 1). This is explained by the different nature of the correlation dependence between the biometric features of each person. In order to take into account the heterogeneity of the curvature/bending of the feature space in the theory a quadratic form of (3), also called the Mahalonobis measure, can be used.

$$y = (\overline{m} - \overline{a})^T \cdot [R]^{-1} \cdot (\overline{m} - \overline{a}), \qquad (3)$$

where $\bar{a}$ is the vector of features values in the normalized coordinate system, $[R]$ is the matrix of pair correlation coefficients between the features. The use of this quadratic form is complicated by the fact that it is required to perform the inversion of the correlation matrix $[R]$, which is not possible in practice. The point is that this operation is ill-conditioned and performed with significant errors [3]. An attempt of such treatment leads to a problem called the 'dimensional curse' [4], at least in the problems of dynamic biometric patterns recognition on training samples in dozens of examples.

The methods for networks configuration of quadratic forms (3) with perceptron training algorithms [5] do not solve the problem of processing highly dependent features.

For these reasons, proximity measures, which are quadratic forms, are focused on processing only independent and weakly dependent data.

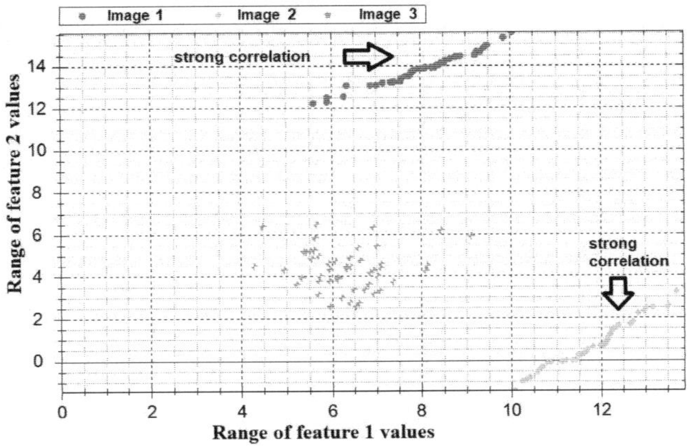

Figure 1. 'Bending' of the two feature space under influence of correlation.

A fundamental scientific problem is the search for effective methods for assessing the closeness of patterns to templates in the space of features, curved by the correlation between them.

3. Theory

A. Proximity measures oriented to processing of dependent features

The results of recent studies indicate that there are functionals (proximity measures) that can take fewer wrong decisions if the correlation between the features is high. The first group of such proximity measures was called multidimensional Bayesian functionals [3, 6, 7, 8]. They evaluate not only the proximity of the pattern to the template, but also the dependence of a certain feature of the pattern on several other features. Besides functionals of the greatest likelihood (Bayesian formulas) there are multidimensional correlation [6], difference [3], Bayesian hyperbolic functionals [7]). The greater value of N and higher the coefficient of equal correlation of features, the more effectively the Bayesian functional works [8].

After the multidimensional Bayesian functionals were proposed, it became clear that the strength and direction of the correlation relationship between the features is a special kind of information about the pattern. Later, other groups of functionals were proposed that could work with dependent data. However, all of them are ineffective working with independent data. Therefore, they should be combined into a network consisting of various measures of proximity, including from quadratic forms.

B. Wide hybrid neural networks

In biometric applications, various methods and approaches to pattern recognition are used: fuzzy extractors [9], artificial neural networks (ANN) [10], artificial immune systems (AIS), methods of mathematical statistics and probability theory [11], k-nearest neighbors method [12], support vector machine [12], and etc. Each of them operates with certain computational elements (CE) and is a variant of generalization of a number of methods, defining the principle of operation and interaction of CE, as well as the general structure of the pattern recognition method. For example, different algorithms of noise-immune coding can be used in fuzzy extractors, in ANN - neurons with different activation functions, in AIS - lymphocytes and antibodies, in the methods of statistics, probability theory and k-nearest neighbors, various measures (proximity metrics) are used to calculate distances in space features (Euclid, Pearson, Mahalonobis, etc.). All approaches have advantages and disadvantages. However, the authors of this article adhere to the ideology of configuration of the so-called 'wide' ANN [13] with a small number of layers based on neurons, which are based on various functionals (analogues of proximity measures). This approach is used in applications where the volume of the training sample is substantially limited. In practice, it is usually required that the biometric authentication system is guaranteed to be trained on 10-20 examples of the user's pattern.

IOP Conf. Series: Journal of Physics: Conf. Series **1050** (2018) 012082 doi:10.1088/1742-6596/1050/1/012082

The training of the 'wide' ANN is layer-by-layer and is carried out without using the backpropagation method (the input of the next layer receives the data of the training sample enriched with the previous layer after its preliminary configuration). Each neuron is trained independently of the rest, weights (or other parameters of the neurons) are calculated determinely, based on the distribution parameters of the characteristic values. Rejection of iterative training procedures allows to remove many disadvantages inherent in multi-layer ('deep') ANN, the most significant of which for biometric applications are the following:

1. to learn ANN classical algorithms in the presence of several layers requires a large training sample (hundreds and thousands of examples).
2. Iterative learning algorithms lose stability when the ANN structure becomes more complex. In particular, it is not possible to configure neurons with a large number of inputs (the 'blindness' effect of the training automaton occurs) [14].
3. The process of learning a multilayer network by an iterative algorithm can be unlimitedly long.
4. The classic ANN configuration algorithms are characterized by the problem of 'retraining'.

It is also has to be noted that the backpropagation method has an exponential computational complexity that does not allow it to be implemented on a weak processor without remote connection to the server. A secret biometric pattern should not be sent over the network, thus it is possible to violate the requirement of Russian standard GOST R 52633.0 not to compromise the biometric template.

All these problems do not apply to 'wide' ANN.

The structure of the artificial neuron of the 'wide' network of this type generally corresponds to the prevailing ideas about their construction. First, the data processes the neuron's functional, then its value is sent to the activation function. However, each neuron can contain several functionals and activation functions and, accordingly, have several outputs (Figure 2). For comparison, in the classical perceptron there is one activation function and one functional, performing a weighted summation of features values. These functionals lose stability processing highly dependent features.

The ability to change the functional of a neuron increases the potential of the ANN in terms of the reliability of the decisions made and generates new benefits. One of them is the ability to process data with varying level of correlations. Commonly, dependent features are tried to be eliminated, and the dimensionality of the feature space is reduced (applying the principal component method, etc.). However, the hybrid 'wide' ANNs is focused on processing all available features, regardless of degree of their mutual correlations. Neurons can consist of different functionals, each of which can give the wrong decision in some cases, but these errors are low-correlated, so their collective solution will be much more powerful. Together they form an effective tool for data enrichment, compensating for the lack of information on the pattern using many ways to process it. The functionals and the number of neuron inputs, as well as the synapses, are selected automatically at the learning stage, depending on mutual correlation and features' informativeness (each neuron must process features that have a comparable amount of information about patterns [10]). Evaluation of information and correlation between the signs is based on the training sample, the volume of which is very limited. At the same time, the increase in the number of features and neurons does not lead to an increase in the required volume of the training sample.

In the simplest case, only one layer of neurons can be configured. This is often enough for most pattern recognition tasks. The increase of the layers number requires justification. A reasonable explanation in this case may be the use of another principle of enriching the data entering the next layer of the network, to take into account additional information about the pattern.

IOP Conf. Series: Journal of Physics: Conf. Series **1050** (2018) 012082 doi:10.1088/1742-6596/1050/1/012082

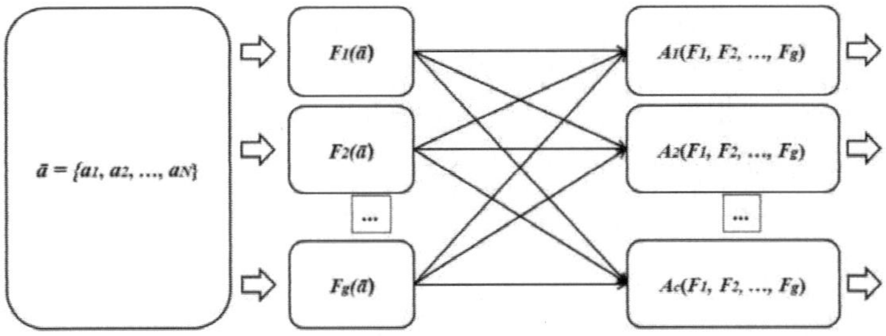

Figure 2. Configuration of a neuron with several functionals and activation functions in the 'wide' ANN.

Thus, the 'wide' ANN considered by the authors are flexible and refer to hybrid neural network algorithms. Each neuron processes the values of those features of the pattern that are best suited for its functionality.

However, this ANN class has its own problems. One of them is the low accuracy of calculating the correlation coefficients and estimates of the features informativeness on small training samples. This makes it difficult to configure many neurons (although it does not reduce the possibility of their training to none).

For the development of 'wide' ANN, it is very important to search for universal functionals that are able to work efficiently both with highly dependent data, and with independent ones, as well as with data of any informativeness. This will make it possible to abandon the assessment of information content and the correlation of features at the training stage.

C. The 'attraction' method of recognizable patterns

The curvature of the feature space by means of a correlation between them seems to be how the real space-time is curved under the influence of gravity. The deformation of space, connected with the presence of mass-energy, was described in the general theory of relativity, proposed by Einstein in 1915. We will not describe the theory in details, but merely note that when patterns are recognized, the effect of the feature space squeezing in the neighborhood of the pattern template is also observed, if the features are dependent. However, in this case, the space is compressed perpendicularly with respect to the direction of the features dependence, i.e. the feature space 'contracts on the sides' (Figure 3), while the gravitational forces cause space-time in the real world to be curved 'around' an object that has a mass (Figure 4).

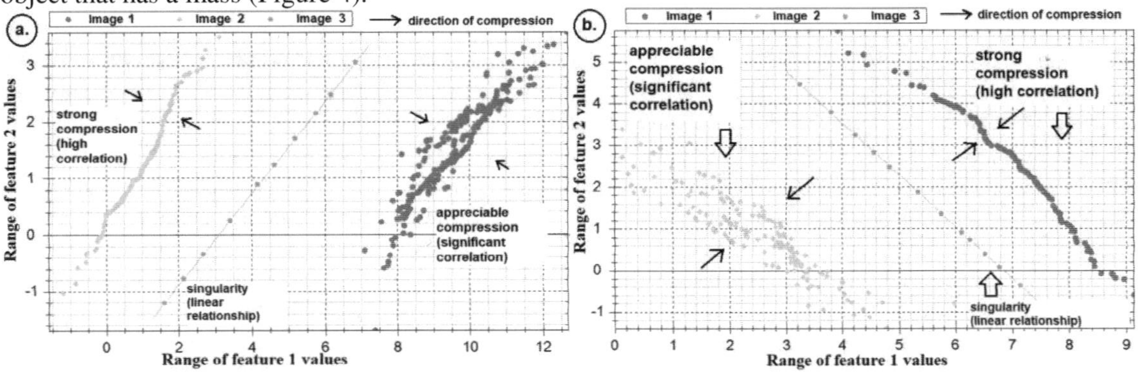

Figure 3. The direction of the space squeezing of two generated features: (a) if there is a positive correlation between them, (b) if there is a negative correlation between them.

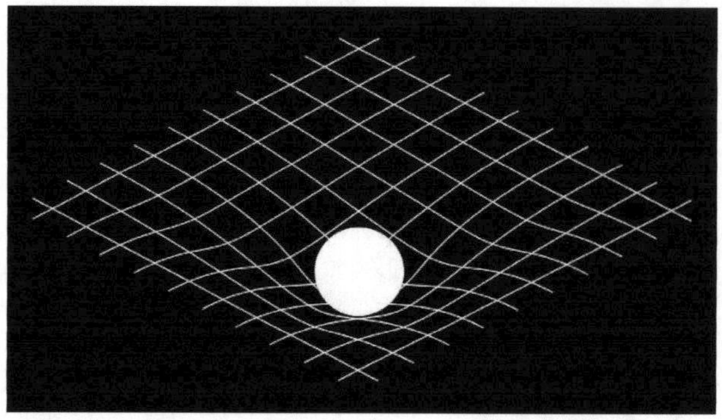

Figure 4. The curvature of the real world space by an object that has mass.

The effect that superdense matter has on space-time (the so-called 'black hole') is widely known. Under the influence of 'black holes', space and time converge to one point, called a singularity. At this point, the usual laws of physics are violated, the space-time dimensions cease to exist (they 'fold' to zero size).

The analogue of the singularity in the two-dimensional feature space is the eigenvalue of the object's template, for which both characteristics are linearly dependent (Figure 3). In fact, in this case, two characteristics turn into one (a singular feature), and the two-dimensional space is transformed into a one-dimensional one. In a multidimensional space for a template, several independent singular signs, each of which can be formed from several linearly dependent features, can exist. For example, in three-dimensional space with two linearly dependent features the template's own scope is converted into a plane, if there are three ones - in a line (in this case one singular feature). Thus, a singularity in feature space can extend to separate dimension and curvature of space under influence of the correlation between variables is heterogeneous and uneven.

Naturally, the described analogy is very conditional, but it makes us to reconsider the classical formulation of the pattern recognition problem. It can be said that the correlation links between the characteristics of templates impart some additional density to it (they increase the template's mass). Suppose that the accumulation of points belonging to the template's own scope is a gravitational field that is generated by a stationary object and propagates in the feature space unevenly in all directions. Then the task of identifying the pattern given by a point in space is reduced to calculating the trajectory of its continuous motion until it reaches the center of one of the templates. We call this approach to pattern recognition by the 'attraction' method.

The idea of using gravity theories in pattern recognition issues is not new. In [15], an edge detector method based on the law of universal gravity is proposed. To construct an edge detector it is assumed that each point of the pattern is a celestial body that has some connection with other points of the pattern within its neighborhood with the help of gravitational forces. For points outside the specified region, all gravitational forces are assumed to be zero. For each point of the pattern, the magnitude and direction of the vector sum of all the gravitational forces that the point exerts on its neighborhood, transmit important information about the edge structure, including magnitude and direction. The algorithm proposed in [15] consists of three stages:

1. The gravitational forces of a pixel exert on every other pixels around it are computed using the law of universal gravity.
2. The vector sum of all gravitational forces is calculated.
3. The magnitude and direction of the vector are used to detect pattern edges.

Gravity not only can distort space-time, but can also be used for data clustering. Samples in each training set can be considered as particles, and if each particle is assigned a mass value, it is possible to form a new equation of attraction. Based on the results of the studies, the edge detector based on the universal gravity theory proved to be more resistant to noise than the methods of Sobel, LOG and Canny [15].

Gravitational search algorithm (GSA) is applied in optimization problems [16]. GSA is a heuristic optimization method. It is proved that this algorithm has a good ability to search for the global optimum, but at the last iteration has a low speed. To address this disadvantage, a hybrid optimization method based on the Particle Swarm Optimization method and GSA was proposed in [17], which was used to configure Feedforward Neural Networks.

In [18] using GSA the search was made to optimize the architecture of the modular neural network (the number of modules, layers and ANN's nodes). Further, ANN, configured using GSA, was used to recognize medical patterns of echocardiograms in order to identify healthy patients. In [19] the advantages of using fuzzy gravitational search in optimization of modular neural networks are shown. The proposed approach provides high reliability of echocardiogram recognition (the number of erroneous decisions was 0.51%).

Algorithm for data clustering (pattern recognition without a teacher), based on Newton's law of gravity is provided in [20]. In the proposed method, the points (samples, patterns) of data and the centers of clusters (templates) are considered fixed and mobile celestial objects, respectively. Fixed objects apply gravity to movable objects and change their position in feature space, and therefore the best positions of cluster centers are determined by applying the law of gravity. To evaluate the effectiveness of the proposed algorithm, a comparative experimental study is performed with some known clustering algorithms using three visual data sets. The experimental results confirm the effectiveness of the proposed algorithm.

In the present work, the described idea is formulated for the first time taking into account the influence of the correlation between features on the properties of the own scope of the template and, thus, an attempt is made to take into account information on the correlation of features for pattern recognition.

D. Gravitational measure of intimacy

Nature has always been an inspiration source for researchers. A large number of methods and algorithms for solving intellectual problems are suggested trying to copy processes from nature: ANN, AIS, genetic algorithms, etc. As a rule, these methods have a number of significant simplifications that, as a result, have little in common between them and the natural prototype. The method of 'attraction' isn't exception. Modeling the process of pattern 'attraction' to the templates can be resource-intensive. Therefore, within the work of this paper, we simplify the problem: assume that the pattern is attracted instantly. Also in the real world, the gravitational attraction is directed in all directions (i.e., in the direction of curvature), in the feature space, the attraction will be stronger in the direction opposite to compression.

A measure (4), that calculates the force of attraction to the template acting on the recognizable pattern, is proposed. Name (4) as the gravitational measure of proximity. The measure does not take into account the gravitational constant and the mass (or attraction force) of the recognized pattern, in this case they are constants that do not affect the recognition result.

$$G_{trig}^2 = \frac{\sum_{j=2}^{N}\left[\sum_{t=1}^{j-1}\left(trig(j,t) = \begin{cases} \cos(\bar{a},\bar{\upsilon}_{j,t})\cdot r_{j,t}, if\ r_{j,t} > 0 \\ -\sin(\bar{a},\hat{v}_{j,t})\cdot r_{j,t}, if\ r_{j,t} < 0 \end{cases}\right)^2\right]}{\Pi(N) = \sqrt{\sum_{j=1}^{N}\frac{(m_j - a_j)^2}{\sigma_j^{\ 2}}}}, \quad (4)$$

where ā is a vector of features' values, $\bar{\upsilon}_{j,t}$ is a vector with coordinates under the numbers j and t, equal to 1, and the remaining coordinates equal to 0, $r_{j,t}$ is the correlation coefficient between the indices j and t, calculated from the training sample data, *trig (j, t)* is the evaluation functional of the attractive force of the template in 2 spatial dimensions (the order of G is 2).The physical meaning of the functional *trig(j,t)* is to increase the attraction more, the closer the location of the recognized pattern to the 'compression line' of space. In fact, this functional takes into account additional information about the position of the pattern and the correlation between the features. The total attraction force is equal to the ratio of the sum of the squares of the component forces in all two-dimensional subspaces of the original feature space to the value of the Pearson's measure. The Pearson's measure plays the role of distance in a normalized coordinate system. Based on a series of computational experiments on pattern recognition, the following empirical formulas for calculating gravitational metrics are also proposed:

$$G_1^2 = \frac{\sum_{j=2}^{N}\left(\sum_{t=1}^{j-1}\cos(\overline{a},\overline{\upsilon}_{j,t})\right)}{\Pi(N)=\sqrt{\sum_{j=1}^{N}\frac{(m_j-a_j)^2}{\sigma_j^{\,2}}}}, r_{j,t}>0 \quad (5)$$

$$G_2^2 = \frac{\sum_{j=2}^{N}\left(\sum_{t=1}^{j-1}\cos(\overline{a},\overline{\upsilon}_{j,t})^2\right)}{\Pi(N)=\sqrt{\sum_{j=1}^{N}\frac{(m_j-a_j)^2}{\sigma_j^{\,2}}}}, r_{j,t}>0 \quad (6)$$

$$G_3^2 = \frac{\sum_{j=2}^{N}\left(\sum_{t=1}^{j-1}(1+\cos(\overline{a},\overline{\upsilon}_{j,t}))\cdot(1-\sin(\overline{a},\overline{\upsilon}_{j,t}))\right)}{\Pi(N)=\sqrt{\sum_{j=1}^{N}\frac{(m_j-a_j)^2}{\sigma_j^{\,2}}}} \quad (7)$$

$$G_4^2 = \frac{\sum_{j=2}^{N}\left(\sum_{t=1}^{j-1}|\cos(\overline{a},\overline{\upsilon}_{j,t})|\cdot|\sin(\overline{a},\overline{\upsilon}_{j,t})|\right)}{\Pi(N)=\sqrt{\sum_{j=1}^{N}\frac{(m_j-a_j)^2}{\sigma_j^{\,2}}}} \quad (8)$$

$$G_5^2 = \frac{\sum_{j=2}^{N}\left(\sum_{t=1}^{j-1}\left(|\cos(\overline{a},\overline{\upsilon}_{j,t})|\cdot|\sin(\overline{a},\overline{\upsilon}_{j,t})|\right)^2\right)}{\Pi(N)=\sqrt{\sum_{j=1}^{N}\frac{(m_j-a_j)^2}{\sigma_j^{\,2}}}} \quad (9)$$

As seen, the above formulas (5) - (9) do not contain computational operations related to the correlation coefficients. The attraction force of the pattern and template is reduced in proportion to the distance, rather than the square of the distance between them (as the law of universal gravity says). Normalizing the force with respect to the square of the distance produces lower results. Newton's law (and other theories of gravitation) help to find an analogy with the problem being solved, but they do not hold for objects which the theory of pattern recognition deals with.

4. Experimental results

An experiment was carried out to recognize patterns in the space of dependent and independent features with different informativeness. Informativeness is the main indicator of the quality of a feature, which is numerically estimated through the intersection area of the feature's probability density function characterizing the recognizable patterns. The informativeness of feature I can be measured in bits or using a relative scale from 0 to 1 (the smaller the I, the more informative the feature). More details on the evaluation of features' informativeness can be found in one of the papers [10, 21, 22]. Patterns and features were generated using the Monte Carlo method for the parameters of the normal distribution law (the most common case for dynamic biometric features). The parameters for generation were chosen so that features were of low-informative ($0.6<I<0.8$). The features values were generated as independent values, then for correlated features a dependence was created, for this value of each feature for each pattern ranged in ascending order (the order of values for each feature varied from the lowest to the highest one). The normal distribution law is the most common for dynamic biometric features.

The reliability of the pattern recognition procedure (as for any biometric authentication) is determined by the probability (or percentage) of errors of the 1st and 2nd kind - False Rejection Rate (FRR) and False Acceptance Rate (FAR), respectively.

In Figure 5, it's shown a comparative evaluation of the effectiveness of the Pearson's measure application (2) and the functional (4) for recognizing the generated patterns in the feature space.

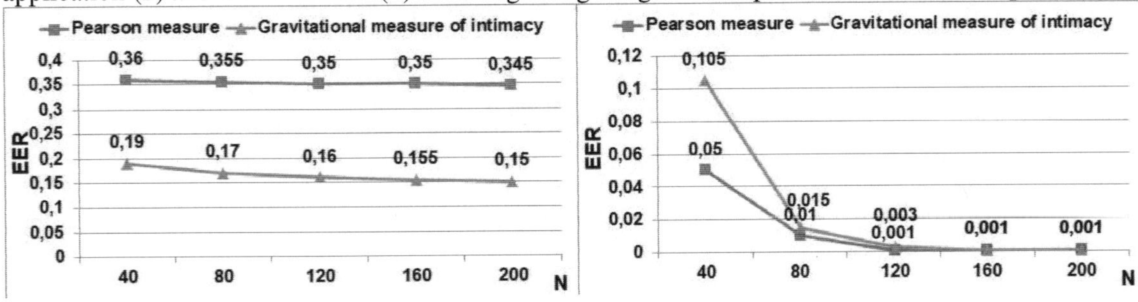

Figure 5. Probability of pattern recognition errors in the space of dependent (above) and independent (below) low-informative features (for I≈0.7).

An experiment was performed comparing functionals of wide networks of neurons formed on considered proximity measures, in the problem of recognizing subjects by the peculiarities of reproduction of handwritten and voice passwords. In the experiment, 90 people took part, each of which during a month entered a secret handwritten pattern using a Wacom graphic tablet (the pen's coordinates and the force of pressing the pen on the tablet were registered) and a voice pattern using a Sony F-V120 microphone. Each week, the subject made at least 20 attempts to introduce each handwritten and voice pattern. As a result, more than 100 biometric samples of voice and handwritten passwords were received from each subject. All samples were subjected to statistical processing to extract the features from them. The methods from [9] were used for analysis of handwritten patterns and for voice patterns – from [22].

After calculating the vectors of features' values from each entered pattern, the formation and training of the ANN on the basis of Pearson's measures and gravitational metrics was carried out. For each subject, a unique network was formed, consisting of two segments (subnets): based on the Pearson's functional and gravitational metrics (Figure 6). The first subnet treated only combinations of weakly dependent or independent features, the second one - combinations of features, the correlation coefficient between which exceeded 0.3. Independent features were divided into categories of informativeness ($0<I<0.9$ for $\Delta I = 0.1$, total 9 categories), the less informative the features are, the more inputs were created for the neuron that processed them. At $I<0.1$, there were 3 inputs of the neuron ($N_1 = 3$), for $0.1 <I<0.2$ - $N_2 = 4$, at $0.2<I<0.3$ - $N_3 = 6$ inputs, and so on. Going to a category with more informative features, the number of neuron's inputs increased ΔN times. It is empirically established that for the types of patterns used, Pearson's network shows good results at $\Delta N = 1.5$. These optimal parameters (N_1, ΔN) were chosen empirically in the process of the experiment.

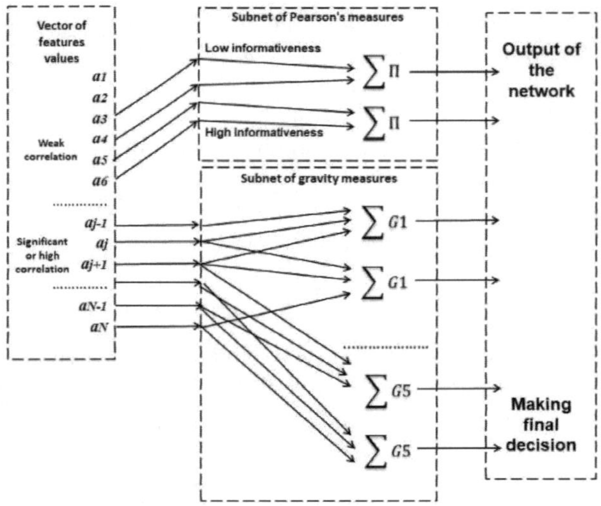

Figure 6. Scheme of a single-layer hybrid 'wide' network based on the Pearson's measure and gravitational metrics.

The subnetwork of gravity proximity measures was formed for processing only those combinations of features, the correlation dependence between which is essential (more then 0.5). In this case, the dimension of the functionals was fixed without considering the informativeness, since dependent features have close informativeness value. The optimal number of inputs of the 'gravitational neuron' (N=20) was chosen empirically in the process of the experiment.

Learning is understood as the computation of mathematical expectations and standard deviations of each feature's value of the subject. 20 biometric samples of the subject were used for training. The remaining samples were used to assess the reliability of subject recognition. As a result (Figure 7), the probabilities of recognition errors of subjects according to the peculiarities of handwritten and voice patterns reproduction were as follows:

– Pearson's subnet: voice – EER=10.2%, handwriting – EER=8%;
– subnet of gravity proximity measures: voice – EER=9,5%, handwriting – EER=6,9%;
– hybrid INS: voice – EER=7.1%, handwriting – EER=6.8%.

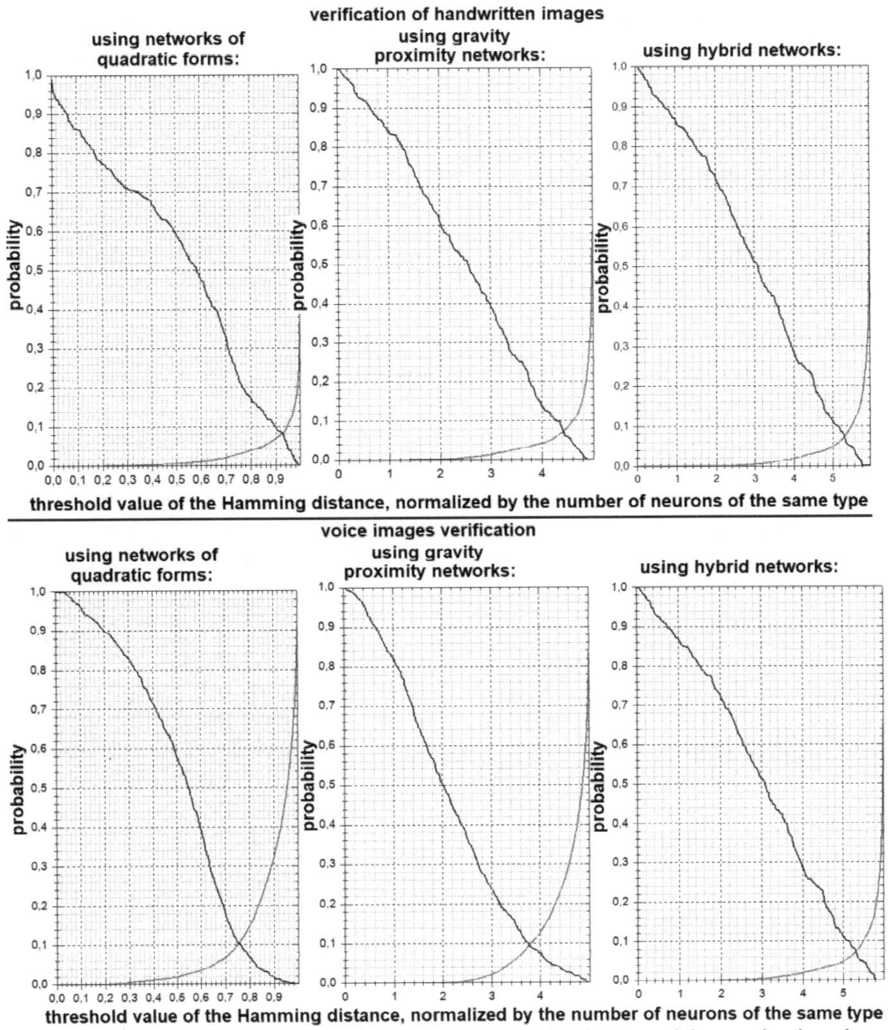

Figure 7. Probabilities of errors of the 1st and 2nd kind of recognition of subjects by handwritten images and voice messages, depending on the threshold value of the Hamming measure from the code generated by the network to the correct one

The positive effect of the integration of quadratic forms with gravitational metrics is most noticeable when recognizing voice images (Figure 7).

5. Discussion of results

As seen from Figure 5, the proximity measure (4) shows similar results with the Pearson's measure (2), if the features are independent, but only for high dimensionality of N. A similar case is observed with respect to the other proposed gravitational metrics. All the proposed proximity measures (4) - (9) give similar indicators of EER in the presence of a positive correlation between the features. However, their error decisions are not strongly correlated, which indicates the possibility of their joint use. When recognizing patterns in the space of correlated features, the proposed gravitational proximity measure shows better results than the Pearson's measure (and Euclid's one respectively).

It is established that the distribution law of the calculated values of the functionals (4)-(9) with received 'Own' input data (corresponding to the subject to whom the template belongs) have a distribution close to lognormal. In the case of the 'Stranger' input data (data from unregistered entities), the values of the functionals (4)-(9) can be described by a normal law (with some reservations). The hypothesis of the distribution law for these quantities was verified by the Chi-square method. Thus, the probability density of the values of gravitational functionals can be used as additional information in the recognition of the pattern. Further enrichment of the input data can be performed by neurons of subsequent layers.

6. Conclusion

A new interpretation of the problem of pattern recognition is proposed, according to which the own scopes of the pattern classes are gravitational fields that are generated by a stationary object and propagate in the feature space unevenly in all directions. Identification of the pattern given by a point in space is reduced to calculating the trajectory of its continuous motion until it reaches the center of one of the templates (classes of patterns).
The measures of proximity are proposed to calculate the 'attractive force' of the pattern to the existing templates in the feature space. This measures of closeness show similar results with measures that are oriented toward the absence of correlation between the features (providing the availability of a large number of processed features), but surpass them when recognizing patterns in the space of correlated features.
Different 'gravitational' functionals can be a big variety, and they can complement each other since their solutions are not completely correlated (at least, it is observed for the proposed metrics). On the basis of each such functional, neurons can be configured for a 'wide' ANN. An important property of 'gravitational' functionals is that they work quite efficiently both in the space of independent features and with strongly correlated ones. In the provided work, one of the possible variants of configuration of hybrid networks of such functionals to solve problems of subjects recognition of by voice and handwritten passwords is proposed. As a result, it was possible to achieve the following indicators of error decisions: voice – EER=7.1%, handwriting – EER=6.8%. These indicators are much lower (by 10-30%), if the system configure individually for each user (estimates of Figure 7 are obtained when setting up networks for the "average user").

References

[1] Boriev Z, Sokolov S, Nyrkov A, Nekrasova A 2016 Mathematical and information maintenance of biometric systems *IOP Conference Series: Materials Science and Engineering* (Saint-Petersburg, Russia) pp 1–4
[2] Sulavko A E, Fedotov A A, Eremenko A V 2017 Users' identification through keystroke dynamics based on vibration parameters and keyboard pressure *Proc. of XI Int. IEEE Scientific and Technical Conf. 'Dynamics of Systems, Mechanisms and Machines' (Dynamics)* (Omsk, Russia, 14–16 Nov. 2017) pp 1–7
[3] Ivanov A I, Lozhnikov P S, Serikova Yu I 2016 Reducing the Size of a Sample Sufficient for Learning Due to the Symmetrization of Correlation Relationships Between Biometric Data *Cybernetics and Systems Analysis* vol 52(3) pp 379–385
[4] Ivanov A I, Lozhnikov P S, Kachajkin E I 2015 Identifikatsiya podlinnosti rukopisnykh avtografov setyami Bayyesa-Khemminga i setyami kvadratichnykh form [Identification of the authenticity of hand-written autographs by Bayes-Hamming networks and networks of quadratic forms] (in Russian) *Voprosy zashchity informatsii* **2** pp 28–34
[5] Lozhnikov P S, Sulavko A E 2017 Usage of quadratic form networks for users' recognition by dynamic biometric patterns *Proc. of XI Int. IEEE Scientific and Technical Conf. 'Dynamics of Systems, Mechanisms and Machines' (Dynamics)* (Omsk, Russia, 14–16 Nov. 2017) pp 1–6
[6] Ivanov A I, Kachajkin E I, Lozhnikov PS 2016 A Complete Statistical Model of a Handwritten Signature as an Object of Biometric Identification *Proc.of 2016 Int. Siberian Conf. on Control and Communications (SIBCON 2016)* (Moscow, Russia, 12–14 May 2016) pp 1–5
[7] Ivanov A I, Lozhnikov P S, Vyatchanin S E 2017 Comparable Estimation of Network Power for Chisquared Pearson Functional Networks and Bayes Hyperbolic Functional Networks while Processing Biometric Data *Proc. of Int. Siberian Conf. on Control and Communications (SIBCON)* (Astana, Kazakhstan, 29–30 June 2017) pp 1–3. DOI: 10.1109/SIBCON.2017.7998435
[8] Ivanov A I, Lozhnikov P S, Sulavko A E 2017 Evaluation of signature verification reliability based on artificial neural networks, Bayesian multivariate functional and quadratic forms *Computer Optics* **5** pp 765–774
[9] Lozhnikov P S, Sulavko A E, Eremenko A V, Volkov D A 2016 Methods of Generating Key Sequences based on Parameters of Handwritten Passwords and Signatures *Information* **7(4)** p 59

[10] Ivanov A I 2016 Mnogomernaya neyrosetevaya obrabotka biometricheskikh dannykh sprogrammnym vosproizvedeniem effektov kvantovoy superpozitsii [Multidimensional neural network processing ofbiometric data with program reproduction of quantum superposition effects] (in Russian) Penza: PNIEI 133 p

[11] Vasilyev V I, Sulavko A E, Eremenko A V, Zhumazhanova S S 2016 Identification potential capacity of typical hardware for the purpose of hidden recognition of computer network users *Proc. of X Int. IEEE Scientific and Technical Conf. 'Dynamics of Systems, Mechanisms and Machines' (Dynamics)* (Omsk, Russia, 15–17 Nov.2016) pp 1–5 DOI: 10.1109/Dynamics.2016.7819106

[12] Pervushin E A 2011 Obzor osnovny'x metodov raspoznavaniya diktorov [Review of the main methods for speakers recognition] (in Russian) *Matematicheskie struktury' i modelirovanie* **24** pp 41–54

[13] Akhmetov B S, Ivanov A I, Funtikov V A, Bezyaev A V, Malygina E A 2014 Tekhnologiya ispol'zovaniya bol'shikh neyronnykh setey dlya preobrazovaniya nechetkikh biometricheskikh dannykh v kod klyucha dostupa [The technology of using large neural networks to convert the fuzzy biometric data in the key code access] (in Russian) Kazakhstan, Almaty: LEM 144 p.

[14] Ivanov A I 2012 Podsoznaniye iskusstvennogo intellekta: programmirovaniye avtomatov neyrosetevoy biometrii yazykom ikh obucheniya [Subconscious of artificial intelligence: programming of neural network biometrics automata by the language of their learning] (in Russian) Penza: PNIEI 125 p

[15] Genyun Sun, Qinhuo Liu,*, Qiang Liu, Changyuan Ji, Xiaowen Li 2007 A novel approach for edge detection based on the theory of universal gravity *Pattern Recognition* **40** pp 2766 – 2775.

[16] Rashedi E, Nezamabadi-pour H, Saryazdi S 2009 GSA: a gravitational search algorithm *Information Science* **179** pp 2232–2248

[17] SeyedAli Mirjalili, Siti Zaiton Mohd Hashim, Hossein Moradian Sardroudi 2012 Training feedforward neural networks using hybrid particle swarm optimization and gravitational search algorithm *Applied Mathematics and Computation* **218 (2012)** pp 11125–11137

[18] González B, Valdez F, Melin P, Prado-Arechiga G 2015 A Gravitational Search Algorithm for Optimization of Modular Neural Networks in Pattern Recognition *In: Castillo O., Melin P. (eds) Fuzzy Logic Augmentation of Nature-Inspired Optimization Metaheuristics. Studies in Computational Intelligence* vol 574. Springer, Cham

[19] Beatriz González, Fevrier Valdez, Patricia Melin, German Prado-Arechiga 2015 Fuzzy logic in the gravitational search algorithm enhanced using fuzzy logic with dynamic alpha parameter value adaptation for the optimization of modular neural networks in echocardiogram recognition *Applied Soft Computing* vol 37 pp 245–254

[20] Abbas Bahrololoum Hossein Nezamabadi-pour Saeid Saryazdi 2015 A data clustering approach based on universal gravity rule *Engineering Applications of Artificial Intelligence* vol 45 pp 415–428

[21] Sulavko A E, Fedotov A A, Eremenko A V 2017 Users' identification through keystroke dynamics based on vibration parameters and keyboard pressure *Proc. of XI International IEEE Scientific and Technical Conference 'Dynamics of Systems, Mechanisms and Machines' (Dynamics)* (Omsk, Russia, 14–16 Nov. 2017) pp 1–7

[22] Sulavko A E, Yeremenko A V, Borisov R V, Inivatov D P 2017 Vliyaniye psikhofiziologicheskogo sostoyaniya diktora na parametry yego golosa i rezul'taty biometricheskoy autentifikatsii po rechevomu parolyu [Influence of speaker's psychophysiological state on the parameters of speaker's voice and the results of biometric authentication on the voice password] (in Russian) *Komp'yuternyye instrumenty v obrazovanii* **4** pp 29–47

Acknowledgments
The research was supported by the Russian Scientific Foundation (project №17-71-10094).

IOP Conf. Series: Journal of Physics: Conf. Series **1050** (2018) 012083 doi:10.1088/1742-6596/1050/1/012083

Low-temperature heat capacity VO2±δ and solid solutions V1-XFeXO2

Vad I Surikov[1], N A Semenyuk[1], Val I Surikov[1], Y V Kuznetsova[2] and S V Yanchij[1]

[1]Omsk State Technical University, 11, Mira ave., Omsk, 644050, Russia
[2]Surgut State University, Surgut, Russia

Abstract. The article reports the study results of the heat capacity of vanadium dioxide compounds (VO_2) (within the homogeneity range) and stoichiometric solid solutions $V_{1-X}Fe_XO_2$ at temperature range from 5 K to 100 K. Separation of various constituents of heat capacity with its additivity assumption has been conducted: crystal lattice heat capacity and defects heat capacity. It is shown that a considerable contribution to the heat capacity is produced by the crystal lattice defects (till ~ 25 K) at low temperatures, and at higher temperatures the heat capacity is determined mainly with the crystal lattice. Using the Debye function table for the heat capacity for all compounds, the characteristic Debye temperatures Θ_D were determined.

1. Introduction

Some compounds 3d -, 4d- and 5f- elements had phase transition metal – dielectric (PTMD) caused with temperature change. These compounds - vanadium dioxide (VO_2) - properties both as in practical as in purely scientific aspects are widely reported [1, 2]. Phase transition temperature of the compound T_{MD} considering hysteresis effects is (340±5) K.

2. Problem statement

However, a number of works noted that naturally, vanadium dioxide is metastable and subject to spontaneous oxidation under the influence of atmospheric oxygen [3]. Certain phase stabilization is achieved with doping the vanadium dioxide (VO_2) by iron [3, 4]. The study of low-temperature heat capacity (when temperature is below 20 K) [5], as well as X-ray diffraction analysis with pycnometric density measurement showed that there is a considerable number of defects both in pure vanadium dioxide and in solid solutions $V_{1-X}Fe_XO_2$ [5]. Obviously, these defects influence various properties including thermophysical properties of these compounds.

3. Theory

The article reported the results of studies of the heat capacity at constant pressure C_P, a number of compounds $VO_{2\pm\delta}$ and solid solutions $V_{1-X}Fe_XO_2$ in the temperature range from 5 K to 100 K. Heat capacity value at constant volume C_V is noted to be used while discussing theoretical positions. It was shown in [6], that at low temperatures the difference between C_P and C_V is not large and at 100 K is no more than 3%. The heat capacity was measured by the method of a vacuum adiabatic calorimeter of the Strelkov type on an installation made in the Khabarovsk VNIIFTRI branch. Measurement of the reference materials heat capacity (in this capacity, benzoic acid and electrolytic copper of Kyshtym copper electrolytic plant were used) showed that the error in determining the heat capacity in the entire investigated temperature range does not exceed 1 %. Liquid helium and liquid hydrogen were used as refrigerants (with evacuation of vapors if necessary). The temperature was registered with a

semiconductor thermometer of resistance (calibrated in Kharkov ILTPE of NASU) and a platinum resistance thermometer attested by VNIIFTRI. The initial sample VO_2 for the studies was prepared from the powdered material of vanadium pentoxide grade HSC by dissociation in vacuum at pressure 7 Pa, at temperature 1300 K. Samples $VO_{2-\delta}$ were obtained from VO_2 with further dissociation in vacuum 0.15 Pa at different temperatures. Samples $VO_{2+\delta}$ were obtained by oxidation at atmospheric pressure at temperature 620 K changing the oxidation time.
To prepare samples containing iron $V_{1-X}Fe_XO_2$, there were mixed necessary amounts of oxides V_2O_3 and V_2O_5 with Fe_2O_3:

$$(1-2X)V_2O_3 + V_2O_5 + 2XFe_2O_3 \xrightarrow{900^0C} 4Fe_XV_{1-X}O_2$$

The samples were tested using a diffractometer Shimadzu Maxima X XRD-7000, studies have shown that all samples are single-phase and belong to the structure VO_2.
The samples element composition was refined on an electronic scanning microscope *JEOL JCM* – 5700 at voltage 15.0 kV.

4. Experimental results

The results of the studies are shown in Fig. 1-3 and in Tables 1–2.

Table 1. Values of heat capacities and characteristic Debye temperatures for VO_2

Temperature, K	C_p, $J(mol\ K)^{-1}$				
	$VO_{1.990}$	$VO_{1.995}$	$VO_{1.998}$	VO_2	$VO_{2.03}$
5	0.0027	0.0032	0.0033	0.0035	0.0060
10	1.16	0.0127	0.0141	0.0154	0.0176
20	0.0334	0.0413	0.0468	0.0577	0.0453
25	0.0562	0.0682	0.0782	0.102	0.0694
50	0.16	0.18	0.17	0.24	0.16
60	0.20	0.26	0.29	0.40	0.23
70	0.31	0.40	0.46	0.63	0.36
80	0.48	0.59	0.68	0.91	0.53
90	0.65	0.84	0.93	1.24	0.75
100	0.87	1.08	1.22	1.59	0.98
Θ_D, K	890	840	780	700	850

Table 2. Values of heat capacities and characteristic Debye temperatures for $V_{1-X}Fe_XO_2$

Temperature, K	C_p, $J(mol\ K)^{-1}$			
	$V_{0.99}Fe_{0.01}O_2$	$V_{0.97}Fe_{0.03}O_2$	$V_{0.95}Fe_{0.05}O_2$	$V_{0.93}Fe_{0.07}O_2$
5	0.0035	0.0027	0.0025	0.0025
10	0.0149	0.0130	0.0116	0.0101
20	0.0547	0.0492	0.0419	0.0396
25	0.0943	0.0840	0.0707	0.0661
50	0.27	0.19	0.21	0.16
60	0.37	0.33	0.27	0.26
70	0.58	0.52	0.43	0.40
80	0.84	0.76	0.63	0.59
90	1.15	1.04	0.87	0.84
100	1.49	1.35	1.15	1.08
Θ_D, K	720	750	800	820

Having information on the heat capacity at temperatures above 50 K, characteristic Debye temperatures Θ_D for all samples in the temperature range from 50 K to 100 K are possible to be determined using Debye function tables for heat capacity. Their values are indicated in Tables 1 and 2. At temperatures $T \sim \Theta_D/12$, within the framework of the Debye model, the lattice heat capacity C_{LET} is related with Debye temperature [7]:

$$C_{LET}=\beta T^3 \tag{1}$$

$$\beta=12\pi^4 VNkn/(5\Theta_D^3) \tag{2}$$

where V is the volume of the unit lattice,
N is the number of cells in the volume under consideration,
k is Boltzmann's constant,
n is the number of atoms in the cell.
Extrapolating certain values of the Debye temperatures to the low-temperature area, the heat capacity of the defects can be determined:

$$C_{DEF}=C_P-C_{LET} \tag{3}$$

Dependences of the heat capacity C_P on temperatures for three samples are presented in Fig.1: VO_2, $V_{0.97}Fe_{0.03}O_2$ and $VO_{1.990}$. A gradual increase in the heat capacity is observed for all samples with increasing temperature. Heat capacity dependences on temperature, both the total and its components are presented in Fig. 2 and Fig. 3: the heat capacity of the crystal lattice and the heat capacity of the defects for two samples (Fig. 1 is for the sample VO_2; Fig. 2 is for the sample $V_{0.97}Fe_{0.03}O_2$). As presented, the heat capacity of crystal lattice defects first increases with increasing temperature (up to $T_K \sim 14$ K), and then with further increasing temperature C_{DEF} does not depend on the temperature. It suggests that the defects in the crystal are as quantum particles, and T_K is degeneracy removal temperature.

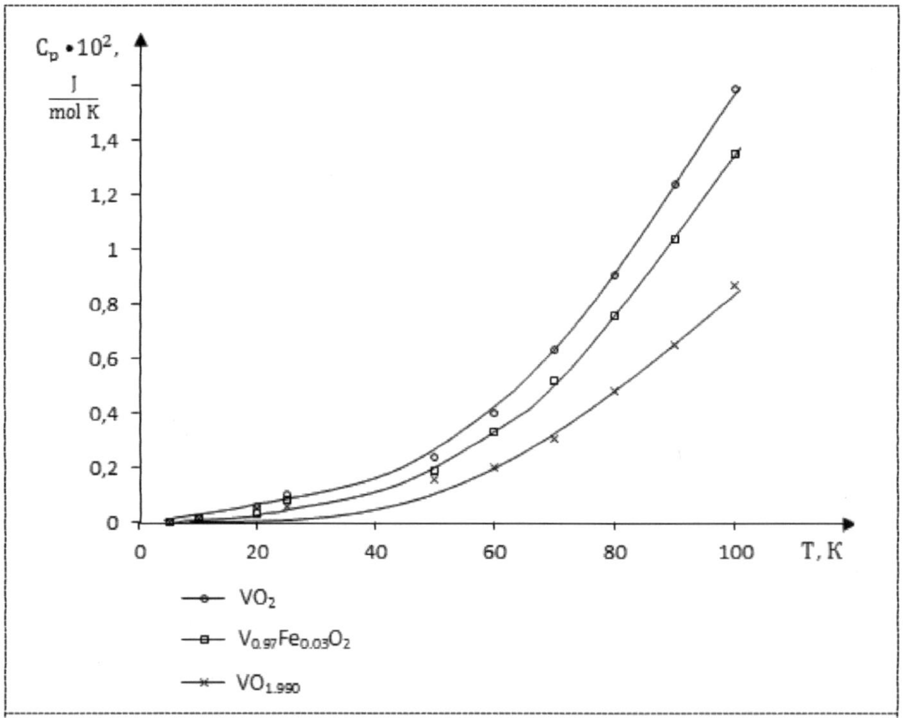

Figure 1. Dependency of heat capacity C_P on temperature.

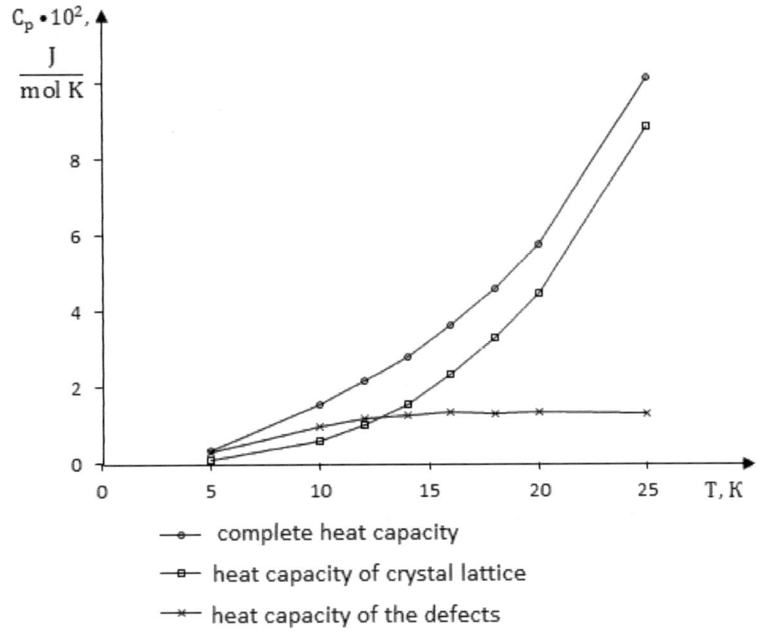

Figure 2. Dependency of heat capacity on temperature for VO_2

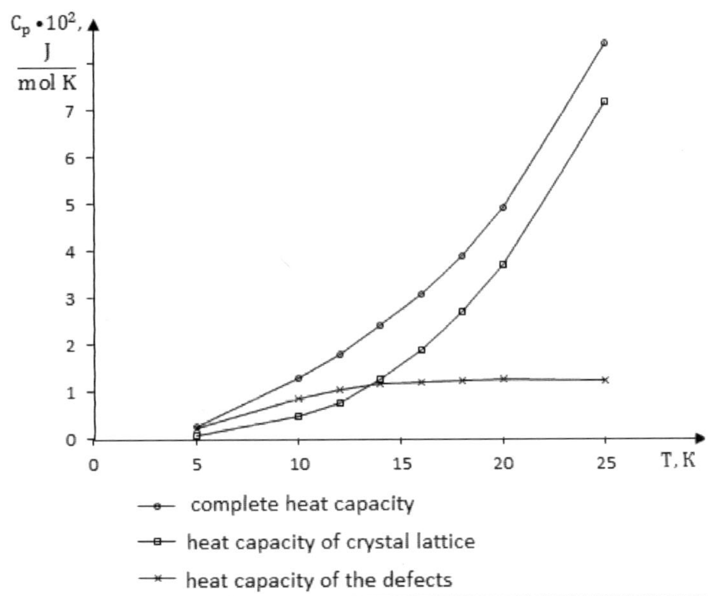

Figure 3. Dependency of heat capacity on temperature for $V_{0.97}Fe_{0.03}O_2$

5. Discussion

As the obtained results prove, at low temperatures the heat capacity of pure vanadium dioxide "VO_2" is minimal for $VO_{1.990}$ (0.0027 J/mol K at 5 K) and it is maximum for $VO_{2.03}$ (0.0060 J/(mol K). The low-temperature heat capacity of other samples "VO_2" varies within these limits. With temperatures close to 100 K (high temperature part of the heat capacity) for the samples "VO_2" minimum values of heat capacity are for $VO_{1.990}$ (0.87 J/(mol K), and maximum are for VO_2 (1.59 J/(mol K).

IOP Conf. Series: Journal of Physics: Conf. Series **1050** (2018) 012083 doi:10.1088/1742-6596/1050/1/012083

At low temperatures, the heat capacity of solid solutions based on vanadium dioxide containing iron $V_{1-X}Fe_XO_2$ is minimal for the samples containing iron 5 % ($V_{0.95}Fe_{0.05}O_2$) and 7 % ($V_{0.93}Fe_{0.07}O_2$) (0.0025 J/mol K at 5K). The heat capacity of a sample containing iron 3 % ($V_{0.97}Fe_{0.03}O_2$) differs a little. The heat capacity $V_{0.99}Fe_{0.01}O_2$ (0.0035 J/mol K) is much larger if to compare. At temperatures close to 100 K, $V_{0.99}Fe_{0.01}O_2$ (1.49 J/(mol K) has the highest values of heat capacity, and the least values are for the heat capacity $V_{0.93}Fe_{0.07}O_2$. In whole, alloying with iron leads to leveling heat capacity samples.

6. Conclusion

Thus, the article reports the results of a study of the heat capacity $VO_{2\pm\delta}$ and solid solutions $V_{1-X}Fe_XO_2$. The heat capacity is established to depend strongly on the composition of the sample. Analysis of the measurement results made possible to isolate the heat capacity of the defects and its dependence on temperature. According to the nature of $C_{DEF}(T)$ the quantum nature of defects is concluded.

References

[1] Mott N F 1974 *Metal – insulator transitions*. Taylr & Francis Ltd., London XVI 278 p
[2] Chudnovskiy F A 1975 *Phase transition metal – semiconductor and its technical application JTP*. vol 45 pp 1561–1583
[3] Kuznetsova Yu V, Surikov Vad I, Surikov Val I, Lyakh O V, Semenuk N A 2016 *Change in the properties of doped vanadium dioxide during long-term storage under natural conditions* vol 3 pp 212–215
[4] Surikov V I, Surikov V I, Semenuk N A, Lyah O V, Kuznetsova Yu V 2016 *Mechanical fracture of vanadium dioxide during thermal cycling* (*Procedia Engineering* vol 152) pp 711–714
[5] Surikov Vad I, Surikov Val I, Lyakh O V, Prokudina N A, Danilov S V 2014 *Crystal structure defects of vanadium dioxide* (Russion Physics Journal) vol 57 № 8 pp 1111–1115
[6] Surikov Vad I, Danilov S V, Surikov Val I, Stoltz A K 1981 *The electronic heat capacity of V_3Ga and V_2O_3 at low temperatures* (Izvestiya VUZov. Physics) pp112–114
[7] Sommerfeld A, Bethe G 1938 *Electronic theory of metals* (M: Foreign Literature) 362 p

IOP Conf. Series: Journal of Physics: Conf. Series **1050** (2018) 012084 doi:10.1088/1742-6596/1050/1/012084

Computer visualization of quality estimation the color reproduction in digital printing systems measuring colour gamut of the color-reproducing system: interpretation aspects

I A Sysuev, L G Varepo S N Litunov, L B Antropova and A A Miroshnichenko

Omsk State Technical University, 11, Mira ave., Omsk, 644050, Russia

E-mail: sia1960@mail.ru

Abstract. Forecasting of the quality estimation the color reproduction for selection of the optimum printing system from the available range and the achievement of the required quality of multicolor products justifies the urgent problem. To solve the problem, a procedure of quality estimation of color reproduction using tools of the CIE L * a * b * – 1976 colorimetric model was used, i. e. the volume of color gamut body and the set of its cross sections. Practical implementation of the approach for complex estimation of the color reproduction quality in digital printing systems and its computer visualization is shown on the example of the "laser (LED) printer-paper" printing system. It is noted that the estimation of color reproduction using the volume index of the gamut body and the set of its sections, as an element of computer graphics for visualizing the result, interprets the accuracy of reproduction of the multicolor image as for the original.

Keywords: color reproduction, computer visualization, geometry of color gamut body section, volume of color gamut body, printing system

1. Introduction

The problem of estimating the color gamut for color reproduction systems, in particular by the use of printing technologies, has been studied in detail since the mid-1970s of the twentieth century. First the investigation was started by Padham and Saunders [1], and then Judd and Vyshetsky [2], whose works have now became classical.

Color gamut of color reproduction systems was carried out with the help of **xy** chromaticity figures (color system CIE (Commission internationale de l'éclairage) XYZ-1931) and **uv** equal-contrast color diagram (CIE uv-1960). In 1976 CIE enacted the color system L*a*b* –1976 and color gamut was evaluated using its parameters, in particular, applying a chromaticity diagram a*b* [3, 4].

With the development of computer technologies, since the 2000s, the estimation of color gamut was carried out in three-dimensional color space, particularly by the body of color gamut in the color system XYZ and in the uniform color space of L * a * b [5–13] color system. However, the evaluation in three-dimensional space is an intricated problem and time-consuming.

Authors [9-12] have developed and proposed an integral performance index of color gamut estimation (IPICGE), further improved by [14–17]. It helped to simplify the technology of estimation and operate with numerical indexes.

Today, the digital printing market is actively developing; the range of printers is increasing every year. Thus, assessing the quality of color reproduction is one of the most important and urgent problems.

2. Statement of the problem

Electrophotographic printing technology, supported in laser and LED printers, is one of the most perspective. To estimate the quality of prints a significant number of indexes are used, including the optical density of background, uniformity of printing, half-toning, resolution of printing, etc.

One of the main tasks is the estimation of color reproduction, i. e. accuracy of reproduction the memorable colors, color scale (gamut), comparison of color reproducing characteristics of printers in different areas of color space.

As noted in [18, 19] printers usually generate a limited number of colors and can't create continuoustone color images. It is noted by the authors that in reproduction of halftone images the blurred edges and textures may appear. To correct the expected errors various solutions are proposed. I. e. the method of using different filters for halftones of different color components [18] and the determination of conditions under which colors reproduce without mistakes [19, 20]. Components of conditions are, for example, resolution of input or output, paper. Herewith, the selected conditions should ensure the adequacy of color images reproduction accuracy.

To evaluate the printing characteristics, a separate model of color areas in halftone printing is proposed by the authors [21]. The result of color separation process is a monochromatic image, which is further processed by segmentation methods for obtaining the exact regions of a certain color or the regions with halftone dots. It should be noted that the accuracy of evaluating the printing characteristics highly depends on the level of segmentation of the color areas involved in printing.

The authors of [22, 23] suppose that modification of paper surface by the use of various chemicals courses the enhancement of pigments fixation on its surface. This result is obtained for inkjet printing.

The aim of this research is a complex estimation of color reproduction and a computer visualization of the result in printing system "laser (LED) printer - paper" as the example.

3. Problem solution

By a printing system it is understood "a combination of technical and technological means of printing setup and directly printing – from the original. This term is equally applicable both to the characteristic of a printing method in general, for example, an inkjet printing system as a mean of obtaining an image, and regarding to a really operated, for example, digital printing system in the configuration of computer – software – settings – coloring materials – printer – printed material "[9].

The system of digital printing simplify if it uses one computer, the same application settings, and in printers – the original consumables (cartridges for electrophotography technology) as "printer - printed material".

Color gamut of any color reproduction system, for example, a printing system, is that the maximum number of colors that this system is capable to reproduce. The method for estimating the color reproduction, considered as standard, consists in estimating the areas of the hexagons of color coverage (Fig. 1), which is represented on the color chart (color diagram) of the color system CIE L * a * b *.

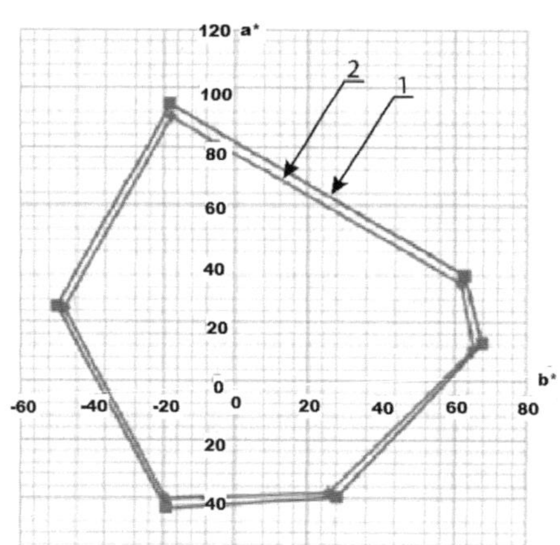

Figure 1. Color chart $a^* b^*$ of printing systems № 3: 1 – mat paper № 2; 2 – gloss paper № 1

In [10, 12], a method, an algorithm and a program for calculating the volume of a color gamut body are developed and tested, and then improved [14–17]. The program (Project 1) is developed to calculate both the volume of the color gamut body and the data for constructing a set of its (body) cross-sections for different values along the L * axis. At that, the more different values along the L * axis are used, the more accurate the configuration will be described. The source data for calculations are the spectral characteristics of reflection (i. e. the dependence of the reflection factor on the wavelength) of the main stimuli (in the case of electrophotography - toners) of subtractive synthesis (Fig. 2).

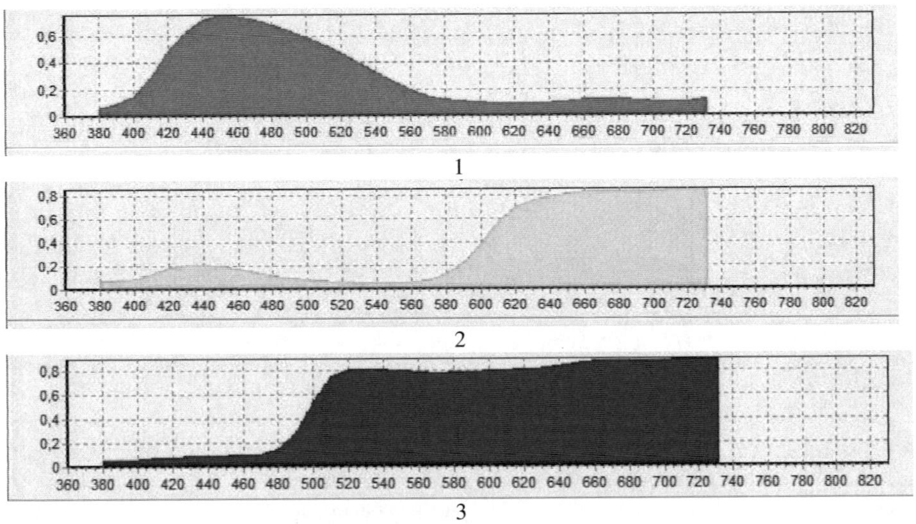

Figure 2. Spectral characteristics of reflection of the main stimuli
(1 – cyan, 2 – magenta, 3 – yellow) for the printing system № 3 (matt paper № 2)

4. Experimental Rerults

The researches of six electrophotographic printing systems of "laser (LED) printer – paper" were carried out for estimating color reproduction in complex.

The components of investigated systems are the modern models of laser (LED) printers by companies that are the world's leading manufacturers (HP Laser Jet Pro 300 Color M351a, Konica Minolta with 364 series PS, OKI C96 55) and paper with different characteristics. The object of the study is the IT-8 test scale.

Measurements of reflection spectra of pure color fields were made and the volumes of color gamut bodies in units of ΔE00 were calculated (Table 1).

Table 1. Characteristics of system colors

Number	**Printing system**				**Volume of color gamut body, units. ΔE_{00}**
		Paper characteristics			
	Printer model	Number	Coating	Whiteness, % CIE	
1	HP Laser Jet Pro 300 Color M351a	1	Gloss	165 ± 2.0	108406
		2	Matt	170 ± 3.0	107399
2	Konica Minolta c 364 series PS	1	Gloss	165 ± 2.0	99592
		2	Matt	170 ± 3.0	93789
3	OKI C96 55	1	Gloss	165 ± 2.0	92582
		2	Matt	170 ± 3.0	84698

According to the received data, in the coordinate system of the color space L * a * b * CIE a color gamut body was visualized on the figure. Particular interest in the solution of the problem assigned in this work was presented in finding the geometric characteristics of color gamut body, primarily its volume. For this purpose, the method of spatial triangulation was used.

To represent the actual color gamma it should be noted that the maximum possible volume of color gamut body is 250,000 units ΔE00. This volume of color gamut body is the maximum possible, i. e. it is the volume of color gamut body for the ideal subtractive synthesis.

Fig. 3 and 4 the examples of computer visualization of a set of cross-sections for color gamut bodies are shown, They were constructed from the data of calculations with a step along the L * axis, equal to 20.

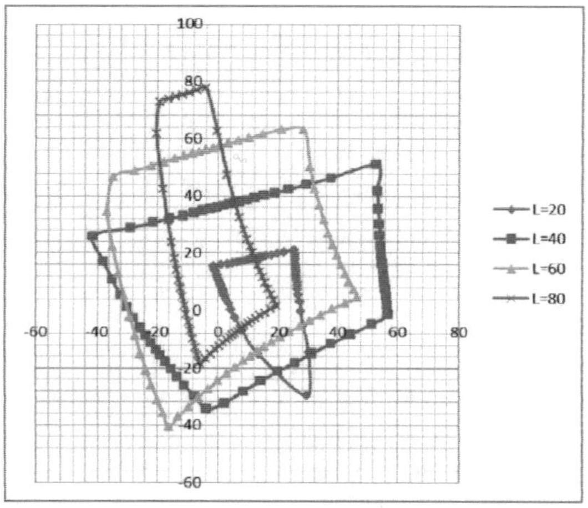

Figure 3. Cross-sections of color gamut body on color diagram a^*b^* for the printing system № 3 (gloss paper 1)

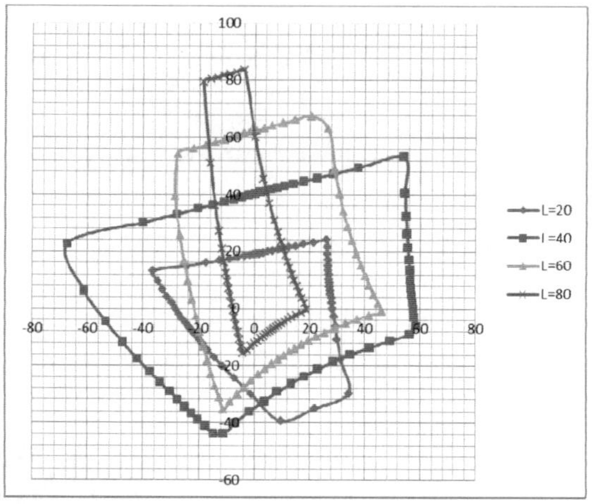

Figure 4. Cross-sections of color gamut body on color diagram a^*b^* for the printing system № 1 (gloss paper 1)

5. Results and Discussion

The results of calculating the volume of color gamut body in the systems "laser (light-emitting) printer - paper" show that the systems reproduce different amounts of colors, which is confirmed by the volume of color gamut body. Differences in the volume of color gamut body of studied printing systems are about 20% (Table 1).

A significant effect of the paper characteristics on the quality of a print in the analyzed printing system was noted. Depending on the paper characteristics changes in the volume of color gamut body ares up to 10%.

Computer visualization of a set of cross-sections of a body for different values along the L * axis clearly shows the reproduction of colors in different areas of color space (Fig. 3, 4). For example, for the HP LaserJet Pro 300 Color M351a printer, the gamma of reproduced shadow and midtone by the luminosity colors (L * = 20 and L * = 40) is about 2 times greater than for the OKI C96 55 printer.

It is estimated that highlight colors (L * = 60 and L * = 80) are better reproduced by the printing system No. 3. On the contrary, the printing system No. 1 better reproduces saturated colors in the yellow, green, cyan, blue, and magenta areas of the color space, only in the red saturation is approximately the same.

6. Conclusions and Summary

On the basis of the analysis the experimental data, it is shown that the improved estimation of the color gamut with the use of the integrated index of color gamut estimation and the set of cross-sections of color gamut body o is more objective, it allows to efficiently estimate the number of reproduced colors and their distribution in the areas of color space.

A computer visual representation of color gamut for the multicolor image reproduction systems interprets the reproduction accuracy of multicolor image according to the original. It allows to compare the potential possibilities of color reproduction for the studying systems and forecast the quality of a multicolor image.

References

[1] Padgham C A, Saunders J E The Perception of Light and Colour. Hardcover, 1975
[2] Judd D B, Wyszecki G (1975) Color in business, science and industry. J. Wiley & Sons New York, London, Toronto, Sydney
[3] Schlicht H-J Bildverarbeitung digital Addison-Wesley Bonn, 1995
[4] Margulis D Makeready: Prepress Resourse. MIS:Press, New York, 1996
[5] Blatner D, Fleishman G, Roth S World Scanning and Halftones. 2nd Edition. An Open House Book, Peachpit Press, 1998
[6] Field G G. (2004) Color and Its reproduction – Fundamentals for the Digital Imaging and Printing industry. Sewickley, PA: GAFT Press
[7] Fraser B, Murphy C, Bunting F World Color Management. Peachpit Press, 2005
[8] Weisberg J Microsoft Windows XP Color management. Peachpit Press, 2006
[9] Pozharsky A O, Sysuev 2005 *J. Omsk Science Herald. Series: Equipment, machinery and technology* **4** (33) p 180-182 (Russian)
[10] Sysuev I A, Pozharsky A O, Zakharenko A A 2006 *J. Omsk Science Herald. Series: Equipment, machinery and technology* **9** (47) p 107–110 (Russian)
[11] Pozharsky A O, Sysuev I A 2006 *Proceedings of the universities. Problems of graphic arts and publishing* **4** p 3–12 (Russia)
[12] Pozharsky A O, Pozharsky T O, Sysuev I A Certif. No 50200601138 of registration of the program Calculation the volume of gamut body by the base points of its surface (Russian Federation); publ. 2016
[13] Maohai Lin, Shisheng Zhou, Mao-hai Lin, Yunhui Luo 2009 *Computer Technology and Development* **1** p 587–589
[14] Varepo L, Golunov A, Golunova A, Trapeznicova O & Nagornova I. 2015 *Testing and Measurement Conference Publications*, 2015 p 69–71
[15] Varepo L G, Trapeznikova O V, Golunov A V, Trapeznikov E V, Nagornova I V 2017 *J. Proceedings of Tula State University. Technical Sciences* **2** p 282–290 (Russian)
[16] Varepo L G, Brazhnikov A Yu, Volinsky A A, Kondratov A P, Nagornova I V 2017 Mechanical Science and Technology Update IOP Publishing. IOP Conf. Series*: Journal of Physics: Conf. Series* **858** (2017) 012038
[17] Varepo L G, Trapeznikova O V, Golunov A V, Panchuk K L, Lyashkov A A 2017 *J. Geometry and Graphics* **3**(5) p 86–91 (Russian)
[18] Jim Z C Lai and Chia-Chi Chen 2003 *Journal of Visual Communication and Image Representation* Vol. **14** Issue 4 p 389–404
[19] Horn B K P 1984 *Computer Vision, Graphics, and Image Processing* Vol. **26** Issue 2 p 135–167
[20] Mohamed R Sh, Yousif G A 2010 *The Egyptian Journal of Remote Sensing and Space Science* Vol. **13** Issue 1 p 75–80
[21] Fribert M 2003 *Computers & Graphics* Vol. **27** Issue 5 p 801–806
[22] Sónia Sousa S, José António Gamelas, Antóniode Oliveira Mendes, Paulo Torrão Fiadeiro, Ana Ramos 2013 *Materials Chemistry and Physics* Vol. **139** Issue 2-3 p 877-884
[23] Kettle J, Lamminmäki T, Gane P 2010 *Surface and Coatings Technology* Vol. **204** Issue 12–13 p 2103–2109

Theory of the piston air spring in the boom machine working equipment

V N Tarasov[1], I V Boyarkina[1] and G N Boyarkin[2]

[1]Siberian State Automobile and Highway University, 5, Mira ave., Omsk, 644080, Russia
[2]Omsk State Technical University, 11, Mira ave., Omsk, 644050, Russia

Abstract. The efficiency coefficient of the boom machine working equipment is around 50%. When payload lifting in the bucket, gravity force of the working equipment performs useless operation. The piston air spring balances the working equipment gravity force and eliminates the indicated defect. The selection problem of the piston air spring parameters has been solved. The theory of the piston air spring and hydraulic power cylinders is supplemented by a new equation connecting the design parameters of kinematic triangles with parameters of hydraulic mechanisms of the boom machine working equipment.

1. Introduction

Technological boom machines have a cyclic nature of the working processes of loading materials and loose cargoes into transport vehicles. The total mass of such machines can be equal to 250–300 tons, and the engine power reaches 1–2 thousand kW. The specific feature of the bucket boom machines consists in the cantilever mounting of the working equipment relative to the machine, at the same time the working equipment mass, reduced to the center of the bucket masses is commensurable in size with the payload mass in the bucket. This means that when lifting the working equipment half of the engine's energy is wasted overcoming the gravity force of the bucket, boom and other elements of the working equipment.

The paper [1] is known, where the authors Yogeshkumar R. Chheta, Rajesh M. Joshi, Krishan Gotewal, M. ManoahStephen pay attention to the negative influence of the gravity force of the manipulator elements on the power characteristics. Passive methods of the gravity force compensation are considered to increase the manipulator efficiency.

In paper [2] the authors Z. Miaofen, S. Shaohui, G. Youping, Z. Dada establish the relationship of mechanical and hydraulic models controlling the working process of a loader. The multiparametric optimization of the engine operation mode have been performed by the numerical integration methods.

In paper [3] the authors Z. Zhihong, W. Yunxin and M. Changxun consider the dynamic modeling of a boom in horizontal and other positions. Mechanical parameters reducing dynamic loads have been established.

In paper [4] the authors S. Kang, J. Park, S. Kim, B. Lee, Y. Kim, P. Kim, H.J. Kim consider the excavator working equipment control using the controller, implementing efficient control processes of the boom, bucket, stick, modeling the experienced operator actions.

In paper [5] the authors H. Xie, G. Zhang determined that in the truck crane upper boom position the hydraulic system pressure increases and vibration increases due to the reduced mass increase and force arm decrease.

IOP Conf. Series: Journal of Physics: Conf. Series **1050** (2018) 012085 doi:10.1088/1742-6596/1050/1/012085

In papers [6, 7] the authors V.N. Tarasov, I.V. Boyarkina proposed a piston-type pneumatic energy-saving drive consisting of a balance cylinder which head end is connected with a gas cylinder charged with compressed air.
The performed review shows that the balancing issue of the gravity force of the working equipment bodies of boom machines is of high priority.

2. Task definition

The article has solved the task of the theory creation relative to the piston air spring mounted in the working equipment of the boom technological machine for balancing the gravity force of the working equipment.
Figure 1 shows the design diagram of the gravity force of excavator working equipment bodies, balanced by the piston air spring *AB*, which is shown in three boom positions: bottom, horizontal and upper. The hydraulic power cylinders are not shown for clarity at this diagram.

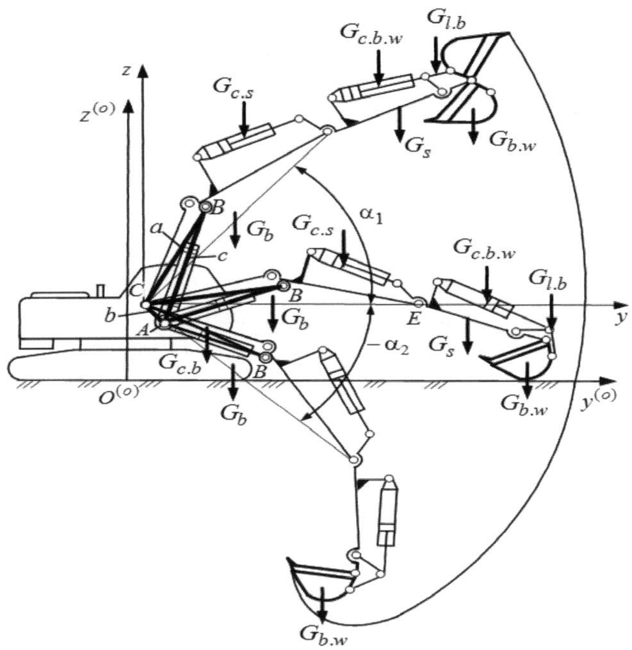

Figure 1. Design diagram of the boom machine working equipment gravity force balancing by the piston air spring

The piston air spring balances the gravity force: $G_{b.w}$ – bucket weight; $G_{l.b}$ – weigh of bucket levers; G_s – stick weight; $G_{c.b.w}$ – bucket cylinder weight; $G_{c.s}$ – stick cylinder weight; G_b – boom weight; $G_{c.b}$ – boom cylinder weight.
Figure 1 selects a kinematic triangle with *CAB* vertices, formed by the sides *a, b, c*, which are connected respectively with the boom, portal and piston air spring. The parameter -*c* of the piston air spring is a variable length link that varies within the limits $c_{\min} \le c \le c_{\max}$.
Figure 2 shows the design scheme for the mechanism research of the hydraulic power cylinders A_bB_b for the boom travel. The kinematic triangle CA_bB_b is selected which is formed by the sides a_b, b_b, c_b [8, 9].

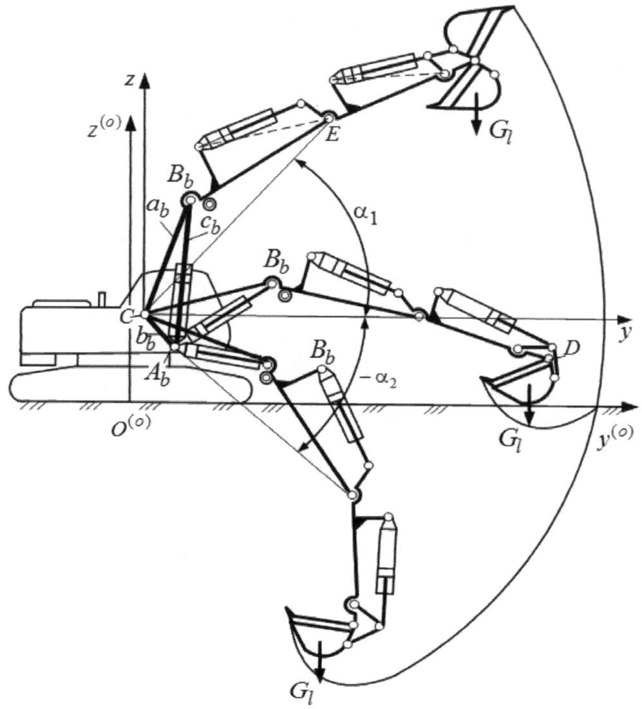

Figure 2. Design diagram of the vertical travel of the gravity force G_l of the payload in the bucket by hydraulic power cylinders

In Figure 2 the hydraulic power cylinders take up only the payload G_l, as all gravity forces are balanced by the piston air spring (see Figure 1).

The idea of the air spring creation is based on the concept of two balanced forces in a mechanical system, one of which is the main vector of the gravity force of the working equipment bodies, reduced to the rod of the air spring, the other is the compressed air pressure force on the piston of the air spring. These balanced forces do not perform operation when moving the boom machine working equipment. The piston air spring ensures the complete balancing of the working equipment gravity force. The piston air spring theory and mechanics of the boom hydraulic power cylinders are based on supporting facts and laws of mechanics [9].

3. Theory

When changing the working equipment position by the boom hydraulic power cylinders (see Figure 2), accordingly the air spring length -*c* is changed. The travel S of the air spring piston and travel S_b of the hydraulic power cylinder piston are determined as the length change of the links AB (Figure 1) and A_bB_b (Figure 2) for different positions of the working equipment. In Figure 1 and Figure 2 the horizontal position with the angle $\varphi_b = 0$ is taken as the initial boom position; upper position has a positive angle $\varphi_b = \alpha_1$ and lower position is characterized by a negative angle $\varphi_b = -\alpha_2$.

In Figure 3,*a* the dependence $S = f(\varphi_b)$ has been receive of the air spring piston travel, and in Figure 3,*b* – dependence $S_b = f(\varphi_b)$ of the piston travel of the hydraulic power cylinders form the boom angle φ_b.

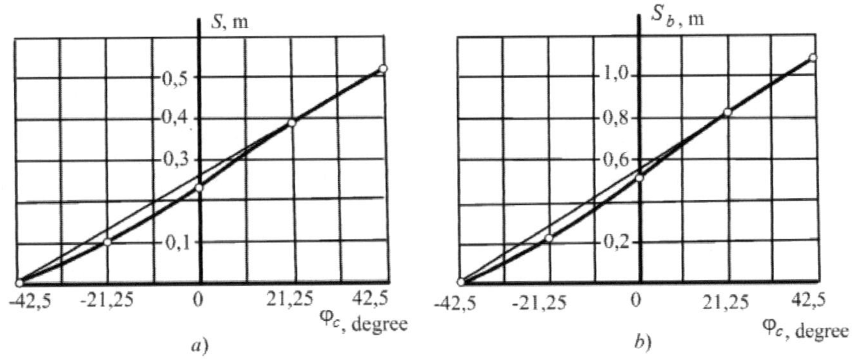

Figure 3. Dependences of the piston travel from the angle of the boom position φ_b : *a*) air spring piston $S = f(\varphi_b)$; *b*) piston of hydraulic power cylinders $S_b = f(\varphi_b)$

The indicated dependences are characterized by polynomial regression equations of third order with high correlation coefficients $R^2 = 0.999$.

For the air spring

$$S = -0.0728\left(\frac{\varphi_b \pi}{180}\right)^3 + 0.0419\left(\frac{\varphi_b \pi}{180}\right)^2 + 0.3958\left(\frac{\varphi_b \pi}{180}\right) + 0.2399. \quad (1)$$

For the boom hydraulic power cylinders

$$S_b = -0.1555\left(\frac{\varphi_b \pi}{180}\right)^3 + 0.0444\left(\frac{\varphi_b \pi}{180}\right)^2 + 0.8303\left(\frac{\varphi_b \pi}{180}\right) + 0.5251. \quad (2)$$

From the theory of mechanics it's known that the derivative of the function $S_b = f(\varphi_b)$ of the hydraulic mechanism piston travel along the boom angle φ_b is the arm of force of the hydraulic cylinder [10, 11].

That's why for the air spring hydraulic mechanism and hydraulic power cylinders the expressions of the derivatives are true

$$\frac{dS}{d\varphi_b} = h; \qquad \frac{dS_b}{d\varphi_b} = h_b, \quad (3)$$

where h, h_b – accordingly, arm of force of the air spring and arm of the hydraulic power cylinders of the boom.

Consequently, it's possible to determine the air spring arm by the numerical differentiation of dependences (Figure 3). We set discrete intervals of the boom angle φ_b and take up the pitch $\Delta\varphi_b = 1^\circ$ of the derivative calculation. According to formula (1) for the given angle φ_b we calculate two current values S_1, S_2, differed by 1° and determine $\Delta S = S_2 - S_1$. We calculate the air spring arm according to the formula $h = \Delta S / \Delta\varphi_b$.

The numerical differentiation operation of the functions $S(\varphi_b)$ and $S_b(\varphi_b)$, shown in Figure 3,*a*,*b*, allows to obtain the dependences of the arm h of the air spring and arm h_b of the hydraulic power cylinders from the boom angle φ_b , which are shown in Figure 4,*a*,*b*.

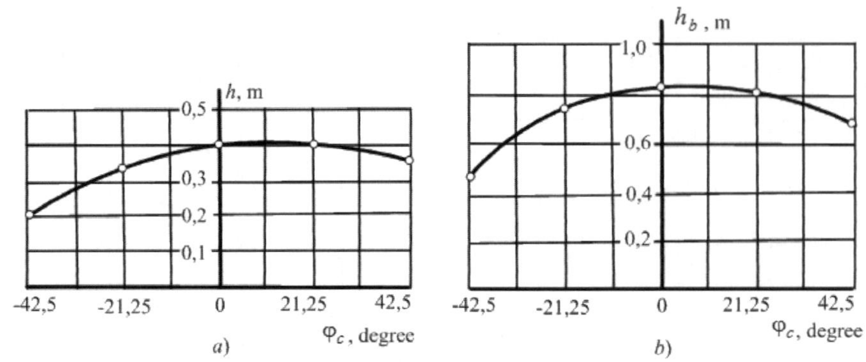

Figure 4. Dependence of the piston air spring arm $h = f(\varphi_b)$ and hydraulic power cylinders $h_b = f(\varphi_b)$ on the boom angle φ_b

The indicated parameters of the air spring and hydraulic power cylinders depend on the working equipment geometric parameters.

In Figure 1 the geometric parameters of mechanisms are represented by the parameters a, b, c, ensuring the air spring reference to the boom machine working equipment, and in Figure 2 – by the parameter a_b, b_b, c_b, ensuring the reference of the hydraulic power cylinders to the boom machine working equipment.

In the article an excavator is considered as an example of the boom machine, which weight is $m_{w.w} = 26000$ kg, working equipment mass is $m_{w.e} = 4051$ kg.

Table 1 lists the working equipment geometric parameters.

Table 1. Geometric parameters of the piston air spring and hydraulic power cylinders of the boom machine

Piston air spring	a, mm	b, mm	$c_{\min}$, mm	$c_{\max}$, mm
	2497	400	2140	2655
Hydraulic power cylinders	a_b, mm	b_b, mm	$c_{b.\min}$, mm	$c_{b.\max}$, mm
	2477	837	1770	2890

In Figure 1 the parameters a, b, c form the arm of force h of the piston air spring. The functional connection [10, 11] of these parameters is determined by the theorem of the height of the triangle vertices according to the equation from paper [10, 11]

$$h^2 = \left(\frac{a \cdot b}{c}\right)^2 - \left(\frac{a^2 + b^2 - c^2}{2c}\right)^2. \tag{4}$$

Similarly for Figure 2 by means of the parameters a_b, b_b, c_b according to expression (4) the arm h_b of the boom hydraulic power cylinders is determined.

The comparison of the values h and h_b, calculated by theorem (4), shows their near-complete coincidence with experimental data for h and h_b, received by the numerical differential method using Figure 3.

The dependences of the arm h of the air spring $h = f(\varphi_b)$ and arm h_b of the hydraulic power cylinders $h_b = f(\varphi_b)$ represent function by which it's possible to determine the piston travel with the final boom turn at an angle of φ_b. Elementary travels of the air spring dS and hydraulic power cylinders dS_b are determined according to the expressions

$$dS = h d\varphi_b; \qquad dS_b = h_b d\varphi_b. \tag{5}$$

That's why the final piston travels of the air spring and hydraulic power cylinders are determined by the integration of expressions (5)

$$S = \int_{-\varphi_b}^{\varphi_b} h d\varphi_b ; \qquad S_b = \int_{-\varphi_b}^{\varphi_b} h_b d\varphi_b . \tag{6}$$

Thus, in Figures 4,*a*,*b* the area, located under the curves $h = f(\varphi_b)$ and $h_b = f(\varphi_b)$, numerically characterize the final travels *S* of the air spring piston and travels S_b of the hydraulic power cylinders.
The considered reference positions consist of the basis of the theory of the air spring and hydraulic power cylinders of the boom machine working equipment.
For the determination of the balancing force *T*, acting at the air spring rod, we apply the principle of virtual displacements. Specify the virtual displacement ds to the piston air spring rod, in the result the boom with the working equipment (see Figure 1) starts virtual angle displacement $d\varphi_b$.
For the system of forces *T* and $\sum G_j$ there is the balance equation in the form of amount elementary works

$$\sum \delta A_i^e = 0 ; \qquad T\delta S - \sum_j G_j y_{j.2} \delta\varphi_b = 0 , \tag{7}$$

where δS, $\delta\varphi_b$ – accordingly the virtual linear and virtual angular displacement of the boom machine working equipment.
The piston air spring force can be defined by equation (7)

$$T = \frac{\sum G_j y_{j.2} \delta\varphi_b}{\delta S} = \frac{\sum G_j y_{j.2}}{h} . \tag{8}$$

where h – arm of the piston air spring.
Thus, obtained expression (8) may be considered as the proof of the provision that the derivative of the function *S* of the piston movement of the air spring along the boom angle φ_b is considered as the arm of force $h = \dfrac{dS}{d\varphi_b}$.
Figure 5 shows the dependence of the force *T* on the piston air spring rod $T = f(S)$ on the piston movement *S*.

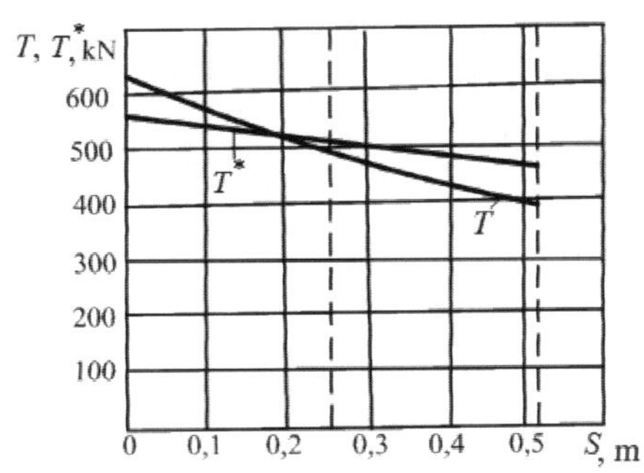

Figure 5. Dependence of the reduced force *T* on the air spring rod on the working equipment gravity force and dependence of the force T^* – compressed air pressure on the air spring piston on the piston movement *S* of the air spring

In Figure 5 the dependence of the force *T*, which represents the reduced force of all gravity forces of the working equipment, it's necessary to balance by the compressed air pressure on the air spring piston. For this purpose it's necessary to assign the maximum force value T^*_{max} on the piston for the boom in the lower position (see Figure 5).
The piston air spring shown in Figure 6 represents a cylindrical case which has the maximum length equal to c_{min} =2140 mm and piston movement in the cylinder *S*=515 mm.

Inside the spring case there is the rod with piston and cylinder. The piston diameter of the air spring $d = 0.16$ m, internal diameter of the gas cavity filled with compressed air $D = 0.25$ m.
The start charging pressure of the air spring for the working equipment in the lower boom position is determined by the formula

$$p_{wH} = p_{w\max} = \frac{4T^*_{\max}}{\pi d^2}. \tag{9}$$

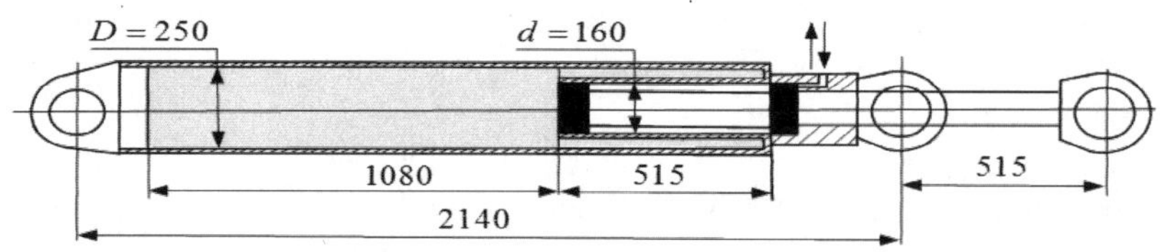

Figure 6. Design of the piston air spring for the gravity force balancing of the boom machine working equipment

The air spring internal cavity volume has the maximum value $V_{\max}$ with the extended rod. While lifting the working equipment the gas chamber volume is increased. The volume change of the air spring internal operating chamber is equal to the volume displaceable while moving the air spring by the operating piston $V = \frac{\pi d^2}{4} S_{\max}$, where $S_{\max}$ – air spring piston movement.

While expanding the compressed air volume in the air spring the pressure is determined by the polytropic process formula [12]

$$p_{w.\min} = p_{w.\max} \left(\frac{V_{\min}}{V_{\max}} \right)^{1,15}. \tag{10}$$

The force T^* of the air spring in the upper boom position is determined by the formula $T^*_{\min} = \frac{\pi d^2}{4} p_{w\min}$.

In Figure 5 the dependence of force is constructed on the air spring piston, created by the air flow for the boom machine $T^* = f(S)$. The obtained dependences allow to determine the work of these forces $T = f(S)$ and $T^* = f(S)$.

The gravity force work of the excavator working equipment is equal to the area, located under the curve $T = f(S)$ (see Figure 5), which is determined by the formula $A(T) = \int_0^S T dS = 261620$ J. The work performed the air spring compressed air is equal to the curve $T^* = f(S)$, which is calculated by the formula $A(T^*) = \int_0^S T^* dS = 261858$ J.

The working process of the hydraulic power cylinders is performed by the similar way. The force in the hydraulic power cylinder rods, implementing the payload lifting G_l (see Figure 2) with the complete balancing of the gravity force is determined by the expression

$$T_b = \frac{G_l y_l \delta\varphi_b}{\delta S} = \frac{G_l y_l}{h_b}, \tag{11}$$

where y_l – bucket gravity center coordinate; h_b – arms of the hydraulic power cylinders.

The dependence of force in the rods of the hydraulic power cylinders on the piston movement $T_b = f(S_b)$ according to formula (11) is given in Figure 7.

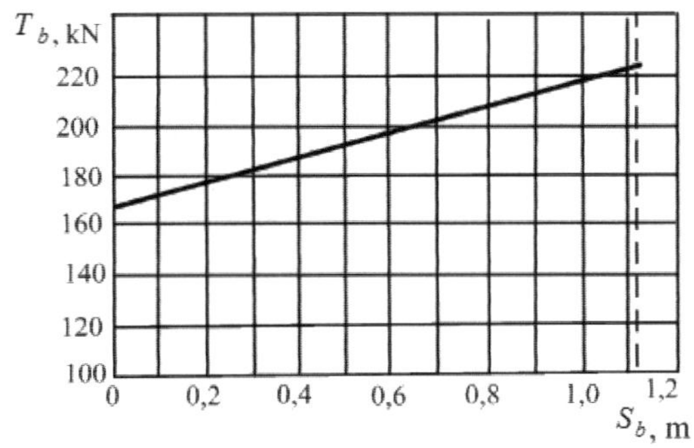

Figure 7. Dependence of force in the rods of the hydraulic power cylinders on the piston movement $T_b = f(S_b)$

The work of the indicated forces of the boom hydraulic cylinders while complete balancing the gravity force is equal to $A(T_b) = 219359$ J. The received values of work of the hydraulic power cylinders $A(T_b^*)$ and piston air spring $A(T^*)$ of the boom machine while complete balancing allow to determine the excavator engine power during the working equipment lifting with cargo in the bucket.

The power of the hydraulic power cylinders reduced to the internal combustion engine shaft is determined by formula [12]

$$N_{b.e} = \frac{A(T_b^*)}{t_b \eta_t}. \quad (12)$$

where t_b – working equipment lifting time; η_t – total efficiency of the excavator working equipment drive.

The total efficiency η_t is determined in the following way $\eta_t = \eta_e \eta_m \eta_{m.w.e} = 0.34$, where η_m – mechanical engine efficiency; $\eta_{m.w.e}$ – mechanical working equipment efficiency; η_e – effective efficiency of an internal combustion engine [12].

The power of the piston air spring reduced to the internal combustion engine shaft is determined by the formula

$$N = \frac{A(T^*)}{t_b \eta_t}. \quad (13)$$

In Table 2 for the discrete values of the working equipment lifting time t_b the power values of the hydraulic power cylinders $N_{b.e}$, air spring power N_e and full power N of the working equipment lifting are given reduced to the engine shaft.

The total capacity of the working equipment while lifting is made up of the power N_{ce} of the hydraulic power cylinders and power N_e of the piston air spring reduced to engine shaft

$$N = N_{be} + N_e. \quad (14)$$

Table 2. Dependence of the hydraulic power cylinder capacity reduced to the engine shaft; power of the piston air spring reduced to the engine shaft on the boom lifting time

Boom lifting time t_b, s.	5	7	9	11	13	15
Average speed of piston movement of hydraulic power cylinders V_b, m/s	0.224	0.16	0.124	0.102	0.086	0.075
Average speed of air spring piston V, m/s	0.103	0.0736	0.0572	0.0468	0.0396	0.0343
Capacity of hydraulic power cylinders	129035	92168	71686	58652	49629	43012

reduced to engine shaft N_{be}, W						
Power of air spring reduced to engine N_e, W	154034	110244	85575	70016	59244	51345
Total capacity of working equipment reduced to engine N, W	283069	202412	157261	128660	108872	94357

4. Research results

The boom lifting time $t_b \approx 15$ s. indicated in the first line of Table 2 corresponds to the installed engine power $N_e = 95.7$ kW indicated in the lower line of Table 2. Such result testifies that there is an insufficient engine power value for this excavator. However, if this boom machine is equipped with the piston air spring, then the boom lifting time will be equal to less than $t_b \approx 7$ s., i.e. working equipment lifting time is reduced by 8 s.

The power value reduction of the hydraulic power cylinders which is equal to the power developed by the engine while the boom lifting, by means of the piston air spring is the important result. The piston air spring balances all the gravity forces of the working equipment by the compressed air pressure in the air spring. The compressed air application as an operating fluid in the piston air spring instead of an expensive working liquid reduces the cost and simplifies the design of the boom machine working equipment.

5. Discussion of results

In this article the results have been received on the basis of power without considering dynamic processes in hydraulic mechanisms. Taking into consideration that the dynamic transient processes in the hydraulic mechanisms are rapidly damping and short-term [13], so their consideration and registration do not exert influence on the excavator boom lifting time (with an error of no more than 3-5%).

That's why the offered method of calculation and theory of the piston air spring are characterized by high accuracy. Errors in the calculation of forces and energy does not exceed 3-5%, as the considered processes of moving huge masses possess low speed of movements (see Table 2, lines 2 and 3).

6. Summary and conclusion

The gravity force balancing of the working equipment bodies by the piston air spring leads to the reduction of the boom lifting time and of the boom machine general operation cycle time. In the considered definite example of the piston air spring application in the context of the excavator with the operating weight of $m_{w.w} = 26$ tons and installed engine power of $N_e = 95.7$ kW, the complete balancing of all the gravity forces appeared to be excessive one, i.e. lifting time of large masses of the working equipment $t_b = 7$ s. is a little bit underestimated. In similar cases it is expedient to fulfill a partial (half) balancing of the gravity forces by the reduction of the maximum charging pressure of the piston air spring by a factor of 2.

References

[1] R Chheta Yogeshkumar, M Joshi Rajesh, Gotewal Krishan and ManoahStephen M 2017 A review on passive gravity compensation *International conference of Electronics Communication and Aerospace Technology (ICECA)* **1** pp 184–189

[2] Miaofen Z, Shaohui S, Youping G and Dada Z 2010 Research on the Multi-domain Modeling and Optimizing Method for Loader Executing System Intelligent System *Design and Engineering Application (ISDEA), International Conference* **2** pp 854–857

[3] Zhihong Z, Yunxin W and Changxun M 2011 Simulation and Optimization of the Driving Forces of Hydraulic Cylinders for Boom of Truck Mounted Concrete Pump *Intelligent Computation Technology and Automation (ICICTA) International Conference* **1** pp 915–919

[4] Kang S, Park J, Kim S, Lee B, Kim Y, Kim P and Kim H J 2014 Path tracking for a hydraulic excavator utilizing proportional-derivative and linear quadratic control *IEEE Conference on Control Applications (CCA)* pp 808–813

[5] Xie H and Zhang G 2013 Research on Characteristics of the Piloted Follow-up Load Control Valve in Automobile Crane Luffing System *Fifth International Conference on Measuring Technology and Mechatronics Automation* pp 887–892

[6] Boyarkina I V 2007 Balanced working equipment as a means of increasing the loading and transporting vehicle efficiency *Construction and road mahines* **12** pp 41–43

[7] Tarasov V N, Boyarkina I V and Kovalenko M V 2007 Energy and resource-saving technology of the gravity force balancing of the boom machine working equipment *Construction and road mahines* **5** pp 46–50

[8] Tarasov V N and Boyarkina I V 2016 Efficiency coefficient of the excavator working equipment lifting mechanism – backhoe *Construction and road mahines* **11** pp 26–32

[9] Tarasov V N and Boyarkina I V [et al.] 2015 Theoretical mechanics (Moscow Publishing House TransLit) p 560

[10] Boyarkina I V 2011 Technological mechanics of single-bucket front-end loaders (Omsk: SibADI) p 336

[11] Boyarkina I V and Tarasov V N 2017 Regularities of the working equipment elements mass reduction to the hydraulic power cylinder piston for the bucket boom machines size standard *Journal of Physics: Conference Series* **858** 1 012006

[12] Tarasov V N and Boyarkina I V 2017 Design procedure of the air spring for the boom machine working equipment in the context of a front loader with the lifting capacity of three tons *Construction and road mahines* **11** pp 9–16

[13] Tarasov V N 1975 Dynamics of control systems of working processes of earth-moving machines (Omsk: SibADI) p 182

IOP Conf. Series: Journal of Physics: Conf. Series **1050** (2018) 012086 doi:10.1088/1742-6596/1050/1/012086

Calculation Of Non-Stationary Magnetic Field Of The Polarized Electromagnet With The External Attracted Anchor

A S Tatevosyan, A A Tatevosyan, N V Zaharova

Omsk State Technical University, 11, Mira ave., Omsk, 644050, Russia

E-mail: karo1@mail.ru

Abstract. Approach to modeling of transition process of the polarized electromagnet of a direct current at a fixed value of a working gap through the solution of a problem of non-stationary magnetic field with the attached electric chain in a complex of the ELCUT 6.0 programs (the professional version) is considered. In the studied design of the polarized electromagnet with the external attracted anchor there are two independent magnetic fluxes. The working magnetic flux arises under the influence of the winding magnetizing forces of the managing director. The polarizing magnetic flux is created by two permanent magnets from magnetic NdFeB alloy (neodymium - an iron - pine forest). In the absence of current in a winding the anchor is affected by the attraction force created by the polarizing magnetic flux. Action of the polarized electromagnet depends on size and the direction of the working magnetic flux created by the magnetizing winding force. Considering wide circulation of electromagnets in various areas of the equipment, including the polarized electromagnets, expansion of opportunities of methods of their research in the established and transition processes represents a relevant task. On the basis of the solution of a problem of non-stationary magnetic field of the polarized electromagnet in a complex of programs ELCUT pictures of magnetic field are calculated and his temporary dependences at connection of a winding of management to a source of constant tension are received.

1. Introduction

The research of transition processes in electromagnets at inclusion of a winding on tension of the power supply is among the widespread tasks solved by students on a lecture and practical training when studying disciplines "Theoretical bases of electrical equipment", "Electrical equipment and electronics" and others. The academism of consideration of a training material on this subject often makes an impression at students that questions of transition processes in electromagnets are rather well studied and there is no need of expansion of knowledge for this area. However in fact the situation is different [1]. Electromagnets differ from each other in a variety of constructive decisions and practical use. Calculation of the established and transition process in electromagnets of different function is a kernel of the problem of their optimum design establishing interrelation of design data of electromagnets with power indicators of their work. In the polarized electromagnets the solution of a problem of optimum design becomes complicated existence in their design of permanent magnets. The offered mathematical model of dynamics of movement of an anchor of an electromagnet uses results of numerical calculation of a magnetic field in a complex of the ELCUT programs at the fixed values of a working gap within a course of an anchor and approximation of dependence of inverse inductance of the coil from an anchor course the straight line equation. Realization of the offered algorithm in a

Matlab/Simulink package gives good coincidence to experimental data that confirms legitimacy of the accepted assumptions at creation of model of dynamics of an electromagnet [2].

2. Fundamentals

Transition process at inclusion of a winding of the polarized electromagnet to a source of constant tension at a fixed value of a working gap happens not only in an electric chain of a winding, but also in the magnetic field distributed in the space surrounding winding rounds, the massive steel core (anchor) and a non laminated magnetic core. The physical processes happening in an electric chain and magnetic field of an electromagnet at transition process are among themselves interconnected. Therefore their separate consideration when calculating transition process with use of the assumptions simplifying a task isn't always justified and can not only exceed an admissible error of calculation, but also yield wrong result if not to consider influence of vortex currents in massive steel parts of a design of a magnetic conductor, steel saturation, true geometry of magnetic system of an electromagnet.

In this article calculation of non-stationary magnetic field of the polarized electromagnet with a motionless anchor with a fixed value of a working gap when giving on a winding of constant tension of the power supply is considered. The modern information technologies provided by a complex of the ELCUT 6.0 programs (the professional version) allow to calculate non-stationary magnetic field of the polarized electromagnet with the attached electric chain (fig. 1). When performing calculation of non-stationary magnetic field in a complex of programs ELCUT[3]the following assumptions are accepted:

- magnetic field of the polarized electromagnet is plane-parallel;
- magnetic permeability and electric conductivity of the used environments are constants;
- in the studied design of the polarized electromagnet the magnetic conductor is made laminate, that is he is taken from stamped sheets of electrotechnical steel therefore in calculation the conductivity of steel is small and other than zero. Calculation of non-stationary magnetic field is made by means of vector magnetic potential. The field of modeling is limited to a rectangle the sizes of 50 mm of 80 mm. Windings of an electromagnet consists of two consistently connected coils. The number of rounds of each coil is equal to 330.

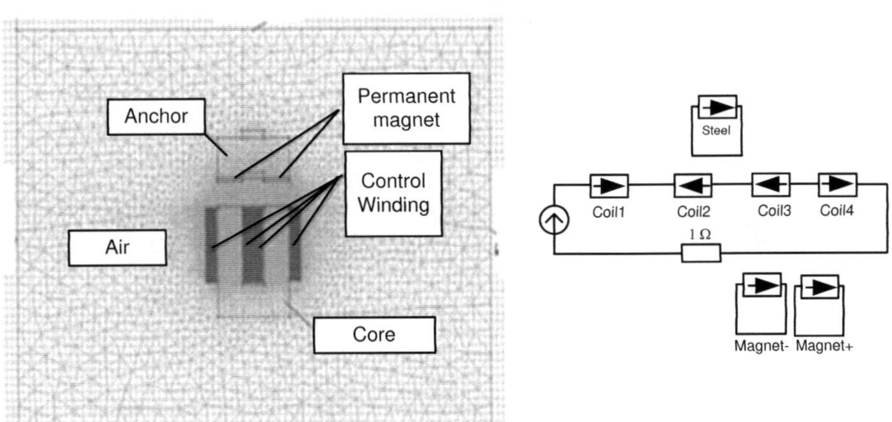

Figure 1. Geometry of model of the polarized electromagnet and the attached electric circuit in ELCUT

3. Theory

Mathematical problem definition is based on the joint solution of the equations describing the magnetic non-stationary field of the polarized electromagnet and an electric chain of a winding at connection her to a source of constant tension under the set entry and boundary conditions in a complex of programs ELCUT [4 – 6]. At the time of switching current is equal in a winding to zero. On borders of settlement field of modeling for all the time of transition process the value of vector magnetic potential is equal to zero. Windows of a task of physical properties of the allocated objects are shown in a complex of programs ELCUT in Table 1.

IOP Conf. Series: Journal of Physics: Conf. Series **1050** (2018) 012086 doi:10.1088/1742-6596/1050/1/012086

Table 1. The Material Property

№	Material	Property	Value
1	Air	Permeability	1
2	Coil 1	Permeability	1
		Electrical Conductivity, Siemens/m	5.8e+7
		Conductor Scheme	Serial
3	Steel	Permeability	1000
		Eelectrical Conductivity, Siemens/m	0.2
4	Magnet -	Permeability	1
		Coercive Force, A/m	8.55e+5
		Direction	90^0
		Electrical Conductivity, Siemens/m	70

4. Results

Results of calculation of non-stationary magnetic field and temporary dependence of current in a winding are shown in fig. 2 and 3.

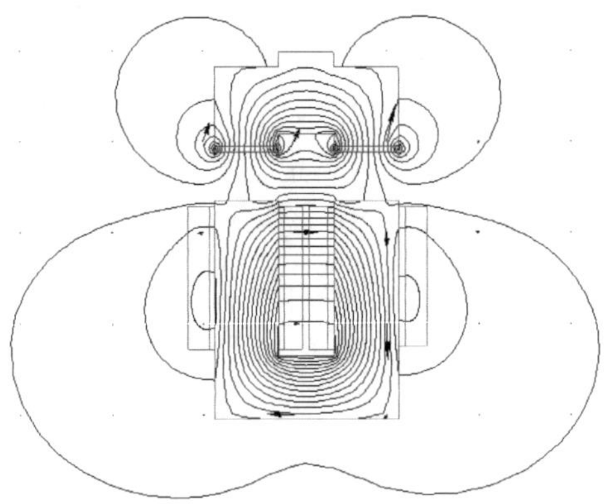

Figure 2. The picture of magnetic field upon termination of transition process for timepoint 0.1 s

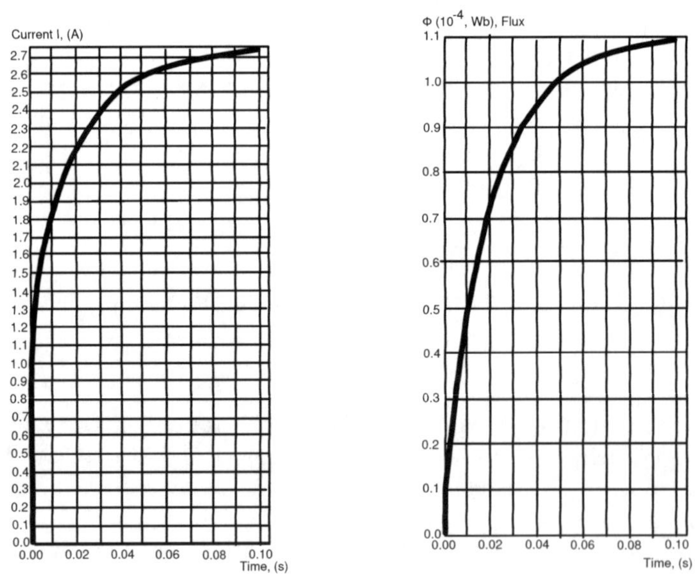

a) b)

IOP Conf. Series: Journal of Physics: Conf. Series **1050** (2018) 012086 doi:10.1088/1742-6596/1050/1/012086

Figure 3. Results of calculation of non-stationary magnetic field of the polarized electromagnet: temporary dependences of current (a) and magnetic flux in the core (b)

5. Conclusions

The developed model of non-stationary magnetic field in a complex of programs ELCUT allows to determine by calculation temporary characteristics of the polarized electromagnet at transition process on the basis of the solution of a chain and field task. The presented mathematical model of the dynamics movements of the armature of an electromagnet, the results of a numerical calculation of the magnetic fields in the complex of ELCUT programs and approximation dependence of inverted coil inductance from the course of the anchor by the equation of a straight line, Possibility to construct an algorithm for solving nonlinear systems of differential equations of dynamics electromagnet during the transient process.

6. References

[1] Kerem K, Afzal S, Edward J P 2016 Analytical modeling of eddy current brakes with the application of time varying magnetic fields, *Applied Mathematical Modelling* (Canada), Vol 40 pp 1168–1179

[2] Tatevosyan A S, Ogorelkov B I, Polyakov D A 2015 Experimental research and mathematical modelling of dynamics of the electromagnet of the direct current *Electrotechnical and data processing facilities and systems*. Vol 1 (11) pp 5–14

[3] *ELCUT program. User's Guide* 2012 – LLC Tor, (St. Petersburg) 356 p

[4] Kurbatov P A 2007 *Modeling Electromechanical Systems of Electromechanically Apparatus* (Moscow, MEI) 110 p

[5] Evstifeev A V 2007 *The Microcontrollers AVR series: User's Guide* (Moscow) 592 p

[6] Zaharova N S, Tatevosyan A S 2013 *Experimental investigation and mathematical modeling of surface effect in current conductor, located in ferromagnetic groove Electrotechnical and Data processing facilities and systems*. (Ufa) Vol 4 (9) pp 5–12

Metallographical researches results of titanium alloy Ti-6Al-4V samples obtained by using traditional and additive technologies

A B Tkhabisimov, A F Mednikov, A Yu Marchenkov, O S Zilova and A A Burmistrov

National Research University «MPEI», Moscow, 111250, Russia

E-mail: TkhabisimovAB@mpei.ru

Abstract. Technologies of additive manufacturing are increasibly used in various industries, including power engineering. It is known that the leading foreign turbo-building companies actively implement such technologies and successfully produce full-size products. Nevertheless, there are many questions related to the operation reliability of such equipment, in particular, resistance to various types of wear.
The purpose of this work was to establish the possibility of titanium alloy Ti-6Al-4V samples obtaining, which is used in turbine construction, with the help of additive technologies to solve problems on the study of their properties and comparison with similar characteristics of samples obtained by the traditional technological method.
As a result of the work, the properties of both types of samples were determined and their comparative analysis was carried out. An almost identical chemical composition of both types of samples was found to comply with the ASTM B265 standard for the Ti-6Al-4V alloy. It is shown that the hardness and microhardness of the investigated both types of samples is the same and amounted to 31 ± 5 HRC and 390 ± 10 HV0.1, respectively. The surface roughness of the «printed» samples is not less than 4 times higher than for the samples manufactured by the traditional technological method, which requires it's additional processing for further research. The microstructure of the alloy obtained by using additive technologies is acicular martensite formed as a result of rapid cooling from the high-temperature beta region.

1. Introduction

Today, one of the promising areas for manufacturing complex parts and components is the method of additive technology or 3D - printing of metal products [1]. Recently, there has been an increased interest in these technologies in the field of mass production, measured in hundreds of units [2]. There are a lot of products made from special materials, often complex geometry, in the aviation industry, the space industry, power engineering and a number of other industries [3].

In these industries there is an increased interest in 3D - printing technologies, as an alternative to traditional technological methods for the production of non-prototypes or prototypes, and products that will be actively used in real objects. The economic feasibility of using this kind of technology is explained by the fact that in some cases, with objective calculations of real costs, additive technologies are less expensive than traditional ones.

At the same time, not only technological quality control is important for the reliable use of additive manufacturing technologies, but also testing the product for resistance to various types of wear, to which the equipment, in particular, solid particle erosion, is exposed. This type of test is regulatory [4] and mandatory for any new manufacturing technology and / or subsequent strengthening of products exposed to gas-abrasive flow. First of all, it concerns the blades of guiding [5] and rotating devices of the powerful power gas-turbine plants compressors first stages [6] and the high and medium pressure cylinders first stages of powerful steam turbines [7].

Carrying out solid particle erosion tests will first of all give the kinetics fundamental dependencies of this process for samples obtained by using innovative additive technologies. To understand the wear resistance of «printed» sample complete physical picture it's important to imagine the process of material manufacture and how great is the difference from the material manufactured by traditional method. Therefore it's necessary to preliminary studies the properties of the sample, the analysis of which can later explain the mechanism of surface destruction during the solid particles impact.

2. Equipment and research methodology specification

Using the traditional technological method (Figure 1) and additive technology (Figure 2) two batches of Ti-6Al-4V titanium alloy samples were manufactured for metallographic researches and subsequent solid particle erosion resistance tests.
The sample manufacturing process by the traditional method included the following stages:
– cutting titanium alloy Ti-6Al-4V bars with a diameter 36 mm into 5 mm long billets;
– material sampling to diameter 34 mm at length 1 mm from the sample edge;
– deburring the sample side edges and flange;
– grinding sample surface on a grinding machine to the required grade of smoothness.
During the sample manufacturing were observed the following technical requirements:
– samples were made of a single material melt with allowances and landings;
– sample heating during mechanical processing in the manufacturing process did not cause structural changes and physicochemical transformations in the metal;
– modes and sequence of the tooling minimized cold-hardening and eliminated the local overheating of the sample working surface during grinding;
– during the sample manufacturing the technological facets have been removed, and sharp edges are blunted up to radius 0.5 mm;
– sample working surface did not contain any corrosion signs, temper colors, etc.

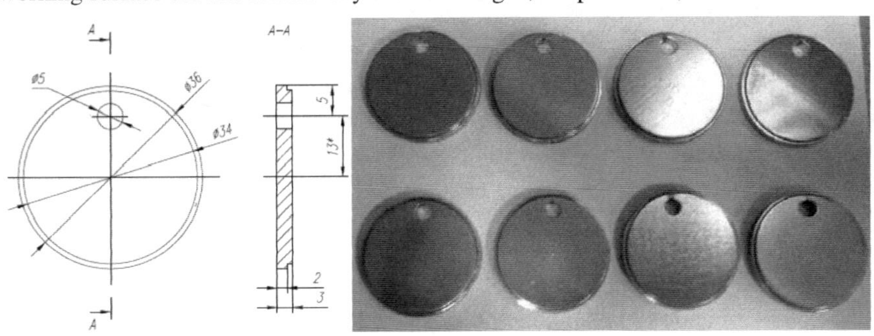

a) b)

Figure 1. Samples made by traditional technological method.

Samples produced by using additive technologies were made of titanium powder Ti-6Al-4V Grade 5. The manufacturing process consisted of the following: device for applying and leveling the layer removed the powder layer from the feeder and distributed it uniformly over the substrate surface. After that, the laser beam scanned this layer surface and formed a product by sintering. At the end of powder layer scanning the platform with manufacturing product has been gone down for deposited layer thickness and the platform with powder has been raised, and the process of applying the powder layer and scanning repeated.

a) b)

Figure 2. Sample's computer model (a)
and samples produced by additive technologies (b).

IOP Conf. Series: Journal of Physics: Conf. Series **1050** (2018) 012087 doi:10.1088/1742-6596/1050/1/012087

After this process consummation unused powder deleted from platform. The chemical compositions of the both materials types are shown in Table 1.

Table 1. Titanium powder and rod Ti-6Al-4V bar.

Element	Ti-6Al-4V Powder	Ti-6Al-4V Bar
	Mass (%)	
Titanium (Ti)	88÷90	89.2
Aluminum (Al)	5.5÷6.5	5.9
Vanadium (V)	3.5÷4.5	4.1
Iron (Fe)	< 0.25	0.2
Oxygen (O)	< 0.13	0.1
Carbon (C)	< 0.08	0.07
Nitrogen (N)	< 0.05	0.03
Hydrogen (H)	< 0.012	0.01

The following measuring equipment URI «Hydroshock rig «Erosion-M» was used to implement the planned metallographic studies:
– TESCAN MIRA 3 LMU scanning field emission electron microscope - to study the both sample types morphology and structure;
– X-Max 50 X-ray energy dispersive spectrometer (EDS) - to obtain the full spectrum of X-ray elements radiation, to carry out qualitative and quantitative elemental analysis of metallographic samples sections with locality on the order of 1 μm;
– Dektak 150 mechanical profilometer - to study the surface morphology, evaluation the material wear of the experimental sample after tribological tests, construction and analysis the surface profiles;
– DuraScan 20 hardness tester for small loads - to determine the surface microhardness in accordance with GOST 9450 76, ISO 6507 and ASTM E384 standards;
– M4C hardness tester - to control the material macrohardness in accordance with the Vickers hardness test method – GOST 2999-75 standard (also ISO 6507, ASTM E-10 standards) at a load of 10 kgf;
– TRB-S-CE-0000 tribometer – to determine the friction coefficient of sample by the «Ball-plane» method in the chosen load values, speed and mileage length.
Following machines by Buehler GmbH company were used to prepare sample for researches: an abrasive cutting machine with a linear movable cutter system PowerMet 3000, an automatic press Simplimet 1000, cold casting apparatus Cast n' Vac 1000 and a grinding machine Beta with an automatic nozzle Vector and a modular system of liquid distribution PriMet 3000.
The structure of both samples types after obtaining the metallographic section and it's subsequent etching was studied at the TESCAN MIRA 3 LMU scanning field emission electron microscope. For metallographic analysis were used a series of standards: GOST 10243-75 «Steel. Methods of testing and macrostructure evaluation»; GOST 8233-56 «Steel. Standards of microstructure»; GOST 9391-80 «Alloys hard sintered. Methods for determination porosity and microstructure»; GOST 5639-82 «Steels and alloys. Methods for the detection and determination the grain size».
Tribological tests of samples were conducted at the following conditions:
– movement scheme - linear reciprocating sample motion relative to a stationary counterbody with a specified displacement amplitude;
– displacement rate – 0.05 m/s;
– applied load– 1 N;
– stroke length (displacement amplitude) – 2 mm;
– mileage length (friction path) – 200 m;
– counterbody – ball ∅6 mm, material – WC.

3. Results and discussion
The metallographical sections of the titanium alloy Ti-6Al-4V samples produced by additive technologies and by traditional technological method are shown in Figure 3 and Figure 4.

IOP Conf. Series: Journal of Physics: Conf. Series **1050** (2018) 012087 doi:10.1088/1742-6596/1050/1/012087

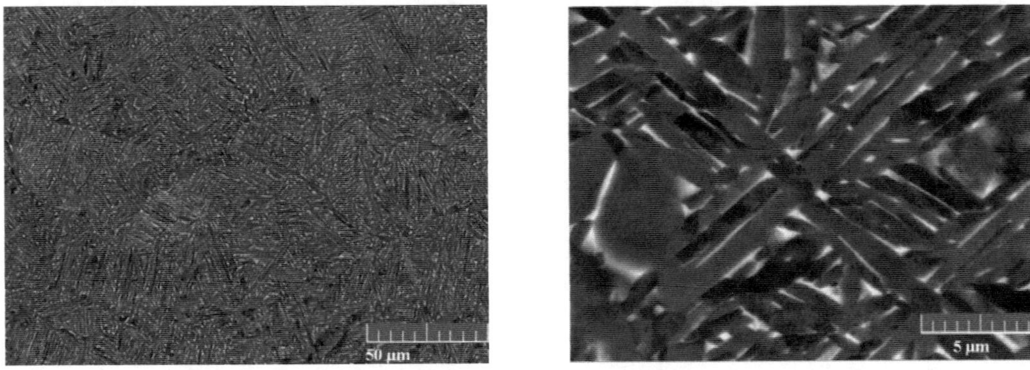

Figure 3. Microstructure image of titanium alloy Ti-6Al-4V sample produced by additive technologies at various multiplications

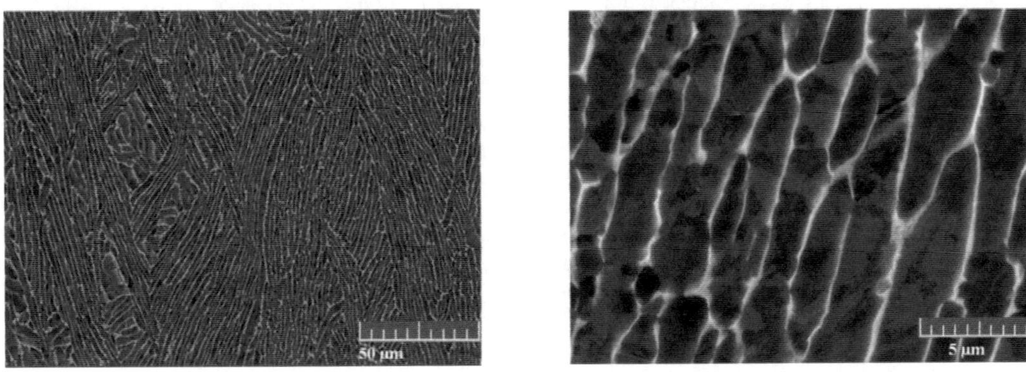

Figure 4. Microstructure image of titanium alloy Ti-6Al-4V sample produced by traditional technological method at various multiplications

The microstructure of the alloy obtained by using additive technologies is acicular martensite formed as a result of rapid cooling from the high-temperature beta region. The microstructure of the alloy manufactured by the traditional process is represented by elongated in the direction of deformation alpha-phase grains with beta-phase streaks located along their boundaries.

After obtaining images of transverse sections and studying their structure, an analysis of the samples elemental composition was performed using the energy-dispersive microanalysis. In this case, spectrum of X-ray radiation were taken in different regions. The spectrum characterizes the relative intensities of the sample characteristic radiation lines, allows qualitatively to reveal the presence of various elements in it, and also estimate their relative content. Thus, on the basis of the obtained spectrum, was determined the elements content in the samples. The results of manufactured samples chemical composition analysis are shown in Table 2.

Table 2. Alloy surface components

No	Samples characteristic	Compositions, %		
		Al	*Ti*	*V*
1	(Ti-6Al-4V, additive technologies)	5.19	91.1	3.74
2	(Ti-6Al-4V, traditional method)	5.97	89.4	4.61

The element analysis carried out showed that the main material of the manufactured samples of both types is titanium, it's content varies in the range 89-92%, the aluminum content is about 5.2-9.9%, the vanadium content is about 3.7-7.6% . The results of tribological tests are shown in Table 3.

IOP Conf. Series: Journal of Physics: Conf. Series **1050** (2018) 012087 doi:10.1088/1742-6596/1050/1/012087

Table 3. The friction coefficient change as a function of the sample's surface mean free path.

Material	
Ti-6Al-4V (traditional method)	**Ti-6Al-4V (additive technologies)**
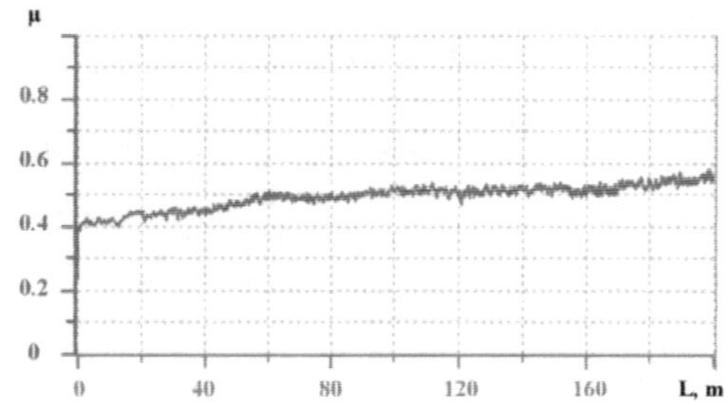	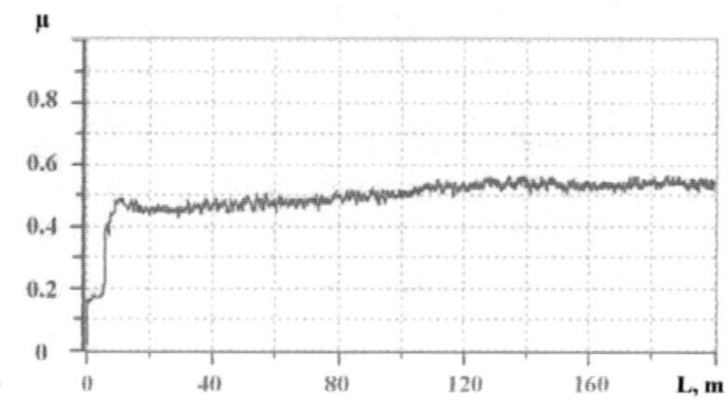
Average friction coefficient $\mu_{av} = 0.5$	Average friction coefficient $\mu_{av} = 0.5$

Conducted researches have shown that both types of samples showed the same value of the average coefficient of friction. This circumstance attests to the high quality of the Ti-6Al-4V powder layers «sintering» and allows us to assert that for this parameter such method of obtaining a sample or billet is not inferior to the well-known hot-rolling process.

The roughness was determined from five profiles, measured at different parts of the each sample surface. The following roughness parameters were calculated: the arithmetic mean deviation of the profile (Ra) and the height of the profile unevenness measured at ten points (Rz). The obtained surface profiles of Ti-6Al-4V titanium alloy samples, manufactured by using additive technologies and the traditional technological method, are shown in Figure 5.

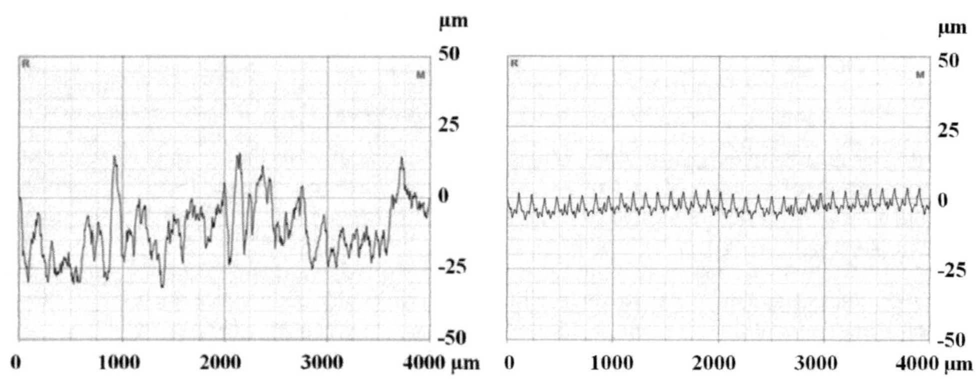

Figure 5. Profilograms of the samples surface:
a) (Ti-6Al-4V, additive technologies); b) (Ti-6Al-4V, traditional method)

A DuraScan 20 (Emco-Test) hardness tester was used to determine the samples microhardness. The measurements were carried out according to the Vickers method at not less than 5 points of each sample surface at a load of 0.05 kgf (0.491 N). The results of measuring surface roughness, hardness and microhardness of the sample surface are shown in Table 4.

Table 4. Results of the roughness, hardness and macrohardness measurement of the test samples surface

No	Samples characteristic	Ra, µm	Rz, µm	HV 0.1	HRC
1	(Ti-6Al-4V, additive technologies)	6.59	45.3	390±10	31±5
2	(Ti-6Al-4V, traditional method)	1.67	9.2	390±40	31±5

From presented profilograms it can be see that the surface roughness of the samples produced by additive technologies with respect to parameters Ra and Rz not less than in 4 times higher than those for samples manufactured by the traditional method.

IOP Conf. Series: Journal of Physics: Conf. Series **1050** (2018) 012087 doi:10.1088/1742-6596/1050/1/012087

4. Conclusions

The complex of metallographic studies showed that the titanium alloy Ti-6Al-4V composition, which is widely used at present for the manufacture of parts by additive technologies, matches the requirements of the ASTM B265 standard.

Manufactured samples have significant differences in microstructure. In the case of additive technologies, this is acicular martensite; in the case of using the traditional method, the microstructure of the alloy is represented by elongated in the direction of deformation alpha-phase grains with beta-phase streaks located along their boundaries.

The values of hardness, microhardness and average coefficient of friction in both types of samples are the same. The results of measuring the roughness of Ra or Rz indicate that the surface of products manufactured by using additive technologies must be further processed.

The obtained data can be used as a necessary knowledge base for carrying out experimental studies of the resistance to hard abrasive particles impact and the surface destruction mechanisms description of samples manufactured by using additive technologies, under different conditions.

Acknowlegements

The study results have been achieved thanks to financial support of the Russian Science Foundation pursuant to the Agreement No. 17-79-10460 of July 28, 2017.

References

[1] Gu D, Hagedorn Y C, Meiners W, Meng G, Santos Batista R J, Wissenbach K, et al. Densification behavior, microstructure evolution, and wear performance of selective laser melting processed commercially pure titanium *Acta Mater* 2012 **60** pp 849–60

[2] Wohlers T T and Caffrey T *Wohlers report 2015*: 3D printing and additive manufacturing state of the industry annual worldwide progress report (Wohlers Associates: Fort Collins)

[3] Baufeld B, Van Der Biest O and Gault R Additive manufacturing of Ti–6Al–4V components by shaped metal deposition: microstructure and mechanical properties *Mater Des* 2010 **31** pp 106–11

[4] RD 153-34.1-17.462-00. Methodical instructions on the procedure for assessing the operability of steam turbines blades in the manufacture process, operation and repair

[5] Reschetnik W, Bruggemann J-P, Aydinoz M E, Grydin O, Hoyer K-P, Kullmer G and Richard H A 2016 Fatigue crack growth behavior and mechanical properties of additively processed EN AW-7075 aluminium alloy *21st European Conference on Fracture* ECF21 20–24 June Catania Italy Procedia Structural Integrity **2** pp 3040–48

[6] Leuders S, Vollmer M, Brenne F, Stroster T and Niendorf T 2015 Fatigue strength prediction for titanium alloy TiAl6V4 manufactured by selective laser melting *Metallurgical and materials transactions* A **46**A pp 3816–23.

[7] Riemer A, Leuders S, Thöne M, Richard H A, Tröster T and Niendorf T 2014 On the fatigue crack growth behavior in 316L stainless steel manufactured by selective laser melting *Engineering Fracture Mechanics* **120** pp 15–25

Testing the Printing Systems for Enlarging Color Reproduction

O V Trapeznikova[1], L G Varepo[1], I A Sysuev[1], O A Novoselskaya[2] and L S Kostuchkova[1]

[1]Omsk State Technical University, 11, Mira ave., Omsk, 644050, Russia
[2]Belarusian State Technological University, 13a, Sverdlova str., Minsk, 220006, Belarus

E-mail: ol-trapeznikova@yandex

Abstract. The process of manufacturing multicolor printed matter is rather complicated as the components of printing system are diverse by the configuration and have a complex of various properties that in aggregate form production quality. Difficulties in solving the problem of ensuring the quality of a multicolor image are determined with the instability of the properties of printing system components, their diversity, both in physicochemical and structural-mechanical properties. All this causes different divergence of the process parameters and does not allow obtaining a qualitatively stable ready-made product. The results of investigation of the possibilities in enlarging color reproduction of a printing system the "inkjet printer – paper" are presented in the work. The role of paper sheet characteristics is noted. It is shown that the priority belongs to paper with a uniform, microporous surface.

1. Introduction

Now in graphic arts the percentage of inkjet printing is increasing due to the operability of obtaining production prints of pictures and the permanently updated assortment of printing systems components. The requirements to improve the quality and competitiveness of Russian multicolor printed matter are increasingly growing due to the constantly rising requirements from the consumers.

The main index of printed matter quality is the color of the print, as a result of interactions between the printing properties of a paper carrier and inks during their spreading and fixing in complex, in a printing press. In current classification of types of printed paper a paper for a contactless printing, i. e. copier, inkjet, electrographic, etc. is marked out in a separate group. To the paper for digital printing, there are requirements that are different from those demanded for printed materials for classical printing methods. The paper for copier must be sufficiently uniform and smooth. The toner fixing in this case will depend on the optimum balance between the surface topography and the thermal properties of the printing material. In inkjet paper a balance between surface topography, porous structure, chemical composition and mechanical strength should be kept. The requirement for such paper are: it must hold color pigments close to the surface, limit and control the spread of the ink, control the amount and speed of its absorption, sustain the ink vehicle until its complete evaporation, form a uniform printed surface. At present, the correlations between the printing and surface properties of paper are not exactly investigated. They could let the forecasting of the printing quality for existing materials for digital printing and are of great practical interest.

2. Statement of the problem

The problem of color reproduction today is successfully solving with the help of color management systems, in particular, CMS. Nevertheless, due to a number of limitations of printing synthesis, the color of the original can be reproduced on the print only identically subjectively, and specified first of all because of the limited range of printing synthesis. Thus it is always actual, to expand the color gamut of the printing system which is achieved in inkjet printing by seeking for optimal printing settings, improving the properties of inks, matching a paper carrier and modifying its surface.

For the optimal choice of printed system components during reproduction of the originals with different complexity groups by the inkjet printing various approaches are known. Improving the quality of inkjet printing, according to [1], can be achieved by modifying the paper surface with surfactants and a cationic polymer. As a result, this treatment has a slight effect on its structural and surface properties, but nevertheless improves the quality of inkjet printing (an increase in gamma area is up to 12 per cent in estimation of color properties). The results of the printing were interpreted by the authors, taking into account the nature of inks and pigments and the molecular structure of the inks, estimated using the IR spectroscopy method. However, despite the positive results achieved, in practice, this type of processing should be carried out directly in the process of paper production.

The key point in the work [2] is the investigation of the fixation of ink colorants in the surface and volume coating layers. The authors of [2] specify that the most important properties of surface coating are the characteristics of the porous structure – the pore size, the amount and the net created by them - and the chemical nature. It, by the authors' opinion, are responsible for ensuring correct fixation of the colorant on paper and quality indexes: high optical density, sharpness, abrasion resistance, etc. In this field of research papers [3–11] are of interest but with the difference that they consider the other methods of printing, mainly offset printing.

A similar opinion is held by the authors of [12]. They believe that coating of the inkjet paper surface should be planned in such a way as to provide good printability with the use of the structural parameters, the pore grid, the chemical composition of the coating, and the properties of the polymer additives. The main direction in this work is to find out how the ions charge and in particular its distribution in the coating layer combines with the binder to influence on the properties of colorant fixing. The aim of the work is to reveal the possibility of enlarging the color reproduction in the printing system "inkjet printer – paper".

3. Materials and Methods

The objects of research are the components of printing systems: samples of paper for digital printing of leading manufacturers in accordance with the assortment of the Russian market, and inkjet systems: Riso (printing system No. 1) and Epson (printing system No. 2).

The measurement of the color coordinates was carried out using a SpectroEye spectrophotometer from Gretag Macbeth. The results were compared using the Adobe PhotoShop CS software. The test chart contains CMYK-colors and their binary overlays (RGB): cyan, magenta, yellow, black and thus red, green, blue. Fillings of the chart fields was made with help of software and contain the dots of 100, 80 and 40% (Figure 1).

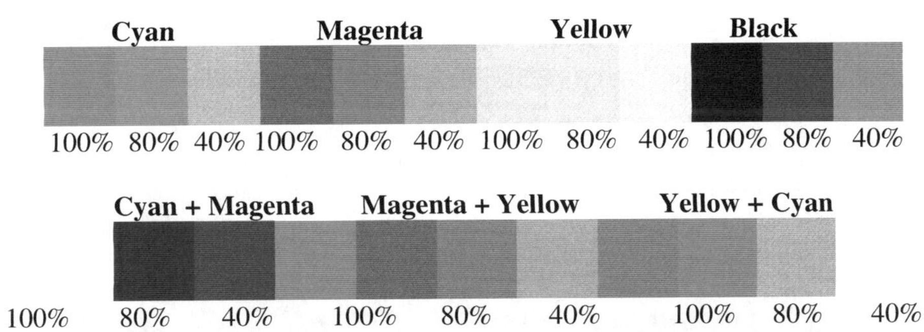

Figure 1. A Test Chart

To determine the surface characteristics of the paper samples tested, a non-contact profilometry method was used using the Micro Measure 3D station [13]. In order to exclude defective samples, their geometric uniformity was preliminarily estimated, the necessary procedures for acclimating the paper were carried out.

IOP Conf. Series: Journal of Physics: Conf. Series **1050** (2018) 012088 doi:10.1088/1742-6596/1050/1/012088

4. Experimental Results

The results of the experiments are presented in Tables 1–5. Table 1 shows the scanned area of the surface profile of the tested paper samples in X and Z coordinates with the calculation of the area of peaks S_{peak} and valleys S_{valley}.

Table 1. Surface Profile of Testing Paper Samples

Number of paper sample	Profile by X and Z with the calculation of peaks S_{peak} and valleys $S_{вп}$ of a surface
1	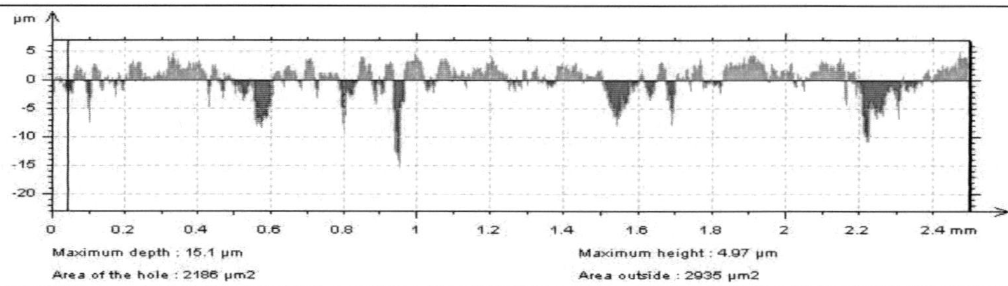
2	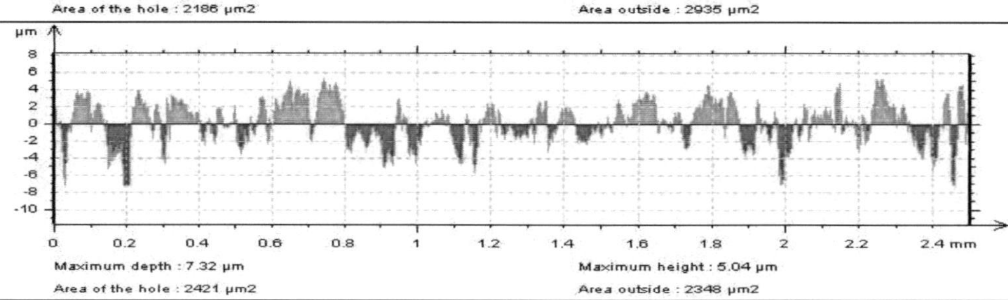
3	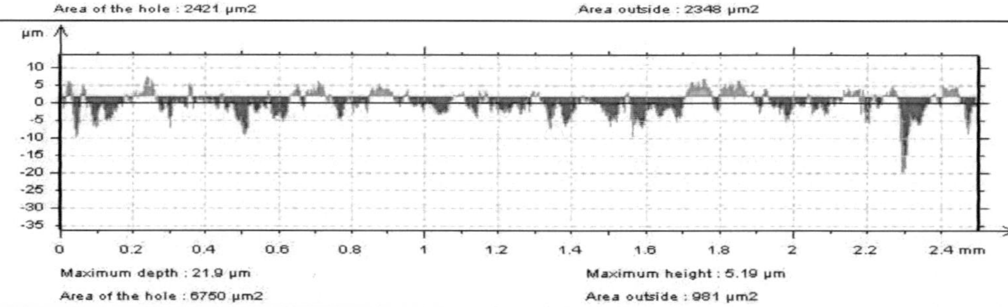
4	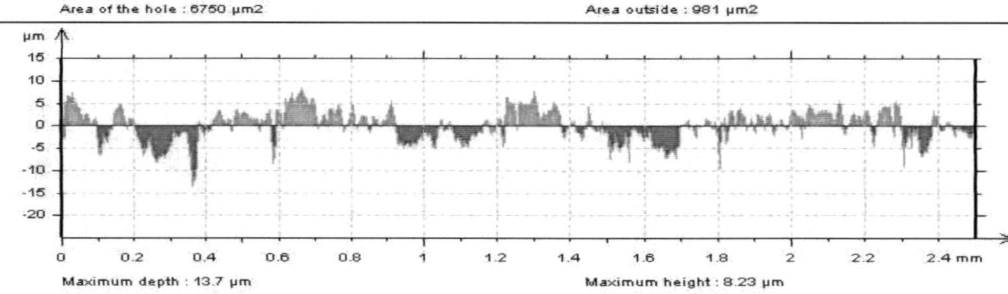
5	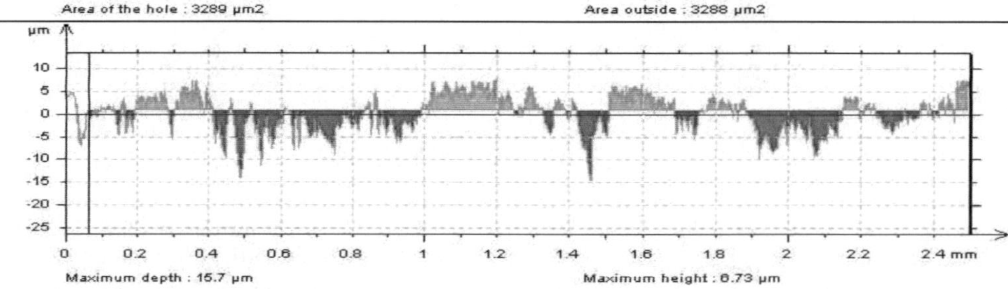

Tables 2–4 show the results of experiments of printing systems.

Table 2. A Test Chart (upper line) and its print (lower layer) by different printing systems

Printing System No.1				**Printing System No.2**			
C	M	Y	K	C	M	Y	K
Paper No. 1							

Paper No.2							
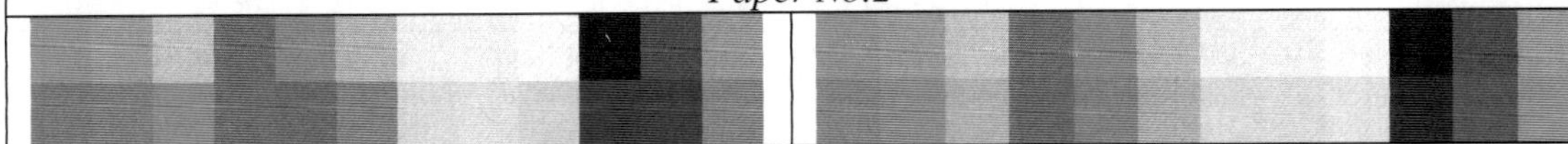							
Paper No. 3							

Paper No.4							

Paper No. 5							

Printing System No.1			**Printing System No.2**		
B	R	G	B	R	G
Paper No.1					

Paper No.2					

Paper No.3					

Paper No.4					

Paper No.5					

Table 2 shows the results of printing a test chart by various printing systems: solid colors are the color fields with 100 % fill; continuous-tone elements – the color fields with a print, representing a certain

IOP Conf. Series: Journal of Physics: Conf. Series **1050** (2018) 012088 doi:10.1088/1742-6596/1050/1/012088

percentage of the solid fill of 80 % and 40 % and are used for measuring the contrast of printing and tone change; binary overlays (RGB fields) – printing one color over another – cyan /magenta, magenta/ yellow and cyan / yellow. Table 3 and 4 show the results of measurements of the optical densities for the obtained printouts sample of the test chart in front.

Table 3. Optical densities of the fields of basic colors obtained by the printing of test chart with different printing systems

Number of paper sample	Field	Printing System No.1				Printing System No.2			
		C	M	Y	K	C	M	Y	K
1	100 %	0.81	0.83	0.84	0.95	0.71	0.99	1.00	1.00
	80 %	0.80	0.82	0.83	0.91	0.69	0.71	0.79	0.71
	40 %	0.72	0.58	0.73	0.46	0.35	0.34	0.34	0.36
2	100 %	0.85	0.86	0.88	0.98	0.76	1.00	1.00	1.00
	80 %	0.84	0.86	0.87	0.90	0.73	0.75	0.75	0.78
	40 %	0.79	0.58	0.73	0.44	0.35	0.34	0.32	0.36
3	100 %	0.82	0.82	0.85	1.00	0.71	1.00	0.98	1.00
	80 %	0.82	0.83	0.86	0.93	0.66	0.71	0.73	0.72
	40 %	0.70	0.56	0.73	0.44	0.33	0.32	0.32	0.35
4	100 %	0.78	0.83	0.83	0.91	0.69	1.00	1.00	1.00
	80 %	0.78	0.82	0.82	0.83	0.65	0.76	0.77	0.71
	40 %	0.67	0.58	0.71	0.44	0.33	0.33	0.34	0.33
5	100 %	0.83	0.84	0.85	0.98	0.69	0.96	0.99	1.00
	80 %	0.83	0.87	0.85	0.92	0.66	0.68	0.76	0.71
	40 %	0.72	0.61	0.74	0.47	0.32	0.30	0.34	0.33

Table 4. Optical densities of the fields of binary overlays of basic colors obtained by the printing of test chart with different printing systems

Number of paper sample	Field	Printing System No.1			Printing System No.2		
		R	G	B	R	G	B
1	100 %	0.86	0.73	0.77	0.89	0.87	0.94
	80 %	0.83	0.73	0.76	0.88	0.77	0.77
	40 %	0.76	0.70	0.71	0.45	0.37	0.41
2	100 %	0.89	0.80	0.83	0.94	0.91	1.00
	80 %	0.85	0.80	0.81	0.91	0.86	0.82
	40 %	0.76	0.72	0.71	0.45	0.37	0.42
3	100 %	0.85	0.76	0.79	0.88	0.87	0.91
	80 %	0.82	0.76	0.78	0.89	0.77	0.75
	40 %	0.74	0.70	0.70	0.46	0.35	0.40
4	100 %	0.84	0.77	0.79	0.89	0.88	0.91
	80 %	0.82	0.76	0.76	0.88	0.78	0.75
	40 %	0.74	0.69	0.68	0.47	0.35	0.39
5	100 %	0.86	0.77	0.89	0.89	0.90	0.92
	80 %	0.82	0.77	0.78	0.87	0.77	0.76
	40 %	0.76	0.72	0.73	0.50	0.38	0.39

Fig. 2, 3 presents the hexagons of color gamut for printing systems No. 1 and No. 2, constructed from the results of measuring the color coordinates of 100% solid pure colors (C, M, Y, R, G, B) in color system L *a* b –1976.

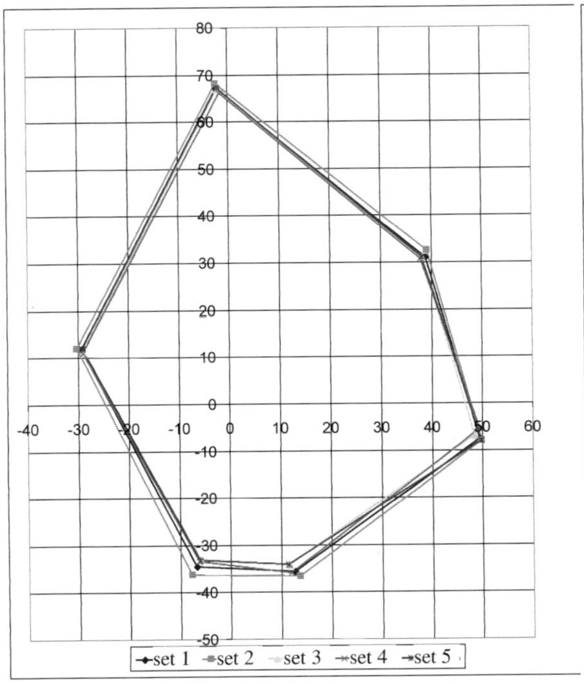

Figure 2. Hexagrams of color gamut of printing system No. 1 on the chromaticity diagram a*b* (the abscissa axis is a^{*M}, the ordinate axis is b*; sets 1–5 of paper No. 1–5, respectively)

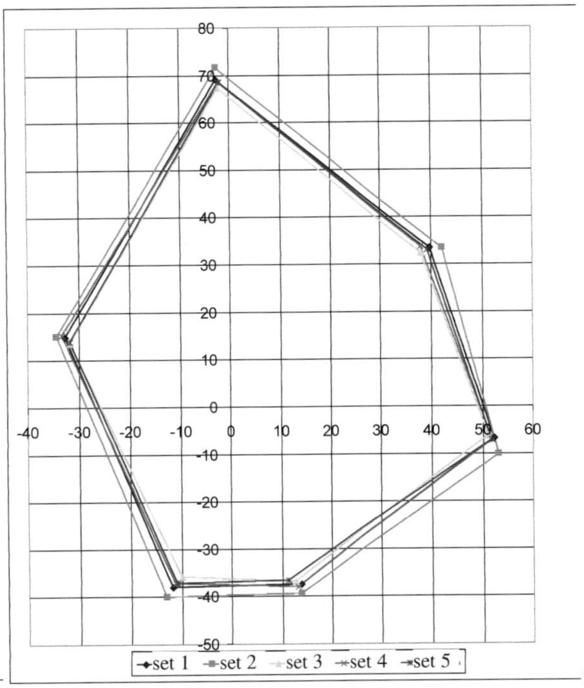

Figure 3. Hexagrams of color gamut of printing system No. 2 on the chromaticity diagram a*b* (the abscissa axis is a^{*M}, the ordinate axis is b*; sets 1–5 of paper No. 1–5, respectively)

5. Results and Discussion

Estimation of the print printed on paper No. 1 by the printing system No. 1 showed that 100 % and 80 % fields practically not differ, and a 40 % field has an optical density much subjects to the digital original. On the contrary, the print obtained with the printing system No. 2 has better characteristics in compare with the printing system No. 1. The colors are much more saturated and closer to the digital original, and the color gamut (Figures 1, 2) is much larger. It is noted that 100 % and 80% fields are sufficiently recognizable, and a 40 % field has an optical density close to the original. At the same time, the yellow color on the paper differs from the digital original. The reproduction of the color with minimal deviation is seen on the the printing system No. 1 with the improved technology of forming ideally round points and drops of variable size for solids and small details of the image.

Prints on paper No. 2, obtained in the same conditions as compared to prints on other test paper samples, have a color rendition that is more identical to the original, and the color gamut on this paper is larger than on the others (Figures 1, 2). This paper sample is characterized by the most uniform closed surface (depth of the depressions is 7.32 μm, height of peaks is 5.04 μm) with an average pore depth of 1.25 μm (Table 1, position 2)

Testing of the prints obtained on paper No. 3 showed that the red and green fields are very brighten, a tinge of magenta hues is noticeable, which is clearly visible on the green field (Table 2, position 3).

On the test chart of the paper sample No. 4 with an average pore depth of 1.96 μm, deviations of fills for 100 and 80 % fields are observed, especially in reproduction of yellow; because of the strong darkening of colors, the 40 % field has a fill close to 80 % (Table 2, position 4).

For the prints on sample paper No. 5, all three fields are practically a single-color solid, this is clearly visible on binary colors, probably due to the lower whiteness of the paper and the less uniform surface of the paper: the depth of the valleys is 15.7 μm, the height of the peaks is 6.73 μm (Table 1, position 5).

6. Conclusion

Analysis of the experimental data made it possible to evaluate the correlation between the quality index of color reproduction and the characteristics of paper surface. The presence of a macroporous structure significantly reduces the quality of reproduction of a multicolour original. The obtained results show that the studied systems reproduce different amounts of colors. The percentage of reproduction of color coordinates for the main colors relatively to the test samples varies from 70 to 98 % that conform to ISO 12647-1-2 - the standard of offset printing (the variability of pure colors should be at least 68 %).

The use of testing techniques for printing system components will provide the correct choice of the printing system, the stability of printing process and forecast the expected quality by achieving the necessary conditions for proper color reproduction.

References

[1] Sónia Sousa, José António Gamelas, Antóniode Oliveira Mendes, Paulo Torrão Fiadeiro, Ana Ramos 2013 *Materials Chemistry and Physics* Vol **139** Is 2–3 pp 877–884

[2] Kettle J, Lamminmäki T, Gane P *2010 Surface and Coatings Technology* Vol **204** Issue pp 2103–2109

[3] Ozaki Y, Kimura M *Appita J.* 2000 (**3**) pp 216–219

[4] Koivula H, Preston J S, Heard P J and Toivakka M 2008 *Colloids and Surfaces A: Physicochemical and Engineering Aspects* Vol **317** Issues 1–3 pp 557–567

[5] Varepo L G, Trapeznikova O V, Nagornova I V 2015 *Procedia Engineering* **113** pp 357–361

[6] Sysuev I A, Pozharsky A O, Zakharenko A A 2006 *Omsk Science Herald* **9** (47) p 107–110 (Russia)

[7] Pozharsky A O, Sysuev I A 2006 *Proceedings of the universities. Problems of graphic arts and publishing* **4** (October–December) pp 3–12

[8] Varepo L, Golunov A, Golunova A, Trapeznicova O & Nagornova I *Testing and Measurement Conference Publications* 2015 pp 69–71

[9] Varepo L G, Brazhnikov A Yu, Volinsky A A, Kondratov A P, Nagornova I V 2017 Mechanical Science and Technology Update *IOP Publishing IOP Conf. Series: Journal of Physics: Conf. Series* **858** (2017) 012038

[10] Varepo L G, Trapeznikova O V, Trapeznikov E V, Nagornova I V 2016 *J. Procedia Engineering* **152** pp 478–481

[11] Gnatjuk S P, Domasjev M V 2008 *Proceedings of the Universities. Problems of Graphic Arts and Publishing* **6** pp 3–19

[12] Lamminmäki T T, Kettle J P, Gane P A C 2011 *Colloids and Surfaces A: Physicochemical and Engineering Aspects* Vol **380** Issues 1–3 pp 79–88

[13] Varepo L G 2010 *Proceedings of the Universities. Problems of Graphic Arts and Publishing* (**3**) pp 3–9 (Russia)

Combined hydraulic power vehicle transmission modes

N N Trushin, G V Shadskii

Tula State University, Tula, Russia

Abstract. Achieving high efficiency is a key objective of optimal vehicle drive train design. Hydraulic power transmissions with hydraulic torque converters are commonly used in wheeled and tracked self-propelled vehicles. Hydraulic torque converters are self-adjusting, and have a relatively low efficiency. The study investigates design options for self-propelled vehicle combined hydraulic power transmission. In order to improve their average efficiency, the engine power is transferred in two combined modes: single- and double-flow. We propose hydraulic power transmission layouts where the hydraulic torque converter operates with regular or summing differentials. Screw and free wheel clutches are used as drive train controls. They enable automated switchover of the torque transmissions modes. The proposed drive trains with switchable hydraulic torque converters are designed for both road and off-road self-propelled vehicles. They can be combined with robotic fixed-ratio gearboxes or as a part of a vehicle drive train consisting of a conventional clutch, and a manual gearbox.

1. Introduction

Hydraulic power transmissions (HPT) are commonly used in wheeled and tracked self-propelled vehicles. HPTs consist of a hydraulic power transmission and a fixed-ratio gearbox [1]. The reasons for extensive use of hydraulic power transmissions with hydraulic torque converters in vehicle/tractor drive trains are as follows:

– automatic ratio adjustment controlled by the drive train load as the engine torque is continuously transferred
– damping drive train torsional vibration that extends service life and ensures smoother operation of the engine and drive train components
– high capacity with relatively compact size and low weight compared to mechanical transmissions
– better drive train controllability and general vehicle handling, specifically in demanding operating conditions [2].

Typically, vehicle/tractor HPTs use a comprehensive hydraulic torque converter with interlocks. An HPT gearbox is a planetary or regular gear with gear shifters: gear clutches, synchromeshes, friction-type clutches, and brakes. An HPT also contains a hydraulic system that supplies fluid to the drive train components, lubricates interfaces, and automatically control the hydraulic torque converter and the gearbox [3].

Hydraulic torque converters are self-adjusting, but their feature low efficiency as compared to mechanical gears. Hydraulic torque converters have the lowest efficiency in a low ratio range. For this reason it is desirable to develop an HPT with a hydraulic torque converter having a high efficiency throughout the entire ratio range.

2. Problem Statement

In an HPT a hydraulic torque converter is installed either in series, or in parallel with the power flow. For the in-series option, the self-propelled vehicle engine power as a single flow is transmitted consequently through the hydraulic torque converter and the gearbox to the output shaft. In this case power losses in the drive train are progressively increased reducing the gearbox/drive train efficiency [4].

For the in-parallel option, the engine power flow is split into two flows. One of them passes through the hydraulic torque converter, and the other one passes through the mechanical gear. In double-flow HPTs, power losses are lower, and efficiency is higher as compared to single-flow drive trains. The disadvantages of double-flow HPTs are reduced converter capacity, and poorer torsional vibration damping.

There are two categories of split-flow HPTs: external and internal flow split designs. In drive trains with internal power flow split the engine power is split in a special (multi-turbine, or multi-stage) hydraulic torque converter. Multi-turbine, or multi-stage hydraulic torque converters have a high conversion ratio, but they are complicated and their application in vehicle drive trains is limited [5].

HPTs with external power flow split are commonly used in self-propelled vehicle drive trains. In such HPTs the engine power flow is split outside of the drive train. A portion of the power is carried over a hydraulic power transmission. and another portion goes through a mechanical gearbox; then the power flows are merged and supplied to the P3E output shaft. In this arrangement a portion of the power goes through a low-efficiency hydraulic torque converter, and another portion is delivered to the output shaft either directly, or through a highly efficient mechanical gearbox. Both power flows are merged at the output shaft. It results in higher overall drive train efficiency.

In double-flow HPTs the engine power is commonly split into two flows or is merges into a single flow with planetary geared differentials. Double-flow HPTs fall into two categories by the mutual arrangement of the differential and the hydraulic torque converter. The first category includes HPTs where the differential is located at the transmission input. In this case, it is a regular differential ("input differential" layout) The second category includes HPTs where the differential is located at the transmission output. In this case, it is a summing differential ("output differential" layout.)

In double-flow HPTs as the max torque ratio is reduced, its efficiency, capacity, transparency increase, and the hydraulic torque converter effective diameter decreases. Also, double-flow HPTs with differentials can be equipped with standard, commercially available hydraulic torque converters.

One way to improve hydraulic power transmission/hydraulic torque converter efficiency in operation is increasing the average transmission efficiency by combining its single- and double-flow modes. It can be done by introducing controllable components into the original hydraulic power transmission design. The extra components would enable two engine power flow modes. In the low ratio range (0...0.6) of the hydraulic torque converter a single-flow transmission/torque conversion is used; in the high hydraulic torque converter ratio range (0.6...0.85) a double-flow arrangement is used.

Figure 1 shows nondimensional properties of single- and double-flow hydraulic power transmissions vs. the hydraulic torque converter **i** ratio: 1) Transmission efficiency curves **η** for single-flow (curve 1) and double-flow (curve 2) modes; 2) Torque ratio **K** curves for single-flow (curve 3) and double-flow (curve 4) modes.

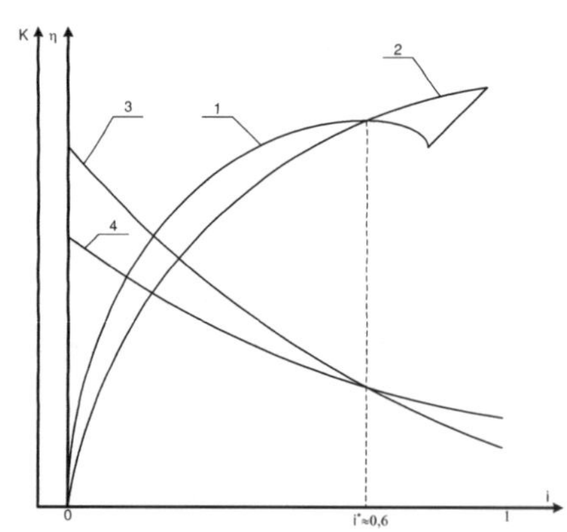

Figure 1. Efficiency (**η**) and torque ratio (**K**) variations in HPTs with single/double-flow modes

The single-flow mode is used for vehicle breakaway and acceleration. The transmission is switched over from the single- to the double-flow mode at the curve intersection (**i*** torque ratio). Besides, the single-flow mode can be utilized for extended vehicle riding in challenging or off-road conditions. While driving in favorable road conditions, the double-flow mode is more efficient, and the transmission operates without switching over to the single-flow mode.

3. Theory

Below we shall consider the original double-flow HPT kinematics used as input data for subsequent study. Dynamic and motion analysis of simple double-flow hydraulic power transmissions with regular and summing differentials are considered in the book [6].

Figure 2 shows a kinematic diagram of a simple double-flow hydraulic power transmission with a regular differential. The HPT consists of a casing (not shown), the input shaft Pos. 1, and the hydraulic torque converter Pos. 2 comprising of the impeller Pos. 3, the runner Pos. 4, and the reactor wheel Pos. 5. The reactor wheel is connected to the HPT casing with the free wheel clutch Pos. 6. The HPT output shaft Pos. 7 is connected to the runner. The input shaft is connected to the carrier Pos. 9 of the differential Pos. 8. The satellite gears Pos. 11 are attached to the carrier. The differential's central (sun) gear Pos. 10 is also connected to the runner and the HPT output shaft. The differential's crown gear Pos. 12 is connected to the hydraulic torque converter's impeller.

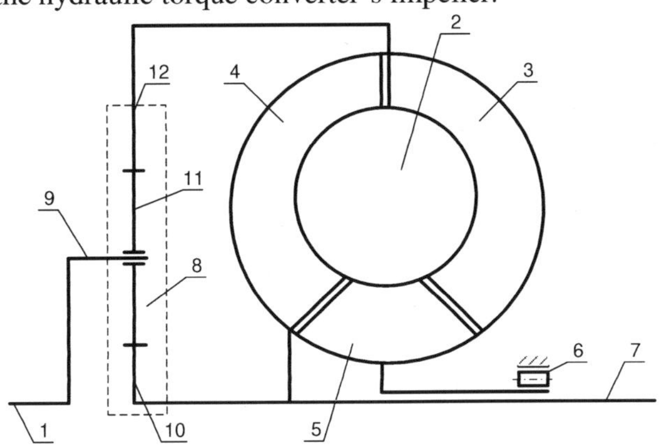

Figure 2. Kinematic diagram of a simple double-flow hydraulic power transmission with a regular differential

The HPT efficiency is higher than that of the original hydraulic torque converter. The transmission also has zero circulating power but features a significantly lower torque ratio. The capacity is increased in the lower engine angular velocity range. The impeller angular velocity exceeds that of the engine. It is controlled by Z_{12}/Z_{10}, where Z_{10} is the teeth number of the central gear, Z_{12} is the teeth number of the planetary ring gear. This may negatively affect the converter performance. A double-flow transmission is also less efficient at torsional vibration damping.

It is possible to arrange a simple double-flow HPT with a regular differential where the central gear is connected to the impeller, while the ring gear is connected to the runner. Such a drive train's properties are similar to that of the drive train shown in Figure 2, but due to even higher impeller angular velocity there are more chances for fluid cavitation in the hydraulic torque converter.

Figure 3 shows a kinematic diagram of a simple double-flow hydraulic power transmission with a summing planetary differential.

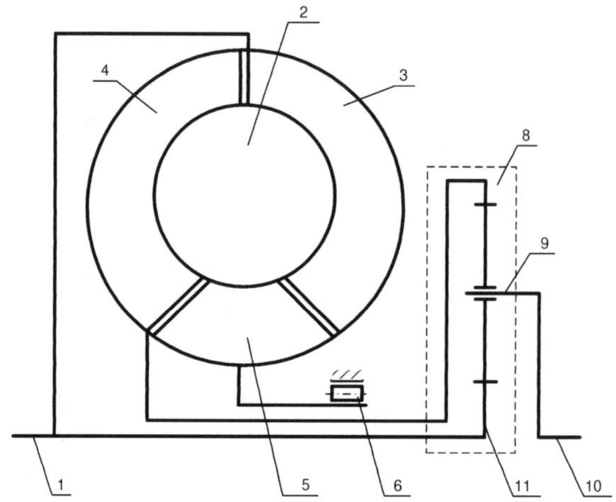

Figure 3. Kinematic diagram of a simple double-flow hydraulic power transmission with a summing differential

This HPT also consists of a casing (not shown), the input shaft Pos. 1, the three-wheel hydraulic torque converter assembly Pos. 2 comprising of the impeller Pos. 3, the runner Pos. 4, and the reactor wheel Pos. 5. The reactor wheel is connected to the HPT casing with the free wheel clutch Pos. 6. The HPT output shaft Pos. 7 is connected to the runner. The output shaft Pos. 10 is connected to the carrier Pos. 9 of the regular differential Pos. 8. The satellite gears Pos. 11 are attached to the carrier. The differential's central (sun) gear Pos. 11 is also connected to the runner and the HPT output shaft. The differential's crown gear Pos. 12 is connected to the hydraulic torque converter's runner.

The HPT efficiency is higher than that of the original hydraulic torque converter. The transmission also has zero circulating power but features a significantly lower torque ratio. The transmission's capacity and transparency are higher. In a certain torque ratio range the hydraulic torque converter operates as a hydraulic brake, i.e. the runner rotates backwards. A double-flow transmission is also less efficient at torsional vibration damping. As compared to double-flow HPTs with summing differentials the impeller angular velocity in this arrangement is always equal to the engine angular velocity.

It is possible to arrange a simple double-flow HPT with a summing differential where the central gear is connected to the input shaft and the impeller, while the ring gear is connected to the runner. The key properties of such a transmission are identical to that of the transmission shown in Figure 3.

4. Results of engineering

We will consider HPT/hydraulic torque converter arrangements that combine single- and double-flow modes.

Figure 4 shows the kinematics of the first two-range HPT option with a regular differential that has a three-wheel hydraulic torque converter assembly and two controllable friction couplings.

IOP Conf. Series: Journal of Physics: Conf. Series **1050** (2018) 012089 doi:10.1088/1742-6596/1050/1/012089

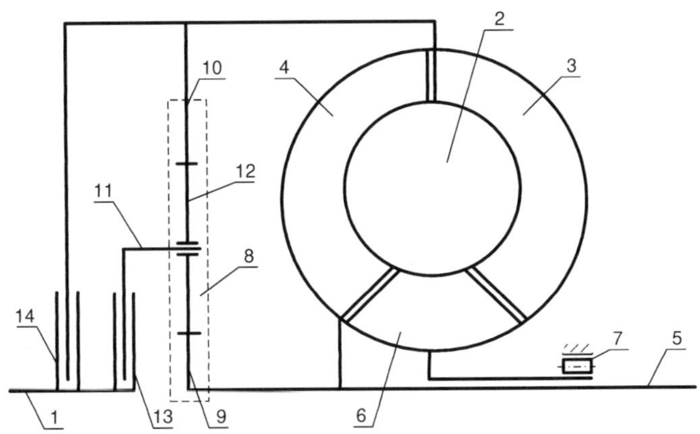

Figure 4. The kinematics of a two-range hydraulic power transmission with a regular differential and two friction clutches

The transmission includes the input shaft Pos. 1. and the hydraulic torque converter Pos. 2 consisting of the pump Pos. 3, and the turbine Pos. 4 connected to the output shaft Pos. 5. The reactor Pos. 6 is attached to the free wheel clutch Pos. 7. The regular differential Pos. 8 consists of the central (sun) gear Pos. 9 connected to the hydraulic torque converter's runner and the transmission output shaft. The crown gear Pos. 10 is connected to the hydraulic torque converter's impeller. The differential carrier Pos. 11 carries the satellite gears Pos. 12. The controllable friction clutch Pos. 13 connects and disconnects the carrier to the input shaft. The second controllable friction clutch Pos. 14 connects the input shaft to the crown gear and the hydraulic torque converter's runner. The friction clutches are controlled by the HPT hydraulic system.

In the first range this HPT operates in the single-flow mode. The friction clutch Pos. 13 is disengaged, and the friction clutch Pos. 14 is engaged. The torque is delivered from the input shaft to the clutch Pos. 14 and to the impeller Pos. 13, and then to the runner Pos. 4, and the output shaft Pos. 5. The carrier Pos. 11 in this mode rotates freely. In the single-flow mode an HPT has an enhanced torque ratio that depends of the hydraulic torque converter specifications.

In the second range the HPT operates as a double-flow transmission with enhanced efficiency. In this mode, the friction clutch Pos. 13 is disengaged, the friction clutch Pos. 14 is engaged and connects the input shaft to the crown gear Pos. 10 and the impeller Pos. 3. The differential splits the engine torque into two parallel flows which are merged at the output shaft.

To ensure the engine power flow continuity as the self-propelled vehicle moves and the HPT ranges are shifted, engagement of one clutch and disengagement of the other clutch shall overlap. Regardless of the HPT mode, both friction clutches shall be sized for the max engine torque.

For the HPT under consideration, a third mode is possible as both Pos. 13 and Pos. 14 friction clutches are engaged. In this case the entire transmission is blocked: both the differential and the hydraulic torque converter are blocked. The ratio in this case is 1. The hydraulic torque converter fluid can be drained, and the engine torque would be transferred from the input to the output shaft with max efficiency. This HPT mode can be used to drive a self-propelled vehicle in favorable road conditions.

Figure 5 shows the second option of a two-range HPT with a regular differential. The friction clutch Pos. 14 is replaced with the friction clutch Pos. 15 [7]. In a single-mode version of this HPT configuration the friction clutch Pos. 13 is disengaged, the differential carrier is disconnected from the input shaft and rotates freely. The torque from the input shaft is transferred to the impeller and to the engaged free wheel clutch Pos. 15. In the double-flow mode the friction clutch Pos. 13 id engaged, the carrier Pos. 14 is connected to the input shaft. Since the impeller and the crown gear Pos. 10 connected to it rotate faster than the input shaft, the free wheel clutch is disengaged.

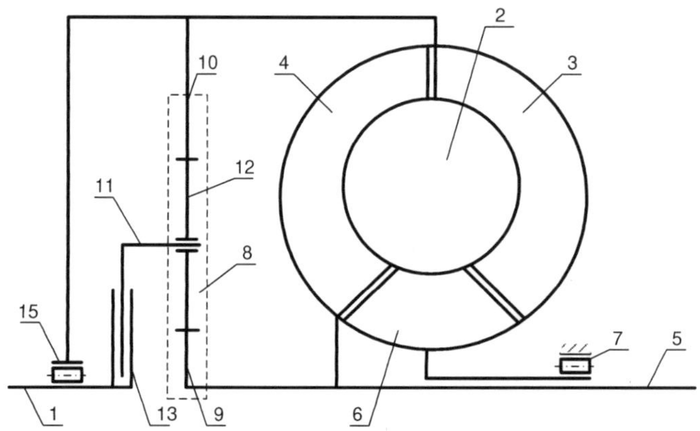

Figure 5. The kinematics of a two-range hydraulic power transmission with a regular differential and one friction clutch

A similar approach to combining a single- and double-flow HPT modes can be applied to a summing differential transmission. Figure 6 shows the kinematics of the first two-range HPT option with a summing differential equipped with a hydraulic torque converter assembly and two controllable friction clutches [3].

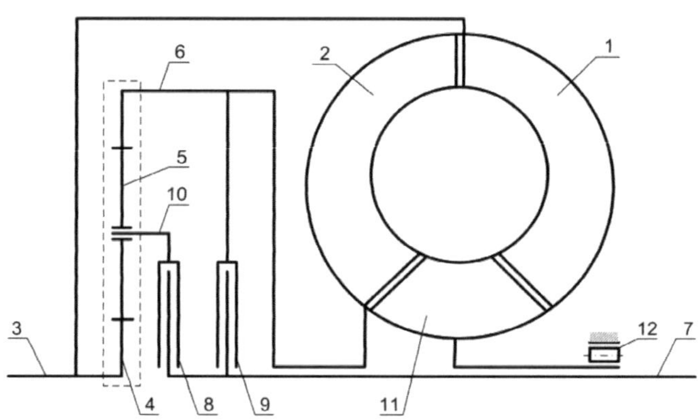

Figure 6. The kinematics of a two-range hydraulic power transmission with a summing differential and two friction clutches

The hydraulic torque converter consists of the impeller Pos. 1 and the runner Pos. 2. The latter is connected to the input shaft Pos. 3. The sun gear Pos. 4 (the differential is highlighted with a dash line) is permanently connected to both the input shaft and the impeller. The runner is connected to the crown gear Pos. 6. The reactor wheel Pos. 11 is attached to the free wheel clutch Pos. 12. The carrier Pos. 10 with the satellite gears Pos. 5 can be engaged/disengaged to the output shaft Pos. 7 using the controllable friction clutch Pos. 8. The second he controllable friction clutch Pos. 9 engages the runner to the output shaft. The Pos. 8 and 9 friction clutches are controlled by the HPT hydraulic system.

In the first range this HPT operates in the single-flow mode. The friction clutch Pos. 8 is disengaged, and the friction clutch Pos. 9 is engaged. The torque is delivered from the input shaft to the the impeller Pos. 13, and then to the runner Pos. 2, and the output shaft Pos. 7. The carrier Pos. 10 rotates freely. In the single-flow mode an HPT has an enhanced torque ratio that depends of the hydraulic torque converter specifications.

In the second range the HPT operates as a double-flow transmission with enhanced efficiency. In this case, the friction clutch Pos. 9 is disengaged, the friction clutch Pos. 8 is engaged and connects the input shaft Pos. 3 to the carrier Pos. 10; the runner Pos. 2 drives the crown gear Pos. 6. The engine torque is split into two parallel flows to be merged at the carrier.

For the HPT under consideration, a third mode is possible as both Pos. 8 and Pos. 9 friction clutches are engaged. In this case the entire transmission is blocked: both the differential and the hydraulic torque converter are blocked. The ratio in this case is 1. The hydraulic torque converter fluid can be drained, and the engine torque would be transferred from the input to the output shaft with max efficiency. This HPT mode can be used to drive a self-propelled vehicle in favorable road conditions.
Figure 7 shows the second option of a two-range HPT with a summing differential. One of the friction clutches is replaced with a friction clutch [8].

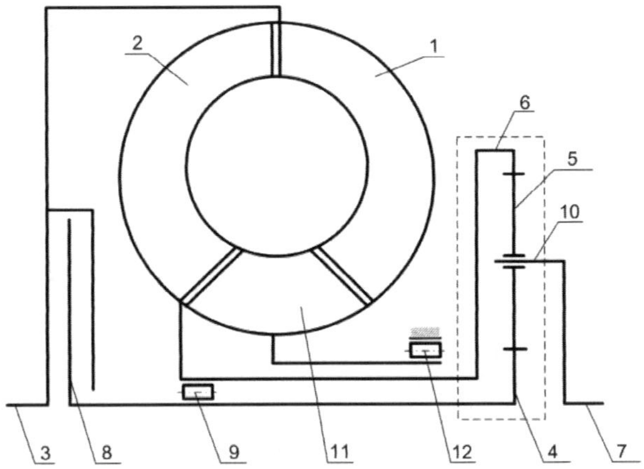

Figure 7. The kinematics of a two-range hydraulic power transmission with a summing differential and one friction clutch

The hydraulic torque converter consists of the impeller Pos. 1 and the runner Pos. 2. The latter is connected to the input shaft Pos. 3. The sun gear Pos. 4 (the differential is highlighted with a dash line) is connected to the input shaft with the clutch Pos. 8. The runner is connected to the crown gear Pos. 6. The reactor wheel Pos. 11 is attached to the free wheel clutch Pos. 12. The carrier Pos. 10 with the satellite gears Pos. 5 are permanently connected to the output shaft Pos. 7. The free wheel clutch Pos. 9 is installed between the differential's sun and crown gears.
In the single-flow range for this HPT configuration the friction clutch Pos. 8 is disengaged. The input shaft Pos. 3 drives the impeller Pos. 1. The runner Pos. 2 drives the crown gear Pos. 6 and the sun gear Pos. 4 through the free wheel clutch Pos. 9. The carrier drives the output shaft. Since the angular velocities of the crown and sun gears are equal, the free wheel clutch Pos. 9 blocks the differential (the differential ratio is 1.)
In the double-flow HPT range the friction clutch Pos. 8 is engaged. It connects the sun gear Pos. 4 to the input shaft Pos. 3. In this case the differential sums the power flows coming from the input shaft and the runner. Since the sun gear angular velocity exceeds that of the crown gear (due to the impeller and the runner slippage) the free wheel clutch Pos. 9 is disengaged (uncoupled.) In this way the free wheel clutch partially automates the HPT range switchover with just one controllable free wheel clutch; it somewhat simplifies the control system. The free wheel clutch and the friction clutch shall be sized for the max engine torque.

5. Conclusion

Hydraulic power transmissions with switchable hydraulic torque converters are designed for both road and off-road self-propelled vehicles. Extra controls enable an extended range for the original hydraulic torque converter ratio. Free wheel clutches automate single/double-flow switchover. The proposed HPTs can be combined with robotic fixed-ratio gearboxes or as a part of a vehicle drive train consisting of a conventional clutch, and a manual gearbox.
The proposed HPT kinematics combines the advantages of a single-flow hydraulic power transmission (higher torque ratios and better efficiency in a lower ratio range) and of a double-flow hydraulic power transmission (better efficiency in a higher ratio range). As a result, the combined HPTs can improve

the average self-propelled vehicle drive train efficiency as compared to a drive train with a single-flow HPT while maintaining high torque ratio in a low ratio range.

Acknowledgments
We thank the personnel of the Research Library, Tula State University, and of the Tula Region Research Library for their assistance with reference and patent search, and innovative design development.

References
[1] Kharitonov S A 2003 *Automatic Transmissions* (Moscow: Astrel) p 479
[2] Naunheimer H, Bertsche B, Ryborz J and Novak W 2011 *Automotive Transmissions Second Edition* (Berlin Heidelberg: Springer-Verlag) p 740
[3] Nanney M J 2007 *Light and Heavy Vehicle Technology Forth Edition* (Oxford: Elsevier) p 671
[4] Fischer R, Küçükay F, Jürgens G, Najork R and Pollak B 2015 *The Automotive Transmission Book* (Heidelberg New York Dordrecht London: Springer) p 355
[5] Heisler H. 2002 *Advanced Vehicle Technologies Second Edition* (Oxford: Butterworth-Heinemann) p 654
[6] Petrov A V 1966 *Planetary and Hydraulic Power Transmissions for Wheeled and Tracked Vehicles* (Moscow: Mashinostroenie) p 383
[7] Shadskii G V, Trushin N N 1992 *A Two-Range Hydraulic Power Transmission* Patent 1754499 USSR
[8] Trushin N N 1992 *A Hydraulic Power Transmission* Patent 1772497 USSR

IOP Conf. Series: Journal of Physics: Conf. Series **1050** (2018) 012090 doi:10.1088/1742-6596/1050/1/012090

Investigation of heat and mass transfer process in the closed volume with different types of impact on liquid

V I Trushlyakov[1], I Y Lesnyak[1], A A Novikov[1] and A V Panichkin[2]

[1]Omsk State Technical University, 11, Mira ave., Omsk, 644050, Russia
[2]Sobolev Institute of Mathematics of the Siberian Branch of Russian Academy of Sciences, 644043, Omsk, Russia

E-mail: lesnyak.ivan@gmail.com

Abstract. Theoretical and experimental investigations of liquid evaporation process under ultrasonic (US) and heat influences (HI) under conditions of reduced pressure are carried out. A mathematical model of the influence of US, HI and pressure of a vapor-gas mixture in a vacuum chamber (VC) on the temperature and rate of fluid evaporation is developed. A program and methods for conducting experiments have been developed, an experimental stand has been created, and an experimental program has been implemented. The experimental dependences of the change in the liquid temperature under combined US and HI under conditions of reduced pressure and under the influence of these factors were obtained separately. A comparative analysis of the calculated and experimental values of liquid temperatures was carried out, which showed a discrepancy of the results to 10%. The areas of parameters of US, HI, and pressure in a vacuum chamber are identified, at which the evaporation rate of the liquid varies significantly, for example, at a pressure of 3 ... 4.5 kPa, a sharp decrease in the liquid temperature is observed, up to the formation of ice crust. The combined use of US and HI in conditions of reduced pressure leads to an increase in the rate of liquid evaporation by 0.7 g / min than when using US and HI separately.

1. Introduction

Unused fuel residues (up to 3 % or more of the initial fuelling) in the tanks of rocket stages with liquid rocket engines provoke explosions in the stages of launch vehicles in orbits and fires in areas of their falls, and drastically deteriorate dynamic characteristics of spent rocket stages at their motion in the atmospheric phase of the descent trajectory.

For avoiding the negative impact of unused residual fuel in the tanks of spent stages of launch vehicles, an on-board system is developed to ensure the removal of these fuel residues fuel residues by evaporation in carrier rockets [1].

Currently, there are various methods for the evaporation of liquids, including by direct heating of the surface on which the liquid is, the convective influence of gas flow on liquid, as well as US, vacuum, electromagnetic influences, etc.

The mechanism of liquid evaporation in US is fairly well studied, such as in [2].

Numerous experiments on the drying of various materials under the action of US confirm the fact that the velocity of the process depends on the magnitude of the US frequency and the pressure, for example, in [2], when the naphthalene evaporates from the surface of metal balls with a diameter of 3 ... 6 mm, the linear dependence of the flow mass on the US.

The dynamics of the liquids cavitation under US was considered in [3, 4], besides, the influence of the ultrasound frequency and amplitude f on the intensity of cavitation on the basis of the Nolting-Nepayras mathematical model was evaluated.

The results of theoretical and experimental investigations of liquid evaporation under various conditions have been obtained, including the following:

– various boundary conditions for the liquid location on a solid surface ("drop" and "mirror") [5, 6];

– gas flow influence on the liquid [7], with certain parameters (input angle, temperature, etc.) [8, 9] by using experimental stands [10, 11];

– use of various types of surfaces on which the liquid is located (micro- and nano-coatings) [12].

Since the existing investigations of liquids evaporation under US [13, 14] and HI [15, 16] are carried out under atmospheric or increased pressure conditions, a special interest is the investigation of US and HI effect on liquid evaporation under conditions of reduced pressure.

It is proposed to investigate the process of heat and mass transfer on liquid evaporation with a free surface under conditions of reduced pressure under combined US and HI, and under the influence of these factors separately.

2. Problem statement

Investigation of the of liquid evaporation process with a free surface in the closed volume under US and HI under conditions of reduced pressure is proposed to be carried out in a limited range of parameters of the evaporated liquid, that are US and HI.

The solution of the task is:

– development of a mathematical model of the influence of US, HI and a vapor-gas mixture (evaporated liquid + gas in a volume) pressure in a closed volume on the process of liquid evaporation, conducting numerical experiments;

– development of a program and methodology for conducting experiments, the formation of requirements for the experimental stand and its development;

– conducting experiments and comparative analysis of the mathematical and physical modeling results.

Restrictions and assumptions are the following:

– distilled water is used as liquid;

– initial mass of the liquid (150 g) at the initial temperature of the liquid (14 ... 16 ^{0}C);

– liquid evaporation is considered up to the point of its freezing;

– change in pressure in the VC is from 101 kPa to 0.8 kPa with a constant pumping speed of 6.5 dm^3 / s;

– US frequency is 40 kHz, sound pressure is 120 dB;

– HI was carried out using an electric heater located in the volume of the evaporated liquid. Electric heater power is 30 watts. The surface area of heating is 3300 mm^2.

3. Mathematical model

The main processes that consume energy with US and HI on the liquid are: a) cavitation; b) micro streams of liquid; c) vortex flows. These processes are manifested in the form of liquid heating [17], by heating the vapor-gas mixture in cavitating bubbles [18] and by dissipating acoustic energy on the inhomogeneities of various fluid layers that form when cavitation occurs and the liquid is stirred.

The total energy of the process is determined by the level of energy consumed and is calculated after measuring the voltage supplying on the ultrasound generator and the current consumed. The calculation takes into account the efficiency of the ultrasonic generator and the coefficient of electro-acoustic conversion. Acoustic energy is expended on the liquid heating and mixing.

For modeling the mixing process, we consider the variant of the turbulent flow and the body moving in the flow. Oscillation of the bottom of the container, in which the liquid is located, creating the effect of motion of the plane in the turbulent flow, allow us to estimate the power required to create this flow.

IOP Conf. Series: Journal of Physics: Conf. Series **1050** (2018) 012090 doi:10.1088/1742-6596/1050/1/012090

In accordance with the well-known formula [19], the power P required for the motion of a body in a turbulent flow can be represented in the form:

$$P = c_r S\rho / 2v^3, \tag{1}$$

where c_r is the coefficient that depends on the shape of the body and is a function of the Reynolds number; S is the the area of the plane in a stream, perpendicular to his direction; ρ is the fluid density; v is the relative velocity of fluid motion. This velocity consists of two components:

1) velocity v_m, which is associated with fluctuations of a bottom of a container at US:

$$v_m = 2\pi f A_m, \tag{2}$$

where f is the frequency of US; A_m is the amplitude of the oscillations;

2) velocity v_k, which is determined by the cavitation process and the flow velocity near the cavitating bubble, is estimated [3] by the following expression:

$$v_k = U_s^2 / 2\pi f R, \tag{3}$$

where U_s is the radial velocity of bubble boundary; R is the average radius of a cavitation bubble.

This expression is associated with the asymmetry of the processes of expansion and contraction of the cavitation bubble [4], and U_s is also determined by the average bubble radius and the time of reduction of the liquid or gas pressure, which, with the asymmetry indicated above, is about 0.75 of the oscillation period.

$$U_s = R / 0{,}5T = fR / 0{,}75, \tag{4}$$

where T is the oscillation period.

$$v_k = 8 fR / 9\pi. \tag{5}$$

The resonance size of the bubble is determined from the known Minnert formula [20], which relates its size to the resonance frequency of the oscillations:

$$f = 1/2\pi R \sqrt{3\gamma / \rho \left(P_0 + 2\sigma / R\right)}, \tag{6}$$

where R is average radius of a cavitation bubble; γ is the air adiabatic index; σ is the surface tension of a liquid; P_0 and ρ is the static pressure and density of liquid, respectively.

For simplification further transformations, we exclude the second member in brackets in the Minnert formula, as a second order quantity of smallness in comparison with the static pressure of liquid. Then the Minnert formula is transformed into an expression for the bubble size:

$$R = 1/2\pi f \sqrt{3P_0\gamma / \rho}. \tag{7}$$

Then the velocity v_k will be determined:

$$v_k = 4/4\pi^2 \sqrt{3P_0\gamma / \rho}. \tag{8}$$

Respectively, the total relative velocity of the liquid:

$$v = v_m + v_k = 2\pi f A_m + 1/\pi^2 \sqrt{3P_0\gamma / \rho}. \tag{9}$$

Power P required to create a turbulent flow:

$$P = c_r S\rho / 2\left(2\pi f A_m + 1/\pi^2 \sqrt{3P_0\gamma / \rho}\right)^3. \tag{10}$$

Knowing the power required for mixing and total acoustic power, it is not difficult to determine the part of the power P_T that determines the heating of the liquid:

$$P_T = S_a \rho c_c 2\pi^2 f^2 A_m^2 - c_r S\rho / 2\left(2\pi f A_m + 1/\pi^2 \sqrt{3P_0\gamma / \rho}\right)^3, \tag{11}$$

IOP Conf. Series: Journal of Physics: Conf. Series **1050** (2018) 012090 doi:10.1088/1742-6596/1050/1/012090

where S_a is the emitter surface area; c_c is the speed of sound in water.

Change of temperature of liquid:

$$\Delta T = P_T t / \left(m_l c_1 + m_c c_c\right), \tag{12}$$

where t is the time; m_l is the mass of liquid; c_l is the specific heat of the liquid; m_c is the container weight; c_c is the specific heat of container material.

On the other hand, the heating process is accompanied by the evaporation of liquid, especially under reduced pressure. In accordance with the laws of thermodynamics, the evaporation process is accompanied by the absorption of energy, which can be determined through the specific heat of vaporization:

$$Q = qm, \tag{13}$$

where q is the specific heat of vaporization; m is the mass of vaporized liquid.

This process leads to a change in the temperature of the liquid:

$$\Delta T = (P_T t - qm) / (m_l c_1 + m_c c_c). \tag{14}$$

The ass of vaporized liquid is as follows:

$$\dot{m} = cS_s(p_n - p) / p_0, \tag{15}$$

where $\dot{m}$ is the amount of liquid converted into steam in 1 second; c is the coefficient of proportionality (hereinafter, determined experimentally); S_s is the evaporation surface area; p_n is the saturated vapor pressure; p is the vapor pressure of a liquid above its free surface; p_0 is the external barometric pressure.

After transformation expression for the velocity of evaporation rate of liquid will look:

$$\dot{m} = cS_s p_n \left(1-\varphi\right) p_0, \tag{16}$$

where φ is the relative humidity.

As experiments are made in the vacuum chamber in the conditions of the decreasing pressure, expression for mass of the liquid passing into steam will be defined as follows:

$$m = \int_0^t \dot{m}\, dt = \int_0^t cS_s p_n \left(1-\varphi\right) / p_0\left(t\right) dt \tag{17}$$

and, accordingly, the expression for changing the temperature of the liquid under US and HI under conditions of decreasing pressure:

$$\Delta T = \left(P_T t - q\int_0^t cS_s p_n \left(1-\varphi\right) / p_0\left(t\right) dt \right) / (m_l c_1 + m_c c_c). \tag{18}$$

At substitution in the received expression formula for the power of US defining liquid heating, we will receive:

$$\Delta T = \left[(S_a \rho c_c 2\pi^2 f^2 A_m^2 - c_r S\rho / 2\left(2\pi f A_m + 1/\pi^2 \sqrt{3P_0\gamma/\rho}\right)^3)t - qcS_s \int_0^t p_n \left(1-\varphi\right) / p_0\left(t\right) dt \right] / \left(m_l c_1 + m_c c_c\right) \tag{19}$$

4. Experimental stand

Experimental investigations for determining the temperature and mass of the vaporized liquid with a free surface in a closed volume under US and HI under conditions of reduced pressure were performed using the developed and created experimental stand shown in Figure 1.

IOP Conf. Series: Journal of Physics: Conf. Series **1050** (2018) 012090 doi:10.1088/1742-6596/1050/1/012090

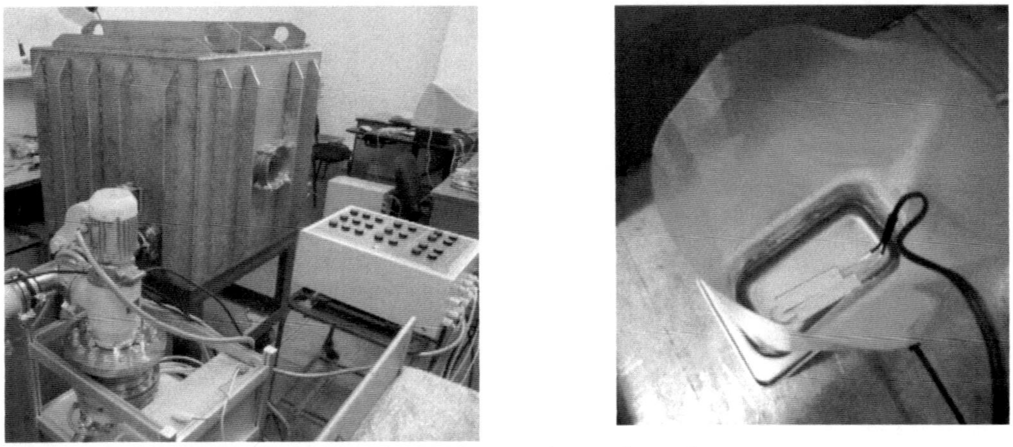

Figure 1. Experimental stand

The experimental stand includes:

– VC with a volume of 0.463 m^3;

– container for placing a liquid with a protective film from splashing drops (length 0.150 m, width 0.085 m, height 0.065 m, weight 762.9 g, material: stainless steel);

– piezoceramic radiator (frequency 40 kHz, sound pressure 120 dB);

– electric heater (power 30 W, surface area of heating 4400 mm^2);

– forvacuum pump (pumping speed 0.0065 m3 / s, ultimate residual pressure 1 Pa);

– system for measuring, recording and processing measurement results (mobile thermocouples, pressure sensor, video camera);

– connecting and shut-off valves (system of hoses, fittings and pneumatic valves).

During the experimental investigation, distilled water was used as the evaporated liquid.

The measurement error for temperatures of the walls, gas, and liquid obtained using a multichannel temperature meter MIT-12 and thermocouples THA is ± 1 °C.

The process of liquid evaporation with a free surface in the VC at US and HI was carried out with the following initial parameters: gas temperature inside the VC is 17 ... 19 ° C; mass of the evaporated liquid is 150 g; water temperature 17 ... 20 ^{0}C; initial pressure in VC 101 kPa.

The experiments were carried out for 7 minutes with the joint US and HI, and under the influence of these factors separately under conditions of constantly decreasing pressure in the VC from 101 kPa to 0.8 kPa.

5. Experimental results

As a result of mathematical modeling and carried out experiments, the dependence of the temperature change of the liquid at US and HI under conditions of reduced pressure was obtained (Fig. 2).

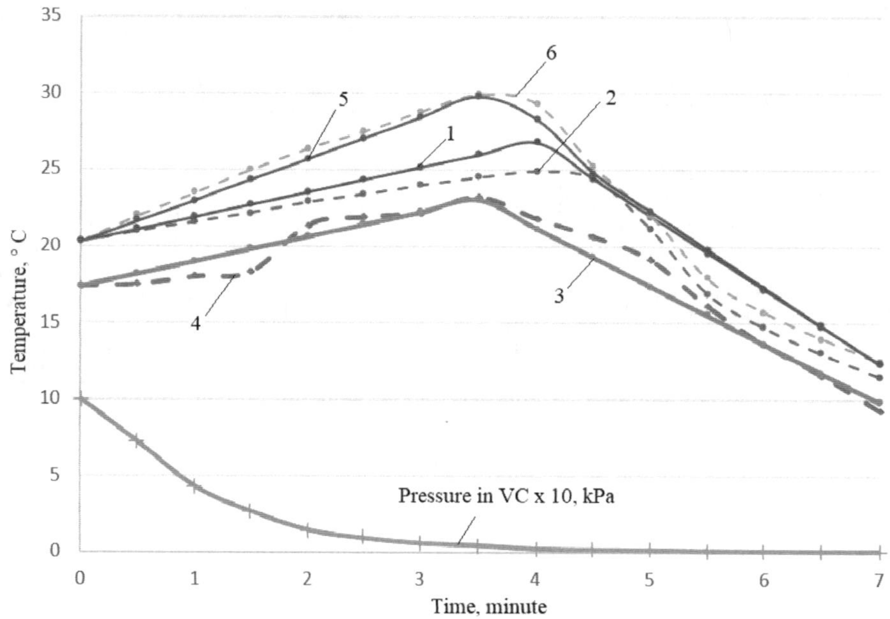

Figure 2. Comparative analysis of the experimental investigations and mathematical modeling results: 1 – US (mathematical model); 2 – US (experiment); 3 – HI (mathematical model); 4 – HI (experiment); 5 – US and HI (mathematical model); 6 – US and HI (experiment)

As can be seen from the results shown in Figure 2, the temperature of the liquid from the initial time to 3.5 ... 4 minutes is smoothly increased by 4.6 ... 9.6 ° C, as in the joint US and HI (graphs 5, 6), and using these factors separately (graphs 1 - 4). Constantly decreasing pressure in the VC leads to active boiling of the liquid and the formation of large collapsing bubbles from the 4th minute of the experiment. At the same time, the temperature of the liquid decreases sharply until the end of the experiment by 13 ... 17 ^{0} C, and the pressure decreases to 0.8 kPa.

US and HI on the liquid lead to an increase in its temperature to a certain pressure value. Consequently, there are areas of the parameters of US, HI, and pressure at which the evaporation rate of the liquid varies substantially.

Table. 1 shows the values of evaporated liquid mass and evaporation rate for separately US and HI, as well as for combined use of US and HI under conditions of reduced pressure.

Table 1. The mass of the evaporated liquid.

Type of action on the liquid	Mass of the evaporated liquid, g	Evaporation rate, g / min
US	7.75	1.1
HI	8.6	1.23
US and HI	12.8	1.83

On the basis on the analysis of the data presented in Figure 2 and in Table 1 the following main research results are formulated:

– US and HI on the liquid lead to an increase in its temperature to a pressure value in the vacuum chamber of 3 ... 4.5 kPa. From 4.5 kPa to 0.8 kPa, the temperature of the liquid and the rate of evaporation of the liquid decrease. Consequently, there are areas of the parameters of US, HI, and pressure at which the evaporation rate of the liquid varies substantially;

– sharing of US and HI in conditions of reduced pressure leads to an increase in the rate of evaporation of the liquid by 0.7 g / min than when using US and HI separately;

– with joint US and HI under conditions of reduced pressure, the mass of the evaporated liquid will be greater than when using US and HI separately.

6. Discussion of the obtained results

The obtained results of theoretical and experimental investigations showed a close compliance of the change in the liquid temperatures at US and HI under conditions of reduced pressure. The discrepancy between the results of mathematical modeling and experimental investigations is no more than 10%.

Investigated US and HI on liquids under reduced pressure, which showed:

– there are areas of parameters of US, HI and pressure at which the evaporation rate of a liquid varies significantly, for example, at a pressure of 3 ... 4.5 kPa, bubbles are observed on the surface of the liquid and the liquid temperature decreases, until the ice crust forms;

– as follows from Table. 1. with the joint US and HI under conditions of reduced pressure, the mass of the evaporated liquid will be greater than when these factors are individually affected;

– at the same energy costs, the use of US leads to an increase in the liquid temperature by 9.6 ° C (Fig. 2, graphs 1, 2), and the use of HI at 5.8 ° C (graphs 3, 4);

– the use of US and HI in conditions of reduced pressure reduces the freezing point of the liquid.

Further recommendations on theoretical and experimental investigations are the following:

– to estimate intensity of US at various frequencies and amplitudes, and also at use of other liquids;

– investigations to determine the optimal combination of factors of US, pressure, temperature, affecting the efficiency of the evaporation process;

– carrying out experimental investigations on the evaporation of a drop of liquid to compare the results obtained with the results presented in [21, 22] with subsequent evaluation of the US and HI use effectiveness.

7. Conclusions

A mathematical and physical model for estimating the influence of US, HI and pressure on the liquid temperature and its evaporation rate is developed.

Experimental investigations of the process of liquid evaporation at US and HI under conditions of low pressure have been carried out. The reliability of the obtained results of the investigated process mathematical modeling is confirmed.

The areas of US, HI and pressure parameters are determined, under which the evaporation rate of the liquid varies substantially.

The recommendations for further theoretical and experimental investigations are formulated.

Acknowledgments

This work was supported by the Russian Federation Ministry of Education and Science within the public contract with subordinate educational organizations, the project "Improvement of environmental safety and economic efficiency of launch-vehicles with cruising liquid rocket engines", application No. 9.1023.2017/PCh.

References

[1] Trushlyakov V I, Kudentsov V Yu, Lesnyak I Yu, Lempert D B and Zarko V E 2016 *Proceedings of the 56th Israel Annual Conference on Aerospace Sciences* (Israel: Tel-Aviv & Haifa) **4.4** L1

[2] Rosenberg L D 1970 *Physics and technology of powerful ultrasound (*Moscow: Science) p 689

[3] Vinogradov B V and Fedin D A 2003 Effect of frequency and amplitude of ultrasonic vibrations on the cavitation intensity *Vopr. Khim. Khim. Tekhnol.* **4** 141–144

[4] Vinogradov B V and Fedin D A 2003 On the development of the cavitation cavity under external influence *Vopr. Khim. Khim. Tekhnol.* **1** 151–153

IOP Conf. Series: Journal of Physics: Conf. Series **1050** (2018) 012090 doi:10.1088/1742-6596/1050/1/012090

[5] Trushlyakov V I, Kudentsov V Yu, Lesnyak I Yu, Rozhaeva K A, Lavruk S A, Dron M M and Zharikov K I 2017 Evaporation of a model liquid *Russian Engineering Research* **10** 14–17
[6] Borodulin V Y, Letushko V N, Nizovtsev M I and Sterlyagov A N 2016 The surface temperature of free evaporating drops *J. Phys.: Conf. Ser.***754** 032018
[7] Trushlyakov V I, Lesnyak I Y and Galfetti L 2017 An experimental investigation of convective heat transfer at evaporation of kerosene and water in the closed volume *Thermophys. and Aeromech.* **24** 771–781
[8] Trushlyakov V and Lavruk S 2015 Theoretical and experimental investigations of interaction of hot gases with liquid in closed volume *Acta Astronautica* **109** 241–247
[9] Lyulinl Y V and Rezanova E V 2017 Convective fluid flows in a horizontal channel with evaporation: analytical and experimental investigations *J. Phys.: Conf. Ser.* **925** 012001
[10] Trushlyakov V I, Kudentsov V Y, Lesnyak I Y, e.a. Sposob modelirovaniya processa gazifikacii ostatkov gidkogo raketnogo topliva v bakah otdelyayusheysya chasti stupeni rakety – nositelya i ustroystvo dlya ego realizacii [A method for modeling the process of gasification of residual liquid rocket fuel in the tanks of the separating part of the rocket stage and the device for its implementation]. Patent RF, no. 2461890, 2012.
[11] Trushlyakov V I, Kudentsov V Y, Lesnyak I Y, e.a. Sposob modelirovaniya processa gazifikacii ostatkov gidkogo raketnogo topliva i ustroystvo dlya ego realizacii [Method for modeling the process of gasification of residual liquid propellant and the device for its implementation]. Patent RF, no. 2474816, 2013.
[12] Semenov A A, Feoktistov D V, Zaitsev D V, Kuznetsov G V and Kabov O A 2015 Experimental investigation of liquid drop evaporation on a heated solid surface *Thermophys. and Aeromech.***22** 771–774
[13] Gubaidullin A A and Pyatkova A V 2016 Acoustic streaming with allowance for heat transfer *Acoustical Physics* **62** 300–305
[14] Perminov S A and Ermakov G V 2010 Boiling-up of superheated water and water solutions under ultrasound influence *Thermophys. and Aeromech.***17** 107–112
[15] Gatapova E Ya, Filipenko R A, LyulinI Yu V, Graur A, Marchuk I V and Kabov O A 2015 Experimental investigation of the temperature field in the gas-liquid two-layer system *Thermophys. and Aeromech.* **22** 729–734
[16] Kuznetsov V V 2011 Heat and mass transfer on a liquid-vapor interface *Fluid Dynamics* **46** 97–107
[17] Beke T 2012 Heat transfer in a thermoacoustic process *European Journal of Physics* **33** 1487–1503.
[18] Shpak O, Stricker L, Versluis M and Lohse D 2013 The role of gas in ultrasonically driven vapor bubble growth *Phys. Med. Biol.* **58** 2523
[19] Kuhling H 1985 *Handbook of Physics* (Fachbuchverlag: Leipzig) p 520
[20] Minnaert M 1933 On musical air-bubbles and the sounds of running water *Philosophical Magazine* **16** 235–248
[21] Kirichenko D P, Zaitsev D V and Kabov O A 2017 Experimental investigation of picoliter liquid drops evaporation on a heated solid surface *J. Phys.: Conf. Ser.* **925** 012026
[22] Borodulin V Y, Letushko V N, Nizovtsev M I and Sterlyagov A N 2016 The surface temperature of free evaporating drops *J. Phys.: Conf. Ser.* **754**

IOP Conf. Series: Journal of Physics: Conf. Series **1050** (2018) 012091 doi:10.1088/1742-6596/1050/1/012091

Optimization of energy consumption at thermal vacuum liquid evaporation in the closed volume

V I Trushlyakov[1], **A V Panichkin**[2]

[1]Omsk State Technical University, 11, Mira ave., Omsk, 644050, Russia
[2]Sobolev Institute of Mathematics, Novosibirsk, Russia
E-mail: vatrushlyakov@yandex.ru

Abstract. The mathematical model (MM) of the process describing the evaporation of model liquid with a free "mirror"-type surface, changes of liquid and vapour-gas mixture (VGM) temperature in the closed volume with application of power consumption optimization at thermal vacuum influence is developed. For MM optimization of power consumption on the basis of the Pontryagin maximum principle with application to process of liquid evaporation from a free surface at joint thermal and vacuum influence was considered. For modeling heat and mass transfer processes with evaporation in the closed volume at thermal vacuum influence, the system of the ordinary differential equations within some assumptions with simplification of a problem definition was used. The MM at some deviations of parameters allows to perform optimization of the cyclorama of pumping out and heating systems works and to define a tendency of power consumption reduction at consecutive inclusions and switching off VGM pumping out and liquid heating. As an optimum condition criterion are accepted power costs of liquid conductive heating and VGM pumping from the closed volume. Based on numerical modeling intervals of liquid heating and VGM pumping time providing minimization of power consumption for evaporation of the set mass of liquid are defined. The test calculations without optimization of power consumption showing satisfactory coordination with experimental data are carried out.

1. Introduction

There is a significant amount of technologies for drying materials which are widely applied in various fields of activity, for example, for drying agricultural products and timber as well as in aviation industry and the missile and space equipment (fuel tanks and highways) after a number of the technological operations connected with their filling with technological liquid (cleaning, calibrating, durability tests, etc.)

Now the task of increasing the efficiency of heat and a mass exchange processes, occurring in various technical drying systems is actual. In this paper, the system of thermal vacuum drying in the closed volume is considered as one of examples of such technical systems.

Numerous theoretical and experimental studies including development of the mathematical models (MM) on materials drying when using various feeders of heat, acoustic influence, change of pressure, etc., which can be divided conditionally into groups, are known, including the following:

- According to various boundary conditions of liquid arrangement on a firm surface ("drop", "mirror"), and also various types of surfaces on which liquid (micro and nanocoverings) is located [1-6].
- According to types of impact on liquid (convective and conductive heating, thermo-capillary, vacuum) [7-17].
- According to optimization of processes and designs of drying systems [18-20].

In this work, the increase in heat and mass exchange efficiency is considered at liquid evaporation from a free surface at joint thermal and vacuum influences with application of the Pontryagin

IOP Conf. Series: Journal of Physics: Conf. Series **1050** (2018) 012091 doi:10.1088/1742-6596/1050/1/012091

maximum principle [21]. The application specificity of the principle which is widely used for problems of optimum control is its applicability for the objects described by systems of the ordinary differential equations, for example, [22].

Thus, to apply the optimization of power consumption on the basis of the Pontryagin maximum principle, it is necessary to develop a simplified MM, describing liquid evaporation process from a free surface at joint thermal and vacuum influence on the basis of the ordinary differential equations that, naturally, will lead to methodical errors, but at the same time will allow to estimate a tendency of power consumption reduction due to the optimization of the pumping out and heating systems operation cyclograma.

2. Definition of liquid evaporation at thermal vacuum influence problem

Based on the literature review carried out above it is possible to formulate the following directions of researches:

- Development of MM of liquid evaporation at thermal vacuum influence based on the ordinary differential equations.
- Development of thermal vacuum process optimum control based on the Pontryagin maximum principle.

For liquid evaporation from the model volume (MV) in the vacuum camera (VC) with pumping of steam-gas mix (VGM) and conductive heating of liquid is considered as criterion of power consumption optimization.

Assumptions:

- The case of limited time (till 10 minutes) interactions of VGM with MV, liquid and the VC metal case when it is possible to present mutual heat exchange between them in the simplified form or to neglect it, allowing some deviations of mathematical modeling results from experimental data is considered;
- The case when in VC there is a constant volume of the air mix V_{V0}, which directly interacts with liquid and in an initial time point is a part of atmospheric volume with the preset parameters is considered .

3. Development of MM of liquid evaporation at thermal vacuum influence

Let us obtain the equations with consideration of VGM pumping from the constant volume of V_{V0} taking into account pumping out with a volume speed of U_V when in the initial equations the current pressure *of P(t)* and temperature of $T_V(t)$ of VGM will be defined from adiabatic expansion with constant k.

The following sizes for the processes will be considered:

V_i, P_i, T_i, m_i, Cp_i are the volumes, pressure, temperatures, masses and heat capacities for a bathtub, liquids, VGM, dry air and steam of liquid at i = *s, w, v,* Ai p, respectively.

Based on the considered assumptions we will further give a MM, describing the process of heat and mass exchange between elements in VC, in the form of the ordinary differential equations.

For MM simplification, describing the process of heat and mass exchange at thermal vacuum influence, the following notations are used [23]:

$W_W(t,T,P)=K_0(P_W(T_W)-P_P(m_P,T_V))(\frac{P_0}{P_V})^q$ is the speed of liquid evaporation from the unit of area, $F_A(t,m_A)=U_V\frac{m_A}{V_{V0}}$ is the function of the rate of dry air decay from VC, $F_P(t,m_P)=U_V\frac{m_P}{V_{V0}}$ is the function of the rate of liquid vapors decay from VC, $F_{mW}(t,T)=\frac{(L_W-Cp_WT_W)}{m_SCp_S+m_WCp_W}$ is the function of change of liquid mass-averaged temperature from its evaporation,

$F_T(t) = \dfrac{Q_T}{m_S Cp_S + m_W Cp_W}$ is the function of change of liquid mass-averaged temperature from the power of thermal influence, $F_V(t) = \dfrac{1}{k-1}\dfrac{T_V}{V_{V0}}$ is the speed of change of VGM temperature at adiabatic expansion, $F_{mV}(t,T) = \dfrac{(L_W - Cp_W T_W)}{m_A Cp_A + m_P Cp_P}$ is the function of change of VGM mass-averaged temperature from liquid evaporation, $F_{m\mu}(t,m,\mu) = (\dfrac{m_A(t)}{\mu_A} + \dfrac{m_P(t)}{\mu_W})$ is the a molar ratio of air mixture mass.

At the considered assumptions in a problem statement, the system of the ordinary differential equations taking into account exchange of masses (liquid evaporation, pumping out of VGM) and heat exchange for liquid and VGM, will have the following form:

$$\frac{\partial m_W}{\partial t} = -SW_W(t,T,P), \tag{1}$$

$$\frac{\partial m_A}{\partial t} = -C_U(t)\frac{\partial V_V}{\partial t}\frac{m_A}{V_{V0}} = -C_U(t)F_A(t,m_A), \tag{2}$$

$$\frac{\partial m_P}{\partial t} = -C_U(t)\frac{\partial V_V}{\partial t}\frac{m_P}{V_{V0}} - \frac{\partial m_W}{\partial t} = -C_U(t)F_P(t,m_P) + SW_W(t,T,P), \tag{3}$$

$$\frac{\partial T_W}{\partial t} = \frac{\partial m_W}{\partial t}F_{mW}(t,T) + C_T(t)F_T(t) = -SW_W(t,T,P)F_{mW}(t,T)\frac{\partial m_W}{\partial t} + C_T(t)F_T(t), \tag{4}$$

$$\frac{\partial T_V}{\partial t} = -F_V(t)C_U(t)U_V + F_{mV}(t,T)SW_W(t,T,P), \tag{5}$$

where $P_V(t) = (\dfrac{m_A(t)}{\mu_A} + \dfrac{m_P(t)}{\mu_W})\dfrac{RT_V(t)}{V_{V0}(t)}$ is the the current pressure in ME (Pa), P_0 is the initial atmospheric pressure (Pa), m_w is the liquid weight (kg), V_V is the volume of VGM (m^3), T_W is the liquid temperature (K), T_V is the VGM temperature (K), S is the area of a free surface liquid (m^2), W_W is the liquid evaporation velocity from a unit of area (kg / ($c \cdot m^2$), L_W is the specific heat of evaporation liquid (J/kg), W_W is the liquid evaporation speed from the unit of area (kg / ($c \cdot m^2$)), $C_T(t)$ is the function of thermal heating on\off control (1 or 0), R= 8.314 is the universal gas constant (J/(mole ·K)), K_0 is the coefficient of a condition of the evaporated liquid surface (undistorted is the $1.39 \cdot 10^{-8}$, wavy is the $4.17 \cdot 10^{-8}$ [24] and the movements of the air medium (at 5 m/s is the $1.67 \cdot 10^{-7}$kg / (Pa · m^2· c)), P_W is the partial pressure of evaporated liquid saturation at its current temperature *of* T_W (Pa), P_P is the partial pressure of liquid in air (Pa), q is the proportionality degree of pressure change for this liquid (0.66÷1.18) [24], m_A is the mass of dry air (kg/mole), m_{P0} is the initial mass of water (liquid) vapors (kg), $m_P = m_{P0} + m_{w0} - m_w$ is the current mass of liquid vapors (kg), μ_W is the liquid molar weight (kg/mole), μ_A is the molar mass of dry air (kg/mole), U_V is the speed of pumping (expansion) of VGM volume (m^3/s),$C_U(t)$ is the function of pumping on/off control (1 or 0), m_{A1} and m_{P1} (kg) is the mass of dry air and liquid vapors in a unit of volume of atmospheric V_{V1} air (m^3), m_s is the mass of MV (kg), Cp_S is the MV heat capacity (J / (kg · K)), Cp_w is the liquid heat capacity (J / (kg · K)), Cp_p is the heat capacity of liquid vapors (J / (kg · K)), Cp_A is the heat capacity of dry air (J / (kg · K)).

At temperatures of liquid condition from 273.1 K and up to 303.1 K and above we shall approximately present the partial pressure of the evaporated liquid saturation in the following form, according to tabular data [24]:

$$P_W(T_W) = 174727.8 - 1312.1T_W + 2.47T_W\,(Pa) \tag{6}$$

Based on Clapeyron-Mendeleyev's equation for mixture gases the partial pressure of liquid in air P_P can be presented in the following form:

$$\mathrm{P}_p(t) = (\frac{m_P(t)}{\mu_W})\frac{RT_V(t)}{V_V(t)}. \tag{7}$$

For the system of equations (1) - (5), initial and boundary conditions can be set from time of m_w, m_A, m_P, T_w, T_V in the following form:

$$t_0 = 0: m_W = m_{W0}, m_A = m_{A0}, m_P = m_{P0}, T_W = T_{W0}, T_V = T_{V0};\; t = t_k: m_W = m_{Wk} \tag{8}$$

where t_k is the undetermined value.

At the constant volume of VC in the system of equations (1)-(5) we suggest that the VGM volume is equal to V_V = to V_{V0}, and at pumping out of air the equation is carried out from VC with change of pressure and temperature on adiabatic process with k = 1.4 (TV^{k-1} = const, PV^k = const) – taking into account the VGM expansion on a differential ratio in the following form:

$$\frac{\partial V_V}{\partial t} = U_V(t) = C_U(t)U_V. \tag{9}$$

4. Criterion of optimization

For formation of optimality criterion for heat and a mass exchange process at thermal vacuum influence we will consider minimization of power costs of a certain liquid mass evaporation with m_0 to m_k for time from $t_0 = 0$ to t_k with change of whole energy in the considered system of elements (MV+ liquid + VGM).

The change of energy of the whole system of material bodies with liquid and VGM have to be corresponded by expenses of the used energy of pumping with expansion of volume on adiabatic process (with pressure decline and temperatures) and the additional conductive energy supplied for heating that is possible to present as integral from a difference of pressure in the form of [25]:

$$E_\Sigma = \int_0^{tk} C_U S_V (P_0 - P_V)\frac{U_V}{S_V}dt + E_T = \int_0^{tk} C_U (P_0 - P_V)U_V dt + \int_0^{tk} C_T Q_T dt, \tag{10}$$

which corresponds to the work on of forces $S_v(P_0 - P_v)$ overcoming on distances $U_v / S_v\,dt$ during t_k, where U_v is the speed of volume expansion (or pumping) on the area S_V, P_0 is the external atmospheric pressure, P_V is the internal pressure of air mixture in the considered volume (in this case, in the constant volume of V_{V0}).

Numerical results of modeling the system of equations (1) – (5) and comparison with experimental studies results [23] are given in paragraph 5

5. Development of optimum control of thermal vacuum process of evaporation based on the principle of a maximum of Pontryagin

Let us consider optimization of power consumption of evaporation of the preset liquid mass by means of the Pontryagin method [21].

According to this theory, in solving the optimization problem with functionality of Hamilton:

$$H = \lambda_1 \frac{\partial m_W}{\partial t} + \lambda_2 \frac{\partial m_A}{\partial t} + \lambda_3 \frac{\partial m_P}{\partial t} + \lambda_4 \frac{\partial T_W}{\partial t} + \lambda_5 \frac{\partial T_V}{\partial t} - f_H \tag{11}$$

it is necessary to introduce the conjugate functions λ_1, λ_2, λ_3, λ_4, λ_5 and the function f_H for power consumption minimization.

The expression for f_H can be written as:

$$f_H = \frac{\partial E_V}{\partial t} = C_U(t)(P_0 - (\frac{m_A(t)}{\mu_A} + \frac{m_P(t)}{\mu_W})\frac{RT_V(t)}{V_{V0}(t)})U_V + C_T Q_T \,. \tag{12}$$

According to equations (1) - (5) and (12) final expression for the function (11) determining the maximizing functional with minimization of expenses of energy by function (12) during time is based $[0, t_k]$:

$$\begin{aligned}
H &= \lambda_1 \frac{\partial m_W}{\partial t} + \lambda_2 \frac{\partial m_A}{\partial t} + \lambda_3 \frac{\partial m_P}{\partial t} + \lambda_4 \frac{\partial T_W}{\partial t} + \lambda_5 \frac{\partial T_V}{\partial t} - \\
&-C_U(t)(P_0 - (\frac{m_A(t)}{\mu_A} + \frac{m_P(t)}{\mu_W})\frac{RT_V(t)}{V_{V0}(t)})U_V - C_T Q_T = \\
&= \lambda_1[-SW_W(t,T,P) + +\lambda_2[-C_U(t)U_V \frac{m_A}{V_{V0}}] + \\
&+\lambda_3[-C_U(t)U_V \frac{m_P}{V_{V0}} + SW_W(t,T,P)] + \\
&+\lambda_4[-SW_W(t,T,P)\frac{(L_W - Cp_W T_W)}{m_S Cp_S + m_W Cp_W} + \frac{C_T Q_T}{m_S Cp_S + m_W Cp_W}] + \\
&+\lambda_5[-\frac{1}{k-1}\frac{T_V}{V_{V0}}\{C_U(t)U_V\} + \frac{(L_W - Cp_W T_W)}{m_A Cp_A + m_P Cp_P} SW_W(t,T,P)] - \\
&-C_U(t)(P_0 - (\frac{m_A(t)}{\mu_A} + \frac{m_P(t)}{\mu_W})\frac{RT_V(t)}{V_{V0}(t)})U_V - C_T Q_T .
\end{aligned}$$

To develop the conjugate system of the equations for definition of functions λ_1, λ_2, λ_3, λ_4, λ_5 separate derivatives of expressions (1) – (5) and (11) are firstly determined by the chosen independent variables (m_W, m_A, m_P, T_W, T_V). After their application in derivative of function H of Hamilton (derivative of H, including f_H) the conjugate system of the equations to (1) - (5) is developed in the following form [21]:

$$\dot{\lambda}_1 = -\frac{\partial H}{\partial m_W} = -\lambda_4[Cp_W SW_W(t,T,P)F_{mW}(t,T) - Cp_W F_T(t)]F_{C1}(t), \tag{13}$$

$$\begin{aligned}
\dot{\lambda}_2 &= -\frac{\partial H}{\partial m_A} = -(\lambda_1 - \lambda_3 + \lambda_4 L_W F_{C1}(t) + \lambda_5 F_{mV}(t,T))[SW_W(t,T,P)\frac{q}{\mu_A F_{m\mu}(t,m,\mu)}] + \\
&+\lambda_2 C_U(t)U_V(t)\frac{1}{V_{V0}} - \lambda_5[-Cp_A F_{mV}(t,T)SW_W(t,T,P)F_{C2}(t)] + C_U(t)F_{RA}(t)U_V,
\end{aligned} \tag{14}$$

$$\dot{\lambda}_3 = -\frac{\partial H}{\partial m_P} = -[\lambda_1 - \lambda_3 + \lambda_4 L_W F_{C1}(t) + \lambda_5 F_{mV}(t,T)] \cdot$$
$$\cdot [SW_W(t,T,P)\frac{q}{\mu_W F_{m\mu}(t,m,\mu)} + SK_0(P_0)F_{RW}(t)F_{Pq}(t)] - \quad (15)$$
$$-\lambda_3 [-C_U(t)U_V(t)\frac{1}{V_{V0}}] - \lambda_5 [-Cp_P F_{mV}(t,T)SW_W(t,T,P)F_{C2}(t)] + [-C_U(t)F_{RW}(t)U_V],$$

$$\dot{\lambda}_4 = -\frac{\partial H}{\partial T_W} = (\lambda_1 - \lambda_3)[SK_0(P_0)[-1312.1 + 4.94\cdot \mathrm{Tw}]F_{Pq}(t)] -$$
$$-\lambda_4 [-L_W F_{C1}(t)\{SK_0(P_0)[-1312.1 + 4.94\cdot \mathrm{Tw}]F_{Pq}(t)\}] - \quad (16)$$
$$-\lambda_5 [-Cp_W F_{C2}(t)SW_W(t,T,P) + F_{mV}(t,T)\cdot \{SK_0(P_0)[-1312.1 + 4.94\cdot \mathrm{Tw}]F_{Pq}(t)\}],$$

$$\dot{\lambda}_5 = -\frac{\partial H}{\partial T_V} = -[\lambda_1 - \lambda_3 + \lambda_4 L_W F_{C1}(t)][SK_0(P_0)F_{Rm}(t)F_{Pq}(t) + SW_W(t,T,P)\frac{q}{T_V(t)}] -$$
$$-\lambda_5 [-\frac{1}{k-1}\frac{1}{V_{V0}}\{C_U(t)U_V(t)\} - F_{mV}(t,T)\{SK_0(P_0)F_{Rm}(t)F_{Pq}(t) + \quad (17)$$
$$+SW_W(t,T,P)\frac{q}{T_V(t)}\}] - C_U(t)F_{m\mu}(t)\frac{R}{V_{V0}(t)}U_V,$$

where the following designations are used

$$F_{C1}(t) = \frac{1}{m_S Cp_S + m_W Cp_W},\ F_{C2}(t) = \frac{1}{m_A Cp_A + m_P Cp_P},\ F_{Pq}(t) = (\frac{P_0}{P_V(t)})^q,\ F_{RA}(t) = \frac{1}{\mu_A}\frac{RT_V(t)}{V_{V0}(t)},$$
$$F_{RW}(t) = (\frac{1}{\mu_W})\frac{RT_V(t)}{V_{V0}(t)},\ F_{Rm}(t) = \frac{m_P(t)}{\mu_W}\frac{R}{V_{V0}(t)}.$$

For equations (13) - (17) initial data $\lambda_{10}, \lambda_{20}, \lambda_{30}, \lambda_{40}, \lambda_{50}$ and final data $\lambda_{1k}, \lambda_{2k}, \lambda_{3k}, \lambda_{4k}$ λ_{5k} are unknown; besides, final values for four functions $m_{Ak}, m_{Pk}, T_{Wk}, T_{Vk}$ are not set yet, therefore, short circuit of boundary conditions in systems of the equations with the unknown final time t_k will require five conditions of transversality at $t = t_k$, which with variables $y_1 = m_{wk}, y_2 = m_{Ak}, y_3 = m_{Pk}, y_4 = T_{wk}, y_5 = T_{Vk}$ the first variation of the optimizing functionality will be:

$$J = \int_0^{tk} H(y_i, \dot{y}_i, t)dt \quad (18)$$

at the requirement of achieving the extremum has to be equal to zero:

$$\delta J\big|_{t_k} = \sum_{i=1}^{5}\frac{\partial H}{\partial \dot{y}_i}\delta y_i(t_k) + H(y_i, \dot{y}_i, t_k)\delta t_k = 0. \quad (19)$$

Thus, for achieving the maximum, the second variation has to be negative:

$$\delta^2 J\big|_{t_k} = [\sum_{i=1}^{5}\frac{\partial H}{\partial \dot{y}_i}\ddot{y}_i + \sum_{i=1}^{5}\frac{\partial^2 H}{\partial \dot{y}_i \partial t}\dot{y}_i + \frac{\partial H}{\partial t}](\delta t_k)^2 < 0. \quad (20)$$

This condition is satisfied from creation of the interfaced system of the equations (13) - (17).

The equation (19), taking into account that already $\delta y_1(t_k) = 0$, will be valid at equalities to zero coefficients at other variations on unknown variables on the right border, i.e. at

$$\frac{\partial H}{\partial \dot{y}_2}=0,\ \frac{\partial H}{\partial \dot{y}_3}=0,\ \frac{\partial H}{\partial \dot{y}_4}=0,\ \frac{\partial H}{\partial \dot{y}_5}=0,\ H(y_i,t_k)=0,$$

or

$$\left.\frac{\partial H}{\partial \dot{m}_A}\right|_{t_k}=0,\left.\frac{\partial H}{\partial \dot{m}_P}\right|_{t_k}=0,\left.\frac{\partial H}{\partial \dot{T}_W}\right|_{t_k}=0,\left.\frac{\partial H}{\partial \dot{T}_V}\right|_{t_k}=0,$$
$$\left.f_H\right|_{t_k}-\dot{m}_W\left.\frac{\partial H}{\partial \dot{m}_W}\right|_{t_k}-\dot{m}_A\left.\frac{\partial H}{\partial \dot{m}_A}\right|_{t_k}-\dot{m}_P\left.\frac{\partial H}{\partial \dot{m}_P}\right|_{t_k}-\dot{T}_W\left.\frac{\partial H}{\partial \dot{T}_W}\right|_{t_k}-\dot{T}_V\left.\frac{\partial H}{\partial \dot{T}_V}\right|_{t_k}=-\left.H\right|_{t_k}=0. \quad (21)$$

These conditions are simplified to the following form:

$$\lambda_2=0,\lambda_3=0,\lambda_4=0,\lambda_5=0,\ \ f_H-\lambda_1\dot{m}_w=0$$

$$C_U(t)(P_0-(\frac{m_A(t)}{\mu_A}+\frac{m_P(t)}{\mu_W})\frac{RT_V(t)}{V_{V0}(t)})U_V\ +C_TQ_T+$$
$$\text{or}\ +\lambda_1[SK_0(P_0)[174727.8\text{-}1312.1\cdot \mathrm{Tw}+2.47\cdot \mathrm{Tw}^2\ -$$
$$-(\frac{m_P(t)}{\mu_W})\frac{RT_V(t)}{V_{V0}(t)}](\frac{P_0}{(\frac{m_A(t)}{\mu_A}+\frac{m_P(t)}{\mu_W})\frac{RT_V(t)}{V_{V0}(t)}})^q]=0. \quad (22)$$

For obtaining the maximum value of functionality from $H(t)$ on an interval $[0,t_k]$ on\off control of VGM pumping of on the $C_U(t)$) function will be carried out from the following conditions:

$$U_U\frac{\partial H}{\partial C_U}\geq 0\ \ at\ \ C_U(t)=1,$$
$$U_U\frac{\partial H}{\partial C_U}<0\ \ at\ \ C_U(t)=0. \quad (23)$$

From here, VGM pumping switch on will be under the condition

$$U_U=-\lambda_2U_V(t)\frac{m_A}{V_{V0}}-\lambda_3U_V(t)\frac{m_P}{V_{V0}}-\lambda_5[\frac{1}{k-1}\frac{T_V}{V_{V0}}U_V(t)\ -$$
$$-(P_0-(\frac{m_A(t)}{\mu_A}+\frac{m_P(t)}{\mu_W})\frac{RT_V(t)}{V_{V0}(t)})U_V\geq 0. \quad (24)$$

For C_T condition of thermal influence switching and off is defined similarly (23):

$$U_T\frac{\partial H}{\partial C_T}\geq 0\ \ at\ \ C_T(t)=1,$$
$$U_T\frac{\partial H}{\partial C_T}<0\ \ at\ \ C_T(t)=0. \quad (25)$$

The requirement of positivity of control functions U_U, U_T indicate an increase of coefficients C_U, C_T (here only with 0 to 1), in order to achieve maximum values for functional J.
On the basis of (11), function of inclusion U_T in the following form is obtained:

IOP Conf. Series: Journal of Physics: Conf. Series **1050** (2018) 012091 doi:10.1088/1742-6596/1050/1/012091

$$U_T = \lambda_4[\frac{Q_T}{m_S Cp_S + m_W Cp_W}] - Q_T. \quad (26)$$

6. Numerical modeling of thermal vacuum evaporation process with power consumption optimization

Numerical modeling of system (1) – (5), with an initial mass of liquid m_{w0} = 10 g to final $m_{wк}$ = 7.1 g from volume with mass m_s = 22 g and area S = 0.01275 m^2 with minimization of energy expenses when using heating with power Q_T = 3.57 W and pumping out with speed U_V = 0.0065m^3/s of VGM from VC volume V_K = 0.463 m^3 was carried out with an initial pressure of VGM P_0 = 105.6 kPas and temperature T_0 = 293.1°K for air, liquid, the camera and the tray from iron with mass m_s = 22 g.

In an initial time point in VC air with humidity of liquid vapors of 50 - 100% that demands, as for the most intense cases, pumping out of VGM from VC is considered.

For the numerical solution of system of the differential equations (1) – (5) and the associated system (13) – (17), the Runge-Kutt method of the 4th order of accuracy on time was used. Calculations were carried out before evaporation of 2.9 g of liquid, and when using optimization conditions, the updating of initial data for variables of the associated system was carried out (at the beginning: λ_{10} = 1, λ_{20} = 1, λ_{30} = –1, λ_{40} = –1, λ_{50} = –1). In optimization calculations, conditions (29) – (32) were used.

By analogy with experimental data given in [23], the following physical quantities were taken: L_w = 2256.0 kJ/kg, R = 8.314 J/(mol · K), K_0= 1.39·10^{-8} kg / (Pas · m^2 · c), q= 1.18, μ_w = 0.0182 kg/mol, μ_A = 0.02897 kg/mol, Cp_S= = 460 J/(kg · K), Cp_w= 4180 J/(kg · K), Cp_p = 2020 J/(kg · K), Cp_A=*1005* J/(kg · K).

For the accounting of heat exchange between VGM, trays with liquid and the VC walls , the following sizes were considered: the tray sizes – 0.150m x 0.085m, the VC walls sizes – 0.755m x 0.785m x 0.78m with thickness 0.018m, a heat capacity of walls of the VC - Cp_K = 460 J/(kg · K), liquid heat conductivity coefficients - λ_W = 0.63 W / (m · K), VGS – λ_V = 0.022 · (P_V/P_0) W/(m · K), VC tray and body - λ_S = λ_K = 92.0 W/(m · J) modeling of heat exchange was carried out according to the methods presented in [23].

The case of joint continuous influence of the heater with VGM pumping out and with control for achievement of total power consumption of E_Σ minimization according to expression (10) was considered.

The calculations results showing the advantage of applying the operating impacts of thermal vacuum drying process (VGM heating and pumping) in comparison with influences on time are given in tables 1 – 4. Thus, the factor of smaller evaporation time is not less important.

Calculations results with initial humidity C_{w0} = 0%, 50% and 100% without heating are given in table 1 when the given mass of liquid in view of the VC limited volume (line 1-3) can not evaporate. In the same place results with heating at the constant power of Q_T = 3.57 W are given when long heating resulted in considerable liquid overheating's (actually it had to be evaporated) which did not correspond to an assumption about small time impact of various factors on process of evaporation (line 4-6).

Table 1. Energy consumption for liquid evaporation without VGM pumping and without optimization

№	Time *t*, s	C_{w0} %	Q_T, W	U_V, m^3/s	λ_{10} λ_{20}	λ_{30} λ_{40}	λ_{50}	P_K, kPa	T_{wK},° K	T_{VK},° K	Δm, g	E_Σ ,kJ
1	*t* = 44187.7	0%	–	–	– –	– –	–	108.611	198.4	299.1	<2.9	0.0
2	*t* = 43050.0	50%	–	–	– –	–	–	107.968	223.7	297.5	<2.2	0.0

3	$t = \infty$	100%	–	–	– –	–	–	105.600	293.1	293.1	0.0	0.0
4	t = 1424.8	0%	3.57	–	– –	–	–	107.317	374.6	296.1	2.9	5.08 66
5	t = 1486.5	100%	3.57	–	– –	–	–	107.268	380.9	295.9	2.9	5.30 68
6	t = 457.2	100%	20.0	–	– –	–	–	106.832	491.9	294.7	2.9	9.14 32

Results of modeling of joint turning on of the thermal heater and VGM pumping out are given in table 2. Parameters ща evaporation of 2.9 g of liquid at continuous impact of VGM pumping out and liquid heating and without optimization are specified in the first line ($\lambda_i = 0$).

Table 2. Comparison of energy consumption for liquid evaporation with pumping and heating optimization

№	Time t, s	Q_T, W	U_V, Dm^3/s	λ_{10}	λ_{20}	λ_{30}	λ_{40}	λ_{50}	E_Σ, kJ
1	t = 143.2	3.57	6.5	–	–	–	–	–	84.7426
2	t = 629.7	3.57 t_0=496.5 t_k=629.7	6.5 t_0=0 t_k=71.5	1182.3	2255.40	5427.80	–216.3	–68.14	35.7568
3	t = 727.7	3.57 t_0=567.2 t_k=727.7	6.5 t_0=0 t_k=68.5	1159.7	2255.10	5428.0	–218.1	–66.85	34.1021
4	t = 3191.9	3.57 t_0=1324.6 t_k=1873.5	6.5 t_0=0 t_k=54.1	1089.0	2250.2	5429.2	–226.4	–58.24	26.0731

Table 2 (line 2-4) shows that by means of controlling the evaporation of the given mass of liquid, power consumption are reduced by three times.

In figures 1-2 cyclogramas of turns-on and turns-off of pumping out and heating to lines 3-4 of table 2 are shown. Besides, the graphs of liquid mass evaporation on time are shown there. As it appears from the graphs, short-term impacts of pumping out and heating lead to liquid evaporation acceleration.

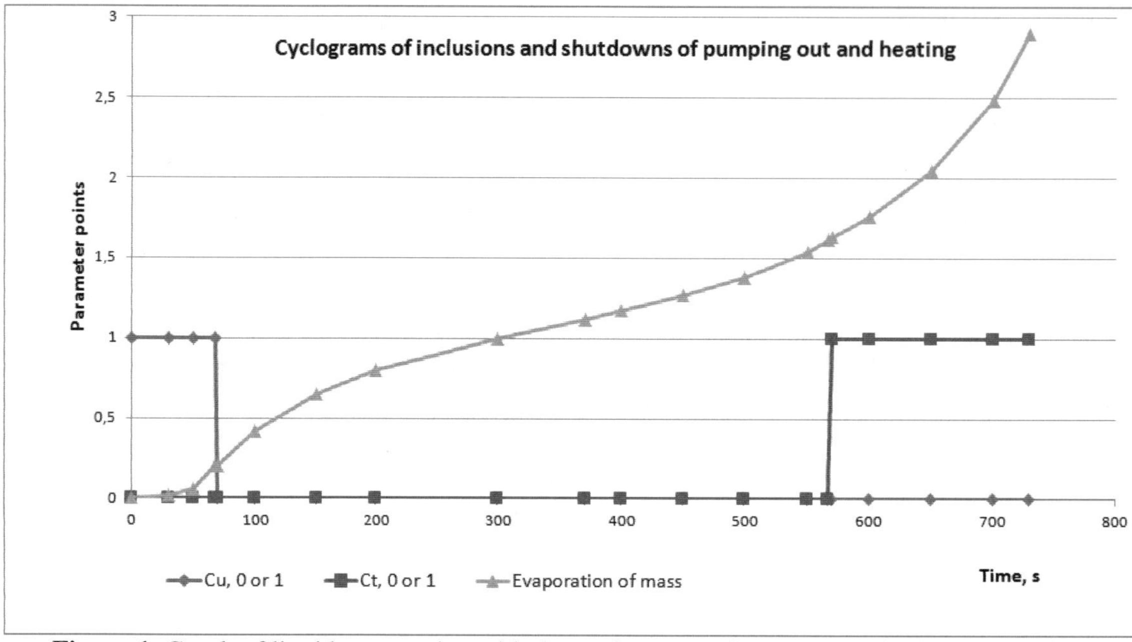

Figure 1. Graph of liquid evaporation with the cyclograma of turns-on - calculation 3, table 2.

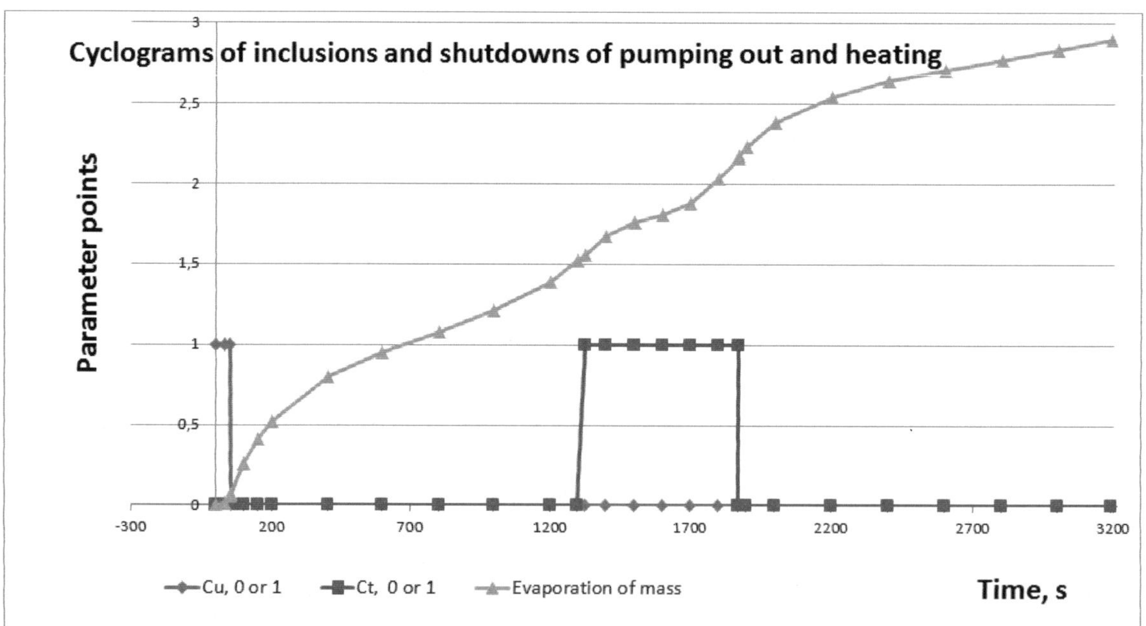

Figure 2. Graph of Liquid Evaporation with the cyclograma of turns-on - calculation 4, table 2.

In table 3 under numbers 2-5 test calculations for liquid evaporation at thermal influence with heat exchange of VGM with liquid in a tray and case walls without optimization are given (in line 2 evaporation at various initial humidity of CwObez of heating is considered). All these results are given as an example since they are in a contradiction with an assumption about a time smallness of thermal vacuum influence (till 10 min.).

Table 3. Energy consumption at liquid evaporation with heat exchange between VGM and liquid in a tray and case walls without optimization and comparison with the experimental results

№	Time *t*, s	C_{w0}	Q_T, W	U_V, Dm^3/s	λ_{10} λ_{20}	λ_{30} λ_{40}	λ_{50}	$P_к$, kPa	T_{WK}, °K	T_{VK}, °K	Δm, g	E_Σ, KJ
1	t=41865.7	50%	–	–	– –	– –	–	106.447	293.0	293.1	2.9	0.0
2	t=23613.9	50%	3.57	–	– –	–	–	106.447	296.3	293.1	2.9	84.301
3	t=41557.1	70%	3.57	–	– –	–	–	106.447	296.5	293.1	2.9	148.358
4	t=74901.8	80%	3.57	–	– –	–	–	106.447	296.7	293.1	2.9	267.391
5	t=77550.5	90%	3.57	–	– –	–	–	106.447	296.9	293.1	2.2	276.889
6	t= 442.8	80%	3.57	6.5	– –	–	–	.0.464	266.9	290.3	2.9	255.726
7	t= 406.1	80%	–	6.5	– –	–	–	0.618	272.6	289.6	2.2	232.099
					Comparison with experiment							
8	t= 435.0	80%	3.57	6.5	– –	–	–	0.573	273.1	282.1	2.9	252.621
9	t= 400.0	80%	–	6.5	– –	–	–	0.613	277.9	279.3	2.2	230.868

Different entry conditions on VGM humidity which showed essential increase of both time for evaporation of 2.9 g of liquid and power consumption at high initial humidity (80% - 90%) were considered

At humidity of C_{w0} = 90% the evaporation process stopped at 2.2 g with continuation of power consumption when heating liquid that is characteristic for the closed camera with heat exchange.

With introduction of VGM pumping out (calculations 6 and 7) at humidity of 80% the essential gain in power consumption for evaporation of 2.9 g and 2.2 g of liquid, is respectively observed. These calculations will well be coordinated with the experimental data [23] given in the table in lines 8 and 9.

In table 4 comparison of energy consumption on liquid evaporation with optimization at the accounting of heat exchange of VGM with liquid is given in a tray and VC case walls.

Table 4. Comparison of energy consumption on evaporation of liquid with optimization and heat exchange of VGM with liquid in a tray and VC case walls.

№	Time *t*, s	Q_T, W	U_V, Dm^3/s	λ_{10}	λ_{20}	λ_{30}	λ_{40}	λ_{50}	E_Σ kJ
1	t= 442.8	3.57	6.5	–	–	–	–	–	255.726
2	t= 435.0	3.57	6.5	–	–	–	–	–	252.621
3	t= 409.5	3.57 t_0=105.7 t_k=409.5	6.5 t_0=0 t_k=409.5	2632.1	13412	8083	–1.918	–70.61	234.035
4	t= 403.7	3.57 t_0=95.3	6.5 t_0=0	2487.2	11380	8903	130.2	–85.01	230.457

		t_k=403.7	t_k=403.7						
5	*t*= 4956.2	3.57 t_0=125.9 t_k=4956.2	6.5 t_0=0 t_k=71.3	2627.2	13120	8131	18.44	–64.72	35.261

In line 1 calculation with evaporation of 2.9 g of liquid without control at constant heating and continuous pumping out is given and corresponds to energy consumption to the experimental data given as No. 2. In lines 3 and 4 two calculations at the same input parameters with optimization which at consecutive approximations show reduction of energy consumption by evaporation of 2.9 g of liquid due to later inclusion of liquid heating are given. Using control parameters with specification of initial data $\lambda_{10} - \lambda_{50}$ provides the decrease of energy consumption by 10% and even more.

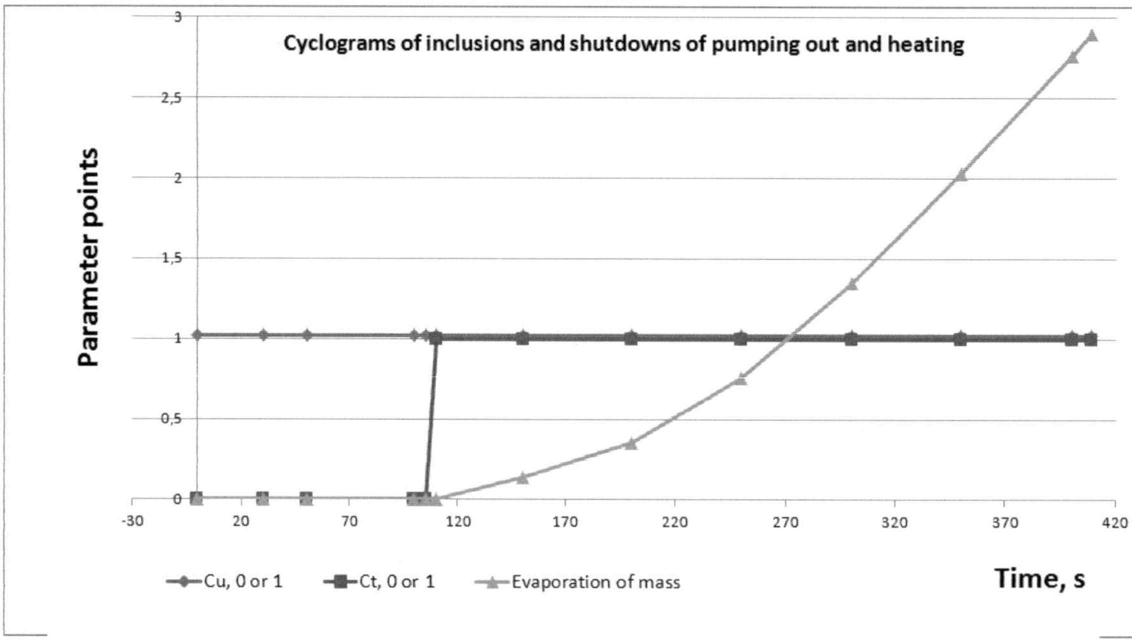

Figure 3. Graph of Liquid Evaporation with the cyclogram of inclusions-calculation 4, table 4.

In figure 3 shown cyclograms of pumping out and heating to line 4 of table 4 and the graph of evaporation of liquid mass on time are shown.

7. Discussion of the results

With optimization and without optimization of energy consumption (table 2) follows from results of modeling of a thermal vacuum drying (VGM heating and pumping) that there appears the gain in optimization due to reduction of an initial interval of time to VGM pumping out and later beginning of liquid heating-up (line 4). Thus, with reduction of energy consumption more optimum control leads to lengthening the time of evaporation of the set mass of liquid. In line 4 more optimum case with heating turn-off with an incomplete mass of evaporation with reduced power consumption is given, but the general time for evaporation of the set weight due to evaporation of liquid in no saturated VGM considerably increases.

The results of modeling of liquid evaporation at thermal influence with heat exchange of VGM with liquid in a tray and case walls at initial humidity> 70% showed the requirement in considerable energy consumption and long time (table 3), at humidity of 90% because of limitation of volume with saturation of VGM evaporation stops (in the considered case at value 2.2 g).

Application of pumping out leads to considerable reduction of time of evaporation of the set mass of liquid and a power prize. Thus comparison with experimental data when the heating and pumping of VGM are constantly involved, gives deviations within 1-3% in different parameters. With application of optimization (table 4) when there is later inclusion of heating, the power prize > 10% can be received.

The further direction of researches provides an assessment of the methodical errors caused by replacement of the differential equations in private derivatives on the ordinary differential equations, and also experimental check of the received results of optimum control of process of thermal vacuum evaporation of liquids with other parameters of system (volumes, masses, speeds of pumping, heat feeders in volume).

8. Conclusion

1. The mathematical model of liquid evaporation at thermal vacuum influence based on the ordinary differential equations is developed.
2. the criterion of optimization of pumping and temperature heating control at the fixed values of the heater power and pumping productivity from a condition of minimization of energy for evaporation of the fixed mass of liquid Is offered.
3. The procedure of the theory of optimum control application based on the Pontryagin principle of maximum Is developed for optimization of thermal vacuum evaporation process.
4. Efficiency of optimum control theory application for the considered problem definition is shown.

Acknowledgements
The research is conducted with financial support of the Ministry of Education and Science of the Russian Federation within the state task to the subordinated educational organizations, the "Increase of Ecological Safety and Economic Efficiency of Carrier Rockets with Mid-flight Liquid Rocket Engines" project task No. 9.1023.2017/PCh.

References

[1] Gatapova E, Filipenkor A, Lyulin Yu, Grauri A, Marchuk I, Kabov O 2015 Experimental research of a temperature field in two-layer system liquid-gas *Thermophysics and aeromechanics* **6** pp 729–734
[2] Gatapova E, Kabov O 2010 Shear-driven flows of locally heated liquid films *Intern. J. Heat and Mass Transfer* **53** pp 2795–2807
[3] Goncharova O, Rezanov E, Tarasov Ya 2013 Mathematical modeling of thermo-capillar currents in a thin layer of liquid taking into account evaporation *The Mathematician and the Mechani,* **3** pp 47–52
[4] Smiths G, Feoktistov D, Orlova E 2016 Evaporation of drops of liquids from a surface of the anodized aluminum *Thermophysics and aeromechanics* **1** pp 17–22
[5] Semenov A, Feoktistov D, Zaytsev D, Kuznetsov G, Kabov O 2015 A pilot study of evaporation of a drop of liquid on the heated firm surface *Thermophysics and aeromechanics* vol **22** pp 801–804
[6] Tsygankov A, Alyoshin A 2016 Modeling of processes of condensation and evaporation in the channel of the regenerative heatutilizer *Messenger of the International academy of cold* **1** pp 82–85
[7] Zudin Yu 2017 Poluempiricheskaya model of intensive evaporation *Thermophysics and aeromechanics* **4** pp 539–552
[8] Trushlyakov V, Lavruk S 2015 Theoretical and experimental investigations of interaction of hot gases with liquid in closed volume *Acta Astronautica* **109** pp 241–247

[9] Zudin Yu 2016 Linear kinetic analysis of evaporation and condensation *Thermophysics and aeromechanics* **3** pp 437–450
[10] Kuznetsov V 2000 Dynamics of locally heated liquid films *Russ. J. Eng. Thermophys* **10** pp 107–120
[11] Kabov O, Kabova Y, Kuznetsov V 2012 Evaporation of not isothermal film of liquid in the microchannel at a cocurrent stream of gas *it is GIVEN* **446** p 5
[12] Kuznetsov V, Andreyev V 2013 The movement of a liquid film and gas stream in the microchannel with evaporation *Thermophysics and aeromechanics* **1** pp 17–28
[13] Iorio G, Goneharova O, Kabov O 2009 Study of evaporative convection in an open cavity under shear stress flow *Microgravity ScL TcchnoL* **21** p 42
[14] Gatapova E, Filipenko R, Lyulin Yu, Graur I, Marchuk I, Kabov O 2015 A pilot study of a temperature field in two-layer system liquid-gas *Thermophysics and aeromechanics* **22** pp 729–734
[15] Kuznetsov V 2011 Heat-mass exchange on an interface liquid-steam *News of the Russian Academy of Sciences Mechanics of liquid and gas* **5** pp 97-107
[16] Trushlyakov V, Lesnyak I, Galfetti L 2017 Pilot studies of process of convective heat exchange at evaporation of kerosene and water in the closed volume *Thermophysics and aeromechanics*, **5** pp 771–781
[17] Semenov G, Budantsev E, Melamed L, Tropkina A 2011 Mathematical modeling and a pilot study of the combined cycles of vacuum drying of thermolabile materials *The bulletin of the international academy of cold* **4** pp 5–11
[18] Kabov O, Kabova Yu 2015 Influence of the sizes of the heater on evaporation of a film of the liquid which is carried away by a gas stream in the microchannel at local heating *Thermophysics and aeromechanics* **22** pp 539–542
[19] Nekrasov S, Volkov V 2017 Computer modeling and optimization of process of thawing of soil by means of energy of the microwave oven *The Engineering and physical journal* **90** pp 55–63
[20] Diligenskaya A, Rapoport E 2016 Metod of minimax optimization in a coefficient return problem of heat conductivity *The engineering and physical journal* **89** pp 1007–1012
[21] Pontryagin L, Boltyansky V, Gamkrelidze R, Mishchenko E 1983 Mathematic theory of optimum processes (Moscow: Science) p 393
[22] Gusev E, Bakulin V 2014 Optimum design of structural and non-uniform materials and designs with the demanded properties *The Engineering and physical journal* **87** pp 250–255
[23] Trushlyakov V, Panichkin A, Prusova O, Zharikov K, Dron' M 2017 Theoretical and pilot studies of process of evaporation of liquid by vacuum drying *Problem of development, production and operation of the missile and space equipment and preparation of engineering shots for aerospace branch: materials XI of the All-Russian scientific conference devoted to memory of the chief designer of "Polyot" A.S. Klinyshkov* pp 100–107
[24] Volkov A, Zharsky I 2005 The big chemical reference book (Munich: Modern school) p 608
[25] Lykov A, Mikhaylov Yu 1963 Theory of heat and mass transfer (Moscow-Leningrad: Gosenergoizdat) p 536

IOP Conf. Series: Journal of Physics: Conf. Series **1050** (2018) 012092 doi:10.1088/1742-6596/1050/1/012092

Dynamic control of tug-debris tethered system after the capturing of the debris

V I Trushlyakov[1], V V Yudintsev[2], R S Pikalov[2]

[1]Omsk State Technical University, 11, ul. Mira, Omsk, 644050, Russian Federation
[2]Samara National Research University, 34, Moskovskoye shosse, Samara, 443086, Russian Federation

Abstract. This paper focuses on the problem of control of the tug-debris space tether system for debris De-orbiting. The motion of the tether system is considered when the space tug and tethered debris have an initial relative velocity, which can affect the tether tension. We investigate two stages of the space tether motion: tether elongation stage and tether retraction stage. Two control laws for the tether length are proposed for the each stage. The control law for the first stage is based on the measurement of the tether tension and is intended to decrease the relative velocity between the space tug and debris object. The control law for the second stage retracts the tether and decrease the distance between the space tug and debris assisting to the process of the docking of the space tug with debris. The proposed control laws are demonstrated by numerical simulations.

1. Introduction

Intensive exploration of the near-Earth space environment since the launch of the Sputnik 1 leads to the fact that now many thousands of space debris have accumulated in near-Earth orbits [1]. Large debris objects are tracked from the Earth and do not carry an immediate threat to functional spacecraft that can adjust its orbit [2,3]. The collisions of large space debris with other small or large pieces can significantly increase the number of small debris in near-Earth orbits. Such events should cause that some orbits become inaccessible [4,5]. The space debris problem is recognized internationally. The countries that have space programs share the common view on the necessity to reduce space debris generation and mitigate risks posed by space objects. A set of International debris mitigation guidelines has been proposed [6,7]. In addition, methods of active debris removal are developing [8–16] many of which involve the use of space tugs for De-orbiting large space debris objects.

The classical docking technique [17] can't be applied to space debris objects. These objects are not controlled and do not cooperate with the space tug for the docking. Space debris objects can rotate at a significant angular velocity, which also makes it impossible for the space tug to dock. Tethered system can be used by the space tug for the capturing rotating debris objects [15]. The tether reduces the impact of the debris motion on the tug when the tether connection is forming. In this case, the debris can be captured using an autonomous docking module [18], net [14] or harpoon [16]. The active debris removal using tethered space tug is one of the promising techniques.

IOP Conf. Series: Journal of Physics: Conf. Series **1050** (2018) 012092 doi:10.1088/1742-6596/1050/1/012092

The control of the space tether system after its formation requires design of new algorithms for the space tug and tether. The quality of this control determines the safety of the active debris removal mission. The control of the tether system during the active stage of a debris removal mission are considered in [19–21]. These works consider the stage of the motion after the space tug burns-up its thruster. These authors investigate the longitudinal oscillations of the tether due to changes of the acceleration of the space tug. These oscillations should be damped to decrease the disturbance to the motion of the space tug and to prevent possible collision of the space tug with the tethered debris.

The control of the tether also is essential before the active stage of the debris removal. When the tethered autonomous module that separates from the space tug is used to capture the debris [18] (figure 1), the capture event may occurs in the presence of relative velocity between the space debris and space tug [22, 23]. This relative velocity can produce a large tension to the tether and to the tether deployment mechanism. So, the tether tension should be controlled after the capture of the space debris.

This work is devoted to the development of the control law for the post-capture stage of the active debris removal mission when the tethered autonomous module is used to capture the debris object that has velocity relative to the space tug.

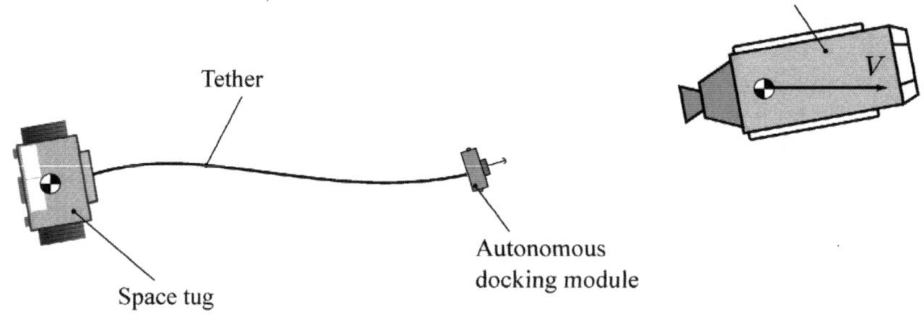

Figure 1. Capturing space debris using the autonomous module

2. Statement of the problem

In this work we investigate the motion of the space tether system consisting of a space tug and tethered space debris with the attached autonomous module. Here and after the term space debris refers to the debris with the connected autonomous module [18, 22, 23]. It is supposed that the debris flyby the space tug with some relative velocity. We consider two stages of the motion of the system:

- tether deployment stage when the tether length is increased with decreasing the relative velocity of the space debris;
- tether retraction stage when the length of the tether is decreased assisting to the docking of the autonomous module and attached debris object with the space tug [24].

At the start of the first stage the tether already has some initial length, which is determined by the relative motion of the autonomous module, space tug and space debris object in the previous stage, when the autonomous docking module separated from the space tug and intercepted the debris object [18].

The space tether system is presented in figure 2. The figure shows the space tug (C_1), space debris with the autonomous module (C_2) and the elastic tether. The space tug has a tether deployment device, which can change length of the tether. The strength of the tether must be taken into account during the tether length control. The tether length is also limited by the capacity of the tether deployment device.

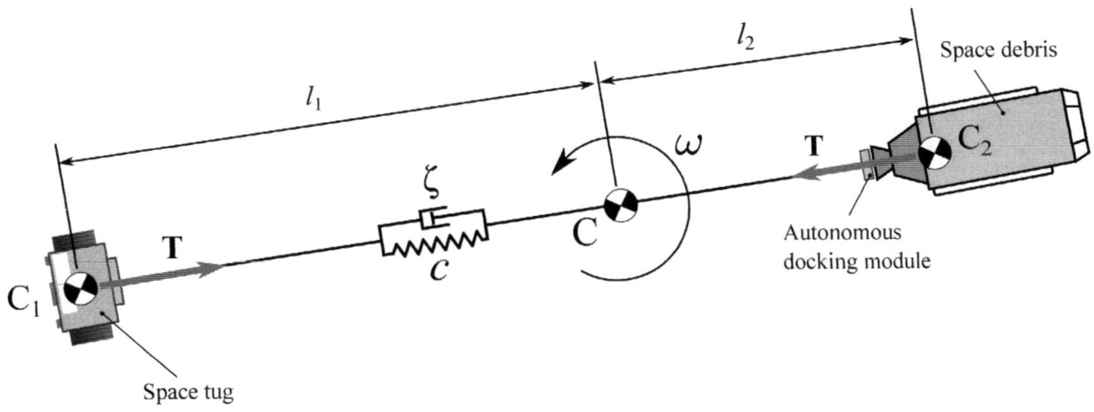

Figure 2. Tug-debris tether system

3. Motion of the tether system and tether control

3.1. Tether deployment stage

The space tug and the space debris are considered as mass points, which both have constant masses m_1 and m_2 respectively. We assume that the time of the considered tether deployment stage is small compared to the orbital period of the system, so the motion of the bodies C_1 and C_2 is considered in a gravityless space. The thrusters of the space tug are not operating. For the motion of two mass points connected by a tether we can write the angular momentum conservation law relative to the center of mass of the system:

$$m_1\omega l_1^2 + m_2\omega l_2^2 = const. \tag{1}$$

Expressing the length l_1 and l_2 in terms of the tether length $l = l_1 + l_2$ (figure 2) we get:

$$l_1 = \frac{m_2}{m_1 + m_2}l, \quad l_2 = \frac{m_1}{m_1 + m_2}l, \tag{2}$$

and rewrite the equation (1) as:

$$\omega l^2 = const. \tag{3}$$

Equation (3) allows to determine the angular velocity of the tether system as a function of the tether length, initial angular velocity ω_0, and initial length l_0 of the tether system:

$$\omega_0 l_0^2 = \omega l^2. \tag{4}$$

Let us consider the motion of the space debris relative to the center of mass of the system. The equation of the motion can be obtained using D'Alembert's principle (figure 2):

$$\omega^2 l_2 m_2 - \ddot{l}_2 m_2 - T = 0, \tag{5}$$

where T is the tether tension force. Substituting (2) into (5) we get:

$$\omega^2 lm - \ddot{l}m - T = 0, \tag{6}$$

where m is the reduced mass of the system:

$$m = \frac{m_1 m_2}{m_1 + m_2}. \tag{7}$$

The angular velocity of the system (ω) is determined by the expression (4), so the equation (6) takes the form:

$$\ddot{l} - \frac{\omega_0^2 l_0^4}{l^3} + \frac{T}{m} = 0 \tag{8}$$

The tether tension force T for its deformation $\delta = l - L$ is determined by the tether stiffness c and damping properties:

$$T = \begin{cases} c\delta + k\dot{\delta}, & l - L > 0, \\ 0, & l - L \leq 0, \end{cases} \tag{9}$$

where L is the tether free length. The stiffness of the tether depends on the modulus of elasticity of the material of the tether (E), cross sectional area of the tether (F) and its free length L:

$$c = \frac{EF}{L}. \tag{10}$$

The damping coefficient k is defined as follows:

$$k = 2\zeta\sqrt{cm}, \tag{11}$$

where ζ is a damping ratio and $2\sqrt{cm}$ is the critical damping of the considered system.

The free length L of the tether can be changed by the tether deployment device installed on the space tug. Let us describe the dynamic of this mechanism as (figure 3):

$$J_r \frac{\ddot{L}}{r} = M_c + Tr, \tag{12}$$

where J_r is the moment of inertia of the reel with the tether of radius r, M_c is the control torque. The equation (12) can be rewritten as

$$M\ddot{L} = F_c + T, \tag{13}$$

where $M = J_r/r^2$ and $F_c = M_c/r$.

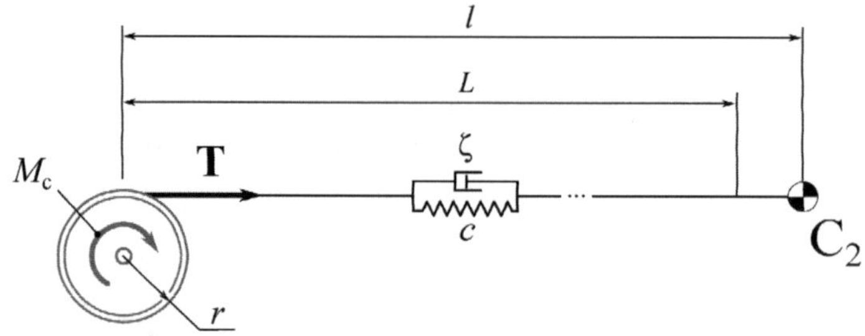

Figure 3. The tether free length (L) control mechanism

Let us consider the following control law for the free length of the tether which depends on the deviation of the tether tension T from some normal operating load T_p:

$$F_c = (T - T_p)\, k_T, \tag{14}$$

where k_T is a positive parameter, T_p is the normal operating tether tension force. The control law (14) ensures that tether tension T is around a certain value T_p, which will lead to the decreasing of the relative velocity of the tug and space debris. The equations for the controlled tether system now can be written as:

$$\begin{cases} \ddot{l} - \dfrac{\omega_0^2 l_0^4}{l^3} + \dfrac{T}{m} = 0, \\ M\ddot{L} = (T - T_p)k_T + T. \end{cases} \tag{15}$$

Equations (15) are to be integrated with the following initial conditions

$$l(0) = l_0,\ \dot{l}(0) = \dot{l}_0,\ L(0) = l_0,\ \dot{L}(0) = \dot{l}_0.$$

The system (15) is nonlinear and its solution can be found numerically but if the initial angular velocity of the system and damping ratio are equal to zero and if the tether stiffness changes slightly during the considered stage of the motion and can be replaced by the constant value, the equation (15) can be integrated analytically. For that case, the equations (15) take the following form:

$$\begin{cases} \ddot{l} + \dfrac{c_0\delta}{m} = 0, \\ M\ddot{L} = (c_0\delta - T_p)k_T + c_0\delta, \end{cases} \tag{16}$$

where $c_0 = EF/l_0$ is the initial stiffness of the tether. The equations (16) has the following solution:

$$\begin{cases} l = l_0 + \dot{l}_0 t - \dfrac{c_0 k_T m^2 M^2 T_p}{A^4}\left(mM\cos\left(\dfrac{At}{mM}\right) + c_0\left(M + (1+k_T)\,m\right)\dfrac{t^2}{2} - mM\right), \\ L = l_0 + \dot{l}_0 t + \dfrac{k_T m^2 M^2 T_p}{A^4}\left(c_d m^2\cos\left(\dfrac{At}{mM}\right) - c_0\left(c_0 M + c_d m\right)\dfrac{t^2}{2} - c_d m^2\right), \end{cases} \tag{17}$$

where

$$A = \sqrt{mM\left(c_0 M + c_d m\right)} \ \text{ and } \ c_d = c_0\left(1 + k_T\right). \tag{18}$$

The tether tension is

$$T = c_0\left(l - L\right) = \frac{1}{A^4} k_T m^3 M^2 T_p \left(c_0 M + c_d m\right)\left(1 - \cos\left(\frac{At}{mM}\right)\right). \tag{19}$$

The tether tension has the maximum when $t = mM/A$:

$$T_{max} = \frac{2c_0 k_T T_p\left(k_T m + m + M\right)}{mM^2\left(c_0 M + c_d m\right)^4}. \tag{20}$$

The expressions (17) and (20) allow to estimate the time needed for the deployment the tether to the maximum length, maximum length of the tether, and maximum tether tension during the deployment stage.

3.2. Tether retraction stage

Let us consider the tether retraction stage. The stage starts with zero velocity of the debris relative to the space tug. We assume that after the tether deployment stage the system is stabilized, and its angular velocity is equal to zero. The tether deployment device in this stage reduces the length of tether assisting to the space tug for docking with the space debris. To avoid collisions of the debris with the space tug and to ensure that the tether is tensioned, the space tug produce a constant thrust force P. The motion equations of the system for the second stage are:

$$\begin{cases} m\ddot{l} = T + P\dfrac{m}{m_1}, \\ M\ddot{L} = F_c + T. \end{cases} \tag{21}$$

We propose the following tether length control law [24]:

$$L = L(t) = l_b + \frac{(l_b - l_e)}{2}\left(1 + \cos\varphi t\right), \tag{22}$$

where $\varphi = \pi/t_e$, l_b is the initial length of the tether, t_e is the duration of the retraction stage, l_e is the final length of the tether (we suppose that the space tug docks with the space debris when the tether length reaches l_e). The control law (22) is the modification of the law, considered in [24]. For the tether length control (22) the control force F_c has the form:

$$F_c = \frac{\pi^2 M\left(l_b - l_e\right)}{2t_e^2}\cos\left(\frac{\pi}{t_e}t\right). \tag{23}$$

4. Simulation

In this section we illustrate the proposed control laws by numerical examples. The parameters of the system are presented in Table 1.

We consider three simulation cases differing in the initial angular velocity of the system to investigate the effect of this velocity on the solution of the system (15) and to compare the results with the estimations (17) and (20). In the first simulation case initial angular velocity of the system is equal to zero, in the second case $\omega_0 = 0.1$ rad/s and in the third case $\omega_0 = 0.1$ rad/s.

Table 1. Parameters of the system

Parameter	Value	Parameter	Value
m, kg	500	T_p, N	1000
M, kg	1	EF, N	$1.35 \cdot 10^5$
$l_0 = L_0$, m	100	ζ	0.005
$\dot{l}_0 = \dot{L}_0$, m/s	20	t_e, s	100
k_T	10	l_e, m	1
P, N	100		

4.1. Tether deployment stage

Figure 4 shows the tether tension (figure 4(a)) and the tether length (figure 4(b)) as functions of time for the first simulation case. These results obtained by numerical integration of the system (15).

Figure 5 shows graphs for the same quantities obtained by the analytical solutions (17) and (19). The maximum tension of the tether obtained by using the expression (20) is equal to 1.8 kN. The numerical solution of the system (15) gives us value of $T_{max} = 1.7$ kN. If the tether length L did not controlled by the tug, the maximum tether tension would be:

$$T^{max} \approx \dot{l}_0\sqrt{c_0 m} \approx 16.4 \text{ kN},$$

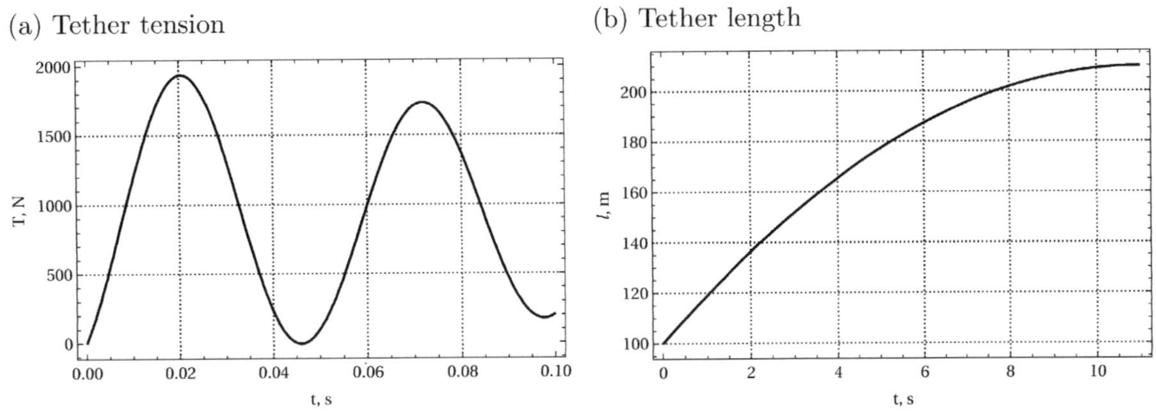

Figure 4. Time history of the tether tension and length for the first case (numerical solution)

which illustrates necessity of the tether control.

The maximum length of the tether, obtained by the numerical solution of the system (15), and by the solution (20) is 210 m.

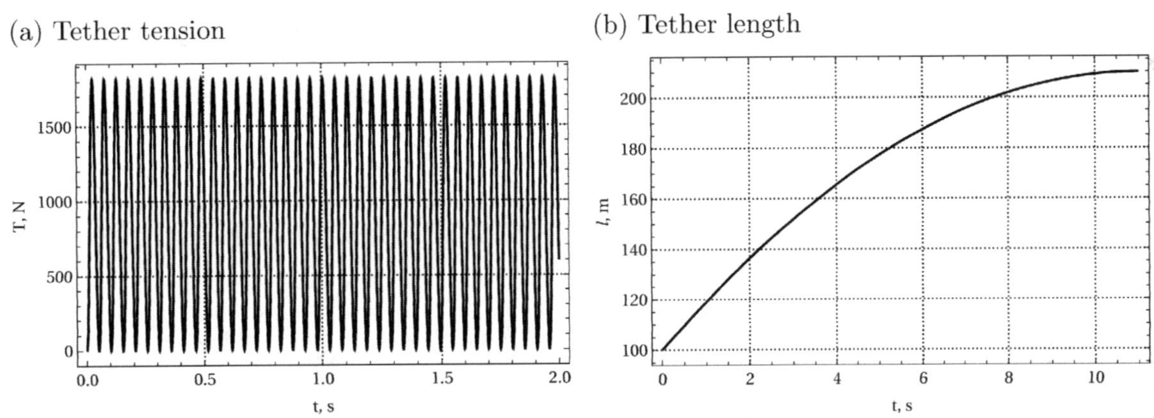

Figure 5. Time history of the tether tension and length for the first case (analytical solution)

Figure 6 shows the elongation of the tether $(l - L)$ and the tether length for the second simulation case ($\omega_0 = 0.1$ rad/s). The graph for the tether tension is not presented here because for the tether tension for the second case does not differ much from the first case (figure 4(a)).

The maximum magnitude of the tether tension is primarily depending on the initial linear relative velocity of the tug and the space debris, which is not changed. In the second simulation case the maximum tether length is about of 232 m (figure 6(a)), which is 22 meters greater than the value obtained by using the approximate solution (17) (figure 5(b)). The tether deployment times also differ from each other.

Figure 7 shows the results for the third case. Figure 7(a) presents the tether elongation and figure 7(b) shows time history of the tether length for the third simulation case. In this case the maximum of length of the tether is about of 308 m, which is 98 meters greater than the value obtained by using the solution presented in figure 5(b). The second and third cases illustrate that when the tether system has initial angular velocity the numerical solution of the system (15) should be used to estimate the final length of the tether.

IOP Conf. Series: Journal of Physics: Conf. Series **1050** (2018) 012092 doi:10.1088/1742-6596/1050/1/012092

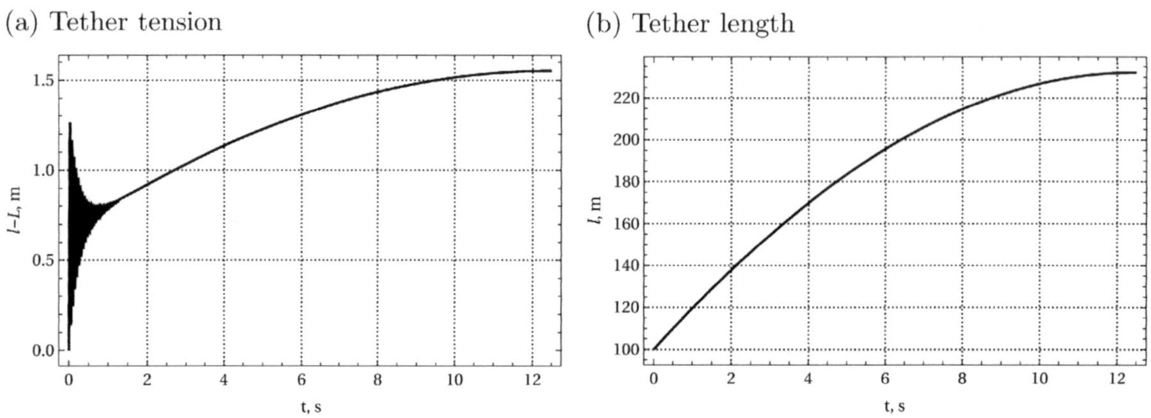

Figure 6. Tether elongation and tether length for the second simulation case (numerical solution)

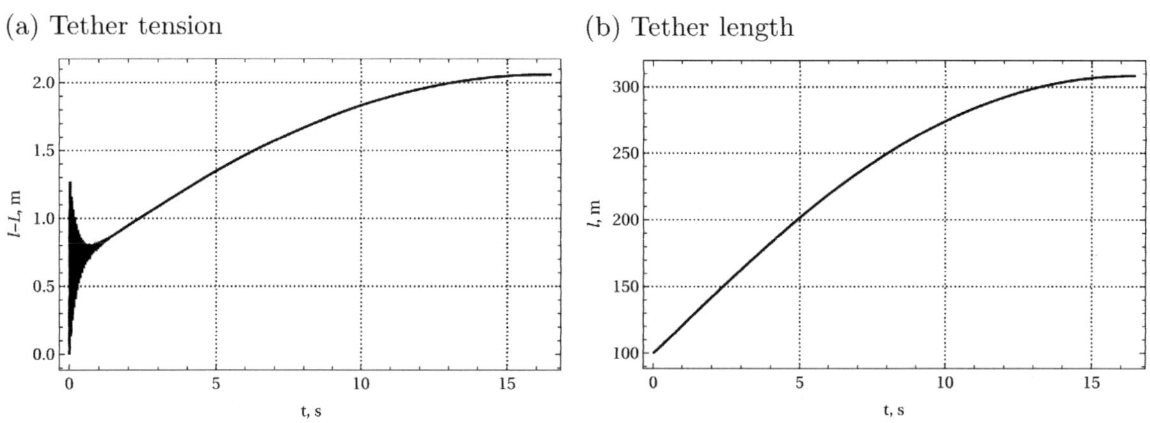

Figure 7. Tether elongation and tether length for the second simulation case (numerical solution)

4.2. Tether retraction stage

Let us consider the second stage of the motion (tether retraction stage). As previously noted, the stage begins after the stabilization of the tether system, so the angular velocity of the system is equal to zero.

Figure 8 shows how the tether tension and tether length change with time. The results show that the influence of the tug's thrust leads to the longitudinal oscillations of the tether. The amplitude of the oscillations of the tether tension can be increased if the damping ratio is small. This case is illustrated by the figure 8(a). The increase of the amplitude of oscillations of the tension leads to the slacking of the tether and uncontrolled relative motion of the space debris. To reduce these oscillations the tether deployment device should damp longitudinal oscillations of the tether.

Figure 9 shows graph for the same quantities with the damping ratio increased by 10 times to a value of 0.05. In this case, the tension amplitude is reduced and at the end of the retraction stage the tether oscillation is absent.

Figure 10 shows the tether retraction velocity. The figure demonstrate that at the end of the tether retraction stage the relative velocity of the tug and the space debris decreases to zero.

(a) Tether tension (b) Tether length

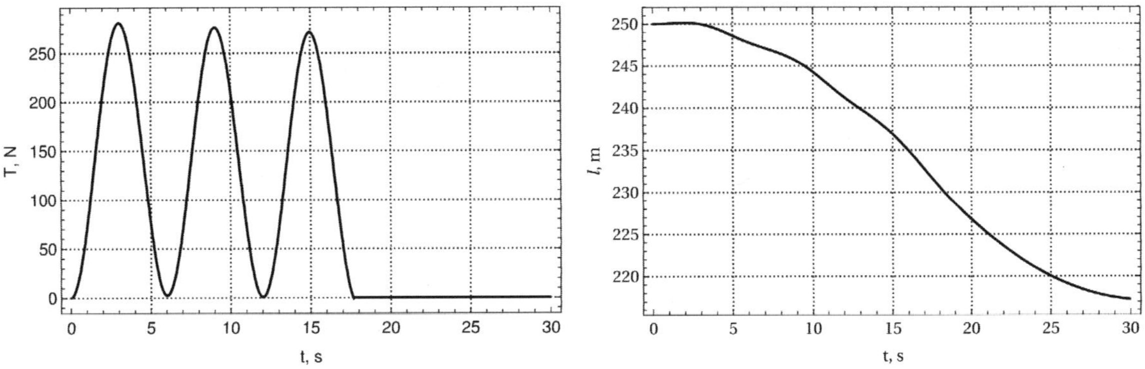

Figure 8. Time history of the tether tension and tether length for the second stage ($\zeta = 0.005$)

(a) Tether tension (b) Tether length

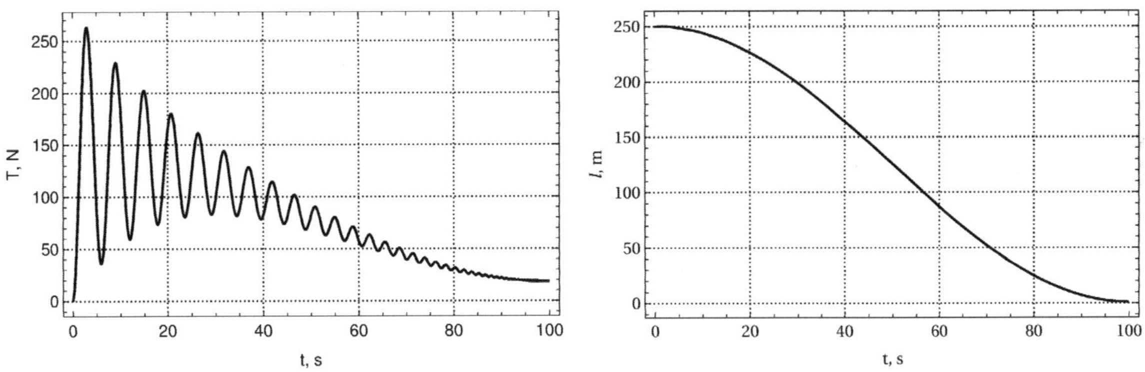

Figure 9. Time history of the tether tension and tether length for the second stage ($\zeta = 0.05$)

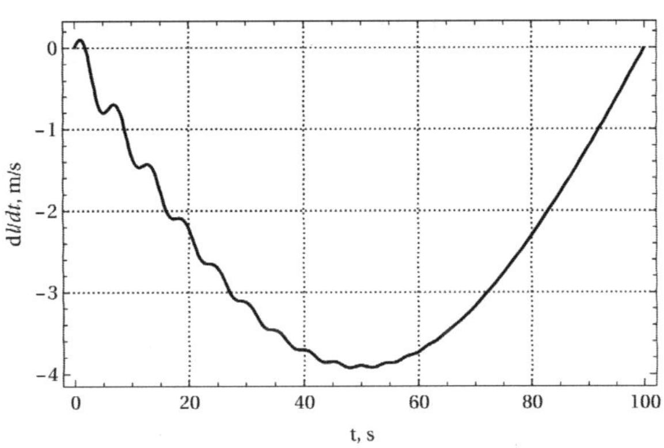

Figure 10. Retraction velocity of the tether

5. Conclusion

In this paper we propose the control laws for the elongation and retraction stages of the space tether system after the capturing of the space debris object by the autonomous module that was separated from the space tug, assuming that the capture event occurs when the space debris have some velocity relative to the space tug.

For the first stage the control law for the tether free length is proposed, which reduce the relative velocity of the bodies to zero and also controls the tension of the tether. For the case of the system motion with the angular velocity equal to zero, the analytical solutions for the tether length and tether tension are obtained. These solutions can help to select the parameters of the proposed control (k_T, T_p) and estimate tether tension and maximum tether length (final length) at the end of the stage. When the tether system has initial angular velocity the numerical solution of the system (15) should be used to estimate the final length of the tether.

In contrast to the tether length control law considered in [22], which depends only on time and initial conditions of the motion of the tether system, the tether length control law proposed in this paper takes into account the tension force of the tether, providing control of this value.

Tether retraction stage should begin after the stabilization of the system for zeroing the initial angular velocity. During this stage the longitudinal oscillations of the tether should be substantially damped to eliminate oscillations of the tether at the end of the stage. The tether length control law for the second stage assists to the space tug for docking with the debris and provides zero relative velocity of the tug and space debris during their rendezvous at the distance l_e. Proposed control requires a constant thrust force acting on the space tug.

Acknowledgments
This work is supported by the Russian Foundation for Basic Research, Project No. 17-41-630274-region.

References

[1] Anselmo L and Pardini C 2016 *Acta Astronaut.* **122** 19–27 ISSN 00945765
[2] Jesus A D C, Sousa R R and Neto E V 2015 *J. Phys. Conf. Ser.* **641** 12021
[3] Ferreira L S, Jesus A D C, Carvalho T C F and Sousa R R 2017 *J. Phys. Conf. Ser.* **911** 12017
[4] Ting Wang 2010 *Sci. Glob. Secur.* **18** 87–118 ISSN 08929882
[5] Johnson N L 2010 Orbital debris: the growing threat to space operations *33rd Annu. Guid. Control Conf.* (Breckenridge)
[6] Ansdell M 2010 *J. Public Int. Aff.* **21** 16 ISSN 1070521X
[7] Johnson N L and Stansbery E G 2010 *Acta Astronaut.* **66** 362–367 ISSN 00945765
[8] Bonnal C, Ruault J M and Desjean M C 2013 *Acta Astronaut.* **85** 51–60 ISSN 00945765
[9] Kumar K, Jankovic M, G N O, Martin J M R, Topputo F, Walker S, Kirchner F and Vasile M 2015 Agora: Mission to demonstrate technologies to actively remove Ariane rocket bodies *Proc. Int. Astronaut. Congr. (IAC 2015). Int. Astronaut. Congr. (IAC-2015), Oct. 12-16, Jerusalem, Isr.* pp 1–16
[10] Shan M, Guo J and Gill E 2015 *Prog. Aerosp. Sci.* **80** 18–32 ISSN 03760421
[11] Udrea B and Nayak M 2015 A cooperative multi-satellite mission for controlled active debris removal from low Earth orbit *IEEE Aerosp. Conf. Proc.* vol 2015-June ISBN 9781479953790 ISSN 1095323X
[12] Huang P, Zhang F, Dongke W, Zhongjie M, Zhengxiong L, Cai J and Wang M 2015 *Prog. Aerosp. Sci.* ISSN 03760421
[13] Schaub H and Sternovsky Z 2014 *Adv. Sp. Res.* **53** 110–118 ISSN 02731177
[14] Liu H, Zhang Q, Yang L, Zhu Y and Zhang Y 2014 *Sci. China Technol. Sci.* **57** 2407–2417 ISSN 1674-7321
[15] Aslanov V S and Yudintsev V V 2015 *Adv. Sp. Res.* **55** 660–667 ISSN 02731177
[16] Dudziak R, Tuttle S and Barraclough S 2015 *Adv. Sp. Res.* **56** 509–527 ISSN 18791948
[17] Hinman E M and Bushman D M 1991 Soviet automated rendezvous and docking system overview *Autom. Rendezvous Capture Rev. Exec. Summ.* vol 1 pp 34–35
[18] Trushlyakov V I, Kudentsov V Y, Shatrov Y T and Makarov Y N 2012 Method of removing space debris from the orbits of payloads using the separated part of the carrier rocket, the upper stage and the device for its realization (In Russian). Pat. 2462399 Russian Federation, IPC B64G1 / 00.
[19] Wen H, Zhu Z H, Jin D and Hu H 2016 *Acta Astronaut.* **119** 110–117 ISSN 00945765
[20] Jasper L E Z, Seubert C R, Schaub H, Trushkyakov V and Yutkin E 2012 Tethered tug for large low Earth orbit debris removal *Adv. Astronaut. Sci.* vol 143 (Charleston) pp 2223–2242 ISBN 9780877035817 ISSN 00653438
[21] Jasper L and Schaub H 2014 *Acta Astronaut.* **105** 373–384 ISSN 00945765

[22] Trushlyakov V and Yudintsev V 2017 Systems engineering design and optimization of an active debris removal mission of a spent rocket body using piggyback autonomous module *3rd IAA Conf. Dyn. Control Sp. Syst. (DYCOSS 2017) 30 May - 1st June 2017* (Moscow) pp 667–681
[23] Lagno O G, Lipatnikova T I, Makarov Y N, Mironova T V, Trushlyakov V I, Shatrov Y T and Yudintsev V V 2017 Parameters design of autonomous docking module and the choice of suitable target and primary payload for ADR *Proc. 7th Eur. Conf. Sp. Debris ESOC 18 - 21 April 2017* (Darmstadt, Germany)
[24] Aslanov V S and Pikalov R S 2017 *Eng. Lett. IAENG* **25** 142–146

Computer visualization of automatized calculation the factors of mass transfer the viscous incompressible liquid with contact width variation

L G Varepo[1], A V Panichkin[2], O V Trapeznikova[1] and I V Nagornova[3]

[1]Omsk State Technical University, 11, Mira ave., Omsk, 644050, Russia
[2]Sobolev Institute of Mathematics, 13, Pevtsova str., Omsk, 644043, Russia
[3]Moscow Polytechnic University, 38, B. Semenovskaya str., Moscow, 107023, Russia

E-mail: larisavarepo@yandex.ru

Abstract. The results of using the finite-difference approximation of combined equations for describing the flow of a viscous incompressible liquid between contacting rotating cylindrical surfaces are presented in the work. A quantitative estimation of mass transfer and splitting factors for a viscous incompressible liquid onto a substrate is obtained depending on a contact width variation. A distinctive feature of the proposed approach is the accounting of the structural and microgeometric parameters of surfaces contacting with a viscous incompressible liquid.

1. Introduction

Various approaches to the mathematical formulation of mass transfer and splitting of viscous incompressible liquids that allow to solve the dynamical problem of transmission a viscous incompressible liquid between the surfaces of two cylinders are given in the works of a number of Russian and foreign scientists [1–3].

Modern computer technologies and wide software potential greatly simplified the modeling of such processes. Among the problem elaborations the works of [4–6] are of great scientific interest

2. Problem definition

Mass flows of an incompressible liquid with mobile barriers are inherent to different technological processes.

An analysis of the previously obtained experimental and theoretical data on the motion of a layer of a viscous liquid between rotating cylinders and its splitting during transfer from one cylinder to another showed that the problem has not been sufficiently studied. Solving such problems is laborious, which is due, first, to the nonlinearity and complexity of the mathematical equations, and second, to the need to determining the free surface when solving the system of partial derivative, which is a characteristic property of flows in layers and films.

Depending on the properties of the viscous incompressible liquid and the rate of layer separation the character of the rupture of the transferred layer also changes. To simplify the concrete problem solving, for example, when transferring a liquid to a porous substrate of one of the contacting cylinders, or a cylinder, it was proposed by many authors to accept conventionally the splitting of the liquid layer divided in half and with a zero absorbency, but there can be an another opinion [7]

The problem solving on the quantitative estimation of mass transfer factors of a viscous incompressible liquid with width variation in the zone of contacting cylinders is, in our opinion, of both practical and scientific interest.

3. The problem solution

To solve the problem, the authors developed an algorithm and its software implementation [8–10].

The most general mathematical model for formulation the flow of a continuous viscous liquid is the Navier-Stokes equations, for the solution of which there are various numerical procedures. Among the most accepted approximation methods, which are of high accuracy, of incompressible flows, the finite elements method is marked out. Nevertheless, finite-difference algorithms are not subject to the above-mentioned method, and in a number of cases they outproduce it in efficiency and accuracy. In this paper, the finite-difference method for solving multidimensional problems is chosen as the main method.

Calculations, for the stated problem situation, were carried out using methods on the finite-difference scheme for equations (1)-(2), using in differential operators of $(\nu\,\nabla^2 - (\bar{V},\nabla))$ the operator Λ in (3) and $\nabla \approx (\frac{\Delta_1+\Delta_{-1}}{2h_x}, \frac{\Delta_2+\Delta_{-2}}{2h_y})$ with usual orientation of the pattern along the nodal curves for the calculation of convective-diffusion transfer by the example of printing ink transfer between two rotating cylinders [8-12].

$$\frac{\partial\bar{V}}{\partial t} + (\bar{V},\nabla)\bar{V} + \frac{1}{\rho}\nabla P = \nu\,\nabla^2\bar{V} + \bar{F}(\mathrm{t},\bar{x},\bar{V},\bar{\omega},\bar{\mathrm{g}}); \tag{1}$$

$$\frac{\partial p}{\partial t} + \nabla\cdot\bar{\boldsymbol{V}} = 0, \tag{2}$$

$$\Lambda = \Lambda_1^0 + \Lambda_2^0, \qquad \Lambda_1^0 = -u\frac{\Delta_1+\Delta_{-1}}{2h_x} + \nu\frac{\Delta_1\Delta_{-1}}{h_x^2};\ \Lambda_2^0 = -\upsilon\frac{\Delta_2+\Delta_{-2}}{2h_y} + \nu\frac{\Delta_2\Delta_{-2}}{h_y^2}, \tag{3}$$

where $\bar{V}$ – velocity; P – presure; ρ – liquid density; t – time; $\bar{x} = (x, y)$ – coordinate of a point in Euclidean space; ∇ – gradient operator; ∇^2– Laplace operator; ν – kinematic viscosity; $\bar{\omega}$ – angular velocity of rotation; $\bar{\mathrm{g}}$ – vector value of acceleration of the gravitational field; $\bar{F}$ – components of acceleration from external forces (gravity forces) and forces from transformation of coordinates; $\Delta_1, \Delta_{-1}, \Delta_2, \Delta_{-2}$ – operators of function shift by step of grid up or down the axes x and y with steps of h_x and h_y.

Specify the conditions of the problem for numerical simulation: ω=*20 rad/s* – angular speed of cylinders rotation, $P = P_{athm} = 10^5\ H/m^2$ – ambient pressure, $\nu = 0.012\ м^2/s$ – kinematic viscosity. Initial dimensions of liquid zone are equal to $\delta_L = 4\ \mu m$, $\delta_L = 5\ \mu m$ (contact width variation) and $\delta_S = 2\ \mu m$ (layer thickness of a liquid), $\delta = 1\ \mu m$ (substrate thickness), cylinder radiuses – $r_1 = r_2 = 0.15\ m$.

A uniform grid with the number of computed nodes by two coordinates N_x and N_y equal to 80 is used. The grid size allows to analyze the convergence and behavior of numerical solutions with an accuracy of 1 % at Reynolds numbers of about 1. The iteration step τ was chosen in the range from 0.0002 to 0.005 for the considered grid in nondimensional quantities. Calculations of controlled mass transfer factors for a viscous incompressible liquid were performed taking into account the filtration of a liquid into a porous substrate on one of the contacting cylinders. The algorithm of numerical simulation and computer visualization of calculations is shown in Fig. 1.

When calculating the fluid flows between the surfaces to be contacted, the deformation of the surface layers of cylinders and substrate, on which the liquid to be transferred, as the result of changes in liquid pressure under the contact with them was taken into consideration.

Taking into account the infinitesimality of the relative velocities in compressed surfaces with the difference between the internal and external pressures $(P_1 - P_0)$ with the value of boundary deformation accumulating over time Δr for the thickness H of the substrate layer or the coating of the first cylinder (layer of rubber-textile sheet) with area S having the elastic modulus E of the center layer thickness changes x_c were calculated by equation (4)

$$\rho HS\ddot{x}_c/2 = -\tfrac{E}{H}\left({}^{\Delta r}/_2 + x_c\right)\cdot S + (P_1 - P_0)\cdot S. \tag{4}$$

IOP Conf. Series: Journal of Physics: Conf. Series **1050** (2018) 012093 doi:10.1088/1742-6596/1050/1/012093

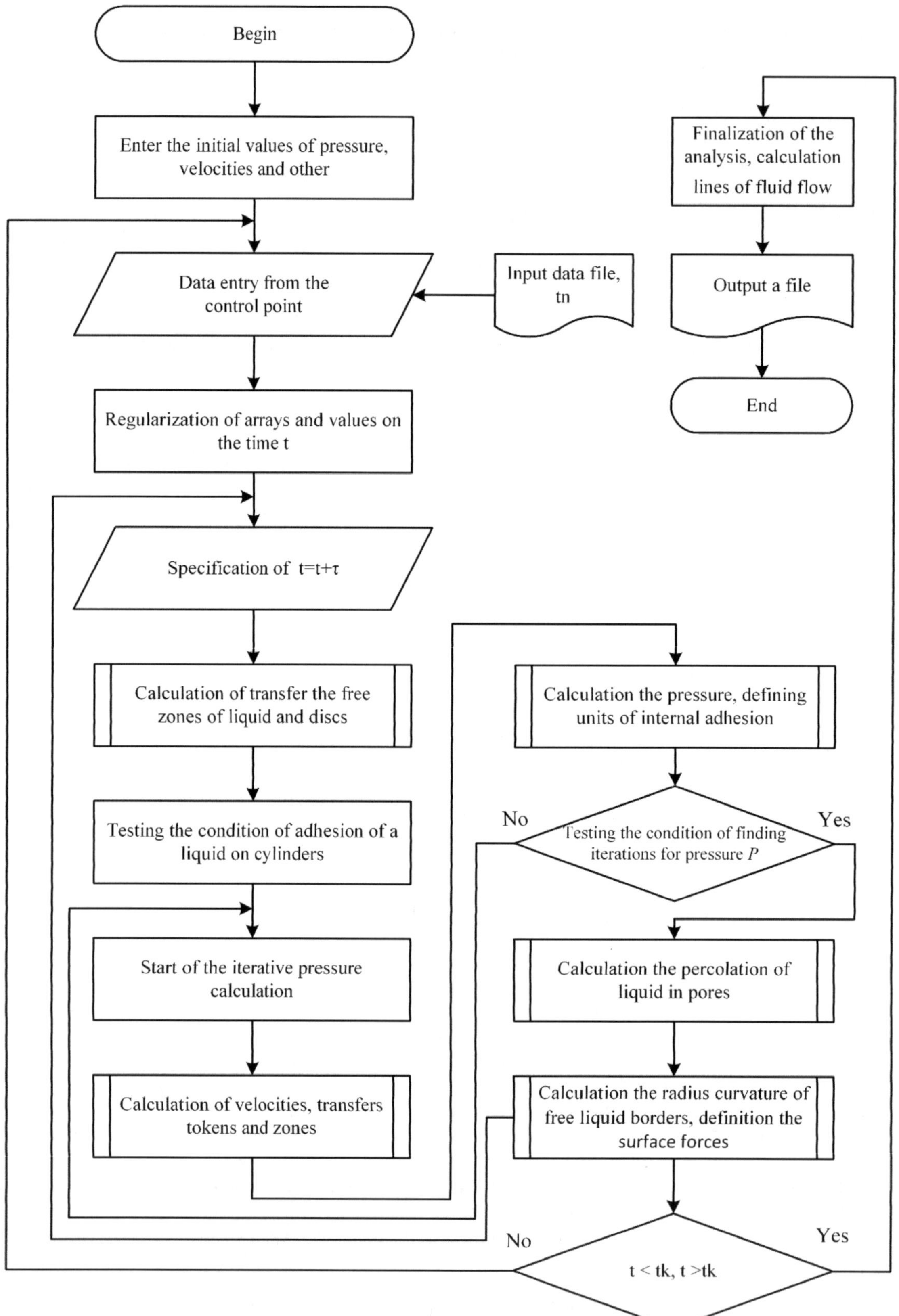

Figure 1. An algorithm for numerical simulation of mass transfer the viscous incompressible liquid onto a substrate of a contacting cylinder

To ensure the automation of the calculation of controlled mass transfer factors the viscous incompressible liquid to a substrate in the zone of two contacting cylinders under the contact width variation, the practical realization of the program on the bases of proposed algorithm is carried out [11].

The results of calculating the controlled factors in given conditions for numerical simulation of studied process are presented (Table 1).

Table 1. Results of numerical simulation

Number of the sample	Contact width variation, μm	Amount of viscous incompressible liquid in a volume layer of a substrate,%	Amount of viscous incompressible liquid in a surface layer of a substrate, %
1	4.0	60.98	3.08
2	5.0	66.96	3.42
3	5.0	59.37	3.48

Computer visualization of numerical simulation of mass transfer the viscous incompressible liquid onto a porous substrate, depending on the changes in contact width variation of two cylinders are in Table 2, depending on the type of substrate – in Table 3.

Table 2. Computer visualization of mass transfer and splitting the viscous incompressible liquid with various contact width of two cylinders

$\delta_L = 4\ \mu m$	$\delta_L = 5\ \mu m$
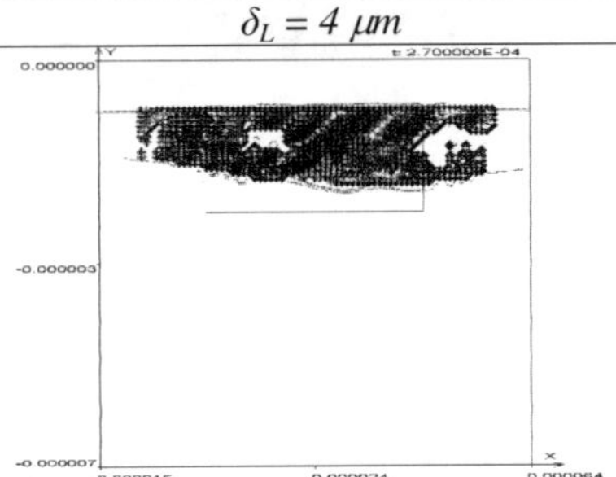	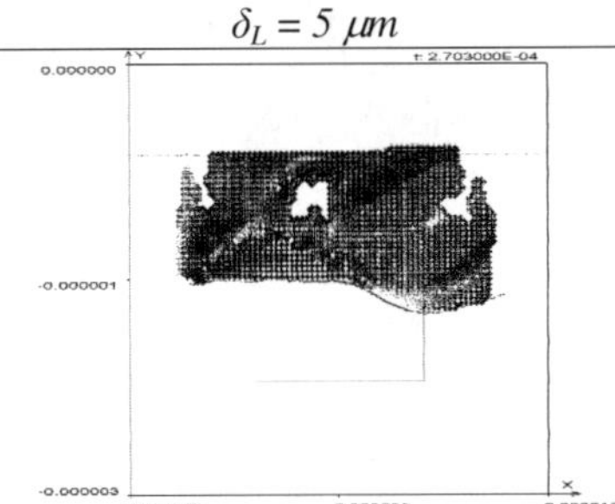
Computer visualization of the behavior of the liquid layer between the contacting surfaces in the center of the contact	
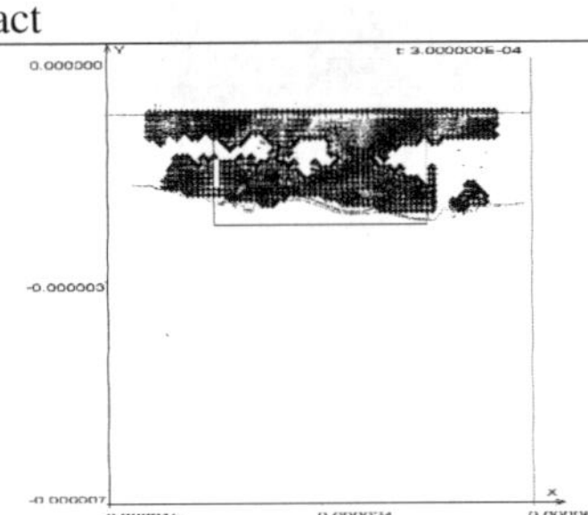	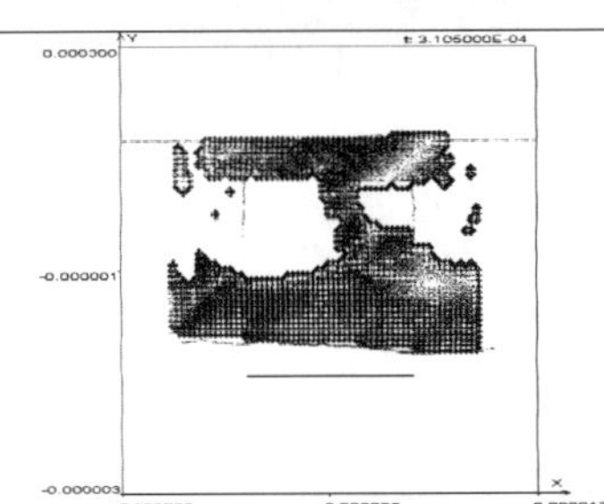
Computer visualization of liquid behavior between contacting surfaces at the end of contact zone	
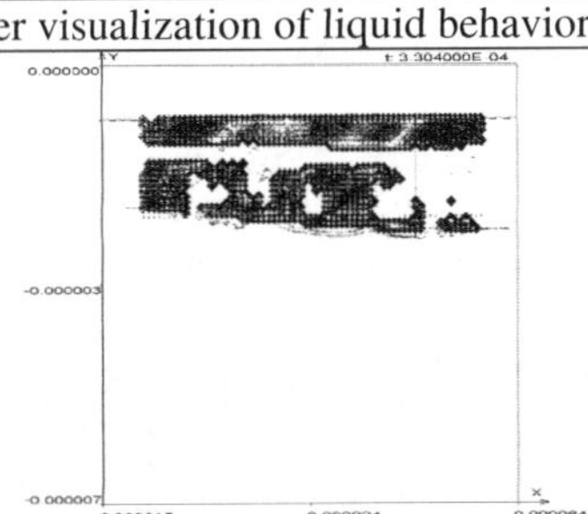	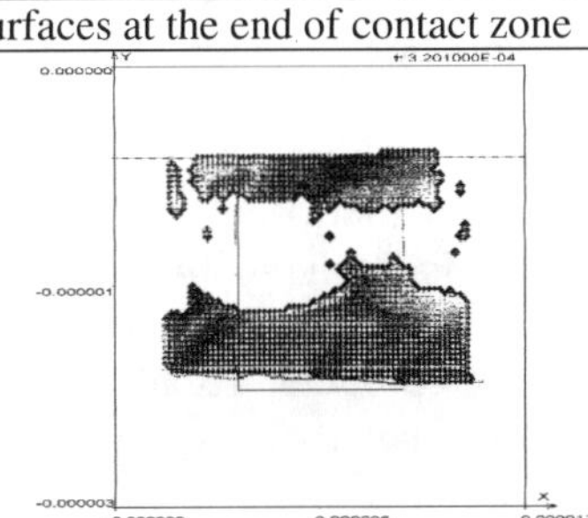
Computer visualization of complete liquid splitting between contacting surfaces at the end of the contact zone	

IOP Conf. Series: Journal of Physics: Conf. Series **1050** (2018) 012093 doi:10.1088/1742-6596/1050/1/012093

Table 3. Computer visualization of mass transfer and splitting the viscous incompressible liquid with varying characteristics of substrate

Microporous substrate (sample 3)	Macroporous substrate (sample 2)
$\delta_L = 5\ \mu m$	
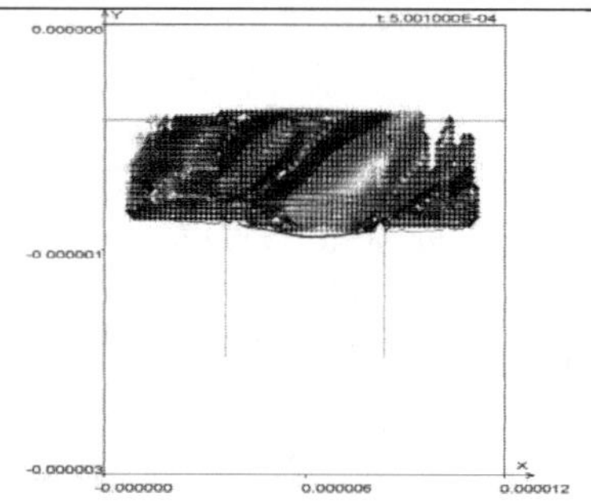	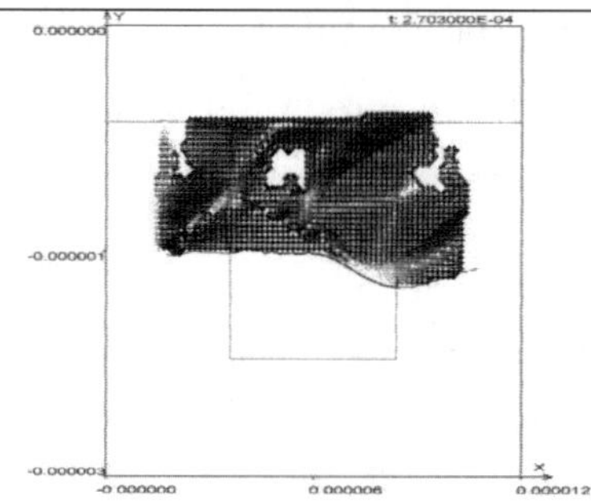
Computer visualization of the behavior of the liquid layer between the contacting surfaces in the center of the contact	
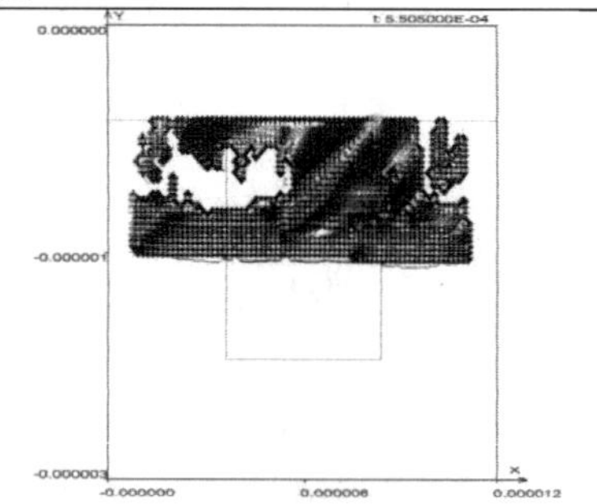	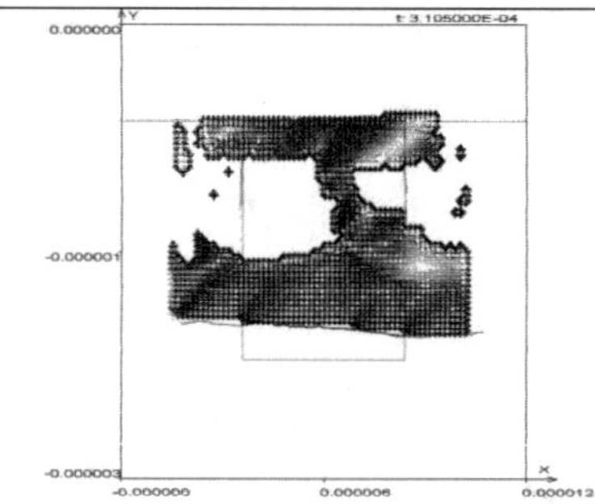
Computer visualization of liquid behavior between contacting surfaces at the end of contact zone	
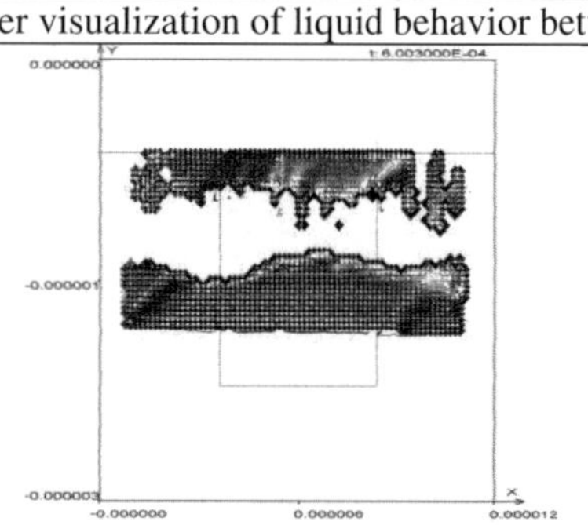	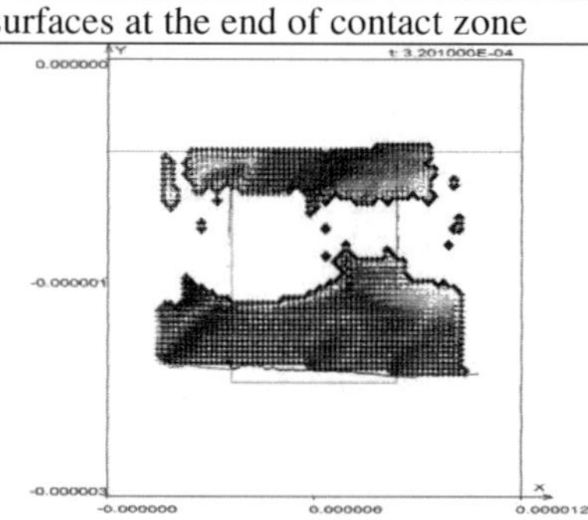
Computer visualization of complete liquid splitting between contacting surfaces at the end of the contact zone	

4. Conclusion

The developed model provides an accurate method for calculating controlled factors of mass transfer the viscous incompressible liquid onto a porous substrate between contacting cylinders, promote the automation of its control.

Figures of computer visualization widen the concept of the mechanism of studying process, allow to see all the changes that occur in the layer of a viscous incompressible liquid when it is transferred onto a substrate in the zone of two contacting cylinders in dependence on conditions being set, that clearly shows the monitoring stages of this process.

In the figures of computer visualization with increasing contact width variation large deformations appear in the liquid layer, i.e. spreading beyond the edges of the contact zone, and an increase of penetration the viscous incompressible liquid into the volumetric layers of the substrate (Sample 1 and Sample 2, respectively). A similar result is observed when estimating the controlled factors of liquid transferring on a macro and microporous substrate surface.

The proposed solution was tested on transferring the ink (viscous incompressible liquid) onto the paper sheet (substrate) by the offset printing method in the contact zone of the cylinders of the press. The obtained result is interpreted as follows. For a substrate with a microporous (Sample 3), the amount of ink in volume layer of paper sheet is much smaller than to a macroporous (Sample 2) surface. There is a higher concentration of ink pigment in the surface layer for the microporous substrate.

References

[1] Calgaro C, Creusé E, Goudon T, Krell S 2017 *Mathematics and Computers in Simulation* **137** pp 201–225
[2] N C Reis Jr, Griffiths R F, Santos J M 2004 *J. Comput.Phys.* **198** pp 747–770
[3] N C Reis Jr, Griffiths R F, Santos J M 2008 *Applied Mathematical Modelling* **32**(3) pp 341–346
[4] Vlachopoulos G, Claypole J, Bould D 2010 Ink mist formation in roller trains: Iarigai proceedings: Advances in printing and media technology. Montreal, Canada **37** pp 227–234
[5] Ozaki Y and Kimura M 2000 *Appita J.* **3** pp 216–219
[6] Koivula H, Preston J S, Heard P J and Toivakka M 2008 *Colloids and Surfaces A: Physicochemical and Engineering Aspects* **317** (1–3) pp 557–567
[7] Fatemeh Ghadiri, Dewan Hasan Ahmed, Hyung Jin Sung, Ebrahim Shirani 2011 *International Journal of Heat and Fluid Flow* **32** (1) pp 308–317
[8] Panichkin A V, Varepo L G 2014 *Springer Proceedings in Physics* **154** pp 79–83
[9] Panichkin A V, Varepo L G, Trapeznikova O V 2016 *IOP: Materials Science and Engineering* **124** pp 1–6
[10] Varepo L G, Panichkin A V 2017 *IOP Conf. Series: Journal of Physics: Conf. Series* (**858**) 012039 pp 1–6
[11] Varepo L G, Panichkin A V Certif. No 2016617873 of state registration of the program Calculation of the transfer coefficients of a viscous incompressible fluid to a substrate between contacting cylindrical surfaces (Russian Federation); publ. 15.07.2016
[12] Varepo L G, Panichkin A V, Panchuk K L 2017 *Proceedings of Computer and Information Technologies* **11** (161) pp 16–22 (Russian)

IOP Conf. Series: Journal of Physics: Conf. Series **1050** (2018) 012094 doi:10.1088/1742-6596/1050/1/012094

Determination of the Internal Combustion Engine Valve Gear Characteristics

Alexander Vasilyev[1] and Yulia Bakhracheva [2]

[1]Department of heat engineering and hydraulics, Volgograd State Technical University, Prosp. Lenin, 28, Volgograd, 400005, Russian Federation
[2]Department of Information Security, Priority Technologies Institute, Volgograd State University, Prosp. Universitetsky, 100, Volgograd, 400062, Russian Federation

E-mail: vasilyev@vstu.ru, bakhracheva@volsu.ru

Abstract. Method of the Internal Combustion Engine (ICE) valve gear characteristics determination by using the numerical differentiation of tappet movement law is submitted. The method of converting these characteristics for different valve types is proposed. Based on the methods of calculation of the motion law of the follower, characteristics of the timing and conversion of the data during the transition from one kinematics to another developed computer software included in the software complex for synthesis of the characteristics of the valve timing. Their use allows to identify resources for improving timing mechanism of an internal combustion engine.

1. Introduction

Availability, reliability and efficiency of the mechanism of gas-distribution (MG) of internal combustion engines greatly depend on the kinematic characteristics [1 - 5]. The movement of the link mechanism is set by the cam profile, which determines the law of motion of the follower. Therefore, for calculation of numerous characteristics of the pairing of the cam-follower (strength, contact-hydrodynamic, tribology, etc.), as well as the laws of motion of all slave units need information about the laws of displacement, velocity and acceleration of the follower on the cam rotation angle [6, 7].

In the most favorable case, a known method of profiling the cam and all the source data defining configuration profiles and the law of motion of the follower, including the above kinematic characteristics of the displacement, velocity and acceleration. If only the dependence of the acceleration of the follower on the cam rotation angle, and the task of subsequent reliable determination of all the required characteristics was facilitated by the fact that the relevant values of the velocity and displacement of the follower is calculated based on the numerical integration of a predetermined law of acceleration (or its equivalent).

However, more often only the law of the pusher displacement depending on the cam rotation angle of can be used as the initial data. In this case, the required characteristics of speed and acceleration of the follower can be determined by numerical differentiation of tabular predetermined law of movement of the pusher on the cam rotation angle, which causes some scatter of data due to insufficient stability of the procedure of the numerical differentiation. However, this calculation is only possible in the case of calculating the characteristics of the real cam, the configuration of which is determined experimentally and tabulated for a given law of movement of the pusher shown in the drawing of the camshaft.

It is also important to note that while calculating the kinematic and other characteristics of the pairing of the cam-follower and valve gear in General is important to consider the type of the used pusher and the kinematic scheme of the mechanism. In this regard, it is often necessary to do the recalculation of characteristics, as the law of movement of the pusher can be defined in one scheme (for example, when the forward movement of the pusher with flat contact surface), and in the timing mechanism of an internal combustion engine implements a different kinematic scheme (for example, the same cam works in conjunction with lever pusher having a cylindrical contact surface). The recalculation of characteristics is also necessary for the manufacture and control of the cam profile, as the scheme of the technological process of processing of the cam profile, as a rule, also different from the kinematic scheme applied in the gas distribution mechanism of internal combustion engines.

2. Mathematical model

In this regard, in the course of the present study was tasked with developing mathematical models, methods and relevant elements of the software system for computers that allows you to calculate the required characteristics of the timing mechanism of an internal combustion engine. And to perform these conversions for different kinematic schemes.

To characterize the velocity and acceleration of the follower (or their analogues) based on the table given the values of the movement used the technique of double numerical differentiation with smoothing. It should be noted that the step angle of the cam driven original table of values of displacement of the pusher is often too large for numerical differentiation. Therefore, the first step is an interpolation function using an interpolation process of the Aitken-Lagrange [8]. At the same time, the subroutine calculates the value of Y uniquely the function $y(x)$ for a given value X of its argument on a given table x_j, y_j (j= 1, 2, ..., n) the corresponding values of argument and function. In this case, the step, the values X and Y get can be quite small.

The first approximation Y_1 of the calculated value Y is the value of the function to be interpolated in X to the nearest node interpolation x_1, i.e., $Y_1= y(x_1)= y_1$. As more accurate approximations are used sequentially computed values of $L_k(X)$ polynomials of Lagrange, the degree of which increases at each new step of the interpolation. The calculations are performed by a known interpolating triangular scheme of Aitken. The interpolation process is terminated when performing, as a rule, the following conditions: the absolute value of the difference between two successive interpolated values is less than the specified ε, that is achieved by the specified accuracy of the iteration.

The selection nearest to a given X value of the argument (on the cam rotation angle) and function (movement of the pusher) from the source table is based on the use of the sub-ordering of the arrays. For a given value of X the function argument $y(x)$ and a given table x_j, y_j (j= 1, 2, ..., n) values of a function and its argument are formed arrays of values x_i, y_i (i= 1, 2, ..., n), are arranged in such a way that the modules of the difference $|x - x_i|$ increased with the growth of index i:

$$|X - x_i| \geq |X - x_k| \text{, if } i > k. \qquad (1)$$

The next step of calculating the characteristics of the velocity and acceleration of the follower is the differentiation of the functions of its displacement, given a table of values at equally spaced points according to equations of Lagrange with five nodes. Used by the routine allows for a function given a set of $y_1, y_2, ..., y_m$ of its values y_k c step change in the argument (on the cam rotation angle) $h= x_k - x_{k-1}$ (k= 2, ..., m), to calculate the set $z_1, z_2, ..., z_m$ values of derivatives in the given m points. To calculate values of derivatives used the Lagrange interpolating polynomial of the fourth degree, built on five successive points x_{k+p}, y_{k+p} (p= -2, -1, 0, 1, 2). The required values of the derivative are calculated by the known formulas [8].

As already mentioned, the operation of numerical differentiation is quite stable and leads to some scatter re-results. In this regard, as shown by numerous calculations, it is advisable after each procedure of numerical differentiation to carry out a two-fold smoothing. For this purpose, a special

IOP Conf. Series: Journal of Physics: Conf. Series **1050** (2018) 012094 doi:10.1088/1742-6596/1050/1/012094

routine smoothing function given a table of values at equally spaced points, with the first-degree polynomial constructed on three consecutive points by the method of least squares. This subroutine computes a set of $y_1, y_2, \ldots, y_m$ smoothed values of the function $z(x)$, given many $z_1, z_2, \ldots, z_m$ of its values at m equally spaced points with step $h = x_k - x_{k-1}$ ($k = 2, \ldots, m$).

Thus, applying the developed methods and software package, it is possible to calculate the characteristics of the velocity and acceleration of the follower (or their analogues) on the basis of the table specified values of its movement. However, the results obtained can be further used to calculate other characteristics of the timing mechanism only when the measured displacement of the follower produced by the same kinematic scheme, which is used in a real test the mechanism. Otherwise, as already noted, there is a need of recalculating the values of displacement, velocity and acceleration of the follower depending on on the cam rotation angle to conform the pattern and size of engine timing.

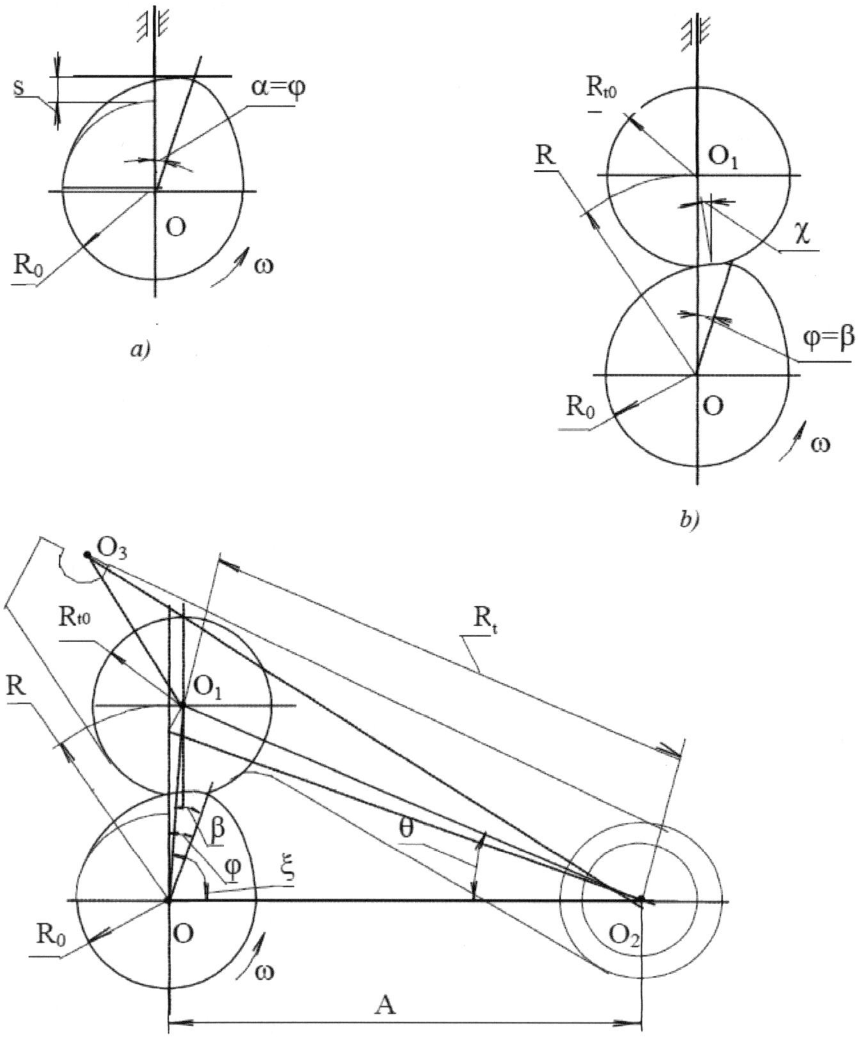

c)

Figure 1. Schematic of cam mechanisms with flat (a), a centrally located cylindrical (b) tappets and a lever having a cylindrical surface of the pusher (c).

Let the scheme of the mechanism corresponds to Fig. 1.(c), and the law of motion of the follower for the cam profile is set depending on the angle β according to the scheme in Fig. 1.(b). Then the analogues of velocity dR/dφ and acceleration $d^2R/d\varphi^2$ pusher can be calculated by the formulas:

$$\frac{dR}{d\varphi}=\frac{dR}{d\beta}\frac{d\beta}{d\varphi} \tag{2}$$

$$\frac{d^2R}{d\varphi^2}=\frac{d^2R}{d\beta^2}\left(\frac{d\beta}{d\varphi}\right)^2+\frac{dR}{d\beta}\frac{d^2\beta}{d\varphi^2} \tag{3}$$

Differentiating the ratio

$$\varphi=\beta\mp(\arccos B_0-\arccos B),\ \text{where} \tag{4}$$

$$B=\frac{A^2-R_t^2+R^2}{2AR};\quad B_0=\frac{A^2-R_t^2+(R_0+R_{t0})^2}{2A(R_0+R_{t0})}, \tag{5}$$

As a result, will receive

$$\frac{d\beta}{d\varphi}=\frac{1}{1\mp\dfrac{dB/d\beta}{\sqrt{1-B^2}}} \tag{6}$$

$$\frac{d^2\beta}{d\varphi^2}=\pm\frac{\dfrac{d\beta}{d\varphi}\left[\dfrac{d^2B}{d\beta^2}\sqrt{1-B^2}+\left(\dfrac{dB}{d\beta}\right)^2\dfrac{B}{\sqrt{1-B^2}}\right]}{\left(\sqrt{1-B^2}\mp\dfrac{dB}{d\beta}\right)^2} \tag{7}$$

In these formulas the upper sign corresponds to a positive sign ξ, and the lower - negative. Thus, substituting (6) and (7) into (2) and (3), we obtain the relations for computing dR/dφ and $d^2R/d\varphi^2$ in terms of the law of motion of the follower, given by the diagram on Fig. 1(b), for the circuit in Fig. 1.(c). The developed software package also provides recalculations for other schemes. For example, participants discussed common case, when the scheme of the mechanism corresponds to Fig. 1.(c), but the movement of the pusher and its derivatives for the cam profile defined by the corner α according to Fig. 1.(a).

3. Discussion

Figure 2 illustrates the differences in the characteristics of the same symmetrical profile of the intake cam tractor engine with the tappets of different types: centrally located (Fig. 1.b) and lever (Fig. 1.C) having a cylindrical contact surface. In the latter case, a clockwise direction of the cam rotation is selected when using the lever pusher engine. The maximum value of the analog acceleration on the side of the lift tappet is 57.3 mm/rad^2, and on the side of lowering – 62.1 mm/rad^2, which is 8.4% more.

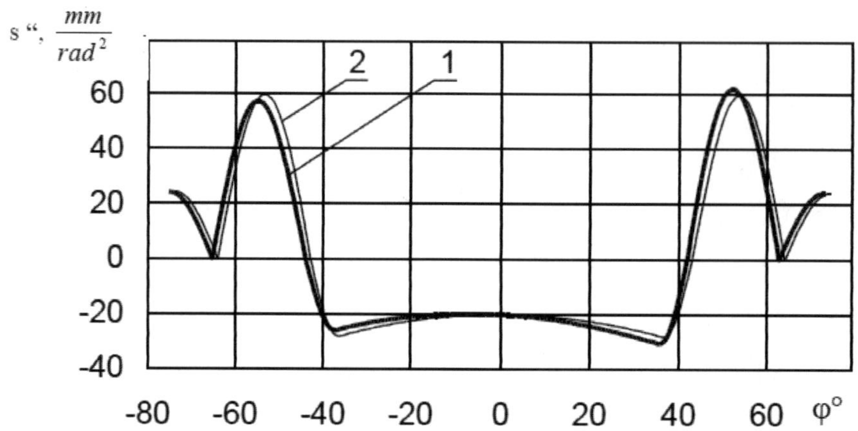

Figure 2. Analogue acceleration of the follower on the cam rotation angle for symmetric kinematic schemes of the lever (1) and centrally located (2) cylindrical tappets

4. Conclusion

Based on the submitted methods of calculation of the follower motion law, timing characteristics and data recalculation for other kinematics schemes computer software is developed and included in the software complex for synthesis of the valve timing characteristics. Their use allows to identify resources for improving timing mechanism of an internal combustion engine.

5. References

[1] Vasilyev A 2006 Simulation of valve gear dynamics using generalized dynamic model *Mechanika* **2** pp 37–43

[2] Vasilyev A and Sidorov D Automobile engine valve train experimental research and dynamics simulation *In:Compillation of Articles of the International Scientific-Technical Symposium «Auto NN 08: Automobile Transport in XXI Century» dedicated to the 45th Anniversary of «Automobile Transport» department of Nizhny Novgorod State Technical University n.a. R.E. Alekseev., Nizhny Novgorod, 18-19 Dec. 2008*

[3] Korchemny L V 1981 Valve gear of the automotive engine: Kinematics and dynamics (Moscow: Mechanical engineering) 353 p

[4] Semenova L M et al 2013 Laws of formation of diffusion layers and solution of the diffusion problem in temperature-cycle carbonitriding of steel *Metal Science and Heat Treatment* 55 pp 34–37

[5] Kosugi T and Seino T 1985 Valve motion simulation method for high-speed internal combustion engines *SAE Techn. Pap* Ser № 850179 pp 1-10

[6] Vasilyev A, Deynichenko E and Popov D 2005 Internal Combustion Engine Valve Gear Cam Wear and Its Influence on Valve Gear and Engine Efficiency *Mekhanika* 54 № 4 pp 44–49

[7] Baron A and Bakhracheva J 2007 A method for impact strength estimation *Mechanika* 66 pp 31–35

[8] Vasilyev A V, Bakhracheva Y S and Storojakov S Y 2016 Dynamics simulation model for the internal combustion engine valve gear *Procedia engineering* "2nd International Conference on Industrial Engineering, ICIE 2016" pp 312-317. doi:10.1016/j.proeng.2016.07.017

[9] Baron A, Bakhracheva J and Osipenko A 2007 Fracture toughness estimation by means of indentation test *Mechanika* 67 pp 33–36

[10] Vasilyev A V et al. 2014 Valve Cam Design Using Numerical Step-by-Step Method *Vestnik Volgogradskogo gosudarstvennogo universiteta. Seriya 10, Innovatsionnaya deyatelnost [Science Journal of Volgograd State University. Innovation Activity]* 10 pp 26–32

IOP Conf. Series: Journal of Physics: Conf. Series **1050** (2018) 012095 doi:10.1088/1742-6596/1050/1/012095

A small spacecraft microengine with SHF impact on working medium: cooling and dilatometric evaporator valve

I S Vavilov, A I Lukyanchik, P S Yachmenev, R N Litau, A V Lysakov

Omsk State Technical University, 11, Mira ave., Omsk, 644050, Russia
E-mail: vava-igg@mail.ru

Abstract. The results of experiments and a theoretical research of a cooling method for SHF-transistor crystal of a correcting small spacecraft (SS) microengine are presented. A distinctive feature of a proposed approach is an efficient use of the heat removed from SHF-transistor in the dilatometric evaporator-valve. Generally, this paper continues to create methodological foundations for a design of a correcting SS microthrust engine with SHF-heating of a working medium. This paper is about cooling of SHF-transistor in vacuum with the application of removed energy in a dilatometric valve and for preliminary heating of a working medium before its input to the SHF-chamber. A mathematical model of SHF transistor crystal cooling by the working medium was verified as a result of studies. In case of certain geometrical and initial thermodynamic conditions are met, the intensity of cooling increases and heating of working medium also increases (micro consumption of air cools the experimental surface by 1 W). The heating of working gas, according to theoretical processing, is suffice to create a jet thrust of 1,2 mN. The application field of studied physical processes is the engine-building for small spacecrafts, for example CubeSat 3U, in particular, building of a microengine with SHF-acceleration of a working medium and a dilatometric throttle valve.

1. Introduction

The authors of this article repeatedly emphasized in previous papers [1, 13] that a low thruster design of very small spacecrafts (VSS) involves a number of restrictions which are:

– low VSS power;
– small VSS weight and dimensions;
– possibilities of an orientation control system.

In this regard the implementation of a «Swiss knife» concept in the development of VSS microengines, when one structural element performs more than one function in the operation of a propulsion system (PS) is relevant for those microengines. The application of this concept reduces the nomenclature of PS elements that will directly lead to reduction of weight, dimensions and power consumption. This poses a question on efficient use of dissipative energy, which, generally, irreversibly passes in the space. The utilization of dissipative energy allows us to increase the PS overall efficiency.

The dissipative energy in an electrothermal microengine (ETME) with SHF-heating of a working medium is a heat flow emerging in a solid-state SHF-generator under its normal operation.

In solid-state SHF-generators basic elements are SHF-transistors which normal operation requires convective cooling. Such devices are usually mounted on massive metal plates providing heat removal onto an external case or in the ambient space by means of convection and emission. For example, the operation of high-power LDMOS-transistors can heat surrounding equipment to 60^0 C [2].

An operating temperature of a transistor's crystal affects greatly the output power, as well as, the device efficiency. Promising carbide-silicon SHF-transistors can provide preset specifications if a crystal temperature is up to 255^0 C [3].

Data on temperatures are given for the research performed "in open air" when a convective heat exchange provides a considerable heat removal from a transistor case. In vacuum the temperature of a case as experimental studies showed, can be $100–120^0$ C higher.

The efficiency of SFH-transistors is in the range of 40 to 70% now. For example, high-power generator SFH-transistor KT9197B with the output power of 5 W has the efficiency of 55% [4], dissipative power losses are 4 W. Regarding to high-power PS (with power consumption of 60-100 W) especially as for large SS, these values can be neglected. However, the propulsion system in case of VSS can be provided only with 5-7 W to work without any negative effects on the main systems. In this case it is necessary to utilize all possible power opportunities of a spacecraft and to minimize losses.

2. Problem statement

The "Swiss knife" concept was embodied in an electrothermal microengine of VSS as a power element of a throttle evaporator-valve representing a rod radiator with an internal gas conduit (fig. 1).

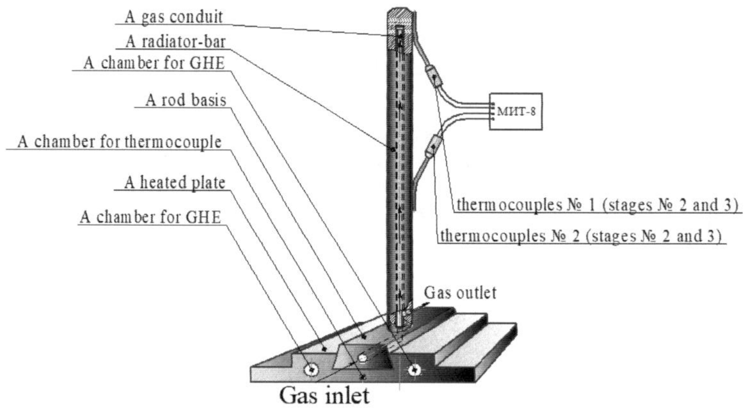

Figure 1. Experimental sample of a rod radiator with an internal gas conduit

Graphite heating elements (GHE) mounted in channels of base-plates created a heat flow of a certain value and a thermocouple installed in a plate registered the value of equilibrium temperature of 250 ^{0}C "in open air" tests without installation of a rod radiator. Thus the electric power required for imitation of heat effects arising during the work of a carbide - silicon SHF-transistor was defined. The basis of a rod and a radiator-rod represent a non-assembled element, in the basis of which there is a gas conduit entering the gas conduit of the rod. Gas in its motion made a double path: from an inlet through a capillary pipe to a dead-end area and back via an annular gap to the rod basis and an outlet in the ambient space. The rod basis was fixed on the heated plate by means of the dovetail lock (joint).

Components and materials' characteristics of the experimental sample, necessary for a theoretical analysis, are given in table 1.

An experimental study was divided into several interrelated stages:

stage 1: a research of heat transfer of a heated plate in vacuum;

stage 2: a research of heat transfer of a heated plate with a rod radiator installed in vacuum;

stage 3: a research of heat exchange of a heated plate with the installed rod radiator and a gas flow in a gas conduit in vacuum.

Table 1. Elements of the experimental sample and the properties of materials

Name / Image/ material	Typical dimensions, mm	Emissivity factor, ε	Thermal conductivity coefficient, λ, W/(m · K)
A heated plate 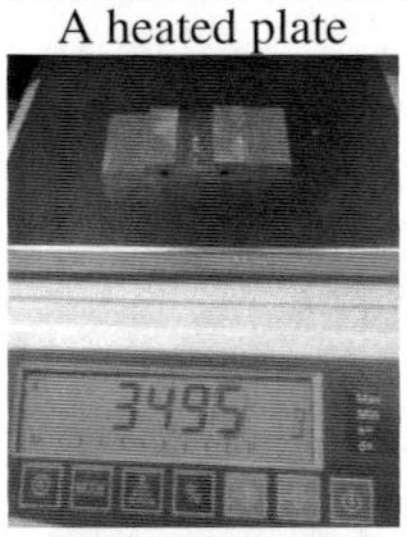 Aluminium	64×33.5×6	0.33	220
A rod-radiator  Duralumin	An external surface Ø10×100 A channel Ø2.5×95	A lateral surface 0.045 An end face 0.03	155
A rod basis  Aluminium	13×33.5×5	0.33	220
GHE 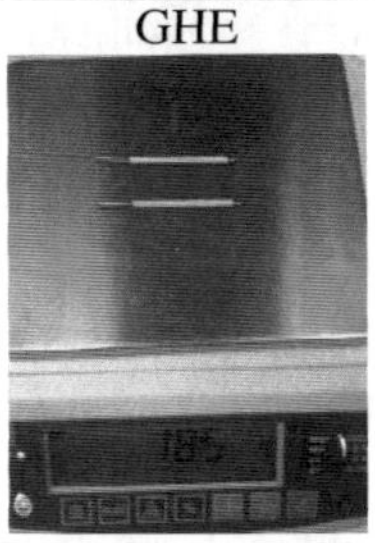 Corundum, graphite	Ø3×35	0.7	3.34

3. Theory

At stage 1 the heated plate was placed in a vacuum chamber, 19.4 W heat flow was fed to the plate from GHE, the plate temperature was measured by three thermocouples, one of them (thermocouple 3 in fig. 1 and 2) was arranged inside the plate, the others were placed on the plate surface. In fig. 2 the temperature dependence of the plate versus heating time according to the results of experiment is shown. We can see that equilibrium temperature of a plate takes value 288 ^{0}C.

The purpose of a theoretical study in stage 1 was the verification of power balance equation which for the "GHE-plate" system looks like:

$$P_{el} = P_{emi.GHE1} + P_{emi.GHE2} + P_{emi.PL}, \quad (1)$$

where P_{el}=19.4 W is instrument electric power brought to the plate from GHE;
$P_{emi.GHE1}$ and $P_{emi.GHE2}$ are respectively, the power emitted by extending areas of GHE and power losses in GHE areas in channels of the heated plate;

$P_{emi.PL}$ is the power emitted by the plate surface in the ambient space.

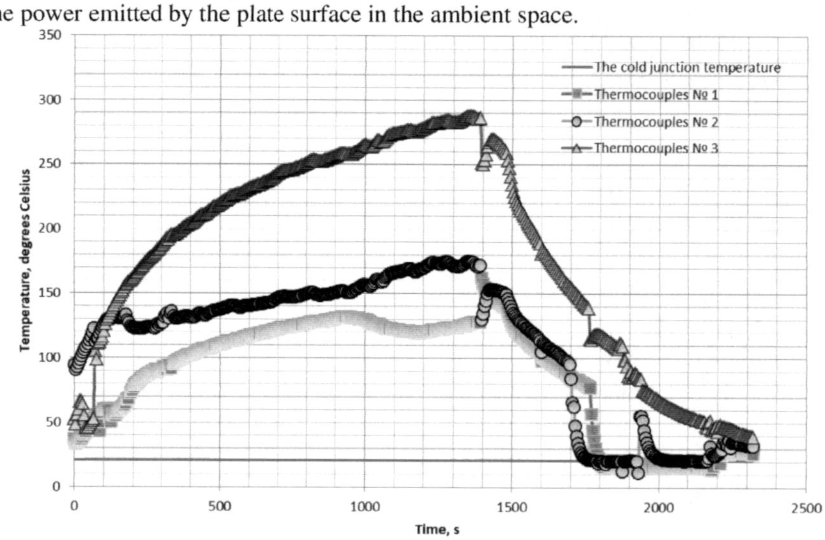

Figure 2. Results of the experiment on plate heat transfer in vacuum

The heat flow emitted by the GHE extending surfaces, with the area F_{Ext} can be expressed in accordance with Stefan-Boltzmann law [5]:

$$P_{emi.GHE1} = \sigma \cdot T^4_{GHE} \cdot \varepsilon_{Al_2O_3} \cdot F_{Ext}, \quad (2)$$

where σ is a Stefan-Boltzmann constant; ε_{Al2O3} is the emissivity factor of the GHE case (the value is determined by [6] and provided in Tab. 1). T_{GHE} is a GHE ceramic surface temperature which is defined by the power balance equation for steady thermal conditions when the entire electric power is used for the radiation from GHE ceramic surface with the total square F_{GHE}:

$$P_{el} = \sigma \cdot T^4_{GHE} \cdot \varepsilon_{Al_2O_3} \cdot F_{GHE}. \quad (3)$$

The value of losses in the transfer by the emission from the GHE ceramic surface to the internal cylindrical surface of plate's channels is defined as the difference between the brought heat flow and the heat flow absorbed by the plate [7]:

$$P_{emi.GHE2} = \sigma \cdot \left(T^4_{GHE} - T^4_{PL}\right) \cdot \varepsilon_{RE} \cdot F_{GHE.PL} \quad (4)$$

where T_{PL} is a temperature of the heated plate;

$$\varepsilon_{RE} = \frac{1}{\dfrac{1}{\varepsilon_{Al_2O_3}} + \dfrac{F_{GHE.PL}}{F_{CHA.PL}} \cdot \left(\dfrac{1}{\varepsilon_{Al}} - 1\right)}$$

is a reduced emissivity factor of the bodies system [7];

$F_{GHE.PL}$ is the area of the GHE emitting surface covered by the internal cylindrical surface of the plate channel $F_{CHA.PL}$;

ε_{Al} is an emissivity factor of a plate material (its value is defined by [8] and provided in table 1).

The heat emission from the whole surface of the heated plate FPL to (ambient) space is defined in accordance with the Stefan-Boltzmann law:

$$P_{emi.PL} = \sigma \cdot T^4_{PL} \cdot \varepsilon_{Al} \cdot F_{PL}. \quad (5)$$

Solving equations (1–5), we obtained the required theoretical heating plate temperature T_{PL}. The experimental and obtained calculated values are given in table 2.

At stage 2 the heated plate with the rod radiator were placed in a vacuum chamber. The heat flow with the power of 19.4 W was brought to the plate from the GHE, the plate temperature was measured by a thermocouple, the rod temperature was measured in the middle point (thermocouple 2, fig. 1) and in the upper point (thermocouple 1, fig. 1).

In fig. 3 the dependence of plate temperature versus heating time according to the experiment results is shown. We can see that the equilibrium temperature of the plate is 195 ^{0}C.

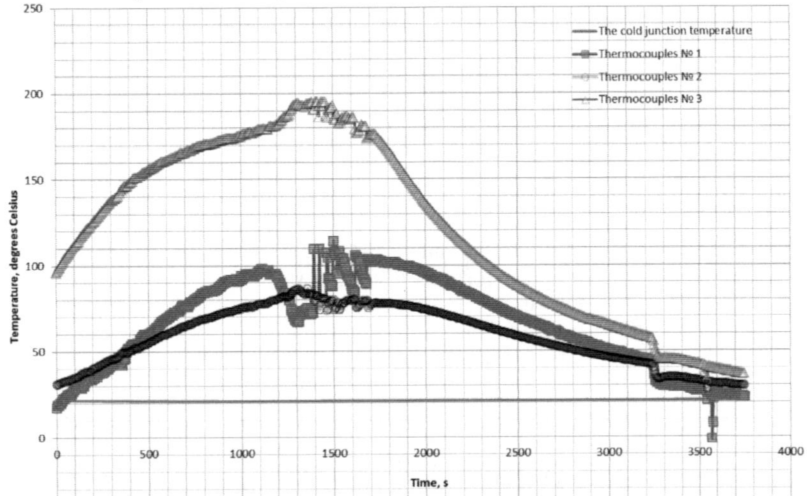

Figure 3. Results of the experiment on the heat transfer of a plate with a rod in vacuum

The purpose of a theoretical research at stage 2 was to define the quantity of heat emitted by a rod to the ambient space in a form of emission.

The quantity of heat transferred by the entire surface of the rod F_{RO} for a finite length L_{RO} and a diameter D_{RO} in heating of one side up to the temperature of $T_{H.RO}$ can be defined, using the formula [9]:

$$P_{em.RO} = \lambda_{DAL} \cdot m \cdot F_{RO} \cdot T_{H.RO} \cdot \frac{\dfrac{\alpha_T}{m \cdot \lambda_{DAL}} + thm \cdot L_{RO}}{1 + \dfrac{\alpha_T}{m \cdot \lambda_{DAL}} + thm \cdot L_{RO}}, \tag{6}$$

where λ_{DAL} is a duralumin thermal conductivity coefficient (it is defined by [10] and provided in table1);

$m = 2 \cdot \sqrt{\dfrac{\alpha_L}{\lambda_{DAL} \cdot D_{RO}}}$ is a coefficient accepted for a round rod;

α_L and α_{EF} are respectively heat-transfer coefficients of a rod lateral surface and of its end face (these values are different since there is an electroplating on a rod surface, but an end face is processed by a lathe. Heat-transfer coefficients in case of radiation heat transfer depend on a body temperature and a surface emissivity factor (values of emissivity factors are defined by [11] and given in table 1).

On the other hand, the heat flow emitted by the rod is a part of heat transferred from the heated plate to the rod basis in a direct contact of surfaces in the dovetail lock. Thus, the same value $P_{em.RO}$ can be expressed by the heat transfer between two flat walls taking into accounting heat emission of power in vacuum [9]:

$$P_{em.RO} = \frac{T_{PL.\exp} - T_{H.RO}}{\dfrac{\delta}{\lambda_{Al}} + \dfrac{T_{H.RO} \cdot F_{BAS.RO}}{P_{em.PL} - \varepsilon_{Al} \cdot \sigma \cdot T_{H.RO}^4 \cdot F_{CON}}} \cdot F_{CON}, \tag{7}$$

where λ_{Al} is an aluminium thermal conductivity (the value is defined by [10] and given in table 1); δ is a basis thickness of a rod; $T_{PL.ex}$ is a temperature of heated plate according to the experiment results (195 ^{0}C); $F_{BAS.RO}$ is an area of a rod in a site of contact with the basis; $F_{BAS.}$ is an emission surface area of a rod basis which is not shaded by its end face surface; $F_{CON.}$ is a contact area of a rod basis with a heated plate.

A simultaneous solution of the equations (6) and (7) gives the value of temperature $T_{H.RO}$ (table 2). From the equation (6) we obtain the rod emission power $P_{em.RO}$ (table 2).

$P_{em.RO}$ value can also be defined in a different way, if to consider the emission power of a rod radiator as the difference between the heat flow brought from GHE and the emission power of a free plate surface (the area of a heated plate minus the area covered with a rod basis):

$$P'_{em.RO} = P_{el} - P_{emPL2}\,. \tag{8}$$

Here $P_{em.PL2}$ is defined according to the formulas (1–5) taking into account a new experimental temperature $T_{PL.exp}$ which is reduced by the value of a plate square covered with its basis. The calculated value is given in table 2.

For verification of calculation results we have to construct a diagram of temperature distribution along the rod length and compare it with experimental values. To construct the temperature diagram we can use the formula [9]:

$$T(x) = T_{H.RO} \cdot \frac{\left(1 - \dfrac{\alpha_T}{m \cdot \lambda_{DAL}}\right) \cdot e^{m \cdot x} \cdot e^{-m \cdot L_{RO}} + \left(1 + \dfrac{\alpha_T}{m \cdot \lambda_{DAL}}\right) \cdot e^{-m \cdot x} \cdot e^{m \cdot L_{RO}}}{e^{m \cdot L_{RO}} + e^{-m \cdot L_{RO}} + \dfrac{\alpha_T}{m \cdot \lambda_{DAL}} \cdot \left(e^{m \cdot L_{RO}} - e^{-m \cdot L_{RO}}\right)}\,. \tag{9}$$

Parameters in the formula (9) are the same, as in the formula (6).

The diagram constructed by the formula (9) is shown in fig. 4. The comparison of calculated values and experimental data is given in table 2.

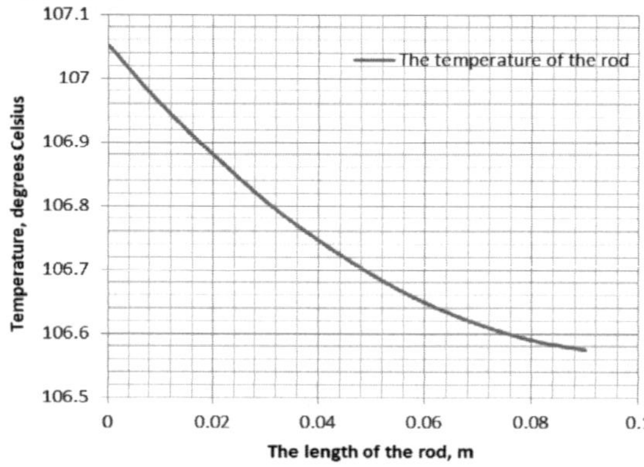

Figure 4. Temperature distribution along the length of the rod

At stage 3 a heated plate and a rod radiator were placed in a vacuum chamber. Gas (air) with the volume flow rate $Q_{air} = 6.007 \cdot 10^{-6}$ m^3/s (it was defined by PM-A-0.063 GUZ rotameter) and inlet pressure $p_{air} = 14665$ Pa in a gas conduit (according to a manometer installed in the vacuum chamber) was fed via the internal gas conduit. The heat flow with the power of 19.4 W was brought to the plate from GHE, the plate temperature was measured by thermocouple 3, the rod temperature was measured in a middle point (thermocouple 2, fig. 1). Fig. 5 shows the dependence of the plate temperature versus heating time, which is constructed on the results of experiments. We can see that the equilibrium temperature of the plate is 173 ^{0}C.

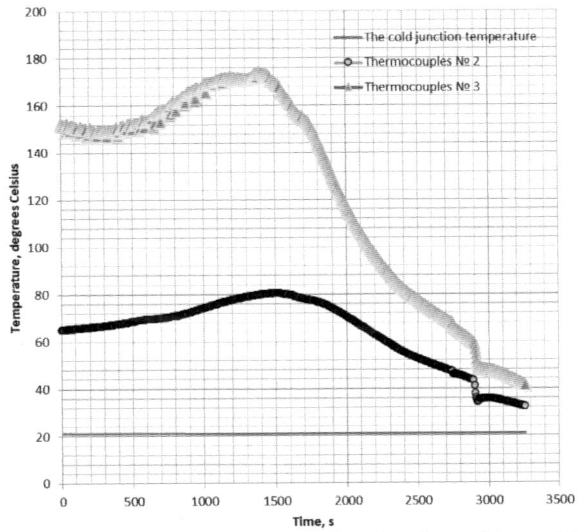

Figure 5. The results of the experiment on the heat transfer of a plate with a rod in vacuum with the of supply gas (the area of attainment of the equilibrium temperature is shown).

The purpose of theoretical research at stage 3 is to determine the value of gas heating when it flows through the internal gas conduit.

Heat exchange between gas and a heated body is described by Newton-Richmann formula [9]:

$$P_{GAS} = \alpha \cdot (T_{AIR} - T_{RO}) \cdot \pi \cdot D_{EQU} \cdot L_{CON} \tag{10}$$

where α is a heat-transfer coefficient from a gas conduit surface to the air; L_{con} is a gas conduit length in a rod; T_{AIR} is a temperature to which air will heat up in its flow in the gas conduit. D_{equ} is an equivalent gas conduit diameter equaled a quadruple square of a gas conduit cross section divided into its entire perimeter [9].

The coefficient α is defined by Nusselt number:

$$\alpha = \frac{Nu \cdot \lambda_{AIR}}{L_{CON}}, \tag{11}$$

where λ_{AIR} is an air thermal conductivity coefficient under pressure in a gas conduit.

The Nusselt number is determined by an empirical dependence for particular flow conditions in a gas conduit. The Reynolds number is defined to establish flow conditions:

$$\mathrm{Re} = \frac{4 \cdot Q_{AIR}}{\pi \cdot D_{EQU} \cdot \eta_{AIR}}, \tag{12}$$

where η_{air} is a dynamic air viscosity under pressure in a gas conduit.

Dynamic viscosity of rarefied polyatomic gas at the gas conduit input η_i and with a wall temperature η_{wa} can be calculated as a function of empirical parameters [12]:

$$\eta_{Ii}(\eta_{WA}) = 2.6693 \cdot 10^{-6} \cdot \frac{\sqrt{M_{AIR} \cdot T_I(T_{WA})}}{\sigma_A^2 \cdot \Omega_\eta}, \tag{13}$$

where M_{AIR} is an air molecular weight; T_I is an air temperature at a gas conduit input and T_{WA} is a wall temperature, K; σ_A is a diameter of molecules, angstrom; Ω_η is a slowly changing function of a dimensionless temperature [12].

The calculation of Re number showed that it is in a range of $10<\mathrm{Re}<1 \cdot 10^4$.

According to [9] an average heat-transfer coefficient can be determined by the formula:

$$Nu = 1.4 \cdot \left(\mathrm{Re} \cdot \frac{D_{EQU}}{L_{EQU}}\right)^{0,4} \cdot \mathrm{Pr}_I^{0,33} \cdot \left(\frac{\mathrm{Pr}_I}{\mathrm{Pr}_{WA}}\right)^{0,25}, \tag{14}$$

where $\mathrm{Pr}_I(\mathrm{Pr}_{WA}) = \frac{\eta_I(\eta_{WA}) \cdot c_{P.I}(c_{P.WA})}{\lambda_I(\lambda_{I.WA})}$ are Prandtl numbers for a gas temperature at the inlet in a gas conduit (Pr_I) and for a temperature of a gas conduit internal wall (Pr_{WA}).

Isobaric heat capacities of air under thermodynamic parameters in a gas conduit (at the inlet in a gas conduit $c_{p.air}$ and with a temperature of a gas conduit wall $c_{p.wa}$), can be expressed by Van der Waals heat capacity of gas:

$$c_{P.I}\left(c_{P.WA}\right)=\left(\frac{R}{\left(1-\frac{2\cdot a_{AIR}\cdot\rho_{AIR}}{T_I\left(T_{WA}\right)\cdot R\cdot M_{AIR}}\cdot\left(1-\frac{b_{AIR}\cdot\rho_{AIR}}{M_{AIR}}\right)^2\right)}+\frac{5\cdot R}{2}\right)\cdot\frac{1}{M_{AIR}}, \tag{15}$$

where R is a universal gas constant; a_{AIR} and b_{AIR} are Van der Waals constants for air; ρ_{AIR} is an air density with pressure p_{air} (which is determined by the Van der Waals equation); T_I is an air temperature at the inlet of a gas conduit and T_{WA} is respectively, a temperature of a gas conduit wall.

Heat conductivities of rarefied air (at the inlet in the gas conduit λ_i and with the temperature of gas conduit wall λ_{wa}) can be estimated by the dependence [12]:

$$\lambda_i\left(\lambda_{WA}\right)=8.322\cdot10^{-2}\cdot\frac{\sqrt{\frac{T_I\left(T_{WA}\right)}{M_{AIR}}}}{\sigma_A^2\cdot\Omega_\lambda}, \tag{16}$$

where Ω_λ is an integral which numerical value is calculated within the model of potential of Lennard-Jones intermolecular interaction [12].

To determine a heat flow value selected from a rod by the air when it flows in a gas conduit it is necessary to define a heat flow value given by the rod to the ambient space. The technique of its definition is described at stage 2. The equations (6) and (7) are solved simultaneously taking into account that at stage 3 $T_{PL.exp}$=173 ^{0}C, this solution gives the temperature value $T_{H.RO3}$ (table 2). According to the equation (6) we obtain the rod emission power $P_{emi.RO3}$ while cooling the rod with the air (table 2).

The heat flow selected by gas can be obtained as the difference of a heat flow of a rod without gas and a heat flow of a rod when the air flows in a gas conduit:

$$P_{GAS}= P_{emi.RO}(P'_{emi.RO}) - P_{emi.RO3}. \tag{17}$$

Here $P`_{emi.RO}$ is defined by the formula (8).

Calculated values of the power for gas heating are given in table 2.

Here $P`_{emi.RO}$ is defined by the formula (8).

Solving the equations (10–17) we obtain the required air temperature T_{AIR} of a rod at the outlet of gas conduit (table 2).

Knowing the gas temperature we can estimate the velocity of particles motion at the outlet of a gas conduit. For this purpose we shall use the equation of average kinetic energy of molecules motion for monoatomic gas which, according to the experimental data, can be used for polyatomic gas (calculation is provided for nitrogen as a predominant component in a mixture):

$$v=\sqrt{\frac{3\cdot k\cdot T_{AIR}}{m_{N2}}} \tag{18}$$

where k is a Boltzmann constant; m_{N2} is a nitrogen molecular mass.

Calculated values of the velocity are given in table 2.

If to consider ETME as a test sample which work results in a creation of a jet thrust then it can be determined by the formula:

$$R_{ETME} = Q_{AIR}\cdot\rho_{AIR}\cdot v. \tag{19}$$

The results of calculation are given in table 2.

It should be noted that the obtained thrust value satisfies the requirements of the orientation control system of VSS. The gas was heated from a radiator-rod which temperature did not exceed 110 ^{0}C.

Table 2. Results of experimental studies and theoretical analysis

Indicator	Experimental value	Calculated value
An equilibrium temperature of a heated plate at stage 1, TPL, K / 0C	561/288	572.507/299.507
An equilibrium temperature of a hot end of a rod at stage 2, TH.ST.RO, K /0C	-	380.05/107.052
A heat flow removed by a radiator-rod at stage 2, $P_{emi.RO}$ / $P'_{emi.RO}$, W	-	7.007/6.481

A temperature of a rod middle at stage 2, ^{0}C	86.41	106.71
A temperature of a rod end at stage 2, ^{0}C	105.63	106.5
An equilibrium temperature of a hot end of a rod at stage 3, $T_{H.RO3}$, K / ^{0}C	-	377.69/104.69
A heat flow removed by a radiator-rod at stage 3, $P_{emi.RO3}$, W	-	5.613
A heat flow removed by the air at stage 3, P_{GAS}, W	-	1.394/0.868
An air temperature at a gas rod outlet at stage 3, K	-	1835.118/1285.76
A velocity of nitrogen molecules in air, ν, m/s	-	1278/1070
A thrust of ETME as an experimental sample, P_{ETME}, mN	-	1.336/1.118

4. Results of experiments

Three experimental stages of studies with theoretical processing of each stage were conducted in a course of problem solving of SHF-transistor cooling with a micro consumption of working gas. The results of experimental study and theoretical analysis are provided in table 2. The data on heating dynamics of sample elements obtained in the course of experimental studies are given in fig. 2,3 and 5.

A mathematical model of cooling for SHF-transistor crystal by a gas working medium have been verified as a result of researches. It is shown that if particular geometrical and initial thermodynamic conditions are met, the intensity of cooling increases, and a heating value of the working medium grows (micro consumption of the air cools the experimental surface by 1 W). The heating of working gas, according to theoretical processing, is suffice to create a jet thrust of 1.2 mN with the gas exit velocity of 1000-1200 m/s.

5. Discussion of results

In the course of experimental and theoretical work the main results essential to further studies have been obtained:

1. The mathematical model describing thermal processes for a gas flow of small volume in a gas conduit of a heating element is verified;
2. The value of a heat quantity selected by gas when it moves in a gas conduit of the sample is obtained experimentally. This value allows us to make a conclusion about a large value of kinetic temperature of gas at the gas conduit outlet and a high value of a specific thrust impulse.

6. Conclusion

Thus we can draw the following conclusions:

1. As for a cooling capacity of a rod radiator and a gas conduit we should note that in case of large rods with a big emitting surface (as in the experiment described) the role of air capillary cooling is not substantial since the gas removes no more than 1.5 W of heat power with 19.4 W fed. Due to the air-cooling, a symbolic crystal of SHF- transistor was "22 ^{0}C colder", than without it. However the rod radiator cooled the symbolic crystal of SHF- transistor by 104 ^{0}C. In case of small emitting surfaces and low power inputs (up to 7–10 W) the cooling effect of gas increases;
2. The first paragraph entails the conclusion that a rod radiator and an internal gas conduit can be efficiently utilized as a working body preheater of a microengine since with small gas consumptions it is possible to obtain a high extent of gas heating and a high specific impulse of the thrust;
3. Concerning a dilatometric effect and a possible use of a rod as a throttle valve plunger we may say that there is no sense to use directly the material of a rod in a dilatometric drive at temperatures up to 110 ^{0}C.

This is due to a small coefficient of thermal extension in metals. We propose to introduce an intermediate element in a form of a polymeric (epoxy resin, ebonite, polyethylene, fluoroplastic) tube in a structure. A thermal expansion coefficient of the tube is 10-15 times higher, than for metals.

6. Source of financing

These studies are carried out within the framework of financial support by the Russian Foundation for Basic Research under Contract No. 31 16-38-60089/15 of 02.12.2015 (Research No. Gr.46-15, registration No. AAAA A161160202100195).

IOP Conf. Series: Journal of Physics: Conf. Series **1050** (2018) 012095 doi:10.1088/1742-6596/1050/1/012095

References

[1] Vavilov I S, Kositsyn V V, Lukyanchik A I [et al] 2016 About a possible decomposition of ammonia by SHF-effect in a correcting microengine of a small spacecraft *Omsk scientific bulletin* **6(150)** pp 58–63
[2] Kucheruk V V, Lebedev N V, Saribekyan A M 2016 The study of output parameters dependence versus temperature with a various type of a heat-conducting substrate on a metal basis exemplified by BLF578XR transistor *Magazine of scientific publications of postgraduate students and postdoctoral students* **4(118)** pp 162–166
[3] Carbide-silicon MESFET SHF-transistors Short catalogue of products PROSOFT URL: http://www.prochip.ru/cms/f/370103.pdf.
[4] KT 9197 B. Powerful SHF generator transistor for work in a meter wave range http://web.vrn.ru/vesta1/transistors/power_lowvoltage/kt_9197_b/kt_9197_b.htm
[5] Sivukhin V N 2005 *General course of physics Optics* (Moscow: FIZMATLIT) **4** pp 544
[6] Martynenko O G 1987 *Spravochnik po teploobmennikam* (Moscow: Energoatomizdat) **1** 352 p.
[7] Kazantsev E I 1975 *Commercial furnaces Reference guide for calculations and design* (Moscow: Metallurgy) p 368
[8] Siegel R and Howell J R 1992 *Thermal Radiation Heat Transfer* (New York: Hemisphere) 1072 p
[9] Mikheyev M A and Mikheyeva I M 1977 *Heat transfer foundations* (Moscow: Energy) p 344
[10] Acherkan N S 1961 *Reference book for mechanical engineers* (Moscow: Mashgiz) **2** 559 p
[11] Babichev A P, Babushkina N A, Bratkovsky A M 1991 *Physical quantities* (Moscow: Energoatomizdat) 1231 p
[12] Bird R B, Stewart W E and Lightfoot E N 2002 Transport Phenomena (John Wiley & Sons, Inc.) 895+3 p
[13] Blinov V N, Shalay V V, Vavilov I S, Kositsin V V, Ruban V I, Lykyanchik A I, Yachmenev P S and Vlasov A S 2017 Gas dynamic model of electrothermal thrusters of small spacecraft and possibility of applying microwave heating of a working *IOP Conf. Series: Journal of Physics: Conf. Series* **858** pp 012004

IOP Conf. Series: Journal of Physics: Conf. Series **1050** (2018) 012096 doi:10.1088/1742-6596/1050/1/012096

Smallsat's dilatometric micromotor: estimation of characteristics

I S Vavilov, A I Lukyanchik, P S Yachmenev, A V Pavlenko, A V Lysakov, R N Litau

Omsk State Technical University, 11, Mira ave., Omsk, 644050, Russia
E-mail: vava-igg@mail.ru

Abstract. The article reflects some results of the work of the authors on the creation of reactive corrective micro-motors for spacecraft of nanoglass. A special feature of these micro-motors is the use of a dilatometric element as a shut-off valve actuator, which is a working gas heater. A prototype of a dilatometric micro-motor was developed and manufactured, on which, under vacuum conditions, efficiency studies were carried out and the output thermodynamic parameters of the gas were obtained. Dilatometric valve engine used thus reducing the nitrogen pressure of 2.8 bar and up to 120 Pa was carried out by the gas flow rate 1.8·10-6 kg/s and heated to a temperature of 470 K. the Specific impulse thrust was 630-650 m/s. Power was 2.6 watts.

1. Introduction

Ultra-small spacecraft (USSC) are used as a means of solving problems in the fields of space research because they can test spacecraft for human safety in an economic way. The USSC class of devices are those weighing up to 100kg (micro-, nano-, pico- and femto satellites).

This class of devices has been made possible by the development and wide distribution of microelectronics, which has been implemented in the miniaturization of the payload and control systems. And the reduction in the cost of USSC components has expanded research in this area to include higher education institutions all over the world, where their development is carried out by undergraduate students and post-graduate students.

The launching of USSC's is carried out by a passing method or by astronauts while in space when they leave their spacecraft (ISS).

The niche of cheap modular USSC is constantly expanding. This is especially noticeable by the increasing frequency of launches of satellites into a near-earth orbit. For example, on 14 July, 2017, Soyuz-2.1a led the launching of 72 small satellites with the launching of the spacecraft "Canopus-V-IR" and in India, with a record number of 104 satellites led by the rocket PSLV-C37.

The next step of development of the USSC class will be the development of guided flights.

For use in large spacecraft, USSC introduce a reactive propulsion system that ensures the correction of the satellite's orbit, the development of apparatus along the orbits of operation in the case of group excavations, and the disposal of the USSC after the end of its active operation period and interplanetary flights.

Unlike microelectronics, the engine building area for USSC is not developing so quickly. The most significant limitation is the amount of excess capacity that the USSC has. The miniaturization of the electronic components of the USSC leads to a reduction in the mass and dimensions of the apparatus and a reduction in the power consumption. The useful surface area of the solar panels is limited by the dimensions of the apparatus. As a consequence, the excess power falls linearly with a decrease in the mass of the apparatus. For example, on the USSC of the CubeSat formfactor of the base standard 1U, the excess power is not more than 1W, while the electromagnetic shut-off valve, mandatory for traditional reactive low-thrust systems, according to some manufacturers, consumes not less than 1W of energy. Because of this, the most common propulsion systems (PS) use "cold gas". Such systems have a flight operation history which has demonstrated that they are easy to implement and reliable, because there are no high temperatures. But a significant disadvantage of PS on "cold gas" is their low profitability.

Another significant disadvantage of remote control on "cold gas" is their low efficiency and high thrust values. The higher thrust has a negative effect on the craft's orientation and stabilization because of the USSC's small mass. For example, the low-thrust engine of the MicroThruster designed by the Marotta engineering company for the ST-5 (25 kg) has a specific impulse of thrust of 65 s for nitrogen operation, a pull of 0.05-2.36 N (depending on the pulse width), and consumes 1 Watt in the pulsed mode [1]. This device is a high-speed impulse valve with pulse duration up to 5 ms. The thrust value is set by the opening time of the flow section. It should be noted that the USSC form factor CubeSat 1U has a mass of up to 1 kg. An interesting device is the correction MEMS company NanoSpace, the working body of which is selected butane. The specific impulse of the motor is 50-75 s, the power consumption is 2Watt at a thrust of 0.01-1 mN [1]. Another micromotor with flight tests is the CNAPS of the Canadian company UTIAS / SFL, operating on sulfur hexafluoride and creating a thrust of 12.5-40 mN at a specific impulse of 40 s.

In PS USSC, resistive micro-motors (electrothermal micromotors) have found application, where the working medium is subjected to preliminary heating, which provides an increase in the specific impulse of thrust. One example is the MRJ Thruster [2] (company Busek) with an energy consumption of 3-15 W, a specific impulse of thrust of 80-150 s and thrust of 2-10 mN. Working body MRJ Thruster-ammonia. Also, it is possible to single out the engine PUC (Propulsion Unit for Cubesats) which is produced by CU Aerospace and VACCO [3], which powers up to 15 W, thrust pulse 70 s and thrust 0.45 N.

The typical scheme of the correcting propulsion system (CPS) USSC is a sequential chain of elements: the fuel tank in the working body - the shut-off electrovalve - the heater of the working body (for resistive motors) - pressure reducer – and the electrothermal micro motor (ETMM). This scheme is used for SSC with a mass of 10 kg with an excess of electrical power from 30 to 110W. The USSC format begins with apparatuses weighing less than 10kg. The traditional layout scheme allows a CPS to operate mainly on "cold gas", i.e. without heating the working fluid.

Structurally, CPS SSC is a sequence of nodes, each of which has a single task. The unavoidable irreversible heat losses in each element reduce the electrical efficiency of the CPS.

Minimization of losses and miniaturization of CPS elements by direct scaling of the elements' constructions does not justify itself, because the boundary conditions for ensuring the operability of electromagnetic and heat-producing devices often conflict with each other. Also, with direct scaling of the construction, consumed powers collide with unscaled gas-dynamic processes occurring in the CPS tract.

From what has been stated above, it follows that at present, there is a problem of developing energy-efficient CPS of the USSC, which ensure the preservation of the given high output characteristics of the ETMM.

The authors proposed the concept of a micromotor, which makes it possible to combine a throttling valve and a working body heater in one technical node; the "Swiss knife" concept. The basis of the mechanics of the valve drive is the dilatometric effect, which is unavoidable in the use of electric heating elements. This approach does not exert an additional energy load on the power supply systems of the USSC. Thus, the increase in the energy efficiency of the CPS of the USSC is made through the intensification of energy exchange processes due to the reduction in the number of intermediate links of the KDU and the use of dissipative energy, using dilatometric effects.

2. Formulation of the problem

The work is based on the results of the experiments carried out with subsequent theoretical processing on the basis of the known laws of physics.

The following tasks have been posed by the authors:

IOP Conf. Series: Journal of Physics: Conf. Series **1050** (2018) 012096 doi:10.1088/1742-6596/1050/1/012096

1. Development and creation of a prototype of an electrothermal micromotor with dilatometric drive of a shut-off throttling valve;
2. Thermal experimental studies of the prototype under vacuum with nitrogen as a working fluid;
3. Theoretical processing of the results of experimental prototype research.

3. Theory

An experimental sample of a dilatometric micromotor (DMM) is shown on Figure 1. The throttling valve 1 is made in the form of a cylindrical sleeve with seals made of vacuum rubber and a side fitting 5 for insertion of the working fluid. The valve has the ability to adjust the flow rate with a screw. This is necessary to adjust to the conditions of the vacuum chamber. The dilatometric element 2 is made in the form of a resistance of the rod, heated by a graphite element, with a diameter of 2 mm. At the end of element 2 is the valve needle. In Figure 1 dilatometric element is shown separately, prior to assembly. The gas path in the engine is an annular channel formed by the element 2 and the motor casing 3. On the tap 4, the heated working medium enters the T-piece, where it enters the rotameter at the fitting 7. The fitting 6 is for connection of a manometer.

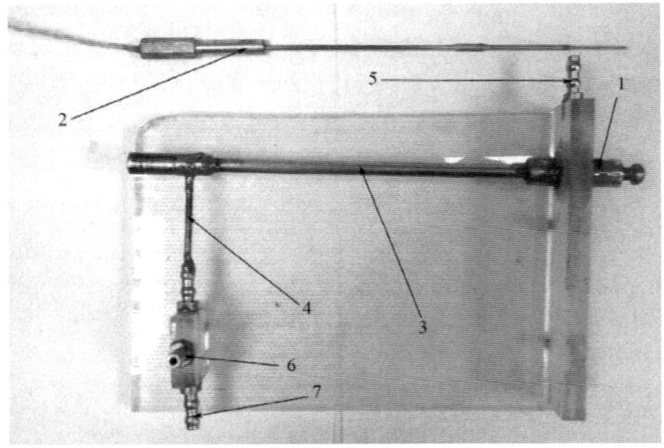

Figure 1. Experimental sample of a micromotor with dilatometric valve drive

The engine elements are made of AISI 321 steel. The fixation and sealing of the elements is with an epoxy resin.

The operation of the DMM is carried out as follows: When the electric current is applied to the element 2, it heats up, the material undergoes a linear expansion, the position of the needle in the valve 1 changes, and the working medium enters the annular gap between the elements 2 and 3, where it heats and expels through the side branch 4. Thus, reactive thrust is created.

Figure 2 shows the experimental setup. In the figure, the items are: 1-vacuum chamber; 2-investigated DMM sample; 3-thermoelectric converters installed on the body and at the side branch of the DMM; 4-rotameter PM-A-0.063 GUZ; 5 – rotameter PM-A-0,1 GUZ; 6-manometer in the mains; 7 – gallon with nitrogen; 8-gas reducer; 9 – Vacuum plate-rotor pump NVR-16D; 10-control panel of the vacuum chamber; 11-lamp PMT-6-3M; 12 – multichannel precision temperature meter MIT-8; 13-PC; 14 – a vacuum gauge Meradat-VIT 16T3 / 2P / 485 / 2M; 15 – valve electromagnetic shutoff; 16 – manometer at the exit from the DMM.

In the course of the experiment, the following values were set: the inlet pressure in the DMM was 2.8 atm. (according to the pressure gauge item 6 in Figure 2) and the electrical voltage on the dilatometer element 3 V (power consumption 2.6 W).

Measured and recorded values in the experiment were: indications of rotameters RM-A-0.063 GUZ and RM-A-0.1 GUZ; temperature of the hull elements of DMM, recorded by thermocouples (item 3, Figure 2); the gauge reading (item 16, Figure 2); The readings of the Meradat-VIT 16T3 / 2P / 485 / 2M vacuum gauge (item 14, Figure 2).

In the course of experimental studies, at the first stage, the fact of DMM operability in vacuum conditions was confirmed: after 25-30 seconds, after applying thermal power to the dilatometric

IOP Conf. Series: Journal of Physics: Conf. Series **1050** (2018) 012096 doi:10.1088/1742-6596/1050/1/012096

element, the float of the rotameter installed in the vacuum chamber came into motion. The initial temperature of the dilatometer element was 18°C.

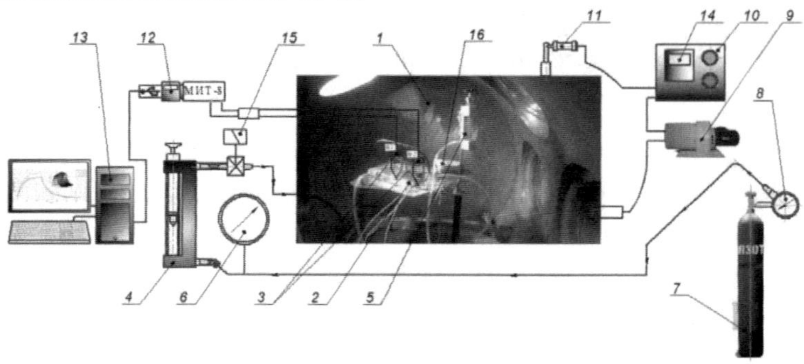

Figure 2. Experimental setup for determining the DMM parameters

To determine the mass flow rate of gas in a high-pressure pipeline, a rotameter was installed (item 4, Figure 2), however, the sensitivity of the instrument was not sufficient, the float of the rotameter 4 did not come into motion with a fluctuating float of the rotameter 5. Thus, the mass flow was determined for the pressure increase in the vacuum chamber (item 1, Figure 2) with the pump off (item 9, Figure 2), the pressure change was recorded with a vacuum gauge 14. With the known volume of the vacuum chamber, the mass flow through the flowing part of the DMM is $\dot{m} = 1.8 \cdot 10^{-6}$ kg/s.

At the second stage of the study, the thermodynamic parameters of the gas in the DMM were determined. An important parameter is the temperature of nitrogen. It was determined in three ways: on the basis of the readings of a rotameter installed in a vacuum chamber; on the basis of the drop in the temperature of the DMM body during the passage of nitrogen; based on the equation of the gas temperature in the heated pipeline, derived from the Newton-Richman equations and the total gas energy.

1. Determination of gas parameters through indications of a rotameter.

The technique is based on two basic equations: the float equation for a rotameter [4] and the van der Waals equation for a real gas.

The equilibrium equation of a float in a rotameter has the form:

$$m_p g = C_X \frac{\rho_{N2}^R \upsilon^2}{2} F_p + \rho_{N2}^R g V_p, \tag{1}$$

where m_p=0.36 grams - the mass of the float of the rotameter (the value is given for the PM-A-0.1 GUZ device, float material D1T); C_X – coefficient of aerodynamic drag of the float; ρ_{N2}^R-density of nitrogen in a rotametric tube; F_p, V_p – the area of the midsection and the volume of the float, respectively. The largest diameter of the float of the PM-A-0.1 GUZ device is d_p=6 mm.

In the tests, the float of the rotameter installed in the vacuum chamber was not in equilibrium in the rotametric tube, but fluctuated, assuming a position from touching the lower stop to 56 divisions on the scale of the rotameter.

The moment of detachment of the float from the lower stop was considered. The diameter of the pipeline (nominal diameter) of the rotameter is d_r=3 mm, hence the flow rate flowing onto the cone of the float can be found from the expression:

$$\upsilon = \frac{4\dot{m}}{\pi d_r^2 \rho_{N2}^R}. \tag{2}$$

Here ρ_{N2}^R – is the unknown nitrogen density at the entrance to the rotameter.

The coefficient of aerodynamic drag of the float was determined by the method of [4]. The final formula for deriving the drag coefficient can be written as:

IOP Conf. Series: Journal of Physics: Conf. Series **1050** (2018) 012096 doi:10.1088/1742-6596/1050/1/012096

$$C_X = \frac{\pi h \cdot tg\alpha \left(2 + \frac{2h \cdot tg\alpha}{d_p}\right) \rho^R_{N2} m_p g d_p}{\dot{m}^2}, \tag{3}$$

where h=2.4-2.6 mm – the amount of lifting of the float at the initial time; tgα=$5\cdot10^{-3}$ – is the tangent of the cone angle of the rotametric tube.

Substituting equations (2) and (3) into equation (1), we obtain the value of the nitrogen density at the input to the rotameter $\rho^R_{N2} = 9.25\cdot10^{-4}$ kg/m^3.

The coefficient of aerodynamic resistance is equal to $C_X = 0.477$, it agrees with the experimental data on flow past cones [5]. For example, for a cone with an angle of 30° $C_X = 0.35$, for an angle of 60° $C_X = 0.61$. The cone angle of the float rotameter RM-A-0.1 GUS is 45°.

The nitrogen temperature at the outlet from the DMM can be determined from the Van der Waals equation:

$$\left(P_{N2} + \frac{a_{N2}\rho^R_{N2}}{M_{N2}}\right)\left(\frac{M_{N2}}{\rho^R_{N2}} - b_{N2}\right) = T_1 R, \tag{4}$$

where $P_{N2} = \frac{4m_p g}{\pi d_p^2} = 124.905$ Pa – minimum float detachment pressure from the stop (pressure in the DMM tract); a_{N2}, b_{N2} –Van der Waals constants for nitrogen; M_{N2} – is the molecular weight of nitrogen; T_1 –is the unknown temperature of nitrogen; R –is the universal gas constant.

2. Determination of gas parameters through the drop in the temperature of the hull elements of the DMM.

The methodology is based on the equation of the balance of the input and output power and the estimation of the gas temperature from the temperature difference between the hull and the gas supply and with the gas fed along the path.

The initial data in this method are the geometric dimensions of the flow section, the physical properties of the materials, the thermocouple readings, the mass of the hull elements.

Thermal loading of DMM occurred under conditions of a vacuum chamber for 5000 s. During this time, the modes of operation of the DMM changed (Figure 3):

A- mode of heating with nitrogen supply (solenoid valve item 15 on Figure 2 is open);

B – mode of gas supply without heating;

C – mode of heating without supply of nitrogen to the pipeline (solenoid valve item 15 on Figure 2 is closed);

D – heating mode with nitrogen supply (solenoid valve item 15 on Figure 2 is open);

E – is the cooling regime with nitrogen supply (there is no heating of the dilatometric element).

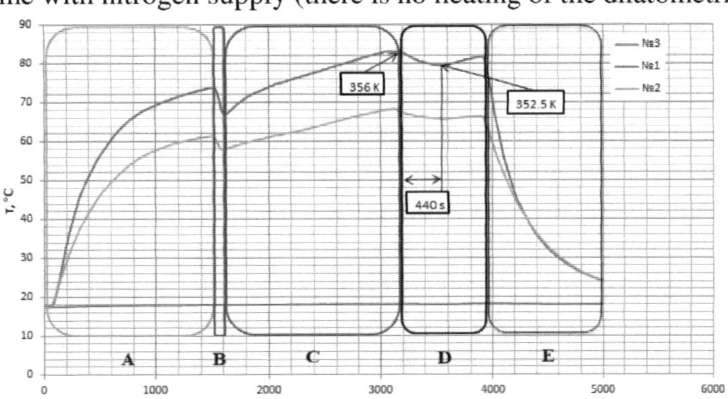

Figure 3. The results of the thermocouple readings mounted on the DMM body (№ 1), at the lateral branch (№ 2), and the cold junction (№ 3)

For the analysis of the thermal regime, mode D was selected (Figure 3), when nitrogen enters the heated DMM path. It can be seen that the transient mode of changing the internal energy of the DMM body material takes 440 s. During this time, the temperature drops by 3.5 K. Curve № 2 shows that in the same section, the temperature changes by the same amount. Hence, it can be assumed that the entire DMM construction, when cooled with nitrogen, loses 3.5–4 K. The mass of DMM elements participating in the heat exchange is 22 grams (as measured directly by electronic scales).
Thus, the thermal power received by nitrogen when it passes through DMM can be expressed:

$$N_{N2} = \sum_{i=1}^{n} c_{Pi} m_i \Delta T \frac{1}{t}, \tag{5}$$

where c_{Pi} -the specific heat of the i-th substance, which is part of the DMM; m_i – mass of the i-th substance, which is part of the DMM; n – is the number of DMM constituents involved in heat transfer and heat transfer; ΔT= 3.5 K – is the temperature drop across the DMM case; t – is the time of the transient process. Because all DMM components involved in heat transfer are made of AISI 321 steel, then in (5) the specific heat of this steel and the weight of 22 grams are taken.

From the power balance equation, it is necessary to determine the temperature of the dilatometric element, since there was no physical possibility to bring thermoelectric converters to it. In the steady state, in the absence of nitrogen, the electrical power supplied to the dilatometric element is expended on heating by the radiation of the inner surface of the DMM body and on the radiation of the body into the surrounding space:

$$N = \sigma\left[\left(T_d^4 - T_k^4\right)\varepsilon_\Sigma \cdot F_1 + T_k^4 \varepsilon F_3\right], \tag{6}$$

where $\varepsilon_\Sigma = \dfrac{1}{\dfrac{1}{\varepsilon} + \dfrac{F_1}{F_2}\left(\dfrac{1}{\varepsilon} - 1\right)}$ – is decreased degree of blackness of the system of bodies from the dilatometric element located inside the shell; $\varepsilon = 0.95$ – the degree of blackness of the surface of AISI 321 steel [6]; σ – the Stefan-Boltzmann constant; T_d – is the unknown temperature of the dilatometric element; T_k=356 K – is the maximum equilibrium temperature of the DMM body when heated without gas; N=2.61 W – electric power input. The surface area of the dilatometric element F_1, the inner surface of the housing F_2 and the outer surface of the housing F_3 are determined by measuring the linear and diametric dimensions of the elements.
To determine the temperature of nitrogen in the annular space of the DMM tract, the Newton-Richman formula [7] is applied:

$$N_{N2} = \alpha\left(T_d - T_{N2}\right)\pi D_\Sigma L_k. \tag{7}$$

In formula (7), the heat transfer coefficient is determined in terms of the Nusselt number:

$$\alpha = \frac{Nu \lambda_{N2}}{L_k}, \tag{8}$$

where λ_{N2} – coefficient of thermal conductivity of nitrogen at thermodynamic parameters of gas in the gap; L_k – the length of the annular gap; D_Σ – equivalent diameter of the annular gap.
It is assumed that the flow regime of the rarefied gas in the gap is laminar. The average heat transfer coefficient for a laminar flow of gas in a pipe under conditions $L_k / D_\Sigma > 10$ and Re >10 can be determined from the formula [7]:

$$Nu = 1.4\left(\mathrm{Re}\frac{D_\Sigma}{L_k}\right)^{0.4} \mathrm{Pr}_{N2}^{0.33}\left(\frac{\mathrm{Pr}_{N2}}{\mathrm{Pr}_k}\right)^{0.25}, \tag{9}$$

where $\mathrm{Pr}_{N2} = \dfrac{\eta_{N2} C_P}{\lambda_{N2}}$ – Prandle number according to the gas temperature at the outlet of the valve when entering the gap; $\mathrm{Pr}_k = \dfrac{\eta_{N2k} C_P}{\lambda_{N2k}}$ – the Prandtl number at the internal wall temperature of the gap.

The isobaric heat capacity of nitrogen, with thermodynamic parameters in the gap, can be expressed as the heat capacity of an ideal diatomic gas:

$$C_p = \frac{7R10^3}{2M_{N2}}, \tag{10}$$

where R – is the universal gas constant; M_{N2} – is the molar mass of nitrogen, kg/kmol.
Reynolds number in the annular gap:

$$\mathrm{Re} = \frac{4\dot{m}}{\pi D_\Sigma \eta_{N2} \rho_{N2}^R}, \tag{11}$$

where η_{N2} – is dynamic viscosity of nitrogen in the gas duct.
The dynamic viscosity of a rarefied polyatomic gas can be found as a function of the empirical parameters [8]:

$$\eta_{N2} = 2.6693 \cdot 10^{-6} \frac{\sqrt{M_{N2} T}}{\sigma^2 \Omega_\eta}, \tag{12}$$

where M_{N2} – is the molecular weight of nitrogen; T – is the temperature in K; σ – is the diameter of the molecules in $\overset{o}{A}$; Ω_η – is the slowly varying function of the dimensionless temperature. The last two quantities are tabular and are given in [8]. Dynamic viscosity is determined for the initial temperature and temperature of the housing.
The thermal conductivity of rarefied nitrogen can be estimated using the dependence [8]:

$$\lambda_{N2} = 0.083 \frac{\sqrt{T / M_{N2}}}{\sigma^2 \Omega_\lambda}, \tag{13}$$

where Ω_λ – is integral, the numerical value of which is calculated in the framework of the Lennard-Jones intermolecular interaction potential model [8].
Solving equations (7) – (13) one can obtain the required gas temperature T_{N2}. The temperature value is given in Table 1.
3. Determination of gas parameters is based on the gas temperature equation.
This equation is derived on the basis of the equations for the change in the total energy and the Newton-Richmann equation and is written taking into account the small difference in the gas pressures before and after heating:

$$T_{\Gamma A3} = \frac{c_V T_0 - \dfrac{21{,}939 \cdot 10^{-3} M_{N2}^{0{,}3} (c_V + R)^{0{,}33}}{\dot{m}^{0{,}6} \sigma^{1{,}2} \Omega^{0{,}6} \left(\rho_{N2}^R\right)^{0.4}} \cdot \dfrac{T_k^{1{,}3} D_\Sigma}{L_k^{0{,}4}}}{c_V - \dfrac{21{,}939 \cdot 10^{-3} M_{N2}^{0{,}3} (c_V + R)^{0{,}33}}{\dot{m}^{0{,}6} \sigma^{1{,}2} \Omega^{0{,}6} \left(\rho_{N2}^R\right)^{0.4}} \cdot \dfrac{T_k^{0{,}3} D_\Sigma}{L_k^{0{,}4}}}. \tag{14}$$

The initial data for equation (14) are the same as for equations (6) and (7). In (14), the molar heat capacity of a gas at a constant volume is defined as $c_V = \frac{i}{2} R$, where i – is the number of degrees of freedom of a nitrogen molecule.

To estimate the velocity of the nitrogen molecules at the exit from the channel, one can use the equation of the average kinetic energy of the molecules for a monatomic gas. This dependence has good convergence with the experimental data for diatomic gases:

$$v = \sqrt{\frac{3RT_{N2}}{M_{N2}}}. \tag{15}$$

The possible thrust is determined by the product of the mass flow rate for the gas velocity. The results of the calculation are given in Table 1.

IOP Conf. Series: Journal of Physics: Conf. Series **1050** (2018) 012096 doi:10.1088/1742-6596/1050/1/012096

4. Research Results

The results of experimental studies are reflected in Table 1.

Table 1. The results of an experimental study of DMM

№	Source data	Intermediate value	Parameters DMM
1.	The mass of the float of the rotameter m_p=0.36 gr. Diameter of the midsection of the float d_p=6 mm The volume of the float V_p=130 mm^3 The oscillation amplitude of the float h=2.5-40 mm Nominal diameter of the flowmeter d_r=3 mm Taper rotometrics tube 1/200 Mass flow of nitrogen $\dot{m}=1.8\cdot10^{-6}$ kg/s.	Aerodynamic drag coefficient $C_X=0.477$ The pressure in the tract DMM $P_{N2}=124.905$ Pa The density of nitrogen in the tract DMM $\rho^R_{N2}=9.25\cdot10^{-4}$ kg/m^3	Gas temperature 455.04 K Specific impulse 636.485 m/s Thrust 1.16 mN
2.	The maximum equilibrium surface temperature during heating without gas 356 K The maximum equilibrium surface temperature when heating with gas 352.5 K (area D, pic. 3) During the transition process t=440 s. Weight DMD m=22 gr. The input electrical power N=2.61 W	Dilatometric element temperature T_d=449.35 K Power, adopted by the gas $N_{N2}=0.096$ W Dynamic viscosity $\eta_{N2}=1.74\cdot10^{-5}$ Pa·s and $\eta_{N2cm}=2.375\cdot10^{-5}$ Pa·s Thermal conductivity $\lambda_{N2}=0.019$ W/(K·m) и $\lambda_{N2k}=0.026$ W/(K·m) Heat transfer coefficient α=4.543 W/(m^2·K)	Gas temperature 477.44 K Specific impulse 651.9 m/s Thrust 1.19 mN
3.	Mass flow of nitrogen $\dot{m}=1.8\cdot10^{-6}$ kg/s Geometric characteristics of the flow part Dilatometric element temperature T_d=449.35 K Density of nitrogen in the tract DMM $\rho^R_{N2}=9.25\cdot10^{-4}$ kg/m^3	-	Gas temperature 553.648 K Specific impulse 702 m/s Thrust 1.28 mN

5. Discussion of results

During the work, a prototype of an electrothermal micromotor with a dilatometric drive of a shut-off valve was built. Experimental research has shown, first of all, that the device operates. It was shown that the gas supply starts in 25-30 seconds after the electric power is applied to the dilatometric element and the gas is cut off after 10-15 s after turning off the power. DMM showed good economy, because it creates a specific thrust impulse of the order of 63-70 s at an energy consumption of 2.61 watts. The heat exchange in this DMM is not effective, because 0.096 W was needed to heat the gas. However, the formula (14) and previous experimental studies of the authors show a strong dependence of the heat transfer rate on the conditional passage diameter of the channel. The study showed that, with a low heat transfer efficiency, the DMM output characteristics slightly exceed the output characteristics of the remote control on «cold gas».

The output characteristics of DMM were obtained by three methods, two of which are based on experimental data, formula (14) gives the ultimate thermal capacity of the gas under the selected conditions.

6. Summary and conclusions

A prototype of an electrothermal micromotor for USSC, in which the dilatometric pair performs the functions of a gas reducer, a heater of the working fluid and a locking organ, cuts off the supply of the working substance to the engine path. A new concept of remote control design for USSC is presented, with low energy consumption while maintaining functioning. The working capacity of the device is shown. The performance of the prototype dilatometric valve when opened is 25-30 seconds, when cut-off-10-15 seconds. Specific impulse thrust power 2.61 W, made up 63 to 70 seconds. These values put the studied prototype of DMM in series with the energy-efficient do the «cold gas» [1, 2, 3].

7. Source of financing

Work is executed at financial support of Council for grants of President of Russian Federation Competition СП-2018, СП-923.2018.1

References

[1] State of the Art of Small Spacecraft Technology URL: https://sst-soa.arc.nasa.gov
[2] Williams D 2012 Propulsion Solutions for CubeSats and Applications URL: http://mstl.atl.calpoly.edu/~workshop/archive/2012/Summer/Day%202/1545-Williams-PropulsionSolutions.pdf (Date of access: 30.10.2017)
[3] Propulsion Unit for CubeSats URL: http://www.vacco.com/images/uploads/pdfs/11044000-01_PUC.pdf (Date of access: 30.10.2017)
[4] MI 1420-86 guidelines. GSE. Flowmeters of constant pressure drop. The conversion of metrological characteristics.
[5] Idelchik I E 1992 *Handbook of hydraulic resistance* (Moscow: Mashinostroenie) 672
[6] Siegel R and Howell J R 1992 *Thermal Radiation Heat Transfer* (USA: Washington) 1072
[7] Mikheev M A and Mikheeva I M 1977 *Basics of heat transfer* (Moscow: Energy) 344
[8] Bird R B, Stewart W E and Lightfoot E N 2002 *Transport Phenomena* (John Wiley & Sons, Inc.) 898
[9] Vavilov I S, Lukyanchik A I and Yachmenev P S [and others] 2017 Study of heat exchange in a capillary gas heater with graphite resistance element *Dynamics of Systems, Mechanisms and Machines* **5** pp 21–31

Data distribution system: preparation of server stations data

E O Vikulov, O V Denisov and L A Denisova

Omsk State Technical University, 11, Mira ave., Omsk, 644050, Russia

e-mail: vikuloveo@gmail.com, denisova@asoiu.com

Abstract. In this article the issue of server stations status data processing for effective distribution of client data on the server stations in high load information systems is discussed. We have analyzed the existing methods for distributing data on server stations of high-loaded web applications, as well as solutions that improve the speed of data delivery to the user. Among the considered methods, we identified a prototype for development and proposed an approach to the design of data distribution systems. We have developed a model that describes the state of the servers, and proposed a structure for representing the parameters of the system. According to the obtained experimental data we have constructed graphs that reflect the dependence of server state parameters on time. In the future, we are planning to use this data about server states, as well as the data about end users (uploading or browsing files), in order to optimize server selection and ensure fast provision of information to the user.

1. Introduction

One of the actual problems of efficient use of computing resources in high-load information systems is to ensure the rapid provision of information to the user by means of the Internet. To increase the speed of data delivery to the user, the following technologies are used: data delivery and distribution networks, data balancing (distribution) systems.

Data delivery and distribution networks (CDN) [1] consist of geographically distributed multifunctional platforms. The interaction of these platforms allows CDN to process data efficiently. This allows CDN to speed up the user's requests for data.

Load balancing systems [2] are technologies based on the data distribution through the server stations. This technology is one of the most successful and is often used in high-load web applications. There are a few algorithms of load balancing through the server stations e.g. DNS Round-robin [3], weighted Round-robin, least and weighted-least connection, agent-based adaptive load balancing, and etc. [3]. Most simple popular algorithm of load balancing is Round-robin. This algorithm allows to system normally distribute tasks between server stations. It helps to optimize resources and decreases processing time of each request [3, 4].

Both methods have drawbacks. The use of data delivery networks significantly increases the cost of the developed application. The cost increases in direct proportion to an increase in the amount of data.

When using the same data balancing systems, the load is distributed between the server stations, which allows us to increase the speed, but does not give a significant increase, as it is often performed by the Round-robin distribution algorithm. Also, the consistent distribution of data does not take into account the information about the data itself, for example, the type of data, the amount of data, and the information about the user who downloads this data. This information helps to determine on which server station balancer should distribute the user file in order to speed up the further operation of the web application. To take into account the current state of the system, you must use adaptive balancing based on the agent. The agent is a software tool that collects and processes data about the current state of the server stations. This type of balancing will allow system to take into account the current state of used server stations and the information about the clients of the system.

2. The problem statement

It is required to develop a data distribution system for server stations, one of the main functions of which is the processing of information about the status of server stations. In addition, the system should implement the following functions: downloading, viewing and issuing data that users work with (images, text, video, etc.). Let's construct the structural scheme of the developed system. This scheme allows us to understand the interaction of users with the system, as well as visualizes the data flows within the application. Here we can see: distributed server stations, application balancer, users who download (on the one side) and scan (on the other side) files using the application (figure 1). The balancer plays a major role in the data transfer chain from the user who has uploaded them to the user downloading or viewing the file. Thus, when downloaded data by the user, the server-balancer must select one of the server stations most suitable for the downloaded data [5, 6]. As a selection criterion, we can use different indicators.

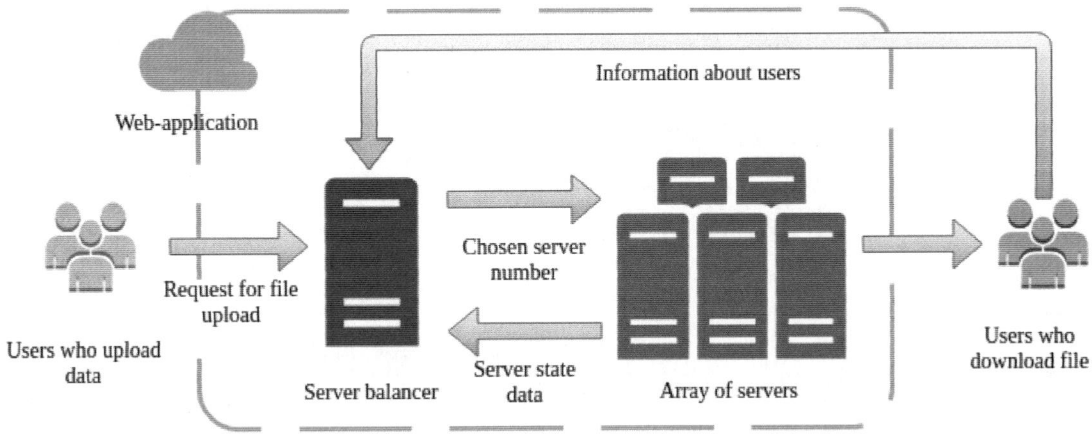

Figure 1. Structural diagram of the system being developed

So, the server-balancer can select the required server station based on the information about the downloaded data (such as type and size) and about the user (such as location, system usage time, previously downloaded files). In order to justify the selection of the quality criterion for the data balancing, it is necessary to carry out the series of experiments confirming the significance of the chosen criterion. As the most simple and affordable, we consider the criterion of the distance between user, downloading files and the server station from the array of servers. Thus, we need to develop a data distribution system for server stations, the tasks of that will be collecting and processing of server stations status data.

3. The development of data distribution system

3.1. System operation algorithm

Let's consider how the developed system should function and imagine the algorithm for data distribution on server stations. At the beginning, the user selects the data that he needs to transfer. After that, the client application sends data to the balancer, which determines the most suitable server for receiving data. Server determination based on server stations status data analysis and user data analysis. After making the decision, the balancer sends the user uploaded data to the selected server and waits for a response. After loading the data, the server responds that the download is successful. The received response is processed and displayed to the user. The block diagram of the algorithm is shown in figure 2.

IOP Conf. Series: Journal of Physics: Conf. Series **1050** (2018) 012097 doi:10.1088/1742-6596/1050/1/012097

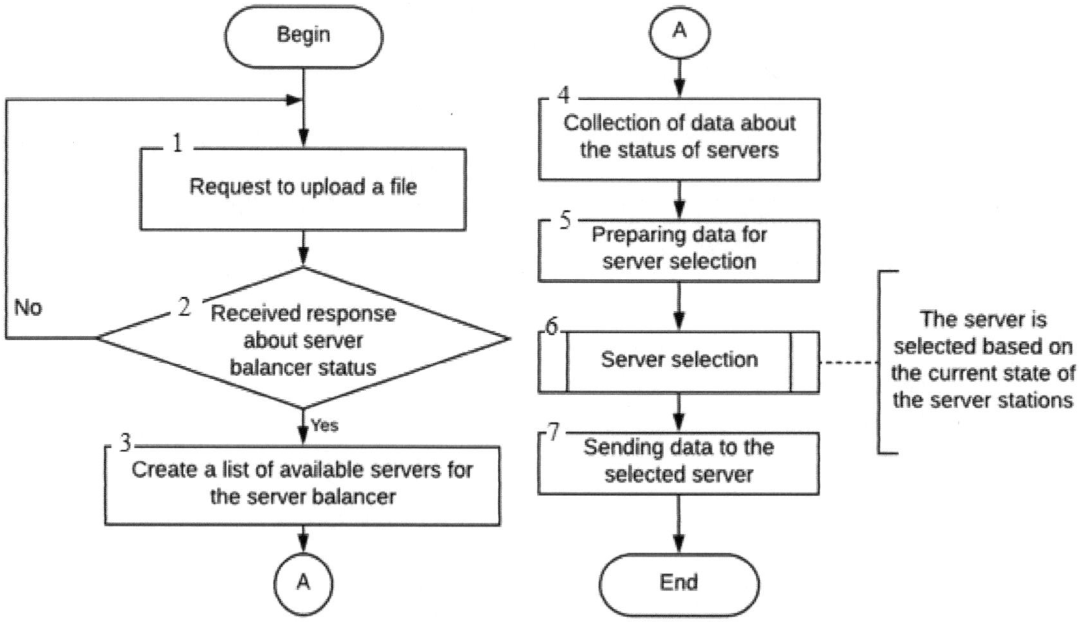

Figure 2. The block diagram of system operation algorithm

Let's consider the operation of the balance server in accordance with blocks 4 and 5 of the algorithm: data collection and data preparation for server selection.

3.2. Preparation data for server selection

Let's consider how we should process the data to make it easy to select a server. We will form an array of data on the status of server stations. Based on this data, the server balancer should make a conclusion about which of the server stations is suitable to transfer of the downloaded file. The data obtained are represented in the form of a multidimensional matrix whose number of columns is equal to the number used in the system parameters and the number of rows is equal to the number of requests sent to the load data. Also we take into account the number of requests for uploading files. To solve the problem of optimizing the choice of the server station, it is important to select the information that will allow the optimal distribution of data on the nodes of the system. The initial data for this problem can be represented in the form of array of size $l \times n \times m$.

$$\boldsymbol{X} = \left[\begin{bmatrix} x_{11}^1 & x_{12}^1 & \dots & x_{1m}^1 \\ x_{21}^1 & x_{22}^1 & x_{2j}^1 & \dots \\ \dots & \dots & x_{ij}^1 & \dots \\ x_{n1}^1 & x_{n2}^1 & \dots & x_{nm}^1 \end{bmatrix} \dots \begin{bmatrix} x_{11}^l & x_{12}^l & \dots & x_{1m}^l \\ x_{21}^l & x_{22}^l & x_{2j}^l & \dots \\ \dots & \dots & x_{ij}^l & \dots \\ x_{n1}^l & x_{n2}^l & \dots & x_{nm}^l \end{bmatrix} \dots \begin{bmatrix} x_{11}^q & x_{12}^q & \dots & x_{1m}^q \\ x_{21}^q & x_{22}^q & x_{2j}^q & \dots \\ \dots & \dots & x_{ij}^q & \dots \\ x_{n1}^q & x_{n2}^q & \dots & x_{nm}^q \end{bmatrix}\right], \quad (1)$$

superscript $l = \overline{1,q}$ is an number of file upload request, q is amount of requests. Subscripts i, j describe the states of the servers, $i = \overline{1,n}$ is an number of server stations, n is amount of server stations, $j = \overline{1,m}$ is an number of operation parameters, m is an amount of parameters. System operation parameters: x_{i1}^l is a hard drive usage of server station number i, x_{i2}^l is a distance between a user and a server station number i, x_{1m}^l is an information about the uploaded file, file size, from a request number l. This data structure will allow us to apply clustering methods to break down subsequent requests to load data into groups and select an appropriate server station.

3.3. Server status data collecting procedure

In order to implement the data collection procedure, a special program was created. This program consists of three main components: agent, model, router. Let's consider each of components. The agent is a subroutine that polls the server about its state. The model generates the data received from the agent in the required form, corresponding to the expression (1). Router provides data to server balancer. Code snippet of each component is shown in figure 3.

```
var schedule = require('node-schedule');
var j = schedule.scheduleJob('*/1 * * * *', function(){
  var os = require('os-utils');
  var freeMemory = os.freememPercentage();
  var systemUptime = os.sysUptime();
  os.cpuUsage(function(v) {
    var cpuUsage = v;
    var diskspace = require('diskspace');
    diskspace.check('/', function (err, result) {
      var stat = new ModelStat();
      stat.set({server : config.get('ip'),
        freemem: freeMemory,
        diskFreeSpace: result.free,
        diskUsage: result.used,
        cpuUsage: cpuUsage,
        upTime: systemUptime,
        createdAt: new Date()
      });
      stat.save()
        .then(function () {})
        .catch(function (error) {
          var message = 'Can not save';
        });
    });
  });
});
```

a)

```
var express = require('express');
var router = express.Router();
var _ = require('lodash');
var http = require('http');
var response = require('../utils/response');
var ModelStat = require('../models/stat');
router.get('/', function (req, res) {
  var request = require('request');
  ModelStat.find(function(){});
  res.send('Hello World!');
});
module.exports = router;
```

b)

```
var http = require('http');
var response = require('../utils/response');
var ModelStat = require('../models/stat');
router.get('/', function (req, res) {
  var request = require('request');
  ModelStat.find(function(){});
  res.send('Hello World!');
});
module.exports = router;
```

c)

Figure 3. Code snippet: a) agent, b) model c) router

Each code snippet is structured to the best performance its function. As you can see, for the agent subroutine, we needed to organize nested calls to the query functions of the server parameters, the model was given a schema with the names of the database fields and their types, an http library for sending and receiving requests was attached to the router.

To develop a program we choose a programming language JavaScript Node.js [7]. Node.js is a runtime environment for executing JavaScript code on the server side. Node.js enables JavaScript to be used for server side scripting, and runs scripts server side to produce dynamic web page content before the page is sent to the user's web browser. Node.js adds to JavaScript ability to interact with input/output devices through the interface (developed on C++). Also Node.js allows us to include another external libraries, written on another programming languages. It provides them JavaScript abilities.

We made a research of different Node.js frameworks and chose ExpressJs. This framework is a modern high performance development tool.

As a database we used MongoDB [8]. This is one of the most developed open-source NoSQL solution. Also this database is a cross-platform, easy to use and provides high performance and availability.

4. The results of the experimental studies

For the experimental research, the developed program on the existing three server stations was installed. After the installation and launch of the program, data on the status of the server stations, in the processed form, was saved in the database of our system. Thus, we got a complete history of the status of each server station after the installation of the program. Figure 4 shows an example of displaying saved parameters (we chose such parameters as the percentage of hard drive usage and the percentage of the central process unit (CPU) usage) for three server stations.

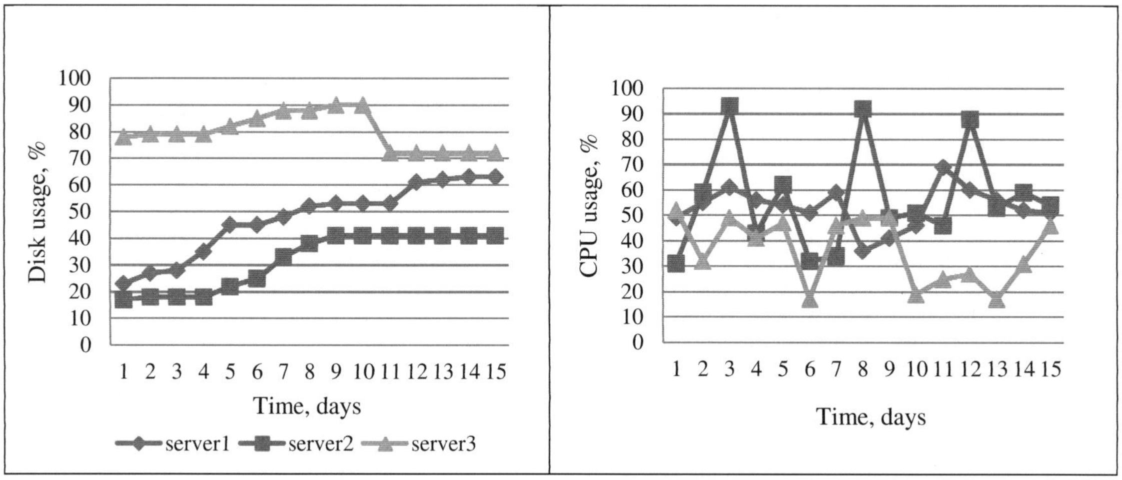

Figure 4. Example of displaying server station parameters

The received data represents the dependences of parameters of functioning of server stations on time. This data can be used to analyze the status of server stations, as well as to perform the distribution of user requests for file downloads. So, we will use the server state data presented in the form (1) at the next stage of the system: at the server selection stage.

5. The discussion of the results

The proposed approach to processing data on the state of server stations differs from the known approaches [9] that the data obtained has a structure suitable for further optimization of server selection. Also the records recorded in the database completely reflect the state of the system at a particular point in time. We believe that it is appropriate to use the multidimensional statistical procedure of cluster analysis [10] for the final processing of data. This procedure will allow us to split the received data into groups with similar parameters and add each new request to one of the groups, thereby determining the appropriate server for it.

6. Conclusion

The article suggests an approach to information processing when designing data distribution systems for highly loaded web applications. A model has been developed and it allows describing the status of server stations at the current time. A structure for presenting the parameters of the current state of the system suitable for further optimization of the server selection has been proposed .As a direction for the further research, we are planning to optimize the distribution of the user data on server stations. To do this, we will use not only the received data on the parameters of the server stations, but also the data about the end users who perform uploading or viewing downloaded files. The task of data distribution will be solved using the methods of cluster analysis.

References

[1] Suherman, Zarlis M, Stirous S And Al-Akaidi M 2016 Applying a rateless code in content delivery networks *IOP Conference Series: Materials Science and Engineering* **237** conf 1
[2] *What is load balancer? How load balancers work.* URL: https://www.nginx.com/resources/glossary/load-balancing/ (reference date 12.10.2017)
[3] Membrey P, Plugge E and Hows D 2013 *Practical Load Balancing: Ride the Performance Tiger (Expert's Voice in Networking)*
[4] *Load Balancing Techniques and Algorithms* URL: https://kemptechnologies.com/load-balancer/load-balancing-algorithms-techniques/ (reference date 12.10.2017)
[5] *Server load balancing | Akamai* URL: https://www.akamai.com/uk/en/resources/server-load-balancing.jsp (reference date 23.10.2017)
[6] Wilson S, Prakash P and Deepalakshmi P 2017 Server-based Dynamic Load Balancing *IEEE Networks & Advances in Computational Technologies (NetACT) International Conference* (Thiruvanthapuram)
[7] *Node.js* URL: https://nodejs.org (reference date 20.09.2017)
[8] *MongoDB for GIANT Ideas | MongoDB* URL: https://mongodb.com (reference date 20.09.2017)
[9] Sahoo J, Salahuddin M and Roch G 2016 A survey on Replica Placement Algorithms for Content Delivery *IEEE Communications Surveys and Tutorials* **19** pp 1002–1026
[10] Knyazev N L and Denisova L A 2018 The detection methods of dynamic objects *J. Phys.: Conf. Ser.* **944** 012051

IOP Conf. Series: Journal of Physics: Conf. Series **1050** (2018) 012098 doi:10.1088/1742-6596/1050/1/012098

High-accuracy finite element

A V Yakovlev, O D Kazakov, E G Ryzhikova and N A Afanasyeva

Bryansk State Technological University of Engineering, Bryansk, Russia

Abstract. The article deals with determining the rigidity matrix of a triangular finite element. Displacement vector components and displacement derivatives are used as nodal degrees of freedom. These finite elements have improved accuracy as they link with continuity conditions not only the displacement field but also derivative fields, which significantly simplifies nodal stress analysis and makes it possible to establish stress boundary conditions. The test calculations verified for a beam were provided.

Key words. Plane elasticity model, finite deformations, displacements, stresses, overhang beam calculations.

1. Introduction

At present a large number of diverse finite elements are used in the finite element method. Triangular finite elements should be considered exhibiting the most promise for membrane solutions of elasticity theory. Firstly, these finite elements provide a more flexible element discretization in the region under study. Secondly, the triangular region has certain advantages in terms of solving the two-dimensional interpolation problem.

Figure 1 features a high-accuracy finite element in which displacements and their derivatives are used as degrees of freedom. This finite element must possess improved accuracy, as it links with continuity conditions not only displacement fields but also deformation fields [1].

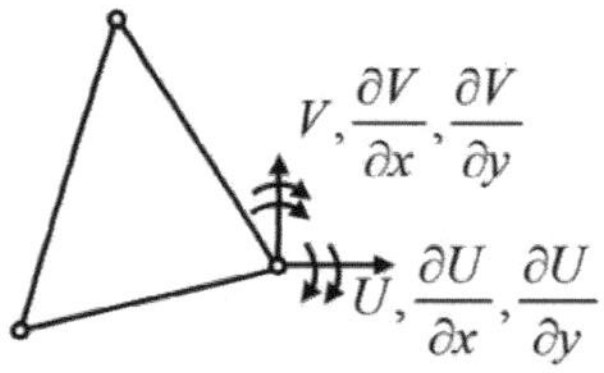

Figure 1. High accuracy finite element

Taking the displacement field differential as a degree of freedom simplifies the nodal stress analysis. Genuinely, deformation tensor components and nodal stress tensor components are expressed with the first-order derivatives of the displacement field. For this very reason, there is an opportunity of setting the stress boundary conditions.

2. Displacement fields

Let us denote the triangle vertices (figure 2) by letters i, j and k. For every triangular node there are coordinates (X_i, Y_i), (X_j, Y_j) and (X_k, Y_k).

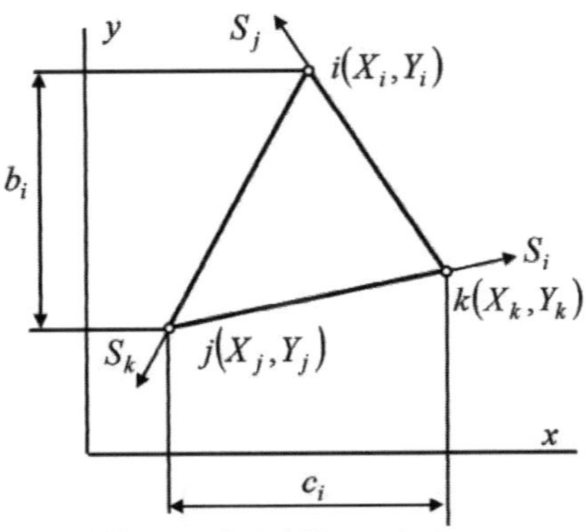

Figure 2. Differentiation

It is more natural to use L-coordinates for a triangular finite element. In this case the displacement field inside the finite element can be described by the pair of homogeneous cubic polynomials:

$$\left.\begin{aligned} u(L)&=\alpha_1 L_i^3+\alpha_2 L_j^3+\alpha_3 L_k^3+\alpha_4 L_i^2 L_j+\alpha_5 L_j^2 L_k+\alpha_6 L_k^2 L_i+\alpha_7 L_j^2 L_i+\alpha_8 L_j^2 L_k^2+\alpha_9 L_i^2 L_k+\alpha_{10} L_i L_j L_k \\ v(L)&=\beta_1 L_i^3+\beta_2 L_j^3+\beta_3 L_k^3+\beta_4 L_i^2 L_j+\beta_5 L_j^2 L_k+\beta_6 L_k^2 L_i+\beta_7 L_j^2 L_i+\beta_8 L_j^2 L_k^2+\beta_9 L_i^2 L_k+\beta_{10} L_i L_j L_k \end{aligned}\right\} \quad (1)$$

where L_i, L_j, L_k $-$ L-coordinates [2].

Dependences (1) in a matrix form:

$$\{u(L)\}=[\alpha_{10}]\{L_{10}\}. \quad (2)$$

At this stage, the derivatives of displacements along triangle sides $S_r (r=i,\ j,\ k)$ will be used as degrees of freedom. To do this, it is necessary to differentiate the dependences (2).

$$\frac{\partial\{u(L)\}}{\partial S_r}=\frac{\partial}{\partial S_r}[\alpha_{10}]\{L_{10}\}=[\alpha_{10}]\frac{\partial\{L_{10}\}}{\partial S_r}. \quad (3)$$

Thus, the vector-valued function differentiation $\{u(L)\}$ reduced to the differentiation of the vector-valued function $\{L_{10}\}$. Let us find the corresponding derivatives. For which purpose, let us assign S_r coordinate starting in the least significant node to each side $l_r (r=i,\ j,\ k)$ (figure 2).

Let us consider side jk of the triangle. The relationship between S_i coordinate and Cartesian system x, y is expressed by the obvious formulae:

$$\left.\begin{aligned} x&=\frac{c_i}{l_i}S_i+X_j \\ y&=-\frac{b_i}{l_i}S_i+Y_j \end{aligned}\right\}. \quad (4)$$

Similar formulae can be written for other sides. In this case the derivative of a composite function with respect to coordinate S_r $(r=i,\ j,\ k)$ will be written as follows:

$$\frac{\partial\{L_{10}\}}{\partial S_r}=\frac{\partial\{L_{10}\}}{\partial L_i}\cdot\frac{\partial L_i}{\partial S_r}+\frac{\partial\{L_{10}\}}{\partial L_j}\cdot\frac{\partial L_j}{\partial S_r}+\frac{\partial\{L_{10}\}}{\partial L_k}\cdot\frac{\partial L_k}{\partial S_r}. \quad (5)$$

Functions $L_p (p=i,\ j,\ k)$ included within (5) are composite functions with respect to coordinates x, y. Thus:

IOP Conf. Series: Journal of Physics: Conf. Series **1050** (2018) 012098 doi:10.1088/1742-6596/1050/1/012098

$$\frac{\partial L_p}{\partial S_r} = \frac{\partial L_p}{\partial x} \cdot \frac{\partial x}{\partial S_r} + \frac{\partial L_p}{\partial y} \cdot \frac{\partial y}{\partial S_r}.$$

Bearing in mind dependence (4)

$$\frac{\partial L_p}{\partial S_r} = \frac{1}{2 \cdot \Delta \cdot l_r}\left(b_p c_r - b_r c_p\right).$$

For a triangle the following correlations are known:

$$c_i b_k - c_k b_i = c_j b_i - c_i b_j = c_k b_i - c_j b_k = a_i + a_j + a_k = 2\Delta .$$

Bearing in mind these correlations, for different combinations p and r from the set i, j, k, we obtain:

$$\frac{\partial L_p}{\partial S_r} = \begin{cases} l_r^{-1} & \text{where} \quad r = i,\ p = k; \quad r = j,\ p = i; \quad r = k,\ p = j \\ -l_r^{-1} & \text{where} \quad r = i,\ p = j; \quad r = j,\ p = k; \quad r = k,\ p = i . \\ 0 & \text{where} \quad r = p \end{cases}$$

Applying these dependences into (5), we obtain the derivatives of particular variables S_i, S_j, and S_k:

$$\left.\begin{aligned} \frac{\partial\{L_{10}\}}{\partial S_i} &= \frac{1}{l_i}\left(-\frac{\partial\{L_{10}\}}{\partial L_j} + \frac{\partial\{L_{10}\}}{\partial L_k}\right) \\ \frac{\partial\{L_{10}\}}{\partial S_j} &= \frac{1}{l_j}\left(-\frac{\partial\{L_{10}\}}{\partial L_k} + \frac{\partial\{L_{10}\}}{\partial L_i}\right) \\ \frac{\partial\{L_{10}\}}{\partial S_k} &= \frac{1}{l_k}\left(-\frac{\partial\{L_{10}\}}{\partial L_i} + \frac{\partial\{L_{10}\}}{\partial L_j}\right) \end{aligned}\right\}.$$

These dependences can be written as the product of some matrix $[T_r](r \in i,\ j,\ k)$, the structure of which is obvious, by vector $\{L_6\}$:

$$\frac{\partial\{L_{10}\}}{\partial S_r} = \frac{1}{l_r}[T_r]\{L_6\}, \text{ where } r \in i,\ j,\ k . \tag{6}$$

Where:

$$\{L_6\}^T = \{L_i^2,\ L_j^2,\ L_k^2,\ L_i L_j,\ L_j L_k,\ L_k L_i\}$$

Applying (6) to (3), we obtain:

$$\frac{\partial\{u(L)\}}{\partial S_r} = \frac{1}{l_r}[\alpha_{10}][T_r]\{L_6\}, \text{ where } r \in i,\ j,\ k . \tag{7}$$

3. Nodal displacement vectors

Let us introduce a nodal displacement vector. With this purpose, let us accomplish some transformations (7). Let us multiply (7) by the side length of a triangle l_r. As a result, we will obtain:

$$\frac{\partial\{u(L)\}}{\partial S_r} l_r = [\alpha_{10}][T_r]\{L_6\}. \tag{8}$$

Let us introduce the definition of function $\{u(L)\}$ along an arbitrary side of a triangle l_r.

$$\{u(l)\}_{,,r} = \frac{\partial\{u(L)\}}{\partial S_r} l_r .$$

In this case, the dependence (8) takes the form:

$$\{u(l)\}_{,,r} = [\alpha_{10}][T_r]\{L_6\}. \tag{9}$$

The new idea of derivative provides the independence of matrix elements $[\alpha_{10}]$ from the triangle side dimension and makes dependence (9) universal for any finite element in the process of differentiation along any of its sides.

Now it is possible to define the nodal displacement values. Let us introduce two local vectors of nodal displacements with the elements ordered along the directions x and y.

$$\left.\begin{aligned}\{U\}_s^T &= \{U_i, U_{i,,j}, U_{i,,k}, U_j, U_{j,,k}, U_{j,,i}, U_k, U_{k,,i}, U_{k,,j}\}\\ \{V\}_s^T &= \{V_i,\ V_{i,,j},\ V_{i,,k},\ V_j, V_{j,,k},\ V_{j,,i},\ V_k,\ V_{k,,i},\ V_{k,,j}\}\end{aligned}\right\}.$$

Where s factor shows that the derivatives are taken along the corresponding sides of a triangle according to formula (9). Taken together, these vectors form a global displacement vector for a triangular finite element with the components ordered in the directions x and y.

$$\{\overline{U}\}_s^T = \{\{U\}_s^T,\ \{V\}_s^T\}.$$

Let us also introduce three local vectors of nodal displacements for node p of a triangular finite element

$$\{U_p\}_s^T = \{U_p, U_{p,,r}, U_{p,,t}, V_p, V_{p,,r}, V_{p,,t}\},$$

where $p \in i, j, k;\quad r \in i, j, k;\quad t \in i, j, k$. Taken together, these three vectors form a global vector of nodal displacements for a finite element whose components are ordered by nodes

$$\{\overline{U}_p\}_s^T = \{\{U_i\}_s^T,\ \{U_j\}_s^T,\ \{U_k\}_s^T\}.$$

Let us relate vectors $\{\overline{U}\}_s$ and $\{\overline{U}_p\}_s$ by matrix $[E_1]$:

$$\{\overline{U}\}_s = [E_1]\{\overline{U}_p\}_s.$$

The structure of matrix $[E_1]$ is obvious.

The vector of nodal displacements with derivatives along the sides of a triangular finite element will be used for defining matrix elements $[\alpha_{10}]$. However, it is impossible to use this vector for rigidity matrix and stress determination. In this instance a displacement vector is necessary, whose components will be derivatives with respect to x and y arguments. Let us introduce the following definition of derivatives of function $\{u(L)\}$ along coordinate axes x and y.

$$\{u(L)\}_{,1} = \frac{\partial\{u(L)\}}{\partial x},\quad \{u(L)\}_{,2} = \frac{\partial\{u(L)\}}{\partial y}. \tag{10}$$

In accordance with (10) let us determine three local vectors of nodal displacements for p node of a triangular finite element:

$$\{U_p\}_x^T = \{U_p,\ U_{p,1},\ U_{p,2},\ V_p,\ V_{p,1},\ V_{p,2}\}, \tag{11}$$

where: $p \in i, j, k$. x factor in (11) shows that the derivatives are taken along coordinate axes x and y.

From (8):

$$\begin{aligned}\{u(L)\}_{,,r} &= \frac{\partial\{u(L)\}}{\partial S_r} l_r = l_r\left(\frac{\partial\{u(L)\}}{\partial x}\frac{\partial x}{\partial S_r} + \frac{\partial\{u(L)\}}{\partial y}\frac{\partial y}{\partial S_r}\right)\\ &= l_r\left(\frac{\partial\{u(L)\}}{\partial x}c_r - \frac{\partial\{u(L)\}}{\partial y}b_r\right).\end{aligned}$$

Now it is possible to find the relationship between vectors $\{U_p\}_s$ and $\{U_p\}_x$ by matrix $\lfloor M_p \rfloor_{sx}$, so that:

$$\{U_p\}_s = \lfloor M_p \rfloor_{sx}\{U_p\}_x \tag{12}$$

where:

$$[M_p]_{sx} = \begin{bmatrix} [M_i]_{sx} & [0_6] & [0_6] \\ [0_6] & [M_j]_{sx} & [0_6] \\ [0_6] & [0_6] & [M_k]_{sx} \end{bmatrix}.$$

Here $[0_6]$ zero matrix has the size of 6×6.

$$[M_i]_{Sx} = \begin{bmatrix} 1 & 0 & 0 \\ 0 & c_j & -b_j \\ 0 & c_k & -b_k \end{bmatrix} \quad [M_j]_{Sx} = \begin{bmatrix} 1 & 0 & 0 \\ 0 & c_k & -b_k \\ 0 & c_i & -b_i \end{bmatrix} \quad [M_k]_{Sx} = \begin{bmatrix} 1 & 0 & 0 \\ 0 & c_i & -b_i \\ 0 & c_j & -b_j \end{bmatrix}$$

Let us find α_l, β_l factors (where $i \in 1, 2, \ldots, 10$) of interpolation polynomials (1). These factors will be obtained from continuity conditions for displacements and derivatives along the triangle sides in its nodes. This procedure is well-known, therefore let us confine ourselves to the ultimate result.

First and foremost, let us write the interpolation polynomial (4) determining the displacement field $u(L)$, and, according to formula (12), let us find two interpolation polynomials describing the fields of derivatives of displacements inside the finite element. Thus, we will obtain:

$$\{u(L)\} = [\Omega(L)]\begin{bmatrix} [Z] & [0] \\ [0] & [Z] \end{bmatrix}\{\overline{U}\}_s.$$

Where $\{u(L)\} = \begin{Bmatrix} u(L) \\ v(L) \end{Bmatrix}$, $[\Omega(L)] = \begin{bmatrix} \{L_{10}\}^T & \{0_{10}\}^T \\ \{0_{10}\}^T & \{L_{10}\}^T \end{bmatrix}$, $\{\overline{U}\}_s = \begin{Bmatrix} \{U\}_s \\ \{V\}_s \end{Bmatrix}$.

Here matrix $[Z]$ contains factors α_l, β_l and links the displacement vector components inside the finite element with nodal displacements.

Let us introduce the shape function matrix:

$$[N(L)] = [\Omega(L)]\begin{bmatrix} [Z] & [0] \\ [0] & [Z] \end{bmatrix}.$$

Then ultimately:

$$\{u(L)\} = [N(L)]\{\overline{U}_p\}_s. \tag{13}$$

In order to simplify the computation of stress tensor, it is worthwhile to employ the nodal displacement vector $\{\overline{U}_p\}_x$, the elements of which are derivatives of displacements along coordinate axes x and y. To do this, let us apply dependence (12) to (13):

$$\{u(L)\} = [N(L)][M]_{sx}\{\overline{U}_p\}_x. \tag{14}$$

4. Deformations and stresses

Let us introduce the deformation vector:

$$\{\varepsilon\}^T = \{\varepsilon_x, \ \tau_{xy}, \ \varepsilon_y\} = \{\varepsilon_{11}, \ \varepsilon_{12}, \ \varepsilon_{22}\}.$$

The Cauchy equations:

$$\{\varepsilon(L)\} = [S]\{u(L)\}. \tag{15}$$

Let us apply (14) to (15):

$$\{\varepsilon(L)\} = [S][N(L)][M]_{sx}\{\overline{U}_p\}_x.$$

The influence of matrix $[A]$ on matrix $[N(L)]$ will give rise to gradient matrix $[B(L)]$:

$$\{\varepsilon(L)\} = [B(L)][M]_{sx}\{\overline{U}_p\}_x. \tag{16}$$

Equation (16) provides the sought dependence between deformations and nodal replacements.

Let us introduce the stress vector:

$$\{\sigma\}^T = \{\sigma_x \quad \tau_{xy}, \quad \sigma_y\} = \{\sigma_{11}, \quad \sigma_{12}, \quad \sigma_{22}\}.$$

Hooke's law:

$$\{\sigma(L)\} = [D]\{\varepsilon(L)\}, \tag{17}$$

where $[D]$ is the elasticity matrix having the size of 3×3. The sought dependence between the stress tensor components and displacement vector can be obtained if equations (16) are applied to (17):

$$\{\sigma(L)\} = [D][B(L)][M]_{sx}\{\overline{U}_p\}_x. \tag{18}$$

5. System of resolving finite element equations

Let us introduce the following notations:

A – work of external forces

Π – potential energy of deformed body

Λ – energy of external and internal forces system

To determine the resolving finite-element equations let us use the Lagrange formula:

$$\delta\Lambda = 0,$$
$$\Lambda = \Pi - A. \tag{19}$$

Potential energy is determined by the Clapeyron formula:

$$\Pi = \frac{1}{2}\int_V \{\varepsilon\}^T \{\sigma\} dV. \tag{20}$$

Let us find potential energy Π^ν for a finite element with the number ν, applying equations (16) and (18) to (20):

$$\Pi^\nu = \frac{1}{2}\int_V \{\overline{U}_p\}_x^T [M^\nu]_{sx}^T [B^\nu(L)]^T [D^\nu][B^\nu(L)][M^\nu]_{sx}\{\overline{U}_p\}_x dV. \tag{21}$$

The work of external forces for node p of a finite element numbered ν:

$$A_p^\nu = \{\widetilde{U}_p^\nu\}^T \{P_p^\nu\},$$

By means of matrix $[F_2]$, let us introduce vector $\{U_p\}_x$ from (11) to dependence (21).

$$A_p^\nu = \{U_p\}_x^T [F_2]\{P_p^\nu\},$$

where: $[F_2] = \begin{bmatrix} 1 & 0 & 0 & 0 & 0 & 0 \\ 0 & 0 & 0 & 1 & 0 & 0 \end{bmatrix}$.

Then, for the finite element numbered ν:

$$A^\nu = \sum_{p=1}^{3} \{U_p^\nu\}^T [F_2]\{P_p^\nu\}.$$

Summating all the finite elements, we will obtain the energy of the system of external and internal forces (19):

$$\Lambda = \frac{1}{2}\sum_{\nu=1}^{E}\int_V \{\overline{U}_p\}_x^T [M^\nu]_{sx}^T [B^\nu(L)]^T [D^\nu][B^\nu(L)][M^\nu]_{sx}\{\overline{U}_p\}_x dV - \sum_{\nu=1}^{E}\sum_{p=1}^{3}\{U_p^\nu\}^T [F_2]\{P_p^\nu\}.$$

Minimization of the obtained functional results in the system of resolving finite-element equations:

$$\sum_{\nu=1}^{E}\int_V [M^\nu]_{sx}^T [B^\nu(L)]^T [D^\nu][B^\nu(L)][M^\nu]_{sx}\{\overline{U}_p\}_x dV = \sum_{\nu=1}^{E}\sum_{p=1}^{3}[F_2]\{P_p^\nu\}. \tag{22}$$

6. Finite element rigidity matrix

The integral on the right side of (22) takes the form of rigidity matrix of finite element $[k]$. Assuming that the finite element width varies linearly, let us write:

$$h(L) = h_i L_i + h_j L_j + h_k L_k \, .$$

where: h_i, h_j, h_k – the finite element width in nodes x i, j, k and assuming that the elements of matrix $\left[M^V\right]_{sx}$ are constant values:

$$\left[k^V\right] = \left[M^V\right]_{sx}^T \int_V \left[B^V(L)\right]^T \left[D^V\right]\left[B^V(L)\right] dV \left[M^V\right]_{sx} \, . \quad (23)$$

The integration in (23) is performed numerically, with 12 points [3].

7. Test calculation

Using the dependences obtained, the authors wrote the program for testing the finite element rigidity matrix and made calculations for a hammer beam displacements (Fig.3). The calculation is of purely mathematical character, therefore dimensions of quantity are not provided.

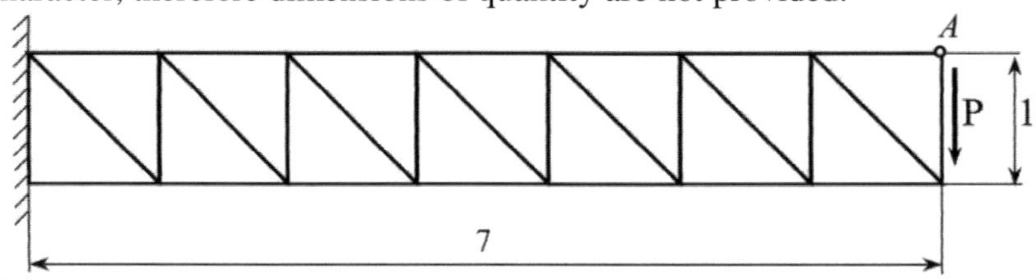

Figure 3. Displacement model of a hammer beam for finite element testing
Elasticity modulus $E = 1000$, Poisson's ratio $\mu = 0.3$

Displacement calculations for point A of a beam (figure. 3), according to exact formulae, amounted to 2.744, but when calculated according to the finite element method, the displacement value amounted to 2.738. Thus, calculation error amounted to 0.22%.

8. Conclusions

In the work presented here the authors examined in detail how to determine the rigidity matrix of a high-accuracy finite element with six nodal degrees of freedom, which is necessary for membrane solutions within the framework of elasticity theory. The test calculation verified the The integration in (23) is performed numerically, with 12 pointsa tetrahedral finite element with twelve nodal degrees of freedom for solving multi-dimensional problems.

References

[1] Gallagher R 1975 Finite element analysis. Fundamentals (Prentice-Hall Publ.) 416 p
[2] Segerlind L 1976 Applied Finite Element Analysis (New York: Wiley)
[3] Strang G and Fix G J 1973 An Analysis of the Finite Element Method (Prentice Hall)

IOP Conf. Series: Journal of Physics: Conf. Series **1050** (2018) 012099 doi:10.1088/1742-6596/1050/1/012099

Growing graphs with addition of communities

E B Yudin

Sobolev Institute of Mathematics, Siberian Branch of the Russian Academy of Sciences, 4 Acad. Koptyug avenue, 630090, Novosibirsk, Russia

The reported study was funded by Russian Foundation for Basic Research, according to the research project No 16-31-60023 mol_a_dk

E-mail: udinev@asoiu.com

Abstract. Paper proposes a model of large networks based on a random preferential attachment graph with addition of complete subgraphs (cliques). The proposed model refers to models of random graphs following the nonlinear preferential attachment rule, and takes into account the possibility of «adding» entire communities of nodes to the network. In the derivation of the relations that determine the vertex degree distribution, the technique of finite-difference equations describing stationary states of a graph is used. The obtained results are tested empirically (by generating large graphs), special cases correspond to known mathematical relations.

1. Introduction

Structures studied by Network Science have different nature, but similar properties, among them the researchers distinguish the following ones:

– vertex degree in network follows a power-law distribution,

– networks have a small diameter, often decreasing with increasing network size («six handshakes» rule),

– there are communities (some groups of nodes connected to each other more than to nodes from other communities) in networks.

As models of large networks, random graphs with preferential attachment are often used [1]. Such models of growing networks allow us to generate graphs that implement:

– a given vertex degree distribution;

– a given vertex degree distribution and a given clustering coefficient [2-5];

– a given vertex degree distribution and a joint distribution of the edge end degrees [6].

It is convenient to consider the scheme of generation of preferential attachment graphs by the example of a random graph following the nonlinear preferential attachment rule (NPA graph). Generation of a NPA graph (figure 1) is an iteration process. At each step of the process an increment is added to the existing graph, an increment is a vertex with a random number of edges, whose free ends are attached to the i-th vertex of the existing graph with a probability proportional to the preference function $f(k_i)$, depending on the vertex degree k_i of the graph $p_i = f(k_i)[\sum_{i=1}^{n} f(k_i)]^{-1}$, $i = 1,\ldots, N$, where N is the number of the vertices.

If the number of edges in the increment is a constant m, and the preference function is linear: $f(k_i) = k_i$, then we obtain a well-known Barabashi-Albert graph (BA) [1], that follows the principle «rich becomes richer» (the more the vertex degree, the greater probability of attaching to it). If the preference function is logarithmic: $f(k_i) = \log(k_i)$, then the law of Weber-Fechner is simulated, according to which the intensity of the perception is proportional to the logarithm the original stimulus intensity.

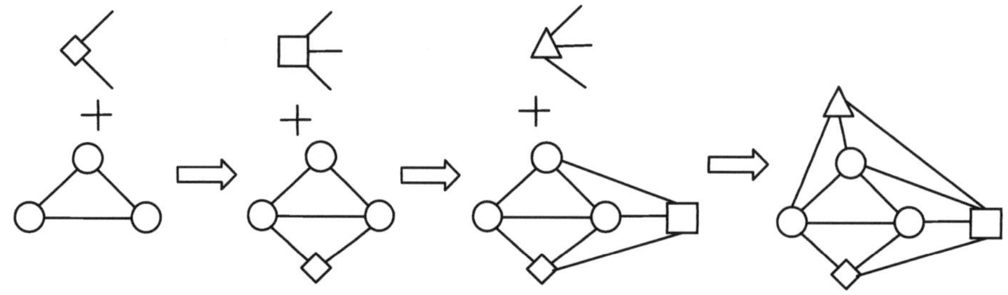

Figure 1. Generation of NPA graphs: vertices with a random number of edges are added to the seed graph. In the figure a vertex with two edges is added, then a vertex with three edges is added twice

However, in the graphs of preferential attachment (especially in the most studied graph of preferential attachment - the BA graph) the communities are absent as a rule, in some ways the graph is homogeneous. Even for small size complete subgraphs (cliques) containing only three vertices, in BA graphs the number of such «triangular» cliques is significantly smaller in comparison to their number in real networks with commensurate number of nodes and edges [3]. A fortiori, it is true for the number of cliques of larger size. This paper proposes preferential attachment graphs, in which this problem is solved. For these graphs it is assumed that the community is added to the network as a whole, thereby the stage of «community increment» is simulated [7]. A community increment is a given number n of nodes, any two nodes of which are linked.

2. Theory

Graphs with the addition of complete subgraphs. A graph with the addition of complete subgraphs is a modified version of a NPA graph. Suppose that at every step of graph growth with a probability γ a complete graph containing n nodes (n-ad, i.e. polyad) is added, with the probability $(1-\gamma)$ – the usual increment of the graph – monad (figure 2). The number of free edges of the monad is determined in accordance with the discrete distribution $\{r_k^{(1)}\}$ and for n-ad – in accordance with the distribution $\{r_k^{(n)}\}$. Free ends of the edges of the monad are joined to the graph independently, according to the probability

$$p_i = \frac{f(k_i)}{\sum_{j=1}^{N} f(k_j)}, \quad i = 1, \ldots, N, \tag{1}$$

where N – is the number of vertices of the graph, the preference function $f(k) > 0$, if $g \le k \le M$, otherwise $f(k) = 0$.

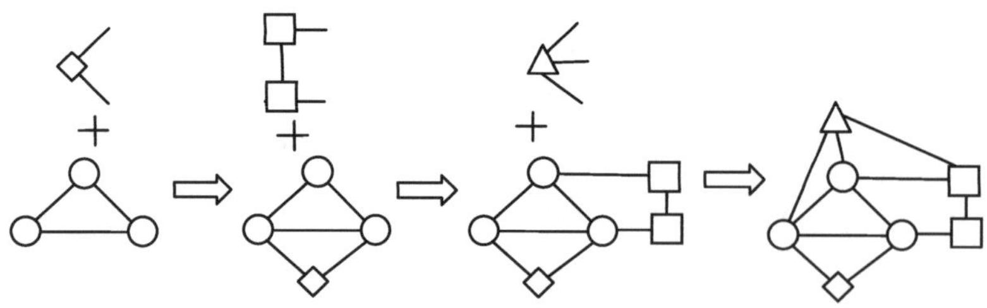

Figure 2. Generation of NPA graphs with communities increment, at each step of the graph generation a two-vertex increment (dyad) is added with probability γ or a single-vertex increment (monad) is added with probability $(1-\gamma)$

Each of the μ free ends of a vertex in n-ad (figure 3) is attached to the same vertex, played according to the formula (1). The other free edge ends of the n-ad attach independently to the existing graph vertices with probability (1).

The following properties can be written for the proposed graph.

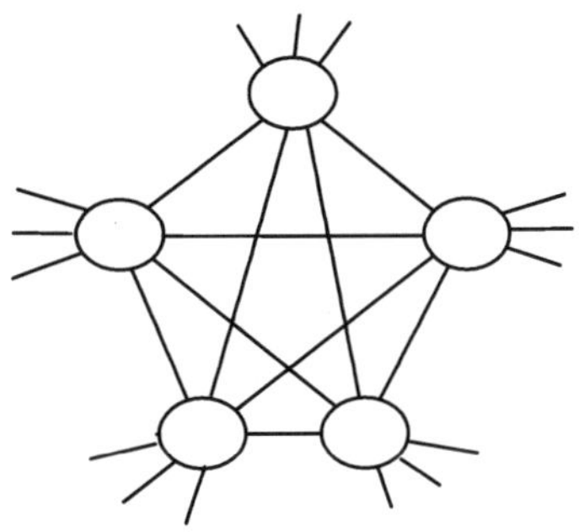

Figure 3. A complete graph with five vertices as an increment (pentad) with 3 played free edges and 10 included in clique.

1) On average at one step $(1+(n-1)\gamma)$ vertices are attached.

2) Denote the average number of edges in a monad as $m^{(1)}$, then $m^{(1)} = \sum_{k=1}^{t} i \cdot r_i^1$, where t is the maximum value of the monad degree.

3) The average number of edges in n-ad $(n \cdot m^{(n)} + n(n-1)/2)$, free (not included in the clique of size n) edges $n \cdot m^{(n)}$, where $m^{(n)}$ is the average number of free edges of each vertices in the n-ad.

4) At one step, on average, $\gamma \cdot (n \cdot m^{(n)} + m^{(n)} \cdot (n-1)/2) + (1-\gamma) \cdot m^{(1)}$ edges are attached to a graph.

The following tasks are set.

1) Within the framework of the proposed modified version of a NPA graph to derive equations that will allow us to calculate the vertex degree distribution (VDD) $\{Q_k\}$ of the generated graph.

2) Within the framework of the proposed modified version of a NPA graph to derive equations that will allow us for a given VDD of real network to select generation parameters that take into account the process of communities increment.

Derivation of the vertex degree distribution. Let us divide the set of all vertices in the existing graph into subsets A_k (layers) of vertices, containing vertices with the same vertex degree k $(k = m, m+1, \ldots)$. The probability Q_k of a layer A_k is defined as the probability of attaching to the layer in a case of random (equiprobable) choice of any of the N vertices in the existing graph: $Q_k = |A_k| / N$, where $|A_k|$ is a number of vertices with the vertex degree k.

The final value P_k of the probability $\sum_{i \in A_k} p_i$ that a vertex of the layer A_k will be chosen to attach to an edge of a new vertex can be obtained from formula (1):

$$\sum_{i \in A_k} p_i = \frac{\sum_{i \in A_k} f(k_i)}{\sum_{j=1}^{N} f(k_j)} = \frac{f(k)|A_k|}{\sum_{l \geq g} f(l)|A_l|} = \frac{f_k |A_k| / N}{\sum_{l \geq g} f_l |A_l| / N} \sim \frac{Q_k f_k}{\sum_{l \geq g} Q_l f_l} = \frac{Q_k f_k}{\langle f \rangle} = P_k .$$

Thus, the probability P_k of selecting a layer A_k can be calculated as

$$P_k = \frac{Q_k f_k}{\langle f \rangle}, \text{ where } \langle f \rangle = \sum_{l \geq g} Q_l f_l \text{ – an average value of the preference function.} \tag{2}$$

Let us consider the factors that influence $|A_k|$ when a monad attaches.

1) If a monad attaches, then $|A_k|$ with probability r_k. increases by 1. On the average $|A_k|$ due to this increases by $1 \cdot r_k^{(1)} + 0 \cdot (1 - r_k^{(1)}) = r_k^{(1)}$.

2) Moreover, on the average $m^{(1)} P_k$ monad edges attach to some vertices of the layer A_k and these vertices go to the layer A_{k+1}, i.e. A_k decreases on average by $m^{(1)} P_k$ and, for a similar reason, A_k increases on average by $m^{(1)} P_{k-1}$.

In total when adding monads

$$\Delta_1 |A_k| = r_k^{(1)} + m^{(1)} P_{k-1} - m^{(1)} P_k. \tag{3}$$

Let us consider the factors that influence $|A_k|$ when a n-ad (figure 4) attaches.

1) The vertices of the n-ad with probability $r_{k+1-n}^{(n)}$ attach to A_k, due to this fact $|A_k|$ increases by $n \cdot r_{k+1-n}^{(n)}$.

2) For each of the vertices in the n-ad it is true that μ outgoing edges attach with the probability P_{k-n} to some vertices of the layer A_{k-n} and these vertices go to the layer A_k. Due to this $|A_k|$ increases by μP_{k-n}.

3) On the average, μ pairs of conjugate edges attach to the vertices of the layer A_k and these vertices go to the layer A_{k+n}. It gives a decrease in $|A_k|$ by the value $-\mu P_k$.

4) The free edges in the n-ad attach to the vertices of the layer A_{k-1} with the probability P_{k-1}. The number of such single edges $n \cdot m^{(n)} - n \cdot \mu$. The increase in $|A_k|$ is equal $(n \cdot m^{(n)} - n \cdot \mu)\, P_{k-1}$.

5) The free edges in the n-ad attach to the vertices of the layer A_k with the probability P_k. A decrease in $|A_k|$ is equal $(n \cdot m^{(n)} - n \cdot \mu)\, P_k$.

In total we obtain

$$\Delta_2|A_k| = n \cdot r_{k+1-n}^{(n)} + \mu P_{k-n} - \mu P_k + (n \cdot m^{(n)} - n \cdot \mu)P_{k-1} - (n \cdot m^{(n)} - n \cdot \mu)P_k . \quad (4)$$

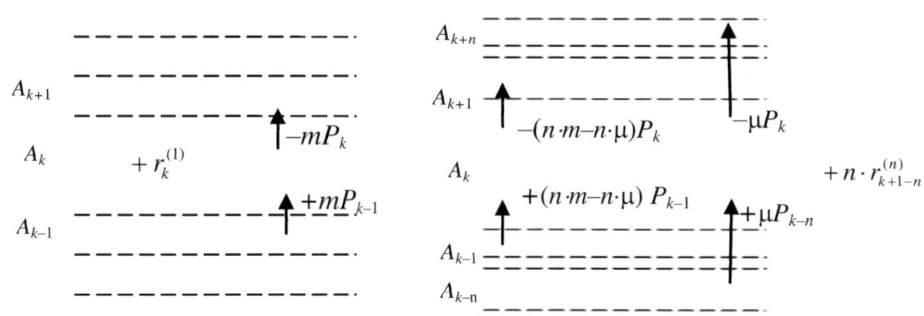

Figure 4. Changing the number of vertices in A_k by adding of monad (left) and n-ad (right)

At each step of the graph growing, n-ad appears with the probability γ, usual increment – with the probability (1 – γ). In view of this

$$\Delta|A_k| = (1-\gamma)\cdot\Delta_1|A_k| + \gamma\cdot\Delta_2|A_k| = (1-\gamma)(r_k^{(1)} + m^{(1)}P_{k-1} - m^{(1)}P_k)+$$
$$+ \gamma\cdot (n \cdot r_{k+1-n}^{(n)} + \mu P_{k-n} - \mu P_k + (n \cdot m^{(n)} - n \cdot \mu)P_{k-1} - (n \cdot m^{(n)} - n \cdot \mu)P_k).$$

Equating the fraction of vertices in the layer A_k before performing the step of graph growing to the corresponding fraction A_{k+1} after this step, we obtain the following balance equation:

$$\frac{|A_k|}{N} = \frac{|A_k| + \Delta|A_k|}{N + ((1-\gamma) + \gamma \cdot n)}.$$

Or

$$\frac{|A_k|}{N} = \frac{|A_k| + r_k^{(1)} + m^{(1)}P_{k-1} - m^{(1)}P_k - \gamma \cdot r_k^{(1)} - \gamma \cdot m^{(1)}P_{k-1} + \gamma \cdot m^{(1)}P_k}{N + ((1-\gamma) + \gamma \cdot n)} +$$
$$+ \frac{\gamma \cdot n \cdot r_{k+1-n}^{(n)} + \gamma \cdot \mu \cdot P_{k-n} - \gamma \cdot \mu \cdot P_k + \gamma \cdot (n \cdot m^{(n)} - n \cdot \mu)P_{k-1} - \gamma \cdot (n \cdot m^{(n)} - n \cdot \mu)P_k}{N + ((1-\gamma) + \gamma \cdot n)}.$$

Rewrite

$$|A_k|N + |A_k| \cdot ((1-\gamma) + \gamma \cdot n) = N|A_k| + N\big(r_k^{(1)} + m^{(1)}P_{k-1} - m^{(1)}P_k - \gamma \cdot r_k^{(1)} - \gamma \cdot m^{(1)}P_{k-1} +$$
$$+ \gamma \cdot m^{(1)}P_k + \gamma \cdot n \cdot r_{k+1-n}^{(n)} + \gamma \cdot \mu \cdot P_{k-n} - \gamma \cdot \mu \cdot P_k + \gamma(n \cdot m^{(n)} - n \cdot \mu)P_{k-1} - \gamma(n \cdot m^{(n)} - n \cdot \mu)P_k\big) =$$
$$= N \cdot |A_k| + N \cdot \big(r_k^{(1)}(1-\gamma) + \gamma \cdot n \cdot r_{k+1-n}^{(n)} + (m^{(1)}(1-\gamma) + \gamma \cdot n \cdot m^{(n)} - \gamma \cdot n \cdot \mu) \cdot P_{k-1} +$$
$$\gamma \cdot \mu \cdot P_{k-n} + (m^{(1)} \cdot (\gamma - 1) - \gamma \cdot n \cdot m^{(n)} + \gamma(n-1) \cdot \mu)P_k\big).$$

Let us reduce the identical terms $|A_k|N$ and divide the equality by N. As a result, the expression $|A_k|((\gamma - 1) + \gamma \cdot n)$ in the left part is transformed into $Q_k((\gamma-1) + \gamma \cdot n)$, and the multipliers N in the right part are reduced. Replacing all P_k in the right part of equality with the expression $P_k = \frac{Q_k f_k}{\langle f \rangle}$, we obtain the balance equation for Q_k:

$$Q_k \cdot \left(1 + \gamma(n-1) - \frac{f_k(m^{(1)}(\gamma - 1) - \gamma \cdot n \cdot m^{(n)} + \gamma(n-1)\mu)}{\langle f \rangle}\right) =$$

$$= \frac{(r_k^{(1)}(1-\gamma) + \gamma \cdot n \cdot r_{k+1-n}^{(n)})\langle f \rangle + (m^{(1)}(1-\gamma) + \gamma \cdot n \cdot m^{(n)} - \gamma \cdot n \cdot \mu - \gamma \cdot m^{(1)})Q_{k-1} + \gamma \cdot \mu \cdot Q_{k-n}}{\langle f \rangle}.$$

In explicit form for Q_k we can write the expression:

$$Q_k = \frac{(r_k^{(1)}(1-\gamma) + \gamma \cdot n \cdot r_{k+1-n}^{(n)})\langle f \rangle + (m^{(1)}(1-\gamma) + \gamma \cdot n \cdot m^{(n)} - \gamma \cdot n \cdot \mu) f_{k-1} Q_{k-1} + \gamma \cdot \mu \cdot f_{k-n} Q_{k-n}}{\langle f \rangle + \langle f \rangle \gamma(n-1) + f_k[m^{(1)} \cdot (1-\gamma) + \gamma \cdot n \cdot m^{(n)} - \gamma(n-1) \cdot \mu]}, \quad (5)$$

where $k = g, g+1, \ldots, M$, $\langle f \rangle = \sum_{k \ge g} f_k Q_k$.

If there is no n-ad, i.e. $\gamma = 0$, then we obtain a well-known formula [8]:

$$Q_k = \frac{r_k^{(1)} \langle f \rangle + m^{(1)} \cdot f_{k-1} Q_{k-1}}{\langle f \rangle + m^{(1)} \cdot f_k}.$$

If there is no monad, i.e. $\gamma = 1$, at $n = 2$ we obtain:

$$Q_k = \frac{2 \cdot r_{k-1}^{(2)} \langle f \rangle + (2m^{(n)} - 2\mu) \cdot f_{k-1} Q_{k-1} + \mu \cdot f_{k-2} Q_{k-2}}{2\langle f \rangle + 2 f_k \cdot m^{(2)} - f_k \mu}.$$

Calibration of the graph by the vertex degree distribution. In the problem of graph calibration it is necessary to find the parameters $\{r_k^{(1)}\}, \{r_k^{(n)}\}$, $\{f_k\}$, γ, μ at which the graph will have a given VDD $\{Q_k\}$. This problem can be solved using equation (5).
Expressing f_k from (5) we obtain the recurrence relation

$$f_k = \frac{[r_k^{(1)}(1-\gamma) + \gamma \cdot n \cdot r_{k+1-n}^{(n)}]\langle f \rangle}{a \cdot Q_k} - \frac{[\langle f \rangle + \langle f \rangle \gamma(n-1)]Q_k}{a \cdot Q_k} + \frac{[(m^{(1)}(1-\gamma) + \gamma \cdot n \cdot m^{(n)} - \gamma \cdot n \cdot \mu) f_{k-1} Q_{k-1} + \gamma \mu f_{k-n} Q_{k-n}]\langle f \rangle}{\langle f \rangle \cdot a \cdot Q_k}, \quad (6)$$

where $a = m^{(1)} \cdot (1-\gamma) + \gamma \cdot n \cdot m^{(n)} - \gamma(n-1) \cdot \mu$.
Relation (6) defines a sequence $\{f_k\}$ defining the preference function, each term of the sequence has a common multiplicative factor $\frac{\langle f \rangle}{a}$. It should be noted that the sequences $\{f_k\}$ and $\{Cf_k\}$, which differ by the factor $C > 0$, are equivalent, taking into account the formula (1). Therefore, we can equate the factor $\frac{\langle f \rangle}{a}$ on the right side of the relation (6) to one (thus determining the equality $\langle f \rangle = a$) and write this ratio in a more compact form:

$$f_k = \frac{r_k^{(1)}(1-\gamma) + \gamma n r_{k+1-n}^{(n)} - (1+\gamma(n-1))Q_k}{Q_k} + \frac{(m^{(1)}(1-\gamma) + \gamma n m^{(n)} - \gamma n \mu) f_{k-1} Q_{k-1} + \gamma \mu f_{k-n} Q_{k-n}}{\langle f \rangle \cdot Q_k}. \quad (7)$$

IOP Conf. Series: Journal of Physics: Conf. Series **1050** (2018) 012099 doi:10.1088/1742-6596/1050/1/012099

Taking into account that as a result of this choice we have equality $\langle f \rangle = a$, and the parameter $a = m^{(1)} \cdot (1-\gamma) + \gamma \cdot n \cdot m^{(n)} - \gamma(n-1) \cdot \mu$, we obtain the following recurrence relation

$$f_k = \frac{r_k^{(1)}(1-\gamma) + \gamma \cdot n \cdot r_{k+1-n}^{(n)} - (1+\gamma(n-1))Q_k}{Q_k} + \\ + \frac{(m^{(1)}(1-\gamma) + \gamma \cdot n \cdot m^{(n)} - \gamma \cdot n \cdot \mu) f_{k-1} Q_{k-1} + \gamma \mu f_{k-n} Q_{k-n}}{[m^{(1)} \cdot (1-\gamma) + \gamma \cdot n \cdot m^{(n)} - \gamma(n-1) \cdot \mu] \cdot Q_k}, \tag{8}$$

where $k = g, g+1, \ldots, M$.

Formula (8) is the solution of graph calibration problem. To realize a given VDD $\{Q_k\}$ it allows to obtain infinitely many weights sequences $\{f_k\}$, which actually set the preference function $f(k)$, depending on which parameters $\{r_k\}$, γ, μ and m are chosen.

3. Numerical experiments

Figure 5 shows a graph calibration using pentads (complete graphs in increments that contain 5 vertices) and monads. The probability of pentad occurrence $\gamma = 0.01$. As a basic vertex degree distribution we get a smoothed distribution, obtained from data on the network of autonomous Internet systems (available at http://www-personal.umich.edu/~mejn/netdata/as-22july06.zip). Figure 5 shows the results of generating a graph with given parameters after adding 50 thousand increments, a complete graph containing 4 vertices was used as a seed graph.

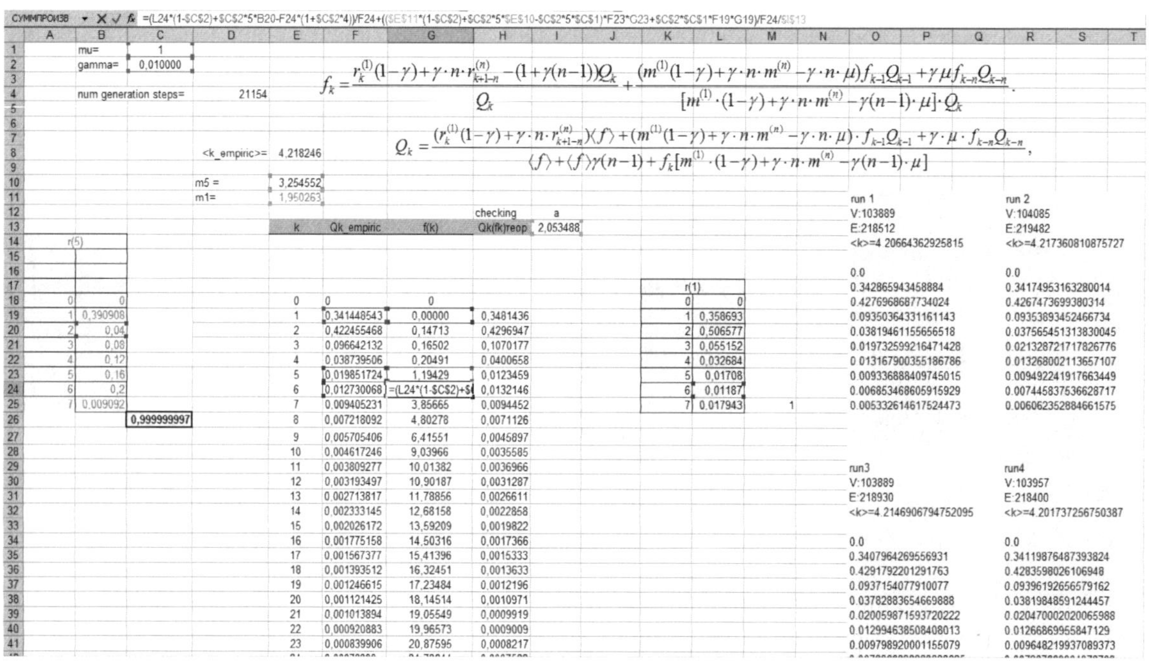

Figure 5. Calibration of a random graph using spreadsheets

4. Conclusion

Despite the numerous publications on the analysis of communities and their dynamics in real networks, the number of papers solving more fundamental questions such as mechanisms of communities emergence modeling is few. For example, in [9, 10] geometric graphs with preferential attachment are proposed (where the geometry determines the degree of similarity, affecting the preference of attachment and defining the form of communities), papers [11, 12] propose to "generate"

graphs as separate weakly connected components (they can also be considered as communities). This paper proposes a random graph that simulates the addition of community as a whole. This process interpretation can be found in social networks. Thus, when the next study group appears, new nodes (representing students) are added to the subnet, the nodes immediately become interconnected (students usually communicate in the subnet with other students from their study group).
Accounting other processes that take place in networks can be referred as a further development of the proposed model; it can be effectively performed with the help of agent-based modeling [13, 14]. By analyzing the dynamics of changes in the network structure papers [15, 16] reveal such processes as destruction of communities (division into several smaller ones or complete destruction of the community), and connection between the age of the community and its size.

Acknowledgments
The reported study was funded by Russian Foundation for Basic Research, according to the research project No 16-31-60023 mol_a_dk.

References
[1] Barabási A L and Albert R 1999 Emergence of scaling in random networks *Science* **286** pp 509–512
[2] Dorogovtsev S N, Mendes J F F and Samukhin A N 2001 Generic scale of the "scale-free" growing networks *Phys. Rev. E* **63** 062101
[3] Klemm K and Eguíluz V 2002 Highly clustered scale-free networks *Phys. Rev. E* **65** 036123
[4] Jackson M O and Rogers B W 2007 Meeting strangers and friends of friends: How random are social networks? *Am. Econ. Rev.* **97(3)** pp 890–915
[5] Zadorozhnyi V N and Yudin E B 2012 Structural properties of the scale-free Barabasi-Albert graph *Automation and Remote Control*, **73 (4)** pp 702-716 doi: 10.1134/S0005117912040091
[6] Zadorozhnyi V N and Yudin E B 2016 Dynamic equations of node degrees in growing networks with connection losses *Dynamics of Systems, Mechanisms and Machines, Dynamics 2016* art 7819111 DOI: 10.1109/Dynamics.2016.7819111
[7] Palla G, Barabási A.-L and Vicsek T 2007 Quantifying social group evolution *Nature* **446** pp 664–667
[8] Zadorozhnyi V, Yudin E 2014 Growing Network: Nonlinear Extension of the Barabasi-Albert Model *Communications in Computer and Information Science* **487** pp 432–439 http://www.springer.com/series/7899
[9] Zuev K, Boguñá M, Bianconi G and Krioukov D 2015 Emergence of Soft Communities from Geometric Preferential Attachment *Scientific Reports* **5** DOI:10.1038/srep09421.
[10] Dorodnykh A, Ostroumova Prokhorenkova L and Samosvat E 2017 Preferential placement for community structure formation *International Workshop on Algorithms and Models for the Web-Graph* (Springer) pp 75–89
[11] Klemm K and Eguiluz V 2002 Growing scale-free network with small world behavior *Phys. Rev. E* **65**
[12] Zadorozhnyi V N, Yudin E B and Yudina M N 2017 Analytical and numerical methods of calibration for preferential attachment randon graphs *2017 International Siberian Conference on Control and Communications, SIBCON 2017* (S. Seifullin Kazakh Agrotechnical UniversityAstana; Kazakhstan; 29 June 2017 30 June 2017) DOI: 10.1109/SIBCON.2017.7998461
[13] Aksyonov K, Antonova A and Goncharova N 2017 Choice of the Scheduling Technique Taking into Account the Subcontracting Optimization. Advances in Signal Processing and Intelligent Recognition Systems *2017 Advances in Intelligent Systems and Computing* **678** (Proceedings of Third International Symposium on Signal Processing and Intelligent Recognition Systems (SIRS-2017, September 13-16, 2017, Manipal, India) pp 297-304 DOI: 10.1007/978-3-319-67934-1_26
[14] Aksyonov K, Bykov E, Aksyonova O, Nevolina A and Goncharova N Architecture of the Multi-agent Resource Conversion Processes Extended with Agent Coalitions *Procedia*

Computer Science **105** (IEEE International Symposium on Robotics and Intelligent Sensors IRIS 2016, Hosei University, Tokyo, Japan, 17-20 December 2016) pp 221–226 DOI: 10.1016/j.procs.2017.01.214

[15] Chi Y, Song X, Zhou D, Hino K and Tseng B .L. 2007 Evolutionary spectral clustering by incorporating temporal smoothness *KDD '07: Proceedings of the 13th ACM SIGKDD International Conference on Knowledge Discovery and Data Mining* (New York: ACM) pp 153–162

[16] Lin Y.-R, Chi Y, Zhu S, Sundaram H and Tseng B L 2008 Facetnet: a framework for analyzing communities and their evolutions in dynamic networks *WWW '08: Proceedings of the 17th International Conference on the World Wide Web* (New York: ACM) pp 685–694

IOP Conf. Series: Journal of Physics: Conf. Series **1050** (2018) 012100 doi:10.1088/1742-6596/1050/1/012100

The kinematics and strength study of indexing bevel gear mechanism in rotating device driver

N V Zakharenkov and I N Kvasov

Omsk State Technical University, 11, Mira ave., Omsk 644050, Russia

Abstract. Indexing spatial gear mechanisms are widely used in automatic machines. The movement in gear mechanisms with stops is defined by complex design shapes that can lose their stiffness at critical loads. To measure the ultimate bearing stress of these mechanisms is possible by both the physical and numerical models tests results. The paper considers the numerical model for the spatial gear mechanism with driven disk indexing at a constant angular velocity of a driving disk. The presented mechanism kinematics and geometry parameters and finite element model were analyzed in the SolidWorks design environment. The structure and kinematics analysis revealed the possible reasons of mechanism failures. The numerical calculations results showing the design performance under the active area contact stresses and teeth bending stresses are represented. The maximal bend of the shaft was determined to avoid engagement failure.
Key words: rotary mechanism, gear mechanism, intermittent motion, orthogonal mechanism

1. Introduction

The consideration is made of bevel gear based spatial mechanism with output member indexing at constant angular speed of the driving member. Similar constructions are used, for example, in advertising installations [1]. A large number of inventions [2, 3] highlights such constructions relevance. Study subject (Fig. 1) includes driving and driven disks and represents a transmission gear with possible stops in three positions.

Bevel gear based mechanism is a transmission with a complex spatial interaction of driving and output members [4]. Stopping in one of the three positions is performed by flat surface sliding, present on mechanism members. Therefore, the numerical simulation in the Solidworks environment was chosen to study kinematics and strength. Examining similar models in the computer environment provides comprehensive information on the study subject [3].

Providing kinematic motion by the given law gives no definite information on the design normal operation. For this cause it is necessary to perform power analysis and field tests [6].

The research objective is to study the mechanism kinematics and to detect the design defects, to assess the strength indices at effective loads and to develop means to eliminate harmful factors arising during operation.

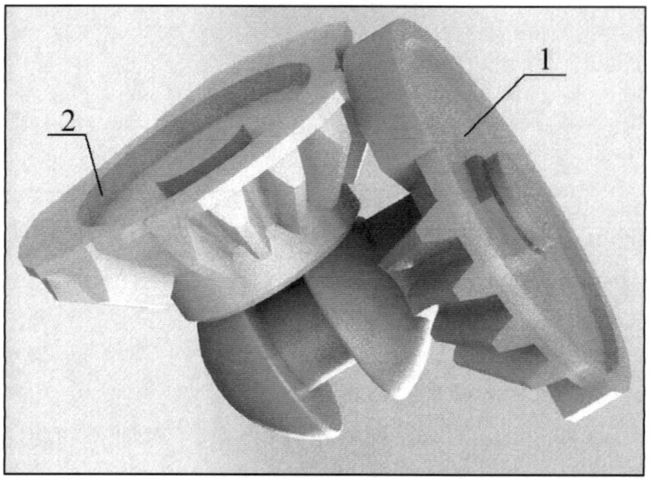

Figure 1. Computer model of spatial mechanism: 1 – driven disk; 2 – driving disk

2. Problem statement

The aims of the study are as follows:

- the design analysis by the performance at high loads, and the mechanism seizure and jamming inability;
- the calculation of loads acting on the mechanism computer model elements;
- the calculation of the design elements strength by contact and bending stresses.

The analysis of the full-scale specimen enabled us to conclude that the design normal operation declines when the first engaging tooth slips at the driving disk. Therefore, it is relevant to develop means of eliminating such a drawback. The slipping can occur: 1) when a tooth breaks; consequently, one should check a tooth by contact stresses; 2) when a tooth changes its shape or bends [7]; therefore, it calls for bending resistance calculation; 3) when the axes of driving and driven disks displace; thus, the threshold displacements are to be determined.

To calculate the loads, we applied the computer model that allowed us to obtain the system behaviour approximate evaluation by performing computational experiments [8]. Moreover, it is also necessary to define missing initial data used in calculations and required for the future model research.

Table 1. Initial data for calculation

Properties	Value
Torque on the shaft with installed driving disk M_1, *N·m*	10
Maximal overload torque at starting M_{imax}, *N·m*	25.33
Rotation frequency of driving disk and driven disk ω, *rad/s*	1.256
Operation time, sec	5
Out-of-service time, sec	15
Lifespan in the stresses variation cycles for a period of one year, N_{k1}	525600
Lifespan in the stresses variation cycles for a period of three years, N_{k3}	1576800

Based on the given initial data in Table 1, it is obvious that at 5 sec operation and 15 sec standstill each tooth of the driven disk experiences the stress once a minute.

Loading cyclogram for operation in two modes looks as shown in Fig. 2. Despite the transmitted torque and idling torque, there is also a peak torque arising at the starting moment (driving and output members engagement). In the following calculations, rating torque M_1 is taken as equal to peak one M_{imax} and effective for whole life cycle.

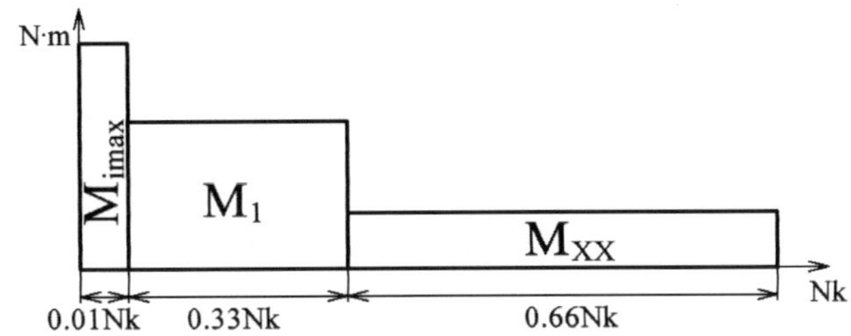

Figure 2. Loading cyclogram

3. Theory

Transmission gear design analysis. The transmission gear design analysis is aimed at identifying the factors influencing the parts operation.

The main working surfaces that are supposed to experience sliding friction are (Fig. 3): ring surface 1 (no load operation provides kinematics), flat surface 2 and 6, spherical surface 3. Working surfaces that are supposed to have rolling friction, with a share of sliding, are teeth surfaces 4.

Flat surface 2 operates under gravity, but it is partly compensated by force $F_a = 200.5$ N, occurring in tooth engagement 4; spherical surface 3 works under axial force from transmission F_a = 200.5 N. Friction may arise on surface 6 as a result of the driven disk axis bend, but only at significant wear of surface 2.

Circumferential speed is defined by the following equation: $V = \dfrac{\pi \cdot 0.857 \cdot d_{e1} \cdot n_1}{60000} = \dfrac{\pi \cdot 0.857 \cdot 68 \cdot 12}{6000} = 0.038$ m/s, where d_{e1} is reference diameter of the driven disk. Consequently, the transmission is low-speed one: lubrication conditions are without lubricant or with grease.

Powers in engagement at peak load $M_{i\text{max}}$ are:

Circumferential $F_t = \dfrac{2000 \cdot M_{u\,\text{max}}}{d_{e1}} = \dfrac{2000 \cdot 25.33}{65} = 779.3$ N;

Radial $F_{r1} = F_{a2} = F_t \cdot tg20° \cdot \cos 45° = 200.5$ N;

Axial $F_{a1} = F_{r2} = F_t \cdot tg20° \cdot \sin 45° = 200.5$ N.

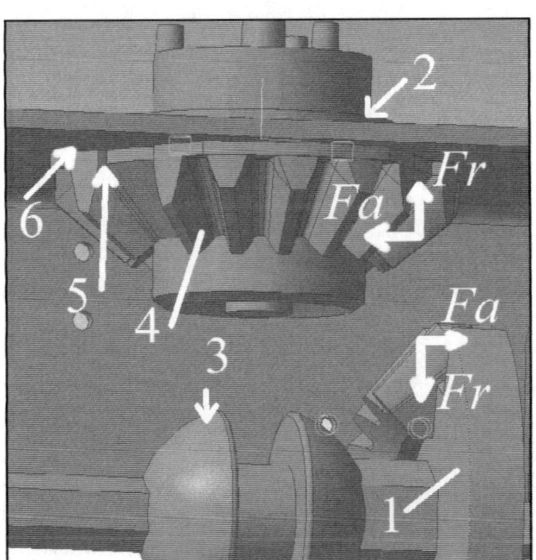

Figure 3. The diagram of forces and friction surface: 1 – ring surface; 2, 6 – flat surface; 3 – spherical surface; 4 – teeth surfaces; 5 – gutter sharp edges

IOP Conf. Series: Journal of Physics: Conf. Series **1050** (2018) 012100 doi:10.1088/1742-6596/1050/1/012100

4. Experimental results

1. Kinematic design analysis

The model kinematics study consists of determining the relevant position of the members during operation and detecting unacceptable operation conditions.

Fig. 4 (on the left) shows the moment of normal gear segment engagement, and Fig. 4 (on the right) demonstrates the first tooth sliding operation moment. The consequence is the mechanism jamming, as shown in Fig. 5.

Fig. 6 shows the motion path in terms of backlashes arising under the load.

Figure 4. Normal gear segment engagement (left) and driving disk tooth slipping (right)

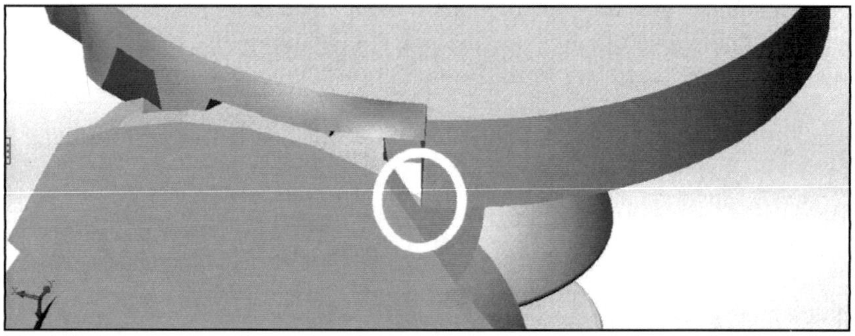

Figure 5. The consequence of jamming at the first engaging tooth slipping

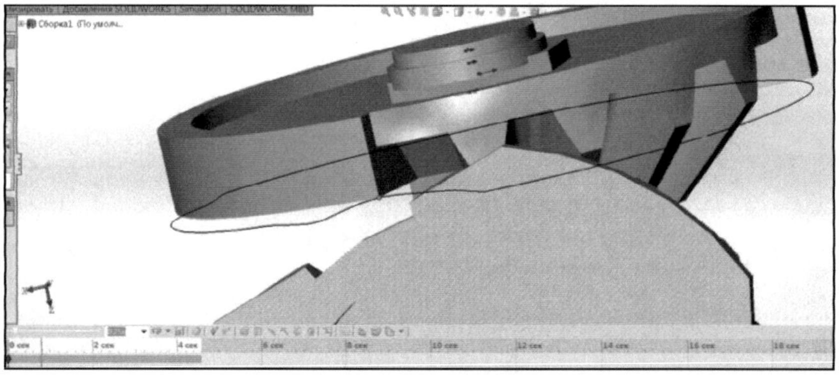

Figure 6. The motion path of the chosen point for one complete revolution of the driving disk

2. Design power analysis

Power analysis is aimed at determining the stresses, displacements and lifespan of the driving and driven disks (Fig. 7, 8, 9).

The force is to be determined in modeling, choosing by the shear absence condition. Maximal power causing no shear for a driving disk is F = 1200 N at 73.3 MPa; and for a driven disk it is F = 1600 N at 77.3 MPa.

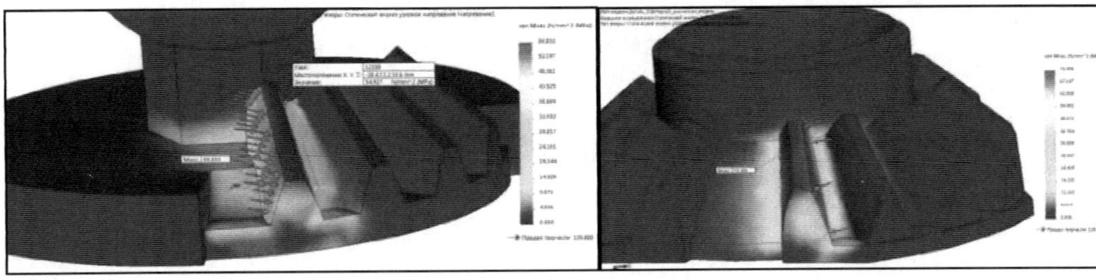

Figure 7. Stress profiles for driving disk (left) and driven disk (right) by maximum distortion energy theory (von Mises theory)

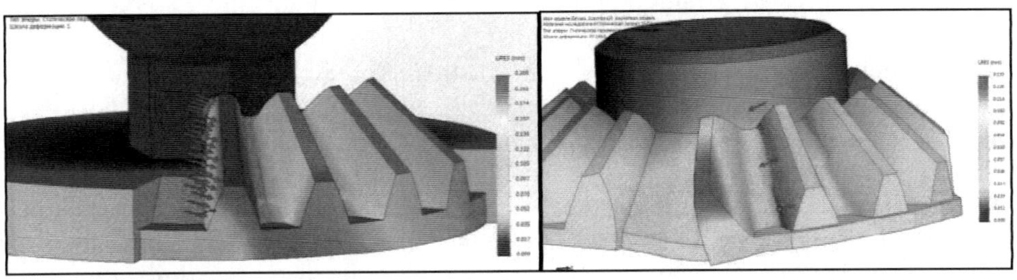

Figure 8. Displacement profiles for driving disk (left) and driven disk (right)

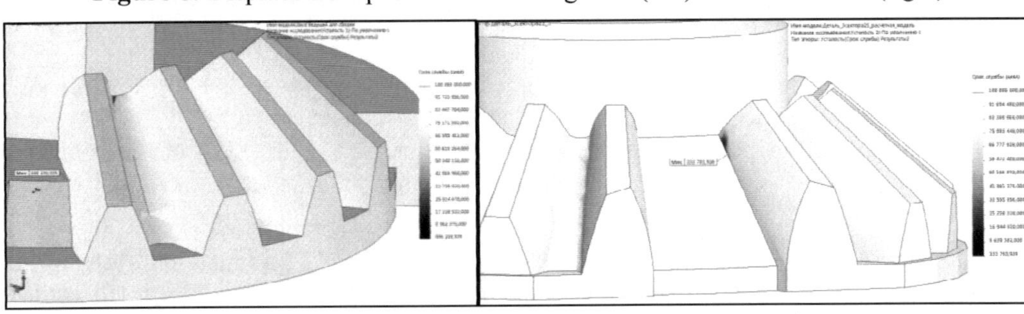

Figure 9. Service lifetime and the place (in black colour) of the most probable failure of the driving disk (left) and the driven disk (right)

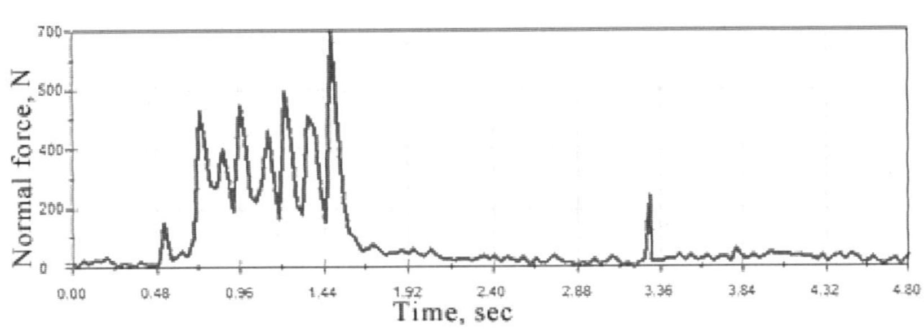

Figure 10. Normal force on the contact surfaces

5. Results discussion

One of the main drawbacks of the gear transmissions is the teeth edge contact at disengagement. This edge creates point load adversely affecting the gear materials wear. The edge contact at the tooth root can be compensated by dulling the contacting edges at the tooth top (Fig. 11).

IOP Conf. Series: Journal of Physics: Conf. Series **1050** (2018) 012100 doi:10.1088/1742-6596/1050/1/012100

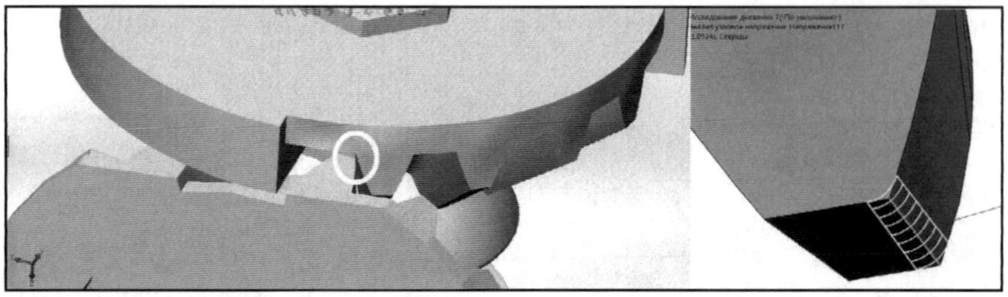

Figure 11. Edge contact (left) and eliminating the effect by rounding the sharp edges (right)

Driving disk material fatigue for the effective loads sets in at 333783 stresses variation cycles (Fig. 9), signifying the lifespan of about 7.5 months.
For the driving disk, fatigue fractions happen after 686223 cycles (Fig. 9), corresponding to the lifespan of 16 moths.
The tooth failure point, under the effect of the contact stresses, coincides with the fatigue fraction area for driving and driven disks.
The buckling failure analysis is not necessary for bulk constructions, since the failure occurs later due to higher stresses.
Since the tooth operates in bending, the stresses are at the tooth root (Fig. 8). Such model behaviour is in agreement with the gear transmission operation practice. The tooth core performs almost no work, which signals about the presence of the surface stresses and irrational material utilization.
Threshold displacements are distributed along the tooth unevenly. They are concentrated on the inner back cone due to less structural stiffness in this area. The maximum deformation value for the driven disk is 0.137 mm (Fig. 8) and for the driving one it is 0.209 mm (Fig. 8).
The maximum shaft bending according to the point tracing path on the driving disk at the teeth engagement in the time moment of 0.8 seconds (Fig. 6) is as follows $\Delta x = 10.19 - 7.26 = 2.93$ mm.
Analytical calculations of the powers in the engagements correspond to the data obtained in computer modelling (Fig. 25). The value of the rating torque M_1 acting upon construction is, however, 30% larger. If circumferential force F_t = 500 N, then $M_1 = \frac{500 \cdot 65}{2000} = 16.25$ N·m. Therefore, we get the improved loading cyclogram (Fig. 12) based on the numerical modelling finding (Fig. 10).

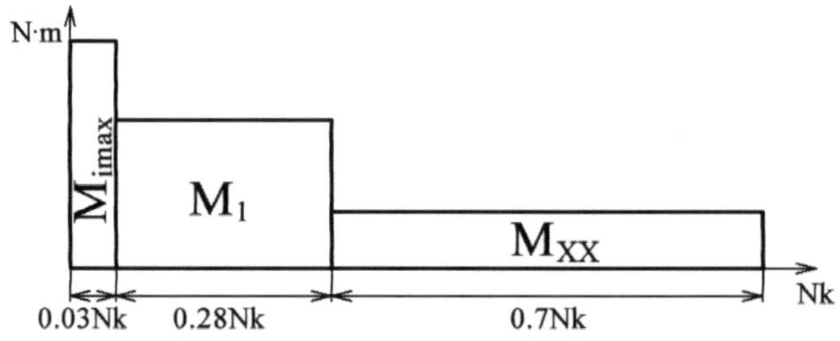

Figure 12. Improved loading cyclogram

6. Conclusions

As a result of the investigation, the mechanism kinematics was studied and the design defects were found. The results of the computer modelling by the contact stresses are obtained, and the test of bending stiffness and components lifespan is performed.
As a result of the undertaken investigation, the model drawbacks were detected, namely, large axial forces arising at engagement. These forces affect not only the teeth engagement but shafts and bearings as well. The maximum shaft bending is limited by 3 mm, to avoid tooth slipping.

The mechanism jamming may occur when the first tooth coming into engagement slips. This is unacceptable and consequently, the driving disk tooth bending stiffness should be improved to lower the bending effect.
The mechanism seizure is possible due to the sharp edges contact; these edges are to be altered in the defined area (Fig. 11) to avoid the mechanism jamming.
The comparison of the components lifespan demonstrates that the driven disk works 48.6% less. Therefore, the model design should be modified to increase the lifespan.
The obtained modelling results show the model application for the given initial data to be acceptable. The marginal stresses for the driving disk are, however, 30% larger than those of the driven disk. It is necessary to create a balanced-life design to distribute whole load along the whole component body and to move the stresses concentration from the most loaded areas. The design balanced life can be achieved: a) by creating stiffening ribs; b) by creating hollows inside the material bulk and by filling it with the stiffening mesh having a shape performing well in bending.
Stresses acting along the line from the tooth to the locking hub centre of the driving and driven disks are distributed unevenly. The strength is excessive in some areas, so balanced-life design is to be found for the model driving and driven disks. It will allow the weight to be lowered and the design balanced life to be increased.

References

[1] Advertising installation. Patent Russia 18792, IPC G09F 11/02 No. 2001108114/20, 30.01.2001, 2001
[2] Buchakov Ju V, Dobrovolskij V P and Kulakov A A 2004 Experimental study of the disc-sprocket rotation mechanism performance of the advertising installation Dynamics Proc. Int. Conf. vol I (Omsk: Omsk State Technical University) pp 428–32
[3] Xu F, Nu N and He W 2009 Modeling and Simulation of Globoidal Indexing Cam Mechanism Based on CATIA and ADAMS Journal of Mechanical Transmission **33** pp 42–3
[4] Ji S, Zhao J and Zhang Y 2015 An application of geodesics to the calculation of the rib-thickness of globoidal cam mechanisms Mechanism and Machine Theory **87** pp 163–76
[5] Li L, Feng X, Zhang Z, Han X and Song Y 2010 3D Modeling and Simulation of a New Type of Globoidal Indexing Cam Mechanism. *Applied Mechanics and Materials* **26-28** pp 931–5
[6] Zhaoa H, Caob Y and Guoc P 2011 Dynamics Simulation of Parallel Indexing Cam Mechanism Based on Rigid-flexible Coupled Modelling *Advanced Materials Research* **156-157** pp 172–76
[7] Hsieh J F 2014 Design and analysis of indexing cam mechanism with parallel axes *Mech. Mach. Theory* **81 pp** 155–65
[8] Guilan T, Haibo F and Weiyi Z 1999 A new method of torque compensation for high speed indexing cam mechanisms *ASME Journal of Mechanical Design* **121** pp 319–23
[9] Gonzalez-Palacios M A and Angeles J 1994 Generation of Contact Surfaces of Indexing Cam Mechanisms a Unified Approach *ASME Journal of Mechanical Design* **116** pp 369–74

IOP Conf. Series: Journal of Physics: Conf. Series **1050** (2018) 012101 doi:10.1088/1742-6596/1050/1/012101

Analysis of strain state prismatic samples for mechanical testing of the biaxial stretching method digital image correlation

E V Zenkov[1], **I P Aistov**[2], **K A Vansovich**[3]
[1]Department of Theoretical Mechanics and Strength of Materials, Irkutsk National Research Technical University, 83, Lermontov, Irkutsk, Russian Federation
[2]Department of Industrial Ecology and Safety, Omsk State Technical University, 11, Mira Pr., Omsk, Russian Federation
[3]Department of Petroleum Engineering, Standardization and Metrology, Omsk State Technical University, 11, Mira Pr., Omsk, Russian Federation

Abstract. The paper deals with the experimental-calculated analysis of the deformed state of a prismatic type sample for mechanical laboratory tests designed to evaluate the strength of a material under a biaxial stress state. A description is given of the advantages of the deformation scheme of this sample, which contributes to a reduction in the technical requirements for the test machines used for performing the test procedures. The distribution of deformation fields of the prismatic sample surface studied is obtained by processing the speckle images of the on-board experiment by the method of correlation of digital images in the Vic-3D system, at the stage of elastic deformation. Their comparison with the results of numerical simulation by the finite element method at control points of the surface of the deformed sample is performed. The discrepancy between the results of the full-scale experiment and the given numerical simulation at the control points did not exceed 14%. Conclusions are formulated for the subsequent use of the results obtained.

1. Introduction

A design estimate of the reliability of assemblies and structural components can be provided if reliable experimental information is available on the strength of their elements. Such an assessment should be carried out for those types of stress-strain state (SSS) in the source of destruction that occur during the action on the part of the operational load. Known methods for modeling the strength characteristics of structural parts are based on the study of the deformation processes of a series of typical laboratory samples tested under standard loading conditions [1, 2]. The results of these tests can then be taken into account as weight factors in the so-called combined strength criteria, for example, Jagn-Buzhinsky, Drucker-Prager, Pisarenko-Lebedev et al. [1-4]. From this point of view, the factor limiting the accuracy of applying the combined strength criteria is the difference between the real type of SSS of the estimated design and the type of SSS of typical samples that give information on the strength of the material with strictly limited types of SSS [3]. In addition, the need for the preliminary determination of the strength characteristics included in the combined strength criteria and the corresponding variety of laboratory test techniques complicate the practical application of a refined assessment of the reliability of assemblies and structural components. In particular, when testing to simulate a biaxial stress state, it is necessary to use cross-shaped patterns [5, 6, 7]. The paper deals with the experimental-calculated analysis of the deformed state of the proposed prismatic-type laboratory sample (prismatic sample) [8], intended to evaluate the strength of a material under a complex stress state, allowing to reproduce the state of biaxial stretching over a wide range of principal stresses under the action of a unidirectional test force.

2. Description of the prismatic sample deformation scheme

When carrying out mechanical tests, the prismatic sample 1 (see Figure 1) is supported by its ends on the end supports *5* and simultaneously by L-shaped projections *2* with support bevels *3* - on the prismatic support *4*. In this case, the contact interaction of the bevel occurs *3* with a prismatic support *4* having the same bevel angles. Under the influence of the test force *7* applied to the central zone *8*, a longitudinal bending of the sample occurs and in the working zone (the lower surface of the horizontal bar of the L-shaped step) positive principal stresses σ_1 arise, which are tensile stresses acting in the longitudinal direction for a prismatic sample direction (Figure 1, view *I*). Simultaneously, during the loading process, the displacement of the lower part of the protrusions *2* also occurs in the direction of the transverse plane of symmetry of the sample. The displacement is caused by the slip of the support bevel *3* of the L-shaped projections *2* along the inclined support surface of the prismatic support *4*. At this displacement, due to the structural shape of the lower part of the L-shaped protrusion, it bends in the transverse plane of symmetry of the sample and positive the main stresses σ_2, the level of which depends on the geometric parameters of the L-shaped projections *2*. The stress σ_3 in the working zone (stresses in the direction normal to the bottom surface of the lateral projection) are close in the test to zero, which creates a biaxial tension.

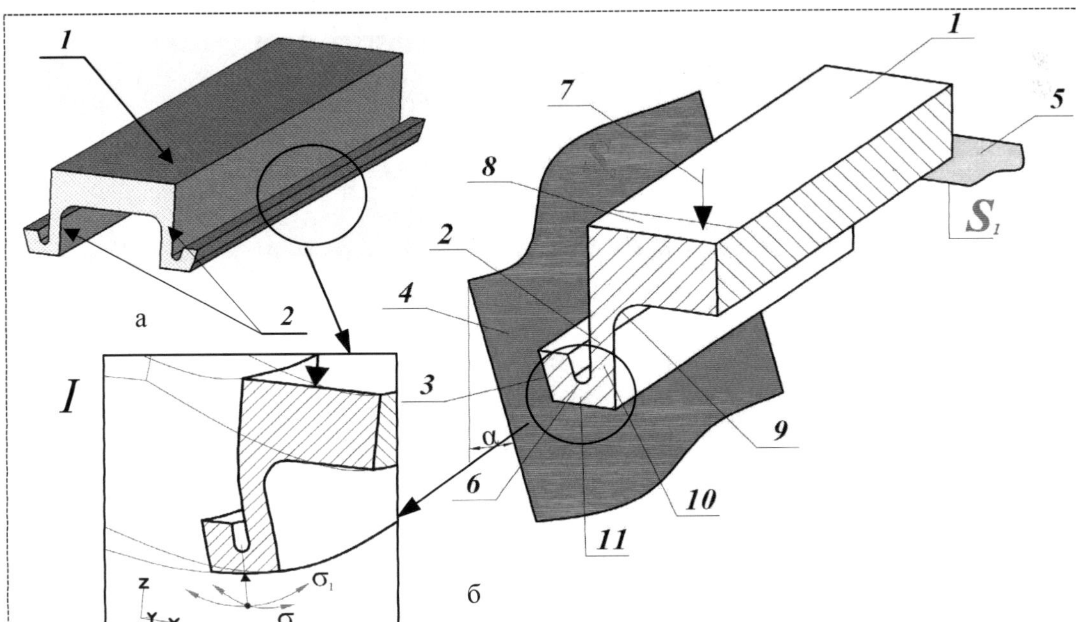

Figure 1. Prismatic sample (a) and its design features (b) (type of quarter of the sample), 1 - prismatic body, 2 - L-shaped projections, 3 - support bevel, 4 - inclined support surface of prismatic support S_1, 5 - bearing surface of the end support S_2, 6 - glancing transition, 7 - test force, 8 - loading surface, 9 - internal radius transition, 10 - the vertical bar of the L-shaped projection, 11 - the horizontal bar of the L-shaped projection, α is the angle of inclination of the support surface S_1, *I* - area of the working area of the sample during deformation

The described scheme of deformation allows to simplify the corresponding equipment used in the testing process and to reduce the requirements for the number of drives of the testing machine. Varying the basic design parameters of the prismatic sample (the width and height of the prismatic part, the height and thickness of the lateral protrusion, the radius of the tilt transition, the angle of the

IOP Conf. Series: Journal of Physics: Conf. Series **1050** (2018) 012101 doi:10.1088/1742-6596/1050/1/012101

lateral support slant), allows creating in the work zone the necessary SSS that simulates the SSS of the original design under study.

3. Experimental research

Together, these elements and the test sample, shown in Figure 1, allowed to develop a laboratory bench for mechanical testing, presented in Figure 2. To create tensile longitudinal stresses σ1 caused by longitudinal bending of the prismatic specimen, monolithic end supports are created (Figure 2), and for creating transverse tensile stresses σ_2 caused by sliding of the support bevels of the L-shaped projections of the sample along the inclined support surface, the support of the prismatic shape is made (Figure 2). This support is a massive prism in the central part of which a longitudinal groove with inclined support surfaces is made, the angles of which are the same as those of the side projections of the test specimen.

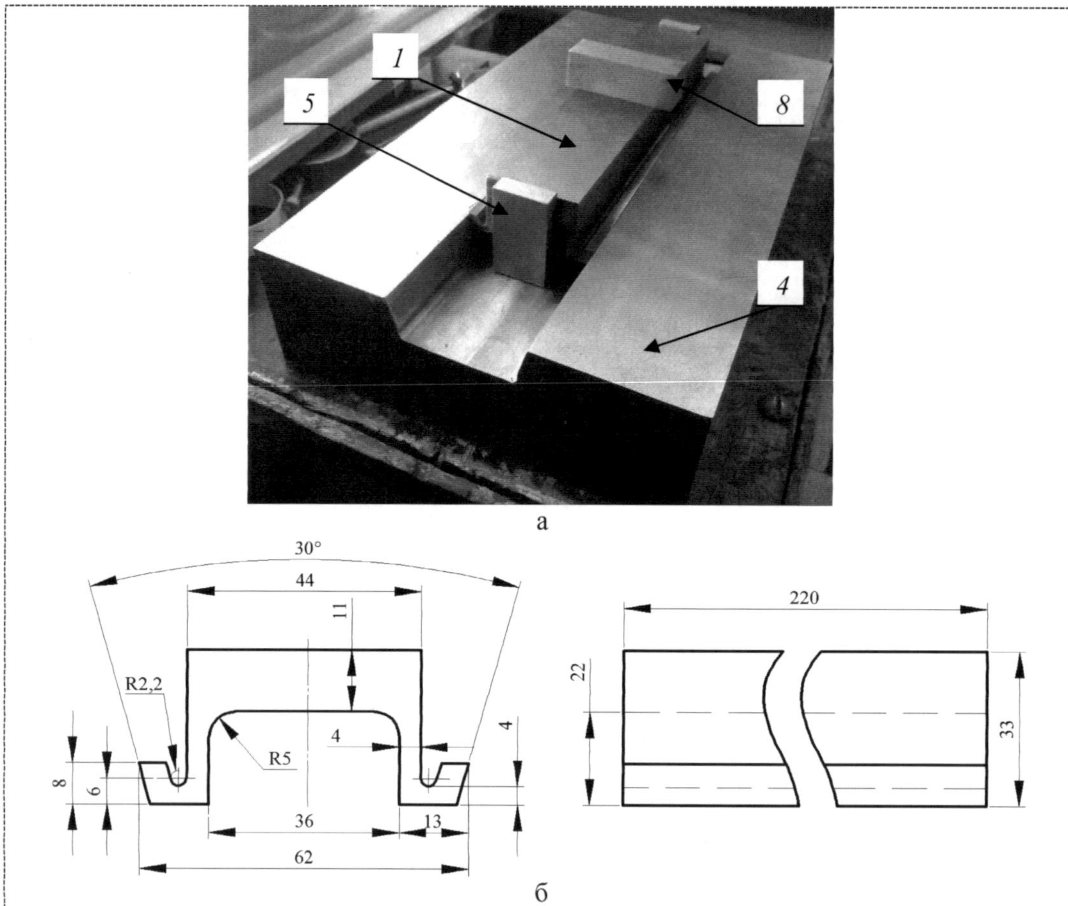

Figure 2. Laboratory bench for mechanical testing (a) and the dimensions (b) of the experimental prismatic sample (the numbering of the elements corresponds to Figure 1

Mechanical tests were carried out on an electromechanical machine Instron 5989. The registration of displacement and deformation fields was carried out using the method of correlating digital images (DIC) [9, 10]. When recording the displacement and deformation fields, the Vic-3D Correlated Solutions digital optical system was used, the mathematical apparatus of which is based on the method of correlation of digital images. The system includes two digital black-and-white cameras with a resolution of 1.4 MPix, a backlight system, a set of calibration tables, specialized software for adjusting and controlling the shooting process (Vic-Snap) and subsequent mathematical image processing (Vic- 3D).

A constructive version of the sample with geometric parameters, shown in Figure 2b. Samples were made from strip rolled bar steel 50CrV4 by milling on 5-coordinate machine tool with numerical control of DMU-80 type. The working and supporting surfaces were subjected to finishing treatment to achieve roughness parameters of these surfaces in accordance with the requirements of the standard [11]. The investigation of the deformation zones of a prismatic sample was carried out at a constant loading rate of 2 mm / min. In Figure 3a shows the appearance of the working body of the Instron 5989 test machine with a test specimen (initial state), figure 3b is the final elastically deformed state of the sample.

Figure 3. The appearance of the Instron 5989 testing machine with the loading unit *1* and the laboratory stand *2* (a), loaded through the pusher *3* (b)

The definition of deformation fields with Vic-3D was as follows. The moment of time corresponding to the test force causing elastic deformation of the material was fixed and then the corresponding number of the speckle image frame was determined from the available per second storyboard of the entire test process in Vic-3D. In Vic-3D, this image is processed - the displacement and deformation components are calculated relative to the original (reference) frame using the DIC [9]. In the course of the experiment, the dynamics of changes in the displacements of the speckle structure were carried out by means of continuous fixation of the images of the lateral surface of the sample. In this case, the fixing rate was set at *1* frame per second, which makes it possible to obtain a detailed picture of the deformation state during the entire test period. The processes of loading and fixing images were synchronized in time. This made it possible to determine the deformation characteristics of the surface under investigation at any instant of time to within a second and correlate them with the current test force obtained as a force / displacement curve of the loading device. The results of processing are displayed on the computer screen in the form of a multicoloured palette, each color of which corresponds to its value of the displacement or deformation field.

On the linear portion of the loading diagram of the prismatic specimen, the value of the test force equal to F = 70 ± 1 kN was recorded and located just above half the length of this section. According to the available second-by-second storyboard of the entire test process in the Vic-3D system, the number of the speckle image frame corresponding to the set F was set equal to t = 130 seconds. In Figure 4 shows the displacement, calculated as a result of the processing in Vic-3D, corresponding to the deformed state of the sample at the stage of elastic deformation, giving an idea of the pattern of the distribution of the longitudinal ε_{xx} deformation fields (see Figure 4a) and axial ε_{yy} deformations (see Figure 4b).

a

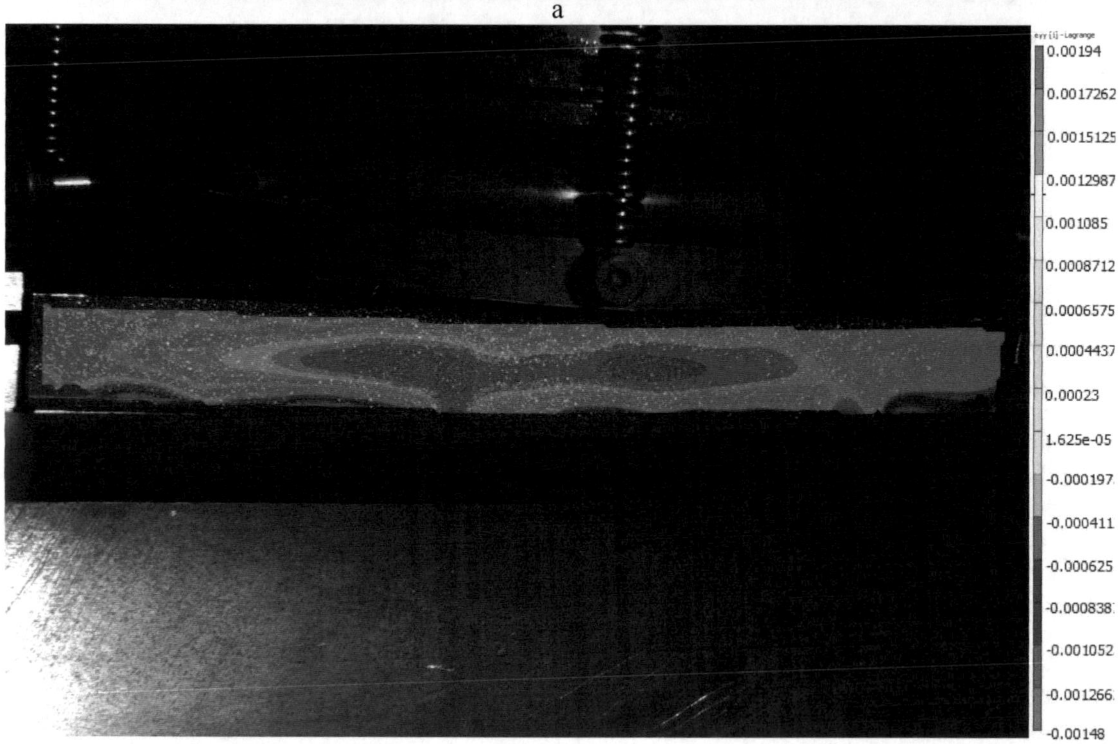

б

Figure 4. The experimental elastically deformed state of a prismatic sample with the distribution of longitudinal ε_{xx} (a) and axial ε_{yy} (b) deformations

4. Calculation research

Analysis of the experimental and calculated data on the deformed state of the sample in question was carried out based on the results of a numerical experiment with the specification of boundary conditions that maximally simulate the testing process. A feature of the process of numerical modelling of the SSS of the samples was the consideration of the contact nature of the elastic interaction of the sample with its supports [14]. The results of a numerical analysis of the computational model of a laboratory bench with boundary conditions for the conjugation of its elements are discussed in detail in [12, 13]. The test force in the numerical experiment corresponded to the indications of the dynamometer of the testing machine (70 kN). In Figure 5 presents the results of finite element modelling of the elastic deformation process performed in the MSC.NASTRAN-MSC.PATRAN software package.

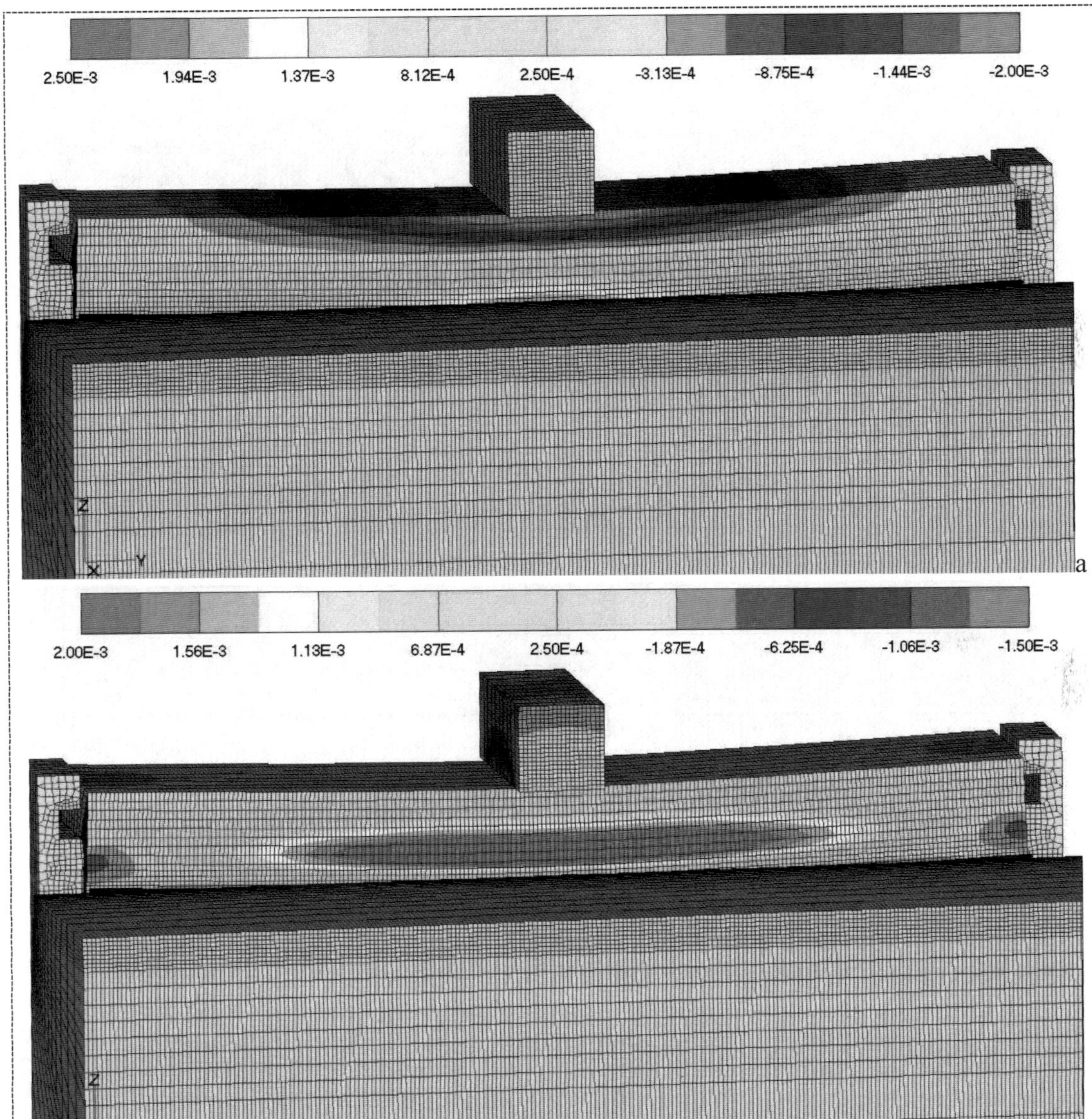

б

Figure 5. The distribution of the fields of the numerical values of the longitudinal ε_{xx} (a) and axial ε_{yy} (b) deformations at the stage of elastic deformations

5. Processing Results

Comparing the obtained results of the full-scale experiment using the Vic-3D system (Figure 4) with the data of numerical simulation (Figure 5) at the stage of elastic deformation, it is seen that the distribution of the field components of longitudinal ε_{xx} and axial ε_{yy} deformations have qualitative similarities. To quantify the reliability of the calculated and experimental deformations available for fixation by the Vic-3D system in the process of elastic deformation of a prismatic sample, the values of these deformations at the reference points of the lateral surface of the deformed sample are compared. The control points are tied to the lateral surface of the sample in a specific way. In the calculated finite element model, the control points are tied to the coordinates of the grid mesh nodes, and on the speckle image processed in the Vic-3D the point coordinate is determined by the shape and size of the speckle image interest area (i.e., the area for which the speckle displacement analysis is assigned). In Figure 6 is a diagram of the arrangement of control points in which the values of the displacement and deformation components were measured and compared.

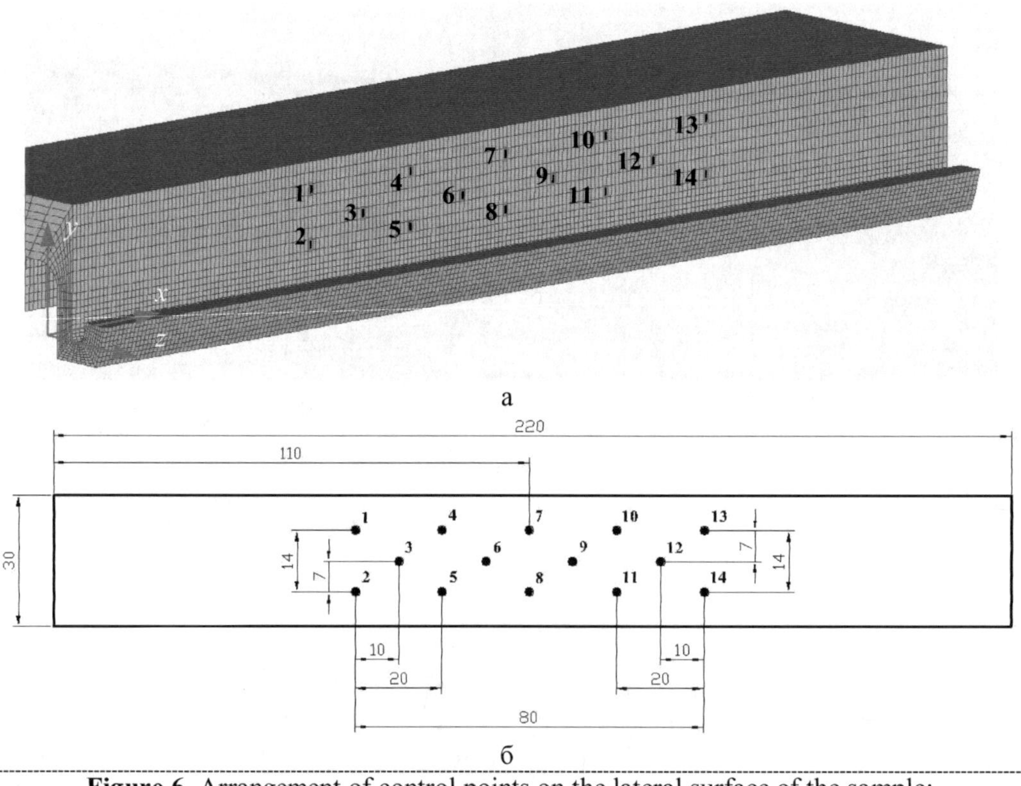

Figure 6. Arrangement of control points on the lateral surface of the sample: a - distribution method, b - location diagram

As a measure of the relative error of the discrepancy between the compared values of η, we considered the ratio

$$\eta = \frac{|\chi_{dic} - \chi_{fem}|}{\chi_{dic}} \cdot 100\%$$

where χ_{dic} - experimental value of the compared value, obtained by means of a method of correlation of digital images, χ_{fem} - the calculated value of the compared value, obtained with the help of FEM. In Figure 7 presents the results of a quantitative comparison of the values of the experimental and calculated displacements and deformations at all 14 control points of the lateral surface of the prismatic sample for various components.

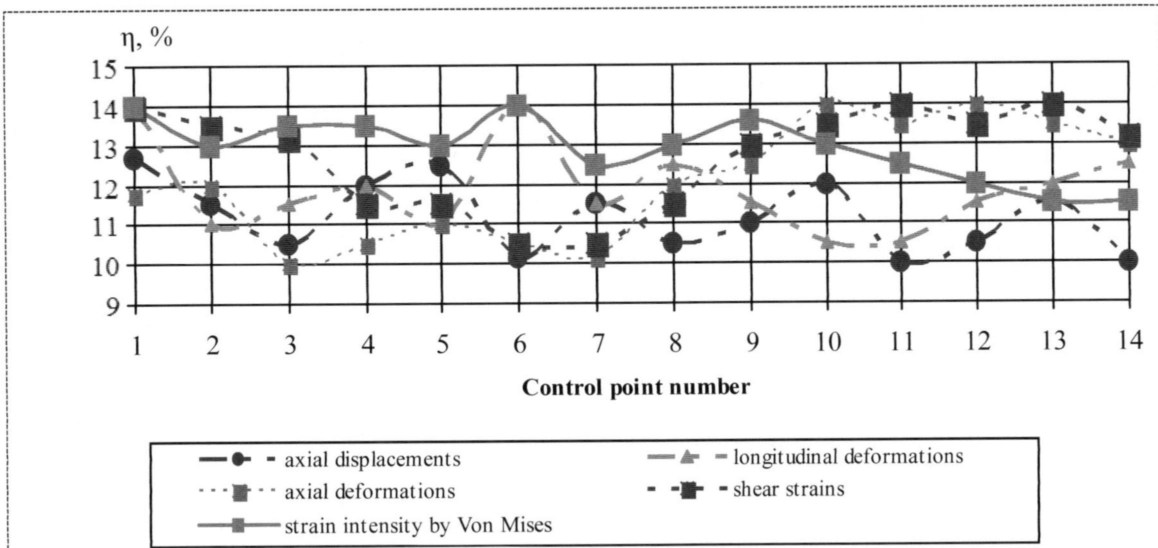

Figure 7. The relative error in the divergence of displacements and deformations in control points of the investigated surface of a prismatic sample

The results shown in Figure 7, in aggregate show that the characteristics of displacements and deformations of the surface of the prismatic sample, obtained by numerical simulation on the FEM and as a result of processing digital images of the full-scale experiment in the Vic-3D system, have a discrepancy not exceeding 14%.

6. Conclusion

It is experimentally shown that the developed mathematical model of elastic deformation of prismatic samples on the basis of FEM allows to estimate the characteristics of SSS in their working zones with an error of no more than 14%. This circumstance makes it possible to apply the developed computational models of the specified samples for the realization of various purposes, in particular, for carrying out variant computational studies to identify patterns of deformation of prismatic samples from their basic geometric parameters. Revealing of the indicated regularities will allow to carry out the reasonable choice of geometrical sizes of samples for reproduction of the required ratio of the main stresses of biaxial stretching and the subsequent determination of the strength characteristics of the material in this state.

7. References

[1] Kogaev V P, Makhutov N A, Gusenkov A P 1985 *The calculations of machine parts and structures for strength and durability* (Moscow: Mechanical engineering) p 224

[2] Pisarenko G S, Yakovlev A P, Matveev V V 2008 *Reference book on the resistance of materials* (Kiev: Publish. house Delta) p 816

[3] Zenkov E V, Tsvik L B 2015 Formation of divergent testing efforts and experimental evaluation of material strength under biaxial stretching *PNRPU Mechanics Bulletin* No 4 pp 110-120

[4] Zenkov E V, Tsvik L B 2013 Stress-Strain State of Prismatic Samples with Hollow Chamfers *Russian Engineering Research* Vol 33 No 10 pp 562-565

[5] Aistov I P, Vansovich K A 2017 Estimation of the growth of nonthrough cracks in the high-pressure cylinder of a long-stroke compressor *Chemical and Petroleum Engineering* Vol 52. pp 1-6

[6] Vansovich K A, Aistov I P, Yadrov V I 2016 The fatigue surface cracks growth model taking into account plastic deformations for cylinder of a low-speed compressor *Procedia Engineering* Vol 152 pp 240-246

[7] Vansovich K A, Yadrov V I 2017 Device for biaxial tests of cruciform specimens *Application No. 2017126593/28 (045818) Russian Federation IPC G01M 13/00*
[8] Zenkov E V, Tsvik L B, Zapolskiy D V 2014 Prismatic sample to estimate the strength of the material *Patent RF No 2516599 publ 27.03.2014 Bulletin of Inventions RF* No 9
[9] Sutton M A, Orteu J-J, Schreier H 2009 *Image Correlation for Shape, Motion and Deformation Measurements* (South Carolina: University of South Carolina) p 364
[10] Zenkov E V, Andreev A A 2013 Methods of experimental study of deformation fields through the use of digital optical system *Problems of transport of Eastern Siberia: Proceedings of the IV All-Russian scientific-practical conference* (Irkutsk: IrGUPS) chapter 1 pp. 95-99
[11] State standard of the Russian Federation 1497-84 1986 *Metals. Methods of tensile testing* (Moscow: Publishing Standards) p 24
[12] Zenkov E V, Tsvik L B, Zapolskiy D V 2012 Mathematical modeling of the processes of deformation of prismatic laboratory samples for fatigue strength testing *Safety of Regions - the Basis of Sustainable Development: Materials of the Third International Scientific and Practical Conference* (Irkutsk: IrGUPS) pp 52-60
[13] Zenkov E V, Tsvik L B 2013 Settlement and experimental evaluation of the stress-strain state of the laboratory sample to fillet *Bulletin of IrSTU* (Irkutsk: IrSTU) No 9 pp 70-78
[14] Zenkevich O 1975 *Finite Element Method in Engineering* (Moscow: Mir) p 572

Numerical modeling of the dynamics of multi-storey buildings with elastoplastic seismic insulators

E V Zenkov, V I Sobolev
Department of Theoretical Mechanics and Strength of Materials, Irkutsk National Research Technical University, 83, Lermontov, Irkutsk, Russian Federation

Abstract. The work is devoted to the problems of reducing the intensity of seismic impacts on multi-storey buildings. The main difficulties in solving the problems of seismic isolation of buildings and structures lie in the field of creating mathematical models of the dynamic interaction of the systems "foundation - seismic isolation device - construction". One of the most effective ways to reduce the level of seismic effects on buildings is to equip them with special devices - seismic isolators, a successful way of implementing which is the use of elastoplastic systems. The method of calculation for seismic actions of multi-storey buildings equipped with elastoplastic seismic insulators is proposed. Numerical modelling of the alternating loading of which uses the rheological model of N N Davidenkov. Modelling of dynamic processes is carried out by means of a mathematical description of the dynamics of two linear multidimensional subsystems that approximate the structures located under seismic insulators and above seismic insulators. The dynamics equations are formed on the basis of the D'Alembert principle. Horizontal seismic effects are considered. The presented technique makes it possible to determine the movements of the nodes of the system at any time of the seismic action specified in the form of a digitized seismogram. The above algorithm is implemented as a software module Proxima. The results of calculations are presented on the example of a model of a 24-storey building designed for the conditions of building in the city of Irkutsk.

1. Introduction

In conditions of dense building of modern cities and increase in the price of land for construction, many designers and customers make bets on increasing the number of storeys of buildings. In areas where there is a danger of earthquakes, an increase in the number of floors of a building can cause difficulties in ensuring the seismic stability of structures [1-3]. One of the possibilities to prevent the destruction of buildings and structures under seismic influences is to equip them with special devices - seismic insulators [7-9], using various physical principles with the formation of hysteresis loops during the alternating loading process [11-13].

Among the most promising are the elastoplastic seismic isolators intended for pumping out the potential energy of the building and reducing the intensity of the stress-strain state of the structures in the process of seismic action. The effect of suppressing seismic influences when using elastoplastic devices is based on the manifestation of residual deformations that decrease the restoring elastic force when deviating from the equilibrium position and, thus, forming a hysteresis loop when the direction of movement is changed [4, 9, 10].

The current practice of calculating buildings and structures for seismic actions uses a predominantly spectral technique, which assumes that the linear elastic-dissipative elements in the

IOP Conf. Series: Journal of Physics: Conf. Series **1050** (2018) 012102 doi:10.1088/1742-6596/1050/1/012102

calculated structures have nonzero rigidity [3, 6]. This condition is necessary to ensure the static stability of the calculated system. At the same time, the introduction of the geometrically variable systems into the design scheme leads to the degeneracy of the stiffness matrices, as a result of which calculation by the spectral technique becomes impossible. The introduction of nonlinear kinematic or elastoplastic elements leads to the inevitable coupling of vibrational modes capable of performing energy exchange. In this case, it is impossible to reduce the initial multiply connected system of equations to separate differential equations using a solution by the spectral method. A numerical method for calculating buildings with elastoplastic seismic insulators is proposed, based on the simulation of nonstationary processes of dynamic interaction of nonlinearly coupled linear subsystems.

2. Method of numerical calculation of building dynamics with seismic insulators

The rheological model of N N Davidenkov [10, 11], which led to the construction of a numerical model of the building on the basis of discrete approximations, is based on the development of the method for calculating the dynamics of a building equipped with elastoplastic seismic insulators [14, 15]. In the general case, the numerical calculation method for building dynamics with seismic insulators can be represented as a sequence of the following stages:

1) Using the known methods [16, 17] and existing computational complexes, finite element sampling of the building's structures is carried out;

2) The discrete dynamic system is divided into two linear parts: the Lower Linear System (LLS) is located below the seismic insulators and the Upper Linear System (ULS) located above the seismic insulators (Figure 1).

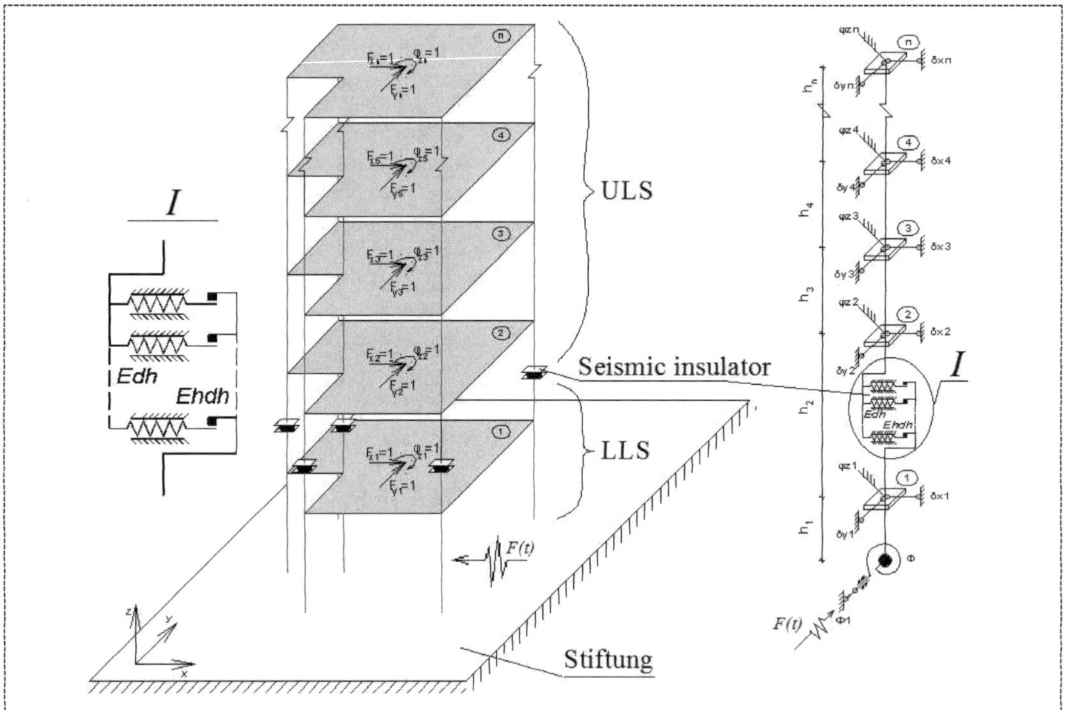

Figure 1. General view of the dynamic system of a building model with seismic insulators

3) On the basis of existing software complexes we obtain the stiffness matrix of two linear systems (LLS and ULS);

4) Determine the inertial parameters of the associated ground and inertial parameters, concentrated in the non-support nodes of the system;

5) We simulate the work of an elastoplastic seismic insulator by introducing an elastoplastic element into the design scheme according to N N Davidenkov's model [10, 11]. A uniaxial stress state is considered. The model of the element (see Figure 1, *I*) consists of an infinite number of arms. In each arm, the spring of rigidity *Edh* is connected in series with the ideal dry-friction damper of the value *Ehdh*. The stiffness characteristics of all the springs *E* are the same, the yield strengths of the springs *H* are different and distributed continuously with a probability density *p(h)*.

The law of deformation of an arbitrary arm has the form:

$$d\sigma = E(\varepsilon - \varepsilon_h)dh$$
$$E(\varepsilon - \varepsilon_h)dh = Ehdhsgn\varepsilon_h^* \tag{1}$$

where $d\sigma$ – the stress in an arbitrary arm *h*, ε – the total deformation of the arm, which is the same for all arms, ε_h – the plastic deformation in the arm *h*. In the equality (1), the function $sgn\varepsilon_h$* is used in the following sense: the function $sgn\varepsilon_h$* is equal to (+1) at a positive plastic deformation rate and $sgn\varepsilon_h$* is equal to (-1) at a negative plastic deformation rate, at zero plastic deformation rate $sgn\varepsilon_h$* in the interval (-1; +1). With this definition of *sgn*, the system of equations (1) is suitable for describing the processes of loading and unloading the arm, both in the absence of plastic deformations and in the presence of them. Summing the efforts in all the arms of the model in accordance with the probability *p(h)*, we obtain

$$\sigma = E\int_0^{\varepsilon}(\varepsilon - \varepsilon_h)p(h)dh$$
$$\varepsilon - \varepsilon_h = hsgn\,\varepsilon_h^* \tag{2}$$

or

$$\sigma = E\varepsilon - E\int_0^{\varepsilon}\varepsilon_h p(h)dh$$
$$\varepsilon = hsgn\varepsilon_h^* + \varepsilon_h \tag{3}$$

In expression (2), the first term represents the elastic force following Hooke's law, the second term describes the dissipation effect. Equation (3) serves to determine the plastic deformation in an arbitrary arm *h*. The system of equations (3) is suitable for describing the processes of loading and unloading a material according to any law in time.

To construct the unloading curve, the Masing principle [10, 18] will be used, which states that the unloading curve can be obtained from the deformation curve by means of coordinate transformation. The shape of the discharge curve does not depend on the strain attained during loading. The form of the bilinear elastoplastic model of the loading-unloading curve is shown in Figure 2.

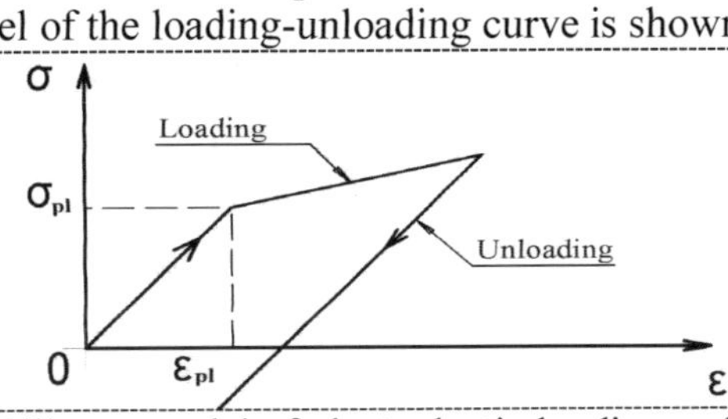

Figure 2. Bilinear model of elastoplastic loading-unloading curve

6) A system of equations of dynamic equilibrium of a discrete model of a building is formed on the basis of the D'Alembert principle [3, 19]. In this case, the state of the elastoplastic elements is uniquely determined by the mutual movement of the upper and lower supports of the seismic isolators using the analytical expressions described. In this discrete version of the model, the inertial parameters (masses and moments of mass inertia relative to the vertical) are considered to be concentrated at the center of mass of each overlap taking into account the masses of the adjacent vertical structures. To form the systems of equations for the dynamic state of a building subject to horizontal influences, we introduce two coordinate systems:

- an absolute coordinate system, rigidly connected with the resting position of the building;
- a relative coordinate system, rigidly connected with the movement of the ground environment.

We denote the coordinates of the position of the model nodes in the absolute coordinate system through x_a, y_a, and the relative coordinates are x, y. The dynamic equilibrium of a building when moving a ground environment can be described in vector form by a system of differential equations:

$$M\frac{d_2V_a}{dt^2}+R(V)=0, \quad (4)$$

where $V_a = (x_{1a}, y_{1a}, \varphi_1, x_{2a}, y_{2a}, \varphi_2, \ldots x_{na}, y_{na}, \varphi_n)^T$ – the vector of displacements of the nodes of the system in the absolute coordinate system, $V = (x_1, y_1, \varphi_1, x_2, y_2, \varphi_2, \ldots x_n, y_n, \varphi_n)^T$ – the displacement vector of the system nodes in the relative coordinate system, $M = diag\,(m_1, m_1, I_1, m_2, m_2, I_2, \ldots m_n, m_n, I_n)$ – the diagonal matrix of inertial parameters, in which m_1, I_1 – the mass and moment of inertia of the solid mass (overlap) with the number i [5, 20].

The vector-valued function $R(V)$ being generally nonlinear, contains linear expressions formed with the help of the previously defined rigidity matrices R_n and R_v for LLS and ULS. For inclusion in the system of equations of dynamics, the matrix R_v expands taking into account the compliance of the support node, which is now a node characterizing the movements of the upper support part of the seismic insulators. Such expansion is carried out on the basis of the equilibrium equations of ULS in projections on the X, Y axis and reactive moments in the horizontal plane. Thus, the matrix R_v increases its order by 3 and becomes triply degenerate. Let R_w be the extended matrix R_v.

If the matrix R_n has the dimension k, then the current number of the reference node of the ULS located on the seismic insulators for V_a and V is $k+1$. In this way

$$V_a=V+V_g, \quad (5)$$

where V_g – the displacement vector of the model reference point in the absolute coordinate system.

Taking into account equality (4), the system of equations (3) will have the form:

$$M\frac{d_2V}{dt^2}+R(V)=-M\frac{d_2V_{гр}}{dt^2} \quad (6)$$

The components of the vector V_g can be specified in the form of earthquake records along the directions of the X, Y axes and rotations of φ with respect to the vertical. In the system of equations (5), the vector function $R(V)$ has the form:

$$R(V)=\begin{bmatrix} R_n\cdot V_n+R_k(\Delta V_{k+1}) & 0 \\ 0 & R_{vr}\cdot V_{Vr}+R_{k+1}(\Delta V_{k+1}) \end{bmatrix}, \quad (7)$$

where $R_k(\Delta V_{k+1}) = (0, 0, 0, \ldots r_{xk}(\Delta V_{k+1}), r_{yk}(\Delta V_{k+1}), r_{\varphi k}(\Delta V_{k+1}))^T$,
$R_{k+1}(\Delta V_{k+1}) = (r_{xk+1}(\Delta V_{k+1}), r_{yk+1}(\Delta V_{k+1}), r_{\varphi k+1}(\Delta V_{k+1}), \ldots 0, 0, 0)^T$,
ΔV_{k+1} – a vector formed from the differences in the linear displacements of the upper and lower reference points of the seismic insulators,
$r_{xk}(\Delta V_{k+1})$, $r_{yk}(\Delta V_{k+1})$, $r_{\varphi k}(\Delta V_{k+1})$ – projections of elastoplastic reactions in the horizontal linear and angular connections of the node with the number k of the LLS,
$r_{xk+1}(\Delta V_{k+1})$, $r_{yk+1}(\Delta V_{k+1})$, $r_{\varphi k+1}(\Delta V_{k+1})$ – projections of elastoplastic reactions in the horizontal linear and angular connections of the node with the number $k+1$ ULS.

The values of $R_k(\Delta V_{k+1})$, $R_{k+1}(\Delta V_{k+1})$ are determined by the algorithms outlined in step 5. The solution of the system of equations (5) is carried out by the Runge-Kutta method [15, 21]. At each step of integration with respect to the independent time variable t_i the components of the vectors V, dV/dt, d_2V/dt^2 are determined.

The presented method of calculating the dynamics of the building makes it possible to determine the movements of the nodes of the system at any time during the action of a real seismogram. However, this is not sufficient to estimate the loss of seismic impact points. It is necessary to determine how much the seismic effect reduces the introduction of an elastoplastic seismic insulator into the building structure. This estimate of the loss of seismic impact points can be performed by the values of the maximum acceleration of the reference node of the ULS, which are considered external seismic influences on the ULS.. The algorithm for calculating the dynamics of the building is implemented in the form of a software module Proxima.

3. Practical implementation of numerical methods for calculating a building with seismic insulators

The Proxima software module is designed to simulate the dynamics of load-bearing structures of a building with elastoplastic seismic insulating supports, with seismic actions.

The structure of the initial data of the Proxima module includes the stiffness and inertial parameters of the bearing structures of the building, the location and parameters of the seismic insulators, the accelerogram of the seismic action, the angle of the direction of action, the correction factor, and the calculation settings. The source data can be imported from a pre-prepared text file, or specified in a table form directly in the program itself.

The results of building dynamics simulation using the Proxima module are presented in the form of trajectories in a tabular form for each degree of freedom of each node in the calculation scheme (Figure 3). The results of the calculation are the acceleration trajectories (*a*), velocities (*V*) and displacements (*u*) of the dynamically calculated building circuit nodes, as well as the trajectory (graphics) of the dynamic forces acting on the nodes of the calculation circuit.

For each trajectory, the maximum and minimum value is reached, which it reaches at a given time interval. The results of the calculation can be exported to a text file, and then used in other systems of engineering analysis.

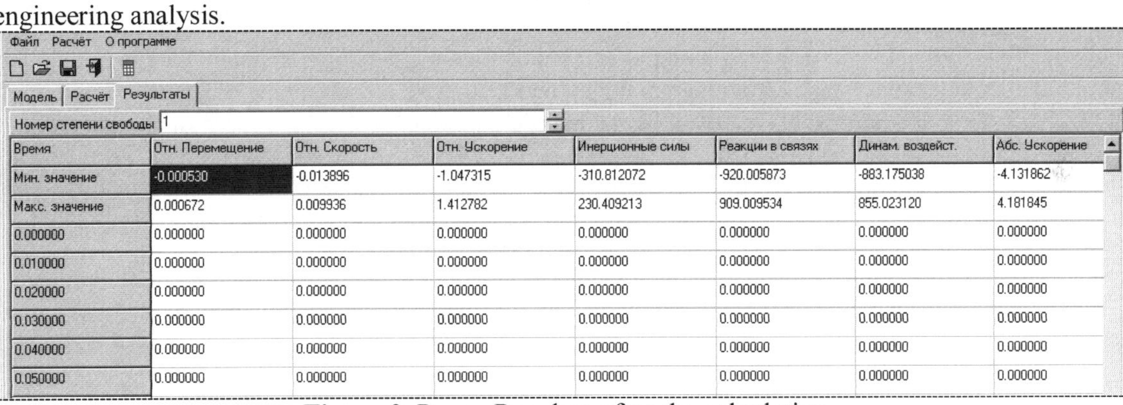

Время	Отн. Перемещение	Отн. Скорость	Отн. Ускорение	Инерционные силы	Реакции в связях	Динам. воздейст.	Абс. Ускорение
Мин. значение	-0.000530	-0.013896	-1.047315	-310.812072	-920.005873	-883.175038	-4.131862
Макс. значение	0.000672	0.009936	1.412782	230.409213	909.009534	855.023120	4.181845
0.000000	0.000000	0.000000	0.000000	0.000000	0.000000	0.000000	0.000000
0.010000	0.000000	0.000000	0.000000	0.000000	0.000000	0.000000	0.000000
0.020000	0.000000	0.000000	0.000000	0.000000	0.000000	0.000000	0.000000
0.030000	0.000000	0.000000	0.000000	0.000000	0.000000	0.000000	0.000000
0.040000	0.000000	0.000000	0.000000	0.000000	0.000000	0.000000	0.000000
0.050000	0.000000	0.000000	0.000000	0.000000	0.000000	0.000000	0.000000

Figure 3. Page «Results» after the calculation

Directly below the field with the counter is a table in which the trajectories and graphs corresponding to the chosen degree of freedom are displayed. The table contains eight columns (see Figure 3). The number of rows in the table depends on the size of the time interval on which the trajectories (the integration segment of the system of differential equations) and the values of the interval for preserving the trajectories are calculated.

The first column contains time marks with a step equal to the interval of preservation of dynamic trajectories. In the second column are the values of the relative displacement of the degree of freedom, corresponding to the time marks. In the third column, the relative velocity is placed, in the fourth column - relative acceleration, in the fifth - the inertial force, in the sixth - the reaction in the connection, in the seventh - the external force acting on the link, in the eighth - the absolute acceleration.

The first line of the table shows the names of physical quantities located in the corresponding columns. In the second and third lines, respectively, the minimum and maximum values of the corresponding quantity are located on the entire time interval.

In Figure 4 shows an accelerogram of seismic effects for modelling the dynamics of a building model consisting of 24 floors, designed for building conditions in the city of Irkutsk.

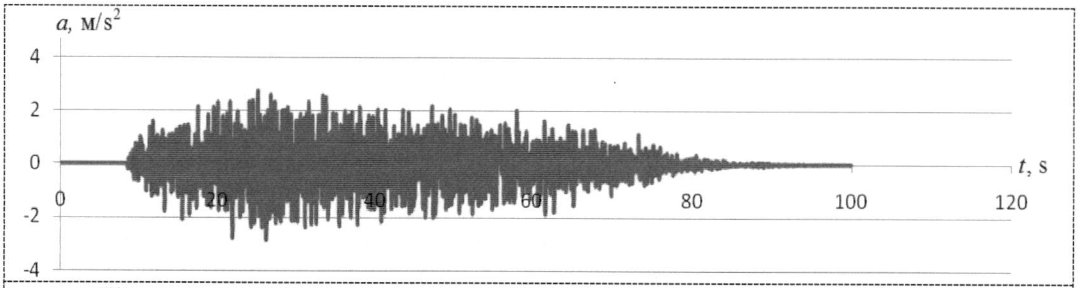

Figure 4. Graphical representation of the earthquake accelerogram (9 points)

The results of calculating the dynamics of the building with the Proxima module are shown in Figure 5 in the form of the trajectory of displacement (u) of the nodes of the dynamic design of the building at a height of 74.4 meters from the base (reference part).

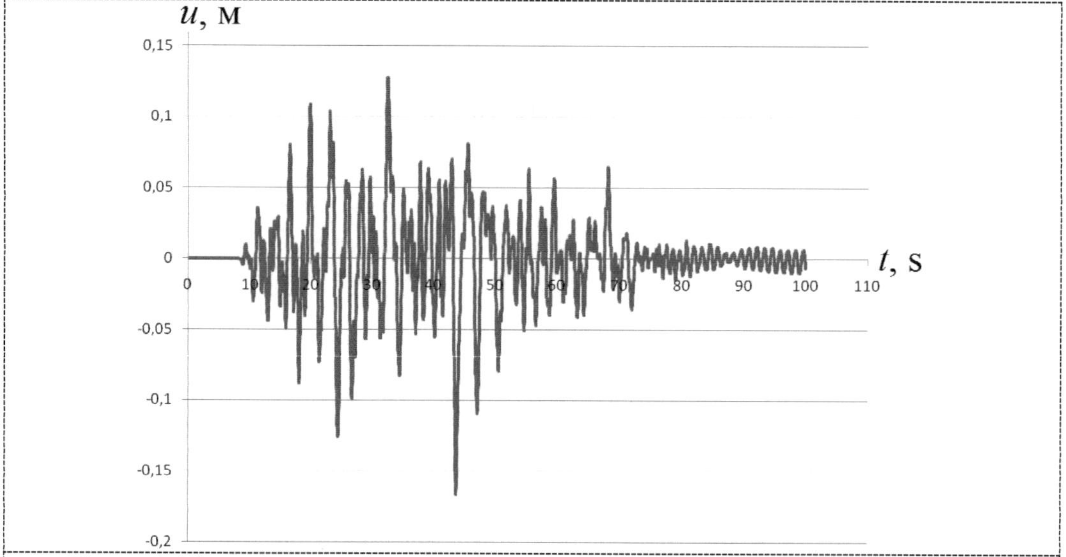

Figure 5. The trajectory of moving the nodes of the building's settlement diagram at the level of 74.4 meters

The results of calculations can be used to evaluate the performance of elastoplastic seismic insulators. For this type of insulator, the main criterion is the pumping out of the earthquake energy due to the so-called plastic effect, that is, the more the seismic insulator works in the plastic stage, the more effectively the earthquake energy is pumped out. On the other hand, there are restrictions on movements beyond which the insulator ceases to satisfy the construction of the bearing capacity.

4. Conclusion

The proposed methodology allows performing calculations of multi-storey buildings equipped with elastoplastic seismic insulators for seismic actions, represented by seismogram records, without using the decomposition procedures of multimass dynamical systems in their own vibrational forms. This circumstance relieves the calculator of doubts connected with the estimation of the errors of such transformations.

Analysis of the movements of the top of the seismic isolator relative to its support part makes it possible to evaluate the effectiveness of using the seismic isolation system. It is obvious that the efficiency of such a system is determined by the measure of the manifestation of plastic deformations, which is caused by hysteresis phenomena and associated energy losses due to vibrations. The latter circumstance makes it possible to perform a quantitative estimation of the level of seismic isolation and to carry out the selection of seismic isolation variants for the given operating conditions.

IOP Conf. Series: Journal of Physics: Conf. Series **1050** (2018) 012102 doi:10.1088/1742-6596/1050/1/012102

5. References

[1] Bershtein S A 1938 *Fundamentals of the Dynamics of Structures* (Moscow: Gosstroyizdat) p 160

[2] Borges D F, Ravara A 1978 *Design of reinforced concrete structures for seismic regions Structures* (Moscow: Stroiizdat) p 135

[3] Claf R, Penzien J 1979 *Dynamics of constructions* (Moscow: Stroiizdat) p 319

[4] Sobolev V I, Gaskin V V 2003 Transformation of seismic influences on multi-storey buildings with vibration isolating foundations *Proceedings of the Vth Russian National Conference on Earthquake-Resistant Construction and Seismic Zoning* (Sochi) p 67

[5] Sobolev V I, Gaskin V V 1999 Numerical studies of buildings under seismic influences given by oscillograms *3rd regional conference on seismic resistance. construction and seismic zoning* (Sochi) p 3

[6] Koreneva B G, Rabinovich I M 1984 Dynamic calculation of buildings and structures *Handbook of the designer* (Moscow: Stroiizdat) p 303

[7] Ishlinsky A Yu 1985 Plasticity *Overview. Mechanics. Ideas. Tasks. Applications* (Moscow: Nauka) pp 258-271

[8] Nashif A, Jones D, Henderson J 1988 *Dampening of oscillations* (Moscow: Mir) p 448

[9] Sobolev V I, Gotovsky S I 2003 Dynamics of seismic manifestations in multi-storey buildings equipped with kinematic foundations *Problems of Mechanics of Modern Machines: Proceedings of the Second International Conference* (Ulan-Ude: GISU) chapter 2 p 6

[10] Palmov V A 1976 *Oscillations of elastoplastic bodies* (Moscow: Science) p 328

[11] Davidenkov N N 1938 *On the scattering of energy in vibrations* (ZhTF) vol 8 pp 43-45

[12] Boevkin V I, Pavlov Yu N *1971 Dynamic absorber with dry friction* (Moscow: MWU) No 153 pp 110-117

[13] Dubkov S V 1988 The method of elastic solutions for transversely isotropic elastoplastic shells *Numerical analysis, mathematical modeling and their application in mechanics* (Moscow: Mosk. University) pp 41 - 42

[14] Sobolev V I, Gaskin V V 1998 Fluctuations in multi-storey buildings with seismic influences given by oscillograms *Bulletin of IrSTU. Series "Construction* (Irkutsk: IrSTU) pp 112-116

[15] Bakhvalov N S 1973 *Numerical methods* (Moscow: Nauka) p 631

[16] Bate K, Wilson E *1982 Numerical Analysis Methods and Finite Element Method* (Moscow: Stroiizdat) p 447

[17] Argyris J H, Boni B, Hinderlang V 1982 Finite element analysis of two- and three dimensional elastoplastic frames - the natural approach *Comp. Meth. Appl. Mech.*, vol 35 No 2 pp 221-248

[18] Annin B D 1988 Development of methods for solving elastoplastic problems *Mechanics and scientific and technical progress. Mechanics of a deformable solid* (Moscow: Science) chapter 3 pp 123-I36

[19] Mandelstam A I 1955 *Lectures on fluctuations* (Moscow: Academy of Sciences of the USSR) p 503

[20] Sobolev V I, Gaskin V V, Snitko A N 1996 Numerical study of multi-storey buildings under seismic influences given by oscillograms *Methods of potential and finite elements in the automation of research of engineering structures* (St. Petersburg) p 1

[21] Golberg S M, Zakharov A Yu, Filippov S S 1976 On some numerical methods for solving nonlinear systems of ordinary differential equations (Moscow: IPM AN USSR) No 12 p 48

The research of the automatic temperature condition control systems for LCD screens of avionics indication equipment

I O Zharinov[1], **O O Zharinov**[2]

[1] Department of Computer Design of On-Board Computer Equipment, Saint Petersburg National Research University of Information Technologies, Mechanics and Optics, 49, Kronverksky ave., Saint Petersburg, 197101, Russia
[2] Department of Problem-Oriented Computing Complexes, Saint Petersburg State University of Aerospace Instrumentation, 69, Bolshaya Morskaia str., Saint Petersburg, 190000, Russia

E-mail: mpbva@mail.ru

Abstract. In the aviation industry the primary modern equipment to display an information is color multi-function displays made on the basis of a plane LCD (liquid crystal display) panel. The problem is that LCD screen performances significantly depend on the ambient temperature. Local automatic temperature control systems are used to keep certain thermal conditions of LCD screens in avionics indication equipment. The problem of development a computer model of the automatic temperature control system of the avionics indication equipment is being studied.

1. Introduction

Exploitation of color multi-function display [1, 2] reveals that the device is affected with some exterior climate factors the most important of which are extremely low (minus 40 Celsius degrees) and extremely high (plus 70 Celsius degrees) ambient temperatures.

After several researches and experiments it is clear that the ambient temperature makes a significant impact on the colorimetric characteristics of an image displayed on the LCD panel. There is a shift of the image (x,y)-chromaticity coordinates which can be explained with the ambient temperature influence on the liquid crystals and on the spectrum reflectance of the backlight lamps (light emitting diodes). The absolute shift value of the (x,y)-chromaticity coordinates was estimated with the level of 0.0003 chromaticity unit per each Celsius degree of ambient temperature [3].

The (x,y)-chromaticity coordinates shift causes change of the position and the area of the LCD panel color gamut triangle on the XY-plane resulting change of the observer image visual perception characteristics. In that way the indication equipment color palette encoded with the RGB software codes (R – Red, G – Green, B – Blue) obtained to reach the maximum image contrast on a lighting equipment under the laboratory conditions and normal ambient temperature may become inappropriate when the ambient temperature varies in extremely wide range. This is the reason why it is an essential task to develop and research an automatic control system (ACS) [4, 5] which is capable to keep certain temperature mode of the LCD of a color multi-function display under different operating conditions.

2. Statement of the automatic control system research and parametric identification problem

Computer simulation of the ACS is one of the most effective ways to obtain the automatic control system quality indices under different operating conditions without long-term full-scale experiments on breadboard models. The most important objective when development an ACS computer model is the model adequacy for describing the real system, compliance with its performances and physical principles which determine the base of the system functionality.

The automatic control system intended for stabilization of the temperature condition of the LCD panel as a main part of the color multi-function display is built as nonlinear automatic system with the relay control law. The classical approach to develop mathematical model of such type of ACS suggests the creation of structural model with feedback, the open circuit consists with serial connection of a relay element (RE) and a linear dynamic unit with a transfer function $W(s)$. The typical structural schematics of the ACS with the relay control law is shown on the figure 1.

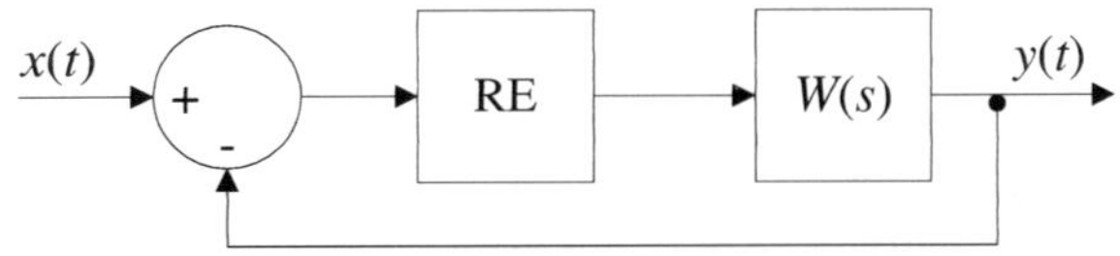

Figure 1. The typical structural schematics of the ACS with the relay control law.

The problem of the ACS research and identification includes identification of the transfer function $W(s)$ type and subsequent definition of the numerical values of the transfer function $W(s)$ parameters. To solve the problem the mathematical expressions which describe the heat exchanging process in an ACS must be defined and the parametric identification procedure must be done. The initial data for the ACS parametric identification procedure is the data obtained from a real experiment in the laboratory with a color multi-function display unit under the predefined ambient temperature conditions.

3. The model of the automatic system of stablilzation of the LCD-panel operating temperature condition

For a comparatively simple ACS of temperature stabilization usually transfer function $W(s)$ is presumed to be first-order aperiodic unit and parameters of the $W(s)$ may be identified using the registered data of the system open loop step response. This simple approach has a disadvantage because heat exchange processes can't be described precisely by the model with constant parameters when ambient temperature varies in the extremely wide range. The temperature mode of the LCD panel as a part of color multi-function display is affected with the following effects [6, 7]:

- the air ambient temperature of the exploitation conditions (in the climatic test chamber in laboratory);
- heat conductivity and convection characteristics of the indication equipment device;
- the level of power which supplies the LCD panel heating element;
- the operational mode of the internal cooling fan which changes the convection characteristics of heat exchange between the LCD panel and the environment.

The development of the temperature stabilization ACS model is based on the second law of thermodynamics and the laws of heat exchange:

- the heat exchange process which uses the thermal conductivity is described with the Fourier`s Law;
- the convection process is described with the Newton-Richmann Law.

Not counting for some differences in the physical processes description which both Laws include there is an experimentally approved fact that the heat quantity Q passing through a surface unit area per time unit t in the direction of normal to the boundary which divide the areas of different temperatures proportionally to those temperatures difference ΔT, in other words the following is true:

$$dQ/dt = \gamma \Delta T \,, \quad (1)$$

where γ – proportionality factor.

And also it is well known that when a heat quantity ΔQ is transferred to a physical body its temperature increases by the value ΔT:

$$\Delta T = \frac{1}{C}\Delta Q, \qquad (2)$$

where C – a physical body heat capacity.

The dynamics heat exchange processes are being considered when the heat quantity is altering in time, taking in mind the formula (2), can be described as:

$$\Delta T(t) = \frac{1}{C}\int_0^t P(\tau)d\tau, \qquad (3)$$

where $P(t) = dQ(t)/dt$ – the heat flow which intensity is changing in time while the system is functioning. The heat flow value is measured with Joule per second.

The expressions (1)-(3) lead to the basic version of the computer model of the ACS of the color multi-function display LCD panel temperature mode stabilization. The ACS computer model is developed in the graphical programming environment Simulink. The proposed model schematic is shown on the figure 2. The output signal of the integrator (*Integrator object*) with the transfer function $W(s) = 1/s$ reproduces the temperature dynamics of the controlled object (i.e. LCD panel) and element *Scope_Temp* visualizes the process. The value applied to the input x_0 defines the integrator initial state, meaning initial temperature of the LCD panel, equal to the ambient temperature value.

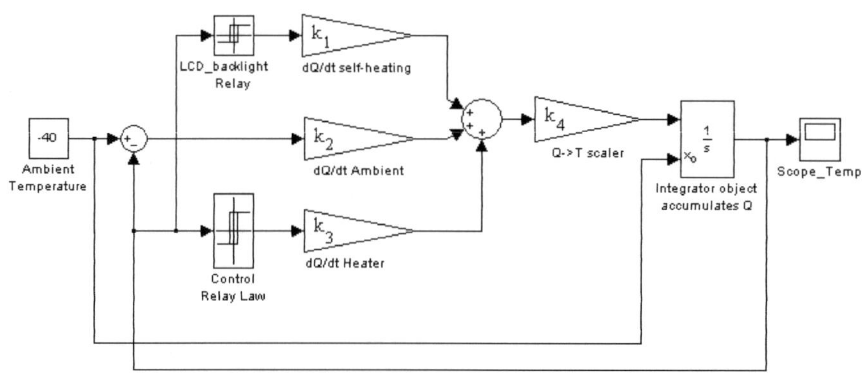

Figure 2. The basic version of the proposed automatic control system computer model.

The proposed ACS computer model is developed taking in mind 3 components of the LCD panel heat exchanging process within the color multi-function display:

- the LCD panel self-heating effect because the LCD illumination lamp functioning (in the scheme this heat exchanging path is represented by serial connection of the element *LCD_Backlight Relay* which controls lamp mode, and the instantaneous element *dQ/dt_self_heating* with a transfer constant k_1 Watts which defines the self-heating source energy equivalent power);

- the LCD heat exchange with the environment (intensity of which obviously depends on the difference between the LCD panel temperature and the environment temperature, this is represented by the proportionality coefficient k_2 Watt/degree of the instantaneous element *dQ/dt_Ambient*);

- the LCD panel heating with the set of built-in resistive heating elements sputtered with the thin-film sputtering technology on the glass (the dissipation power is defined with the coefficient k_3 Watts and is modelled as an instantaneous element *dQ/dt_Heater* in the scheme, element *Control Relay Law* controls this heating source mode).

The transfer constant $k_4 = 1/C$ of the instantaneous element $Q \rightarrow T\_Scaler$ connected to the input of the integrator in the schematic model on the figure 2 has a reciprocal value to the equivalent heat capacity C of the LCD panel which is a part of a color multi-function display. The relay element

Control_Relay_Law is a device with the hysteresis of the following functionality logic:

$$x_{out} = \begin{cases} 0, & x_{in} > L_{off} \\ 1, & x_{in} < L_{on} \end{cases}, \tag{4}$$

where x_{in} – the relay element input signal, x_{out} – the relay element output signal. The trigger threshold levels L_{on} and L_{off} of the element relay *Control_Relay_Law* are defined by the ACS developer and correspond to the heater element on-threshold and off-threshold. In the closed-loop ACS it is necessary to set $L_{on} > L_{off}$.

The relay element *LCD_backlight_Relay* included in the ACS to take into consideration the restriction established by color multi-function display developers, preventing the first switching-on of the illumination lamp of the LCD panel if its temperature is below zero. The necessity to apply such decision became clear after the researches showed that the LCD panel illumination lamp resource decreases significantly if the screen is used often without warming up when the ambient temperature is below zero. The exact numerical values of the ACS model could be obtained after some test measurements of the LCD panel temperature alteration while it is being tested inside the temperature stabilized chamber of tranquil air with any preset temperature. In the temperature test of avionic equipment there is an excessively low ambient temperature preset at the level minus 40 Celsius degrees. The ACS relay control law presumes alternate switching-on and switching-off of the LCD panel heating element when the temperature reaches the low or high preset thresholds of the relay element. The ACS equivalent structural schemes which correspond to the switched-on and switched-off modes of the heater are shown on the figure 3.

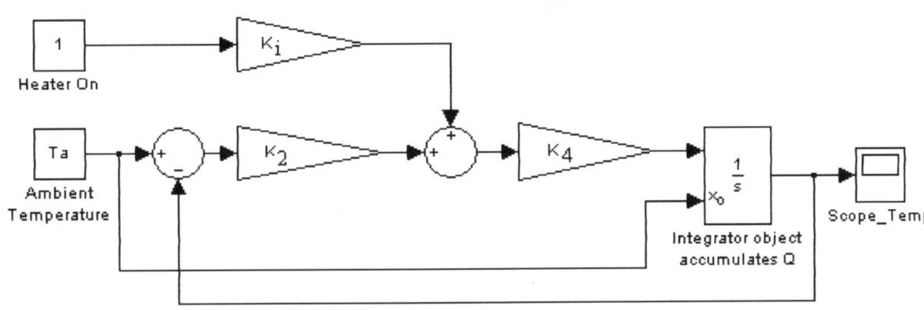

a)

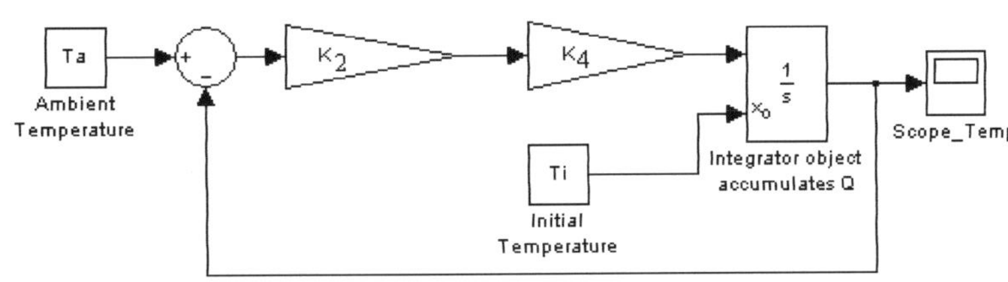

b)

Figure 3. Linear models of the automatic control system in test modes:
a) in the warming-up mode when the ambient temperature is excessively low,
б) in the cooling mode when device was previously warmed up.

Using the linear system analysis methods it can be shown that for the ACS model which scheme is shown on the figure 3,a the transient process when the initial value of the temperature of the LCD panel is equal to the ambient temperature T_a, may be described with the formula:

$$T(t) = T_a + \frac{k_i}{k_2}\left(1 - e^{-k_2 k_4 t}\right), \tag{5}$$

where for the parameter k_i either the value k_3 (heater is on, LCD panel illumination lamp is off) or the value k_1 (heater is off, LCD panel illumination lamp is on) must be used according to the color multi-function display selected test mode.

The model shown on the figure 3,b corresponds to the situation when a color multi-function display was warmed up to the temperature T_i and now it is cooling down when the ambient temperature is excessively low T_a. In this case the ACS transient process must be expressed as:

$$T(t) = T_a + (T_i - T_a)e^{-k_2 k_4 t}. \tag{6}$$

Numerical values of the parameters k_1, k_2 and k_4 of the basic model shown on the figure 2 may be estimated using the least-squares method for the expressions (5) and (6) together with known test measurement data. The numerical value of the coefficient k_3 (heater power) can be found in the color multi-function display technical documentation.

Any further improvement of the offered basic model of the LCD panel temperature stabilization automatic control system must be done keeping in mind the additional specific features of a color multi-function display, in particular when it is exploited in the excessively low temperatures: in the moment when the display reaches the above zero area of temperatures and the LCD panel is switched on for the first time the power of the heater must be reduced.

The final version of the ACS Simulink model is shown on the figure 4. To implement alteration of the heater power, there is an element *D*-Flip-Flop included in the model, output signal of the *D*-Flip-Flop controls the *Switch* element, so the heating element of the LCD panel gets reduced level of the power right after the LCD panel illumination lamp is switched on (30 Watts from the element *dQ/dt_Heater_Low* against 90 Watts from the element *dQ/dt_Heater_Hi* in the nominal mode).

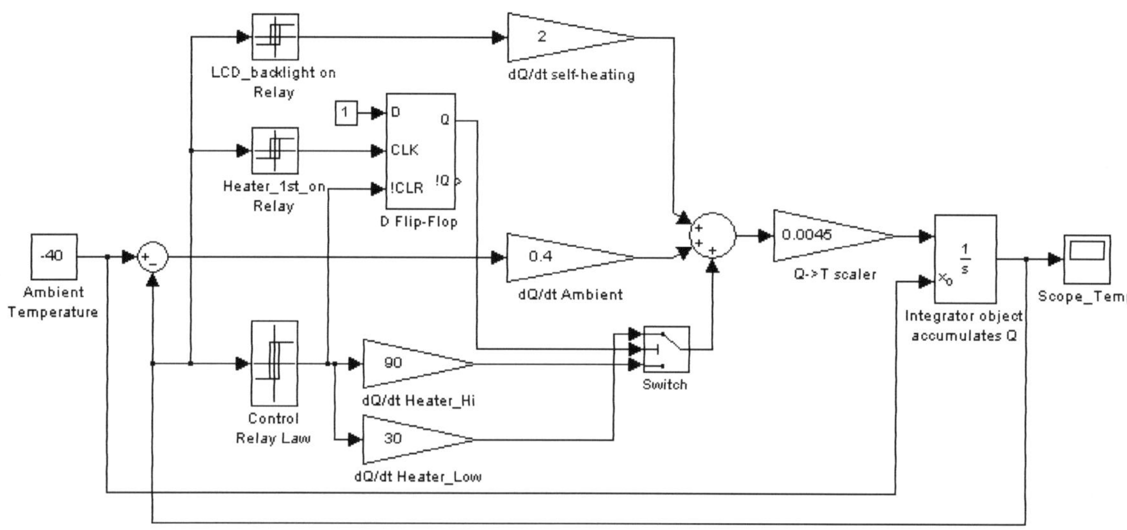

Figure 4. Final version of the computer model of the automatic control system.

After the procedure of the parametric identification of the automatic control system of the LCD panel operational mode temperature stabilization as a part of a mass production color multi-function display sample the following numerical values of parameters are obtained: $k_1 = 2$ Watts, $k_2 = 0.4$ Watts/Degree, $k_3 = 90$ Watts (in the heater full power mode), $k_4 = 0.0045$ Degrees/Joules.

The ACS computer model adequacy for the real device unit can be evaluated through comparison of the dynamic processes in the real system and in the model. The graphs for the both dynamic processes are shown on the figure 5. The relay element switching levels are L_{on}=25, L_{off}=10 Celsius degrees. Advantage of the proposed computer model of the ACS of the LCD panel temperature stabilization is that parameters non-stationarity may be implemented. Particularly non-stationarity can be included with the parameter k_2 to simulate the altering conditions of the convection heat exchange process. The heat exchange conditions depend on the atmospheric pressure and there is a wide range of variation while a color multi-function display is exploited as a part of aviation equipment.

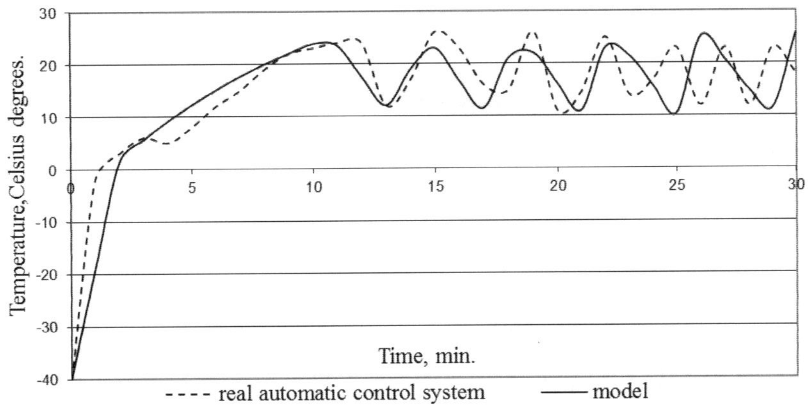

Figure 5. The dynamic process graph of the real automatic control system and its model.

To simulate the dynamics of the k_2 parameter changing an additional input for exterior disturbing influence can be added in the automatic control system model. The exterior disturbing influence on the parameter k_2 can be expressed with a two-input multiplier model. The similar approach to modelling the ACS parameters alternating dynamics can be used for the coefficients k_1, k_3 when necessary.

4. Conclusion

The proposed computer model of the LCD panel temperature conditions stabilization system is a universal one. The model parameters have a clear physical meaning and can be estimated on the basis of the simple temperature test data [8-10]. It is important to notice that the proposed model is not so complicated being compared with known kind of mathematical models of thermodynamic systems described with a system of non-linear differential equations in partial derivatives which are more accurate in the theory but doubly applicable for the real manufacturing necessities.

The approaches used to develop the proposed temperature conditions stabilization system computer model can be productive too when they are used for non-linear models of other automatic systems which are used in avionics.

References

[1] Livada B 2012 *Scientific Technical Review* **62** 70-79

[2] Zharinov I O, Zharinov O O 2017 *IOP Conference Series: Materials Science and Engineering* **286** 012015

[3] Chhajed S, Xi Y, Li Y-L, Gessmann Th and Schubert E F 2005 *Journal of applied physics* **97** 1-8

[4] Wang Ch-Fu, Tang Ch-W, Huang B-J 2010 *IEEE Trans. on power electronic* **25** 417-428

[5] Chiu T-L, Lee J-H 2010 *Optics communications* **283** 373-378

[6] Kihara K, Kawahara J, Takeda Y 2010 *Behavior research methods* **42** 1105-13

[7] Lagroix H E, Yanko M R, Spalek T M 2012 *Attention, perception & psychophysics* **74** 1033-41

[8] Menesatti P, Angelini C, Pallottino F, Antonucci F, Aguzzi Y and Costa C 2012 *Sensors* **12** 7063-79

[9] Thomas J-B, Hardeberg J Y, Foucherot I and Gouton P 2008 *Color research and application* **33** 449-460

[10] Huang W, Li J-M, Yang L-M, Jin Zh-L, Zhong Zh-G and Liu Y 2011 *Optics & Laser technology* **43** 214-217

IOP Conf. Series: Journal of Physics: Conf. Series **1050** (2018) 012104 doi:10.1088/1742-6596/1050/1/012104

The research of statistical distribution of the image tristimulus values displayed on the avionics indication equipment

I O Zharinov[1], **O O Zharinov**[2]

[1] Department of Computer Design of On-Board Computer Equipment, Saint Petersburg National Research University of Information Technologies, Mechanics and Optics, 49, Kronverksky ave., Saint Petersburg, 197101, Russia
[2] Department of Problem-Oriented Computing Complexes, Saint Petersburg State University of Aerospace Instrumentation, 69, Bolshaya Morskaia str., Saint Petersburg, 190000, Russia

E-mail: mpbva@mail.ru

Abstract. The problem of evaluation of statistical distribution of the image tristimulus values displayed on the avionics indication equipment screen is being studied. There are some mathematical expressions and simulation results which describe the behavior of the tristimulus values distribution probability density function. There are some histograms for tristimulus values statistical distribution in various combinations of the primary color codes which are used in the avionics indication equipment software.

1. Introduction
The development and exploitation experience of a color multi-function displays [1] made on the base of a LCD (liquid crystal display) panels by foreign manufacturers revealed that the same *RGB* program code (*R* – Red, *G* – Green, *B* – Blue) which defines the displayed image color causes different *XYZ*-tristimulus values and (x,y)-chromaticity coordinates [2, 3] in various samples of color multi-function display.

Difference in the values of the *XYZ*-tristimulus values and (x,y)-chromaticity coordinates can be explained by technological dispersion of the LCD panel parameters. The experiments with colorimeter show clearly that the color difference in the values of (x,y)-coordinates displayed on the different samples of the color multi-function display with the same *RGB* code may reach the marker of 0.01-0.03 units what can be perceived by the observer and the measuring instruments.

In that way a group of color multi-function displays manufactured with the same software and construction documentation when the mass production LCD panels are used and arranged in a row in the flight instruments panel of a civilian plane (up to 6 units) differ in the values of the tristimulus values (chromaticity coordinates) of the same displayed image. And that is unacceptable.

In that case the problem of research of the tristimulus values statistical distribution laws of the synthesizing images in order to establish the requirements for manufacturing quality to provide the acceptable level of the technological dispersion of the avionics mass production screens is an essential task.

2. Color spaces in avionics
Tristimulus values of the image displayed on a flight instruments LCD panel screen are defined by the

RGB code values of the color multi-function display software with the Grassmann Laws of additive color mixture [3]:

$$\begin{bmatrix} X \\ Y \\ Z \end{bmatrix} = \begin{bmatrix} X_r\, X_g\, X_b \\ Y_r\, Y_g\, Y_b \\ Z_r\, Z_g\, Z_b \end{bmatrix} \begin{bmatrix} R \\ G \\ B \end{bmatrix},\ \begin{bmatrix} R \\ G \\ B \end{bmatrix} = \begin{bmatrix} X_r\, X_g\, X_b \\ Y_r\, Y_g\, Y_b \\ Z_r\, Z_g\, Z_b \end{bmatrix}^{-1} \begin{bmatrix} X \\ Y \\ Z \end{bmatrix}, \quad (1)$$

where X, Y, Z – tristimulus values; X_r, X_g, X_b, Y_r, Y_g, Y_b, Z_r, Z_g, Z_b – screen profile coefficients; R, G, B – displayed color programed codes.

The LCD panel profile coefficients define the color gamut triangle on the *XY*-plane. The color gamut triangle includes the locus of all possible (x,y)-chromaticity coordinates of the image displayed on the LCD panel. The values of the (x,y)-chromaticity coordinates are defined through the *XYZ*-tristimulus values:

$$x = \frac{X}{X+Y+Z},\ y = \frac{Y}{X+Y+Z}. \quad (2)$$

3. The LCD panel parameters technological dispersion model

The LCD panel colorimetric parameters technological dispersion model which takes into account the screen color reproduction stochastic effects may be represented as [4]:

$$\begin{bmatrix} X \\ Y \\ Z \end{bmatrix} = \begin{bmatrix} X_r + \Delta\cdot\xi_{X_r} & X_g + \Delta\cdot\xi_{X_g} & X_b + \Delta\cdot\xi_{X_b} \\ Y_r + \Delta\cdot\xi_{Y_r} & Y_g + \Delta\cdot\xi_{Y_g} & Y_b + \Delta\cdot\xi_{Y_b} \\ Z_r + \Delta\cdot\xi_{Z_r} & Z_g + \Delta\cdot\xi_{Z_g} & Z_b + \Delta\cdot\xi_{Z_b} \end{bmatrix} \begin{bmatrix} R \\ G \\ B \end{bmatrix} \Rightarrow$$

$$\Rightarrow \begin{cases} X = R\left(X_r + \Delta\cdot\xi_{X_r}\right) + G\left(X_g + \Delta\cdot\xi_{X_g}\right) + B\left(X_b + \Delta\cdot\xi_{X_b}\right) \\ Y = R\left(Y_r + \Delta\cdot\xi_{Y_r}\right) + G\left(Y_g + \Delta\cdot\xi_{Y_g}\right) + B\left(Y_b + \Delta\cdot\xi_{Y_b}\right) \\ G = R\left(Z_r + \Delta\cdot\xi_{Z_r}\right) + G\left(Z_g + \Delta\cdot\xi_{Z_g}\right) + B\left(Z_b + \Delta\cdot\xi_{Z_b}\right) \end{cases}, \quad (3)$$

where

$$\xi_{X_r} \in \left[-X_r/2; +X_r/2\right],\ \xi_{X_g} \in \left[-X_g/2; +X_g/2\right],\ \xi_{X_b} \in \left[-X_b/2; +X_b/2\right],$$
$$\xi_{Y_r} \in \left[-Y_r/2; +Y_r/2\right],\ \xi_{Y_g} \in \left[-Y_g/2; +Y_g/2\right],\ \xi_{Y_b} \in \left[-Y_b/2; +Y_b/2\right],$$
$$\xi_{Z_r} \in \left[-Z_r/2; +Z_r/2\right],\ \xi_{Z_g} \in \left[-Z_g/2; +Z_g/2\right],\ \xi_{Z_b} \in \left[-Z_b/2; +Z_b/2\right]$$

— random values distributed with the uniform distribution law, Δ – the model parameter which includes the presence of the screen colorimetric characteristics technological dispersion while the screens are being manufactured.

4. The tristimulus values distribution probability density function theoretical estimate

After the analysis of the equations (1) and (3) it is clear that the tristimulus values could be defined in the three combinations of the *RGB* code components conditions:

1. The color displayed on the LCD panel screen is formed with one of the *RGB* code components when the other two are equal to zero. In other words those are situations when $R \neq 0, G = 0, B = 0$, $R = 0, G \neq 0, B = 0$ or $R = 0, G = 0, B \neq 0$.

In this case the tristimulus values are distributed [5, 6] according to the uniform law with boundaries defined by the absolute value of the *RGB* code non-zero component, corresponding LCD panel profile coefficient and the technological dispersion parameter Δ.

The expressions to evaluate the probability density functions (PDF) $f_X(t)$, $f_Y(t)$, $f_Z(t)$ according to the X, Y, Z tristimulus values distribution are designated in the table 1. If the variable t is out of the intervals designated in the table 1 the PDF $f(t)$ is equal to zero.

Table 1. Expressions to evaluate the PDF of the *X*, *Y*, *Z* tristimulus values distribution.

The *RGB* code components	The tristimulus values PDF $f_X(t)$	$f_Y(t)$	$f_Z(t)$
$R \neq 0$, $G = 0$, $B = 0$	$1/X_r R\Delta$, $t \in [RX_r - RX_r\Delta/2; RX_r + RX_r\Delta/2]$	$1/Y_r R\Delta$, $t \in [RY_r - RY_r\Delta/2; RY_r + RY_r\Delta/2]$	$1/Z_r R\Delta$, $t \in [RZ_r - RZ_r\Delta/2; RZ_r + RZ_r\Delta/2]$
$R = 0$, $G \neq 0$, $B = 0$	$1/X_g R\Delta$, $t \in [GX_g - GX_g\Delta/2; GX_g + GX_g\Delta/2]$	$1/Y_g R\Delta$, $t \in [GY_g - GY_g\Delta/2; GY_g + GY_g\Delta/2]$	$1/Z_g R\Delta$, $t \in [GZ_g - GZ_g\Delta/2; GZ_g + GZ_g\Delta/2]$
$R = 0$, $G = 0$, $B \neq 0$	$1/X_b R\Delta$, $t \in [BX_b - BX_b\Delta/2; BX_b + BX_b\Delta/2]$	$1/Y_b R\Delta$, $t \in [BY_b - BY_b\Delta/2; BY_b + BY_b\Delta/2]$	$1/Z_b R\Delta$, $t \in [BZ_b - BZ_b\Delta/2; BZ_b + BZ_b\Delta/2]$

2. The color is formed with two non-zero components of the *RGB* code when the third component is equal to zero. In other words those are situations when $R \neq 0, G \neq 0, B = 0$, $R = 0, G \neq 0, B \neq 0$ or $R \neq 0, G = 0, B \neq 0$. In this case the PDF *f*(*t*) of the tristimulus values distribution is a mathematical function of trapezoidal type [7, 8]. The function *f*(*t*) must be calculated as convolution of two PDF. For example for *X*-coordinate $f_X(t) = f_{\xi_{X_r}}(t-s) \otimes f_{\xi_{Xg}}(s)$ when $R \neq 0, G \neq 0, B = 0$:

$$f_X(t) = f_{\xi_{X_r}}(t-s) \otimes f_{\xi_{Xg}}(s) = \int_{-\infty}^{+\infty} f_{\xi_{X_r}}(t-s) f_{\xi_{Xg}}(s) ds = \begin{cases} 0, t < a_1 + a_2, b_1 + b_2 \leq t \\ \dfrac{t - a_1 - a_2}{(b_2 - a_2)(b_1 - a_1)}, a_1 + a_2 \leq t < b_1 + a_2 \\ \dfrac{1}{b_2 - a_2}, b_1 + a_2 \leq t < a_1 + b_2 \\ \dfrac{b_1 + b_2 - t}{(b_2 - a_2)(b_1 - a_1)}, a_1 + b_2 \leq t < b_1 + b_2 \end{cases} . \quad (4)$$

The system (4) is true when $b_2 - a_2 > b_1 - a_1 > 0$, if the inequality expression is not valid than the coefficient indices a_i, b_i must be redefined. It could be shown that expressions of the system (4) for the PDF lead to different kind of functions in depends on the ratio of the interval lengths $[a_1; b_1]$ and $[a_2; b_2]$. In the table 2 there are some particular cases of all possible interval boundaries to evaluate the PDF distribution of the *X*, *Y* and *Z* tristimulus values. The cases $b_1 - a_1 = b_2 - a_2 > 0$ correspond to equal interval lengths $[a_1; b_1]$ and $[a_2; b_2]$ in (4) when the PDF *f*(*t*) of the trapezoidal type becomes of the triangular type. In the mathematical statistics when $[a_1 = a_2 = 0; b_1 = b_2 = 1]$ the function *f*(*t*) defines the Simpson triangular distribution which is an Irwin-Hall distribution special case:

$$f_n(t) = \frac{1}{2(n-1)!} \sum_{k=0}^{n} (-1)^k \binom{n}{k} (t-k)^{n-1} sign(t-k) ,$$

for the sum *n* of the uniformly distributed [0;1] random values when *n*=2.

3. The color is formed with a combination of all three non-zero codes of the primary color components. In other words it is general case when $R \neq 0, G \neq 0, B \neq 0$. In this case the probability density function *f*(*t*) of the tristimulus values distribution is a mathematical function of piecewise parabolic type [9].

Table 2. Tristimulus values distribution PDF parameters ratio.

The *RGB* code components		Tristimulus values distribution PDF $f_X(t)$	$f_Y(t)$	$f_Z(t)$
$R\neq 0$, $G\neq 0$, $B=0$	a_1	$RX_r - RX_r\Delta/2$	$RY_r - RY_r\Delta/2$	$RZ_r - RZ_r\Delta/2$
	b_1	$RX_r + RX_r\Delta/2$	$RY_r + RY_r\Delta/2$	$RZ_r + RZ_r\Delta/2$
	a_2	$GX_g - GX_g\Delta/2$	$GY_g - GY_g\Delta/2$	$GZ_g - GZ_g\Delta/2$
	b_2	$GX_g + GX_g\Delta/2$	$GY_g + GY_g\Delta/2$	$GZ_g + GZ_g\Delta/2$
$R=0$, $G\neq 0$, $B\neq 0$	a_1	$GX_g - GX_g\Delta/2$	$GY_g - GY_g\Delta/2$	$GZ_g - GZ_g\Delta/2$
	b_1	$GX_g + GX_g\Delta/2$	$GY_g + GY_g\Delta/2$	$GZ_g + GZ_g\Delta/2$
	a_2	$BX_b - BX_b\Delta/2$	$BY_b - BY_b\Delta/2$	$BZ_b - BZ_b\Delta/2$
	b_2	$BX_b + BX_b\Delta/2$	$BY_b + BY_b\Delta/2$	$BZ_b + BZ_b\Delta/2$
$R\neq 0$, $G=0$, $B\neq 0$	a_1	$RX_r - RX_r\Delta/2$	$RY_r - RY_r\Delta/2$	$RZ_r - RZ_r\Delta/2$
	b_1	$RX_r + RX_r\Delta/2$	$RY_r + RY_r\Delta/2$	$RZ_r + RZ_r\Delta/2$
	a_2	$BX_b - BX_b\Delta/2$	$BY_b - BY_b\Delta/2$	$BZ_b - BZ_b\Delta/2$
	b_2	$BX_b + BX_b\Delta/2$	$BY_b + BY_b\Delta/2$	$BZ_b + BZ_b\Delta/2$

The function $f(t)$ has to be calculated again through the convolution of probability density functions. For the X-coordinate it is $f_X(t)=f_{\xi_{X_r}+\xi_{X_g}}(t-s)\otimes f_{\xi_{X_b}}(s)$, where a probability density function of trapezoidal type for example (4) is defined on the interval $[a_1+a_2;b_1+b_2]$ and a probability density function $f_{\xi_{X_b}}(s)$ of uniform distribution is defined on its own interval $[a_3;b_3]$ (see the table 1). Statistically the function $f(t)$ is a PDF of the sum of three uniformly distributed random values where each value is defined on its own interval $[a_i;b_i]$, $b_i>a_i>0$, i=1, 2, 3 and in general case $a_1\neq a_2\neq a_3$, $b_1\neq b_2\neq b_3$. Generally a piecewise parabolic type probability density function $f(t)$ of the tristimulus value distribution (as in the example of the X-coordinate) must be represented as:

$$f(t)=\begin{cases} f_1(t), a_1+a_2+a_3\le t<a_1+a_2+b_3 \\ f_2(t), a_1+a_2+b_3\le t<a_1+b_2+a_3 \\ f_3(t), a_1+b_2+a_3\le t<a_1+b_2+b_3 \\ f_4(t), a_1+b_2+b_3\le t<b_1+a_2+a_3 \\ f_5(t), b_1+a_2+a_3\le t<b_1+a_2+b_3 \\ f_6(t), b_1+a_2+b_3\le t<b_1+b_2+a_3 \\ f_7(t), b_1+b_2+a_3\le t<b_1+b_2+b_3 \\ 0, 0<t<a_1+a_2+a_3, t\ge b_1+b_2+b_3 \end{cases}, \tag{5}$$

where

$$f_1(t)=\underbrace{\frac{1}{b_3-a_3}\int_{t-b_3}^{a_1+a_2} f_{\xi_{X_r}+\xi_{X_g}}(s)ds}_{0}+\frac{1}{b_3-a_3}\int_{a_1+a_2}^{t-a_3} f_{\xi_{X_r}+\xi_{X_g}}(s)ds=\frac{\left(t-(a_1+a_2+a_3)\right)^2}{2(b_1-a_1)(b_2-a_2)(b_3-a_3)},$$

$$f_2(t)=\frac{1}{b_3-a_3}\int_{t-b_3}^{t-a_3}\frac{s-(a_1+a_2)}{(b_1-a_1)(b_2-a_2)}ds=\frac{(t-a_3)^2+2(a_1+a_2)(a_3-b_3)-(t-b_3)^2}{2(b_1-a_1)(b_2-a_2)(b_3-a_3)},$$

$$f_3(t)=\frac{1}{b_3-a_3}\int_{t-b_3}^{a_1+b_2}\frac{s-(a_1+a_2)}{(b_1-a_1)(b_2-a_2)}ds+\frac{1}{b_3-a_3}\int_{a_1+b_2}^{t-a_3}\frac{1}{b_1-a_1}ds=$$

$$=\frac{2(b_2-a_2)(t-(a_1+b_2+a_3))+(a_1+b_2)^2}{2(b_1-a_1)(b_2-a_2)(b_3-a_3)}+\frac{2(a_1+a_2)(t-(a_1+b_2+b_3))-(t-b_3)^2}{2(b_1-a_1)(b_2-a_2)(b_3-a_3)},$$

$$f_4(t)=\frac{1}{b_3-a_3}\int_{t-b_3}^{t-a_3}\frac{1}{b_1-a_1}ds=\left.\frac{s}{(b_3-a_3)(b_1-a_1)}\right|_{t-b_3}^{t-a_3}=\frac{1}{b_1-a_1},$$

$$f_5(t)=\frac{1}{b_3-a_3}\int_{t-b_3}^{a_2+b_1}\frac{1}{b_1-a_1}ds+\frac{1}{b_3-a_3}\int_{a_2+b_1}^{t-a_3}\frac{(b_1+b_2)-s}{(b_1-a_1)(b_2-a_2)}ds=$$

$$=\frac{2(b_2-a_2)((b_1+a_2+b_3)-t)+(a_2+b_1)^2}{2(b_1-a_1)(b_2-a_2)(b_3-a_3)}+\frac{2(b_1+b_2)(t-(b_1+a_2+a_3))-(t-a_3)^2}{2(b_1-a_1)(b_2-a_2)(b_3-a_3)},$$

$$f_6(t)=\frac{1}{b_3-a_3}\int_{t-b_3}^{t-a_3}\frac{(b_1+b_2)-s}{(b_1-a_1)(b_2-a_2)}ds=\frac{(t-b_3)^2+2(b_1+b_2)(b_3-a_3)-(t-a_3)^2}{2(b_1-a_1)(b_2-a_2)(b_3-a_3)},$$

$$f_7(t)=\frac{1}{b_3-a_3}\int_{t-b_3}^{b_1+b_2}f_{\xi_{X_r}+\xi_{X_g}}(s)ds+\underbrace{\frac{1}{b_3-a_3}\int_{b_1+b_2}^{t-a_3}f_{\xi_{X_r}+\xi_{X_g}}(s)ds}_{0}=\frac{(t-(b_1+b_2+b_3))^2}{2(b_1-a_1)(b_2-a_2)(b_3-a_3)}.$$

The value of the coefficients a_i, b_i in the system (5) are different for the probability density function of the *X*-coordinate $f_X(t)$, *Y*-coordinate $f_Y(t)$ and *Z*-coordinate $f_Z(t)$. There are some exact expressions in the table 3 to calculate the coefficients a_i, b_i of the system (5).

Table 3. The expressions to calculate the tristimulus values distribution PDF coefficients.

Coefficients	The tristimulus value distribution PDF		
	$f_X(t)$	$f_Y(t)$	$f_Z(t)$
a_1	$RX_r-RX_r\Delta/2$	$RY_r-RY_r\Delta/2$	$RZ_r-RZ_r\Delta/2$
b_1	$RX_r+RX_r\Delta/2$	$RY_r+RY_r\Delta/2$	$RZ_r+RZ_r\Delta/2$
a_2	$GX_g-GX_g\Delta/2$	$GY_g-GY_g\Delta/2$	$GZ_g-GZ_g\Delta/2$
b_2	$GX_g+GX_g\Delta/2$	$GY_g+GY_g\Delta/2$	$GZ_g+GZ_g\Delta/2$
a_3	$BX_b-BX_b\Delta/2$	$BY_b-BY_b\Delta/2$	$BZ_b-BZ_b\Delta/2$
b_3	$BX_b+BX_b\Delta/2$	$BY_b+BY_b\Delta/2$	$BZ_b+BZ_b\Delta/2$

It is important to notice that the system (5) is true if: $b_1-a_1\geq b_2-a_2\geq b_3-a_3>0$. To comply with this condition the distribution intervals could be previously ranked with the indices coefficients a_i, b_i redefinition.

On the figure 1 are presented some PDF $\hat{f}_X(t)$, $\hat{f}_Y(t)$, $\hat{f}_Z(t)$ evaluations for the *XYZ*-tristimulus values distribution obtained through simulation and also some theoretical estimations of probability density functions $f_X(t)$, $f_Y(t)$, $f_Z(t)$ of the *XYZ*-tristimulus values distribution which are built according to the system (5) when the *RGB* code components values are non-zero. The analysis of (4) and (5) shows that function $f(t)$ depends on technological dispersion parameter Δ of the LCD panel profile coefficients [10] and the *RGB* code components values. It means that under some non-zero values of the *RGB* code components the $f(t)$ distribution of the *XYZ*-tristimulus values is a trapezoidal (triangular) function and under the others non-zero values it is a piecewise parabolic function.

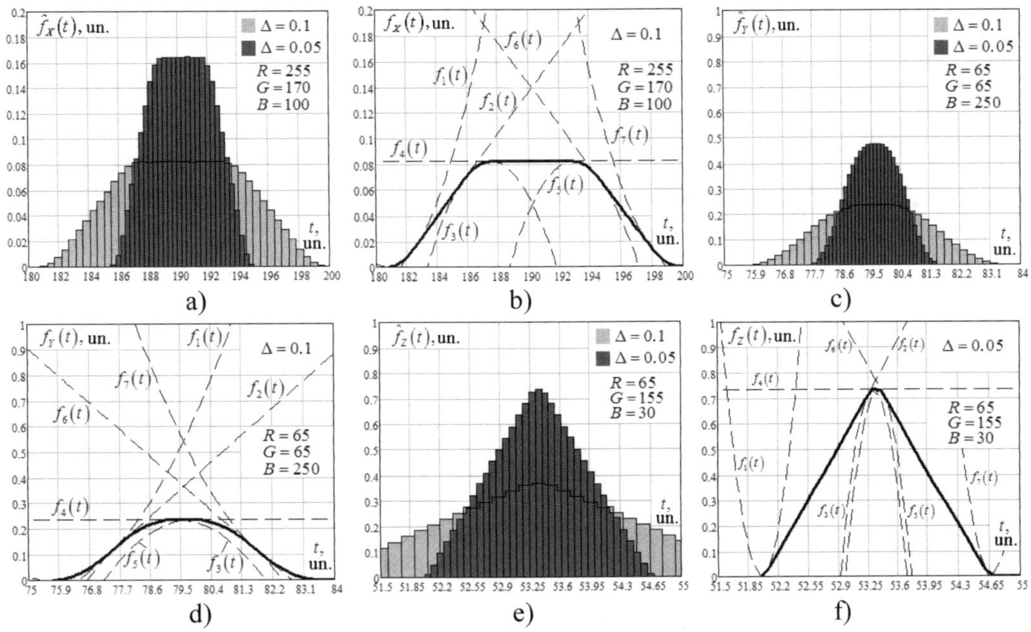

Figure 1. The evaluation of the PDF: a) $\hat{f}_X(t)$, c) $\hat{f}_Y(t)$, e) $\hat{f}_Z(t)$ of the *XYZ*-tristimulus values distribution and the piecewise parabolic PDF: b) $f_X(t)$, d) $f_Y(t)$, f) $f_Z(t)$.

5. Conclusion

The expressions from the tables 1 to 3 and systems (4) and (5) are presented analytically and can be used to define the tristimulus values distribution probability density functions for both the International Commission on Illumination standards of 1931 and 1964 with different white balance standards: D-75, D-65, D-55, D-50 etc. It's only need to define corresponding values [10] of screen profile components X_r, X_g, X_b, Y_r, Y_g, Y_b, Z_r, Z_g, Z_b of the basic conversions (1) and (2).

The obtained analytical expressions for tristimulus values distribution probability density functions may be used not only in colorimetry. The mathematical expressions of the system (5) expand the applicability boundaries of the Irwin-Hall distribution law dealing with sum of uniformly distributed random values $\xi_i \in [0;1]$ on the situation of sum of three uniformly distributed random values with different boundaries $\xi_i \in [a_i;b_i]$, $a_1 \neq a_2 \neq a_3$, $b_1 \neq b_2 \neq b_3$, $b_i > a_i > 0$.

References

[1] Livada B 2012 *Scientific Technical Review* **62** 70-79
[2] Ibraheem N A, Hasan M M, Khan R Z and Mishra P K 2012 *ARPN Journal of science and technology* **3** 265-275
[3] Masaoka K 2010 *Optics Letter* **12** 2031-33
[4] Zharinov I O, Zharinov O O 2017 *IOP Conference Series: Materials Science and Engineering* **286** 012015
[5] Bradley D M, Gupta R C 2002 *Annals of the Institute of Statistical Mathematics* **54** 689-700
[6] Sadooghi-Alvandi S M, Nematollahi A R, Habibi R 2009 *Statistical papers* **50** 171-175
[7] Gard M, Choudhary S, Kalla S L 2009 *International journal of optimization: theory, methods and applications* **1** 279-290
[8] Buonocore A, Pirozzi E, Caputo L 2009 *Statistics & Probability Letters* **79** 2092-97
[9] Potuschak H, Muller W G 2009 *Statistical papers* **50** 177-183
[10] Zharinov I O, Zharinov O O 2017 *IOP Conference Series: Materials Science and Engineering* **286** 012014

IOP Conf. Series: Journal of Physics: Conf. Series **1050** (2018) 012105 doi:10.1088/1742-6596/1050/1/012105

Nonlinear effects in materials obtained by the method of selective laser melting

A S Zhukov, B K Barakhtin, P A Kuznetsov
NRC "Kurchatov Institute" – CRISM "Prometey", Saint-Petersburg, 191015, Russia

Abstract. The analysis of the structure of composite materials, produced by additive technology from powders of steels of different chemical composition, is performed. On the basis of the obtained data, a kinetic model of the organization of the process according to the self-organizing (oscillatory) scenario is formulated.

1. Introduction

In the methods of obtaining composite materials with the required level of performance properties, the high-temperature sintering method of powder compositions has been known and applied for several decades [1]. The improvement of technological equipment and the possibility of industrial use of concentrated energy fluxes in the form of electronic, plasma, laser, and other beams initiated the development of new high-temperature selective alloying methods, called "Additive Technologies" [2-4] as ways to create bulk products from powder materials by a given (computer) 3-D model.

Practical implementation of new approaches has shown that the use of concentrated energy flows not only substantially expands the possibilities of material processing, but also requires the consideration of the features of kinetics in heat and mass transfer in the zones of melting of the powder compositions and subsequent crystallization of the melt [5–6]. It is known [7] that in the interaction volume of a powerful energy beam with the material surface, it is possible to activate self-oscillating processes that affect both the quality of the received bulk material and the efficiency (productivity) of the technology.

The aim of the work is to identify structurally conditioned and technological parameters in a volumetric metallic composite material that allow to realize an additive technological process according to a self-organizing (oscillatory) scenario.

2. Materials and experimental methods

Table 1. Chemical composition of powders

Steel grades	Basic alloying elements,% by weight						
	Cr	Ni	Mn	Si	Mo	Cu	Nb
410L	12.039	–	0.907	0.713	–	–	–
17-4 PH	15.539	3.113	0.715	0.383	–	3.407	0.175
316L	16.5	10.47	1.764	–	2.196	–	–

Monolithic composite materials in the form of rods and plates of different sizes and orientations with respect to the direction of the laser beam were obtained from dispersed powders of AISI 316L, 410L, 17-4 PH structural steels (Tab. 1) at the specialized EOSint M270 plant.

The appearance of the powder used in selective laser melting technologies is shown in Fig. 1, and the grain size distribution of steel powders is shown in Fig. 2 The laser worked in a continuous mode and allowed to vary the speed of the spot moving up to 0.8 m/s and energy up to 195 W.

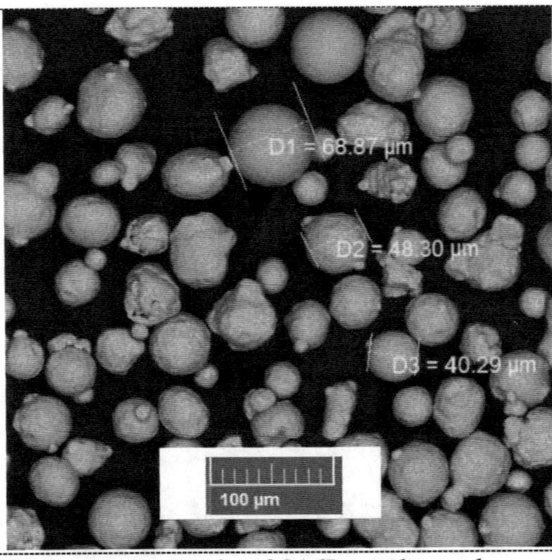

Figure 1. Example of 316L steel powder

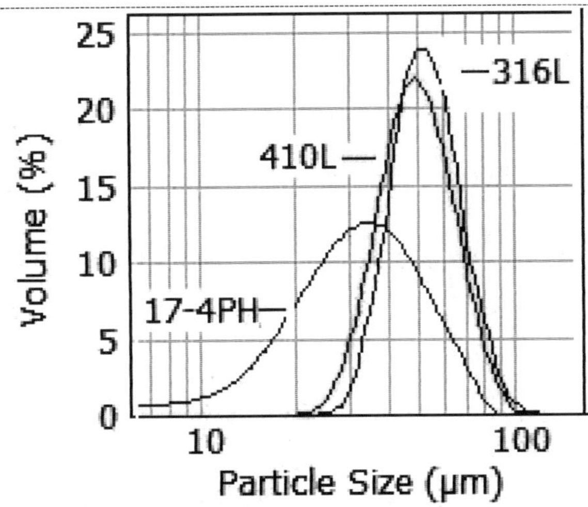

Figure 2. The grain size distribution of steel powders

The quality of the resulting composite was judged by the shape of the crystallization zones, the presence of discontinuities, and the data on the disorientation of crystallites that were observed on the thin sections of representative samples using a light (Leica DM 2500) and electronic raster (Quanta 200 3D FEG) microscopes.

3. Experimental results and discussion

Visual inspection of the produced rods and plates, as well as thin sections prepared on the analytical cross-sections of steel composites, showed that the power and velocity of the laser beam affect the reproduction of the given 3D shape of the sample and the quality of the metal crystallization zones (Fig. 3).

For example, in the sequence of acts of melting, mixing and crystallization of a metal, the increase in laser power is accompanied by a concentration of temperature and pressure, which leads to an increase in the volume of the alloy being melted. With increasing speed of moving the laser beam, the volume decreases, but the width of the processing zone remains. In such heterogeneous "temperature-geometric" conditions, crystallization starts from the melted powders and the epitaxy mechanism extends perpendicular to the isotherms. The trajectories of the isotherms determine the direction and curvature of the growing crystallites. With increasing speed of laser movement, the solidification front is stretched in the direction of its motion and the probability of formation of equiaxed grains of recrystallization appears.

IOP Conf. Series: Journal of Physics: Conf. Series **1050** (2018) 012105 doi:10.1088/1742-6596/1050/1/012105

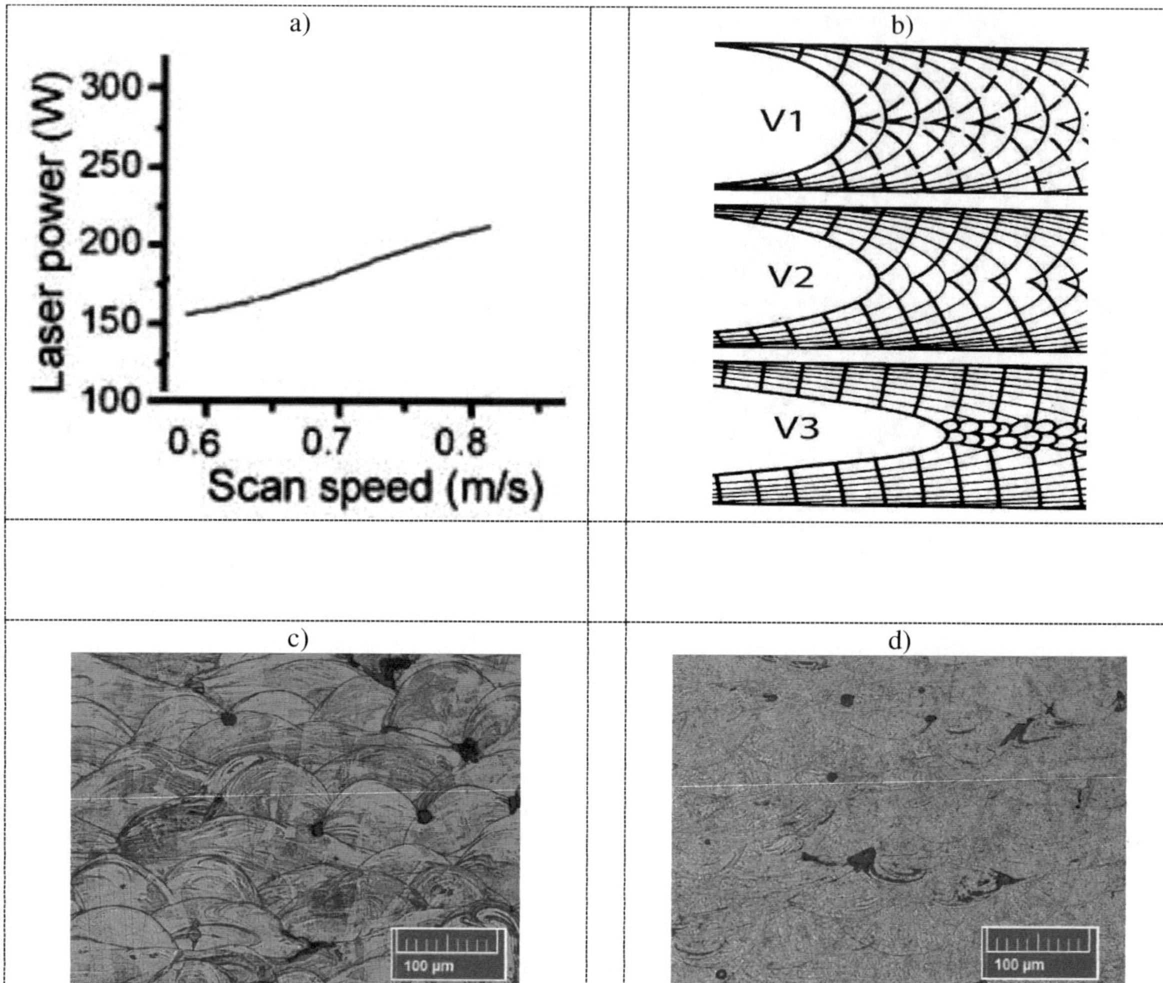

Figure 3. (a) Change in power and speed of the laser beam. **(b)** Scheme for changing isotherm lines (fine lines) and crystallite growth directions (thickened lines) when the beam moves in the direction of the arrow with velocities $V_3>V_2>V_1$. **(c, d)** Quasiregular structure of the obtained composites 316L and 17-4 PH with nonmetallic inclusions and pores

The composite obtained by selective laser melting is characterized by a quasiregular lamellar structure, caused by fluctuations in the chemical composition in the volume of melting and crystallization. On the thin sections of the samples in EBSD electrons, arched periodic traces of stopping of the crystallization fronts are observed with steps of 20±10 μm, due to temperature pulsations and concentration inhomogeneities of impurities from low-melting or silicate eutectics (Fig. 4)

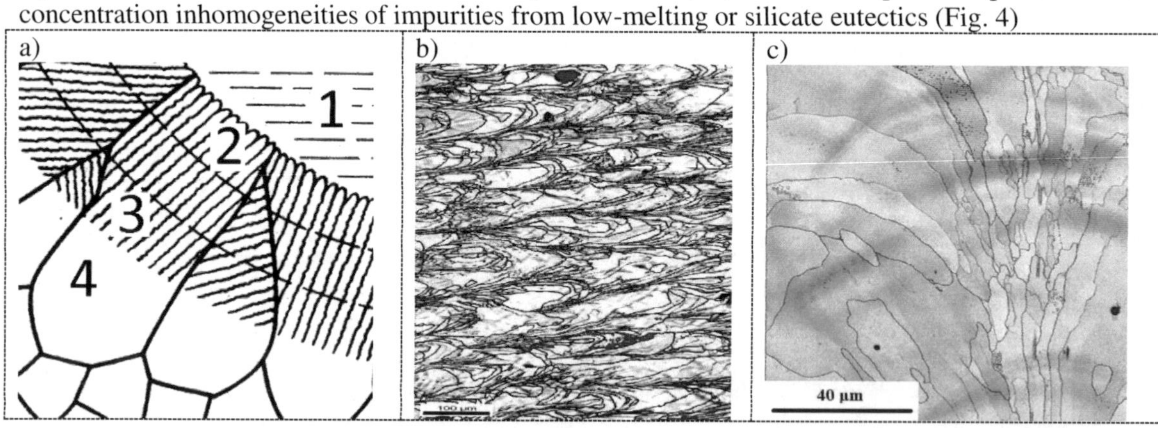

Figure 4. (a) The crystallization sequence along the direction of the temperature field gradient from the fusion zone (1) through the epitaxy layer (2) and the primary crystallites (3) to the quasiregular structure (4). **(b, c)** In the rays of EBSD electrons, an example of typical images of thin sections of the obtained 410L steel composite at different optical magnifications

When studying crystal orientations, a transition zone consisting of an epitaxy layer (2) and primary crystallites (3) is not detected. The orientation lines indicate only the contours of the melting zones of the powder particles. Due to the epitaxial growth of the crystals formed during solidification from the melting zone, the size of the primary crystallites approaches the size of the powder particles of the base metal.

The results obtained did not contradict the published data, for example, [8-9]. At the same time, it was noted that the shape of the grains that crystallized along the direction of the temperature field gradient was periodically reproduced in steps of 150±50 μm.

Since the motion of the laser beam was carried out at a constant rate, possible pulsations of the thermal effects, by analogy with [10], can be related to the features of the interaction of the beam with the molten powder in the sequence of the "laser beam – metal vapor over the melt zone – melt". In this formulation of the physical model of the process, the metal vapor above the melt is a "control valve" in the path of the introduced thermal energy.

Under the condition that the parameters of the laser beam are constant, the nonlinear process of exchanging thermal energy and matter in the form of vapors and melt in its simplest form can be represented by a system of kinetic equations:

$$\begin{cases} \frac{dT}{dt} = \alpha T - \beta T P \\ dP/dt = \gamma T P - \delta \eta \end{cases} \quad (1)$$

where T is the temperature in the interaction zone of the ray with matter, t is the time, P is the vapor density of the melted substance, α, β, γ, δ are the coefficients depending on the physical properties (viscosity η) and the chemical composition of the melted substance.

The meaning of the above equations is that the changes in energy (temperature field) and aggregate quality of the substance (vapor density) in the path of the input energy are determined by their parameters at each moment in time, taking into account possible losses. For example, (β T P) is the energy expenditure for evaporation, and (δ η) is the "escape" of steam to the condensate (epitaxy layer).

Indirect signs of the validity of the formulated scheme for the exchange of energy and matter are the nonlinear (with maximum) ratio between the power and the speed of the beam during the formation of the melted object, as well as the influence of the speed of the beam on the quality of the crystallization of the metal.

The determination of the coefficients and the verification of the operability of the model are a task of subsequent experiments. Referring to the results of the analytical studies presented in [7, 10], one can be sure that in the production of parts by additive technology from the assigned powder substance it will be possible to determine experimentally or to select the numerical values of the coefficients at which the solution of the proposed system of equations will correspond to the oscillatory scenario. In this case, an adequate technological process will satisfy the conditions of self-organization, in which the quality of consolidation of the substance will be the best, and the productivity will be going to the maximum.

4. Conclusions

Based on the results of the work performed, it can be stated that the dominant parameters of the technology for these powders' parameters are the power of the input energy (laser power), the speed of the beam and the physical properties of the evaporated substance (viscosity of the melt vapor).

Acknowledgments

This work was supported by the grant of Russian Scientific Fund №15-19-00210. Experimental studies were performed on the equipment of the collective use center "Composition, Structure and Properties of Structural Materials" NRC "Kurchatov Institute" – CRISM "Prometey", Saint-

Petersburg, Russia, supported by the Ministry of Education and Science of the Russian Federation (unique subsidy identifier – RFMEF159517X0004).

References

[1] Samokhin A V, Alekseev N V, Tsvetkov Yu V 2006 *Plasma-assisted processes for manufacturing nanosized powder materials.* High Energy Chemistry **2(40)** pp 120–126
[2] Barakhtin B K, Bobyr V V, VoznIuk A V et al. 2017 *Optimization of the technological parameters and determination of selective laser melting modes of 316L-based composition powder.* Problems of Materials Science **2(90)** pp 146–152
[3] de Lima M S F, Sankaré S 2014 *Microstructure and mechanical behavior of laser additive manufactured AISI316 stainless steel stringers*. Materials and Design **55** pp 526–532
[4] Bobkova T I, Bystrov R Yu, Farmakovsky B V et al. 2014 *Plasma-chemical synthesis of nanoscale aluminum oxide powders and their use as a reinforcing components in the microplasma spraying.* Problems of Materials Science **1(77)** pp 73–78
[5] Akhmanov S A, Emelyanov V I, Koroteev N I, Seminogov V N 1985 *Effects of powerful laser radiation on the surface of semiconductors and metals: nonlinear optical effects and nonlinear optical diagnostics.* Successes of Physical Sciences **4(147)** pp 675–745
[6] Barakhtin B K, VoznIuk A V, Deev A A, Zhukov A S 2017 *Structure-mechanical condition of additively sintered material under conditions of hot plastic deformation.* Deformation and Destruction of Materials **2** pp 8–14
[7] Uglov A A, Selischev S V 1987 *Self-oscillatory processes under the action of concentrated energy flows* (Moscow: Nauka) 148 p
[8] Hanzl P, Zetek M, Bakša T, Kroupa T 2015 *The Influence of Processing Parameters on the Mechanical Properties of SLM Parts.* Procedia Engineering **100** pp 1405–1413
[9] Qiu C, Yue S, Adkins N J E et al. 2015 *Influence of processing conditions on structure and compressive properties of cellular lattice structures fabricated by selective laser melting.* Materials Science & Engineering **A628** pp 188–197
[10] Zuev A V 1998 *Treatment of materials with concentrated energy flows* (Moscow: Moscow Power Engineering Institute) 162 p

Diagnostics and evaluation of the residual life of an induction motor according to energy parameters

Y Zhukovskiy, N Koteleva

St. Petersburg Mining University, 2, 21 Line of Vasilyevsky Island, St. Petersburg, 199106, The Russian Federation

Email: zhukovskiy_yul@pers.spmi.ru; Koteleva_NI@ pers.spmi.ru

Abstract. The various devices based on vibration, thermal, electromagnetic and other methods are used to diagnose an asynchronous motor today. The use of vibration diagnostics is common for large machines. The emergency stop of the large machines leads to significant downtime and non-produced products meaning that the enterprise incurs significant material losses. However, for low-power engines, vibration diagnostics are used only as a periodic assessment of the state without accumulation of data and their subsequent analysis. It should also be noted that the vibration signals measured from the rolling bearing are affected by varying operating conditions and noise that distort the signal and complicate the faults identification for further evaluation of the residual life. In a number of cases, such as submersible pumps, nuclear power plant safety aggregates, mining equipment, it is practically impossible to use vibration, thermal and other methods for diagnosing asynchronous motors. Therefore, the actual task of this research is the development of a method for diagnosing and estimating the residual life in conditions of the systems without any maintenance, based on the electric motor energy parameters.

1. Introduction

Electric motors, like all electromechanical equipment, are subject to wear and tear, and their operation often causes malfunctions, breakdowns or operation with parameters that differ from the nominal values. Since electric power is converted into mechanical energy in the electric motor, it is obvious that the malfunctions of electric motors can be caused both by malfunctions in electric and electromagnetic systems, and by defects in mechanisms[1,2]. In the manufacture and operation of asynchronous motors, there are deviations in the engine parameters from the nominal values [3]. These deviations can be caused both by technological errors in the manufacture of the engine, by incorrect operating conditions, and by wear during operation.

One of the most common defects are those associated with the electric motor bearings, the rotor cage, and also with the change of the amount of looseness between the rotor and the stator. Defects of rolling bearing are one of the main causes of the engine disintegration of electrical machines, so diagnostics and early detection of designing defects are necessary. It makes possible to prevent more catastrophic consequences of a failure. Operation of the electric motor with defects of the rolling bearing does not lead to its immediate failure, but reduces reliability of an operation, a durability and other technical and economic indicators. The magnetic field in the air gap distorts, the efficiency is reduces by 1–2% [7], additional ultra-harmonics of the consumed current appear, the starting torque reduces by 10-12% [3, 4], the heat loss by 4–5% [6].

2. The essence of the diagnostic method

The essence of the proposed method is that the basis for the application of electrical equipment and associated devices is the transform of electrical energy [1]. At the same time, a part of the energy is lost in the electrical equipment and associated devices, so any thier damage, one way or another, leads to increased losses and lower coefficient of efficiency. For technically sound electrical equipment, the dependence of losses (a coefficient of efficiency) from load is known, and in the event of a malfunction this dependence changes. The method provides registration and accumulation of values of electrical quantities (currents, voltage and power) of the controlled object and creation of a pattern set of calibration values corresponding to the resource of the testing equipment. If there is the sufficient statistical data, it is possible to predict the dynamics of these changes in the future, i.e. estimate the residual resource. More detailed information on the causes of the loss is carried by oscillography charts of instantaneous power and by power constructed for individual frequency components, after the Fourier transform.

An integral assessment of the state of the equipment is formed as a set of estimates of all nodes and subsystems, therefore for an asynchronous motor as a whole, the state estimation will be formed on the basis of parallel chains of diagnostic parameters for each detected defect. The resulting residual resource will be determined by the most rapidly developing and the least manageable defect [2]. Owning full information about the detected defects and the possible influence of each of them on the residual life of the product, it is possible to solve the task of determining the amount of services work necessary to bring the life of the product to the required level. Knowing the causes of each defect and the factors that affect its development, it is possible, by influencing the causes and factors, to suspend or slow down the development of the most critical defects, thus saving the performance capabilities. For this, the use of IoT technology is proposed. This technology allows an accumulating, aggregating and analyzing large amounts of data "Big Data" to identify the causes of defects using artificial neural networks and machine learning.

3. The electrical parameters analyzing

The system receives electric power from the industrial network with power P_0 , an electric transducer transforms the electrical energy parameters, giving out at the output energy with power P_1 , In the process of transformation, part of the energy is lost as a loss $\Delta P_1 = P_0 - P_1$. Similarly, the energy receives on the input of electromechanical transducer P_1 , part of which $\Delta P_2 = P_1 - P_2$ is lost during the transform process, on the output remains P_2 . For mechanical transducer, accordingly, on the input is received P_2 , in the transform process is lost $\Delta P_3 = P_2 - P_3$, on the output remains P_3 .

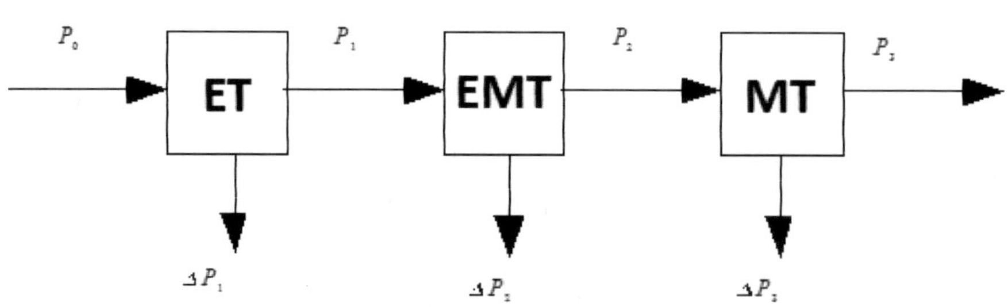

Figure 1. Structural circuit of energy transform in the electric motor

In the static mode, the conversion factor of each of the sides of the circuit is characterized by a constant transmission factor, the coefficient of efficiency. Accordingly, for the power on the output of the i-th transducer, expression

$$P_{i-1} = \eta_i \cdot P_i ,$$

where P_{i-1} – the power on the output of the transducer, W.; P_i - the power on the input of the block, η_i - the coefficient of efficiency of the transducer, it equals

$$\eta_i = \frac{P_{i-1}}{P_i} = \frac{P_i - \Delta P_i}{P_i} = 1 - \frac{\Delta P_i}{P_i} ,$$

where ΔP_i – value of losses in the transducer, W.

It should be noted that the summand $\frac{\Delta P_i}{P_i}$ in essence represents the amount of losses in the transducer, referred to the consumed power of the transducer. So, the new designation for this summand have to be introduced

$$\rho_i = \frac{\Delta P_i}{P_i} . \tag{1}$$

With different loads of the transducer, that is, with different values of the output power P_{i-1}, the efficiency is different. The dependence $\eta_i = \eta_i(P_{i-1})$ is usually given in the technical data of the corresponding transducer, or can be obtained experimentally, for a technically sound transducer. It should be taken into account that this dependence, as a rule, is nonlinear. If this relationship is known, then it is possible to determine the dependence of the parameter ρ_i on the output power P_i .

$$\rho_i = 1 - \eta_i(P_i) = \rho_i(P_i). \tag{2}$$

The electric motor contains three transducers connected in series, therefore, one can write

$$P_3 = \eta_3 \cdot \eta_2 \cdot \eta_1 \cdot P_0 ,$$

or,

$$P_3 = [1 - \rho_1(P_1)] \cdot [1 - \rho_2(P_2)] \cdot [1 - \rho_3(P_3)] \cdot P_0 . \tag{3}$$

During operation of the electric motor, the transducers accumulate damage that leads to a deterioration in their energy characteristics. The consumption power is increasing with present the constant output power, and for various types of damage, the additional component of loss will depend on the output power in different ways.

We denote additional losses in the i-th transducer $\xi_i(P_i)$, then

$$P_3 = [1 - \rho_1(P_1) - \xi_i(P_i)] \cdot [1 - \rho_2(P_2) - \xi_i(P_i)] \cdot [1 - \rho_3(P_3) - \xi_i(P_i)] \cdot P_0 . \tag{4}$$

Losses in each transducer using the function of the consumed power can be determined for state of unit in a technically sound and for the state of unit in the real time. Comparison of these dependencies allows us to estimate the value of additional losses caused by the development of damage, which is used as an assessment of the technical condition of the electric motor (Fig. 2).

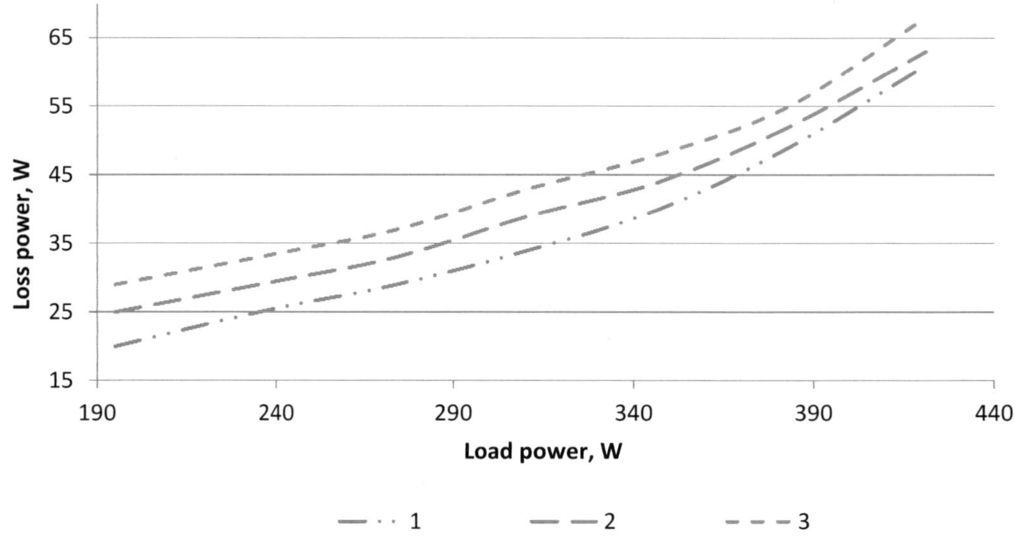

Figure 2. Energy losses for various faults and loads: 1 – technically sound, 2 – damaged 2 rotor winding rods, 3 – damaged 6 rods and defects in foundation installation

As the electric motor generates a resource, the amount of power loss in the electric motor at the same load increases. In this case, the coefficient of change in losses equal to the difference of the relative power consumed by the electric motor at a given load value can serve as an indicator of its technical condition.

$$\Delta\rho(P_2)=\frac{(P_{1b}-P_{1m})}{P_{1b}}, \tag{5}$$

where – P_{1b} initial value of power consumption of the electric motor under load P_2, measured on a technically sound motor in the initial period of its operation, kW; P_{1m} – the current value of the motors power consumption under load P_2, measured during its operation, kW.

In this article the method for determining the energy losses in the electric drive of mining machines is proposed. At the beginning of operation, on a known serviceable motor, during the shift, the power consumption P_{1i} of the electric motor is measured N times in the time interval

$$\Delta t_i=\frac{T_{sh}}{N}, \tag{6}$$

Where T_{sh} – the time of a shift, hr.

The shorter the time interval Δt_i, the higher the reliability of the measurements

Further, the average value of the power consumption $\overline{P_1}$ of the electric motor is determined and its standard deviation $S_{\overline{P_1}}$ according to the formulas:

$$\overline{P_1}=\sum_{i=1}^{N}\frac{P_{1i}}{N}, \tag{7}$$

$$S_{\overline{P_1}}=\sqrt{\sum_{i=1}^{N}\frac{\left(\overline{P_1}-P_{1i}\right)^2}{N-1}}. \tag{8}$$

IOP Conf. Series: Journal of Physics: Conf. Series **1050** (2018) 012106 doi:10.1088/1742-6596/1050/1/012106

According to the "Three sigma rule" with a probability of not less than 99.7%, the value of the consumed power should fall within the interval

$$P_1 \in \left(P_{1\max}, P_{1min}\right), \tag{9}$$

where

$$P_{1\min} = \overline{P_1} - 3 \cdot S_{\overline{P_i}}, \tag{10}$$

$$P_{1\max} = \overline{P_1} + 3 \cdot S_{\overline{P_i}}. \tag{11}$$

Further, at intervals of time T_{diag}, the value of which should be determined on the basis of statistical data on motor failures, the power consumption is measured and the average $\overline{P_1}'$, minimum P'_{1min}, maximum values P'_{1max} of the power consumption during the shift are determined. Next, the average value of the coefficient of loss variation is determinates

$$\Delta\rho_{av} = \frac{\overline{P_1}' - \overline{P_1}}{P_{1b}}, \tag{12}$$

Then, the value of the coefficient of loss variation at the minimum and maximum load

$$\Delta\rho_{min} = \frac{P'_{1min} - P_{1min}}{P_{2nom}}, \tag{13}$$

$$\Delta\rho_{max} = \frac{P'_{1max} - P_{1max}}{P_{2nom}}. \tag{14}$$

where P_{2nom} – nominal value of mechanical power of electric motor, kW.

The results of calculation of losses according to formulas (12) - (14) characterize the technical state of the motor, if these values exceed the permissible limits, the value of which should be determined on the basis of statistical data on the failures of electric drives of this type, it should be concluded that a malfunction is likely.

The residual life evaluation

The technical resource T_p is usually expressed in units of time and characterizes the time interval during which, with a given probability, the diagnosed object does not go over to the limit state. It should be noted that the reasons for the failure of these or other parts of the electric drive are the energy losses in them, in particular, the life of the insulation of the windings of the electric machine is determined by its heating and is proportional to the square of the current consumed and, accordingly, to the losses of power consumption.

The concept of an energy resource Q is added. It equals to the reduced value of losses in a motor during the time before its transition to the limit state

$$Q = \rho T_p. \tag{15}$$

The nominal energy resource can be defined as

$$Q_{nom} = \rho_{nom} T_{p.nom}. \tag{16}$$

The theoretical and experimental dependences of the energy losses and the efficiency of the drives of mining machines as functions of the consumed electric power, current, as well as parameters characterizing the operating modes of the mining machine, in particular, on the feed rate, are presented in [2, 6, 7]. These dependences are non-linear, especially with high loads. To account for energy losses during overloads above nominal, it is proposed to introduce an overload coefficient K_{ov} characterizing the increment of losses in the motor during overload and determined by the performance characteristics of this motor.

Thus, by monitoring the power consumption at certain intervals, it is possible to estimate the amount of spent energy resource from expression

$$Q_0 = \sum_{i=1}^{N} K_{ovi} \cdot \rho_i \Delta t_i \,, \tag{17}$$

where K_{ovi} – coefficient of overload in *i*-th measurement; ρ_i – reduced losses during the *i*-th measurement, determined by the performance characteristics: Δt_i – duration of the time interval between measurements, during which the load is assumed unchanged, h;

Using the concept of the resource, expressed by the formula (17), the expected residual resource can be determined:

$$T_{p.ex} = \left(1 - \frac{Q_0}{Q_{nom}}\right) T_{nom} \,. \tag{18}$$

With a polyharmonic composition of the consumed power:

$$\rho_i = \sum_{k=1}^{M} \rho_{ik}(\omega_k) \,, \tag{19}$$

where $\rho_{ik}(\omega_k)$ – *k*-th harmonic component in the loss spectrum, M – is the number of harmonics of the spectrum analyzed.

It is suggested in the expression to take into account that individual components of the spectrum have a greater impact on energy resources, introduction of weighting factors

$$\rho_i = \sum_{k=1}^{M} K_{hhk} \rho_{ik}(\omega_k) \,, \tag{20}$$

where K_{hhk} – weight coefficients that take into account the significance of the *k*-th harmonic of the spectrum.

Taking into account the above, for the expected residual resource the expression (17) takes the form:

$$Q = \sum_{i=1}^{N} K_{ovi} \left[\sum_{k=1}^{M} K_{hh.\kappa}(\omega_k) \rho(\omega_k) \right] \Delta t_i \,. \tag{21}$$

Substituting expression (20) in (18), we obtain for the remaining resource:

$$T_{p.ex} = \left(1 - \frac{Q_0}{Q_{nom}}\right) T_{p.nom} \sum_{i=1}^{N} K_{ovi} \left[\sum_{k=1}^{M} K_{hh.k}(\omega_k) \rho(\omega_k) \right] \Delta t_i \,.$$

4. Conclusions

To solve the problems of diagnostics of electric motors located in unattended, hard-to-reach technological zones, information-measuring methods are needed that allow estimating the remaining resource by electrical parameters. The proposed methodology is based on the collection, the aggregation, the classification and the analyzation of statistical data of electrical quantities and control of the consumed electric power in order to determine the remaining life of the electric motor. Comparison of load loss diagrams allows you to obtain knowledge about the current state of the machine and can be used to predict the residual resource. If it is necessary to determine the type of damage, oscillography charts of the power transformed in a Fourier series can be used. Monitoring of the power change of losses at characteristic frequencies will determine the type and level of the defect. For the practical application of this method, a large array of statistical data is needed that will allow us to extract knowledge of the changes in the frequency components of power and to fix a diagnostic picture of the initial and critical state

5. Acknowledgments

The presented results were obtained as a part of scientific researches according to the contract № 13.3746.2017/8.9 within the scope of the State task "The designing on the base of systematic and logic probability evaluations of rational and economically proved structure of centralized, autonomous and combined power supply systems with high reliability and stability level with usage of alternative and renewable power sources for uninterrupted power supply of enterprises with continuous technological cycle"

References

[1] Korolev N A and Solovev S V 2017 *IOP Conference Series: Materials Science and Engineering* **177(1)** 012007
[2] Zhukovskiy Yu L, Korolev N A, Babanova I S and Boikov A V 2017 *IOP Conf. Series: Earth and Environmental Science* **87** 032055
[3] Haidong Shao, Hongkai Jiang, Xun Zhang and Maogui Niu Published 2015 *IOP Publishing Ltd Measurement Science and Technology* **26(**11)
[4] Zhukovskiy Y and Koteleva N (2017) *IOP Conference Series: Materials Science and Engineering* **177(1)** 012014
[5] Belsky A, Skamyin A and Iakovleva E 2016 *International Journal of Applied Engineering Research* **11(1)** pp 233–238
[6] Kozyaruk A 2011 *Notes of the Mining Institute* **192** pp 100–104
[7] Leonidovich Z Y, Urievich V B 2015 *International Journal of Applied Engineering Research* **10 (20)** 41150–41155
[8] Pashkov E N, Martyushev N V, Ponomarev A V 2014 *IOP Conference Series: Materials Science and Engineering* **66 (1)** 012014
[9] Pashkov E N, Martyushev N V, Masson I A 2014 *IOP Conference Series: Materials Science and Engineering* **66 (1)** 012011
[10] Shevchuk V A, Muravlev O P, Stolyarova O O, & Shevchuk V P (2016) *MATEC Web of Conferences* **91** 01033

The influence of ultrasound on physico- mechanical properties of composite materials reinforced with carbonaceous fibers in the formation process

I V Zlobina, N V Bekrenev and G K Muldasheva

Yuri Gagarin State Technical University of Saratov, 77 Politechnicheskaya street, Saratov, Russia, 410054

Abstract. The process of reinforcement in the ultrasonic field of structural elements from ABS plastic composite structures containing carbon fibers was studied. The fact of formation of an adhesive-strong interface of "composite-plastic" under the action of ultrasound on the sample during the curing of the composite is established. It is shown that ultrasonic action causes the formation of a homogeneous morphology of the surface and structure of the composite layer, as well as a change in the ratio of the strength parameters of the material. In tensile tests, an increase in strength of 18 % was found with a certain decrease in the elongation.

1. Introduction

The development of modern technology and its production is characterized by a wide use of non-metallic composite materials [1] and digital technologies, one of the directions of which are three-dimensional printing technologies, or additive technologies [2]. Among the various methods of three-dimensional printing, the most common are the technologies of selective laser sintering of products from metal powders and layer-by-layer fusing with polymer filament. However, composite materials are characterized by pronounced anisotropy of physicomechanical characteristics, determined by the type and orientation of the reinforcing components [3]. Objects made of polymeric materials, obtained on the basis of additive technologies, are characterized along with significant heterogeneity of the structure by instability of properties from product to product and from installation to installation [2], which limits the scope of this technology by rapid prototyping and making models for foundry production. At the same time, the possibilities of additive technologies make it possible to strengthen reinforcement of objects from polymeric materials by introducing composite structures containing, for example, carbon fibers. However, to increase the strength of such products, it will be of a great importance to have an adhesive interaction between the reinforcing composite and the polymer base. At the same time, due to the insignificant strength of the polymer structure for compression, which does not allow the use of significant loads in pressing the composite, it is very difficult to obtain a uniform impregnation of the composite with a binder. It is known that ultrasonic vibrations generated in liquid media contribute not only to the development of cavitation processes, but also to a significant intensification of capillary effects [4, 5].

2. Formulation of the problem

The purpose of our research was to study the possibility of improving the uniformity of impregnation of a composite based on an epoxy matrix filled with carbon fibers, as well as improving its contact interaction with a plastic substrate in an ultrasonic field.

3. Research methods and equipment

In the experiments we used samples of dimensions 120x40x4.5 mm from the thermoplastic ABS, obtained by three-dimensional printing using FDM technology from a 1.75 mm thick thread with a depth of 3 mm filled with carbon plastic. We used the 3D printer Felix 3.1 Single Extruder for printing. The cavity was layer-by-layer filled with carbon fiber produced by LLC Balakovo Carbon Production (Balakovo, Saratov Region) and epoxy resin ED-20 with PEP hardener. 10 samples were prepared and 5 samples after filling with a composite were subjected to ultrasound through the lower surface of the sample using a piezoceramic transducer (Figure 1).

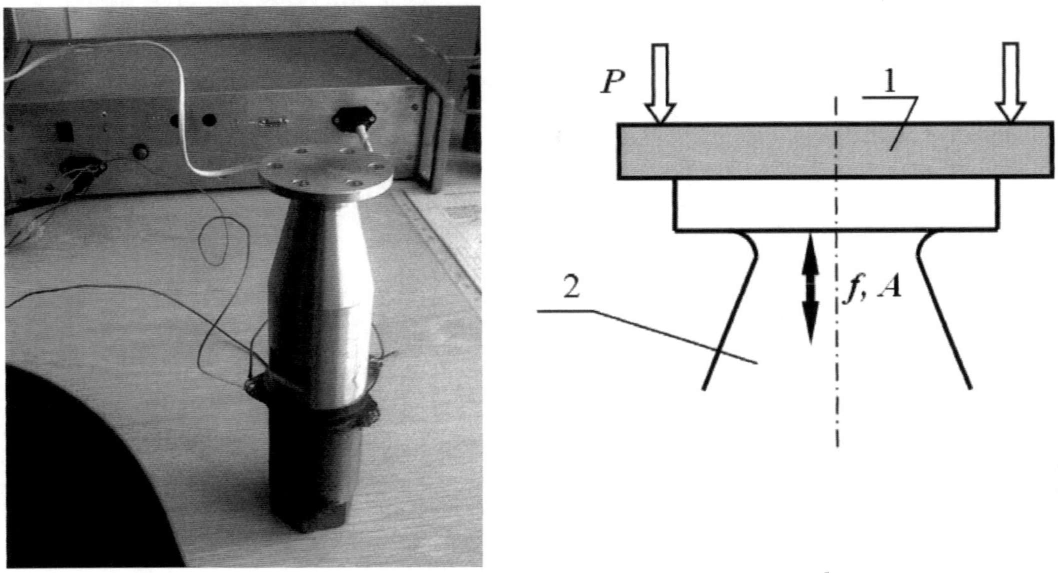

a *b*

Figure 1. Experimental setup (a) and the loading scheme of the sample (b) 1 – sample filled with a composite, 2 – source of ultrasonic vibrations

The converter was supplied with power from the designed by our technical task in the SSTU ultrasonic generator with a programmable frequency range from 20 to 60 kHz with the ability to fine-tune the joystick with an accuracy of 10 Hz. The resonant frequency of the converter was 23600 Hz, the amplitude of the oscillations was regulated from 5 to 15 μm by varying the generator voltage from 120 to 200 V. The processing was carried out at the attained maximum ultrasound intensity in the demonstration mode at an amplitude of 12 μm. After curing, the control and ultrasonically generated samples were subjected to tensile tests on a tearing machine IR-5082-100 with computer processing of the results. The surface of the composite layer and its contact area with plastic was examined with a digital microscope Digital Microscope 2.0 MP 1000X (GAOSUO, China) with an increase in x500, displaying on the laptop screen.

4. Results and discussion

The test results are shown in FIG. 2 and in the table. It can be seen that the curing of the composite structure with ultrasound influence causes a change in the ratio of the force parameters characterizing the process of deformation and destruction of the sample. With an insignificant increase in the maximum destructive load, loads corresponding to the elastic limit of the material increase significantly (by 14–18 %). In this case, the yield strength is reduced by 70 % and the duration of plastic deformation of the sample decreases. This corresponds to a twofold decrease in the relative elongation.

a

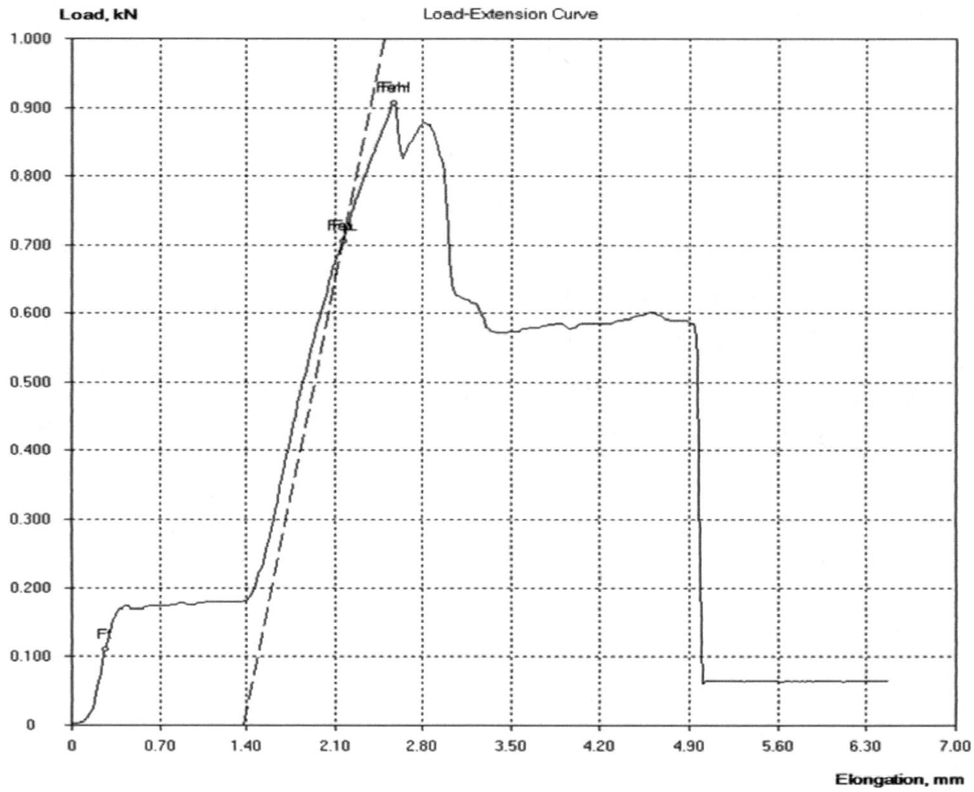

b

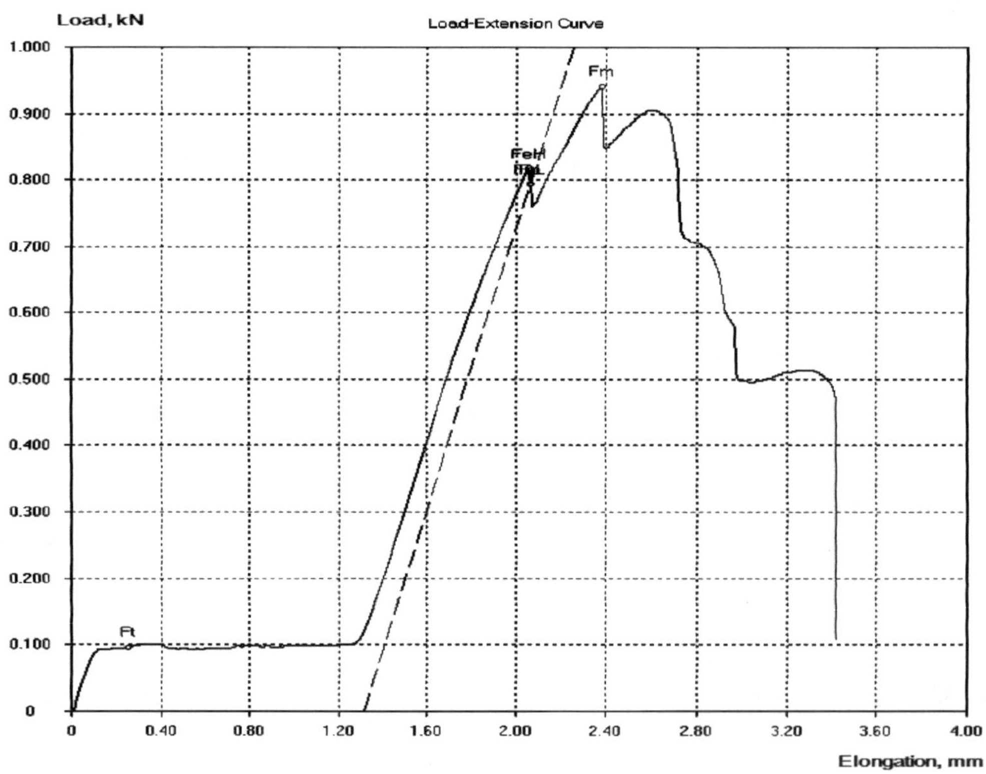

Figure 2. Loading schedules of samples: with a composite structure (a) cured under normal conditions, with a composite structure cured by ultrasonic vibrations (b)

Table 1. Mechanical characteristics of the control and ultrasound-generated samples obtained during the tensile tests.

Parameter / Sample	Fm, kN	Feh, kN	Fel, kN	Fp, kN	Ft, kN	A,%
Control sample	0.9	0.7	0.7	0.7	0.17	7.5
After ultrasound processing	0.95	0.83	0.8	0.8	0.1	5.0
Parameter change	+5 %	+18 %	+14 %	+14 %	- 70 %	- 50 %

At the same time, the maximum modulus of elasticity increases by almost 20 %. The spread of the values of all the listed parameters for the control sample was from 25 to 35 %, for samples formed with ultrasound – did not exceed 12 %.

The obtained results can be explained by analyzing the micrographs of the formed composite material (Figures 3 and 4). On all control samples, the separation of the reinforcing structure from the thermoplastic substrate is clearly visible. Apparently, almost separate perception of the tensile load by the substrate and the composite caused a decrease in some force parameters of the deformation of the control sample and especially its elastic modulus in comparison with the treated one, since the peeling of the reinforcing structure is of a random nature and the number of bond points is different for different samples. In the structure of the reinforcing composite,there are seen separate voids , not filled with a binder. The surface is characterized by a highly heterogeneous morphology. These results are apparently explained by the poor wettability of the carbon fibers and the incomplete flow of the binder into the gaps between them in the absence of evacuation or surface pressure.

The contact zone "composite-plastic", formed under conditions of ultrasonic influence, is solid. Clear penetration of the binder into the structure of the plastic substrate is seen. The local penetration of the binder through the entire layer of the substrate to the external surface contacted with the ultrasonic radiator was visually determined. The structure of the composite is more uniformly filled with a binder, there are practically no voids. The morphology of the surface has a smooth, wavy, repetitive appearance, which is connected to acoustic currents in the uncured composite.

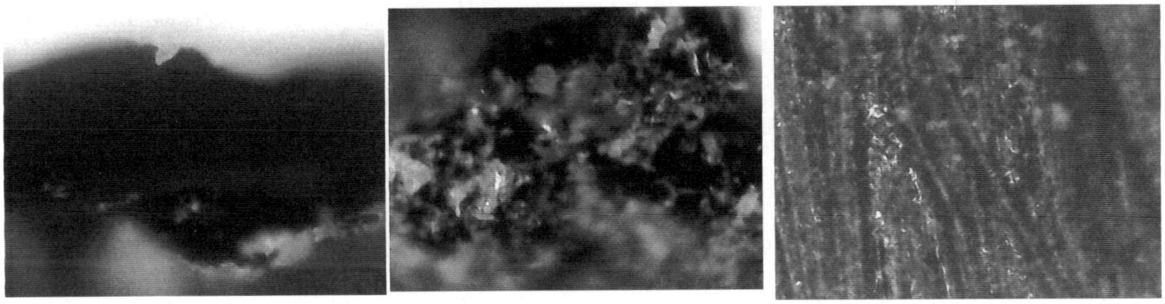

a *b* *c*

Figure 3. The contact zone "composite-plastic" (a), the microstructure of the composite layer (b) and the morphology of the surface (c) of the control sample

a *b* *c*

Figure 4. Contact zone "composite-plastic" (a), microstructure of the composite layer (b) and morphology of the surface (c) of the sample formed under influence of ultrasound

5. Conclusion

The noted features may be the reason of an increase in the elasticity modulus of the samples and a decrease in the relative elongation due to the joint perception of the load by both the plastic of the substrate and the composite, as well as an increase in the uniformity of the strength characteristics and an increase in the parameters determining the increase in the elastic limit. To optimize the elastoplastic characteristics of composite materials formed under conditions of ultrasonic influence, it is advisable to conduct studies on the selection of the frequency range and intensity of ultrasound, taking into account the wave properties of the substrate materials and reinforcing fibers, as well as the cavitation threshold of the binder.

6. Acknowledgments

The studies were carried out with the support of RFBR grant No. 17-03-00720 "Methodology of optimizing micro construction of composite materials for objects of complex shape with increased dynamic strength, formed in layers by electrotechnological methods".

References

[1] Mirny M 2016 *The world market of carbon plastics will reach $ 23 billion by 2022* (Electronic Resource)

[2] Bourell D L, Beaman J J, Leu M C and Rosen D W 2009 *A Brief History of Additive Manufacturing and the 2009 Roadmap for Additive Manufacturing* Looking Back and Looking Ahead, *U. S. Turkey Workshop on Rapid Technologies*

[3] Kablov E N 2012 *Materials and chemical technologies for aviation equipment* Vestnik Rossiiskoi Akademii Nauk T 82 Vol 6 pp 520–530

[4] Rosenberg D 2012 *Physics and Technology of High-Power Ultrasound* Vol 3 *Physical fundamentals of ultrasonic technology* Moscow *Book on Demand* 689 p

[5] Prikhodko V M, Medelyaev I A and Fatyukhin D S 2015 *Formation of operational properties of machine parts by ultrasonic methods: monograph* Moscow *MADI* 264 p

IOP Conf. Series: Journal of Physics: Conf. Series **1050** (2018) 012108 doi:10.1088/1742-6596/1050/1/012108

Strengthening microwave modification of structural elements of composite materials reinforced with synthetic fibers

I V Zlobina[1], N V Bekrenev[1] and D P Teterin[2]

[1]Yuri Gagarin State Technical University of Saratov, 77 Politechnicheskaya street, Saratov, Russia, 410054
[2]JSC "Air Automatics" named after V.V. Tarasova, Kursk, Zapolnaya street, 47, Russia

Abstract. The strength characteristics of composite carbon and fiberglass plastics have been studied, after their modification by microwave radiation of various power input. It has been established that the microwave effect on the finally shaped object under various conditions of the parameters increases the ultimate bending stresses of the carbon fiber reinforced plastic by 47–53 %, the GRP by 60 %. At the same time, significant changes in the microstructure are noted, which are manifested in the increase in its homogeneity and density as the contact points of the matrix agglomerates and reinforcing fibers increase.

1. Introduction

Analysis of scientific and technical literature, conference materials and exhibitions testifies to the intensive development of production of composite materials based on carbon fibers and glass fabrics and their wide application in the aviation, automotive, shipbuilding, wind power, rocketry and space technology. According to the research company Grand View Research, the world market of carbon plastic will reach $ 23.55 billion by 2022. According to another research company, MarketsandMarkets, the global market for carbon–fiber reinforced plastics will rise to $ 35.74 billion by 2020 [1, 2]. The aerospace industry accounted for more than 30 % of the consumption of these materials in 2014. The foregoing testifies to the existence and preservation in the near and distant future of the needs of high-tech industries in the use of high-quality composite materials.

However, due to a number of reasons, composite materials and products made of them have a pronounced anisotropy of properties [3]. This causes the need for additional reinforcement in some hazardous areas, leading to weight gain. This is undesirable for aerospace products, especially for highly maneuverable and high-speed objects.

There are different ways of forming a given set of material properties. One of them is the modification of its structure. An effective method of action is the use of microwave radiation (microwave electromagnetic field). This method is applicable to non-metallic composite materials.

We performed experimental studies [4] of the effect of the microwave radiation on the strength of the pultruded carbon and carbon plastics with a quasi–isotropic structure. It is established that the following changes in strength characteristics are noted for the specimens exposed to optimal regimes. The strength of the bending stresses increases by 11–16 %, the shear stresses by 13–21 %, while the interlayer shift tests increase the stress values by 14–15 % compared to the control samples. It should be noted that the temperature of the sample did not exceed 39–40 °C after the treatment was completed.

IOP Conf. Series: Journal of Physics: Conf. Series **1050** (2018) 012108 doi:10.1088/1742-6596/1050/1/012108

The foregoing indicates a promising continuation of the work on the investigation of the mechanisms of strengthening under the influence microwave radiation of finally formed products made of composite materials.

2. Formulation of the problem

The analysis of the materials of domestic and foreign scientific publications [5–7] shows that the greatest attention in the development of research in the field of ultrahigh–frequency processing of materials is devoted to the search for fundamentally new solutions for the application of microwave technologies with the aim of replacing existing heat treatment technologies. Recently Russian and foreign scientists V.A. Kolomeitsev, G.A. Morozov, O.G. Morozov, Y.S. Arkhangelsky, E.V. Kolesnikov, V.A. Tsarev, G. Piushner, J.R.Cannon, W.C. Chew, A. Razek, T.L. White and others made a significant contribution to the development of the theory of microwave heating of dielectric materials. The thermal effects of microwave radiation are manifested mainly in the stages of material formation, in the synthesis of reinforcing fibers, in the curing of the matrix. This does not solve the problem of the appearance of structural heterogeneity in the process of product formation, curing and finishing machining. It, in turn, leads to an increase in residual stresses and a decrease in strength characteristics. However, the thermal effects of microwave radiation are manifested mainly in the stages of material formation, in the synthesis of reinforcing fibers, when the matrix is cured. At the same time, the problem of the appearance, in the process of product formation, curing and finish machining of structural heterogeneities, leading to an increase in residual stresses and a decrease in strength characteristics, is not solved. Most studies of these electrotechnological processes concerned effects on the process of synthesis of components or curing of the composition [5]. There is almost no information about the mechanism of the effect of microwave radiation on the cured structure of the finally formed product. At the same time, the application of this influence on composite materials during their formation is not always expedient because of the need to introduce changes into the smooth and complex technological process of synthesizing the components, laying out the composite layers, curing and finishing.

The aim of our studies is the experimental determination of possible reinforcing microwave modifying of finally formed and cured structure of constructional composite materials such as glass and carbon plastics.

3. Research methods and equipment

We have studied samples from layered glass, carbon plastics in the form of plates with dimensions of 80.0×15.0×2.0 mm. In the experiments we used a microwave installation “Zhuk-2-02” manufactured by LLC "AgroEcoTech", Obninsk, Kaluga region with a radiating horn antenna. The installation generates an electromagnetic field of 2450 MHz with an output power of 1200 W. Three microwave power modes were used: small P_I, medium P_{II} and high P_{III}. Specific technological regimes are not described due to the procedure for patenting this method of modification.

The processing time was set to 1, 2, 3 minutes at different power levels. Tests of the samples before and after treatment were carried out on a machine equipped with strain gage force sensors and a worm loading mechanism. Processing of the results of measuring the build-up of the load applied to the sample was carried out using a special program (LabVIEW, Orel) and allowed to obtain load graphs (torque on the drive) in dynamics from the moment of application to destruction. Based on the known dependences of the material mechanics, bending stresses were calculated. The surface of the deformation zone was studied using a digital microscope Digital Microscope 2.0 MP 1000X (GAOSUO, China) with an increase in x500 with displaying on the laptop screen.

4. Results and discussion

Based on the results of processing the data of the experimental setup, graphical dependencies were obtained (Figure 1).

IOP Conf. Series: Journal of Physics: Conf. Series **1050** (2018) 012108 doi:10.1088/1742-6596/1050/1/012108

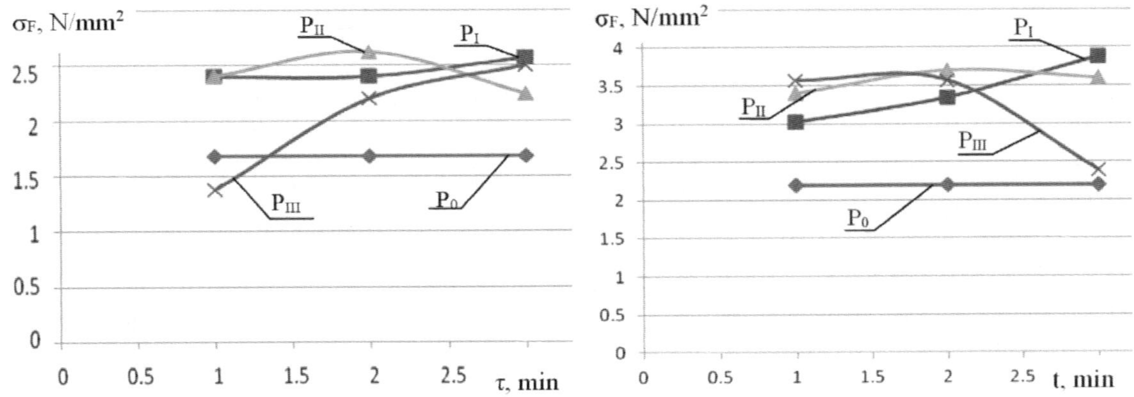

a *b*

Figure 1. Influence of time of microwave treatment of CFRP (a) and GRP (b) on flexural stresses σ_F at different power input

At low microwave radiation power, bending stresses increase by an average of 41 % and are practically independent of the exposure time. While testing the samples after low-power microwave radiation treatment, the flexural stresses increase by an average of 41 % and practically doesn`t depend on the exposure time. While examining samples processed at medium power, the dependence of the hardening effect on time is also insignificant. The hardening effect is unstable and ranges from 29 % to 53 %. Analysis of the test results of samples treated with the use of a high-power electromagnetic field shows an explicit dependence of the hardening of the material on the time of exposure. When it increases from 1.5 to 3 min, bending stresses increase by 46 %. However, at low values of time, a decrease in strength is observed, which requires an explanation.

The nature of the dependencies for fiberglass differs from the ones shown in Figure 1a: at short exposure time with an increase in the input power of radiation, a stable increase in the strength of the sample is observed, determined from the limiting bending stresses. Finally, at the maximum time of the investigated range of exposure, the maximum hardening effect is achieved at low power, but with its increase the effect decreases sharply. At maximum power, the strength of the sample remains practically comparable with the control one. In this case, obviously, another mechanism of the electromagnetic field`s influence is manifested. With the passage of time, at a small power level, the processes of formation of additional cross links – in the matrix, between the matrix and reinforcing fibers, as well as between the fibers themselves, which form conglomerates (braided bundles). As a result the material acquires maximum strength. The effect of hardening on bending stresses in this case is more than 60 % and exceeds that achieved one for CFRP.

A study of the microstructure of the deformation zone of the samples (Figures 2 and 3) showed the following.

a *b* *c*

Figure 2. Microstructure of a control sample made of CFRP (a) and processed with medium (b) and large (c) microwave power x500

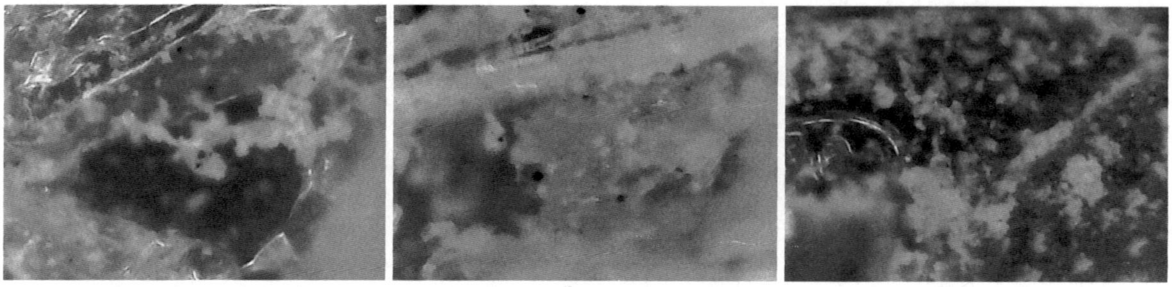

a *b* *c*

Figure 3. Microstructure of a control sample made of fiberglass (a) and processed with medium (b) and large (c) microwave power x500

In the zone of destruction of the control sample made from CFRP (Figure 2a), breaks of the crumbled matrix are seen, which led to a weakening of bonds between the reinforcing fibers. Though they mostly retained uniformity, they did not preserve the shape of the sample due to separate functioning. The structure of a sample processed at low power is characterized by greater monolithicity. One can note the presence of minor cracks, fibers, interconnected structures of the matrix, which have a more developed contact surface. Studying the sample processed at medium power for 3 minutes (Figure 2b), cracks in the matrix are visible, they divide the structure into blocks, although the blocks consist of fiber conglomerates. The microstructure of samples treated at high electromagnetic field power (Figure 2c) has the appearance of “alloyed” blocks and microblocks having a very developed surface. As a result large number of matrix agglomerates with a highly developed surface appear. In consequence of the described effect, flexibility, elasticity of the fiber-matrix system and its strength increase.

In the fracture zone of a control glass made of fiberglass (Figure 3a), breaks in the fibers and particles of the crumbled matrix are seen. The structure of the sample processed at a low power is monolithic, the fibers are linked together by “star” structures of the matrix, which have a developed contact surface. When processing at medium power for 3 minutes (Figure 3c), individual fibers are seen. They are separated from each other, although with the remaining “star” formations on the surface. After exposure for 3 minutes at high power, the sample looks completely destructured.

5. Conclusion

The obtained results can be explained as follows. The investigated carbon fiber plastic has a relatively loose (untight) packing of carbon fibers and a smaller content of reinforcing components. Accordingly, the effect of microwave radiation, which is mainly manifested at the level of the formation of additional cross-linking of the matrix molecules, leads to a greater effect on a material with a smaller content of the carbon reinforcing component. As the radiation power supplied to the GRP increases, dehydration and embrittlement of the matrix and reinforcing fibers are noted. At maximum power, this effect is manifested to a greater extent, therefore an increase in the exposure time causes heating of the material, which adversely affects the adhesion of the components and also the cohesive interaction of the structural components of the matrix. Accordingly, the resistance to the action of loads decreases and reaches a minimum with prolonged processing at the maximum of the investigated power range.

References

[1] Mirny M 2016 *The world market of carbon plastics will reach $ 23 billion by 2022* (Electronic Resource)

[2] 2011 *Share of application carbon plastic in the domestic the aviation industry will increase* Composite World *Special Issue "Aviation and Space"* pp 14–15

[3] Kablov E N 2012 *Materials and chemical technologies for aviation equipment* Vestnik Rossiiskoi Akademii Nauk **82** Vol 6 pp 520–530

[4] Zlobina I V and Bekrenev N V 2016 *The effect of electromagnetic field on CFRP (carbon fiber reinforced polymer) structural* Solid State Phenomena **870** pp 101–106

[5] Arkhangelsky Y S 2011 *A reference book on microwave electrothermia: reference book* (Saratov: Scientific Book) 560 p

[6] Kolomeitsev V A, Kuzmin Y A, Nikuiko D N and Semenov A E 2013 *Experimental studies of the level of uneven heating of dielectric materials and absorbed power in microwave devices of the resonator type* Electromagnetic waves and electronic systems **18**(12) pp 25–31

[7] Estel L, Lebaudy Ph, Ledoux A, Bonnet C and Delmotte M 2004 *Microwave assisted blow molding of polyethylene-terephthalate (PET) bottles* Proceedings of the Fourth World Congress on Microwave and Radio Frequency Applications **11** pp 33

Acknowledgments

The results presented in this article were received with partial financing from the project 9.2108.2017 / IF "Development and experimental working off of theoretical bases for the use of complexes with helicopter-type unmanned aerial vehicles with a take-off weight of up to 500 kg while performing search and rescue operations on water" carried out on the instructions of the Ministry of Education and Science of the Russian Federation.

IOP Conf. Series: Journal of Physics: Conf. Series **1050** (2018) 012109 doi:10.1088/1742-6596/1050/1/012109

Iteration of database relations for queries of special form

Sergey Zykin, Andrey Poluyanov

Sobolev Institute of Mathematics, Novosibirsk, Russia

Abstract. In this paper, the material for compiling a plan of performing "join-selection-projection" queries to relational databases is systematized. This is necessary for counting the number of iterations in the query execution algorithms. The presented topic is relevant for a considerable period of time. However, every time there are new requirements for queries, new methods of data presentation, etc. This leads to the necessity of resolving old problems. Many papers were devoted to various aspects of the optimization of queries to relational databases. Thus, it is possible to define two main directions. They are logical optimization of the queries and physical optimization of the queries execution. The basis of logical optimization is heuristics about the order of relational operations. However, more effective solution to the optimization problem is minimization of input-output operation quantity during the query execution. The approach to the development of the query execution plan on the basis of estimators for input-output operations for conventional and specialized methods of data storage is offered in this paper. The expediency of transformations, use of indexes etc. are determined by estimators.
Keywords: database, joint storage, query, estimator

1. Introduction

The single iteration usually used as a unit of measurement of algorithm difficulty. However, the difficulty of the iteration itself usually is not considered. In this paper the iteration scheme of relations with using results obtained earlier is offered: the basic properties of queries logical optimization [1]; methods for transforming relational queries [2]; optimization and design infrastructure for efficient and correct SQL transformations [3]. The physical storage design methods of data and the calculation of the corresponding estimates are presented in the following works: the basics of query optimization in relational system [4]; development of database structures [5]; creating an optimal query execution plan [6]; the use of statistical estimates when performing queries [7]. (the references relate only to the summarized and systematized publications).

The necessity to study the problem of optimizing database queries periodically arises in the analysis of ad hoc queries, or special ways of storing data as well as in solving unusual application problems. Let us consider only some of them, which are the closest to the subject of this paper. An area of query optimization such as e-commerce is considered in [8]. In this work at researching the message system in the instant messaging system based on the multi-language e-commerce platform in order to design the instant messaging system in multi-language environment and exhibit the national characteristics based information as well as applying national languages to e-commerce. Oracle memory database is adopted as user message buffer, so the messages transmitted and accepted in user communication process can be directly read from the memory and there is no need to read and write in the magnetic disk for each time, thus improving system throughput capacity and user concurrency.

Currently considerable interest is taken in the methods of using the cache [9, 10] to reduce the execution time of the query. Consideration of the specific problem in this case allows more efficient use of the cache contents. In work [9] the use of cache in a distributed database is analyzed. The problem of the time required to receive information about the cache contents from multiple sources is investigated in this paper. The model of distributed query optimization with a cache location between the local optimizer and the local database is proposed. Whenever a request is given to the local optimizer, the optimizer first checks the local cache. The proposed model to optimize the cost of work is based on four different factors: the distance to the server, server capacity, server load and the current length of the queue at the node where the query should be executed. In our work on the basis of these results we propose the method of using estimators in a distributed system.

In paper [10] a new method of client-side data caching for relational databases with a central server and distant clients is proposed. Data are downloaded to the client cache, in which the results of previous queries are stored. The query result is a special form called a "universal relational query". The basis of such queries is three main operations of relational algebra: selection, projection, and natural join. A subsequent query may be executed in a client's local cache. For this truth spaces of the logical restrictions in a new user's query and the results of the queries execution in the cache are compared. Such a comparison can be performed analytically without the need of additional queries on a server. This method may be used to define the lack of data in the cache and execute the query on the server only for these data. The solution to this problem with the use of analytical methods which distinguishes the research results shown in this paper from those ones obtained before. Our research is complemented by estimators of cache update if necessary: all contents of the table or its part will be updated in cache.

The problem of managing complex database systems will not lose its relevance in the foreseeable future. Such systems include a set of different hardware and software resources of a large number of configuration options that require a lot of experience, and it is time consuming. In work [11] a multilevel control system "MAG" is presented. This system can be used to monitor, analyze, predict configuration parameters, and it allows adjusting these parameters of database management. This system has the following features: a prediction of system performance depending on the set parameters; analysis and identification of the source of the problem. The method of estimation proposed in this paper can be embedded in "MAG" control system. In addition, there is a necessity for peculiarities of the storage structure in the system to be taken into account.

In work [12] the data integration for distributed query execution is considered. The main focus is on the algorithms for processing heterogeneous and distributed data. The conclusion is that the well-known universal algorithms are ineffective in such environment. This work considers the possibility of joint storage of data in the form of clusters. In our paper we consider a generalization of the clustered storage environment which allows using the projection operations instead of the operations of natural join of an arbitrary set of database relations being stored together.

As the initial form of the query representation we shall consider a "universal" relational query:

$$\pi_X(\sigma_C(R_1 \bowtie R_2 \bowtie \ldots \bowtie R_k) \qquad (1)$$

where π, σ, $\bowtie$ are the projection, selection and join operators; X is the set of attributes in the resulting relation; C is the logic expression. The attribute names in (1) have an extended format: <relation name>.<attribute name>. This allows an unambiguous determination of attribute affiliation in X and in C, and the natural join operation is converted into Cartesian product. Expression (1) covers a wide class of relational queries, but not all possible queries.

In [1, 2] the graph decomposition algorithm requiring transformation of expression $\boldsymbol{C}$ to the conjunctive normal form is considered. The heuristics proposed in [2] leads to the need of iterative procedures over two not small relations that requires more input-output operations than a pair-by-pair iterative procedure. Besides, opportunities of physical optimization including the use of indexes are not taken into account. In [13, 14] queries are optimized with the use of the index, but the lack of rules for determining the usefulness of the index could lead to increased costs. The formulated problems can be effectively resolved with the use of estimators.

First of all we shall consider the environment of storage and data processing. It is supposed that there is a distributed database on several servers connected by reasonably high-speed channels. The offered technique can be used without changes for one database server and one workstation.

2. Algorithmic basis of the method

For further treatment it is necessary to consider the basic algorithm of iterating (BAI). Let the sequence of the relations R_1, R_2,..., R_k is fixed by some rules; D_i is the viewing domain for R_i in compliance with the restrictions C, and current restrictions from the relations R_j, $j<i$. D_i contains all R_i tuples in the worst case, however, the use of indexes and links between relations or clustering can essentially reduce D_i. The domain D_i for a single restriction and for one index is defined by the set of initial and final interval values. The number of intervals is determined by the rating: the cost of the search of the next interval by the index should not exceed the cost of viewing the relation before the start of the interval. The domain D_i for the set of restrictions with the use of inverted file is the set of references on database records. In addition to the need of splitting on the intervals there is a need to determine the feasibility of using the indexes of fields. The calculation algorithms of D_i in BAI on the expression C and the data access method (indexations, links, clustering etc.) are assumed to be given.
Let us consider the basic algorithm of iteration:

```
<calculation D1>
<search of the first tuple t[R1] ∈ D1>
i=1
r=Ø
<cycle>
  if t[Ri] ∈ Di then
    if i=k then
      if C(t)=TRUE then
        r=r+t
        skip Di
      end if
    else
      i=i+1
      <calculation Di>
      <search of the first tuple t[Ri] ∈ Di>
    end if
  else
    if i=1 then
      <exit from cycle>
    else
      i=i–1
      skip Di
    end if
  end if
<the end of a cycle>
<the end of the algorithm>,
```

where r is the resulting relation; t is the tuple that is defined on attributes X (1) and on attributes C; skip D_i is the transition from the current tuple to the next tuple.
Remark 1. The relation R_i in the algorithm refers to as the leading for the relations R_j, $j>i$.
Remark 2. The submitted algorithm is reasonably simple and can be transformed into recursive algorithm. However, this will bring to an increase of operative memory costs by the system software.
Let $\varphi(1, 2,..., k; l)$ is the iteration cost of relations R_1, R_2,..., R_k sequence on server with number l, $\psi(1, 2,..., k; l)$ is the cost of result formation of iteration on server l. The general costs of iteration procedure are defined by the sum: $\varphi + \psi$, where φ includes the quantity of input operations and costs of initial data transfer on server l, ψ is the quantity of output operations on server l.
Let us consider the common schema of construction the quazyoptimal query realization plan:
1. Pair of relations R_i, R_j and server l, for which:

$$\varphi(i, j; l) + \psi(i, j; l) \rightarrow min\{(i, j; l)\}$$

is chosen.
1.1. For each pair the leading relation is defined. If $\varphi(i, j; l) \leq \varphi(j, i; l)$, the leading relation is R_i; otherwise the leading one is R_j.

1.1.1. For the driven relation R_j a preliminary iteration is necessary if:

$$\varphi(j;\, l) + \psi(j;\, l) + \varphi(\, i,\, m;\, l) < \varphi(i,\, j;\, l),$$

where R_m is the resulting relation after the R_j iteration.
1.2. $k = 0$.
2. For the relations not having been chosen, the relation R_n is determined such that:

$$\varphi\ (i, j, i_1, i_2,\ldots, i_k, n;\, l) \le \varphi(i, j, i_1, i_2,\ldots, i_k, n;\, l) + $$
$$+\ \psi\ (i, j, i_1, i_2,\ldots, i_k, n;\, l) + \varphi(m, n;\, l),$$

where R_m is the result of the chosen relation iteration.
2.1. For R_n a preliminary iteration is necessary if:

$$\varphi\ (n;\, l) + \psi(n;\, l) + \varphi(i, j, i_1, i_2,\ldots, i_k, m;\, l) < \varphi(i, j, i_1, i_2,\ldots, i_k, n;\, l),$$

where R_m is the result of the R_n iteration.
2.2. If in item 2.1 R_n is found, then $k = k+1$, $i_k = n$, and we go to item 2; otherwise we go to item 3.
3. The BAI for the chosen relations will be executed. An intermediate iteration will be executed if the necessity is defined in items 1.1.1 and 2.1. If there has been non-used relations, then the statistics for calculation ~~and ψ is~~ formed, and we go to item 1; otherwise there is the end of the algorithm.
Remark. More optimum sequences of the query execution (1) are possible in some cases. The choice of the submitted schema is stipulated by the following:

- The complete formation of optimum schema is an NP–complex problem.
- Estimator calculations for the completely optimum schema by the estimator ψ are executed. This estimator contains errors that will cause an increase of calculation errors.
- The costs of the quazyoptimal schema search have the second orders which are quite acceptable in comparison of the processor speed and the data transfer speed.

3. Estimators

The estimators φ and ψ on the statistics basis are calculated. The effective technique [4, 6] for the interval estimation of distribution function of the attributes in relations is based on:

$$(N, k, d, \{a_i, b_i, N_i\}, i = 1,\ldots, n), \tag{2}$$

where N is the capacity of the relation (quantity of tuples in R or a cardinal number), k is the capacity of the attribute (quantity of various values) in R, d is the minimum difference between various values of the attribute, n is the quantity of intervals, a_i is the values of the lower interval limit, b_i is the values of the upper interval limit, N_i is the attribute value quantity in the interval (a_i, b_i).
The values k and d can be estimated by an expert. The quantity of various values on the interval can be determined by the formula:

$$k_i = kN_i/N.$$

The updating of the given value can be executed by using the value d:

$$k_i \le 1+(\, b_i - a_i)/d.$$

With $n = 1$ the expression (2) corresponds to the assumption of an evenly distributed attribute.
For the calculation of φ and ψ on several attributes of one relation the assumption about independence of these attributes may be used. In some cases the estimation results will be substantially improved, if the covariation function of the attributes [4, 6] is used. The dynamic evaluation of the covariation functions is inexpedient due to essential costs either on memory, or on time. However, covariation and more qualitative statistics (2) can be received by a system in the background mode, when the loading of any server is minimal.
We shall consider the way of expression C represented in (1). For a problem of sequential iteration under the estimates control and for multiple access realization to the data the most expedient is the disjunctive normal form (DNF) of C representation:

$$C = \vee_{i=1}^{n} \wedge_{j=1}^{k_i} C_{i_j}\,, \tag{3}$$

where C_{i_j} = <attribute><operation><attribute> or C_{i_j} = <attribute><operation><constant>.
C_{i_j} can contain expressions on attributes, if they allow the normalization [3], and the functions φ and ψ can be calculated for them.

It is necessary to expand expression C by constraints before its transformation to DNF. Expression C must be expanded by the equivalent conditions, for example:

$$A_1 > A_2 \wedge A_2 = A_3 \equiv A_1 > A_2 \wedge A_2 = A_3 \wedge A_1 > A_3.$$

These operations will allow essential reduction of the intermediate computational complexity.
Let us consider the realization schema of restrictions C in the iteration process of the relations. Restrictions C_{i_j} in (3) are active under the iteration procedure, if all their attributes belong to the considered relations. All other C_{i_j} are replaced by the value *TRUE*. The active restrictions are used for the estimators φ and ψ computation and for tuples filtration in the resulting relation. When the current iteration terminated, the active restrictions C_{i_j} from C should be deleted, if they have the *TRUE* value on all tuples of the iteration result. Other C_{i_j} will participate in the next iterations as active.

Remark. If all C_{i_j} in any conjunctive group are deleted, then the expression C will be equivalent to the *TRUE* value and on the next iterations the Cartesian products execution will be needed. If the C_{i_j} restriction is deleted, then the matching attributes can be deleted, if the set X does not contain these attributes.

The iteration result for query (1) will be specified by r, $r_2 = R_1 \times R_2 \times \ldots \times R_k$, $r_1 = \sigma_C(r_2)$, $r_0 = \pi_X(r_1)$.

***Theorem* 1**. The iterative algorithm and the correct realization of the restriction schema form the resulting relation: $r = r_0$.

Proof. 1. Let a tuple $t \in r$ and $t \notin r_0$ are existed. The tuples $\{t[R_i]\}$ in each relation are existed, and they are used for tuple $t \in r$ generation. Let $t_1 \in \{t[R_1]\} \times \{t[R_2]\} \times \ldots \times \{t[R_k]\}$, hence $t=\pi_X(t_1)$. Under the assumption of $C(t_1) = FALSE$. Hence, in each conjunctive group $C_{ij(i)}(t_1) = FALSE$ exists, for each $i \in \{1, 2, \ldots, n\}$, $j(i)$ is the component number in the group. The restrictions $C_{ij(i)}$ under the condition after each iteration cannot be deleted. Hence, the tuple t_1 at least on the last iteration will be deleted, and this contradicts the initial assumption.

2. Let a tuple $t \notin r$ and $t \in r_0$ are existed. Let C_j is the restriction of j-s iteration, where the unused restrictions are replaced by the value *TRUE*. Obviously, that is executed by the following identity:

$$C(u) \rightarrow C_j(u) \equiv TRUE$$

for arbitrary iteration j and arbitrary tuple u, including tuple t. Let us assume that $C(t) = TRUE$. Hence, there are no iterations on which the tuples join $\{t[R_i]\}$, i = 1, 2, …, k would be removed, and this contradicts the initial assumption. This completes the proof.

4. Model of data representation

As the implementation of the context we use data representation in the form of the Table of Joins, which is a modification of relation data representation [15].

Let us consider the transformation of relational database: $R = \{R_1, R_2, ..., R_k\}$ to the Table of Joins (*TJ*) has a scheme (S,g), where S is a scheme that defined on the set of attributes $U=\{A_1, A_2, ..., A_n\}$, g is a vector length k. We consider the method of forming tuples $t \in s$, where s is the implementation of the *TJ* table. Consider all possible combinations without repetition of relations R_1, R_2, ..., R_k satisfying lossless join property [1]. This set of relations will be called a "context". Let the set of relations $C'=\{R_{m(1)}, R_{m(2)}, ..., R_{m(p)}\}$ be the context, where m is an integer array of p numbers of relations of the current combination and c' is the implementation of context constrained by selection σ_F with logical formula F:

$$c'=\sigma_F(R_{m(1)}[W_{m(1)}] \bowtie R_{m(2)}[W_{m(2)}] \bowtie \ldots \bowtie R_{m(p)}[W_{m(p)}]) \quad (4)$$

For each tuple $u \in c'$ we form a tuple t by the following rules: if the attribute A_j belongs to at least one relation of current join, then $t[A_j]=u[A_j]$, otherwise $t[A_j]=emp$, where emp is an empty value. To each tuple we assign a bit vector $g(t) = (g_1(t), g_2(t), \ldots, g_k(t))$. If the relation R_j is involved in the current join, then $g_j(t) = 1$, otherwise $g_j(t) = 0$.

We define a partial order on the tuples $t \in s$.

IOP Conf. Series: Journal of Physics: Conf. Series **1050** (2018) 012109 doi:10.1088/1742-6596/1050/1/012109

***Definition* 1.** A tuple $t \in s$ is less definite or equal to a tuple $t' \in s$ if for any attribute A_i the following condition holds: if $t[A_i] \neq t'[A_i]$, then $t[A_i]=emp$ and $g_j(t') \geq g_j(t)$, $j=1,\ldots,k$. In this case we write $t \prec t'$, the tuple t is said to be subordinate to the tuple t' and both of these tuples are called comparable.
Note. If we assume that A_i is *NULL* in both tuples, then $t[A_i]=t'[A_i]$.
In the representation s, it is sufficient to store only the tuple t', because t' contains all subordinate tuples. On the next step we remove all subordinate tuples. Equality of *NULL* values in definition 1 gets rid of the subordinate tuples t, which are derived from the same database tuples that t'. The difference between *NULL* and *emp* is that the first refers to indefinite attribute value, and the second refers to the absence of a corresponding tuple in the current join. It is obvious that the comparison operation $\prec$ is transitive.
The final stage of constructing a representation s is to remove tuples for which $F(t)=FALSE$. In this case, the tuples t', for which $t' \prec t$ and $F(t')=TRUE$ are excessive, because indefinite data in t' are defined in t and the user rejects them.
Suppose that $R(L) = (R_{m(1)}, R_{m(2)}, \ldots, R_{m(p)})$, where $L = (m(1), m(2), \ldots, m(p))$, and m is an integer array of numbers of p relations. Let us define the operation of the projection on the s.
***Definition* 2**. Projection $\pi_{R(L)}(s)$ is a set of tuples $u[R(L)]$, defined on the set of all attributes of relations $R(L)$, where for each $u[R(L)]$ there is a tuple $t \in s$ such that $u[R(L)]=t[R(L)]$ and $g_{m(i)}(t)=1$, $i = 1, 2,\ldots, p$.
Logical constraint $F(t)$ for each dimension is represented as a disjunctive normal form:

$$F=F_1 \vee F_2 \vee \ldots \vee F_m,$$

where each formula F_i is a conjunction:

$$F_i=F_{i,1} \wedge F_{i,2} \wedge \ldots \wedge F_{i,q},$$

where $F_{i,j}$ is predicate comparisons of SQL. If any predicate $F_{i,j}$ is not defined on a tuple t, then it is replaced with the value *TRUE* in case this conjunct still has some predicates, otherwise it is replaced with *FALSE*. This substitution allows us to remove from s the tuples for which some attributes are not defined yet or there are no tuples with coinciding values in other relations. This situation is also the subject of data analysis. After such substitution formula F can be either *TRUE* or *FALSE*.
Implementation of this property requires specific way of building s. If it is generated starting from all relations of the context, continued by relation subsets and so on up to a single relation with tuples filtering by using formula F, then in s there are tuples subordinate to those deleted in previous iterations. In addition, different sequences of deleting subordinate tuples may yield different results, as far as a specified tuple order is a partial order, not a strict linear one. Consequently, the shaping of Table of Join should be executed on the considered rule.

5. Scheme of queries realization

We shall consider a situation, when any R_i in one Table of Joins or one cluster [13] is represented. Probably, the same relation is represented in (1) for several times under various extended names. Let us name such relations by the joint stored.
The one pass from Table of Joins or cluster or duplicated relation in this case is more advantageous. The BAI modification is necessary so that resulting relations in BAI could be formed.
Let R_j is the next relation when generating scheme of iterations. The relations are stored jointly with it

$$R_{m(1)}, R_{m(2)}, \ldots, R_{m(p)}$$

which has not been iterated yet. For each $R_{m(q)}$, $q = 1, 2, \ldots, p$, we realize checking. If R_j and $R_{m(q)}$ in the Table of Joins or the cluster are jointly stored and natural join conditions are specified for them, then $R_{m(q)}$ should be added to iterative schema without generating a new resulting relation.
Otherwise, generation of a new resulting relation for $R_{m(q)}$ is necessary, if:

$$\varphi(n;l) < \varphi(m(q);l)$$

where R_n is the iteration result of $R_{m(q)}$. Further R_n instead of $R_{m(q)}$ in (1) is used.
***Remark* 1**. The Table of Joins, unlike the cluster, can contain any relations which have different common attributes. Hence, the relation number in one iteration can be larger for the Table of Joins.
***Remark* 2**. If attributes of the restriction C_{i_j} belong to different resulting relations, then this restriction is not activated.

6. Conclusion

We shall consider a situation, when any R_i in one Table of Joins or one cluster [13] is represented. Probably, the same relation is represented in (1) for several times under various extended names. Let us name such relations by the joint stored.

The representation of C in DNF permits one to unite several queries execution. The simultaneous occurrence of the several queries is usually due to preparation of one document, where the same data should be processed at various restrictions. In this case queries are united in groups, having common relations in (1). Groups of the queries with common relations are optimized and executed at first:

$$X = X_1 \cup X_2 \cup \ldots \cup X_n$$

and

$$C = C_1 \vee C_2 \vee \ldots \vee C_n,$$

where n is the quantity of the queries in group, X_i and C_i are the attributes set and restriction of i's query. Groups are separated after the exhaustion of the common relations. The process of disintegration is finished by forming the resulting relations, each of them being the response to a separate query.

The queries do not arrive simultaneously at the multiple data access. Then the system can suspend the group of queries execution and finish a new query execution at the same stage so that the total time of queries execution is reduced. In particular case:

$$C \rightarrow C_i \equiv TRUE\ ,\ X_i \subseteq X,$$

where C and X are current restriction and attributes set at the stage of queries group execution, C_i and X_i are restriction and attributes set of a new query. Then a new query is connecting to the group and its cooperative execution is resumed.

For groups of queries their parallel execution on various servers will be realized [16]. Within the one group the parallelism is possible for various iterations, but it will cause an increase of intermediate results.

7. References

[1] Ullman J 1980 *Principles of database systems* (Stanford: Computer Science Press) p 379

[2] Maier D 1983 *The theory of relational databases* (Stanford: Computer Science Press) p 637

[3] Chaudhuri S 1998 An overview of query optimization in relational system *ACM SIGACT-SIGMOD-SIGART symp. on Principles of database systems (PODS '98)* pp 34–43 URL http://dx.doi.org/10.1145/275487.275492

[4] Hong J, He Z, Bell D 2010 An overview of query optimization in relational system *Information Systems* **35** pp 140–148 URL http://dx.doi.org/10.1016/j.is.2009.08.004

[5] Teorey T, Fry J 1982 *Design of database structures* (NJ: Prentice Hall, Upper Saddle River)

[6] Kuznetsov S D 1990 Optimal query plan generation in relational DBMSs *Programming and Computer Software* **2**

[7] Chen L, Dobra A 2013 Histograms as statistical estimators for aggregate queries *Information Systems* **38** pp 213-230 URL https://doi.org/10.1016/j.is.2012.08.003

[8] Yang H, Chen X, Li J, Cao Y 2017 Design of Instant Messaging System of Multi-language E-commerce Platform *IOP Conference Series: Materials Science and Engineering* **231** 012023 URL http://dx.doi.org/10.1109/ICDCS.2013.21

[9] Kumar M, Batra N, Aggarwal H 2012 Cache Based Query Optimization Approach in Distributed Database *International Journal of Computer Science Issues* **9** pp 389–395

[10] Mosin S V, Zykin S V 2015 Truth Space Method for Caching Database Queries *Modeling and Analysis of Information Systems* **22** pp 248–258

[11] Khattab A, Algergawy A, Sarhan A 2015 MAG: A performance evaluation framework for database systems *Knowledge-Based Systems* **85** pp 245–255 URL http://dx.doi.org/10.1016/j.knosys.2015.05.010

[12] Arebi P, Gonbadipoor N 2011 A Genetic Algorithm for Query Optimization in Database Grid by Dynamic Cost Estimation *Computer Modelling and Simulation (UKSim) UkSim 13th International Conference* pp 81–86 URL http://dx.doi.org/10.1109/UKSIM.2011.25

[13] Hursch J, Hursch C 1989 *Working with ORACLE version 6.0* (TAB Books) p 464
[14] Nield T 2016 *Getting Started with SQL* (O'Reilly Media) p 134
[15] Zykin S V 2010 Automation of the interface formation between multidimensional and relational representation of the data *In Book: Relational Databases and Open Source Software Development* ed Taylor J (New York: Nova Science Publishers) chapter 2 pp 43–66
[16] Dobra A, Garofalakis M, Gehrke J, Rastogi R 2009 Multi-query optimization for sketch-based estimation *Information Systems* **34** pp 209–230 URL https://doi.org/10.1016/j.is.2008.06.002

Acknowledgments

The work was supported by the program of fundamental scientific researches of the SB RAS № I.5.1., project № 0314-2016-0020.

Institute of Physics
Dirac House, Temple Back
Bristol BS1 6BE UK

ISSN: 1742-6588
ISBN 978-1-5108-6822-9